Total Stress Tensor $\tilde{\underline{\underline{\Pi}}} = -p\underline{\underline{I}} + \underline{\underline{\tau}}$

$$\begin{pmatrix} \tilde{\Pi}_{11} & \tilde{\Pi}_{12} & \tilde{\Pi}_{13} \\ \tilde{\Pi}_{21} & \tilde{\Pi}_{22} & \tilde{\Pi}_{23} \\ \tilde{\Pi}_{31} & \tilde{\Pi}_{32} & \tilde{\Pi}_{33} \end{pmatrix}_{123} = \begin{pmatrix} \tilde{\tau}_{11} - p & \tilde{\tau}_{12} & \tilde{\tau}_{13} \\ \tilde{\tau}_{21} & \tilde{\tau}_{22} - p & \tilde{\tau}_{23} \\ \tilde{\tau}_{31} & \tilde{\tau}_{32} & \tilde{\tau}_{33} - p \end{pmatrix}_{123}$$

Dynamic Pressure $\mathcal{P} \equiv p + \rho g h$

Newtonian Constitutive Equation

$$\tilde{\underline{\underline{\tau}}} = \mu\left(\nabla\underline{v} + (\nabla\underline{v})^T\right)$$

$$= \mu \begin{pmatrix} 2\frac{\partial v_1}{\partial x_1} & \frac{\partial v_2}{\partial x_1} + \frac{\partial v_1}{\partial x_2} & \frac{\partial v_3}{\partial x_1} + \frac{\partial v_1}{\partial x_3} \\ \frac{\partial v_2}{\partial x_1} + \frac{\partial v_1}{\partial x_2} & 2\frac{\partial v_2}{\partial x_2} & \frac{\partial v_2}{\partial x_3} + \frac{\partial v_3}{\partial x_2} \\ \frac{\partial v_3}{\partial x_1} + \frac{\partial v_1}{\partial x_3} & \frac{\partial v_2}{\partial x_3} + \frac{\partial v_3}{\partial x_2} & 2\frac{\partial v_3}{\partial x_3} \end{pmatrix}_{123}$$

Total Molecular Fluid Force on a Finite Surface $\mathcal{S}$

$$\underline{\mathcal{F}} = \iint_{\mathcal{S}} [\hat{n} \cdot \tilde{\underline{\underline{\Pi}}}]_{\text{at surface}}\ dS$$

Stationary Fluid $\quad [\hat{n} \cdot \tilde{\underline{\underline{\Pi}}}] = -p\hat{n}$

Moving Fluid $\quad [\hat{n} \cdot \tilde{\underline{\underline{\Pi}}}] = -p\hat{n} + \hat{n} \cdot \tilde{\underline{\underline{\tau}}}$

Total Fluid Torque on a Finite Surface $\mathcal{S}$

$$\underline{\mathcal{T}} = \iint_{\mathcal{S}} [\underline{R} \times (\hat{n} \cdot \tilde{\underline{\underline{\Pi}}})]_{\text{at surface}}\ dS$$

Total Flow Rate Out Through a Finite Surface $\mathcal{S}$

$$Q = \dot{V} = \iint_{\mathcal{S}} [\hat{n} \cdot \underline{v}]_{\text{at surface}}\ dS$$

Average Velocity Across a Finite Surface $\mathcal{S}$

$$\langle v \rangle = \frac{Q}{S}$$

AN INTRODUCTION TO FLUID MECHANICS

This is a modern and elegant introduction to engineering fluid mechanics enriched with numerous examples, exercises, and applications. The goal of this textbook is to introduce the reader to the analysis of flows using the laws of physics and the language of mathematics. The approach is rigorous, but mindful of the student. Emphasis is on building engagement, competency, and problem-solving confidence that extends beyond a first fluids course.

This text delves deeply into the mathematical analysis of flows, because knowledge of the patterns fluids form and why they are formed and the stresses fluids generate and why they are generated is essential to designing and optimizing modern systems and devices. Inventions such as helicopters and lab-on-a-chip reactors would never have been designed without the insight brought by mathematical models.

Faith A. Morrison is Professor of Chemical Engineering at Michigan Technological University, where she has taught for 22 years. Morrison's expertise is in polymer rheology, in particular focusing on materials with structure, including high-molecular-weight polymers, block copolymers, hydrogels, and composites. She is the Past President of the Society of Rheology and Editor of the *Rheology Bulletin*. Morrison is the author of *Understanding Rheology* (2001).

AN INTRODUCTION TO FLUID MECHANICS

Faith A. Morrison

Department of Chemical Engineering
Michigan Technological University

CAMBRIDGE
UNIVERSITY PRESS

CAMBRIDGE UNIVERSITY PRESS
Cambridge, New York, Melbourne, Madrid, Cape Town,
Singapore, São Paulo, Delhi, Mexico City

Cambridge University Press
32 Avenue of the Americas, New York, NY 10013-2473, USA

www.cambridge.org
Information on this title: www.cambridge.org/9781107003538

First published 2013

Printed in the United States of America

A catalog record for this publication is available from the British Library.

Library of Congress Cataloging in Publication data

Morrison, Faith A.
An introduction to fluid mechanics / Faith A. Morrison.
    p.   cm.
ISBN 978-1-107-00353-8 (hardback)
1. Fluid mechanics.   I. Title.
QA901.M67   2012
532–dc23   2011049511

ISBN 978-1-107-00353-8 Hardback

Additional resources for this publication at www.cambridge.org/us/knowledge/isbn/item6684157/.

Cover photo: The Naruto Whirlpools, Japan, as seen from a tourist cruise boat. Photo taken by Hellbuny.

*This book is dedicated to my mother Frances P. Morrison,
my father Philip W. Morrison, and my elder brother
Professor Philip W. Morrison, Jr.*

Contents

Preface

This book forms the basis of a one-semester introductory course in fluid mechanics for engineers and scientists. Students working with this text are expected to have a background in multivariable calculus, linear algebra, and differential equations; review of these topics as applied to fluid mechanics is provided in Chapter 1. Problem solving is taught by example throughout the text. We include numerous solved examples and end-of-chapter problems, and a complete solution manual is available for instructors.

Fluid mechanics can be a difficult subject. Nonlinear physics governs flow, and thus we often resort to a variety of simplifications to obtain solutions. Different simplifications are used under different conditions, making fluid mechanics intimidating, at least to a beginner. *An Introduction to Fluid Mechanics* presents the topic through a discovery process, as described in this preface, that mimics engineering practice. The process used seeks solutions by answering the following questions:

1. *What is the problem?*
2. *What do we need to know, and do, to address the problem?*
3. *What is the solution to the problem?*
4. *What other problems/opportunities may be addressed now that we have solved this problem?*

This organizational choice builds critical thinking skills by emphasizing the thought processes that lead to model development. The book is divided into four parts that answer these four questions for the study of fluid mechanics.

1. **What is the problem?** [*Part I: Preparing to Study Flow*]
   Chapter 1: Why Study Fluid Mechanics
   Chapter 2: How Fluids Behave

The problem addressed in this book is how to bring readers to an understanding of flow behavior and to mastery of flow-modeling calculations. To accomplish this objective, students must come to the task with skills in mathematics and simple flow calculations. In Chapter 1 we introduce the problem, cover needed background calculations (i.e., the macroscopic mass balance and the mechanical engineering balance), and review mathematics that is prerequisite to the study of fluid mechanics (i.e., calculus and differential equations). In Chapter 2, we showcase the diversity and complexity of fluid behaviors—showing readers that the mechanical energy balance is insufficient to explain flow patterns and making the case that effort spent learning fluid mechanics is worth it. The presentation in Chapter 2 is at the survey level and spans from the introduction of viscosity to discussions of magnetohydrodynamics and vorticity. Overall, the text follows a path inspired by the spiral learning curve [Bruner, 1966], with the topics of Chapter 2 revisited at the end of the book (*Chapter 10: How Fluids Behave (Redux)*). That final chapter demonstrates how the intervening presentation leads to the ability to solve complex flow problems.

2. **What do we need to know, and do, to address the problem?** [*Part II: The Physics of Flow*]
   Chapter 3: Modeling Fluids
   Chapter 4: Molecular Fluid Stresses
   Chapter 5: Stress-Velocity Relationships

Having clarified our objectives in Part I, we seek methods to address the objectives in Part II. The continuum and the control volume are introduced in Chapter 3, and the stress components, fluid statics, and surface tension are presented in Chapter 4. To apply momentum conservation to a continuum, we need the stress constitutive equations, developed in Chapter 5 (Newtonian and non-Newtonian). These three chapters introduce the complete continuum model.

It can be a challenge to maintain student focus when covering background material, and we address this issue in a unique way: we provide a storyline. At the end of Chapter 3 we introduce two flow calculations and follow them longitudinally throughout Part II. These two problems (flow down an incline plane and flow in a 90-degree bend) are addressed in a just-in-time format, beginning before readers know enough fluid mechanics to be able to solve them. The solution develops gradually, incorporating new model pieces as they are covered. The repeated appearance of the two highlighted problems focuses readers on new developments, demonstrating the utility of the most recent step. Both highlighted problems are completed in Chapter 5, and Part II closes with the continuum model in place.

3. **What is the solution to the problem?** [*Part III: Flow Field Calculations*]
   Chapter 6: Microscopic Balance Equations
   Chapter 7: Internal Flows
   Chapter 8: External Flows

Model in hand, we turn to flows of interest. In Chapter 6 we develop the microscopic momentum balance (i.e., the Navier-Stokes equation), which represents an adaptation of the methods of Part II to the general case. We introduce the expressions for flow rates, fluid forces on walls, and fluid torques and show how to use these. In Chapter 7 a range of internal flows is discussed (pipes and ducts); in Chapter 8 external flows and boundary-layer flows are presented in detail (drag and lift).

The reader's path through Chapters 7 and 8 follows once again a storyline of a pair of highlighted flow problems. Chapter 7 begins with the quest to determine the extent of a home flood. Although not transparently related to the continuum model, the home flood problem is readily associated with pipe flow and motivates the examination of pressure drop/flow rate relationships, laminar and turbulent flow, and other internal-flow topics. We repeat this structure in Chapter 8, asking about a skydiver, which raises the question of flow past an obstacle in general, leading to discussion of drag, lift, and boundary layers.

Throughout Part III we employ dimensional analysis when the models we develop are too difficult to solve. Dimensional analysis is presented as a natural step in a problem-solving methodology that begins with addressing simplified versions of a real problem (because those are the problems we can solve and they give us insight), progresses to solving mathematically complex models, and turns ultimately to obtaining practical data correlations.

4. **What other problems/opportunities may be addressed now that we have solved this problem?** [*Part IV: Advanced Flow Calculations*]
   Chapter 9: Macroscopic Balance Equations
   Chapter 10: How Fluids Behave (*Redux*)

The final two chapters of *An Introduction to Fluid Mechanics* guide readers through advanced modeling calculations on a variety of flows. In Chapter 9 the macroscopic balances, including the mechanical energy balance and the macroscopic momentum balance, are derived and applied. Although simple uses of the mechanical energy balance are covered in Chapter 1, in Chapter 9 the applications are more involved, including pump sizing and open-channel flow. Applying the macroscopic momentum balance is generally considered to be a difficult topic; we systemize macroscopic momentum solutions, making them more accessible. In Chapter 10, the learning spiral returns us to the more complex flows introduced in Chapter 2, and we apply the now-familiar continuum model to begin to understand these flows. Chapter 10 discusses numerical solutions, statistical aspects of turbulence, lift, circulation, vorticity, and supersonic flow.

The text includes reference materials provided to aid the student. The appendices contain a glossary of terms and mathematical tables. There is additional mathematical assistance available on the Internet in the Web Appendix. Finally, key equations are presented on the inside covers as an aid to problem solving.

REFERENCE

Bruner, Jerome S., *The Process of Education* (Harvard University Press: Cambridge, MA, 1966).

Acknowledgments

My path to choosing this presentation method for my fluids class began in 1998 when I was first asked to teach fluid mechanics. I looked at the texts available, and, given the goals of both my course and my students, I had difficulty choosing a text. Although I did not find a book that satisfied my needs, I did find notes from a colleague, Professor Davis W. Hubbard, that got me started in the right direction. Professor Hubbard passed away in 1994, before this text was conceived, but his contribution to pedagogy lives on through his influence on this book.

I would like to thank many colleagues, friends, and family members for their assistance, encouragement, and support during the time spent working on this project. A partial list includes Tomas Co, Susan Muller, Scott Chesna, Denise Lorson, Pushpalatha Murthy, Madhukar Vable, Frances Morrison, Rosa Co, Tommy Co, and my colleagues and students in the Department of Chemical Engineering at Michigan Technological University and in the Society of Rheology.

In 2005–6 I spent a sabbatical year at Korea University in Seoul, Korea, teaching and working on this text. Many thanks to my hosts and colleagues for their welcome and for creating such a productive atmosphere in which to work. I would like to thank particularly Jae Chun Hyun, Chongyoup Kim, Joung Sook Hong, Jun Hee Sung, Kwan Young Lee, Jae Sung Lee (Postech), and my students in CBE614 Rheology, especially Yang Soo Son, Wun-gwi Kim, and Seoung Hyun Park. Final edits on this manuscript were prepared during another sabbatical year in 2012–13 as the William R. Kenan, Jr., Visiting Professor for Distinguished Teaching at Princeton University. I would like to thank my hosts at Princeton for this opportunity, especially Robert K. Prud'homme and Richard Register.

PART I

PREPARING TO STUDY FLOW

1 Why Study Fluid Mechanics?

1.1 Getting motivated

Flows are beautiful and complex. A swollen creek tumbles over rocks and through crevasses, swirling and foaming. A child plays with sticky taffy, stretching and reshaping the candy as she pulls and twists it in various ways. Both the water and the taffy are fluids, and their motions are governed by the laws of nature. Our goal is to introduce readers to the analysis of flows using the laws of physics and the language of mathematics. On mastering this material, readers can harness flow to practical ends or create beauty through fluid design.

In this text we delve into the mathematical analysis of flows; however, before beginning, it is reasonable to ask if it is necessary to make this significant mathematical effort. After all, we can appreciate a flowing stream without understanding why it behaves as it does. We also can operate machines that rely on fluid behavior—drive a car, for example—without understanding the fluid dynamics of the engine. We can even repair and maintain engines, piping networks, and other complex systems without having studied the mathematics of flow. What is the purpose, then, of learning to mathematically describe fluid behavior?

The answer is quite practical: Knowing the patterns that fluids form and why they are formed, and knowing the stresses that fluids generate and why they are generated, is essential to designing and optimizing modern systems and devices. The ancients designed wells and irrigation systems without calculations, but we can avoid the waste and tedium of the trial-and-error process by using mathematical models. Some inventions, such as helicopters and lab-on-a-chip reactors, are sufficiently complex that they never would have been designed without mathematical models. Once a system is modeled accurately, it is then straightforward to calculate operating variables such as flow rates and pressures or to evaluate proposed design or operating changes. A mathematical understanding of fluids is important in fields such as airplane and space flight, biomedicine, plastics processing, volcanology, enhanced oil recovery, pharmaceuticals, environmental remediation, green energy, and astrophysics. Although a trial-and-error approach can get us started in fluids-related problems, significant progress requires formal mathematical analysis.

We seek, then, to understand and model flows. As we begin, one advantage we have is that we already know much about flow: We interact daily with fluids, from throwing balls through the air to watering the lawn (Figure 1.1). We can

Figure 1.1 Reducing the cross-sectional area of the nozzle of a garden hose increases the fluid velocity, causing the water to travel farther before gravity pulls the stream to the ground. The upstream pressure is approximately constant at the pressure supplied by the municipal water system.

build on this familiarity (Chapter 2) and add tools from calculus and physics (Chapters 3–6) to arrive at sensible modeling and engineering results and insights (Chapters 7–10).

We cover the basics, one of which is the use of the continuum model to describe flow. The continuum model treats fluids not as molecules but rather as a deformable whole with properties that can be described by continuous functions of space and time (Chapter 3). Another basic we must master is understanding how molecular stress is generated and diffused in flowing materials. This is a complex topic, and we use two chapters to discuss it (Chapters 4 and 5). We will see that a systematic approach to fluid-stress modeling can make this challenging topic accessible. The stress constitutive equation (Chapter 5) connects fluid stress and motion in a way that leads directly to predictions of flow behavior in subsequent chapters.

We ultimately solve flow problems with momentum balances, which we introduce in Chapter 3 and learn to apply to flows in subsequent chapters. The flows we consider are divided into internal and external flows (Chapters 7 and 8). In both internal and external flows, we consider two regimes of flow: laminar and turbulent. As shown in a water jet in Figure 1.2, in the slow-flow regime, called *laminar flow*, small pieces of fluid move in an orderly fashion in smooth and more-or-less straight lines. At higher flow rates (or at other times when conditions are right), the flow becomes disordered and fluid particles move along seemingly random paths, causing substantial mixing; this is called *turbulent flow*. Another classic behavior exhibited by fluids is the formation of *boundary layers* in rapid flows (Figure 1.3). Boundary layers, both laminar and turbulent (Chapter 8), form in rapid flows as a result of the interaction of fluid momentum with solid boundaries. Knowledge of the mechanisms of laminar flow, turbulent flow, and boundary layers provides the background we need to understand the intricate momentum exchanges in complex flows.

Once the basics are established, we move to a more advanced study of fluids (Chapters 9 and 10). The purpose of advanced study varies among individuals, but the ability to innovate and invent new technologies rests on having an advanced understanding of physical systems, including flowing systems

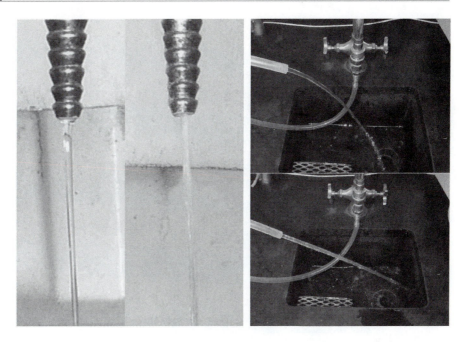

Figure 1.2 There are two basic flow regimes: a smooth slow flow-rate regime (i.e., laminar flow) and a rough, rapid flow-rate regime (i.e., turbulent flow).

(Figures 1.4 and 1.5). Advanced study may take the form of exploring: hemo-dynamics (i.e., the study of blood flow) [53]; non-Newtonian fluid mechanics, also called rheology [12, 104]; aeronautics [11, 76]; magnetohydrodynamics, which is important in astrophysics and metallurgy [35]; and microfluidics, a new field that explores the behavior of liquids confined in small spaces (Figure 1.5)

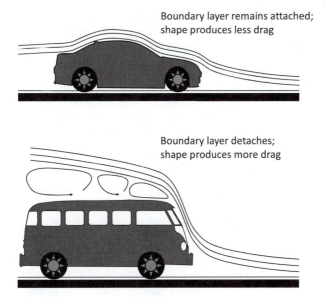

Boundary layer remains attached;
shape produces less drag

Boundary layer detaches;
shape produces more drag

Figure 1.3   Schematic of an attached boundary layer flowing over a streamlined vehicle versus a detached boundary layer flowing over a blunt object such as a van.

Figure 1.4 The human body relies on fluid flow to provide the necessary functions of life. The circulatory system, with blood as the transport medium, keeps nutrients and oxygen flowing to every part of the body as needed and also transports waste back to the lungs and kidneys for disposal. Blood responds as a Newtonian fluid when flowing in arteries and larger veins but, in smaller regions, it displays non-Newtonian behavior [53]. Different flow behaviors are covered in Chapter 4. Shown here is an artificial heart. Detailed knowledge of blood-flow dynamics (i.e., hemodynamics) is required to contribute to the design and manufacture of such devices. Photo courtesy of Abiomed.

[2, 51, 75]. The last of these, microfluidics, is contributing to the development of new biological processing devices (e.g., sensors or lab-on-a-chip devices) that carry out molecular separations in microscopic channels. In this text we touch briefly on some advanced topics of fluid mechanics but, more important, we lay the groundwork needed for the study of such subjects.

The equations that govern flow are nonlinear, second-order, partial differential equations (PDEs); thus, they are complex. In this text we study solutions of PDEs, but we also study simple algebraic equations based on mass, energy, and momentum conservation that tell us a great deal about flows. In fact, the first step in a detailed system analysis usually is to perform algebraic macroscopic balances. In the next section, we introduce the macroscopic mass and energy balances for flow; we use these balances throughout the text, especially in the analysis of pumps and other fluid-driven machinery (Figure 1.6 and Chapter 9).

For detailed flow analysis, we must set up and solve partial differential equations (Chapters 6–8). For complex flows, although we know the PDEs that govern the flow, we cannot always solve them, even with modern methods and computers. When the complete solution of flow equations is not possible, an effective approach is to divide the flow domain into separate regions, where the equations may be simplified and therefore solved. This "divide-and-conquer" approach to

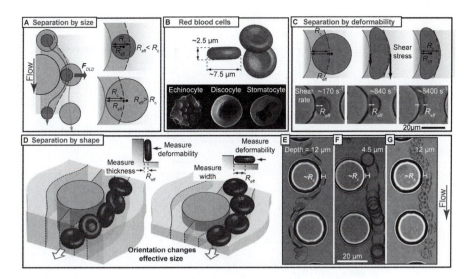

Figure 1.5 *Deterministic lateral displacement* (DLD) is a fluid-mechanics based mechanism for separating blood cells (RBC, see (B)) by (A) size, (C) deformability, or (D) by shape. In (A) particles with effective size R_{eff} smaller than a critical size R_c follow the flow streamlines which pass close to the obstacle, while larger particles cannot approach the obstacle and are forced onto a new path. In (C) shear forces deform particles, and flow at various shear rates is used to measure deformability. In (D) variation of the channel geometry allows researchers to investigate particle shape since different shapes respond to the geometry in specific ways. In (E), (F), and (G) cells are shown in the DLD device. From J. P. Beech et al. *Lab on a Chip*, vol. 12, 1048 (2012). Reproduced by permission of the authors and The Royal Society of Chemistry.

fluid mechanics includes the boundary-layer approach, in which regions close to solid boundaries are handled separately from the main flow (Chapter 8). Dimensional analysis, discussed throughout the text, helps to quantify which forces dominate in which regions of complex flow, thereby helping to address such problems. At the end of the book, we introduce *vorticity*, a physical quantity associated with a flow field that helps track momentum exchange in rapid, curling, twisting flows.

In this book, we explain fluid mechanics. The subjects and type of discussion presented here have been chosen to bring you to a real understanding of how fluids work. We explain the techniques that experts have discovered to model flows. More than just teaching students to pass a fluids course, our goal is to produce a competency with fluid-mechanics modeling that will allow students to contribute to the field and to apply their knowledge to engineering applications. We present many examples that build this understanding as well as competence and confidence in solving problems in fluid mechanics. End-of-chapter problems are provided, and we also direct readers to several published volumes of solved problems to supplement their efforts with this text [46, 56].

We proceed now to the study of elementary fluid mechanics. We begin with a quick-start section in which we show what type of fluid mechanics can be understood with a simple energy balance, without the detailed understanding of momentum exchange that is the primary topic of this text. We introduce

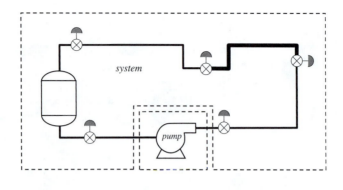

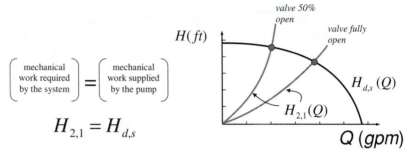

Figure 1.6 The performance of a centrifugal pump may be understood through pumping-head curves (see Section 9.2.4.1). These curves of head (i.e., mechanical energy per unit weight) versus capacity (i.e., flow rate) give the operating point of a pump as the intersection between the curve that is characteristic of the pump and the curve that is characteristic of the system through which the fluid is moving. When the system changes (e.g., a valve is closed somewhat), the system curve shifts as does the operating point. Both system and pump curves are derived from the mechanical energy balance.

the mechanical energy balance (MEB) and its no-friction, no-work version—the macroscopic Bernoulli equation—and we solve some basic problems. To proceed beyond the mechanical energy balance to an understanding of the patterns that fluids create and the stresses that fluids generate, we must consider momentum balances. Momentum balances concern us for the majority of this book.

The last section of this chapter discusses mathematical methods used in fluid mechanics. This overview connects mathematics in the abstract to the specific topic of fluid mechanics.

1.2 Quick start: The mechanical energy balance

In flowing systems, the laws of conservation of mass, momentum, and energy allow us to calculate how systems behave. For a detailed understanding of flows, we study the versions of conservation laws that apply to microscopic systems called control volumes (Figure 1.7, top). The equations that result from microscopic balances are nonlinear partial differential equations. It is an involved process to develop these equations and to learn to apply them; we start this task in Chapter 2.

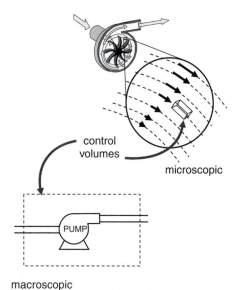

control volumes

microscopic

macroscopic

Figure 1.7

Conservation equations applied to small regions in a flow result in partial differential equations that can provide detailed information about the flow field. Conservation equations applied to entire devices or piping systems result in algebraic equations that give relationships among process variables such as average velocity, pressure, and frictional losses.

If a detailed understanding is not required, the conservation laws can be applied to larger-scaled systems rather than microscopic control volumes. Flow systems studied with macroscopic equations can be an entire pumping flow loop, for example (Figure 1.6; Figure 1.7, bottom), or a power station generating electricity at a waterfall. The balance equations in these cases are algebraic rather than differential equations, making them easier to apply and to solve. The drawback to macroscopic analysis is that we must make many assumptions and, because the assumptions sacrifice accuracy, we must supplement theoretical calculations with experiments. Another drawback of macroscopic analysis is that many of the flow details are not determined using such methods. Both microscopic and macroscopic analyses are useful, depending on the information that is sought.

We derive the macroscopic conservation laws later in the book (Chapter 9). In this quick-start section, we present the macroscopic conservation equations without derivation, and we show how they sometimes may be used to calculate and relate flow rates, pressure drops, frictional losses, and work. Practice with these elementary macroscopic calculations is good background for our primary task, which is the detailed study of fluid patterns and fluid stresses in complex flows.

The topic of this section is the mechanical energy balance (MEB), an energy balance applicable to a narrow class of flows that nevertheless are common and practical. We consider the special case of a single-input, single-output flow system such as a liquid pushed through a piping system by a pump (Figure 1.8). The fluid moves through the system at a mass flow rate, m, which corresponds to a particular volumetric flow rate Q and average velocity $\langle v \rangle$

$$\text{Volumetric flow rate:} \quad Q = \frac{m}{\rho} \tag{1.1}$$

$$\text{Average fluid velocity:} \quad \langle v \rangle = \frac{Q}{A} \tag{1.2}$$

where ρ is the density and A is the cross-sectional area of the pipe (see following discussion). There are pressure changes along the flow path as well as velocity and elevation changes. In addition, friction due to fluid contact with the wall or jumbled flow through fittings or other apparatuses causes energy to be converted

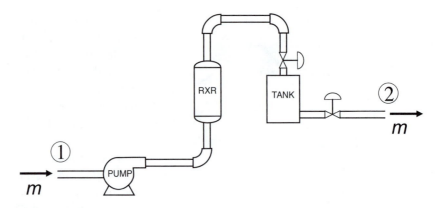

Figure 1.8 A very common system is one with a single-input stream (1), a single-output stream (2), and in which an incompressible (ρ = constant), nonreacting, nearly isothermal fluid is flowing.

to heat and essentially lost. Finally, mechanical devices put energy into or extract energy from the system in the form of *shaft work*, which refers to work associated with devices such as pumps, turbines, and mixers that interact with the fluid through a rotating shaft (see Chapter 9).

A macroscopic energy balance that may be applied to a single-input, single-output system with no reaction, no phase change, and little heat loss or heat generation is the mechanical energy balance, which is derived in Chapter 9 (Figure 1.9).

Mechanical energy balance
(single-input, single-output,
steady, no phase change,
incompressible,
$\Delta T \approx 0$, no reaction)

$$\frac{\Delta p}{\rho} + \frac{\Delta \langle v \rangle^2}{2\alpha} + g\Delta z + F = -\frac{W_{s,by}}{m} \quad (1.3)$$

$$\frac{p_2 - p_1}{\rho} + \frac{\langle v \rangle_2^2 - \langle v \rangle_1^2}{2\alpha} + g(z_2 - z_1) + F_{2,1} = -\frac{W_{s,by,21}}{m} \quad (1.4)$$

Definition of Terms in the
Mechanical Energy Balance

Δ out–in

F friction in system (always positive)

$W_{s,by}$ shaft work done, by fluid (negative for pumps
          and mixers; positive for turbines)

α
(velocity profile
shape parameter) $\begin{cases} \alpha = \frac{1}{2} & \text{laminar flow} \\ \alpha \approx 1 & \text{turbulent flow} \end{cases}$

Figure 1.9 The mechanical energy balance relates changes in key energy properties to the friction and work associated with the fluid in the system.

Table 1.1. Requirements for using the MEB

- Single-input, single-output (no branching)
- Steady state
- Constant density (incompressible fluid)
- Temperature approximately constant
- No phase changes or other chemical changes occur
- Only insignificant amounts of heat transferred

where p is pressure in the fluid, $\langle v \rangle$ is the average velocity in the pipe, z is the elevation, 1 and 2 refer to two locations in the flow, g is the acceleration due to gravity, ρ is fluid density, and m is the mass flow rate of fluid. The Δ in the MEB refers to the difference between the value of a quantity (p, $\langle v \rangle$, or z) at an outlet position minus the value of that quantity at an inlet position (out$-$in). The term $F = F_{2,1}$ accounts for all frictional losses in the system between the chosen outlet (2) and inlet (1) positions, and $W_{s,by} = W_{s,by,21}$ accounts for all the shaft work done by the fluid in the system between the chosen outlet and inlet positions. The quantity $W_{s,by}$ is positive for devices such as turbines, in which the fluid works on the surroundings, and $W_{s,by}$ is negative for pumps and mixers. The work done *by* the fluid equals the negative of the work done *on* the fluid, $W_{by} = -W_{on}$.

The quantity α in Equation 1.3 is a constant that depends on the type of flow pattern—that is, laminar or turbulent. We discuss the differences between laminar and turbulent flow in Section 2.4 and throughout the text. Here, we simply recall the discussion around Figure 1.2: Laminar flow is an organized flow with straight flow lines, and turbulent flow is a more rapid, disorganized flow with a jumbled structure. The quantity α in Equation 1.3 is approximately equal to 1 for turbulent flow and is exactly equal to 1/2 for laminar flow (see Chapter 9 for a derivation of α).

We provide no detailed justification of Equation 1.3 here because our purpose is to dive in and attempt some basic flow calculations. It is important, however, to know the assumptions involved in deriving Equation 1.3 so that we apply this result appropriately. The mechanical energy balance is limited to systems for which all of the following requirements hold: single-input, single-output, steady state, constant density and temperature, no reaction, no phase change, and negligible heat transferred (Table 1.1). The mechanical energy balance may be used only on systems that meet the requirements listed in Table 1.1.

To apply the mechanical energy balance to a flow system (e.g., the system shown in Figure 1.8), we first choose locations to designate as the inflow (1) and the outflow (2) locations. Strategically, they should be chosen so that some of the quantities in the MEB (e.g., pressure, average fluid velocity, and elevation) are measured easily at the chosen points. Shaft work often is the quantity to be calculated with the MEB. The friction term sometimes may be neglected; when it cannot be neglected, it must be calculated from experimental results—that is, from data correlations (see Section 1.2.3 and the Glossary). Care must be taken when using the MEB because the natural units of each term are not automatically the same and unit conversions are necessary. In the sections that follow, we show how to apply the MEB to situations of interest.

One relationship that we will need for the mechanical-engineering-balance calculations is the one between volumetric flow rate Q or $\dot{V}$[1] and average velocity in the pipe $\langle v \rangle$. We already mentioned this relationship and we discuss it in detail later (see Equations 3.71 and 6.255). To obtain average velocity from volumetric flow rate, we proceed as follows:

Average velocity
through pipe of
cross section A

$$\langle v \rangle = \frac{Q}{A}$$

(1.5)

where A is the pipe cross-sectional area and Q is the volumetric flow rate. For a circular pipe of diameter $D = 2R$, this becomes:

Average velocity
through circular pipe
of diameter D

$$\langle v \rangle = \frac{Q}{\pi R^2} = \frac{4Q}{\pi D^2}$$

(1.6)

The mass flow rate is just the volumetric flow rate multiplied by fluid density ρ:

$$\left(\frac{\text{mass}}{\text{time}} \right) = \left(\frac{\text{mass}}{\text{volume}} \right) \left(\frac{\text{volume}}{\text{time}} \right)$$

(1.7)

Mass flow rate
through pipe of
cross section A

$$m = \rho Q = \rho A \langle v \rangle$$

(1.8)

Mass flow rate
through circular pipe of
inner diameter D

$$m = \rho Q = \rho \pi R^2 \langle v \rangle = \frac{\rho \pi D^2}{4} \langle v \rangle$$

(1.9)

Thus, a measurement of mass flow rate can be converted to average velocity in a circular pipe as:

Average velocity
through circular pipe
of diameter D

$$\langle v \rangle = \frac{4m}{\rho \pi D^2}$$

(1.10)

We will have numerous occasions to use the relationships given in Equations 1.5–1.10.

In the sections that follow, we show how to use the mechanical energy balance (Equation 1.3) to solve for flow variables. The method that we discuss follows the steps listed in Table 1.2. We conclude this section with examples of flow calculations employing the relationships previously introduced. Note how the units are converted in these examples. In Section 1.2.1, we begin our work with the MEB using the simplest applications: those in which friction and shaft work are both zero.

---

[1] The symbol Q is used conventionally for both volumetric flow rate and a quantity of heat (see the energy-balance discussion in Chapter 6). Because of this dual use, we sometimes use the symbol $\dot{V}$ for volumetric flow rate, especially when the use of Q for flow rate would cause confusion.

Table 1.2. Method for applying the MEB

1. Choose inlet (1) and outlet (2) points (choose points where much is known).
2. Evaluate the pressure, average velocity, and elevation at these chosen points, if possible.
3. Calculate the frictional losses F or neglect them, if appropriate.
4. If there are no moving parts, $W_{s,by} = 0$ and the missing pressures, velocities, or elevations may be calculated or related.
5. If there are moving parts, the shaft work $W_{s,by}$ may be calculated from the MEB.

EXAMPLE 1.1. *Water is flowing in a $1/2 = $ in. Schedule 40 pipe at 3.0 gallons per minute (gpm). What is the average velocity in the pipe?*

SOLUTION. The average velocity in a pipe is equal to the volumetric flow rate, Q, divided by the cross-sectional area, A. We are given the flow rate and the rating of the pipe, from which we can find the cross-sectional area. Using Equation 1.5, we therefore can find the average velocity $\langle v \rangle$.

The nomenclature "Schedule 40 pipe" refers to a standard-size steel pipe as rated by the American National Standards Institute (ANSI). The true dimensions of this piping is found in tables published in the literature; a useful reference is *Perry's Chemical Engineers' Handbook* [132], which is available online. Consulting the literature, we find that the inner diameter of Schedule 40 1/2-in. pipe is 0.620 in., which corresponds to a cross-sectional area of $A = \pi D^2/4 = 0.3019$ in.^{2}. Using Equation 1.5 and performing the necessary unit conversions, we arrive at the average fluid velocity in the pipe. A link to a table of common unit conversions and physical property data is on the inside front cover of this book.

$$\langle v \rangle = \frac{Q}{A} \tag{1.11}$$

$$= \left(\frac{3.0 \text{ gpm}}{0.3019 \text{ in.}^2} \right) \left(\frac{35.3145 \text{ ft}^3/\text{s}}{15,850.2 \text{ gpm}} \right) \left(\frac{144 \text{ in.}^2}{\text{ft}^2} \right) \tag{1.12}$$

$$= 3.1882 \text{ ft/s} \tag{1.13}$$

$$= \boxed{3.2 \text{ ft/s} = 0.97 \text{ m/s}} \tag{1.14}$$

A flow rate of 3 gpm is a typical household-water flow rate. It is worth memorizing the order of magnitude of these numbers:

$$Q \approx 3 \text{ gpm}$$

Typical household flows: $\quad \langle v \rangle \approx 3$ ft/s (nominal half-inch pipe)
$$\langle v \rangle \approx 1 \text{ m/s (nominal half-inch pipe)}$$

EXAMPLE 1.2. *Water flows steadily through a converging section of piping shown in Figure 1.10. The pipe diameter at the inlet to the contraction is D_1 and the pipe diameter at the exit of the contraction is D_2. What is the relationship between the average velocity at the inlet $\langle v \rangle_1$ and the average velocity at the exit $\langle v \rangle_2$?*

Figure 1.10   Steady flow through a converging section of pipe causes the flow to accelerate. With a mass balance, we can relate inlet and outlet average velocities.

SOLUTION. The flow we consider here is through a contraction (Figure 1.10). The flow through the contraction is steady; thus, the mass flow rate is the same everywhere throughout the device. The mass balance between Points 1 and 2 may be written as:

$$\text{Macroscopic mass balance} \quad \begin{pmatrix} \text{mass} \\ \text{in} \end{pmatrix} = \begin{pmatrix} \text{mass} \\ \text{out} \end{pmatrix} \qquad (1.15)$$
$$\text{(steady state)}$$

$$m_1 = m_2$$

where m_1 is the mass flow rate at Point 1 and m_2 is the mass flow rate at Point 2. Using Equation 1.9, we write mass flow rate in terms of average velocity and solve for the relationship between the average velocities in the contraction. Note that the density of water ρ is a constant:

$$m_1 = m_2$$

$$\rho Q_1 = \rho Q_2$$

$$Q_1 = Q_2$$

$$\langle v \rangle_1 \frac{\pi D_1^2}{4} = \langle v \rangle_2 \frac{\pi D_2^2}{4}$$

$$\boxed{\langle v \rangle_1 = \left(\frac{D_2}{D_1} \right)^2 \langle v \rangle_2} \qquad (1.16)$$

At a constant flow rate, the average velocity depends inversely on the square of the pipe diameter.

1.2.1 MEB with no friction, no work: Macroscopic Bernoulli equation

When the friction term and the shaft work are zero, the mechanical energy balance simplifies to a form known as the macroscopic Bernoulli equation:

$$\frac{\Delta p}{\rho} + \frac{\Delta \langle v \rangle^2}{2\alpha} + g\Delta z = 0$$

Macroscopic Bernoulli equation (single-input, single-output, steady, no phase change, incompressible, $\Delta T \approx 0$, no reaction, no friction, no shaft work)   (1.17)

$$\frac{p_2 - p_1}{\rho} + \frac{\langle v \rangle_2^2 - \langle v \rangle_1^2}{2\alpha} + g(z_2 - z_1) = 0 \qquad (1.18)$$

Recall that Δ refers to the change in the property from the inlet to the outlet (out–in). Although the Bernoulli equation seems constrained, it has proven useful because the assumptions listed in Equation 1.17 are met in certain important flows. The Bernoulli equation is one of the most widely used equations in fluid mechanics (for advanced uses, see Chapters 8–10); unfortunately, it also is one of the most widely misused equations. To show how the Bernoulli equation may be used properly, we present three examples. Note that the assumptions listed in Equation 1.17 must be met to permit the use of the Bernoulli equation.

EXAMPLE 1.3. *A Venturi meter is a flow-rate measuring device in which a pressure drop is measured and flow rate is inferred (Figure 1.11). For flow through a Venturi meter, what is the relationship between measured pressure change and flow rate?*

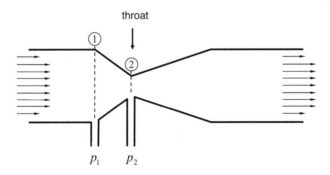

Figure 1.11   The relationship between the measured pressures and the fluid velocity in Venturi meters may be deduced from the mechanical energy balance (for systems in which friction may be neglected) or from the mechanical energy balance and a calibration specific to the device (if friction effects are considered).

SOLUTION. The device in Figure 1.11 is a flow meter. We begin our solution with some background on flow measurement.

In a piping system, we often want to know the average flow rate of a fluid in a pipe. One sure way to measure this flow is the pail-and-scale method (Figure 1.12),

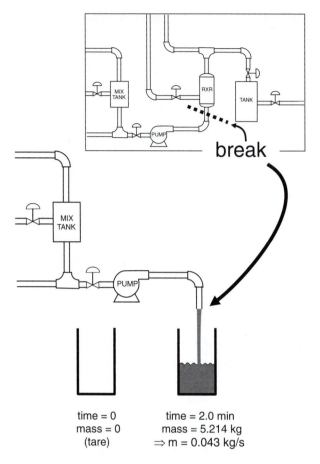

time = 0   time = 2.0 min
mass = 0   mass = 5.214 kg
(tare)   ⇒ m = 0.043 kg/s

Figure 1.12   An accurate way to measure flow rate is the pail-and-scale method, in which we break into the flow loop and measure the amount of fluid that accumulates in a pail over a set time interval. For an operating chemical plant, this is not a convenient method.

in which we break into the flow loop and measure the time it takes for an amount of fluid to fill a pail or another container. From these data, we can calculate the mass flow rate and, subsequently, the average fluid velocity (Equation 1.10):

$$m \equiv \frac{(\text{mass collected})}{(\text{collection time})} \tag{1.19}$$

$$\langle v \rangle = \frac{4m}{\rho \pi D^2} \tag{1.20}$$

where m is the mass flow rate, ρ is the fluid density, $\langle v \rangle$ is the average fluid velocity in the pipe, and D is the pipe diameter. The pail-and-scale method is accurate for measuring time-averaged flow rate, but it is highly undesirable to break into flow streams in functioning chemical plants or in many other operations. Also, the pail-and-scale method takes some time and therefore does not provide an instantaneous value of flow rate. Thus, engineers invented a wide variety of devices with which the flow rate may be inferred from measurement of a process variable such as pressure [118]; the Venturi meter is one of these devices.

A Venturi meter allows for the calculation of flow rate in pipes from a measurement of a particular pressure difference (see Figure 1.11). The design of a Venturi meter is of a converging section of pipe followed by a diverging section; the changes in the cross-sectional area are gradual to minimize the frictional losses within the device. Pressure measurements are taken at the points indicated in Figure 1.11; through application of the macroscopic mass and energy balances, we can relate this pressure difference to the instantaneous flow rate in the tube. Venturi meters allow for an accurate measurement of flow rate without significantly disturbing the flow.

Because the flow in a Venturi meter is a steady, single-input, single-output system with no reaction or phase change occurring and little heat generated or lost, we analyze this flow using the mechanical energy balance, neglecting at first the frictional contribution ($F = 0$). There are no moving parts and no shafts; therefore, $W_{s,by} = 0$, and we can use the macroscopic Bernoulli equation, as follows:

$$\boxed{\frac{\Delta p}{\rho} + \frac{\Delta \langle v \rangle^2}{2\alpha} + g\Delta z = 0}$$

Macroscopic Bernoulli equation (single-input, single-output, steady, no phase change, incompressible, $\Delta T \approx 0$, no reaction, no friction, no shaft work)   (1.21)

$$\frac{p_2 - p_1}{\rho} + \frac{\langle v \rangle_2^2 - \langle v \rangle_1^2}{2\alpha} + g(z_2 - z_1) = 0 \tag{1.22}$$

where Subscript 1 indicates the value of that variable at the inlet position and Subscript 2 indicates the value of that variable at the outlet position.

If we carefully choose Points 1 and 2 for our problem, it is straightforward to relate pressure and average velocity with the MEB. In the Venturi meter, we choose Point 1 as the point of the upstream pressure measurement and Point 2 is at the throat, the location of the other pressure measurement. Venturi meters are installed horizontally; thus, $z_2 - z_1 = 0$. The Bernoulli equation simplifies in this case to:

$$\frac{p_2 - p_1}{\rho} + \frac{\langle v \rangle_2^2 - \langle v \rangle_1^2}{2\alpha} + g(z_2 - z_1) = 0 \tag{1.23}$$

$$\frac{p_2 - p_1}{\rho} + \frac{\langle v \rangle_2^2 - \langle v \rangle_1^2}{2\alpha} = 0 \tag{1.24}$$

In an example in the previous section, we related $\langle v \rangle_1$ and $\langle v \rangle_2$ through the mass balance over a converging section of pipe. The result was as follows:

From the mass balance: (Equation 1.16):
$$\langle v \rangle_1 = \left(\frac{D_2}{D_1} \right)^2 \langle v \rangle_2 \tag{1.25}$$

where D_1 is the pipe diameter at Point 1 and D_2 is the pipe diameter at Point 2. Substituting this result into the macroscopic Bernoulli equation (Equation 1.24), we obtain the final relationship between the volumetric flow rate through the

Venturi meter and the measured pressure drop ($p_1 - p_2$):

$$\frac{p_2 - p_1}{\rho} + \frac{\langle v \rangle_2^2 - \langle v \rangle_1^2}{2\alpha} = 0$$

$$\frac{p_2 - p_1}{\rho} + \frac{1}{2\alpha}\left[\langle v \rangle_2^2 - \left(\frac{D_2}{D_1}\right)^4 \langle v \rangle_2^2\right] = 0$$

$$\langle v \rangle_2 = \sqrt{\frac{\frac{2\alpha(p_1 - p_2)}{\rho}}{\left[1 - \left(\frac{D_2}{D_1}\right)^4\right]}}$$

$$Q = \left(\begin{array}{c}\text{cross-sectional}\\\text{area}\end{array}\right)\left(\begin{array}{c}\text{average}\\\text{velocity}\end{array}\right) = \frac{\pi D_2^2}{4}\langle v \rangle_2$$

$$\boxed{Q = \frac{\pi D_2^2}{4}\sqrt{\frac{\frac{2\alpha(p_1 - p_2)}{\rho}}{\left[1 - \left(\frac{D_2}{D_1}\right)^4\right]}}}\qquad \begin{array}{c}\text{Flow rate}\\\text{measured by a}\\\text{Venturi meter}\\\text{(no friction)}\end{array} \qquad (1.26)$$

When the flow is sufficiently rapid as measured by a quantity called the *Reynolds number* (Re) (Re $= (\rho \langle v \rangle D)/\mu > 10^4$; see Equation 1.62) where μ is fluid viscosity, the no-friction relationship in Equation 1.26 accurately describes the pressure-drop/flow-rate relationship for many Venturi meters. For slower flows, friction is more important to the total energy, and experiments should be conducted to determine the neglected friction. In the experiments needed to calibrate a Venturi meter for frictional losses, we measure the time-averaged flow rate Q by an independent method (e.g., pail-and-scale) and we measure the pressure drop $p_1 - p_2$; finally, we deduce an empirical friction correction factor C_V that makes Equation 1.27 correct according to the measured data:

$$\boxed{Q = C_V\left(\frac{\pi D_2^2}{4}\right)\sqrt{\frac{\frac{2\alpha(p_1 - p_2)}{\rho}}{\left[1 - \left(\frac{D_2}{D_1}\right)^4\right]}}}\qquad \begin{array}{c}\text{Flow rate}\\\text{measured by a}\\\text{Venturi meter}\\\text{(with friction)}\end{array} \qquad (1.27)$$

C_V must be determined experimentally by either the user or the manufacturer of the Venturi meter. Venturi meters typically are used in turbulent flow for which $\alpha = 1$.

With only the Bernoulli equation and a simple mass balance, we can completely describe the operation of a Venturi meter. The complexities of the flow are swept up into the friction coefficient C_V, which is determined experimentally. The strategy of the mechanical energy balance and other macroscopic balances is: Perform balances on macroscopically sized control volumes, make reasonable assumptions, improve accuracy by making experimental measurements, and adjust the equations to match the experiments. The examples that follow discuss additional situations in which the macroscopic Bernoulli equation may be applied.

EXAMPLE 1.4. *Water drains from a tank as shown in Figure 1.13. The level of water in the tank is maintained at a constant height through control of flow in the overhead pipe. What is the drain flow rate in terms of the height of the fluid in the tank?*

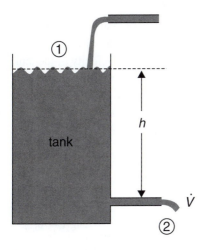

Figure 1.13 Water drains from a tank that is maintained at a constant level. This type of arrangement is known as a constant-head tank.

SOLUTION. The system of water in the tank flowing out the bottom drain is a single-input, single-output, steady flow of an incompressible fluid. There is no heat transfer and no chemical reaction or phase change; therefore, all requirements of the mechanical energy balance are met.

$$\frac{\Delta p}{\rho} + \frac{\Delta \langle v \rangle^2}{2\alpha} + g\Delta z + F = -\frac{W_{s,by}}{m}$$

Mechanical energy balance (single-input, single-output, steady, no phase change, incompressible, $\Delta T \approx 0$, no reaction)   (1.28)

We choose as our two points (1) the surface of the fluid in the tank and (2) the point at the lower exit where the fluid emerges into the air. These are good choices because we know much about the pressure, average velocity, and elevation at these points, as we now discuss.

There are no moving parts in the chosen system and therefore no shaft work. The flow in the tank is tranquil and little friction is generated. The flow through the exit pipe may have a frictional contribution, but the exit pipe is short; thus, it seems reasonable to entirely neglect friction. The prediction from this frictionless calculation can be checked experimentally to see if this last assumption is valid.

The mechanical energy balance thus simplifies to the Bernoulli equation:

$$\frac{\Delta p}{\rho} + \frac{\Delta \langle v \rangle^2}{2\alpha} + g\Delta z = 0$$

Macroscopic Bernoulli equation (single-input, single-output, steady, no phase change, incompressible, $\Delta T \approx 0$, no reaction, no friction, no shaft work)   (1.29)

$$\frac{p_2 - p_1}{\rho} + \frac{\langle v \rangle_2^2 - \langle v \rangle_1^2}{2\alpha} + g(z_2 - z_1) = 0 \tag{1.30}$$

At Point 1, the surface of the water in the tank, the pressure is atmospheric. At Point 2, the discharge of the pipe, the pressure also is atmospheric; therefore, $p_2 - p_1 = 0$. The expression $\langle v \rangle_1$ refers to the velocity of the tank water surface, which is zero. The average velocity of the water at the exit $\langle v \rangle_2$ is the quantity in which we are interested. Finally, z_1 and z_2 refer to the elevations of the two chosen points. We may choose the elevation of the discharge as the reference level for measuring elevation; thus, $z_2 = 0$ and $z_1 = h$. The mechanical energy balance becomes:

$$\frac{p_2 - p_1}{\rho} + \frac{\langle v \rangle_2^2 - \langle v \rangle_1^2}{2\alpha} + g(z_2 - z_1) = 0 \tag{1.31}$$

$$\frac{\langle v \rangle_2^2}{2\alpha} - gh = 0 \tag{1.32}$$

The value of α to use depends on whether the flow is laminar or turbulent (see Figure 1.9), and the established way to infer this is discussed in the next section. For now, we assume a turbulent flow; thus, $\alpha = 1$.

We now can solve for $\langle v \rangle_2$ and the volumetric flow rate Q:

$$\frac{\langle v \rangle_2^2}{2} - gh = 0 \tag{1.33}$$

Torricelli's law: discharge velocity from a constant-head tank (no friction)

$$\langle v \rangle_2 = \sqrt{2gh} \tag{1.34}$$

$$Q = A\langle v \rangle_2 = \frac{\pi D_2^2}{4} \langle v \rangle_2 \tag{1.35}$$

Discharge flow rate from a constant-head tank (no friction)

$$Q = \frac{\pi D_2^2}{4} \sqrt{2gh} \tag{1.36}$$

Equation 1.34 is known as Torricelli's law, named for Evangelista Torricelli, who invented the barometer and discovered in 1643 the equation for discharge velocity from a constant-head tank. Experiments verify Torricelli's law for tanks with short exit pipes.

EXAMPLE 1.5. *Water is siphoned from a tank as shown in Figure 1.14. What is the flow rate of water in the siphon tube (inner diameter =1.5 cm)? What is the limit in the wall height that the siphon can overcome?*

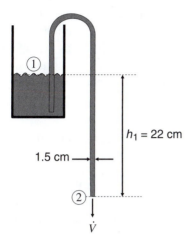

$h_1 = 22$ cm

1.5 cm

$\dot{V}$

Figure 1.14   A siphon works because liquids prefer to form unbroken streams, and the weight of the fluid below the tank level is sufficient to draw the trailing stream over a barrier. The siphon breaks when the pressure in the stream is low enough to allow the fluid to boil, breaking the liquid stream.

SOLUTION. The system of water flowing in the siphon is a single-input, single-output, steady flow of an incompressible fluid. There is no heat transfer and no chemical reaction or phase change; therefore, all requirements of the mechanical energy balance are met.

$$\boxed{\frac{\Delta p}{\rho} + \frac{\Delta \langle v \rangle^2}{2\alpha} + g\Delta z + F = -\frac{W_{s,by}}{m}}$$

Mechanical energy balance (single-input, single-output, steady, no phase change, (1.37) incompressible, $\Delta T \approx 0$, no reaction)

We choose our two points as locations for which we know a great deal: (1) the free surface in the tank, and (2) the exit point of the siphon. There are no moving parts in the chosen system and therefore no shaft work. The flow in the tank and siphon is tranquil and little friction is generated. The mechanical energy balance simplifies to the Bernoulli equation:

$$\boxed{\frac{\Delta p}{\rho} + \frac{\Delta \langle v \rangle^2}{2\alpha} + g\Delta z = 0}$$

Macroscopic Bernoulli equation (single-input, single-output, steady, no phase change, (1.38) incompressible, $\Delta T \approx 0$, no reaction, no friction, no shaft work)

$$\frac{p_2 - p_1}{\rho} + \frac{\langle v \rangle_2^2 - \langle v \rangle_1^2}{2\alpha} + g(z_2 - z_1) = 0 \qquad (1.39)$$

At Point 1, the surface of water in the tank, the pressure is atmospheric. At Point 2, the discharge of the tube, the pressure also is atmospheric; therefore, $p_2 - p_1 = 0$. The expression $\langle v \rangle_1$ refers to the velocity of the tank water surface, which is approximately zero if we confine our analysis to the initial stages of the flow. The average velocity of water at the exit $\langle v \rangle_2$ is the quantity in which we are interested. Finally, z_1 and z_2 refer to the elevations of the two chosen points. We may choose the elevation of the discharge as our reference level for measuring elevation; thus, $z_2 = 0$ and $z_1 = h_1$. In all important ways, this calculation is identical to the previous example. The mechanical energy balance thus becomes:

$$\frac{p_2 - p_1}{\rho} + \frac{\langle v \rangle_2^2 - \langle v \rangle_1^2}{2\alpha} + g(z_2 - z_1) = 0 \tag{1.40}$$

$$\frac{\langle v \rangle_2^2}{2\alpha} - gh_1 = 0 \tag{1.41}$$

Again, we assume a turbulent flow and $\alpha = 1$. We now solve for $\langle v \rangle_2$ and the volumetric flow rate Q:

$$\frac{\langle v \rangle_2^2}{2} - gh_1 = 0 \tag{1.42}$$

Discharge velocity from
a siphon
(no friction)

$$\boxed{\langle v \rangle_2 = \sqrt{2gh_1}} \tag{1.43}$$

$$Q = A \langle v \rangle_2 = \frac{\pi D_2^2}{4} \langle v \rangle_2 \tag{1.44}$$

Discharge flow rate from
a siphon
(no friction)

$$\boxed{Q = \frac{\pi D_2^2}{4} \sqrt{2gh_1}} \tag{1.45}$$

For the dimensions shown in the water siphon in Figure 1.14, the initial discharge flow rate is:

$$Q = \frac{\pi (0.015 \text{ m})^2}{4} \sqrt{(2)(9.8066 \text{ m/s}^2)(0.22 \text{ m})}$$

$$= 3.6708 \times 10^{-4} \text{ m}^3/\text{s} \left(\frac{15,850.2 \text{ gpm}}{\text{m}^3/\text{s}} \right)$$

$$= 5.8183 \text{ gpm}$$

$$= \boxed{5.8 \text{ gpm}} \tag{1.46}$$

This is the fluid volumetric flow rate as the siphon starts up and little water has drained from the tank and for which no friction is accounted.

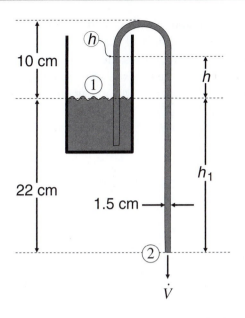

Figure 1.15 The height of the barrier in a siphon is limited by the creation of subatmospheric pressures near the top of the barrier. When the pressure at the top is low enough, vapor forms and the continuity of the fluid is interrupted. This interruption causes the siphon to break.

The second part of the problem asks for the maximum height of the barrier at which the siphon stops working. To determine this height, we consider the status of the fluid within the siphon (Figure 1.15). The siphon functions well as long as fluid pressure never drops below its vapor pressure. At the vapor pressure, fluid boils, and the vapor produced causes the liquid stream to break. We can determine the pressure at any point in the siphon using the mechanical energy balance (MEB).

To calculate pressure in the siphon, we perform a mechanical energy balance between Point 1 at the tank surface and a second point somewhere in the pipe flow. We call the second point h, which indicates the elevation of our chosen point above the water level in the tank. A mechanical energy balance between Points 1 and h (assuming no friction, no shaft work, and a turbulent flow) indicates the pressure at h, and we can compare that pressure to the vapor pressure to determine the value of h at which the liquid boils.

$$\boxed{\frac{\Delta p}{\rho} + \frac{\Delta \langle v \rangle^2}{2\alpha} + g\Delta z = 0}$$

Macroscopic Bernoulli equation (single-input, single-output, steady, no phase change, incompressible, $\Delta T \approx 0$, no reaction, no friction, no shaft work)   (1.47)

$$\frac{p_2 - p_1}{\rho} + \frac{\langle v \rangle_2^2 - \langle v \rangle_1^2}{2\alpha} + g(z_2 - z_1) = 0 \qquad (1.48)$$

Recall that Δ means out–in.

The pressure at the tank surface is atmospheric, $p_1 = p_{atm}$, and the velocity of the tank surface is approximately zero. We choose the reference elevation as the water surface in the tank; thus, $z_1 = 0$. The height z_h is h.

$$\frac{p_h - p_1}{\rho} + \frac{\langle v \rangle_h^2 - \langle v \rangle_1^2}{2} + g(z_h - z_1) = 0 \qquad (1.49)$$

$$\frac{p_h - p_{atm}}{\rho} + \frac{\langle v \rangle_h^2}{2} + gh = 0 \qquad (1.50)$$

We previously solved the siphon-discharge velocity for this problem as $\langle v \rangle_2 = \sqrt{2gh_1}$. Because the cross-sectional area of the tube is constant, the average velocity of the fluid throughout the tube is the same as at the discharge $\langle v \rangle_2 = \langle v \rangle_h$ (from a mass balance). Substituting the previous solution for discharge velocity, Equation 1.43, into the MEB, Equation 1.50, we now obtain an expression for the pressure at point h as a function of h:

$$\frac{p_h - p_{atm}}{\rho} + \frac{\langle v \rangle_h^2}{2} + gh = 0 \tag{1.51}$$

$$\frac{p_h - p_{atm}}{\rho} + gh_1 + gh = 0 \tag{1.52}$$

Pressure within a
working siphon
at a point elevated
a distance h above
the tank fluid surface

$$\boxed{p_h = p_{atm} - \rho g \left(h_1 + h \right)} \tag{1.53}$$

Readers may verify that if we had written a mechanical energy balance between Points h and the discharge Point 2, we would have arrived at the same result.

Equation 1.53 indicates the pressure in the siphon and shows that it may be less than atmospheric. At a point elevated by an amount h above the tank water level, the pressure in the siphon is less than p_{atm} by an amount $\rho g(h_1 + h)$. Recall that h_1 reflects the height of the section of siphon that drops below the tank water level (see Figure 1.15), which is constant. Note that the pressure is lowest when h is large, such as when h is at the highest point in the siphon.

With Equation 1.53 we can calculate when the pressure in the siphon becomes so low that the fluid boils and vapor fills the tube. This is called vapor-lock (Figure 1.16). We obtain the maximum height to which an intermediate point of the siphon may be raised by equating the pressure at height h, p_h, to the vapor pressure of the liquid being siphoned, p_v^*.

Vapor-lock occurs
when p_h drops
to fluid vapor pressure

$$\boxed{p_v^* = p_h} \tag{1.54}$$

$$p_v^* = p_{atm} - \rho g \left(h_1 + h_{max} \right) \tag{1.55}$$

Maximum siphon
height above
tank fluid level

$$\boxed{h_{max} = \frac{p_{atm} - p_v^*}{\rho g} - h_1} \tag{1.56}$$

Vapor pressures p_v^* for various fluids are found in the literature. From *Perry's Handbook* [132], we find that water at 25°C has a vapor pressure of 23.756 mmHg, or 3.167206×10^3 N/m^2, and a density of 997.08 kg/m^3. For the dimensions in

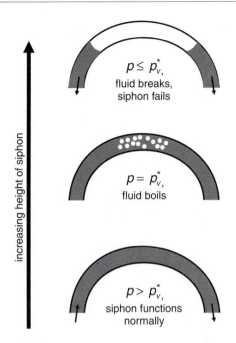

Figure 1.16 Schematic of the formation of a vapor gap in a siphon. This occurs when the barrier over which the fluid travels rises above a critical height; the critical height may be calculated as shown in the example.

the water siphon in Figure 1.15, the maximum height over which the siphon can operate is:

$$h_{max} = \frac{p_{atm} - p_v^*}{\rho g} - h_1$$

$$= \frac{\left(1.01325 \times 10^5 - 3.167206 \times 10^3\right) \text{N/m}^2 \left(\frac{\text{kg m/s}^2}{1 \text{ N}}\right)}{\left(\frac{997.08 \text{ kg}}{\text{m}^3}\right)\left(\frac{9.80066 \text{ m}}{\text{s}^2}\right)} - (0.22 \text{ m})$$

$$= 9.825 \text{ m}$$

$$\boxed{h_{max} = \quad 9.8 \text{ m}}$$

This large value for h_{max} becomes smaller as the temperature increases because the vapor pressure increases. Also, if the flow exit elevation is lowered (i.e., h_1 increases), Equation 1.56 indicates that h_{max} decreases.

For the systems described in this section, the macroscopic Bernoulli equation applies because there is no friction and no shaft work. In the next section, we consider systems that are slightly more complicated: those that include shaft work.

1.2.2 MEB with shaft work

The Bernoulli equation does not apply to systems that include turbines, pumps, or other devices that produce or consume shaft work; such systems must be analyzed using the full mechanical energy balance (MEB). In this section, we first analyze a pumping loop with the MEB and calculate the shaft work necessary to move fluid at a given flow rate. Second, we analyze the conversion of flow energy to electrical energy in a hydroelectric power plant. We do not yet consider the effect of frictional losses on the required work in these systems; frictional losses in fluid networks are addressed in Section 1.2.3.

EXAMPLE 1.6. *What is the work required to pump 6.0 gpm of water in the piping network shown in Figure 1.17? You may neglect the effect of friction.*

SOLUTION. When a flow problem involves the amount of shaft work required to bring about a flow, the mechanical energy balance is the place to start. The system of water in the flow loop is a single-input, single-output, steady flow of an incompressible fluid. There is no heat transfer and no chemical reaction or phase change; therefore, all requirements of the mechanical energy balance are met.

We choose Points 2 and 1 to be where we know the most about the problem. We choose Location 2 to be where the fluid exits the pipe; Location 1 is the liquid free-surface in the tank. For both locations, we know the pressure, the velocity of

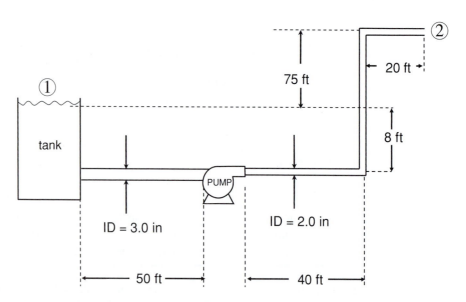

Figure 1.17   A common problem in engineering involves pumping a fluid from a tank at atmospheric pressure through a piping system. The amount of work required to pump at a chosen flow rate may be calculated using the mechanical energy balance.

the fluid, and the elevation, which is all of the information we need to calculate $W_{s,by}$ from the friction-free MEB.

Mechanical energy balance
(single-input, single-output,
steady, no phase change,
incompressible,
$\Delta T \approx 0$, no reaction)

$$\frac{\Delta p}{\rho} + \frac{\Delta \langle v \rangle^2}{2\alpha} + g\Delta z + F = -\frac{W_{s,by}}{m}$$

$$\frac{p_2 - p_1}{\rho} + \frac{\langle v \rangle_2^2 - \langle v \rangle_1^2}{2\alpha} + g(z_2 - z_1) + F_{2,1} = -\frac{W_{s,by,21}}{m} \quad (1.57)$$

At Position 1, $p_1 = 1.0$ atm, $z_1 = 0$ (Position 1 is chosen as the reference elevation), and $\langle v \rangle_1 \approx 0$. At Position 2, $p_2 = 1.0$ atm, $z_2 = 75$ ft, and the velocity $\langle v \rangle_2$ may be calculated from the volumetric flow rate and the cross-sectional area of the pipe. The density of water at 25°C is 62.25 lb_m/ft^3 (from *Perry's Handbook* [132]). The frictional term $F = F_{2,1}$ is assumed to be zero, as indicated in the problem statement. A table of unit-conversion factors is available at the link provided on the inside cover of this text.

$$Q = \left(\frac{6.0 \text{ gal}}{\text{min}}\right) \left(\frac{1 \text{ ft}^3}{7.4805 \text{ gal}}\right) \left(\frac{\text{min}}{60 \text{ s}}\right)$$

$$= 0.013368 \text{ ft}^3/\text{s} = \boxed{1.3 \times 10^{-2} \text{ ft}^3/\text{s}}$$

$$m = Q\rho = \frac{0.013368 \text{ ft}^3}{\text{s}} \left(\frac{62.25 \text{ lb}_m}{\text{ft}^3}\right)$$

$$= 0.83216 \text{ lb}_m/\text{s} = \boxed{8.3 \times 10^{-1} \text{ lb}_m/\text{s}}$$

$$\langle v \rangle_2 = \frac{Q}{\pi R^2} = \frac{0.013368 \text{ ft}^3}{\text{s}} \left(\frac{1}{\pi (1.0 \text{ in.})^2}\right) \frac{(12 \text{ in.})^2}{(1 \text{ ft})^2}$$

$$= 0.612744 \text{ ft/s} = \boxed{6.1 \times 10^{-1} \text{ ft/s}}$$

Note that significant figures should be considered when reporting values for $W_{s,by,21}$, Q, m, and $\langle v \rangle_2$ (e.g., $\langle v \rangle_2 = 6.1 \times 10^{-1}$ ft/s). However, when the numbers are needed to carry forward the calculation, the complete number (i.e., all digits; e.g., $\langle v \rangle_2 = 0.612744$ ft/s) should be used to minimize calculator or computer roundoff error (see the Glossary).

The average velocity of fluid in the 3-inch inner-diameter (ID) pipe may be calculated from the macroscopic mass balance:

$$\begin{array}{l} \text{Steady-state} \\ \text{macroscopic} \\ \text{mass balance} \end{array} \quad \begin{pmatrix} \text{mass flow} \\ \text{2-inch pipe} \end{pmatrix} = \begin{pmatrix} \text{mass flow} \\ \text{3-inch pipe} \end{pmatrix} \qquad (1.58)$$

$$\rho \langle v \rangle_2 \frac{\pi D_2^2}{4} = \rho \langle v \rangle_1 \frac{\pi D_1^2}{4}$$

$$\langle v \rangle_1 = \left(\frac{D_2}{D_1} \right)^2 \langle v \rangle_2 \qquad (1.59)$$

$$\langle v \rangle_1 = (0.612744 \text{ ft/s}) \left(\frac{2.0 \text{ in.}}{3.0 \text{ in.}} \right)^2 \qquad (1.60)$$

$$= 0.272331 \text{ ft/s} \qquad (1.61)$$

$$= \boxed{2.7 \times 10^{-1} \text{ ft/s}}$$

To choose α in the mechanical energy balance, we need to determine if the flow is laminar or turbulent. As discussed in Chapter 2 and derived in detail in Chapter 7, we can determine if the flow is laminar or turbulent based on a quantity known as the Reynolds number:

$$\begin{array}{c} \text{Reynolds number} \\ \text{(dimensionless flow rate,} \\ \text{ratio of inertial} \\ \text{to viscous forces)} \end{array} \quad \boxed{\text{Re} \equiv \frac{\rho \langle v \rangle D}{\mu}} \qquad (1.62)$$

where ρ is the fluid density, $\langle v \rangle$ is the fluid average velocity, D is the pipe diameter, and μ is the fluid viscosity. Viscosity is the property of a fluid that quantifies how easily it flows; we discuss viscosity from many angles in this text. From *Perry's Handbook* [132], we find that the viscosity of water at 25°C is 0.8937 centipoise (abbreviated cp), where 1 poise = 1 g/(cm · s). In American engineering units, the viscosity of water is 6.005×10^{-4} lb$_{\text{m}}$/(ft · s). The Reynolds number indicates whether the flow in the pipe is laminar (Re < 2,100) or turbulent (Re > 4,000).

$$\begin{array}{c} \text{Observed transition} \\ \text{from laminar flow to} \\ \text{turbulent flow in pipes} \\ \text{(see Chapters 2 and 7)} \end{array} \quad \boxed{\begin{array}{l} \text{laminar tube flow: } \text{Re} < 2,100 \\ \text{turbulent tube flow: } \text{Re} > 4,000 \end{array}} \qquad (1.63)$$

The Reynolds number indicates the ratio of inertial to viscous forces in the flow. The Reynolds number is discussed later in the text.

For the flow in our system, the Reynolds number depends on whether the flow is in the 2-inch or 3-inch pipe because average velocity and D differ for those two pipe sections. The Reynolds number is calculated as:

$$\text{Re}_{2\,in\,pipe} = \left.\frac{\rho\langle v\rangle D}{\mu}\right|_{2\,in\,pipe} = \frac{\left(\frac{62.25\ \text{lb}_\text{m}}{\text{ft}^3}\ \frac{0.612744\ \text{ft}}{\text{s}}\ \frac{2.0\ \text{in.}}{12\ \text{in./ft}}\right)}{(0.8937\ \text{cp})\left(\frac{6.7197\times10^{-4}\ \text{lb}_\text{m}}{\text{ft}\cdot\text{s}\cdot\text{cp}}\right)}$$

$$= 10{,}586 = 1.1 \times 10^4 > 4{,}000 \Rightarrow \text{turbulent}$$

$$\text{Re}_{3\,in\,pipe} = \left.\frac{\rho\langle v\rangle D}{\mu}\right|_{3\,in\,pipe} = \frac{\left(\frac{62.25\ \text{lb}_\text{m}}{\text{ft}^3}\ \frac{0.272331\ \text{ft}}{\text{s}}\ \frac{3.0\ \text{in.}}{12\ \text{in./ft}}\right)}{(0.8937\ \text{cp})\left(\frac{6.7197\times10^{-4}\ \text{lb}_\text{m}}{\text{ft}\cdot\text{s}\cdot\text{cp}}\right)}$$

$$= 7{,}057 = 7.1 \times 10^3 > 4{,}000 \Rightarrow \text{turbulent}$$

Note that the Reynolds number is dimensionless. From the values of Re, we conclude that the flow in both pipe sections is turbulent; therefore, $\alpha = 1$ for our calculations.

Now we assemble the mechanical energy balance and calculate the shaft work. Warning: It is always important to carefully consider the units in engineering calculations. Problems using the mechanical energy balance are particularly tricky because fundamentally different properties are being related (e.g., pressure, velocity, and work). The units of the American engineering system (both pounds-mass and pounds-force are used) cause initial confusion.[2] The best approach is to work carefully when using numbers and explicitly show all unit conversions. Note that there are 32.174 ft $\text{lb}_\text{m}/\text{s}^2$ per lb_f.

$$\frac{p_2 - p_1}{\rho} + \frac{\langle v\rangle_2^2 - \langle v\rangle_1^2}{2\alpha} + g(z_2 - z_1) + F_{2,1} = \frac{-W_{s,by,21}}{m} \tag{1.64}$$

$$\frac{\langle v\rangle_2^2}{2\alpha} + gz_2 = \frac{-W_{s,by,21}}{m} \tag{1.65}$$

$$\left[\frac{(0.612744\ \text{ft/s})^2}{2(1)} + \frac{32.174\ \text{ft}}{\text{s}^2}(75\ \text{ft})\right]\frac{\text{s}^2\cdot\text{lb}_\text{f}}{32.174\ \text{ft}\cdot\text{lb}_\text{m}} = \frac{-W_{s,by,21}}{0.83216\ \text{lb}_\text{m}/\text{s}} \tag{1.66}$$

$$-W_{s,by,21} = (5.83484 \times 10^{-3} + 75)\frac{\text{ft}\ \text{lb}_\text{f}}{\text{lb}_\text{m}}\left(0.83216\ \frac{\text{lb}_\text{m}}{\text{s}}\right) \tag{1.67}$$

$$= 62.417\ \text{ft}\cdot\text{lb}_\text{f}/\text{s}\left(\frac{1.341 \times 10^{-3}\ \text{hp}}{0.7376\ \text{ft}\cdot\text{lb}_\text{f}/\text{s}}\right)$$

$$-W_{s,by,21} = 0.1135\ \text{hp} = \boxed{1.1 \times 10^{-1}\ \text{hp}} = W_{s,pump} \tag{1.68}$$

---

[2] The unit conversion 32.174 ft $\text{lb}_\text{m}/(\text{s}\ \text{lb}_\text{f})$ is given the symbol g_c. For more on g_c, see the Glossary. In any equation from the literature with the symbol g_c included, the g_c can be omitted safely with no effect on the equation, provided that all units are reconciled with appropriate unit conversions.

The work done by the fluid $W_{s,by,21}$ is negative, which is correct because the fluid is not producing work but instead is experiencing the effects of work done on it by the pump. The work done by the pump, $W_{s,pump}$, is the negative of the work done by the fluid, $W_{s,pump} = -W_{s,by,21}$.

It is interesting that the kinetic-energy contribution (i.e., the velocity term) in this problem (5.8×10^{-3} ft lb$_f$/lb$_m$; Equation 1.67) is small compared to the potential energy contribution (i.e., the gravity term, 75 ft lb$_f$/lb$_m$).

Fluids are worked on by pumps, and performing the mechanical energy balance on the fluid yields a calculation of a negative amount of work done by the fluid. An example of the fluid doing positive work on a piece of machinery is water flowing through the turbine in a hydroelectric power plant. Following is an example on this topic.

EXAMPLE 1.7. *A tropical town is located next to a 40.0-m waterfall in a river that has a 1,000.0 m³/s average volumetric flow rate during the rainy season and an average flow rate of 300.0 m³/s during the dry season. What is the maximum amount of hydroelectric power that can be produced by this waterfall? If operating a laptop computer consumes approximately 30.0 W, estimate the number of computers that could be run by the waterfall.*

SOLUTION. Hydroelectric power is produced by channeling falling water through large turbines in a hydroelectric power plant (Figure 1.18). The spinning water vanes inside the turbines turn electromagnets through a wire coil and generate electricity through electromagnetic induction [167]. The turbine thereby creates usable electrical power from shaft work performed by the water ($W_{s,by} > 0$). A typical commercial hydroelectric plant produces between 1 and 1,300 megawatts (MW) of electrical power. The system of water flowing through the turbine is a single-input, single-output, steady flow of an incompressible fluid. There is no heat transfer and no chemical reaction or phase change; therefore, all requirements of the mechanical energy balance are met, and we can

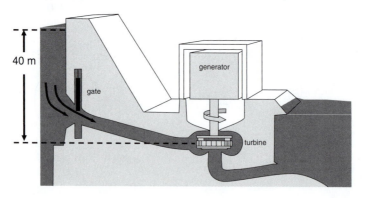

Figure 1.18 The potential energy of water at the top of a waterfall can be used to generate electricity by channeling gravity-driven flow through a turbine. The rushing water rotates the turbine, which in turn rotates an electromagnet through a coil, producing electricity by induction.

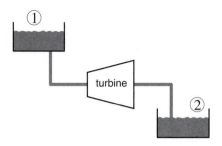

Figure 1.19 Schematic of the hydroelectric power plant analyzed in Example 1.7.

calculate the shaft work produced by water passing through the hydroelectric plant.

$$\frac{\Delta p}{\rho} + \frac{\Delta \langle v \rangle^2}{2\alpha} + g\Delta z + F = -\frac{W_{s,by}}{m}$$

Mechanical energy balance
(single-input, single-output,
steady, no phase change,   (1.69)
incompressible,
$\Delta T \approx 0$, no reaction)

The operations of a hydroelectric plant are illustrated in Figure 1.19. We choose Input Point 1 as the slow-moving water above the falls and Output Point 2 as the slow-moving water below the falls. Our choice is driven, as usual, by our ability to evaluate terms in the mechanical energy balance at these locations. At both points, the pressure is atmospheric and the velocity is negligible. Thus, the MEB becomes:

$$\frac{p_2 - p_1}{\rho} + \frac{\langle v \rangle_2^2 - \langle v \rangle_1^2}{2\alpha} + g(z_2 - z_1) + F_{2,1} = -\frac{W_{s,by,21}}{m} \qquad (1.70)$$

$$g(z_2 - z_1) + F_{2,1} = -\frac{W_{s,by,21}}{m} \qquad (1.71)$$

The shaft work is work done by the water on the turbine, and this is the quantity that we seek to calculate (i.e., $W_{s,by,21} > 0$ for fluid in a turbine). The term $F_{2,1}$ is the friction between Points 2 and 1 and includes the friction in the turbine and the frictional losses associated with the flow before and just after the turbine.

Reflecting on the likely flow pattern in the hydroelectric plant, we surmise that the largest frictional loss in the system is inside the turbine. We therefore split the 2,1-system friction $F_{2,1}$ into the friction in the turbine and all other losses. Then, as a first calculation, we neglect all of the frictional losses outside of the turbine:

$$g(z_2 - z_1) + F_{2,1} = -\frac{W_{s,by,21}}{m} \qquad (1.72)$$

$$g(z_2 - z_1) + F_{turbine} + F_{other} = -\frac{W_{s,by,21}}{m} \qquad (1.73)$$

Neglecting the losses other than those in the turbine and grouping the turbine losses with the turbine shaft work, we arrive at an expression for work performed by the fluid on the turbine:

$$g(z_2 - z_1) + F_{turbine} = -\frac{W_{s,by,21}}{m} \tag{1.74}$$

$$\frac{W_{s,by,21}}{m} + F_{turbine} = g(z_1 - z_2) > 0 \tag{1.75}$$

We define η as the turbine efficiency. The turbine efficiency reflects the fraction of the fluid energy delivered to the turbine that is actually extracted as shaft work, omitting energy being dissipated as frictional losses:

$$\text{Turbine efficiency} \quad \eta = \frac{\left(\dfrac{\text{useful energy}}{\text{mass fluid}}\right)}{\left(\dfrac{\text{total fluid energy}}{\text{mass fluid}}\right)} \tag{1.76}$$

$$\eta \equiv \frac{\dfrac{W_{s,by,21}}{m}}{\left(\dfrac{W_{s,by,21}}{m} + F_{turbine}\right)} \tag{1.77}$$

Substituting turbine efficiency η from Equation 1.77 into Equation 1.75, we obtain:

$$\frac{W_{s,by,21}}{m} + F_{turbine} = g(z_1 - z_2) \tag{1.78}$$

$$\frac{1}{\eta} \frac{W_{s,by,21}}{m} = g(z_1 - z_2) \tag{1.79}$$

For the 40-m waterfall under consideration and for a turbine that is 80 percent efficient (an estimate derived from a literature search), we calculate the amount of electricity that can be generated:

$$\frac{1}{\eta} \frac{W_{s,by,21}}{m} = g(z_1 - z_2) \tag{1.80}$$

$$\frac{W_{s,by,21}}{m} = \eta g(z_1 - z_2) \tag{1.81}$$

$$= (0.80)(9.8066 \text{ m/s}^2)(40 \text{ m}) = \left(313.8112 \frac{\text{m}^2}{\text{s}^2}\right)\left(\frac{1 \text{ N}}{\text{kg m/s}^2}\right)$$

$$\frac{W_{s,by,21}}{m} = 313.8112 \frac{J}{kg} = \boxed{310 \frac{J}{kg}}$$

We now calculate the power produced under low-flow-rate (300 m³/s) and high-flow-rate (1,000 m³/s) conditions. The density we need comes from the literature (see *Perry's Handbook* [132]):

$$\begin{matrix} \text{Power generated} \\ \text{by turbine} \end{matrix} = W_{s,by,21} \qquad (1.82)$$

$$= \left(\frac{W_{s,by,21}}{m}\right)(\text{m}) = \left(\frac{W_{s,by,21}}{m}\right)(\rho Q) \qquad (1.83)$$

$$\begin{matrix} \text{Low-flow} \\ \text{power:} \end{matrix} \quad W_{s,by,21} = \left(313.8112 \,\frac{J}{kg}\right)\left(997.08 \,\frac{kg}{m^3}\right)\left(300 \,\frac{m^3}{s}\right)\left(\frac{W}{J/s}\right)$$

$$\boxed{W_{s,by,21}(\text{low flow}) = 9.4 \times 10^7 \ W = 94 \ \text{MW}}$$

$$\begin{matrix} \text{High-flow} \\ \text{power:} \end{matrix} \quad W_{s,by,21} = \left(313.8112 \,\frac{J}{kg}\right)\left(997.08 \,\frac{kg}{m^3}\right)\left(1,000 \,\frac{m^3}{s}\right)\left(\frac{W}{J/s}\right)$$

$$\boxed{W_{s,by,21}(\text{high flow}) = 3.1 \times 10^8 \ W = 310 \ \text{MW}}$$

The actual amount of electricity generated by the turbine is less than either of these results because there are frictional losses other than those in the turbine. Turbine efficiency is a number that must be measured (by independently measuring the amount of electrical power produced by the turbine; see Chapter 9). Manufacturers of turbines supply experimental data on their products' efficiencies.

For computers consuming power at a rate of 30 W, our calculations indicate that 3 million computers could be powered during the dry season and 10 million during the rainy season. Hydroelectric power is an economic and renewable resource because the rainfall cycle replenishes the upstream water supply. Care must be taken in designing and operating hydroelectric power plants, however, because fish and other species in a river are disrupted by the diversion of water through the turbine.

For fluid systems with rotating machinery such as pumps and turbines, the mechanical energy balance is an essential tool in quantifying shaft work. To apply the MEB to problems, we must learn to be strategic in choosing inlet (Point 1) and outlet (Point 2) points. As in the previous examples, the free surface of an open tank is a good location to choose as an inlet or outlet because we know the pressure (i.e., atmospheric) and the velocity (i.e., approximately zero). Another location about which we know much is the discharge point of a pipe ($\langle v \rangle = Q/\pi R^2$, $p = p_{atm}$). When we seek information about the operating capabilities of a device, the points immediately before and after the device are appropriate choices for the mechanical energy balance. Remember that the steady-state macroscopic mass balance (i.e., mass in = mass out) provides an essential relationship between flows at different points in an apparatus.

We turn now to using the mechanical energy balance in systems in which friction is important.

1.2.3 MEB with friction

The friction term often makes an important contribution to the mechanical energy balance (MEB). In piping systems, this is true when there are changes in pipe diameter, twists and turns in the pipe, flow obstructions such as an orifice plate, or when there are long runs of piping. When friction is important, the F term in the MEB must be determined experimentally—just as the friction coefficient C_V for Venturi meters and turbine efficiency η are determined experimentally, as discussed previously. To quantify the friction, we first apply the MEB to the system to determine which measurable quantities are of interest; we subsequently conduct experiments to obtain those quantities. In practice, the experiments already have been performed for common devices, and we use published experimental results to calculate F. The study of friction begins by considering frictional losses in the steady flow in a long, straight run of horizontal pipe.

EXAMPLE 1.8. *For household water in steady flow in a 1/2-inch Schedule 40 horizontal pipe at 3.0 gpm (Figure 1.20), what are the frictional losses over a 100.0-foot run of pipe? The flow may be laminar or turbulent.*

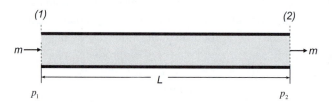

Figure 1.20 A mechanical energy balance on a pipe section yields the expression for the frictional losses in a straight pipe.

SOLUTION. The system of water flowing in a tube is a single-input, single-output, steady flow of an incompressible fluid. There is no heat transfer and no chemical reaction or phase change. Therefore, all requirements of the mechanical energy balance are met.

Mechanical energy balance (single-input, single-output, steady, no phase change, incompressible, $\Delta T \approx 0$, no reaction)

$$\frac{\Delta p}{\rho} + \frac{\Delta \langle v \rangle^2}{2\alpha} + g\Delta z + F = -\frac{W_{s,by}}{m} \qquad (1.84)$$

We choose as our two points (1) a point upstream where the pressure p_1 is measured and (2) a point downstream where the pressure p_2 is measured. There are no pumps or moving parts in the chosen system, which means $W_{s,by} = 0$. The pipe has a constant flow rate and a constant cross-sectional area; therefore, from the mass balance, $\rho A \langle v \rangle_2 - \rho A \langle v \rangle_1 = 0$; and, therefore, $\langle v \rangle$ does not change between Points 1 and 2. The pipe is horizontal; therefore, $z_2 - z_1 = 0$. The MEB becomes:

$$\frac{p_2 - p_1}{\rho} + \frac{\langle v \rangle_2^2 - \langle v \rangle_1^2}{2\alpha} + g(z_2 - z_1) + F_{2,1} = -\frac{W_{s,by,21}}{m} \qquad (1.85)$$

$$\frac{p_2 - p_1}{\rho} + F_{2,1} = 0$$

The frictional term is found to be:

$$\boxed{F_{\text{straight pipe}} = F_{2,1} = \frac{p_1 - p_2}{\rho}} \qquad \begin{array}{c} \text{Friction} \\ \text{in steady flow} \\ \text{in pipes} \end{array} \qquad (1.86)$$

Thus, to characterize friction in straight pipes, data can be obtained about pressure drop for a variety of flow rates and tube geometries (e.g., length and diameter) and for a variety of fluids (e.g., with different densities ρ and viscosities μ), and the data can be tabulated and published. The published data are then used to calculate frictional losses in future MEB analyses of straight lengths of pipe.

$$\text{Data needed:} \qquad \boxed{\Delta p(Q) \text{ for various } \rho, \mu, D, L} \qquad (1.87)$$

With these data, we can use Equation 1.86 to calculate frictional losses in the pipe. The needed data correlations for $\Delta p(Q)$ are discussed next.

As discussed previously, to determine F for flows in pipes, we need data on pressure drop as a function of velocity or volumetric flow rate, $\Delta p(Q)$. The problem of pressure drop as a function of flow rate in pipe flow has been studied in depth and, with the help of momentum balance, it largely has been solved (discussed in subsequent chapters). To keep moving forward with this quick-start section, we summarize the practical results of the analyses. These equations are derived in Chapter 7.

From an in-depth analysis of pipe flow, we find that a useful defined quantity in pipe flow is the Fanning friction factor, f, which is a dimensionless wall force that may be used to correlate friction in pipe flows with the Reynolds number (i.e., the dimensionless flow rate introduced in Equation 1.62). The Fanning friction factor f is defined as:

$$\begin{array}{c} \text{Fanning friction factor} \\ \text{(dimensionless} \\ \text{fluid force on pipe wall)} \end{array} \qquad \boxed{f \equiv \frac{\text{Wall force}}{(\text{area})(\text{kinetic energy})}} \qquad (1.88)$$

$$f = \frac{\mathcal{F}_{\text{drag}}}{(2\pi R L)\left[\frac{1}{2}\rho \langle v \rangle^2\right]} \qquad (1.89)$$

where R is pipe radius, ρ is fluid density, $\langle v \rangle$ is average fluid velocity, and L is the length of the pipe. For flows in straight pipes, $\mathcal{F}_{\text{drag}}$ may be shown as given by $\mathcal{F}_{\text{drag}} = (p_1 - p_2)\pi R^2$ (see Chapter 9); thus, from Equation 1.89, the Fanning friction factor for straight pipes is:

$$\begin{array}{c}\text{Wall drag}\\\text{in straight pipes}\\(\text{see Equation 9.236})\end{array} \qquad \mathcal{F}_{\text{drag}} = (p_1 - p_2)\pi R^2 \qquad (1.90)$$

$$\boxed{f = \frac{(p_1 - p_2)D}{2L\rho\langle v \rangle^2}} \qquad \begin{array}{c}\text{Fanning friction factor}\\\text{in terms of}\\\text{experimental variables}\\(\text{straight pipes})\end{array} \qquad (1.91)$$

where $D = 2R$ is the pipe inner diameter.

The methods used in this text allow us to show that for steady flow of any Newtonian fluid in any smooth tube, the Fanning friction factor is a function of only the Reynolds number (see Chapter 7).

$$\boxed{f = f(\text{Re}) \text{ only}; \ \text{Re} \equiv \frac{\rho\langle v \rangle D}{\mu}} \qquad \begin{array}{c}\text{Dimensional analysis result}\\\text{for Fanning friction factor}\\\text{in pipe flow}\\(\text{see Chapter 7})\end{array} \qquad (1.92)$$

This powerful result simplifies data reporting for frictional losses in pipes. The literature for flow in tubes reports a single plot of $f(\text{Re})$, which determines f for smooth pipes of all sizes, in all flow regimes, and for normal fluids of all densities and viscosities. Once we have f for our flow, we can use Equations 1.91 and 1.86 to obtain friction loss for straight pipes.

$$\boxed{F_{\text{straight pipe}} = \frac{p_1 - p_2}{\rho} = \frac{2fL}{D}\langle v \rangle^2} \qquad \begin{array}{c}\text{MEB friction term}\\\text{in steady flow in a}\\\text{straight section of pipe}\end{array} \qquad (1.93)$$

Alternatively, the friction loss in straight pipes may be expressed in terms of head loss, h_f:

$$\text{Head loss} \left(\frac{\text{energy}}{\text{unit weight}}\right): \quad h_f \equiv \frac{F_{\text{straight pipe}}}{g} = \frac{2fL}{gD}\langle v \rangle^2$$

This equation is known as the Darcy-Weisbach equation [178].

This discussion illustrates the power of the analytical methods that we are studying. From general considerations (see Chapter 7), we can deduce simple equations that allow us to make practical calculations of pressure drop and shaft work in flows (see Equations 1.84 and 1.86). Engineers who master the fluid-mechanics methods in this text have a distinct advantage in designing, optimizing, and inventing devices that employ fluids.

The data correlations for f are well established; they are developed in Chapter 7. For laminar flow, we can use direct theoretical calculations to determine f

Table 1.3. Surface roughness for various materials

Material	ε (mm)
Drawn tubing (e.g., brass, lead, glass)	1.5×10^{-3}
Commercial steel or wrought iron	0.05
Asphalted cast iron	0.12
Galvanized iron	0.15
Cast iron	0.46
Wood stave	0.2–0.9
Concrete	0.3–3.0
Riveted steel	0.9–9.0

Source: Perry's Handbook [132]

as a function of the Reynolds number.

$$f_{\text{laminar flow}} = \frac{16}{\text{Re}}$$

Fanning friction factor
in steady laminar flow in pipes
(analytical result;
see Equation 7.155) (1.94)

Experiments show that laminar flow takes place in straight pipes with a circular cross section for $\text{Re} < 2,100$ and that a fully turbulent flow occurs for $\text{Re} > 4,000$. Between $\text{Re} = 2,100$ and $\text{Re} = 4,000$, the flow is called transitional flow, which is neither stable laminar flow nor fully turbulent flow. Operating devices in the transitional-flow regime generally is avoided for stability reasons.

For turbulent flow, the correlations of friction factor as a function of the Reynolds number cannot be obtained analytically but have been found through careful experiments (see Chapter 7). A useful empirical equation that fits the data for turbulent flow is the Colebrook formula [43], which gives f as a function of the Reynolds number and ε, a surface roughness parameter relevant for commercial pipe.

$$\frac{1}{\sqrt{f}} = -4.0 \log \left(\frac{\varepsilon}{D} + \frac{4.67}{\text{Re}\sqrt{f}} \right) + 2.28$$

Colebrook formula
Fanning friction factor
in steady turbulent (1.95)
flow in pipes
(See equation 7.161)

where D is pipe diameter. Values of ε for various materials are listed in Table 1.3 and the Colebrook correlation is graphed in Figure 1.21 on a log-log plot. Because the friction factor appears twice in Equation 1.95, the Colebrook equation requires an iterative solution. For smooth pipes, an explicit correlation that works for all Reynolds numbers is given in Equation 7.158.

In the following example, we use the data correlations in Equations 1.94 and 1.95 to predict losses in long straight pipes. The household pipe-flow problem in Example 1.8 also can be solved now that we have $f(\text{Re})$ (see Problem 8).

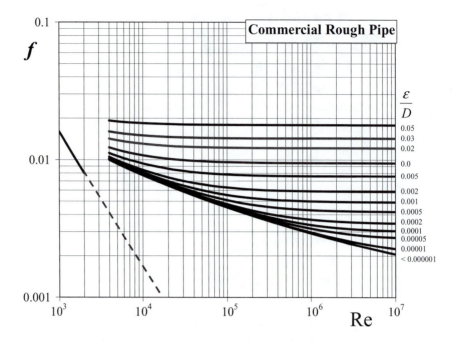

Figure 1.21 Fanning friction factor versus Reynolds number from the Colebrook formula; see Equation 1.95. For Re < 2,100, $f = 16/\text{Re}$, which on a log-log graph is a line of slope -1. This is the Moody plot [103].

EXAMPLE 1.9. *An oil pipeline carries crude oil from Northern Alaska to the year-round port in Valdez, Alaska, for shipment to refineries for processing. For one horizontal section of straight pipe that is 10.0 miles long (16.1 km) and 4.0 feet (1.22 m) ID, what is the pressure drop for oil traveling at 7.0×10^5 barrels per day (42 U.S. gallons per barrel) (Figure 1.22)? The pipe walls may be assumed to be smooth ($\varepsilon = 0$). The kinematic viscosity (i.e., ratio of viscosity to density) of the crude oil at the flow temperature is $\mu/\rho = v = 7.0$ centistokes (1 stoke = 1 cm^2/s) and the density of the crude oil is 800.0 kg/m^3.*

SOLUTION. The system of oil in the pipeline is a single-input, single-output, steady flow of an incompressible fluid. There is no heat transfer and no chemical reaction or phase change. Therefore, all requirements of the mechanical energy balance are met.

$$\frac{\Delta p}{\rho} + \frac{\Delta \langle v \rangle^2}{2\alpha} + g\Delta z + F = -\frac{W_{s,by}}{m}$$

Mechanical energy balance
(single-input, single-output,
steady, no phase change,   (1.96)
incompressible,
$\Delta T \approx 0$, no reaction)

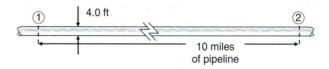

① ↓ 4.0 ft   ②

10 miles
of pipeline

Figure 1.22 A long pipeline generates frictional losses that cannot be ignored in the design of a pumping system.

We choose Points 1 and 2 to be two points separated by 10 miles of straight horizontal pipe (see Figure 1.22). There are no moving parts in the chosen system and therefore no shaft work. The pipe is horizontal ($\Delta z = 0$) and, because the pipe cross section is constant, there is no change in velocity from one end to the other ($\Delta \langle v \rangle^2 = 0$). The mechanical energy balance simplifies to:

$$\frac{p_2 - p_1}{\rho} + \frac{\langle v \rangle_2^2 - \langle v \rangle_1^2}{2\alpha} + g(z_2 - z_1) + F_{2,1} = -\frac{W_{s,by,21}}{m} \qquad (1.97)$$

$$\frac{p_2 - p_1}{\rho} + F_{2,1} = 0 \qquad (1.98)$$

$$p_1 - p_2 = \rho F_{2,1} \qquad (1.99)$$

To calculate the pressure drop $p_1 - p_2$, we need $F_{2,1}$ friction loss in straight pipe. As discussed previously in this section, we can calculate $F_{2,1}$ for straight pipes using Equation 1.93:

$$\text{Friction in straight pipe} \quad F_{2,1} = \frac{2fL}{D} \langle v \rangle^2 \qquad (1.100)$$

where f is the Fanning friction factor. We know the pipe ID (4.0 feet) and the pipe length (10 miles); we can calculate the average velocity from the volumetric flow rate (10^6 barrels/day [bpd]), and we obtain f from Re, ε, and the Colebrook equation (see Equation 1.95). We show the calculation here; as always when performing MEB calculations, we must be mindful of the unit conversions.

$$\text{Flow rate:} \quad Q = \left(\frac{700,000 \text{ barrels}}{\text{day}}\right)\left(\frac{42 \text{ US gal}}{\text{barrel}}\right)\left(\frac{\text{day}}{24 \text{ h}}\right)\left(\frac{\text{h}}{60 \text{ min}}\right)$$

$$= 20,416 \text{ gal/min} \left(\frac{\text{m}^3/\text{s}}{15,850 \text{ gpm}}\right)$$

$$= 1.288 \text{ m}^3/\text{s} = \boxed{1.3 \text{ m}^3/\text{s}} \qquad (1.101)$$

$$\text{Average velocity:} \quad \langle v \rangle = \frac{4Q}{\pi D^2}$$

$$= \frac{4(1.288 \text{ m}^3/\text{s})}{\pi \left((4.0 \text{ ft})(0.3048 \frac{\text{m}}{\text{ft}})\right)^2}$$

$$= 1.10336 \text{ m/s} = \boxed{1.1 \text{ m/s}} \qquad (1.102)$$

Reynolds number:

$$\text{Re} = \frac{\rho \langle v \rangle D}{\mu} = \frac{\langle v \rangle D}{\mu / \rho} \tag{1.103}$$

$$= \frac{(1.10336 \text{ m/s}) \left((4.0 \text{ ft})(0.3048 \text{ m/ft}) \right)}{0.070 \text{ cm}^2/\text{s} \left(\frac{\text{m}^2}{10{,}000 \text{ cm}^2} \right)}$$

$$= 192{,}174 = \boxed{190{,}000} \Rightarrow \text{turbulent} \tag{1.104}$$

Colebrook correlation (smooth pipe, $\varepsilon = 0$)

$$\frac{1}{\sqrt{f}} = -4.0 \log \left(\frac{4.67}{\text{Re}\sqrt{f}} \right) + 2.28 \tag{1.105}$$

To solve Equation 1.105 for f, we consult Figure 1.21 for $\text{Re} = 190{,}000$ to estimate a first guess of $f_{(1)} = 0.004$ and use that in an iterative solution:

$$\frac{1}{\sqrt{f_{(2)}}} = -4.0 \log \left(\frac{4.67}{\text{Re}\sqrt{f_{(1)}}} \right) + 2.28 \tag{1.106}$$

$$\frac{1}{\sqrt{f_{(2)}}} = -4.0 \log \left(\frac{4.67}{192{,}174\sqrt{0.004}} \right) + 2.28 \tag{1.107}$$

$$f_{(2)} = 0.003935 \tag{1.108}$$

Substituting this next guess into the righthand side of Equation 1.109, we iterate as shown here until the final solution is found. We stop our calculations when there is no change within the accuracy of the calculation:

$$\frac{1}{\sqrt{f_{(n)}}} = -4.0 \log \left(\frac{4.67}{\text{Re}\sqrt{f_{(n-1)}}} \right) + 2.28 \tag{1.109}$$

$$f_{(1)} = 0.004$$

$$f_{(2)} = 0.003935$$

$$f_{(3)} = 0.003942$$

Final result: $$\boxed{f = 0.0039} \tag{1.110}$$

Substituting the appropriate values into Equation 1.100, we now calculate the friction and then the pressure drop $p_1 - p_2$ in the oil pipeline. To reduce the impact of roundoff error in intermediate calculations, we use all of the digits we have for f, $\langle v \rangle$, L, and D and all unit conversions.

$$F_{2,1} = \frac{2fL}{D}\langle v \rangle^2 \tag{1.111}$$

$$= \frac{(2)(0.003941)(10 \text{ miles})(1609.344 \frac{m}{mile})(1.10336 \frac{m}{s})^2}{(4.0 \text{ ft})(0.3048 \text{ m/ft})}\left(\frac{1 \text{ N}}{\text{kg m/s}^2}\right) \tag{1.112}$$

$$= 126.657 \frac{\text{N m}}{\text{kg}} = \boxed{130 \text{ J/kg}} \tag{1.113}$$

$$p_1 - p_2 = \rho F_{2,1} \tag{1.114}$$

$$= \left(\frac{800 \text{ kg}}{\text{m}^3}\right)\left(\frac{126.657 \text{ N m}}{\text{kg}}\right)\left(\frac{\text{Pa}}{\text{N/m}^2}\right) \tag{1.115}$$

$$= 101,326 \text{ Pa} \tag{1.116}$$

$$= \boxed{100 \text{ kPa}} = 1.0 \text{ atm} = 15 \text{ psi} \tag{1.117}$$

As introduced previously, we often report friction results in terms of *head loss*, which is defined as $h_f = F_{2,1}/g$—that is, energy per unit fluid weight (see Equation 1.93 and Section 9.2.2). The units of head loss are feet or meters. In units of head loss, the friction result is:

$$\text{Head loss:} \quad h_f = \frac{F_{2,1}}{g} = \frac{126.657 \frac{\text{N m}}{\text{kg}}}{9.8066 \text{ m/s}^2} \tag{1.118}$$

$$= \boxed{13 \text{ m} = 42 \text{ ft}} \tag{1.119}$$

The frictional loss in the 10-mile pipe is the equivalent of the energy per unit weight that a pump needs to expend in order to raise the fluid 13 m (see Chapter 9).

In addition to wall drag in straight pipes, many other sources of friction exist in piping systems: valves, fittings, pumps, expansions, and contractions (Figure 1.23). To quantify the amount of fluid friction generated in these devices as a function of fluid velocity, we use the same procedure as for deducing the result for straight pipes: We apply the mechanical energy balance to the valve, fitting, or other friction-generating segment of the piping system; we simplify the resulting equation by using mass and momentum balances as appropriate; and we conduct experiments to find any needed data correlations. For valves, fittings, expansions, and contractions, the data correlations that result from such analyses (for derivation, see the steps leading to Equation 9.318) may be written in the

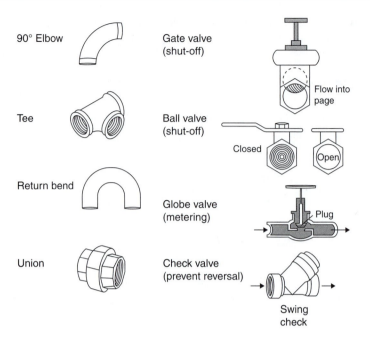

90° Elbow

Tee

Return bend

Union

Gate valve
(shut-off)

Flow into
page

Ball valve
(shut-off)

Closed   Open

Globe valve
(metering)   Plug

Check valve
(prevent reversal)

Swing
check

Figure 1.23   Sketches of common pipe fittings and valves. Ball valves and gate valves are two-position valves—open and closed—and are designed for minimum frictional loss during continuous flow. Globe valves are designed to vary the flow through the valve (i.e., metering valves). The ability to meter the flow, however, introduces frictional losses as the flow moves around the obstruction of the valve's moving parts [132]. The design differences in the valves are reflected in the frictional-loss coefficients.

following form:

$$F_{\text{fitting}} = K_f \frac{\langle v \rangle^2}{2} \qquad \begin{array}{l}\text{Friction from}\\ \text{fittings}\end{array} \qquad (1.120)$$

where K_f is friction coefficient for the valve or fitting. The empirical friction coefficients K_f are different for each type of valve or fitting, and they are different for laminar and turbulent flows. Values of K_f are listed in Tables 1.4 and 1.5.

The values of K_f for expansions and contractions are listed in Tables 1.4 and 1.5 and as follows for both laminar ($\alpha = 0.5$) and turbulent ($\alpha = 1$) flows:

$$\text{Expansion loss} \qquad K_{exp} = \frac{1}{\alpha}\left(1 - \frac{A_1}{A_2}\right)^2 \qquad (1.121)$$

$$\text{Contraction loss} \qquad K_{cont} = \frac{0.55}{\alpha}\left(1 - \frac{A_2}{A_1}\right) \qquad (1.122)$$

where A_1 is the upstream cross-sectional area and A_2 is the downstream cross-sectional area. These expressions are derived in Chapter 9. The $\langle v \rangle$ to be used in Equations 1.121 and 1.122 for expansions and contractions, by convention, is the faster average velocity (i.e., the upstream velocity for an expansion and the downstream velocity for a contraction). Frictional coefficients in the literature sometimes also are given in terms of equivalent pipe lengths [132].

Table 1.4. Published friction-loss factors for turbulent flow
through valves, fittings, expansions, and contractions

Fitting	Friction-loss factor, K_f
Standard elbow, 45°	0.35
Standard elbow, 90°	0.75
Tee used as ell	1.0
Tee, branch blanked off	0.4
Return bend	1.5
Coupling	0.04
Union	0.04
Gate valve, wide open	0.17
Gate valve, half open	4.5
Globe valve, bevel seat, wide open	6.0
Globe valve, bevel seat, half open	9.5
Check valve, ball	70.0
Check valve, swing	2.0
Water meter, disk	7.0
Expansion from A_1 to A_2	$\left(1 - \dfrac{A_1}{A_2}\right)^2$
Contraction from A_1 to A_2	$0.55\left(1 - \dfrac{A_2}{A_1}\right)$

Source: Perry's Handbook [132]

Table 1.5. Friction-loss factors K_f for laminar flow through selected valves, fittings, expansions
and contractions

	K_f					
Fitting	$Re_i = 50$	100	200	400	1,000	Turbulent
Elbow, 90°	17	7	2.5	1.2	0.85	0.75
Tee	9	4.8	3.0	2.0	1.4	1.0
Globe valve	28	22	17	14	10	6.0
Check valve, swing	55	17	9	5.8	3.2	2.0
Expansion from A_1 to A_2	$2\left(1 - \dfrac{A_1}{A_2}\right)^2$					$\left(1 - \dfrac{A_1}{A_2}\right)^2$
Contraction from A_1 to A_2	$\dfrac{0.55}{0.5}\left(1 - \dfrac{A_2}{A_1}\right)$					$0.55\left(1 - \dfrac{A_2}{A_1}\right)$

Source: Perry's Handbook [132]

Table 1.6. Calculating piping friction from published correlations

1. Count and identify valves, bends, and couplings in the system.
2. Find the published friction coefficients K_f in the literature (e.g., Tables 1.4 and 1.5).
3. Measure lengths of all straight-pipe segments and total them (separate different diameters).
4. Calculate f for each pipe section using f (Re) (i.e., Colebrook formula or another correlation).
5. Calculate friction F_{piping} from Equation 1.124. Use F_{piping} as needed in the mechanical energy balance to calculate quantities of interest.

The friction for a complete piping system is equal to the friction caused by the straight-pipe sections (see Equation 1.93) plus the friction caused by each of the valves, fittings, expansions, and contractions present in the flow loop (see Equation 1.120).

$$F_{piping} = \sum \left(\begin{array}{c} \text{friction of} \\ \text{straight-pipe sections} \end{array} \right) + \sum \left(\begin{array}{c} \text{friction of} \\ \text{fittings and valves} \end{array} \right) \quad (1.123)$$

Friction in a piping system

$$F_{piping} = \sum_{\substack{j,\, straight \\ pipe \\ segments}} \left[4 f_j \frac{L_j}{D_j} \frac{\langle v \rangle_j^2}{2} \right] + \sum_{i,\, fittings} \left[n_i K_{f,i} \frac{\langle v \rangle_i^2}{2} \right] \quad (1.124)$$

where n_i is the number of each type of fitting or valve. Note that in the correlations there are different values of K_f depending on whether the flow is laminar or turbulent. Also, the $\langle v \rangle_j$ used in the summation over the straight-pipe segments is the average velocity in the straight pipe, which is different for different values of D_j.

With the development of Equation 1.124 for the friction term in piping systems, we now are ready to calculate a mechanical energy balance with friction. The procedure for using published correlations to calculate the friction term for piping systems is outlined in Table 1.6. The following example uses this procedure.

EXAMPLE 1.10. *What is the work required to pump 6.0 gpm of water in the piping network shown in Figure 1.17? Do not neglect the effect of friction. The piping may be considered to be smooth pipe.*

SOLUTION. We previously solved this problem without friction. Now we perform the same calculation with the addition of the frictional contribution $F_{2,1}$. We begin with the mechanical energy balance (Equation 1.57):

$$\frac{\Delta p}{\rho} + \frac{\Delta \langle v \rangle^2}{2\alpha} + g \Delta z + F = -\frac{W_{s,by}}{m}$$

Mechanical energy balance
(single-input, single-output,
steady, no phase change,   (1.125)
incompressible,
$\Delta T \approx 0$, no reaction)

As before, we choose Point 2 as the exit of the pipe and Point 1 as the free surface of the tank.

$$\frac{p_2 - p_1}{\rho} + \frac{\langle v \rangle_2^2 - \langle v \rangle_1^2}{2\alpha} + g(z_2 - z_1) + F_{2,1} = -\frac{W_{s,by,21}}{m} \qquad (1.126)$$

In the previous example, we obtained Equation 1.65 for the current system without friction; going back one step in the previous solution to Equation 1.64, the mechanical energy balance for this problem with friction included is:

$$\frac{\langle v \rangle_2^2}{2} + g z_2 + F_{2,1} = \frac{-W_{s,by,21}}{m} \qquad (1.127)$$

Substituting values for this problem, we obtain:

$$\left[\frac{(0.612744 \text{ ft/s})^2}{2(1)} + \frac{32.174 \text{ ft}}{s^2}(75 \text{ ft}) + F_{2,1} \right] \frac{s^2 \cdot \text{lb}_f}{32.174 \text{ ft} \cdot \text{lb}_m} = \frac{-W_{s,by,21}}{0.83216 \text{ lb}_m/s}$$

$$(1.128)$$

To make the units consistent on both sides of the equation, we converted the lefthand units (ft^2/s^2) to $\text{ft} \cdot \text{lb}_f/\text{lb}_m$. To calculate $F_{2,1}$, we use Equation 1.124:

$$F_{2,1} = F_{\text{piping}} = \sum_{\substack{j, \text{ straight} \\ \text{pipe} \\ \text{segments}}} \left[4 f_j \frac{L_j}{D_j} \frac{\langle v \rangle_j^2}{2} \right] + \sum_{i, \text{fittings}} \left[K_{f,i} \frac{\langle v \rangle_i^2}{2} \right]$$

We have two types of straight-pipe segments: one that is 50 feet long with an ID of 3.0 inches, and one that is a total of $40 + 8 + 75 + 20 = 143$ feet long with an ID of 2.0 inches. The average velocities in the pipes were calculated in the previous example to be as follows (all digits included):

$$\langle v \rangle_{2in \text{ } pipe} = 0.612744 \text{ ft/s}$$

$$\langle v \rangle_{3in \text{ } pipe} = 0.272331 \text{ ft/s}$$

We retain all digits because this is an intermediate calculation.

The Fanning friction factors f for each of the two types of straight-pipe segments are different. The Fanning friction factor is a function of the Reynolds number, which depends on $\langle v \rangle$. The friction factor may be obtained from the appropriate correlations—that is, $f = 16/\text{Re}$ (see Equation 1.94) for laminar flow and the Colebrook formula (see Equation 1.95) for turbulent flow. We previously calculated the Reynolds numbers for the two pipe sizes:

$$\text{Re}_{2in \text{ } pipe} = \left. \frac{\rho \langle v \rangle D}{\mu} \right|_{2in \text{ } pipe} = 10,586 = 1.1 \times 10^4 \qquad (1.129)$$

$$\text{Re}_{3in \text{ } pipe} = \left. \frac{\rho \langle v \rangle D}{\mu} \right|_{3in \text{ } pipe} = 7,077 = 7.1 \times 10^3 \qquad (1.130)$$

and the flow is everywhere turbulent (Re > 4,000). The Fanning friction factors are found from an iterative solution of the Colebrook formula (see the technique in Example 1.9), and the results are $f = 0.007603$ for the 2-inch pipe and

$f = 0.00848$ for the 3-inch pipe. Again, we retain extra digits for these intermediate calculations to avoid roundoff error in subsequent calculations.

The fittings for our flow loop are two 90° elbows and two contractions—one from the tank to the inlet of the 3-inch pipe and one immediately upstream of the pump. For the contraction from the tank to the 3-inch pipe, the velocity is the same as in the 3-inch pipe (i.e., the larger velocity). For the contraction to 2 inches and for the two elbows, the velocity is the same as in the 2-inch pipe. For the fittings in our system, the friction-loss factors K_f obtained from Table 1.4 are listed here:

Fitting	K_f
Contraction (tank to 3-inch pipe, $A_1/A_2 = \infty$)	0.55
Contraction (3 inches to 2 inches), $A_2/A_1 = 4/9$	0.305556
90° elbow	0.75

We now calculate the friction contribution to the mechanical energy balance for this system:

$$
F_{(2,1)} = \left[\sum_{\substack{j,\,straight \\ pipe \\ segments}} 4f_j \frac{L_j}{D_j} \frac{\langle v \rangle_j^2}{2} \right] + \left[\sum_{i,\,fittings} K_{f,i} \frac{\langle v \rangle_i^2}{2} \right]
$$

$$
= \left[(4)(0.00848) \left(\frac{50 \text{ ft}}{3.0 \text{ in}} \frac{12 \text{ in}}{\text{ft}} \right) \frac{(0.272331 \text{ ft/s})^2}{2} \right.
$$

$$
+ (4)(0.007603) \left(\frac{143 \text{ ft}}{2.0 \text{ in}} \frac{12 \text{ in}}{\text{ft}} \right) \frac{(0.612744 \text{ ft/s})^2}{2} \right]
$$

$$
+ \left[0.55 \frac{(0.272331 \text{ ft/s})^2}{2} \right.
$$

$$
+ (0.305556 + (2)0.75) \frac{(0.612744 \text{ ft/s})^2}{2} \right]
$$

$$
= (0.252 + 4.899 + 0.020 + 0.057 + 0.282) \text{ ft}^2/\text{s}^2
$$

$$
= 5.50946 \frac{\text{ft}^2}{\text{s}^2} \left(\frac{1 \text{ lb}_f}{32.172 \text{ ft lb}_m/\text{s}} \right)
$$

$$
= 0.17124 \frac{\text{ft lb}_f}{\text{lb}_m} \tag{1.131}
$$

$$
= \boxed{ 0.2 \frac{\text{ft lb}_f}{\text{lb}_m} } \tag{1.132}
$$

Note that the dominant term is the friction from the flow in the 2-inch pipe (i.e., the smaller pipe). Finally, we combine this result with Equation 1.66 from the previous example to arrive at the value for the shaft work. For the final answer, we convert all terms from ft lb_f/s to horsepower (hp).

$$\frac{-W_{s,by,21}}{0.83216 \, \text{lb}_{m/s}} = \left[\frac{(0.612744 \, \text{ft/s})^2}{2} + \frac{32.174 \, \text{ft}}{s^2}(75 \, \text{ft}) + F_{2,1}\frac{\text{ft}^2}{s^2}\right]\frac{s^2 \cdot \text{lb}_f}{32.174 \, \text{ft} \cdot \text{lb}_m}$$

$$W_{s,by,21} = -62.55935 \, \text{ft} \cdot \text{lb}_f/s \left(\frac{1.341 \times 10^{-3} \, \text{hp}}{0.7376 \, \text{ft} \cdot \text{lb}_f/s}\right) = -0.1137366 \, \text{hp}$$

$$W_{s,pump} = -W_{s,by,21} = 1.1 \, \text{hp} \qquad\qquad (1.133)$$

The work done by the fluid is negative because it receives an infusion of energy from the pump (i.e., the pump works on the fluid, not the other way around). This is the final answer.

To separate individual contributions to the total friction, we calculate the friction for each fitting separately. The answers in the following table are expressed as both energy per unit mass (both units ft lb_f/lb_m and ft^2/s^2 are shown) and energy per unit weight (ft), also called fluid-head units.

Fitting	Energy/Mass		Energy/Weight	Percent of Total Friction Losses
	$K_f\frac{\langle v\rangle_i^2}{2}$ $\frac{\text{ft}^2}{s^2}$	$K_f\frac{\langle v\rangle_i^2}{2}$ $\frac{\text{ft lb}_f}{\text{lb}_m}$	$K_f\frac{\langle v\rangle_i^2}{2g}$ ft	
50 ft of 3-in. pipe	0.252	0.0078	0.0078	4.6
43 ft of 2-in. pipe	4.899	0.1520	0.1520	88.9
contraction, tank to 3 in.	0.020	0.0006	0.0006	0.4
contraction, 3 to 2 in.	0.057	0.0017	0.0017	1.0
2 90° elbows	0.282	0.0087	0.0087	5.1
Total	5.509	0.1712	0.1712	100.0

Note that the numerical values are the same in the second and third columns; however, the second column is energy per unit mass, $K_f \langle v\rangle^2/2$, whereas the third column is energy per unit weight, $K_f \langle v\rangle^2/2g$ (see Section 9.2.2). The major frictional loss is the turbulent flow in a small-diameter, long pipe, followed by the two 90° elbows and the bigger pipe. In this example, however, friction losses in the fittings are small compared to the Δz term (i.e., potential energy, also called elevation head). In a problem with a less significant elevation rise (e.g., less than 10 feet) or for a system with longer runs of a narrower pipe, the frictional losses comprise a more important part of the problem.

The result calculated in the previous example was the same—to two significant figures—as the calculation without friction (compare Equations 1.133 and 1.68). If we examine the contributions to the shaft work, we see that in this flow loop, the $\Delta z = 75$-foot elevation rise (i.e., potential energy) dominates the kinetic-energy change $\Delta\langle v\rangle^2/2$ and the frictional losses F. If we convert the kinetic energy and frictional contributions into energy per weight (i.e., fluid head) in units of feet, we start to intuit how the various types of energy contribute to the load on a pump. We write each contribution in terms of equivalent feet of elevation change by dividing the terms of the mechanical energy balance by the acceleration due to gravity ($g = 32.174$ ft/s^2 = 980 cm/s^2), as follows:

$$-\frac{W_{s,by}}{m} = \frac{\Delta p}{\rho} + \frac{\Delta \langle v \rangle^2}{2\alpha} + g\Delta z + F \tag{1.134}$$

$$\frac{-W_{s,by,21}}{m} = \frac{(1-1)\,\text{atm}}{\rho} + \frac{(0.612744\,\text{ft/s})^2 - 0^2}{2} + \frac{32.174\,\text{ft}}{s^2}(75\,\text{ft} - 0\,\text{ft})$$

$$+ 5.50946\,\frac{\text{ft}^2}{s^2}$$

$$\frac{-W_{s,by,21}}{mg} = 0 + \frac{(0.612744\,\text{ft/s})^2}{(2)(32.174\,\text{ft/s}^2)} + 75\,\text{ft} + \frac{(5.50946\,\text{ft}^2/s^2)}{(32.174\,\text{ft/s}^2)}$$

$$\frac{-W_{s,by,21}}{mg} = 0\,\text{ft} + 0.006\,\text{ft} + 75\,\text{ft} + 0.17\,\text{ft} \tag{1.135}$$

$$\frac{-W_{s,by,21}}{mg} = \left(\begin{matrix}\text{pressure}\\ \text{head}\end{matrix}\right) + \left(\begin{matrix}\text{velocity}\\ \text{head}\end{matrix}\right) + \left(\begin{matrix}\text{elevation}\\ \text{head}\end{matrix}\right) + \left(\begin{matrix}\text{friction}\\ \text{head}\end{matrix}\right) \tag{1.136}$$

The four contributions on the right hand side of Equation 1.135 are called *pressure head*, *velocity head*, *elevation head*, and *friction head*. The elevation head dominates in this example. Because head has units of length, it is intuitive to compare the various quantities in Equation 1.136 using head. Each contribution (in feet or meters) is the same amount of energy per unit weight as is stored in a column of fluid of height given by the head. Because we can visualize these heights, it is convenient to use these units rather than less intuitive units such as ft^2/s^2 or ft lb$_f$/lb$_m$ or their metric equivalents. We discuss the concept and utility of fluid head (i.e., energy per unit weight) in more detail in Chapter 9, which also discusses pumps and the shaft work of pumps as well as pumping efficiency.

Thus far in this chapter we present reasons to study fluid mechanics and we describe the strategy used in this book. We also discuss algebraic energy-balance techniques based on the mechanical energy balance (MEB), and we find them to be useful for several flow situations. Advanced mathematics are not needed for the MEB, but it is applicable only in single-input, single-output systems that meet the criteria listed in Equation 1.3 and Table 1.1. Also, to complete MEB calculations, we need additional empirical data in the form of the device coefficient C_V, pump or turbine efficiency η, friction factor $f(\text{Re})$, or fitting friction coefficient K_f.

We are now ready to proceed to the detailed analyses that lead to both the equations used in this chapter and more complex equations and calculations that deepen our understanding of fluid systems. The more intensive study of fluid mechanics begins in Chapter 2 with a quantitative discussion of observed fluid behaviors. A continuing discussion of the mechanical energy balance, including the derivation of the balance equations, is in Chapter 9. The final section of this chapter describes the mathematical techniques used throughout this text.

1.3 Connecting mathematics to fluid mechanics

In mathematics classes, students comment that they cannot see how their studies can be applied. In engineering classes, students comment that they cannot make the connection between the abstract mathematics they study and the concrete problems they face. The difficulty for engineering professors is that we cannot teach an engineering subject (e.g., fluid-mechanics modeling) until the students know sufficient mathematics (e.g., manipulating vectors and matrices; and differentiating, integrating, and solving differential equations). Students thus spend years studying mathematics outside the engineering context, not knowing how it relates to engineering. When the mathematics is finally needed in engineering courses, students find it hard to recall and apply.

Our goal as engineers is to be able to design, build, operate, and optimize equipment and systems in modern society, and mathematics is essential to these engineering tasks. The era of trial-and-error is fading fast—high-tech fields are not amenable to random tinkering, and the financial, environmental, and safety risks involved in unproven designs are too high for most applications to support. We must learn to use modeling tools and our knowledge of how the physical world operates to carry out engineering tasks.

The physical world, however, is complex, and this is why mathematics is important to engineers. It has taken centuries to organize scientific observations of how the world works into the body of knowledge that we know as the engineering curriculum. One breakthrough that allowed this to happen was the development of *calculus* in the 1600s (Figure 1.24). Calculus is the field of mathematics that deals with rates of change, and the flows of fluids, heat, and mass (i.e., the so-called transport phenomena) are governed by transport laws that involve rates of change. Thus, when transport phenomena are important, rates of change are important, and we need calculus.

We need calculus not only in the sense that integrations and differentiations appear in the problems we solve; we also need the concepts of calculus to develop the governing equations of fluid mechanics, which involve rates of change and summations over infinitesimal regions of space. Studying fluid mechanics, therefore, requires students to reexamine the concepts of calculus—having already mastered the mechanics of integration and differentiation—and to deepen their understanding of the rate-of-change processes presented abstractly in calculus class. With a physical system to consider—liquid flow—those rates of change have a concrete name and a physical situation. Rates of change and integration also may make more sense when studying flow than when first studied abstractly in a mathematics course.

In this section, we review aspects of calculus that are directly applicable to fluid mechanics, including the calculus of tensors. Tensor mathematics is not a standard component of the undergraduate introduction to calculus, and here we cover those aspects that are useful for the study of fluid mechanics. The

Date	Field	Contributors
1666–84	Calculus	Newton, Leibniz
1656–1859	Thermodynamics	Boyle, Hooke, Joule, Thompson
1687	Laws of motion	Newton
1704	Optics	Newton
1738	Bernoulli equation	Bernoulli
1750	Coordinate systems	Euler
1751	Electricity	Franklin
1769	Steam engine	Watt
1822–50	Motion of liquids, solids	Navier, Stokes
1839–40	Tube flow	Hagen, Poiseuille
1855	Diffusion	Fick
1873	Electricity and magnetism	Maxwell
1870–95	Laminar and turbulent flow	Reynolds
1903	Controlled flight	Wright and Wright
1904	Boundary layers	Prandtl
1920–50	Rocketry	Goddard, von Braun
1951	Heart bypass surgery	Dennis
1958	Integrated circuit	Kilby
1965	Moore's law	Moore
1969	Supersonic commercial aviation	Several
1978	Commercial mobile phones	Several
1981	Personal computer	
2007	Human geonome sequenced	Levy, Venter

Figure 1.24 The development of fluid mechanics and the other transport fields depended on the invention of calculus.

use of tensors may be avoided in an elementary study of fluid mechanics; since tensors make that study easier, we include and use them for readers who find them helpful. Studying the mathematics review in this chapter prepares students to learn fluid-mechanics modeling [17, 179, 184].

1.3.1 Calculus of continuous functions

Calculus is the mathematics that allows us to quantify concepts that deal with rates of change (i.e., derivatives) and summations over infinitesimal regions of space (i.e., integrals). We will use the defining equations of derivatives and integrals in our fluid-mechanics discussions, and they are presented here. We also use derivatives and integrals to calculate engineering quantities of interest (see Section 6.2.3); therefore, examples of these types of calculations are presented here. Problems at the end of the chapter are provided so that students can practice working with these mathematics tools. A rigorous and general treatment of calculus is found in standard textbooks [166].

1.3.1.1 DERIVATIVES

When differentiation is introduced in first-semester calculus courses, it is in the context of finding the slope of a line tangent to a curve. An arbitrary curve is

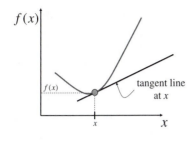

shown in Figure 1.25. At position x, a tangent line is drawn, and we write an expression for the slope of this tangent line.

Slope of a line is defined as rise over run but, for the tangent line, which only touches $f(x)$ at one point, we can write neither rise nor run in terms of the function $f(x)$. A secant line (see the topmost line in Figure 1.26) intersects the function $f(x)$ in two places; for such a line, it is easy to write an expression for slope in terms of rise over run and the values of the function $f(x)$:

Figure 1.25   A line tangent to a one-dimensional function $f(x)$ may be drawn at any point x.

$$\begin{array}{c}\text{Slope of a}\\\text{secant line}\\\text{(from geometry)}\end{array} = \frac{f(x + \Delta x) - f(x)}{\Delta x} \qquad (1.137)$$

If the interval Δx is made smaller, the secant lines approach the tangent line and Equation 1.137 becomes a better approximation for the slope of the tangent line at x. In the limit that Δx goes to zero, the ratio in Equation 1.137 becomes arbitrarily close to the slope of the tangent line at x; this limit serves as the definition of a derivative:

$$\text{Derivative defined} \qquad \boxed{\frac{df}{dx} \equiv \lim_{\Delta x \longrightarrow 0} \left[\frac{f(x + \Delta x) - f(x)}{\Delta x} \right]} \qquad (1.138)$$

Shown here is an alternative notation for the same quantity:

$$\text{Derivative defined} \qquad \boxed{\frac{df}{dx} \equiv \lim_{\Delta x \longrightarrow 0} \left[\frac{f|_{x+\Delta x} - f|_x}{\Delta x} \right]} \qquad (1.139)$$

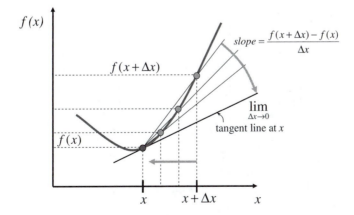

Figure 1.26   For a simple one-dimensional function $f(x)$, the limit in the definition of the derivative (see Equation 1.138) represents the slope of the tangent to the curve at a point. This expresses the instantaneous rate of change of the function $f(x)$ with respect to the variable x [166].

The expression $f|_{x+\Delta x}$ is read as "f evaluated at $x + \Delta x$." Because the definition of a derivative requires the limit as Δx goes to zero, the function f must be continuous for a derivative to be meaningful. Calculus is well suited for making calculations in fluid mechanics since we use the continuous variables ρ (density), $\underline{v}$ (velocity), and $\underline{\underline{\tilde{\tau}}}$ (molecular stress) to describe systems.

Although the derivative df/dx usually is discussed in terms of being the slope of the tangent line of the curve $f(x)$, that is only one visualization of this quantity. The expression in Equation 1.138 is the fundamental definition of a derivative; thus, in any analysis when such a limit of a ratio appears, that limit may be replaced with a derivative, and all of the properties of derivatives as sorted out by mathematicians may be invoked in subsequent calculations. In Chapter 3, we use a formulation like Equation 1.138 to keep track of momentum transfers in fluids. The basic physics allows us to write the property of interest, and calculus allows us to write this physics in differential form and to proceed to the solution.

An engineering task for which differentiation is useful is calculating the maximum value of a function. Following is an example of such a calculation.

EXAMPLE 1.11. *The function in Equation 1.140 represents the z-direction velocity of a flow between two vertical parallel plates. At what position in the flow does the velocity reach a maximum?*

$$v_z(y) = \frac{\bar{\rho}g\bar{\beta}(T_2 - T_1)b^2}{12\mu}\left[\left(\frac{y}{b}\right)^3 - \left(\frac{y}{b}\right)\right] \tag{1.140}$$

SOLUTION. The flow between two vertical plates shown in Figure 1.27 is the result of fluid-density differences driven by a temperature difference in the y-direction. All of the following quantities are constant: $\bar{\rho}$, average density; $\bar{\beta}$, average coefficient of thermal expansion; $T_2 - T_1$, temperature difference; b, gap; and μ, viscosity. The methods in this text lead to the ability to obtain Equation 1.140. Here, we have the simpler task of determining from the solution the location and magnitude of the maximum and minimum in velocity.

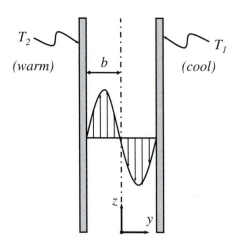

Figure 1.27   Temperature difference generates a flow between two long wide plates (i.e., hot air rises). We obtain the velocity profile in Equation 1.140 by using the methods in this book in conjunction with energy-balance equations (see Problem 40 in Chapter 7).

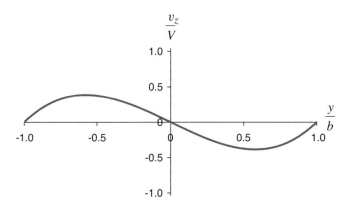

Figure 1.28
The velocity profile that develops when a fluid is trapped between two plates held at two different temperatures is given by the cubic equation in Equation 1.140. We use calculus to find the maximum and minimum values of the function.

The equation provided has many different quantities in it; for now, all of those different constants are simply confusing. Note that the combination of quantities in front of the square brackets must have units of velocity (m/s); thus, for simplicity, we call that combination of variables V. Our equation to work with is then:

$$v_z(y) = V \left[\left(\frac{y}{b} \right)^3 - \left(\frac{y}{b} \right) \right] \tag{1.141}$$

which is plotted in Figure 1.28 in dimensionless form as v_z/V versus y/b.

The location of the maximum value of a function can be determined from the slope of the tangent line as a function of position, which can be calculated from the derivative (Figure 1.29). When the value of a function at a point is increasing, the slope of the tangent line at that point is positive. When the value of a function at a point is decreasing, the slope of the tangent line at that point is negative. When the slope of the tangent line at a point is zero, the value of the function is neither increasing nor decreasing but rather has reached a maximum or a minimum. To find the location of the maximum (or minimum) of a function, we calculate the

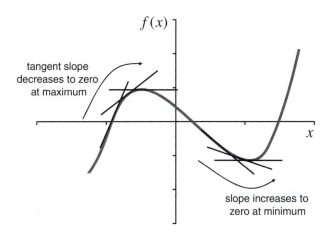

Figure 1.29
The locations of extrema are found using the derivative. For both maxima and minima, the slope of the tangent line (i.e., the derivative at that point) is zero.

derivative of the function, set the derivative equal to zero, and solve for the values of location that satisfy the resulting equation:

$$\text{Location of extrema:} \qquad \frac{dv_z}{dy} = 0 \tag{1.142}$$

$$\frac{d}{dy}\left(\frac{Vy^3}{b^3} - \frac{Vy}{b}\right) = 0 \tag{1.143}$$

$$\frac{3Vy^2}{b^3} - \frac{V}{b} = 0 \tag{1.144}$$

$$y = \pm\frac{b}{\sqrt{3}} \tag{1.145}$$

$$\text{Location of extrema:} \qquad \boxed{\frac{y}{b} = \pm 0.577} \tag{1.146}$$

Substituting these two values into the function for velocity (Equation 1.141), we obtain the maximum and minimum values of velocity, which are located at $y/b = \pm 0.577$:

$$v_z\left(\frac{b}{\sqrt{3}}\right) = V\left[\left(\frac{y}{b}\right)^3 - \left(\frac{y}{b}\right)\right] \tag{1.147}$$

$$= V\left[\left(\frac{1}{\sqrt{3}}\right)^3 - \left(\frac{1}{\sqrt{3}}\right)\right] = \frac{-2V}{3\sqrt{3}} = \boxed{-0.39V} \quad \text{(minimum)} \tag{1.148}$$

$$v_z\left(\frac{-b}{\sqrt{3}}\right) = V\left[\left(\frac{-1}{\sqrt{3}}\right)^3 - \left(\frac{-1}{\sqrt{3}}\right)\right] = \frac{2V}{3\sqrt{3}} = \boxed{0.39V} \quad \text{(maximum)} \tag{1.149}$$

Analogous derivatives on multivariable functions—partial derivatives—are useful in calculations on continuous functions of two, three, or more variables (see the Web appendix [108] for a review). The fluid-mechanics variables ρ, $\underline{v}$, and $\underline{\underline{\tau}}$ are all multivariable, continuous functions.

1.3.1.2 INTEGRALS

We turn now to the other key concept of calculus: the integral. When integration is introduced in calculus courses, it is usually in the context of finding areas. The area under the positive function $f(x)$ between $x = a$ and $x = b$, depicted in Figure 1.30, may be approximated by the sum of the areas of appropriately chosen rectangles. First, the interval between a and b is divided into N equally sized intervals. Second, the areas of the N rectangles are summed to approximate

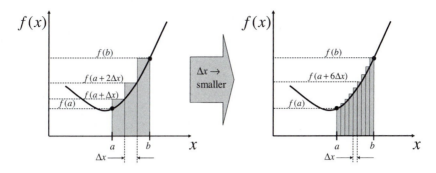

Figure 1.30 For a simple one-dimensional function $f(x)$, the limit in the definition of the integral (Equation 1.151) represents the area under the curve between the chosen limits.

the total area under the curve. The interval size Δx is arbitrary:

$$\begin{array}{c} \text{Area between } f(x) \\ \text{and } x\text{-axis} \\ \text{(from geometry)} \end{array} \approx \sum_{i=1}^{N} f(a + i\Delta x)\Delta x \qquad (1.150)$$

$$\Delta x \equiv \frac{b - a}{N}$$

If the interval Δx is made smaller, Equation 1.150 becomes a better approximation for the area under $f(x)$. In the limit that Δx goes to zero, the summation in Equation 1.150 becomes arbitrarily close to the area under $f(x)$, and this limit serves as the definition of an integral:

Integral defined
$$\boxed{I = \int_{a}^{b} f(x)dx \equiv \lim_{N \to \infty} \left[\sum_{i=1}^{N} f(a + i\Delta x)\Delta x \right]} \qquad (1.151)$$

$$\Delta x = \frac{b - a}{N}$$

Because the definition requires the limit as Δx goes to zero, the function f must be continuous for an integral to be meaningful. Many properties of interest in fluid mechanics are calculated from limits of summations.

Integrals may be used whenever a calculation can be put into the form of Equation 1.151.[3] In Chapter 3, we use a form of Equation 1.151 to sum various mass and momentum transfer effects in deforming liquids. Another task for which integration is useful is calculating the average of a function, discussed in the following example (see also Section 6.2.3).

EXAMPLE 1.12. *The shape of the velocity profile for a steady flow in a narrow slit between two plates is given by* $f(y)$ *(see Section 7.1.1 and Figure 1.31), where* $f = v_x/v_{max}$ *is dimensionless and* y *and the number 10 have units of mm.*

---

[3] There are rigorous mathematical rules that restrict which types of functions are integratable. See the mathematical literature for more on this subject [166].

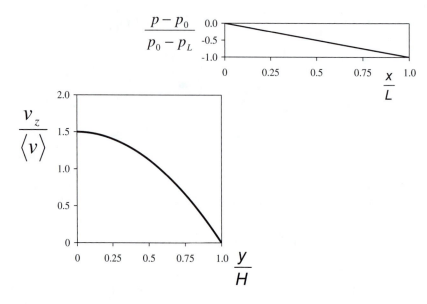

Figure 1.31 Methods in this book allow us to calculate the velocity profile for laminar flow in a narrow slit.

Over the range $0 \le y \le 10$ mm, what is the average value of the velocity in the slit? (See Figure 1.32.)

$$\frac{v_x}{v_{max}} = f(y) = \left[1 - \left(\frac{y}{10} \right)^2 \right] \tag{1.152}$$

SOLUTION. As discussed here, we can calculate the average value of a function over a range by integrating the function between the endpoints of the range and dividing by the range:

$$\text{Average of } f(y) = \langle f \rangle = \frac{\displaystyle\int_{y_{min}}^{y_{max}} f(y)\, dy}{(y_{max} - y_{min})} \tag{1.153}$$

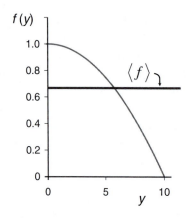

Figure 1.32   This example requests the average of the function $f(y)$ over the range $0 \le y \le 10$.

For the current problem, we obtain:

$$\langle f \rangle = \frac{1}{(10-0)} \int_0^{10} \left[1 - \left(\frac{y}{10} \right)^2 \right] \, dy$$

$$= \frac{1}{10} \left[y - \frac{y^3}{300} \right]\Bigg|_0^{10} = \boxed{\frac{2}{3}}$$

Equation 1.153 may be shown rigorously to hold by breaking up the interval $y_{min} \le y \le y_{max}$ into smaller intervals, averaging the values of the function at each position, and taking the limit of this average as the size of the interval between values goes to zero. We demonstrate this calculation for a function of a single variable.

Consider the function $f(y)$; we seek to derive the expression for the average of a function, Equation 1.153. To calculate $\langle f \rangle$, the average of the function between limits y_{min} and y_{max}, we choose a sampling of N evenly spaced points and assemble the average of these values. Later, we take the limit as N goes to infinity:

Average
of a function
(definition of $\approx \dfrac{1}{N} \Big(f(y_1) + f(y_2) + \ldots + f(y_k) + \ldots + f(y_{N-1}) + f(y_N) \Big)$
arithmetic
mean)

$$= \frac{1}{N} \sum_{k=1}^{N} f(y_{min} + k\Delta y) \tag{1.154}$$

$$\Delta y = \frac{y_{max} - y_{min}}{N} \tag{1.155}$$

We now solve Equation 1.155 for N and substitute this into Equation 1.154. Finally, we take the limit as N goes to infinity:

$$N = \frac{y_{max} - y_{min}}{\Delta y}$$

Average
of a function $\approx \left[\dfrac{\Delta y}{y_{max} - y_{min}} \right] \displaystyle\sum_{k=1}^{N} f(y_{min} + k\Delta y)$

Average
of a function $= \dfrac{\displaystyle\lim_{N \to \infty} \left[\displaystyle\sum_{k=1}^{N} f(y_{min} + k\Delta y)\Delta y \right]}{y_{max} - y_{min}} \tag{1.156}$

Comparing the limit in Equation 1.156 to the definition of a single integral (see Equation 1.151), we obtain the expression we seek:

$$\text{Average of } f(y) = \langle f \rangle = \frac{\int_{y_{min}}^{y_{max}} f(y)\, dy}{(y_{max} - y_{min})} \tag{1.157}$$

We will see limits of sums in fluid mechanics being equated to integrals in Chapter 2. Analogous double and triple integrals are useful in calculations on continuous functions of two, three, or more variables (see the Web appendix [108]).

1.3.2 Vector calculus

The mathematics of fluid mechanics is vector calculus, which is a calculation system that allows us to keep track of not only the magnitude of interactions but also the character of the interactions: how forces are applied to a body, for example, or how bodies move in space. In fluid mechanics, important vectors include velocity and force. Vector calculus relies on fundamental definitions such as for *scalars* and *vectors*. We begin with these definitions; once this background is established, we introduce *tensors*, a more complex entity related to scalars and vectors. In this section we also review how to express vectors and tensors in coordinate systems, both Cartesian and curvilinear (i.e., cylindrical and spherical). In addition, we cover differential operations as applied to vectors and tensors. The core equations of fluid mechanics are partial differential equations that express vectors and tensors in those coordinate systems. The mathematics in this section is relied upon throughout this text.

The term scalar refers to a constant or variable function that conveys magnitude. Numbers in the usual sense are scalars. Examples of scalars are fluid density, the speed of a bullet, or the number of molecules in a vessel. Scalar variables can be manipulated through the usual mathematical methods. We summarize here the rules of algebra for scalars (e.g., α, β, and γ):

$$\text{Rules of algebra for scalars} \begin{cases} \text{commutative law} & \alpha\beta = \beta\alpha \\ \text{associative law} & (\alpha\beta)\gamma = \alpha(\beta\gamma) \\ \text{distributive law} & \alpha(\beta + \gamma) = \alpha\beta + \alpha\gamma \end{cases}$$

A vector is a constant or variable function that conveys magnitude and direction. The directional property of vectors is what separates them from scalars. Examples of vectors are the velocity of a baseball (i.e., not just its speed, but also its direction of travel), the force due to gravity, and the momentum of a fluid particle. Two vectors that have the same magnitude can have drastically different

Figure 1.33 Schematic representation of forces acting on a table. If the same magnitude of force, f, is applied in different directions, the vectors describing those forces are different in the two cases.

effects. For example, a downward force on a table will not move it, while a force to the side will cause the table to slide (Figure 1.33).

The two characteristics of a vector, magnitude and direction, can be written separately. For a vector $\underline{f}$, the magnitude is written $f = |\underline{f}|$ and the direction is expressed by using a unit vector in the direction of $\underline{f}$. In this text we write vectors with an bar under the symbol and unit vectors with a caret ($\hat{\ }$) over the symbol.

$$\text{Vector magnitude:} \quad |\underline{f}| = f \tag{1.158}$$

$$\text{Vector direction:} \quad \hat{f} = \frac{\underline{f}}{f} \tag{1.159}$$

$$|\hat{f}| = 1 \tag{1.160}$$

When adding or subtracting vectors we line up the vectors head to tail and calculate the sum as the new vector that joins the first tail with the last head (Figure 1.34). When multiplying a vector by a scalar, the rules of algebra are the same as the rules for multiplying scalars.

$$\text{Rules of algebra for scalars with vectors} \quad \begin{cases} \text{commutative law} & \alpha \underline{a} = \underline{a}\alpha \\ \text{associative law} & (\alpha \underline{a})\beta = \alpha(\underline{a}\beta) \\ \text{distributive law} & \alpha(\underline{a} + \underline{b}) = \alpha \underline{a} + \alpha \underline{b} \end{cases}$$

When multiplying two vectors, there are two different operations defined, the *scalar product* (also called the dot product or inner product) and the *vector product* (i.e., cross product or outer product). They are defined as follows

$$\text{Scalar product:} \quad \underline{a} \cdot \underline{b} = ab \cos \psi \tag{1.161}$$

$$\text{Vector product:} \quad \underline{a} \times \underline{b} = ab \sin \psi \, \hat{n} \tag{1.162}$$

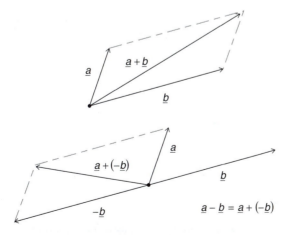

Figure 1.34 Pictorial representation of the addition and subtraction of two vectors.

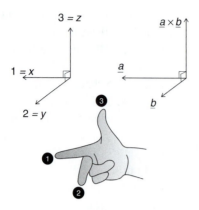

Figure 1.35 Definition of a righthanded coordinate system and the righthand rule for cross products.

where ψ is the angle between the vectors and $\hat{n}$ is a unit vector perpendicular to both $\underline{a}$ and $\underline{b}$ subject to the righthand rule (Figure 1.35). From geometry (Figure 1.36) we see that the dot product of a vector with a unit vector results in a quantity that equals the projection of the first vector in the direction of the unit vector.

$$\begin{array}{c}\text{Projection of } \underline{b} \\ \text{in the direction } \hat{n}\end{array} \qquad \underline{b} \cdot \hat{n} = (b)(1)\cos\psi = b\cos\psi \qquad (1.163)$$

This is an important operation in determining the quantity of flow through a surface.

The rules of algebra for the dot and cross products are summarized here:

$$\begin{array}{l}\text{Rules of algebra for} \\ \text{the vector dot product:}\end{array} \left\{ \begin{array}{ll} \text{commutative} & \underline{a} \cdot \underline{c} = \underline{c} \cdot \underline{a} \\ \text{associative} & \text{not possible} \\ \text{distributive} & \underline{a} \cdot (\underline{c} + \underline{w}) = \underline{a} \cdot \underline{c} + \underline{a} \cdot \underline{w} \end{array} \right.$$

$$\begin{array}{l}\text{Rules of algebra for} \\ \text{the vector cross product:}\end{array} \left\{ \begin{array}{ll} \text{NOT commutative} & \underline{a} \times \underline{c} \neq \underline{c} \times \underline{a} \\ \text{NOT associative} & (\underline{a} \times \underline{c}) \times \underline{w} \neq \underline{a} \times (\underline{c} \times \underline{w}) \\ \text{distributive} & \underline{a} \times (\underline{c} + \underline{w}) = \underline{a} \times \underline{c} + \underline{a} \times \underline{w} \end{array} \right.$$

The dot product provides a way to calculate the magnitude of a vector.

$$\underline{a} \cdot \underline{a} = (a)(a)\cos(0) = a^2 \qquad (1.164)$$

$$|\underline{a}| = +\sqrt{\underline{a} \cdot \underline{a}} \qquad (1.165)$$

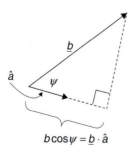

$$b\cos\psi = \underline{b} \cdot \hat{a}$$

Figure 1.36 The projection of a vector in a chosen direction is equal to the dot product of the vector with a unit vector in the chosen direction.

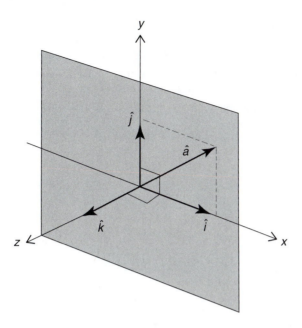

Figure 1.37 Schematic of the Cartesian coordinate system (xyz) and the Cartesian basis vectors ($\hat{i}, \hat{j}, \hat{k}$), also called $\hat{e}_x, \hat{e}_y,$ $\hat{e}_z$ or $\hat{e}_1, \hat{e}_2, \hat{e}_3$. The vector $\hat{a}$ is in the xy-plane and may be written as the sum of its x- and y-components: $a_x\hat{i} + a_y\hat{j}$ or $a_x\hat{e}_x + a_y\hat{e}_y$. The z-component of $\hat{a}$ is zero.

By convention, the magnitude of a vector is taken to be positive; any negative signs are associated with the vector direction.

1.3.2.1 COORDINATE SYSTEMS

Making calculations with vectors requires us to choose a coordinate system for reference. The most familiar coordinate system is the Cartesian coordinate system (xyz), but we begin with general considerations first, because we use non-Cartesian coordinate systems as well.

A coordinate system is composed of three non-coplanar basis vectors. Any vector may be expressed as the linear combination of any three basis vectors. If $\underline{a}, \underline{b},$ and $\underline{c}$ form a basis, then any vector $\underline{v}$ may be written as

$$\begin{array}{cc} \underline{v} \text{ expressed in terms} & \underline{v} = v_a\,\underline{a} + v_b\,\underline{b} + v_c\,\underline{c} \\ \text{of basis vectors } \underline{a},\underline{b},\underline{c} & \end{array} \qquad (1.166)$$

where v_a, v_b, and v_c are the coefficients of $\underline{v}$ with respect to the coordinate system $\underline{a}\ \underline{b}\ \underline{c}$. For the Cartesian coordinate system, the basis vectors are $\hat{i}, \hat{j},$ and $\hat{k}$ or $\hat{e}_x, \hat{e}_y,$ and $\hat{e}_z$ (this is our preferred nomenclature). We may also use $\hat{e}_1, \hat{e}_2,$ and $\hat{e}_3$ for a Cartesian coordinate system (Figure 1.37). In the Cartesian system the basis vectors are unit vectors, and $\hat{i} = \hat{e}_x = \hat{e}_1$ points parallel to the x-axis, $\hat{j} = \hat{e}_y = \hat{e}_2$ points parallel to the y-axis, and $\hat{k} = \hat{e}_z = \hat{e}_3$ points parallel to the z-axis. At every point in space $\hat{i}$ is parallel to the x-axis and points in the direction of increasing x, and likewise $\hat{j}$ is parallel to the y-axis and $\hat{k}$ is parallel to the z-axis, and they point in the directions of increasing y and z, respectively. The Cartesian basis vectors are constant. This is an advantage when integrating vectors, as we see in this text. In fluid mechanics we often use coordinate systems that vary with position in order to simplify boundary conditions, for example. We discuss the non-Cartesian coordinate systems in Section 1.3.2.4.

We indicate at the beginning of this discussion that it is desirable to express vectors in a common coordinate system so that we can manipulate them. To see how this works, we write an arbitrary vector in the Cartesian coordinate system and apply the rules of algebra for vectors to simplify the results.

$$\underline{u} = u_1\hat{e}_1 + u_2\hat{e}_2 + u_3\hat{e}_3 \tag{1.167}$$

$$\underline{v} = v_1\hat{e}_1 + v_2\hat{e}_2 + v_3\hat{e}_3 \tag{1.168}$$

$$\underline{w} = w_1\hat{e}_1 + w_2\hat{e}_2 + w_3\hat{e}_3 \tag{1.169}$$

Adding $\underline{u}$ and $\underline{v}$ together and factoring out the basis vectors yields

$$\underline{w} = \underline{u} + \underline{v} = (u_1 + v_1)\hat{e}_1 + (u_2 + v_2)\hat{e}_2 + (u_3 + v_3)\hat{e}_3 \tag{1.170}$$

Comparing Equations 1.170 and 1.169 we find:

$$w_1 = u_1 + v_1 \tag{1.171}$$

$$w_2 = u_2 + v_2 \tag{1.172}$$

$$w_3 = u_3 + v_3 \tag{1.173}$$

This is easy to remember: when adding two vectors expressed in the same coordinate system, add the coefficients of each basis vector to obtain the coefficients of the sum.

We find it convenient in this text to use matrix representation for vectors; It is arbitrary whether to write a vector as a column vector or a row vector.

$$\underline{v} = \begin{pmatrix} v_1 \\ v_2 \\ v_3 \end{pmatrix}_{123} = (v_1 \quad v_2 \quad v_3)_{123} \tag{1.174}$$

Note that in this text we write the subscript 123 on the matrix version of $\underline{v}$ to remind us that the coordinate system $\hat{e}_1\hat{e}_2\hat{e}_3$ was used to define v_1, v_2, and v_3. We can write equation 1.170 in matrix form as follows:

$$\begin{pmatrix} w_1 \\ w_2 \\ w_3 \end{pmatrix}_{123} = \begin{pmatrix} u_1 \\ u_2 \\ u_3 \end{pmatrix}_{123} + \begin{pmatrix} v_1 \\ v_2 \\ v_3 \end{pmatrix}_{123} = \begin{pmatrix} u_1 + v_1 \\ u_2 + v_2 \\ u_3 + v_3 \end{pmatrix}_{123} \tag{1.175}$$

To express a dot product between two vectors using basis-vector notation, we write each vector with respect to the basis and apply the distributive law of the dot product.

$$\underline{v} \cdot \underline{u} = (v_1\hat{e}_1 + v_2\hat{e}_2 + v_3\hat{e}_3) \cdot (u_1\hat{e}_1 + u_2\hat{e}_2 + u_3\hat{e}_3) \tag{1.176}$$

$$= v_1u_1\hat{e}_1 \cdot \hat{e}_1 + v_2u_1\hat{e}_2 \cdot \hat{e}_1 + v_3u_1\hat{e}_3 \cdot \hat{e}_1 + v_1u_2\hat{e}_1 \cdot \hat{e}_2$$

$$+ v_2u_2\hat{e}_2 \cdot \hat{e}_2 + v_3u_2\hat{e}_3 \cdot \hat{e}_2 + v_1u_3\hat{e}_1 \cdot \hat{e}_3 + v_2u_3\hat{e}_2 \cdot \hat{e}_3 + v_3u_3\hat{e}_3 \cdot \hat{e}_3 \tag{1.177}$$

The basis vectors of the Cartesian coordinate system are orthonormal, and therefore the dot products of unlike vectors are 0, while the dot products of a vector with itself yields 1. Equation 1.177 therefore simplifies to

$$\underline{v} \cdot \underline{u} = v_1 u_1 + v_2 u_2 + v_3 u_3 \tag{1.178}$$

We obtain the same result by using matrix notation and linear algebra:

$$\begin{pmatrix} v_1 & v_2 & v_3 \end{pmatrix}_{123} \cdot \begin{pmatrix} u_1 \\ u_2 \\ u_3 \end{pmatrix}_{123} = v_1 u_1 + v_2 u_2 + v_3 u_3 \tag{1.179}$$

Likewise, we can write the cross product of two vectors in terms of their coefficients in an orthonormal coordinate system. For vectors $\underline{u}$ and $\underline{v}$:

$$\underline{u} \times \underline{v} = (u_1 \hat{e}_1 + u_2 \hat{e}_2 + u_3 \hat{e}_3) \times (v_1 \hat{e}_1 + v_2 \hat{e}_2 + v_3 \hat{e}_3) \tag{1.180}$$

$$= u_1 v_1 \hat{e}_1 \times \hat{e}_1 + u_1 v_2 \hat{e}_1 \times \hat{e}_2 + u_1 v_3 \hat{e}_1 \times \hat{e}_3 + u_2 v_1 \hat{e}_2 \times \hat{e}_1 + u_2 v_2 \hat{e}_2 \times \hat{e}_2$$

$$+ u_2 v_3 \hat{e}_2 \times \hat{e}_3 + u_3 v_1 \hat{e}_3 \times \hat{e}_1 + u_3 v_2 \hat{e}_3 \times \hat{e}_2 + u_3 v_3 \hat{e}_3 \times \hat{e}_3$$

Because the basis vectors are orthonormal, each cross product is either 1, -1, or 0 (see Equation 1.162) and several of these terms are zero. Therefore, we write:

$$\underline{u} \times \underline{v} = u_1 v_2 \hat{e}_3 - u_1 v_3 \hat{e}_2 - u_2 v_1 \hat{e}_3 + u_2 v_3 \hat{e}_1 + u_3 v_1 \hat{e}_2 - u_3 v_2 \hat{e}_1 \tag{1.181}$$

$$= \begin{pmatrix} u_2 v_3 - u_3 v_2 \\ u_3 v_1 - u_1 v_3 \\ u_1 v_2 - u_2 v_1 \end{pmatrix}_{123} \tag{1.182}$$

This result is equivalent to the calculation implicit in the following determinate:

$$\underline{u} \times \underline{v} = \det \begin{vmatrix} \hat{e}_1 & \hat{e}_2 & \hat{e}_3 \\ u_1 & u_2 & u_3 \\ v_1 & v_2 & v_3 \end{vmatrix} \tag{1.183}$$

We provide practice with coordinate-system–based vector calculations in the following several examples.

EXAMPLE 1.13. *What is $\underline{u} \cdot \underline{v}$ for the following vectors?*

$$\underline{u} = \begin{pmatrix} 1 \\ 1 \\ 2 \end{pmatrix}_{123} \qquad \underline{v} = \begin{pmatrix} 1 \\ 3 \\ 0 \end{pmatrix}_{123} \tag{1.184}$$

SOLUTION. We can calculate $\underline{u} \cdot \underline{v}$ by matrix multiplying the coefficients of $\underline{u}$ and $\underline{v}$ in the orthonormal coordinate system $\hat{e}_1 \hat{e}_2 \hat{e}_3$.

$$\underline{u} \cdot \underline{v} = \begin{pmatrix} 1 & 1 & 2 \end{pmatrix}_{123} \cdot \begin{pmatrix} 1 \\ 3 \\ 0 \end{pmatrix}_{123} = 4 \tag{1.185}$$

Alternatively, we can use the formula in Equation 1.179:

$$v_1 u_1 + v_2 u_2 + v_3 u_3 = (1)(1) + (1)(3) + (2)(0) = 4 \tag{1.186}$$

Both methods are correct when the two vectors are expressed in the same orthonormal coordinate system.

EXAMPLE 1.14. *What is the component of the velocity vector $\underline{v}$ in the $\hat{e}_1$ direction? (See the inside cover for equations employing the dot product.)*

SOLUTION. When introducing the dot product of two vectors, we noted that the projection of a vector in a certain direction can be found by dotting the vector with a unit vector in the desired direction. For an orthonormal basis, the basis vectors are the unit vectors, and we can solve for the components of a vector with respect to the orthonormal basis by taking the following dot products:

$$\underline{v} \cdot \hat{e}_1 = v_1 \tag{1.187}$$

$$\underline{v} \cdot \hat{e}_2 = v_2 \tag{1.188}$$

$$\underline{v} \cdot \hat{e}_3 = v_3 \tag{1.189}$$

This also may be confirmed by dotting Equation 1.168 with each of the unit vectors in turn and remembering that we are assuming the three basis vectors $\hat{e}_i$ ($i = 1, 2, 3$) to be mutually perpendicular and of unit length. For example:

$$\hat{e}_1 \cdot \underline{v} = \hat{e}_1 \cdot (v_1\hat{e}_1 + v_2\hat{e}_2 + v_3\hat{e}_3) \tag{1.190}$$

$$= \hat{e}_1 \cdot v_1\hat{e}_1 + \hat{e}_1 \cdot v_2\hat{e}_2 + \hat{e}_1 \cdot v_3\hat{e}_3 \tag{1.191}$$

$$= v_1 \tag{1.192}$$

EXAMPLE 1.15. *What is the component of the force vector $\underline{u}$ in the $\underline{a}$ direction for $\underline{u}$ and $\underline{a}$ given here? (Finding components of vectors appears in drag calculations.)*

$$\underline{u} = \begin{pmatrix} 1 \\ 5 \\ -1 \end{pmatrix}_{123} \qquad \underline{a} = \begin{pmatrix} 1 \\ 0 \\ -1 \end{pmatrix}_{123} \tag{1.193}$$

SOLUTION. The solution method for this example is the same as for the previous example. We dot the vector (this time it is $\underline{u}$) with a unit vector in the direction of $\underline{a}$.

$$u_a = \underline{u} \cdot \frac{\underline{a}}{|\underline{a}|} \tag{1.194}$$

To calculate $|\underline{a}|$, we dot $\underline{a}$ with itself and take the square root:

$$|\underline{a}| = \sqrt{\underline{a} \cdot \underline{a}} \tag{1.195}$$

$$= \sqrt{\begin{pmatrix} 1 & 0 & -1 \end{pmatrix}_{123} \cdot \begin{pmatrix} 1 \\ 0 \\ -1 \end{pmatrix}_{123}} = \sqrt{2} \tag{1.196}$$

Our final answer is calculated as:

$$u_a = \underline{u} \cdot \frac{\underline{a}}{|\underline{a}|} \tag{1.197}$$

$$= \frac{\begin{pmatrix} 1 & 5 & -1 \end{pmatrix}_{123} \cdot \begin{pmatrix} 1 \\ 0 \\ -1 \end{pmatrix}_{123}}{\sqrt{2}} \tag{1.198}$$

$$= \frac{2}{\sqrt{2}} = \boxed{\sqrt{2}} \tag{1.199}$$

EXAMPLE 1.16. *What is $\underline{u} \times \underline{v}$ for the two vectors given here? (Cross products appear in torque calculations; see Example 1.17.):*

$$\underline{u} = \begin{pmatrix} 3 \\ 0 \\ 0 \end{pmatrix}_{123} = 3\hat{e}_1 \tag{1.200}$$

$$\underline{v} = \begin{pmatrix} 1 \\ -2 \\ 0 \end{pmatrix}_{123} = \hat{e}_1 - 2\hat{e}_2 \tag{1.201}$$

SOLUTION. To solve for $\underline{u} \times \underline{v}$, we write the expression in a Cartesian coordinate system and follow the rules of algebra for the cross product:

$$\underline{u} \times \underline{v} = 3\hat{e}_1 \times (\hat{e}_1 - 2\hat{e}_2) \tag{1.202}$$

$$= 3(\hat{e}_1 \times \hat{e}_1) - 6(\hat{e}_1 \times \hat{e}_2) \tag{1.203}$$

$$= -6\hat{e}_3 = \begin{pmatrix} 0 \\ 0 \\ -6 \end{pmatrix}_{123} \tag{1.204}$$

Alternatively, we can use Equation 1.182:

$$\underline{u} \times \underline{v} = \begin{pmatrix} u_2 v_3 - u_3 v_2 \\ u_3 v_1 - u_1 v_3 \\ u_1 v_2 - u_2 v_1 \end{pmatrix}_{123} \tag{1.205}$$

$$= \begin{pmatrix} (0)(0) - (0)(-2) \\ (0)(1) - (3)(0) \\ (3)(-2) - (0)(1) \end{pmatrix}_{123} \tag{1.206}$$

$$= \begin{pmatrix} 0 \\ 0 \\ -6 \end{pmatrix}_{123} \tag{1.207}$$

Another method of solving for this cross product is to carry out the determinant in Equation 1.183.

$$\underline{u} \times \underline{v} = \det \begin{vmatrix} \hat{e}_1 & \hat{e}_2 & \hat{e}_3 \\ 3 & 0 & 0 \\ 1 & -2 & 0 \end{vmatrix} \tag{1.208}$$

$$= -6e_3 \tag{1.209}$$

All three methods arrive at the same answer.

EXAMPLE 1.17. *What is the torque on a lever attached to the shaft shown in Figure 1.38?*

top view:

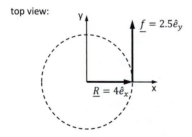

Figure 1.38   A shaft is turned by application of a torque a distance R from the axis of the shaft. The vector from the axis of rotation to the point of application of force is the lever arm. Force is applied at the circumference of an imaginary circle made by the projected rotation of the shaft and lever.

SOLUTION. Torque is the amount of effort to produce a rotation in a body; the definition of torque is the cross product of the lever arm and the force [167] (see Section 6.2.3.2). The lever arm is the distance from the point of application of the force to the axis of rotation:

$$\underline{T} = (\text{lever arm}) \times (\text{force}) \tag{1.210}$$

$$= \underline{R} \times \underline{f} \tag{1.211}$$

Writing f and $\underline{R}$ in the coordinate system shown, we use Equation 1.182 to carry out the cross product:

$$\underline{R} = \begin{pmatrix} 4 \\ 0 \\ 0 \end{pmatrix}_{xyz} \qquad \underline{f} = \begin{pmatrix} 0 \\ 2.5 \\ 0 \end{pmatrix}_{xyz} \tag{1.212}$$

$$\underline{T} = \underline{R} \times \underline{f} = \begin{pmatrix} R_2 f_3 - R_3 f_2 \\ R_3 f_1 - R_1 f_3 \\ R_1 f_2 - R_2 f_1 \end{pmatrix}_{123} \tag{1.213}$$

$$= \begin{pmatrix} 0 \\ 0 \\ 10 \end{pmatrix}_{123} \tag{1.214}$$

In Chapter 6, we discuss the many engineering quantities of interest that may be calculated from the modeling described in this text. For example, in machinery that employs axles lubricated by fluids or mixing shafts turning in fluids, the torque on the shaft is a quantity of interest.

1.3.2.2 TENSORS

Molecular stress in a moving fluid is best described as a *tensor*, defined as a mathematical entity related to vectors and scalars but somewhat more complicated. It is possible to skirt most details of tensor analysis and still understand fluid mechanics. It is sufficient for our purposes to think of tensors as 3×3 matrices that hold the information about fluid stresses (see Equation 1.223). For those who want to understand tensors more fully, we provide a brief overview; many texts are available in the literature for comprehensive covereage [6, 13, 14].

A tensor is a mathematical entity related to vectors, but it is not easy to graphically represent a tensor. For our purposes, a tensor is a mathematical machine that works through the dot product to transform vectors in a convenient way. To make tensors work, we write them as 3×3 matrices, as we discuss here.

The simplest tensor is the *dyad* or *dyadic product*. The dyadic product is formed by writing two vectors side by side.

$$\text{Tensor:} \quad \underline{\underline{A}} = \underline{a}\,\underline{b} \tag{1.215}$$

There is no dot or cross symbol in the dyadic product; this type of product is called the *indeterminate vector product*. When we write a tensor with a single symbol we use two underlines as shown for $\underline{\underline{A}}$ in Equation 1.215.

The indeterminate vector product has its rules of algebra, and these are listed here. We draw the reader's attention to the first rule of the indeterminate vector product: this type of product is not commutative.

$$\begin{array}{l} \text{Laws of algebra} \\ \text{for the indeterminate} \\ \text{vector product:} \end{array} \left\{ \begin{array}{ll} NOT \text{ commutative} & \underline{a}\,\underline{b} \neq \underline{b}\,\underline{a} \\ \text{associative} & (\underline{a}\,\underline{b})\underline{c} = \underline{a}(\underline{b}\,\underline{c}) \\ \text{distributive} & \underline{a}(\underline{b} + \underline{c}) = \underline{a}\,\underline{b} + \underline{a}\,\underline{c} \\ & (\underline{a} + \underline{b})(\underline{c} + \underline{d}) = \underline{a}\,\underline{c} + \underline{a}\,\underline{d} + \underline{b}\,\underline{c} + \underline{b}\,\underline{d} \end{array} \right.$$

Scalars may be placed anywhere within an expression containing the indeterminate vector product:

$$\gamma \underline{c}\,\underline{d} = \underline{c}\gamma\underline{d} = \underline{c}\,\underline{d}\gamma \tag{1.216}$$

where γ is a scalar and $\underline{c}$ and $\underline{d}$ are vectors.

To use tensors in fluid-mechanics calculations we write them relative to a coordinate system. For the Cartesian coordinate system and using the rules of tensor algebra, we obtain:

$$\underline{\underline{A}} = \underline{a}\,\underline{b} \tag{1.217}$$

$$= (a_1\hat{e}_1 + a_2\hat{e}_2 + a_3\hat{e}_3)(b_1\hat{e}_1 + b_2\hat{e}_2 + b_3\hat{e}_3) \tag{1.218}$$

$$= a_1b_1\hat{e}_1\hat{e}_1 + a_1b_2\hat{e}_1\hat{e}_2 + a_1b_3\hat{e}_1\hat{e}_3 + a_2b_1\hat{e}_2\hat{e}_1$$
$$+ a_2b_2\hat{e}_2\hat{e}_2 + a_2b_3\hat{e}_2\hat{e}_3 + a_3b_1\hat{e}_3\hat{e}_1 + a_3b_2\hat{e}_3\hat{e}_2 + a_3b_3\hat{e}_3\hat{e}_3 \tag{1.219}$$

The indeterminate vector product does not commute, and therefore terms with $\hat{e}_2\hat{e}_1$, for example, are not equivalent to terms with $\hat{e}_1\hat{e}_2$. There are 9 distinct dyads of the coordinate-system basis. The scalar pre-factors of each term are called the *coefficients* of the tensor. We can write the tensor coefficients of $\underline{\underline{A}}$ as a 3×3 matrix.

$$\underline{\underline{A}} = \underline{a}\,\underline{b} \tag{1.220}$$

$$= a_1b_1\hat{e}_1\hat{e}_1 + a_1b_2\hat{e}_1\hat{e}_2 + a_1b_3\hat{e}_1\hat{e}_3 + a_2b_1\hat{e}_2\hat{e}_1$$
$$+ a_2b_2\hat{e}_2\hat{e}_2 + a_2b_3\hat{e}_2\hat{e}_3 + a_3b_1\hat{e}_3\hat{e}_1 + a_3b_2\hat{e}_3\hat{e}_2$$
$$+ a_3b_3\hat{e}_3\hat{e}_3 \tag{1.221}$$

$$= \begin{pmatrix} a_1b_1 & a_1b_2 & a_1b_3 \\ a_2b_1 & a_2b_2 & a_2b_3 \\ a_3b_1 & a_3b_2 & a_3b_3 \end{pmatrix}_{123} \tag{1.222}$$

$$\underline{\underline{A}} = \begin{pmatrix} A_{11} & A_{12} & A_{13} \\ A_{21} & A_{22} & A_{23} \\ A_{31} & A_{32} & A_{33} \end{pmatrix}_{123} \tag{1.223}$$

The first index on the coefficient A_{ij} indicates the row number of the term and the second index indicates the column number.

There is a dot product between two tensors. For tensors expressed with respect to an orthonormal coordinate system, the tensor dot product works exactly like 3×3 matrix multiplication:

$$\underline{\underline{C}} = \underline{\underline{A}} \cdot \underline{\underline{B}} = \begin{pmatrix} A_{11} & A_{12} & A_{13} \\ A_{21} & A_{22} & A_{23} \\ A_{31} & A_{32} & A_{33} \end{pmatrix}_{123} \cdot \begin{pmatrix} B_{11} & B_{12} & B_{13} \\ B_{21} & B_{22} & B_{23} \\ B_{31} & B_{32} & B_{33} \end{pmatrix}_{123} \tag{1.224}$$

$$C_{11} = A_{11}B_{11} + A_{12}B_{21} + A_{13}B_{31}$$

$$C_{12} = A_{11}B_{12} + A_{12}B_{22} + A_{13}B_{32}$$

$$C_{13} = A_{11}B_{13} + A_{12}B_{23} + A_{13}B_{33}$$

$$C_{21} = A_{21}B_{11} + A_{22}B_{21} + A_{23}B_{31}$$

$$C_{22} = A_{21}B_{12} + A_{22}B_{22} + A_{23}B_{32}$$

$$C_{23} = A_{21}B_{13} + A_{22}B_{23} + A_{23}B_{33}$$

$$C_{31} = A_{31}B_{11} + A_{32}B_{21} + A_{33}B_{31}$$

$$C_{32} = A_{31}B_{12} + A_{32}B_{22} + A_{33}B_{32}$$

$$C_{33} = A_{31}B_{13} + A_{32}B_{23} + A_{33}B_{33}$$

$$C_{ij} = \sum_{k=1}^{3} A_{ik}B_{kj} \tag{1.225}$$

The Equation 1.225 is a compact way of writing all nine relationships. Dot multiplication of a vector and a tensor written with respect to the same orthonormal basis works like the matrix multiplication of a 1×3 matrix with a 3×3 matrix. The result is a vector:

$$\underline{w} = \underline{v} \cdot \underline{\underline{A}} \tag{1.226}$$

$$= (v_1 \quad v_2 \quad v_3)_{123} \cdot \begin{pmatrix} A_{11} & A_{12} & A_{13} \\ A_{21} & A_{22} & A_{23} \\ A_{31} & A_{32} & A_{33} \end{pmatrix}_{123} \tag{1.227}$$

$$= (w_1 \quad w_2 \quad w_3)_{123} \tag{1.228}$$

where:

$$w_1 = v_1 A_{11} + v_2 A_{21} + v_3 A_{31} \tag{1.229}$$

$$w_2 = v_1 A_{12} + v_2 A_{22} + v_3 A_{32} \tag{1.230}$$

$$w_3 = v_1 A_{13} + v_2 A_{23} + v_3 A_{33} \tag{1.231}$$

We use matrix algebra to carry out the dot product on components of vectors with tensors written with respect to orthonormal bases. The inside cover shows some fluid-mechanics equations involving the dot product of vectors and tensors.

EXAMPLE 1.18. *For the vectors given here, what are the coefficients in the 123-coordinate system of the tensor* $\underline{\underline{B}} = \underline{u}\,\underline{v}$?

$$\underline{u} = \begin{pmatrix} 1 \\ 1 \\ 2 \end{pmatrix}_{123} \qquad \underline{v} = \begin{pmatrix} 1 \\ 3 \\ 0 \end{pmatrix}_{123} \tag{1.232}$$

SOLUTION. We can form $\underline{\underline{B}}$ by following the rules of algebra. We begin by writing $\underline{u}$ and $\underline{v}$ explicitly in terms of the basis vectors. The final result is obtained by distributing the indeterminate vector product:

$$\underline{\underline{B}} = \underline{u}\,\underline{v} \tag{1.233}$$

$$= (\hat{e}_1 + \hat{e}_2 + 2\hat{e}_3)(\hat{e}_1 + 3\hat{e}_2) \tag{1.234}$$

$$= \hat{e}_1\hat{e}_1 + 3\hat{e}_1\hat{e}_2 + \hat{e}_2\hat{e}_1 + 3\hat{e}_2\hat{e}_2 + 2\hat{e}_3\hat{e}_1 + 6\hat{e}_3\hat{e}_2 \tag{1.235}$$

We write this result in matrix form as follows:

$$\underline{\underline{B}} = \begin{pmatrix} 1 & 3 & 0 \\ 1 & 3 & 0 \\ 2 & 6 & 0 \end{pmatrix}_{123} \tag{1.236}$$

EXAMPLE 1.19. *For the tensor* $\underline{\underline{B}} = 2\hat{e}_1\hat{e}_1 + \hat{e}_1\hat{e}_2 - \hat{e}_1\hat{e}_3 + 2\hat{e}_2\hat{e}_2 + 1\hat{e}_3\hat{e}_1 - 2\hat{e}_3\hat{e}_2$, *what is* $\underline{\underline{B}} \cdot \underline{\underline{B}}$?

SOLUTION. We can calculate $\underline{\underline{B}} \cdot \underline{\underline{B}}$ by matrix multiplying the coefficients of $\underline{\underline{B}}$ in the orthonormal coordinate system 123:

$$\underline{\underline{B}} \cdot \underline{\underline{B}} = \begin{pmatrix} 2 & 1 & -1 \\ 0 & 2 & 0 \\ 1 & -2 & 0 \end{pmatrix}_{123} \cdot \begin{pmatrix} 2 & 1 & -1 \\ 0 & 2 & 0 \\ 1 & -2 & 0 \end{pmatrix}_{123} \tag{1.237}$$

$$= \begin{pmatrix} 3 & 6 & -2 \\ 0 & 4 & 0 \\ 2 & -3 & -1 \end{pmatrix}_{123} \tag{1.238}$$

In fluid-mechanics modeling, molecular stress in a fluid is a tensor. Tremendous simplification is achieved when matrix–tensor calculations are used to keep track of fluid motion and force transmission. Molecular stress is discussed in detail in Chapter 4.

1.3.2.3 DIFFERENTIAL OPERATIONS

Three of the most important equations in fluid mechanics are those of conservation of mass, momentum, and energy. They are differential equations and they are derived in Chapter 6:

Mass conservation $\qquad \left(\dfrac{\partial \rho}{\partial t} + \underline{v} \cdot \nabla \rho\right) = -\rho\,(\nabla \cdot \underline{v})$ (1.239)

Momentum conservation $\qquad \rho\left(\dfrac{\partial \underline{v}}{\partial t} + \underline{v} \cdot \nabla \underline{v}\right) = -\nabla p + \nabla \cdot \underline{\underline{\tau}} + \rho \underline{g}$ (1.240)

Energy conservation $\quad \rho\left(\dfrac{\partial \hat{E}}{\partial t} + \underline{v} \cdot \nabla \hat{E}\right) = -\nabla \cdot \underline{q} - \nabla \cdot (p\underline{v})$

$$+ \nabla \cdot \underline{\underline{\tau}} \cdot \underline{v} + S_e \tag{1.241}$$

(Chapter 6 defines the variables in these equations.) These equations contain both time derivatives $(\partial/\partial t)$ and spatial derivatives $(\partial/\partial x_1, \partial/\partial x_2,$ and $\partial/\partial x_3)$; the spatial derivatives are hidden in the symbol ∇, as we now discuss. Chapters 6 through 8 describe how to apply these differential conservation equations to situations of interest.

In the previous equations, the vector differential operator, ∇ (called *del* or *nabla*), expresses differentiation operations in physical space (i.e., three dimensions). Equations written with vectors in terms of a letter with an underbar ($\underline{a}$) and spatial differentiation written with the symbol ∇ are said to be written in *Gibbs notation* (see Glossary). Nabla is an operator that operates on scalars, vectors, or tensors. For example, the term ∇p, which appears in Equation 1.240, is a vector and may be defined in Cartesian coordinates as follows (see [146] for a more physical treatment):

$$\nabla p = \begin{pmatrix} \dfrac{\partial p}{\partial x} \\[2mm] \dfrac{\partial p}{\partial y} \\[2mm] \dfrac{\partial p}{\partial z} \end{pmatrix}_{xyz} = \begin{pmatrix} \dfrac{\partial p}{\partial x_1} \\[2mm] \dfrac{\partial p}{\partial x_2} \\[2mm] \dfrac{\partial p}{\partial x_3} \end{pmatrix}_{123} \qquad (1.242)$$

To use the conservation equations, we also must evaluate the expressions $\nabla \cdot \underline{v}$, $\nabla \cdot \underline{\underline{\tau}}$, $\underline{v} \cdot \nabla \rho$, and $\underline{v} \cdot \nabla \underline{v}$. The details of differential operations on vectors and tensors are discussed in Appendix B.1. Differential operations in the Cartesian coordinate system are carried out in Table B.2. In our study of fluid mechanics, we rely on the tables in Appendix B.1 to translate expressions in Gibbs notation to the equivalent matrix or component notation with respect to a chosen coordinate system. Some vector identities that apply to operations with del are provided in the inside front cover of this book and in Appendix B.

EXAMPLE 1.20. *If the pressure p in a fluid varies with position (x, y, z) according to the following equation, what is the gradient field of the pressure, ∇p? The answer is a vector. Note that ∇p appears in the microscopic momentum balance, the central equation of fluid mechanics.*

$$p(x, y, z) = 16x^2 + 4y \qquad (1.243)$$

SOLUTION. To calculate ∇p for the given pressure distribution, we follow Equation 1.242:

$$\nabla p = \begin{pmatrix} \dfrac{\partial p}{\partial x} \\[2mm] \dfrac{\partial p}{\partial y} \\[2mm] \dfrac{\partial p}{\partial z} \end{pmatrix}_{xyz} = \begin{pmatrix} \dfrac{\partial(16x^2 + 4y)}{\partial x} \\[2mm] \dfrac{\partial(16x^2 + 4y)}{\partial y} \\[2mm] \dfrac{\partial(16x^2 + 4y)}{\partial z} \end{pmatrix}_{xyz} \qquad (1.244)$$

$$\nabla p = \begin{pmatrix} 32x \\ 4 \\ 0 \end{pmatrix}_{xyz} \qquad (1.245)$$

Note that pressure is a scalar field (i.e., varies with position), and ∇p is a vector field.

EXAMPLE 1.21. *For the following fluid velocity field, what is $\nabla \cdot \underline{v}$? Note that the term $\nabla \cdot \underline{v}$ appears in the mass conservation equation, Equation 1.239.*

$$\underline{v} = \begin{pmatrix} -0.06x_1 \\ 0 \\ 0.06x_3 \end{pmatrix}_{123} \tag{1.246}$$

SOLUTION. The expression for $\nabla \cdot \underline{v}$ in Cartesian coordinates 123 is given in Table B.2 and repeated here:

$$\nabla \cdot \underline{v} = \frac{\partial v_1}{\partial x_1} + \frac{\partial v_2}{\partial x_2} + \frac{\partial v_3}{\partial x_3} \tag{1.247}$$

For the velocity field given, we obtain:

$$\nabla \cdot \underline{v} = \frac{\partial}{\partial x_1}(-0.06x_1) + \frac{\partial}{\partial x_2}(0) + \frac{\partial}{\partial x_3}(0.06x_3) \tag{1.248}$$

$$= -0.06 + 0 + 0.06 \tag{1.249}$$

$$\boxed{\nabla \cdot \underline{v} = 0} \tag{1.250}$$

Note that $\underline{v}$ is a vector field and $\nabla \cdot \underline{v}$ is a scalar.

EXAMPLE 1.22. *Using Table B.2 to write $\nabla \underline{w}$, what is $\underline{v} \cdot \nabla \underline{w}$? Note that a term like this appears in the momentum conservation equation, Equation 1.240.*

SOLUTION. We calculate the result of the dot product of $\underline{v}$ and $\nabla \underline{w}$ using matrix multiplication when both expressions are written in the same Cartesian coordinate system. From Table B.2, we write:

$$\underline{v} = \begin{pmatrix} v_1 \\ v_2 \\ v_3 \end{pmatrix}_{123} \tag{1.251}$$

$$\nabla \underline{w} = \begin{pmatrix} \dfrac{\partial w_1}{\partial x_1} & \dfrac{\partial w_2}{\partial x_1} & \dfrac{\partial w_3}{\partial x_1} \\[2mm] \dfrac{\partial w_1}{\partial x_2} & \dfrac{\partial w_2}{\partial x_2} & \dfrac{\partial w_3}{\partial x_2} \\[2mm] \dfrac{\partial w_1}{\partial x_3} & \dfrac{\partial w_2}{\partial x_3} & \dfrac{\partial w_3}{\partial x_3} \end{pmatrix}_{123} \tag{1.252}$$

Now, taking the dot product of the two:

$$
\underline{v} \cdot \nabla \underline{w} = \begin{pmatrix} v_1 & v_2 & v_3 \end{pmatrix}_{123} \cdot \begin{pmatrix} \dfrac{\partial w_1}{\partial x_1} & \dfrac{\partial w_2}{\partial x_1} & \dfrac{\partial w_3}{\partial x_1} \\[2mm] \dfrac{\partial w_1}{\partial x_2} & \dfrac{\partial w_2}{\partial x_2} & \dfrac{\partial w_3}{\partial x_2} \\[2mm] \dfrac{\partial w_1}{\partial x_3} & \dfrac{\partial w_2}{\partial x_3} & \dfrac{\partial w_3}{\partial x_3} \end{pmatrix}_{123} \tag{1.253}
$$

$$
= \left(\left[v_1 \tfrac{\partial w_1}{\partial x_1} + v_2 \tfrac{\partial w_1}{\partial x_2} + v_3 \tfrac{\partial w_1}{\partial x_3} \right] \left[v_1 \tfrac{\partial w_2}{\partial x_1} + v_2 \tfrac{\partial w_2}{\partial x_2} + v_3 \tfrac{\partial w_2}{\partial x_3} \right] \left[v_1 \tfrac{\partial w_3}{\partial x_1} + v_2 \tfrac{\partial w_3}{\partial x_2} + v_3 \tfrac{\partial w_3}{\partial x_3} \right] \right)_{123} \tag{1.254}
$$

which is a 1×3 matrix. We can introduce summation signs to write this result more compactly. The ability to use summation notation is facilitated by the use of $\hat{e}_1, \hat{e}_2, \hat{e}_3$ notation instead of $\hat{e}_x, \hat{e}_y, \hat{e}_z$ or $\hat{i}, \hat{j}, \hat{k}$ notation.

$$
\underline{v} \cdot \nabla \underline{w} = \left(\sum_{k=1}^{3} \frac{\partial w_1}{\partial x_k} v_k \quad \sum_{k=1}^{3} \frac{\partial w_2}{\partial x_k} v_k \quad \sum_{k=1}^{3} \frac{\partial w_3}{\partial x_k} v_k \right)_{123} \tag{1.255}
$$

$$
\underline{v} \cdot \nabla \underline{w} = \begin{pmatrix} \sum_{k=1}^{3} \dfrac{\partial w_1}{\partial x_k} v_k \\[3mm] \sum_{k=1}^{3} \dfrac{\partial w_2}{\partial x_k} v_k \\[3mm] \sum_{k=1}^{3} \dfrac{\partial w_3}{\partial x_k} v_k \end{pmatrix}_{123} \tag{1.256}
$$

Notice that the answer is a vector and we change from row to column notation when convenient.

EXAMPLE 1.23. *Using Table B.2 to write $\nabla \underline{v}$, what is $\underline{v} \cdot \nabla \underline{v}$ for the velocity vector $\underline{v}$ given here?*

$$
\underline{v} = -6.0 x_1 \hat{e}_1 + 6.0 x_3 \hat{e}_3 \tag{1.257}
$$

SOLUTION. We calculate the result of the dot product using matrix multiplication when both expressions are written in the same Cartesian coordinate system. From Table B.2, we write:

$$
\underline{v} = \begin{pmatrix} -6.0 x_1 \\ 0 \\ 6.0 x_3 \end{pmatrix}_{123} \tag{1.258}
$$

$$\nabla \underline{v} = \begin{pmatrix} \dfrac{\partial v_1}{\partial x_1} & \dfrac{\partial v_2}{\partial x_1} & \dfrac{\partial v_3}{\partial x_1} \\[2mm] \dfrac{\partial v_1}{\partial x_2} & \dfrac{\partial v_2}{\partial x_2} & \dfrac{\partial v_3}{\partial x_2} \\[2mm] \dfrac{\partial v_1}{\partial x_3} & \dfrac{\partial v_2}{\partial x_3} & \dfrac{\partial v_3}{\partial x_3} \end{pmatrix}_{123} = \begin{pmatrix} -6.0 & 0 & 0 \\ 0 & 0 & 0 \\ 0 & 0 & 6.0 \end{pmatrix}_{123} \tag{1.259}$$

Now, taking the dot product of the two:

$$\underline{v} \cdot \nabla \underline{w} = \begin{pmatrix} -6.0x_1 & 0 & 6.0x_3 \end{pmatrix}_{123} \cdot \begin{pmatrix} -6.0 & 0 & 0 \\ 0 & 0 & 0 \\ 0 & 0 & 6.0 \end{pmatrix}_{123} \tag{1.260}$$

$$= \begin{pmatrix} 36x_1 \\ 0 \\ 36x_3 \end{pmatrix}_{123} \tag{1.261}$$

1.3.2.4 CURVILINEAR COORDINATES

So far, we use the Cartesian coordinate system to express vectors and tensors in terms of scalar coefficients. Because vector and tensor quantities are independent of the coordinate system, we also use the convenient Cartesian system to express vector–tensor relations. The goal of this text is to show how to mathematically model flows. In flow modeling, the Cartesian system is a natural choice for solving problems if the flow boundaries are straight lines. This is the case for straight flows in rectangular ducts or in wide straight-line flows. In both cases, the flow boundaries coincide with coordinate surfaces (e.g., at $x_2 = H$, $v_1 = 0$, for all x_1 and x_3; Figure 1.39a). It is convenient to choose a coordinate system that makes the boundaries easy to specify because we must mathematically specify the boundaries in the solutions. When the coordinate system makes the boundaries easy to specify, the entire problem is easier to solve. However, when the boundaries are curved—for example, flow in a pipe or around a falling sphere or rising bubble—it is mathematically awkward to use the Cartesian system (Figure 1.39b). To solve problems with cylindrical and spherical symmetry, we use coordinate systems that share these symmetries.

The cylindrical and spherical coordinate systems are shown in Figures 1.40 and 1.41. The position of a point in space may be specified by its Cartesian coordinate position (x, y, z) or by its location in terms of cylindrical coordinates (r, θ, z), as shown in Figure 1.40. The cylindrical coordinate variables, r, θ, and z, may be written in terms the Cartesian coordinate variables, x, y, and z, as follows:

Cylindrical coordinate variables:
(from geometry)

$$x = r \cos \theta \tag{1.262}$$
$$y = r \sin \theta \tag{1.263}$$
$$z = z \tag{1.264}$$

Also associated with each point are three basis vectors. In the Cartesian system, these basis vectors are $\hat{e}_x$, $\hat{e}_y$, and $\hat{e}_z$, as discussed previously. The directions of $\hat{e}_x$,

(a)

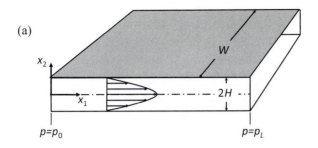

(b)

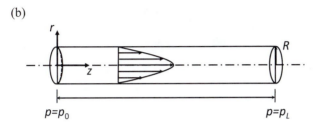

Figure 1.39   When the important surfaces of a flow are rectangular (a), Cartesian coordinates are convenient. When the important surfaces of a flow are cylindrical or spherical, one of the curvilinear coordinate systems is more convenient (b).

$\hat{e}_y$, and $\hat{e}_z$ are the same no matter which Point P is considered: If, for example, $\hat{e}_z$ points upward toward the sky from any one point, then $\hat{e}_z$ points upward toward the sky at every point considered. In the cylindrical coordinate system, the three basis vectors associated with a Point P are $\hat{e}_r$, $\hat{e}_\theta$, and $\hat{e}_z$ (see Figure 1.41). The vector $\hat{e}_z$ at P is the same as the vector of the same name in the Cartesian coordinate system. The vector $\hat{e}_r$ at P is a vector that points radially outward from the nearest point on the z-axis in the direction of increasing r; thus, $\hat{e}_r$ is perpendicular to the z-axis. Furthermore, $\hat{e}_r$ is defined to make an angle θ with the positive x-axis of the Cartesian system. The last cylindrical basis vector, $\hat{e}_\theta$, is defined as perpendicular to $\hat{e}_r$ and $\hat{e}_z$ and points in the direction counterclockwise to the x-axis—that is, in the direction of increasing θ. Both $\hat{e}_r$ and $\hat{e}_\theta$ vary with position (see Figure 1.41). For an arbitrary point at coordinates (x, y, z) or (r, θ, z), the cylindrical basis vectors are related to the constant Cartesian basis

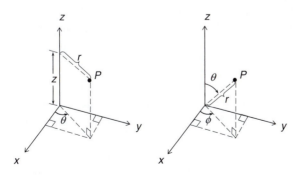

Figure 1.40   Schematic of the geometries of the cylindrical and spherical coordinate systems.

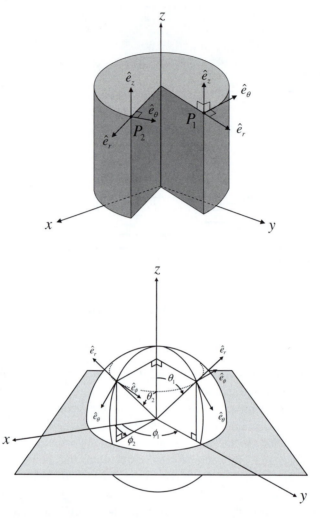

Figure 1.41 Pictorial representation of the basis vectors associated with the cylindrical (top) and spherical (bottom) coordinate systems. The directions of the curvilinear basis vectors at two positions are highlighted above, demonstrating that the directions of the basis vectors vary with position.

vectors, as follows:

$$\hat{e}_r = \cos\theta\,\hat{e}_x + \sin\theta\,\hat{e}_y \qquad (1.265)$$

Cylindrical basis vectors:
(from geometry)

$$\hat{e}_\theta = -\sin\theta\,\hat{e}_x + \cos\theta\,\hat{e}_y \qquad (1.266)$$

$$\hat{e}_z = \hat{e}_z \qquad (1.267)$$

These relationships result from careful consideration of the geometry in Figures 1.40 and 1.41. The cylindrical coordinate system is an orthonormal basis system, which means that at any chosen position, the basis vectors are mutually perpendicular and of unit length.

The cylindrical basis vectors vary with position, and this affects how spatial derivatives are written in the cylindrical coordinate system. To perform operations with the spatial derivative operator ∇, we write del and the other quantities in Cartesian coordinates, subsequently converting the result to cylindrical coordinates. Differential operations expressed in cylindrical coordinates are summarized in Table B.3 in Appendix B. (We refer to Appendix B when we need components in cylindrical coordinates.)

For systems with spherical symmetry, we use a spherical coordinate system. The position of a point in space may be specified by its spherical coordinates (r, θ, ϕ), as shown in Figure 1.40. The spherical coordinate variables (r, θ, ϕ) may be written in terms of the Cartesian coordinates as follows:

$$
\begin{array}{ll}
\text{Spherical coordinate variables:} & x = r \sin\theta \cos\phi \qquad (1.268) \\
& y = r \sin\theta \sin\phi \qquad (1.269) \\
\text{(from geometry)} & z = r \cos\theta \qquad\qquad (1.270)
\end{array}
$$

In the spherical coordinate system, all three basis vectors associated with a Point P vary with position (see Figure 1.41). The three unit vectors are $\hat{e}_r$, $\hat{e}_\theta$, and $\hat{e}_\phi$. The vector $\hat{e}_r$ points radially from the origin toward a point of interest in the direction of increasing r. The vector $\hat{e}_\theta$ is perpendicular to $\hat{e}_r$ and points in the direction that rotates away from the positive z-axis; this is the direction of increasing θ. The vector $\hat{e}_\phi$ is perpendicular to $\hat{e}_r$ and $\hat{e}_\theta$ and points counterclockwise from the x-axis. The definitions of r and θ and $\hat{e}_r$ and $\hat{e}_\theta$ are different in the cylindrical and spherical coordinate systems. The spherical basis vectors $(\hat{e}_r, \hat{e}_\theta, \hat{e}_\phi)$ may be written in terms of the Cartesian coordinates as follows:

$$
\begin{array}{ll}
& \hat{e}_r = (\sin\theta \cos\phi)\hat{e}_x + (\sin\theta \sin\phi)\hat{e}_y \\
& \qquad + (\cos\theta)\hat{e}_z \qquad\qquad\qquad\qquad (1.271) \\
\text{Spherical basis vectors:} & \hat{e}_\theta = (\cos\theta \cos\phi)\hat{e}_x + (\cos\theta \sin\phi)\hat{e}_y \\
\text{(from geometry)} & \qquad + (-\sin\theta)\hat{e}_z \qquad\qquad\qquad (1.272) \\
& \hat{e}_\phi = (-\sin\phi)\hat{e}_x + (\cos\phi)\hat{e}_y \qquad (1.273)
\end{array}
$$

Operating with the spatial derivative operator ∇ in spherical coordinates has the same difficulties described for cylindrical coordinates: Because the basis vectors vary with position in space, spatial derivatives must be carefully evaluated when this coordinate system is used. The solution to this problem when working in spherical coordinates is the same as the solution when using cylindrical coordinates: Write ∇ and the vectors in the Cartesian system and carefully carry out the operations. This already has been done with the results shown in Table B.4 in Appendix B. The extra difficulty caused by definitions in the curvilinear coordinate systems is offset by the mathematical simplifications that result when cylindrically or spherically symmetric flow problems are expressed in these coordinate systems (see Chapters 7–10). Several examples are presented for practice with curvilinear coordinates and vectors.

EXAMPLE 1.24. *For the following vectors, what is* $\underline{a} \cdot \underline{m}$? *Note that the two vectors are not written in the same coordinate system.*

$$\underline{a} = \begin{pmatrix} 2 \\ -1 \\ 1 \end{pmatrix}_{xyz} \qquad \underline{m} = \begin{pmatrix} 1 \\ 1 \\ 3 \end{pmatrix}_{r\theta z} \tag{1.274}$$

SOLUTION. Because the two vectors are not written in the same coordinate system, we must convert them before carrying out the dot product. Alternatively, we can write the vectors with the basis vectors explicitly shown and use the distributive law:

$$\underline{a} \cdot \underline{m} = \left(2\hat{e}_x - \hat{e}_y + \hat{e}_z\right) \cdot \left(\hat{e}_r + \hat{e}_\theta + 3\hat{e}_z\right) \tag{1.275}$$

$$= 2\hat{e}_x \cdot (\hat{e}_r + \hat{e}_\theta + 3\hat{e}_z) - \hat{e}_y \cdot (\hat{e}_r + \hat{e}_\theta + 3\hat{e}_z) + \hat{e}_z \cdot (\hat{e}_r + \hat{e}_\theta + 3\hat{e}_z) \tag{1.276}$$

To evaluate the individual dot products, we use Equations 1.265–1.267:

$$\underline{a} \cdot \underline{m} = 2\hat{e}_x \cdot \hat{e}_r + 2\hat{e}_x \cdot \hat{e}_\theta + 2\hat{e}_x \cdot 3\hat{e}_z - \hat{e}_y \cdot \hat{e}_r$$

$$- \hat{e}_y \cdot \hat{e}_\theta - \hat{e}_y \cdot 3\hat{e}_z + \hat{e}_z \cdot \hat{e}_r + \hat{e}_z \cdot \hat{e}_\theta + \hat{e}_z \cdot 3\hat{e}_z \tag{1.277}$$

$$= 2\cos\theta - 2\sin\theta - \sin\theta - \cos\theta + 3 \tag{1.278}$$

$$\underline{a} \cdot \underline{m} = \cos\theta - 3\sin\theta + 3 \tag{1.279}$$

Alternatively, we convert $\underline{m}$ from the cylindrical to the Cartesian coordinate system first:

$$\underline{m} = \begin{pmatrix} 1 \\ 1 \\ 3 \end{pmatrix}_{r\theta z} = 1\hat{e}_r + 1\hat{e}_\theta + 3\hat{e}_z \tag{1.280}$$

$$= 1\left(\cos\theta\hat{e}_x + \sin\theta\hat{e}_y\right) + 1\left(-\sin\theta\hat{e}_x + \cos\theta\hat{e}_y\right) + 3\left(\hat{e}_z\right) \tag{1.281}$$

$$= \begin{pmatrix} \cos\theta - \sin\theta \\ \sin\theta + \cos\theta \\ 3 \end{pmatrix}_{xyz} \tag{1.282}$$

and the dot product is formed by summing the products of the x-, y-, and z-coefficients. (Matrix multiplication of coefficients is allowed when the vectors or tensors are written with respect to the same orthonormal basis.)

$$\underline{a} \cdot \underline{m} = \begin{pmatrix} 2 & -1 & 1 \end{pmatrix}_{xyz} \cdot \begin{pmatrix} \cos\theta - \sin\theta \\ \sin\theta + \cos\theta \\ 3 \end{pmatrix}_{xyz} \tag{1.283}$$

$$= \cos\theta - 3\sin\theta + 3 \tag{1.284}$$

We obtain the same result with both methods.

EXAMPLE 1.25. *A flow is produced in the gap between two cylinders by turning the inner cylinder. The device is tall and the gap between the cylinders is small; thus, the effect of the flow at the bottom of the device is negligible. For the flow in the gap, the velocity field is* $\underline{v}$. *If* $\underline{v}$ *is expressed in Cartesian coordinates, what are the nonzero components of the velocity vector? If the cylindrical coordinate system is used, what are the nonzero components? Comment on the solution.*

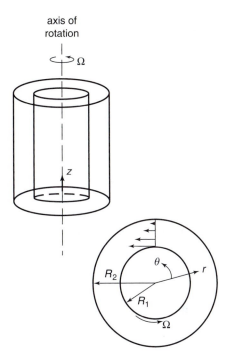

Figure 1.42   A flow is produced in the gap between two cylinders by turning the inner cylinder. The device is tall and the gap between the cylinders is small.

SOLUTION. The flow shown in Figure 1.42 is in the azimuthal direction, and the paths followed by fluid particles are circular. In Cartesian coordinates, there would be no z-component of the velocity but there would be both x- and y-components.

$$\underline{v} = \begin{pmatrix} v_x \\ v_y \\ v_z \end{pmatrix}_{xyz} = \begin{pmatrix} v_x \\ v_y \\ 0 \end{pmatrix}_{xyz} = v_x \hat{e}_x + v_y \hat{e}_y \qquad (1.285)$$

The vector $\underline{v}$ is a vector in a plane with unit normal $\hat{e}_z$.

If we describe the flow using cylindrical coordinates, the three components we seek are v_r, v_θ, and v_z. The component v_r is the component in the direction of increasing r; this component is zero. The component v_θ is the component of velocity in the direction of increasing θ; this is the component that we are seeking. The component v_z is the component of velocity in the direction of increasing z;

this component is zero. Thus, there would be only one nonzero component, v_θ.

$$\underline{v} = \begin{pmatrix} v_r \\ v_\theta \\ v_z \end{pmatrix}_{r\theta z} = \begin{pmatrix} 0 \\ v_\theta \\ 0 \end{pmatrix}_{r\theta z} = v_\theta \hat{e}_\theta \qquad (1.286)$$

If we use Equations 1.262, 1.263, and 1.266, we can convert between these two coordinate systems (remember that $\underline{v}$ is independent of the coordinate system). Beginning with $\underline{v}$ in the cylindrical coordinate system and using Equation 1.266 for $\hat{e}_\theta$ and Equations 1.262–1.264 to convert $\sin\theta$, we obtain:

$$\underline{v} = v_\theta \hat{e}_\theta = v_\theta \left(-\sin\theta \hat{e}_x + \cos\theta \hat{e}_y \right) \qquad (1.287)$$

$$= \begin{pmatrix} -v_\theta \sin\theta \\ v_\theta \cos\theta \\ 0 \end{pmatrix}_{xyz} = \begin{pmatrix} -v_\theta \frac{y}{\sqrt{x^2+y^2}} \\ v_\theta \frac{x}{\sqrt{x^2+y^2}} \\ 0 \end{pmatrix}_{xyz} = \begin{pmatrix} v_x \\ v_y \\ 0 \end{pmatrix}_{xyz} \qquad (1.288)$$

Equation 1.286 for $\underline{v}$ in the $r\theta z$ coordinate system is much simpler than Equation 1.288 for $\underline{v}$ written in the xyz coordinate system. By choosing the cylindrical coordinate system, we reduce the number and complexity of the velocity coefficients that we must solve for. In addition, the boundary conditions are simpler in the cylindrical coordinate system. The boundary conditions are no-slip at the two cylindrical surfaces (see Chapter 6). The no-slip conditions require the velocity of the fluid at the surface to be the same as the velocity of the surface:

$$\text{Boundary conditions:} \quad \begin{cases} r = R_2 \ v_\theta = 0 \\ r = R_1 \ v_\theta = R_1\Omega \end{cases} \qquad (1.289)$$
$$\text{(cylindrical coordinates)}$$

If we use a Cartesian coordinate system, the same boundary conditions are written as:

$$\text{Boundary conditions:} \quad \begin{cases} x^2 + y^2 = R_2^2 \ v_x^2 + v_y^2 = 0 \\ x^2 + y^2 = R_1^2 \ v_x^2 + v_y^2 = R_1^2\Omega^2 \end{cases} \qquad (1.290)$$
$$\text{(Cartesian coordinates)}$$

The boundary conditions written in this way are more difficult to work with than those written in the cylindrical coordinate system.

EXAMPLE 1.26. *Water flows in a horizontal pipe. We want to calculate the flow in the cylindrical coordinate system centered along the pipe axis. In the cylindrical coordinate system, what is the vector expression for the acceleration due to gravity?*

SOLUTION. The two coordinate systems of interest are shown in Figure 1.43. The acceleration due to gravity is given most naturally by the Cartesian vector $\underline{g} = -g\hat{e}_2$. To convert this expression to the cylindrical coordinate system, we must relate the two sets of basis vectors and then use algebra to convert $\underline{g}$. From Figure 1.43, we see that:

$$\hat{e}_1 = \cos\theta \hat{e}_r - \sin\theta \hat{e}_\theta \qquad (1.291)$$

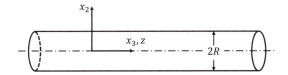

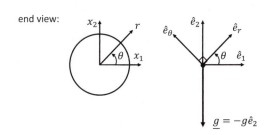

end view:

Figure 1.43 Two coordinate systems are of interest for flow in a horizontal pipe: (1) the cylindrical coordinate system centered along the axis of the tube, and (2) the Cartesian coordinate system containing the axis of the tube as $\hat{e}_3$ and the vertical direction as $\hat{e}_2$.

We also can obtain this result from solving Equations 1.265–1.266 for $\hat{e}_x = \hat{e}_1$. Thus, $\underline{g}$ becomes:

$$\underline{g} = -g\hat{e}_2 = \begin{pmatrix} 0 \\ -g \\ 0 \end{pmatrix}_{123} \tag{1.292}$$

$$= \begin{pmatrix} -g\cos\theta \\ g\sin\theta \\ 0 \end{pmatrix}_{r\theta z} \tag{1.293}$$

These two ways of expressing g are completely equivalent; a vector is independent of the coordinate system in which we express it. When using matrix notation we identify the coordinate system we are using by writing a subscript on the vector or tensor.

EXAMPLE 1.27. *In a liquid of density ρ, what is the net fluid force on a submerged sphere (i.e., a ball or a balloon) (Figure 1.44)? What is the direction of the force and how does the magnitude of the fluid force vary with fluid density?*

SOLUTION. We are not ready to solve this problem at this stage in the text, but when it is solved in Chapter 4, we arrive at the following expression for $\underline{f}$ in terms of an integral in the spherical coordinates (see Figure 4.23):

$$\underline{f} = -\rho g R^2 \int_0^{2\pi} \int_0^{\pi} (H_0 - R\cos\theta)\,\hat{e}_r \sin\theta\,d\theta\,d\phi \tag{1.294}$$

$$= -\rho g R^2 \int_0^{2\pi} \int_0^{\pi} (H_0 - R\cos\theta) \begin{pmatrix} 1 \\ 0 \\ 0 \end{pmatrix}_{r\theta\phi} \sin\theta\,d\theta\,d\phi \tag{1.295}$$

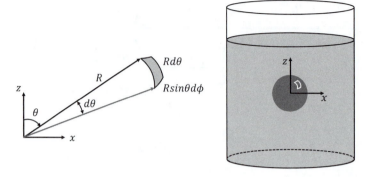

Fluid exerts a net force on a submerged sphere. If the sphere is light, the force from the fluid pressure acts to float the sphere. If the sphere is heavy, the fluid sinks in the fluid but is decelerated by the fluid force.

The basis vector $\hat{e}_r$ varies with θ and ϕ and therefore must be treated as a variable in the integration. The simplest way to proceed is to convert $\hat{e}_r$ to constant Cartesian coordinates before attempting to integrate. The basis vector $\hat{e}_r$ is expressed in Cartesian coordinates in Equation 1.271:

$$\hat{e}_r = \sin\theta\cos\phi\hat{e}_x + \sin\theta\sin\phi\hat{e}_y + \cos\theta\hat{e}_z \tag{1.296}$$

Substituting this into Equation 1.294, we obtain:

$$\underline{f} = -\rho g R^2 \int_0^{2\pi}\int_0^{\pi}(H_0 - R\cos\theta)\,\hat{e}_r\,\sin\theta d\theta d\phi \tag{1.297}$$

$$= -\rho g R^2 \int_0^{2\pi}\int_0^{\pi}(H_0 - R\cos\theta)\sin\theta\,[\sin\theta\cos\phi\hat{e}_x$$
$$+ \sin\theta\sin\phi\hat{e}_y + \cos\theta\hat{e}_z]\,d\theta d\phi \tag{1.298}$$

The equation for $\underline{f}$ is a vector equation, and there are three Cartesian components in Equation 1.298, as emphasized here:

$$\underline{f} = -\rho g R^2 \int_0^{2\pi}\int_0^{\pi}(H_0 - R\cos\theta)\sin\theta\begin{pmatrix}\sin\theta\cos\phi\\\sin\theta\sin\phi\\\cos\theta\end{pmatrix}_{xyz}d\theta d\phi \tag{1.299}$$

Each vector component is integrated separately.

For the x-component ϕ-integration, we integrate $\cos\phi$ from zero to 2π. The result of this definite integral is zero.

$$\int_0^{2\pi}\left[(H_0 - R\cos\theta)\sin^2\theta\right]\cos\phi d\phi = \int_0^{2\pi}[\text{function of }\theta]\ \cos\phi d\phi \tag{1.300}$$

$$= [\text{function of }\theta]\ \sin\phi|_0^{2\pi} = 0 \tag{1.301}$$

For the y-component ϕ-integration, we integrate $\sin \phi$ from zero to 2π. The result of this definite integral also is zero.

$$\int_0^{2\pi} \left[(H_0 - R\cos\theta)\sin^2\theta \right] \sin\phi d\phi = \int_0^{2\pi} [\text{function of } \theta] \; \sin\phi d\phi \quad (1.302)$$

$$= [\text{function of } \theta] \; (-\cos\phi)|_0^{2\pi} = 0$$

$$(1.303)$$

For the z-component ϕ-integration, we integrate an expression independent of ϕ from zero to 2π. This integral is 2π times the expression.

$$\int_0^{2\pi} [(H_0 - R\cos\theta)\sin\theta\cos\theta] \; d\phi = \int_0^{2\pi} [\text{function of } \theta] \; d\phi \quad (1.304)$$

$$= [\text{function of } \theta] \, 2\pi \quad (1.305)$$

Substituting these results into Equation 1.299, we obtain:

$$\underline{f} = -\rho g R^2 \int_0^{\pi} \begin{pmatrix} 0 \\ 0 \\ 2\pi \, (H_0 - R\cos\theta)\sin\theta\cos\theta \end{pmatrix}_{xyz} d\theta \quad (1.306)$$

The last step is to carry out the remaining θ-integral. For the x- and y-components, the θ-integral is the integral of zero, which is zero; the θ-integral for the z-component is straightforward:

$$\underline{f} = -\rho g R^2 \int_0^{\pi} \begin{pmatrix} 0 \\ 0 \\ 2\pi \, (H_0 - R\cos\theta)\sin\theta\cos\theta \end{pmatrix}_{xyz} d\theta \quad (1.307)$$

$$\frac{f_z}{-2\pi R^2 \rho g} = H_0 \int_0^{\pi} \sin\theta\cos\theta d\theta - R \int_0^{\pi} \sin\theta\cos^2\theta d\theta \quad (1.308)$$

The first definite integral is zero (confirm for yourself), indicating that the absolute depth of the sphere, H_0, has no effect on the magnitude of the force. The second definite integral gives a nonzero result that carries forward. The final result is:

$$f_z = \frac{4\pi R^3}{3}\rho g \quad (1.309)$$

$$\underline{f} = \begin{pmatrix} 0 \\ 0 \\ \frac{4\pi R^3}{3}\rho g \end{pmatrix}_{xyz} \quad (1.310)$$

The net fluid force on the sphere is an upward force $\underline{f} = f_z\hat{e}_z$ equal in magnitude to the weight of a sphere-shaped quantity of fluid. Thus, the fluid exerts an upward force (i.e., a force in the $+z$-direction) on the sphere equal in magnitude to the weight of the fluid displaced by the sphere (i.e., Archimedes' principle). This is the *buoyancy effect*, which is why objects float. When the weight of the fluid displaced by an object is higher than the weight of the object itself, the object floats. When the weight of the fluid displaced is less than the weight of the object, the object sinks. Chapter 4 discusses these forces in fluids in more detail.

EXAMPLE 1.28. *For the tensor $\underline{\underline{A}} = 2\hat{e}_1\hat{e}_1$, what is $\underline{\underline{A}}$ written in the cylindrical coordinate system?*

$$\underline{\underline{A}} = 2\hat{e}_1\hat{e}_1 = \begin{pmatrix} 2 & 0 & 0 \\ 0 & 0 & 0 \\ 0 & 0 & 0 \end{pmatrix}_{123} \tag{1.311}$$

SOLUTION. To translate $\underline{\underline{A}}$ written in the 123-coordinate system to the same tensor written in the $r\theta z$-coordinate system, we begin with the tensor written explicitly in terms of the basis vectors $\hat{e}_1$, $\hat{e}_2$, $\hat{e}_3$. We use the expressions in Equations 1.265–1.267 to algebraically convert the basis vectors:

$$\underline{\underline{A}} = 2\hat{e}_1\hat{e}_1 \tag{1.312}$$

To write $\hat{e}_1$ in terms of $\hat{e}_r$, $\hat{e}_\theta$, and $\hat{e}_z$, we explicitly solve Equations 1.265 and 1.266 for $\hat{e}_1$:

$$\hat{e}_r = \cos\theta\hat{e}_1 + \sin\theta\hat{e}_2 \tag{1.313}$$

$$\hat{e}_\theta = -\sin\theta\hat{e}_1 + \cos\theta\hat{e}_2 \tag{1.314}$$

Solving for $\hat{e}_1$:

$$\hat{e}_1 = \cos\theta\hat{e}_r - \sin\theta\hat{e}_\theta \tag{1.315}$$

Substituting this result into Equation 1.312 twice and carrying out the distributive law, we obtain:

$$\underline{\underline{A}} = 2\hat{e}_1\hat{e}_1 \tag{1.316}$$

$$= (\cos\theta\hat{e}_r - \sin\theta\hat{e}_\theta)(\cos\theta\hat{e}_r - \sin\theta\hat{e}_\theta) \tag{1.317}$$

$$= \cos^2\theta\hat{e}_r\hat{e}_r - \sin\theta\cos\theta\hat{e}_r\hat{e}_\theta - \sin\theta\cos\theta\hat{e}_\theta\hat{e}_r + \sin^2\theta\hat{e}_\theta\hat{e}_\theta \tag{1.318}$$

$$\underline{\underline{A}} = \begin{pmatrix} \cos^2\theta & -\sin\theta\cos\theta & 0 \\ -\sin\theta\cos\theta & \sin^2\theta & 0 \\ 0 & 0 & 0 \end{pmatrix}_{r\theta z} \tag{1.319}$$

The same tensor $\underline{\underline{A}}$ is expressed in Equations 1.311 and 1.319—the two versions are expressed with respect to different coordinate systems.

1.3.3 Substantial derivative

The mass, momentum, and energy conservation equations introduced in Section 1.3.2 are written in Equations 1.239–1.241 in a way that emphasizes the similarity of the lefthand terms. Notice that on the lefthand side of those equations, the following pattern recurs:

$$\frac{\partial f}{\partial t} + \underline{v} \cdot \nabla f \tag{1.320}$$

where, depending on which equation we look at, f is density, velocity, or energy. This pattern is called the *substantial derivative*. The notation for a substantial derivative is a derivative written with a capital D:

Substantial derivative (Gibbs notation)

$$\frac{Df}{Dt} \equiv \frac{\partial f}{\partial t} + \underline{v} \cdot \nabla f$$

(1.321)

Cartesian coordinates (see Table B.2)

$$\frac{Df}{Dt} \equiv \frac{\partial f}{\partial t} + \frac{\partial f}{\partial x_1} v_1 + \frac{\partial f}{\partial x_2} v_2 + \frac{\partial f}{\partial x_3} v_3$$

(1.322)

The substantial derivative has a physical meaning: the rate of change of a quantity (i.e., mass, energy, or momentum) as experienced by an observer that is moving along with the flow. The observations made by a moving observer are affected by the stationary time rate of change of the property ($\partial f/\partial t$); however, what is observed also depends on where the observer goes as it floats along with the flow ($\underline{v} \cdot \nabla f$). If the flow takes the observer into a region where, for example, the local energy is higher, then the observed amount of energy will be higher due to this change in location. The rate of change from the perspective of an observer floating along with a flow appears naturally in the equations of change.

The physical meaning of the substantial derivative is discussed more completely in the sidebar and in National Committee for Fluid Mechanics Films (NCFMF) available on the Internet [120]. This chapter concludes with practical mathematical advice in Section 1.3.4. Chapter 2 describes fluid behavior as a first step to fluid-mechanics modeling.

Substantial Derivative in Fluid Mechanics

In fluid mechanics and other branches of physics, we often deal with properties that vary in space and change with time. Thus, we must consider the differentials of multivariable functions. Consider a multivariable function, $f(t, x_1, x_2, x_3)$, associated with a particle of fluid, where t is time and x_1, x_2, and x_3 are the three spatial coordinates. The function f might be, for example, the density of flowing material as a function of time and position. The expression Δf is the change in f when comparing the value of the function f at two nearby points, (t, x_1, x_2, x_3) and $(t + \Delta t, x_1 + \Delta x_1, x_2 + \Delta x_2, x_3 + \Delta x_3)$.

$$f = f(t, x_1, x_2, x_3)$$

(1.323)

$$\Delta f = f(t + \Delta t, x_1 + \Delta x_1, x_2 + \Delta x_2, x_3 + \Delta x_3) - f(t, x_1, x_2, x_3)$$

(1.324)

In the limit that the two points are close together, Δf becomes the differential df:

$$df = \lim_{\substack{\Delta x_1 \longrightarrow 0 \\ \Delta x_2 \longrightarrow 0 \\ \Delta x_3 \longrightarrow 0 \\ \Delta t \longrightarrow 0}} \Delta f$$

(1.325)

(continued)

Substantial Derivative in Fluid Mechanics *(continued)*

We can write Δf in terms of partial derivatives, which are functions that give the rates of change of f (i.e., slopes) in the three coordinate directions x_1, x_2, and x_3 (see Web appendix [108] for a review):

$$\Delta f = \frac{\partial f}{\partial t}\Delta t + \frac{\partial f}{\partial x_1}\Delta x_1 + \frac{\partial f}{\partial x_2}\Delta x_2 + \frac{\partial f}{\partial x_3}\Delta x_3 \qquad (1.326)$$

Because the differential df is the limit of Δf as all changes of variable go to zero, we can take the limit of Equation 1.326 to obtain df in terms of dx_1, dx_2, and dx_3:

$$df = \lim_{\substack{\Delta x_1 \to 0 \\ \Delta x_2 \to 0 \\ \Delta x_3 \to 0 \\ \Delta t \to 0}} \Delta f \qquad (1.327)$$

$$df = \lim_{\substack{\Delta x_1 \to 0 \\ \Delta x_2 \to 0 \\ \Delta x_3 \to 0 \\ \Delta t \to 0}} \frac{\partial f}{\partial t}\Delta t + \frac{\partial f}{\partial x_1}\Delta x_1 + \frac{\partial f}{\partial x_2}\Delta x_2 + \frac{\partial f}{\partial x_3}\Delta x_3 \qquad (1.328)$$

$$df = \frac{\partial f}{\partial t}dt + \frac{\partial f}{\partial x_1}dx_1 + \frac{\partial f}{\partial x_2}dx_2 + \frac{\partial f}{\partial x_3}dx_3 \qquad (1.329)$$

This is the familiar chain rule. The direction in going from (t, x_1, x_2, x_3) to $(t + \Delta t, x_1 + \Delta x_1, x_2 + \Delta x_2, x_3 + \Delta x_3, t + \Delta t)$ is not specified in the definition of df; Equation 1.329 applies to any path between any two nearby points.

There is a particular path and set of neighboring particles that are of recurring interest in fluid mechanics: the path that fluid particles take. Fluid particles are discussed in detail in Chapter 3 but, briefly, a fluid particle is an infinitesimally small amount of fluid. For a chosen particle, its motion describes a path through three-dimensional space (Figure 1.45). These paths are called pathlines of the flow.

Consider variation in the function f along a particular path—that is, the path that a fluid particle traces out as it travels through a flow. The function f might be density as a function of position and time for example, or temperature as a function of position and time. Beginning at an arbitrary point in the flow, we compare the value of f at the original point and at the nearby point $f + \Delta f$. For an arbitrary path as just discussed, Δf is given by Equation 1.328 repeated below:

$$\Delta f|_{along\ ANY\ path} = \frac{\partial f}{\partial t}\Delta t + \frac{\partial f}{\partial x_1}\Delta x_1 + \frac{\partial f}{\partial x_2}\Delta x_2 + \frac{\partial f}{\partial x_3}\Delta x_3 \qquad (1.330)$$

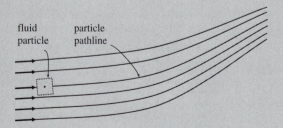

fluid   particle
particle   pathline

Figure 1.45   A fluid particle consists of the same molecules at all times. The path that a particle follows through a flow is called a pathline.

Substantial Derivative in Fluid Mechanics *(continued)*

If we now follow fluid particles along a particular path, the particle pathline, then we can relate the directions Δx_1, Δx_2, and Δx_3 to the local fluid velocity components v_1, v_2, and v_3:

$$\text{Along a flow pathline:} \quad \begin{cases} \Delta x_1 = v_1 \Delta t \\ \Delta x_2 = v_2 \Delta t \\ \Delta x_3 = v_3 \Delta t \end{cases} \tag{1.331}$$

Substituting these expressions into Equation 1.330, we obtain:

$$\Delta f|_{\text{along particle pathline}} = \frac{\partial f}{\partial t} \Delta t + \frac{\partial f}{\partial x_1} v_1 \Delta t + \frac{\partial f}{\partial x_1} v_2 \Delta t + \frac{\partial f}{\partial x_1} v_3 \Delta t \tag{1.332}$$

$$= \Delta t \left(\frac{\partial f}{\partial t} + \frac{\partial f}{\partial x_1} v_1 + \frac{\partial f}{\partial x_1} v_2 + \frac{\partial f}{\partial x_1} v_3 \right) \tag{1.333}$$

Dividing through by Δt and taking the limit as Δt goes to zero, we arrive at the following expression, which is the substantial derivative:

$$\frac{\Delta f}{\Delta t}\bigg|_{\text{along particle pathline}} = \frac{\partial f}{\partial t} + \frac{\partial f}{\partial x_1} v_1 + \frac{\partial f}{\partial x_1} v_2 + \frac{\partial f}{\partial x_1} v_3 \tag{1.334}$$

$$\frac{Df}{Dt} \equiv \frac{df}{dt}\bigg|_{\text{along particle pathline}} = \lim_{\Delta t \to 0} \frac{\Delta f}{\Delta t}\bigg|_{\text{along particle pathline}} \tag{1.335}$$

Substantial derivative or rate of change of f along a particle pathline

$$\boxed{\frac{Df}{Dt} = \frac{\partial f}{\partial t} + \frac{\partial f}{\partial x_1} v_1 + \frac{\partial f}{\partial x_2} v_2 + \frac{\partial f}{\partial x_3} v_3} \tag{1.336}$$

Thus, the substantial derivative gives the time rate of change of a function f as an observer floats along a pathline in a flow, attached to a fluid particle. Why does this matter in fluid mechanics? One reason is that sometimes measurements are made in just this way, by floating an instrument in a flow—for example, a weather balloon (Figure 1.46). The density, velocity, or temperature as a function of time recorded this way is the substantial derivative along the pathline traveled. In meteorology and oceanography, it is common to take measurements of the substantial derivative.

However, the main reason that the substantial derivative is important is that it appears in the mass, momentum, and energy-conservation equations (Equations 1.239–1.241):

$$\text{Mass conservation} \quad \frac{D\rho}{Dt} = -\rho \left(\nabla \cdot \underline{v} \right) \tag{1.337}$$

$$\text{Momentum conservation} \quad \rho \frac{D\underline{v}}{Dt} = -\nabla p + \mu \nabla^2 \underline{v} + \rho \underline{g} \tag{1.338}$$

$$\text{Energy conservation} \quad \rho \frac{D\hat{E}}{Dt} = -\nabla \cdot \underline{q} - \nabla \cdot (p\underline{v}) + \nabla \cdot \underline{\underline{\tau}} \cdot \underline{v} + \mathcal{S}_e \tag{1.339}$$

The substantial derivative appears because each equation is written in terms of the properties of a *field* (written in terms of the field variables ρ, $\underline{v}$, and $\hat{E}$) rather than of a single isolated body (Figure 1.47). To understand the difference, consider mass, momentum, and energy conservation

(continued)

Substantial Derivative in Fluid Mechanics *(continued)*

Figure 1.46　Weather balloons float along at the average velocity of the fluid and measure velocity, temperature, and other variables of interest to weather forecasters. This is an example of a measurement of properties along a pathline. (Meteorologist Jeff DeRosa launches a weather balloon. © Russ Durkee, 2005, NSF, USAP Photo Library)

	rate of change associated with a body	*rate of change of a field variable as recorded by an observer moving with the flow*
mass	$\dfrac{d(m)_{body}}{dt}$	$\dfrac{D\rho}{Dt}$
momentum	$\dfrac{d(m\underline{v})_{body}}{dt}$	$\rho\dfrac{D\underline{v}}{Dt}$
energy	$\dfrac{d(E)_{body}}{dt}$	$\rho\dfrac{D\hat{E}}{Dt}$

Figure 1.47　The balance equations for mass, momentum, and energy may be written for a body or a position in space—that is, for a field.

Substantial Derivative in Fluid Mechanics *(continued)*

of a body. The mass of a body is conserved in the sense that if the mass changes—if a piece is shaved off, for example—it is not the same body. The momentum of a body is conserved (i.e., Newton's second law; see Chapter 3), and the energy of a body is conserved (i.e., first law of thermodynamics; see Chapter 6). For a body, the conservation laws contain the usual time rates of change of mass (dm/dt), momentum ($d(m\underline{v})/dt$), and energy (dE/dt).

When we are concerned with the properties characteristic of a location in a field rather than of a chosen body, the correct expression for the rate of change of the field variable at a fixed point is shown to be the substantial derivative (see Chapters 3 and 6). The rate of change of a property—mass, momentum, and energy—for a given position in a field depends on the instantaneous rate of change of the property at that location ($\partial/\partial t$) as well as the rate at which the property is convected to that location by the fluid motion ($\underline{v} \cdot \nabla$). In Chapter 6, we derive the mass, momentum, and energy balances for a position in a field, and the substantial derivative appears naturally. The concepts outlined here are discussed fully in Chapters 3 and 6. We present two examples to build familiarity with the substantial derivative.

EXAMPLE 1.29. *Using Equation 1.242 to write ∇f, use matrix multiplication to verify the equality of the following two expressions for the substantial derivative:*

$$\begin{matrix} \text{Substantial} \\ \text{derivative} \\ \text{(Gibbs notation)} \end{matrix} \qquad \frac{Df}{Dt} \equiv \frac{\partial f}{\partial t} + \underline{v} \cdot \nabla f \qquad (1.340)$$

$$\begin{matrix} \text{Cartesian coordinates} \\ \text{(see Table B.2)} \end{matrix} \quad \frac{Df}{Dt} \equiv \frac{\partial f}{\partial t} + \frac{\partial f}{\partial x_1}v_1 + \frac{\partial f}{\partial x_2}v_2 + \frac{\partial f}{\partial x_3}v_3 \qquad (1.341)$$

SOLUTION. To show the equality of these two equations, we write the Gibbs notation expressions $\underline{v}$ and ∇f in Cartesian coordinates and matrix multiply:

$$\underline{v} = \begin{pmatrix} v_1 \\ v_2 \\ v_3 \end{pmatrix}_{123} \qquad (1.342)$$

$$\nabla f = \begin{pmatrix} \dfrac{\partial f}{\partial x_1} \\[2mm] \dfrac{\partial f}{\partial x_2} \\[2mm] \dfrac{\partial f}{\partial x_3} \end{pmatrix}_{123} \qquad (1.343)$$

$$\underline{v} \cdot \nabla f = \begin{pmatrix} v_1 & v_2 & v_3 \end{pmatrix}_{123} \cdot \begin{pmatrix} \dfrac{\partial f}{\partial x_1} \\[2mm] \dfrac{\partial f}{\partial x_2} \\[2mm] \dfrac{\partial f}{\partial x_3} \end{pmatrix}_{123} \qquad (1.344)$$

$$= v_1\frac{\partial f}{\partial x_1} + v_2\frac{\partial f}{\partial x_2} + v_3\frac{\partial f}{\partial x_3} \qquad (1.345)$$

(continued)

Substantial Derivative in Fluid Mechanics *(continued)*

Thus:

$$\frac{Df}{Dt} \equiv \frac{\partial f}{\partial t} + \underline{v} \cdot \nabla f = \frac{\partial f}{\partial t} + \frac{\partial f}{\partial x_1} v_1 + \frac{\partial f}{\partial x_2} v_2 + \frac{\partial f}{\partial x_3} v_3 \qquad (1.346)$$

EXAMPLE 1.30. *What is the substantial derivative $D\underline{v}/Dt$ of the steady-state velocity field represented by the following velocity vector? Note that the answer is a vector.*

$$\underline{v}(x, y, z, t) = \begin{pmatrix} -3.0x \\ -3.0y \\ 6z \end{pmatrix}_{xyz} \qquad (1.347)$$

SOLUTION. We begin with the definition of the substantial derivative in Equation 1.321 and substitute $\underline{v}$ for f:

$$\frac{D\underline{v}}{Dt} = \frac{\partial \underline{v}}{\partial t} + \underline{v} \cdot \nabla \underline{v} \qquad (1.348)$$

We now consult Table B.2 to determine the components of $\underline{v} \cdot \nabla \underline{v}$ in Cartesian coordinates, and we construct the Cartesian expression for $D\underline{v}/Dt$:

$$\underline{v} \cdot \nabla \underline{v} = \begin{pmatrix} v_x \frac{\partial v_x}{\partial x} + v_y \frac{\partial v_x}{\partial y} + v_z \frac{\partial v_x}{\partial z} \\ v_x \frac{\partial v_y}{\partial x} + v_y \frac{\partial v_y}{\partial y} + v_z \frac{\partial v_y}{\partial z} \\ v_x \frac{\partial v_z}{\partial x} + v_y \frac{\partial v_z}{\partial y} + v_z \frac{\partial v_z}{\partial z} \end{pmatrix}_{xyz} \qquad (1.349)$$

$$\frac{D\underline{v}}{Dt} = \frac{\partial \underline{v}}{\partial t} + \underline{v} \cdot \nabla \underline{v} \qquad (1.350)$$

$$= \begin{pmatrix} \frac{\partial v_x}{\partial t} \\ \frac{\partial v_y}{\partial t} \\ \frac{\partial v_z}{\partial t} \end{pmatrix}_{xyz} + \begin{pmatrix} \frac{\partial v_x}{\partial t} + v_x \frac{\partial v_x}{\partial x} + v_y \frac{\partial v_x}{\partial y} + v_z \frac{\partial v_x}{\partial z} \\ \frac{\partial v_y}{\partial t} + v_x \frac{\partial v_y}{\partial x} + v_y \frac{\partial v_y}{\partial y} + v_z \frac{\partial v_y}{\partial z} \\ \frac{\partial v_z}{\partial t} + v_x \frac{\partial v_z}{\partial x} + v_y \frac{\partial v_z}{\partial y} + v_z \frac{\partial v_z}{\partial z} \end{pmatrix}_{xyz} \qquad (1.351)$$

Finally, we carry out the partial derivatives on the various terms of the velocity field and substitute them into equation 1.351:

$$\frac{D\underline{v}}{Dt} = \begin{pmatrix} 0 + (-3)(-3x) + 0 + 0 \\ 0 + 0 + (-3)(-3y) + 0 \\ 0 + 0 + 0 + 6(6z) \end{pmatrix}_{xyz} \qquad (1.352)$$

$$\frac{D\underline{v}}{Dt} = \begin{pmatrix} 9x \\ 9y \\ 36z \end{pmatrix}_{xyz} \qquad (1.353)$$

1.3.4 Practical advice

The analysis of flows often means solving for density, velocity, and stress fields. The equations that we encounter in these analyses are ordinary differential equations (ODEs) and partial differential equations (PDEs) [17]. The solutions of differential equations give the complete density, velocity, and stress fields for a problem, from which many engineering quantities can be calculated. In this text, it is assumed that students have taken multivariable calculus, linear algebra, and a first course in solving differential equations; we apply these and other mathematics skills in our study of fluid mechanics.

To prepare students to study fluid mechanics, the Web appendix [108] contains a review of solution methods for differential equations. Also, several exercises provide problem-solving practice that may be helpful. For instructional videos on mathematics through differential equations, see [73]. For more on solving ODEs and PDEs, see the Web appendix [108] and [61]. We move on to modeling flows in general in Chapter 2.

EXAMPLE 1.31. *In fluid mechanics, we encounter the following equation:*

$$\frac{p_L - p_0}{L} = \frac{\mu}{r} \frac{d}{dr} \left(r \frac{dv_z}{dr} \right) - \rho g \sin \alpha \qquad (1.354)$$

This equation appears in the analysis of pressure-driven flow in a tilted tube in cylindrical coordinates. Solve the differential equation for $v_z(r)$; note that p_L, p_0, L, μ, ρ, g, and α are all constants.

SOLUTION. In this example and the one that follows, we show the details of integration for problems related to fluid mechanics. At first, the differential equations to solve appear to be complex; in this case, however, only the most elementary integrations are required. Strategies for recognizing and carrying out the solution are discussed.

The first step in solving an equation—once it has been derived from the physics—is to take careful stock of it. Is it an algebraic or a differential equation? If it is a differential equation, is it an ODE (i.e., function of a single independent variable) or a PDE (i.e., function of two or more independent variables)? Which expressions in the equation are constant and which are variable?

To clarify the structure of Equation 1.354, we group the constants together and rename that group:

$$\frac{1}{r} \frac{d}{dr} \left(r \frac{dv_z}{dr} \right) = \left[\frac{p_L - p_0}{\mu L} + \frac{\rho g \sin \alpha}{\mu} \right] \equiv B \qquad (1.355)$$

$$\frac{d}{dr} \left(r \frac{dv_z}{dr} \right) = B r \qquad (1.356)$$

where B is a constant equal to the quantity in the square brackets of Equation 1.355. Equation 1.356 is a cleaner representation of Equation 1.355 and therefore is easier to solve. This is further clarified when we recognize that the lefthand side of Equation 1.356 is written as the derivative of a grouped quantity. We can simplify the appearance of Equation 1.356 if we define the quantity in parentheses as a new variable Ψ:

$$\text{Define } \Psi: \quad \Psi \equiv \left(r \frac{dv_z}{dr} \right) \tag{1.357}$$

$$\text{Substitute } \Psi \text{ into the} \atop \text{differential equation:} \quad \frac{d}{dr} \left(r \frac{dv_z}{dr} \right) = B \, r \tag{1.358}$$

$$\frac{d\Psi}{dr} = B \, r \tag{1.359}$$

Equation 1.359 is clear and simple to solve. Integrating once:

$$\int d\Psi = \int B \, r \, dr \tag{1.360}$$

$$\Psi = B \frac{r^2}{2} + C_1 \tag{1.361}$$

where C_1 is an integration constant. Substituting the definition of Ψ, we now rearrange and integrate:

$$\Psi = \left(r \frac{dv_z}{dr} \right) = B \frac{r^2}{2} + C_1 \tag{1.362}$$

$$\frac{dv_z}{dr} = \frac{B}{2} r + \frac{C_1}{r} \tag{1.363}$$

$$\int dv_z = \int \left(\frac{B}{2} r + \frac{C_1}{r} \right) dr \tag{1.364}$$

$$v_z = \frac{B}{4} r^2 + C_1 \ln r + C_2 \tag{1.365}$$

where C_2 is a second integration constant. This is as far as we can go. Because Equation 1.354 is a second-order ODE, we need two boundary conditions on r to determine the two integration constants, C_1 and C_2.

EXAMPLE 1.32. *In pressure-driven flow in a tube (Poiseuille flow; see Section 7.1), the z-component of the momentum balance simplifies to the equation shown here (see also Equation 7.16). Solve for $v_z(r)$ and $p(z)$.*

$$\frac{\partial p(z)}{\partial z} = \frac{\mu}{r} \frac{\partial}{\partial r} \left(r \frac{\partial v_z(r)}{\partial r} \right) + \rho g \tag{1.366}$$

SOLUTION. Although Equation 1.366 is a PDE, it is among the simplest PDEs to solve because it is separable. A separable PDE of two variables is one that can be completely separated into two independent equations to solve, as we now demonstrate.

The pressure $p(z)$ in Equation 1.366 is given as only a function of z and the velocity $v_z(r)$ is given as only a function of r. Gravity (g), density (ρ), and viscosity (μ) are constant. If we rearrange Equation 1.366, we can collect all of the z-dependent terms on the left and all of the r-dependent terms on the right. The constant terms can go on either side; we arbitrarily group the constant terms with the pressure:

$$\frac{\partial p}{\partial z} - \rho g = \frac{\mu}{r} \frac{\partial}{\partial r} \left(r \frac{\partial v_z}{\partial r} \right) \tag{1.367}$$

The lefthand side is only a function of z and the righthand side is only a function of r. We have succeeded in separating the two variables, r and z. Thus, both sides must be equal to the same constant, which we call λ [58]:

$$\frac{\partial p}{\partial z} - \rho g = \frac{\mu}{r} \frac{\partial}{\partial r} \left(r \frac{\partial v_z}{\partial r} \right) = \lambda \tag{1.368}$$

We separated the z and r parts of Equation 1.367 into two independent equations that we can solve directly:

$$\frac{\partial p}{\partial z} - \rho g = \lambda \tag{1.369}$$

$$\frac{\mu}{r} \frac{\partial}{\partial r} \left(r \frac{\partial v_z}{\partial r} \right) = \lambda \tag{1.370}$$

Because Equations 1.369 and 1.370 are now ODEs, we change the differentiation from partial differentiation $\partial/\partial r$, $\partial/\partial z$ to total differentiation d/dr, d/dz. The remaining steps are straightforward:

$$\text{Pressure ODE:} \quad \frac{dp}{dz} - \rho g = \lambda \tag{1.371}$$

$$\frac{dp}{dz} = (\lambda + \rho g) \tag{1.372}$$

$$\int dp = \int (\lambda + \rho g)\, dz \tag{1.373}$$

$$p = (\lambda + \rho g)z + C_3 \tag{1.374}$$

where C_3 is an integration constant.

$$z\text{-velocity ODE:} \quad \frac{\mu}{r} \frac{d}{dr} \left(r \frac{dv_z}{dr} \right) = \lambda \tag{1.375}$$

$$\frac{d}{dr} \left(r \frac{dv_z}{dr} \right) = \left(\frac{\lambda}{\mu} \right) r \tag{1.376}$$

The solution of Equation 1.376 is discussed in the previous example (compare to Equation 1.356).

1.4 Problems

1. Create a list of five real engineering problems or societal challenges that can be addressed with the modeling introduced in this chapter and studied in fluid mechanics.

2. The green hose fills a swimming pool in 4 hours, the red hose fills the same pool in 6 hours, and the yellow hose fills it in 8 hours. With all three hoses running at those rates, how long will it take to fill the pool?

3. What is a typical volumetric flow rate (in gpm and lpm (liters per minute) for household plumbing? What is a typical value of average velocity in a pipe? Assume half-inch type-K copper tubing (see *Perry's Chemical Engineering Handbook* [132] for dimensions).

4. Compare typical values of velocity head, pressure head, elevation head, and friction head. What is a good rule of thumb for velocity differences that are significant in the flow of household water? Assume that the relevant piping is half-inch type K copper tubing (see *Perry's Chemical Engineering Handbook* [132] for dimensions).

5. What are the viscosity and density of glycerin at room temperature? A useful reference for physical-property data is *Perry's Chemical Engineering Handbook* [132].

6. How do the viscosity of sugar–water solutions vary with concentration and temperature? (Find the answer in the literature.) Provide a plot that shows how the data vary; consider carefully how to plot the data so that the trend is displayed meaningfully.

7. Examine the friction factor/Reynolds number relationship for turbulent flow in pipes (see Figure 1.21). Calculate the pressure drop versus the flow rate for turbulent flow in a rough pipe in an existing apparatus at a chemical plant. List the information needed about the pipe to make the calculation. Which factors are the most critical?

8. For household water in steady flow in a half-inch Schedule 40 horizontal pipe at 3.0 gpm (see Figure 1.20), what are the frictional losses over a 100-foot run of pipe? The flow may be laminar or turbulent. (This problem was proposed originally as Example 1.8; on completion of this chapter, we now can solve it.)

9. What is the range of the friction factor for turbulent flow in smooth and rough pipes? What is the range of the friction factor for laminar flow?

10. Water at 25°C flows at 6.3×10^{-3} m^3/s through the irregularly shaped container in Figure 1.48. What is the average fluid velocity at the exit? The apparatus is open to the atmosphere at the entrance and the exit.

11. At a Reynolds number of 10,000, flow in a pipe is turbulent and it is not possible to produce a laminar flow. What is the friction factor for a flow in smooth pipe at this Reynolds number? If somehow we could produce a laminar flow at this Reynolds number, what would the friction factor be? Repeat for Re $= 10^5$. Compare the two answers and discuss.

12. *Piping* and *tubing* are names for conduits of fluids, but the two terms differ in that the outer diameter (OD) of piping is standardized to allow pipefitters to mount pipes into standard-size holders. The tubing OD is not standardized. What are the ID and OD of nominal 1/2-inch, 3/4-inch, and 1-inch Schedule 40 pipes? Give dimensions in both inches and mm. What are the closest metric standard pipe sizes to these three sizes? Search for these answers in the literature.

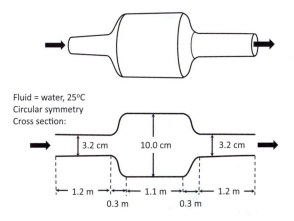

Fluid = water, 25°C
Circular symmetry
Cross section:

| **Figure 1.48** | Flow through an irregular container (Problem 10). |

13. Piping is rated by its nominal size—for example, 1/2-inch or 3/8-inch pipe—but the true ID is not the same as the nominal size. For water flowing in 1/2-inch, Schedule 40 PVC (smooth) pipe at 3.0 gpm, calculate the average velocity and the Reynolds number using the correct, true ID of the pipe. Calculate the average velocity and Reynolds number using 0.5 inch (i.e., the nominal size) as the diameter. Calculate the friction factor based on these two numbers (e.g., using the Colebrook equation or Equation 7.158). Calculate the predicted pressure drop per unit length $\Delta p/L$ in the two cases. How much error in pressure drop is generated for 100 feet of pipe when the wrong diameter is used?

14. Glycerin at room temperature is made to flow through a pipe (the ID is 1.2 mm) at a Reynolds number of 1.00×10^2. What is the average velocity of the glycerin? What is the average velocity if the fluid is water instead? Which flow generates more friction? Be quantitative in your answer and explain.

15. A 30-gallon bathtub takes about 8.0 minutes to fill. What is the flow rate of water in the pipes (1/2-inch type-K copper tubing) in gpm? What is the flow rate in cm^3/s?

16. Water (25°C) flows through 1-inch Schedule 40 steel pipe at 2.0 gpm. What is the Reynolds number of the flow? What is the friction factor? Is the flow laminar or turbulent?

17. Water (25°C) flows through 1-1/2-inch Schedule 40 pipe at 2.0 gpm. What is the pressure drop along 5,000 feet of smooth pipe? If the pipe is not smooth but rather commercial steel, what is the pressure drop?

18. Room temperature water comes out of a spigot at 3.0 gpm. How long would it take to fill a 5-gallon bucket?

19. Water at room temperature comes out of a spigot at the maximum speed possible for the flow to still be laminar. What is the flow rate in gpm and in liters/minute? The flow line is 1/2-inch, Schedule 40 smooth pipe.

20. Water (25°C) flows through DN40 (metric pipe size) Schedule 40 smooth pipe at 8.0 liters/minute. What is the pressure drop along 1,500 meters of pipe? If the flow rate doubles to 16 liters/minute, what is the pressure drop?

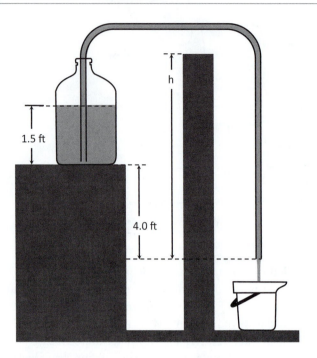

1.5 ft

h

4.0 ft

Figure 1.49   Schematic for Problem 23.

21. A Venturi meter with a 1.00-inch diameter throat is to be installed in a 2-inch line (i.e., Schedule 40 piping, smooth) with water flowing at 25°C. If the flow is turbulent and the range of expected flow rates is 0–200 gpm, what is the expected range of pressure drop in the Venturi meter? You may neglect frictional losses.

22. A Venturi meter with a 4.00 mm ID throat is installed in a 25DN line (metric pipe size, Schedule 40, piping, smooth) with water flowing at 25°C. If the flow is turbulent and the maximum flow rate is 40.0 liter/min, what is the pressure drop in the Venturi meter? You may neglect friction.

23. A gasoline tank is connected to a 25-foot hose (ID $= 1.50$ cm) as shown in Figure 1.49. The ambient temperature is 38°C. What is the maximum height of the barrier over which the gasoline may be siphoned? You may neglect frictional losses. Note the following physical property data: density of gasoline $= 5.6$ lb$_m$/gal and vapor pressure at 38°C is 12.3 psia.

24. A water tank is connected to a 100-foot hose (ID $= 1.50$ cm), as shown in the top of Figure 1.50. The height h is 1.8 meters. Calculate the average velocity of water in the hose. Do not neglect friction; you may assume turbulent flow.

25. For the flow setup in Problem 24 ($h = 1.8$ meters), if we elevate the center of the hose, the flow will continue unabated. At some elevation, however, the pressure inside the elevated part will drop to the vapor pressure of water at 25°C and the water will boil, breaking the siphon. At what height, H, will the siphon break?

26. A pipeline of diameter d connects the fluid (density $= \rho$) in an elevated open tank and a closed tank (Figure 1.51). The fluid is motionless. Determine the pressure in the lower tank in terms of the labeled heights.

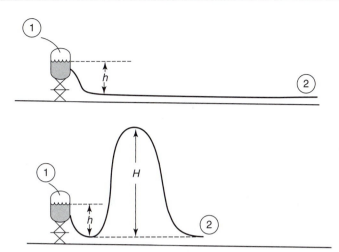

Figure 1.50   Schematic for Problems 24 and 25.

27. Water at 25°C fills the irregularly shaped container in Figure 1.52. What is the absolute pressure P in psia at the position noted? The apparatus is open to the atmosphere at the top. The apparatus is 100.0 cm thick into the page.

28. A tall scaffolding is erected next to a lake where a pump is operating. The maximum head deliverable by the pump is $W_{pump}/mg = 70$ ft. A long hose is connected to the pump exit, and the pump draws water from the lake. The

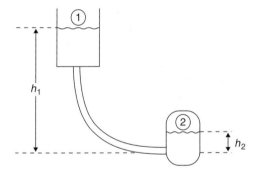

Figure 1.51   Schematic of apparatus for Problem 26.

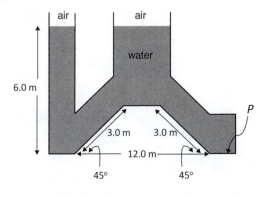

Figure 1.52   Schematic for Problem 27.

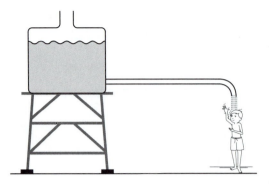

Figure 1.53   Schematic of a rustic shower arrangement in the woods (Problem 29).

pump is running and water is coming out of the hose. You grab the end of the hose and start climbing the scaffolding. How high do you have to climb before the water stops coming out of the hose? Justify your answer using the mechanical energy balance.

29. At a vacation camp in the woods, the owner collects rainwater for washing. She plans to construct a cold-water shower by mounting the collection tank (i.e., 150-gallon, 36-inch diameter) on a platform and using gravity to provide the flow through piping attached to a hole in the side near the bottom of the tank (Figure 1.53). She easily can obtain PEX tubing (i.e., cross-linked polyethylene) in nominal 1/2-inch and 1-inch sizes. What is the flow rate at the pipe exit at the beginning of the shower if she connected 10 feet of the 1/2-inch PEX (ID = 0.632 inches) to a full tank of water? What is the flow rate if the tank were only half full? Do not neglect friction.

30. Your grandfather has a cottage at the lake and wants to install a pump to deliver water to the house. He plans to pump water at night to fill a storage tank that he installed next to the cottage (Figure 1.54). The pipes and fittings he chose to use for the installation are listed in the table given. The pumps in the catalog your grandfather consulted are rated by their value of horsepower (hp). What is the minimum hp rating of a pump capable of providing a flow

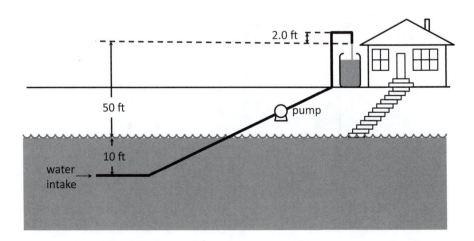

Figure 1.54   Schematic of the water system at cottage (Problem 30).

rate of 5 gpm of water at the tank? Assume that the pump is 65 percent efficient (i.e., of the energy put out by the pump, only 65 percent goes toward work on the fluid) and that the pipe is PVC (i.e., polyvinyl chloride, a polymeric material that is assumed to be smooth).

Fitting	Number of fittings
straight pipe, 1 inch, Schedule 40	95 feet
coupling	8
globe valve	1
gate valve	4
disk water meter	1

31. Pressure-drop versus flow-rate data were taken on water (room temperature) flowing in a 30.0 m section of old 1-inch Schedule 40 pipe (Table 1.7). Calculate the friction factor versus the Reynolds number for these data. How do the results compare to the standard correlation for the friction factor (i.e., the Colebrook equation)? Be quantitative. If we assume that there has been some scaling (i.e., deposition of hard deposits on the inner walls) that has decreased the effective pipe ID, can we improve the correspondence between the data and the literature correlation? Discuss.

32. A pump is connected between two tanks as shown in Figure 1.55. Calculate the pressure head, the velocity head, the elevation head, and the friction head

Table 1.7. Data for flow in a pipe
for Problem 31

Δp (kPa)	Q (cm^3/s)
8.0	350
20	560
45	880
88	1,400
230	2,200
470	3,500

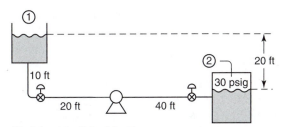

All piping 1 in. Schedule 40
2 gate valves
1 90° bend

Figure 1.55   Schematic of flow between an open and a closed tank (Problem 32).

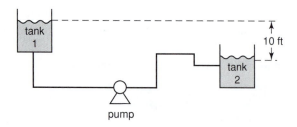

Figure 1.56 Schematic of flow for Problem 34.

between the outlet and the inlet for a flow at 5.0 gpm. Calculate the pumping head W_{pump}/mg for this flow.

33. For the flow loop shown in Figure 1.55, develop an equation that gives the friction head loss $h_f \equiv F_{21}/g$ in feet as a function of flow rate Q in gpm. The answer is an approximately quadratic equation. Plot your answer as friction head versus capacity (i.e., flow rate) for turbulent flow rates up to 10 gpm.

34. Pumps are rated in terms of fluid head (i.e., energy per unit weight of the fluid that they are pumping). A pump is connected between two open tanks as shown in Figure 1.56. The shaft work delivered by the pump at 6.0 gpm is measured at $W_{pump}/mg = 75$ feet, where W_{pump} is the shaft work done by the pump, m is the mass flow rate of the fluid being pumped, and g is the acceleration due to gravity. What is the friction loss of the system between Points 1 and 2? Give your answer in feet of head. The frictional losses of the pump already have been accounted for and should not be included in the calculations.

35. A run of water piping crosses a field where a road is to be built. The piping will be routed temporarily over the road as shown in Figure 1.57. How is the load on the pump affected by the temporary change? Estimate the additional load on the pump as a function of flow rate for the dimensions and fittings shown in Figure 1.57. Both new valves are ball valves.

36. For the piping system shown in Figure 1.58, what is the average fluid velocity at the pipe discharge? Write the answer in terms of the variables defined in the figure. You may neglect friction in the solution. The tank is not open to the atmosphere; the pipe discharges fluid to the atmosphere. P is the absolute pressure inside the vapor space over the fluid in the tank, and P is held constant.

37. Modify the solution for the discharge velocity of a siphon (see Example 1.5) by accounting for the friction term. Assume that the friction factor is approximately constant and that flow is in the turbulent regime ($0.002 < f < 0.010$; see the Moody chart, Figure 1.21 [103]). What is the error involved in neglecting friction in a siphon?

38. Water at 25°C flows at 3.2 gpm through the multipath pipeline in Figure 1.59. Calculate the volumetric flow rate in each branch and the pressure drop between points (a) and (b). Note: the pressure drop across each branch is the same and is equal to the pressure drop from (a) to (b). Equation 1.93 shows us that since ΔP is the same, then the head loss $h_f = \frac{\Delta P}{\rho g} = \frac{2fLV^2}{Dg}$ in each branch is the same. The mass balance provides a second relationship between the two velocities, allowing the problem to be solved.

Pipeline crosses desert road:

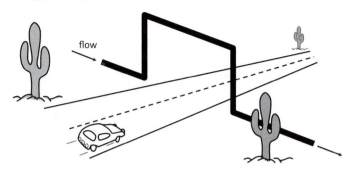

40 ft 2 in. Schedule 40 steel
2 new ball valves
4 new 90° bends

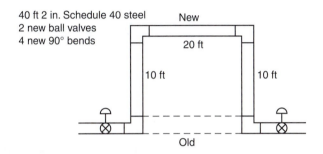

Schematic of circumstances described in Problem 35.

Inner Diameter = D
Density ρ
Viscosity μ
Temperature T

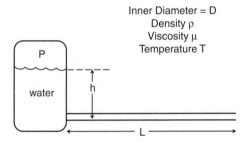

Schematic for Problem 36.

All piping is 1/2 inch Schedule 40
Length branch (1) = 245 ft
Length branch (2) = 540 ft

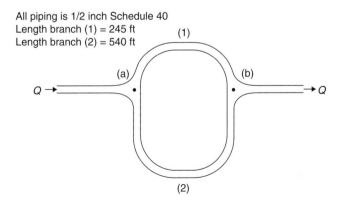

Schematic for Problem 38.

39. What is the effect of viscosity on the operation of a siphon?

40. Sketch the cylindrical coordinate system basis vectors $\hat{e}_r$, $\hat{e}_\theta$, and $\hat{e}_z$ at the following points (r, θ, z): $(3, 0, 0)$, $(3, \frac{\pi}{2}, 0)$, $(3, \frac{3\pi}{4}, 0)$, $(6, 0, 0)$, $(6, \frac{\pi}{2}, 0)$, and $(6, \frac{3\pi}{4}, 0)$. Sketch the Cartesian basis vectors $\hat{e}_x$, $\hat{e}_y$, $\hat{e}_z$ at the same locations. Comment on your sketches.

41. For the vector $\underline{v} = U\hat{e}_\theta$ written in the cylindrical coordinate system, what is the component of $\underline{v}$ in the $\hat{e}_x$ direction? U is a constant.

42. For the vector $\underline{v} = Ur\hat{e}_\theta$ written in the cylindrical coordinate system, what is the component of $\underline{v}$ in the $\hat{e}_y$ direction? U is a constant and r is the coordinate variable of the cylindrical coordinate system.

43. For the following vectors $\underline{v}$ and $\underline{a}$, what is the component of the velocity $\underline{v}$ (m/s) in the direction of vector $\underline{a}$?

$$\underline{v} = 3\hat{e}_x + 2\hat{e}_y + 7\hat{e}_z = \begin{pmatrix} 3 \\ 2 \\ 7 \end{pmatrix}_{xyz}$$

$$\underline{a} = 6\hat{e}_z = \begin{pmatrix} 0 \\ 0 \\ 6 \end{pmatrix}_{xyz}$$

44. For the following vector and tensor (matrix), what is $\hat{n} \cdot \underline{\underline{\tilde{\tau}}}$? Both expressions are written in the cylindrical coordinate system.

$$\hat{n} = \begin{pmatrix} 1 \\ 1 \\ 0 \end{pmatrix}_{r\theta z} \qquad \underline{\underline{\tilde{\tau}}} = \begin{pmatrix} 0 & 12 & 0 \\ 12 & 0 & 0 \\ 0 & 0 & 0 \end{pmatrix}_{r\theta z}$$

45. What is the dot product of the following two vectors? Both vectors represent properties at the point $(1, 0, 0)_{xyz}$. Note: The expressions here are written in two different coordinate systems.

$$\begin{pmatrix} 1 \\ 0 \\ 2 \end{pmatrix}_{xyz} \qquad \begin{pmatrix} 1 \\ 0 \\ 2 \end{pmatrix}_{r\theta z}$$

46. What is the dot product of the following two vectors? Both vectors represent properties at the point $(0, 1, 0)_{xyz}$. Note: The expressions here are written in two different coordinate systems.

$$\begin{pmatrix} 1 \\ 2 \\ 0 \end{pmatrix}_{xyz} \qquad \begin{pmatrix} 1 \\ -1 \\ 2 \end{pmatrix}_{r\theta z}$$

47. What is the cross product of the following two vectors? Both vectors represent properties at the point $(1, 1, 0)_{xyz}$. Note: The expressions here are written in

two different coordinate systems.

$$\begin{pmatrix} 1 \\ 0 \\ 1 \end{pmatrix}_{xyz} \qquad \begin{pmatrix} 1 \\ 0 \\ 0 \end{pmatrix}_{r\theta z}$$

48. Write the vector $\underline{w} = 4\hat{e}_1 - \hat{e}_2 + \hat{e}_3$ in cylindrical coordinates.
49. Write the vector $\underline{v} = (1 - 4y^2)\hat{e}_x$ in cylindrical coordinates.
50. Write the vector $\underline{w} = -3\hat{e}_1 - \hat{e}_2 + \hat{e}_3$ in spherical coordinates.
51. Write the vector $\underline{v} = (1 - 2y^2)\hat{e}_y$ in spherical coordinates.
52. The solution for the velocity field for steady, pressure-driven flow in a tube is provided in Chapter 7 (see Equation 7.23). Convert this solution, which is given in cylindrical coordinates, to Cartesian coordinates, $x, y, z, \hat{e}_x, \hat{e}_y,$ and $\hat{e}_z$.
53. The solution for the velocity field for steady, uniform flow around a sphere is provided in Chapter 8 (see Equation 8.23). Convert this solution, which is given as a vector written in the spherical coordinate system, to a vector written in the Cartesian coordinate system. You may leave your answer in terms of spherical coordinate variables r, θ, π. What relationships between r, θ, ϕ and x, y, z do we need to complete the conversion to the Cartesian coordinate system?
54. The solution for the velocity field for steady, pressure-driven flow in a slit is provided in Chapter 7 (see Equation 7.188). Convert this solution, which is given in Cartesian coordinates centered in the middle of the slit, to Cartesian coordinates anchored on the bottom wall.
55. What is a boundary condition? Why are boundary conditions needed when solving differential equations?
56. How many boundary conditions on x are needed for the following partial differential equation? How many boundary conditions are needed on y?

$$\alpha \frac{\partial \underline{v}}{\partial x} = \frac{\partial^2 \underline{v}}{\partial y^2}$$

57. For the steady laminar flow of water through a long pipe, calculate the flow rate Q from the velocity profile, which is given here. Show your work. The following quantities are constants: $R, L, \rho, g, P_o, P_L, \mu; r$ is the coordinate variable in the cylindrical coordinate system.

$$\underline{v} = v_z \hat{e}_z$$

$$v_z = \frac{R^2(L\rho g + p_0 - p_L)}{4\mu L} \left(1 - \frac{r^2}{R^2}\right)$$

58. Using a computer program (i.e., spreadsheet or other), plot the velocity profile given in Equation 1.140, which represents the velocity profile between two long vertical plates separated by a narrow gap. The flow is caused by natural convection: One plate is hotter than the other.
59. For the natural-convection velocity profile given in Equation 1.140, calculate the second derivative of the velocity-profile function and evaluate the second

derivative at the extrema of the function. What does the second derivative tell us about the extrema?

60. Which of the following expressions is $\underline{v} \cdot \nabla \underline{v}$? Explain how you arrive at your answer.

$$
\begin{pmatrix} v_x \frac{\partial v_x}{\partial x} + v_y \frac{\partial v_x}{\partial y} + v_z \frac{\partial v_x}{\partial z} \\ v_x \frac{\partial v_y}{\partial x} + v_y \frac{\partial v_y}{\partial y} + v_z \frac{\partial v_y}{\partial z} \\ v_x \frac{\partial v_z}{\partial x} + v_y \frac{\partial v_z}{\partial y} + v_z \frac{\partial v_z}{\partial z} \end{pmatrix}_{xyz}
\quad \text{or} \quad
\begin{pmatrix} v_x \frac{\partial v_x}{\partial x} + v_x \frac{\partial v_y}{\partial y} + v_x \frac{\partial v_z}{\partial z} \\ v_y \frac{\partial v_x}{\partial x} + v_y \frac{\partial v_y}{\partial y} + v_y \frac{\partial v_z}{\partial z} \\ v_z \frac{\partial v_x}{\partial x} + v_z \frac{\partial v_y}{\partial y} + v_z \frac{\partial v_z}{\partial z} \end{pmatrix}_{xyz}
$$

61. What is the substantial derivative $D\underline{v}/Dt$ of the steady-state velocity field represented by the following velocity vector? Note that the answer is a vector. Explain how you arrive at your answer.

$$
\underline{v}(x_1, x_2, x_3, t) = \begin{pmatrix} 1 - 9x_2^2 \\ 0 \\ 0 \end{pmatrix}_{123}
$$

62. Under the pull of gravity, a Newtonian fluid drains from a cylindrical tank through a small hole in the center of the bottom of the tank. The tank has radius R and is of height H. Which coordinate system do you choose for solving for the flow field in this problem? In your chosen coordinate system, what is the general expression for the velocity field $\underline{v}$? Are any of the components of $\underline{v}$ zero in your coordinate system? If so, why? Of what variables is $\underline{v}$ a function?

63. A Newtonian fluid flows past a stationary sphere. Upstream of the sphere, the flow is uniform; that is, the velocity is constant in both magnitude and direction. The radius of the sphere is $D/2$. Which coordinate system do you choose for solving for the flow velocity field in this problem? In the chosen coordinate system, what is the general expression for the velocity field $\underline{v}$? Are any of the components of $\underline{v}$ zero in your coordinate system? If so, why? Of what variables is $\underline{v}$ a function?

64. A Newtonian fluid flows under a driving pressure gradient and down the axis of a duct with a rectangular cross section. The width of the duct is $2W$ and the height is $2H$. The duct has a length of L. Which coordinate system do you choose for solving for the flow in this problem? In the chosen coordinate system, what is the general expression for the velocity field $\underline{v}$? Are any of the components of $\underline{v}$ zero in your coordinate system? If so, why? Of what position variables is $\underline{v}$ a function?

65. A Newtonian fluid flows under a driving pressure gradient down the axis of a duct with a circular cross section. The radius of the duct is $D/2$ and the length is L. Which coordinate system do you choose for solving for the flow in this problem? In the chosen coordinate system, what is the general expression for the velocity field $\underline{v}$? Are any of the components of $\underline{v}$ zero in your coordinate system? If so, why? Of what variables is $\underline{v}$ a function?

66. A Newtonian fluid flows under a driving pressure gradient down the axis of a duct with an elliptical cross section. The longer axis of the ellipse is a and the shorter axis is b. The length of the duct is L. Which coordinate system do you choose for solving for the flow in this problem? In the chosen coordinate

system, what is the general expression for the velocity field $\underline{v}$? Are any of the components of $\underline{v}$ zero in your coordinate system? If so, why? Of what variable is $\underline{v}$ a function?

67. A Newtonian fluid flows past a three-dimensional stationary object that is a simplified version of a modern automobile. Upstream of the object, the flow is uniform; that is, the velocity is constant in both magnitude and direction. The object presents a cross section to the flow of A_p. Which coordinate system do you choose for solving for the flow velocity field in this problem? In the chosen coordinate system, what is the general expression for the velocity field $\underline{v}$? Are any of the components of $\underline{v}$ zero in your coordinate system? If so, why? Of what variable is $\underline{v}$ a function?

2 How Fluids Behave

Our task is to learn to model flows. To set up the models, we draw on our intuition of how fluids behave; for example, we often can guess the direction that a flow takes under the influence of particular forces. Intuition also may enable us to identify symmetries in a flow field. Intuition comes from experience, however, and for introductory students, experience may be in short supply.

One solution to a lack of experience is to experiment with fluids. Unfortunately, not all of us have access to pumps, flow meters, and piping systems; therefore, it is worthwhile to take a laboratory course in fluid mechanics, if possible. Another way to build experience with fluid behavior is to view flow-visualization videos. Between 1961 and 1969, a group of experts in fluid mechanics (the National Committee for Fluid Mechanics Films [NCFMF]) produced a series of flow-demonstration films [112] that introduce fluid behavior; the films and film notes are now available on the Internet. There also are books [170] and other media [65] that catalog fluid behavior, as well as Web sites on which researchers have posted flow-visualization videos, including the *Gallery of Fluid Motion* [133], and elsewhere [182]. These sites bring to life all types of fluid behavior, from the mundane to the esoteric.

In addition to these sources of intuition on fluid behavior, there are experiments that we conduct in our daily life. We wash, cook, eat, water the lawn, and drive and maintain automobiles, all activities that involve interaction with one or more fluids. In this chapter, we discuss several qualitative effects observed in flows that we may encounter daily. The intuition built by these descriptions serves us well in the chapters that follow as we are required to make inferences about unknown flows. We also introduce simple mathematical relations based on the concepts discussed here, and we revisit the balance equations introduced in Chapter 1. The goal of this chapter is to make an initial pass through the entire range of fluid behavior. We refer back to these phenomena throughout the remainder of the text as we develop the appropriate models and techniques to describe the behavior introduced here. In Chapter 10, we formally revisit the topics of this chapter to consolidate the understanding of fluid mechanics that we achieve through our study.

2.1 Viscosity

Not all liquids flow in the same way, as we know from handling foods and other household fluids. Honey or syrup poured from a container flows more slowly

Figure 2.1 Force is required to make water (left) and honey (right) flow. Honey flows more slowly under the pull of gravity than water; therefore, honey has a higher viscosity than water.

than water. In engineering terms, honey resists the pull of gravity more than water resists the same pulling force; honey is said to be more viscous (Figure 2.1). Viscosity measures the tendency of a fluid to resist flow. If honey and water were made to flow at the same volumetric flow rate—by squeezing both liquids from plastic bottles, for example—it would take more effort to produce the flow of honey than the flow of water.

In the garage, we encounter another viscous fluid—motor oil—for which viscosity is a particularly important property. Motor oil lubricates an engine's moving parts. An effective lubricant must not flow off the moving parts during operation of the engine; thus, the viscosity of an effective oil must be above a specified minimum value when the engine is warm and running. High viscosity is not an advantage when changing the oil, however, because the old oil must flow out of the engine casing under the pull of gravity. Also, when a cold engine first is started, low viscosity is desired so that less torque is required to start the engine. We see then that the viscosity of engine oil must be neither too high nor too low. The design process for engine oil is complicated further by the fact that its viscosity decreases rapidly with increasing temperature—thus, as an engine heats up, the viscosity of engine oil drops, which is the exact opposite of the desired effect.

A solution to the motor-oil dilemma is to formulate oils differently for different engines, operating conditions, and uses. Multigrade motor oils are graded for at least two viscosities [109]. A typical automotive oil (e.g., 15W 30 motor oil) is designed to have a viscosity of at most 7,000 centipoise (cp) at $-20°C$ (1 poise = g/cm s). At 100°C, the same oil is required to have a viscosity high enough so that the ratio of viscosity to density (called the *kinematic viscosity*) is above 9.3 centistokes (cs). In the SAE rating (i.e., 15W30 in the previous example), the first number is the cold-temperature performance (W = winter) and the second number is the high-temperature performance. For both numbers, the higher the number, the higher is the viscosity of the oil.

The main property of viscous liquids is that forces can be transferred through them. Consider the experiment of spreading honey on a piece of toast or, as shown in Figure 2.2, on a piece of parchment paper. If you try to spread honey without holding the paper, the honey will not spread. Instead, the knife, honey, and paper

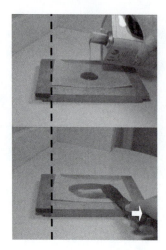
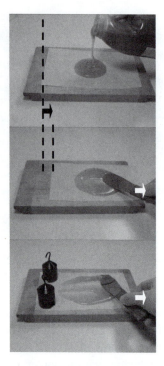

Figure 2.2   When oil is poured on a sheet of parchment paper and then spread with a knife, no restraining force is necessary to prevent the paper from sliding along. Oil has a low viscosity and transmits little stress. When the experiment is repeated with honey, the paper slides along to the right with the spreading honey. A weight of 400 g is necessary to keep the paper from sliding when honey is spread. Honey is high viscosity and transmits a great amount of stress from the knife to the paper.

all move together. If you hold onto the paper, the honey spreads, but you must exert a force to hold the paper in place. What is happening? The force you are exerting on the honey with the knife is being transferred from the knife to a layer of honey, to another layer of honey, and so on until it is transferred to the paper. In the process, the honey flows. This is a property of viscous liquids: Frictional forces are transferred through liquids causing deformation of the liquid. The force required to hold the paper or bread depends on the viscosity of the spreading fluid. To spread honey takes a larger force than to spread oil or water. Another way of thinking about it is that more force is transferred by honey than by oil or water. The equation that relates the viscosity, the force per area generated in the deforming fluid, and the relative speed of the object (or knife or hand) is called Newton's law of viscosity:

$$
\tilde{\tau}_{21} = \mu \frac{\partial v_1}{\partial x_2} \tag{2.1}
$$

Newton's law of viscosity:
(force-deformation relationship;
see Chapter 5)

where μ is the viscosity, $\tilde{\tau}_{21}$ is the molecular shear stress, v_1 is the fluid velocity in the x_1 coordinate direction, and x_2 is the coordinate direction orthogonal

Figure 2.3 Honey is a Newtonian fluid—that is, it flows even under the mild force imposed by gravity. After only seconds, the surface of the honey is level and smooth. Paint is a yield-stress fluid. When spread on a wall it stays where it is placed, resisting the pull of gravity, allowing it to dry in place. Photos courtesy of (left) Silva/AGE Photostock, and (right) Steeger/AGE Photostock.

to x_1. Newton's law of viscosity is one of the founding equations of fluid mechanics.

We discuss viscosity and Newton's law of viscosity in more detail in Chapter 5. For now, we associate viscosity with the tendency to resist flow. There are subtleties, however, to our experience with fluids that resist flow. Mayonnaise is a fluid that resists flow—so much so that it does not flow out of a jar when poured. Yet, mayonnaise spreads easily with a knife—in fact, it spreads on a piece of toast with less effort than honey, and thus appears to have a lower viscosity. Why is it, then, that mayonnaise does not flow when poured?

Honey, water, and other fluids that flow when poured from a container belong to the class of fluids called Newtonian fluids (Figure 2.3). They have a constant viscosity and respond to all attempts to deform them, regardless of how small the applied effort. Newtonian fluids follow Equation 2.1. Mayonnaise is not a Newtonian fluid because it can resist small efforts to deform it, such as the small tug of gravity that seeks to level out fluid in a jar. When a material does not flow until a certain amount of stress is applied, it is called a yield-stress or Bingham fluid [104]. Non-Newtonian fluids like mayonnaise are common in both the kitchen (e.g., peanut butter and ketchup) and engineering (e.g., paint, slurries, asphalts, and suspensions). In addition to yield-stress fluids, other types of fluids are non-Newtonian because they have variable viscosities depending on how fast they are stirred. Another type of non-Newtonian fluid is a memory fluid like Silly Putty, which stretches when pulled, partially recoils when released, yet flows into a puddle with enough time. The study of non-Newtonian fluid flow is called *rheology* [104] (see Chapter 5).

In summary, viscosity describes the ability of a fluid to resist flow, and viscous fluids can transmit forces from one surface to another. The viscous behavior of fluids can be simple (Newtonian) or complex (non-Newtonian). The following two examples get us started with viscosity-related calculations.

EXAMPLE 2.1. *What are the units of viscosity in the metric system and in the American engineering unit system?*

SOLUTION. Because all equations must be dimensionally consistent, we can use Newton's law of viscosity (see Equation 2.1) to deduce the units of viscosity. In the metric system, the units of stress are Pa and the units of velocity are m/s. Solving Equation 2.1 for viscosity and substituting the units, we determine the units of viscosity:

$$\mu = \frac{(\tau_{21})}{\left(\dfrac{\partial v_1}{\partial x_2}\right)} \tag{2.2}$$

$$[=] \left(\frac{\mathrm{Pa}}{}\right)\left(\frac{\mathrm{m}}{\mathrm{m/s}}\right)\left(\frac{\mathrm{N/m^2}}{\mathrm{Pa}}\right)\left(\frac{\mathrm{kg\ m}}{\mathrm{N\ s^2}}\right) \tag{2.3}$$

$$[=] \frac{\mathrm{kg}}{\mathrm{m\ s}} = \mathrm{Pa\ s} \tag{2.4}$$

If centimeters and grams are used, the unit becomes g/cm s, which is called a *poise*. The viscosity of water at room temperature is about one centipoise (*cp*), or one milli-pascal-second (mPa s).

In the American engineering system of units, the same manipulation yields:

$$\mu = \frac{(\tau_{21})}{\left(\dfrac{\partial v_1}{\partial x_2}\right)} \tag{2.5}$$

$$[=] \left(\frac{\mathrm{lb_f}}{}\right)\left(\frac{\mathrm{ft}}{\mathrm{ft/s}}\right)\left(\frac{32.174\ \mathrm{ft\ lb_m}}{\mathrm{lb_f\ s^2}}\right) \tag{2.6}$$

$$[=] \frac{\mathrm{lb_m}}{\mathrm{ft\ s}} \tag{2.7}$$

In these units, the viscosity of water at room temperature is about 6×10^{-4} $\mathrm{lb_m/ft\ s}$. Note that the factor 32.174 is attached to the conversion of ft $\mathrm{lb_m/s^2}$ to $\mathrm{lb_f}$. With American engineering units, be sure to include this factor when converting force units to units of mass times acceleration (see Glossary under g_c).

EXAMPLE 2.2. *How much force does it take to slowly inject a water-like solution through a 16-gauge needle?*

SOLUTION. A syringe with a needle attached is shown in Figure 2.4. It takes force to move the plunger through the barrel of the syringe even if the syringe is empty of fluid. This is a small force, however; a much larger resistance can develop when fluid fills the barrel of the syringe and a small needle is attached. In our calculation, we are concerned with the contribution to force on the plunger that is due to flow resistance in the needle.

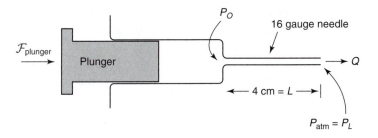

Figure 2.4
The flow through the needle of a syringe can be modeled as flow through a tube. Analysis of slow tube flow (see Chapter 7) results in the Hagen-Poiseuille equation.

In slow flows in a tube, the effects of viscosity dominate, and we can show by using a momentum balance (see Chapter 7) and Newton's law of viscosity (see Equation 2.1 and Chapter 4) that the pressure drop from the upstream point to the downstream point in a tube is related to the flow rate in the tube according to the Hagen-Poiseuille equation:

$$\begin{array}{l}\text{Hagen-Poiseuille equation}\\\text{(flow-rate/pressure-drop}\\\text{for laminar tube flow)}\end{array}\qquad \boxed{Q = \frac{\pi(p_0 - p_L)R^4}{8\mu L}} \qquad (2.8)$$

where Q is the volumetric flow rate in the tube, $p_0 - p_L$ is the pressure drop across a tube length L, μ is the viscosity of the fluid, and R is the radius of the tube. This equation is derived in Chapter 7 for laminar flow.

To know how much force is needed to make an injection from the syringe, we need the force on the plunger, which is related to the pressure inside the syringe reservoir:

$$\text{Magnitude of force on plunger:} \quad \mathcal{F}_{\text{plunger}} = \left(\frac{\text{force}}{\text{area}}\right)\left(\begin{array}{c}\text{plunger}\\\text{cross-sectional}\\\text{area}\end{array}\right) \quad (2.9)$$

The pressure in the reservoir is the same as p_0, the pressure in the fluid at the beginning of the flow through the narrow needle. Therefore, we write:

$$\mathcal{F}_{\text{plunger}} = p_0 \left(\pi R_p^2\right) \qquad (2.10)$$

where R_p is the radius of the plunger and p_0 is the force per unit area on the plunger, which is equal to the gauge pressure in the fluid in the syringe reservoir. For a slow injection, we can obtain p_0 from the Hagen-Poiseuille equation (Equation 2.8):

$$Q = \frac{\pi(p_0 - p_L)R^4}{8\mu L} \qquad (2.11)$$

$$p_0 - p_L = \frac{8Q\mu L}{\pi R^4} \qquad (2.12)$$

$$p_0 = p_L + \frac{8Q\mu L}{\pi R^4} \qquad (2.13)$$

where R is the inner radius of the needle and p_L is the pressure at the exit of the needle, which is equal to zero in terms of gauge pressure (i.e., gauges read zero when exposed to atmospheric pressure).

A review of the literature reveals that water at room temperature has a viscosity of 0.8937×10^{-2} poise [132]. We also learn from the literature that a 16-gauge needle has an ID of 1.194 mm $= 1.194 \times 10^{-3}$ m. The piston of a typical syringe is about 1 cm in diameter and the speed of the piston when injecting is about 0.5 cm/s. We can convert piston speed to flow rate in the needle as follows:

$$\text{Flow rate} = \left(\begin{array}{c} \text{average} \\ \text{velocity} \end{array} \right) \left(\begin{array}{c} \text{cross-sectional} \\ \text{area} \end{array} \right) \tag{2.14}$$

$$Q = \langle v \rangle \pi R_p^2 \tag{2.15}$$

We used this relationship in Chapter 1 (Equation 1.2) and it is derived formally in Chapter 3. For this problem, with the values of $\langle v \rangle = 0.5$ cm/s and $R_p = 0.5$ cm assumed previously, we calculate the volumetric flow rate as:

$$Q = \langle v \rangle \pi R_p^2 \tag{2.16}$$

$$= \left(\frac{0.5 \text{ cm}}{\text{s}} \right) (\pi) (0.5 \text{ cm})^2 \tag{2.17}$$

$$= 0.3927 \text{ cm}^3/\text{s} = \boxed{0.4 \text{ cm}^3/\text{s}} \tag{2.18}$$

We now can calculate the upstream pressure from the rearranged Hagen-Poiseuille equation in Equation 2.13. We assume the needle length to be 4 cm:

$$p_0 = \frac{8 Q \mu L}{\pi R^4} \tag{2.19}$$

$$= \frac{8 \left(\frac{0.3927 \text{ cm}^3}{\text{s}} \right) \left(\frac{0.8937 \times 10^{-2} \text{ g}}{\text{cm s}} \right) (4 \text{ cm})}{\pi \left(\frac{0.1194 \text{ cm}}{2} \right)^4} \tag{2.20}$$

$$= \frac{2814 \text{ g}}{\text{cm s}^2} \left(\frac{\text{kg}}{1{,}000 \text{ g}} \right) \left(\frac{100 \text{ cm}}{\text{m}} \right) \left(\frac{\text{N}}{\text{kg m/s}^2} \right) \left(\frac{\text{Pa}}{\text{N/m}^2} \right) \tag{2.21}$$

$$= 281.4 \text{ Pa} \tag{2.22}$$

Now that we know the pressure in the barrel of the syringe, we can calculate the force on the plunger:

$$\mathcal{F}_{\text{plunger}} = p_0 \left(\pi R_p^2 \right) \tag{2.23}$$

$$= (281.4 \text{ Pa}) (\pi) \left(0.5 \text{ cm} \frac{\text{m}}{100 \text{ cm}} \right)^2 \left(\frac{\text{N/m}^2}{\text{Pa}} \right) \tag{2.24}$$

$$= 0.022 \text{ N} \tag{2.25}$$

This force is slightly less than the weight of a U.S. penny coin (after 1982, the U.S. penny's mass was 2.5 g, which weighs $F = mg = 0.0245$ N). For a more viscous fluid or for a syringe of different geometry, we can adjust the quantities in Equation 2.19 and calculate the appropriate result.

A final comment: The equation used to solve this problem (i.e., Equation 2.8) was for laminar flow only. We can check whether the flow is laminar by calculating the Reynolds number:

$$\text{Re} = \frac{\rho \langle v \rangle D}{\mu} \tag{2.26}$$

$$= \frac{\left(\frac{1.0 \text{ g}}{\text{cm}^3}\right)\left(\frac{0.3927 \text{ cm}^3}{\text{s}}\right)\left(\frac{4}{\pi(1.194\times 10^{-3} \text{ m})^2}\right)(1.194 \times 10^{-3} \text{ m})\left(\frac{\text{m}}{100 \text{ cm}}\right)}{0.8937 \times 10^{-2}\frac{\text{g}}{\text{cm s}}}$$

$$= 468$$

Because the Re is less than 2,100, we confirm that the flow is laminar (see Equation 1.63).

2.2 Drag

In discussing viscosity, we have a fluid-centered view—that is, we ask what is the effect on a fluid if a force were applied to it from the outside. The fluid deforms and flows (e.g., honey pushed by a knife or medicine pushed by a syringe), and the source of the stress is the motion of the solid boundary (e.g., knife or plunger). The principal issue in this view is: How does the fluid deform?

We also can have a solid-centered view and ask which forces act on solids when fluids move around them. This is an intuitive perspective when there is a large amount of fluid and a small solid object moving through it, such as when a ball is thrown through the air or a child swims in the ocean (Figure 2.5). It also is natural to have a solid-centered view when fluid rushes by a stationary object, such as when wind blows on a building or molten plastic is forced over integrated circuit chips in an encapsulation process.

The force transferred from a fluid to a solid opposing the object's motion is called drag. This is the same force that transfers from a knife to honey; only the point of view has changed. We encounter both points of view depending on whether we are more concerned with the deformation taking place in the fluid

Fluid on inside: Fluid on outside:

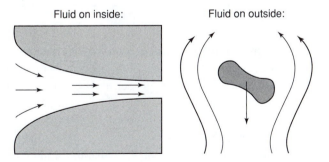

Figure 2.5   Although they both represent situations in which fluids and solids interact, we consider two cases: (1) a fluids-centered view in which a small amount of fluid is trapped between solid-bounding surfaces (i.e.,internal flow; see Chapter 7); and (2) a solids-centered view in which there is a large amount of fluid and isolated bodies move through the fluid (i.e., external flow; see Chapter 8).

(i.e., fluid-centered view) or with the forces on objects moving through fluids (i.e., solids-centered view).

The role of viscosity in creating air drag may not seem obvious at first because we usually associate viscosity with thick fluids such as water and honey. Like water and honey, however, air is a viscous fluid—even a Newtonian fluid. An important difference among water, honey, and air is that the viscosity of air is 50 times smaller than that of water and a half-million times smaller than that of honey. As shown in Equation 2.1 and discussed in more detail in Chapter 5, stress is generated in viscous fluids when there are velocity differences. Fluids with high viscosities develop high stresses, but even low-viscosity fluids can develop high stresses if the velocity gradients dv_1/dx_2 are high enough. In air flows, often the speed of air is quite high (e.g., hurricane wind speed past a house or relative speed between air and an airplane); therefore, forces caused by air drag can be significant.

EXAMPLE 2.3. *Fluids with higher viscosity produce more drag. How much difference is there among the viscosities of air, alcohol, water, olive oil, and honey? How much do other material properties (e.g., the density) of these materials vary? Comment on the differences.*

SOLUTION. We can find the viscosities of common fluids in the literature [87], some of which are listed in Table 2.1. Included is the viscosity of pitch, a highly viscous material derived from wood.

Table 2.1. The viscosity of familiar materials

Fluid	T (°C)	μ (Pa s)	μ lb$_m$/(ft s)	ρ (kg/m^3)	ρ lb$_m$/ft^3
air	25	18.6×10^{-6}	12.50×10^{-6}	1.20	74.9×10^{-3}
water	25	0.8937×10^{-3}	0.6005×10^{-3}	997	62.2
n-propyl alcohol	25	1.96×10^{-3}	1.32×10^{-3}	804	50.2
olive oil	25	69×10^{-3}	46×10^{-3}	918	57.3
honey	25	9	6	1360	84.9
pitch	25	1×10^6	0.67×10^6	1100	69

Note: The range of viscosity is 10 orders of magnitude; the range of density is only 3 orders of magnitude.

The striking feature about the values of viscosity is that they range over 10 orders of magnitude (Figure 2.6 and Table 2.1). The density of air is significantly less than densities of liquids; however, slight density differences among the liquids do not explain viscosity variations of nine orders of magnitude.

Viscosity is the material parameter that determines how much stress is generated in a given flow (recall Newton's law of viscosity; see Equation 2.1). Thus, based on these values of viscosity, we conclude that the amount of stress generated by different fluids can vary widely. Because of this strong variation among materials, researchers often resort to using logarithmic scales when plotting data related to viscosity [104].

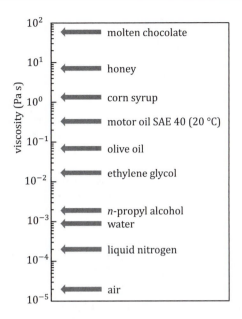

Figure 2.6 Viscosities of familiar materials compared on a logarithmic scale.

Drag can have a confounding effect even at low speeds if it is forgotten or if it is not accounted for properly. Consider the reported experiments[1] of Galileo Galilei (1564–1642) that contributed to the discovery of the nature of gravity. In the 300s BCE, Aristotle postulated the view that heavier objects fall faster to the Earth than lighter objects. What was Aristotle's evidence? Aristotle compared the gentle floating of dropped feathers to the rapid descent of stones. Viewed with modern hindsight, Aristotle's experiments were of bodies moving through a fluid (i.e., air) under the action of a force (i.e., gravity). Galileo's experiments tested Aristotle's hypothesis by proposing to drop stones from a great height, such as the Leaning Tower of Pisa. If Aristotle were right, two stones, one twice the weight of the other, would land at different times when released from the top of the tower. In Galileo's experiments, the differently weighted stones landed simultaneously, proving that the speed of the falling stones was independent of their weight.

What was wrong with Aristotle's observations and conclusions? The problem was precisely a failure to understand the effect of the viscous drag due to the presence of air. In Aristotle's observations, the falling stone or feather exerted a force on the air through which it fell (recall the moving knife transferring force through honey to the bread). This exertion of force slightly decelerates the falling stone and severely decelerates the lightweight feather. If Aristotle had dropped the stone and feather in a vacuum chamber or on the moon, both objects would have fallen at the same rate, and he would have reached a different conclusion (Figure 2.7).[2]

[1] Historians now believe that these experiments never actually were carried out and more likely were only "thought" experiments.

[2] Aristotle would have been unable to consider experiments in a vacuum chamber; the first practical experiments on a vacuum were conducted in the 17th century.

Figure 2.7   Astronaut David Scott conducted Galileo's experiment on the moon [159] during Apollo 15 in 1971 and verified that in the absence of air drag, a falcon feather and a geology hammer land simultaneously when dropped (art credit: Tomas Co).

In many applications, effort is made to minimize drag—for example, swimmers wear specially designed clothing and shave their body to reduce drag, bicyclists hunch over or draft one another to reduce drag during a race (Figure 2.8), and automobiles and airplanes are designed with smooth curves to ease the flow of air around the moving body to increase fuel efficiency (Figure 2.9; see also Section 8.2.3). We track drag through the drag coefficient, a quantity that is constant for blunt objects moving at high flow speeds (see Chapter 8):

$$\text{Drag coefficient} \qquad C_D = \frac{\mathcal{F}_{\text{drag}}}{\frac{1}{2}\rho\langle v\rangle^2 A_p} \qquad (2.27)$$

where C_D is the drag coefficient (unitless); $\mathcal{F}_{\text{drag}}$ is the drag, which is a force magnitude; ρ is the density of the fluid; $\langle v\rangle$ is the average velocity of the object or the velocity of the fluid as it flows past the object; and A_p is the reference area for drag coefficient—often the area presented by the object to the oncoming flowstream. The following example is a problem that can be addressed with knowledge of drag coefficient as a function of system geometry and the Reynolds number.

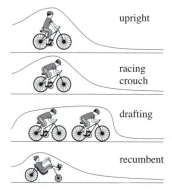

upright

racing
crouch

drafting

recumbent

Figure 2.8   Bicycle racers gain an edge by adopting a more streamlined shape or by drafting—that is, riding in the wake produced by another cyclist. The drag coefficient (a measure of drag generated) for an upright bicycle driven in air ($C_D = 1.1$) is significantly larger compared to that of a cyclist in the racing crouch (0.88), in the drafting position (0.50), or riding a streamlined bicycle (0.12) [183]. The cross-sectional area presented by the cyclist in these four positions also varies: upright, 5.5 feet^2; racing and drafting, 3.9 feet^2; and recumbent, 5.0 feet^2.

Figure 2.9 Automobile manufacturers devote significant effort to reducing drag caused by the shape of cars. Modern automobiles with smooth lines experience less drag than the boxy cars of yesteryear. Computational techniques can be used to accurately predict drag on automobiles before they are even constructed. The streamlines shown above for flow over a race car were calculated with computational fluid dynamics (CFD) software. Image courtesy NASA.

EXAMPLE 2.4. *How much faster will a bicycle racer traveling at 40 mph go if she adopts a racing crouch rather than riding upright?*

SOLUTION. Changing one's posture on a bicycle from upright to a racing crouch reduces the amount of area presented to the oncoming air from 5.5 feet^2 to 3.9 feet^2 (see Figure 2.8), but it also changes the drag coefficient because of the change in the shape of flow around the bicyclist. We can determine the effect of the posture change on the bicyclist's speed by using the drag expression in Equation 2.27 and the experimental values of the drag coefficient in Figure 2.8.

$$\text{Drag coefficient} \quad C_D = \frac{\mathcal{F}_{\text{drag}}}{\frac{1}{2}\rho \langle v \rangle^2 A_p} \tag{2.28}$$

We find the density of air for the conditions of interest to be $\rho = 0.0766 \text{ lb}_{\text{m}}/\text{ft}^3$ [87]. The drag coefficient is a variable at some speeds; however, at the cyclist's speed, the drag coefficient is constant.

A bicycle racer, traveling at 40 mph while upright ($C_D = 1.1$), generates a drag of:

$$\mathcal{F}_{\text{drag}} = \left(\frac{C_D \rho A_p}{2}\right) \langle v \rangle^2 \tag{2.29}$$

$$= \frac{(1.1)}{} \left(\frac{0.0766 \text{ lb}_{\text{m}}}{\text{ft}^3}\right) \frac{(5.5 \text{ ft}^2)}{} \left(\frac{1}{2}\right) \left((40 \text{ mph}) \frac{1.46667 \text{ ft/s}}{\text{mph}}\right)^2 \tag{2.30}$$

$$= \frac{797.5 \text{ lb}_{\text{m}} \text{ ft}}{\text{s}^2} \left(\frac{\text{lb}_{\text{f}}}{32.174 \text{ ft lb}_{\text{m}}/\text{s}^2}\right) \tag{2.31}$$

$$= 24.8 \text{ lb}_{\text{f}} \tag{2.32}$$

Again, conversion of mass-acceleration units to force units in the American engineering system requires a unit conversion of 32.174 ft lb_m/s^2 per lb_f.

For the same cyclist traveling with the same drag but now in the crouching position (from Figure 2.8, $C_D = 0.88$, and $A_p = 3.9$ ft^2), the speed is:

$$C_D = \frac{\mathcal{F}_{\text{drag}}}{\frac{1}{2}\rho\langle v\rangle^2 A_p} \tag{2.33}$$

$$\langle v\rangle^2 = \frac{\mathcal{F}_{\text{drag}}}{\frac{1}{2}\rho C_D A_p} \tag{2.34}$$

$$= \frac{(24.8\ lb_f)}{\left(\frac{1}{2}\right)\left(\frac{0.0766\ lb_m}{ft^3}\right)(0.88)(3.9\ ft^2)}\left(\frac{32.174\ ft\ lb_m}{s^2\ lb_f}\right) \tag{2.35}$$

$$= 6{,}070.3\ ft^2/s^2 \tag{2.36}$$

$$\langle v\rangle = \frac{77.9\ ft}{s}\left(\frac{mph}{1.4667\ ft/s}\right) \tag{2.37}$$

$$= 53\ mph \tag{2.38}$$

A cyclist rides 33 percent faster in the crouching position than when she rides in the upright position. From this calculation, we see how important an athlete's posture can be to performance. Calculations like this can be used to motivate drag-reducing changes in clothing and bicycle architecture and technique.

In Chapter 8, we derive the drag-coefficient equation and model flows in which drag is the dominant engineering concern.

2.3 Boundary layers

Drag is a straightforward consequence of bodies moving through viscous fluids and, because all fluids have viscosity, drag is always present. In many flows, however, there are locations in the flow where drag is negligible. In boundary-layer analysis, introduced here, researchers simplify their calculations by using their knowledge of how viscous effects are distributed throughout a flow (Figure 2.10). After the boundary-layer concept was introduced in the early 20th century, the field of aeronautics developed rapidly. Boundary-layer concepts also are important in heat-transfer and mass-transfer analyses [15].

Because relative speeds in flows (i.e., dv_1/dx_2 in Equation 2.1) can be enormous (e.g., airplanes, bullets, and spacecraft), the viscous drag on surfaces touched by rapid flow can be significant. When large forces are generated in a flow, the flow around an obstacle rearranges to localize the effect of viscosity. In 1904, Ludwig Prandtl [134] identified two distinct regions in rapid flow: (1) a narrow layer near the surfaces in which the fluid's viscosity dictates the flow pattern and stresses; and (2) a region away from the surfaces in which viscous effects are negligible. The thin layer in which viscosity is important is called the boundary layer; outside of the boundary layer, the fluid (often air or water) behaves as if it had zero viscosity.

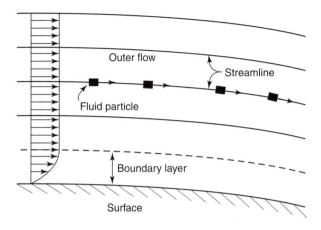

Figure 2.10 Viscosity is important in the part of the flow near surfaces (i.e., the boundary layer), but it is insignificant in the part of the flow far from solid surfaces. By separately considering the two regions—the regions near to and far from the surface—the analysis is greatly simplified.

The importance of boundary-layer study is illustrated with an example from sports. The flight of a golf ball is dominated by the structure of its boundary layer. Because manufacturers and golfers did not understand the science behind their flight, golf balls initially were manufactured to be smooth [173]. Golfers noticed, however, that old dented balls flew farther than brand-new balls. Golfers started roughing up new balls before playing them. If we examine the structure of the flow of air around a ball (i.e., the flow field), we can understand why a rough golf ball flies farther than a completely smooth ball.

Figure 2.11b shows the flow of air around a ball from the point of view of the ball. If the ball moves very slowly, the air creeps around the ball, forming a smooth flow pattern (Figures 2.11b and 2.12a). This type of flow is called creeping flow (see Chapter 8), and there is no boundary layer; viscosity is important throughout this flow. As the ball moves faster (or, from the point of view of the ball, as the air rushes by more rapidly), a boundary layer forms (Figure 2.12d). At high flow rates, the fluid outside the boundary layer moves at a uniform speed, and our attention shifts to the boundary layer. Because all of the viscous or friction effects take place in the boundary layer, the character of the boundary layer determines how much decelerating drag is felt by the ball and, therefore, how far the ball will fly.

In the flight of a smooth ball, the boundary layer appears as shown in Figure 2.12d. The details of the flow depend on geometry and speed of the air rushing by the ball, but the flow has several general characteristics. On the face of the ball that parts the flow

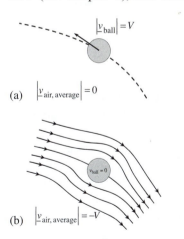

Figure 2.11 We can visualize a ball as an object moving through the air or as a ball with air moving around it. We choose to observe the flow from the moving ball; thus, the flow field appears as shown here.

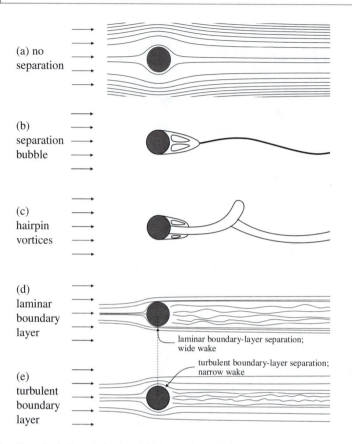

(a) no separation

(b) separation bubble

(c) hairpin vortices

(d) laminar boundary layer

laminar boundary-layer separation; wide wake

turbulent boundary-layer separation; narrow wake

(e) turbulent boundary layer

Figure 2.12   At very low ball speeds (i.e., low airflow rates), the air passing around the ball exhibits a flow in which the streamlines hug the sphere and form an orderly flow pattern. As the flow rate increases, separation occurs, and a recirculation region forms around the downstream stagnation point. At higher Reynolds numbers, complex three-dimensional hairpin vortices form. At still higher Reynolds numbers, a laminar boundary layer forms, which separates from the sphere surface near the equator, and a wide turbulent wake trails the sphere. At the highest Reynolds numbers, the boundary layer becomes thick and turbulent but separates from the sphere surface at a position downstream of the sphere equator. As a result of this delayed separation, the turbulent wake behind the turbulent boundary layer is narrower than the wake behind the laminar boundary layer. For more details, see Chapter 8.

(i.e., the leading face), a thin boundary layer hugs the ball's surface. At some position, for a smooth ball this position is about halfway from the front to the back, the boundary layer separates from the ball and forms the wake region. The wake region is a complex flow region with vortices and curvy flow lines, and significant drag is generated by the presence of the wake. The total amount of drag on the ball can be reduced by redesigning the ball's surface so that the boundary-layer separation occurs farther back from the leading face of the ball (Figures 2.12e and 2.13). By delaying the boundary-layer separation, we reduce the size of the wake and thereby reduce drag. The dimples on the surface of a golf ball do exactly this: They delay boundary-layer separation.

The aerodynamic shape of an airplane wing also is designed to delay or eliminate boundary-layer separation. Boundary-layer separation is caused by an adverse pressure gradient—which means that the pressure downstream is higher than the pressure upstream. The adverse pressure gradient in flow around a sphere

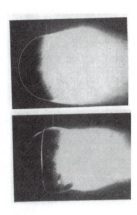

Figure 2.13 Roughening the surface of a sphere can trip the turbulent boundary layer and delay separation (Source: [34], original source U.S. Naval Ordinance Test Station, Pasadena Annex). On the left, the dropped ball is smooth and the boundary layer is laminar. On the right, the tip has sand grains cemented to its nose, and the sand trips the boundary layer, delaying separation. The rightmost series shows flow past a sphere both with and without a thin wire ring placed before the widest part of the sphere (Source: [147]; original reference Wieselsberger, ZFM, vol. 5, 140 (1914). The wire serves the same purpose as the sand; that is, the turbulent boundary layer is tripped.

is established by the uniform, viscosity-free flow outside the boundary layer. The boundary layers that form on a smooth ball are divided into two types: laminar and turbulent. Laminar boundary layers form at low speeds and are fairly regular in their flow patterns; however, laminar boundary layers are less able to withstand adverse pressure gradients without separation. Laminar boundary layers separate from the ball surface at the equator (90 degrees from the stagnation point, the centerline point of impact with the sphere). Turbulent boundary layers form at higher speeds or when something disrupts the flow. Turbulent boundary layers have a disorganized internal flow structure, but they are more able to withstand adverse pressure gradients without separation [154] due to their ability to borrow energy from the outer flow. Turbulent boundary layers separate behind the equator, about 110 degrees from the stagnation point. On golf balls, dimples or dents on the surface trip the boundary layer from a laminar boundary layer to a turbulent boundary layer, delaying separation, reducing drag, and making the balls fly farther (see Figure 2.13 and Section 8.2).

Another advantage of the boundary-layer picture is that it tells us for which situations we can ignore viscosity altogether in our calculations. We want to ignore viscosity because the flow outside the boundary layer where viscous effects can be neglected is much easier to analyze than the viscous boundary-layer flow. In the equations that govern the calculation of velocity in flows, the Navier-Stokes equations (see Chapter 6), the viscous term has second derivatives of velocity. If we can avoid including this term, solving the Naiver-Stokes equation is much easier. Solutions to the Navier-Stokes equations when viscosity is neglected are called potential flow solutions.[3]

For steady flow in the potential-flow limit, pressure calculations from velocity are particularly simple, as now discussed. Consider the outer region of a steady

---

[3] The flow of an inviscid fluid is called potential flow because of the similarity between equations for this flow and those for electrical potentials [167].

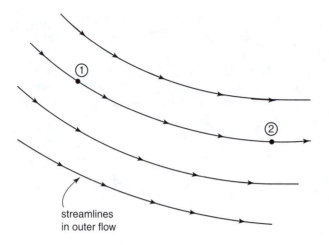

streamlines
in outer flow

Figure 2.14   In the outer flow region (away from walls), there is no effect of viscosity and the Bernoulli equation holds. We can choose Points 1 and 2 as any two points on the same streamline. The two points must be on the same streamline so that the Bernoulli equation requirement of single-input, single-output is satisfied.

flow shown in Figure 2.10. The individual particles of fluid follow paths that are called streamlines. The system of the fluid traveling along a streamline is a single-input, single-output, steady flow of an incompressible fluid. Viscosity is not important along a streamline that is far from a surface because there is little relative motion ($\partial v_1/\partial x_2 = 0$); with no effect of viscosity, there is no heat generated. The system considered has no chemical reaction or any phase change. Thus, along the streamline and in the outer flow away from a surface, all requirements of the mechanical energy balance (MEB) are met.

Furthermore, because there is no effect of viscosity in the region far from any surface, the friction term of the MEB is zero. There is, of course, no pump or any moving shafts in this flow along a streamline. The α quantity in the MEB is related to the distribution of velocity across the inlet and outlet cross sections (see Chapter 9); because we are following a single streamline, there is no velocity distribution across the inlet or outlet, and $\alpha = 1$ and $\langle v \rangle = v$. Thus, for flow along a streamline when all of the assumptions discussed are valid, the mechanical energy balance reduces to the Bernoulli equation with the average velocity now equal to the velocity on the streamline $\langle v \rangle = v$ (Figure 2.14):

$$\boxed{\frac{\Delta p}{\rho} + \frac{\Delta v^2}{2} + g\Delta z = 0}$$

Bernoulli equation along a streamline, in steady, rapid flow, far from any surface, with no phase change, incompressible, no velocity distribution, $\Delta T \approx 0$, no reaction, no friction, no shaft work

(2.39)

$$\frac{p_2 - p_1}{\rho} + \frac{v_2^2 - v_1^2}{2} + g(z_2 - z_1) = 0 \qquad (2.40)$$

We can write Equation 2.40 for any streamline in the outer flow because each streamline is a single-input, single-output system. This is a powerful application of the Bernoulli equation.

To put Equation 2.40 in traditional form, we can move all of the properties of Point 1 to one side of the equation and all of the properties of Point 2 to the other side, yielding:

$$\frac{p_2 - p_1}{\rho} + \frac{v_2^2 - v_1^2}{2} + g(z_2 - z_1) = 0 \tag{2.41}$$

$$\frac{p_1}{\rho} + \frac{v_1^2}{2} + z_1 = \frac{p_2}{\rho} + \frac{v_2^2}{2} + z_2 \tag{2.42}$$

The choice of Points 1 and 2 along the streamline is completely arbitrary, however; we can keep Point 1 the same and change the choice of Point 2 to be any point along the streamline. Because the choice of Points 1 and 2 to use in the Bernoulli equation along a streamline is arbitrary, the combination of pressure, velocity, and elevation terms on the lefthand and righthand sides of Equation 2.42 must be equal to the same scalar constant for every point on the streamline:

$$\boxed{\left(\frac{p}{\rho} + \frac{v^2}{2} + z\right) = \begin{array}{l} \text{constant along a streamline} \\ \text{in inviscid flow (away from surfaces)} \end{array}} \tag{2.43}$$

This is a powerful result for flow along a streamline in the region of a flow in which viscosity is not important—that is, in the outer region of a boundary-layer flow. Equation 2.43 allows us to relate pressures, velocities, and elevations for rapid flows away from surfaces. This result is derived more formally in Example 8.13.

Chapter 8 discusses solutions of the momentum-balance equations for the case of outer flows in which viscosity may be neglected. These potential-flow solutions are useful in aeronautics and other applications in which flow speeds are very high. The potential-flow solutions of the governing equations give the velocity distribution in the outer flow, and application of the Bernoulli equation along a streamline gives the pressure distribution from the velocity result. Knowing the pressure distribution in the outer flow then permits us to solve for the flow field in the inner region—that is, within the boundary layer (see Chapter 8). Potential-flow results also are useful in problems when only the outer flow is of interest, as in the following example. The key contribution of boundary-layer analysis is to clarify the existence of the two regions—the inner region where viscosity is important and the outer region where viscosity is not important—which then allows us to solve for the flows in both regions, if desired. Boundary layers figure into the development of airplanes, projectiles (e.g., bullets, torpedos, and missiles), and fuel-efficient automobiles and trucks (Figure 2.15).

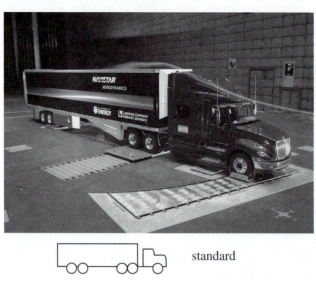

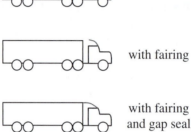

Figure 2.15 The amount of drag experienced by an 18-wheeler can be reduced by adding a piece above the truck cab that allows a smooth boundary layer to develop over the cab, the extra piece, and the roof of the trailer. Without this piece, the flow develops recirculation zones behind the cab and near the front of the trailer. These recirculation zones increase drag on the vehicle [65, 183]. Photo courtesy of Lawrence Livermore National Laboratory.

EXAMPLE 2.5. *A new tower hotel, cylindrical in shape and 100. feet in diameter, was built in a resort town near the sea on the windward side of an island (Figure 2.16). Residents complained that there often are uncomfortably high winds near several of the tower entrances. How does the wind speed vary with position around the tower and with onshore wind speed?*

SOLUTION. The air flow around the tower is a complex flow, particularly near its circular walls where a boundary layer forms and drag is produced. The question is about wind speed for someone standing a little distance away from the walls, however, and this question is about the flow outside the boundary layer. To address this question, we must evaluate carefully what the flow structure is like in the various locations under consideration.

Flow transverse to a long circular cylinder has been researched thoroughly and, at high flow speeds (i.e., high Reynolds numbers), we review the flow structures in Figure 2.17. A uniform high-speed wind approaches the cylinder at velocity v_∞; divides around the cylinder forming a boundary layer on the leading face of the cylinder; and at a position immediately past the equator of the circular cross section, the boundary layer separates from the cylinder surface and a turbulent wake is observed.

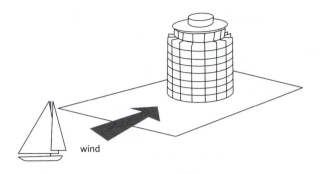

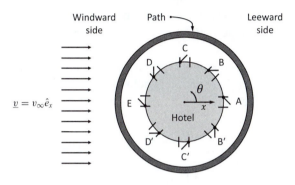

Figure 2.16   The new resort tower is cylindrical in shape with eight entrances equally spaced around the circumference.

Creeping flow (streamlines)

A flow with separation (streamlines)

Oscillating flow with Karman vortex street (pathlines)

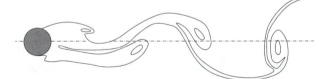

Boundary-layer flow with separation (turbulent BL, streamlines)

(laminar BL, streamlines)

Figure 2.17   For different values of the Reynolds number, $\rho v_\infty D/\mu$ different flow regimes are observed for flow around a long cylinder [149].

For the three door positions on the leeward side of the building, the wind velocity is the wind speed in the wake, which is somewhat chaotic and small compared to the wind on the windward side. For the position directly facing the oncoming wind, we expect some deceleration of the wind due to the presence of the building. For this position and others on the windward side of the building, we can estimate the wind speed as a function of onshore wind speed v_∞ by using the velocity solution for the outer flow—that is, the potential-flow solution for flow around a long cylinder (see Chapter 8). The potential-flow solution does not consider viscosity, but rapid flows form boundary layers and outside of the boundary layer, viscosity is not important. The case we consider is a rapid flow; for the windward side where the boundary layer is still attached to the cylinder, the potential-flow solution should give a reasonable result for flow away from the walls.

The velocity as a function of position for potential flow around a long cylinder is given in the literature [9] (for this solution the x-axis points in the wind direction, perpendicular to the cylinder; the z-axis of the $r\theta z$ system points along the cylinder axis):

$$\text{Potential flow around a long cylinder} \quad \underline{v} = \begin{pmatrix} v_\infty \left(1 - \frac{R^2}{r^2}\right) \cos\theta \\ -v_\infty \left(1 + \frac{R^2}{r^2}\right) \sin\theta \\ 0 \end{pmatrix}_{r\theta z} \quad (2.44)$$

$$\underline{v} = v_\infty \hat{e}_x$$

$$= v_\infty \left(1 - \frac{R^2}{r^2}\right) \cos\theta \, \hat{e}_r - v_\infty \left(1 + \frac{R^2}{r^2}\right) \sin\theta \, \hat{e}_\theta \quad (2.45)$$

We can calculate the speed of the fluid as a function of position from the magnitude of $\underline{v}$:

$$|\underline{v}| = \sqrt{\underline{v} \cdot \underline{v}} = \sqrt{v_r^2 + v_\theta^2 + v_z^2} \quad (2.46)$$

$$= v_\infty \sqrt{\left(1 + \left(\frac{R}{r}\right)^4 - 2\left(\frac{R}{r}\right)^2 \cos 2\theta\right)} \quad (2.47)$$

We now examine wind speeds along a path around the building. We choose a path that is 10 feet from the wall, which we assume is outside the boundary layer. With this choice, the path position is coordinate value $r = 60$ feet; the radius of the cylinder is $R = 50$ feet; and, for the eight doors, the values of θ are listed here with the predicted potential-flow speeds from Equation 2.47.

Door Location	θ	$\dfrac{v}{v_\infty}$	
behind the cylinder, center, A	$\theta = 0$	0.31	
behind the cylinder, to the side, B, B'	$\theta = \pm\frac{\pi}{4}$	1.22	(2.48)
at the equator, C, C'	$\theta = \pm\frac{\pi}{2}$	1.69	
in front of the cylinder, to the side, D, D'	$\theta = \pm\frac{3\pi}{4}$	1.22	
in front of the cylinder, center, E	$\theta = \pi$	0.31	

As discussed previously, the boundary layer in rapid flow around a cylinder is observed to detach downstream of the cylinder equator. Thus, for positions A, B, and B', we do not use the potential-flow results but rather estimate that the cylinder shields the doors ($v \longrightarrow 0$). For the other five doors where the boundary layer is attached, however, the potential-flows results are a good estimate. We therefore find that the wind velocity along a path around the building 10 feet from the wall of the hotel will vary as follows:

$$
\begin{array}{cc}
\text{door} & \dfrac{v}{v_\infty} \\
\hline
A & 0 \\
B, B' & \approx 0 \\
C, C' & 1.7 \\
D, D' & 1.2 \\
E & 0.3 \\
\hline
\end{array}
\tag{2.49}
$$

These calculations show that the windiest spot is half way around the building from where the wind first hits. At this location the wind speed is 70 percent higher than the speed of the offshore breeze.

The key knowledge needed in this example is an awareness of the existence and impact of the boundary layer and the meaning of potential-flow solutions. We study boundary layers in Chapter 8.

2.4 Laminar versus turbulent flow: Reynolds number

The introduction to boundary layers in the previous section is concerned with flows in which a large amount of fluid is moving past a surface: so-called external flows. Boundary layers also are present in internal flows: flows inside fixed boundaries such as within pipes, reactors, or blood vessels (see Chapter 7). When water enters an intake pipe at low flow rates (Figure 2.18), the flow in the entry region is uniform in the core with a boundary-layer structure near the walls that thickens and grows as the flow adapts to the presence of the pipe walls. Once the flow fully develops inside the pipe, the region outside the boundary layer disappears, and we observe a well-defined flow throughout the pipe, called

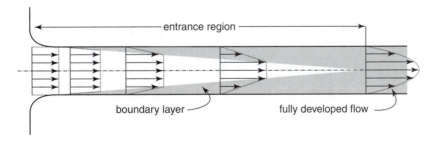

entrance region

boundary layer

fully developed flow

Figure 2.18 The flow near the entrance of the pipe is different than the flow in the rest of the pipe. A boundary layer forms on the inner pipe surface and friction effects are concentrated there. The boundary layer grows rapidly; soon, the core region outside of the boundary layer disappears and a well-developed pipe flow appears, which is dominated by frictional effects. In this illustration, the fluid is assumed to be incompressible; thus, the flow rate is constant throughout the pipe, including in the entrance region.

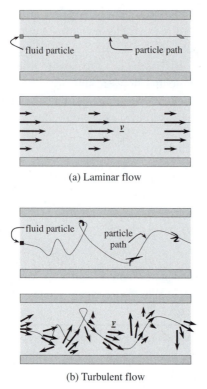

(a) Laminar flow

(b) Turbulent flow

Figure 2.19   Laminar flow (a) is a flow in which fluid particles move in layers, one layer sliding over the other (the word *laminar* comes from the Latin word for *layer*). In laminar flow, particles move along straight paths and the velocity along those paths is constant at steady state. The fluid particles deform in a well-defined manner. In turbulent flow (b), the detailed motion of fluid particles is not well defined and much mixing occurs. Particles move along tortuous trajectories, one of which is shown here, and are deformed in ways that are difficult to quantify. The velocity field, even at steady state, is a wildly fluctuating function of space and time.

laminar flow. In laminar flow, viscous effects dominate throughout and cylindrical layers of fluid slide over one another, transferring stress from the flow to the walls of the pipe.

Laminar pipe flow is similar to flow in a laminar boundary layer discussed in the previous section—that is, the flow is organized, with fluid layers sliding over one another, transferring stress in an orderly manner (Figure 2.19, top). Steady laminar flow in a pipe is a simple flow for which we can fully calculate all aspects of the flow—pressure field, velocity field, and stress field—using the methods in this book (see Chapter 7). We introduced one laminar-flow result, the Hagen-Poiseulle equation, in Equation 2.8. Although it is a simple flow, steady laminar flow has practical applications in real-world situations, such as in the analysis of blood flow in arteries, in studies of the flow of high-viscosity liquids (e.g., polymers, foods, and slurries), and in viscosity measurements.

If the flow rate of a laminar flow in a tube is increased (e.g., by increasing the driving pressure), the flow eventually becomes unstable. By "unstable," we mean that the flow no longer moves in well-defined layers from upstream to downstream but rather breaks up into many small eddies swirling over one another, tumbling in the flow direction (Figure 2.19, bottom). We encountered turbulent flow in the previous section when discussing turbulent boundary layers (see Figure 2.12). For both turbulent pipe flow and turbulent boundary layers, the flow is disorganized, and significant energy is churned up in the motions of the flow.

In turbulent pipe flow, there is a dominant flow direction; however, on a small lengthscale, the flow is jumbled and mixed with small eddies and whirls that are impossible to predict and difficult even to characterize mathematically. The distinction between laminar and turbulent flow was elucidated by Osborne Reynolds in 1883 [139]. In his experiments, a dye was injected into the center of pipe flow and observed through the transparent walls of the pipe (Figure 2.20). At low flow rates, the dye moves downstream in a straight line that mixes slightly with the main fluid due to molecular diffusion. At high flow rates (Figure 2.20c), the dye stream breaks up soon after injection and spreads across the cross section of the

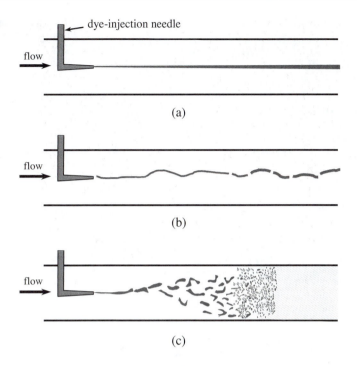

(a)

(b)

(c)

Figure 2.20 Reynolds demonstrated the fundamental difference between laminar and turbulent flow by injecting dye into water flowing in a pipe. At low flow rates (a), the dye moves downstream in a straight line. At high flow rates (c), the dye stream breaks up soon after injection and spreads across the cross section of the pipe, ultimately resulting in a stream that is homogeneously colored with dye. In transitional flow, the dye stream distorts and elongates but mixing is incomplete [14].

pipe, resulting in a stream that is homogeneously colored with dye. There is much cross-stream mixing in high-flow-rate turbulent flow. By carefully increasing the flow rate from low (laminar) to high (turbulent), the transition to turbulence can be captured. In transitional flow (Figure 2.20b), the dye stream distorts and elongates but mixing is incomplete. These pipe-flow regimes were discussed in the quick-start section of Chapter 1. Reynolds's dye-tracing experiments established that the Reynolds number, $\text{Re} = \rho \langle v \rangle D / \mu$, is the parameter that distinguishes the three flow regimes in a pipe: He found that laminar flow occurs for $\text{Re} < 2{,}100$; between $\text{Re} = 2{,}100$ and $\text{Re} = 4{,}000$, the flow is transitional; and fully turbulent flow occurs for $\text{Re} > 4{,}000$ (see Equation 1.63).

Turbulent flow is very common. Flows are turbulent in most industrial process units (see Example 1.6) and in the air around us; some blood flow in the human body is turbulent as well (Table 2.2 and Figure 2.21). For example, the narrowing of arteries characteristic of advanced heart disease can be detected by a physician listening with a stethoscope for turbulent blood flow in constricted arteries [53]. Narrowing of and obstructions in the arteries increase the blood average velocity, causing turbulence; the rapid pressure fluctuations associated with turbulence produce a noise (i.e., the Korotkov sound) that can be heard with a stethoscope. The higher drag associated with turbulent flow produces wear and tear on the arteries and is one of the dangers of heart disease. Prosthetics designed to treat heart disease (e.g., artificial valves, artificial hearts, and stents; see Figure 2.22) must be designed to minimize turbulence.

Table 2.2. Reynolds numbers in the circulatory system vary from 0.0007 to almost 6,000 [145]

Location	Diameter (cm)	$\langle v \rangle$ (cm/s)	Re
Ascending aorta	2.0–3.2	63	3,600–5,800
Descending aorta	1.6–2.0	27	1,200–1,500
Large arteries	0.2–0.6	20–50	110–850
Capillaries	0.0005–0.001	0.05–0.1	0.0007–0.003
Large veins	0.5–1.0	15–20	210–570
Vena cavae	2.0	11–16	630–900

Note: Original reference is Whitmore, R.L., *Rheology of the Circulation*, Oxford, 1968.

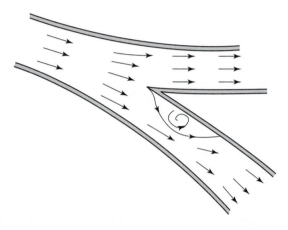

Figure 2.21   Schematic of turbulent blood flow in arteries after a bifurcation.

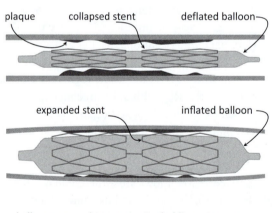

plaque   collapsed stent   deflated balloon

expanded stent   inflated balloon

balloon removed; stent remains holding artery open

Figure 2.22   A stent is a tiny expandable stainless-steel tube that holds heart arteries open following angioplasty [21]. In angio-plasty, a small balloon is used to force open blocked arteries. The stent is placed around the balloon and used to prop open the artery after the balloon is deflated. Buildup of cholesterol plaque on artery walls—the cause of narrowing of the arteries (i.e., atherosclerosis)—usually is found near branching points in the blood vessels. Researchers believe that flow disturbances near these branches or near obstructions like plaque deposits or the stent itself may encourage atherosclerosis [53]. Minimizing such flow disturbances thus becomes a matter of life and death.

Because of the considerable mixing in turbulent flow, more drag is produced compared to laminar flow.[4] Although we cannot predict the detailed velocity, pressure, and stress fields of turbulent flow, extensive study of turbulent flow since its description by Reynolds has shown how to predict turbulence and how to design practical equipment around average values of velocity, pressure, and stress in turbulent flow. We discuss turbulent flow in Chapters 7 and 10.

In the following example, we illustrate the power of the Reynolds number in analyzing pipe flow.

EXAMPLE 2.6. *When choosing the pump for a flow application, it is essential to know how much pressure is needed to produce the flow. Experiments with three different fluids in three different pipes show that a wide range of pressure drops are needed to bring about flows at modest flow rates (Table 2.3). The flows were carried out in clear pipes and visually inspected; some were smooth and laminar, some patterns were chaotic-looking and therefore deemed turbulent, and some were difficult to evaluate for flow type; these were designated as transitional. What are the key factors that determine flow type for these fluids? How could we have predicted the pattern of flow type as a function of pressure drop?*

Table 2.3. Three fluids pumped at fixed flow rates through 2.0-m pipes (6.6 ft) of various sizes

| Fluid | Q gpm | Q cm³/s | Pressure drop (Pa) in various pipes | | |
			1/4 in.	3/8 in.	1/2 in.
water 4°C	0.5	32	**660**	*170*	64
water 4°C	1.0	63	3,600	800	**210**
water 4°C	2.0	126	12,000	2,800	930
water 4°C	3.0	189	24,000	5,500	1,800
water 25°C	0.5	32	980	**190**	**66**
water 25°C	1.0	63	2,900	700	240
water 25°C	2.0	126	9,400	2,400	780
water 25°C	3.0	189	20,000	4,700	1600
blood 37°C	0.5	32	*1,000*	*310*	*120*
blood 37°C	1.0	63	**3,700**	**590**	*250*
blood 37°C	2.0	126	14,500	3,700	**1,000**
blood 37°C	3.0	189	28,700	7,200	2,400

Notes: Schedule 40 pipes with nominal size given.
The difference between the pressure at the inlet and at the outlet is given in *Pa*. All flows were observed to be turbulent except those indicated with *italics* and **boldface**, which were laminar and transitional, respectively.

SOLUTION. There is much data in Table 2.3; we begin by plotting it to see which trends are revealed (Figure 2.23). Inspecting the graph, we see that the highest pressures are generated in the smallest pipes. Also, higher flow rates generated the highest pressures. The pressure data vary over several orders of magnitude; because of the large pressure values at high flow rate and small diameter, the

---

[4]More drag is produced in turbulent flow than in a hypothetical laminar flow at the same flow rate. However, it is not possible to produce such laminar flows in most high-flow-rate situations (see Chapter 7).

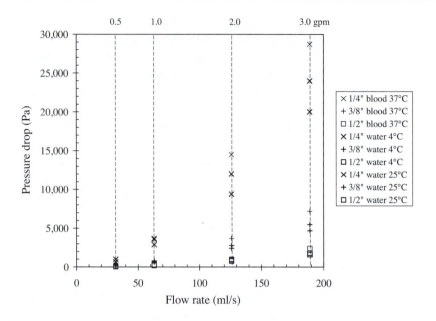

Figure 2.23 Pressures generated in steady flow of various fluids through 2-m pipes of various diameters.

lower pressure data are difficult to see in Figure 2.23. We fix this problem by changing the y-scale on that plot to be logarithmic (Figure 2.24).

In the log-linear view of the pressure data in Figure 2.24, we see that pipe size makes an important difference among the observed pressure drops. The data are in three groups, with the smallest-pipe data in the topmost trend and the largest-pipe data along a trend at the bottom. Therefore, it appears that determining factors for laminar or turbulent flow may be critical values of pressure or pipe size.

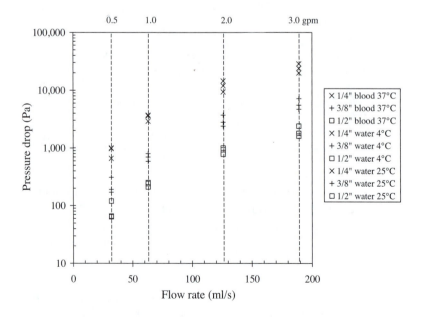

Figure 2.24 Pressures generated in steady flow of various fluids through 2-m pipes of various diameters; log-lin plot.

Table 2.4. Physical property data for fluids in the example

Fluid	ρ kg/m^3	μ Pa s ($\frac{kg}{m\,s}$)
water 4°C [132]	1.000×10^3	1.57×10^{-3}
water 25°C [132]	0.99708×10^3	0.894×10^{-3}
blood 37°C [145]	1.060×10^3	3.0×10^{-3}

However, comparing the flow types given in Table 2.3 with the figures does not reveal a pattern. For example, at 1.0 gpm in the 1/4-inch pipe, both water flows are turbulent but blood flow, which has the highest pressure drop, is transitional. At 1.0 gpm in the 3/8-inch pipe, blood again is transitional whereas water is turbulent but, in this case, blood has the lowest pressure drop, $\Delta p = 590$ Pa.

It turns out there is a simple way to correlate the data in Table 2.3, and it was used in Chapter 1. The methods in this text led researchers of a previous generation (including Osborne Reynolds) to discover that all of the data could be correlated if flow rate were written in dimensionless form as the Reynolds number, Re, and if pressure drop along a pipe were written in terms of a dimensionless wall force, called the Fanning friction factor f:

$$\text{Reynolds number:} \quad \text{Re} \equiv \frac{\rho \langle v \rangle D}{\mu} \tag{2.50}$$

$$\text{Fanning friction factor:} \quad f = \frac{D \Delta p}{2 \rho \langle v \rangle^2 L} \tag{2.51}$$

where ρ is the density of the fluid, $\langle v \rangle$ is the average velocity of the fluid, μ is the fluid's viscosity, Δp is pressure drop, and L is length of the pipe. A review of the literature allows us to find the densities and viscosities of our fluids (Table 2.4). The average velocity in the pipe may be calculated from the experimental flow rates using the usual expression (i.e., Equation 1.2):

$$\langle v \rangle = \frac{Q}{\pi R^2} = \frac{4Q}{\pi D^2} \tag{2.52}$$

where $R = D/2$ is the inner radius of the tube. For Schedule 40 piping, again from the literature [132], we obtain the precise values of the pipe IDs (Table 2.5); note that the values in the table are quite different from the nominal sizing values. We now can convert the data in Table 2.3 to friction factor versus Reynolds number, as shown in the following calculation:

Table 2.5. Inner diameter for pipes in the example [132]

Nominal pipe size [132]	ID inches	ID meters
1/4	0.364	0.925×10^{-2}
3/8	0.493	1.252×10^{-2}
1/2	0.622	1.580×10^{-2}

For water at 4°C in the 1/4-inch pipe at 0.5 gpm, we calculate:

$$\langle v \rangle = \frac{4Q}{\pi D^2} \tag{2.53}$$

$$= \frac{(4)\left(\frac{32 \text{ cm}^3}{\text{s}}\right)}{\pi (0.925 \text{ cm})^2} \tag{2.54}$$

$$= 47.62 \text{ cm/s} = \boxed{48 \text{ cm/s}} \tag{2.55}$$

$$\text{Re} \equiv \frac{\rho \langle v \rangle D}{\mu} \tag{2.56}$$

$$= \frac{\left(\frac{1{,}000 \text{ kg}}{\text{m}^3}\right)\left(\frac{0.4763 \text{ m}}{\text{s}}\right)(0.925 \times 10^{-2} \text{ m})}{1.57 \times 10^{-3} \frac{\text{kg}}{\text{m s}}} \tag{2.57}$$

$$= 2{,}806 = \boxed{2{,}800 \text{ (unitless)}} \tag{2.58}$$

$$f = \frac{D \Delta p}{2 \rho \langle v \rangle^2 L} \tag{2.59}$$

$$= \frac{(0.00925 \text{ m})(660 \text{ Pa})}{(2)\left(\frac{1{,}000 \text{ kg}}{\text{m}^3}\right)\left(\frac{0.4762 \text{ m}}{\text{s}}\right)^2 (2.0 \text{ m})} \tag{2.60}$$

$$= \boxed{0.0067 \text{ (unitless)}} \tag{2.61}$$

When all the data in Table 2.3 are converted to friction factor versus Reynolds number, we plot these quantities on a log-log plot (Figure 2.25).

The friction factor/Reynolds number plot is striking in its simplicity compared to the same data plotted in either Figure 2.23 or 2.24. First, there is one single curve for all three fluids. All of the differences in the experiments due to choice of fluid are captured by including the viscosity and the density in the Reynolds number. Second, there is no longer any evidence in the plot of a dependence on pipe diameter. The recasting of the data into dimensionless pressure drop and dimensionless flow rate fully captures the effect of pipe diameter on flow rate and pressure drop. By comparing the observed flow types listed in Table 2.3, we also see that there is a clear separation of flow types by Reynolds number in the data of this example: Laminar flow is observed for $\text{Re} < 2{,}100$, turbulent flow is observed for $\text{Re} > 4{,}000$, and unstable flow is observed between them. This is precisely Osborne Reynolds's observation that flow type depends on only the combined variable Reynolds number, not individually on the parameters ρ, $\langle v \rangle$, D, and μ.

The problem statement asked what are the key factors that determine flow type for the fluids studied. The answer is that flow type is determined by the Reynolds

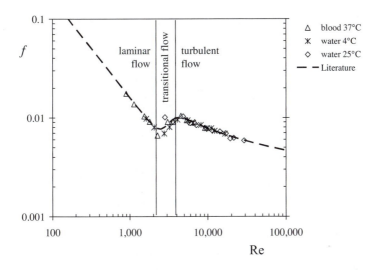

Figure 2.25 The data in this example are rendered into friction factor versus Reynolds number. Laminar flow is observed when the Reynolds number is less than about 2,100; turbulent flow is observed when the Reynolds number is above 4,000.

number of the flow. We were asked how we could have predicted the observed variations in pressure drop instead of carrying out involved experiments. We see now that because the friction factor/Reynolds number curve is available from prior experiments by Reynolds and others (see Figure 1.21 and the Colebrook equation, Equation 1.95 in Chapter 1), we could have proceeded as follows: Calculate the Reynolds number from the flow rates of interest (using Equations 2.50 and 2.52), read friction factor from the published correlation plot (see Figure 1.21), and calculate Δp for each datapoint using Equation 2.51. Because $\Delta p(Q)$ in pipes in the form of $f(Re)$ already is well known, we can predict with confidence many important quantities in pipe flows of all types (see Chapter 8 for more details).

The incredible simplicity of Figure 2.25 was not luck. Dimensional analysis of the governing equations tells us directly to expect that plotting friction factor versus Reynolds number would collapse the data for pipe flow. The dimensional analysis of pipe flow is described in Chapter 7.

EXAMPLE 2.7. *For the flow of water at 3.0 gpm in $\frac{1}{2}$-inch type-L copper tubing, is the flow laminar or turbulent? What is the highest flow rate for laminar flow in this tubing?*

SOLUTION. To determine if the flow is laminar or turbulent, we calculate the Reynolds number. The viscosity and density of water are available from the literature, as is the true ID of $\frac{1}{2}$-inch copper tubing [132].

$$\rho = 62.25 \ \text{lb}_\text{m}/\text{ft}^3 \tag{2.62}$$

$$\mu = 6.005 \times 10^{-4} \ \text{lb}_\text{m}/\text{ft s} \tag{2.63}$$

$$D = 0.545 \ \text{in} \tag{2.64}$$

First, we calculate the average velocity from the flow rate:

$$\langle v \rangle = \frac{4Q}{\pi D^2} \tag{2.65}$$

$$= 4\,(3\text{ gpm})\left(\frac{2.228 \times 10^{-3}\text{ ft}^3/\text{s}}{\text{gpm}}\right)\left(\frac{4}{\pi\,(0.545/12\text{ ft})^2}\right) \tag{2.66}$$

$$= 16.503\text{ ft/s} = \boxed{17\text{ ft/s}} \tag{2.67}$$

Second, we calculate the Reynolds number:

$$Re = \frac{\rho\langle v \rangle D}{\mu} \tag{2.68}$$

$$= \frac{(62.25\text{ lb}_\text{m}/\text{ft}^3)\,(16.503\text{ ft/s})\left(\frac{0.545}{12}\text{ ft}\right)}{6.005 \times 10^{-4}\text{ lb}_\text{m}/\text{ft s}} \tag{2.69}$$

$$= \boxed{78{,}000} \tag{2.70}$$

Because the Reynolds number is higher than 4,000, the flow is turbulent.

To determine the highest flow rate that gives laminar flow, we seek the flow that corresponds to a Reynolds number of 2,100:

$$Re = 2{,}100 \tag{2.71}$$

$$2100 = \frac{\rho\langle v \rangle D}{\mu} \tag{2.72}$$

$$\langle v \rangle = \frac{(2{,}100)\left(6.005 \times 10^{-4}\text{ lb}_\text{m}/\text{ft s}\right)}{(62.25\text{ lb}_\text{m}/\text{ft}^3)\left(\frac{0.545}{12}\text{ ft}\right)} \tag{2.73}$$

$$= 0.4460\text{ ft/s} = \boxed{0.45\text{ ft/s}} \tag{2.74}$$

Third, we calculate the flow rate in gpm that corresponds to this average velocity:

$$Q = \frac{\pi D^2}{4}\langle v \rangle \tag{2.75}$$

$$= \frac{\pi\left(\frac{0.545}{12}\text{ ft}\right)^2(0.4460\text{ ft/s})}{4}\left(\frac{\text{gpm}}{2.228 \times 10^{-3}\text{ ft}^3/\text{s}}\right) \tag{2.76}$$

$$= \boxed{0.3\text{ gpm}} \tag{2.77}$$

We calculated that the maximum flow rate for laminar flow in the context of a $\frac{1}{2}$-inch tubing is about 0.3 gpm, which is a very low flow rate in a household or industrial application. For comparison, a low-flow bathroom showerhead has a flow rate between 1 and 3 gpm through a $\frac{1}{2}$-inch pipe. Laminar flow is used rarely in plumbing and industrial piping unless conditions are designed deliberately to achieve laminar flow.

2.5 Aerodynamics: Lift

In Section 2.2, we discuss the concept of fluid drag, a force that slows down objects that move through a fluid. Drag is a consequence of viscosity—which is a measure of the ability of a fluid to transfer stress. Drag is a force that acts counter to the principal flow direction. Lift is another component of force created when objects move through fluids. Lift tends to move objects in a direction perpendicular to the main flow direction (Figure 2.26), and it is lift that gets an airplane off the ground.

How does an airplane fly? This is not an easily answered question despite the existence of many published explanations. The technology of heavier-than-air flight involves a discussion of viscosity, pressure, boundary layers, and boundary-layer separation and boundary-layer attachment (recall the complex golf-ball discussion). To give an idea of how flight depends on complex flow phenomena, we outline the reasons for flight. The physics of flight is better understood by

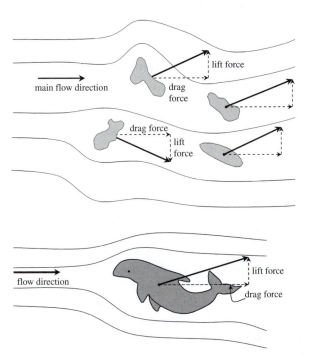

Figure 2.26 Lift is the force that allows airplanes and helicopters to resist the pull of gravity and to fly, but it also is a lateral force that affects any body that moves through a fluid such as particles in a water stream or dolphins in the ocean.

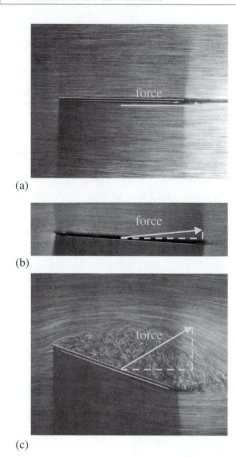

(a)

(b)

(c)

Figure 2.27   Anisotropic bodies experience lift when placed at an inclined angle relative to a uniform flow. (a) If the body is placed at zero inclination—that is, facing head on into the flow, there is no lift (i.e., no vertical component of force); (b) at a 2.5-degree angle of attack, there is a small amount of lift; and (c) at a 20-degree angle of attack, there is more lift. Notice the change in the flow around the body as the angle of attack increases [170]. Images Copyright © 1974 ONERA.

studying aeronautics after completing this first course in fluid mechanics [11, 76] (see also Section 10.4).

Anisotropic bodies[5] experience lift when placed at an inclined angle relative to a uniform flow field. The type of object orientation that produces lift is shown in Figure 2.27. A body placed at an inclined angle relative to a uniform upstream flow field splits the flow, pushing part of the fluid down and past the object and part of the fluid up and over the object. The object does not need to be an airfoil to experience lift. We discuss airfoils in more detail later.

One part of lift is caused by the force that the object uses to push down the portion of the stream that flows down the underside of the object. Imagine your hand to be the object and you are holding it in a strong oncoming air jet (Figure 2.28). To do this, you are imposing a force on the air that is pushing the air down. Following Newton's law of motion, the air pushes back on your hand with an equal and opposite force. Your hand is inclined relative to the

[5] Highly symmetric bodies such as spheres also can experience lift in some flows; see Chapter 8 for a discussion.

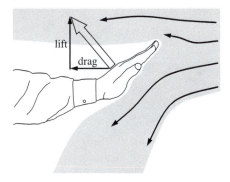

A hand held up to deflect an oncoming air jet experiences lift.

velocity of the incoming air jet; to divert the air jet, you must exert a vector force that has two components: one opposite in direction to that of the incoming air jet and one downward. The upward force your hand feels is part of the lift.

Lift due to the upward component of force on the bottom of an object in a stream is not the only contribution to total lift, however, and here the explanation becomes more complicated. There also is a component of lift from the fluid that pours over the top of the object. The angle that the object makes with the

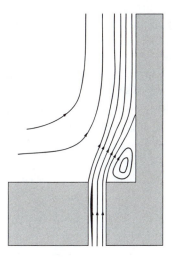

Figure 2.29   The Coanda effect is the tendency of a fluid moving over the top of an object to turn and flow along the surface. The Coanda effect occurs because less energy is lost by deflecting the stream toward the solid surface than if the stream continued to flow in a straight line. The free jet moving in a straight line tends to entrain fluid that it flows past. The presence of the surface prevents this inward flow. Nevertheless, the inward forces of that flow are present and they redirect the jet toward the surface.

oncoming fluid-velocity direction is called the angle of attack (see Figure 2.27). For a modest angle of attack, the air flowing over the top of the object does not simply rush past in a straight line; rather, it turns and flows down the surface (see Figure 2.27b). This effect is particularly pronounced if the top surface of the object is smoothly curved (e.g., an airplane wing). The smoothly varying surface of an airplane wing is designed to prevent boundary-layer detachment, thereby enhancing the tendency of the flow to cling to the wing. The tendency of a fluid jet in some geometries to attach to a nearby surface is called the Coanda effect, which is caused by the inertia of the uniform outer flow and the tendency of fluid jets to entrain bystanding fluid on either side of the jet [168] (Figure 2.29). The net effect is that the object in the airstream not only directs downward the air that flows under the object; due to the Coanda effect, it also directs downward the air that flows over the object (Figure 2.30). Thus, the lift experienced by the object is the equal and opposite reaction force generated by the downward-forced airstreams passing under and over the object. Airplane wings are objects that move rapidly relative to air, and they are set at an angle to push and pull down the

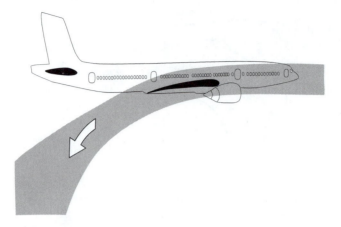

An airplane wing moving rapidly deflects air downward by pushing on it. Due to the Coanda effect, the air moving over the top of the wing also is pushed downward by the motion of the wing.

air. Airplane engines are designed to give horizontal thrust, which is used to direct the air downward and also to overcome the horizontal drag caused by the air.

The angle of attack is an important parameter in flight. At a zero angle of attack—that is, with an airfoil such as an airplane wing facing squarely into onrushing air—the airfoil experiences little lift. As the angle of attack increases, the lift increases. There is a limit to this effect, however, due to the dynamics of boundary layers. As the angle of attack increases, the boundary layer on the top surface of the airfoil is increasingly unstable; it eventually detaches near the trailing edge and drag increases (Figure 2.31). If the angle of attack is increased further, the flow completely separates from the top surface of the airfoil, which now is said to be stalled. During a stall, the lift decreases and the severe loss in lift causes the airplane to drop. A midflight stall can be extremely dangerous, although a stunt pilot who understands a stall can control it and recover by adjusting the angle of attack to a more acceptable value.

The lifting characteristics of a well-designed airfoil are quantified in the lift coefficient, C_L (compare with drag coefficient; Equation 2.27):

$$\text{Lift coefficient} \qquad \boxed{C_L = \frac{\mathcal{F}_{\text{lift}}}{\frac{1}{2}\rho\langle v\rangle^2\, A_p}} \qquad (2.78)$$

where C_L is the lift coefficient (unitless); $\mathcal{F}_{\text{lift}}$ is the lift, which is a force magnitude; ρ is the density of the fluid; $\langle v\rangle$ is the average velocity of the object or the velocity of the fluid as it flows past the object; and A_p is the reference area for lift coefficient, often the planform area, which is the projected area of the object in the direction of lift (i.e., perpendicular to the oncoming flow stream). The lift coefficient, like the drag coefficient, is a function of the Reynolds number. For Reynolds numbers associated with airplane flight, the lift coefficient of an airfoil is primarily a function of the angle of attack. Figure 2.32 shows the measured C_L as a function of the angle of attack α for a typical airfoil [64]. The lift coefficient rises linearly with α up to a maximum where boundary-layer separation causes stall; the lift

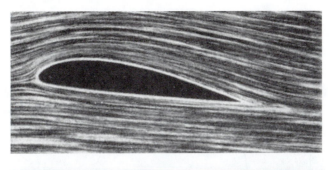

angle of attack

Figure 2.31 Increasing the angle of attack increases lift on an airfoil. When the flow completely separates from the top surface, the flow is stalled [147]. Image from L. Prandtl and O. Tietjens, Hydro- und Aeromechanik, Springer, Berlin, (1929).

coefficient decreases with further increases in α. The maximum value of C_L can be obtained readily from such data and is a strong function of airfoil shape as well as the Reynolds number (Figure 2.33). Advanced airfoil designs incorporate flaps, slots, and other types of boundary-layer control allowing C_L to increase up to values of 4 or higher [121, 176].

The mathematical complexities of the fluid mechanics of airplane design are beyond the scope of this text, but the previous discussion describes the richness of fluid phenomena and the practicality of the study of fluid mechanics, especially drag, lift, and boundary layers (see Chapters 8 and 10). The concept of lift is applicable to more than airplane flight—lift affects the settling of anisotropic particles in a suspension, wind stresses on structures, propulsion in sailboats, and racecar aerodynamics (e.g., the front and rear wings on Formula One racing machines are designed to counteract the tendency of lift to raise the vehicle off the pavement). This section concludes with an example of a lift calculation.

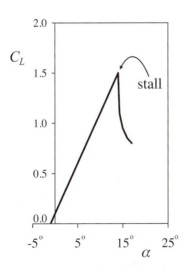

Figure 2.32 Lift coefficient for an airfoil as a function of the angle of attack. For symmetric airfoils, there is zero lift at zero angle of attack. The airfoil corresponding to these data is not symmetric and has zero lift at a downward angle of attack [64].

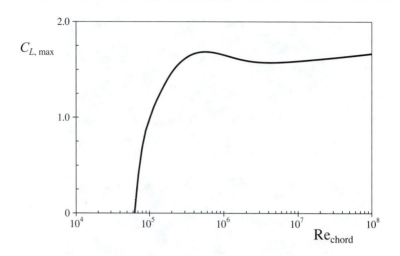

Maximum lift coefficient as a function of the Reynolds number for a typical simple airfoil. Reynolds number is based on the chord length, which is the distance from the airfoil tip to the tail. At takeoff, Reynolds numbers are about 10^5, increasing to higher than 10^7 at cruising [64].

EXAMPLE 2.8. *The takeoff speed of an aircraft is roughly 1.2 times the stall speed. What is the takeoff speed of an aircraft with a mass of 74,000 lb_m and a planform area of 2,600 ft^2? Use the lift data in Figure 2.32 for the calculations. The density of air is 0.07625 lb_m/ft^3.*

SOLUTION. To lift the aircraft into the air, we must generate a vertical force that equals the weight of the plane. Thus, we need a lift of magnitude:

$$\mathcal{F}_{\text{lift}} = mg \tag{2.79}$$

where m is the mass of the plane and g is the acceleration due to gravity. For the example plane, then:

$$\mathcal{F}_{\text{lift}} = mg \tag{2.80}$$

$$= (74{,}000 \text{ lb}_\text{m}) \left(\frac{32.174 \text{ ft}}{\text{s}^2} \right) \left(\frac{\text{s}^2 \text{ lb}_\text{f}}{32.174 \text{ ft lb}_\text{m}} \right) \tag{2.81}$$

$$= 74{,}000 \text{ lb}_\text{f} \tag{2.82}$$

Note the conclusion that a mass of 74,000 lb_m weighs 74,000 lb_f; this is the logic behind the otherwise confounding American engineering units.

Because the takeoff speed is related to the stall speed, we begin by calculating the stall speed. The stall speed is the speed at which an aircraft reaches the stall point at the maximum of the C_L versus α curve in Figure 2.32. The lift coefficient at the maximum from the curve in Figure 2.32 is 1.5. The lift coefficient and the plane stall speed are related in Equation 2.78 when C_L is set equal to this maximum value:

$$C_L = \frac{\mathcal{F}_{\text{lift}}}{\frac{1}{2}\rho \langle v \rangle^2 A_p} \tag{2.83}$$

For $C_L = 1.5$ and $\langle v \rangle = v_{stall}$, we obtain:

$$v_{stall} = \sqrt{\frac{\mathcal{F}_{lift}}{\frac{1}{2} C_L \rho A_p}} \tag{2.84}$$

$$v_{stall} = \sqrt{\frac{(74{,}000 \text{ lb}_f) \left(\frac{32.174 \text{ ft lb}_m}{s^2 \text{ lb}_f} \right)}{\frac{1}{2}(1.5)(0.07625 \text{ lb}_m/\text{ft}^3)(2{,}600 \text{ ft}^2)}} \tag{2.85}$$

$$= 127 \text{ ft/s} = \boxed{130 \text{ ft/s}} \tag{2.86}$$

The problem stated that the takeoff speed is 1.2 times the stall speed:

$$v_{takeoff} = 1.2 \, v_{stall} = 152 \text{ ft/s} \tag{2.87}$$

$$= \boxed{150 \text{ ft/s}} \tag{2.88}$$

In the remainder of this chapter, we introduce specialty flows that produce fascinating behaviors that merit study once fluid basics are understood. Readers who prefer to begin their own modeling efforts may proceed to Section 2.11, which is a summary of the chapter and a launching point for the remainder of the text.

2.6 Supersonic flow

When fluids move extremely rapidly—such as when a gas flows through a relief valve on an overpressurized tank or when air passes through a jet engine—the flow can become so fast that its fundamental nature changes. These high-speed flows are called supersonic because the speed of sound is the critical speed that marks when the change in physics occurs.

Sound is the result of forces on a gas, liquid, or solid causing a disturbance that then propagates through the matter as a longitudinal compressive wave. An example of sound propagation is a hammer striking a bell (Figure 2.34) causing the bell to vibrate. The vibration of the bell causes the air around the bell to move, and the information that the bell is vibrating travels through the air at a speed called the speed of sound in air. Our ears pick up and interpret this vibration through our physiology as sound. From the perspective of physics, what happened is that forces at the source of the sound (i.e., in our example, the forces between the hammer and the bell) caused the bell to vibrate, which in turn causes a disturbance in the fluid near the bell. These disturbances cause subsequent disturbances in neighboring fluid particles, and the process repeats as the wave propagates.

Striking a bell creates sound waves that propagate through the air.

The propagation of sound waves is so rapid and of such low amplitude that it typically is not mentioned in the discussion of fluid mechanics; however, wave propagation of disturbances in fluids has a role "behind the scenes" in everyday fluid mechanics. For example, a hydraulic lift is a device used to amplify forces using a clever geometry and a quiescent fluid through which pressure propagates. When forces act on a liquid at one side of a hydraulic lift (Figure 2.35; see also Section 4.2.4.2), they affect the nearest layer of fluid, which contacts and affects the next nearest layer, and so on. The information that a force has been applied at Point (a) travels throughout the fluid reservoir as a longitudinal pressure wave, and the speed of the propagation of that wave is the speed of sound in that fluid. This happens so rapidly that it usually is considered to have occurred instantaneously, and it is not necessary to discuss the transmission of this information. Thus, when analyzing the hydraulic lift (see Chapter 4), we state simply that the pressure applied at one location in a quiescent fluid spreads instantly to all locations. In a hydraulic lift, the fluid moves slowly (or not at all), and the information on pressure change moves rapidly.

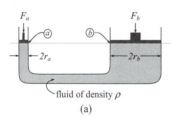

(a)

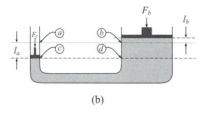

(b)

When a fluid is moving rapidly or when an object moves rapidly through a fluid, the speed of the fluid and the speed of the information waves may be similar. When this occurs, we cannot ignore the time that it takes for information about forces to travel

A hydraulic lift is used to amplify forces. The forces applied at surface (a) move the piston down, thereby affecting the fluid beneath the piston. The forces on the fluid are transmitted through the entire fluid reservoir at the speed of sound. The net effect is to raise the pressure in the reservoir, and the raised pressure applied to the larger surface (b) creates a force large enough to lift an automobile (see Section 4.2.4.2).

within the fluid. This is the regime of supersonic or near-supersonic flows. A parameter called the Mach number (Ma) delineates whether a flow is below, near, or above the speed of sound, which is the speed of information:

$$\text{Mach number} \quad \text{Ma} \equiv \frac{v_0}{v_{\text{sound}}} \tag{2.89}$$

where v_0 is the speed of an object in the flow and v_{sound} is the speed of sound in the fluid. At a temperature of $15°C$ and at sea level, the speed of sound in air is 340.3 m/s (761.2 mph). An object traveling at the speed of sound is traveling at Mach 1.

Supersonic flows are important in space travel and ballistics. In process engineering, supersonic flows occur in relief valves and, in this application, it is critical that their special physics be considered when analyzing the valves (see Chapter 10). Complete consideration of supersonic flows requires the incorporation of fluid compressibility into the modeling equations of fluid mechanics and, therefore, involves issues related to the fluid thermodynamics. These topics are summarized in Chapter 10; more information on supersonic flows, including the development and use of compressible flow models, is in the literature [3].

2.7 Surface tension

The flows discussed so far involve a single fluid phase: either a fluid producing drag on an obstacle or fluid filling a tube or channel. When two fluids are present (e.g., both air and water), an interface forms between the two phases and new phenomena appear. To understand flows in which one of the boundaries is another fluid, we must consider the properties of the phase boundary, known as the free surface of the flow.

For many, an early introduction to science was learning the distinctions among the three basic states of matter;[6] solid, liquid, and gas. In the solid state, matter holds its own shape; whereas in the liquid state, matter conforms to the shape of its container; and in the gaseous state, matter expands to fill all available space. Because it does not expand to fill all space, a quantity of material in the liquid or solid state creates an interface or phase boundary between the material and its surroundings. For both liquids and solids, there can be interesting properties associated with the phase boundaries because the molecules near the free surface do not experience the same environment as those deep inside the material. The unique surface properties of solids are exploited in fields such as catalysis, in which chemically active groups on the surface can accelerate chemical reactions. For liquids, the existence of a free surface often leads to motion of the interface and subsequently to the creation of interesting surface shapes and phenomena (Figures 2.36 and 2.37).

To account for free-surface effects within flow models, we introduce an additional material parameter beyond density and viscosity. The unbalanced molecular

---

[6]In the 21st century, children are taught about the five states of matter: solid, liquid, and gas, Bose-Einstein condensate (a phase that appears at absolute zero), and plasma (high-temperature ionized gas).

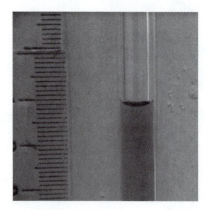

Figure 2.36   Surface forces cause the curvature of interfaces in small tubes, which is called the meniscus effect. The scale on the left is marked in millimeters.

effects at free surfaces may be accounted for by defining a fluid property called the surface tension (see Section 4.4):

$$
\begin{matrix}
\text{Surface tension} \\
\text{(extra tension/length} \\
\text{in a surface due to} \\
\text{unbalanced molecular forces)}
\end{matrix}
\qquad \sigma\,[\text{N/m}]
\qquad (2.90)
$$

We can understand surface tension by considering what is unique about fluid surfaces. The fluid properties at a free surface are exceptional because the environment faced by the fluid molecules at a free surface is different from the environment experienced by them away from the free surface (Figure 2.38). In a liquid, there are attractive forces between the molecules that constitute the liquid (see Section 4.2). In the center of a container of liquid at rest, a given molecule experiences a cohesive pull from every direction, the different pulls balance one another, and the molecule experiences no net force. At the interface

Figure 2.37   Surface tension allows engineers and designers to create interesting effects with water, such as the curving water sheet in this fountain (*National Museum of Contemporary Art, Gyeonggi-do, South Korea*).

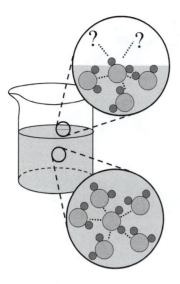

Figure 2.38 Fluid particles deep in a container experience forces from the fluid particles surrounding them. For the particles at the surface, however, the forces on one side are different: They are due to the presence of air or whatever fluid is on the other side of the interface. This difference causes an imbalance in the forces for particles at the surface.

between the liquid and a surrounding gas, however, the attractive liquid forces pull only from one side. Molecules at the free surface experience the downward pull of attraction to neighboring liquid particles, but there is a negligible balancing upward pull from the gas molecules above them.

The molecular-force imbalance at the free surface is not captured by bulk fluid properties such as the density or viscosity. Because the downward force on surface particles is not balanced by an upward force of fluid on the other side of the interface, the net effect is that the free surface behaves like a thin massless film under tension. In other words, the free surface is like a piece of a balloon that has been stretched in all directions (Figure 2.39). When force balances (i.e., momentum balances) are performed on systems the boundaries of which cross the free surface, there is an additional force that must be included to account for the free-surface physics. Observations show that this force is tangent to the surface and normal to the line where the system boundaries and the free surface intersect.

The idea of the free surface as a massless membrane under tension helps us to understand why some insects and small particles that are heavier than water do not sink when they walk on water. The water strider is a common example (Figure 2.40); it is heavier than water and should sink when it steps out onto the water. Instead, however, the water strider produces dimples in the fluid free surface, as if it were walking on a stretched balloon. The survival of these insects depends on the surface tension of the water. Pollutants that reduce the surface tension jeopardize the existence of the water strider.

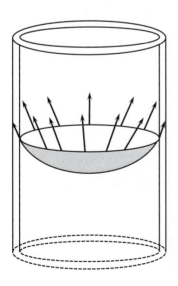

Figure 2.39 Unbalanced forces at the free surface of a fluid must be accounted for by including the surface tension in fluid models. The surface tension is the tension per unit length present in an imaginary stretched film coincident with the free surface.

Surface tension has a role in capillary action, or capillarity, in which liquids climb up narrow tubes or narrow gaps between surfaces (Figure 2.41). Capillarity, which is important in the flow of water through soils as well as in flows in the human body, is the result of free-surface forces and fluid-solid attractive forces. In space travel, where the pull of gravity is small, capillarity causes liquids to crawl out of open containers.

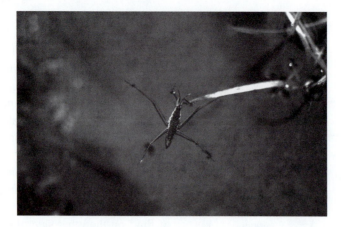

Figure 2.40 The legs of the water strider make impressions on the water surface as it walks across the free surface. The free surface acts like a membrane under tension that supports the insect. Photo courtesy the U.S. National Park Service photographer Rosalie LaRue.

Therefore, space travelers must drink with special straws that clamp shut when not in use to prevent snacks from climbing up the straw and floating freely throughout the cabin.

Surface forces are important in a wide variety of technical applications, including the breakup of jets, processes involving thin films, and foams [122]. Wicking, the drawing of fluid up into a fabric or wick as in a candle or away from the body as in the design of exercise clothing, is another process that works by capillary action. The opposite effect, waterproofing, is a manipulation of surface forces to prevent wicking. Surface tension causes striking effects that are exploited to make engaging fountain displays (see Figure 2.37). In soap and water solutions, for example, variation of the concentration of the solute can cause the surface tension to vary, which in turn causes flow. Flow driven by surface-tension gradients—called the Marangoni effect [112]—stabilizes soap bubbles, among other effects (Figure 2.42). Finally, the emerging field of micromechanics creates machinery that works on nearly molecular-size scales. The properties of any liquids involved in micromachines are dominated by interfacial forces.

Interfacial forces are not always important, however, even when a large amount of free surface is present. In an ocean, for example, wave motion depends on viscous forces and gravity forces, but the contribution of surface-tension forces

Figure 2.41 The surface forces between the glass capillary walls and the fluid (i.e., water and food coloring) are attractive, and this attractive force is sufficient to draw liquid into the capillary. The capillary is open at the top.

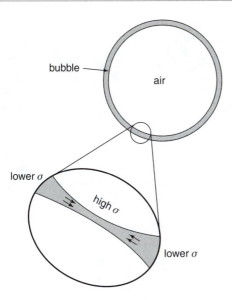

Soap bubbles are composed of thin fluid layers sandwiched between two free surfaces. Surfactant molecules occupy the free surfaces and reduce the surface tension of the bubble surface compared to the surface tension of pure water. If an external force deforms or inflates the bubble, more surface is generated, reducing the concentration of surfactant molecules at the bubble surfaces. Lower surfactant concentration implies higher surface tension, however, and this locally higher surface tension pulls fluid into the thinning layer, stabilizing the film and preventing bubble rupture.

to the momentum balance in oceanic flows is negligible. One goal in studying fluid mechanics is to map out when different types of forces are important and when they are not. Our tool in this endeavor is dimensional analysis, which we study in Chapters 7, 8, and 10.

2.8 Flows with curved streamlines

This chapter discusses flows that form many different patterns, and the same analysis techniques apply to mostly straight flows as to those that are strongly curving. Flows that are strongly curving, however, present a particular challenge to our intuition because rotational motion is more complex and can lead to counterintuitive results. For this reason, flows with curved streamlines typically are covered only in advanced courses in fluid mechanics. An introduction to flows with curved streamlines is in Section 10.5.

Many important flows have curved streamlines. A tornado is an extreme example of such a flow, and understanding their velocity and pressure distributions can be of great humanitarian importance. Other curved flows include fluids stirred in a vessel, water flowing in curving rivers or through pipe bends, blood flowing throughout the human body, vessels being drained of their contents, plastic flowing into a mold, smoke rings, and vortices formed at the tips of airplane wings or in the wake following a propeller (Figure 2.43).

An unusual phenomenon associated with curved flow is the development of secondary flows near boundaries. We experience this type of flow when stirring

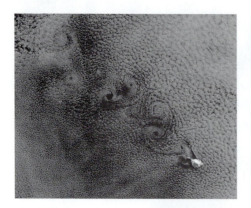

Figure 2.43 Various flows with curved streamlines are observed: (a) vortices shed by a stationary object in a flow; (b) tornadoes; and (c) whirlpools. Images courtesy of the National Science Foundation (nsf.gov), the National Oceanic and Atmospheric Administration's National Severe Storm Laboratory, and tippecanoe.in.gov.

loose tea leaves in a cup (Figure 2.44).[7] After stirring, the circular flow dies slowly, and the brewed tea comes to rest. It is interesting to observe that the tea leaves collect in the center of the cup. Because of the inertia[8] of the spinning tea leaves, it seems intuitive that the leaves would be thrown to the outer perimeter of the cup rather than collect in the center. They collect in the center rather than at the periphery because of a weak radial flow in the boundary layer near the bottom of the cup. We use this example to frame a brief discussion of secondary flows.

The strong circular flow in the teacup experiment is called the primary flow, and the weaker radial flow that takes place at the bottom is called the secondary flow [154]. In the teacup, the secondary flow occurs because the fluid near the bottom is slowed down by its proximity to the motionless bottom surface. Away from the bottom, in the strong primary flow, a pressure distribution builds up, resulting in a larger pressure near the outer edges of the cup compared to the center. Near the bottom wall of the teacup, the slowed fluid is unable to maintain this pressure gradient and becomes subject to it instead, and fluid is pushed toward

---

[7]In some cultures, the teabag has replaced the practice of brewing loose tea in a cup, so this phenomenon may not be familiar; a little fieldwork therefore may be required to observe the secondary flow discussed.

[8]Recall that inertia is the tendency of a body once in motion to remain in motion unless an outside force acts on it. Thus, inertial forces in a circular flow refer to the tendency of fluid particles to experience an outward force pushing them toward larger radial positions.

Figure 2.44   After azimuthal stirring, tea leaves tend to gravitate to the center of a cup. This effect is due to the secondary flow near the bottom of the cup.

the center of the cup. The tea leaves, which are heavier than water and therefore settle to the bottom of the cup, are dragged along in this inward flow and collect in the center of the flow (see Figure 2.44).

A second example of secondary flow induced by curved streamlines occurs near the bottom of a riverbed. This flow is partially responsible for the tendency of rivers and streams to develop exaggerated bends and turns. If a mild bend develops in a river or stream, the induced secondary flow drags silt and other sediments from the outer bank of the river and deposits them on the inner bank, accentuating the bend and strengthening the secondary flow [112].

Secondary flows can be beneficial in applications that require good mixing, such as in a heart–lung machine (HLM). The HLM, or pump oxygenator [54], is an instrument used in surgery when the heart must be stopped to allow a surgeon to perform repairs. A body cannot survive without a heart; thus, the duties of the heart are taken over by the HLM. An important function of the heart is to pump blood to the lungs, where carbon dioxide is removed from the blood and oxygen is replenished. The heart also pumps the newly restored blood to the rest of the body, where it is needed. When the HLM takes over for the heart, it pumps blood to an external device in which oxygen is added to the blood and carbon dioxide is removed (Figure 2.45). The transfer of gases to and from the blood in a membrane oxygenator is effected through gas-permeable circular tubing arranged in coils (Figure 2.46). The primary flow is down the length of the tube, but the tube is curved intentionally to induce a secondary flow. The streamlines for this

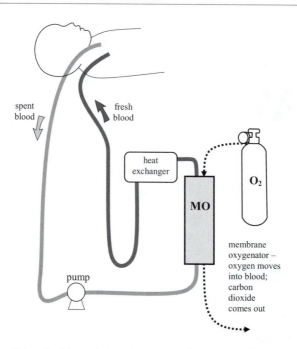

Figure 2.45 Schematic of the surgical use of a membrane oxygenator, a type of heart–lung machine. In a HLM, blood returning to the heart is pumped outside of the body and through the membrane oxygenator. In the membrane oxygenator, oxygen diffuses through membranes and dissolves into the blood, and carbon dioxide diffuses back through the membranes and exits the oxygenator. The oxygen-laden blood exiting the membrane oxygenator returns to the body.

secondary flow are shown in Figure 2.46. The vortices in a helical tube were first described by W. R. Dean [38, 39] and are called Dean vortices. This secondary flow in the HLM moves blood from the walls to the center of the tube and back again as the fluid progresses downstream [21, 53, 110, 140]. Thus, the secondary flow stirs up the blood and results in an improvement of a factor between two and four in the blood oxygenation that occurs [110].

The subtle nature of curved flow makes these flows a challenge to study. Flows with curvature are usually analyzed with the help of the concept of *vorticity*, which is a vector quantity related to the amount of rotational character in a flow

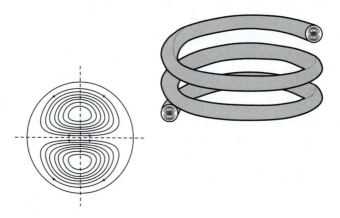

Figure 2.46 The oxygen that diffuses into the blood near the tube surface mixes efficiently with the rest of the blood because of a secondary flow that takes place in the curved tubes.

field at a particular point. In terms of vector calculus, vorticity $\underline{\omega}$ is defined as $\underline{\omega} \equiv \nabla \times \underline{v}$, where $\underline{v}$ is the vector that describes the local direction and magnitude of fluid velocity, and ∇ is the spatial differentiation operator (see Section 1.3). Like velocity, vorticity forms a field, and we speak of a pattern of vortex lines in a flow that map out the local vorticity vector by tracing lines that are everywhere tangent to the vorticity (see Section 8.3). Vortex lines drawn through every point on a closed curve form a vortex tube. Various mathematical theorems based on momentum, mass, and energy conservation apply to vorticity and vortex tubes and can be helpful in understanding fluid motions involving strong amounts of curvature. For example, the product of the magnitude of the vorticity and the cross-sectional area of a vortex tube must be constant for a vortex tube [72]. Vorticity is introduced in Section 8.3 [114], and flows with curved streamlines are discussed in Section 10.5 [123]. Many resources in the literature [9, 79, 154, 168] can guide further study of highly rotational flows once the basics in this text have been mastered.

2.9 Magnetohydrodynamics

The fluid behaviors described in the preceding sections are exhibited by normal fluids including air, water, oils, and foods. In addition to these behaviors, there are specialized types of fluid behaviors characteristic of more esoteric fluids, such as the molten core of the Earth. Research fields have arisen around unusual fluids, and basic fluid mechanics is the entry point to the study of these advanced topics, one of which is the field of magnetohydrodynamics (MHD), which helps us to understand flows in the core of the Earth or on the surface of the sun.

As discussed in this chapter, flow and deformation of fluids is caused by the imposition of forces such as a knife spreading peanut butter or gravity pulling water over Niagara Falls. Three types of forces cause most flows: pressure differences, imposed forces that act on the boundaries of a fluid, and gravity (see Chapter 6). A more unusual source of flow driving force is a magnetic field. When a fluid is electrically conductive, forces are induced in the fluid by an external electric field. These forces cause fluid motion; in turn, the fluid motion alters the magnetic field. To understand the effect of magnetic field on the motion of a conductive fluid, the electromagnetic and the fluid-mechanics equations must be considered simultaneously. The electromagnetic equations are the Maxwell differential equations [167] and are taught in physics and chemistry courses. The fluid-mechanics equations are those that are discussed in this book (see Chapter 6). Both types of equations are vector-field differential equations and are best described with vector calculus.

The phenomenon of MHD is due to the mutual interaction of a magnetic field $\underline{B}$ and a fluid velocity field $\underline{v}$ [35]. For convenience, we divide the process into three parts. In the first part, relative movement of a conducting fluid and a magnetic field causes an electromotive force (e.m.f.) to develop. This is a consequence of Faraday's law of induction [167], and when a conducting fluid moves in a magnetic field, a current begins to flow in the conducting fluid. The induced current in the fluid must itself create a magnetic field, in accordance with Ampère's law. In the second part, the induced magnetic field adds to the

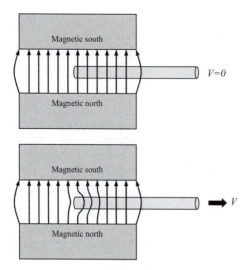

Figure 2.47 When a conductive rod is drawn through a vertical magnetic field, the induced current in the rod creates an induced magnetic field [35]. The effect on the magnetic field lines is that they bend. The visual effect is as if the rod is dragging the magnetic field lines in the direction of its motion. The effect is similar for a conductive fluid, although more complicated because the moving conductor is deformable in that case. Reprinted with the permission of Cambridge University Press.

original magnetic field, altering the field lines. The change is usually such that the fluid appears to drag the magnetic field lines in the direction of the flow (Figure 2.47). The third step in this simplified explanation of MHD is when the modified magnetic field interacts with the induced current density to give rise to a Lorentz force. This is a force exerted on moving charged particles, and it acts perpendicular to both the direction of the motion of the charged particles and the magnetic field lines [167]. In MHD, the Lorentz force is directed so that it inhibits the relative movement of the magnetic field and the fluid [35].

MHD figures prominently in astronomy, geology, and metallurgy (Figure 2.48). The Earth's magnetic field is a result of fluid motion in its core, and the solar magnetic field generates sunspots and solar flares due to MHD. Because liquid metals are conductive, MHD is used in the metallurgical industry to heat, pump, stir, and levitate liquid metals. MHD also is used to damp surface motion in metallurgical processing [35]. MHD flows are highly rotational, and vorticity is an important tool in their study.

This discussion is only a summary of an advanced application of fluid mechanics, but it demonstrates that a mathematical understanding of basic fluid flow is essential before attempting to master complex fluid motions, such as those induced in conducting fluids by magnetic fields. Investing in the study of basic fluid mechanics opens up a wide variety of avenues for advanced fluid applications.

2.10 Particulate flow

This chapter on fluid behavior concludes with a cautionary note. Not everything that flows may be considered by using the continuum assumptions discussed in

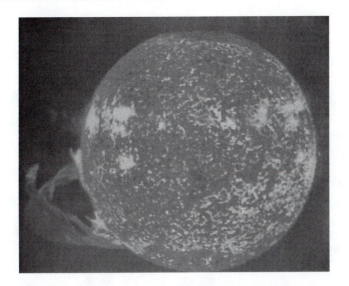

Figure 2.48   Many cosmic bodies are composed of hot ionized gases or plasmas. These conducting fluids interact with magnetic fields to display a wide variety of MHD-induced motions, including the solar flares shown here. Photo courtesy of NOAA.

this book. The continuum assumption (see Chapter 3) is a model that states that fluids are continuous and everywhere characterized by average properties—for example, locally averaged values of density, viscosity, and concentration. An important class of materials that is not continuous, and therefore beyond the scope of this book, is particulate solids.

Particulate solids are systems composed of small solid grains of matter—for example, sand, salt, sugar, baby powder, corn starch, gravel, dirt, and polymer pellets. Particulates can flow, and they move and deform in ways that sometimes are similar to the flow of continuous fluids; however, there are many ways in which they are different from continuous fluids.

Consider the flow of two fluids through a funnel (Figure 2.49). A Newtonian fluid poured into a funnel always flows out. A viscous fluid takes longer to flow out than a less viscous material, but the flow does not stop moving until all of it has passed through the funnel. Consider what happens, however, if a particulate fluid is loaded into a funnel. The particulates (e.g., tapioca pearls) flow but, at some point, they jam. To restart the flow, it is necessary to tap the funnel or to lift it and set it down again. After the distribution of particles is disturbed, the flow starts again; however, several iterations of tapping and jarring may be necessary to make the tapioca to pass completely through the funnel.

Tapioca pearls flow sporadically in a funnel due to the formation of particle bridges or arches over the bottom opening of the funnel (Figure 2.50). Functioning like the stones that comprise archways in buildings, the particles form a sturdy structure that is able to support the weight of the material above it. This structure stops the flow until the structure itself is disrupted externally by either tapping on the funnel, blasting the arch with high speed air, or otherwise jostling the flow [153].

Particulate solids exhibit properties that are a combination of liquid-like behavior, solid-like behavior, and particle-interface–dominated behavior [163].

flowing
liquid

flowing
granular
material

jammed
granular
material

Figure 2.49   When simple fluids like milk flow through a funnel, the flow is predictable and reproducible (left). However, when particulate solids such as tapioca pearls flow, sometimes the flow moves (center) and sometimes it jams due to a bridging effect (right). The flow varies depending on particle characteristics and initial conditions.

Particulates flow through openings, take the basic shape of the container they occupy, and exert pressure on the walls of the container. Liquids cannot sustain shear stresses without flow (see Chapter 4), but particulates can. An example of particulates sustaining a shear stress is a sand pile—a liquid cannot form a pile. Another property of liquids is that when a load is applied to a liquid, an isotropic pressure distribution is observed throughout the liquid (see Chapter 4). For particulate solids, even if a uniform load is applied to the mass of particulates, the stress may not be isotropically distributed—there may be a buildup of stress at points in the sample, as occurs during bridging (see, e.g., Figure 4.27). The flow stresses for particulates are proportional to the normal load—that is, to the magnitude of force directed perpendicular to the flow direction—whereas the flow stresses for Newtonian liquids are proportional to the deformation rate (Equation 2.1, see also Chapter 5)—the list of differences is long. Perhaps the most vexing property

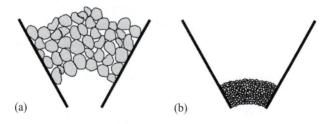

(a)   (b)

Figure 2.50   Particulates can form bridges or arches that block the flow and support large stresses. Two types of arches form: (a) mechanical, caused by the interlocking of large (>3,000 μm diameter) free-flowing particles; and (b) cohesive, caused by interparticle attractions in fine powders [153].

of particulate flows is that the magnitude of shearing stress is generally indeterminate: For two particulate flows with the same apparent velocity field, different stresses can be generated [163].

The study of particulate flows is diverse and growing [30]. Numerous engineering applications involve particulate flows, including agriculture, food processing, polymer processing, geology, construction, pharmaceuticals, and chemical manufacturing. The principles of continuum fluid mechanics are not useful for these systems, although some nomenclature and concepts are used in common.

2.11 Summary

This chapter is an overview of fluid behavior including an introduction to viscosity, drag, boundary layers, laminar flow, turbulent flow, lift, supersonic flow, surface tension, curving flow, and magnetohydrodynamics. The ultimate task is to make engineering calculations on flows. To assess our current knowledge of fluid mechanics, we now attempt to carry out a fluid mechanics calculation. Simply "jumping in" is a tactic that we endorse: Even if we are ill-prepared to complete the task, the attempt often leads to greater understanding, which then identifies and motivates the background study needed to truly solve the problem.

EXAMPLE 2.9. *In this chapter, we presented Newton's law of viscosity, the force-deformation relationship for fluids. For the various flow phenomena introduced in this chapter, how do we solve Newton's law of viscosity to obtain the fluid velocity field?*

SOLUTION. Newton's law of viscosity relates $\tilde{\tau}_{21}$, the force per area generated in the deforming fluid; $\partial v_1/\partial x_2$, the local relative speed of the object; and μ, the viscosity (Equation 2.1):

$$\text{Newton's law of viscosity:} \quad \boxed{\tilde{\tau}_{21} = \mu \frac{\partial v_1}{\partial x_2}} \quad (2.91)$$

(force-deformation relationship)

To solve any equation, we need to know what is constant, what is variable, and the meaning of the variables.

In Equation 2.91, the viscosity is a constant, and $\tilde{\tau}_{21}$ and v_1 are variables that are different in every location in a flow. The variable v_1 is the magnitude of the velocity in some direction; however, in many of the flows discussed in this chapter, the velocity is a fully three-dimensional function that varies in every direction. The general expression for the velocity is a vector:

$$\underline{v} = \begin{pmatrix} v_1 \\ v_2 \\ v_3 \end{pmatrix}_{123} = \begin{pmatrix} v_1(x_1, x_2, x_3) \\ v_2(x_1, x_2, x_3) \\ v_3(x_1, x_2, x_3) \end{pmatrix}_{123} \quad (2.92)$$

The stress $\tilde{\tau}_{21}$ also is a function of x_1, x_2, x_3, as well as only one of several stresses associated with a complex flow. In the tensor discussion in Chapter 1, we learned that there are nine components of stress.

$$\underline{\underline{\tilde{\tau}}} = \begin{pmatrix} \tilde{\tau}_{11} & \tilde{\tau}_{12} & \tilde{\tau}_{13} \\ \tilde{\tau}_{21} & \tilde{\tau}_{22} & \tilde{\tau}_{23} \\ \tilde{\tau}_{31} & \tilde{\tau}_{32} & \tilde{\tau}_{33} \end{pmatrix}_{123} \qquad (2.93)$$

Although Equation 2.91 gives an important local relationship between one particular stress $\tilde{\tau}_{21}$ and the local distribution of velocity, it alone does not contain sufficient information about the flow to allow us to calculate the velocity field $\underline{v}$ or the stress field $\underline{\underline{\tilde{\tau}}}$ for the entire flow. As discussed in subsequent chapters, we need mass and momentum balances, properly applied for a given situation, and we must incorporate force-deformation information such as Newton's law of viscosity to solve for velocity and stress fields. The momentum balance for fluids is introduced in Chapter 3, the stress tensor in Chapter 4, the complete stress-velocity relationship in Chapter 5, and the techniques for solving the modeling equations in Chapters 6–10.

This is the beginning of our study of fluid mechanics. After completing the course, readers will be able to calculate fluid velocity and stress fields for many flows. Therefore, the solution of this problem is postponed until more is known about how fluids work.

In this text, we present fluid-mechanics modeling in a step-by-step manner. We introduce fluid physics by tying together the familiar physics of rigid bodies (e.g., blocks sliding down a hill) with the physics of deforming systems—that is, fluids. In the next chapter, we choose our model for quantifying fluid behavior, which is called the continuum model.

2.12 Problems

1. When you have the oil changed in your car, the service attendant asks, "Would you like 10W40 or 10W30?" How should you decide?
2. What is the density of acetone? What is the viscosity of acetone? Compared to water, does acetone generate more or less stress in flow? Be quantitative in your answer.
3. What is the density of blood? What is the viscosity of blood? When doctors give a "blood thinner," is it the viscosity, the density, or something else that they are changing?
4. When medical technicians draw blood for laboratory tests, they first insert a needle attached to a tubeholder into a vein. The second step is to push a tube onto the needle, causing blood to flow into the tube (i.e., the needle penetrates a septum covering the top of the tube). Why does the blood flow into the tube? It may be necessary to search the device on the Internet to determine the answer.
5. In addition to solid, liquid, and gas, another common state of matter is foam. Foams occur in food processing (e.g., whipped cream and frothed milk), consumer products (e.g., hair mousse and shaving cream), and industrial

applications (e.g., wall insulation and fire-extinguishing fluid). Describe the structure of foam from a scientific perspective. Do the foams flow and deform like Newtonian liquids (e.g., water and oil), Bingham plastics (e.g., mayonnaise and paint), or do they comprise their own class of materials? Describe the flow behavior of foams.

6. Honey is trapped between two long wide plates (plate area $= 9.0$ cm^2) and the top plate is moved at 1.0 cm/s. The gap between the plates is 0.50 mm. What is the velocity gradient $\partial v_1/\partial x_2$, where 1 indicates the flow direction and 2 indicates the direction perpendicular to the plates?

7. If water is trapped between two long wide plates and subjected to a velocity gradient of 10.0 s^{-1} in the 2-direction, what is the magnitude of the shear stress τ_{21} that is generated? If the area of the top plate in contact with the water is 25 cm^2, what is the force needed to maintain the motion of the plate?

8. If water is trapped between two long wide plates and subjected to a velocity gradient of 5.0 s^{-1} in the 2-direction, what is the magnitude of the shear stress τ_{21} that is generated? If the fluid between the plates is changed from water to honey, how much shear stress is generated?

9. Olive oil is placed between two long wide plates (plate area $= 97.5$ in^2) and the top plate is moved at 0.25 in/s. The gap between the plates is 0.0126 in. What is the force that it takes to maintain the motion?

10. Two fluids are examined with a parallel-plate apparatus like Newton used to study fluids. The two plates have the same area A; and with the test fluid in the gap, a constant gap of H is maintained as the top plate is dragged in a uniform direction, causing the fluid to deform. When the two fluids are tested, it takes twice as much force to move the plate at a fixed velocity V with Fluid 2 as with Fluid 1. What is the ratio of the viscosities of the two fluids?

11. A tree in the wind is an object subjected to a uniform flow (a flow that everywhere has the same speed and direction). How much drag is a tree subjected to by modest winds and by hurricane-force winds? Search the literature for air speeds and drag coefficients to answer this question.

12. A bicycle racer in a racing crouch is traveling at 50 mph. How much faster will she go if her teammate drafts her by riding immediately in front of her?

13. How much wind force is a flag subjected to on a typical day? Search the literature for air speeds and drag coefficients to answer this question.

14. A disk (i.e., radius is R and thickness is H) is dropped from a great height. How much faster does the disk fall when dropped edge first versus dropped with the large circular surface perpendicular to the fall direction? Search the literature for drag coefficients for the disk in these two orientations.

15. When you stir water (or coffee or tea) in a cup, how does the shape and position of the fluid surface change compared to the fluid at rest? Sketch the quiescent and steady-state fluid interfaces. Note: The sketch should be consistent with the principle of conservation of mass.

16. The viscosity of water is about 1 cp. Showing the unit conversions, what does 1 cp translate into in American engineering units (involving lb$_f$, ft, s)? What does this quantity translate into in SI units (Système international d'unités, the metric system, involving kg, m, s)?

17. In this chapter, we discuss the force it takes to push fluid through a needle attached to a syringe. If the fluid ejected from the needle is glycerin rather than water, how much force would it take? Use all of the same assumptions as in the text example.

18. The rate of blood circulation in the body is 5.0 lpm (liters per minute) [145]. How much blood passes through the heart in a day?

19. Laminar flow in a tube is described by the Hagen-Poiseuille equation, introduced in this chapter. (a) For water ($25°C$) flowing through a 10-foot section of 1/2-inch pipe (Schedule 40) in laminar flow, what is the maximum flow rate (in gpm) through the pipe before the flow becomes transitional? (b) What is the pressure drop (in psi) across the pipe at this maximum flow rate?

20. Water ($25°C$) is pushed through a pipe (ID 4.0 mm and length 1.5 m) and laminar flow is produced at a Reynolds number of 800. For the same fluid subjected to the same pressure drop in a pipe of the same length, at what pipe diameter will it no longer be possible to produce laminar flow?

21. Blood travels through the large arteries of a human body (internal radius 12 mm) at an average velocity of about 50 cm/s. What is the flow rate of blood through these arteries? Is the flow laminar or turbulent? The viscosity of blood is 3.0 cp and the density of blood is 1,060 kg/m^3 [145].

22. Blood travels through the human heart's ascending aorta (diameter = 3.2 cm) at an average velocity of about 63 cm/s [145]. What is the flow rate of blood through this vessel? See the previous problem for viscosity and density of blood.

23. A carbon-dioxide bubble rises in a glass of soda. From the perspective of an observer sitting on the bubble, sketch the flow lines as the liquid parts and flows around the rising bubble. Is there a flow (i.e., motion) in the carbon dioxide inside the bubble? Discuss why or why not.

24. How does drinking from a straw work? In your answer, use scientific terms such as pressure, continuum, and flow.

25. An open container of fluid has a hole in the side and the fluid leaks out under the force of gravity. If a tight-fitting but movable piston is placed on top of the fluid and a 10-kg weight is placed on top of the piston, how would the flow out the hole change? Sketch your answer. Why is there a change?

26. Many teapots dribble. Why? Which forces mentioned in this chapter influence teapot dribble?

27. Consider the following (admittedly improbable) two ways to make an open-faced peanut-butter-and-jelly sandwich: (1) Spread a layer of peanut butter on a slice of bread; then top this layer with a layer of jelly. (2) First spread a layer of jelly on a slice of bread; then spread peanut butter over the jelly layer. Discuss the pros and cons of the two methods. If forced to choose one of these two methods, which would you choose? Give a fluid-mechanics explanation of your choice.

28. Why do ice cubes float in water? Why do olives sink in water? Which physical property is important to the answer: viscosity, density, surface tension, or something else?

29. You are standing facing a strong wind and you are cold. Will any of the following actions reduce how cold you feel? Explain your reasoning using

Figure 2.51 Water towers are visible in many towns in the United States and Europe, and they serve an important role in daily life (Problem 32).

fluid mechanics: (a) turning sidewise to the wind; (b) laying flat on the ground; and (c) crouching down on the ground.

30. What is a boundary layer? Give an example of a boundary-layer effect that you have experienced.

31. Players at the 2010 Football World Cup in South Africa complained that the ball had an erratic flight path. Discuss possible fluid-mechanics reasons for problems with the ball.

32. What is the purpose of water towers built in many towns (Figure 2.51)? What determines how high the water tank should be?

33. Fill a straw by placing it in a liquid. If we place a finger over the top and remove the straw from the liquid, the straw remains full. Why does the water not flow out of the straw (Figure 2.52)?

Figure 2.52 Liquid can be captured by submerging a straw in liquid and then plugging the top of the straw with a finger before withdrawing the straw from the liquid (Problem 33).

Plumbing vents are visible on housing rooflines.

An important design element of an indoor plumbing system is the presence of vents (Problem 34).

34. Why do home plumbing systems have vents (Figure 2.53)?

35. In an experiment showed to schoolchildren, a colored liquid is placed in a 2-liter soda bottle that is subsequently connected at the neck to a second 2-liter bottle. When all the liquid is in one bottle and the contraption is inverted, the liquid flows slowly and haltingly from top to bottom. If the fluid is swirled, however, it drains rapidly from top to bottom. What is happening in this experiment? Use fluid-mechanics concepts in your explanation.

36. Many adventure and horror movies feature quicksand. What is quicksand? Does it really exist? How does quicksand work? Is quicksand a Newtonian fluid?

37. In some homes, residents learn that when someone is showering, no one should flush the toilet or otherwise use water lest the person in the shower receives a scalding from hot water. What is happening in this circumstance? What is wrong with the plumbing design to cause this effect?

38. Trees need water to live, and they get much of the required water from the ground through their roots. How does water flow up a tree trunk against the downward pull of gravity? Use scientific principles in your answer.

39. In the living space in a spacecraft in orbit around planet Earth, Earth's gravitational pull is not very strong. How are the following processes affected by a zero-gravity working environment?

(a) Drinking water from the lip of an open glass.
(b) Drinking water with a straw from an open glass.
(c) Drinking water with a straw from a closed box.
(d) Flushing a toilet.
(e) Brewing coffee with an automatic-drip coffeemaker.
(f) The human digestive system.
(g) Swallowing food.
(h) Blood circulating in the human body.

40. Why do helium balloons float in air?

41. What is the definition of the Fanning friction factor? What is the definition of the Darcy or Moody friction factor? How can the friction factor be measured for a given piping system?

42. Water flows through a smooth pipe at a Reynolds number of 53,000. What is the Fanning friction factor for this flow? For glycerin and acetone flowing at the same Reynolds number, what are the friction factors?

43. When a balloon inflated with air is released, it accelerates and flies around. Where does the kinetic energy of the balloon originate?

44. Can you suck foam (e.g., frothed milk from a cappuccino or whipped cream from a milkshake) up a straw? Why or why not?

45. What is a tornado? How does it form? How does it dissipate?

46. How does water-repellant fabric work?

47. When the flow rate in a water faucet is high, water emerges as an unbroken column of fluid. When the flow rate is decreased, the faucet eventually begins to drip. Why does the fluid stream break up into droplets?

48. What is vorticity? For what types of flow is vorticity important?

49. How fast is an aircraft going in km/hr if it is traveling at Mach 1.4?

50. An aircraft has a mass of 35,000 kg and a planform area of 250 m^2. How much lift must the aircraft generate to fly?

51. An aircraft (mass $= 25,000$ kg; planform area $= 203$ m^2) has a lift coefficient of $C_L = 1.8$ at stall. What is the stall speed of the aircraft?

PART II

THE PHYSICS OF FLOW

3 Modeling Fluids

Chapter 2 describes fluid behaviors, and Chapters 1 and 2 introduce basic fluids calculations. We turn now to developing a modeling method that allows us to understand fluids behavior in detail.

Fluids move and deform in predictable ways that are governed by the laws of physics. To apply the laws of physics to fluids, we must develop a mathematical picture or model of fluid motion. With an effective model, we can predict fluid patterns and stresses and apply these predictions to engineering calculations.

To build up the fluid model that we use, we begin with a reminder about how to calculate the motions of individual rigid bodies. To apply these methods to fluids, we then introduce the continuum model, a mathematical picture of fluids in which we consider small packets of fluid to be individual bodies. We discuss how we apply the laws of physics to these small fluid packets or particles to deduce velocities and forces for the fluid particles. Finally, we introduce the control volume, a point of view used for fluid modeling that focuses our calculations on a physical region in space rather than on individual bodies in motion. This difference in strategy—that is, considering a control volume rather than individual bodies—is a key difference between the modeling techniques of fluid mechanics and those of solid-body mechanics.

3.1 Motion of rigid bodies

Many of the classical laws of physics are conservation laws, which hold that some property may neither be created nor destroyed but may interconvert only between various forms. Mass is conserved [47],[1] and so are energy and momentum [157, 167]. Motion is governed by the momentum balance.

The fundamental expression of conservation of momentum for a body is given in Newton's second law of motion [167]: "The time rate of change of momentum of a body is equal to the resultant external force acting on the body." In mathematical symbols, this becomes:

$$\begin{pmatrix} \text{resultant of} \\ \text{external forces} \\ \text{on a body} \end{pmatrix} = \begin{pmatrix} \text{time rate of change} \\ \text{of momentum} \\ \text{on the body} \end{pmatrix} \qquad (3.1)$$

---

[1] Mass is conserved if a nuclear reaction does not occur. In a nuclear reaction, mass is converted to energy [167].

Newton's second law:
$$\sum_{\substack{\text{all forces} \\ \text{acting on body}}} \underline{f} = \frac{d(m\underline{v})_{\text{body}}}{dt} \qquad (3.2)$$

where $\underline{f}$ represents the various forces on the body, m is the mass of the body, $\underline{v}$ is the velocity of the body, and t is time. The derivative $d(m\underline{v})/dt$ is the rate of change of momentum for the body. Note that if the mass of the system is constant and there is a single force, Newton's second law becomes the familiar $\underline{f} = m\underline{a}$, where $\underline{a} = d\underline{v}/dt$ is the acceleration of a body of constant mass m. Engineering and science students spend considerable time learning how to apply Newton's second law to rigid bodies in their physics and mechanics classes.

How can we apply Newton's second law to flowing systems? This is a difficult question, but we arrive at a method by building on what we know about Newton's second law as applied to individual solid bodies. As a refresher problem, consider the motion of a block sliding down an incline—first without friction, then with friction considered.

EXAMPLE 3.1. *A block of mass m slides down an inclined plane as shown in Figure 3.1. The surface of the plane is smooth, and the friction between the block and the plane may be neglected. What is the velocity of the block at steady state?*

SOLUTION. We solve for the motion of the block by applying Newton's second law, the momentum balance:

Newton's second law:
$$\sum_{\substack{\text{all forces} \\ \text{acting on body}}} \underline{f} = \frac{d(m\underline{v})_{\text{body}}}{dt} = m\underline{a} \qquad (3.3)$$

where $\underline{f}$ represents the forces on the block, m is the mass of the block, $\underline{v}$ is the velocity of the block, t is time, and $\underline{a}$ is the acceleration of the block.

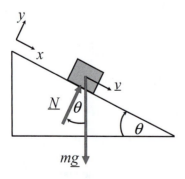

Figure 3.1   A block of mass m slides down a smooth incline. The motion of the block may be solved for by applying the principle of conservation of momentum (i.e., Newton's second law).

There are two forces on the block: the downward force $m\underline{g}$ due to gravity and the normal force $\underline{N}$ exerted by the surface of the incline (see Figure 3.1). We must choose a coordinate system in which to solve the problem; we choose a Cartesian coordinate system where the x-direction points down the incline, the y-direction points normal to the inclined surface, and the z-direction is perpendicular to the xy-plane such that a righthand coordinate system is formed (i.e., z-direction is out of the page). In this coordinate system, $\underline{g}$ and $\underline{N}$ may be written as:

$$\underline{g} = \begin{pmatrix} g\sin\theta \\ -g\cos\theta \\ 0 \end{pmatrix}_{xyz} \tag{3.4}$$

$$\underline{N} = \begin{pmatrix} N_x \\ N_y \\ N_z \end{pmatrix}_{xyz} = \begin{pmatrix} 0 \\ N_y \\ 0 \end{pmatrix}_{xyz} \tag{3.5}$$

where we have incorporated the fact that $\underline{N}$ is normal to the surface and therefore in the y-direction only. We also can simplify the expression for $\underline{v}$ in this coordinate system because $\underline{v}$ is in the x-direction only:

$$\underline{v} = \begin{pmatrix} v_x \\ v_y \\ v_z \end{pmatrix}_{xyz} = \begin{pmatrix} v_x \\ 0 \\ 0 \end{pmatrix}_{xyz} \tag{3.6}$$

We substitute these vectors into Equation 3.3 and solve for $\underline{v}$ and $\underline{N}$:

$$\frac{d(m\underline{v})}{dt} = m\underline{a} = \sum_{\substack{\text{all forces} \\ \text{acting on body}}} \underline{f} = m\underline{g} + \underline{N} \tag{3.7}$$

$$\begin{pmatrix} m\dfrac{dv_x}{dt} \\ 0 \\ 0 \end{pmatrix}_{xyz} = \begin{pmatrix} mg\sin\theta \\ -mg\cos\theta \\ 0 \end{pmatrix}_{xyz} + \begin{pmatrix} 0 \\ N_y \\ 0 \end{pmatrix}_{xyz} \tag{3.8}$$

Equating the y-components of each vector in Equation 3.8, we obtain:

$$y\text{-component:} \qquad N_y = mg\cos\theta \tag{3.9}$$

Equating the x-components of each vector in the same equation, we obtain:

$$x\text{-component:} \qquad m\frac{dv_x}{dt} = mg\sin\theta \tag{3.10}$$

Now we solve for v_x. Note that g and $\sin\theta$ are constant:

$$\frac{dv_x}{dt} = g\sin\theta \equiv B \tag{3.11}$$

Integrating:

$$\int dv_x = \int B\, dt \tag{3.12}$$

$$v_x = Bt + C_1 \tag{3.13}$$

$$v_x(t) = (g\sin\theta)t + C_1 \tag{3.14}$$

The constant C_1 is an arbitrary constant of integration. If we assume that at time $t = 0$ the velocity is zero, then $C_1 = 0$ and $v_x = (g\sin\theta)t$. Thus, the velocity of the block is:

$$\underline{v}(t) = \begin{pmatrix} (g\sin\theta)t \\ 0 \\ 0 \end{pmatrix}_{xyz} = (g\sin\theta)t\hat{e}_x \tag{3.15}$$

As a result of gravity, the block moves down the slippery surface of the incline with a constant acceleration $d\underline{v}/dt = \underline{a} = g\sin\theta\,\hat{e}_x$ (Equation 3.11) and a linearly increasing, time-dependent velocity $\underline{v} = (g\sin\theta)t\,\hat{e}_x$. In the solution to this problem, we also obtained an expression for the normal force vector acting on the block:

$$\underline{N} = \begin{pmatrix} 0 \\ mg\cos\theta \\ 0 \end{pmatrix}_{xyz} = mg\cos\theta\,\hat{e}_y \tag{3.16}$$

EXAMPLE 3.2. *A block of mass m slides down an inclined plane as shown in Figure 3.2. The surface of the plane is rough, and the contact between the block and the plane creates a frictional force on the block that retards its motion down the plane. The frictional force $\underline{F}$ is proportional to the magnitude of the normal force exerted by the inclined plane on the block. What is the velocity of the block at steady state?*

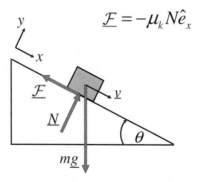

Figure 3.2
A block of mass m slides down a rough incline. The frictional force that slows the block is proportional to the magnitude of the normal force imposed on the block by the surface of the incline. The constant of proportionality is called the coefficient of sliding friction [167]. Within a reasonable range of velocities, the coefficient of sliding friction is independent of the block velocity.

SOLUTION. The solution procedure is the same as that used in the previous example with the addition of a force, $\underline{\mathcal{F}}$, which acts on the block in the negative x-direction. We solve for the motion of the block by applying Newton's second law, the momentum balance. Newton's second law is:

$$\text{Newton's second law:} \quad \sum_{\substack{\text{all forces}\\\text{acting on body}}} \underline{f} = \frac{d(m\underline{v})_{\text{body}}}{dt} = m\underline{a} \quad (3.17)$$

where $\underline{f}$ represents the forces on the block, m is the mass of the block, $\underline{v}$ is the velocity of the block, t is time, and $\underline{a}$ is the acceleration of the block.

Experiments on sliding friction show that the magnitude of the retarding frictional force is proportional to the magnitude of the normal force exerted on the block. The constant of proportionality is called the coefficient of sliding friction μ_k [167]. In our solution, we write the frictional force $\underline{\mathcal{F}}$ in terms of μ_k:

$$\underline{\mathcal{F}} = \begin{pmatrix} -\mu_k N \\ 0 \\ 0 \end{pmatrix}_{xyz} = -\mu_k N \hat{e}_x \quad (3.18)$$

where N is the magnitude of the normal-force vector $\underline{N}$. The negative sign in Equation 3.18 reflects that friction slows the block; that is, it acts in the opposite direction to the direction of the velocity of the block.

To solve for the velocity of the block, we substitute $\underline{\mathcal{F}}$ and the expressions for the other two forces (i.e., gravity and the normal force) into Equation 3.17 and simplify:

$$\frac{d(m\underline{v})}{dt} = m\underline{a} = \sum_{\substack{\text{all forces}\\\text{acting on body}}} \underline{f} = m\underline{g} + \underline{N} + \underline{f}_{friction} \quad (3.19)$$

$$\begin{pmatrix} m\dfrac{dv_x}{dt} \\ 0 \\ 0 \end{pmatrix}_{xyz} = \begin{pmatrix} mg\sin\theta \\ -mg\cos\theta \\ 0 \end{pmatrix}_{xyz} + \begin{pmatrix} 0 \\ N_y \\ 0 \end{pmatrix}_{xyz} + \begin{pmatrix} -\mu_k N \\ 0 \\ 0 \end{pmatrix}_{xyz} \quad (3.20)$$

Equating the y-components of each vector in Equation 3.20, we again obtain $N_y = mg\cos\theta$ for the y-component of normal force. Note that N, the magnitude of $\underline{N}$, therefore is given by:

$$\underline{N} = \begin{pmatrix} 0 \\ mg\cos\theta \\ 0 \end{pmatrix}_{xyz} \quad (3.21)$$

$$N = |\underline{N}| = +\sqrt{\underline{N} \cdot \underline{N}} = mg\cos\theta \quad (3.22)$$

Equating the x-components of each vector in Equation 3.20, we obtain:

$$x\text{-component} \quad m\frac{dv_x}{dt} = mg\sin\theta - \mu_k N \quad (3.23)$$

Substituting $N = mg \cos\theta$ into this equation and solving for v_x yields the result for v_x:

$$m\frac{dv_x}{dt} = mg \sin\theta - \mu_k mg \cos\theta \tag{3.24}$$

$$\frac{dv_x}{dt} = g \sin\theta - g\mu_k \cos\theta \tag{3.25}$$

$$v_x = (g \sin\theta - g\mu_k \cos\theta)\, t + C_1 \tag{3.26}$$

where C_1 is an arbitrary constant of integration. If we assume that at time $t = t_0$ the velocity is v_0,[2] then we can evaluate $C_1 = v_0 - (g \sin\theta - g\mu_k \cos\theta)\, t_0$. After substituting this into Equation 3.26, we obtain the final result for v_x:

$$v_x = v_0 + (g \sin\theta - g\mu_k \cos\theta)(t - t_0) \tag{3.27}$$

Thus, the steady-state velocity of the block in this example is:

$$\underline{v}(t) = \begin{pmatrix} v_0 + (g \sin\theta - g\mu_k \cos\theta)(t - t_0) \\ 0 \\ 0 \end{pmatrix}_{xyz} \tag{3.28}$$

When analyzed with friction present, the block moves with constant acceleration $\underline{a} = (g \sin\theta - g\mu_k \cos\theta)\, \hat{e}_x$ (Equation 3.25), although the acceleration is less than when the block slides without friction. With friction present (Equation 3.28) or not present (Equation 3.15), the steady-state velocity is a linear function of time.

As shown in these two examples, the general method for calculating the motion of solid bodies is to write expressions for forces on a chosen body in a chosen coordinate system and solve the components of the momentum balance for unknown quantities.

The motions of fluids likewise are governed by the balance of momentum. In flow, momentum transfers from one part of a fluid to another part or from fluids to solids (e.g., walls, paddles, and suspended particles) and vice versa. Our challenge in applying the momentum balance to fluids is to learn how to interpret the momentum-balance law when the system of interest is not an easily recognizable solid body but is, instead, a deformable medium.

3.2 Motion of deformable media

In elementary momentum-balance problems such as those discussed in the previous section, the bodies of interest are discrete (i.e., countable) and rigid. When forces are applied to a rigid body, the body retains its shape and moves with a resultant acceleration that depends on the forces applied and the body's mass.

---

[2] We cannot use the same initial condition as for the previous example because of the problem of static friction. When a block starts up from rest, it first must overcome the static-friction forces, which are higher than the sliding-friction forces [167]. Thus, the modeling of the block velocity is more complex if the block is assumed to start from rest. We can avoid this by choosing a time boundary-condition that gives the velocity at some time during the sliding of the block.

Figure 3.3 In rigid-body mechanics, individual bodies (e.g., a rock) comprise the system on which balances are made. For fluids, deformation makes the concept of a system somewhat more complicated.

The body has an associated location ($\underline{x}$), velocity ($\underline{v} = d\underline{x}/dt$), and acceleration ($\underline{a} = d\underline{v}/dt$), and these properties as a function of time can be calculated by applying the momentum balance ($\sum \underline{f} = m\underline{a}$) to the body.

When forces are applied to fluids, they move in ways totally unthinkable for a rigid solid. Compare, for instance, dropping a rock (i.e., rigid body) to pouring water from a pitcher (i.e., fluid) (Figure 3.3). A rock dropped from your hand accelerates to the ground under the force of gravity and perhaps rolls before stopping. The motion of water draining from a pitcher onto the ground also is caused by the force of gravity and, in response to this force, the water accelerates toward the ground in a stream. In the course of becoming that stream, the water in the pitcher deformed and accelerated; when the water hit the ground, it deformed again and divided into separate pieces of fluid and moved off in many directions before coming to rest.

The motion of the dropped rock can be analyzed by straightforward application of momentum-conservation laws using methods reviewed in the previous section. For flowing, deforming systems, the situation is different. Our task is to find a mathematical method whereby we can apply the conservation laws of nature to flowing, deforming systems.

The solution to this problem is to introduce the concept of the continuum, which is a mathematical idea that allows us to describe fluid motion by (1) defining a small number of continuous functions to account for material behavior; and (2) applying the laws of physics to infinitesimally small regions of a fluid, called fluid particles. The fluid particles together comprise the whole of the fluid; however, by dividing the fluid into many particles (i.e., bodies), we transform the complex deforming-fluid system into many simpler microscopic systems. Once the fluid is divided into infinitesimally small fluid particles, we can apply familiar methods from rigid-body mechanics to calculate the overall motion.

The other tool used in our analysis of fluid motion is the control volume, which is a concept that frees us from having to follow individual fluid particles throughout a flow and instead allows us to monitor the stream of fluid particles that pass through a chosen volume in space. The control-volume approach used in fluid mechanics and the mass-body-motion approach favored in rigid-body mechanics are equally correct implementations of the laws of physics. We will see, however, that for most fluid problems, balances on control volumes are easier to compute than those on individual fluid particles.

A fluids problem that is geometrically similar to a box sliding down an incline is introduced in the following example.

EXAMPLE 3.3. *What is the velocity field in a wide, thin film of water that runs steadily down an inclined surface under the force due to gravity? The fluid has a constant density ρ.*

SOLUTION. The flow we are considering may be the water running down a car's windshield or part of the industrial operation shown in Figure 3.4. The flow is driven by an external force (i.e., gravity), and the velocity is different at different points throughout the film thickness. The steady-state velocity distribution in the film depends on momentum exchanges within the fluid.

As with any type of problem solving in physics or mechanics, the first step is to reflect on the nature of the problem. We must use our judgment to determine

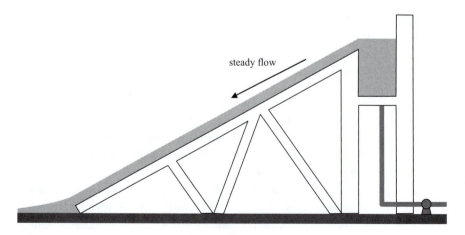

steady flow

Figure 3.4   A film of constant thickness flowing down an incline may be produced by flow through a weir, as shown here. Away from the edges and away from the top and bottom of the flow, the flow can be sufficiently idealized that we can solve for the velocity field.

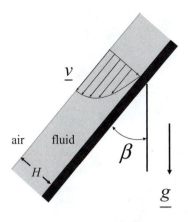

air   fluid

v

H

β

g

Figure 3.5

The idealized version of flow down an incline is a film of constant thickness such that the velocity is everywhere in the same direction but speed varies with position in the film. We seek to calculate the velocity as a function of position.

which properties are constant, which are variable, and how they vary. We have many choices about how to proceed; some choices make the problem easier to solve, others make the problem more difficult.

To solve for the velocity field, we must idealize the situation. Because the flow is wide, we consider only a two-dimensional cut near the center of the flow, as shown in Figure 3.5. The film is assumed to be of uniform thickness H and the water is isothermal.

We are asked to calculate the fluid velocity as a function of position for the situation shown in Figure 3.5. The water is flowing down the surface because gravity is acting on it. Looking at the situation, it seems reasonable to assume that the flow occurs parallel to the surface, with layers of fluid sliding over one another. The fluid in direct contact with the surface does not move; fluid at other locations moves parallel to the surface under the pull of gravity.

We need a way to identify small pieces of the flow as individual items so that we somehow can work out how they interact and create the final flow. We need a way to keep track of the motion of these pieces of fluid as well as of the forces on the fluid pieces and the forces that the fluid exerts on the walls and other boundaries of the flow.

In this chapter, we address these requirements. In the next section, we introduce the continuum model of fluids, field variables to describe particle motion and forces, and an overall approach—the control-volume approach—that allows us to organize the balances we need to apply to this situation. With these tools, we can return to this problem and address the issues in the problem statement.

3.2.1 The continuum model

For the solid sliding block in Section 3.1, the ideas of mass, force, and acceleration were easy to apply ($\sum \underline{f} = m\underline{a}$). For the example of water flowing from a pitcher, however (see Figure 3.3), mass, force, and velocity are more difficult to translate from ideas into hard equations. Is the m in $\sum \underline{f} = m\underline{a}$ the mass of all of the water or the mass of only the water in motion? What is meant by *velocity* or *acceleration* when one part of the system is moving but other parts are not or are moving in different directions? The forces are different in different parts of the water, leading to a confusing situation.

To address this complex situation, we divide the deforming system into multiple subsystems and apply momentum conservation to them. This strategy allows us to account for the exchanges of momentum between different parts of the original

Figure 3.6   Water flowing in a channel is not characterized by a single velocity but rather by a velocity field. (This photograph is a portion of the Cheonggye Stream, a restored waterway in Seoul, South Korea.)

system. It is these internal momentum exchanges that make fluid mechanics so complicated and interesting.

To implement the continuum approach, first we need to adopt a method for quantifying system properties such as mass, velocity, and force for different regions of the fluid. We use the concepts of the density field, the velocity field, and the stress field. Once we have a way of writing mass, motion, and force for a deforming medium, we can move on to applying Newton's laws to these systems.

3.2.1.1 FIELD VARIABLES

Consider a flow such as that shown in Figure 3.6. Water flows in a stream and falls over an edge under the pull of gravity. What is the velocity of the water? For fluids, there is not a single value of velocity that can describe completely the motion. For a fluid, the velocity is a property that varies with position in the flow. The function that gives the magnitude and direction of the velocity in a flow for every location (x, y, z) and time (t) is called the *velocity field*, $\underline{v}(x, y, z, t)$.

How can we measure the velocity field for moving water in a stream? If we drop a ping-pong ball in the stream, we can infer that the velocity of the fluid in contact with the ball is the same as the velocity of the ball. If we then measured the position of the ping-pong ball as a function of time,[3] we would know the fluid velocity at the surface of the stream in various locations. If the flow in the stream is steady, meaning that the flow patterns and speeds at every position do not change with time, we could repeat the experiment hundreds of times from different starting locations and obtain enough data to map out the surface velocity in the stream as a function of position. To obtain the velocity field below the surface, we must figure out another way to mark and follow the fluid that is below the surface, and there are techniques to accomplish this.[4] The results of

---

[3] This could be done with a camera taking timed photographs of the position of the ping-pong ball, for example.

[4] This can be done with neutrally buoyant particles—that is, particles whose density is the same as the density of water. Such particles would not sink or float but rather would follow the local

these experiments are data that correspond to the three-dimensional velocity field in the stream.

The velocity field measured in this way would be a map of velocities found in chosen locations in the stream. If we assume that the velocity is smoothly varying from point to point, and if we have taken a sufficient number of datapoints, we are justified in fitting a smooth function to the data that we gathered. After performing such a fit on steady-flow data, we obtain a continuous three-dimensional function $\underline{v}(x, y, z)$ that describes the velocity at every point in the stream. More sophisticated methods are needed to measure a time-dependent velocity field, but the general principle of the experiment is the same.

Similarly, we can use a continuous function to describe mass in a fluid. If we measure the mass of different samples of fluid throughout the flow, we can map the mass per volume or density as a three-dimensional function of position and time. The resulting function $\rho(x, y, z, t)$ is called the *density field*. Force per area or stress at different locations $\underline{\underline{\tilde{\Pi}}}(x, y, z, t)$[5] likewise can be expressed as a field variable, although it is more complicated, and we postpone a discussion of the details of stress until Chapter 4. The continuous functions fluid density ρ, fluid velocity $\underline{v}$, and fluid stress $\underline{\underline{\tilde{\Pi}}}$ are the field variables that we use to describe the physics of fluids.

EXAMPLE 3.4. *The density field $\rho(x, y, z)$ is one of the continuous variables of fluid mechanics. In the ocean, the density of the water varies with position due to salt concentration and temperature. At a latitude of 35 degrees south in the Pacific Ocean, density measurements as a function of depth z were made, and the data are shown in Table 3.1. What is the water-density function $\rho(z)$ for this location? What is the gradient of the density $\nabla \rho$ as a function of depth? Calculate the gradient both numerically and by fitting a function to $\rho(z)$.*

Table 3.1. Pacific Ocean water density as a function of vertical distance from the surface at a position 30 degrees south latitude

Depth m	ρ g/cm^3
0	1.0250
250	1.0260
500	1.0274
1,000	1.0280
1,500	1.0280
2,500	1.0280
3,500	1.0280
4,500	1.0280

SOLUTION. The data in Table 3.1 represent the function $\rho(z)$ in digital form; to see what the function look likes, we plot it in Figure 3.7. To deduce an equation that fits the data, we first nondimensionalize and scale the data to vary between zero and one. If we call the data $\rho(z)$, we can shift the data to zero by plotting $\rho(z) - \rho_{min}$. We can scale the data further by dividing this shifted data by the range of the data, $\rho_{max} - \rho_{min}$:

$$\text{Scaled variable:} \quad \frac{\rho(z) - \rho_{min}}{\rho_{max} - \rho_{min}} \text{ versus } z \qquad (3.29)$$

---

velocity. This technique, in conjunction with high-speed video and data processing, is used in advanced fluid-mechanics studies to measure complex velocity fields [1, 138].

[5] There are two related stress variables, $\underline{\underline{\tilde{\Pi}}}$ and $\underline{\underline{\tilde{\tau}}}$, with $\underline{\underline{\tilde{\Pi}}} = \underline{\underline{\tilde{\tau}}} - p\underline{\underline{I}}$. We define these variables in Chapter 4.

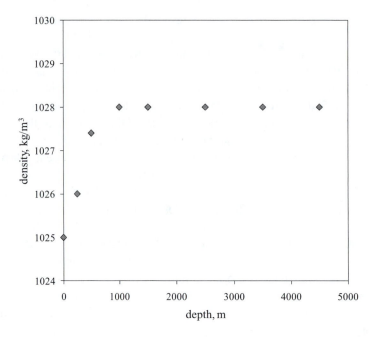

Figure 3.7　A pycnocline is a layer of water in which the water density changes rapidly with depth. These data show measured seawater densities as a function of depth.

The scaled variable is plotted versus z in Figure 3.8. The resulting curve rises to an asymptote. This type of curve often can be fit to a function of the following form:

$$\text{Smooth rise to asymptote:} \quad y = 1 - e^{-\frac{x}{\alpha}} \tag{3.30}$$
$$\text{(first-order response)}$$

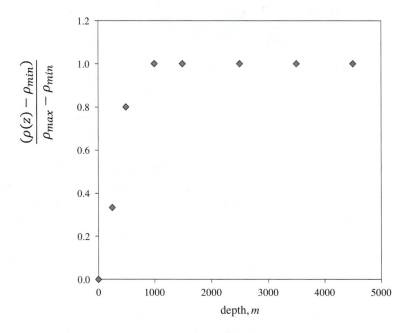

Figure 3.8　To find the functional form that best fits the data, we scale the variable and nondimensionalize as a first step.

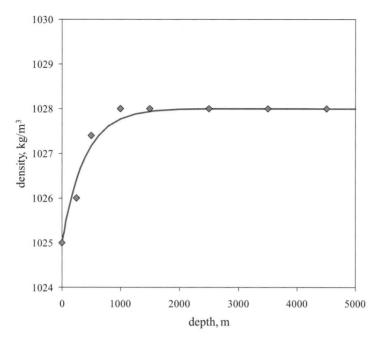

density, kg/m³ (y-axis, values 1024–1030)

depth, m (x-axis, values 0–5000)

Figure 3.9 The final fit given by Equation 3.33 is not perfect but the general trend of the data is captured. Limitations of the fit should be considered when the model is applied to any calculations.

where the value of the parameter α is adjusted to achieve the fit. Using a numerical program to minimize the error between the model and the data,[6] we obtain a reasonable agreement for $\alpha = 390$ m. The raw data with the fit plotted for comparison are shown in Figure 3.9.

$$\frac{\rho(z) - \rho_{min}}{\rho_{max} - \rho_{min}} = 1 - e^{-z/z_0} \tag{3.32}$$

$$\frac{\rho(z) - 1.025 \text{ g/cm}^3}{0.003 \text{ g/cm}^3} = 1 - e^{-z/390 \text{ m}} \tag{3.33}$$

To calculate the gradient of the function (see Equation 1.242), we must differentiate the data relative to the direction z:

$$z\text{-component of the} \quad \frac{\partial \rho}{\partial z} \tag{3.34}$$
$$\text{gradient of the function:}$$

Note that we are not given any information about how the density changes in the other two Cartesian directions; thus, nothing can be said about the x- and y-components of the gradient $\nabla \rho$.

[6] To do this, create columns of model predictions and experimental data and then calculate an error vector from:

$$\text{error} = \frac{(\text{model} - \text{data})^2}{(\text{data})^2} \tag{3.31}$$

The sum of the errors then is minimized by manipulating the model parameters; for the current case, we minimize the sum of the errors by manipulating α.

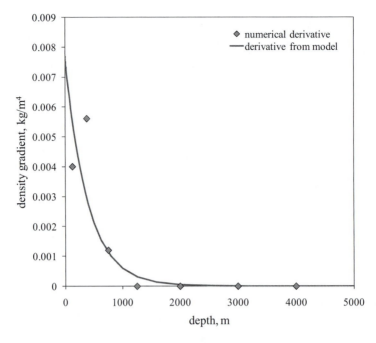

Figure 3.10 We calculate the gradient function in two ways: The first method used the data and estimated the derivative as $\Delta\rho/\Delta z$ (numerical). The second method fit the density data to a function and then analytically took the derivative of that function; that curve is shown as a smooth line. The two methods agree but noise was introduced to the numerical calculation that is avoided with the model method.

We can calculate $\partial\rho/\partial z$ in two ways. One way is to differentiate the function that we fit to the data. From the result of the differentiation, we can calculate $\partial\rho/\partial z$ at each depth given in Table 3.1:

$$\frac{\rho(z) - \rho_{min}}{\rho_{max} - \rho_{min}} = 1 - e^{-\frac{z}{\alpha}} \tag{3.35}$$

$$\rho(z) = \rho_{min} + (\rho_{max} - \rho_{min})\left(1 - e^{-\frac{z}{\alpha}}\right) \tag{3.36}$$

$$= \rho_{max} - (\rho_{max} - \rho_{min})e^{-\frac{z}{\alpha}} \tag{3.37}$$

$$\frac{\partial\rho}{\partial z} = -(\rho_{max} - \rho_{min})\left(\frac{-1}{\alpha}\right)e^{-\frac{z}{\alpha}} \tag{3.38}$$

We know from Figure 3.9 that our function does not exactly fit the data; thus, we also can estimate the gradient function by estimating it directly from the data in Table 3.1. We calculate $\Delta\rho/\Delta z$ for each neighboring set of datapoints and associate the calculated slope with the midpoint between the two neighboring points. The gradient calculated in these two ways is plotted in Figure 3.10. The two calculation methods agree in trend, but the numerical calculation has a great amount of scatter. The estimate of $[\nabla\rho]_z$ that comes from the curve fit uses information from near-neighbor points but also uses data from points that are farther apart than near neighbors; thus, a more smoothly varying function is obtained. The continuous function $\rho(z)$ for a fluid is useful in mass- and momentum-balance calculations, as discussed in subsequent chapters.

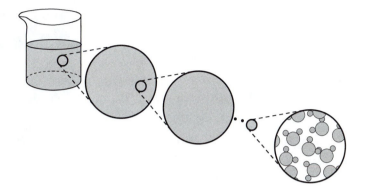

Figure 3.11   Liquids, like all matter, consist of molecules. The discrete nature of molecular structure is important when we examine a system at the nanometer lengthscale. At macroscopic lengthscales, however, it is sufficient to look at average properties as reflected in the continuous field variables of density, velocity, and stress.

In summary, to apply the laws of physics to deformable media, we define continuously varying field functions ρ, $\underline{v}$, and $\underline{\underline{\tilde{\Pi}}}$ to express mass/volume, motion, and molecular stress. In the next section, we explain how these continuous functions are used with the continuum picture to express fluid physics.

3.2.1.2 THE CONTINUUM HYPOTHESIS

The laws of physics—mass, momentum, and energy conservation—require that we quantify mass, material velocity, and molecular forces for the systems considered. For the physics of solid bodies, we simply write variables for these properties of the body. For fluids, we use continuous functions to quantify these properties throughout space and time. This is the continuum approach.

Using continuous functions to describe mass, velocity, and stress is not entirely consistent with what we know about the physical world. Although liquids appear continuous—they have no visible sharp boundaries between particles like powders—we know that fluids are made of individual pieces of mass—that is, molecules—and that there is empty space between molecules in a fluid (Figure 3.11). The molecular nature of matter is relevant in chemical studies—for example, studies of reactions. Because we are interested in fluid properties on a macroscopic lengthscale, however, we ignore molecular details and instead model systems in a more average way.

The continuum picture is an artificial model of the physical world that is convenient to use for making calculations on fluids. When using the continuum picture, we do not consider the motions of and forces on individual molecules; instead, we apply the laws of physics to the continuous functions that describe the density, velocity, and stress fields. Working at the level of continuum particles instead of at the molecular level is convenient because it involves fewer details—the behavior of billions of billions of billions of molecules can be summarized in three or four continuous functions. Details of molecular arrangement and motion are lost, however, when using the artificial continuum model rather than dealing directly with real molecules; fortunately, these details usually turn out to be unimportant to macroscopic observations of flow.

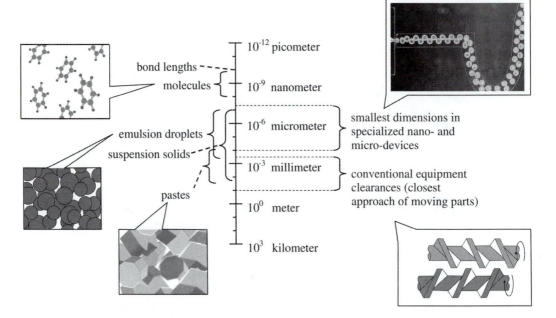

Figure 3.12 The continuum picture is applicable to many systems as long as the smallest dimension of the flow (e.g., channel width or gap between rotating screws) is much larger than the largest dimension of the material structure of the fluid. Homogeneous chemicals are understood readily to produce continuous liquids (e.g., water and benzene), but mixtures such as liquid–liquid mixtures (e.g., emulsions) or liquid–solid mixtures (e.g., suspensions and pastes) also may be modeled as continua as long as the lengthscale of the flow is large enough. If the lengthscale of the flow is very small, such as in modern micro- and nano-scale devices [44], then even simple pure liquids may cease to behave as continua, and the methods in this chapter cannot be used. (The bubbles in the microchannel shown here are from the research work of Shelley Anna and collaborators [4] used with permission.)

In the continuum model, the properties of a material vary continuously on any lengthscale, even the smallest possible that we can imagine. The continuum model is applicable for most fluids, from water to molten plastics, and can be acceptable for modeling heterogeneous systems such as emulsions and suspensions, depending on which properties are being calculated (Figure 3.12). In general, as long as equipment dimensions are much larger than the largest fluid structural dimension (e.g., particle or droplet size), the continuum model is effective in predicting fluid behavior (Figure 3.13).

In addition to simplicity, a major advantage of the continuum description for fluids is that calculus may be used in problem solving with this model. The interrelations among fluid density, velocity, and stress in flowing liquids are intricate and can be baffling, but with calculus—which was invented for this purpose—the physics can be organized into equations that may be solved with what are now well-known methods. Expressions for both rate of change and integration appear naturally when we apply conservation principles to fluid motion represented by the field variables ρ, $\underline{v}$, and $\underline{\tilde{\Pi}}$. Specifically, the fundamental definitions of derivative and integral from calculus (see Section 1.3) appear when mass, momentum, and energy-conservation principles are applied to a continuum (see Section 3.2.2

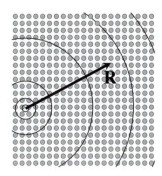

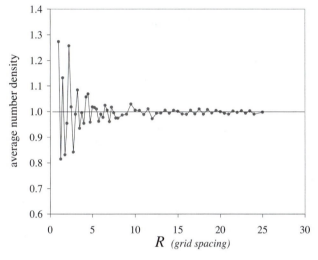

The density of any material becomes meaningless if the lengthscales considered are very small. This shows the result of a calculation of the average number of dots/area enclosed by circles of varying diameter (top figure). For large diameters, the ratio (area of dots)/(unit area) in the figure is a constant equal to one. As the size of the measurement area (R) decreases, however, the discrete nature of the dot distribution begins to be seen in the density measurement, and the calculated density is no longer a meaningful number.

and Equation 3.126):

Derivative defined

$$\frac{df}{dx} \equiv \lim_{\Delta x \to 0}\left[\frac{f|_{x+\Delta x} - f|_{x}}{\Delta x}\right] \tag{3.39}$$

Integral defined

$$\int_a^b f(x)dx \equiv \lim_{N \to \infty}\left[\sum_{i=1}^{N} f(a + i\Delta x)\Delta x\right] \tag{3.40}$$

$$\Delta x = \frac{b - a}{N}$$

Thus, our picture of a fluid is a mathematical continuum described by a set of field variables that capture the fluid's motion ($\underline{v}$) and other properties of the fluid (i.e., density and molecular stress). To analyze the behavior of a fluid continuum and to solve for the field variables, we use concepts from calculus. After we model the physics of flow by applying the conservation equations in terms of our continuous field variables, we obtain the final results for the field variables ρ, $\underline{v}$, and $\underline{\underline{\tilde{\Pi}}}$ by applying our prior knowledge of how to integrate and differentiate expressions and how to solve differential equations.

To demonstrate the type of calculations that the continuum model and calculus allow, in the following example we integrate the continuous function for density $\rho(z)$ to obtain a desired system property: the average density.

EXAMPLE 3.5. *What is the average density of seawater in the layer within 2,000 m of the ocean surface? Refer to the previous example for information on seawater density as a function of position.*

SOLUTION. Calculating averages of functions is a classic task of calculus. We begin with the expression for the average of a function, Equation 1.157:

$$\text{Average of } f(z) = \langle f \rangle = \frac{\displaystyle\int_{z_{min}}^{z_{max}} f(z)\, dz}{(z_{max} - z_{min})} \tag{3.41}$$

The equation we fit to the seawater density data now makes the calculation of the average density straightforward:

$$\langle \rho \rangle = \frac{\displaystyle\int_0^{2000\text{ m}} \rho(z)\, dz}{(2{,}000\text{ m} - 0\text{ m})} \tag{3.42}$$

$$\rho(z) = \rho_{min} + (\rho_{max} - \rho_{min})\left(1 - e^{-\frac{z}{\alpha}}\right) \tag{3.43}$$

$$\langle \rho \rangle = \frac{1}{2{,}000\text{ m}} \int_0^{2000\text{ m}} \rho_{min} + (\rho_{max} - \rho_{min})\left(1 - e^{-\frac{z}{\alpha}}\right) dz$$

$$= \left(\frac{\rho_{min}}{2{,}000\text{ m}}\right) z \Big|_{0\text{ m}}^{2000\text{ m}} + \frac{(\rho_{max} - \rho_{min})}{2{,}000\text{ m}} \left(z + \alpha e^{-\frac{z}{\alpha}}\right)\Big|_{0\text{ m}}^{2000\text{ m}}$$

$$= 1{,}025\,\frac{\text{kg}}{\text{m}^3} + \frac{3\,\frac{\text{kg}}{\text{m}^3}}{2{,}000\text{ m}} \left(2{,}000\text{ m} + 390\text{ m}\left(e^{-\frac{2{,}000\text{ m}}{390\text{ m}}} - 1\right)\right)$$

$$\langle \rho \rangle = \boxed{1{,}027.4\,\frac{\text{kg}}{\text{m}^3}} \tag{3.44}$$

3.2.1.3 FLUID PARTICLES

We choose to analyze flow patterns in terms of the continuum model and to quantify flows through the field variables of density, velocity, and stress. It remains for us to subject fluid motion to the laws of nature, particularly the law of conservation of momentum. When that connection is made, we can begin to

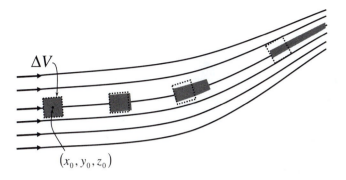

(x_0, y_0, z_0)

Figure 3.14 A fluid particle is defined at some initial time. The particle always contains the same molecules that it contained when it was defined (marked red or shaded). At later times, as the particle moves with the flow, the shape, speed, and direction of motion change according to the constraints imposed on it by the flow. Shown for comparison is the undeformed shape of the particle convected along with the flow.

understand why fluids make the patterns they do and why flows generate their associated forces.

Newton's second law, $\sum \underline{f} = m\underline{a}$, applies to individual bodies. As discussed in Section 3.1, when Newton's law is applied to a body, we can calculate the motion of the body. For a fluid, not all parts of it have the same position, velocity, and acceleration; therefore, to describe fluid motion with Newton's laws, we must divide the fluid into smaller entities that can be followed as a function of time. To obtain an accurate description of the flow, we divide the fluid into very small fluid particles.

A fluid particle is defined as a small quantity of mass occupying a volume ΔV. The mass inside this small volume at a chosen point is given by $\rho \Delta V$, where ρ is the fluid density at that point. The small mass $\rho \Delta V$ constitutes a fluid particle to which we can apply Newton's laws of motion and other laws of physics.

Consider a two-dimensional portion of a flow as shown in Figure 3.14. At some initial time t_0, we define one particular fluid particle as all of those molecules that are contained in a cube of volume ΔV in one chosen location (x_0, y_0, z_0). We imagine the molecules within the chosen particle as colored red. As time moves forward, the red particle moves forward with the flow and the shape, speed, and direction of its motion all change. Our task is to apply $\sum \underline{f} = m\underline{a}$ to this body.

The mass of the particle is straightforward: $m = \rho \Delta V$. Acceleration of the particle is the average acceleration of all of the molecules within the particle. We choose ΔV to be very small to increase the accuracy of using the average acceleration:

$$\sum_{\substack{\text{all forces} \\ \text{acting on particle}}} \underline{f}_i = m\underline{a} \tag{3.45}$$

$$= (\rho \Delta V) \langle \underline{a} \rangle_{\substack{\text{average for all} \\ \text{molecules}}} \tag{3.46}$$

The forces on the red particle are gravity and molecular surface forces imposed by neighboring fluid particles. It is simple to account for the force of gravity on the particle: It is equal to particle mass multiplied by acceleration due to gravity.

$$t_0 \qquad\qquad t$$

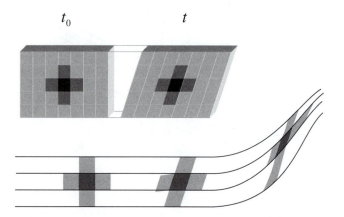

Figure 3.15   When we first divide the flow into small, cubical particles, the arrangement is orderly and it is easy to express mathematically the forces on such systems. In a complex flow, however, convection and deformation move and drastically change the shape of the particles as a function of time. When the particles become highly irregular in shape, it is difficult to model the forces that act on the particles, especially the surface forces.

Thus, the force on the particle due to gravity is $\rho \Delta V \underline{g}$:

$$\sum_{\substack{\text{all forces} \\ \text{acting on particle}}} \underline{f}_i = (\rho \Delta V) \, \langle \underline{a} \rangle_{\substack{\text{average for all} \\ \text{molecules}}} \qquad (3.47)$$

$$\underline{f}_{\text{gravity}} + \underline{f}_{\text{surface}} = (\rho \Delta V) \, \langle \underline{a} \rangle_{\substack{\text{average for all} \\ \text{molecules}}} \qquad (3.48)$$

$$\rho \Delta V \underline{g} + \underline{f}_{\text{surface}} = (\rho \Delta V) \, \langle \underline{a} \rangle_{\substack{\text{average for all} \\ \text{molecules}}} \qquad (3.49)$$

The molecular surface forces on the red particle due to the neighboring particles are the difficult part of this problem (Figure 3.15). If we divide the entire flow field into identical particles at t_0, then at that initial moment, the red particle has six neighboring particles of the same size and shape. As the flow progresses in time, all of these particles and the red particle move and deform. At each time of the flow, the red particle has oddly shaped and possibly different neighbors imposing different forces on it. Calculating the motion and shape of the red particle at any time depends on an accurate modeling of the interactions with its neighbors. In some flows, the deformation of particles is severe (Figure 3.16) and it is a challenging problem to account for the forces on each small deforming fluid particle; such calculations can be attempted only with the help of powerful computers and advanced numerical techniques [49].

Thus, we find ourselves at a deadend. The law of conservation of momentum requires that we consider the effects of forces on individual bodies. The individual bodies in a deforming medium, however, are constantly changing shape, making the application of the conservation laws difficult and, for the moment, impractical. We need a new approach: a method that allows us to consider a more fixed type of system, a method that makes it easier to account for the molecular surface forces imposed by neighboring particles.

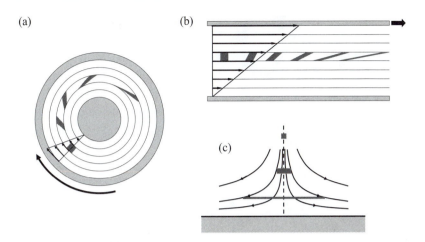

(a)

(b)

(c)

We seek a method for calculating velocity and stress fields for all types of flows. In the flows pictured here, small fluid particles deform and adopt shapes that are difficult to track. In some cases (c), the velocities of the two ends of the particle go in opposite directions.

3.2.2 Control-volume approach

When we reach a deadend, a good way to restart is to return to the beginning and remind ourselves of what we want to do. We want to determine an effective way to model the flow patterns and forces associated with fluid motion. Fluid motion can be complex (see Chapter 2), but it makes sense to start with the simplest systems. A simple system to consider is flow in a straight section of a river or stream (Figure 3.17) or, even simpler, flow in a straight open channel in the laboratory,

We look for simple flows, such as the gentle flowing of a stream, to learn to model fluid motion. (The photograph is a portion of the Cheonggye Stream, a restored waterway in Seoul, South Korea.)

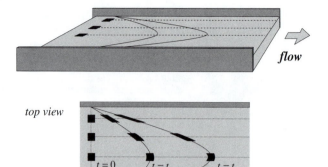

Figure 3.18 We begin again, considering a very simple flow: steady flow in a straight channel. Even in this simple flow, however, fluid particles undergo considerable deformation.

where the channel walls have a regular shape (Figure 3.18). In such a channel, the flow can be made to be steady, meaning that the velocity at any chosen point is constant in time.

Steady flow in a straight channel is simple, but it still has the problem outlined in the previous section; that is, fluid particles deform into awkward shapes as the flow progresses, and applying momentum balances to these shapes is difficult (see Figure 3.18).

Another approach is suggested by watching a steady stream of fluid particles move past from a fixed position on the shore. Rather than follow individual particles of fluid over the course of time, is it possible to calculate a momentum balance by accounting for the forces on fluid that moves through a fixed region within the flow? If we choose balances on a rigid (i.e., nondeforming), motionless volume, we would not have the fluid-particle-tracking problem.

It is an appealing idea but there remains the problem of Newton's second law, which is written with respect to individual bodies. If we want to consider a fixed volume in space with different bodies flowing through the volume, we must adapt Newton's second law to this new circumstance, if possible. This is what we do in Sections 3.2.2.1 and 3.2.2.2.

The result we need is Newton's second law written for a control volume (CV), which is called the *Reynolds transport theorem* and is derived as Equations 3.135 and 3.136; the final equations are as follows.

Reynolds transport theorem (momentum balance on CV derived in this section)

$$\sum_{\substack{\text{on} \\ \text{CV}}} \underline{f} = \frac{d\mathbf{P}}{dt} + \iint_{\text{CS}} (\hat{n} \cdot \underline{v})\, \rho \underline{v}\, dS \qquad (3.50)$$

$$\begin{pmatrix} \text{sum of} \\ \text{forces} \\ \text{on CV} \end{pmatrix} = \begin{pmatrix} \text{rate of} \\ \text{increase of} \\ \text{momentum of} \\ \text{fluid in CV} \end{pmatrix} + \begin{pmatrix} \text{net outflow of} \\ \text{momentum} \\ \text{through bounding} \\ \text{surfaces of CV} \end{pmatrix} \qquad (3.51)$$

Momentum-Conservation Equations

Individual Bodies
Newton's Second Law of Motion

$$\sum_{\substack{\text{on} \\ \text{body}}} \underline{f} = m\underline{a} = \frac{d(m\underline{v})_{\text{body}}}{dt}$$

Control Volumes
Reynolds Transport Theorem

$$\sum_{\substack{\text{on} \\ \text{CV}}} \underline{f} = \frac{d(m\underline{v})_{\text{CV}}}{dt} + \iint_{\text{CS}} (\hat{n} \cdot \underline{v})\rho\,\underline{v}\,dS$$

Figure 3.19 Momentum is conserved. For individual bodies, Newton's second law is a convenient equation for making calculations of motion and forces. For fluids, it is often more convenient to use an equivalent expression: the Reynolds transport theorem with the momentum balance written on a control volume.

where CV is the control volume, CS is the control surface, $\underline{f}$ represents various forces on the control volume, $\underline{\mathbf{P}}$ is the momentum in the control volume, and the integral expresses the net outflow of momentum through the bounding surface of the control volume. The Reynolds transport theorem, named for Osborne Reynolds, states that the sum of forces on a control volume is equal to the rate of change of momentum of the fluid in the CV plus the net outward flux of momentum through the surfaces bounding the CV (Figure 3.19). Newton's second law, by comparison, states that the sum of the forces on a body is equal to the rate of change of momentum of the body. When the momentum-balance calculation is performed on a control volume (i.e., a stationary, rigid, imaginary volume in our usage), an extra term is needed compared to the body case because the net forces on the CV can affect more than just the rate-of-change term: Material can cross the boundaries of the control volume, bringing along momentum. This extra term is called the *convective term*. For momentum balances on bodies (i.e., Newton's second law), there is no issue of momentum being carried into the system by another body; the balance always is carried out on a chosen body or bodies. Once the system of interest is chosen, no other bodies enter the picture.

We derive the Reynolds transport theorem in the next section and discuss the convective term in Section 3.2.2.2. We show how to use this equation in Section 3.2.3. Readers who want to begin with solving problems using the Reynolds transport theorem may proceed to Section 3.2.3 and subsequently return to this derivation section as desired.

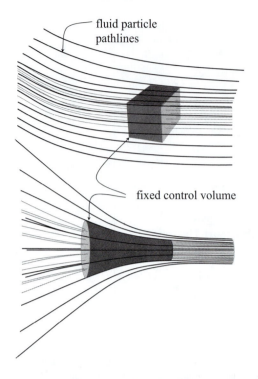

fluid particle
pathlines

fixed control volume

Figure 3.20   A control volume is an imagined region in space through which fluid moves. In our discussion, we assume the control volume to be fixed in shape and position. The shape of the control volume is arbitrary, and we usually choose a shape that mimics the flow pattern because this choice simplifies mass, momentum, and energy-balance calculations. The paths of the particles that pass through the control volume are emphasized here.

3.2.2.1 MOMENTUM BALANCE ON A CONTROL VOLUME

The volume on which we do our balances is called the control volume (CV), which is an imaginary container through which fluid particles move (Figure 3.20). For the derivation of the momentum balance on a control volume, we consider an arbitrarily shaped control volume fixed in position and shape in an arbitrary flow (Figure 3.21).

At chosen time t, the control volume contains certain fluid particles. These fluid particles are a body in the sense of Newton's laws. We imagine that the fluid

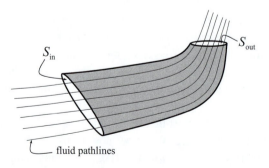

S_{in}

S_{out}

fluid pathlines

Figure 3.21   For the derivation of the momentum balance as applied to a control volume, we do not assume any special shape, but we do assume the control volume to be fixed in shape and position.

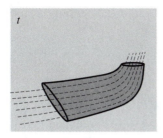 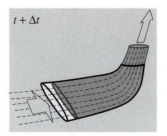

Figure 3.22 At time t, the fluid in the control volume is imagined to be colored red and all of the fluid outside of the control volume is colored blue. At a slightly later time $t + \Delta t$, some of the red fluid has exited the control volume and some of the new (blue) fluid has entered through the inlet surface(s).

in the control volume at time t is colored red (Figure 3.22, left). The red fluid is subject to forces on it, and the relationship between the net forces on and the momentum of the red fluid is given by Newton's second law:

$$\sum_{\substack{\text{on} \\ \text{body}}} \underline{f} = m\underline{a} = \frac{d(m\underline{v})_{\text{body}}}{dt} \tag{3.52}$$

$$\sum_{\substack{\text{on} \\ \text{body}}} \underline{f} = \begin{pmatrix} \text{net force} \\ \text{on red fluid at } t \\ = \text{net force} \\ \text{on CV at } t \end{pmatrix} = \begin{pmatrix} \text{rate of change} \\ \text{of momentum of} \\ \text{red fluid at } t \end{pmatrix} \tag{3.53}$$

We must work on this equation to see how the momentum of the fluid in the CV changes with time.

We can use the definition of derivative (see Equation 3.39) to rewrite the derivative that appears on the righthand side of Equation 3.52 as a limit of a rate of change of momentum over the interval between time t and a slightly later time $t + \Delta t$:

$$\sum_{\substack{\text{on} \\ \text{body}}} \underline{f} \bigg|_t = \sum_{\substack{\text{on} \\ \text{CV}}} \underline{f} \bigg|_t = \frac{d(m\underline{v})}{dt} \bigg|_t \tag{3.54}$$

$$= \lim_{\Delta t \to 0} \left[\frac{(m\underline{v})|_{t+\Delta t} - (m\underline{v})|_t}{\Delta t} \right] \tag{3.55}$$

To fill in the terms on the righthand side of Equation 3.55, we must think about the momentum of the red fluid at t and at $t + \Delta t$. Our goal is to relate these quantities to the forces on the control volume.

Returning to our picture of the control volume (see Figure 3.22), we can visualize the process of the red fluid passing through the CV between times t and $t + \Delta t$. At time t, all of the red fluid is in the CV. At time $t + \Delta t$, some of the red fluid has left the CV and some of the upstream fluid has entered it. For simplicity, we call the upstream fluid the blue fluid. We divide the red fluid into the red fluid that stays in the CV between t and $t + \Delta t$ and the red fluid that leaves during that interval. Dropping the limit symbol, Equation 3.55 becomes:

$$
\Delta t \left. \sum_{\substack{\text{on} \\ \text{CV}}} \underline{f} \right|_t = \left(\begin{array}{c} \text{momentum of} \\ \text{red fluid} \end{array} \right) \Bigg|_{t+\Delta t} - \left(\begin{array}{c} \text{momentum of} \\ \text{red fluid} \end{array} \right) \Bigg|_t \tag{3.56}
$$

$$
= \left[\left(\begin{array}{c} \text{momentum of} \\ \text{red fluid} \\ \text{that stays} \end{array} \right) + \left(\begin{array}{c} \text{momentum of} \\ \text{red fluid} \\ \text{that exits} \end{array} \right) \right]_{t+\Delta t}
$$

$$
- \left[\left(\begin{array}{c} \text{momentum of} \\ \text{red fluid} \\ \text{that stays} \end{array} \right) + \left(\begin{array}{c} \text{momentum of} \\ \text{red fluid} \\ \text{that exits} \end{array} \right) \right]_t \tag{3.57}
$$

Although we temporarily omitted the limit symbol, at the end of this derivation, we again take the limit as Δt goes to zero. Because this separation is convenient in a later step in the derivation, we distinguish here between red fluid that ultimately stays and red fluid that ultimately exits.

Newton's second law relates the net forces on a body (i.e., the red fluid) to the rate of change of momentum of the body. We now are trying to relate forces in a fluid to the rate of change of momentum of the fluid in the control volume. The fluid in the CV is different fluid at different times, which is the complicating factor. Beginning with the red-fluid momentum balance as written in Equation 3.57, we make definitions and rearrangements that allow us to isolate the rate of change of momentum of the fluid in the CV at a time of interest.

We define a variable $\underline{P}$ to represent the momentum of the fluid in the control volume at any time:

$$
\left(\begin{array}{c} \text{momentum} \\ \text{of fluid} \\ \text{in the CV} \end{array} \right) \equiv \underline{P} \tag{3.58}
$$

Because the fluid in the CV at time t is different fluid from that in the CV at time $t + \Delta t$, the momentum of the fluid in the CV is different at these two times, and we write it in terms of red and blue fluid, as follows.

First, at time t, the red fluid fills the CV so that $\underline{P}|_t$ is the momentum of all of the red fluid at t:

$$
\left(\begin{array}{c} \text{momentum} \\ \text{of fluid} \\ \text{in CV} \end{array} \right) \Bigg|_t = \underline{P}|_t = \left(\begin{array}{c} \text{momentum} \\ \text{of red fluid} \\ \text{that stays} \end{array} \right) \Bigg|_t + \left(\begin{array}{c} \text{momentum} \\ \text{of red fluid} \\ \text{that exits} \end{array} \right) \Bigg|_t \tag{3.59}
$$

Second, we write the momentum in the CV at time $t + \Delta t$. At this time, the fluid in the CV is the red fluid that stayed and the new blue fluid that entered:

$$
\left(\begin{array}{c} \text{momentum} \\ \text{of fluid} \\ \text{in CV} \end{array} \right) \Bigg|_{t+\Delta t} = \underline{P}|_{t+\Delta t} = \left(\begin{array}{c} \text{momentum} \\ \text{of red fluid} \\ \text{that stays} \end{array} \right) \Bigg|_{t+\Delta t}
$$

$$
+ \left(\begin{array}{c} \text{momentum} \\ \text{of blue fluid} \\ \text{that enters} \end{array} \right) \Bigg|_{t+\Delta t} \tag{3.60}
$$

We now combine the two previous equations with Equation 3.57, which is the momentum balance on the red fluid; this yields a new relationship between forces and the fluid in the CV.

First, we solve Equation 3.59 for the momentum at time t of red fluid that stays:

$$\left(\begin{array}{c} \text{momentum} \\ \text{of red fluid} \\ \text{that stays} \end{array}\right)\Bigg|_{t} = \underline{P}|_{t} - \left(\begin{array}{c} \text{momentum} \\ \text{of red fluid} \\ \text{that exits} \end{array}\right)\Bigg|_{t} \tag{3.61}$$

Second, we solve Equation 3.60 for the momentum at time $t + \Delta t$ of red fluid that stays:

$$\left(\begin{array}{c} \text{momentum} \\ \text{of red fluid} \\ \text{that stays} \end{array}\right)\Bigg|_{t+\Delta t} = \underline{P}|_{t+\Delta t} - \left(\begin{array}{c} \text{momentum} \\ \text{of blue fluid} \\ \text{that enters} \end{array}\right)\Bigg|_{t+\Delta t} \tag{3.62}$$

Combining these two expressions with Equation 3.57 results in:

$$\Delta t \sum_{\substack{\text{on} \\ \text{CV}}} \underline{f}\Bigg|_{t} = \left(\begin{array}{c} \text{momentum} \\ \text{of red fluid} \\ \text{that stays} \end{array}\right)\Bigg|_{t+\Delta t} + \left(\begin{array}{c} \text{momentum of} \\ \text{red fluid} \\ \text{that exits} \end{array}\right)\Bigg|_{t+\Delta t}$$
$$- \left(\begin{array}{c} \text{momentum} \\ \text{of red fluid} \\ \text{that stays} \end{array}\right)\Bigg|_{t} - \left(\begin{array}{c} \text{momentum} \\ \text{of red fluid} \\ \text{that exits} \end{array}\right)\Bigg|_{t} \tag{3.63}$$

$$= \underline{P}|_{t+\Delta t} - \left(\begin{array}{c} \text{momentum} \\ \text{of blue fluid} \\ \text{that enters} \end{array}\right)\Bigg|_{t+\Delta t} + \left(\begin{array}{c} \text{momentum} \\ \text{of red fluid} \\ \text{that exits} \end{array}\right)\Bigg|_{t+\Delta t}$$
$$- \underline{P}|_{t} + \left(\begin{array}{c} \text{momentum} \\ \text{of red fluid} \\ \text{that exits} \end{array}\right)\Bigg|_{t} - \left(\begin{array}{c} \text{momentum} \\ \text{of red fluid} \\ \text{that exits} \end{array}\right)\Bigg|_{t} \tag{3.64}$$

The final two terms cancel, yielding:

$$\Delta t \sum_{\substack{\text{on} \\ \text{CV}}} \underline{f}\Bigg|_{t} = \underline{P}|_{t+\Delta t} - \underline{P}|_{t} - \left(\begin{array}{c} \text{momentum} \\ \text{of blue fluid} \\ \text{that enters} \end{array}\right)\Bigg|_{t+\Delta t} + \left(\begin{array}{c} \text{momentum} \\ \text{of red fluid} \\ \text{that exits} \end{array}\right)\Bigg|_{t+\Delta t}$$

$$\tag{3.65}$$

We have made considerable progress in our quest to relate red-fluid momentum changes to momentum changes of the fluid in the control volume. To proceed, we write mathematical expressions for the two quantities expressed in words on the righthand side of Equation 3.65. These two quantities are entering and exiting fluid momenta at $t + \Delta t$—that is, momenta of fluid that crosses the CV boundaries. Both expressions can be written following the same approach; the calculation results in a double integral over the control-volume bounding surfaces.

The final mathematical expression for the terms in Equation 3.65 are derived in the next section. The final results, derived as Equation 3.132, are as follows.

The two integrals are called the convective terms.

$$\sum_{\substack{on \\ CV}} \underline{f}\,\Bigg|_t = \frac{\mathbf{P}|_{t+\Delta t} - \mathbf{P}|_t}{\Delta t} + \left(\iint_{S_{in}} (\hat{n} \cdot \underline{v})\, \rho \underline{v}\, dS \right)\Bigg|_{t+\Delta t}$$

$$+ \left(\iint_{S_{out}} (\hat{n} \cdot \underline{v})\, \rho \underline{v}\, dS \right)\Bigg|_{t+\Delta t} \qquad (3.66)$$

3.2.2.2 THE CONVECTIVE TERM

To convert the word expressions in Equation 3.65 to mathematical terms, we must consider how to use the continuum model to keep track of mass or momentum flow through a surface. We begin by considering the simplest case of direct mass and momentum flow through a flat surface. We derive key mathematical tools in the next two examples.

EXAMPLE 3.6. *Liquid passes through a chosen area A as shown in Figure 3.23. The velocity is perpendicular to the surface A at every point and does not vary across the cross section. What are the volumetric flow rate (volume liquid/time), mass flow rate (mass/time), and momentum flow rate (momentum/time) through A?*

SOLUTION. Figure 3.23 shows that for the case under consideration, the velocity of the fluid is perpendicular to the surface A and is constant (i.e., it does not vary with position). Consider the fluid that passes through A during a short time interval Δt (Figure 3.24). The volume of fluid that passes through A during the interval Δt forms a solid, the volume of which is given by:

$$\begin{pmatrix} \text{volume of fluid} \\ \text{passing through } A \\ \text{in time } \Delta t \end{pmatrix} = \begin{pmatrix} \text{height} \\ \text{of solid} \end{pmatrix} \begin{pmatrix} \text{cross section} \\ \text{of solid} \end{pmatrix} \qquad (3.67)$$

$$= \Delta x\, A \qquad (3.68)$$

where Δx is the change in location of fluid that started at A and has moved in the x-direction for time Δt. The magnitude of the fluid velocity, v, can be written as:

$$\begin{matrix} \text{Magnitude of} \\ \text{fluid velocity} \end{matrix} \qquad |\underline{v}| = v = \frac{\Delta x}{\Delta t} \qquad (3.69)$$

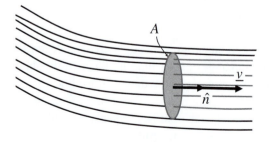

Figure 3.23   In this example, we consider the flow through a surface A. The velocity of the fluid is perpendicular to the surface A.

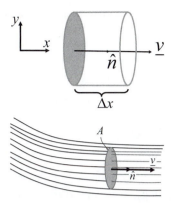

Figure 3.24 During the time interval Δt, a volume of fluid of height Δx and of cross-sectional area A passes through area A.

With these two expressions, we can calculate all of the quantities of interest. The volumetric flow rate is the volume of fluid divided by the time interval:

$$Q = \frac{\text{fluid volume}}{\text{time interval}} = \frac{\Delta x \, A}{\Delta t} = v \, A \qquad (3.70)$$

Volumetric flow
of liquid through A
(velocity perpendicular to A;
$\underline{v}$ does not vary across A)

$$\boxed{Q = v \, A} \qquad (3.71)$$

The mass flow rate can be calculated from the volumetric flow rate and the density:

$$m = \left(\frac{\text{mass}}{\text{volume}}\right)\left(\frac{\text{volume}}{\text{time}}\right) \qquad (3.72)$$

$$= (\rho)(v \, A) \qquad (3.73)$$

Mass flow
of liquid through A
(velocity perpendicular to A;
$\underline{v}$ does not vary across A)

$$\boxed{m = (\rho)(v \, A)} \qquad (3.74)$$

Finally, the momentum flow rate (a vector quantity) can be calculated from the definition of momentum and the previous results:

$$\left(\begin{array}{c}\text{momentum flow} \\ \text{of liquid through } A\end{array}\right) = \left(\frac{\text{momentum}}{\text{volume}}\right)\left(\frac{\text{volume}}{\text{time}}\right) \qquad (3.75)$$

$$= \frac{(\text{mass})(\text{velocity})}{\text{volume}}\left(\frac{\text{volume}}{\text{time}}\right) \qquad (3.76)$$

$$= \left(\frac{\text{mass}}{\text{volume}}\right)(\underline{v})\left(\frac{\text{volume}}{\text{time}}\right) \qquad (3.77)$$

$$= \rho \, \underline{v} \, (v A) \qquad (3.78)$$

Note that for this example, the velocity of the fluid was perpendicular to the surface A and $\underline{v}$ does not vary across A.

$$
\boxed{
\begin{array}{c}
\text{Momentum flow} \\
\text{of liquid through } A \\
\text{(velocity perpendicular to } A; \\
\underline{v} \text{ does not vary across } A)
\end{array}
= \rho \, \underline{v} \, (vA)
}
\tag{3.79}
$$

The previous example shows how powerful the continuum approach is. With simple logic (essentially, unit matching), we can express volume, mass, and momentum flows for a chosen system in terms of two field variables: density and velocity. For more complex systems, we build on these relationships and use vector tools, as shown in the next example.

EXAMPLE 3.7. *Liquid passes through a chosen area A as shown in Figure 3.25. The velocity of the fluid makes an angle θ with the unit normal to A, which is called $\hat{n}$. The velocity does not vary across the surface A. What are the volumetric flow rate (volume liquid/time), mass flow rate (mass/time), and momentum flow rate (momentum/time) through A?*

SOLUTION. The logic of the solution is the same for this case as in the previous example; there is, however, a difference in the volume of fluid that passes through A in time interval Δt.

Consider the fluid that passes through A during the short time interval Δt (Figure 3.26). The x-direction is the direction of flow. In time interval Δt, fluid that started on the surface A moved along x a distance Δx. The volume of fluid that passed through A in this time interval is the volume of the mathematical solid shown. The height of the solid is $\Delta x \cos \theta$. The volume of fluid that passes through A during the interval Δt thus is given by:

$$
\begin{pmatrix}
\text{volume of fluid} \\
\text{passing through } A \\
\text{in time } \Delta t
\end{pmatrix}
=
\begin{pmatrix}
\text{height} \\
\text{of solid}
\end{pmatrix}
\begin{pmatrix}
\text{cross section} \\
\text{of solid}
\end{pmatrix}
\tag{3.80}
$$

$$
= (\Delta x \cos \theta) \, A
\tag{3.81}
$$

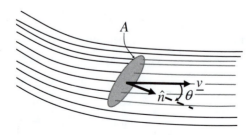

Figure 3.25 In this example, we consider the flow through a surface A. The velocity of the fluid is not perpendicular to the surface A; instead, it makes an angle θ with the surface unit normal $\hat{n}$.

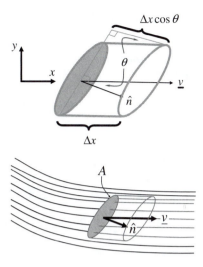

Figure 3.26 During the time interval Δt, a volume of fluid of height $\Delta x \cos \theta$ and of cross-sectional area A passes through area A.

The magnitude of the fluid velocity, v, can be written as before as follows:

$$\text{Magnitude of fluid velocity} \quad |\underline{v}| = v = \frac{\Delta x}{\Delta t} \tag{3.82}$$

With these two expressions, we can calculate all of the quantities of interest:

$$\text{Volumetric flow of liquid through } A \quad Q = \frac{\text{fluid volume}}{\text{time interval}} \tag{3.83}$$

$$= \frac{\Delta x \cos \theta \ A}{\Delta t} \tag{3.84}$$

$$= v \cos \theta \ A \tag{3.85}$$

$$= (\hat{n} \cdot \underline{v})A \tag{3.86}$$

Volumetric flow
of liquid through A
(general orientation case;
$\underline{v}$ does not vary across A)
$$\boxed{Q = v \cos \theta \ A = (\hat{n} \cdot \underline{v})A} \tag{3.87}$$

We use the definition of the dot product to write the final result (Equation 3.85) in vector notation ($\hat{n} \cdot \underline{v} = |\hat{n}||\underline{v}| \cos \theta = v \cos \theta$; see Equation 1.161). As before, the mass flow rate can be calculated from the volumetric flow rate and the density:

$$\text{Mass flow of liquid through } A \quad m = \left(\frac{\text{mass}}{\text{volume}}\right)\left(\frac{\text{volume}}{\text{time}}\right) \tag{3.88}$$

$$= (\rho)(v \cos \theta \ A) = \rho (\hat{n} \cdot \underline{v}) \ A \tag{3.89}$$

Mass flow
of liquid through A
(general orientation case;
$\underline{v}$ does not vary across A)
$$\boxed{m = \rho (\hat{n} \cdot \underline{v}) \ A} \tag{3.90}$$

Finally, the momentum flow rate can be calculated as before from the definition of momentum and the previous results:

$$\left(\begin{array}{c} \text{momentum flow} \\ \text{of liquid through } A \end{array} \right) = \left(\frac{\text{momentum}}{\text{volume}} \right) \left(\frac{\text{volume}}{\text{time}} \right) \tag{3.91}$$

$$= \frac{(\text{mass})(\text{velocity})}{\text{volume}} \left(\frac{\text{volume}}{\text{time}} \right) \tag{3.92}$$

$$= \left(\frac{\text{mass}}{\text{volume}} \right) (\underline{v}) \left(\frac{\text{volume}}{\text{time}} \right) \tag{3.93}$$

$$= \rho\, \underline{v}\, (v \cos \theta\, A) = \rho\, \underline{v}\, (\hat{n} \cdot \underline{v}) A \tag{3.94}$$

This is the general result when $\underline{v}$ is not necessarily perpendicular to A:

$$\left(\begin{array}{c} \text{momentum flow} \\ \text{of liquid through } A \\ \text{(general orientation case;} \\ \underline{v} \text{ does not vary across } A) \end{array} \right) = \rho \underline{v}\, (\hat{n} \cdot \underline{v}) A \tag{3.95}$$

We recover the case of velocity perpendicular to A (see Equation 3.79) when $\theta = 0$ ($\cos 0 = 1$, $\hat{n} \cdot \underline{v} = v$).

The relationship obtained in Equation 3.87 for volumetric flow rate through an area as a function of the locally constant velocity $\underline{v}$ ($Q = (\hat{n} \cdot \underline{v})A$) is similar to the equation introduced in Chapter 1 that relates overall volumetric flow rate through a pipe to the *average* velocity in the pipe $\langle v \rangle$ (see Equation 1.2). If we write Equation 3.87 on a microscopic piece of cross-sectional area in a pipe flow with varying $\underline{v}$ and integrate over the pipe cross section (recall Equation 1.157), we obtain Equation 1.2; this calculation is shown in Chapter 6 (see Equation 6.254). In the following example, we practice with the relationships just developed.

EXAMPLE 3.8. *Consider a control volume in the shape of the square pyramid as shown in Figures 3.27 and 3.28. The square pyramid is a pentahedron with a square for a base and four triangles for sides; the one in Figure 3.27 has four equilateral triangles for sides (i.e., a Johnson solid). The pyramid is a control volume placed in a uniform flow (i.e., velocity $\underline{v}$ in the flow is constant at every position in space). The flow direction is parallel at all points to a vector in the plane of the pyramid's base that bisects two opposite sides of the base. Calculate the mass flow rate of fluid of density ρ through each of the five sides of the pentahedron. Write the answer in terms of the speed of the fluid v and the pyramid edge length α.*

SOLUTION. The use of a pentahedron as a control volume is unusual, but the calculations involved in solving this problem are not unusual when making calculations of the convective contribution to the momentum balance. This problem provides an opportunity to practice with angles, geometry, the dot product, and the relationships in this section.

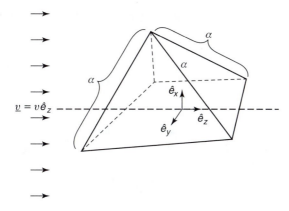

The control volume is a square pyramid that has five sides, four of which are equilateral triangles.

The mass flow through a surface is given by Equation 3.90:

$$\begin{matrix} \text{Mass flow of liquid} \\ \text{through surface } A \end{matrix} \qquad m = \rho \, (\hat{n} \cdot \underline{v}) \, A \qquad (3.96)$$

For each of the five surfaces of the control volume, we need the unit normal $\hat{n}$ and the area A. The density ρ is constant, and the velocity vector $\underline{v} = v\hat{e}_z$ is the same at all locations for uniform flow.

xz-section through center:

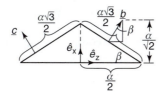

xy-section through center:

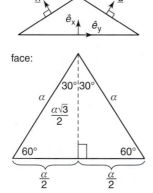

The unit normals needed for the calculations in this example can be determined through the geometry of sections cut through the center of the control volume.

We choose a Cartesian coordinate system with the flow direction as the z-direction:

$$\underline{v} = \begin{pmatrix} 0 \\ 0 \\ v \end{pmatrix}_{xyz} = v\hat{e}_z \qquad (3.97)$$

The outwardly pointing unit normal vectors for each surface of the control volume are shown in Figure 3.28. For the bottom of the pyramid, the outwardly pointing unit vector $\underline{a}$ points downward, $\underline{a} = -\hat{e}_x$. The dot product of $\underline{a}$ and $\underline{v} = v\hat{e}_z$ is therefore zero, and the mass flow rate through the bottom is zero:

$$m = \rho\,(\hat{n} \cdot \underline{v})\,A \qquad (3.98)$$

$$m|_a = \rho(\underline{a} \cdot \underline{v})\alpha^2 \qquad (3.99)$$

$$= \rho\alpha^2\,(1\ 0\ 0)_{xyz} \cdot \begin{pmatrix} 0 \\ 0 \\ v \end{pmatrix}_{xyz} \qquad (3.100)$$

$$= 0 \qquad (3.101)$$

For surface b, the geometry in Figure 3.28 shows that the outwardly pointing unit normal vector $\underline{b}$ is:

$$\text{From geometry:} \qquad \hat{n}|_b \equiv \underline{b} = \begin{pmatrix} \frac{1}{\sqrt{3}} \\ 0 \\ \sqrt{\frac{2}{3}} \end{pmatrix}_{xyz} \qquad (3.102)$$

and the area of the equilateral triangle that comprises the face is $A = (1/2)(\alpha)(\alpha\sqrt{3}/2)$. The mass flow rate through surface b is therefore:

$$m = \rho\,(\hat{n} \cdot \underline{v})\,A \qquad (3.103)$$

$$m|_b = \rho(\underline{b} \cdot \underline{v})\frac{\alpha^2\sqrt{3}}{4} \qquad (3.104)$$

$$= \frac{\rho\alpha^2\sqrt{3}}{4}\left(\frac{1}{\sqrt{3}}\ 0\ \sqrt{\frac{2}{3}}\right)_{xyz} \cdot \begin{pmatrix} 0 \\ 0 \\ v \end{pmatrix}_{xyz} \qquad (3.105)$$

$$= \frac{\rho v\alpha^2}{2\sqrt{2}} \qquad (3.106)$$

For surface c, also shown in Figure 3.28, the outwardly pointing unit normal vector $\underline{c}$ is similar to $\underline{b}$, but the z-component points in the opposite direction:

$$\text{From geometry:} \qquad \hat{n}|_c \equiv \underline{c} = \begin{pmatrix} \frac{1}{\sqrt{3}} \\ 0 \\ -\sqrt{\frac{2}{3}} \end{pmatrix}_{xyz} \qquad (3.107)$$

The mass flow rate through surface c is therefore:

$$m|_c = \rho(\underline{c} \cdot \underline{v}) \frac{\alpha^2 \sqrt{3}}{4} \tag{3.108}$$

$$= \frac{\rho \alpha^2 \sqrt{3}}{4} \left(\tfrac{1}{\sqrt{3}} \; 0 \; -\sqrt{\tfrac{2}{3}} \right)_{xyz} \cdot \begin{pmatrix} 0 \\ 0 \\ v \end{pmatrix}_{xyz} \tag{3.109}$$

$$= -\frac{\rho v \alpha^2}{2\sqrt{2}} \tag{3.110}$$

The mass flow rates out through surfaces b and c are the same, but one is positive, indicating that the flow is outward (i.e., surface $\underline{b}$); and one is negative, indicating that the flow is inward (i.e., surface c).

For surfaces d and h, the two side faces of the pyramid, the unit normal vectors are in the xy-plane. Thus, when the outwardly pointed unit normal $\hat{n}$ is dotted with $\underline{v} = v\hat{e}_z$ in each case, the result is zero; there is no mass flow out of the control volume through either surface:

$$\hat{n}|_d = \underline{d} = \begin{pmatrix} d_x \\ d_y \\ 0 \end{pmatrix}_{xyz} \tag{3.111}$$

$$\hat{d} \cdot \underline{v} = (d_x \hat{e}_x + d_y \hat{e}_y) \cdot v\hat{e}_z = 0 \tag{3.112}$$

$$\hat{n}|_h = \underline{h} = \begin{pmatrix} h_x \\ h_y \\ 0 \end{pmatrix}_{xyz} \tag{3.113}$$

$$\hat{h} \cdot \underline{v} = (h_x \hat{e}_x + h_y \hat{e}_y) \cdot \underline{v}\hat{e}_z = 0 \tag{3.114}$$

Finally, notice that the sum of all of the mass flow rates is zero, which is in accord with the mass balance that at steady state the net outflow of mass from the control volume is zero:

$$\begin{pmatrix} \text{net outflow} \\ \text{of mass from} \\ \text{CV} \end{pmatrix} = m|_a + m|_b + m|_c + m|_d + m|_h \tag{3.115}$$

$$= 0 + \frac{\rho v \alpha^2}{2\sqrt{2}} - \frac{\rho v \alpha^2}{2\sqrt{2}} + 0 + 0 \tag{3.116}$$

$$= 0 \tag{3.117}$$

We return now to Equation 3.65 and seek to convert the two word expressions in that equation to mathematical terms. Both of the word expressions under consideration account for momentum flows through the surfaces that bound the control volume. In Example 3.8, we practiced writing momentum flows through a surface (see Equation 3.95), and we now turn to applying this technique to the control volume.

Beginning with the blue fluid that enters the control volume, consider the surface area S_{in} through which blue fluid enters (Figure 3.29). We choose a surface

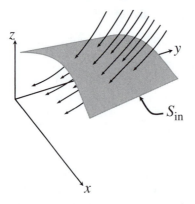

Figure 3.29 The momentum carried by fluid moving across a curved surface is calculated with a surface integral.

with an arbitrary shape and orientation for this derivation. In a general flow, fluid velocity varies with position; therefore, care must be taken when calculating the momentum entering the control volume through S_{in}. We must divide the surface S_{in} in some way and sum the contributions from various regions. In addition, the surface S_{in} generally is not flat; therefore, the task of dividing S_{in} is a challenge. This problem was addressed in the development of integral calculus (for a review, see the Web appendix [108]), and we can apply these methods directly to the calculation of the flow of momentum through S_{in}.

Our approach is to project S_{in} onto a plane that we arbitrarily call the xy-plane (Figure 3.30). The area of the projection is $\mathcal{R}$. Because $\mathcal{R}$ is in the xy-plane, the unit normal to $\mathcal{R}$ is $\hat{e}_z$. We divide the projection $\mathcal{R}$ into areas $\Delta A = \Delta x \Delta y$ and seek to write the momentum flow rate in different regions of S_{in} associated with their projections ΔA_i. By focusing on $\mathcal{R}$ and equal-sized divisions of $\mathcal{R}$ (rather than directly dividing the curvy surface S_{in}), we can arrive at the appropriate integral expression.

Figure 3.30 shows the area S_{in} and its projection $\mathcal{R}$ in the xy-plane. The area $\mathcal{R}$ is divided into rectangles of area ΔA_i, and we consider only the ΔA_i that are wholly contained within the boundaries of $\mathcal{R}$. For each ΔA_i in the xy-plane, we choose a point within ΔA_i and call it $(x_i, y_i, 0)$. The point (x_i, y_i, z_i) is located on the surface S_{in} directly above $(x_i, y_i, 0)$. If we draw a plane tangent to S_{in} through the point (x_i, y_i, z_i), we can construct an area ΔS_i that is a portion of the tangent plane whose projection onto the xy-plane is ΔA_i (see Figure 3.30). We soon take a limit as ΔA_i becomes infinitesimally small; therefore, it is not important which point $(x_i, y_i, 0)$ is chosen as long as it is in ΔA_i.

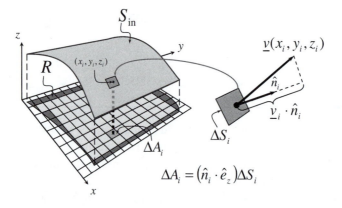

Figure 3.30 For a surface that is not flat, we first project the surface onto the xy-plane. We then divide the projection and proceed to write and sum the momentum flow rate through each small piece. The surface differential ΔS_i can be related to ΔA_i—its projection onto the xy-plane—by $\Delta S_i = \Delta A_i / (\hat{n}_i \cdot \hat{e}_z)$.

Each tangent-plane area ΔS_i approximates a portion of the surface S_{in}, and we write an estimate of the total momentum flow through S_{in} as a sum of the momentum flows through all of the tangent planes ΔS_i. The momentum entering the control volume between t and $t + \Delta t$ through one such ΔS_i can be calculated as follows:

$$\begin{pmatrix} \text{momentum} \\ \text{entering CV} \\ \text{through } i^{th} \\ \text{tangent plane } \Delta S_i \end{pmatrix} = \left(\frac{\text{momentum}}{\text{volume}} \right) \left(\frac{\text{volume flow inward}}{\text{time}} \right) \Delta t \quad (3.118)$$

Volumetric flow inward may be written using Equation 3.87:

$$\begin{pmatrix} \text{momentum} \\ \text{entering CV} \\ \text{through } i^{th} \\ \text{tangent plane } \Delta S_i \end{pmatrix} = \left(\frac{\text{mass} \cdot \text{velocity}}{\text{volume}} \right) \begin{pmatrix} \text{inflow} \\ \text{velocity} \\ \text{magnitude} \end{pmatrix} \cdot \text{area} \Delta t \quad (3.119)$$

$$= (\rho_i \, \underline{v}|_i)(-(\hat{n}_i \cdot \underline{v}|_i)\Delta S_i) \, \Delta t \quad (3.120)$$

where ρ_i and $\underline{v}|_i$ are the density and velocity at (x_i, y_i, z_i) and $\hat{n}_i$ is the outwardly pointing unit normal vector at (x_i, y_i, z_i) (compare with Equation 3.95). Note that we have a choice for unit normal vector $\hat{n}_i$ because any surface has two unit normal vectors: one pointing into and one pointing out of the control volume. The fluid-mechanics convention is to choose the outwardly pointing unit normal. The negative sign in Equation 3.120 is a consequence of this choice, and the expression $\hat{n}_i \cdot \underline{v}|_i$ corresponds to the outwardly moving component of the velocity. Because we are interested in the inwardly moving flow in Equation 3.120, we must include a negative sign.

Equation 3.120 gives the contribution of momentum passing through each ΔS_i. To approximate the total momentum flow through S_{in}, we now sum over all tangent-planes ΔS_i. Note that we are including only the ΔS_i associated with those projections ΔA_i that are fully contained within $\mathcal{R}$. Subsequently, we take the limit as ΔA becomes small to make the calculation exact:

$$\begin{pmatrix} \text{momentum} \\ \text{of blue fluid} \\ \text{that enters CV} \end{pmatrix} \approx \sum_{i=1}^{N} \begin{pmatrix} \text{momentum} \\ \text{entering CV} \\ \text{through } i^{th} \\ \text{tangent plane } \Delta S_i \end{pmatrix} \quad (3.121)$$

$$= -\sum_{i=1}^{N} (\rho_i \, \underline{v}|_i)((\hat{n}_i \cdot \underline{v}|_i)\Delta S_i) \, \Delta t \quad (3.122)$$

$$= -\Delta t \sum_{i=1}^{N} ((\hat{n}_i \cdot \underline{v}|_i)\rho_i \, \underline{v}|_i \, \Delta S_i) \quad (3.123)$$

$$\begin{pmatrix} \text{momentum} \\ \text{of blue fluid} \\ \text{that enters CV} \end{pmatrix} = -\Delta t \lim_{\Delta A \to 0} \left[\sum_{i=1}^{N} ((\hat{n}_i \cdot \underline{v}|_i)\rho_i \, \underline{v}|_i \, \Delta S_i) \right] \quad (3.124)$$

where N is the number of projections ΔA_i that are wholly within $\mathcal{R}$. We can relate the tangent-plane area ΔS_i and the projected area ΔA_i through geometry (see the Web appendix [108]). The result is:

$$\Delta A_i = (\hat{n}_i \cdot \hat{e}_z) \Delta S_i \tag{3.125}$$

where $\hat{e}_z$ is the unit normal of the ΔA_i and $\hat{n}_i$ is the unit normal of ΔS_i. Substituting this relationship, Equation 3.124 becomes:

$$\begin{pmatrix} \text{momentum} \\ \text{of blue fluid} \\ \text{that enters CV} \end{pmatrix} = -\Delta t \lim_{\Delta A \to 0} \left[\sum_{i=1}^{N} \frac{(\hat{n}_i \cdot \underline{v}|_i) \rho_i \, \underline{v}|_i}{\hat{n}_i \cdot \hat{e}_z} \Delta A_i \right] \tag{3.126}$$

The limit of the sum on the righthand side of Equation 3.126 is related to the definition of a double integral [108]:

$$\begin{array}{c} \text{Double integral} \\ \text{of a function} \\ \text{(general version)} \end{array} \quad \boxed{I = \iint_{\mathcal{R}} f(x, y) \, dA \equiv \lim_{\Delta A \to 0} \left[\sum_{i=1}^{N} f(x_i, y_i) \Delta A_i \right]}$$

$$\tag{3.127}$$

where $\mathcal{R}$ is the region in the xy-plane over which f is being integrated (i.e., summed). Comparing Equations 3.126 and 3.127, we write:

$$\begin{pmatrix} \text{momentum} \\ \text{of blue fluid} \\ \text{that enters CV} \end{pmatrix} = -\Delta t \iint_{\mathcal{R}} \frac{(\hat{n} \cdot \underline{v}) \rho \underline{v}}{\hat{n} \cdot \hat{e}_z} \, dA \tag{3.128}$$

If we define $dS \equiv dA/(\hat{n} \cdot \hat{e}_z)$, then Equation 3.128 becomes [108]:

$$\begin{pmatrix} \text{momentum} \\ \text{of blue fluid} \\ \text{that enters CV} \end{pmatrix} = -\Delta t \iint_{S_{in}} (\hat{n} \cdot \underline{v}) \, \rho \underline{v} \, dS \tag{3.129}$$

This is the expression we need to finish writing the convective terms in Equation 3.65. Our calculations show that the momentum of blue fluid that enters the CV is equal to the surface integral of the crossing momentum per unit volume $(\hat{n} \cdot \underline{v}) \, \rho \underline{v}$ over the inlet surface S_{in}.

The momentum of the red fluid that exits the control volume may be written similarly, resulting in an analogous integral over the outflow surface S_{out}:

$$\begin{pmatrix} \text{momentum} \\ \text{of red fluid} \\ \text{that exits CV} \end{pmatrix} = \Delta t \iint_{S_{out}} (\hat{n} \cdot \underline{v}) \, \rho \underline{v} \, dS \tag{3.130}$$

Notice that there is no negative sign in Equation 3.130 (recall the discussion related to Equation 3.120) because in this case, we are accounting for fluid that is exiting, and the outwardly pointing normal dotted with the velocity vector gives the component of velocity corresponding to outflow. We now substitute the results in Equations 3.129 and 3.130 into Equation 3.65 to replace the word expressions:

$$\Delta t \left. \sum_{\substack{on \\ CV}} \underline{f} \right|_t = \left. \mathbf{P} \right|_{t+\Delta t} - \left. \mathbf{P} \right|_t - \left. \begin{pmatrix} \text{momentum} \\ \text{of blue fluid} \\ \text{that enters} \end{pmatrix} \right|_{t+\Delta t} + \left. \begin{pmatrix} \text{momentum} \\ \text{of red fluid} \\ \text{that exits} \end{pmatrix} \right|_{t+\Delta t}$$

(3.131)

$$\left. \sum_{\substack{on \\ CV}} \underline{f} \right|_t = \frac{\left. \mathbf{P} \right|_{t+\Delta t} - \left. \mathbf{P} \right|_t}{\Delta t} + \left. \left(\iint_{S_{in}} (\hat{n} \cdot \underline{v}) \rho \underline{v} \, dS \right) \right|_{t+\Delta t}$$

$$+ \left. \left(\iint_{S_{out}} (\hat{n} \cdot \underline{v}) \rho \underline{v} \, dS \right) \right|_{t+\Delta t}$$

(3.132)

The two integrals in Equation 3.132 may be combined because the first is over all inlet surfaces and the second is over all outlet surfaces. All CV surfaces are either inlet or outlet surfaces or those through which no fluid passes. Surfaces through which no fluids pass would have $\hat{n} \cdot \underline{v} = 0$ because $\underline{v} = 0$ there. We therefore can write these two integrals together as the integral over the entire enclosing surface of the control volume, CS:

$$\left. \left(\iint_{S_{in}} (\hat{n} \cdot \underline{v}) \rho \underline{v} \, dS \right) \right|_{t+\Delta t} + \left. \left(\iint_{S_{out}} (\hat{n} \cdot \underline{v}) \rho \underline{v} \, dS \right) \right|_{t+\Delta t}$$

$$= \left. \left(\iint_{CS} (\hat{n} \cdot \underline{v}) \rho \underline{v} \, dS \right) \right|_{t+\Delta t}$$

(3.133)

Making this change in Equation 3.132 and taking the limit as Δt goes to zero, we arrive at the final relationship we seek: between the forces on the CV and the rate of change of momentum of the fluid in the CV:

$$\left. \sum_{\substack{on \\ CV}} \underline{f} \right|_t = \lim_{\Delta t \to 0} \left(\frac{\left. \mathbf{P} \right|_{t+\Delta t} - \left. \mathbf{P} \right|_t}{\Delta t} \right) + \lim_{\Delta t \to 0} \left. \left(\iint_{CS} (\hat{n} \cdot \underline{v}) \rho \underline{v} \, dS \right) \right|_{t+\Delta t}$$

(3.134)

Reynolds transport theorem (momentum balance on CV)

$$\boxed{\sum_{\substack{on \\ CV}} \underline{f} = \frac{d\mathbf{P}}{dt} + \iint_{CS} (\hat{n} \cdot \underline{v}) \rho \underline{v} \, dS}$$

(3.135)

$$\begin{pmatrix} \text{sum of} \\ \text{forces} \\ \text{on CV} \end{pmatrix} = \begin{pmatrix} \text{rate of} \\ \text{increase of} \\ \text{momentum of} \\ \text{fluid in CV} \end{pmatrix} + \begin{pmatrix} \text{net outflow of} \\ \text{momentum} \\ \text{through bounding} \\ \text{surfaces of CV} \end{pmatrix}$$

(3.136)

Going from Equation 3.134 to Equation 3.135, we again have used the fundamental definition of a derivative (see Equation 3.39).[7] The integral term is called the convective term.

Equation 3.135, called the Reynolds transport theorem, gives the equivalent of Newton's second law ($\sum f = m\underline{a}$) for a control volume. The Reynolds transport theorem states that the sum of forces on a control volume is equal to the rate of increase of momentum of the fluid in the control volume plus the net outward flux of momentum through the surfaces bounding the control volume (see Figure 3.19). In the next section, we learn how to apply this equation to control volumes that interest us in fluid mechanics.

3.2.3 Problem solving with control volumes

With development of the Reynolds transport theorem, we have the main tool needed to solve a wide variety of flow problems:

Reynolds transport theorem
(momentum balance on CV)

$$\sum_{\substack{\text{on} \\ \text{CV}}} \underline{f} = \frac{d\mathbf{P}}{dt} + \iint_{CS} (\hat{n} \cdot \underline{v})\, \rho \underline{v}\, dS \qquad (3.137)$$

$$\begin{pmatrix} \text{sum of} \\ \text{forces} \\ \text{on a CV} \end{pmatrix} = \begin{pmatrix} \text{rate of} \\ \text{increase of} \\ \text{momentum of} \\ \text{fluid in CV} \end{pmatrix} + \begin{pmatrix} \text{net outflow of} \\ \text{momentum} \\ \text{through bounding} \\ \text{surfaces of CV} \end{pmatrix} \qquad (3.138)$$

The Reynolds transport theorem gives the equivalent of Newton's second law ($\sum f = ma$) for a control volume. This expression states that the sum of forces on a CV is equal to the rate of change of momentum of the fluid in the CV plus the net outward flux of momentum through the surfaces bounding the CV (see Figure 3.19). When properly applied to a flow situation and solved, the momentum balance gives the velocity field and information on how forces interact in a fluid. The Reynolds transport theorem is a powerful tool, and it solves the problem of the difficulty in applying Newton's laws to fluids. Now the challenge becomes to learn how to apply this tool to problems of interest.

We turn to two problems that represent those we seek to solve. The first is a fluids version of the sliding-block problem—the flow of a thin film of fluid down an inclined plane (introduced previously)—in which we seek a detailed prediction of the velocity field. The second example applies our CV approach to a macroscopic scale, enabling us to calculate forces on the bend of a pipe with fluid flowing inside.

For both problems, our approach is qualitatively the same. First, we interpret the situation as a problem involving the field variables of density and velocity. Second, we choose a CV. In the first example, we choose a microscopic CV because we seek to calculate the velocity at every point in the fluid; in the second

---

[7] The momentum of the fluid in the control volume $\underline{\mathbf{P}}$ is a function only of time; for more discussion on this point, see Deen [40] and the supplemental web materials [108].

example, we choose a macroscopic CV because a macroscopic force is sought. Third, we apply the Reynolds transport theorem and solve.

3.2.3.1 MICROSCOPIC CONTROL-VOLUME PROBLEM

In Section 3.2, we introduced the problem of calculating the velocity field in flow down an incline. We can make more progress on this problem now that we understand the Reynolds transport theorem.

EXAMPLE 3.9 (Incline, continued). *What is the velocity field in a wide, thin film of water that runs steadily down an inclined surface under the force due to gravity? The fluid has a constant density ρ.*

SOLUTION. The flow considered is driven by an external force (i.e., gravity), and the velocity is different at different points in the flow. The velocity distribution in the film depends on momentum exchanges within the fluid. This problem is the type for which we derived the Reynolds transport theorem:

Reynolds transport theorem
(momentum balance on CV)

$$\sum_{\substack{\text{on} \\ \text{CV}}} \underline{f} = \frac{d\underline{\mathbf{P}}}{dt} + \iint_{CS} (\hat{n} \cdot \underline{v}) \, \rho \underline{v} \, dS \qquad (3.139)$$

When correctly applied, the Reynolds transport theorem allows us to calculate the velocity distribution in the flow down an incline.

To apply the Reynolds transport theorem and solve for the velocity field, we idealize the situation. Because the flow is wide, we consider only a two-dimensional cut near the center of the flow (see Figure 3.5). The film is assumed to be of uniform thickness H and the water is isothermal. We assume that the flow occurs parallel to the surface with layers of fluid sliding over one another. The fluid in direct contact with the surface does not move; fluid at other locations moves parallel to the surface under the pull of gravity.

We quantify the situation within the continuum model. The velocity is a field variable that indicates the speed and direction of travel of bits of fluid at every possible location. Velocity is a vector; therefore, in the most general case, velocity may be written in a chosen coordinate system in terms of the three components relative to that coordinate system:

Fluid
velocity field

$$\underline{v} = \begin{pmatrix} v_1 \\ v_2 \\ v_3 \end{pmatrix}_{123} \qquad \text{(arbitrary coordinates)} \qquad (3.140)$$

Our first task is to choose the coordinate system in which we solve the coefficients of $\underline{v}$. The choice of coordinate system is arbitrary—that is, the meaning of a vector is independent of the coordinate system in which it is expressed. Although the choice of coordinate system is arbitrary, this does not render the choice unimportant. If we choose wisely, we simplify the problem; if we choose unwisely, we may be unable to solve the problem. We choose the Cartesian coordinate system in Figure 3.31 with the z-direction parallel to the flow direction, the x-direction perpendicular to the wall, and the y-direction following the righthand rule and

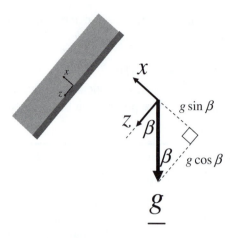

Figure 3.31 Because we chose a coordinate system that simplifies the variable velocity vector, the gravity vector is slightly more complicated than it might be with another choice of coordinate system.

into the paper. By choosing a coordinate system that aligns with the flow direction, we reduce to one the number of nonzero velocity components. Recall that in the sliding-block problem, we also chose a coordinate system parallel to the surface and in the direction of motion of the block. In Chapter 6, we examine the issue of choosing a coordinate system and discuss the impact of this choice on boundary conditions.

In our chosen coordinate system, the velocity vector is given by:

$$\underline{v} = \begin{pmatrix} 0 \\ 0 \\ v_z \end{pmatrix}_{xyz} = v_z \hat{e}_z \tag{3.141}$$

In this coordinate system, the boundaries of the problem are at $x = 0$ (i.e., the surface of the incline) and at $x = H$ (i.e., the top surface of the film; also called the free surface). A disadvantage of our choice is that the acceleration due to gravity in this system does not line up with any coordinate direction in our coordinate system. We can write the two nonzero components of $\underline{g}$ in terms of the angle β that the incline makes with the vertical (see Figure 3.31):

$$\begin{matrix} \text{Gravity field} \\ \text{(from geometry)} \end{matrix} \qquad \underline{g} = \begin{pmatrix} g_x \\ g_y \\ g_z \end{pmatrix}_{xyz} = \begin{pmatrix} -g\sin\beta \\ 0 \\ g\cos\beta \end{pmatrix}_{xyz} \tag{3.142}$$

$$= -g\sin\beta\,\hat{e}_x + g\cos\beta\,\hat{e}_z \tag{3.143}$$

We now are ready to apply the momentum balance to our problem. The momentum balance for a control volume is the Reynolds transport theorem (see Equation 3.139). All of the terms in the Reynolds transport theorem relate to the momentum associated with a chosen CV. The next step, therefore, is to choose the CV to which we will apply the momentum balance.

Like the choice of coordinate system, the choice of CV is arbitrary; there are choices that make the problem easy and those that make it nearly impossible to solve. We seek to calculate how the velocity component v_z varies with position in

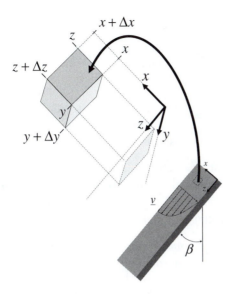

Figure 3.32   The control volume we chose is small so that v_z does not vary much within it. Its shape reflects the symmetries of the problem.

the flow. We therefore choose a CV that is small enough to characterize a single position in the flow: the general position (x, y, z).

Choosing control volumes becomes easier with practice. Because this is the first CV that we are choosing, the process may seem mysterious at this point. As we study more and different problems, however, the concerns that go into making a good choice of CV become clearer (see also Chapter 4, Figure 4.2, and Chapter 9).

For flow of a thin film down an incline, we choose a small CV, the shape of which reflects the symmetries of the problem. Because the flow is rectilinear, we choose a small rectangular parallelepiped (i.e., a box; Figure 3.32). The chosen CV allows us to write the forces that act on a little packet of fluid at a point within the flow of interest.

Having chosen the CV, the next step is to apply the momentum balance on a CV, the Reynolds transport theorem (see Equation 3.139), to our chosen control volume:

$$\frac{d\mathbf{P}}{dt} + \iint_{CS} (\hat{n} \cdot \underline{v})\, \rho \underline{v}\, dS = \sum_{\substack{\text{on} \\ \text{CV}}} \underline{f} \qquad (3.144)$$

The first term on the lefthand side of Equation 3.144 is the rate of change of momentum of the CV. The flow we are considering is at steady state; therefore, the rate of change of the momentum in our CV is zero.

The integral is the net outflow of momentum from the CV (recall that $\hat{n}$ is the outwardly pointing normal). Momentum flows in or out through a surface only when mass crosses that surface—that is, when the velocity component $(\hat{n} \cdot \underline{v})$ is nonzero at that surface. For our CV, no mass crosses the top, bottom, or in-page or out-of-page surfaces, leaving two terms to be evaluated:

$$\iint_{CS} (\hat{n} \cdot \underline{v}) \, \rho \underline{v} \, dS = \begin{pmatrix} \text{net momentum} \\ \text{out of CV} \end{pmatrix} \tag{3.145}$$

$$= \begin{pmatrix} \text{momentum} \\ \text{out through} \\ \text{upstream} \\ \text{side of CV} \end{pmatrix} + \begin{pmatrix} \text{momentum} \\ \text{out through} \\ \text{downstream} \\ \text{side of CV} \end{pmatrix} \tag{3.146}$$

The same amount of momentum enters and leaves our CV; thus, the net momentum out of the CV is zero. We formally obtain this result by writing Equation 3.146 in terms of our variables and simplifying with the mass balance. Momentum is mass multiplied by velocity; thus, Equation 3.146 becomes:

$$\iint_{CS} (\hat{n} \cdot \underline{v}) \, \rho \underline{v} \, dS = \begin{pmatrix} \text{momentum} \\ \text{out through} \\ \text{upstream} \\ \text{side of CV} \end{pmatrix} + \begin{pmatrix} \text{momentum} \\ \text{out through} \\ \text{downstream} \\ \text{side of CV} \end{pmatrix} \tag{3.147}$$

$$= -\hat{e}_z \cdot (v_z|_z \, \hat{e}_z) \rho \, v_z|_z \, \hat{e}_z \Delta x \, \Delta y$$
$$+ \hat{e}_z \cdot (v_z|_{z+\Delta z} \, \hat{e}_z) \rho \, v_z|_{z+\Delta z} \, \hat{e}_z \Delta x \, \Delta y \tag{3.148}$$

$$= \rho \left(v_z|_{z+\Delta z} \right)^2 \Delta x \, \Delta y \begin{pmatrix} 0 \\ 0 \\ 1 \end{pmatrix}_{xyz}$$

$$- \rho \left(v_z|_z \right)^2 \Delta x \, \Delta y \begin{pmatrix} 0 \\ 0 \\ 1 \end{pmatrix}_{xyz} \tag{3.149}$$

The magnitude of the velocity at the upstream surface $v_z|_z$ and the magnitude of the velocity at the downstream surface $v_z|_{z+\Delta z}$ are related through the mass balance, $dM_{CV}/dt = 0$:

Steady-state mass balance on CV
$$\frac{dM_{CV}}{dt} = \begin{pmatrix} \text{mass} \\ \text{in} \end{pmatrix} - \begin{pmatrix} \text{mass} \\ \text{out} \end{pmatrix} = 0 \tag{3.150}$$

$$0 = (\rho)(v_z|_z \, \Delta x \, \Delta y) - (\rho)(v_z|_{z+\Delta z} \, \Delta x \, \Delta y) \tag{3.151}$$

$$0 = v_z|_z - v_z|_{z+\Delta z} \tag{3.152}$$

The result, $v_z|_z = v_z|_{z+\Delta z}$, allows us to conclude that the convective term in Equation 3.149 is zero. Note that if we divide Equation 3.152 by Δz and take the limit as Δz goes to zero, we obtain:

$$0 = v_z|_z - v_z|_{z+\Delta z} \tag{3.153}$$

$$0 = \lim_{\Delta z \to 0} \left[\frac{v_z|_{z+\Delta z} - v_z|_z}{\Delta z} \right] \tag{3.154}$$

mass balance result: $$0 = \frac{dv_z}{dz} \tag{3.155}$$

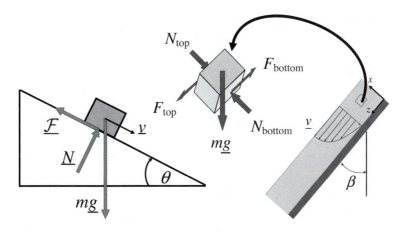

Figure 3.33   The molecular contact forces affecting our control volume are analogous to the contact forces on a sliding solid block: There are shear stresses due to friction and normal stresses that counteract the downward pull of gravity.

where in Equation 3.155 we used the definition of a derivative, Equation 3.39. In Chapter 6, we arrive at this same result through a different path.

Thus, from the mass balance and the assumption of steady state, two of the three terms in the Reynolds transport theorem (see Equation 3.144) are equal to zero for the chosen CV, and we are left with a simple force balance to solve:

$$0 = \sum_{\substack{\text{on} \\ \text{CV}}} \underline{f} \qquad (3.156)$$

To continue, we now must write the forces that act on the control volume, one of which is gravity. The force due to gravity on the mass in the CV is the acceleration due to gravity $\underline{g}$ (a vector) multiplied by the mass of the fluid in the CV:

$$\begin{pmatrix} \text{force due} \\ \text{to gravity} \\ \text{on CV} \end{pmatrix} = (M_{CV}) \, \underline{g} \qquad (3.157)$$

$$= \left(\frac{\text{mass}}{\text{volume}} \right) (\text{volume}) \, \underline{g} \qquad (3.158)$$

$$= \rho \, (\Delta x \Delta y \Delta z) \, \underline{g} \qquad (3.159)$$

$$= \rho \Delta x \Delta y \Delta z \begin{pmatrix} -g \sin \beta \\ 0 \\ g \cos \beta \end{pmatrix}_{xyz} \qquad (3.160)$$

What are the other forces on the CV? If we think of the sliding-block problem in Section 3.1, gravity was one of the forces; the other forces were the retarding force due to friction and the normal force from the incline surface, which supported the block.

Within our falling film, there should be forces of this type on our control volume as well (Figure 3.33). The sliding-friction force acts on the top $\underline{f}_{\text{top}}$ and bottom $\underline{f}_{\text{bottom}}$ of the CV and comes from the sliding of neighboring fluid particles. The normal force supporting the CV likewise acts on the top N_{top} and bottom N_{bottom} of the CV and again comes from the fluid particles that are in contact with it.

In the sliding-block example, when we needed to write expressions for these forces, we brought in observations from experiments—namely, Equation 3.18, $\underline{\mathcal{F}} = -\mu_k N \hat{e}_x$, which recorded for us the relationship between sliding friction and normal force for that sliding solid block. To continue with this falling-film example, we need to learn how frictional and normal forces act in liquids.

We find that we cannot complete this example at this time. We first must discuss intermolecular forces in liquids, which are the subject of Chapter 4. We return to this example after we investigate this subject well enough to fill in the required molecular forces. The momentum balance so far on the chosen microscopic control volume in the flow down an incline is summarized as follows:

$$0 = \sum_{\substack{\text{on} \\ \text{CV}}} \underline{f} \tag{3.161}$$

$$0 = \begin{pmatrix} \text{force due} \\ \text{to gravity} \\ \text{on CV} \end{pmatrix} + \begin{pmatrix} \text{molecular} \\ \text{sliding} \\ \text{surface forces} \\ \text{on CV} \end{pmatrix} + \begin{pmatrix} \text{molecular} \\ \text{normal} \\ \text{surface forces} \\ \text{on CV} \end{pmatrix} \tag{3.162}$$

$$0 = \rho \Delta x \Delta y \Delta z \begin{pmatrix} -g \sin \beta \\ 0 \\ g \cos \beta \end{pmatrix}_{xyz} + \begin{pmatrix} \text{molecular} \\ \text{sliding} \\ \text{surface forces} \\ \text{on CV} \end{pmatrix} + \begin{pmatrix} \text{molecular} \\ \text{normal} \\ \text{surface forces} \\ \text{on CV} \end{pmatrix} \tag{3.163}$$

3.2.3.2 MACROSCOPIC CONTROL-VOLUME PROBLEM

The previous example used a microscopic control volume (volume $= \Delta x \Delta y \Delta z$ at point x, y, z) because we sought to calculate the velocity at a point in the fluid. Because the velocity varies from place to place in that problem, we needed a small CV (infinitely small, in fact) so that a single value of the velocity or, at most, a minutely changing velocity is captured by the CV.

In some situations, we seek a more macroscopic engineering variable, in which case an infinitesimal control volume is not necessary. An example of a macroscopic-engineering variable is the total restraining force on a piece of equipment in which liquids flow. Following is such a calculation.

EXAMPLE 3.10. *What is the direction and magnitude of the force needed to support the 90-degree pipe bend shown in Figure 3.34? An incompressible (i.e., constant-density) liquid enters the pipe at volumetric flow rate Q_a and exits at volumetric flow rate Q_b. The cross-sectional area of the pipe bend is πR^2 throughout.*

SOLUTION. The flow is driven by external forces (i.e., the upstream pressure, perhaps provided by a pump) and is affected by another external force (i.e., gravity). The direction of velocity is different at different points in the flow

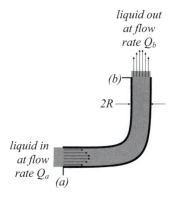

*liquid out
at flow
rate Q_b*

(b)

2R

*liquid in
at flow
rate Q_a*   *(a)*

Figure 3.34   A liquid flowing in a pipe bend of circular cross section exerts forces on the pipe. To calculate the forces on the pipe, we perform a momentum balance on a macroscopic control volume.

(i.e., inlet/outlet). We seek to calculate a net force, which depends on momentum exchanges within the system. This problem is the type for which we derived the Reynolds transport theorem.

Reynolds transport theorem
(momentum balance on CV)

$$\sum_{\substack{\text{on} \\ \text{CV}}} \underline{f} = \frac{d\mathbf{P}}{dt} + \iint_{\text{CS}} (\hat{n} \cdot \underline{v}) \, \rho \underline{v} \, dS \qquad (3.164)$$

The Reynolds transport theorem correctly applied allows us to calculate the net force on the bend.

We first examine this problem by imagining the situation. Fluid entering the bend travels horizontally to the right, but the shape of the pipe causes the flow direction to change. The fluid exits the pipe bend traveling vertically upward.

If the pipe bend were suspended by a light string and the fluid were directed into the bend by a high-capacity firehose, surely the string would break because there would be a large horizontal component to the force coming from the fluid, and nothing restrains the pipe in the horizontal direction. If we turned down the flow rate and stabilized the bend with our hands, we would have to exert a horizontal component of force to counterbalance the horizontal momentum of the incoming fluid; we also would have to exert a vertically upward force to keep the momentum of the upwardly flowing exiting fluid from breaking the supporting string and jamming the bend into the ground.

By first thinking about the problem, we conclude that there must be a restraining force on the pipe bend that is directed approximately as shown in Figure 3.35. Our task is to use the momentum balance to calculate the vector restoring force $\mathcal{R}$ that is needed to keep the pipe stationary.

To calculate the force on the bend, we perform a momentum balance using the Reynolds transport theorem, and we begin

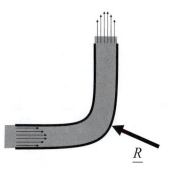

R

Figure 3.35   Using only our imagination and intuition, we reason that the solution to this problem must be a vector directed approximately as shown here.

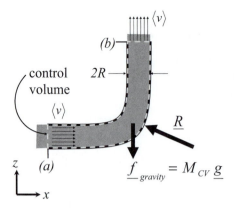

Figure 3.36   The force that flowing liquid exerts on a pipe bend can be calculated by performing a momentum balance on a macroscopic control volume such as the one outlined with a dotted line. The y-direction is into the page.

by choosing a coordinate system and a control volume. Unlike in the previous example of flow down an incline, in this problem, the CV need not be microscopic because here we do not seek to know the details of velocity distribution inside the pipe. We choose instead a CV that encloses all of the fluid inside the pipe (Figure 3.36). We seek to calculate the net force exerted by the inside walls of the pipe on the fluid in this CV.

To choose the coordinate system for our calculations, we again consider the quantities that we seek to calculate. A horizontal Cartesian coordinate system is a reasonable choice for this problem because it is easy to express the incoming and exiting velocities. Although the pipe is circular in cross section, a cylindrical coordinate system does not make the calculations easier because of the bend in the pipe. We choose the xyz coordinate system shown in Figure 3.36. The y-direction is into the page.[8]

Having chosen the control volume and the coordinate system, we proceed with writing the terms of the momentum balance as they apply to these choices:

Reynolds transport theorem (momentum balance on CV)
$$\frac{d\underline{\mathbf{P}}}{dt} + \iint_{CS} (\hat{n} \cdot \underline{v}) \rho \underline{v} \, dS = \sum_{\substack{\text{on} \\ \text{CV}}} \underline{f} \quad (3.165)$$

The flow is steady; therefore, the rate of change of momentum in the CV is zero, $d\underline{\mathbf{P}}/dt = 0$. The surface integral term gives the net momentum flow out of the CV due to convection. In our CV, momentum is convected in and out through the surfaces labeled (a) and (b) in Figure 3.36. The momentum convection term does not sum to zero for flow in a 90-degree bend because of the change in flow direction. We calculate this convective term by carefully evaluating the integral in Equation 3.165.

_____

[8] We use a cylindrical coordinate system later in this problem when calculating forces on the ends of the CV.

At the inlet and outlet surfaces, we write the two unit vectors $\hat{n}$ in our chosen coordinate system as:

$$\hat{n}|_a = \begin{pmatrix} -1 \\ 0 \\ 0 \end{pmatrix}_{xyz} \qquad \hat{n}|_b = \begin{pmatrix} 0 \\ 0 \\ 1 \end{pmatrix}_{xyz} \tag{3.166}$$

Recall that $\hat{n}$ is the outwardly pointing unit normal of the CV surfaces.

We can write versions of the velocity vectors at (a) and (b) in our coordinate system as well. At surface (a), the velocity points in the x-direction; at surface (b), the velocity points in the z-direction. Because our CV is macroscopic, we use average velocities to characterize the momentum flowing in and out of it. With this choice, we ignore the variations in velocity profile across the inlet and outlet surfaces.

$$\underline{v}|_a = \begin{pmatrix} \langle v \rangle|_a \\ 0 \\ 0 \end{pmatrix}_{xyz} \tag{3.167}$$

$$\underline{v}|_b = \begin{pmatrix} 0 \\ 0 \\ \langle v \rangle|_b \end{pmatrix}_{xyz} \tag{3.168}$$

$$\tag{3.169}$$

To relate inlet and outlet average velocities, we first perform a mass balance. For a constant-density system, we obtain:

$$\frac{dM_{CV}}{dt} = \begin{pmatrix} \text{rate of} \\ \text{mass flow} \\ \text{into the CV} \end{pmatrix} - \begin{pmatrix} \text{rate of} \\ \text{mass flow} \\ \text{out of the CV} \end{pmatrix} = 0 \tag{3.170}$$

$$0 = \rho Q_a - \rho Q_b \tag{3.171}$$

$$Q_a = Q_b \equiv Q \tag{3.172}$$

where M_{CV} is the mass of fluid in the CV. The cross-sectional area of the pipe πR^2 is the same for the entrance and the exit; because the flow rate also is constant, the magnitude of the average velocity at the entrance and at the exit is the same. For the inlet and exit surfaces, we therefore write:

$$\langle v \rangle|_a = \langle v \rangle|_b = \langle v \rangle \tag{3.173}$$

$$\underline{v}|_a = \begin{pmatrix} \langle v \rangle \\ 0 \\ 0 \end{pmatrix}_{xyz} \qquad \underline{v}|_b = \begin{pmatrix} 0 \\ 0 \\ \langle v \rangle \end{pmatrix}_{xyz} \tag{3.174}$$

With these expressions for $\underline{v}$ at the inlet and outlet surfaces, we now can calculate the surface integral in the momentum balance—Equation 3.165 (the convective term)—for our control volume. The surface integral may be broken into two parts—an integral over the inlet and an integral over the outlet:

$$\iint_{CS} (\hat{n} \cdot \underline{v}) \, \rho \underline{v} \, dS = \iint_{S_a} (\hat{n} \cdot \underline{v}) \, \rho \underline{v} \, dS + \iint_{S_b} (\hat{n} \cdot \underline{v}) \, \rho \underline{v} \, dS \tag{3.175}$$

Because $\hat{n}$, $\underline{v}$, and ρ are constant across both the inlet and outlet surfaces, we can remove them from the integrals:

$$\iint_{CS} (\hat{n} \cdot \underline{v}) \rho \underline{v} \, dS = (\hat{n} \cdot \underline{v})|_a \, \rho \, \underline{v}|_a \iint_{S_a} dS + (\hat{n} \cdot \underline{v})|_b \, \rho \, \underline{v}|_b \iint_{S_b} dS \quad (3.176)$$

Both surface integrals now are evaluated and give the tube cross-sectional area πR^2. We also know the various vectors $\hat{n}$ and $\underline{v}$ at the inlet and outlet; thus, we can proceed to the final answer for this term:

$$\iint_{CS} (\hat{n} \cdot \underline{v}) \rho \underline{v} \, dS = (\hat{n} \cdot \underline{v})|_a \, \rho \, \underline{v}|_a \, \pi R^2 + (\hat{n} \cdot \underline{v})|_b \, \rho \, \underline{v}|_b \, \pi R^2 \quad (3.177)$$

$$= \rho \pi R^2 \left[\begin{pmatrix} -1 \\ 0 \\ 0 \end{pmatrix}_{xyz} \cdot \begin{pmatrix} \langle v \rangle \\ 0 \\ 0 \end{pmatrix}_{xyz} \right] \begin{pmatrix} \langle v \rangle \\ 0 \\ 0 \end{pmatrix}_{xyz}$$

$$+ \rho \pi R^2 \left[\begin{pmatrix} 0 \\ 0 \\ 1 \end{pmatrix}_{xyz} \cdot \begin{pmatrix} 0 \\ 0 \\ \langle v \rangle \end{pmatrix}_{xyz} \right] \begin{pmatrix} 0 \\ 0 \\ \langle v \rangle \end{pmatrix}_{xyz} \quad (3.178)$$

$$= \langle v \rangle^2 \rho \pi R^2 \begin{pmatrix} -1 \\ 0 \\ 1 \end{pmatrix}_{xyz} \quad (3.179)$$

To assess our progress, we substitute the convective term and the fact that $d\underline{P}/dt = 0$ into the momentum balance (i.e., the Reynolds transport theorem):

$$\frac{d\underline{P}}{dt} = \iint_{CS} -(\hat{n} \cdot \underline{v}) \rho \underline{v} \, dS + \sum_{\substack{\text{on} \\ \text{CV}}} \underline{f} \quad (3.180)$$

$$0 = -\langle v \rangle^2 \rho \pi R^2 \begin{pmatrix} -1 \\ 0 \\ 1 \end{pmatrix}_{xyz} + \sum_{\substack{\text{on} \\ \text{CV}}} \underline{f} \quad (3.181)$$

All that remains is to write the forces on the control volume and solve for the desired restraining-force vector.

One of the forces on the CV is gravity, which is the CV mass multiplied by the acceleration due to gravity. For our chosen coordinate system, this becomes:

$$\begin{matrix} \text{Force on CV} \\ \text{due to gravity} \end{matrix} = M_{CV} \begin{pmatrix} 0 \\ 0 \\ -g \end{pmatrix}_{xyz} \quad (3.182)$$

where M_{CV} is the mass of fluid in the CV.

A second force on the control volume is the force exerted on the fluid by the walls of the pipe. To sort out what these are, we can do a "thought experiment." If the pipe were straight and the water were directed through the straight pipe without touching the walls, there would be no force between the fluid and the walls. Because the fluid touches the walls, however, there is a molecular contact force between the fluid and the walls. The force exerted by the walls on the fluid is the negative of the force exerted by the fluid on the walls:

$$\begin{pmatrix} \text{force on CV} \\ \text{due to contact} \\ \text{with pipe walls} \end{pmatrix} = -\begin{pmatrix} \text{force } on \text{ walls} \\ \text{due to contact} \\ \text{with fluid} \end{pmatrix} = \begin{pmatrix} \text{force } by \text{ walls} \\ \text{due to contact} \\ \text{with fluid} \end{pmatrix} \equiv \begin{pmatrix} \mathcal{R}_x \\ \mathcal{R}_y \\ \mathcal{R}_z \end{pmatrix}_{xyz}$$

(3.183)

The vector $\underline{\mathcal{R}}$ is the restoring force we seek.

The contact forces between the walls and the fluid are accounted for but the walls of the pipe are not the only boundaries of the CV: The surfaces at (a) and (b) also are bounding surfaces. Will there be molecular contact forces on these surfaces? The fluid upstream of the CV is moving at volumetric flow rate Q and is pushing the fluid ahead of it. This pushing force is a molecular contact force on surface (a). Likewise, the fluid in the CV immediately inside surface (b) is pushing on the fluid outside of the CV. The molecular forces on the CV include the forces on (a) and (b), as well as the force on the walls $\underline{\mathcal{R}}$:

$$\begin{pmatrix} \text{molecular} \\ \text{force} \\ \text{on CV} \end{pmatrix} = \begin{pmatrix} \text{force on CV} \\ \text{due to contact} \\ \text{with walls} \end{pmatrix} + \begin{pmatrix} \text{molecular} \\ \text{force on CV} \\ \text{at (a)} \end{pmatrix} + \begin{pmatrix} \text{molecular} \\ \text{force on CV} \\ \text{at (b)} \end{pmatrix}$$

(3.184)

$$= \begin{pmatrix} \mathcal{R}_x \\ \mathcal{R}_y \\ \mathcal{R}_z \end{pmatrix}_{xyz} + \begin{pmatrix} \text{molecular} \\ \text{force on CV} \\ \text{at (a)} \end{pmatrix} + \begin{pmatrix} \text{molecular} \\ \text{force on CV} \\ \text{at (b)} \end{pmatrix}$$ (3.185)

The molecular forces on surfaces (a) and (b) are due to the forces between the molecules in the flowing liquid. If the liquid is a simple one, such as water, these forces are straightforward (see Chapter 4). If the liquid is complex, such as a high-molecular-weight polymer, the intermolecular forces are complicated. In either case, to write an expression for these forces, we need to know how intermolecular forces for various types of liquids can be accounted for in the continuum model.

As in the previous example, we must postpone a solution because we do not yet know how to handle the molecular contact forces in fluids. We return to finish this problem after we have the proper tools. Following is the momentum balance thus far for the flow of water in a 90-degree pipe bend:

$$0 = -\langle v \rangle^2 \rho \pi R^2 \begin{pmatrix} -1 \\ 0 \\ 1 \end{pmatrix}_{xyz} + \sum_{\substack{\text{on} \\ \text{CV}}} \underline{f}$$ (3.186)

$$0 = \langle v \rangle^2 \rho \pi R^2 \begin{pmatrix} 1 \\ 0 \\ -1 \end{pmatrix}_{xyz} + M_{CV} \begin{pmatrix} 0 \\ 0 \\ -g \end{pmatrix}_{xyz} + \begin{pmatrix} \mathcal{R}_x \\ \mathcal{R}_y \\ \mathcal{R}_z \end{pmatrix}_{xyz}$$

$$+ \begin{pmatrix} \text{molecular} \\ \text{force on CV} \\ \text{at (a)} \end{pmatrix} + \begin{pmatrix} \text{molecular} \\ \text{force on CV} \\ \text{at (b)} \end{pmatrix}$$ (3.187)

3.3 Summary

In this chapter, we take the first steps toward developing a problem-solving method for two types of flow problems: microscopic and macroscopic. We now summarize our progress.

The continuum model is a way of viewing fluids using a set of continuous functions to keep track of fluid behavior, ignoring molecular details. The continuous functions of fluid mechanics include the density field, the velocity field, and the molecular-stress field, which is discussed in Chapter 4. Calculus is the mathematics of continuous functions and we use it extensively to make our calculations of fluid motion and fluid forces.

Fluid motion is governed by mass, momentum, and energy balances. We choose to use balances on control volumes instead of on individual bodies. The control-volume method is more convenient to use in fluid mechanics because fluids are not individual rigid bodies like those with which we deal in introductory physics and mechanics courses. The control-volume method is well suited for use with the continuum picture, as shown in the final two examples in this chapter. We continue study of these two problems in Chapters 4 and 5 and consider more problems of this type in Chapters 7–10.

The appropriate momentum balance to use with a control volume is given by the Reynolds transport theorem:

$$\text{Reynolds transport theorem} \atop \text{(momentum balance on CV)} \qquad \frac{d\mathbf{P}}{dt} + \iint_{CS} (\hat{n} \cdot \underline{v})\, \rho \underline{v}\, dS = \sum_{\substack{\text{on} \\ \text{CV}}} \underline{f} \quad (3.188)$$

Recall that $\hat{n}$ is the outwardly pointing normal to the CV enclosing surface CS; thus, the integral in Equation 3.188 is net outflow of momentum from the CV.

To apply the Reynolds transport theorem to a problem, we must be able to identify the forces that are acting on the CV, including molecular forces. In this chapter, we discuss one force—gravity—that acts on a CV. Chapter 4 introduces molecular stress, the source of a second significant force that acts on a CV. In Chapter 5, we discuss the link between molecular stress and fluid motion. When these topics have been covered, we can complete our flow calculations on the inclined plane and the 90-degree bend, and we will be ready to tackle a wide variety of problems in fluid mechanics.

3.4 Problems

1. What is a control volume? Why does the field of fluid mechanics introduce this concept?
2. What is a fluid particle? How big is a fluid particle?
3. How is the concept of a continuum different from your understanding of matter from chemistry studies?
4. What is meant by the term *velocity field*? What other "fields" are there in fluid mechanics and physics?

Table 3.2. Data of $y(t)$ for Problem 11

t (s)	y (m/s)
3.6	3.9
4.0	12.1
4.4	22.7
4.8	33.0
5.2	41.9
5.6	49.7
6.0	55.6
6.4	61.0
6.8	65.5
7.4	71.0
8.2	76.7
9.2	82.0
10.4	86.2
12.0	90.2
13.4	92.7
15.4	94.6
17.4	95.9
19.4	97.1
21.4	98.0
22.8	98.4
24.4	98.5
26.0	98.7
27.6	99.0
29.4	100.0
31.2	100.2
33.6	100.7

5. What are the principal forces that cause flow?

6. What is Newton's second law $\sum \underline{f} = m\underline{a}$ when written on a control volume V with bounding surfaces CS?

7. We derived the Reynolds transport theorem for the momentum balance. What is it for the mass balance?

8. Why are we unable to use the momentum balance $\sum \underline{f} = m\underline{a}$ (i.e., Newton's second law) directly in fluid-flow calculations?

9. What is the difference between the rate of change of momentum terms $\frac{d(m\underline{v})}{dt}$ and $\frac{d\underline{P}}{dt}$ in Newton's second law (Equation 3.52) and the Reynolds transport theorem (Equation 3.135)?

10. In Equation 3.126 in the development of the convective term of the momentum balance, an indeterminate vector product $(\underline{v}\,\underline{v})$ appears. How did that expression come to include a dyadic product? What is the meaning of the tensor $\rho\underline{v}\,\underline{v}$?

11. For the data given in Table 3.2 (i.e., arbitrary time-dependent quantity y), find a function $y(t)$ that fits the data well. What is your estimate of $y\,(7.0)$?

12. For the experimental data given in Table 3.3 (i.e., viscosity of aqueous sugar solutions as a function of concentration), find a function $\mu(c)$ that fits the data well. What is your estimate of μ (28.2 wt%) and μ (50.0 wt%)?

Table 3.3. Experimental data of viscosity as a function of concentration $\mu(c)$ of aqueous sugar solutions for Problem 12

c (wt% sugar)	μ (cp)
10	0.62
10	0.87
10	0.88
10	0.89
20	1.0
20	1.2
20	1.2
20	1.2
20	1.2
20	1.3
30	2.0
30	2.1
30	2.1
30	2.3
30	3.0
40	3.8
40	4.3
40	4.3
40	4.4
40	4.6
45	5.2
45	5.3
45	5.3
50	8.4
50	9.3
50	9.5
50	9.7
50	14
60	28
60	30
60	30
60	32
65	63
65	64
65	65
65	69

13. For the experimental data given in Table 3.4 (i.e., pumping head as a function of volumetric flow rate [102]), find a function $H_{pump}(Q)$ that fits the data well. How much head does the pump develop at 2.2 gpm?

14. A uniform flow $\underline{v} = U\hat{e}_z$ of an incompressible fluid of density ρ passes through a volume that is in the shape of a half sphere of radius R. The outwardly pointing unit normal of the flat surface of the half sphere is $\hat{n} = -\hat{e}_z$. What is the mass flow rate through the hemispherical surface of this volume? Show that you can obtain the correct answer by integrating the formal expression for Q (Equation 3.87).

15. What is the flow of momentum through the hemispherical surface described in Problem 14?

Table 3.4. Experimental data of pumping head as a function of volumetric flow rate $H_{pump}(Q)$ for a laboratory pump [102] (Problem 13)

Q gpm	Head ft
0.88	72.5
1.00	68.1
1.38	70.5
1.87	67.2
1.99	70.4
2.37	63.6
2.86	58.9
3.23	57.3
3.36	52.7
3.85	46.2

16. A uniform flow $\underline{v} = U\hat{e}_x$ of an incompressible fluid of density ρ passes through a volume that is in the shape of a block (i.e., rectangular parallelepiped). The sides of the block are lengths $a < b < c$. The unit normal to the cb surface is $\hat{n} = (\hat{e}_x - \hat{e}_y)/\sqrt{2}$. What is the mass flow rate through the cb surface? What is the momentum flow rate through the cb surface?

17. For the volume described in Problem 16, what are the unit normals to the other two surfaces?

18. For the volume described in Problem 16, what is the mass flow rate through the ac surface? What is the momentum flow rate through the ac surface?

19. For the function $f(x)$ given here, what is the average value $\langle f \rangle$ of the function between $x = 0$ and $x = 2$?

$$f(x) = 2x^2 + 3$$

20. For the velocity-profile function $v_y(x)$ given here (equation uses Cartesian coordinates xyz, $\underline{v} = v_y\hat{e}_y$), what is the average value $\langle v_y \rangle$ of the function between $x = 0$ and $x = 2$? The units of velocity are m/s and the units of x are m.

$$v_y(x) = 3\left(\frac{x}{6}\right)^2 + 1.5$$

21. The y-component of a velocity field in flow through a slit (equation uses Cartesian coordinates) is given here. What is the average value of the velocity? $2H$ is the gap between the plates. At what location is the velocity a maximum? The units of velocity and A are m/s and the units of x and H are m.

$$\underline{v} = \begin{pmatrix} 0 \\ v_y(x) \\ 0 \end{pmatrix}_{xyz}$$

$$v_y(x) = A\left(1 - \frac{(x-H)^2}{H^2}\right)$$

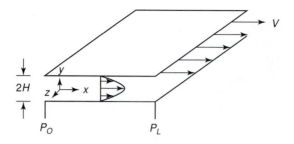

Figure 3.37   Pressure-driven flow (i.e., Poiseuille flow) through a slit with a superimposed drag flow due to the motion of the top plate (Problem 24).

22. The z-component of a velocity field in flow through a tube (given in cylindrical coordinates) is shown here. What is the average value of the velocity? R is the radius of the tube. The units of velocity and A are m/s and the units of r and R are m.

$$\underline{v} = \begin{pmatrix} 0 \\ 0 \\ v_z(r) \end{pmatrix}_{r\theta z}$$

$$v_z = A\left(1 - \frac{r^2}{R^2}\right)$$

23. For the velocity-profile function $v_z(r)$ given here, what is the average value $\langle v_z \rangle$ of the function between $r = 5$ and $r = 10$? The function is written in cylindrical coordinates. The units of velocity are m/s and the units of r are m.

$$v_z(r) = 8\ln\left(\frac{r}{3}\right)$$

24. The x-component of a velocity field is given here (expressed in Cartesian coordinates). This velocity profile results from pressure-driven flow through a slit with the top wall moving at velocity V (Figure 3.37). What is the average value of the velocity? $2H$ is the gap between the plates, a pressure gradient $\Delta P/L$ is imposed, and the fluid viscosity is μ. At what location is the velocity a maximum?

$$\underline{v} = \begin{pmatrix} v_x(y) \\ 0 \\ 0 \end{pmatrix}_{xyz}$$

$$v_x(y) = \left(\frac{H^2(\Delta P)}{2\mu L}\right)\left(1 - \frac{y^2}{H^2}\right) + \frac{V}{2}\left(1 + \frac{y}{H}\right)$$

25. What is the wetted surface area of water flowing in a tube? Show that you can obtain the answer by performing an integration in cylindrical coordinates.

26. What is the wetted surface area of a sphere dropping in a fluid? Show that you can obtain the answer by integrating an appropriate quantity.

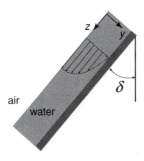

Figure 3.38   Flow coordinate system for Problem 31.

27. What is the wetted surface area of an open, semicircular channel (i.e., half pipe) of length L and pipe radius R, in which the fluid height in the center is h. Show that you can obtain the answer by integrating an appropriate quantity.

28. For a pipe that is only 80 percent full (i.e., occupied volume = 80 percent of the total pipe volume), what is the wetted surface area? The pipe is of length L and radius R.

29. For the two vectors given here, what is $|\underline{w}|$? What is $|\underline{v}|$? What is $(\underline{w} \cdot \underline{v})$? What is the angle between the two vectors?

$$\underline{w} = \begin{pmatrix} 1 \\ 1 \\ \sqrt{2} \end{pmatrix}_{123} \qquad \underline{v} = \begin{pmatrix} 1 \\ 6 \\ 3 \end{pmatrix}_{123}$$

30. For the two vectors given here, what is $|\underline{w}|$? What is $|\underline{v}|$? What is $(\underline{w} \cdot \underline{v})$? What is the angle between the two vectors? Note that the two vectors are not written relative to the same coordinate system.

$$\underline{w} = \begin{pmatrix} 1 \\ 1 \\ 0 \end{pmatrix}_{r\theta z \, | \, r=1, \theta=\pi, \phi=0} \qquad \underline{v} = \begin{pmatrix} 1 \\ 6 \\ 3 \end{pmatrix}_{123}$$

31. For the Cartesian coordinate system shown in Figure 3.38, what is a unit vector in the direction of gravity? What is the component of gravity in the flow direction?

32. For the cylindrical coordinate system shown in Figure 3.39 for the axial flow in a wire-coating operation, what is a unit vector in the direction of gravity? What is the component of gravity in the flow direction?

33. For the horizontal flow around a sphere in a wind tunnel, the top view of the geometry is shown in Figure 3.40. Relative to the spherical coordinate system shown, what is a unit vector in the direction of gravity? What is the component of gravity in the flow direction?

34. For a particular problem, the control volume is chosen to be a rectangular parallelepiped of dimensions length L, width W, and height H. What is the total surface area of the control volume? What is the volume of the control volume? Choose a coordinate system and write formal surface integrals over the surfaces and verify your answer for total surface area. Write a formal volume integral and verify your answer for volume.

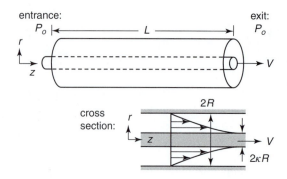

Figure 3.39　Axial annular flow that occurs in wire coating (Problem 32).

35. For a particular problem, the control volume is chosen to be a right-circular cylinder of radius R and height H. What is the total surface area of the control volume? What is the volume of the control volume? Choose a coordinate system and write formal surface integrals over the surfaces and verify your answer for total surface area. Write a formal volume integral and verify your answer for volume.

36. For a particular problem, the control volume is chosen to be a cone of height H and widest radius R. What is the total surface area of the control volume? What is the volume of the control volume? Choose a coordinate system and write formal surface integrals over the surfaces and verify your answer for total surface area. Write a formal volume integral and verify your answer for volume.

37. For a particular flow problem, the control volume is chosen to be a rectangular parallelepiped with dimensions of length L, width W, and height H. The Cartesian coordinate system chosen is located at one corner of the control volume ($0 \leq x \leq L, 0 \leq y \leq W, 0 \leq z \leq H$). For each enclosing control

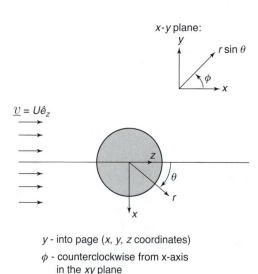

Figure 3.40　Flow around a sphere in a wind tunnel (Problem 33).

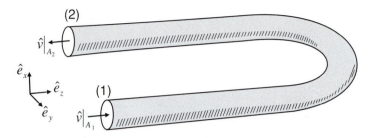

Figure 3.41   When fluid flows in a U-shaped tube, the momentum changes direction and forces are required to restrain the tube (Problems 41 and 44).

surface of this control volume, what are the outwardly pointing unit normal vectors $\hat{n}$ for each control surface? For a uniform flow $\underline{v} = U_o(\hat{e}_x + \hat{e}_y)$ through the control volume, what is the mass flow rate through each control surface? The fluid has constant density ρ and U_0 is constant.

38. For a particular flow problem, the control volume is chosen to be a vertical-right circular cylinder of radius R and height H. Choose a cylindrical coordinate system for flow down the cylindrical axis of this control volume. For each enclosing control surface of the control volume, write the unit vectors that are normal to each control surface. For a uniform flow $\underline{v} = U\hat{e}_z$ through the control volume (U is constant), what is the flow rate through each control surface? The fluid has variable density ρ. For a flow $\underline{v} = \left(U\frac{1}{r}\right)\hat{e}_r$ through the control volume, what is the mass flow rate through each control surface?

39. For a particular flow problem, the control volume is chosen to be a truncated cone of height H, bottom widest radius R_1, and top smaller radius R_2. The cone is truncated a distance l from the tip and the cone angle is $\theta = \alpha$, where θ is the coordinate variable for a spherical coordinate system with origin at the core tip. For each enclosing control surface, write the unit vectors that are normal to each control surface. For a uniform flow $\underline{v} = -U\hat{e}_z$ down the axis of the control volume, what is the mass flow rate through each control surface? The fluid has constant density ρ and the flow first passes through the bottom of the control volume.

40. An incompressible fluid (i.e., density is constant) enters a rectangular duct flowing at a steady flow rate of Q gpm. The width of the duct is W, the height of the duct is H, and the length of the duct is L. What is the average velocity of fluid entering the duct in terms of these variables? What is the average velocity of fluid exiting the duct?

41. An incompressible fluid (i.e., density is constant) enters a U-shaped conduit flowing at a steady flow rate of Q gpm (Figure 3.41). The conduit has a circular cross section all along its length and the radius of the conduit is R. What is the average velocity of fluid entering the conduit in terms of these variables? What is the average velocity of fluid exiting the conduit?

42. An incompressible fluid enters a converging bend flowing at a steady flow rate of Q gpm (Figure 3.42). The bend makes a 20-degree turn and has a circular cross section all along its length. At the inlet to the bend, the radius of the conduit is R_1; at the exit, the radius is a smaller value, R_2. What is

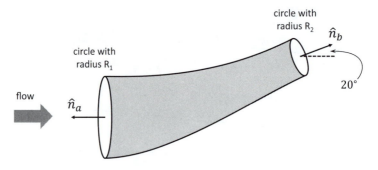

circle with
radius R_2

$\hat{n}_b$

circle with
radius R_1

flow

$\hat{n}_a$

20°

Figure 3.42   Schematic of a converging fitting (Problems 42 and 47).

the average velocity of fluid entering the conduit in terms of these variables? What is the average velocity of fluid exiting the conduit?

43. An incompressible fluid enters a horizontal, diverging conduit flowing at a steady flow rate of Q gpm. The conduit has a circular cross section all along its length. At the inlet, the radius of the conduit is R_1; at the exit, the radius is a larger value, R_2. What is the average velocity of fluid entering the conduit in terms of these variables? What is the average velocity of fluid exiting the conduit?

44. In this chapter, we introduced the Reynolds transport theorem:

$$\text{Reynolds transport theorem} \atop \text{(momentum balance on CV)} \quad \sum_{\substack{\text{on} \\ \text{CV}}} \underline{f} = \frac{d\mathbf{P}}{dt} + \iint_S (\hat{n} \cdot \underline{v}) \, \rho \underline{v} \, dS$$

The convective term is the integral in the Reynolds transport theorem, and this term accounts for the net loss of momentum from the control volume through its bounding surfaces. Consider two cases of flow with an average inlet velocity of $\langle v \rangle$: (a) steady flow through a straight tube of radius R, and (b) steady flow through a U-shaped tube of radius R (see Figure 3.41). For Case (a), the convective term is zero; for Case (b), the convective term is not zero. Perform each calculation and explain the results.

45. In Equation 3.181 for the problem of flow in a right-angle bend, the convective term of the macroscopic momentum balance is not equal to zero, even though an equal magnitude of momentum enters and exits the control volume. Explain why this is so.

46. Evaluate the convective term of the Reynolds transport theorem for the 162-degree bend-reducing fitting shown in Figure 3.43. The flow is into the wider cross section.

47. Evaluate the convective term of the Reynolds transport theorem for the 20-degree bend-reducing fitting shown in Figure 3.42.

48. Set up the problem of steady flow of a Newtonian fluid down an inclined plane using a Cartesian coordinate system in which gravity is in the $(-z)$-direction.

49. Set up the problem of steady flow of a Newtonian fluid through a right-angle bend using a cylindrical coordinate system with the z-direction as the inlet flow direction. What is the velocity vector like at the exit for this

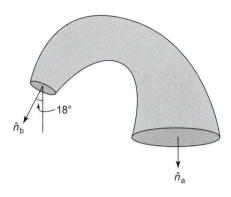

Figure 3.43   Schematic of a reducing fitting (Problem 46).

chosen coordinate system? What is the gravity vector? Comment on your observations.

50. The definition of a derivative is given in Chapter 1 (see Equation 1.138):

$$\frac{df}{dx} \equiv \lim_{\Delta x \to 0} \left[\frac{f(x + \Delta x) - f(x)}{\Delta x} \right]$$

What is the derivative (df/dx) of $f(x) = x^2$? Formally verify your answer by plugging in $f(x)$ and $f(x + \Delta x)$ into the definition and carrying out the limit.

4 Molecular Fluid Stresses

In our presentation thus far, we are seeking an effective way to model the flow patterns and forces associated with fluid motion. Our picture of a fluid is a mathematical continuum described by a set of field variables, including density $\rho(x, y, z, t)$ and velocity $\underline{v}(x, y, z, t)$, that capture the fluid's motion. We mentioned but have not explained yet the field variable fluid molecular stress $\underline{\underline{\tilde{\Pi}}}$ or $\underline{\underline{\tilde{\tau}}}(x, y, z, t)$, which describes molecular surface forces in a fluid.

In Chapter 3, we introduced the control volume (CV), a fixed region in space through which fluid particles move and on which we perform balances. Using CVs in fluids calculations frees us from having to follow individual particles from place to place. The momentum balance (i.e., Newton's second law), written with respect to a control volume, is given by the Reynolds transport theorem:

$$\text{Reynolds transport theorem (momentum balance on CV)} \qquad \boxed{\frac{d\mathbf{P}}{dt} + \iint_{CS} (\hat{n} \cdot \underline{v})\, \rho \underline{v}\, dS = \sum_{\substack{\text{on} \\ \text{CV}}} \underline{f}} \qquad (4.1)$$

where $d\mathbf{P}/dt$ is the rate of increase of momentum in the CV, the integral represents the net flow of momentum out of the CV, and $\sum \underline{f}$ is a sum of the forces on the CV. The Reynolds transport theorem requires that we write an expression for the forces on a control volume, including molecular forces; how to do this is the topic of this chapter.

There are two types of forces on a control volume in a fluid: noncontact forces such as gravity; and contact forces, which in fluids arise from molecular forces. This chapter discusses the fact that molecular force in liquids is quantified with stress, force per area; and that molecular stress is divided into two types, *isotropic* and *anisotropic*. Isotropic molecular stress, or pressure, figures into flow and also is important in static-fluid applications such as in manometers (see Section 4.2.4.1) and hydraulic lifts (see Section 4.2.4.2). Anisotropic molecular stress, which includes shear stress, is present only when fluids are in motion.

The stress tensor is a field variable used to write the molecular forces on a control volume. The stress-tensor concept makes molecular stresses easier to handle. Although it is a complicated subject, we do not need to understand all of the details of tensor mathematics to use the stress-tensor components in our study of fluids in motion. In practical calculations, fluid forces may be obtained from the stress tensor by using 3×3 matrix operations.

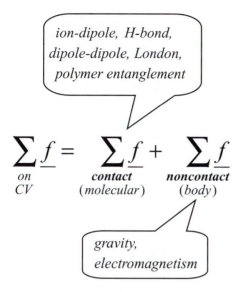

$$\sum_{\substack{on \\ CV}} f = \sum_{\substack{contact \\ (molecular)}} f + \sum_{\substack{noncontact \\ (body)}} f$$

Figure 4.1 Two types of forces act: contact forces and noncontact forces. The contact forces in a fluid are molecular forces due to the nature of the chemicals that comprise the fluid. Depending on the material that is flowing, contact forces include dipole-dipole, ion-dipole, London, hydrogen-bond, and entanglement forces. Noncontact forces include gravity and electromagnetism.

In this chapter, we discuss the origin of molecular forces on a control volume, describe stress in static fluids, and introduce the stress tensor $\tilde{\underline{\underline{\Pi}}}$ for stresses in static and moving fluids. At the conclusion of this chapter, we return to the two examples in Chapter 3: flow down an incline (i.e., microscopic balance) and forces on a right-angle bend (i.e., macroscopic balance) and incorporate the stress components into those solutions.

In Chapter 5, we discuss the connection between stress $\tilde{\underline{\underline{\Pi}}}$ and fluid velocity $\underline{v}$. The relationship between $\tilde{\underline{\underline{\Pi}}}$ and $\underline{v}$ is called the *stress constitutive equation*, which is the final piece of information needed to model flows with the continuum model and control volumes. The stress constitutive equation of Chapter 5 also is the final piece of physics needed to complete the two example calculations.

4.1 Forces on a control volume

The general momentum-balance equation derived in Chapter 3, the Reynolds transport theorem, requires expressions for all of the forces f acting on a control volume. To use the Reynolds transport theorem to solve for flow patterns and to solve for other flow properties of interest, we must determine which forces act on the CV.

Fundamentally, there are two types of forces in nature: contact forces and noncontact or body forces [167] (Figure 4.1):

$$\sum_{\substack{on \\ CV}} f = \sum_{\substack{on \\ CV}} f \bigg|_{contact} + \sum_{\substack{on \\ CV}} f \bigg|_{noncontact} \tag{4.2}$$

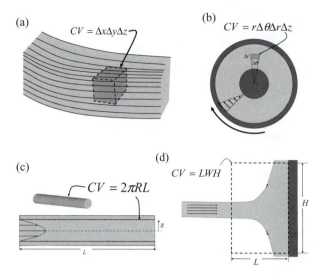

Figure 4.2   Control volumes are arbitrary volumes; thus, we choose them for convenience. Control-volume boundaries can be fluid-fluid (a, b) or fluid-solid (parts of c, d). When calculating velocity fields, microscopic control volumes embedded within the flow (a, b) allow us to calculate the velocity field. Macroscopic control volumes (c, d) are convenient when calculating, for example, the total force on the walls of a piece of process equipment.

Gravity and electromagnetism are familiar noncontact forces. Gravity is an important force in fluid mechanics, and it is straightforward to express the effect of gravity on a CV through Newton's law—that is, as mass multiplied by the acceleration due to gravity:

$$\sum \underline{f}\Bigg|_{\substack{\text{on} \\ \text{CV}}}^{noncontact} = \begin{pmatrix} \text{force due to} \\ \text{gravity} \\ \text{on CV} \end{pmatrix} = \begin{pmatrix} \text{mass of fluid} \\ \text{in CV} \end{pmatrix} \underline{g} \qquad (4.3)$$

$$= M_{CV}\, \underline{g} \qquad (4.4)$$

where M_{CV} is the mass of fluid in the CV and $\underline{g}$ is the acceleration due to gravity. If electromagnetic forces are important in a flow (e.g., in a conducting liquid; see Section 2.9), the electromagnetic force on the CV can be written analogously [35]. Electromagnetic forces need be considered only in specialized applications that involve magnetic fields and conductive fluids; we omit further discussion of these types of forces.

Contact forces on a chosen CV act through the control surface (CS) that bounds the CV. To identify contact forces on a CV, we must choose one (example CVs are shown in Figure 4.2) and then ask what touches the surfaces of the CV.

The forces that act on a CV surface in a flow are intermolecular forces–either the forces between molecules in the fluid or those between molecules in the fluid and molecules in a solid, such as a wall. To model these forces in detail, we must specify which molecules are present in our system. For example, if the fluid is polar (e.g., water and ethanol are both polar) (Figure 4.3), dipole-dipole attractions and perhaps hydrogen bonding contribute to forces on the control surface. In a nonpolar oil or a polymer melt, electrostatic attractions are

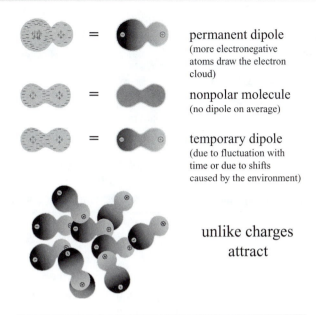

permanent dipole
(more electronegative
atoms draw the electron
cloud)

nonpolar molecule
(no dipole on average)

temporary dipole
(due to fluctuation with
time or due to shifts
caused by the environment)

unlike charges
attract

An extreme case of charge polarization:

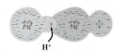

hydrogen bond
(strong permanent dipole
donates proton to make a
bridge between molecules)

Figure 4.3 Fluids are held together by a variety of forces, which are responsible for flow behavior. Most intermolecular forces are electrostatic in nature, a result of the attraction between positive and negative charges between molecules. Polar molecules have permanent dipoles that attract one molecule to another. A nonpolar molecule exhibits a temporary dipole when its electron cloud shifts due to, for example, the presence of a nearby positive or negative charge. London-dispersion forces are those intermolecular forces that result from the instantaneous temporary dipoles formed in all molecules due to the continuous motion of the electron cloud. One of the strongest intermolecular forces is associated with the hydrogen bond, in which a proton (i.e., positive charge) is donated from one atom to another and the resulting dipole is strongly polarized.

not of concern; however, other forces contribute (e.g., London dispersion forces or polymer entanglement forces; Figure 4.4). Different fluids are affected by different intermolecular forces (Table 4.1):

$$\sum f \Big|_{\substack{\text{on} \\ \text{CV}}}^{contact} = \begin{pmatrix} \text{force on CV} \\ \text{due to} \\ \text{dipole-dipole} \\ \text{interactions} \end{pmatrix} + \begin{pmatrix} \text{force on CV} \\ \text{due to} \\ \text{London-dispersion} \\ \text{interactions} \end{pmatrix} + \begin{pmatrix} \text{force on CV} \\ \text{due to} \\ \text{ion-dipole} \\ \text{interactions} \end{pmatrix}$$

$$+ \begin{pmatrix} \text{force on CV} \\ \text{due to} \\ \text{H-bond} \\ \text{interactions} \end{pmatrix} + \begin{pmatrix} \text{force on CV} \\ \text{due to} \\ \text{entanglement} \end{pmatrix} \tag{4.5}$$

Thus, writing an expression for molecular contact forces on a control volume is a complex problem. We need to know which molecules are located at the control surface and which types of intermolecular forces (i.e., dipole-dipole, hydrogen

Table 4.1. Liquids held together by intermolecular forces [95]

Type of force	Strength of force, kJ/mole	Present between
Dipole-dipole	3–4	Polar molecules (e.g., acetone and glycerol)
London dispersion	1–10	All molecules; depends on polarizability of the electron cloud
Ion-dipole	10–50	Ions and polar solvents (e.g., salt solutions)
Hydrogen bond	10–40	O-H, N-H, and F-H bonds (e.g., water, ammonia, and strands of DNA)
Polymer entanglement	Unknown	Polymer chains over a critical molecular weight

Note: Different forces are more important for different types of fluids, and the strength of the forces depends on temperature. As temperatures rise, intermolecular forces lessen and liquids evaporate and form gases.

bonding, or other) are important for those molecules. Furthermore, we must write this information for every portion of the bounding surface of the control volume so that we can evaluate the summation in Equation 4.1.

The situation described here is complex, and at this point in our modeling it would be convenient if we did not have to be so specific about molecular behavior and mechanisms. Fortunately, in the continuum approach, it is possible to sidestep the details of molecular structure and interactions and still account for molecular forces on a control volume. The continuum approach views the fluid as a field characterized by position-dependent density and velocity functions, ignoring the existence of individual molecules. Consistent with this approach, we now seek a method that allows us to quantify molecular forces without addressing molecular details. We seek a field variable that can capture the effect of molecular forces

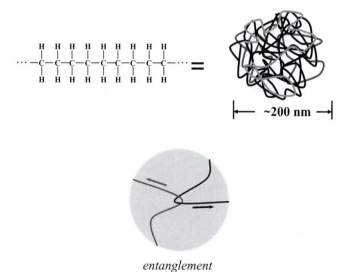

entanglement

Figure 4.4 Polymers and other large molecules experience intermolecular forces such as entanglement forces that are due to the physical size and complicated shape of molecules. Entanglement can hold polymeric liquids together even when they are subjected to strong forces pulling them apart.

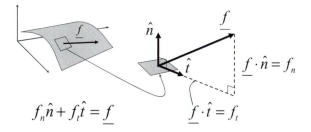

Figure 4.5 The forces at a location on an arbitrary surface include both normal forces and tangential or shear forces.

acting on control surfaces in a fluid. The position-dependent field variable that makes this possible is the stress tensor—although, at this point, it is not obvious as to how we can use a single variable to account for the numerous effects that comprise molecular contact forces in a fluid:

$$\sum \underline{f}\Bigg|_{\substack{\text{on} \\ \text{CV}}}^{contact} = \iint_{CS} \left(\begin{array}{c} \text{a continuum expression} \\ \text{that works for} \\ \text{all types of contact forces} \end{array} \right) dS \qquad (4.6)$$

To develop a continuum variable for molecular forces, we begin with the most basic characteristics of forces in fluids. In general, the molecular force $\underline{f}$ on a tiny surface at a given location in a fluid may point in any direction. Such a force vector can be resolved into two components: one that is normal to the surface and one that is tangent to the surface (Figure 4.5). The component of the force vector that is normal to the surface is called the *normal force*, whereas the component of force that is tangent to the surface is called the *shear force*.

Shear and normal forces affect fluids differently. A material, in fact, is classified as a fluid based on how it responds to shear forces: *A fluid is a substance that cannot withstand a shear force without continuously deforming* (Figures 4.6 and 4.7). If we attempt to shear a substance and if it continuously deforms under shearing forces, the substance is a fluid. Fluids behave this way because the intermolecular forces that hold liquid and gas molecules together are not strong enough in a fluid to prevent continuous lateral sliding if a tangential force is imposed. This formal description of a fluid is consistent with our intuitive

	At rest	In motion
Solid	shear and normal forces	shear and normal forces
Fluid (gases and liquids)	normal forces only (shear = 0)	shear and normal forces

Figure 4.6 Solids at rest may support both shear and normal forces. Fluids at rest have no shear forces. When shear forces are applied to a fluid, the fluid deforms continuously.

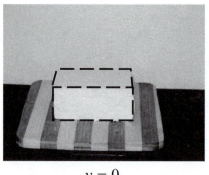 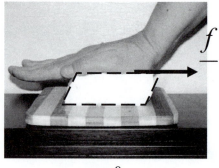

$$\underline{v} = 0 \qquad\qquad\qquad \underline{v} = 0$$

Figure 4.7 A fluid is a substance that cannot withstand a shear force without continuously deforming. Tofu is not a fluid because it can withstand a shear stress when it is at rest.

understanding that a fluid is a type of matter that moves and deforms easily. The fact that fluids continuously deform under shear forces distinguishes them from soft elastic solids such as gelatin and tofu, which can sustain a shear force at rest. When tofu for example (see Figure 4.7), is under shear stress, it deforms but eventually holds a final deformed shape. Fluids, by contrast, cannot do this: Fluids cannot be at rest when a shear force is applied.

Because of the difference between how fluids respond to shear and normal forces, it makes sense to divide our study of molecular contact forces in fluids into two parts: (1) fluids at rest where shear forces are zero (see Section 4.2); and (2) fluids in motion (see Section 4.3). Our discussion of stationary and moving fluids leads to an efficient mathematics that simplifies and organizes the task of accounting for the contact forces within a fluid, both normal and tangential. The result is Equation 4.285, presented here:

$$\begin{array}{c}\text{Total molecular fluid force}\\ \text{on a finite surface } \mathcal{S}\end{array} \qquad \boxed{\mathcal{F} = \iint_{\mathcal{S}} [\hat{n} \cdot \underline{\underline{\tilde{\Pi}}}]_{\text{at surface}} \, dS} \qquad (4.7)$$

To explain this mathematics, we begin in Section 4.2 with the simplest situation involving forces in fluids—that is, the case of a stationary fluid. The case of forces in a moving fluid is discussed in Section 4.3, in which we introduce the stress tensor $\underline{\underline{\tilde{\Pi}}}$. In the following example, we practice dividing forces into shear and normal components.

EXAMPLE 4.1. *Flow in the vicinity of a sphere produces a molecular force on the sphere. The forces at Points (a) and (b) (Figure 4.8) are given by the following two vectors (arbitrary force units):*

$$\underline{f}\Big|_{(a)} = \begin{pmatrix} 2 \\ 0 \\ -4 \end{pmatrix}_{xyz} \qquad \underline{f}\Big|_{(b)} = \begin{pmatrix} 1 \\ 1 \\ -3 \end{pmatrix}_{xyz} \qquad (4.8)$$

What is the normal force on the sphere at Point (a)? What is the tangential force on the sphere at Point (a)? Point (a) is located at coordinate point $(R, \pi/2, 0)$

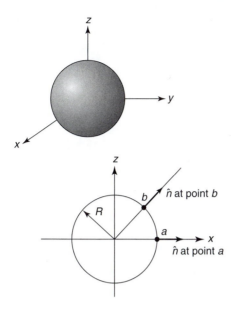

Figure 4.8
A flow field causes forces on a sphere. With force vectors given, the example asks for the normal and tangential forces at Points (a) and (b). In this chapter, we learn how to calculate the force vector at a point in a flow from the stress tensor.

in the $r\theta\phi$ coordinate system. What are the normal and tangential forces on the sphere at Point (b)? Point (b) is located at coordinate point $(R, \pi/4, 0)$ in the $r\theta\phi$ coordinate system.

SOLUTION. To find the component of $\underline{f}$ on the sphere in the normal direction, we dot $\underline{f}$ with the unit normal at Point (a). The unit normal at all points on the surface of a sphere is $\hat{e}_r$ of the spherical coordinate system evaluated at the chosen point. The unit vector $\hat{e}_r$ of the spherical coordinate system is as follows (Cartesian coordinates; see the inside cover of this book):

$$\hat{e}_r = (\sin\theta \cos\phi)\,\hat{e}_x + (\sin\theta \sin\phi)\,\hat{e}_y + (\cos\theta)\,\hat{e}_z \tag{4.9}$$

Point (a) is located at $(R, \pi/2, 0)$ of the $r\theta\phi$ coordinate system; thus, $\hat{n}$ at Point (a) is:

$$\hat{n} = \hat{e}_r|_{(a)} = \hat{e}_x = \begin{pmatrix} 1 \\ 0 \\ 0 \end{pmatrix}_{xyz} \tag{4.10}$$

The magnitude of the normal force component of $\underline{f}$ is the component of $\underline{f}$ in the $\hat{n}$ direction, $f_n = \hat{n} \cdot \underline{f} = \hat{e}_x \cdot \underline{f}|_{(a)}$, which is 2. Thus:

$$(f_n\hat{n})|_{(a)} = 2\hat{e}_x \tag{4.11}$$

$$= \begin{pmatrix} 2 \\ 0 \\ 0 \end{pmatrix}_{xyz} \tag{4.12}$$

The normal and tangential forces add up to the total force; thus, we can calculate the tangential force from the difference:

$$\underline{f} = f_n\hat{n} + f_t\hat{t} \tag{4.13}$$

$$(f_t\hat{t})\big|_{(a)} = \underline{f}\big|_{(a)} - (f_n\hat{n})\big|_{(a)} \tag{4.14}$$

$$= \begin{pmatrix} 2 \\ 0 \\ -4 \end{pmatrix}_{xyz} - \begin{pmatrix} 2 \\ 0 \\ 0 \end{pmatrix}_{xyz} \tag{4.15}$$

$$= \begin{pmatrix} 0 \\ 0 \\ -4 \end{pmatrix}_{xyz} = -4\hat{e}_z \tag{4.16}$$

Similar calculations enable us to find the normal and tangential components of $\underline{f}\big|_{(b)}$. Point (b) is located at $(R, \pi/4, 0)$; thus, $\hat{n}$ at Point (b) is:

$$\hat{n} = \hat{e}_r\big|_{(b)} = \frac{1}{\sqrt{2}}\hat{e}_x + \frac{1}{\sqrt{2}}\hat{e}_z = \begin{pmatrix} \frac{1}{\sqrt{2}} \\ 0 \\ \frac{1}{\sqrt{2}} \end{pmatrix}_{xyz} \tag{4.17}$$

$$f_n\big|_{(b)} = \hat{n}\cdot\underline{f} = \frac{1}{\sqrt{2}} + 0 - \frac{3}{\sqrt{2}} = -\sqrt{2} \tag{4.18}$$

$$(f_n\hat{n})\big|_{(b)} = (-\sqrt{2})\begin{pmatrix} \frac{1}{\sqrt{2}} \\ 0 \\ \frac{1}{\sqrt{2}} \end{pmatrix}_{xyz} = \begin{pmatrix} -1 \\ 0 \\ -1 \end{pmatrix}_{xyz} \tag{4.19}$$

The normal and tangential forces add up to the total force; thus, we calculate the tangential force as:

$$\underline{f} = f_n\hat{n} + f_t\hat{t} \tag{4.20}$$

$$(f_t\hat{t})\big|_{(b)} = \begin{pmatrix} 1 \\ 1 \\ -3 \end{pmatrix}_{xyz} - \begin{pmatrix} -1 \\ 0 \\ -1 \end{pmatrix}_{xyz} \tag{4.21}$$

$$= \begin{pmatrix} 2 \\ 1 \\ -2 \end{pmatrix}_{xyz} \tag{4.22}$$

4.2 Stationary fluids: Hydrostatics

We now begin development of an expression for fluid contact forces by considering fluids at rest. Stationary fluids cannot support tangential forces; thus, in this section, we need be concerned only with normal forces.

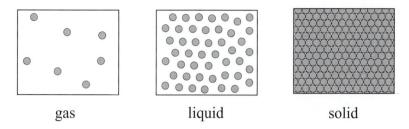

gas   liquid   solid

Figure 4.9 Gas molecules have little or no attraction for one another, whereas liquid molecules are held together by intermolecular attractions. Solids have strong intermolecular attractions that hold molecules in nearly fixed positions.

Fluids may be gases or liquids, and these two types of systems respond differently to normal forces due to differences in their fundamental natures [95] (Figure 4.9). In gases, molecules have little or no attraction for one another, are widely spaced, and are free to move about in the volume available. In liquids, molecules have strong attractive forces for one another, and these forces hold the molecules in proximity. Both gases and liquids respond to normal forces.

In the last two centuries, scientists discovered the basic nature of stationary gases and liquids and developed models that explain normal forces in these systems. We first discuss gases and the *kinetic-molecular theory of gases*, which is a model that explains how molecular motions result in normal forces on surfaces in stationary gases. Second, we discuss a simple liquid model that uses a potential-energy function to describe the relationship between intermolecular structure and normal forces in a stationary liquid. These two models allow us to connect molecular behavior of fluids with continuum functions, which then are used to quantify molecular normal contact force in static fluids.

4.2.1 Gases

The fundamental nature of gases is discussed in elementary science and chemistry courses. Gas molecules have little or no attraction for one another and are free to move about in the volume available. The behavior of simple gases is captured by the ideal gas law, which relates pressure, volume, temperature, and the number of moles present [95].

Ideal gas law
$$p = \frac{N}{V}RT \tag{4.23}$$

In this equation, p is pressure (force/area), V is volume, N is the number of moles, T is absolute temperature, and $R = 0.08206\ l\ atm/mol\ K$ is the ideal gas constant. We are interested in forces in a gas at rest, and the ideal gas law states that the force per area (i.e., pressure) on a surface in a gas is proportional to both the number of moles per unit volume (N/V) in a container and to the absolute temperature T of the system.

The ideal gas law is a consequence of molecular behavior that can be summarized by five modeling assumptions that together are known as the kinetic-molecular theory of gases [62, 95] (Table 4.2). According to this theory, gas

Table 4.2. The Kinetic-Molecular theory of gases [62]

1. A gas consists of tiny particles, either atoms or molecules, moving around at random.
2. The volume of individual particles is negligible compared with the total volume occupied by the gas. Thus, most of the volume occupied by a gas is empty space.
3. The gas particles act independently. There are neither attractive nor repulsive forces among particles.
4. Collisions involving gas particles are elastic, which means that no kinetic energy is lost by particles when they collide.
5. The average kinetic energy of gas particles is proportional to the absolute temperature of the gas. At constant temperature, the kinetic energy is constant.

pressure on a surface results from the collisions of gas particles with the surface in question (Figure 4.10), and pressure is higher when there are more collisions (i.e., N/V increases) or when particle momentum is higher (i.e., higher kinetic energy or temperature). The first three assumptions of kinetic theory—that the gas molecules move around rapidly at random in a vast empty space and do not interact—predict that pressure is the same on any surface at any location in an isothermal stationary gas.

$$
\begin{array}{c}
\text{Kinetic-molecular} \\
\text{theory of gases}
\end{array}
\left(
\begin{array}{c}
\text{pressure on} \\
\text{any surface} \\
\text{in an ideal gas} \\
\text{at rest}
\end{array}
\right)
= p = \text{constant}
\qquad (4.24)
$$

This is a consequence of the fact that rapidly moving gas molecules mix readily, resulting in the same frequency of collisions on any possible surface.

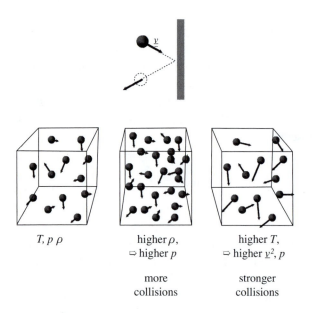

$T, p\ \rho$ higher ρ,   higher T,
⇨ higher p ⇨ higher $\underline{v}^2, p$

more   stronger
collisions   collisions

Figure 4.10   In gases, pressure on a surface is produced by the collision of gas molecules on that surface. As gas density increases, pressure increases because the number of collisions per unit time increases. As gas temperature increases, pressure again increases—in this case, because temperature is proportional to kinetic energy or the square of the molecular speed. When the speeds of molecules are higher, the forces of the collisions also are higher and pressure is increased.

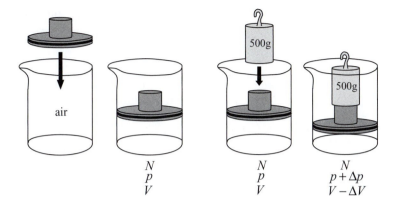

Figure 4.11 A gas is confined in a container by a piston. There is no friction between the walls of the container and the piston. A weight subsequently is placed on top of the piston, causing the gas to compress. The volume change of the gas causes the pressure in the gas to increase throughout the container.

External forces on a gas have an effect on pressure, as shown in the following thought experiment. Consider a fixed quantity of gas at a constant temperature T confined in a piston-container arrangement (Figure 4.11). The piston moves freely (i.e., no friction at the wall) but makes a tight seal such that no gas escapes. We can apply an external force to the gas by placing a weight on the piston. Before the weight is introduced, the forces on the piston are in balance; that is, the weight of the piston is balanced by the upward force on the piston due to the gas pressure. When the weight is placed on the piston, the forces are no longer in balance and the piston descends, compressing the gas. This deformation of the gas is called bulk deformation and occurs in compressible fluids when they are subjected to normal forces. As the volume occupied by the gas shrinks in size, the gas molecules crowd together and the density of the gas increases. The increase in density increases the frequency of collisions on any surface and therefore the pressure in the gas. The gas pressure rises continuously as the piston descends until the upward force on the piston due to the gas pressure again balances the downward force due to the combined mass of the piston and the extra weight. When the piston comes to rest, the new pressure inside the container is given by the ideal gas law (Equation 4.23) with the new volume $(V - \Delta V)$ inserted for V.

Thus, forces on surfaces in stationary gases are well understood in terms of the kinetic-molecular theory of gases: The force on any surface within a stationary gas is given by the gas pressure multiplied by the area of the surface. The pressure force acts normal to the surface. We write this mathematically as:

$$\left. \underline{f} \right|_{on\ \Delta A} = p\ \Delta A\ (-\hat{n}) \tag{4.25}$$

Force on any
small surface ΔA
of unit normal $\hat{n}$
in a stationary ideal gas
$$\boxed{\left. \underline{f} \right|_{on\ \Delta A} = \frac{NRT}{V}\ \Delta A\ (-\hat{n})} \tag{4.26}$$

where ΔA is the flat area of the surface, p is the pressure, and $\hat{n}$ is the unit normal vector for the flat surface ΔA. The negative sign changes the direction of the

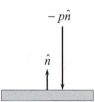

$-p\hat{n}$

$\hat{n}$

Figure 4.12 The pressure is a compressive force per unit area. Pressure on a surface acts in the direction opposite to the direction of the outwardly pointing unit normal vector $\hat{n}$.

outwardly pointing unit normal $\hat{n}$ to the appropriate pushing direction of pressure (Figure 4.12).

In summary, the molecular contact force on any surface in a gas at rest is purely a normal force because stationary fluids cannot support shear forces. Gases are compressible fluids; thus, they undergo bulk deformation and become more dense under application of external normal forces. Pressure and density in simple gases are related through the ideal gas law, and the force on any small surface in a stationary ideal gas is a vector given in Equation 4.26. A problem of calculating force in a stationary gas is worked out in the following example.

EXAMPLE 4.2. *A shelter is created by leaning a hard plastic sheet up against a wall as shown in Figure 4.13. Severe weather can cause very low pressures to exist on the outside of the shelter. For an outside atmospheric pressure of 720 torr, what is the force (provide magnitude and direction) due to the air pressure on the outside of the sheet? What is the component of this force in the downward (i.e., gravity) direction? The inside pressure is 760 torr.*

SOLUTION. The air on the outside of the shelter is in motion due to the storm and cannot be modeled with static-fluid equations, but we can examine the effect of pressure differences and postpone the determination of those pressures. Consider the static case of low pressure on the outside of the shelter compared to fixed

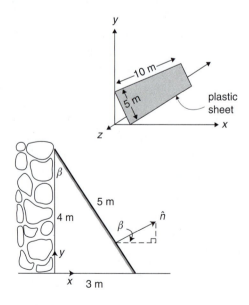

Figure 4.13 A shelter is created by leaning a hard plastic sheet up against a wall, as shown.

atmospheric pressure (760 torr) inside the shelter. The force on the outside of the sheet may be calculated from Equation 4.26:

Force on any
flat surface ΔA
of unit normal $\hat{n}$
in a stationary gas

$$\left. \underline{f} \right|_{on\ \Delta A} = p\,\Delta A\,(-\hat{n}) \qquad (4.27)$$

The pressure on the outside of the shelter p_{out} exerts a force on the area of the sheet ($\Delta A = (5\ m)(10\ m)$) in the direction $-\hat{n}$, where we use geometry to write $\hat{n}$:

$$\hat{n} = \begin{pmatrix} \cos \beta \\ \sin \beta \\ 0 \end{pmatrix}_{xyz} \qquad (4.28)$$

$$\left. \underline{f} \right|_{\Delta A} = (720\ torr)\left(\frac{1.01325 \times 10^5\ N/m^2}{760\ torr}\right)(50\ m^2)(-1)\begin{pmatrix} \frac{4}{5} \\ \frac{3}{5} \\ 0 \end{pmatrix}_{xyz} \qquad (4.29)$$

Outside force:

$$\left. \underline{f} \right|_{\Delta A} = \begin{pmatrix} -3.8 \\ -2.9 \\ 0 \end{pmatrix}_{xyz} MN \qquad (4.30)$$

where $1\ MN = 10^6\ N$.

To calculate the component of $\left. \underline{f} \right|_{\Delta A}$ in the direction of gravity, we dot this vector with the unit vector in the downward direction, $-\hat{e}_y$:

$$\left. \underline{f} \right|_{\Delta A} \cdot (-\hat{e}_y) = \begin{pmatrix} -3.8 & -2.9 & 0 \end{pmatrix}_{xyz} \cdot \begin{pmatrix} 0 \\ -1 \\ 0 \end{pmatrix}_{xyz} \qquad (4.31)$$

$$= 2.9 \times 10^6\ N \qquad (4.32)$$

This downward force is positive. Other forces on the sheet include the vertical component of pressure on the inside of the shelter and the downward force due to gravity (i.e., the weight of the sheet). A force balance determines whether the shelter roof would blow away due to the low outside pressure.

4.2.2 Liquids

In Section 4.2, we arrive at Equation 4.26 for $\left. \underline{f} \right|_{on\ \Delta A}$, the molecular contact forces on a small flat surface ΔA in gases at rest. We obtained this expression by considering the fundamental nature of gases beginning with the kinetic theory.

Liquids are similar to gases in many ways. Referring to the list of assumptions that characterize the kinetic theory of gases (see Table 4.2), the first assumption

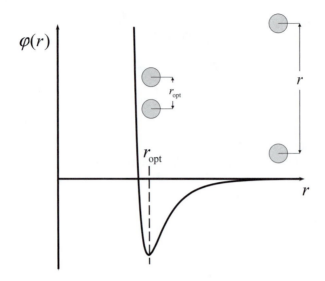

This potential energy function describes qualitatively the attraction of two molecules in a liquid. There is an optimum spacing for the molecules, which is the spacing corresponding to the minimum in the potential energy curve. Pressing the molecules to be closer than the optimum spacing requires significant force [62].

holds for liquids as well as for gases: A liquid consists of tiny particles, either atoms or molecules, moving around at random.

The second assumption of kinetic theory, however, does not hold for liquids because the volume of the individual liquid molecules is not negligible compared with the total volume occupied by the liquid. For a gas at reasonable temperatures and pressures, less than 0.1 percent of the volume is taken up by the molecules, whereas approximately 70 percent of a liquid's volume is taken up by molecules [95].

The most important feature of liquids, however, is their violation of the third assumption of the kinetic theory of gases: Particles act independently. In liquids, molecules do not act independently. Rather, the molecules in a liquid are constantly subjected to the attractive and repulsive forces of their neighbors and, if pressed together by outside forces, the molecules strongly repel one another to preserve optimum molecular spacing.

The behavior of liquids can be understood by reference to an intermolecular potential-energy function shaped like that shown in Figure 4.14. The potential-energy function $\phi(r)$ describes the energy penalty if molecules approach to within a distance r. For example, at large intermolecular spacings r, there is no attractive or repulsive force felt by the molecules, and the value of the potential-energy function ϕ is zero. As the molecules approach one another, however, the liquid molecules attract (i.e., there is a negative energy penalty). The attraction increases as two molecules get closer; eventually, the electron orbitals of the molecules begin to overlap and this conflict results in a large positive energy penalty. The repulsive force is strong at small spacings ($r < r_{opt}$), and the potential-energy function increases steeply as the two molecules are forced together. The optimum average spacing for molecules r_{opt} is the spacing at which the potential-energy curve reaches its minimum.

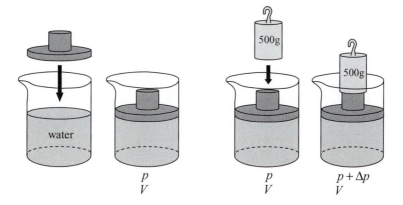

Figure 4.15 A liquid is held in a container and a piston sits on top. There is no friction between the walls of the container and the piston. A weight subsequently is placed on top of the piston, causing the pressure in the liquid to increase throughout the container.

External forces have a different effect when applied to liquids rather than to gases. Imagine a container (Figure 4.15) in which the top surface is a piston that moves freely; however, this time a liquid such as water is placed in the container. The liquid and the piston are both motionless; thus, the forces on the piston are in balance. Subsequently, as with the gas in the previous section, we add a weight to the piston. In the case of the gas, the piston moved and compressed the gas, increasing the pressure ($p = NRT/(V - \Delta V)$). In the case of a liquid, the molecules already are close together and no reduction in liquid volume is observed.

Although no appreciable volume change occurs in the liquid, on a molecular level, something changes after the weight is added. The weight placed on the piston increases the downward force on the liquid, and this force attempts to squeeze the molecules closer together than r_{opt}. Any attempt to squeeze the molecules closer together, however, is resisted by intermolecular repulsion (see Figure 4.14). Intermolecular repulsion acts among all of the molecules in the liquid, and when the weight is added to the top of the liquid, all the molecules in the container are raised to the higher energy state of being slightly squeezed closer together (Figure 4.16).

When a weight is applied to a confined liquid, what increases throughout the liquid—as is true with a gas—is the pressure. The pressure increased in the case of the gas due to a volume decrease; the pressure change in a liquid has a different cause. In a gas, pressure is due to collisions among molecules and between molecules and surfaces. In a liquid, pressure also is due to collisions but, more important, it is a result of the repulsive and attractive intermolecular forces described by the potential-energy function in Figure 4.14. In our example, when the weight is placed on the piston, the force of its weight is immediately transferred to the piston and to the liquid in contact with the piston. The liquid does not compress—at least not significantly. Instead, the force on the piston is transmitted to the liquid in contact with the piston, which transfers the force to the next nearest layer of liquid, and so on, until all of the fluid in the container is affected. This transference occurs rapidly. To accommodate the applied force, all molecules are nudged slightly closer together, raising their intermolecular

no external
force

external force
applied

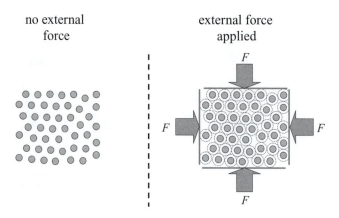

Figure 4.16   When an external compressive force attempts to squeeze a confined liquid, the molecules cannot easily be made to get closer. Instead, an electrostatic repulsion acts as a restoring force to resist the external deformation.

potential energy. The net result of the application of the external normal force on a confined liquid is that the pressure rises nearly instantaneously throughout the container.

This is qualitative information, which we now seek to turn into a quantitative rule describing how pressure in a liquid is related to external forces. The molecular mechanism for pressure in a liquid, as discussed previously, is repulsion based on intermolecular forces. Because intermolecular forces vary among chemicals, however, we face a more complex problem with liquids than with gases. For liquids, there is no single "ideal liquid law," analogous to the ideal gas law, that relates liquid pressure to molecular variables. To write a law that relates liquid pressure to liquid properties such as density or molecular structure, we must specify the type and intensity of the intermolecular forces acting in the liquid and causing the pressure rise. That is, we need the exact curve of the potential function in Figure 4.14 for the liquid in question. We can know this curve only if we first specify which liquid we are considering.

It is undesirable, however, to be too specific about the liquid being considered because choosing one chemical or one class of chemicals for the analysis severely limits the results. We remind ourselves of our goal: We seek an expression for molecular contact forces on a surface in a stationary liquid. In a gas, we arrived at the expression we needed (Equation 4.26) through a comprehensive ideal gas law, but this appears to be a difficult path for liquids. Perhaps there is a different way to reach our goal.

We abandon the idea of relating liquid forces to molecular parameters with a comprehensive equation of state and turn instead to writing an expression for stress on a surface that requires the fewest assumptions about intermolecular forces within the liquid. As we show herein, it turns out that liquids at rest can be understood without specifying much about intermolecular forces. We now pursue a general method for describing the effect of external forces on liquids, although we revisit specific intermolecular forces in Section 4.3 and in Chapter 5, where we discuss liquids in motion.

We seek a liquid equation equivalent to Equation 4.26, the fluid-force equation for ideal gases, and we want to assume as little as possible about the molecules that comprise our liquid. We begin with the fact that the forces on surfaces in liquids at rest are normal forces. Thus, we can immediately write:

$$\underline{f}\Big|_{on\ \Delta A} = \left(\frac{force}{area}\right) (\text{flat area}) \left(\begin{array}{c} \text{unit vector} \\ \text{specifying} \\ \text{direction} \end{array}\right) \quad (4.33)$$

Force at a point on a plane of unit normal $\hat{n}$ in a stationary liquid

$$\boxed{\underline{f}\Big|_{on\ \Delta A} = p(x, y, z, \hat{n})\, \Delta A\, (-\hat{n})} \quad (4.34)$$

Recall that pressure is a pushing force/area; thus, the negative sign changes the direction of $\hat{n}$ so that $\underline{f}$ pushes on ΔA. Until we prove otherwise, pressure p in a liquid may depend on position (x, y, z) and orientation $(\hat{n})$ of the chosen measurement surface ΔA. Now our task is to see whether there is anything general we know about liquids that might help us to be more specific in Equation 4.34.

Although the details of how forces are generated vary among liquids, all forces are subjected to the laws of physics—specifically, the law of conservation of momentum. We can apply momentum balances to a portion of a stationary fluid to determine whether momentum conservation places any constraints on the function p in a liquid. Our calculations result in the discovery of two pieces of information that engineers find extremely useful: (1) pressure at a point in a static fluid is isotropic (i.e., does not depend on $\hat{n}$; Equation 4.60); and (2) in a static fluid subjected to gravity, the pressure on a surface varies linearly with liquid depth and is independent of horizontal position (see Equation 4.68).

To see what can be learned from a momentum balance, we begin by choosing a control volume (CV). Consider a small wedge of fluid within a liquid at rest (Figure 4.17). The wedge has triangular faces in the planes parallel to the xy-plane with sides of lengths Δx, Δy, and Δl; the uniform height of the wedge is Δz. Gravity acts in the negative z-direction, $\underline{g} = -g\hat{e}_z$. The dimensions Δx, Δy, and Δz are small enough that the pressure does not vary significantly across the faces of the wedge. Let the pressures on the faces be $p|_x$, $p|_y$, p, $p|_{z+\Delta z}$, and $p|_z$, as shown in Figure 4.17. We call the surface of area $\Delta l \Delta z$ the "l-surface," and the pressure on the l-surface is p.

The momentum balance on the control volume is given by the Reynolds transport theorem—that is, Newton's second law as applied to a CV:

Reynolds transport theorem (momentum balance on CV)

$$\frac{d\underline{P}}{dt} + \iint_{CS} (\hat{n} \cdot \underline{v})\, \rho \underline{v}\, dS = \sum_{\substack{on \\ CV}} \underline{f} \quad (4.35)$$

where $\underline{P}$ is the momentum in the CV, $\underline{v}$ is local fluid velocity, ρ is local fluid density, and $\hat{n}$ is the outwardly pointing unit normal of a small portion of the control surface dS. The wedge CV is in a fluid at rest ($\underline{v} = 0$, $\underline{P} = 0$); therefore, from Equation 4.35, the sum of the forces on the CV is zero. The forces on the

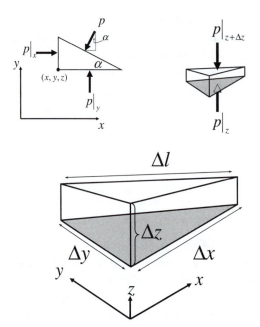

A small wedge of fluid with forces on the sides is shown. By writing the force balance on this system, we can show that pressure is isotropic in a fluid at rest. The balance is carried out in a Cartesian coordinate system in which gravity is in the negative z direction.

CV are gravity and the contact forces on the five faces of the CV:

$$0 = \sum_{\substack{on \\ CV}} \underline{f} \tag{4.36}$$

$$0 = \begin{pmatrix} \text{contact} \\ \text{forces} \end{pmatrix} + \begin{pmatrix} \text{noncontact} \\ \text{forces} \end{pmatrix} \tag{4.37}$$

$$0 = \begin{pmatrix} \text{contact forces} \\ \text{on 5 sides} \end{pmatrix} + \begin{pmatrix} \text{force due} \\ \text{to gravity} \end{pmatrix} \tag{4.38}$$

We now write the forces on the CV. The force due to gravity is given by the mass of fluid in the CV multiplied by the acceleration due to gravity. The contact forces on the five faces are equal to the pressures on them multiplied by their areas A. Because the fluid is stationary, all of the contact forces on the control surfaces act in directions perpendicular to the surfaces (i.e., normal forces); stationary fluids cannot sustain shear forces. The momentum balance in Equation 4.38 requires that the vector combination of the contact forces and gravity force on the control volume sum to zero. Writing this information mathematically, we obtain:

$$0 = \sum_{\substack{on \\ CV}} \underline{f} \tag{4.39}$$

$$0 = \begin{pmatrix} \text{contact forces} \\ \text{on 5 sides} \end{pmatrix} + \begin{pmatrix} \text{force due} \\ \text{to gravity} \end{pmatrix} \tag{4.40}$$

$$0 = (-pA\hat{n})|_l + (-pA\hat{n})|_x + (-pA\hat{n})|_y + (-pA\hat{n})|_z + (-pA\hat{n})|_{z+\Delta z}$$
$$+ \rho \underline{g} \frac{1}{2} \Delta x \Delta y \Delta z \tag{4.41}$$

where $(1/2)\Delta x \Delta y \Delta z$ is the volume of the wedge. From the geometry and coordinate system in Figure 4.17, we can specify the various pressures, areas, and unit normals in Equation 4.41 as follows:

$$
\begin{pmatrix} 0 \\ 0 \\ 0 \end{pmatrix}_{xyz} = \begin{pmatrix} -p \sin \alpha \, \Delta l \Delta z \\ -p \cos \alpha \, \Delta l \Delta z \\ 0 \end{pmatrix}_{xyz} + \begin{pmatrix} p|_x \, \Delta y \Delta z \\ 0 \\ 0 \end{pmatrix}_{xyz} + \begin{pmatrix} 0 \\ p|_y \, \Delta x \Delta z \\ 0 \end{pmatrix}_{xyz}
$$

$$
+ \begin{pmatrix} 0 \\ 0 \\ p|_z \frac{1}{2} \Delta x \Delta y \end{pmatrix}_{xyz} + \begin{pmatrix} 0 \\ 0 \\ -p|_{z+\Delta z} \frac{1}{2} \Delta x \Delta y \end{pmatrix}_{xyz}
$$

$$
+ \begin{pmatrix} 0 \\ 0 \\ -\rho g \frac{1}{2} \Delta x \Delta y \Delta z \end{pmatrix}_{xyz} \tag{4.42}
$$

From geometry, $\Delta x = \Delta l \cos \alpha$ and $\Delta y = \Delta l \sin \alpha$.

$$
\begin{pmatrix} 0 \\ 0 \\ 0 \end{pmatrix}_{xyz} = \begin{pmatrix} -p \Delta y \Delta z \\ -p \Delta x \Delta z \\ 0 \end{pmatrix}_{xyz} + \begin{pmatrix} p|_x \, \Delta y \Delta z \\ 0 \\ 0 \end{pmatrix}_{xyz} + \begin{pmatrix} 0 \\ p|_y \, \Delta x \Delta z \\ 0 \end{pmatrix}_{xyz}
$$

$$
+ \begin{pmatrix} 0 \\ 0 \\ p|_z \frac{1}{2} \Delta x \Delta y \end{pmatrix}_{xyz} + \begin{pmatrix} 0 \\ 0 \\ -p|_{z+\Delta z} \frac{1}{2} \Delta x \Delta y \end{pmatrix}_{xyz}
$$

$$
+ \begin{pmatrix} 0 \\ 0 \\ -\rho g \frac{1}{2} \Delta x \Delta y \Delta z \end{pmatrix}_{xyz} \tag{4.43}
$$

Equation 4.43 is three equations, one each for the x-, y-, and z-directions, and we can solve them for the pressures. The x-component of the momentum balance gives us:

$$
0 = -p \, \Delta y \Delta z + p|_x \, \Delta y \Delta z \tag{4.44}
$$

$$
\boxed{p = p|_x} \tag{4.45}
$$

Similarly, in the y-direction:

$$
0 = -p \, \Delta x \Delta z + p|_y \, \Delta x \Delta z \tag{4.46}
$$

$$
\boxed{p = p|_y} \tag{4.47}
$$

We have shown that the pressures on the x-, y-, and l-surfaces are equal. The orientation of the coordinate system in the xy-plane was arbitrary, however, as is the angle α. We conclude that the pressure in a static fluid at a point is the same on all planes drawn perpendicular to the xy-plane (i.e., horizontal relative to gravity).

In the z-direction, the momentum balance gives:

$$p|_z \frac{1}{2} \Delta x \Delta y - p|_{z+\Delta z} \frac{1}{2} \Delta x \Delta y - \rho g \frac{1}{2} \Delta x \Delta y \Delta z = 0 \qquad (4.48)$$

$$p|_{z+\Delta z} - p|_z = -\rho g \Delta z \qquad (4.49)$$

$$\frac{p|_{z+\Delta z} - p|_z}{\Delta z} = -\rho g \qquad (4.50)$$

In the limit as Δz goes to zero, the left side of Equation 4.50 becomes the definition of the derivative of a function—specifically, the derivative of the pressure in the z-direction (see Equation 1.139):

$$\lim_{\Delta z \longrightarrow 0} \left[\frac{p|_{z+\Delta z} - p|_z}{\Delta z} \right] = -\rho g \qquad (4.51)$$

$$\boxed{\frac{dp}{dz} = -\rho g} \qquad (4.52)$$

Equation 4.52 tells us that in a static liquid, there is a nonzero pressure gradient in the direction of gravity.

The momentum balance on the wedge-shaped control volume yielded much information about the state of stress at a point in a stationary fluid. The x- and y-components of the momentum balance told us that when evaluating pressure at a point, we may choose any plane through our point as long as the chosen plane is perpendicular to the xy-plane. We also learned from the z-component how pressure varies in the z-direction when gravity is present; this is an important observation to which we subsequently return.

It turns out that we need not choose a measurement plane perpendicular to the xy-plane in our analysis because the pressure on *any* plane through a chosen point is the same, as we now show. Consider the stress in a static fluid at a chosen point (Figure 4.18). We again choose our coordinate system so that gravity is in the $(-z)$-direction, $(\underline{g} = -g\hat{e}_z)$, and p represents the pressure on any plane perpendicular to the xy-plane. We choose an arbitrary, infinitesimally small surface ΔS that is not perpendicular to the xy-plane. The unit normal vectors of ΔS are $\hat{n}$ and $-\hat{n}$, and we choose the x-direction so that $\hat{n}$ and $-\hat{n}$ are in the xz-plane. Because the fluid is stationary, the forces on the two sides of ΔS are equal and opposite in direction, $f\hat{n}$ and $-f\hat{n}$ (i.e., balanced normal forces). The area of ΔS is $\Delta l \Delta y$, and the magnitude of pressure on either side of ΔS is $f/\Delta l \Delta y$.

The projection of ΔS in the x-direction is a rectangular piece of surface ΔA, of area $\Delta l \cos \theta \Delta y$ (see Web appendix [108] for details), shown in Figure 4.18. The projection of a surface in a direction gives the effective size of a surface in that direction. The magnitude of the force on either side of ΔA is equal to the portion of $f\hat{n}$ that acts in the x-direction; that is, the x-component of the force

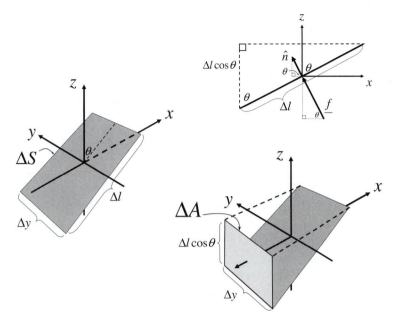

Figure 4.18 We consider a plane that is not perpendicular to the xy-plane. We can show that the pressure on any such plane is the same as on planes perpendicular to the xy-plane.

vector $f\hat{n}$. We calculate this as follows:

$$\hat{n} = \begin{pmatrix} \cos\theta \\ 0 \\ \sin\theta \end{pmatrix}_{xyz} \tag{4.53}$$

$$f\hat{n} = \begin{pmatrix} f\cos\theta \\ 0 \\ f\sin\theta \end{pmatrix}_{xyz} \tag{4.54}$$

$$\begin{pmatrix} \text{magnitude of force} \\ \text{on } \Delta S \text{ in positive} \\ x\text{-direction} \end{pmatrix} = f\cos\theta = \begin{pmatrix} \text{force on} \\ \Delta A \end{pmatrix} \tag{4.55}$$

However, the pressure on either side of ΔA is the ratio of this force to the area of ΔA:

$$\begin{pmatrix} \text{pressure} \\ \text{on } \Delta A \end{pmatrix} = \frac{\text{force on } \Delta A}{\text{area of } \Delta A} \tag{4.56}$$

$$= \frac{f\cos\theta}{\Delta l\cos\theta\,\Delta y} \tag{4.57}$$

$$= \frac{f}{\Delta l\,\Delta y} \tag{4.58}$$

ΔA is perpendicular to the xy-plane, and the pressure on *any plane* perpendicular to the xy-plane is given by p. Thus, the pressure on ΔA, given by Equation 4.58, is equal to p:

$$p = \begin{pmatrix} \text{pressure} \\ \text{on } \Delta A \end{pmatrix} = \frac{f}{\Delta l\,\Delta y} \tag{4.59}$$

Pressure at a point is
independent of $\hat{n}$
(isotropic)
in a stationary liquid

Figure 4.19 Stress is isotropic in a stationary fluid. This means that at a point in a fluid, the pressure on a surface through that point does not depend on the orientation of the surface chosen.

We have shown that $f/\Delta l \Delta y$, the pressure on the original surface ΔS, is equal to p, the pressure on any surface perpendicular to the xy-plane. Because the choice of surface ΔS was arbitrary, we conclude that the pressure on any plane through the point is equal to $p = f/\Delta l \Delta y$. Thus, at a point in a static fluid, the pressure on all planes is the same; that is, the function p is independent of $\hat{n}$. The pressure is isotropic (i.e., independent of position) in a stationary liquid (Figure 4.19).

Our discussion so far allows us to conclude that pressure is isotropic and varies only in the direction of gravity, z (Equation 4.52). We can summarize these results with the following equation (compare to Equation 4.34):

Force at a point on a
plane of unit normal $\hat{n}$
in a stationary liquid

$$\left. \underline{f} \right|_{on\ \Delta A} = p(z)\,\Delta A\,(-\hat{n}) \tag{4.60}$$

Returning to the momentum balance on the wedge CV, the z-component of the momentum balance (Equation 4.52) has enough information to solve for $p(z)$. Integrating Equation 4.52 for a fluid with constant density, we obtain:

$$\frac{dp}{dz} = -\rho g \tag{4.61}$$

$$\int dp = \int -\rho g dz \tag{4.62}$$

$$p(z) = -\rho g z + C_1 \tag{4.63}$$

where C_1 is an arbitrary constant of integration. If we use the boundary condition that $p = p_0$ at $z = 0$, we can solve for C_1:

Pressure in a
stationary fluid
at elevation z
(z-direction upward)

$$p(z) = -\rho g z + p_0 \tag{4.64}$$

Note that the z-direction points upward; that is, gravity is given by $\underline{g} = -g\hat{e}_z$. We also can write this as:

$$p_o = p(z) + \rho g z \tag{4.65}$$

Pressure at the
bottom of a
column of fluid

$$p_{bottom} = p_{top} + \rho g h \tag{4.66}$$

where p_{bottom} is the pressure at the bottom, p_{top} is the pressure at the top, and h is the height of the column of fluid.

We summarize our results by writing an equation for the force on a surface in a stationary liquid (compare to Equation 4.34):

$$\left. \underline{f} \right|_{on\ \Delta A} = p(z)\,\Delta A\,(-\hat{n}) \tag{4.67}$$

Force at a point
on a plane of area ΔA
with unit normal $\hat{n}$
in a stationary liquid

$$\left. \underline{f} \right|_{on\ \Delta A} = (-\rho g z + p_0)\,\Delta A\,(-\hat{n}) \tag{4.68}$$

As before, $\hat{n}$ is the unit normal vector of the surface ΔA, ρ is the density of the fluid, g is the magnitude of the acceleration due to gravity, and p_0 is the pressure at $z = 0$. The force $\underline{f}$ depends on the z-position of the point in question because the weight of the fluid above a chosen surface affects the pressure on that surface. The molecular force on a plane ΔA at a point in a stationary fluid is independent of the orientation of the surface ΔA and is independent of the x- and y-positions of the point.

We did not succeed in deducing $\underline{f}$ in terms of molecular parameters like temperature and volume as for gases (Equation 4.26), but we developed a useful equation. Having established that pressure is isotropic and a function only of the z-position, we now can solve a wide variety of problems in stationary fluids, including pressure effects on nonflat surfaces (Example 4.6). We can best understand Equations 4.68 and 4.64 by applying these results to several examples. In Section 4.2.3, we apply the static-fluid equation to explain the functioning of manometers and hydraulic lifts.

EXAMPLE 4.3. *Consider the water-filled device in Figure 4.20. What are the pressures at the points indicated in the schematic?*

SOLUTION. Pressure in a stationary liquid depends on only the elevation of the point, Equation 4.64, repeated here:

Pressure in a
stationary fluid
at elevation z

$$p(z) = -\rho g z + p_0 \tag{4.69}$$

where p_0 is the pressure at the position $z = 0$ and the z-coordinate direction points upward (see Figure 4.17). Pressure in a stationary liquid depends on neither the shape of its container nor the lateral (x- or y-) position of the point.

To analyze the pressures in the unusual device in Figure 4.20, we first choose our coordinate system. We locate our coordinate system at the top surface (Point A). The three top surfaces are open to the air and have the same fluid level; thus, the pressure at the top free surface ($z = 0$) is 1.0 atm $= 1.01325 \times 10^5$ Pa:

$$p(z) = -\rho g z + p_0 \tag{4.70}$$

$$p(0) = p_0 \tag{4.71}$$

$$= p_A = p_B = p_C = 1.0 \text{ atm} = 1.01325 \times 10^5 \text{ Pa} \tag{4.72}$$

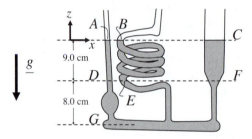

Figure 4.20 The pressure exerted at a point at the bottom of a quantity of fluid depends on only the vertical height of the fluid above the point. Thus, the pressure is the same at the bottom of each tube, no matter the shape. If more fluid is poured into any tube in the apparatus shown, flow occurs until the levels are even and the pressure is equilibrated. Device constructed by Eugenijus Urnezius and Timothy Gasperich.

Below the top surface, the pressures are higher due to the weight of the fluid above. Points D, E, and F have the same elevation ($z = -h$) and therefore the same pressure. The density of water at room temperature is about 995 kg/m^3. We apply Equation 4.64 for $z = -h = -9$ cm:

$$p(z) = -\rho g z + p_0 \tag{4.73}$$

$$p(-h) = \rho g h + p_0 \tag{4.74}$$

$$= \left(995\frac{\text{kg}}{\text{m}^3}\right)\left(9.80\frac{\text{m}}{\text{s}^2}\right)(9.0 \times 10^{-2}\,\text{m})\left(\frac{\text{Pa s}^2\,\text{m}}{\text{kg}}\right)$$

$$+\,1.01325 \times 10^5\,\text{Pa}$$

$$p_D = p_E = p_F = 1.02203 \times 10^5\,\text{Pa} \tag{4.75}$$

$$= 1.02 \times 10^5\,\text{Pa} \tag{4.76}$$

The pressure at Point G is calculated the same way for $z = -17$ cm:

Pressure at Point G: $p(h_G) = \rho g h_G + p_0$

$$= \left(995\frac{\text{kg}}{\text{m}^3}\right)\left(9.80\frac{\text{m}}{\text{s}^2}\right)(17.0 \times 10^{-2}\ \text{m})\left(\frac{\text{Pa s}^2\ \text{m}}{\text{kg}}\right)$$

$$+ 1.01325 \times 10^5\ \text{Pa}$$

$$p_G = 1.02983 \times 10^5\ \text{Pa} \tag{4.77}$$

$$= 1.03 \times 10^5\ \text{Pa} \tag{4.78}$$

These two results are different by only a small amount because the heights of fluid considered are quite small.

EXAMPLE 4.4. *Write the value of the pressure and the equation for the force vector acting on the following surface: a flat surface of area 6.00 cm² facing upward, 10.0 m below the surface of the ocean and located 120 km due south of New Orleans, Louisiana, USA. The density of seawater near the ocean surface is 1,025 kg/m³.*

SOLUTION. The pressure in a liquid at rest does not depend on the location of the point in the xy-plane—that is, on how far it is from New Orleans; the pressure depends on only how much fluid is above the surface of interest. This relationship is codified in Equation 4.64, repeated here:

Pressure in a
stationary fluid at $\boxed{p(z) = -\rho g z + p_0}$ (4.79)
elevation z

where p_0 is the pressure at the position $z = 0$ and the z-coordinate direction points upward (Figure 4.21). For our problem, all points on the surface of interest are 10.0 m below the ocean surface. We designate $z = 0$ to be the surface of the ocean; thus, the location we are interested in is $z = -h = -10.0$ m. At $z = 0$, $p = p_0 = 1.0$ atm $= 1.01325 \times 10^5$ Pa:

$$p(z) = -\rho g z + p_0 \tag{4.80}$$

$$p(-h) = \rho g h + p_0 \tag{4.81}$$

$$p(10\ \text{m}) = \left(1,025\frac{\text{kg}}{\text{m}^3}\right)\left(9.80\frac{\text{m}}{\text{s}^2}\right)(10.0\ \text{m})\left(\frac{\text{N s}^2}{\text{kg m}}\right)\left(\frac{\text{Pa m}^2}{\text{N}}\right)$$

$$+ 1.01325 \times 10^5\ \text{Pa}$$

$$p(10\ \text{m}) = 2.01775 \times 10^5\ \text{Pa} \tag{4.82}$$

$$= 2.02 \times 10^5\ \text{Pa} \tag{4.83}$$

Because the pressure is the same across the surface, to calculate the vector force on the surface, we multiply the pressure by the area over which it acts. The

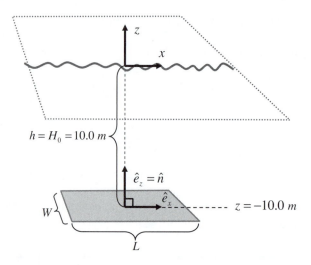

Figure 4.21 The pressure on a horizontal surface in a liquid is calculated using Equation 4.79 in terms of the coordinate system shown.

direction of the force is toward the surface and perpendicular to it:

$$\underline{f} = [p(h = 10 \text{ m})]\,(\text{area})(-\text{unit normal vector}) \tag{4.84}$$

$$\underline{f} = \left(2.01775 \times 10^5 \,\frac{\text{N}}{\text{m}^2}\right)(6.00 \text{ cm}^2)\left(\frac{\text{m}^2}{10^4 \text{ cm}^2}\right)(-\hat{e}_z) \tag{4.85}$$

$$= -121 N\hat{e}_z \tag{4.86}$$

This is the final answer.

Although we calculated correctly the force on the surface, the methods used were informal. Using more formal mathematics points to how to approach complex problems such as the examples that follow. The more formal approach is to write first the local pressure on a small piece of the surface, ΔA. We then can sum that expression over the entire surface, take the limit as $\Delta A \to 0$, and obtain an integral. Because the pressure is constant on the surface in which we are interested (i.e., the surface is oriented horizontally and all points are at the same elevation), the pressure comes out of the integral, and we obtain the same result as before, as we must. Beginning with Equation 4.60:

$$\underline{f}\Big|_{\text{on}\Delta A} = p(z)\,\Delta A(-\hat{n}) \tag{4.87}$$

$$\underline{f} = \lim_{\Delta A \to 0}\left[\sum_{i=1}^{N}\underline{f}\Big|_{\text{on}\Delta A}\right] \tag{4.88}$$

$$= \lim_{\Delta A \to 0}\left[\sum_{i=1}^{N}(p(z)\,\Delta A(-\hat{n}))_i\right] \tag{4.89}$$

We recognize this as the definition of a two-dimensional integral (see Web appendix [108]) for details):

$$\underline{f} = \iint_S (-n)p \, dS \tag{4.90}$$

For our problem $dS = dxdy$, $\hat{n} = \hat{e}_z$, and incorporating the limits, we obtain:

$$\underline{f} = \int_0^W \int_0^L p(-\hat{e}_z) \, dxdy \tag{4.91}$$

where W is the width and L is the length of the surface. Because p and $\hat{e}_z$ are constant, we move them out of the integral:

$$\underline{f} = (-\hat{e}_z)p \int_0^W \int_0^L dxdy \tag{4.92}$$

$$\underline{f} = p(WL)(-\hat{e}_z) \tag{4.93}$$

$$\underline{f} = -201N\hat{e}_z \tag{4.94}$$

This second approach to the solution (Equation 4.90) is helpful when solving problems in which the pressure varies across the surface (see the next example) or when the surface is not flat (Example 4.6).

EXAMPLE 4.5. *What is the total vector force on a 0.500 m × 1.00 m rectangular plate submerged 12.0 m below the surface of a water tank and oriented as shown in Figure 4.22 (tilted $\alpha = 30$ degrees from the vertical)?*

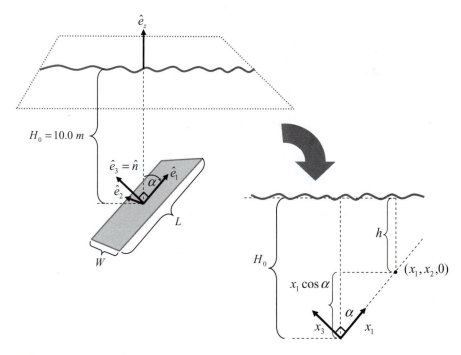

Figure 4.22 A tilted surface submerged 12 m below the surface of a tank experiences a total pressure that is calculated in this example.

SOLUTION. We choose a Cartesian coordinate system on the plate such that the 3-direction is normal to the plate and the surface of the plate is in the 12-plane. If z is the vertical direction ($\underline{g} = -g\hat{e}_z$) and we take $z = 0$ at the air–water interface, we can write p as:

<table>
<tr><td>Pressure in a stationary fluid at elevation z</td><td>$$p(z) = -\rho g z + p_0$$</td><td>(4.95)</td></tr>
</table>

$$p(x_1, x_2) = p(-h) = \rho g h + p_0 \tag{4.96}$$

where h is the distance below the surface of the point $(x_1, x_2, 0)$. We can relate $z = -h$ and x_1 through geometry (see the inset in Figure 4.22):

$$h = H_0 - x_1 \cos\alpha \tag{4.97}$$

where H_0 is the depth of the plate center of gravity, $H_0 = 12.0$ m. Combining this result with Equation 4.96, we obtain:

$$p(x_1, x_2) = \rho g (H_0 - x_1 \cos\alpha) + p_0 \tag{4.98}$$

$$= (\rho g H_0 + p_0) + (-\rho g \cos\alpha) x_1 \tag{4.99}$$

The force is calculated from Equation 4.90 with $\hat{n} = \hat{e}_3$, $dS = dx_1 dx_2$, and p given by Equation 4.98:

$$\underline{f} = \iint_S (-\hat{n}) p \, dS \tag{4.100}$$

$$= \int_{x_2} \int_{x_1} (p)(-\hat{e}_3) dx_1 dx_2 \tag{4.101}$$

where the $x_1 x_2 x_3$ coordinate system is centered on the surface of the plate. The unit vector $\hat{e}_3$ is constant and can come out of the integral; however, p is a function of x_1 as given in Equation 4.99 and the function must be integrated:

$$\underline{f} = \int_{x_2} \int_{x_1} [p(x_1)](-\hat{e}_3) dx_1 dx_2 \tag{4.102}$$

$$= -\hat{e}_3 \int_{-W/2}^{W/2} \int_{-L/2}^{L/2} [(\rho g H_0 + p_0) + (-\rho g \cos\alpha) x_1] \, dx_1 dx_2 \tag{4.103}$$

$$= -\hat{e}_3 W \int_{-L/2}^{L/2} [A x_1 + B] \, dx_1 \tag{4.104}$$

where we have carried out the x_2 integration, and $A = -\rho g \cos\alpha$ and $B = \rho g H_0 + p_0$. Integrating over x_1, we obtain:

$$\underline{f} = -\hat{e}_3 W \left[A \frac{x_1^2}{2} + B x_1 \right] \Bigg|_{-L/2}^{L/2}$$

$$= -\hat{e}_3 W B L$$

$$= -\hat{e}_3 W L (p_0 + \rho g H_0) \tag{4.105}$$

Notice that the total force is equal to the pressure at the center of mass (the center of mass is at $x_1 = 0$) multiplied by the total area WL. The force is directed normal to the plate and toward the plate. For the numbers given in this example, the final result is:

$$\begin{pmatrix} \text{Force} \\ \text{on plate} \end{pmatrix} = -\hat{e}_3 WL (p_0 + \rho g H_0)$$

$$= -\hat{e}_3(0.500 \text{ m}^2) \left[1.01325 \times 10^5 \text{ Pa} \right.$$

$$\left. + \left(995 \frac{\text{kg}}{\text{m}^3} \right) \left(9.80 \frac{\text{m}}{\text{s}^2} \right) (12.0 \text{ m}) \right]$$

$$= -1.09169 \times 10^5 \text{ N } \hat{e}_3 \tag{4.106}$$

$$\boxed{= \quad -1.09 \times 10^5 \text{ N } \hat{e}_3} \tag{4.107}$$

EXAMPLE 4.6. *In a liquid of density ρ, what is the net fluid force on a submerged sphere (e.g., a ball or a balloon)? What is the direction of the force and how does the magnitude of the fluid force vary with fluid density?*

SOLUTION. The problem again asks for the net force on a surface, but this time the surface is the surface of a sphere, which means that force varies with position because p varies with z. To calculate the net force on the sphere, we write the force on a small portion of the sphere surface and then integrate over the entire surface to obtain the net force (Figure 4.23).

In our usual coordinate system for pressure problems (i.e., $z = 0$ at the liquid interface), the pressure at a point in the fluid is a function of the vertical distance of the point from the surface of the fluid:

Pressure in a stationary fluid at elevation z

$$\boxed{p(z) = -\rho g z + p_0} \tag{4.108}$$

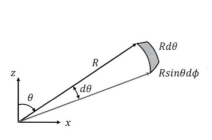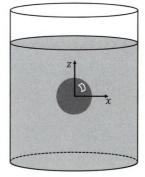

Figure 4.23 Fluid exerts a net force on a submerged sphere. If the sphere is light, the force from the fluid pressure acts to float the sphere. If the sphere is heavy, the fluid sinks in the fluid, but the sphere is decelerated by the fluid force.

where p_0 is atmospheric pressure. Let H_0 be the distance from the center of the sphere to the surface of the liquid in the tank. We carry out our force integration in a spherical coordinate system centered on the sphere. In this coordinate system, the vertical position z is given by:

$$\text{From geometry:} \qquad -z = H_0 - R\cos\theta \tag{4.109}$$

The pressure at points on the sphere surface is therefore:

$$p(z) = -\rho g z + p_0 \tag{4.110}$$

$$p|_{surface} = \rho g\,(H_0 - R\cos\theta) + p_0 \tag{4.111}$$

$$p|_{surface} - p_0 = p_{gauge} = \rho g\,(H_0 - R\cos\theta) \tag{4.112}$$

We show in Example 4.4 that the force on a finite surface S in a stationary fluid is given by:

$$\underline{f} = \iint_S (-n)\,p\,dS \tag{4.113}$$

For the sphere surface:

$$p \longrightarrow \rho g\,(H_0 - R\cos\theta) \tag{4.114}$$

$$dS \longrightarrow (R\sin\theta\,d\phi)(R\,d\theta) \tag{4.115}$$

$$\hat{n} = \hat{e}_r \tag{4.116}$$

Substituting these and adding appropriate limits yields:

$$\underline{f} = -\rho g R^2 \int_0^{2\pi} \int_0^{\pi} (H_0 - R\cos\theta)\,\hat{e}_r \sin\theta\,d\theta\,d\phi \tag{4.117}$$

The basis vector $\hat{e}_r$ varies with position and, thus, we convert to Cartesian coordinates centered on the sphere before attempting to integrate. The basis vector $\hat{e}_r$ is expressed in Cartesian coordinates in Equation 1.273:

$$\hat{e}_r = \sin\theta\cos\phi\,\hat{e}_x + \sin\theta\sin\phi\,\hat{e}_y + \cos\theta\,\hat{e}_z \tag{4.118}$$

$$\underline{f} = -\rho g R^2 \int_0^{2\pi} \int_0^{\pi} (H_0 - R\cos\theta)\,\hat{e}_r \sin\theta\,d\theta\,d\phi \tag{4.119}$$

$$= -\rho g R^2 \int_0^{2\pi} \int_0^{\pi} (H_0 - R\cos\theta)\sin\theta\,[\sin\theta\cos\phi\,\hat{e}_x$$
$$+ \sin\theta\sin\phi\,\hat{e}_y + \cos\theta\,\hat{e}_z]\,d\theta\,d\phi \tag{4.120}$$

We carry out this integration in an example in Chapter 1 (see Example 1.27). The equation is a vector equation, and there are three nonzero Cartesian components of $\underline{f}$, as emphasized here:

$$\underline{f} = -\rho g R^2 \int_0^{2\pi} \int_0^{\pi} (H_0 - R\cos\theta)\sin\theta \begin{pmatrix} \sin\theta\cos\phi \\ \sin\theta\sin\phi \\ \cos\theta \end{pmatrix}_{xyz} d\theta\,d\phi \tag{4.121}$$

The x- and y-components integrate to zero, indicating that the net force due to the fluid is only in the z-direction; that is, net force is either upward or downward.

The z-component integrates to give the magnitude of the force, as discussed in Chapter 1. The final result is as follows (see Equation 1.310):

$$\underline{f} = \begin{pmatrix} 0 \\ 0 \\ \frac{4\pi R^3}{3}\rho g \end{pmatrix}_{xyz} \tag{4.122}$$

The final result is an upward force on the sphere equal in magnitude to the weight of a sphere-shaped hunk of fluid. Thus, the fluid exerts an upward force on the sphere equal in magnitude to the weight of the fluid displaced by the sphere. This is the buoyancy effect, also known as Archimedes' principle, which was articulated by Archimedes in the third century BCE.

In Example 4.6, we derive Archimedes' principle from our result for forces on finite surfaces in stationary liquids (see Equation 4.90, repeated here):

$$\begin{array}{c} \text{Force on a plane} \\ \text{of finite area } S \\ \text{in a stationary liquid} \end{array} \quad \boxed{\underline{f} = \iint_S (-\hat{n}) p \, dS} \tag{4.123}$$

where the pressure p is only a function of elevation with respect to gravity; for $\underline{g} = -g\hat{e}_z$, $p = -\rho g z + p_0$. Archimedes arrived at his principle without calculus, but the advantage of our methods is that we are building a systematic modeling protocol that, so far, is yielding correct results. We seek methods that work in applications that are far more complex than those addressed by Archimedes. The discovery of this systematic, correct protocol for stationary fluids contributes to our ability to model the more complex and less intuitive problems of modern engineering.

In the next example, we apply our methods to a slightly more complex case, that of finding the pressure distribution in the atmosphere, a problem in which the density is not constant.

EXAMPLE 4.7. *What is the effect of gravity on the pressure distribution in a compressible fluid such as air in Earth's atmosphere?*

SOLUTION. In our discussion of gases in Section 4.2.1, we did not consider the effect of gravity on density. Gravity acts on all masses, including gas molecules, although the effect is negligible except when great distances are considered, such as in Earth's atmosphere.

The application of the momentum balance to a wedge of liquid allows us to conclude that pressure at a point is isotropic. We also saw that the effect of gravity is to produce a gradient of pressure in the direction of gravity (see Equation 4.52):

$$\frac{dp}{dz} = -\rho g \tag{4.124}$$

We integrated Equation 4.124 for constant-density fluids to obtain the hydrostatic-pressure Equation 4.64. For gases, density is not constant, and we must modify our derivation. For a gas at modest pressures, gas density is given by the ideal-gas equation (see Equation 4.23):

$$\text{Ideal gas law } \rho = \frac{pM}{RT} \tag{4.125}$$

where p is pressure, M is molecular weight, T is absolute temperature, and R is the ideal-gas constant. Combining these two relationships, we can solve for the pressure as a function of elevation z for an ideal gas:

$$\frac{dp}{dz} = -\rho g \tag{4.126}$$

$$= -\frac{pMg}{RT} \tag{4.127}$$

$$\int \frac{dp}{p} = -\int \frac{Mg}{RT} dz = -\frac{Mg}{RT} \int dz \tag{4.128}$$

$$\ln p = -\frac{Mg}{RT} z + C_1 \tag{4.129}$$

where C_1 is an arbitrary constant of integration. Note that we have assumed that temperature is not a function of z—that is, that temperature does not vary with elevation. If we know the pressure at one elevation—for example, at $z = 0$, $p = p_0$—we obtain the final result for pressure as a function of elevation in an ideal gas. Applying this boundary condition:

$$\text{BC:} \quad z = 0 \quad p = p_0 \tag{4.130}$$

$$\ln p_0 = C_1 \tag{4.131}$$

$$\ln p = -\frac{Mg}{RT} z + \ln p_0 \tag{4.132}$$

$$\ln \left(\frac{p}{p_0} \right) = -\frac{Mg}{RT} z \tag{4.133}$$

$$\frac{p}{p_0} = e^{-\frac{Mg}{RT} z} \tag{4.134}$$

Pressure variation due to gravity in an isothermal ideal gas

$$\boxed{p = p_0 e^{-\frac{Mg}{RT} z}} \tag{4.135}$$

The result in Equation 4.135 predicts that for air ($M = 29$ g/mol) at standard temperature (T $= 300$K), a height difference of about a kilometer produces a 10 percent change in pressure. Thus, the variations in pressure experienced by a stationary ideal gas due to gravity are not severe except when large distances are considered. Our analysis assumes that temperature is constant in a gas, which is not true in the atmosphere. If measurements of $T(z)$ are available, we can include that effect in the integration in Equation 4.128 and obtain a more accurate equation for pressure variation in Earth's atmosphere.

From these examples, we see that even without an "ideal liquid law," we can make meaningful calculations of forces in static liquids. The force on a surface in a static liquid is calculated from pressure. Pressure in a static liquid or gas is isotropic (i.e., the same in all directions) and depends on only elevation in a gravity field (see Equation 4.68). With these facts and the momentum balance (see Equation 4.1), we can analyze static-fluid devices.

The next two sections discuss the application of our new understanding of forces in stationary fluids to engineering devices that contain static fluids. We treat the more complicated subject of forces in moving fluids in Section 4.3.

4.2.3 Pascal's principle

In the previous section, we establish that the pressure on a stationary liquid is given by:

$$\begin{array}{c}\text{Pressure in a}\\ \text{stationary fluid}\\ (z\text{-direction upward})\end{array} \qquad \boxed{p(z) = -\rho g z + p_0} \qquad (4.136)$$

$$\text{Alternate expression:} \quad p_{bottom} = p_{top} + \rho g h \qquad (4.137)$$

where ρ is the density of the liquid, g is the magnitude of the acceleration due to gravity, p_0 is the pressure at $z = 0$, and z points upward ($\underline{g} = -g\hat{e}_z$). In that discussion, we consider the unconventional device shown in Figure 4.20, in which the three branches of the vessel are open to the atmosphere. We used Equation 4.136 to explain the pressure distribution in the device.

The fluid in the device in Figure 4.20 was unconfined—that is, open to the atmosphere. In many practical uses of liquids in engineering devices, external forces are imposed on confined liquids. An important reason that confined liquids are used in engineering designs is the way they transmit external forces. To see how stationary, confined liquids transmit external forces, consider the same unconventional device but modified such that we can impose an external force on the top surface of the liquid.

EXAMPLE 4.8. *The device shown in Figure 4.24 is pressurized using a pump until the gas pressure in the device is 2.00 atm. What are the pressures at the points indicated in the figure?*

SOLUTION. To analyze the new device, we follow the same procedure as when considering the original device. We apply the results of this chapter: Pressure at a point in a stationary liquid depends on only the elevation of the point (Equation 4.136). Pressure in a stationary liquid depends on neither the shape of the container nor the lateral (x- or y-) position of the point.

The three top surfaces are open to the pressurized gas; thus, the pressure at the top free surface ($z = 0$) is 2.00 atm $= 2.03 \times 10^5$ Pa:

$$p(z) = -\rho g z + p_0 \qquad (4.138)$$

$$p_A = p_B = p_C = p(0) = 2.00 \text{ atm} = 2.03 \times 10^5 \text{ Pa} \qquad (4.139)$$

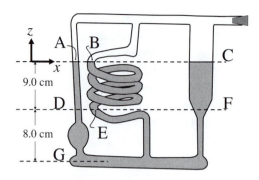

Figure 4.24 A device similar to the one shown in Figure 4.20 is constructed but in this device, the fluid is pressurized by an external source and then sealed.

Below the top surface, the pressures are higher. Points D, E, and F have the same elevation ($z = -h_D$) and, hence, the same pressure:

Pressure at D, E, F: $p(z) = -\rho g z + p_0$ (4.140)

$$p(-h_D) = \rho g h_D + p_0 \tag{4.141}$$

$$= \left(995\frac{kg}{m^3}\right)\left(9.80\frac{m}{s^2}\right)(9.0 \times 10^{-2}\,m)\left(\frac{Pa\ s^2\ m}{kg}\right)$$

$$+\ 2.02650 \times 10^5\,Pa$$

$$p_D = p_E = p_F = 2.03528 \times 10^5\,Pa \tag{4.142}$$

$$= 2.04 \times 10^5\,Pa \tag{4.143}$$

The pressure at Point G is calculated the same way:

$$p(z) = -\rho g z + p_0 \tag{4.144}$$

Pressure at G: $p(-h_G) = \rho g h_G + p_0$

$$= \left(995\frac{kg}{m^3}\right)\left(9.80\frac{m}{s^2}\right)(17.0 \times 10^{-2}\,m)\left(\frac{Pa\ s^2\ m}{kg}\right)$$

$$+\ 2.02650 \times 10^5\,Pa$$

$$p_G = 2.04308 \times 10^5\,Pa \tag{4.145}$$

$$= 2.04 \times 10^5\,Pa \tag{4.146}$$

Note that to three significant figures, the later two answers are the same because the fluid heights are modest. As the imposed gas pressure increases, the contribution to the total pressure that is made by $\rho g h$ decreases and becomes insignificant.

Note that the extra pressure applied to the device in the previous example was distributed equally throughout the liquid; that is, the pressure at every point after the extra pressure was applied is equal to its previous pressure plus the newly applied extra pressure. This ability of confined liquids to distribute pressure equally throughout a device has important engineering applications.

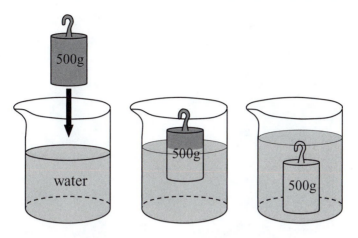

Figure 4.25 Unconfined gases and liquids move when pushed by a normal force. If the fluid is unconfined, the imposed normal force transmits through the fluid and finds a location where a shearing motion is possible. Molecules subjected to a shearing stress deform because fluids cannot sustain a shear stress without continuous deformation.

Gases and liquids are both fluids (i.e., by definition, unable to sustain shear forces without deforming continuously) but, as we have seen, these two types of fluid respond differently to applied external normal forces. Under the action of external normal forces, confined gases are made to occupy smaller volumes (see Figure 4.11). Confined liquids under the action of normal forces cannot reduce in volume, and they resist normal forces by building up internal pressure through intermolecular repulsion (see Figure 4.16). Key to this picture of liquid or gas response is that the fluid is confined. Unconfined fluids—both gases and liquids—move when pressed on by a normal force (Figure 4.25).

The response of a confined incompressible liquid to a normal force is summarized in a compact statement known as *Pascal's principle: Pressure exerted on a confined liquid is transmitted equally to every part of the liquid and to the walls of the container.*

We describe Pascal's principle (without naming it) in Section 4.2.2 when we discuss how a confined incompressible fluid responds to a normal force: Force applied to the top liquid layer transmits through that layer to the next layer, and so on, until all molecules in the container are affected. Each molecule pushes against its neighbors or against the confining walls and—finding no relief through motions of the neighbors or of the wall—all molecules in the container share the burden of the normal force and enter into the elevated energy state of being slightly closer together. The rigidity of the container walls is essential to forcing the liquid molecules to attempt to compress and therefore to store energy by adopting a slightly deformed molecular-orbital state. If the molecules are not trapped, they would rather move than be compressed (see Figure 4.25).

Pascal's principle is unique to confined incompressible liquids at rest. Confined compressible fluids reduce in volume under the action of an external force, and the final pressure of a compressible fluid in a device such as shown in Figure 4.11 depends on the final volume through the equation of state ($pV = NRT$). Incompressible solids do not follow Pascal's principle because in solids, it is forces that transmit directly, not pressure (see the sidebar). A solid with a

Figure 4.26 A solid stored in a container does not occupy the container in the same way that a liquid does. When a foam brick sits in a bag (the container), it barely touches the walls and little stress is transferred. When a weight is placed on the brick, little or no force (i.e., normal force or shear force) is transferred to the walls of the container. A liquid, by contrast, must transmit forces to the walls of its container.

rectangular shape, for example, housed in a rectangular container can sustain normal forces and shear forces without imparting them to the container walls (Figure 4.26). Granular solids, by contrast, transmit some forces to the walls of their containers; however, not all forces are transmitted and the transmitted forces are not distributed equally to all locations. Depending on how a container is loaded with a granular material, many different physical configurations are possible. Figure 4.27 shows three different configurations of similar amounts of

Figure 4.27 A granular material such as foam gems stored in a container may or may not transmit applied normal forces to the walls of the container. The configuration adopted by a stored granular material depends on the shape and size of the grains as well as how a container is loaded [30].

granular material; the amount of load transferred to the walls is different in each case.

Pascal's principle has been exploited for centuries in engineering devices. In the next two sections, we conclude our discussion of static fluids by considering the operation of two important engineering devices: the manometer and the hydraulic lift. These devices depend on Pascal's principle for their operation. In Section 4.3, we discuss forces in moving fluids.

How do solids transmit external forces from one surface to another?

Pressure is defined as isotropic normal force per unit area. Thus, if we know the magnitude of normal force on a surface and the area of the surface, the pressure is the ratio of the two:

$$p = \left(\frac{\text{normal force}}{\text{area}} \right) \tag{4.147}$$

This definition is straightforward; however, because forces transmit differently in solids and liquids, we must apply carefully the concept of pressure for different systems. Consider the pressure at the bottom of a cylindrical column made of a homogeneous solid such as gold (Figure 4.28). The bottom surface has an area πR^2, where R is the radius of the column. The magnitude of normal force exerted by the column on the surface on which it stands is

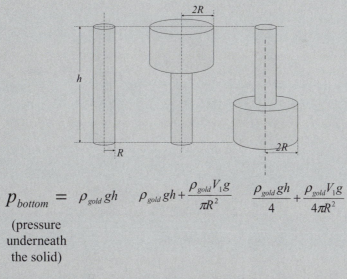

solid

$$p_{bottom} = \rho_{gold} g h \qquad \rho_{gold} g h + \frac{\rho_{gold} V_1 g}{\pi R^2} \qquad \frac{\rho_{gold} g h}{4} + \frac{\rho_{gold} V_1 g}{4\pi R^2}$$

(pressure underneath the solid)

Figure 4.28 Solids and liquids transmit forces differently, which we can understand by considering objects of various shapes. For solids, all of the weight of the solid exerts force on the bottom surface.

(continued)

How do solids transmit external forces from one surface to another? *(continued)*

the force due to the weight:

$$\begin{pmatrix} \text{magnitude of} \\ \text{force exerted} \\ \text{by solid-gold} \\ \text{column} \end{pmatrix} = (\text{mass}) \begin{pmatrix} \text{acceleration} \\ \text{due to} \\ \text{gravity} \end{pmatrix} \tag{4.148}$$

$$= \left[\left(\frac{\text{mass}}{\text{volume}} \right) (\text{volume}) \right] (\text{gravity}) \tag{4.149}$$

$$= \rho_{gold} \, \pi R^2 h \, g \tag{4.150}$$

The table on which the column stands exerts an equal and opposite force upward on the column. The gauge pressure[1] at the bottom of the column is the weight divided by the area:

$$p = \left(\frac{\text{force}}{\text{area}} \right) = \frac{\rho_{gold} g h \pi R^2}{\pi R^2} \tag{4.151}$$

Gauge pressure at bottom of solid-gold rod

$$\boxed{p_{bottom} = \rho_{gold} g h} \tag{4.152}$$

If we consider a solid object of a less regular shape, the calculation is the same. Consider the middle solid-gold object in Figure 4.28. This object is the same height as the original rod, but at the top there is now a ledge of radius $2R$. Let V_1 be the volume of gold added to make the second object from the first object. The total mass of the new object is then:

$$\begin{pmatrix} \text{mass of} \\ \text{irregular rod} \end{pmatrix} = \rho_{gold} (\text{original volume} + V_1) \tag{4.153}$$

$$= \rho_{gold} \left(\pi R^2 h + V_1 \right) \tag{4.154}$$

The pressure on the bottom of the irregular rod is force due to gravity divided by the contact area:

$$p = \frac{\rho_{gold} g \left(\pi R^2 h + V_1 \right)}{\pi R^2} \tag{4.155}$$

Gauge pressure at bottom of irregular rod (small side down)

$$\boxed{p_{bottom} = \rho_{gold} g h + \frac{\rho V_1 g}{\pi R^2}} \tag{4.156}$$

We see that for a solid object, the pressure at the bottom increases when the mass of the material increases.

We also can turn the irregular column upside down and recalculate the pressure at the bottom. The mass is the same (Equation 4.154) but the rod sits on the wide bottom; thus, the

---

[1] A pressure gauge usually reads zero when opened to atmospheric pressure. To convert gauge pressure to absolute pressure, we add the atmospheric pressure.

How do solids transmit external forces from one surface to another? *(continued)*

force is distributed over a larger area. The pressure at the bottom is calculated in the same way as before:

$$p = \frac{\rho_{gold}g\left(\pi R^2 h + V_1\right)}{\pi(2R)^2} \tag{4.157}$$

Pressure at
bottom of
irregular rod
(larger side down)

$$p_{bottom} = \frac{\rho_{gold}gh}{4} + \frac{\rho V_1 g}{4\pi R^2} \tag{4.158}$$

When oriented large side down, the irregular rod exerts a different pressure on the table that supports it. Because the contact area is larger by a factor of four, the pressure is smaller by that same factor.

Now consider a fluid system. Figure 4.29 is a cylindrical container holding water. The volume of water in the cylindrical container is the same as the volume of the first gold rod considered previously. The pressure at the bottom of the column of water is given by Equation 4.64 applied to this system. Note that we are measuring pressure at the bottom of the fluid, *not between the solid container and the table*:

$$p(z) = -\rho gz + p_0 \tag{4.159}$$

where the z-direction points upward. If we choose $z = 0$ at the base of the column, then p_0 is the pressure at the bottom. The pressure at the top, $p(z) = p(h)$, is atmospheric:

$$p(h) = -\rho gh + p_0 \tag{4.160}$$

$$p(h) = p_{atm} \tag{4.161}$$

liquid

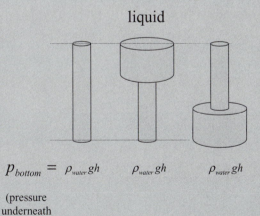

$$p_{bottom} = \rho_{water}gh \qquad \rho_{water}gh \qquad \rho_{water}gh$$

(pressure
underneath
the liquid)

Figure 4.29   Solids and liquids transmit forces differently, which we can understand by considering objects of various shapes. For liquids, not all of the weight exerts its force on the bottom liquid surface; some of the force due to the weight is exerted on the walls as well as any ledges in the container. Note that the pressure at the bottom discussed in this figure is the pressure inside the container, measured near the bottom of the liquid.

(continued)

How do solids transmit external forces from one surface to another? *(continued)*

$$p_0 = p_{atm} + \rho_{water}gh \qquad (4.162)$$

Gauge pressure
at the bottom
of the fluid column

$$p_{bottom} = p_0 - p_{atm} = \rho_{water}gh \qquad (4.163)$$

where ρ is the density of the fluid, water, and p_{bottom} is given in gauge pressure. This result is the same (except for the difference in density) as Equation 4.152, which was the calculation of gauge pressure at the bottom of the solid-gold column of the same shape.

To determine the effect of irregular shapes on pressure in liquids, we imagine a container with the same shape as the irregular gold rod with the ledge at the top. If we fill this irregularly shaped container with water, the shape of the water in the container is the same as the shape of the irregular gold rod (see Figure 4.29, center). For this container, what is the fluid gauge pressure at the bottom of the water?

For the solid gold, we note that the force on the bottom is the mass of the gold multiplied by gravity; the gauge pressure on the bottom is this force divided by the area in contact with the surface on which it sits. When we add extra weight at the top of the rod, the gauge pressure exerted on the bottom increases.[2]

For water, however, force transmits differently than it does for solids. In deriving Equation 4.159, we made no mention of the size or shape of the container holding the liquid, and it was not necessary. Equation 4.159 resulted from carrying out a momentum balance on a microscopic wedge-shaped control volume within a mass of fluid. This equation is equally valid for fluid in the odd-shaped container as in the cylindrical container. Equation 4.64 states that the pressure in a fluid depends on any imposed pressure on the top of the column and the vertical height of fluid above the point in question. Because the water in the irregular vessel rises to the same elevation as the water in the cylindrical container, we calculate the same pressure at the bottom of the irregular container as for the cylindrical container:

$$p_0 = p_{atm} + \rho_{water}gh \qquad (4.164)$$

Gauge pressure
at the bottom
of the fluid column
(irregular shape,
small side down)

$$p_{bottom} = p_0 - p_{atm} = \rho_{water}gh \qquad (4.165)$$

where ρ is the density of water and p_{bottom} again is given in gauge pressure. If we repeat the calculation for a third container with the shape of the irregular rod but this time with the wide

---

[2] These were gauge pressures because we did not consider the one atmosphere pressure that acts on the rod by virtue of it being present in the atmosphere.

How do solids transmit external forces from one surface to another? *(continued)*

part at the bottom, the result would be the same:

$$
\begin{pmatrix}
\text{pressure} \\
\text{at the bottom} \\
\text{of the fluid column} \\
\text{of height } h \\
\text{(cylindrical)}
\end{pmatrix}
=
\begin{pmatrix}
\text{pressure} \\
\text{at the bottom} \\
\text{of the fluid column} \\
\text{of height } h \\
\text{(irregular shape,} \\
\text{small side down)}
\end{pmatrix}
=
\begin{pmatrix}
\text{pressure} \\
\text{at the bottom} \\
\text{of the fluid column} \\
\text{of height } h \\
\text{(irregular shape,} \\
\text{large side down)}
\end{pmatrix}
= \rho_{water} g h
$$

(4.166)

We see from this discussion that solids and liquids are different in how they transmit forces. The difference is due to intermolecular forces. The strong intermolecular forces that hold solids together do not allow a solid's molecules to move under application of a force. Thus, solids retain their shape under the pull of gravity, and the molecules in a solid sample move as one.

Liquids have intermolecular forces that hold the liquid's molecules together, but these forces are not strong. Liquids deform under the pull of gravity (adopting the shape of their container) and, when forces are applied to a liquid, the liquid usually deforms.

The difference between the gold rods and the water containers in our example is the role of the container. For the solid, there is no container. All of the force due to the weight of the gold is transmitted—because of the strong intermolecular forces holding the gold together—to the bottom surface in contact with the table. For the liquid, there is a container. If there were no container, the liquid would spread out on the table under the force of gravity. In the cylindrical container (Figure 4.30), the shape of the container directs the effect of gravity on the liquid. Each layer of water exerts a force due to gravity on the layer below it. The cumulative effect of all of this mass, stacked up vertically by the cylindrical shape of the container, is to exert the total gravitational force of the water on subsequent liquid layers and finally on the bottom surface of the container (of area πR^2). The cylindrical container holds the water in the shape of a rod; as a rod, the water exerts the same force as a rod of gold (the only difference is the density of the two materials).

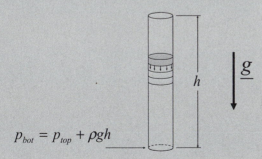

$$p_{bot} = p_{top} + \rho g h$$

Figure 4.30   The liquid in the cylindrical container is vertically stacked up and aligned by the shape of the container. Thus, the entire weight of the fluid is directed by the strong cohesive intermolecular forces of the solid toward the bottom of the column of fluid; the pressure at the bottom of the column is the same as exerted by a solid of the same shape.

(continued)

How do solids transmit external forces from one surface to another? *(continued)*

What about irregular shapes? For an irregular solid bar, the mass is larger than the mass of a cylindrical rod, and all of the force of gravity on this mass is directed by cohesive forces in the solid toward the area of contact with the table. We calculated the pressure due to the irregular rod as the new mass divided by the contact area. For the water, the situation is different. Again, the container serves to hold the liquid in the shape of the container but, because of its irregular shape, the container also supports some of the weight of the water.

Consider the water in the region $R < r < 2R$ in the irregular vessel (Figure 4.31). This liquid is sitting on a shelf created by the shape of the container, and the glass container wall below this fluid is supporting the liquid. A portion of the force due to gravity on the liquid thus is transmitted to the solid walls of the container and subsequently is transmitted by the solid walls to the table. The only pressure effect *in the liquid* is the effect that is present in the column of fluid directly above the bottom surface: The pressure in a liquid varies with elevation:

$$\begin{array}{c}\text{Pressure}\\\text{at the bottom}\\\text{of a column}\\\text{of height } h\end{array} \qquad \boxed{p_{bottom} = p_{top} + \rho g h} \qquad (4.167)$$

We reiterate that we are discussing the pressure inside the container at the bottom: the pressure in the liquid. The net result is that the pressure exerted by solids of the same height vary as shown in Figure 4.28, whereas the pressure exerted by liquids of the same height is constant (see Figure 4.29).

The way that forces transmit in solids gives rise to the mechanical advantage of levers, pulleys, and other simple machines [167]. The way that pressure transmits in liquids gives rise to the mechanical advantage of the hydraulic lift and other hydraulic devices, as discussed in the next section.

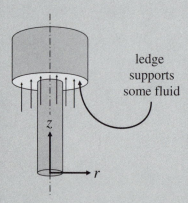

ledge
supports
some fluid

| **Figure 4.31** | In containers that are not cylindrical, some of the force due to gravity on the liquid is transmitted to the walls of the container. The support by the walls is directed to the table through the solid walls of the container. Thus, not all of the mass of the fluid contributes to the pressure at the bottom of the column of liquid. |

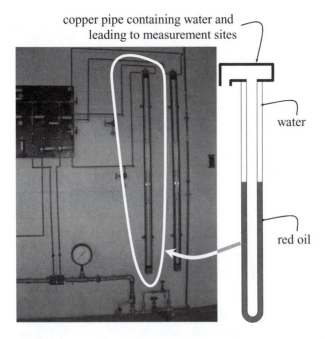

copper pipe containing water and
leading to measurement sites

water

red oil

Manometers are a reliable way to measure pressure because they do not depend on calibration of any sort. As long as the density of the measurement fluid is known, the pressure difference between the two sides of a manometer may be measured accurately. Two manometers are shown but only one is highlighted. The measurement fluid is red oil and the process fluid is water; the reading on the manometer is zero pressure difference between the two sides. (Photograph courtesy of David Caspary, Michigan Technological University)

4.2.4 Static fluid devices

A manometer is a simple device that can be used to accurately measure pressure differences. Manometers are used rarely in an industrial setting, but they often are found in a laboratory (Figure 4.32). Manometers work through the principles we elucidate in this chapter: Pressure in a stationary fluid is isotropic, independent of horizontal position, and a linear function of elevation (Equation 4.64).

4.2.4.1 MANOMETERS

The operation of a manometer exploits the z-dependence of pressure in a liquid. We learned in the previous section that $p(z)$ can be written as:

$$p(z) = -\rho g z + p_0 \qquad (4.168)$$

where gravity is in the $-\hat{e}_z$-direction and p_0 is the pressure at $z = 0$. Solving Equation 4.168 for p_0:

Pressure at the
bottom of a column
of fluid of height z

$$\boxed{p_0 = p(z) + \rho g z} \qquad (4.169)$$

$$\boxed{p_{bottom} = p_{top} + \rho g h} \qquad (4.170)$$

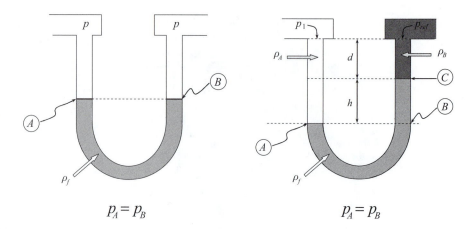

$$p_A = p_B \qquad\qquad p_A = p_B$$

Figure 4.33 General schematic of a U-tube manometer. Fluid of density ρ_f fills the manometer. Different fluids fill in above the manometer fluid on the two sides of the instrument.

where $p(z)$ is the pressure at the top of the column of fluid (position z). This is the main equation needed to understand the functioning of manometers.

A typical schematic of a U-tube manometer is shown in Figure 4.33. A heavy measuring fluid of density ρ_f is trapped in a U-shaped glass tube. The left end of the tube is connected to a fluid at a point where it is desired to measure the pressure. The right end of the manometer tube is exposed to atmospheric pressure or to some other pressure that serves as a reference. When the pressures on the two sides of the manometer are equal (left), the measuring fluid levels out on both sides of the manometer (imagine holding a flexible tube with some water trapped in it (Figure 4.34)). Recall that pressure in a liquid depends on only elevation, not on lateral position (x, y). The two sides of the manometer are at the same elevation; therefore, they are at the same pressure. Conversely, because the two sides are at the same pressure, they rise to the same elevation.

When the pressure on the left is higher than the pressure on the right side of a manometer, the fluid level on the left depresses and the fluid level on the

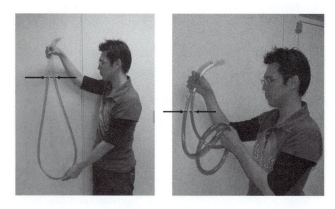

Figure 4.34 A flexible tube demonstrates the principle that drives all manometer calculations. When a fluid is trapped in a flexible tube, both ends are open to the same pressure (i.e., atmospheric), and the fluid levels are the same. Changing the shape of the tube by coiling it does not change the fluid levels.

right rises (see Figure 4.33, right). The difference in levels in the manometer under pressure can be analyzed using Equation 4.169 to yield a measurement of the pressure difference between the two locations. Because it takes very little pressure difference to produce measurable fluid-height differences, manometers are sensitive measuring devices.

If we draw a dotted horizontal line from Point A to Point B in the schematic on the right side of Figure 4.33, below this line is an equilibrium state that we understand—a continuous plug of one type of fluid with equal fluid levels on both sides. The conclusion drawn from Equation 4.169 is that the pressures at Points A and B are the same:

$$p_A = p_B \tag{4.171}$$

To determine the unknown pressure p_1, we apply Equation 4.169 to each side of the manometer, calculate p_A and p_B, and solve Equation 4.171.

The pressure at Point A is obtained from a straightforward application of Equation 4.169: p_1 is the pressure at the top and $z = (h + d)$ is the height of the fluid:

$$p_0 = p(z) + \rho g z \tag{4.172}$$

$$p_A = p_1 + \rho_A g(h + d) \tag{4.173}$$

The pressure at Point B has contributions from two different columns of fluid. We apply Equation 4.169 to the two different fluid columns sequentially:

$$p_C = p_{\text{ref}} + \rho_B g d \tag{4.174}$$

$$p_B = p_C + \rho_f g h \tag{4.175}$$

$$= p_{\text{ref}} + \rho_B g d + \rho_f g h \tag{4.176}$$

Substituting Equations 4.173 and 4.176 into Equation 4.171, we obtain:

$$p_A = p_B \tag{4.177}$$

$$p_1 + \rho_A g(h + d) = p_{\text{ref}} + \rho_B g d + \rho_f g h \tag{4.178}$$

U-tube
manometer
equation
$$\boxed{p_1 - p_{\text{ref}} = (\rho_f - \rho_A)g h + (\rho_B - \rho_A)g d} \tag{4.179}$$

Thus, the pressure difference between p_1 and a reference pressure may be calculated by measuring h and d and knowing the densities of the fluids in the manometer. If fluids A and B are the same (e.g., both air), then the term involving d vanishes ($\rho_A = \rho_B$) and we obtain a simple result:

U-tube
manometer equation
(same fluid above both sides)
$$\boxed{p_1 - p_{\text{ref}} = (\rho_f - \rho_A)g h} \tag{4.180}$$

The relationships we explored to arrive at the U-tube manometer equations can be used in more complex manometers, as shown in the following examples.

The key to analyzing manometers is to remember the principles used to derive the equations: For a section of tube filled with only one type of fluid, the pressure is the same at two points at the same elevation; pressure at the bottom of a column of fluid is equal to the pressure at the top plus ρg(height).

EXAMPLE 4.9. *A manometer is configured as shown in Figure 4.35. A heavy fluid has been placed in the bottom of the manometer. A light fluid has been added to the left side of the manometer only. Both sides of the manometer are connected to a process stream, and the manometer is used to measure the pressure difference $p_1 - p_2$. The fluid densities and the manometer readings are indicated in Figure 4.35. What is the pressure difference $p_1 - p_2$ in terms of fluid heights and fluid densities?*

SOLUTION. The pressures must be equal at the two points labeled (a) and (b) in Figure 4.35. We can relate these two pressures to the unknown pressures p_1 and p_2 using the manometer principles described previously. Before we use numbers, we first derive the equation to use in the calculation.

In a continuous fluid, fluid at the same elevation has the same pressure. This condition is met for Points a and b in Figure 4.35:

$$p_a = p_b \tag{4.181}$$

The other principle of static fluids is that the pressure at the bottom of a column of fluid is equal to the pressure at the top, plus density multiplied by gravity multiplied by the height of the column of fluid. When different columns of fluid stack on top of one another as in this example, the pressures due to each column simply add up. Thus, the pressure at Point a is:

$$\text{pressure at } a = \begin{pmatrix} \text{pressure} \\ \text{at} \\ \text{top} \end{pmatrix} + \begin{pmatrix} \text{pressure} \\ \text{due to} \\ \text{fluid C} \end{pmatrix} + \begin{pmatrix} \text{pressure} \\ \text{due to} \\ \text{fluid A} \end{pmatrix} \tag{4.182}$$

$$= p_1 + \rho_C g h + \rho_A g d \tag{4.183}$$

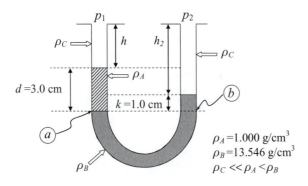

Figure 4.35 Manometers come in various shapes and use a variety of fluids. In this example, we consider a manometer that contains two different measurement fluids.

where h is the unknown height of the fluid above the ρ_A fluid in the lefthand side. In the figure, it is given that the density of fluid C is very small; therefore, the $\rho_C gh$ term may be neglected. Similarly, the pressure at Point b is:

$$\text{pressure at } b = p_2 + \rho_C gh_2 + \rho_B gk \qquad (4.184)$$

$$= p_2 + \rho_B gk \qquad (4.185)$$

where h_2 is the height of ρ_C fluid on the righthand side; in Equation 4.185, we neglect this contribution. Equating the pressures at Points a and b allows us to solve for $p_1 - p_2$:

$$(\text{pressure at } a) = (\text{pressure at } b) \qquad (4.186)$$

$$p_1 + \rho_A gd = p_2 + \rho_B gk \qquad (4.187)$$

$$p_1 - p_2 = g(\rho_B k - \rho_A d) \qquad (4.188)$$

Substituting the numerical values, we obtain:

$$p_1 - p_2 = 980\frac{\text{cm}}{\text{s}^2}\left[\left(13.546\frac{\text{g}}{\text{cm}^3}\right)(1.0 \text{ cm}) - \left(1.000\frac{\text{g}}{\text{cm}^3}\right)(3.0 \text{ cm})\right]$$

$$= 10{,}335 \text{ g/(cm} \cdot \text{s}^2) \qquad (4.189)$$

$$= \boxed{1.0 \times 10^4 \text{ dynes/cm}^2}$$

EXAMPLE 4.10. *A double-well manometer is an instrument that can be used to measure small pressure differences. The double well is a U-tube–type manometer; however, at the top of each side, there is a well with a larger cross-sectional area. There are two types of fluid in the double well: the heavier bottom fluid and the lighter top fluid. The same amount of top fluid is placed on both sides of the manometer, as shown in Figure 4.36a. What is the expression for the pressure difference between the two sides of the double-well manometer in Figure 4.36b? The cross-sectional area of both wells is A, and the cross-sectional area of the U-tube is a.*

SOLUTION. When the pressure is equal on both sides of this device, the top levels of the two fluids are equal (Figure 4.36a). When pressure is increased on the lefthand side, the levels change as shown in Figure 4.36b. We can analyze this problem if we recognize the relationship between the pressures at Points 1 and 2. In a continuous fluid, fluid at the same elevation has the same pressure. This condition is met for the two points marked 1 and 2 in Figure 4.36.

$$p_1 = p_2 \qquad (4.190)$$

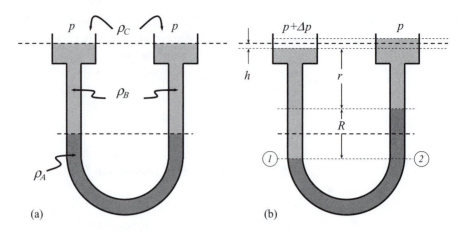

Figure 4.36 The double-well manometer can be used to magnify small changes in pressure. (a) When the pressure is the same on both sides, the level of the manometer fluid is the same. (b) When the pressure is higher on the left side, a small change in fluid height in the well translates to a much larger change in fluid height within the U-tube portion of the manometer.

We write the pressures at these two points in terms of Δp and the various fluid heights using Equation 4.169, first applied to the left side and then to the right side:

$$p_1 = p_2 \tag{4.191}$$

$$p + \Delta p + \rho_C g(2h) + \rho_B g(r + R) = p + \rho_B g(r + 2h) + \rho_A g R \tag{4.192}$$

$$\Delta p + \rho_C g(2h) + \rho_B g R = \rho_B g(2h) + \rho_A g R \tag{4.193}$$

$$\Delta p = (\rho_B - \rho_C) g(2h) + (\rho_A - \rho_B) g R \tag{4.194}$$

Equation 4.194 contains h, which is not measured. We can eliminate this height from the equation by relating h to other distances in the manometer. Imagine the process of applying the extra pressure Δp to the lefthand well in Figure 4.36a. The two wells start out at the same height. When Δp is applied, ρ_B-fluid moves from the wide well into the narrow tube. On the righthand side, ρ_B fluid moves from the narrow tube into the wide well above it. By invoking mass conservation, we can relate the changes in height that take place.

When an additional pressure Δp is applied to the lefthand side of the well in Figure 4.36a, the fluid that leaves the left well is of volume hA and of mass $\rho_B h A$. In the manometer lefthand tube, this same quantity of fluid takes up a volume $(R/2)a$ and mass $\rho_B a(R/2)$. Equating the two masses, we find $hA = aR/2$ or $2h = R(a/A)$. Substituting this into Equation 4.194, we arrive at our final result for Δp:

$$\Delta p = g R \left[(\rho_B - \rho_C) \left(\frac{a}{A} \right) + (\rho_A - \rho_B) \right] \tag{4.195}$$

The pressure difference Δp thus is determined by measurement of R alone; all of the densities as well as the geometric terms a and A must be known.

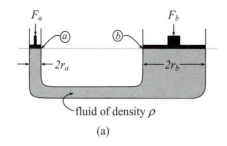

(a)

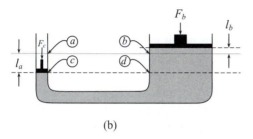

(b)

Figure 4.37   A hydraulic lift works on the same principle as a manometer. Pressure applied to the side with a small diameter is transmitted through the fluid and acts on the larger diameter on the other side. Although the pressure is the same, because the area is larger, the force is proportionally larger. When the pressure is the same on both surfaces, the liquid levels are the same (a). When the hydraulic lift operates, fluid is displaced from the left side to the right side, and a large force is generated (b).

Note that in solving these two problem, we did not attempt to adapt the U-tube manometer formula previously derived but rather applied the principles of fluid statics directly to the device under consideration. These principles are as follows:

Principles of Fluid Statics

1. In a continuous fluid, fluid at the same elevation has the same pressure.
2. The pressure at the bottom of a column of fluid is equal to the pressure at the top, plus density multiplied by gravity multiplied by the height of the column of fluid.

A consequence of the second principle is that when different columns of fluid stack on top of one another, the pressures due to each column simply add up.

4.2.4.2 HYDRAULIC LIFTS

We can show that there is a significant mechanical advantage implicit in Pascal's principle. Consider the distorted manometer in the top of Figure 4.37. The left side of the apparatus has a small radius r_a, whereas the right side has a much larger radius r_b. Because the same fluid fills both sides of the apparatus to the same levels like a manometer, the pressures on the liquid surfaces a and b are equal in this first configuration and $p_a = p_b$. From this observation, we can relate F_a and F_b:

$$p_a = p_b \tag{4.196}$$

$$\frac{F_a}{\pi r_a^2} = \frac{F_b}{\pi r_b^2} \tag{4.197}$$

where F_a and F_b are the forces on the left and right pistons, respectively. Because $r_a^2/r_b^2 < 1$, the force on surface b is larger than the force on surface a:

$$F_b = \frac{r_b^2}{r_a^2} F_a \tag{4.198}$$

Equation 4.198 indicates that a relatively small force applied to the left side of the apparatus in Figure 4.37 can be translated into a larger force on the right side. The apparatus in Figure 4.37 is called a hydraulic lift, and the hydraulic-lift principle is used by automobile mechanics to raise vehicles off the shop floor. The same principle is used in automotive braking systems and pneumatic actuators for lifting or other applications requiring large forces.

The physics involved in a hydraulic lift is Pascal's principle. The force on the small surface is transmitted to the confined incompressible liquid in the lift. The liquid cannot compress or escape, so the force is transmitted to all of the molecules in the container and acts in all directions. The molecules are forced closer together than their equilibrium position, which affects all of the molecules in the container. The molecules in contact with the righthand side of the lift exert the same pressure on that piston as is exerted among all of the molecules at the same elevation. That pressure applied over the larger piston surface area results in a larger cumulative force. Note that because high pressure exists everywhere in the device, the entire device must be designed to withstand high pressure.

The force amplification in a hydraulic lift may appear to violate the principle of conservation of energy, but this is not the case, as seen in the following example.

EXAMPLE 4.11. *Show that the hydraulic lift does not violate conservation of energy.*

SOLUTION. Consider Figure 4.37b. To calculate the work done when the lift operates, let l_a be the downward distance traveled by the piston between a and c when a force F_c is applied. The downward displacement of fluid on the left side of the lift must be matched with an upward displacement of fluid on the right side, but the distance upward traveled by the larger piston on the right will not be l_a; rather, it will be much smaller due to the difference in cross-sectional area. We can use a mass balance to calculate l_b, the distance upward that larger piston travels:

$$\begin{pmatrix} \text{mass displaced} \\ \text{downward} \\ \text{on left side} \end{pmatrix} = \begin{pmatrix} \text{mass displaced} \\ \text{upward} \\ \text{on right side} \end{pmatrix} \tag{4.199}$$

$$\rho \pi r_a^2 l_a = \rho \pi r_b^2 l_b \tag{4.200}$$

$$\frac{r_a^2}{r_b^2} = \frac{l_b}{l_a} \tag{4.201}$$

where ρ is the density of the incompressible fluid in the device, called the hydraulic fluid. We can calculate the relationship between F_c and F_b as well. The principles used to analyze manometers apply; that is, pressure is equal at equal elevations of fluid. For the operating hydraulic lift, the pressures are equal at Points c and d in Figure 4.37b. The pressure at Point c is due to the downward force F_c, whereas the pressure at Point d is due to the downward force F_b plus the weight of the hydraulic fluid above the cd level:

$$p_c = p_d \tag{4.202}$$

$$\frac{F_c}{\pi r_a^2} = \frac{F_b}{\pi r_b^2} + \rho g(l_a + l_b) \tag{4.203}$$

Substituting Equation 4.201 into Equation 4.203 and performing some algebra, we obtain:

$$F_c l_a = F_b l_b + [\pi r_a^2 l_a \rho] g (l_a + l_b) \tag{4.204}$$

The quantity in square brackets in Equation 4.204 is the mass of fluid that transfers from the left side to the right side. Calling this quantity m, we obtain:

$$F_c l_a - mg l_a = F_b l_b + mg l_b \tag{4.205}$$

Equation 4.205 can be interpreted as an energy balance. Energy is the potential to do work. Equation 4.205 states that the energy expended on the left side of the hydraulic lift is equal to the energy expended on the right side. The energy expended on the left side is the work pushing down the small piston (i.e., force times displacement, $F_c l_a$) plus the work pushing down the mass of hydraulic fluid ($-mg l_a$, which is negative because work is recovered rather than expended when a weight is pushed down in a gravity field). The work performed on the right side is equal to the work done lifting the large piston a distance l_b ($F_b l_b$) plus the work done to lift the mass m of hydraulic fluid ($+mg l_b$). An equal amount of work is done on both sides of the apparatus, even though a larger force is generated on the right side compared to the left. The mechanical advantage obtained in a hydraulic lift is similar to the mechanical advantages gained in other elementary machines, such as a lever and a block and tackle [67].

EXAMPLE 4.12. *For a hydraulic lift with dimensions given in Figure 4.38 and using a maximum force of 20 lb$_f$, what diameter is needed on the small-diameter side to lift a large pickup truck that weighs 5.5 tons (11,000 lb$_f$)? How much vertical displacement on the left is needed to lift the vehicle 87 inches?*

SOLUTION. The numbers we calculate in this example show us that the simplistic design discussed in Figure 4.38 is not the actual design of a hydraulic lift. The calculations also lead us to a better design.

The hydraulic lift in Figure 4.38 is designed to lift an 11,000 lb$_f$ vehicle 87 in. (220 cm) upward. The piston under the truck is 4.0 feet in diameter with an area of:

$$\text{piston area} = \pi R^2 = \pi (2.0 \text{ ft})^2 = 1810 \text{ in.}^2 \tag{4.206}$$

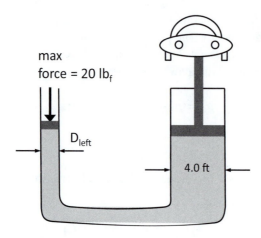

max
force = 20 lb$_f$

D_{left}

4.0 ft

Figure 4.38

A hydraulic lift is to be designed to lift an 11,000-lb$_f$ vehicle 87 inches in the air. The piston is 4.0 feet in diameter; on the left side, a maximum of 20 lb$_f$ is to be used to do the lifting.

The maximum displacement of the piston is 87 inches, from which we can calculate the volume of hydraulic fluid that must enter the righthand side of the lift:

$$\begin{array}{l} \text{volume of hydraulic fluid} \\ \text{displaced from left side} \\ \text{to right side} \end{array} = \pi R^2 L = \left(1810 \text{ in.}^2 \right) (87 \text{ in.}) \quad (4.207)$$

$$= 157{,}431 \text{ in.}^3 (682 \text{ gal}) \quad (4.208)$$

The main operating principle of the lift is that the pressure is the same on both sides (variation of pressure with elevation is negligible in these high-pressure devices). With the 11,000-lb$_f$ truck sitting on the 4-foot diameter piston on the right, a pressure of 6.1 psi is generated:

$$\text{Pressure on the right:} \quad p_{right} = \frac{11{,}000 \text{ lb}_f}{1810 \text{ in.}^2} \quad (4.209)$$

$$= 6.1 \text{ psi} \quad (4.210)$$

This is also the pressure on the left side. If the maximum force on the left side is 20 lb$_f$, we can calculate the diameter on the left:

$$p_{right} = p_{left} \quad (4.211)$$

$$6.1 \text{ } psi = \frac{20 \text{ lb}_f}{\pi \left(D_{left}/2 \right)^2} \quad (4.212)$$

$$D_{left} = 2.0 \text{ in.} \quad (4.213)$$

This is a small diameter, but the situation becomes worse when we consider the amount of vertical displacement required to lift the truck the desired amount. We calculated that a volume of 157,431 in.^{3} (682 gal) must pass from the left side to the right side to bring about the lifting. From this number, we can calculate the

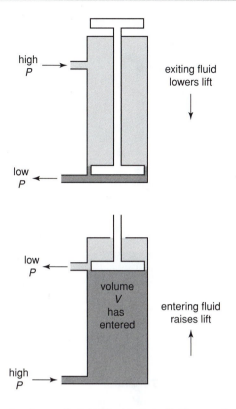

P →

exiting fluid
lowers lift
↓

low
P ←

low
P ←

volume
V
has
entered

entering fluid
raises lift
↑

high
P →

Figure 4.39 If we focus on the right side of the hydraulic lift, we see that what is required is for fluid to be fed below the piston to raise the vehicle. To lower the vehicle, fluid needs to be allowed to drain from under the piston.

vertical displacement on the left side:

$$V_{right} = V_{left} \tag{4.214}$$

$$157{,}431 \text{ in.}^3 = \pi \left(D_{left}/2\right)^2 L \tag{4.215}$$

$$L_{left} = 50{,}112 \text{ in.} = 4176 \text{ ft} = 0.79 \text{ mile} \tag{4.216}$$

These calculations are somewhat discouraging. If we want the force on the left to be cut down to the reasonable 20 lb$_f$, we must tradeoff the force with a large displacement. If we change the design to allow for a larger force (perhaps 100 lb$_f$), we still cannot obtain a reasonable displacement for the left (try this calculation for practice).

The problem is with the left side. If we can feed the fluid to the right side in little chunks—drawing the fluid from a reservoir, for example—we could accomplish the lift with the desired force reduction. Figure 4.39 shows the implementation of this idea. A pump is used to feed hydraulic fluid to the "right side" of the hydraulic lift. In actual practice, there is no longer a left side, only the right side where the vehicle is being lifted. The pump feeds fluid against the operating pressure of the lift (constant) and, as the fluid enters the lift, the truck rises. The force that the pump must use to put a small amount of fluid into the lift against the reservoir pressure is much smaller than the 11,000 lb$_f$ weight of the truck because each stroke of the pump is like the action of a left side of the hydraulic lift.

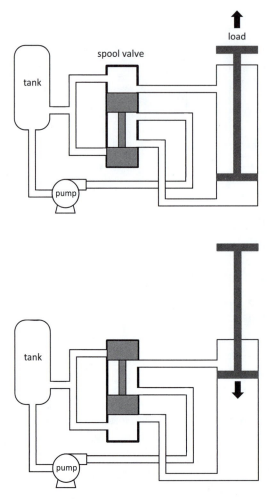

Figure 4.40 An actual hydraulic lift uses a pump to feed fluid below the piston of the lift. A spool valve switches how the suction and discharge ports on the pump are connected to the lift.

Figure 4.40 shows the conventional design of a pump-enabled hydraulic lift. In this design, hydraulic fluid is on both sides of the piston. A special valve called a spool valve is used to switch how the pump is connected to the lift. When the lift is operating to raise a vehicle, the discharge side of the pump is connected below the lift piston and the suction side is connected above the lift piston. Hydraulic fluid moves into the chamber and the vehicle rises. When the spool valve shifts into the alternate position, the discharge side of the pump is connected above the lift piston and the suction side is connected to the chamber below the piston. In this configuration, the piston descends. Note that once the vehicle is lifted, other valves may be closed to hold the lift in place and the pump may be turned off.

Our study of stationary fluids has familiarized us with some of the intrinsic properties of fluids. Fluids and solids differ in their responses to shear forces, as follows:

Properties of Fluids

1. Solids can remain deformed but stationary when subjected to shear forces, whereas fluids must move. Therefore, fluids at rest are experiencing only normal forces.
2. Normal forces in gases cause volume change, and the magnitude of normal force on a surface in a gas is related to volume, temperature, and number of moles through the equation of state; for example, the ideal gas law.
3. In stationary liquids, pressure is produced by intermolecular forces not present in gases.
4. For both liquids and gases, pressure is isotropic.
5. In both gases and liquids, pressure depends on elevation, although the density of gases is so small that no elevation dependence is observed unless the distances are very large.

The properties of static fluids may be exploited in engineering applications such as manometers and hydraulic lifts and must be accounted for in constructing weirs, dams, water towers, and other structures, as well as in the manufacture of boats, piers, plumbing networks and other systems involving stationary liquids.

We now discuss moving and deforming fluids in which shear forces are present; here, the situation is considerably more complex than in stationary fluids.

4.3 Fluids in motion

In this chapter, we discuss fluid contact forces, which are forces due to intermolecular effects. The major issue to address is how to account for intermolecular forces in our continuum model of fluid behavior—specifically, how to write intermolecular contact forces on a control volume (CV) so that we can carry out a momentum balance and solve for flow fields and other quantities of interest in moving fluids.

$$\text{Reynolds transport theorem} \atop \text{(momentum balance on CV)} \qquad \frac{d\mathbf{P}}{dt} + \iint_{CS} (\hat{n} \cdot \underline{v}) \, \rho \underline{v} \, dS = \sum_{\substack{\text{on} \\ \text{CV}}} \underline{f} \quad (4.217)$$

Our first topic was stationary fluids for which the Reynolds transport theorem reduces to a force balance $0 = \sum \underline{f}$. There are no shear forces in stationary fluids because shear forces always cause motion in fluids. For both stationary gases and liquids, the force on a fluid surface can be expressed in terms of pressure, a normal force per unit area. In stationary fluids, both gases and liquids, pressure at a point is isotropic—that is, pressure is independent of the orientation of the measurement surface. On a small flat surface ΔA with unit normal vector $\hat{n}$, the force on that surface in a stationary fluid is given by:

$$\text{Force at a point} \atop {\text{on a surface } \Delta A \atop \text{in a stationary fluid}} = [p(z)] \Delta A \, (-\hat{n}) \qquad (4.218)$$

The field p is a function of elevation if gravity is important. In the previous section, we discussed several applications of momentum balances (i.e., force balances) in stationary liquids.

In this section, we address the more complicated case of contact forces in moving fluids. A major complication in determining contact forces on surfaces in moving fluids is that shear forces may be present. In a stationary fluid, once a surface is chosen, we know the direction of the force on that surface—all contact forces in stationary fluids are normal forces. For example, in Equation 4.218 for force in a stationary fluid, knowing that the forces were normal forces allowed us to write $\underline{f} \propto \hat{n}$. When we seek to write the force on a surface in a moving fluid, we do not know the direction of the force—there may be a normal component to the force but there also may be a tangential (i.e., shear) component. In addition, both the magnitude and the direction of the force are a function of the fluid velocity:

$$\text{For fluid in motion:} \quad \underline{f}_{surface} = \begin{pmatrix} \text{expression} \\ \text{giving magnitude} \\ \text{of force} \\ (\text{function of } \underline{v}) \end{pmatrix} \begin{pmatrix} \text{unit vector} \\ \text{giving direction} \\ \text{for force} \\ (\text{function of } \underline{v}) \end{pmatrix}$$

$$(4.219)$$

Our challenge then is to find a modeling method and a mathematics that allows us to write the force on an arbitrary surface in a moving fluid as a function of the fluid properties and as a function of the velocity. Once this is obtained, we can proceed with solutions for fluids variables using the Reynolds transport theorem.

An additional challenge is that surface contact forces in a moving liquid are due to intermolecular attractions, which vary greatly from one material to another. Our preference in continuum modeling is to specify as little as possible about the chemical details of the fluid so that the result will be general. We succeeded in obtaining such a simple expression for stationary flows; we now face the challenging situation of wanting a general, elegant solution to the complex problem of expressing molecular forces in moving fluids. In the next section, we introduce the fluid-mechanics community's solution to this problem: the nine stress components $\tilde{\Pi}_{ij}$, also known as the *stress tensor*.

As we explain in the remainder of this chapter and text, we do not need to understand all of the details of tensor mathematics to use the stress tensor in our study of fluids in motion. For our purposes, the stress tensor functions as a 3×3 matrix that allows us to calculate magnitude and direction of molecular contact forces on a surface in a moving fluid. The deeper meaning and implications of the stress tensor are important in advanced fluid-mechanics study, and readers are referred to additional information on the stress tensor and tensor mathematics in the literature [6, 12, 88, 89, 104].

4.3.1 Total molecular stress

We seek an expression to use in the momentum balance that represents all possible molecular contact forces that may be present in a moving fluid. We must think in very general terms.

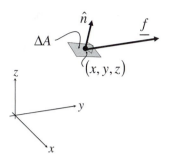

Figure 4.41

The force vector acts on a surface through Point (x, y, z). The surface has unit normal $\hat{n}$ and area ΔA. A different surface through the same point generally has a different stress vector, both in magnitude and direction.

Molecular stress is contact force per unit area on a chosen surface. In a moving fluid, the stress has both normal and tangential components. We can specify a surface by specifying the unit normal to and the area of the surface. If a force vector $\underline{f}$ acts on a surface with unit normal $\hat{n}$ and area ΔA, the stress vector on that surface is just $\underline{f}/\Delta A$ (Figure 4.41). The stress vector $\underline{f}/\Delta A$ on a surface in a moving fluid depends on position in the fluid (x, y, z) and time. $\underline{f}/\Delta A$ also depends on which surface we choose at the location under consideration; that is, $\underline{f}/\Delta A$ depends on $\hat{n}$ as well:

$$\begin{array}{c}\text{Stress vector} \\ \text{in a fluid on } \Delta A \\ \text{at } (x, y, z)\end{array} = \underline{f}(x, y, z, \hat{n})/\Delta A \qquad (4.220)$$

Instead of the vector function $\underline{f}(x, y, z, \hat{n})/\Delta A$, which points in an unknown direction, we prefer to define a new type of stress-field variable that allows us to express contact stress magnitude and direction in terms of a simple dot product between the stress field variable and the surface unit normal vector $\hat{n}$. The mathematical entity that makes this possible is called the second-order tensor [6, 12, 93, 104].

Initially, the formal derivation for the stress tensor can be intimidating. The idea of stress in a moving fluid is unfamiliar, and the need for nine different stress components can make the entire concept seem impossible to understand. For this reason, some introductory textbooks on fluid mechanics avoid mentioning the stress tensor and instead focus on problems and methods that use the components of the stress tensor but do not call them by that name.

Avoiding the expression *stress tensor* is possible but, beyond the simplest flows, accounting for intermolecular contact forces in flows is more difficult without the stress tensor than with it. In engineering devices and in high-tech fluids applications, the fluids from a region are constantly exchanging momentum with both fluids from other regions and with the boundaries by exerting forces. These momentum exchanges determine the flow field and the stress distribution; therefore, they determine how the devices operate. Without the stress tensor, developing enough knowledge to understand the stresses in such systems is an enormous task—and it is an unnecessary task because the stress tensor was developed to do the accounting for us.

Nevertheless, the stress tensor is a difficult subject to understand fully. Fortunately, we do not need to understand it fully to use it at the beginner level. The principal relationship we need is given in Equation 4.221, and its meaning is fairly straightforward:

$$\begin{array}{c}\text{Tension on} \\ \text{a surface in a fluid} \\ \text{with unit normal } \hat{n} \\ \text{(Gibbs notation)}\end{array} \qquad \boxed{\underline{f}(x, y, z, \hat{n}) = \Delta A \; \left[\hat{n} \cdot \underline{\underline{\Pi}}\right]\big|_{\Delta A}} \qquad (4.221)$$

The tension on a surface ΔA with unit normal $\hat{n}$ in a fluid may be calculated by the dot product of $\hat{n}$ and the stress tensor $\underline{\underline{\tilde{\Pi}}}$ at that location. The fact that stress is a tensor can be ignored, and we can focus on the calculation that is implied by Equation 4.221, which in Cartesian coordinates or other orthonormal coodinates is simply a matrix multiplication:

$$
\begin{array}{c}
\text{Tension on} \\
\text{a surface in a fluid} \\
\text{with unit normal } \hat{n}
\end{array}
\qquad
\boxed{\underline{f} = \Delta A \begin{pmatrix} n_1 & n_2 & n_3 \end{pmatrix}_{123} \cdot \begin{pmatrix} \tilde{\Pi}_{11} & \tilde{\Pi}_{12} & \tilde{\Pi}_{13} \\ \tilde{\Pi}_{21} & \tilde{\Pi}_{22} & \tilde{\Pi}_{23} \\ \tilde{\Pi}_{31} & \tilde{\Pi}_{32} & \tilde{\Pi}_{33} \end{pmatrix}_{123}}
$$

$$(4.222)$$

The multiplication of $\hat{n}$ and $\underline{\underline{\tilde{\Pi}}}$ in Equation 4.222 is carried out following the usual conventions of matrix multiplication:

$$
\underline{f} = \Delta A \begin{pmatrix} n_1 & n_2 & n_3 \end{pmatrix}_{123} \cdot \begin{pmatrix} \tilde{\Pi}_{11} & \tilde{\Pi}_{12} & \tilde{\Pi}_{13} \\ \tilde{\Pi}_{21} & \tilde{\Pi}_{22} & \tilde{\Pi}_{23} \\ \tilde{\Pi}_{31} & \tilde{\Pi}_{32} & \tilde{\Pi}_{33} \end{pmatrix}_{123} \tag{4.223}
$$

$$
= \Delta A \left(n_1 \tilde{\Pi}_{11} + n_2 \tilde{\Pi}_{21} + n_3 \tilde{\Pi}_{31} \right.
$$
$$
+ n_1 \tilde{\Pi}_{12} + n_2 \tilde{\Pi}_{22} + n_3 \tilde{\Pi}_{32}
$$
$$
\left. + n_1 \tilde{\Pi}_{13} + n_2 \tilde{\Pi}_{23} + n_3 \tilde{\Pi}_{33} \right) \tag{4.224}
$$

$$
= \Delta A \sum_{p=1}^{3} \sum_{m=1}^{3} n_p \tilde{\Pi}_{pm} \tag{4.225}
$$

where the stress coefficients $\tilde{\Pi}_{pm}$ are evaluated at the surface ΔA.

To arrive at Equation 4.221, we first define the stress tensor and subsequently show how this quantity is used to express molecular forces on an arbitrary surface in both stationary and moving fluids. The derivation of Equation 4.221 follows; the discussion of how to use Equation 4.221 begins after Equation 4.261.

4.3.1.1 STRESS TENSOR

To describe the stress at a point in a moving fluid, we choose three planes as the standard reference planes for stress. These three planes are perpendicular to the three basis vectors $\hat{e}_1$, $\hat{e}_2$, and $\hat{e}_3$ associated with a Cartesian coordinate system $x_1x_2x_3$ (Figure 4.42). At an arbitrary Point P in a fluid, the stress vector on the plane through Point P perpendicular to $\hat{e}_1$ (the 1-plane) we call $\underline{a}$; the stress vector on the plane through Point P perpendicular to $\hat{e}_2$ (the 2-plane) we call $\underline{b}$; and the stress vector on the plane through Point P perpendicular to $\hat{e}_3$ (the 3-plane) we call $\underline{c}$. The vectors $\underline{a}$, $\underline{b}$, and $\underline{c}$ in general do not point in any special direction; for instance, they are not necessarily perpendicular to the plane with which they are associated. In deriving a general expression for the stresses at Point P, we make no assumptions about the directions of the stress vectors associated with these three planes.

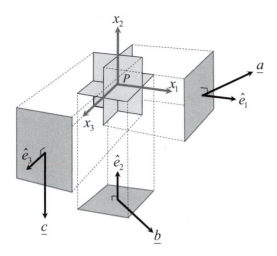

Figure 4.42 Schematic of the state of stress at Point P in a flowing system. The vectors shown indicate the stresses on three mutually perpendicular planes passing through P.

We begin by examining $\underline{a}$ in the chosen coordinate system:

$$\underline{a} = a_1\hat{e}_1 + a_2\hat{e}_2 + a_3\hat{e}_3 = \begin{pmatrix} a_1 \\ a_2 \\ a_3 \end{pmatrix}_{123} \tag{4.226}$$

The vector $\underline{a}$ is the stress on a 1-surface—a 1-surface being a surface with unit normal $\hat{e}_1$. The quantity a_1 is the coefficient of $\underline{a}$ in the 1-direction; thus, a_1 is the stress on a "1" plane in the "1" direction. We now define a scalar quantity $\tilde{\Pi}_{11}$ to be equal to a_1. By writing the coefficients of $\underline{a}$ (and, subsequently, $\underline{b}$ and $\underline{c}$) in terms of these double-subscripted quantities, we can organize the different stress components at Point P that act on planes perpendicular to the three coordinate surfaces at Point P:

$$\tilde{\Pi}_{11} = a_1 = \begin{pmatrix} \text{stress at Point } P \\ \text{on a 1-surface} \\ \text{in the 1-direction} \end{pmatrix} \tag{4.227}$$

$$\tilde{\Pi}_{12} = a_2 = \begin{pmatrix} \text{stress at Point } P \\ \text{on a 1-surface} \\ \text{in the 2-direction} \end{pmatrix} \tag{4.228}$$

$$\tilde{\Pi}_{13} = a_3 = \begin{pmatrix} \text{stress at Point } P \\ \text{on a 1-surface} \\ \text{in the 3-direction} \end{pmatrix} \tag{4.229}$$

$$\underline{a} = (\text{stress on a 1-surface at } P)$$

$$= \tilde{\Pi}_{11}\hat{e}_1 + \tilde{\Pi}_{12}\hat{e}_2 + \tilde{\Pi}_{13}\hat{e}_3 = \begin{pmatrix} \tilde{\Pi}_{11} \\ \tilde{\Pi}_{12} \\ \tilde{\Pi}_{13} \end{pmatrix}_{123} \tag{4.230}$$

Following the same logic for $\underline{b}$ and $\underline{c}$, we obtain analogous results. The vector $\underline{b}$ is a force at Point P on a 2-surface (i.e., surface with unit normal $\hat{n} = \hat{e}_2$). The three components are the forces on a 2-surface in the 1-, 2-, and 3-directions:

$$\tilde{\Pi}_{21} = b_1 = \begin{pmatrix} \text{stress at Point } P \\ \text{on a 2-surface} \\ \text{in the 1-direction} \end{pmatrix} \tag{4.231}$$

$$\tilde{\Pi}_{22} = b_2 = \begin{pmatrix} \text{stress at Point } P \\ \text{on a 2-surface} \\ \text{in the 2-direction} \end{pmatrix} \tag{4.232}$$

$$\tilde{\Pi}_{23} = b_3 = \begin{pmatrix} \text{stress at Point } P \\ \text{on a 2-surface} \\ \text{in the 3-direction} \end{pmatrix} \tag{4.233}$$

$$\underline{b} = (\text{stress on a 2-surface at Point } P)$$

$$= \tilde{\Pi}_{21}\hat{e}_1 + \tilde{\Pi}_{22}\hat{e}_2 + \tilde{\Pi}_{23}\hat{e}_3 = \begin{pmatrix} \tilde{\Pi}_{21} \\ \tilde{\Pi}_{22} \\ \tilde{\Pi}_{23} \end{pmatrix}_{123} \tag{4.234}$$

For a 3-surface, the result is:

$$\underline{c} = (\text{stress on a 3-surface at Point } P)$$

$$= \tilde{\Pi}_{31}\hat{e}_1 + \tilde{\Pi}_{32}\hat{e}_2 + \tilde{\Pi}_{33}\hat{e}_3 = \begin{pmatrix} \tilde{\Pi}_{31} \\ \tilde{\Pi}_{32} \\ \tilde{\Pi}_{33} \end{pmatrix}_{123} \tag{4.235}$$

In general, $\tilde{\Pi}_{ik}$ is the stress on an i-plane in the k-direction. Remember that an i-plane means a plane perpendicular to the $\hat{e}_i$-direction. There are nine stress quantities, $\tilde{\Pi}_{ik}$.

The quantities $\tilde{\Pi}_{ik}$ are the coefficients of the total stress tensor at Point P. A tensor is a mathematical entity related to vectors but of higher order or complexity. Whereas vectors have three components when written in Cartesian coordinates, tensors have nine. Tensors and vectors can be multiplied in carefully proscribed ways (see Section 1.3), but an in-depth discussion of tensors is beyond the scope of this book [6, 93, 104]. In vector Gibbs notation, tensor variables are written with double underlines; thus, the total stress tensor is written as $\underline{\underline{\tilde{\Pi}}}$. The nine coefficients of a tensor relative to a coordinate system may be written in a 3×3 matrix:

$$\underline{\underline{\tilde{\Pi}}} = \begin{pmatrix} \tilde{\Pi}_{11} & \tilde{\Pi}_{12} & \tilde{\Pi}_{13} \\ \tilde{\Pi}_{21} & \tilde{\Pi}_{22} & \tilde{\Pi}_{23} \\ \tilde{\Pi}_{31} & \tilde{\Pi}_{32} & \tilde{\Pi}_{33} \end{pmatrix}_{123} \tag{4.236}$$

where the subscript '123' refers to the Cartesian coordinate system $\hat{e}_1$, $\hat{e}_2$, and $\hat{e}_3$.

The total stress tensor $\underline{\underline{\tilde{\Pi}}}$ is more powerful than a naming convention, and it has meaning for more than on the three planes discussed so far. The total stress tensor $\underline{\underline{\tilde{\Pi}}}$ contains all of the information about the state of stress at a point in a stationary or moving fluid [6]. In the next example, we use rules of tensor algebra to show how $\underset{=}{\tilde{\Pi}}$, written as we have in Equation 4.236, can be used to calculate stress on *any plane* through Point P. The final matrix result for how this is done is given in Equation 4.261.

EXAMPLE 4.13. *What is the molecular contact force at a point on an arbitrary plane of unit normal $\hat{n}$ in a moving fluid?*

SOLUTION. Our task is to calculate the force vector at a point on ΔA, a small surface in a fluid. Let P be our chosen point and $\hat{n}$ be the unit vector normal to ΔA (Figure 4.43). The vector $\hat{n}$ can be written relative to a chosen Cartesian coordinate system as shown here:

$$\hat{n} = n_1\hat{e}_1 + n_2\hat{e}_2 + n_3\hat{e}_3 = \begin{pmatrix} n_1 \\ n_2 \\ n_3 \end{pmatrix}_{123} \tag{4.237}$$

In general, $\hat{n}$ does not line up with any of the coordinate-basis vectors $\hat{e}_i$, for $i = 1, 2, 3$.

In this section, we defined the stress tensor $\underset{=}{\tilde{\Pi}}$. In our chosen Cartesian coordinate system, $\underset{=}{\tilde{\Pi}}$ may be written as:

$$\underset{=}{\tilde{\Pi}} = \begin{pmatrix} \tilde{\Pi}_{11} & \tilde{\Pi}_{12} & \tilde{\Pi}_{13} \\ \tilde{\Pi}_{21} & \tilde{\Pi}_{22} & \tilde{\Pi}_{23} \\ \tilde{\Pi}_{31} & \tilde{\Pi}_{32} & \tilde{\Pi}_{33} \end{pmatrix}_{123} \tag{4.238}$$

This is shorthand for the more complete tensor representation shown here (see Section 1.3.2.2):

$$\begin{aligned} \underset{=}{\tilde{\Pi}} = {} & \hat{e}_1\hat{e}_1\tilde{\Pi}_{11} + \hat{e}_1\hat{e}_2\tilde{\Pi}_{12} + \hat{e}_1\hat{e}_3\tilde{\Pi}_{13} \\ & + \hat{e}_2\hat{e}_1\tilde{\Pi}_{21} + \hat{e}_2\hat{e}_2\tilde{\Pi}_{22} + \hat{e}_2\hat{e}_3\tilde{\Pi}_{23} \\ & + \hat{e}_3\hat{e}_1\tilde{\Pi}_{31} + \hat{e}_3\hat{e}_2\tilde{\Pi}_{32} + \hat{e}_3\hat{e}_3\tilde{\Pi}_{33} \end{aligned} \tag{4.239}$$

$$= \sum_{p=1}^{3}\sum_{k=1}^{3} \tilde{\Pi}_{pk}\hat{e}_p\,\hat{e}_k \tag{4.240}$$

Later in this example, we need to reference the version of $\underset{=}{\tilde{\Pi}}$ in Equation 4.239.

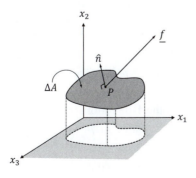

Figure 4.43 Consider a surface ΔA in a flowing fluid. At any point in the fluid, we can draw an infinite number of planes through that point. If we arbitrarily choose one such plane with unit normal $\hat{n}$, what is the force on that surface?

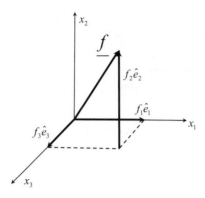

Figure 4.44

Force on the surface ΔA has components in three coordinate directions.

In Equation 4.239, an unusual type of vector multiplication appears, written as two vectors sitting side by side (e.g., $\hat{e}_1\hat{e}_2$ or $\hat{e}_2\hat{e}_3$). This type of product is called the *indeterminate vector product*, which is explained in Section 1.3.2.2. All of the possible indeterminate vector products of the unit vectors $\hat{e}_1$, $\hat{e}_2$, and $\hat{e}_3$ appear in Equation 4.239, for a total of nine terms. The indeterminate vector product of two vectors produces a tensor, as discussed in Section 1.3.2.2. The terms $\hat{e}_1\hat{e}_2$ and $\hat{e}_2\hat{e}_1$ and similar mirror-image pairs are not equal to one another; thus, the positions of individual unit vectors (i.e., which vector is first, which is second) are important within the indeterminate multiplication of vectors.

We want to write the force on ΔA in terms of the nine stress coefficients $\tilde{\Pi}_{pk}$. As discussed, the nine components $\tilde{\Pi}_{pk}$ at our Point P are defined relative to the stresses on the surfaces through P that are perpendicular to the coordinate-basis vectors, $\hat{e}_1$, $\hat{e}_2$, and $\hat{e}_3$. Because ΔA is not lined up with any of these reference planes, to carry out our task, we must find a way to relate the forces on the 1-, 2-, and 3-surfaces with the force on ΔA.

We define the vector $\underline{f}$ as the force (not stress) on ΔA at Point P (see Figure 4.43). In our chosen coordinate system, $\underline{f}$ is written as:

$$\underline{f} = \begin{pmatrix} f_1 \\ f_2 \\ f_3 \end{pmatrix}_{123} = f_1\hat{e}_1 + f_2\hat{e}_2 + f_3\hat{e}_3 \tag{4.241}$$

This way of expressing $\underline{f}$ shows that $\underline{f}$ is composed of the sum of three vectors: a force in the 1-direction, $f_1\hat{e}_1$; a force in the 2-direction, $f_2\hat{e}_2$; and a force in the 3-direction, $f_3\hat{e}_3$ (Figure 4.44):

$$\underline{f} = \begin{pmatrix} f_1 \\ 0 \\ 0 \end{pmatrix}_{123} + \begin{pmatrix} 0 \\ f_2 \\ 0 \end{pmatrix}_{123} + \begin{pmatrix} 0 \\ 0 \\ f_3 \end{pmatrix}_{123} \tag{4.242}$$

The magnitude of the force on ΔA in the 1-direction is f_1. Examining the nine coefficients $\tilde{\Pi}_{pk}$, we see that there are three $\tilde{\Pi}_{pk}$ that describe stresses in the 1-direction. These components refer to stresses on three specific surfaces, none of which is the surface of interest, ΔA:

$$\tilde{\Pi}_{11} = \left(\begin{array}{c} \text{stress at Point } P \\ \text{on a 1-surface} \\ \text{in the 1-direction} \end{array} \right) \tag{4.243}$$

$$\tilde{\Pi}_{21} = \left(\begin{array}{c} \text{stress at Point } P \\ \text{on a 2-surface} \\ \text{in the 1-direction} \end{array} \right) \tag{4.244}$$

$$\tilde{\Pi}_{31} = \left(\begin{array}{c} \text{stress at Point } P \\ \text{on a 3-surface} \\ \text{in the 1-direction} \end{array} \right) \tag{4.245}$$

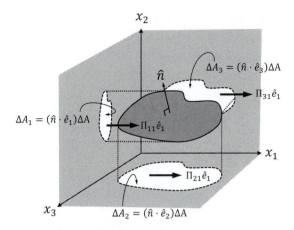

The stress vector on an arbitrary surface ΔA can be expressed in terms of the nine components of the stress tensor $\tilde{\Pi}_{ik}$. For example, the 1-component of the force is related to the three $\tilde{\Pi}_{i1}$ components that concern forces in the 1-direction. The projections of ΔA in the three coordinate directions are required to complete the expressions for f_1, f_2, and f_3, as discussed in this chapter.

We can decompose f_1 into the sum of three contributions that incorporate the three stresses $\tilde{\Pi}_{11}$, $\tilde{\Pi}_{21}$, and $\tilde{\Pi}_{31}$. The component f_1, which is the stress on ΔA in the 1-direction, is the sum of the 1-direction forces that act on projections of the surface ΔA in the 1-, 2-, and 3-directions (Figure 4.45).

$$f_1 = \begin{pmatrix} \text{the force on} \\ \Delta A \text{ in the} \\ \text{1-direction} \end{pmatrix} \tag{4.246}$$

$$f_1 = \begin{pmatrix} \text{stress acting} \\ \text{on a 1-surface} \\ \text{in the 1-direction} \end{pmatrix} \begin{pmatrix} \text{projection} \\ \text{of } \Delta A \text{ in the} \\ \text{1-direction} \end{pmatrix}$$

$$+ \begin{pmatrix} \text{stress acting} \\ \text{on a 2-surface} \\ \text{in the 1-direction} \end{pmatrix} \begin{pmatrix} \text{projection} \\ \text{of } \Delta A \text{ in the} \\ \text{2-direction} \end{pmatrix}$$

$$+ \begin{pmatrix} \text{stress acting} \\ \text{on a 3-surface} \\ \text{in the 1-direction} \end{pmatrix} \begin{pmatrix} \text{projection} \\ \text{of } \Delta A \text{ in the} \\ \text{3-direction} \end{pmatrix} \tag{4.247}$$

$$= \tilde{\Pi}_{11} \Delta A_1 + \tilde{\Pi}_{21} \Delta A_2 + \tilde{\Pi}_{31} \Delta A_3 \tag{4.248}$$

The projection of a surface in a direction gives the effective size of a surface in the chosen direction. Thus, Equation 4.247 sums the product of the effective size of ΔA in three directions multiplied by the 1-direction forces on those effective surfaces. The net sum is the total force on ΔA in the 1-direction.

Needed in Equation 4.247 are the stress tensor components $\tilde{\Pi}_{11}$, $\tilde{\Pi}_{21}$, and $\tilde{\Pi}_{31}$. Each projection needed is the projection in a coordinate direction $\hat{e}_i$ of a surface of area ΔA; this projection is given by (for justification, see the Web appendix [108]):

$$\Delta A_i = \begin{pmatrix} \text{projection of } \Delta A \\ \text{in direction of } \hat{e}_i \end{pmatrix} = (\hat{n} \cdot \hat{e}_i) \, \Delta A \tag{4.249}$$

where $\hat{n}$ is the unit normal vector to ΔA. Applying Equation 4.249, the area over which $\tilde{\Pi}_{11}$ acts is the projection of ΔA in the 1-direction, $\hat{n} \cdot \hat{e}_1 \, \Delta A$. Likewise, the areas over which $\tilde{\Pi}_{21}$ and $\tilde{\Pi}_{31}$ act are $\hat{n} \cdot \hat{e}_2 \, \Delta A$ and $\hat{n} \cdot \hat{e}_3 \, \Delta A$, respectively. We now substitute these expressions into Equation 4.248:

$$f_1 = \begin{pmatrix} \text{the force on} \\ \Delta A \text{ at } P \text{ in the} \\ \text{1-direction} \end{pmatrix}$$

$$= \tilde{\Pi}_{11} \Delta A_1 + \tilde{\Pi}_{21} \Delta A_2 + \tilde{\Pi}_{31} \Delta A_3 \tag{4.250}$$

$$= (\tilde{\Pi}_{11}(\hat{n} \cdot \hat{e}_1) \Delta A) + (\tilde{\Pi}_{21}(\hat{n} \cdot \hat{e}_2) \Delta A) + (\tilde{\Pi}_{31}(\hat{n} \cdot \hat{e}_3) \Delta A) \tag{4.251}$$

We can simplify the expression for f_1 by collecting terms that are in common:

$$f_1 = \Delta A \, (\hat{n} \cdot \hat{e}_1 \tilde{\Pi}_{11} + \hat{n} \cdot \hat{e}_2 \tilde{\Pi}_{21} + \hat{n} \cdot \hat{e}_3 \tilde{\Pi}_{31}) \tag{4.252}$$

$$= \Delta A \, \hat{n} \cdot (\hat{e}_1 \tilde{\Pi}_{11} + \hat{e}_2 \tilde{\Pi}_{21} + \hat{e}_3 \tilde{\Pi}_{31}) \tag{4.253}$$

We used the distributive law of the vector dot product to factor $(\hat{n} \cdot)$ out to the front of Equation 4.253. The result in Equation 4.253 for f_1 relates the 1-component of $\underline{f}$ to the stress components $\tilde{\Pi}_{i1}$ and the unit normal of the surface ΔA, given by $\hat{n}$. We follow the same logic to arrive at expressions for f_2, the 2-component of force on ΔA; and f_3, the 3-component of force on ΔA:

$$f_2 = \Delta A \, \hat{n} \cdot (\hat{e}_1 \tilde{\Pi}_{12} + \hat{e}_2 \tilde{\Pi}_{22} + \hat{e}_3 \tilde{\Pi}_{32}) \tag{4.254}$$

$$f_3 = \Delta A \, \hat{n} \cdot (\hat{e}_1 \tilde{\Pi}_{13} + \hat{e}_2 \tilde{\Pi}_{23} + \hat{e}_3 \tilde{\Pi}_{33}) \tag{4.255}$$

Having related f_1, f_2, and f_3 to the $\tilde{\Pi}_i$, the three expressions for f_1, f_2, and f_3 in terms of the $\tilde{\Pi}_{pk}$ can be substituted into Equation 4.241, yielding the complete vector expression for $\underline{f}$ in terms of the $\tilde{\Pi}_{pk}$. We now assemble $\underline{f}$ and simplify using the rules of algebra and factoring:

$$\underline{f} = f_1 \hat{e}_1 + f_2 \hat{e}_2 + f_3 \hat{e}_3 \tag{4.256}$$

$$= \left[\Delta A \, \hat{n} \cdot (\hat{e}_1 \tilde{\Pi}_{11} + \hat{e}_2 \tilde{\Pi}_{21} + \hat{e}_3 \tilde{\Pi}_{31}) \right] \hat{e}_1$$

$$+ \left[\Delta A \, \hat{n} \cdot (\hat{e}_1 \tilde{\Pi}_{12} + \hat{e}_2 \tilde{\Pi}_{22} + \hat{e}_3 \tilde{\Pi}_{32}) \right] \hat{e}_2$$

$$+ \left[\Delta A \, \hat{n} \cdot (\hat{e}_1 \tilde{\Pi}_{13} + \hat{e}_2 \tilde{\Pi}_{23} + \hat{e}_3 \tilde{\Pi}_{33}) \right] \hat{e}_3 \tag{4.257}$$

$$= \Delta A \, \hat{n} \cdot (\hat{e}_1 \hat{e}_1 \tilde{\Pi}_{11} + \hat{e}_2 \hat{e}_1 \tilde{\Pi}_{21} + \hat{e}_3 \hat{e}_1 \tilde{\Pi}_{31}$$

$$+ \hat{e}_1 \hat{e}_2 \tilde{\Pi}_{12} + \hat{e}_2 \hat{e}_2 \tilde{\Pi}_{22} + \hat{e}_3 \hat{e}_2 \tilde{\Pi}_{32}$$

$$+ \hat{e}_1 \hat{e}_3 \tilde{\Pi}_{13} + \hat{e}_2 \hat{e}_3 \tilde{\Pi}_{23} + \hat{e}_3 \hat{e}_3 \tilde{\Pi}_{33}) \tag{4.258}$$

The final expression in Equation 4.258 is a quantity (ΔA) multiplied by the dot product of a vector $(\hat{n})$ with a new, more complex quantity: a tensor $(\underline{\tilde{\Pi}} \equiv$ sum of indeterminate vector products; compare with Equation 4.239 and see also Section 1.3.2.2 and [6]).

The terms of $\underline{\underline{\tilde{\Pi}}}$ are not in the usual order that we take when we write a tensor in matrix form, but all of the terms are there. Rearranging, we obtain:

$$\underline{f} = \Delta A \, \hat{n} \cdot (\hat{e}_1 \hat{e}_1 \tilde{\Pi}_{11} + \hat{e}_1 \hat{e}_2 \tilde{\Pi}_{12} + \hat{e}_1 \hat{e}_3 \tilde{\Pi}_{13}$$
$$+ \hat{e}_2 \hat{e}_1 \tilde{\Pi}_{21} + \hat{e}_2 \hat{e}_2 \tilde{\Pi}_{22} + \hat{e}_2 \hat{e}_3 \tilde{\Pi}_{23}$$
$$+ \hat{e}_3 \hat{e}_1 \tilde{\Pi}_{31} + \hat{e}_3 \hat{e}_2 \tilde{\Pi}_{32} + \hat{e}_3 \hat{e}_3 \tilde{\Pi}_{33}) \tag{4.259}$$

$$= \Delta A \, \hat{n} \cdot \left[\sum_{p=1}^{3} \sum_{k=1}^{3} \tilde{\Pi}_{pk} \hat{e}_p \hat{e}_k \right] \tag{4.260}$$

The rules for dot multiplying a vector with a tensor using coefficients in an orthonormal coordinate system are the same as those for matrix multiplication of a 1×3 matrix with a 3×3 matrix [6]. Therefore, we can write Equation 4.259 as:

Tension on a surface of area ΔA with unit normal $\hat{n}$ (matrix notation)

$$\underline{f} = \Delta A \, \begin{pmatrix} n_1 & n_2 & n_3 \end{pmatrix}_{123} \cdot \begin{pmatrix} \tilde{\Pi}_{11} & \tilde{\Pi}_{12} & \tilde{\Pi}_{13} \\ \tilde{\Pi}_{21} & \tilde{\Pi}_{22} & \tilde{\Pi}_{23} \\ \tilde{\Pi}_{31} & \tilde{\Pi}_{32} & \tilde{\Pi}_{33} \end{pmatrix}_{123} \tag{4.261}$$

This is the final result in a form that clarifies how to calculate the force on a surface using matrix multiplication. In terms of tensor notation, the same result may be written as:

$$f_m = \Delta A \sum_{p=1}^{3} n_p \tilde{\Pi}_{pm} \tag{4.262}$$

Force on a surface of area ΔA with unit normal $\hat{n}$ (Gibbs notation)

$$\underline{f} = \Delta A \, [\hat{n} \cdot \underline{\underline{\tilde{\Pi}}}]|_{\Delta A} \tag{4.263}$$

The total stress tensor $\underline{\underline{\tilde{\Pi}}}$ contains all of the information about the state of molecular stress at a point in a fluid. To calculate the contact force acting on any specific surface, we simply dot the unit normal to the surface at the point with $\underline{\underline{\tilde{\Pi}}}$ at the point and multiply by the area of the surface (Equation 4.261 or Equation 4.263).

Equation 4.263 is the key result of the discussion in this section. The stress tensor is an elegant solution to the problem of how to mathematically express the state of stress at a point on a chosen surface in a fluid. Equation 4.263 (or Equation 4.261) allows us to calculate the magnitude and direction of force on a surface in a stationary or moving fluid. To see the power of this expression, we have three examples in which we calculate the force vector on a surface when the stress tensor is known. In Chapter 5, we discuss how to find the coefficients of $\underline{\underline{\tilde{\Pi}}}$ for a chosen flow.

EXAMPLE 4.14. *For a fluid whose stress tensor is given by the following general expression:*

$$\tilde{\underline{\underline{\Pi}}} = \begin{pmatrix} \tilde{\Pi}_{11} & \tilde{\Pi}_{12} & \tilde{\Pi}_{13} \\ \tilde{\Pi}_{21} & \tilde{\Pi}_{22} & \tilde{\Pi}_{23} \\ \tilde{\Pi}_{31} & \tilde{\Pi}_{32} & \tilde{\Pi}_{33} \end{pmatrix}_{123} \tag{4.264}$$

what is the force on a surface of area S whose unit normal is $\hat{e}_2$? All quantities are written relative to the same Cartesian coordinate system.

SOLUTION. We already know the answer to this question because we know $\underline{b}$, the stress on a 2-surface: The answer is the area of the surface multiplied by the stress on the surface, given by $\underline{b}$ (see Equation 4.234). We gain some practice with matrix-tensor notation, however, by using Equation 4.261 to calculate this result.

$$\underline{f} = \Delta A \, \hat{n} \cdot \tilde{\underline{\underline{\Pi}}} \tag{4.265}$$

$$\begin{pmatrix} \text{force on surface} \\ \text{of area } S \\ \text{whose unit normal} \\ \text{vector is } \hat{e}_2 \end{pmatrix} = S \, \hat{e}_2 \cdot \tilde{\underline{\underline{\Pi}}}$$

$$= S \begin{pmatrix} 0 & 1 & 0 \end{pmatrix}_{123} \cdot \begin{pmatrix} \tilde{\Pi}_{11} & \tilde{\Pi}_{12} & \tilde{\Pi}_{13} \\ \tilde{\Pi}_{21} & \tilde{\Pi}_{22} & \tilde{\Pi}_{23} \\ \tilde{\Pi}_{31} & \tilde{\Pi}_{32} & \tilde{\Pi}_{33} \end{pmatrix}_{123}$$

$$= S \begin{pmatrix} \tilde{\Pi}_{21} \\ \tilde{\Pi}_{22} \\ \tilde{\Pi}_{23} \end{pmatrix}_{123}$$

$$= S(\tilde{\Pi}_{21}\hat{e}_1 + \tilde{\Pi}_{22}\hat{e}_2 + \tilde{\Pi}_{23}\hat{e}_3)$$

$$= S\underline{b}$$

This is the result we expected (compare with Equation 4.234).

EXAMPLE 4.15. *The stress tensor $\tilde{\underline{\underline{\Pi}}}$ at Point P in a fluid is given in matrix form in Equation 4.266. For a flat square surface of area 3.1 mm² submerged in the fluid oriented perpendicular to the 13-plane as shown in Figure 4.46, what is the force on that surface?*

$$\tilde{\underline{\underline{\Pi}}} = \begin{pmatrix} -2.0\,Pa & 0 & 0 \\ 0 & -2.0\,Pa & 0 \\ 0 & 0 & -2.0\,Pa \end{pmatrix}_{123} \tag{4.266}$$

SOLUTION. We showed in this section that the force vector $\underline{f}$ on a surface with unit normal $\hat{n}$ is given by:

$$\underline{f} = \Delta A \left[\hat{n} \cdot \tilde{\underline{\underline{\Pi}}} \right] \tag{4.267}$$

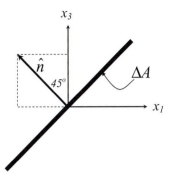

The unit normal of a chosen surface is given by the vector $\hat{n}$, which lies in the 13-plane at an angle 45 degrees counterclockwise from the 3-axis, as shown.

where ΔA is the area of the surface and $\underline{\underline{\tilde{\Pi}}}$ is the stress tensor at that point. When all of the vectors are written in the same Cartesian coordinate system, this expression may be evaluated using matrix multiplication:

$$
\begin{pmatrix} f_1 & f_2 & f_3 \end{pmatrix}_{123} = \Delta A \begin{pmatrix} n_1 & n_2 & n_3 \end{pmatrix}_{123} \cdot \begin{pmatrix} \tilde{\Pi}_{11} & \tilde{\Pi}_{12} & \tilde{\Pi}_{13} \\ \tilde{\Pi}_{21} & \tilde{\Pi}_{22} & \tilde{\Pi}_{23} \\ \tilde{\Pi}_{31} & \tilde{\Pi}_{32} & \tilde{\Pi}_{33} \end{pmatrix}_{123} \quad (4.268)
$$

We have $\underline{\underline{\tilde{\Pi}}}$ and we know $\Delta A = 3.1$ mm^2. The coefficients of the unit normal vector $\hat{n}$ in the $x_1 x_2 x_3$ coordinate system can be worked out by geometry:

$$
\hat{n} = \begin{pmatrix} \frac{-1}{\sqrt{2}} \\ 0 \\ \frac{1}{\sqrt{2}} \end{pmatrix}_{123} \quad (4.269)
$$

The final step is to substitute these values into Equation 4.268 and evaluate the matrix multiplications:

$$
\begin{pmatrix} f_1 & f_2 & f_3 \end{pmatrix}_{123} = \Delta A \begin{pmatrix} n_1 & n_2 & n_3 \end{pmatrix}_{123} \cdot \begin{pmatrix} \tilde{\Pi}_{11} & \tilde{\Pi}_{12} & \tilde{\Pi}_{13} \\ \tilde{\Pi}_{21} & \tilde{\Pi}_{22} & \tilde{\Pi}_{23} \\ \tilde{\Pi}_{31} & \tilde{\Pi}_{32} & \tilde{\Pi}_{33} \end{pmatrix}_{123}
$$

$$
= 3.1 \text{ mm}^2 \begin{pmatrix} \frac{-1}{\sqrt{2}} & 0 & \frac{1}{\sqrt{2}} \end{pmatrix}_{123} \cdot \begin{pmatrix} -2.0 \text{ Pa} & 0 & 0 \\ 0 & -2.0 \text{ Pa} & 0 \\ 0 & 0 & -2.0 \text{ Pa} \end{pmatrix}_{123}
$$

$$
= \begin{pmatrix} (3.1)\frac{2.0}{\sqrt{2}} \\ 0 \\ (3.1)\frac{-2.0}{\sqrt{2}} \end{pmatrix}_{123} (\text{Pa})(\text{mm}^2) \left(\frac{\text{m}}{10^3 \text{ mm}} \right)^2 \frac{\text{N/m}^2}{\text{Pa}}
$$

$$
= \begin{pmatrix} (3.1)\frac{2.0}{\sqrt{2}} \\ 0 \\ (3.1)\frac{-2.0}{\sqrt{2}} \end{pmatrix}_{123} 10^{-6} \text{ N}
$$

Note that this final result is parallel to the unit normal vector $\hat{n}$:

$$\underline{f} = \begin{pmatrix} 2(3.1)\frac{1}{\sqrt{2}} \\ 0 \\ 2(3.1)\frac{-1}{\sqrt{2}} \end{pmatrix}_{123} \mu N = (-6.2\ \mu N)\hat{n}$$

where a μN is a microNewton.[3]

The force vector turned out to be parallel to $\hat{n}$ because the stress given by Equation 4.266 is isotropic. As discussed in Section 4.3.2 isotropic stress gives a diagonal stress tensor with the three coefficients on the diagonal equal to one another.

EXAMPLE 4.16. *The stress tensor $\tilde{\underline{\underline{\Pi}}}$ at Point P in a viscous, moving fluid is given in matrix form here. For the same submerged flat square surface discussed in the previous example, what is the force on that surface using this new stress tensor?*

$$\tilde{\underline{\underline{\Pi}}} = \begin{pmatrix} -2.0\text{Pa} & 4.0\text{Pa} & 0 \\ 4.0\text{Pa} & -2.0\text{Pa} & 0 \\ 0 & 0 & -2.0\text{Pa} \end{pmatrix}_{123} \tag{4.270}$$

SOLUTION. The solution to this problem follows the same steps as those in the previous problem:

$$\underline{f} = \Delta A\, \hat{n} \cdot \tilde{\underline{\underline{\Pi}}}$$

$$\left(f_1\ \ f_2\ \ f_3 \right)_{123} = \Delta A \left(n_1\ \ n_2\ \ n_3 \right)_{123} \cdot \begin{pmatrix} \tilde{\Pi}_{11} & \tilde{\Pi}_{12} & \tilde{\Pi}_{13} \\ \tilde{\Pi}_{21} & \tilde{\Pi}_{22} & \tilde{\Pi}_{23} \\ \tilde{\Pi}_{31} & \tilde{\Pi}_{32} & \tilde{\Pi}_{33} \end{pmatrix}_{123}$$

$$= 3.1\ \text{mm}^2 \left(\frac{-1}{\sqrt{2}}\ \ 0\ \ \frac{1}{\sqrt{2}} \right)_{123} \cdot \begin{pmatrix} -2.0\ \text{Pa} & 4.0\ \text{Pa} & 0 \\ 4.0\ \text{Pa} & -2.0\ \text{Pa} & 0 \\ 0 & 0 & -2.0\ \text{Pa} \end{pmatrix}_{123}$$

$$= 3.1 \begin{pmatrix} \frac{2.0}{\sqrt{2}} \\ \frac{-4.0}{\sqrt{2}} \\ \frac{-2.0}{\sqrt{2}} \end{pmatrix}_{123} (\text{Pa})(\text{mm}^2) \left(\frac{m}{10^3\ \text{mm}} \right)^2 \frac{\text{N/m}^2}{\text{Pa}}$$

$$= 3.1 \begin{pmatrix} \frac{2.0}{\sqrt{2}} \\ \frac{-4.0}{\sqrt{2}} \\ \frac{-2.0}{\sqrt{2}} \end{pmatrix}_{123} 10^{-6}\ \text{N}$$

---

[3] The placement of the coefficients of a physical vector in column vectors (3×1) or in row vectors (1×3) is arbitrary. These vectors have their meaning as coefficients that multiply the basis vectors with which they are associated. How they are displayed is a matter of convenience, and we switch between the representations as needed.

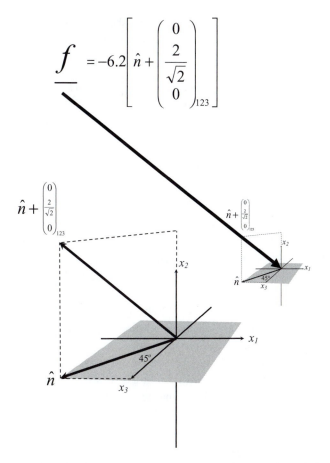

$$\underline{f} = -6.2\left[\hat{n} + \begin{pmatrix} 0 \\ \frac{2}{\sqrt{2}} \\ 0 \end{pmatrix}_{123}\right]$$

Figure 4.47 When a vector dot multiplies a stress tensor that is not diagonal, the tensor rotates and stretches the vector to produce the output vector. In this example, the output vector is related to the input vector by a rotation and a stretch.

Note that this final result is not parallel to the unit normal vector $\hat{n}$:

$$\underline{f} = -6.2\begin{pmatrix} \frac{-1}{\sqrt{2}} \\ \frac{2}{\sqrt{2}} \\ \frac{1}{\sqrt{2}} \end{pmatrix}_{123} \mu N$$

$$= (-6.2\ \mu N)\left[\hat{n} + \begin{pmatrix} 0 \\ \frac{2}{\sqrt{2}} \\ 0 \end{pmatrix}_{123}\right] \tag{4.271}$$

We can tell that the stress tensor from this example (Equation 4.270) is not isotropic because it has off-diagonal elements. The isotropic stress tensor of the previous example stretched the unit normal vector to give the force vector but did not rotate it, and $\underline{f}$ is parallel to $\hat{n}$ for Example 4.15. The stress tensor from this example stretched and rotated the unit normal vector to produce the final force vector (Figure 4.47). The stress tensor is usually anisotropic when the fluid is in motion. We discuss both isotropic and anisotropic stress tensors in subsequent sections of this chapter.

Experimental studies on the stress tensor show that $\underline{\underline{\tilde{\Pi}}}$ is symmetric for the majority of fluids; that is, $\tilde{\Pi}_{rs} = \tilde{\Pi}_{sr}$ and rows and columns if $\underline{\underline{\tilde{\Pi}}}$ may be exchanged with no impact:

$$
\begin{matrix} \text{Stress tensor} \\ \text{is symmetric} \end{matrix} \quad
\begin{pmatrix} \tilde{\Pi}_{11} & \tilde{\Pi}_{12} & \tilde{\Pi}_{13} \\ \tilde{\Pi}_{21} & \tilde{\Pi}_{22} & \tilde{\Pi}_{23} \\ \tilde{\Pi}_{31} & \tilde{\Pi}_{32} & \tilde{\Pi}_{33} \end{pmatrix}_{123} =
\begin{pmatrix} \tilde{\Pi}_{11} & \tilde{\Pi}_{21} & \tilde{\Pi}_{31} \\ \tilde{\Pi}_{12} & \tilde{\Pi}_{22} & \tilde{\Pi}_{32} \\ \tilde{\Pi}_{13} & \tilde{\Pi}_{23} & \tilde{\Pi}_{33} \end{pmatrix}_{123}
\qquad (4.272)
$$

For a symmetric stress tensor, there are six independent stresses at a Point P in a fluid:

$$
\begin{matrix} \text{Six independent} \\ \text{stress components} \\ \text{in a fluid} \end{matrix} \quad
\begin{matrix} \tilde{\Pi}_{11} & \tilde{\Pi}_{22} & \tilde{\Pi}_{33} \\ \tilde{\Pi}_{21} = \tilde{\Pi}_{12} & \tilde{\Pi}_{31} = \tilde{\Pi}_{32} & \tilde{\Pi}_{23} = \tilde{\Pi}_{32} \end{matrix}
\qquad (4.273)
$$

Note that in both examples, $\underline{\underline{\tilde{\Pi}}}$ is symmetric. The total stress tensor can be shown rigorously to be symmetric for nonpolar fluids [6]. For polar fluids, if body moments[4] to couple stresses [85, 169] are absent, we can show that $\underline{\underline{\tilde{\Pi}}}$ is symmetric [40]. We always assume $\underline{\underline{\tilde{\Pi}}}$ to be symmetric.

In terms of the reference Cartesian coordinate system, $\tilde{\Pi}_{11}$, $\tilde{\Pi}_{22}$, and $\tilde{\Pi}_{33}$ are all normal stresses: The stress is in the same direction as the unit normal to the surface on which it acts (e.g., $\tilde{\Pi}_{11}$ is the stress on a 1-surface in the 1-direction). An alternative notation for the normal stresses is $\sigma_{11} = \Pi_{11}$, $\sigma_{22} = \Pi_{22}$, and $\sigma_{33} = \Pi_{33}$. The remaining off-diagonal terms of $\underline{\underline{\tilde{\Pi}}}$ are all pure shear stresses (e.g., $\tilde{\Pi}_{31}$ is the stress on a 3-surface in the 1-direction).

4.3.1.2 STRESS SIGN CONVENTION

A final unresolved surface force issue is the sign of the total stress tensor $\underline{\underline{\tilde{\Pi}}}$. The expression $\Delta A \, \hat{n} \cdot \underline{\underline{\tilde{\Pi}}}$ gives the force vector on a surface with unit normal $\hat{n}$ and area ΔA. Force can be either a push (i.e., compression) or a pull (i.e., tension). In this text, we follow the standard engineering convention and choose $\underline{\underline{\tilde{\Pi}}}$ to express tension on the surface (Figure 4.48). This convention implies that stress on a surface $\underline{\underline{\tilde{\Pi}}}$ and pressure p on a surface have opposite signs. The choice of tension or compression is arbitrary, and the sign convention affects only expressions that contain the total stress tensor $\underline{\underline{\tilde{\Pi}}}$ or the extra stress tensor, $\underline{\underline{\tilde{\tau}}}$, which is introduced in Section 4.3.2.

A warning to readers: Several chemical-engineering textbooks use the opposite convention [14, 15, 104]. The choice is arbitrary, and there are good reasons for both choices. We choose the tension-positive convention here to match the majority of fluid-mechanics textbooks. The tension-positive convention implies that the forces generated *by* a control volume will be negative and forces acting *on* the control volume will be positive. The alternate convention is desirable in the study of transport phenomena, in which the compression-positive choice

---

[4]Body moments are torques experienced by particles of fluid due to some intrinsic property of the fluid; that is, not due to the usual body forces (i.e., gravity) or surface forces (molecular action at the surface of the particle). Ferrofluids, which are suspensions of magnetic particles, experience body moments when they flow in the presence of a magnetic field [85]. For these fluids, it is inappropriate to assume that the stress tensor is symmetric.

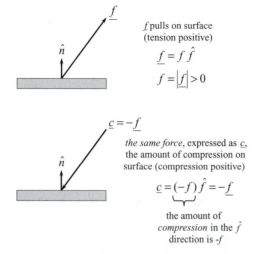

f pulls on surface
(tension positive)

$$\underline{f} = f\,\hat{f}$$

$$f = |\underline{f}| > 0$$

$\underline{c} = -\underline{f}$

the same force, expressed as $\underline{c}$,
the amount of compression on
surface (compression positive)

$$\underline{c} = (-f)\,\hat{f} = -\underline{f}$$

the amount of
compression in the $\hat{f}$
direction is $-f$

There are two common stress conventions in the chemical-engineering fluid-mechanics community. The one used here is used by most engineers—that is, the stress tensor expresses tension stress. This convention affects only expressions that contain the total stress tensor or the extra stress tensor.

makes the stress-velocity constitutive law (see Chapter 5 and the following discussion) have parallel construction to the flux/temperature and flux/concentration constitutive laws (see Equations 5.67 and 5.68 for identification of symbols):

$$
\begin{array}{c}
\text{stress/velocity law} \\
\text{(compression positive)}
\end{array}
\qquad
-\tilde{\tau}_{ik} = \tau_{ik} = -\mu\frac{\partial v_i}{\partial x_k}
\qquad (4.274)
$$

$$
\text{flux/temperature law}
\qquad
\frac{q_k}{A} = -k\frac{\partial T}{\partial x_k}
\qquad (4.275)
$$

$$
\text{stress/velocity law}
\qquad
J^*_{Ak} = -D_{AB}\frac{\partial c_A}{\partial x_k}
\qquad (4.276)
$$

We refer readers to Bird et al. [14, 15] for a complete discussion of the reasons for choosing the opposite convention. Also, in some textbooks [98], the meaning of the subscripts of $\underline{\tilde{\Pi}}$ are reversed from our usage; that is, in some textbooks $\tilde{\Pi}_{rs}$ is the stress on an s-plane in the r-direction. This convention is not as common as what we are using; when reading other sources, it is important to note which convention is being followed.

Our goal in this chapter is to develop an expression for molecular contact forces that we can use in the momentum balance on a control volume. We seek to represent all of the possible molecular contact forces that may be present in a moving fluid. We developed such an expression, Equation 4.263, repeated here:

$$
\begin{array}{c}
\text{Molecular fluid tension} \\
\text{on a submerged surface} \\
\text{of area } \Delta A \\
\text{with unit normal } \hat{n} \\
\text{(Gibbs notation)}
\end{array}
\qquad
\boxed{\underline{f}\Big|_{\Delta A} = \Delta A\ \left[\hat{n} \cdot \underline{\tilde{\Pi}}\right]\Big|_{\Delta A}}
\qquad (4.277)
$$

Tension on
a surface of area ΔA
with unit normal $\hat{n}$
(matrix notation)

$$\underline{f}\Big|_{\Delta A} = \Delta A \; \begin{pmatrix} n_1 & n_2 & n_3 \end{pmatrix}_{123} \cdot \begin{pmatrix} \tilde{\Pi}_{11} & \tilde{\Pi}_{12} & \tilde{\Pi}_{13} \\ \tilde{\Pi}_{21} & \tilde{\Pi}_{22} & \tilde{\Pi}_{23} \\ \tilde{\Pi}_{31} & \tilde{\Pi}_{32} & \tilde{\Pi}_{33} \end{pmatrix}_{123}$$

(4.278)

Equation 4.277 gives the force on a small surface ΔA in a flowing fluid. The contact force at a point in a fluid is written in terms of the product of an appropriate vector stress $(\hat{n} \cdot \tilde{\underline{\Pi}})$ and an area (ΔA). The stress at a point on a particular surface is calculated from the stress tensor, $\tilde{\underline{\Pi}}$, which is a convenient mathematical construct that allows us to calculate the stress on a surface if we know the unit normal to the surface $\hat{n}$. Equation 4.277 can be evaluated most easily by using matrix calculations with the components of $\hat{n}$ and $\tilde{\underline{\Pi}}$ when they are written in Cartesian or other convenient coordinate systems. As yet, we do not know how to obtain $\tilde{\underline{\Pi}}$ for a flow of interest; this is addressed in Chapter 5.

The expression in Equation 4.277 for force on a surface is applicable only to infinitesimal surfaces. If the surface ΔA is too large, then the unit normal vector $\hat{n}$ and the stress tensor $\tilde{\underline{\Pi}}$ may vary for different locations within ΔA. We can calculate force on a surface of finite size by applying Equation 4.277 to small pieces of the surface of interest and summing the forces on the various pieces. We addressed this type of sum over a surface in Chapter 3; the result is a surface integral. If we divide a surface into small tangent planes ΔS_i at various points on the surface (see Figure 3.30), we can write the force due to the fluid at one piece of the wall tangent plane ΔS_i using Equation 4.277:

Molecular fluid force
on surface ΔS_i
with unit normal $\hat{n}$
at Point (x_i, y_i, z_i)

$$\underline{f}\Big|_{\Delta S_i} = \left[\hat{n} \cdot \tilde{\underline{\Pi}}\right]_{(x_i y_i z_i)} \Delta S_i \qquad (4.279)$$

where $\tilde{\underline{\Pi}}$ is the total stress tensor and $[\hat{n} \cdot \tilde{\underline{\Pi}}]_{x_i y_i z_i}$ is the stress on ΔS_i at x_i, y_i, z_i. To obtain the force on an entire finite surface S, we sum all of the pieces that comprise the surface and take the limit as $\Delta S = \Delta A/(\hat{n} \cdot \hat{e}_z) \to 0$ (see web Appendix [108]). The result is a two-dimensional surface integral:

Total molecular fluid force
on a finite surface S

$$= \sum_{i=1}^{N} \underline{f}\Big|_{\Delta S_i} \qquad (4.280)$$

$$= \sum_{i=1}^{N} \left[\hat{n} \cdot \tilde{\underline{\Pi}}\right]_{(x_i y_i z_i)} \Delta S_i \qquad (4.281)$$

$$= \lim_{\Delta A \to 0} \left[\sum_{i=1}^{N} \frac{\left[\hat{n} \cdot \tilde{\underline{\Pi}}\right]_{(x_i y_i z_i)}}{\hat{n}_i \cdot \hat{e}_z} \Delta A_i \right] \qquad (4.282)$$

$$= \iint_{R} \frac{\left[\hat{n} \cdot \tilde{\underline{\Pi}}\right]_{\text{at surface}}}{\hat{n} \cdot \hat{e}_z} \, dA \qquad (4.283)$$

$$= \iint_{S} \left[\hat{n} \cdot \tilde{\underline{\Pi}}\right]_{\text{at surface}} \, dS \qquad (4.284)$$

Total molecular fluid force
on a finite surface $\mathcal{S}$

$$\mathcal{F} = \iint_{\mathcal{S}} \left[\hat{n} \cdot \underline{\tilde{\Pi}} \right]_{\text{at surface}} dS \qquad (4.285)$$

These steps come directly from calculus and are similar to the development of the convective term in the Reynolds transport theorem (see Equation 3.129).

The stress tensor is a powerful tool. For beginners, the key relationships are: Equation 4.277; the matrix version of that equation given in Equation 4.261; and Equation 4.285 for force on a finite surface. The power of the tensor notation is that all we need is the unit normal to a surface and, through simple matrix calculations, the force due to the fluid is predicted by Equation 4.261 or for a finite surface by Equation 4.285. The use of tensors simplifies a complex situation to straightforward matrix manipulations. The following is an example of how to calculate force on a finite surface in a simple flow.

EXAMPLE 4.17. *A Newtonian fluid is placed between two long, wide plates. The top plate is made to slide at a constant speed in the x_1–direction of the coordinate system shown (Figure 4.49). What is the force needed to move the plate? The stress tensor $\underline{\tilde{\Pi}}$ for the flow is given here (see Chapter 5 for the discussion of how to obtain the stress tensor for this flow):*

$$\underline{\tilde{\Pi}} = \begin{pmatrix} 0 & \dot{\gamma}_0 x_2 & 0 \\ \dot{\gamma}_0 x_2 & 0 & 0 \\ 0 & 0 & 0 \end{pmatrix}_{123} \qquad (4.286)$$

SOLUTION. The quantity $\dot{\gamma}_0$ in Equation 4.286 (read as "gamma-dot-naught") is a constant called the *shear rate*, which is a quantity that characterizes the intensity of the deformation produced by the flow. The total fluid force on a finite surface is given by Equation 4.285.

Total molecular fluid force
on a finite surface $\mathcal{S}$

$$\mathcal{F} = \iint_{\mathcal{S}} \left[\hat{n} \cdot \underline{\tilde{\Pi}} \right]_{\text{at surface}} dS \qquad (4.287)$$

We are interested in the force on the top plate of our flow cell, and this plate touches the top of the fluid. In the chosen coordinate system, the fluid surface of interest is a plane in the fluid at uniform position $x_2 = B$ with constant unit normal

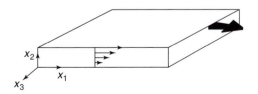

Figure 4.49 Fluid is deformed between two long, wide plates separated by a gap B. The top plate moves at a constant speed to the right. We seek to calculate the force needed to move the plate with this motion.

$\hat{n} = -\hat{e}_2$. Making these substitutions into Equation 4.287, we can calculate the force:

$$\begin{matrix} \text{Total molecular fluid force} \\ \text{on a finite surface } S \end{matrix} \qquad \underline{\mathcal{F}} = \iint_S \left[\hat{n} \cdot \underline{\tilde{\Pi}} \right]_{\text{at surface}} dS \qquad (4.288)$$

$$= \int_0^L \int_0^W \left[-\hat{e}_2 \cdot \underline{\tilde{\Pi}} \right]_{x_2 = B} dx_3 dx_1 \quad (4.289)$$

$$\underline{\mathcal{F}} = \int_0^L \int_0^W \begin{pmatrix} 0 & -1 & 0 \end{pmatrix}_{123} \cdot \begin{pmatrix} 0 & \dot{\gamma}_0 B & 0 \\ \dot{\gamma}_0 B & 0 & 0 \\ 0 & 0 & 0 \end{pmatrix}_{123} dx_3 dx_1 \quad (4.290)$$

$$= \int_0^L \int_0^W \begin{pmatrix} -\dot{\gamma}_0 B & 0 & 0 \end{pmatrix}_{123} dx_3 dx_1 \qquad (4.291)$$

$$\underline{\mathcal{F}} = \begin{pmatrix} -\dot{\gamma}_0 B L W \\ 0 \\ 0 \end{pmatrix}_{123} = -\dot{\gamma}_0 B L W \hat{e}_1 \qquad (4.292)$$

The final calculated force on the wall by the fluid is in the negative 1-direction of magnitude $\mathcal{F} = \mathcal{F}_x = B L W \dot{\gamma}_0$. To overcome the fluid force, a force of equal magnitude in the positive 1-direction must be applied to the plate.

4.3.2 Isotropic and anisotropic stress

It is useful to separate molecular fluid stresses into two categories: isotropic and anisotropic. When stress is isotropic at a point, the same magnitude of stress is exerted on any surface through the point and the stress always acts normally to the surface. In moving fluids, the molecular stress is not isotropic; rather, shear stresses—which are anisotropic stresses—are usually present, along with isotropic and anisotropic normal stresses. When stress is anisotropic, there are different values of stress on different surfaces through a chosen point, and anisotropic stresses are not limited to acting normally to the surface. Anisotropic molecular stresses are related straightforwardly to velocity gradients in a flow (see Chapter 5). In preparation for that discussion, we separate the isotropic and anisotropic parts of the stress tensor $\underline{\tilde{\Pi}}$ into two pieces.

To separate the isotropic part of $\underline{\tilde{\Pi}}$ from the anisotropic part, we simply subtract the isotropic part. The isotropic part of $\underline{\tilde{\Pi}}$ is easy to identify because it must be independent of surface orientation and always must act normally to any surface through a point. In terms of the components of the stress tensor $\underline{\tilde{\Pi}}$, the normal stresses are on the diagonal, $\tilde{\Pi}_{11}$, $\tilde{\Pi}_{22}$, and $\tilde{\Pi}_{33}$. For a stationary fluid, we know that the normal stresses at a point are equal to one another and have a value of $-p$. Thus, for a stationary fluid, we can write:

$$\begin{matrix} \text{Total stress tensor} \\ \text{(stationary fluid)} \end{matrix} \qquad \underline{\tilde{\Pi}} = \begin{pmatrix} -p & 0 & 0 \\ 0 & -p & 0 \\ 0 & 0 & -p \end{pmatrix}_{123} \qquad (4.293)$$

$$\underline{\tilde{\Pi}} = -p \underline{I} \qquad (4.294)$$

Recall from linear algebra that $\underline{I}$ is the identity tensor.

Equation 4.293 is a general expression for an isotropic stress in both stationary and moving fluids. It may seem odd to arrive at this expression by considering only one special coordinate system, the reference Cartesian system. We expect the stress tensor $\tilde{\underline{\underline{\Pi}}}$ to describe the stress on any surface, not only the reference surfaces with unit normals $\hat{e}_1$, $\hat{e}_2$, and $\hat{e}_3$. In fact, Equation 4.293 expresses the meaning we intend, the fact that pressure is isotropic and always acts normally no matter which surface is chosen, and we can show this by making a simple matrix calculation.

EXAMPLE 4.18. *Show that writing the total stress tensor at a point as $\tilde{\underline{\underline{\Pi}}} = -p\underline{\underline{I}}$, where p is pressure, implies that the stress acts equally in all directions and acts normally to any chosen surface at that point. $\underline{\underline{I}}$ is the identity tensor.*

$$\underline{\underline{I}} = \begin{pmatrix} 1 & 0 & 0 \\ 0 & 1 & 0 \\ 0 & 0 & 1 \end{pmatrix}_{123} \tag{4.295}$$

$$\tilde{\underline{\underline{\Pi}}} = \begin{pmatrix} -p & 0 & 0 \\ 0 & -p & 0 \\ 0 & 0 & -p \end{pmatrix}_{123} = -p\underline{\underline{I}} \tag{4.296}$$

SOLUTION. Let $\hat{n}$ be a unit normal vector to an arbitrary surface at the point of interest. As described in the previous section (Equation 4.263), we can calculate force on a surface of area ΔA with unit normal $\hat{n}$ by matrix multiplying $\hat{n}$ with $\tilde{\underline{\underline{\Pi}}}$:

$$\begin{array}{c} \text{Tension on} \\ \text{a surface of area } \Delta A \\ \text{with unit normal } \hat{n} \\ \text{(Gibbs notation)} \end{array} \qquad \boxed{\underline{f}\Big|_{\Delta A} = \Delta A \left[\hat{n} \cdot \tilde{\underline{\underline{\Pi}}}\right]\Big|_{\Delta A}} \tag{4.297}$$

For $\tilde{\underline{\underline{\Pi}}}$ given in Equation 4.296 and $\hat{n}$ written in the 123-coordinate system, we calculate the stress vector at the point of interest:

$$\frac{\underline{f}}{\Delta A} = \hat{n} \cdot \tilde{\underline{\underline{\Pi}}} \tag{4.298}$$

$$= \begin{pmatrix} n_1 & n_2 & n_3 \end{pmatrix}_{123} \cdot \begin{pmatrix} -p & 0 & 0 \\ 0 & -p & 0 \\ 0 & 0 & -p \end{pmatrix}_{123}$$

$$= -\begin{pmatrix} n_1 p & n_2 p & n_3 p \end{pmatrix}_{123}$$

$$= -p \begin{pmatrix} n_1 & n_2 & n_3 \end{pmatrix}_{123}$$

$$= -p\hat{n}$$

Thus, we see that, for $\tilde{\underline{\underline{\Pi}}} = -p\underline{\underline{I}}$, $\underline{f}/\Delta A = -\hat{n}p$; that is, the stress vector for any $\hat{n}$ is parallel to $\hat{n}$ of magnitude p. Thus, for the stress tensor given, the stress vector is stress normal to the surface and of a magnitude equal to the pressure, no matter which surface is chosen.

From the previous example, we see that the pressure contribution to the stress tensor may be written as a diagonal tensor $-p\underline{I}$. Thus far, we considered pressure in the context of stationary fluids; there also is a pressure contribution to the stress in flowing fluids. However, pressure is not the only stress in flowing fluids; neither is pressure the only normal stress in flowing fluids. In moving fluids, in addition to the isotropic normal stresses, $-p\underline{I}$, there are anisotropic stresses, that is off-diagonal elements of $\underline{\underline{\tilde{\Pi}}}$ and unequal diagonal terms in $\underline{\underline{\tilde{\Pi}}}$.

To separate the isotropic normal stresses from the anisotropic stresses, we choose to subtract the pressure contribution from the total stress tensor $\underline{\underline{\tilde{\Pi}}}$ and define a new stress tensor, called the *extra stress tensor* $\underline{\underline{\tilde{\tau}}}$.

$$\underline{\underline{\tilde{\tau}}} = \begin{pmatrix} \tilde{\tau}_{11} & \tilde{\tau}_{12} & \tilde{\tau}_{13} \\ \tilde{\tau}_{21} & \tilde{\tau}_{22} & \tilde{\tau}_{23} \\ \tilde{\tau}_{31} & \tilde{\tau}_{32} & \tilde{\tau}_{33} \end{pmatrix}_{123} \tag{4.299}$$

$$\equiv \begin{pmatrix} \tilde{\Pi}_{11} & \tilde{\Pi}_{12} & \tilde{\Pi}_{13} \\ \tilde{\Pi}_{21} & \tilde{\Pi}_{22} & \tilde{\Pi}_{23} \\ \tilde{\Pi}_{31} & \tilde{\Pi}_{32} & \tilde{\Pi}_{33} \end{pmatrix}_{123} - \begin{pmatrix} -p & 0 & 0 \\ 0 & -p & 0 \\ 0 & 0 & -p \end{pmatrix}_{123} \tag{4.300}$$

Extra stress
tensor
defined

$$\underline{\underline{\tilde{\tau}}} = \begin{pmatrix} \tilde{\Pi}_{11} + p & \tilde{\Pi}_{12} & \tilde{\Pi}_{13} \\ \tilde{\Pi}_{21} & \tilde{\Pi}_{22} + p & \tilde{\Pi}_{23} \\ \tilde{\Pi}_{31} & \tilde{\Pi}_{32} & \tilde{\Pi}_{33} + p \end{pmatrix}_{123} \tag{4.301}$$

$$\underline{\underline{\tilde{\tau}}} = \underline{\underline{\tilde{\Pi}}} + p\underline{I} \tag{4.302}$$

Like the total stress tensor $\underline{\underline{\tilde{\Pi}}}$, the extra stress tensor $\underline{\underline{\tilde{\tau}}}$ is symmetric, and there are six components of $\underline{\underline{\tilde{\tau}}}$ that characterize the stress at a point: the three normal stresses, $\tilde{\tau}_{11}, \tilde{\tau}_{22}, \tilde{\tau}_{33}$; and the three shear stresses, $\tilde{\tau}_{21} = \tilde{\tau}_{12}, \tilde{\tau}_{31} = \tilde{\tau}_{13}, \tilde{\tau}_{32} = \tilde{\tau}_{23}$.

We have accomplished much in our quest to write molecular stress on a control volume. We know that molecular tension $\underline{f}$ on a surface ΔA is given by:

Molecular fluid force
on a surface
of area ΔA
with unit normal $\hat{n}$

$$\underline{f}\Big|_{\Delta A} = \Delta A \, [\hat{n} \cdot \underline{\underline{\tilde{\Pi}}}]\Big|_{\Delta A} \tag{4.303}$$

where $\hat{n}$ is the unit normal to the surface and $\underline{\underline{\tilde{\Pi}}}$ is the total stress tensor. For a finite surface S, we also know that:

Total molecular fluid force
on a finite surface S

$$\mathcal{F} = \iint_S [\hat{n} \cdot \underline{\underline{\tilde{\Pi}}}]_{\text{at surface}} \, dS \tag{4.304}$$

In any problem we tackle, we choose the control volume and, therefore, we know the area on which we are calculating molecular stress and the associated $\hat{n}$ for that area. The only missing ingredient is $\underline{\underline{\tilde{\Pi}}}$. Thus far, we still do not know the nine $\tilde{\Pi}_{ij}$, but we know that $\underline{\underline{\tilde{\Pi}}}$ is symmetric, and we have seen how to separate out the pressure contribution $-p\underline{\underline{I}}$ from the anisotropic contribution $\underline{\underline{\tilde{\tau}}}$:

$$\text{Stress tensor} \quad \boxed{\underline{\underline{\tilde{\Pi}}} = \underline{\underline{\tilde{\tau}}} - p\underline{\underline{I}}} \qquad (4.305)$$

We can see the power and organization of $\underline{\underline{\tilde{\Pi}}}$ and Equations 4.303 and 4.304 in the trial calculations that follow.

EXAMPLE 4.19. *When we study stationary fluids at the beginning of this chapter, we calculated the buoyancy effect on a sphere submerged in a fluid of density ρ (see Figure 4.23, Example 4.6). Reexamine this problem using the stress tensor $\underline{\underline{\tilde{\Pi}}}$ to obtain the stress expressions needed.*

SOLUTION. In the previous solution to the problem of calculating the fluid force on a sphere submerged in a liquid, we did not have the stress tensor to work with and we had to carefully reason the role of pressure. Now that we have introduced the stress tensor and have calculated it for stationary fluids (Equation 4.293), we can calculate the force on the sphere surface directly from Equation 4.304.

$$\begin{array}{c}\text{Total stress tensor} \\ \text{(stationary fluid)}\end{array} \quad \underline{\underline{\tilde{\Pi}}} = \begin{pmatrix} -p & 0 & 0 \\ 0 & -p & 0 \\ 0 & 0 & -p \end{pmatrix}_{123} = -p\underline{\underline{I}} \quad (4.306)$$

$$\begin{array}{c}\text{Total molecular fluid force} \\ \text{on a finite surface } S\end{array} \quad \underline{\mathcal{F}} = \iint_S [\hat{n} \cdot \underline{\underline{\tilde{\Pi}}}]_{\text{at surface}} \, dS \quad (4.307)$$

We are calculating the fluid force on a sphere. The unit normal at every location on the sphere surface is $\hat{e}_r$ at that location. For a static fluid, the stress tensor at every location in the fluid is $\underline{\underline{\tilde{\Pi}}} = -p\underline{\underline{I}}$, where p is a function of elevation due to the hydrostatic effect. To solve the problem, we form the dot product in Equation 4.307 and carry out the integration on the surface of the sphere:

$$[\hat{n} \cdot \underline{\underline{\tilde{\Pi}}}]|_{r=R} = \left[(1 \ \ 0 \ \ 0)_{r\theta\phi} \cdot \begin{pmatrix} -p & 0 & 0 \\ 0 & -p & 0 \\ 0 & 0 & -p \end{pmatrix}_{r\theta\phi} \right]\Bigg|_{r=R} \quad (4.308)$$

$$= (-p|_R \ \ 0 \ \ 0)_{r\theta\phi} \quad (4.309)$$

We can directly carry out this matrix multiplication because both vectors are written in the same orthonormal coordinate system; now we substitute the result into Equation 4.307.

$$\begin{array}{c}\text{Total molecular fluid force} \\ \text{on a submerged sphere}\end{array} \quad \underline{\mathcal{F}} = \iint_S [\hat{n} \cdot \underline{\underline{\tilde{\Pi}}}]_{r=R} \, dS \quad (4.310)$$

$$= \int_0^{2\pi} \int_0^\pi \begin{pmatrix} -p|_R \\ 0 \\ 0 \end{pmatrix}_{r\theta\phi} R^2 \sin\theta \, d\theta \, d\phi$$

$$(4.311)$$

The pressure is a function of elevation due to the hydrostatic effect; we previously worked out the function for $p(\theta)$ using geometry (see Equation 4.112):

$$p|_R = \rho g \left(H_0 - R \cos \theta\right) \qquad (4.312)$$

where H_0 is the distance from the fluid surface to the center of the sphere. Combining this result with Equation 4.311, we obtain:

$$\begin{array}{c}\text{Total fluid force} \\ \text{on a sphere} \\ \text{in static fluid}\end{array} \qquad \underline{\mathcal{F}} = R^2 \int_0^{2\pi} \int_0^{\pi} \begin{pmatrix} -\rho g \left(H_0 - R \cos \theta\right) \\ 0 \\ 0 \end{pmatrix}_{r\theta\phi} \sin \theta\, d\theta\, d\phi$$

$$(4.313)$$

This is the same equation we obtained in the earlier example, and the steps to the final result are given there (see Equation 4.117 and the equations that follow). We see that the stress tensor makes it straightforward to set up the calculation of forces on surfaces in fluids.

EXAMPLE 4.20. *A cup-and-bob apparatus is widely used to measure viscosities for fluids. For the apparatus in Figure 4.50, what is the torque needed to turn the inner cylinder (called the bob) at an angular speed of Ω? The stress tensor $\underline{\tilde{\Pi}}$ for the flow is given here (see Chapter 6 for discussion of how to calculate the*

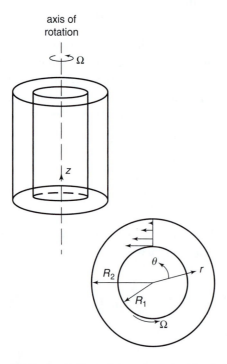

Figure 4.50 Fluid is placed in the gap between two concentric cylinders. The inner cylinder (i.e., the bob) is made to turn at an angular speed of Ω; the outer cylinder (i.e., the cup) remains stationary. We seek to calculate the torque needed to move the bob with the proscribed motion.

stress tensor for this flow):

$$\tilde{\underline{\underline{\Pi}}} = \begin{pmatrix} 0 & \mu r \frac{df}{dr} & 0 \\ \mu r \frac{df}{dr} & 0 & 0 \\ 0 & 0 & 0 \end{pmatrix}_{r\theta z} \tag{4.314}$$

where $f = \frac{v_\theta}{r} = \alpha\Omega\left(1 - \frac{R_2^2}{r^2}\right)$ and $\alpha = \left(\frac{R_1^2}{R_2^2}\right) / \left(\frac{R_1^2}{R_2^2} - 1\right)$. (The effect of pressure has been neglected).

SOLUTION. We are interested in the total torque exerted by the fluid on the inner cylinder, which has a radius of R_1. Torque is the vector expressing the amount of effort needed to produce a rotation in a body; the definition of torque is the cross product of lever arm and the force [167]. We calculate the torque on the entire bob beginning with the torque generated by the force on an infinitesimal surface in contact with the fluid (see also Section 6.2.3.2).

$$\begin{array}{c} \text{Molecular fluid force} \\ \text{on surface } \Delta S_i \\ \text{with unit normal } \hat{n} \\ \text{at point } (x_i, y_i, z_i) \end{array} \qquad \underline{f}\Big|_{\Delta S_i} = \left[\hat{n} \cdot \tilde{\underline{\underline{\Pi}}}\right]_{(x_i y_i z_i)} \Delta S_i \tag{4.315}$$

On a small piece of the surface of the inner cup (surface at $r = R_1$ with unit normal $\hat{n} = \hat{e}_r$), the stress is given by $\tilde{\underline{\underline{\Pi}}}$ over an area $R_1 d\theta dz$; thus, the torque is:

$$d\underline{\mathcal{T}} = \text{(lever arm)} \times \text{(force)} \tag{4.316}$$

$$= \text{(lever arm)} \times [\hat{n} \cdot \tilde{\underline{\underline{\Pi}}}]_{\Delta S} \Delta S \tag{4.317}$$

$$= (R_1 \hat{e}_r) \times \left(\hat{e}_r \cdot \tilde{\underline{\underline{\Pi}}}\big|_{r=R_1} R_1 d\theta dz\right) \tag{4.318}$$

$$= \begin{pmatrix} R_1 \\ 0 \\ 0 \end{pmatrix}_{r\theta z} \times \left(\begin{pmatrix} 0 \\ \left[\mu r \frac{df}{dr}\right]\Big|_{r=R_1} (R_1 d\theta dz) \\ 0 \end{pmatrix}_{r\theta z} \right) \tag{4.319}$$

$$= \begin{pmatrix} 0 \\ 0 \\ 2\mu\alpha\Omega R_2^2 d\theta dz \end{pmatrix}_{r\theta z} \tag{4.320}$$

where the dot product $\hat{e}_r \cdot \tilde{\underline{\underline{\Pi}}}$ was evaluated using Equation 4.314 and the cross product using Equation 1.182. The total torque on the bob is this quantity integrated over the entire bob surface (see Section 6.2.3.2):

$$\underline{\mathcal{T}} = \int_0^L \int_0^{2\pi} \begin{pmatrix} 0 \\ 0 \\ 2\mu\alpha\Omega R_2^2 \end{pmatrix}_{r\theta z} d\theta dz \tag{4.321}$$

$$= \int_0^L \int_0^{2\pi} 2\mu\alpha\Omega R_2^2 \hat{e}_z d\theta dz \tag{4.322}$$

Because all of the quantities in the integral are constant, including the basis vector $\hat{e}_z$, the integration is straightforward:

$$\underline{\mathcal{T}} = 4\pi R_2^2 L \mu \alpha \Omega \hat{e}_z \tag{4.323}$$

$$\underline{\mathcal{T}} = \left(\frac{4\pi R_1^2 L \mu \Omega}{\frac{R_1^2}{R_2^2} - 1} \right) \hat{e}_z = \begin{pmatrix} 0 \\ 0 \\ \frac{4\pi R_1^2 L \mu \Omega}{\frac{R_1^2}{R_2^2} - 1} \end{pmatrix}_{r\theta z} \tag{4.324}$$

The result is a clockwise fluid torque (since the inner radius/outer radius < 1); the negative of this result is the counterclockwise torque that must be applied to the bob to produce the flow in the gap between the two cylinders. Measurement of the torque and rotational speed Ω for an apparatus of known geometry (known R_1 and R_2) allows us to determine the viscosity of the fluid in the gap.

EXAMPLE 4.21. *What is the force on an airfoil subjected to a uniform flow* $\underline{v} = U\hat{e}_x$ *(Figure 4.51)? To make this calculation, numerical results for the stress field* $\underline{\tilde{\Pi}}$ *as a function of position* (x, y, z) *are available. Also, the shape and orientation of the airfoil are given.*

SOLUTION. The techniques in this textbook lead to equations that can be solved numerically for $\underline{\tilde{\Pi}}(x, y, z)$ for flows such as the one discussed here. Once the stress field is known, it often is our goal to calculate a concrete engineering quantity such as the net force on the airfoil. The force vector on an airfoil in a flow reveals the drag and the lift that the airfoil experiences (see Chapter 8):

$$\text{Drag} = \begin{pmatrix} \text{Component of force} \\ \text{on an object} \\ \text{parallel to} \\ \text{the incident flow direction} \end{pmatrix} \tag{4.325}$$

$$\text{Lift} = \begin{pmatrix} \text{Component of force} \\ \text{on an object} \\ \text{normal to} \\ \text{the incident flow direction} \end{pmatrix} \tag{4.326}$$

$$\underline{f} = \begin{pmatrix} \mathcal{F}_{\text{drag}} \\ \mathcal{F}_{\text{lift}} \\ 0 \end{pmatrix}_{xyz} \tag{4.327}$$

The y-direction has been chosen to be parallel to the direction of lift.

$\underline{V} = U\hat{e}_x$

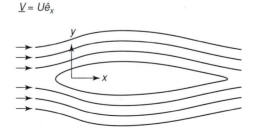

Figure 4.51 Uniform flow approaches the airfoil from the left. The flow splits as it moves around the object.

To calculate the force on the airfoil, we must implement the integral in Equation 4.304 over the airfoil surface:

$$\text{Total molecular fluid force} \atop \text{on a surface } \mathcal{S} \qquad \underline{\mathcal{F}} = \iint_{\mathcal{S}} \left[\hat{n} \cdot \underline{\underline{\tilde{\Pi}}} \right]_{\text{at surface}} dS \qquad (4.328)$$

The integral is carried out numerically. To revert Equation 4.328 to a form that may be carried out numerically, we go back to the steps used in this chapter to develop this integral, ending with Equation 4.282:

$$\text{Total molecular fluid force} \atop \text{on airfoil} \qquad \underline{f} = \begin{pmatrix} \mathcal{F}_{\text{drag}} \\ \mathcal{F}_{\text{lift}} \\ 0 \end{pmatrix}_{xyz} \qquad (4.329)$$

$$= \sum_{i=1}^{N} \underline{f} \Big|_{\Delta S_i} \qquad (4.330)$$

$$= \sum_{i=1}^{N} \left[\hat{n} \cdot \underline{\underline{\tilde{\Pi}}} \right]_{(x_i y_i z_i)} \Delta S_i \qquad (4.331)$$

$$= \lim_{\Delta A \to 0} \left[\sum_{i=1}^{N} \frac{\left[\hat{n} \cdot \underline{\underline{\tilde{\Pi}}} \right]_{(x_i y_i z_i)}}{\hat{n}_i \cdot \hat{e}_z} \Delta A_i \right] \qquad (4.332)$$

Thus, to calculate $\underline{f}$ on the airfoil, for each point $(x_i y_i z_i)$ located on the surface of the airfoil, we carry out the following calculations:

1. Calculate the outwardly pointing unit vector $\hat{n}|_i$ at the point.
2. Calculate the dot product of $\hat{n}|_i$ with the value of $\underline{\underline{\tilde{\Pi}}}$ at the point.
3. Calculate the area ΔA_i of the airfoil surface that is associated with the point. The value of ΔA_i depends on how close together the points $(x_i y_i z_i)$ are available.
4. Calculate the sum in Equation 4.332.
5. Confirm that the chosen ΔA_i are sufficiently small for the results to be accurate. This is usually accomplished by carrying out multiple calculations with successively smaller ΔA_i until the size of this area has no effect on the calculated answer (i.e., mesh refinement; see Chapter 10).

Numerical simulators available commercially [27] have built-in routines that can make these calculations. In that case, it is sufficient for the user of the software to understand which integration is needed to address the question at hand.

In this chapter, we develop a powerful tool, the stress tensor. The stress tensor is an entity that allows us to calculate the force on a small surface in a fluid or, with integration, the force on a finite surface in a fluid:

$$\text{Fluid force on} \atop \text{a flat surface of area } \Delta A \atop \text{with unit normal } \hat{n} \qquad \boxed{\underline{f} = \Delta A \left[\hat{n} \cdot \underline{\underline{\tilde{\Pi}}} \right] \Big|_{\Delta A}} \qquad (4.333)$$

Tension Stress Tensor in Flow

- Molecular contact stress
- Symmetric; i.e., $\widetilde{\Pi}_{ik} = \widetilde{\Pi}_{ki}$, $\widetilde{\tau}_{pm} = \widetilde{\tau}_{mp}$
- $\widetilde{\Pi}_{jn}$ is stress on a j-surface in the n-direction
- Infinitesimal surface: $\underline{f} = \Delta A\,\hat{n}\cdot\underline{\underline{\widetilde{\Pi}}}$
- Finite surface:

$$\underline{f} = \iint_S \left[\hat{n}\cdot\underline{\underline{\widetilde{\Pi}}}\right]dS$$

$$\boxed{\underline{\underline{\widetilde{\Pi}}} = \underline{\underline{\widetilde{\tau}}} - p\underline{\underline{I}}}$$

$$\begin{pmatrix} \widetilde{\Pi}_{11} & \widetilde{\Pi}_{12} & \widetilde{\Pi}_{13} \\ \widetilde{\Pi}_{21} & \widetilde{\Pi}_{22} & \widetilde{\Pi}_{23} \\ \widetilde{\Pi}_{31} & \widetilde{\Pi}_{32} & \widetilde{\Pi}_{33} \end{pmatrix}_{123} = \begin{pmatrix} \widetilde{\tau}_{11}-p & \widetilde{\tau}_{12} & \widetilde{\tau}_{13} \\ \widetilde{\tau}_{21} & \widetilde{\tau}_{22}-p & \widetilde{\tau}_{23} \\ \widetilde{\tau}_{31} & \widetilde{\tau}_{32} & \widetilde{\tau}_{33}-p \end{pmatrix}_{123}$$

Figure 4.52 The stress tensor in matrix form can be used to calculate the tension on a surface in a flow.

$$\text{Total molecular fluid force} \atop \text{on a finite surface } \mathcal{S} \qquad \underline{\mathcal{F}} = \iint_{\mathcal{S}} \left[\hat{n}\cdot\underline{\underline{\widetilde{\Pi}}}\right]_{\text{at surface}} dS \qquad (4.334)$$

The stress tensor may be thought of as composed of an isotropic part $-p\underline{\underline{I}}$ and an anisotropic part $\underline{\underline{\widetilde{\tau}}}$, both of which are a function of velocity in a moving fluid (Figure 4.52). In a stationary fluid, there is no anisotropic part ($\underline{\underline{\widetilde{\tau}}}\big|_{stationary} = 0$) and $\underline{\underline{\widetilde{\Pi}}}\big|_{stationary} = -p\underline{\underline{I}}$. To proceed to modeling moving fluids from this point, we need the components of $\underline{\underline{\widetilde{\tau}}}$ as a function of velocity (see Chapter 5).

To see the progress made in the development of problem-solving methods, we return to the two unfinished problems in Chapter 3—flow down an inclined plane and flow in a right-angle bend—and see how the stress-tensor components figure into them. Section 4.4 addresses a related but independent fluid-stress subject: free-surface molecular-stress effects, including surface tension.

EXAMPLE 4.22 (Incline, continued). *What is the velocity field in a wide, thin film of water that runs steadily down an inclined surface under the force of gravity? The fluid has a constant density ρ.*

SOLUTION. We began this problem in Chapter 3, and the geometry is shown in Figure 4.53. After choosing a coordinate system, we chose a rectangular solid shape of volume $\Delta x\,\Delta y\,\Delta z$ as the control volume (Figure 4.54). To solve for the

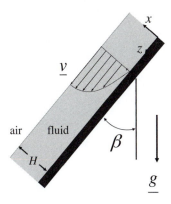

Figure 4.53 The idealized version of flow down an incline is a film of constant thickness where the velocity is everywhere in the same direction but varies in magnitude with position in the film. We seek to calculate the velocity as a function of position relative to the wall (i.e., as a function of x).

velocity field, we applied the momentum balance to this CV:

Reynolds transport theorem (momentum balance on CV)

$$\frac{d\underline{\mathbf{P}}}{dt} + \iint_{CS} (\hat{n} \cdot \underline{v}) \, \rho \underline{v} \, dS = \sum_{\substack{\text{on} \\ \text{CV}}} \underline{f} \quad (4.335)$$

For steady flow, the rate of accumulation of momentum in the CV is zero ($d\underline{\mathbf{P}}/dt = 0$). As discussed in Chapter 3, for steady, unidirectional flow, the net outflow of momentum from the control surface (the integral in Equation 4.335) is zero. There are two types of forces that we identified as acting on the CV: gravity

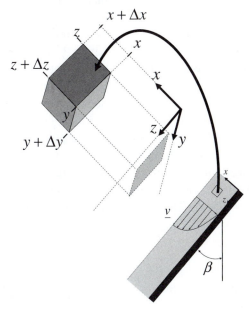

Figure 4.54 The control volume we chose for this problem is a microscopic parallelepiped of volume $\triangle x \triangle y \triangle z$. It is located at an arbitrary location $x\,y\,z$ within the flow. By choosing this type of control volume, we can derive the differential equations that relate fluid stresses.

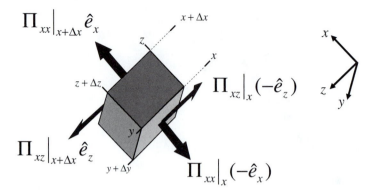

Figure 4.55 The molecular forces in the inclined flow problem are analogous to the forces that previously acted on a block sliding down an incline studied. All stress components are positive when there is a pull (i.e., tension) in the positive *k*-direction.

(noncontact) and molecular (contact) forces. Thus, as we showed in Chapter 3, the momentum balance for this problem becomes:

$$0 = \sum_{\substack{\text{on} \\ \text{CV}}} \underline{f} \tag{4.336}$$

$$= \Delta x \Delta y \Delta z \begin{pmatrix} -\rho g \sin \beta \\ 0 \\ \rho g \cos \beta \end{pmatrix}_{xyz} + \begin{pmatrix} \text{molecular} \\ \text{sliding} \\ \text{forces} \\ \text{on CV} \end{pmatrix} + \begin{pmatrix} \text{molecular} \\ \text{normal} \\ \text{forces} \\ \text{on CV} \end{pmatrix} \tag{4.337}$$

We are now in a position to write the two molecular-force contributions in terms of $\underline{\underline{\tilde{\Pi}}}$.

When identifying the molecular forces in this example, we take inspiration from the sliding-block example discussed in Chapter 3. In that situation, as in this problem, there was a downward pull due to gravity. The surface of the incline imposed a retarding frictional force and an upward, supporting normal force on the block; in the current flow example, we see that the situation is similar. Tangential frictional forces act on the top and the bottom of the CV (Figure 4.55). The layers of fluid above the CV are sliding past the top, like a block sliding down an inclined plane. These forces act in the z-direction on the top control surface, which is a surface with unit normal $\hat{e}_x$. In the stress tensor written in our coordinate system, the stress in the flow direction on this surface is the component $\tilde{\Pi}_{xz}$ evaluated at the top surface. The top surface is at location $x + \Delta x$ and has an area of $\Delta y \Delta z$. The force on the top surface due to sliding is then:

$$\begin{pmatrix} \text{molecular} \\ \text{sliding force} \\ \text{on top of CV} \\ z\text{-component} \end{pmatrix} = \left[\left(\frac{\text{force}}{\text{area}} \right) (\text{area}) \begin{pmatrix} \text{unit vector} \\ \text{indicating} \\ \text{direction} \end{pmatrix} \right]_{top} \tag{4.338}$$

$$= \tilde{\Pi}_{xz}\big|_{x+\Delta x} \, \Delta y \Delta z \, \hat{e}_z \tag{4.339}$$

$$= \tilde{\tau}_{xz}\big|_{x+\Delta x} \, \Delta y \Delta z \, \hat{e}_z \tag{4.340}$$

For shear terms, $\tilde{\Pi}_{ij} = \tilde{\tau}_{ij}$; the only differences between $\underline{\underline{\tilde{\Pi}}}$ and $\underline{\underline{\tilde{\tau}}}$ occur in the normal stress terms along the diagonal (see Equation 4.301).

This way of arriving at the molecular friction forces on the top of the CV relies on thought and evaluation. We also can arrive at this result by automatically applying Equation 4.263, which is repeated here:

$$\begin{array}{c} \text{Force on a surface} \\ \text{of area } \Delta A \\ \text{with unit normal } \hat{n} \end{array} \qquad \boxed{\underline{f}\Big|_{\Delta A} = \Delta A \, [\hat{n} \cdot \underline{\underline{\tilde{\Pi}}}]|_{\Delta A}} \qquad (4.341)$$

For the top control surface, $\hat{n} = \hat{e}_x$ and $\Delta A = \Delta y \Delta z$. We write $\underline{\underline{\tilde{\Pi}}}$ in matrix form and then simplify the calculation:

$$\begin{pmatrix} \text{molecular} \\ \text{sliding force} \\ \text{on top of CV} \end{pmatrix} = \left[\Delta A \, \hat{n} \cdot \underline{\underline{\tilde{\Pi}}}\right]\Big|_{x+\Delta x} \qquad (4.342)$$

$$= \Delta y \Delta z \, \hat{e}_x \cdot \underline{\underline{\tilde{\Pi}}}\Big|_{x+\Delta x} \qquad (4.343)$$

$$= \Delta y \Delta z \, \begin{pmatrix} 1 & 0 & 0 \end{pmatrix}_{xyz} \cdot \begin{pmatrix} \tilde{\Pi}_{xx} & \tilde{\Pi}_{xy} & \tilde{\Pi}_{xz} \\ \tilde{\Pi}_{yx} & \tilde{\Pi}_{yy} & \tilde{\Pi}_{yz} \\ \tilde{\Pi}_{zx} & \tilde{\Pi}_{zy} & \tilde{\Pi}_{zz} \end{pmatrix}_{xyz}\Bigg|_{x+\Delta x} \qquad (4.344)$$

$$= \Delta y \Delta z \, \begin{pmatrix} \tilde{\Pi}_{xx} & \tilde{\Pi}_{xy} & \tilde{\Pi}_{xz} \end{pmatrix}_{xyz}\Big|_{x+\Delta x} \qquad (4.345)$$

$$= \Delta y \Delta z \, \begin{pmatrix} \tilde{\Pi}_{xx}|_{x+\Delta x} \\ \tilde{\Pi}_{xy}|_{x+\Delta x} \\ \tilde{\Pi}_{xz}|_{x+\Delta x} \end{pmatrix}_{xyz} \qquad (4.346)$$

We write this result in terms of the components of $\underline{\underline{\tilde{\tau}}}$ by writing $\underline{\underline{\tilde{\Pi}}} = \underline{\underline{\tilde{\tau}}} - p\underline{\underline{I}}$:

$$\begin{pmatrix} \text{molecular} \\ \text{sliding force} \\ \text{on top of CV} \end{pmatrix} = \underline{f}_{top} = \Delta y \Delta z \begin{pmatrix} \tilde{\tau}_{xx}|_{x+\Delta x} - p|_{x+\Delta x} \\ \tilde{\tau}_{xy}|_{x+\Delta x} \\ \tilde{\tau}_{xz}|_{x+\Delta x} \end{pmatrix}_{xyz} \qquad (4.347)$$

The z-component of Equation 4.347 is the result obtained in Equation 4.340. The x- and y-components of Equation 4.347 are force components that have not yet been considered. They are the forces on the top of the CV in the x- and y-directions. The matrix calculation shows how pressure enters the problem for this surface—pressure is a normal force that adds to the other normal-force component on this surface, $\tilde{\tau}_{xx}$. The mathematics conveniently guides us to include these terms.

To write all of the molecular forces on the CV, we must carry out the equivalent calculation (Equation 4.341) for each of the remaining five surfaces of the CV. The calculations can be made systematically, as shown in Table 4.3 and as follows:

$$\underline{f}_{k^{th} \, side} = \left[\Delta A \, \hat{n} \cdot \underline{\underline{\tilde{\Pi}}}\right]\Big|_{k^{th} \, side} \qquad (4.348)$$

Table 4.3. *The molecular stresses acting on the chosen control volume calculated individually for six sides*

Side	$\hat{n}$	ΔA	$\hat{n} \cdot \underline{\underline{\tilde{\Pi}}}$	Simplified
Top	$\hat{e}_x$	$\Delta y \Delta z$	$\left. \begin{pmatrix} \tilde{\tau}_{xx} - p \\ \tilde{\tau}_{xy} \\ \tilde{\tau}_{xz} \end{pmatrix}_{xyz} \right|_{x+\Delta x}$	$\begin{pmatrix} -p\vert_{x+\Delta x} \\ 0 \\ \tilde{\tau}_{xz}\vert_{x+\Delta x} \end{pmatrix}_{xyz}$
Bottom	$-\hat{e}_x$	$\Delta y \Delta z$	$-\left. \begin{pmatrix} \tilde{\tau}_{xx} - p \\ \tilde{\tau}_{xy} \\ \tilde{\tau}_{xz} \end{pmatrix}_{xyz} \right|_{x}$	$\begin{pmatrix} p\vert_{x} \\ 0 \\ -\tilde{\tau}_{xz}\vert_{x} \end{pmatrix}_{xyz}$
Neutral(1)	$\hat{e}_y$	$\Delta x \Delta z$	$\left. \begin{pmatrix} \tilde{\tau}_{yx} \\ \tilde{\tau}_{yy} - p \\ \tilde{\tau}_{yz} \end{pmatrix}_{xyz} \right|_{y+\Delta y}$	$\begin{pmatrix} 0 \\ -p\vert_{y+\Delta y} \\ 0 \end{pmatrix}_{xyz}$
Neutral(2)	$-\hat{e}_y$	$\Delta x \Delta z$	$-\left. \begin{pmatrix} \tilde{\tau}_{yx} \\ \tilde{\tau}_{yy} - p \\ \tilde{\tau}_{yz} \end{pmatrix}_{xyz} \right|_{y}$	$\begin{pmatrix} 0 \\ p\vert_{y} \\ 0 \end{pmatrix}_{xyz}$
Downstream	$\hat{e}_z$	$\Delta x \Delta y$	$\left. \begin{pmatrix} \tilde{\tau}_{zx} \\ \tilde{\tau}_{zy} \\ \tilde{\tau}_{zz} - p \end{pmatrix}_{xyz} \right|_{z+\Delta z}$	$\begin{pmatrix} \tilde{\tau}_{zx}\vert_{z+\Delta z} \\ 0 \\ -p\vert_{z+\Delta z} \end{pmatrix}_{xyz}$
Upstream	$-\hat{e}_z$	$\Delta x \Delta y$	$-\left. \begin{pmatrix} -\tilde{\tau}_{zx} \\ \tilde{\tau}_{zy} \\ \tilde{\tau}_{zz} - p \end{pmatrix}_{xyz} \right|_{z}$	$\begin{pmatrix} \tilde{\tau}_{zx}\vert_{z} \\ 0 \\ p\vert_{z} \end{pmatrix}_{xyz}$

Note: Also shown are the same terms simplified by knowledge of the stress-velocity constitutive equation (see Chapter 5).

We now substitute these expressions back into the momentum balance, Equation 4.337:

$$0 = \sum_{\substack{\text{on} \\ \text{CV}}} \underline{f} \tag{4.349}$$

$$= \begin{pmatrix} \text{gravity} \\ \text{force} \\ \text{on CV} \end{pmatrix} + \begin{pmatrix} \text{molecular} \\ \text{sliding forces} \\ \text{on CV} \end{pmatrix} + \begin{pmatrix} \text{molecular} \\ \text{normal forces} \\ \text{on CV} \end{pmatrix} \tag{4.350}$$

$$= \underline{f}_{gravity} + \sum_{k=1}^{6} \underline{f}_{k^{th} side} \tag{4.351}$$

$$= \Delta x \Delta y \Delta z \begin{pmatrix} -\rho g \sin \beta \\ 0 \\ \rho g \cos \beta \end{pmatrix}_{xyz}$$

$$+ \Delta y \Delta z \begin{pmatrix} \tilde{\tau}_{xx}\vert_{x+\Delta x} - p\vert_{x+\Delta x} \\ \tilde{\tau}_{xy}\vert_{x+\Delta x} \\ \tilde{\tau}_{xz}\vert_{x+\Delta x} \end{pmatrix}_{xyz} - \Delta y \Delta z \begin{pmatrix} \tilde{\tau}_{xx}\vert_{x} - p\vert_{x} \\ \tilde{\tau}_{xy}\vert_{x} \\ \tilde{\tau}_{xz}\vert_{x} \end{pmatrix}_{xyz}$$

$$+ \Delta x \Delta z \begin{pmatrix} \tilde{\tau}_{yx}\vert_{y+\Delta y} \\ \tilde{\tau}_{yy}\vert_{y+\Delta y} - p\vert_{y+\Delta y} \\ \tilde{\tau}_{yz}\vert_{y+\Delta y} \end{pmatrix}_{xyz} - \Delta x \Delta z \begin{pmatrix} \tilde{\tau}_{yx}\vert_{y} \\ \tilde{\tau}_{yy}\vert_{y} - p\vert_{y} \\ \tilde{\tau}_{yz}\vert_{y} \end{pmatrix}_{xyz}$$

$$+ \; \Delta x \, \Delta y \begin{pmatrix} \tilde{\tau}_{zx}|_{z+\Delta z} \\ \tilde{\tau}_{zy}|_{z+\Delta z} \\ \tilde{\tau}_{zz}|_{z+\Delta z} - p|_{z+\Delta z} \end{pmatrix}_{xyz} - \Delta x \, \Delta y \begin{pmatrix} \tilde{\tau}_{zx}|_z \\ \tilde{\tau}_{zy}|_z \\ \tilde{\tau}_{zz}|_z - p|_z \end{pmatrix}_{xyz} \qquad (4.352)$$

This is a complex expression, but it is systematic. We simplify this momentum balance by combining like terms, dividing through every term by the volume $\Delta x \, \Delta y \, \Delta z$, and rearranging:

$$0 = \begin{pmatrix} -\rho g \sin \beta \\ 0 \\ \rho g \cos \beta \end{pmatrix}_{xyz}$$

$$+ \frac{1}{\Delta x} \begin{pmatrix} \tilde{\tau}_{xx}|_{x+\Delta x} - \tilde{\tau}_{xx}|_x \\ \tilde{\tau}_{xy}|_{x+\Delta x} - \tilde{\tau}_{xy}|_x \\ \tilde{\tau}_{xz}|_{x+\Delta x} - \tilde{\tau}_{xz}|_x \end{pmatrix}_{xyz} - \frac{1}{\Delta x} \begin{pmatrix} p|_{x+\Delta x} - p|_x \\ 0 \\ 0 \end{pmatrix}_{xyz}$$

$$+ \frac{1}{\Delta y} \begin{pmatrix} \tilde{\tau}_{yx}|_{y+\Delta y} - \tilde{\tau}_{yx}|_y \\ \tilde{\tau}_{yy}|_{y+\Delta y} - \tilde{\tau}_{yy}|_y \\ \tilde{\tau}_{yz}|_{y+\Delta y} - \tilde{\tau}_{yz}|_y \end{pmatrix}_{xyz} - \frac{1}{\Delta y} \begin{pmatrix} 0 \\ p|_{y+\Delta y} - p|_y \\ 0 \end{pmatrix}_{xyz}$$

$$+ \frac{1}{\Delta z} \begin{pmatrix} \tilde{\tau}_{zx}|_{z+\Delta z} - \tilde{\tau}_{zx}|_z \\ \tilde{\tau}_{zy}|_{z+\Delta z} - \tilde{\tau}_{zy}|_z \\ \tilde{\tau}_{zz}|_{z+\Delta z} - \tilde{\tau}_{zz}|_z \end{pmatrix}_{xyz} - \frac{1}{\Delta z} \begin{pmatrix} 0 \\ 0 \\ p|_{z+\Delta z} - p|_z \end{pmatrix}_{xyz} \qquad (4.353)$$

We can split this vector equation into three components in the x-, y-, and z-directions. We begin with the x-component:

$$0 = -\rho g \sin \beta + \left(\frac{\tilde{\tau}_{xx}|_{x+\Delta x} - \tilde{\tau}_{xx}|_x}{\Delta x} \right) - \left(\frac{p|_{x+\Delta x} - p|_x}{\Delta x} \right)$$

$$+ \left(\frac{\tilde{\tau}_{yx}|_{y+\Delta y} - \tilde{\tau}_{yx}|_x}{\Delta y} \right) + \left(\frac{\tilde{\tau}_{zx}|_{z+\Delta z} - \tilde{\tau}_{zx}|_z}{\Delta z} \right) \qquad (4.354)$$

If we now take the limits as Δx, Δy, and Δz go to zero, we see that the expressions in parentheses in Equation 4.354 are equal to first derivatives (see Equation 1.138):

$$\rho g \sin \beta = \lim_{\Delta x \to 0} \left[\frac{\tilde{\tau}_{xx}|_{x+\Delta x} - \tilde{\tau}_{xx}|_x}{\Delta x} \right] - \lim_{\Delta x \to 0} \left[\frac{p|_{x+\Delta x} - p|_x}{\Delta x} \right]$$

$$+ \lim_{\Delta y \to 0} \left[\frac{\tilde{\tau}_{yx}|_{y+\Delta y} - \tilde{\tau}_{yx}|_y}{\Delta y} \right] + \lim_{\Delta z \to 0} \left[\frac{\tilde{\tau}_{zx}|_{z+\Delta z} - \tilde{\tau}_{zx}|_z}{\Delta z} \right] \qquad (4.355)$$

$$\rho g \sin \beta = \frac{\partial \tilde{\tau}_{xx}}{\partial x} - \frac{\partial p}{\partial x} + \frac{\partial \tilde{\tau}_{yx}}{\partial y} + \frac{\partial \tilde{\tau}_{zx}}{\partial z} \qquad (4.356)$$

x-component, momentum balance

$$\boxed{0 = -\frac{\partial p}{\partial x} + \left[\frac{\partial \tilde{\tau}_{xx}}{\partial x} + \frac{\partial \tilde{\tau}_{yx}}{\partial y} + \frac{\partial \tilde{\tau}_{zx}}{\partial z} \right] - \rho g \sin \beta} \qquad (4.357)$$

The result is a differential equation in terms of the stress components $\tilde{\tau}_{xx}$, $\tilde{\tau}_{xx}$, $\tilde{\tau}_{xx}$, and the pressure p. We arrive at analogous expressions for the y- and z-components of the momentum balance by following similar steps:

y-component, momentum balance

$$0 = -\frac{\partial p}{\partial y} + \left[\frac{\partial \tilde{\tau}_{xy}}{\partial x} + \frac{\partial \tilde{\tau}_{yy}}{\partial y} + \frac{\partial \tilde{\tau}_{zy}}{\partial z} \right]$$ (4.358)

z-component, momentum balance

$$0 = -\frac{\partial p}{\partial z} + \left[\frac{\partial \tilde{\tau}_{xz}}{\partial x} + \frac{\partial \tilde{\tau}_{yz}}{\partial y} + \frac{\partial \tilde{\tau}_{zz}}{\partial z} \right] + \rho g \cos \beta$$ (4.359)

By writing the stresses in terms of the components of the stress tensor $\tilde{\Pi}_{jm} = \tilde{\tau}_{jm} - p\underline{I}$, we progress from a word-equation balance of momentum in Equation 4.337 to three partial differential equations. Our goal is to solve for the velocity field. To proceed further, we must find out how the stress components are related to the velocity field $\underline{v}$. This subject was studied many years ago by several scientists, including Isaac Newton (1643–1727). In Chapter 5, we discuss Newton's results and apply them to finish this example.

The momentum-balance results in Equations 4.357, 4.358, and 4.359 are more complicated than we were likely to arrive at through an ad hoc procedure of imagining which stresses act on the various faces of a fluid control volume. The introduction of the stress tensor made systematic the expression of molecular stresses. We are now confident that we are properly accounting for the contributions of molecular forces to the momentum balance. In Chapter 6, we relate these terms to the velocity and solve the differential equation for velocity and pressure as a function of position.

The second unfinished example in Chapter 3 used a macroscopic control volume on the flow in a right-angle bend. We now continue our work on this problem, using what we know about molecular stress to advance it.

EXAMPLE 4.23 (90 Degree bend, continued). *What is the direction and magnitude of the force needed to support the 90 degree pipe bend shown in Figure 4.56? An incompressible (i.e., constant density) liquid enters the pipe at volumetric flow rate Q_a and exits at volumetric flow rate Q_b at steady state. The cross section of the pipe bend is πR^2.*

SOLUTION. In this problem, we use a control volume that includes all of the fluid in the 90 degree bend (see Figure 3.36) and perform a momentum balance (see equation 3.187). The flow is incompressible and at steady state; thus, $d\underline{P}/dt = 0$ and $Q_a = Q_b$. The average velocity was used in Example 3.10 to quantify the convection of momentum into and out of the control volume through surfaces (a) and (b). The forces on the CV were identified as gravity and three molecular contact forces: the restoring force on the sides $\underline{R}$ for which we are solving, and

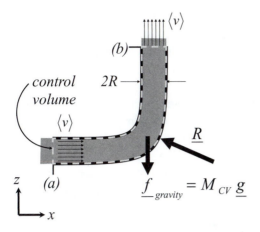

Figure 4.56 In Chapter 3, we carried out a momentum balance on a macroscopic control volume such as the one outlined with a dotted line here. In this chapter, we specify the molecular stresses.

the as-yet-unspecified molecular forces on inlet and outlet surfaces (a) and (b). With these forces included, the momentum balance was found to be:

$$\frac{d\mathbf{P}}{dt} = - \iint_S (\hat{n} \cdot \underline{v})\, \rho \underline{v}\, dS + \sum_{\substack{\text{on} \\ \text{CV}}} \underline{f} \tag{4.360}$$

$$0 = -\langle v \rangle^2 \rho \pi R^2 \begin{pmatrix} -1 \\ 0 \\ 1 \end{pmatrix}_{xyz} + \sum_{\substack{\text{on} \\ \text{CV}}} \underline{f} \tag{4.361}$$

$$= \langle v \rangle^2 \rho \pi R^2 \begin{pmatrix} 1 \\ 0 \\ -1 \end{pmatrix}_{xyz} + M_{CV} \begin{pmatrix} 0 \\ 0 \\ -g \end{pmatrix}_{xyz} + \begin{pmatrix} \mathcal{R}_x \\ \mathcal{R}_y \\ \mathcal{R}_z \end{pmatrix}_{xyz}$$

$$+ \begin{pmatrix} \text{molecular} \\ \text{force on CV} \\ \text{at } (a) \end{pmatrix} + \begin{pmatrix} \text{molecular} \\ \text{force on CV} \\ \text{at } (b) \end{pmatrix} \tag{4.362}$$

We are now in a position to specify those molecular forces in terms of the stress tensor $\underline{\tilde{\Pi}} = \underline{\tilde{\tau}} - p\underline{I}$.

The fluid force $\mathcal{F}$ on a finite surface S is given by:

$$\text{Total molecular fluid force} \atop \text{on a surface } S \qquad \mathcal{F} = \iint_S \left[\hat{n} \cdot \underline{\tilde{\Pi}} \right]_{\text{at surface}} dS \tag{4.363}$$

We can calculate the molecular forces on the inlet surface (a) and the outlet surface (b) by applying Equation 4.363 to each surface in turn. The unit normal vectors of the surfaces (a) and (b) are given in our chosen 90 degree-bend Cartesian coordinate system by:

$$\hat{n}|_a = \begin{pmatrix} -1 \\ 0 \\ 0 \end{pmatrix}_{xyz} \qquad \hat{n}|_b = \begin{pmatrix} 0 \\ 0 \\ 1 \end{pmatrix}_{xyz} \tag{4.364}$$

When we previously worked on this problem, we decided that the variation of the velocity across the cross section was not of interest and that we would characterize the velocity by its average value $\langle v \rangle$, which is the same at the inlet and at the outlet. Now we also seek to express the molecular stress in terms of a quantity averaged over surfaces (a) and (b).

The surface at (a) is a circle, and integration over a circle most easily is performed in a cylindrical coordinate system. At the (a) surface, we choose for convenience to carry out the integration in a cylindrical coordinate system $r\theta\bar{z}$ with the $\bar{z}$-direction of the cylindrical system parallel to the x-direction of our 90-degree-bend Cartesian system. We adopt the $\bar{z}$ nomenclature to avoid confusion with the Cartesian z-direction. The outwardly pointing unit normal to (a) in the chosen cylindrical system is $\hat{n} = -\hat{e}_{\bar{z}}$:

$$\begin{matrix} \text{Total molecular fluid force} \\ \text{on a surface } \mathcal{S} \end{matrix} \qquad \underline{\mathcal{F}} = \iint_{\mathcal{S}} \left[\hat{n} \cdot \underline{\underline{\tilde{\Pi}}} \right]_{\text{at surface}} dS \quad (4.365)$$

$$\underline{\mathcal{F}}|_a = \int_0^{2\pi} \int_0^R -\hat{e}_{\bar{z}} \cdot \underline{\underline{\tilde{\Pi}}}\big|_a r\, dr\, d\theta \qquad (4.366)$$

$$= \int_0^{2\pi} \int_0^R -\hat{e}_{\bar{z}} \cdot \left[(\underline{\underline{\tilde{\tau}}} - p\underline{\underline{I}}) \right]_a r\, dr\, d\theta \qquad (4.367)$$

$$= \int_0^{2\pi} \int_0^R \left(-\hat{e}_{\bar{z}} \cdot \underline{\underline{\tilde{\tau}}}\big|_a \right) r\, dr\, d\theta + \int_0^{2\pi} \int_0^R (p|_a \, \hat{e}_{\bar{z}}) r\, dr\, d\theta \qquad (4.368)$$

where we have written $\underline{\underline{\tilde{\Pi}}} = \underline{\underline{\tilde{\tau}}} - p\underline{\underline{I}}$ and $\hat{e}_{\bar{z}} \cdot p\underline{\underline{I}} = p\hat{e}_{\bar{z}}$ (see Example 4.18). The pressure integral is the integral of a constant (i.e., both p and the unit vector $\hat{e}_{\bar{z}} = \hat{e}_x$ are constant); thus, we easily carry out that integral. The extra-stress integral is more problematic. We cannot carry out the integral with $\underline{\underline{\tilde{\tau}}}$ until we know more about $\underline{\underline{\tilde{\tau}}}$. We define $\underline{f}_\mu$ as $\hat{n} \cdot \underline{\underline{\tilde{\tau}}}$ at a surface of interest, which allows us to write the unknown stress integral term as the spatial average of $\underline{f}_\mu$:

$$\underline{\mathcal{F}}|_a = \left[\int_0^{2\pi} \int_0^R \left(-\hat{e}_{\bar{z}} \cdot \underline{\underline{\tilde{\tau}}}\big|_a \right) r\, dr\, d\theta \right] + p|_a \, \pi R^2 \hat{e}_x \qquad (4.369)$$

$$= \pi R^2 \left\langle \underline{f}_\mu\big|_a \right\rangle + p|_a \, \pi R^2 \hat{e}_x \qquad (4.370)$$

where:

$$\left\langle \underline{f}_\mu\big|_b \right\rangle = \frac{1}{\pi R^2} \left[\int_0^{2\pi} \int_0^R \left(-\hat{e}_{\bar{z}} \cdot \underline{\underline{\tilde{\tau}}}\big|_a \right) r\, dr\, d\theta \right] \qquad (4.371)$$

We revert to using the global Cartesian coordinate system in the pressure term in Equation 4.370.

For the (b) side, the outwardly pointing unit vector in our Cartesian system is $\hat{n} = \hat{e}_z$; for the cylindrical coordinate system convenient for this integral, we choose $\hat{e}_{\bar{z}} = \hat{e}_z$. Thus, we obtain for the (b) surface:

$$\underline{\mathcal{F}}|_b = \int_0^{2\pi} \int_0^R \hat{e}_{\bar{z}} \cdot \left[(\underline{\underline{\tilde{\tau}}} - p\underline{\underline{I}}) \right]_b r \, dr \, d\theta \tag{4.372}$$

$$= \left[\int_0^{2\pi} \int_0^R \left(\hat{e}_{\bar{z}} \cdot \underline{\underline{\tilde{\tau}}} \Big|_b \right) r \, dr \, d\theta \right] - p|_b \, \pi R^2 \hat{e}_z \tag{4.373}$$

$$= \pi R^2 \left\langle \underline{f}_\mu \Big|_b \right\rangle - p|_b \, \pi R^2 \hat{e}_z \tag{4.374}$$

where:

$$\left\langle \underline{f}_\mu \Big|_b \right\rangle = \frac{1}{\pi R^2} \left[\int_0^{2\pi} \int_0^R \left(\hat{e}_{\bar{z}} \cdot \underline{\underline{\tilde{\tau}}} \Big|_b \right) r \, dr \, d\theta \right] \tag{4.375}$$

Substituting these results into Equation 4.362 yields:

$$0 = \langle v \rangle^2 \rho \pi R^2 \begin{pmatrix} 1 \\ 0 \\ -1 \end{pmatrix}_{xyz} + M_{CV} \begin{pmatrix} 0 \\ 0 \\ -g \end{pmatrix}_{xyz} + \begin{pmatrix} \mathcal{R}_x \\ \mathcal{R}_y \\ \mathcal{R}_z \end{pmatrix}_{xyz}$$

$$+ \pi R^2 \left\langle \underline{f}_\mu \Big|_a \right\rangle + \pi R^2 \begin{pmatrix} p|_a \\ 0 \\ 0 \end{pmatrix}_{xyz} + \pi R^2 \left\langle \underline{f}_\mu \Big|_b \right\rangle - \pi R^2 \begin{pmatrix} 0 \\ 0 \\ p|_b \end{pmatrix}_{xyz} \tag{4.376}$$

From this point, we are again stuck. We replaced words in Equation 4.362 with the stress variables $\underline{\underline{\tilde{\tau}}}$ and p, and we now see how pressure is accounted for in the problem. We cannot evaluate the two terms with $\underline{f}_\mu$ in them without knowing how $\underline{\underline{\tilde{\tau}}}$ is related to velocity, especially how $\underline{\underline{\tilde{\tau}}}$ is related to flow direction. We return to this example in Chapter 5 after the discussion of stress-velocity constitutive relationships.

To move forward in our calculations, we need concrete values for the stress tensor and we must know how stress is related to velocity $\underline{v}$. To specify these required relationships, we must discuss the behavior of molecules in motion. Different types of fluids with different molecular forces give different functions $\underline{\underline{\tilde{\tau}}}(\underline{v})$:

<table>
<tr><td>Stress-velocity
constitutive equation
(different for different fluids)</td><td>$\underline{\underline{\tilde{\tau}}} = f(\underline{v})$</td><td>(4.377)</td></tr>
</table>

For simple fluids, the stresses $\tilde{\tau}_{ij}$ and the velocity field v_p are connected through a relationship identified by Newton—the Newtonian constitutive equation—which is introduced in Chapter 5. As discussed there, the Newtonian constitutive equation reflects the effects of *Brownian motion* on momentum transfer among fluid layers. Because the behavior of many fluids is dominated by Brownian motion, the random thermal motion associated with all molecules, the Newtonian equation is found to be widely applicable—it works for water, oil, solvents, and even air. The Newtonian constitutive equation relates derivatives of velocity to the extra stress components $\tilde{\tau}_{ij}$. Once we understand the relationships between $\tilde{\tau}_{ij}$ and the velocity components v_p, we can substitute these stress functions into the momentum balance and subsequently solve the balance equations. For a microscopic CV, the solutions of the balance equations provide the velocity and stress fields; once we know these, we have complete knowledge of the flow pattern and the distribution of forces in the flow. For a macroscopic balance, the solutions of the balance equation provide the missing quantity—often a force vector or the pressure.

Readers may proceed directly to Chapter 5. Before leaving the subject of stresses in fluids, we present a discussion of stress-related phenomena that occur at the interfaces between phases. Near an interface, fluids behave differently than within a bulk fluid. In Section 4.4, we introduce a new continuum function—the surface tension—to account for these effects. In some flows, surface tension enters as a boundary condition. Surface tension also can drive flows, particularly when the dimensions of the flow are small such as in microfluidics (see Figure 1.5) [75].

4.4 Free-surface stress effects

Interesting effects occur at fluid-gas interfaces due to the intermolecular attractions that hold liquids together (see Section 2.7). These attractions act between neighboring molecules; when a molecule is located deep within the fluid in a container, it feels attractive pulls from every direction (Figure 2.38). The different pulls offset one another and the molecule feels a balanced force. At the interface between a liquid and a gas, however, the gas exerts a negligible attractive force on the liquid molecules at the interface. Molecules at the surface feel the downward pull of their attraction to the liquid molecules beneath them, but there is no balancing upward pull from gas molecules above them. Liquid molecules at the surface are therefore pulled toward the bulk of the liquid, and the surface layer compresses until the liquid's natural resistance to compression allows the interface to attain equilibrium. The state of stress at the interface is different from that in the bulk and must be treated differently in our continuum model.

In liquid–solid–gas systems, a related phenomenon occurs. For example, when a liquid is contained in a solid vessel but is open to a gas, the location and shape of the contact line where solid, liquid, and gas meet are determined by intermolecular forces (Figure 4.57). The liquid molecules near the liquid–gas interface experience the same unbalanced forces as described previously. The liquid molecules in contact with the solid surface experience intermolecular forces of two types: (1) the attraction of their neighboring liquid molecules; and (2) either an attraction to or a repulsion from the molecules on the surface of the solid. If the liquid molecules are attracted to the solid, the contact line will climb the wall of the container due to these attractions. A balance between the intermolecular attractions between liquid and solid molecules and the downward force due to gravity ultimately determines the location of the contact line. If the liquid molecules are repelled by the solid, the contact line drops below the surface of

Table 4.4. Surface Tension for Several Liquids

Liquid	σ (dyne/cm)
Water	73
Salt water	75
Soapy water	20-30
Ether	17
Alcohol	23
Carbon tetrachloride	27
Lubricating oil	25-35
Mercury	480

Source: reference [154]

the bulk liquid in the container, and a balance between intermolecular forces and gravity again determines the final position of the contact line. This interfacial effect is magnified in small tubes or in narrow gaps where the attraction of a liquid for a solid surface can cause liquid to climb several centimeters in height above the bulk-liquid level (Figure 2.41).

Liquid–gas and liquid–solid–gas interfaces occur in engineering flows, and the effects of surface forces often must be taken into account. Because free-surface phenomena depend on intermolecular forces, however, the effect is different for every liquid/gas/solid combination. Our challenge is to determine how to account for these intermolecular effects within the continuum model.

The overall effect of the unbalanced molecular forces at an interface can be modeled with a continuum property known as the *surface tension*. The unbalanced forces at a liquid–gas interface make the interface act as if it were covered with a massless membrane that is in a state of tension. The tension in this imaginary membrane is expressed as an amount of force per unit length acting along any line drawn in the interface. This tension/length is given the symbol σ and is called the surface tension:

$$\text{Surface tension} \quad \sigma\,[\text{N/m}] \tag{4.378}$$

Surface tension manifests as a force that acts tangent to the interface and perpendicular to the chosen line [40].

Representative values of surface tension are given in Table 4.4. As shown, water has a higher surface tension than most other fluids; liquid metals such as mercury are the exception to this rule [154]. When other molecules are added to water, the surface tension usually decreases: for example, when soap is added to water, the surface tension decreases by more than 70 percent. Certain salts slightly

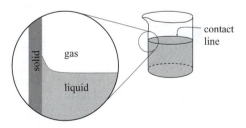

Figure 4.57 When a solid, liquid, and a gas meet, the intermolecular forces between the solid and the liquid and between the liquid and the gas are not usually the same. Thus, liquid molecules have a preference for being in contact with either the solid or the gas. The differences in intermolecular attractions between the phases cause the surface to curve as shown.

increase the surface tension of water. Surface tension decreases with increasing temperature for all liquids.

To use the surface tension σ in making calculations, we apply the idea of the interface as a massless membrane in a state of tension. When performing momentum balances on control volumes that intersect a fluid interface, surface tension acts along the closed line of intersection of the interface with the surface of the control volume. This method can be illustrated with the following examples.

EXAMPLE 4.24. *What is the pressure inside a spherical water droplet in air (Figure 4.58)? What is the pressure difference across interfaces when the shape is not spherical?*

SOLUTION. Consider the momentum balance on a control volume that consists of half of the droplet (see Figure 4.58). Because the droplet is motionless ($\underline{v} = 0$, $\mathbf{P} = 0$), the sum of the forces on the CV is zero. The forces on the CV are gravity, molecular-contact forces, and surface-tension forces. In this analysis, we neglect the pull of gravity:

$$\frac{d\mathbf{P}}{dt} + \iint_{CS} (\hat{n} \cdot \underline{v}) \, \rho \underline{v} \, dS = \sum_{\substack{\text{on} \\ \text{CV}}} \underline{f} \qquad \begin{array}{l} \text{Reynolds transport theorem} \\ \text{(momentum balance on CV)} \end{array} \qquad (4.379)$$

$$0 = \sum_{\substack{\text{on} \\ \text{CV}}} \underline{f} \qquad (4.380)$$

$$0 = \underline{f}_{gravity} + \underline{f}_{surface} + \underline{f}_{\sigma} \qquad (4.381)$$

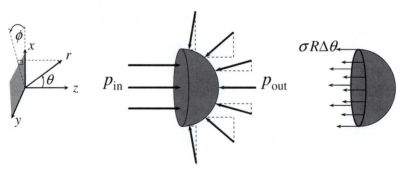

Figure 4.58 A water droplet in air has a higher pressure inside compared to outside due to the surface tension. We can calculate the pressure difference by considering the forces on a control volume consisting of half of the spherical droplet. The surface tension acts like a force along the rim and the pressure acts normally to the control-volume surface at every point.

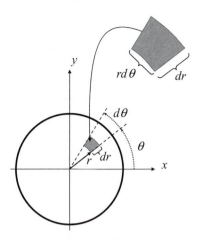

Figure 4.59 The surface element for the calculation of the force on the flat circular surface of the control volume is shown here. The cylindrical coordinate system is used.

The molecular force on a CV is evaluated by calculating the net force on all of the surfaces that enclose it. The surface forces on the chosen hemispherical CV can be calculated in two pieces: force on the flat surface, $\underline{f}_{flat}$, and force on the hemispherical surface, $\underline{f}_{hemisph}$. The integral for force on a finite surface S is:

$$
\begin{array}{c}
\text{Total molecular fluid force} \\
\text{on a finite surface } \mathcal{S}
\end{array}
\qquad
\underline{\mathcal{F}} = \iint_{\mathcal{S}} \left[\hat{n} \cdot \tilde{\underline{\tilde{\Pi}}}\right]_{\text{at surface}} dS \qquad (4.382)
$$

We begin with $\underline{f}_{flat}$, the force on the circular surface portion of the CV bounding surface. For this surface, the outwardly pointing unit normal vector $\hat{n} = -\hat{e}_z$. The stress tensor in the stationary fluid inside the droplet is due entirely to isotropic stress, $\tilde{\underline{\tilde{\Pi}}}|_{flat} = -p_{in}\underline{\underline{I}}$ (see Equation 4.293). The surface element dS in the cylindrical coordinate system of the circular control surface is given by $(dr)(r\,d\theta)$ (Figure 4.59). Combining these results in Equation 4.382, we obtain:

$$
\begin{array}{c}
\text{Total molecular fluid force} \\
\text{on a finite surface } \mathcal{S}
\end{array}
\qquad
\underline{\mathcal{F}} = \iint_{\mathcal{S}} \left[\hat{n} \cdot \tilde{\underline{\tilde{\Pi}}}\right]_{\text{at surface}} dS \qquad (4.383)
$$

$$
\underline{f}_{flat} = \int_0^{2\pi} \int_0^R \left[-\hat{e}_z \cdot (-p_{in}\underline{\underline{I}})\right] (dr)(r\,d\theta) \qquad (4.384)
$$

$$
= 2\pi \int_0^R \begin{pmatrix} 0 & 0 & 1 \end{pmatrix}_{r\theta z} \cdot \begin{pmatrix} p_{in} & 0 & 0 \\ 0 & p_{in} & 0 \\ 0 & 0 & p_{in} \end{pmatrix}_{r\theta z} r\,dr \qquad (4.385)
$$

$$
= 2\pi \int_0^R \begin{pmatrix} 0 & 0 & p_{in} \end{pmatrix}_{r\theta z} r\,dr \qquad (4.386)
$$

$$
= \begin{pmatrix} 0 & 0 & \pi R^2 p_{in} \end{pmatrix}_{r\theta z} \qquad (4.387)
$$

$$
\underline{f}_{flat} = p_{in}\pi R^2 \hat{e}_z = \begin{pmatrix} 0 \\ 0 \\ p_{in}\pi R^2 \end{pmatrix}_{r\theta z} = \begin{pmatrix} 0 \\ 0 \\ p_{in}\pi R^2 \end{pmatrix}_{xyz} \qquad (4.388)
$$

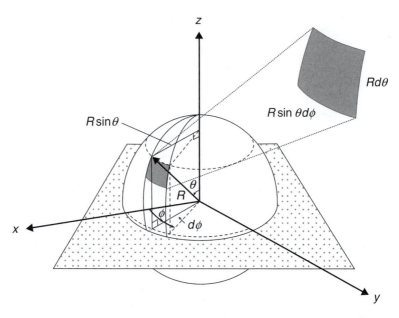

The surface element for the calculation of the force on the spherical surface of the control volume is shown here.

where p_{in} is the pressure on the inside of the droplet and R is the radius of the droplet. Note that Equation 4.388 contains the result written in two different coordinate systems.

Next, we calculate $\underline{f}_{hemisph}$. The fluid on the outside of the bubble is different from the fluid inside; thus, $\underset{=}{\tilde{\Pi}}$ is different. For the stationary outer fluid:

$$\begin{matrix} \text{Total molecular fluid force} \\ \text{on a finite surface } \mathcal{S} \end{matrix} \qquad \underline{\mathcal{F}} = \iint_{\mathcal{S}} \left[\hat{n} \cdot \underset{=}{\tilde{\Pi}} \right]_{\text{at surface}} dS \qquad (4.389)$$

$$\underline{f}_{hemisph} = \iint_{\mathcal{S}} \left[\hat{n} \cdot -p_{ext}\underset{=}{I} \right]\Big|_{r=R} dS \quad (4.390)$$

The result of this calculation is $\underline{f}_{hemisph} = p_{ext}\pi R^2$ in the $-\hat{e}_z$ direction because the radially inward and radially outward components of the force due to pressure on the hemispherical surface of our CV exactly balance out, leaving only the z-contribution. We show this formally in the following discussion.

To calculate the force on the outside of the droplet from Equation 4.390, it is easiest to use a spherical coordinate system rather than the cylindrical system used on the other control surface. The definition of the spherical coordinate system relative to the Cartesian system is given in equations 1.268–1.273. For the hemispherical surface of our CV, $\hat{n} = \hat{e}_r$, where $\hat{e}_r$ is a basis vector of the spherical coordinate system. The stress tensor in the stationary fluid outside the droplet is $\underset{=}{\tilde{\Pi}}\big|_{external} = -p_{ext}\underset{=}{I}$. The surface element dS of the spherical coordinate system of the hemispherical control surface is given by $(Rd\theta)(R\sin\theta d\phi)$ (Figure 4.60). Substituting these results into Equation 4.390, we obtain:

$$\underline{f}_{hemisph} = \iint_{\mathcal{S}} \left[\hat{n} \cdot \underset{=}{\Pi} \right]\big|_{surface} dS \qquad (4.391)$$

$$= \int_0^{2\pi} \int_0^{\pi/2} \left[\hat{e}_r \cdot \left(-p_{ext}\underset{=}{I} \right) \right]\big|_{r=R} (Rd\theta)(R\sin\theta d\phi) \quad (4.392)$$

$$= -\int_0^{2\pi} \int_0^{\pi/2} \begin{pmatrix} 1 & 0 & 0 \end{pmatrix}_{r\theta\phi} \cdot \begin{pmatrix} p_{ext} & 0 & 0 \\ 0 & p_{ext} & 0 \\ 0 & 0 & p_{ext} \end{pmatrix}_{r\theta\phi} R^2 \sin\theta d\theta d\phi$$

(4.393)

$$= -\int_0^{2\pi} \int_0^{\pi/2} \begin{pmatrix} p_{ext} & 0 & 0 \end{pmatrix}_{r\theta\phi} R^2 \sin\theta d\theta d\phi$$

(4.394)

$$= -\int_0^{2\pi} \int_0^{\pi/2} p_{ext} \hat{e}_r \, R^2 \sin\theta d\theta d\phi$$

(4.395)

The unit vector $\hat{e}_r$, which is one of the basis vectors of the spherical coordinate system, is not a constant; rather, it is a function of both θ and ϕ, the variables over which we are integrating. To finish the integration in Equation 4.395, we write $\hat{e}_r$ in the Cartesian coordinate system and carry out the two integrations. The expression for $\hat{e}_r$ in the Cartesian coordinate system was given in Equation 1.271:

$$\underline{f}_{hemisph} = -p_{ext} R^2 \int_0^{2\pi} \int_0^{\pi/2} \hat{e}_r \, \sin\theta d\theta d\phi$$

(4.396)

$$= -p_{ext} R^2 \int_0^{2\pi} \int_0^{\pi/2} (\sin\theta \cos\phi \, \hat{e}_x + \sin\theta \sin\phi \, \hat{e}_y$$

$$+ \cos\theta \, \hat{e}_z) \, \sin\theta d\theta d\phi$$

(4.397)

$$= -p_{ext} R^2 \int_0^{2\pi} \int_0^{\pi/2} \begin{pmatrix} \sin^2\theta \cos\phi \\ \sin^2\theta \sin\phi \\ \sin\theta \cos\theta \end{pmatrix}_{xyz} d\theta d\phi$$

(4.398)

Integrating each component separately and applying the appropriate integration limits, the ϕ integrals of the x- and y-components give zero. Thus, the force on the outside has no x or y component, which was expected. The ϕ-integral of the third term yields 2π. To calculate the final result, we now carry out the θ-integration of the remaining term:

$$\underline{f}_{hemisph} = -p_{ext} R^2 \int_0^{2\pi} \int_0^{\pi/2} \sin\theta \cos\theta \, \hat{e}_z \, d\theta d\phi$$

(4.399)

$$= -p_{ext} 2\pi R^2 \hat{e}_z \int_0^{\pi/2} \cos\theta \sin\theta \, \cos\theta d\theta$$

(4.400)

$$= -p_{ext} 2\pi R^2 \hat{e}_z \left[\frac{\sin^2\theta}{2} \right] \Big|_{\theta=0}^{\theta=\pi/2}$$

(4.401)

$$= -p_{ext} \pi R^2 \hat{e}_z = \begin{pmatrix} 0 \\ 0 \\ -p_{ext} \pi R^2 \end{pmatrix}_{xyz}$$

(4.402)

This is the result we reasoned out previously (see the discussion following Equation 4.390).

The final force acting on our CV is the force due to surface tension $\underline{f}_\sigma$. To visualize how surface tension works, imagine that the surface of the droplet is actually a spherical balloon. We impose a pressure on the inside of the balloon so that it inflates exactly as a sphere. If we cut the sphere in half but somehow keep it inflated, the balloon is elastic and wants to contract and collapse. To secure the edges of the balloon and contain its contents in place, we must hold onto the circular rim of the half balloon with a certain amount of force that appropriately balances the inside pressure (see Figure 4.58, right). The magnitude of the force/length that we must apply along the rim of the circular balloon is equal to the surface tension, σ. The force we use to restrain the balloon is tangential to the balloon surface and is directed in the $(-z)$-direction of the coordinate system shown, perpendicular to the circular line of intersection of the CV and the interface. The net surface-tension force is given by:

$$\begin{pmatrix} \text{force on CV} \\ \text{due to} \\ \text{surface} \\ \text{tension} \end{pmatrix} = \underline{f}_\sigma = \begin{pmatrix} \text{force/length} \\ \text{along} \\ \text{circumference} \end{pmatrix} (\text{length}) \begin{pmatrix} \text{unit vector} \\ \text{indicating} \\ \text{direction} \end{pmatrix} \quad (4.403)$$

$$= \sigma(2\pi R)(-\hat{e}_z) = \begin{pmatrix} 0 \\ 0 \\ -2\pi R\sigma \end{pmatrix}_{xyz}$$

We return to Equation 4.380 and assemble the force balance:

$$0 = \sum_{\substack{\text{on} \\ \text{CV}}} \underline{f} = \underline{f}_{surface} + \underline{f}_\sigma \quad (4.404)$$

$$= \underline{f}_{flat} + \underline{f}_{hemisph} + \underline{f}_\sigma \quad (4.405)$$

$$= \begin{pmatrix} 0 \\ 0 \\ p_{in}\pi R^2 \end{pmatrix}_{123} + \begin{pmatrix} 0 \\ 0 \\ -p_{ext}\pi R^2 \end{pmatrix}_{123} + \begin{pmatrix} 0 \\ 0 \\ -2\pi R\sigma \end{pmatrix}_{123} \quad (4.406)$$

$$= p_{in}\pi R^2 \hat{e}_z - p_{ext}\pi R^2 \hat{e}_z - 2\pi R\sigma \hat{e}_z \quad (4.407)$$

Solving for the pressure difference, we find that Δp across the bubble interface is proportional to the surface tension:

$$\boxed{\Delta p = p_{in} - p_{ext} = \frac{2\sigma}{R}} \quad (4.408)$$

The pressure inside the water drop is greater than the pressure outside the drop by the amount $2\sigma/R$. This extra pressure is due to the unbalanced molecular forces at the droplet surface, which lead to an extra inward pull on the surface-water molecules. Note that the pressure difference increases as the size of the droplet decreases.

For surfaces other than spheres, we define two local radii of curvature, R_1 and R_2 (Figure 4.61), and derive the pressure difference across the surface with a similar although geometrically more complicated calculation (see Web appendix [108] for details). The CV we choose is a thin volume that envelopes the piece of surface shown in Figure 4.61. For steady state $(d\underline{P}/dt = 0)$ and zero velocity, the

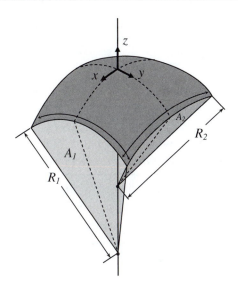

Figure 4.61 For a surface of arbitrary shape, we relate the local pressures on the two sides of the surface to the surface tension using this illustration. The local radii of curvature on two sides are R_1 and R_2 with included angles $2\theta_1$ and $2\theta_2$. (See the Web appendix [108] for details.)

momentum balance becomes:

$$\text{Reynolds transport theorem}\quad \frac{d\underline{\mathbf{P}}}{dt} + \iint_S (\hat{n} \cdot \underline{v})\, \rho \underline{v}\, dS = \sum_{\substack{\text{on} \\ \text{CV}}} \underline{f} \tag{4.409}$$

$$0 = \sum_{\substack{\text{on} \\ \text{CV}}} \underline{f} = \underline{f}_{surface} + \underline{f}_\sigma \tag{4.410}$$

$$= \begin{pmatrix} \text{force on CV} \\ \text{due to} \\ \text{inside fluid} \end{pmatrix} + \begin{pmatrix} \text{force on CV} \\ \text{due to} \\ \text{outside fluid} \end{pmatrix}$$

$$+ \begin{pmatrix} \text{force on CV} \\ \text{due to} \\ \text{surface tension} \\ \text{on 2 arcs of} \\ \text{length } 2\theta_1 R_1 \end{pmatrix} + \begin{pmatrix} \text{force on CV} \\ \text{due to} \\ \text{surface tension} \\ \text{on 2 arcs of} \\ \text{length } 2\theta_2 R_2 \end{pmatrix} \tag{4.411}$$

$$= \Big(p_{in}(2\theta_1 R_1)(2\theta_2 R_2) - p_{ext}(2\theta_1 R_1)(2\theta_2 R_2)$$

$$-2\,\sigma(2\theta_1 R_1)\sin\theta_1 - 2\,\sigma(2\theta_2 R_2)\sin\theta_2 \Big)\, \hat{e}_z \tag{4.412}$$

where $2\theta_1 R_2$ is the arc length swept out in an xz-plane through the point of interest, and $2\theta_2 R_2$ is the arc length swept out in a yz-plane through the point of interest (see Figure 4.61). The sine terms originate in the geometry of the z-component of the surface-tension force (see Web appendix [108] for details). For $\theta_1 = \theta_2 = \theta = $ small, we approximate $\sin\theta \approx \theta$ and obtain the final result:

$$\Delta p = p_{in} - p_{ext} = \sigma \left(\frac{1}{R_1} + \frac{1}{R_2} \right) \tag{4.413}$$

This equation is known as the Young–Laplace equation. Note that for $R_1 = R_2$, the Young–Laplace equation gives the spherical droplet result (Equation 4.408).

Another classic surface-tension effect is the rise of liquids in capillary tubes, known as capillary action. This effect can be understood by invoking the idea of a membrane under tension at the interface, as shown in the next example.

EXAMPLE 4.25. *How is the height of a fluid in a capillary tube related to the surface tension σ?*

SOLUTION. The phenomenon of capillary action (see Figure 2.41) was known in ancient times. In small tubes, fluid is observed to rise upward, defying gravity. Capillary action is caused by the molecular forces between the liquid molecules and those that comprise the surface of the capillary tube. Water, for example, has favorable molecular interactions with glass; thus, the water is attracted to the glass. These attractions are sufficiently favorable to allow the water to climb several centimeters into capillary tubes of the appropriate diameter.

In the continuum model, the phenomenon of capillary action can be quantified by the surface tension. Consider the schematic of a capillary tube immersed in a fluid shown in Figure 4.62. The fluid is shown to rise in the capillary tube, and

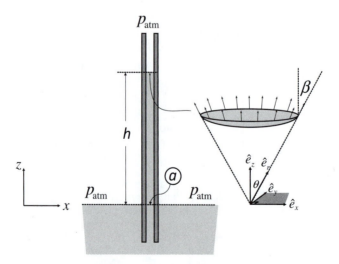

Figure 4.62   In cross section, we see the forces at work that bring about capillary action. The force holding up the column of liquid is the surface tension. Two coordinate systems are shown: the Cartesian system is used for the overall calculation and a spherical coordinate system is used for the calculation of the net surface-tension force. We also used a cylindrical coordinate system (not shown) in this problem.

the shape of the interface is concave upward. The interface meets the wall of the capillary tube at an angle designated as β.

The capillary tube is open to the atmosphere; therefore, the pressure on the upper liquid surface is equal to p_{atm}. We also know that the pressure at Point (a) is equal to p_{atm}, because Point (a) is at the same elevation as the outside liquid level and locations in a fluid at the same elevation have the same pressure (i.e., Pascal's principle).

It seems contradictory that the pressure is p_{atm} at both the top of the capillary and near the bottom of the capillary. The contradiction is resolved, however, if we examine the shape and behavior of the interface at the top of the column in terms of surface tension.

Consider the interface at the top of the column of fluid in the capillary to be a massless membrane under tension. To relate the forces on this membrane to the forces in the fluid, we follow the same process used in the last example: we choose a control volume and perform a momentum balance. The control volume we choose is a cylindrical column of height h, the height of the fluid in the capillary.

As in the previous example, the velocity into and out of the CV is zero, the time rate of change of the momentum $d\mathbf{P}/dt$ is zero, and the momentum balance on the CV reduces to the sum of the forces equal to zero. The forces are gravity, the contact forces on all surfaces, and the force due to surface tension:

$$\frac{d\underline{\mathbf{P}}}{dt} + \iint_{CS} (\hat{n} \cdot \underline{v})\,\rho\underline{v}\,dS = \sum_{\substack{\text{on} \\ \text{CV}}} \underline{f} \qquad \begin{array}{l}\text{Reynolds transport theorem} \\ \text{(momentum balance on CV)}\end{array} \qquad (4.414)$$

$$0 = \sum_{\substack{\text{on} \\ \text{CV}}} \underline{f} = \underline{f}_{gravity} + \underline{f}_{surface} + \underline{f}_{\sigma} \qquad (4.415)$$

$$0 = \underline{f}_{gravity} + \underline{f}_{bottom} + \underline{f}_{top} + \underline{f}_{sides} + \underline{f}_{\sigma} \qquad (4.416)$$

On the bottom of the CV (Point (a), $\hat{n} = -\hat{e}_z$), the molecular force is the force due to pressure in the liquid at that location. This force acts normally to the bottom surface—that is, in the positive z-direction. The steps to arrive at the needed expression (see Equation 4.388, beginning with Equation 4.382) are the same as in Example 4.24.

$$\begin{array}{l}\text{Force on} \\ \text{the bottom} \\ \text{of the CV}\end{array} \quad \underline{f}_{bottom} = \iint_{S_{bot}} [\hat{n} \cdot \underline{\underline{\Pi}}]\big|_{bot}\,dS \qquad (4.417)$$

$$= \int_0^{2\pi} \int_0^R [-\hat{e}_z \cdot (-p_{atm}\underline{\underline{I}})]\,r\,dr d\theta \qquad (4.418)$$

$$= p_{atm}\,\pi R^2\,\hat{e}_z \qquad (4.419)$$

where r and θ are the coordinate variables of the cylindrical coordinate system. Similarly, the force on the flat top of the cylindrical CV ($\hat{n} = \hat{e}_z$) is due to air pressure. The force on the CV at this location is in the $(-z)$-direction:

$$\begin{matrix} \text{Force on} \\ \text{the top} \\ \text{of the CV} \end{matrix} \quad \underline{f}_{top} = \iint_{S_{top}} [\hat{n} \cdot \underline{\underline{\Pi}}]\big|_{top} \, dS \tag{4.420}$$

$$= \int_0^{2\pi} \int_0^R [\hat{e}_z \cdot (-p_{atm}\underline{\underline{I}})] \, (dr)(r\,d\theta) \tag{4.421}$$

$$= -p_{atm} \, \pi \, R^2 \, \hat{e}_z \tag{4.422}$$

For this portion of the calculation, we modeled the top surface of the CV as a flat circle, neglecting curvature.

The force on the side walls is calculated from the same starting point as the calculation for the top and bottom shown previously. Again, we use a cylindrical coordinate system (for sides $\hat{n} = -\hat{e}_r$ of the cylindrical coordinate system):

$$\begin{matrix} \text{Force on} \\ \text{the sides} \\ \text{of the CV} \end{matrix} \quad \underline{f}_{sides} = \iint_{S_{sides}} [\hat{n} \cdot \underline{\underline{\Pi}}]\big|_{sides} \, dS \tag{4.423}$$

$$= \int_0^L \int_0^{2\pi} [-\hat{e}_r \cdot (-p\underline{\underline{I}})]\big|_R \, R\,d\theta \, dz \tag{4.424}$$

Writing $\hat{e}_r$ and $\underline{\underline{I}}$ in Cartesian coordinates and $p = p(z)$ as the wall pressure, we now carry out the matrix multiplication and integrate:

$$\underline{f}_{sides} = \int_0^L \int_0^{2\pi} \left(\cos\theta \;\; \sin\theta \;\; 0 \right)_{xyz} \cdot \begin{pmatrix} p(z) & 0 & 0 \\ 0 & p(z) & 0 \\ 0 & 0 & p(z) \end{pmatrix}_{xyz} R\,d\theta dz \tag{4.425}$$

$$= \int_0^L \int_0^{2\pi} \left(p(z)\cos\theta \;\; p(z)\sin\theta \;\; 0 \right)_{xyz} R\,d\theta dz \tag{4.426}$$

$$= R \int_0^L p(z) \left[\int_0^{2\pi} \begin{pmatrix} \cos\theta \\ \sin\theta \\ 0 \end{pmatrix}_{xyz} d\theta \right] dz \tag{4.427}$$

$$= R \int_0^L p(z) \left[\begin{pmatrix} \sin\theta \\ -\cos\theta \\ 0 \end{pmatrix}_{xyz} \Bigg|_0^{2\pi} \right] dz \tag{4.428}$$

$$= \begin{pmatrix} 0 \\ 0 \\ 0 \end{pmatrix}_{xyz} \tag{4.429}$$

This calculation indicates that there is no net force on the CV due to fluid contact with the sides; the forces between the fluid and the vertical walls are symmetrical, so they integrate out when we integrate all the way around the circumference.

The force on the CV due to gravity is given by the mass of liquid multiplied by the acceleration due to gravity:

$$\text{Force on CV due to gravity} = \underline{f}_{gravity} = \rho \left(\pi R^2 h \right) g \left(-\hat{e}_z \right) \tag{4.430}$$

where ρ is the density of the liquid, g is the acceleration due to gravity, and h is the height of the column of fluid. For the calculation of the force due to gravity, we again assume that the fluid volume is cylindrical, which means we neglect the mass of the liquid that would fill in the concave cap of the CV. Because the capillary is narrow, this should be a good assumption. We include a negative sign in Equation 4.430 to reflect that gravity acts in the $(-z)$-direction.

The remaining molecular force is due to surface tension. Following the same procedure used in the previous example, we must calculate the force due to surface tension. The surface-tension force is the force at the top of the CV applied along the contact line between the CV and the interface—that is, along the line between the liquid and the inside of the column. If we imagine that the top of the CV is a massless, stretched membrane, then a force applied along the arrows shown in Figure 4.62 must be applied to hold the membrane in place as the weight of the column of fluid pulls downward on the membrane.

The net surface-tension force may be calculated by an integral along the contact line at the top of the column. In the spherical coordinate system, this is along the line $\theta = \beta$. The surface tension σ is the tension per unit length along that line. This force is directed at angle β relative to the vertical (see Figure 4.62). For an arc along this circle of length $Rd\phi$, where ϕ is the other angular coordinate of the spherical coordinate system and R is the radius of the capillary, the magnitude of the surface-tension force is:

$$\left(\frac{\text{force}}{\text{length}} \right) (\text{length}) = \sigma \, R d\phi \tag{4.431}$$

This force is directed at an angle $\theta = \beta$ from the vertical. The vector force due to this small arc is:

$$\text{Force due to surface tension along a small arc of length } Rd\phi \text{ at } \theta = \beta \qquad d\underline{f}_{\sigma} = \sigma R d\phi \; \hat{e}_r|_{\theta=\beta} \tag{4.432}$$

where $\hat{e}_r$ is the r-direction unit vector of the spherical coordinate system shown in Figure 4.62. The total surface-tension force is obtained by integrating this expression around the circle—that is, for ϕ varying from 0 to 2π:

$$\text{Total force due to surface tension} \qquad \oint d\underline{f}_{\sigma} = \underline{f}_{\sigma} = \sigma R \int_0^{2\pi} \hat{e}_r|_{\theta=\beta} \, d\phi \tag{4.433}$$

If we write the spherical-coordinate-system basis vector $\hat{e}_r$ in the constant Cartesian coordinate system, we can integrate around ϕ to obtain the net surface-tension force:

$$
\text{(Equation 1.271)} \qquad \hat{e}_r = \begin{pmatrix} \sin\theta\cos\phi \\ \sin\theta\sin\phi \\ \cos\theta \end{pmatrix}_{xyz} \tag{4.434}
$$

$$
\begin{array}{c} \text{Total force due to} \\ \text{surface tension} \end{array} \qquad \underline{f}_\sigma = \sigma R \int_0^{2\pi} \hat{e}_r|_{\theta=\beta} \, d\phi \tag{4.435}
$$

$$
= \sigma R \int_0^{2\pi} \begin{pmatrix} \sin\beta\cos\phi \\ \sin\beta\sin\phi \\ \cos\beta \end{pmatrix}_{xyz} d\phi \tag{4.436}
$$

The angle β is a constant in the integration, and the ϕ-integral is straightforward for each component of the vector:

$$
\underline{f}_\sigma = \sigma R \begin{pmatrix} (\sin\beta)\sin\phi \\ -(\sin\beta)\cos\phi \\ (\cos\beta)\phi \end{pmatrix}_{xyz}\Bigg|_0^{2\pi} \tag{4.437}
$$

$$
= \sigma R \begin{pmatrix} 0 \\ 0 \\ (\cos\beta)2\pi \end{pmatrix}_{xyz} = 2\pi R\sigma\cos\beta \, \hat{e}_z \tag{4.438}
$$

We now assemble the force balance:

$$
0 = \left(\underline{f}_{bottom} + \underline{f}_{top} + \underline{f}_{sides} \right) + \underline{f}_{gravity} + \underline{f}_\sigma \tag{4.439}
$$

$$
\begin{pmatrix} 0 \\ 0 \\ 0 \end{pmatrix}_{xyz} = \begin{pmatrix} 0 \\ 0 \\ p_{atm}\pi R^2 \end{pmatrix}_{xyz} - \begin{pmatrix} 0 \\ 0 \\ p_{atm}\pi R^2 \end{pmatrix}_{xyz}
$$

$$
+ \begin{pmatrix} 0 \\ 0 \\ 0 \end{pmatrix}_{xyz} - \begin{pmatrix} 0 \\ 0 \\ \rho g\pi R^2 h \end{pmatrix}_{xyz} + \begin{pmatrix} 0 \\ 0 \\ 2\pi R\sigma\cos\beta \end{pmatrix}_{xyz} \tag{4.440}
$$

$$
\begin{pmatrix} 0 \\ 0 \\ 0 \end{pmatrix}_{xyz} = \begin{pmatrix} 0 \\ 0 \\ -\rho g R h + 2\sigma\cos\beta \end{pmatrix}_{xyz} \tag{4.441}
$$

We can solve the z-component of Equation 4.441 for the height h of the fluid in the capillary in terms of the surface tension, the shape of the meniscus, and the geometry of the capillary tube:

$$
\boxed{ h = \frac{2\sigma\cos\beta}{\rho g R} } \tag{4.442}
$$

Note that as the radius of the capillary tube R increases, the height h of the column of fluid goes to zero.

Surface-tension gradients caused by concentration, temperature, or electrical gradients can drive flows. This effect, called the Marangoni effect, is demonstrated in the NCFMF film that highlights surface-tension effects [112]. Surface tension is important in slow flow through porous media. In most macroscopic engineering flows, surface-tension effects are negligible. There are two dimensionless numbers that can be used to determine whether surface tension is important in a flow—the Bond number and the Weber number:

$$
\begin{array}{l}
\text{Ratio of} \\
\text{gravity forces and} \quad \text{Bond number} \quad Bo = \dfrac{\rho g L^2}{\sigma} \quad (4.443) \\
\text{surface-tension forces}
\end{array}
$$

$$
\begin{array}{l}
\text{Ratio of} \\
\text{inertial forces and} \quad \text{Weber number} \quad We = \dfrac{\rho V^2 L}{\sigma} \quad (4.444) \\
\text{surface-tension forces}
\end{array}
$$

The topic of surface tension highlights a significant aspect of the continuum model. Because the continuum picture is a model and not physical reality, it does not reflect all of the physics in a system that it approximates. In this chapter, the idea of a continuous field of matter is valuable for calculations on bulk fluids, but this picture cannot capture the boundary effects that result from the real physics at interfaces. We must think again about the real system and adjust our model to account for this newly appreciated aspect of the system. The cost of this adjustment is that we have a new material parameter—the surface tension—that we must measure and consider in continuum modeling to correctly capture the true behavior of our system.

This circumstance—the need to adjust or to complicate a chosen model—will recur as we seek to apply our models to complex systems. At every new juncture where new or neglected physics intrudes, we revisit and adjust our initial model. This does not mean that the model is wrong; it means only that any model is limited to the circumstances under which it was developed. Dimensional analysis is a tool that helps us to quantify when we need to switch from one description of a physical situation to another.

In Chapter 5, we return to the task of incorporating molecular physics into our description of stress so that we can relate $\underline{\underline{\tilde{\tau}}}$ and $\underline{v}$ in a moving fluid. Once we know the relationship between $\underline{\underline{\tilde{\tau}}}$ and $\underline{v}$, we can complete our balance calculations on moving fluids.

4.5 Problems

1. What is a control volume? Can fluid pass through the walls of a control volume?
2. Thinking about a fluid as a chemist would—as a collection of molecules—which properties of molecules generate contact forces on a control volume in a fluid? Which properties of molecules generate noncontact forces on a control volume in a fluid?

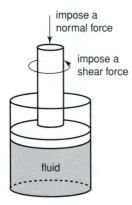

impose a
normal force

impose a
shear force

fluid

Figure 4.63 Olive oil is confined in the device shown (see Problems 6 and 7).

3. How does the continuum picture differentiate between the flow behavior of a polar molecule like water compared to a nonpolar molecule like methane?

4. How does the continuum picture differentiate between the flow behavior of a long-chain polymer that entangles with other long-chain molecules and a short-chain molecule that is not capable of entangling?

5. Describe how a stiff solid like steel responds to shear forces. Describe for normal forces. How does a soft solid like a block of tofu respond to shear forces? Normal forces?

6. Describe how a confined fluid, such as olive oil in the device shown in Figure 4.63, responds to shear forces imposed by the rotation of the lid. How does the fluid respond to normal forces imposed by pressing down on the lid?

7. For the fluid-filled device in Figure 4.63, we can write the forces imposed on the fluid in a cylindrical coordinate system, with $\hat{e}_z$ vertically upward and parallel to the axis of rotation. In this coordinate system with a pure normal force imposed, what is the vector representation of the normal force $\underline{N}$ on the lid? Which components are zero? Give your answer in both matrix form and component/basis-vector form (with the $\hat{e}_r$, $\hat{e}_\theta$, $\hat{e}_z$ basis vectors). In the case of the lid rotating, write the tangential force on the lid in the cylindrical coordinate system. Which components are zero? If the lid is subjected to a force that has both normal and tangential components, what does the force vector look like in the cylindrical coordinate system?

8. River flow in the vicinity of a vertical bridge support induces a molecular force on the rod-shaped support. The forces at Points (a) and (b) (Figure 4.64) and are given by the following two vectors (i.e., arbitrary force units):

$$\underline{f}\Big|_{(a)} = \begin{pmatrix} 320 \\ 210 \\ 0 \end{pmatrix}_{xyz} \qquad \underline{f}\Big|_{(b)} = \begin{pmatrix} 310 \\ -200 \\ -3 \end{pmatrix}_{xyz}$$

What is the normal force on the support at (a)? What is the tangential force on the support at (a)? Point (a) is located at coordinate point $(R, 3\pi/4, 5)$ in the $r\theta z$-coordinate system. What are the normal and tangential forces on

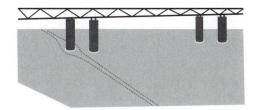

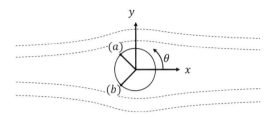

Figure 4.64 A bridge support is subject to normal and tangential forces by the river (Problem 8).

the support at (b)? Point (b) is located at coordinate point $(R, 5\pi/4, 5)$ in the $r\theta z$-coordinate system.

9. A rectangular parallelepiped is subjected to the force $\underline{f}$ below on its top surface (perpendicular to z-axis, arbitrary units). What is the shear force? What is the normal force?

$$\underline{f} = 3\hat{e}_x + 2\hat{e}_y - 0.5\hat{e}_z$$

10. A cylinder of height L is subjected to the force $\underline{f}$ on its side surface at location $(R, \pi/4, L/2)$. What is the shear force on the cylinder surface? What is the normal force? The units of $\underline{f}$ are arbitrary.

$$\underline{f} = 2\hat{e}_r + 2\hat{e}_\theta + \hat{e}_z$$

11. An ideal gas fills a balloon that has a spherical shape. The coordinate system chosen for the problem is a Cartesian system located at the center of the sphere with x pointing east, y pointing north, and z pointing vertically upward. The temperature of the gas is 305 K, and the molar volume is 12.5 l/mol. What are the forces (your answers should be vectors) on the 1.0 cm^2 areas of balloon surface located as follows:

(a) At the equator of the balloon and centered where the balloon intersects the x-axis?

(b) At the equator of the balloon and centered where the balloon intersects the y-axis?

(c) At the equator of the balloon and centered halfway between the first two areas?

12. A cubical box 0.10 m on a side contains 0.121 moles of ideal gas at 403 K. What is the force on each side? The effect of gravity may be neglected. The answer should be a vector; choose a convenient coordinate system.

13. A standard manometer is used to calibrate a digital pressure meter. For the manometer shown in Figure 4.65, one side is open to atmospheric pressure ($p_{\text{atm}} = 76.2$ cm Hg). What is the unknown pressure P? The manometer

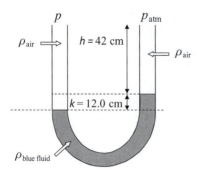

Figure 4.65 Manometer for Problem 13.

fluid is Blue Fluid 175 with a density of 1.75 g/cm³, and the gas on both sides of the manometer is air at 25°C.

14. A closed manometer (pressure is zero absolute on the closed end) contains a dense nonvolatile liquid. The open end is connected by tubing to a gas process stream at pressure p. The manometer fluid has density $\rho = 13.6$ g/cm³, and the height difference between the two sides of the manometer is 5.4 mm. What is the pressure in the process stream?

15. A tilted manometer is used to measure very small pressure differences. For the tilted manometer shown in Figure 4.66, what is the pressure difference between the two sides in terms of the variables defined in the figure?

16. A manometer is configured as shown in Figure 4.67. A heavy fluid (Fluid B) has been placed in the bottom of the manometer; a light fluid (Fluid A) has been added to the left side only. The left side of the manometer is connected to a process stream in which water (25°C, Fluid C) is flowing; the right side is open to the atmosphere (Fluid D). The manometer is being used to measure the pressure difference $p_1 - p_2$. The density of the two fluids in the manometer are 1.75 and 13.6 g/cm³. What is the pressure difference $p_1 - p_2$ in terms of fluid heights and fluid densities? For $h_1 = 2.3$ cm, $h_2 = 2.3$ cm, $h_3 = 1.0$ cm, what is the pressure difference in psi?

17. The pressure in the vapor space of a tank (Figure 4.68) is measured with a mercury manometer that is open to air. The manometer is isolated from the water in the tank by an intermediate section of piping in which oil is trapped.

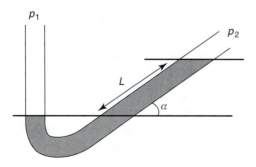

Figure 4.66 Schematic of a tilted manometer (Problem 15).

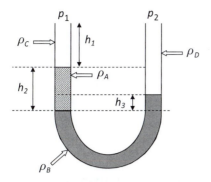

Figure 4.67 A schematic of a manometer that contains two different measurement fluids (Problem 16).

What is the pressure in the head space above the water for the heights and fluids shown in Figure 4.68?

18. In the double-well manometer discussed in this chapter, the same amount of top fluid is placed on each side of the manometer (see Figure 4.36); this is difficult to achieve in practice, however. If slightly more fluid is placed on the lefthand side than on the righthand side, there will be a change in the reading of the manometer, even when the pressures on both sides are the same (Figure 4.69). Show how Equation 4.195 is modified if the initial reading of the double-well manometer is R_0 rather than zero.

19. *Will it float?* In the 1990s on American television, David Letterman's comedy show had a nonsense segment called *Will It Float?* in which Letterman and his bandleader Paul Shafer guessed whether an item would float or sink when dropped into a container of water. For an object that occupies a volume of 4.0 liters, what is the maximum weight that can float?

20. What is the pressure at the bottom of the fluid in the container shown in Figure 4.70 if the angle between the container wall and the horizontal is α? Most of the fluid in this container is not vertically above the bottom surface of the container. Explain how hydrostatic pressure is transferred to that bottom surface in this container.

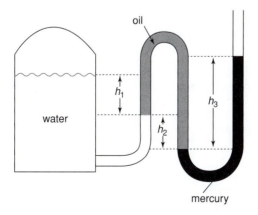

Figure 4.68 We use the static-pressure relationships to relate pressure in a tank to the various heights of fluids as shown (Problem 17).

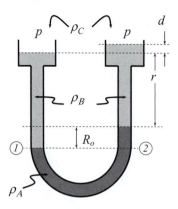

Figure 4.69 If more top fluid is present on the left than on the right, the initial reading of a double-well manometer is R_0 (Problem 18).

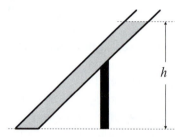

Figure 4.70 A tilted container holds a liquid (Problem 20).

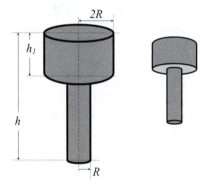

Figure 4.71 An irregularly shaped container holds a liquid (Problem 21).

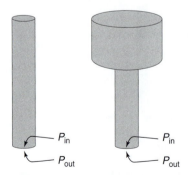

Figure 4.72 Two vessels are constructed and filled with water. We compare the pressure (force/area) near the bottom inside the vessel and the pressure on the solid surface at the bottom of the vessel (Problem 22).

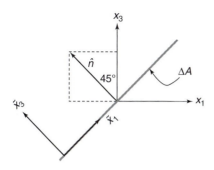

Figure 4.73   Pressure can be calculated in any coordinate system. Problem 23 asks for a calculation in the coordinate system shown.

21. In terms of the dimensions shown in Figure 4.71, what is the pressure on the bottom surface of the container (area $= \pi R^2$)? What is the pressure on the ledge-surface in Figure 4.71? Check whether all of the force due to gravity on the liquid in the container is accounted for by the pressure on these two surfaces.

22. Consider two vessels of different shapes but that have the same height (Figure 4.72). What is the pressure at the bottom of the fluid in each case? What is the pressure at the bottom of the vessel (outside) in each case? Are they the same or different? Explain.

23. What is the total vector force on a 0.550 m $\times$ 1.00 m rectangular plate submerged 13.0 m below the surface of a water tank and oriented as shown in Figure 4.73? Express your answer in the coordinate system shown.

24. As a result of a flood, air is trapped in a room by water of depth 10.0 feet pressing down on a metal sheet as shown in Figure 4.74. How much force (a vector) is the water exerting on the metal sheet? How much force is the weight of the sheet exerting on the portions of the walls that are holding it in place? Is it possible for someone trapped in the room to push up the sheet? The sheet is 4 feet by 6 feet by 0.10 inch thick and made of steel (density $=$ 0.286 $\text{lb}_\text{m}/\text{in.}^3$).

25. What is the force on a six inch cube of balsa wood held 5 feet below the surface of water in a tank? The answer must be a vector.

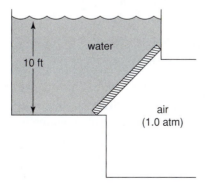

Figure 4.74   A flood creates an air pocket that is protected by a sheet of steel leaning at a 45 degree angle (Problem 24).

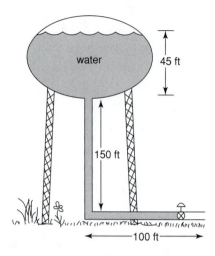

Figure 4.75 A municipal water tank stores water 150 feet in the air (Problem 30).

26. What is the force on a box of polymer foam (density 330 kg/m^3) held 3 m below the surface of water in a tank? The answer must be a vector. The dimensions of the box are 45 cm by 18.0 cm by 6.0 cm.

27. A ball of radius 20.0 cm is submerged 5.00 m below the surface of the ocean. What is the vector force on this object?

28. An object the shape of half a sphere of radius 20 cm is submerged 2.00 m below the surface of the ocean. What is the vector force on this object?

29. A box made of polymeric foam (density $= 330$ kg/m^3) is to be weighted so that it will float 4.5 m below the surface of a pool filled with water at 25°C. The dimensions of the box are 20 cm by 25 cm by 6.5 cm. What is the target weight of the box so that it neither sinks nor rises to the surface? If the pool is filled with seawater, what should the new weight be?

30. Figure 4.75 shows a water tower and some piping. For the dimensions shown and assuming that all pipes are Schedule 40 nominal 1.5-inch pipe, calculate the flow rate Q.

31. Pressure is an isotropic normal stress, meaning that it acts perpendicularly to any chosen surface and has the same magnitude for all surfaces. Consider the stress tensor given here, which has only normal stress components. Is this stress tensor isotropic? The stress is in Pa.

$$\underset{\approx}{\tilde{\Pi}} = \begin{pmatrix} 5 & 0 & 0 \\ 0 & 7 & 0 \\ 0 & 0 & 1 \end{pmatrix}_{123}$$

32. Consider the stress tensor given here, which has only normal stress components. For a cube subjected to this same stress on its six surfaces, calculate the vector force on each surface. The outwardly pointing normal to the top surface is in the x_3-direction and the cube is 20 cm on a side. The stress is in Pa.

$$\underset{\approx}{\tilde{\Pi}} = \begin{pmatrix} 3 & 0 & 0 \\ 0 & -2 & 0 \\ 0 & 0 & 1 \end{pmatrix}_{123}$$

33. The stress tensor $\underset{\sim}{\tilde{\Pi}}$ at Point P is given in the following matrix form. For a flat square surface centered at P of area 3.1 mm^2 oriented perpendicular to the 13-plane as shown in Figure 4.46, what is the stress on that surface at Point P? Demonstrate that the stress in this problem is not isotropic.

$$\underset{\sim}{\tilde{\Pi}} = \begin{pmatrix} -2.0Pa & 0 & 0 \\ 0 & -4.0Pa & 0 \\ 0 & 0 & -5.0Pa \end{pmatrix}_{123}$$

34. Consider the stress tensor given here (spherical coordinates). For a sphere subjected to this stress field, calculate the vector stress at the following locations (all written in the r, θ, ϕ coordinate system): (R,0,0), (R,$\pi/2$, 0), (R, π,0), and (R, $3\pi/2$, 0). Comment on the results. The stress is in Pa.

$$\underset{\sim}{\tilde{\Pi}} = \begin{pmatrix} -3 & 0 & 0 \\ 0 & -3 & 0 \\ 0 & 0 & -3 \end{pmatrix}_{r\theta\phi}$$

35. Consider the stress tensor given here (Cartesian coordinates, Pa): For a cube (side length is L; located in the first quadrant of the 123-coordinate system with a vertex at the origin) subjected to this stress field, calculate the vector force on the surface at the following locations (all written in the 123-coordinate system): $(\frac{L}{2}, \frac{L}{2}, 0)$, $(\frac{L}{2}, \frac{L}{2}, L)$, $(L, \frac{L}{2}, \frac{L}{2})$, and $(\frac{L}{2}, L, \frac{L}{2})$. Comment on the results.

$$\underset{\sim}{\tilde{\Pi}} = \begin{pmatrix} -3 & 2x_2 & 0 \\ 2x_2 & -3 & 0 \\ 0 & 0 & -3 \end{pmatrix}_{123}$$

36. For a flow in which the pressure $p = 7$ (all units arbitrary) and the extra-stress tensor is given here, what is the total stress tensor $\underset{\sim}{\tilde{\Pi}}$ in matrix form?

$$\underset{\sim}{\tilde{\tau}} = \begin{pmatrix} 1 & 5 & 2 \\ 5 & 1 & 3 \\ 2 & 3 & 0 \end{pmatrix}_{123}$$

37. For a flow in which the pressure $p = -7x_1 + 3$ (all units arbitrary) and the extra-stress tensor is given here, what is the total stress tensor $\underset{\sim}{\tilde{\Pi}}$ in matrix form?

$$\underset{\sim}{\tilde{\tau}} = \begin{pmatrix} 0 & 5\left(1 - \frac{x_2^2}{9}\right) & 0 \\ 5\left(1 - \frac{x_2^2}{9}\right) & 0 & 0 \\ 0 & 0 & 0 \end{pmatrix}_{123}$$

38. For a flow in which the pressure $p = 3$ (arbitrary units) and the extra-stress tensor $\underset{\sim}{\tilde{\tau}} = 6x_2 \left(\hat{e}_1\hat{e}_2 + \hat{e}_2\hat{e}_1\right)$, what is the total stress tensor $\underset{\sim}{\tilde{\Pi}}$ in matrix form?

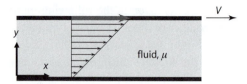

Drag flow is the classic geometry for determining viscosity (Problem 40).

39. Consider the stress tensor given here (Cartesian coordinates, Pascals):

$$\underset{=}{\tilde{\Pi}} = \begin{pmatrix} -3 & 7x_2 & 0 \\ 7x_2 & -3 & 0 \\ 0 & 0 & -3 \end{pmatrix}_{123}$$

For a sphere subjected to this stress field, calculate the vector stress at the following locations (all written in the $r\theta\phi$ coordinate system with θ the angle between the x_3-axis and r): (R, 0, 0), (R, $\pi/2$, 0), (R, π, 0), and (R, $3\pi/2$, 0). Comment on the results.

40. For the flow shown in Figure 4.76, a fluid is trapped between two long, wide plates and the upper plate is made to move at a speed V in the x-direction. This is Newton's experiment, called *drag flow*. The pressure is everywhere atmospheric. It takes some force to move the upper plate. Does the pressure (normal force/area) have a role in determining the force that it takes to move the top plate? State your reasoning.

41. For the flow shown in Figure 4.77, a fluid jet impinges on a wall. The pressure around the jet is everywhere atmospheric. The jet produces a force on the wall. Does the fluid pressure p (normal force/area) have a role in determining the force on the wall? State your reasoning.

42. A surface of interest in a flow has a unit normal of $\hat{n} = \frac{1}{\sqrt{6}}\left(\hat{e}_x + 2\hat{e}_y - \hat{e}_z\right)$ and an area of 8.0 cm^2. If the pressure is 1.02×10^6 Pa and the extra-stress tensor $\underset{=}{\tilde{\tau}}$ in mega Pa is given here, what is the force on the surface? Give the

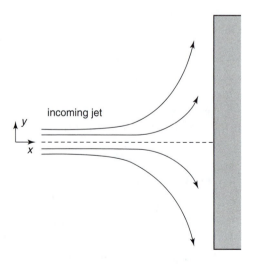

incoming jet

A jet produces a force on a wall (Problem 41).

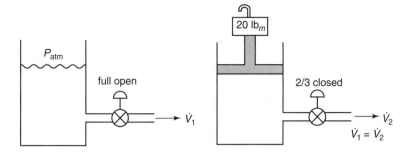

Figure 4.78 Water drains from two tanks at the same rate. However, the tank is open to the atmosphere in one case; in the other case, an external pressure is imposed as shown (Problem 43).

complete force vector.

$$\tilde{\underline{\underline{\tau}}} = \begin{pmatrix} 0 & 2 & 0 \\ 2 & 0 & 0 \\ 0 & 0 & 0 \end{pmatrix}_{xyz}$$

43. Figure 4.78 is a tank with a spout from which water can drain in a controlled manner. When the water surface is open to the atmosphere, the water drains at flow rate $\dot{V}$ liters/min. If a piston that seals with the sides of the tank is added to the top and a 20-kg weight is added, the flow rate increases. If we close the valve, we can reduce the flow rate until it is again $\dot{V}$. We have two situations, then, in which water drains from the tank at flow rate $\dot{V}$; what is the difference in the state of the fluid between the two situations? If we poke a hole in the side of the tank, will the flow from the hole be the same in the two different situations? How?

44. We can use a technique called *quasi-steady-state problem solving* to estimate the time it takes a tank to drain completely. Consider the tank in Figure 4.79.

   (a) When the drain is first opened, what is the instantaneous flow rate Q from the tank in terms of the fluid height h?
   (b) The height of the fluid in the tank is changing as the tank drains. What is the speed of the fluid surface in terms of h? As the tank drains, what is the flow rate through a cross section at the middle of the tank?

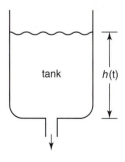

Figure 4.79 Schematic of a tank draining. This problem is solved with the mechanical energy balance and quasi-steady-state methods (Problem 44).

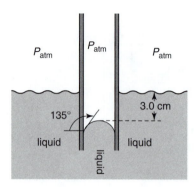

Figure 4.80   When liquids have a repulsive interaction with solids, capillary depression is observed (Problem 49).

(c) Equating the instantaneous flow rate from Part (a) with the flow rate in the tank from Part (b), obtain a differential equation for h and solve. Note that the initial height of the fluid is h_0.

45. In the water-droplet example (see Example 4.24), we concentrated on the z-component of the momentum balance. What information do the x- and y-components of the momentum balance convey?

46. In Example 4.24, we examined the pressure inside a spherical water droplet. For water droplets of various sizes between 0.02 and 3.0 mm, what is the pressure inside the droplet? What is the pressure inside droplets of mercury for this range of sizes?

47. How high will acetone rise in a glass capillary of diameter 0.03 mm? The angle between the vertical and the meniscus is unknown. Make a reasonable estimate for this angle.

48. A carbon-dioxide bubble is motionless at the bottom of a glass of carbonated beverage. Estimate the pressure inside the bubble.

49. Intermolecular forces at solid–liquid–gas interfaces can cause capillary rise or capillary depression. Consider a glass capillary tube submerged in a fluid that does not wet glass—that is, a fluid that has repulsive intermolecular interactions with glass. This situation creates capillary depression as shown in Figure 4.80. If the depression $h = -0.3$ cm and the angle of the meniscus is 135 degrees, what is the surface tension of the liquid? The fluid density is 1200 kg per cubic meter and the capillary is 1.0 mm in diameter.

50. A laminar jet of an oil inside a bath of a second oil forms a cylindrical column of fluid. Because the two oils are not the same material, there is an effect

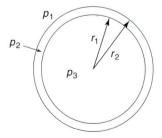

Figure 4.81   A soap bubble can be modeled as shown; the film thickness has been greatly exaggerated (Problem 51).

of surface tension in the formation of the interface. What is the pressure difference due to surface tension between the inside and the outside of a cylindrical column of fluid? Hint: Think of the cylindrical column as a fluid shape with two radii of curvature.

51. What is the pressure difference between the inside and the outside of a soap bubble? Hint: We can draw a soap bubble as shown in Figure 4.81. We write the liquid pressure within the film p_2 in terms of the atmospheric pressure p_1. We subsequently can write the inside air pressure p_3 in terms of the pressure in the film. Taking the limit that $r_1 = r_2$ gives the final result.

5 Stress-Velocity Relationships

In previous chapters, we have been developing a mathematical model for fluid behavior. The task is almost complete: The missing piece in our continuum model is the relationship between molecular stresses and fluid velocity; that connection is made in this chapter.

First, we review our modeling efforts thus far. Our fluids calculations begin with the momentum balance on a control volume (CV)—the Reynolds transport theorem:

$$
\text{Reynolds transport theorem} \atop \text{(momentum balance on CV)} \qquad \boxed{\frac{d\mathbf{P}}{dt} + \iint_{CS} (\hat{n} \cdot \underline{v})\, \rho \underline{v}\, dS = \sum_{\substack{\text{on} \\ \text{CV}}} \underline{f}} \qquad (5.1)
$$

This equation is Newton's second law ($\sum \underline{f} = m\underline{a}$) written on a control volume. The three terms in the equation are the rate of change of momentum on a CV $d\mathbf{P}/dt$, the sum of forces on the CV, and the convective term (i.e., the integral), which accounts for the net momentum added to the CV by flow through the bounding control surface (CS).

Chapter 3 describes how to evaluate the convective term of the momentum balance. In Chapter 4, the forces in the summation term are written as gravity ($\underline{f}_{gravity}$) plus the molecular surface forces ($\underline{f}_{surface}$). We also introduced the stress tensor $\underline{\underline{\tilde{\Pi}}}$ to write the molecular surface forces in terms of the surface area and the unit normal of a chosen surface:

$$
\text{Force in a fluid on} \atop {\text{a flat surface of area } \Delta A \atop {\text{with unit normal } \hat{n} \atop \text{(Gibbs notation)}}} \qquad \underline{f}(\Delta A, \hat{n}) = \Delta A\ [\hat{n} \cdot \underline{\underline{\tilde{\Pi}}}]_{surface} \qquad (5.2)
$$

For a finite surface, this expression becomes an integral:

$$
\text{Total molecular fluid force} \atop \text{on a finite surface } \mathcal{S} \qquad \underline{\mathcal{F}} = \iint_{\mathcal{S}} [\hat{n} \cdot \underline{\underline{\tilde{\Pi}}}]_{\text{at surface}}\ dS \qquad (5.3)
$$

The expressions in Equations 5.2 and 5.3 require that we know the stress tensor $\underline{\underline{\tilde{\Pi}}}$ for the fluid. Chapters 1 and 4 explain that a tensor is a mathematical entity characterized by nine scalar components; we also learned that tensors work like 3×3 matrices in practical calculations in Cartesian coordinates.

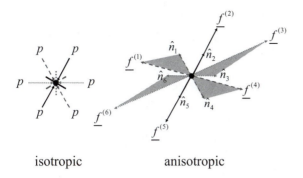

isotropic   anisotropic

Figure 5.1 When the stress is isotropic at a point, as it is in a stationary fluid, any chosen surface ΔA experiences the same force, $-p\hat{n}\,\Delta A$ (left). When a fluid is in motion (right), the force on a chosen surface with normal $\hat{n}$ depends on $\hat{n}$. Six possible choices of $\hat{n}_i$ are shown with corresponding force vectors.

The expression for the stress tensor in a fluid $\underset{=}{\tilde{\Pi}}$ depends on whether the fluid is stationary or in motion. The stress tensor in stationary fluids is simple because the stress is isotropic, as described in Chapter 4. Molecular stress in a stationary fluid is a function of elevation in a gravity field, but the molecular force acts equally in all directions at a given point. Writing pressure as a diagonal stress tensor with $-p$ along the diagonal, as shown here, encodes both the isotropic nature of pressure stress and the trait that pressure acts normally on any surface:

$$\begin{matrix}\text{Total stress tensor} \\ \text{for stationary fluids} \\ \text{(isotropic stress)}\end{matrix} \quad \underset{=}{\tilde{\Pi}} = \begin{pmatrix} -p & 0 & 0 \\ 0 & -p & 0 \\ 0 & 0 & -p \end{pmatrix}_{xyz} = -p\underset{=}{I} \quad (5.4)$$

Chapter 4 describes how to make calculations with $\underset{=}{\tilde{\Pi}} = -p\underset{=}{I}$ when dealing with stationary fluids.

Fluids in motion are different than stationary fluids because the stresses in moving fluids are not isotropic; therefore, the stress tensor is not equal to $-p\underset{=}{I}$. In moving fluids, we can choose to separate $\underset{=}{\tilde{\Pi}}$ into isotropic $(-p\underset{=}{I})$ and anisotropic $(\underset{=}{\tilde{\tau}})$ stresses, as discussed in Chapter 4 (Figure 5.1):

$$\begin{matrix}\text{Total stress tensor} \\ \text{in moving fluids}\end{matrix} \quad \underset{=}{\tilde{\Pi}} = \begin{pmatrix}\text{isotropic} \\ \text{stress}\end{pmatrix} + \begin{pmatrix}\text{anisotropic} \\ \text{stress}\end{pmatrix} \quad (5.5)$$

$$\begin{pmatrix} \tilde{\Pi}_{xx} & \tilde{\Pi}_{xy} & \tilde{\Pi}_{xz} \\ \tilde{\Pi}_{yx} & \tilde{\Pi}_{yy} & \tilde{\Pi}_{yz} \\ \tilde{\Pi}_{zx} & \tilde{\Pi}_{zy} & \tilde{\Pi}_{zz} \end{pmatrix}_{xyz} = \begin{pmatrix} -p & 0 & 0 \\ 0 & -p & 0 \\ 0 & 0 & -p \end{pmatrix}_{xyz} + \begin{pmatrix} \tilde{\tau}_{xx} & \tilde{\tau}_{xy} & \tilde{\tau}_{xz} \\ \tilde{\tau}_{yx} & \tilde{\tau}_{yy} & \tilde{\tau}_{yz} \\ \tilde{\tau}_{zx} & \tilde{\tau}_{zy} & \tilde{\tau}_{zz} \end{pmatrix}_{xyz} \quad (5.6)$$

$$\underset{=}{\tilde{\Pi}} = -p\underset{=}{I} + \underset{=}{\tilde{\tau}} \quad (5.7)$$

It remains to relate these stresses to the velocity field in a moving fluid.

In continuum modeling, the stress-velocity constitutive equation—a function that is introduced in this chapter—relates the anisotropic stress field in a moving fluid to the velocity field.

$$\tilde{\underline{\underline{\tau}}} = f(\underline{v}) \qquad (5.8)$$

Stress-velocity constitutive equation (different for different fluids; discussed in this chapter)

We deduce the stress-velocity constitutive equation from observations of how fluids behave. The isotropic stress $-p\underline{\underline{I}}$ in a moving fluid does not have its own constitutive equation; rather, it is related to the velocity through the momentum balance. This means that once all of the other terms of the momentum balance are expressed correctly, a pressure that is consistent with momentum conservation can be calculated. Alternatively, if the pressure field is known, then solving the momentum balance leads to the velocity field.

A final note: The relationship we seek in Equation 5.8 between anisotropic stress and velocity is a tensor equation. In this chapter, we treat it as such but we also present a simpler scalar version of the stress/velocity relationships. For both those who are comfortable with tensors and for those who want to avoid them, the relationship between stress and velocity begins with observations of fluids in a simple flow—that is, shear flow. Once the stress-velocity equation is obtained, we can complete the momentum balance. Tables of vector and tensor components (see Appendix B) provide expressions to use in momentum-balance calculations.

At the end of this chapter, we carry to completion the two momentum balances that have been in progress: (1) calculating the velocity distribution in a film flowing down an incline, and (2) calculating the force on a 90-degree bend. In Chapter 6, we strengthen our understanding of the continuum-modeling method by discussing the general microscopic-balance equations and the methods used for solving them and for making engineering calculations with the results.

5.1 Simple shear flow

In flow, velocity and stress are related, which we can prove by thinking about simple experiences we have had with flow. When stirring a thick cake batter, it takes more effort to stir the fluid rapidly than slowly. In terms of fluid-mechanics variables, the total force required to sustain the flow of the batter increases with the average velocity of the flow. Another example is the flow produced when a tube such as a honey bottle is squeezed, expelling liquid. During squeezing, a small force expels a small amount of honey; if a larger squeezing force is applied, the honey emerges faster (Figure 5.2).

To quantify these effects, we must be guided by experiments. However, stirring a cake batter and squeezing a fluid from a bottle are fairly complicated

Mixing
thick
cake

Squeezing
honey from
a bottle

Figure 5.2   Larger stresses are associated with higher flow rates, as we know from working with household fluids. The stress-constitutive equation puts this relationship in mathematical form.

situations to model. When mixing cake batter, the stirring motions of the baker vary with every stroke; therefore, the velocity profile in the batter has three nonzero components (e.g., v_x, v_y, and v_z), which depend on time and vary in three dimensions. The flow from the honey bottle also is three-dimensional and complex (Figure 5.3).

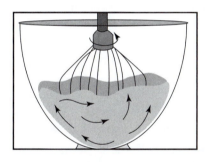

Figure 5.3   The flow from a bottle of honey and the stirring of cake batter are both interesting flows, but the velocity field in both cases is complex. Fluid particles move along three-dimensional paths and accelerate and decelerate depending on how the flow is maintained.

The stress-velocity relationship for a fluid would be clearer if we studied less complicated cases than these two. To simplify the situation, we consider a flow in which the velocity at every position points in the same direction. We can produce such a unidirectional flow by putting a fluid in the narrow gap between two parallel plates (Figure 5.4). If we move the top plate at a constant speed V in a straight line, the flow between the plates should be unidirectional. We choose the plates to be long and wide to minimize effects due to the edges. By carefully choosing the flow geometry and conditions, we make the physics of the situation simple enough to model.

The sliding flow described in Figure 5.4 is called *simple shear flow*. The flow domain has a rectangular shape; therefore, we choose a three-dimensional rectangular Cartesian coordinate system xyz in which to analyze the experiments. Because we have selected a unidirectional flow and lined up the x-axis with the flow direction, the

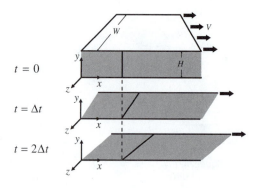

$t = 0$

$t = \Delta t$

$t = 2\Delta t$

Figure 5.4 A simple flow apparatus is produced by confining a fluid in the narrow gap between two large flat plates. Moving the top plate parallel to the bottom plate produces shear flow. A vertical line drawn on the fluid rotates and stretches as the flow proceeds in time from top to bottom.

velocity vector in this flow can be written as a simple vector:

$$\underline{v} = \begin{pmatrix} v_x \\ v_y \\ v_z \end{pmatrix}_{xyz} = \begin{pmatrix} v_x \\ 0 \\ 0 \end{pmatrix}_{xyz} = v_x \hat{e}_x \tag{5.9}$$

We want to know the relationship between the stresses in this flow and the velocity field. We begin by looking into the form of the velocity field; that is, how does v_x vary with position in the flow?

5.1.1 Velocity field

Because our chosen flow has straight streamlines and otherwise is not complicated, we can deduce the velocity field in shear flow from simple arguments. Except near the edges of the plates, the velocity in all planes of constant z should look like the velocity profile shown in the top of Figure 5.5. The fluid in contact with the bottom plate is moving at the speed of the bottom plate, which is zero. The fluid in contact with the top plate is moving at the speed of that plate, which means that the speed of fluid at that top surface is equal to V. Because we chose a particularly simple flow situation, we can deduce the likely flow field as a linear interpolation between these two points, as shown in the bottom of Figure 5.5:

$$\text{Linear velocity profile:} \quad v_x = (\text{slope})\, y + (\text{intercept}) \tag{5.10}$$

$$\text{Known points:} \quad y = 0 \quad v_x = 0 \quad \Rightarrow \quad \text{intercept} = 0$$

$$y = H \quad v_x = V \quad \Rightarrow \quad \text{slope} = V/H$$

$$\begin{aligned} &\text{Velocity profile:} \\ &\text{Simple shear:} \end{aligned} \quad \boxed{v_x = \frac{V}{H}y} \tag{5.11}$$

$$\frac{v_x}{V} = \frac{y}{H} \tag{5.12}$$

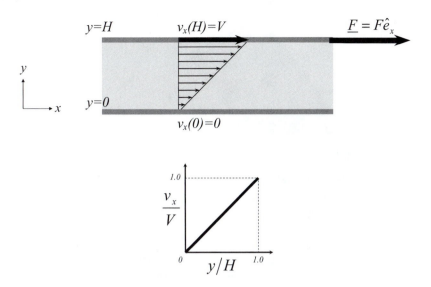

The simplicity of the shear-flow experiment leads to a simple velocity profile.

Many researchers have carried out experiments in the narrow-gap, parallel-plate shear apparatus, confirming that simple fluids and modest plate velocities produce the linear velocity profile deduced by linear interpolation, $v_x = (V/H)y$. The simple-shear velocity profile is plotted as a function of y and z in Figure 5.6. The flow at values of y and z near the edges of the plates ($x = L, z = W$) do not follow this idealized picture of the flow due to the finite extent of the flow and the presence of an air interface. We limit our discussion to wide and long plates so that these contributions to the overall forces driving the flow are negligible. The complete velocity vector for our chosen flow is as follows (combining Equations 5.11 and 5.9):

$$\begin{matrix}\text{Velocity profile}\\ \text{in simple shear flow}\\ \text{(experimental result)}\end{matrix} \qquad \underline{v} = \begin{pmatrix} (V/H)y \\ 0 \\ 0 \end{pmatrix}_{xyz} = \frac{V}{H}y\hat{e}_x \qquad (5.13)$$

By choosing a simple flow situation to study as test flow, we can deduce the complete velocity field from simple arguments and observations.

5.1.2 Stress field

After carefully designing a test flow and guessing and then experimentally verifying the velocity field for this flow, the next step is to look at the flow force measurements. We are interested in the forces within the bulk fluid. The force that it takes to move the top plate can be measured, and we start there and call that vector force $\underline{F}$. The force F is applied in the x-direction; thus, $\underline{F} = F\hat{e}_x$ (see Figure 5.5). The force pulling the plate acts on the top layer of the fluid, and we want to relate this force to our stress variables $\tilde{\Pi}_{ij}$ from Chapter 4, the stress in

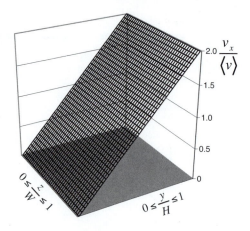

Figure 5.6 The velocity profile for steady drag flow between two long, wide plates is shown in a three-dimensional view. The average velocity $\langle v \rangle = V/2$; thus, the wall velocity is twice the average velocity of this flow. The data are plotted versus normalized coordinates y/H and z/W, which both range from 0 to 1.

the fluid on an $\hat{e}_i$-surface in the $\hat{e}_j$-direction. Our goal is to discover the function $\tilde{\underline{\underline{\Pi}}}(v)$.

We calculate the force to move the top plate in terms of the $\tilde{\Pi}_{ij}$ by using informal arguments. The tangential force exerted by the moving plate on a patch of fluid in the top layer is a force on a $\hat{e}_y$-surface in the $\hat{e}_x$-direction at location $y = H$ (Figure 5.7); in terms of $\tilde{\underline{\underline{\Pi}}}$, this force on a small area $\Delta x \, \Delta z$ is:

$$\begin{array}{l} \text{Tangential force} \\ \text{on surface } \Delta x \, \Delta z = (\text{stress})(\text{area}) \left(\begin{array}{l} \text{unit vector} \\ \text{indicating} \\ \text{direction} \end{array} \right) \qquad (5.14) \\ \text{in top fluid layer} \end{array}$$

$$= \left. \tilde{\Pi}_{yx} \right|_{y=H} \Delta x \, \Delta z \, \hat{e}_x \qquad (5.15)$$

where $\left. \tilde{\Pi}_{yx} \right|_{y=H} = \left. \tilde{\Pi}_{yx} \right|_{H}$ is $\tilde{\Pi}_{yx}$ evaluated at the position (x, H, z). We sum these contributions over the entire surface of the plate to arrive at an expression

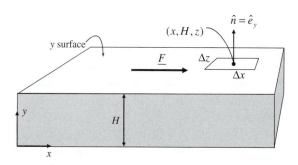

Figure 5.7 The stress on the top layer of fluid is the stress on a surface with outward unit normal $\hat{e}_y$ in the $\hat{e}_x$ direction, $\tilde{\Pi}_{yx}$.

that gives the total force on the top plate:

$$\text{Total tangential force on top fluid layer} \qquad \underline{F} = \lim_{\substack{\Delta x \to 0 \\ \Delta z \to 0}} \left[\sum_{p=1}^{N} \tilde{\Pi}_{yx}\big|_H \, \Delta x_p \Delta z_p \, \hat{e}_x \right] \qquad (5.16)$$

$$\underline{F} = \int_0^W \int_0^L \tilde{\Pi}_{yx}\big|_H \, \hat{e}_x \, dx\,dz \qquad (5.17)$$

We used the definition of integral in two dimensions in going from Equation 5.16 to Equation 5.17 (see Web appendix [108] for details).

Because all of the fluid in the top layer is moving at the same velocity, $\tilde{\Pi}_{yx}\big|_H$ is constant throughout the top layer of fluid; $\hat{e}_x$ also is constant (i.e., always points in the same direction). Moving these quantities out of the integral and integrating what remains, we obtain:

$$\text{Total tangential force on top fluid layer} \qquad = \tilde{\Pi}_{yx}\big|_H \, \hat{e}_x \int_0^W \int_0^L dx\,dz \qquad (5.18)$$

$$\underline{F} = \tilde{\Pi}_{yx}\big|_H \, L\,W \, \hat{e}_x = \begin{pmatrix} \tilde{\Pi}_{yx}\big|_H \, L\,W \\ 0 \\ 0 \end{pmatrix}_{xyz} \qquad (5.19)$$

The force on the top surface is of magnitude $\tilde{\Pi}_{yx}\big|_H \, L\,W$ in the x-direction.

We arrive at the same result more formally by beginning with the general expression from Chapter 4 for fluid force on a surface with unit normal $\hat{n} = \hat{e}_y$ (see Equation 4.263):

$$\text{Total molecular fluid force on a finite surface } \mathcal{S} \qquad \mathcal{F} = \iint_{\mathcal{S}} [\hat{n} \cdot \underline{\tilde{\Pi}}]_{\text{at surface}} \, dS \qquad (5.20)$$

$$= \int_0^W \int_0^L \hat{e}_y \cdot \underline{\tilde{\Pi}}|_H \, dx\,dz \qquad (5.21)$$

Carrying out this integration, we obtain:

$$\begin{array}{c} \text{Force on} \\ \text{surface of area } L\,W \\ \text{in top fluid layer} \\ \text{at } y = H \end{array} \quad \underline{F} = \int_0^W \int_0^L (0\ \ 1\ \ 0)_{xyz} \cdot \begin{pmatrix} \tilde{\Pi}_{xx}\big|_H & \tilde{\Pi}_{xy}\big|_H & \tilde{\Pi}_{xz}\big|_H \\ \tilde{\Pi}_{yx}\big|_H & \tilde{\Pi}_{yy}\big|_H & \tilde{\Pi}_{yz}\big|_H \\ \tilde{\Pi}_{zx}\big|_H & \tilde{\Pi}_{zy}\big|_H & \tilde{\Pi}_{zz}\big|_H \end{pmatrix}_{xyz} dx\,dz$$

$$= \int_0^W \int_0^L \begin{pmatrix} \tilde{\Pi}_{yx}\big|_H \\ \tilde{\Pi}_{yy}\big|_H \\ \tilde{\Pi}_{yz}\big|_H \end{pmatrix}_{xyz} dx\,dz \qquad (5.22)$$

Because the stress components on the surface do not depend on positions x or z, we can pull the entire vector out of the integral and carry out the remaining area integration. The result for $\underline{F}$ is:

$$\underline{F} = LW \begin{pmatrix} \tilde{\Pi}_{yx}|_H \\ \tilde{\Pi}_{yy}|_H \\ \tilde{\Pi}_{yz}|_H \end{pmatrix}_{xyz} \tag{5.23}$$

The x-component of this force vector is the same as the result at which we arrived previously using ad hoc arguments (see Equation 5.15). The other two components of Equation 5.23 tell us how to calculate the y- and z-components of the force $\underline{F}$ needed to move the top plate. We stated previously that we determined experimentally that creating this flow required only an x-direction force $\underline{F} = F\hat{e}_x$; thus, these two components must be zero or nearly zero for this flow:[1]

$$\underline{F} = LW \begin{pmatrix} \tilde{\Pi}_{yx}|_H \\ \tilde{\Pi}_{yy}|_H \\ \tilde{\Pi}_{yz}|_H \end{pmatrix}_{xyz} = LW \begin{pmatrix} \tilde{\Pi}_{yx}|_H \\ 0 \\ 0 \end{pmatrix}_{xyz} \tag{5.24}$$

Returning to Equation 5.19 or Equation 5.24 and noting that by definition $\tilde{\Pi}_{yx} = \tilde{\tau}_{yx}$ (Equation 5.7), we can relate the magnitude $F = |\underline{F}|$ to the extra-stress component $\tilde{\tau}_{yx}$:

$$\underline{F} = F\hat{e}_x = \tilde{\Pi}_{yx}|_H \, LW \, \hat{e}_x \tag{5.25}$$

$$\begin{pmatrix} F \\ 0 \\ 0 \end{pmatrix}_{xyz} = \begin{pmatrix} \tilde{\Pi}_{yx}|_H \, LW \\ 0 \\ 0 \end{pmatrix}_{xyz} = \begin{pmatrix} \tilde{\tau}_{yx}|_H \, LW \\ 0 \\ 0 \end{pmatrix}_{xyz} \tag{5.26}$$

$$F = \tilde{\tau}_{yx}|_H \, LW \tag{5.27}$$

$$\boxed{\tilde{\tau}_{yx}|_H = \frac{F}{LW}} \tag{5.28}$$

Our analysis and experiments yield a value for $\tilde{\tau}_{yx}$ at the top plate, which represents stress at a single value of y, $y = H$. To determine the relationship between velocity and stress throughout the flow, we need values for stress at other locations, such as in the bulk of the fluid, away from the wall. We have no stress measurements within the flow; thus, we appear to be prevented from obtaining information about the stresses in the bulk.

Because we chose a simple flow, however, there is a way out of this apparent dead end, and the path is through the momentum balance. In the example that follows, we show that the stress $\tilde{\tau}_{yx}$ is constant throughout the flow and equal to the value that we calculated in Equation 5.28. Because we found this particularly

[1] Some materials exist that are observed to require a force $\underline{F} = F_x\hat{e}_x + F_y\hat{e}_y$ to produce simple shear flow (see Section 5.3).

simple flow—a flow with a linear velocity profile and constant stress throughout—we can deduce, in the next section, the stress-velocity relationship for a fluid subjected to the flow.

EXAMPLE 5.1. *Calculate the value of $\tilde{\tau}_{yx}$ as a function of position in steady simple shear flow between wide, long parallel plates.*

SOLUTION. We calculate $\tilde{\tau}_{yx}$ at all values of y by performing a momentum balance on the control volume outlined in cross section by the dashed line in Figure 5.8. The chosen control volume is a rectangular parallelepiped of height $H - y$ and cross-sectional area LW.

The momentum balance on a CV is given by the Reynolds transport theorem:

$$
\begin{array}{ll}
\text{Reynolds transport theorem} \\
\text{(momentum balance on CV)}
\end{array}
\qquad
\boxed{\dfrac{d\mathbf{P}}{dt} + \iint_{CS} (\hat{n} \cdot \underline{v})\, \rho \underline{v}\, dS = \sum_{\substack{\text{on} \\ \text{CV}}} \underline{f}}
\qquad (5.29)
$$

The flow is steady; thus, $d\mathbf{P}/dt = 0$. The convective term (i.e., the term containing the integral) turns out to be zero as well, as we show by calculating the integral on the six bounding surfaces of the CV:

$$
\iint_{CS} (\hat{n} \cdot \underline{v})\, \rho \underline{v}\, dS = \sum_{\substack{\text{6 surfaces} \\ S_i}} \iint_{S_i} (\hat{n} \cdot \underline{v})\, \rho \underline{v}\, dS
\qquad (5.30)
$$

The convection term contains the velocity $\underline{v}$. For the bottom surface, the velocity $\underline{v} = 0$ and there is no contribution to the convective term. For the two side surfaces with $\hat{n} = \pm \hat{e}_z$ and for the top, the velocity $\underline{v} = V\hat{e}_x$ is parallel to the surface; therefore, $\hat{n} \cdot \underline{v} = 0$ and there also is no contribution to the convective term. For the upstream bounding surface (for which $\hat{n} = -\hat{e}_x$) and the down-stream bounding surface ($\hat{n} = \hat{e}_x$), we previously determined that $\underline{v} = (V/H)y\hat{e}_x$ (see Equation 5.13); thus, $\hat{n} \cdot \underline{v} \neq 0$. The convective term therefore is calculated as the sum of contributions from the upstream and downstream surfaces.

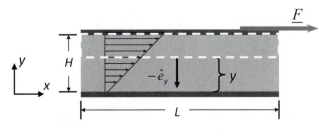

Figure 5.8   Drag flow shown in cross section; the cell is of length L in the x-direction, of height H in the y-direction, and of width W in the z-direction (out of the page). The CV outlined with a dashed line is used to calculate how stress varies with position in simple shear flow.

The calculation is straightforward:

$$\iint_S (\hat{n} \cdot \underline{v}) \rho \underline{v} \, dS = \left[\iint (\hat{n} \cdot \underline{v}) \rho \underline{v} \, dS \right]\bigg|_{upstream}$$

$$+ \left[\iint (\hat{n} \cdot \underline{v}) \rho \underline{v} \, dS \right]\bigg|_{downstream} \tag{5.31}$$

$$= \int_0^W \int_0^H \left(-\hat{e}_x \cdot \frac{Vy}{H}\hat{e}_x \right) \rho \frac{Vy}{H}\hat{e}_x \, dydz$$

$$\int_0^W \int_0^H \left(\hat{e}_x \cdot \frac{Vy}{H}\hat{e}_x \right) \rho \frac{Vy}{H}\hat{e}_x \, dydz \tag{5.32}$$

$$= \int_0^W \int_0^H -\left(\frac{Vy}{H} \right)^2 \rho\hat{e}_x \, dydz$$

$$\int_0^W \int_0^H \left(\frac{Vy}{H} \right)^2 \rho\hat{e}_x \, dydz \tag{5.33}$$

$$= 0 \tag{5.34}$$

Thus, the convective term is zero, and the momentum balance on our CV becomes:

$$0 = \sum_{\substack{on \\ CV}} \underline{f} \tag{5.35}$$

The forces on the chosen CV are gravity and the molecular contact forces on the six bounding surfaces. Gravity is given by the volume of the CV multiplied by density, and it acts in the $-y$-direction (Figure 5.9):

$$\underline{f}_{gravity} = LW(H - y)(\rho)(-\hat{e}_y) \tag{5.36}$$

The forces on the upstream, downstream, and side surfaces are small because the areas of these surfaces are small (H small). The forces on the top ($\hat{n} = \hat{e}_y$) and bottom ($\hat{n} = -\hat{e}_y$) of the CV are the dominant forces. Neglecting the contact forces on the upstream, downstream, and side surfaces, we complete the momentum balance:

$$\begin{array}{c} \text{Momentum balance} \\ \text{on CV} \end{array} \qquad 0 = \sum_{\substack{on \\ CV}} \underline{f} \tag{5.37}$$

$$0 = \underline{f}_{gravity} + \underline{f}_{top} + \underline{f}_{bottom} \tag{5.38}$$

We measured the force on the top as a force in the x-direction, $\underline{f}_{top} = \underline{F} = F\hat{e}_x$. The force on the bottom is evaluated using Equation 5.3:

$$\begin{array}{c} \text{Total molecular fluid force} \\ \text{on a finite surface } \mathcal{S} \end{array} \qquad \underline{\mathcal{F}} = \iint_{\mathcal{S}} [\hat{n} \cdot \underline{\tilde{\Pi}}]_{\text{at surface}} \, dS \tag{5.39}$$

$$\underline{f}_{bottom} = \int_0^W \int_0^L (-\hat{e}_y) \cdot \underline{\tilde{\Pi}}|_y \, dxdz \tag{5.40}$$

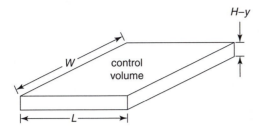

Figure 5.9 The control volume for this example is a rectangular parallelepiped of height $(H - y)$ and cross section WL.

As before, the quantities inside the integral do not depend on the x or z position; thus, we carry out the integral to obtain:

$$\underline{f}_{bottom} = LW(-\hat{e}_y) \cdot \underline{\underline{\tilde{\Pi}}}\big|_y \tag{5.41}$$

We use matrix notation to carry out the dot product:

$$\underline{f}_{bottom} = -LW \begin{pmatrix} \tilde{\Pi}_{yx}\big|_y \\ \tilde{\Pi}_{yy}\big|_y \\ \tilde{\Pi}_{yz}\big|_y \end{pmatrix} \tag{5.42}$$

With this and previous results for $\underline{f}_{top}$ and $\underline{f}_{gravity}$, the momentum balance becomes:

$$0 = \sum_{\substack{on \\ CV}} \underline{f} \tag{5.43}$$

$$0 = \underline{f}_{gravity} + \underline{f}_{top} + \underline{f}_{bottom} \tag{5.44}$$

$$\begin{pmatrix} 0 \\ 0 \\ 0 \end{pmatrix}_{xyz} = \begin{pmatrix} 0 \\ -\rho g L W (H - y) \\ 0 \end{pmatrix}_{xyz} + \begin{pmatrix} F \\ 0 \\ 0 \end{pmatrix}_{xyz} - LW \begin{pmatrix} \tilde{\Pi}_{yx}\big|_y \\ \tilde{\Pi}_{yy}\big|_y \\ \tilde{\Pi}_{yz}\big|_y \end{pmatrix} \tag{5.45}$$

$$\begin{pmatrix} 0 \\ 0 \\ 0 \end{pmatrix}_{xyz} = \begin{pmatrix} 0 \\ -\rho g L W (H - y) \\ 0 \end{pmatrix}_{xyz} + \begin{pmatrix} F \\ 0 \\ 0 \end{pmatrix}_{xyz} - \begin{pmatrix} LW\, \tilde{\tau}_{yx}\big|_y \\ LW\left(\tilde{\tau}_{yy}\big|_y - p\big|_y\right) \\ LW\, \tilde{\tau}_{yz}\big|_y \end{pmatrix} \tag{5.46}$$

The x-component of Equation 5.46 tells us what we wanted to know about the shear-stress distribution in the flow: The shear stress at any value of y in simple shear flow is equal to F/LW (Figure 5.10).

$$\tilde{\tau}_{yx}\big|_y = \frac{F}{LW} \tag{5.47}$$

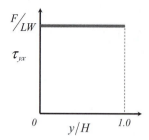

Figure 5.10 The shear stress $\tilde{\tau}_{yx}$ in drag flow is constant and equal to F/LW.

Shear stress in
simple shear flow
(verified experimentally)

$$\tilde{\tau}_{yx}(y) = \text{constant} = \frac{F}{LW} \qquad (5.48)$$

Experiments with optical techniques that allow measurements of stresses within the gap of shear flow verify the result in Equation 5.48. With this result, we can deduce the stress-velocity relationship for steady simple shear flow.

Before leaving this example, we note that Equation 5.46 provides information about two other stress components, $\tilde{\tau}_{yy}$ and $\tilde{\tau}_{yz}$. The z-component of Equation 5.46 indicates that $\tilde{\tau}_{zy} = 0$ for all values of y. The y-component relates gravity to stress components; as a practical matter, gravity is negligible in this horizontal flow, and we therefore can conclude that $\tilde{\Pi}_{yy} = \tilde{\tau}_{yy} - p = 0$ for all values of y in simple shear.

To recap, we seek the relationship between the anisotropic stress $\underset{=}{\tilde{\tau}}$ and the velocity in moving fluids. We plan to deduce this relationship through experiments, but we need a simple flow on which to conduct the experiments. We chose to investigate one such simple flow—shear flow—and we find that it is appropriate for our purposes. The velocity distribution in shear flow is a linear function and the shear stress is constant throughout the flow domain:

Steady simple shear flow
(drag flow,
experimentally verified)

$$\underline{v} = \begin{pmatrix} v_x \\ 0 \\ 0 \end{pmatrix}_{xyz} = \begin{pmatrix} \frac{V}{H}y \\ 0 \\ 0 \end{pmatrix}_{xyz} \qquad (5.49)$$

$$\tilde{\tau}_{yx}(y) = \frac{F}{LW} = \text{constant}$$

Simple shear flow may be characterized by its geometry (L, W, H), and just two measurements: the velocity of the plate V and the force needed to move the top plate, F. Steady simple shear flow created in this way also is called *drag flow*.

Having identified an appropriate test flow that is simple enough to provide clear information on fluid behavior, we now are prepared to examine the behavior of various fluids in drag flow to see how the flow parameters V, F, L, W, and H are related for different fluids subjected to flows of this type. First we try a numerical example that employs Equation 5.49.

EXAMPLE 5.2. *A fluid is placed in the apparatus shown in Figure 5.5. The gap between the plates is 1.000 mm; the plate size is 10.0 cm by 10 cm. The sample completely fills the plates. When the top plate is in motion in the x-direction at 1.000 mm/s, the force required to maintain the motion is 1.00 mN. What is the shear stress at the moving plate? What is the magnitude of the velocity at the moving plate? What is the shear stress at the stationary plate? What is the magnitude of the velocity at the stationary plate? To make the numbers more real, calculate the mass (in grams) that weighs 1.00 mN.*

SOLUTION. The answers to the many questions of this problem are obtained from Equation 5.49. The velocity at the top plate ($y = H$) and at the bottom plate ($y = 0$) are $V = 1.000$ mm/s and zero, respectively. These quantities were used to derive Equation 5.49. The shear stress is constant throughout the gap in drag flow, as shown thus, the shear stress is F/LW at both locations, which for the values given is:

$$\tilde{\tau}_{yx}(y) = \frac{F}{LW} = \text{constant} \tag{5.50}$$

$$= \left(10^{-3}\,\text{N}\right)\left(\frac{1}{(0.10\,\text{m})(0.10\,\text{m})}\right)\left(\frac{\text{Pa}}{\text{N}/\text{m}^2}\right) \tag{5.51}$$

$$= 1.00 \times 10^{-5}\,\text{Pa} \tag{5.52}$$

To calculate the driving force 1.00 mN in an intuitive sense, we use Newton's second law in a gravity field:

$$F = mg \tag{5.53}$$

$$m = \frac{F}{g} = \left(\frac{10^{-3}\,\text{N}}{9.80\,\text{m}/\text{s}^2}\right)\left(\frac{\text{kg m}/\text{s}^2}{\text{N}}\right)\left(\frac{10^3\,\text{g}}{\text{kg}}\right) \tag{5.54}$$

$$= 0.10\,\text{g} \tag{5.55}$$

This small amount of force—equivalent to the weight of a tenth of a gram—indicates that the measuring device for force on the apparatus in Figure 5.8 must be sensitive and accurate to be of any practical use. Alternatively, we can increase the area of the plates to increase the signal from the apparatus (i.e., larger plates require a larger force to pull at the indicated speed).

The forces to deform water, oil, and honey in the shear-flow apparatus would be different. In the next section, we define viscosity—the material parameter that quantifies the differences between the flow behaviors of different fluids.

5.1.3 Viscosity

A great deal of research has been conducted on liquids in a parallel-plate apparatus such as the one described in Figure 5.5. In this device, a fluid is trapped between two large plates and one plate is moved at a constant speed while the gap is maintained constant. For many fluids, including water, oil, milk, and solvents, measurements of the force on the plate F as a function of plate speed V, area $A = LW$, and gap H reveal a simple pattern. As noted with the batter-stirring and honey-squeezing examples (see Figure 5.2), as V increases, F increases. When quantitative measurements are made in parallel-plate flow, we find that F is directly proportional to V:

$$\begin{matrix} \text{Experimental result} \\ \text{in parallel-plate apparatus:} \end{matrix} \quad F \propto V \tag{5.56}$$

That is, when velocity doubles, the force doubles.

We also find that changes in geometry affect the measurements. For a given speed V, if gap H is reduced, the required shear force F increases. Also, for a given speed V, if the area of plate A is increased, the required shear force F increases. The relationships among these four shear-flow variables are well described by the single equation shown here:

$$\begin{matrix} \text{Experimental results:} \\ \text{(steady shear flow)} \end{matrix} \quad \frac{F}{LW} \propto \frac{V}{H} \tag{5.57}$$

We introduce a constant of proportionality μ between F/LW and V/H and write:

$$\frac{F}{LW} = \mu \frac{V}{H} \tag{5.58}$$

We recognize $F/A = F/LW$ as the shear stress $\tilde{\tau}_{yx}$ (Equation 5.49), which is constant in steady shear flow. The ratio V/H is equal to the slope of the v_x-versus-y curve; that is, to the y-derivative of the velocity v_x (see equation 5.49):

$$v_x = \frac{V}{H} y \tag{5.59}$$

$$\frac{dv_x}{dy} = \frac{V}{H} \tag{5.60}$$

Equation 5.58 thus becomes:

$$\begin{matrix} \text{Newton's law of viscosity} \\ \text{in steady simple shear flow} \\ \text{(shear flow in } x\text{-direction,} \\ \text{gradient in } y\text{-direction)} \end{matrix} \quad \boxed{\tilde{\tau}_{yx} = \mu \frac{dv_x}{dy}} \tag{5.61}$$

Table 5.1. Viscosity of familiar materials [132]

Fluid	T (°C)	μ (Pa s)	μ $lb_m/(ft\ s)$	ρ (kg/m^3)	ρ lb_m/ft^3
Air	25	18.6×10^{-6}	12.50×10^{-6}	1.20	74.9×10^{-3}
Water	25	0.8937×10^{-3}	0.6005×10^{-3}	997	62.2
n-Propyl alcohol	25	1.96×10^{-3}	1.32×10^{-3}	804	50.2
Olive oil	25	69×10^{-3}	46×10^{-3}	918	57.3
Honey	25	9	6	1360	84.9
Pitch	25	1×10^{6}	0.67×10^{6}	1100	69

Note: The range of viscosity is ten orders of magnitude; the range of density is only three orders of magnitude.

This equation, deduced from the carefully chosen test flow and found to hold for a wide variety of fluids subjected to it, provides the missing link between the velocity field and the stress field. This equation was first introduced in Chapter 2.

We have arrived at a shear stress–velocity relationship for simple fluids in a simple flow. This equation is called *Newton's law of viscosity*, after Isaac Newton, who performed many early experiments in shear flow [125]. The proportionality constant μ in Equation 5.61 is called the *viscosity* or the *steady-shear viscosity*. A different value of viscosity μ is found for every fluid (Table 5.1), and viscosity is observed to be a sensitive function of temperature.

Viscosity is a material parameter that states how much shear stress $\tilde{\tau}_{yx}$ is generated by a given velocity gradient dv_x/dy in simple shear flow. The units of viscosity are Pa s = kg/(m·s). The *poise* = cm/(g·s), is named after Jean Marie Poiseuille, a physician who made important contributions to the study of pressure-driven flow in arteries (see Chapter 7). The magnitude of viscosity can vary over a wide range (see Table 5.1). For air, viscosity is approximately 2×10^{-4} poise; for water, $\mu \approx 0.01P = 1$ cp; for oil, $\mu \approx 1P = 100$ cp; and for polymer melts, 10^5 poise or higher.

Newton's law of viscosity is a relationship that describes molecular forces in flow ($\tilde{\tau}_{yx}$ is molecular contact stress), but we did not arrive at this law by invoking any particular molecular behavior. We arrived at the need for viscosity by making observations on a simple test flow. Viscosity, like density and surface tension, is a necessary parameter of the continuum model so that it correctly reflects the actual behavior of molecules. By inventing these parameters, we build up the ability of the continuum model to capture a wide variety of fluid behavior. As is true for density and surface tension, however, there also is a molecular interpretation of viscosity (see the sidebar).

In the next section, we describe how Newton's law of viscosity may be used in the momentum-balance calculations of Chapters 3 and 4 to finish those analyses and to calculate flow properties of interest.

Molecular Interpretation of Viscosity

The material-property density is the average mass per volume of the material, for a large sample size. We formally write this definition as:

$$\text{Molecular interpretation of } \rho \quad \rho = \lim_{N \to \infty} \left[\frac{\text{mass of } N \text{ molecules}}{\text{volume occupied by } N \text{ molecules}} \right] \tag{5.62}$$

The material and continuum property surface tension is a measure of the unbalance of inter-molecular forces in fluid near a phase boundary. This unbalance has the effect of creating a tension per unit length along a line drawn in an interface:

$$\text{Molecular interpretation of } \sigma \quad \sigma = \frac{\left(\begin{array}{c} \text{extra molecular} \\ \text{tension} \end{array} \right)}{\left(\begin{array}{c} \text{length along line} \\ \text{in interface} \end{array} \right)} \tag{5.63}$$

The material and continuum property viscosity also has a molecular mechanism associated with it. Viscosity is a measure of the tendency in a material of momentum to transport down a velocity gradient due to Brownian motion, which is the random motion of molecules first observed by Robert Brown in 1827. Brown looked through his microscope at pollen floating on water drops and saw pollen in constant motion, being pushed by the random thermal motions of water molecules. Brownian motion is a fundamental behavior exhibited by liquids and gases: Gases and liquids are composed of molecules that move about at random.

Brownian motion can be a source of momentum transport if there is a gradient of momentum in a fluid [62]. The basics of the effect are illustrated in Figure 5.11. Consider a fluid in motion in the x-direction, with a gradient of velocity in the z-direction:

$$\underline{v} = \begin{pmatrix} v_x \\ v_y \\ v_z \end{pmatrix}_{xyz} = \begin{pmatrix} v_x(z) \\ 0 \\ 0 \end{pmatrix}_{xyz} \tag{5.64}$$

A positive gradient dv_y/dz means that as we look at larger values of z, the average velocity of the molecules is higher:

$$\frac{dv_x}{dz} > 0 \tag{5.65}$$

In addition to the average macroscopic velocity, molecules have a Brownian contribution to their velocity, which is random in magnitude and direction. If we choose a plane in the flow as a reference plane, we can watch the effect of random Brownian motion on the transport of momentum through this plane.

We choose a reference plane through the origin with unit normal equal to $\hat{e}_z$ (Figure 5.11). At a distance l above the plane, the molecules have a slightly higher average velocity v_x than the molecules a distance l below the plane. This difference in average bulk velocity is a reflection of the positive velocity gradient dv_x/dz. In addition to the bulk motion, Brownian motion moves the molecules around at random. We know, therefore, that due to Brownian motion, equal numbers of molecules from above and from below the reference plane pass through the plane per unit time.

Molecular Interpretation of Viscosity *(continued)*

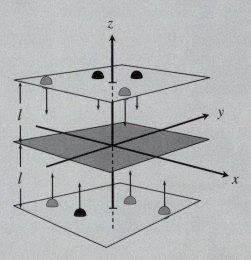

Figure 5.11 Particles cross the z-plane through the origin from above and from below due to Brownian motion. Some molecules are faster than others; in this figure, the faster molecules are black. There is a gradient in velocity in the z-direction; thus, there are more black (i.e., fast) molecules above the plane than below it. When Brownian motion causes equal numbers of molecules to cross the z-plane from above and below, there is a net flux of black molecules (i.e., faster ones) in the $(-z)$-direction.

Although equal numbers of molecules from above and below our chosen plane end up crossing the plane, because of the velocity gradient, the net effect of these crossings on momentum in the region is not zero. The molecules that cross the reference plane from above have a slightly higher bulk momentum, on average, than the molecules that cross it from below. Before and after the crossing, there are the same numbers of molecules on both sides of the plane, but there is a net transfer of momentum from the faster region to the slower region.

It is the random Brownian motion acting in the nonrandom velocity gradient that causes a net momentum flux down a velocity gradient. If these ideas are pursued more formally and if we identify $-\tilde{\tau}_{zx}$ as the flux of x-momentum in the z-direction, we obtain Newton's law of viscosity written in the current coordinate system:[2]

$$
\begin{array}{c}
\text{Newton's law of viscosity} \\
\text{(flow in } x\text{-direction,} \\
\text{gradient in } z\text{-direction)}
\end{array}
\qquad
\boxed{(-\tilde{\tau}_{zx}) = -\mu \frac{dv_x}{dz}}
\qquad (5.66)
$$

The effect of Brownian motion in spatially inhomogeneous materials accounts for more than Newton's law of viscosity. If the same ideas are applied to materials in which there are concentration or temperature gradients, Brownian motion explains important laws related to

(continued)

---

[2] The negative sign is needed in the term $-\tilde{\tau}_{zx}$ because of our choice of convention for the sign of the stress tensor. We chose tension as positive; if we had chosen the other convention, compression is positive, the negative sign would not be needed here [15].

Molecular Interpretation of Viscosity *(continued)*

diffusion (i.e., Fick's law) and heat transfer (i.e., Fourier's law). For diffusion:

$$\text{Fick's law of diffusion} \qquad \boxed{J_{Az}^* = -D_{AB}\frac{dc_A}{dz}} \qquad (5.67)$$

where J_{Az}^* is the molar flux of species A, D_{AB} is the diffusion coefficient, and c_A is the concentration of species A. For heat transfer:

$$\text{Fourier's law of heat conduction} \qquad \boxed{q_z = -k\frac{dT}{dz}} \qquad (5.68)$$

where q_z is the heat flux in the z-direction, k is the thermal conductivity, and T is temperature. For more on the similarities and differences in the transport equations for mass, heat, and momentum, see the literature [15].

5.2 Newtonian fluids

In Section 5.1, we discussed simple experiments in shear flow from which we deduced Newton's law of viscosity. Newton's law of viscosity gives a relationship between shear stress and velocity gradient that holds for many fluids when they are subjected to steady, simple shear flow between very wide and very long parallel plates:

$$\begin{matrix}\text{Newton's law} \\ \text{of viscosity} \\ \text{(steady, simple shear)}\end{matrix} \qquad \boxed{\tilde{\tau}_{yx} = \mu\frac{dv_x}{dy}} \qquad (5.69)$$

The flow down an inclined plane considered previously was a flow not too dissimilar from steady, simple shear flow. With the insight provided by Equation 5.69, we return to the incline problem and attempt to apply Newton's law to this new situation with the goal of calculating the velocity and stress distributions in the flow.

EXAMPLE 5.3 (Incline: continued). *What is the velocity field in a wide, thin film of water that runs steadily down an inclined surface under the force of gravity? The fluid has a constant density ρ (continued from Chapters 3 and 4).*

SOLUTION. We started this problem in Chapter 3 (Figure 5.12). We performed a momentum balance on a microscopic control volume of size $\Delta x\,\Delta y\,\Delta z$; in Chapter 4, we wrote the molecular contact forces in terms of the extra-stress tensor components $\tilde{\tau}_{jk}$ and the pressure. We incorporated the stress components into the momentum balance (see Equations 4.357–4.359), but we paused our development there because we did not know how the stress components varied with velocity.

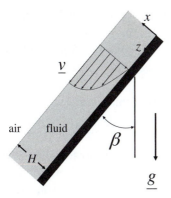

Figure 5.12 The idealized version of flow down an incline is a film of constant thickness where the velocity is everywhere in the same direction but the magnitude varies with position in the film. We seek to calculate the velocity as a function of position relative to the wall (i.e., as a function of x).

Now that we have deduced a relationship between stress and velocity in simple shear flow—Newton's law of viscosity—we can proceed with the solution.

Newton's law of viscosity is given in Equation 5.69 and repeated here:

$$\begin{matrix} \text{Newton's law of viscosity} \\ \text{(drag-flow coordinate} \\ \text{system; see Figure 5.5)} \end{matrix} \qquad \tilde{\tau}_{yx} = \mu \frac{dv_x}{dy} \qquad (5.70)$$

This equation was inferred from experiments in drag flow in which the flow direction was the x-direction and the gradient direction was the y-direction. We adjust Newton's law for our current flow by changing the first subscript on $\tilde{\tau}_{ij}$ to the gradient direction of our flow, and the second subscript of $\tilde{\tau}_{ij}$ to the flow direction. We also make the corresponding changes in the velocity gradient:

$$\begin{matrix} \text{Newton's law} \\ \text{of viscosity} \\ \text{(current coordinate system)} \end{matrix} \qquad \tilde{\tau}_{xz} = \mu \frac{dv_z}{dx} \qquad (5.71)$$

The steps taken to adapt Newton's law of viscosity seem reasonable, although we may hesitate to apply this equation to a flow other than drag flow—that is,

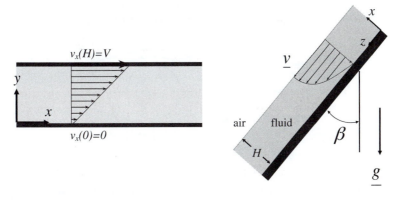

Figure 5.13 We developed Newton's law of viscosity in drag flow (left), and we now seek to apply it to the falling-film flow. The two flows are similar but there are differences that may be significant.

the one on which Newton's law of viscosity is based. The falling-film flow we are considering is different from drag flow in that the falling film has a nonlinear (Figure 5.13) rather than a linear velocity profile. Also, we know little about the shear-stress distribution in the falling-film problem. How do we know if the adaptation in Equation 5.71 is correct?

Again we must postpone completion of this example to clarify the questions raised by changes in coordinate system, velocity field, and stress field.

One approach to the dilemma of how to adapt Newton's law of viscosity to the falling-film problem is to assume that our approach is correct and then check the answer obtained for the velocity profile against measurements. If we do this with the falling-film example, we find that the adaptation works—Equation 5.71 is the correct version of Newton's law of viscosity to use in this flow. We do not always find it so easy to adapt Newton's law of viscosity, however, as the next example shows.

EXAMPLE 5.4. *What is the correct form for Newton's law of viscosity for the two flows shown in Figure 5.14?*

SOLUTION. The top schematic in Figure 5.14 depicts a flow produced by a sheet of fluid impinging on a wall. To make the flow simpler, we consider a

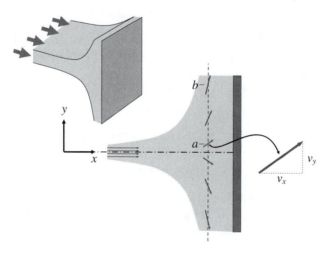

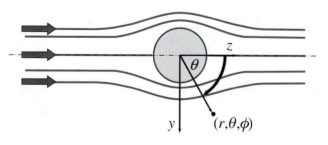

Figure 5.14 When the flow is not unidirectional, it is not clear how to adapt Newton's law of viscosity. The top flow is analyzed in Cartesian coordinates, and the bottom flow uses spherical coordinates.

very wide sheet of fluid so that changes in the width direction z are negligible. The curved streamlines of the flow make it a challenge to analyze the flow. The flow is symmetric from top to bottom, so we choose to model the flow in a Cartesian coordinate system with the x-direction toward the wall and the y-direction upward, as shown in Figure 5.14.

Our choice of coordinate system takes advantage of the symmetry of the problem and allows us to eliminate the z-component of the velocity profile:

$$\text{Planar jet flow} \quad \underline{v} = \begin{pmatrix} v_x \\ v_y \\ v_z \end{pmatrix}_{xyz} = \begin{pmatrix} v_x \\ v_y \\ 0 \end{pmatrix}_{xyz} \tag{5.72}$$

We now want to adapt Newton's law of viscosity for this flow. Newton's law tells us that the shear stress is proportional to the velocity gradient. We begin, therefore, by looking for the velocity gradients. We have two nonzero components of velocity, v_x and v_y. As previously when analyzing shear flow in Section 5.1.1, we look at the velocity boundary conditions to see how the velocity components change with position in the flow.

Along the centerline of the flow, at the inflow boundary of the flow, v_x is large; but, at the wall, v_x goes to zero. Thus, v_x is a function of the x-position. The component v_x also is a function of y, although this is more difficult to see. If we consider the variation of v_x along the vertical dotted line in Figure 5.14, we see that at Point (b), the x-component of velocity is slightly smaller than at Point (a) because fluid particles are losing speed in the x-direction as they move through the flow. Hence, v_x also is a function of y. If we now consider the component v_y in the same way, we see that v_y also is a function of both x and y. Neither component is a function of z due to the large width of the flow.

To adapt Newton's law of viscosity for the jet flow, we now have four nonzero velocity gradients that may affect stress in this flow:

$$\frac{\partial v_x}{\partial x} \quad \frac{\partial v_x}{\partial y} \quad \frac{\partial v_y}{\partial x} \quad \frac{\partial v_y}{\partial y} \tag{5.73}$$

It is not easy to see where we should go from here. One option is to create four different Newton's laws of viscosity with the four velocity gradients:

$$\tilde{\tau}_{xx} \overset{?}{=} \mu \frac{\partial v_x}{\partial x}$$

$$\tilde{\tau}_{yx} \overset{?}{=} \mu \frac{\partial v_x}{\partial y}$$

$$\tilde{\tau}_{xy} \overset{?}{=} \mu \frac{\partial v_y}{\partial x} \tag{5.74}$$

$$\tilde{\tau}_{yy} \overset{?}{=} \mu \frac{\partial v_y}{\partial y}$$

It turns out this would be incorrect. One way we can tell that this approach is incorrect is to notice that such a system results in $\tilde{\tau}_{xy} \neq \tilde{\tau}_{yx}$. Because the stress tensor is symmetric, there is a problem with this guess.

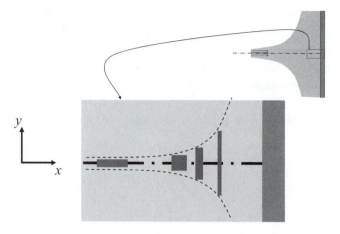

In the flow of the planar jet into the wall, the fluid particles experience an extreme compressional deformation near the wall; that is, v_x varies in the x-direction, $\partial v_x / \partial x \neq 0$.

Another problem with our guess is that Newton's law of viscosity was developed in a sliding flow, and we have two velocity gradients that are not the sliding type; $\partial v_x / \partial x$ and $\partial v_y / \partial y$. These velocity gradients describe compressional and extensional aspects of the flow (Figure 5.15). It does not seem reasonable to directly adapt the shear-stress–velocity relationship developed in shear flow to flows that are so different from simple shear. This flow with curved streamlines shows that we have more work before we can understand how stress and velocity are related in general flows—even for materials that follow Newton's law of viscosity in simple shear.

As a second example, consider the axisymmetric flow around a sphere (see Figure 5.14, bottom). This flow is related to the settling of spherical particles and to the flow around bubbles. Because the flow is around a sphere, the flow has a symmetry that is easiest to describe using the spherical coordinate system, $r\theta\phi$. In that coordinate system, the velocity vector of the flow around a sphere has two nonzero coordinates, v_r and v_θ. The v_ϕ component is zero because there is no spiral or spinning component to the flow:

$$\underline{v} = \begin{pmatrix} v_r \\ v_\theta \\ v_\phi \end{pmatrix}_{r\theta\phi} = \begin{pmatrix} v_r \\ v_\theta \\ 0 \end{pmatrix}_{r\theta\phi} \tag{5.75}$$

In addition, the flow is symmetrical in the ϕ-direction (i.e., if we slice the flow at different values of ϕ, we see the same flow pattern). Both v_r and v_θ, however, are functions of both r and θ:

$$\begin{matrix} \text{Flow around} \\ \text{a sphere} \end{matrix} \qquad \underline{v} = \begin{pmatrix} v_r(r, \theta) \\ v_\theta(r, \theta) \\ 0 \end{pmatrix}_{r\theta\phi} \tag{5.76}$$

We do not attempt to write an ad hoc adaptation of Newton's law of viscosity for this flow. We know from Chapter 1 to expect complications from the change of coordinates from Cartesian to spherical coordinates, which must be done

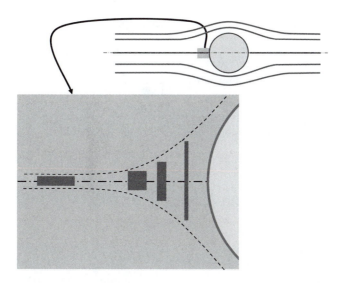

Figure 5.16 In the flow around a sphere, the fluid particles moving along the centerline impact the sphere and experience an extreme compressional deformation.

rigorously. In addition, this flow has both shear and elongation/compression aspects, as did the planar-jet flow considered. For example, the fluid particles that travel along the axis of symmetry go from traveling at the mean velocity to traveling at zero velocity when they encounter the sphere (Figure 5.16).

The complications due to curved streamlines and the change of coordinates convince us that we must reconsider the mathematical form of Newton's law of viscosity. We must seek a general formulation for the stress-velocity relationship. The formulation we need must work in the widest possible variety of flows and in any coordinate system.

The previous example shows that we need a stress–velocity relationship that is more general than Newton's law of viscosity. As stated in the introduction to this chapter, the true relationship between stress and velocity is a tensor relationship. Tensors are independent of coordinate system; thus, if we formulate a correct tensor relationship between the stress tensor and velocity, we satisfy the requirement for an equation that works in any coordinate system. The second requirement is that the tensor relationship between stress and velocity should correctly capture material behavior. The relationship that does this correctly for many fluids is called the *Newtonian constitutive equation*.

To rigorously derive a tensor version of Newton's law of viscosity, we need to know more about tensors. Instead, we present simple arguments to justify the final result, but this is not a proof. Interested readers may pursue more information and a complete proof in the literature [6].

5.2.1 The constitutive equation

The tensor that contains all of the information about how the velocity vector varies in space is called the *velocity gradient tensor* $\nabla \underline{v}$. In a Cartesian coordinate system, $\nabla \underline{v}$ is given by the following (see Appendix B):

$$
\begin{array}{c}
\text{Velocity} \\
\text{gradient} \\
\text{tensor}
\end{array}
\quad
\nabla \underline{v} =
\begin{pmatrix}
\frac{\partial v_x}{\partial x} & \frac{\partial v_y}{\partial x} & \frac{\partial v_z}{\partial x} \\[6pt]
\frac{\partial v_x}{\partial y} & \frac{\partial v_y}{\partial y} & \frac{\partial v_z}{\partial y} \\[6pt]
\frac{\partial v_x}{\partial z} & \frac{\partial v_y}{\partial z} & \frac{\partial v_z}{\partial z}
\end{pmatrix}_{xyz}
\tag{5.77}
$$

We now guess a relationship between $\underline{\underline{\tilde{\tau}}}$ and $\nabla \underline{v}$. The simplest relationship is that $\underline{\underline{\tilde{\tau}}}$ is proportional to $\nabla \underline{v}$, $\underline{\underline{\tilde{\tau}}} = a \nabla \underline{v}$. In Cartesian coordinates, this becomes:

$$
\underline{\underline{\tilde{\tau}}} \overset{?}{=}
\begin{pmatrix}
a\frac{\partial v_x}{\partial x} & a\frac{\partial v_y}{\partial x} & a\frac{\partial v_z}{\partial x} \\[6pt]
a\frac{\partial v_x}{\partial y} & a\frac{\partial v_y}{\partial y} & a\frac{\partial v_z}{\partial y} \\[6pt]
a\frac{\partial v_x}{\partial z} & a\frac{\partial v_y}{\partial z} & a\frac{\partial v_z}{\partial z}
\end{pmatrix}_{xyz}
\tag{5.78}
$$

Notice that if we make this choice, the yx-component of $\underline{\underline{\tilde{\tau}}} = a \nabla \underline{v}$ is the equivalent of Equation 5.61, Newton's law of viscosity, if $a = \mu$. This simplest equation is not appropriate, however, because we know that $\underline{\underline{\tilde{\tau}}}$ is symmetric. Because $\nabla \underline{v}$ is not symmetric (e.g., the xy-component is not equal to the yx-component), a simple proportionality between $\underline{\underline{\tilde{\tau}}}$ and $\nabla \underline{v}$ cannot be correct, and $\underline{\underline{\tilde{\tau}}} = \mu \nabla \underline{v}$ can not be the correct stress–velocity equation.

It is straightforward to show that the tensor is $\nabla \underline{v} + (\nabla \underline{v})^T$ is symmetric. The notation $\underline{\underline{A}}^T$ means the transpose of the tensor is $\underline{\underline{A}}$, and $\underline{\underline{A}}^T$ is the new tensor obtained by switching the rows and columns of the matrix representation of $\underline{\underline{A}}$:

$$
(\nabla \underline{v})^T =
\begin{pmatrix}
\frac{\partial v_x}{\partial x} & \frac{\partial v_x}{\partial y} & \frac{\partial v_x}{\partial z} \\[6pt]
\frac{\partial v_y}{\partial x} & \frac{\partial v_y}{\partial y} & \frac{\partial v_y}{\partial z} \\[6pt]
\frac{\partial v_z}{\partial x} & \frac{\partial v_z}{\partial y} & \frac{\partial v_z}{\partial z}
\end{pmatrix}_{xyz}
\tag{5.79}
$$

$$
\nabla \underline{v} + (\nabla \underline{v})^T =
\begin{pmatrix}
\frac{\partial v_x}{\partial x} & \frac{\partial v_y}{\partial x} & \frac{\partial v_z}{\partial x} \\[6pt]
\frac{\partial v_x}{\partial y} & \frac{\partial v_y}{\partial y} & \frac{\partial v_z}{\partial y} \\[6pt]
\frac{\partial v_x}{\partial z} & \frac{\partial v_y}{\partial z} & \frac{\partial v_z}{\partial z}
\end{pmatrix}_{xyz}
+
\begin{pmatrix}
\frac{\partial v_x}{\partial x} & \frac{\partial v_x}{\partial y} & \frac{\partial v_x}{\partial z} \\[6pt]
\frac{\partial v_y}{\partial x} & \frac{\partial v_y}{\partial y} & \frac{\partial v_y}{\partial z} \\[6pt]
\frac{\partial v_z}{\partial x} & \frac{\partial v_z}{\partial y} & \frac{\partial v_z}{\partial z}
\end{pmatrix}_{xyz}
\tag{5.80}
$$

$$
=
\begin{pmatrix}
2\frac{\partial v_x}{\partial x} & \frac{\partial v_y}{\partial x} + \frac{\partial v_x}{\partial y} & \frac{\partial v_z}{\partial x} + \frac{\partial v_x}{\partial z} \\[6pt]
\frac{\partial v_x}{\partial y} + \frac{\partial v_y}{\partial x} & 2\frac{\partial v_y}{\partial y} & \frac{\partial v_z}{\partial y} + \frac{\partial v_y}{\partial z} \\[6pt]
\frac{\partial v_x}{\partial z} + \frac{\partial v_z}{\partial x} & \frac{\partial v_y}{\partial z} + \frac{\partial v_z}{\partial y} & 2\frac{\partial v_z}{\partial z}
\end{pmatrix}_{xyz}
\tag{5.81}
$$

Considering the symmetry requirement, we propose that a possible relationship between the stress tensor $\underline{\underline{\tilde{\tau}}}$ and $\underline{v}$ is a simple proportionality between $\underline{\underline{\tilde{\tau}}}$ and $(\nabla \underline{v} + \nabla \underline{v})^T$:

$$
\begin{array}{c}
\text{Hypothesis: This is the} \\
\text{general stress–viscosity} \\
\text{relationship for Newtonian fluids}
\end{array}
\qquad
\underline{\underline{\tilde{\tau}}} \overset{?}{=} \mu \left(\nabla \underline{v} + (\nabla \underline{v})^T \right)
\tag{5.82}
$$

The first test of our hypothesis is shear flow. In shear flow, we know that $\underline{v} = v_x \hat{e}_x$:

$$\text{Shear flow:} \quad \underline{v} = \begin{pmatrix} v_x \\ 0 \\ 0 \end{pmatrix}_{xyz} = \begin{pmatrix} \frac{Vy}{H} \\ 0 \\ 0 \end{pmatrix}_{xyz} \tag{5.83}$$

and from Newton's law of viscosity, $\tilde{\tau}_{yx} = \mu(dv_x/dy)$. First, we write Equation 5.82 in Cartesian component form:

$$\begin{matrix} \text{Hypothesis test:} \\ \text{shear flow} \end{matrix} \quad \underset{=}{\tilde{\underline{\tau}}} \overset{?}{=} \mu \left(\nabla \underline{v} + (\nabla \underline{v})^T \right) \tag{5.84}$$

$$= \mu \begin{pmatrix} 2\frac{\partial v_x}{\partial x} & \frac{\partial v_y}{\partial x} + \frac{\partial v_x}{\partial y} & \frac{\partial v_z}{\partial x} + \frac{\partial v_x}{\partial z} \\ \frac{\partial v_x}{\partial y} + \frac{\partial v_y}{\partial x} & 2\frac{\partial v_y}{\partial y} & \frac{\partial v_z}{\partial y} + \frac{\partial v_y}{\partial z} \\ \frac{\partial v_x}{\partial z} + \frac{\partial v_z}{\partial x} & \frac{\partial v_y}{\partial z} + \frac{\partial v_z}{\partial y} & 2\frac{\partial v_z}{\partial z} \end{pmatrix}_{xyz} \tag{5.85}$$

Note that the predicted $\underset{=}{\tilde{\underline{\tau}}}$ is symmetric. We can simplify Equation 5.85 for simple shear flow because $\underline{v} = v_x \hat{e}_x$, which means that $v_y = v_z = 0$. With these velocity components equal to zero, Equation 5.85 becomes:

$$\underset{=}{\tilde{\underline{\tau}}} \overset{?}{=} \mu \begin{pmatrix} 2\frac{\partial v_x}{\partial x} & \frac{\partial v_x}{\partial y} & \frac{\partial v_x}{\partial z} \\ \frac{\partial v_x}{\partial y} & 0 & 0 \\ \frac{\partial v_x}{\partial z} & 0 & 0 \end{pmatrix}_{xyz} \tag{5.86}$$

We know from the experiments discussed previously in this chapter that in simple shear flow, $v_x = (V/H)y\hat{e}_x$; therefore, v_x is not a function of x or of z and, thus, $\partial v_x/\partial z = 0$ and $\partial v_x/\partial y = 0$. Including these two facts in Equation 5.86, the final result for $\underset{=}{\tilde{\underline{\tau}}}$ is:

$$\begin{matrix} \text{Stress tensor} \\ \text{steady simple shear} \\ \text{(testing proposed relationship)} \end{matrix} \quad \underset{=}{\tilde{\underline{\tau}}} \overset{?}{=} \mu \begin{pmatrix} 0 & \frac{\partial v_x}{\partial y} & 0 \\ \frac{\partial v_x}{\partial y} & 0 & 0 \\ 0 & 0 & 0 \end{pmatrix}_{xyz} \tag{5.87}$$

This tensor for simple shear flow has only two nonzero components, and they give Newton's law of viscosity. With the additional information that in simple shear flow the velocity is a function only of y, we can change the partial derivatives in Equation 5.87 to total derivatives. Thus, the guessed tensor relationship between stress and velocity works for the simple shear-flow case.

$$
\begin{matrix}
\text{Stress tensor} \\
\text{steady simple shear} \\
\text{(testing proposed relationship)}
\end{matrix}
\qquad
\underline{\underline{\tilde{\tau}}} = \mu
\begin{pmatrix}
0 & \dfrac{dv_x}{dy} & 0 \\[2mm]
\dfrac{dv_x}{dy} & 0 & 0 \\[2mm]
0 & 0 & 0
\end{pmatrix}_{xyz}
\qquad (5.88)
$$

$$
\tilde{\tau}_{yx} = \tilde{\tau}_{xy} = \mu \frac{dv_x}{dy}
$$

In fact, it may be shown through rigorous experimental testing on many shear and non-shear flows that our guessed constitutive equation, Equation 5.82, is the correct constitutive equation for many fluids. The tensor constitutive equation we propose captures the full three-dimensional nature of the flow of simple fluids and is called the Newtonian constitutive equation:

$$
\begin{matrix}
\text{Newtonian} \\
\text{constitutive equation} \\
\text{(stress-velocity relationship)}
\end{matrix}
\qquad
\underline{\underline{\tilde{\tau}}} = \mu \left(\nabla \underline{v} + (\nabla \underline{v})^T \right)
\qquad (5.89)
$$

The Newtonian constitutive equation in Cartesian, cylindrical, and spherical coordinates are given here:

$$
\begin{pmatrix}
\tilde{\tau}_{xx} & \tilde{\tau}_{xy} & \tilde{\tau}_{xz} \\
\tilde{\tau}_{yx} & \tilde{\tau}_{yy} & \tilde{\tau}_{yz} \\
\tilde{\tau}_{zx} & \tilde{\tau}_{zy} & \tilde{\tau}_{zz}
\end{pmatrix}_{xyz}
= \mu
\begin{pmatrix}
2\dfrac{\partial v_x}{\partial x} & \dfrac{\partial v_y}{\partial x} + \dfrac{\partial v_x}{\partial y} & \dfrac{\partial v_z}{\partial x} + \dfrac{\partial v_x}{\partial z} \\[3mm]
\dfrac{\partial v_x}{\partial y} + \dfrac{\partial v_y}{\partial x} & 2\dfrac{\partial v_y}{\partial y} & \dfrac{\partial v_z}{\partial y} + \dfrac{\partial v_y}{\partial z} \\[3mm]
\dfrac{\partial v_x}{\partial z} + \dfrac{\partial v_z}{\partial x} & \dfrac{\partial v_y}{\partial z} + \dfrac{\partial v_z}{\partial y} & 2\dfrac{\partial v_z}{\partial z}
\end{pmatrix}_{xyz}
\qquad (5.90)
$$

$$
\begin{pmatrix}
\tilde{\tau}_{rr} & \tilde{\tau}_{r\theta} & \tilde{\tau}_{rz} \\
\tilde{\tau}_{\theta r} & \tilde{\tau}_{\theta\theta} & \tilde{\tau}_{\theta z} \\
\tilde{\tau}_{zr} & \tilde{\tau}_{z\theta} & \tilde{\tau}_{zz}
\end{pmatrix}_{r\theta z}
= \mu
\begin{pmatrix}
2\dfrac{\partial v_r}{\partial r} & r\dfrac{\partial}{\partial r}\left(\dfrac{v_\theta}{r}\right) + \dfrac{1}{r}\dfrac{\partial v_r}{\partial \theta} & \dfrac{\partial v_r}{\partial z} + \dfrac{\partial v_z}{\partial r} \\[3mm]
r\dfrac{\partial}{\partial r}\left(\dfrac{v_\theta}{r}\right) + \dfrac{1}{r}\dfrac{\partial v_r}{\partial \theta} & 2\left(\dfrac{1}{r}\dfrac{\partial v_\theta}{\partial \theta} + \dfrac{v_r}{r}\right) & \dfrac{1}{r}\dfrac{\partial v_z}{\partial \theta} + \dfrac{\partial v_\theta}{\partial z} \\[3mm]
\dfrac{\partial v_r}{\partial z} + \dfrac{\partial v_z}{\partial r} & \dfrac{1}{r}\dfrac{\partial v_z}{\partial \theta} + \dfrac{\partial v_\theta}{\partial z} & 2\dfrac{\partial v_z}{\partial z}
\end{pmatrix}_{r\theta z}
$$

$$
(5.91)
$$

$$
\begin{pmatrix}
\tilde{\tau}_{rr} & \tilde{\tau}_{r\theta} & \tilde{\tau}_{r\phi} \\
\tilde{\tau}_{\theta r} & \tilde{\tau}_{\theta\theta} & \tilde{\tau}_{\theta\phi} \\
\tilde{\tau}_{\phi r} & \tilde{\tau}_{\phi\theta} & \tilde{\tau}_{\phi\phi}
\end{pmatrix}_{r\theta\phi}
$$

$$
= \mu
\begin{pmatrix}
2\dfrac{\partial v_r}{\partial r} & r\dfrac{\partial}{\partial r}\left(\dfrac{v_\theta}{r}\right) + \dfrac{1}{r}\dfrac{\partial v_r}{\partial \theta} & \dfrac{1}{r\sin\theta}\dfrac{\partial v_r}{\partial \phi} + r\dfrac{\partial}{\partial r}\left(\dfrac{v_\phi}{r}\right) \\[3mm]
r\dfrac{\partial}{\partial r}\left(\dfrac{v_\theta}{r}\right) + \dfrac{1}{r}\dfrac{\partial v_r}{\partial \theta} & 2\left(\dfrac{1}{r}\dfrac{\partial v_\theta}{\partial \theta} + \dfrac{v_r}{r}\right) & \dfrac{\sin\theta}{r}\dfrac{\partial}{\partial \theta}\left(\dfrac{v_\phi}{\sin\theta}\right) + \dfrac{1}{r\sin\theta}\dfrac{\partial v_\theta}{\partial \phi} \\[3mm]
\dfrac{1}{r\sin\theta}\dfrac{\partial v_r}{\partial \phi} + r\dfrac{\partial}{\partial r}\left(\dfrac{v_\phi}{r}\right) & \dfrac{\sin\theta}{r}\dfrac{\partial}{\partial \theta}\left(\dfrac{v_\phi}{\sin\theta}\right) + \dfrac{1}{r\sin\theta}\dfrac{\partial v_\theta}{\partial \phi} & 2\left(\dfrac{1}{r\sin\theta}\dfrac{\partial v_\phi}{\partial \phi} + \dfrac{v_r}{r} + \dfrac{v_\theta\cot\theta}{r}\right)
\end{pmatrix}_{r\theta\phi}
$$

$$
(5.92)
$$

These components also are given in Appendix B, Table B.8.

Newton published his one-dimensional law of viscosity in 1687, and G. G. Stokes generalized it to three dimensions in 1845 [90]. Important experimental verification of the Newtonian constitutive equation did not occur until the work of Poiseuille (1856) [90, 91] and Couette (1890) [29] later in the 19th century. For a discussion of the development of the Newtonian constitutive equation from Newton's law of viscosity, see Bird et al. [15].

Having the Newtonian constitutive equation, we now are ready to complete the problem of the film flowing down the incline. Before we return to complete that example, however, we try again to evaluate the stress-velocity laws for the two complex flows recently considered: the planar-jet flow and flow around a sphere.

EXAMPLE 5.5 (Stress-velocity relationship: concluded). *What are the correct forms of the stress-velocity relationship for the two flows shown in Figure 5.14?*

SOLUTION. The solution to this problem is to look up the Newtonian constitutive equation in our chosen coordinate system and then simplify it based on what we know about the velocity field and how it varies in space.

For the planar jet hitting the wall, we chose the Cartesian coordinate system; the constitutive equation therefore is given by Equation 5.90:

$$
\begin{pmatrix} \tilde{\tau}_{xx} & \tilde{\tau}_{xy} & \tilde{\tau}_{xz} \\ \tilde{\tau}_{yx} & \tilde{\tau}_{yy} & \tilde{\tau}_{yz} \\ \tilde{\tau}_{zx} & \tilde{\tau}_{zy} & \tilde{\tau}_{zz} \end{pmatrix}_{xyz} = \mu \begin{pmatrix} 2\frac{\partial v_x}{\partial x} & \frac{\partial v_y}{\partial x} + \frac{\partial v_x}{\partial y} & \frac{\partial v_z}{\partial x} + \frac{\partial v_x}{\partial z} \\ \frac{\partial v_x}{\partial y} + \frac{\partial v_y}{\partial x} & 2\frac{\partial v_y}{\partial y} & \frac{\partial v_z}{\partial y} + \frac{\partial v_y}{\partial z} \\ \frac{\partial v_x}{\partial z} + \frac{\partial v_z}{\partial x} & \frac{\partial v_y}{\partial z} + \frac{\partial v_z}{\partial y} & 2\frac{\partial v_z}{\partial z} \end{pmatrix}_{xyz} \quad (5.93)
$$

The velocity vector for the planar-jet flow has $v_z = 0$; thus, we may eliminate the terms of Equation 5.93 that contain v_z:

$$
\underline{v} = \begin{pmatrix} v_x \\ v_y \\ 0 \end{pmatrix}_{xyz} \quad (5.94)
$$

As discussed previously, in this flow, v_x and v_y are both functions of x and y; however, because the flow is wide, they are not functions of z. We therefore also eliminate the terms in the constitutive equation that contain derivatives with respect to z. Incorporating these observations in Equation 5.93, we obtain:

$$
\begin{pmatrix} \tilde{\tau}_{xx} & \tilde{\tau}_{xy} & \tilde{\tau}_{xz} \\ \tilde{\tau}_{yx} & \tilde{\tau}_{yy} & \tilde{\tau}_{yz} \\ \tilde{\tau}_{zx} & \tilde{\tau}_{zy} & \tilde{\tau}_{zz} \end{pmatrix}_{xyz} = \mu \begin{pmatrix} 2\frac{\partial v_x}{\partial x} & \frac{\partial v_y}{\partial x} + \frac{\partial v_x}{\partial y} & 0 \\ \frac{\partial v_x}{\partial y} + \frac{\partial v_y}{\partial x} & 2\frac{\partial v_y}{\partial y} & 0 \\ 0 & 0 & 0 \end{pmatrix}_{xyz} \quad (5.95)
$$

This is the final result. There are four nonzero stresses in this flow and two of them are equal to one another, $\tilde{\tau}_{yx} = \tilde{\tau}_{xy}$:

$$
\begin{aligned}
\tilde{\tau}_{yx} = \tilde{\tau}_{xy} &= \mu \left(\frac{\partial v_y}{\partial x} + \frac{\partial v_x}{\partial y} \right) \\
\tilde{\tau}_{xx} &= 2\mu \frac{\partial v_x}{\partial x} \\
\tilde{\tau}_{yy} &= 2\mu \frac{\partial v_y}{\partial y}
\end{aligned}
\tag{5.96}
$$

This result is different from our simplistic guess in Equation 5.74. If we perform a mass balance on a microscopic control volume, we discover other relationships among the velocity derivatives that also simplify the problem.

For the flow around a sphere, we begin with the Newtonian constitutive equation written in spherical coordinates, Equation 5.92:

$$
\begin{pmatrix} \tilde{\tau}_{rr} & \tilde{\tau}_{r\theta} & \tilde{\tau}_{r\phi} \\ \tilde{\tau}_{\theta r} & \tilde{\tau}_{\theta\theta} & \tilde{\tau}_{\theta\phi} \\ \tilde{\tau}_{\phi r} & \tilde{\tau}_{\phi\theta} & \tilde{\tau}_{\phi\phi} \end{pmatrix}_{r\theta\phi}
$$

$$
= \mu \begin{pmatrix} 2\frac{\partial v_r}{\partial r} & r\frac{\partial}{\partial r}\left(\frac{v_\theta}{r}\right) + \frac{1}{r}\frac{\partial v_r}{\partial \theta} & \frac{1}{r\sin\theta}\frac{\partial v_r}{\partial \phi} + r\frac{\partial}{\partial r}\left(\frac{v_\phi}{r}\right) \\[2mm]
r\frac{\partial}{\partial r}\left(\frac{v_\theta}{r}\right) + \frac{1}{r}\frac{\partial v_r}{\partial \theta} & 2\left(\frac{1}{r}\frac{\partial v_\theta}{\partial \theta} + \frac{v_r}{r}\right) & \frac{\sin\theta}{r}\frac{\partial}{\partial \theta}\left(\frac{v_\phi}{\sin\theta}\right) + \frac{1}{r\sin\theta}\frac{\partial v_\theta}{\partial \phi} \\[2mm]
\frac{1}{r\sin\theta}\frac{\partial v_r}{\partial \phi} + r\frac{\partial}{\partial r}\left(\frac{v_\phi}{r}\right) & \frac{\sin\theta}{r}\frac{\partial}{\partial \theta}\left(\frac{v_\phi}{\sin\theta}\right) + \frac{1}{r\sin\theta}\frac{\partial v_\theta}{\partial \phi} & 2\left(\frac{1}{r\sin\theta}\frac{\partial v_\phi}{\partial \phi} + \frac{v_r}{r} + \frac{v_\theta \cot\theta}{r}\right) \end{pmatrix}_{r\theta\phi}
$$

$$
\tag{5.97}
$$

The stress tensor written in spherical coordinates has many extra terms that result from the coordinate transformation from Cartesian to spatially varying spherical coordinates. These nonintuitive terms can be arrived at only by performing rigorous algebraic calculations of the transformation of the Cartesian coordinates to the spherical coordinates using Equations 1.268–1.273. Because these calculations are complex, we use tables that provide the final results needed.

For the flow around a sphere, $v_\phi = 0$, and the flow is symmetric relative to ϕ, which means that derivatives with respect to ϕ are zero. These two facts allow us to simplify the expression for $\tilde{\underline{\tau}}$ in Equation 5.97:

$$
\underline{v} = \begin{pmatrix} v_r(r, \theta) \\ v_\theta(r, \theta) \\ 0 \end{pmatrix}_{r\theta\phi}
\tag{5.98}
$$

$$
\begin{pmatrix}
\tilde{\tau}_{rr} & \tilde{\tau}_{r\theta} & \tilde{\tau}_{r\phi} \\
\tilde{\tau}_{\theta r} & \tilde{\tau}_{\theta\theta} & \tilde{\tau}_{\theta\phi} \\
\tilde{\tau}_{\phi r} & \tilde{\tau}_{\phi\theta} & \tilde{\tau}_{\phi\phi}
\end{pmatrix}_{r\theta\phi}
$$

(5.99)

$$
= \mu
\begin{pmatrix}
2\frac{\partial v_r}{\partial r} & r\frac{\partial}{\partial r}\left(\frac{v_\theta}{r}\right) + \frac{1}{r}\frac{\partial v_r}{\partial \theta} & 0 \\
r\frac{\partial}{\partial r}\left(\frac{v_\theta}{r}\right) + \frac{1}{r}\frac{\partial v_r}{\partial \theta} & 2\left(\frac{1}{r}\frac{\partial v_\theta}{\partial \theta} + \frac{v_r}{r}\right) & 0 \\
0 & 0 & 2\left(\frac{v_r}{r} + \frac{v_\theta \cot\theta}{r}\right)
\end{pmatrix}_{r\theta\phi}
$$

This is as far as we can simplify the constitutive equation. There are four unique, nonzero stress–velocity relations to consider in this flow:

$$
\tilde{\tau}_{rr} = 2\mu\frac{\partial v_r}{\partial r}
$$

$$
\tilde{\tau}_{r\theta} = \tilde{\tau}_{\theta r} = \mu\left(r\frac{\partial}{\partial r}\left(\frac{v_\theta}{r}\right) + \frac{1}{r}\frac{\partial v_r}{\partial \theta}\right)
$$

$$
\tilde{\tau}_{\theta\theta} = 2\mu\left(\frac{1}{r}\frac{\partial v_\theta}{\partial \theta} + \frac{v_r}{r}\right)
$$

$$
\tilde{\tau}_{\phi\phi} = 2\mu\left(\frac{v_r}{r} + \frac{v_\theta \cot\theta}{r}\right)
$$

(5.100)

These components are more complex than any guess we are likely to make for the relationship between stresses and velocity gradients. It was essential to use the correct tensor transformation of $\underset{\approx}{\tilde{\tau}}$ to the chosen coordinate system. It was straightforward to obtain those terms from a reference table (see Equation 5.92 and Table B.8) as long as we knew the constitutive equation in Gibbs notation (Equation 5.89).

The stress-velocity laws obtained in Equations 5.96 and 5.100 can be used in the momentum balances on control volumes appropriate to the particular flow. Complex problems such as those in Figure 5.14 generally are solved with the aid of the general microscopic-momentum-balance equation, which is derived for balances on arbitrary control volumes in Chapter 6. Once the microscopic-momentum-balance equation is solved for $\underline{v}$ with the help of the constitutive equation, Equations 5.96 and 5.100 are used again to calculate stresses in the two flows.

The Newtonian constitutive equation is a powerful result. This simple equation, which contains a single material-flow parameter, the viscosity μ, gives the stress tensor as a function of velocity at any position in a flow. With the Newtonian constitutive equation, we now know $\underset{\approx}{\tilde{\tau}}$ and $\underset{\approx}{\tilde{\Pi}}$ and can calculate forces or surfaces in flows with the usual integral of $\hat{n} \cdot \underset{\approx}{\tilde{\Pi}}|_{surface}$ (Equation 5.3). In Section 5.3, we explore constitutive equations for materials that do not follow Newton's law of viscosity. Non-Newtonian fluids are common, and the field of study that focuses on non-Newtonian fluids is called rheology [12, 90, 104].

The next two examples provide practice with calculations using the Newtonian constitutive equation. In Section 5.2.2, we deploy the Newtonian constitutive equation to allow us to finish the two example problems from Chapters 3 and 4.

EXAMPLE 5.6. *Calculate the stress field in the steady upward flow of an incompressible Newtonian fluid around a stationary solid sphere of diameter $2R$ (velocity and pressure fields given here). The fluid approaches the sphere with a uniform upstream velocity of v_∞ (Figure 5.17).*

SOLUTION. We are not ready to carry out such a calculation from scratch, but if we are given the velocity solution, we now know how to calculate the stress tensor from the velocity solution.

The solution for the velocity field in creeping flow around a sphere is given here and discussed in detail in Chapter 8. The problem is solved and reported in spherical coordinates:

Solution [43], creeping flow (Stokes flow) around a sphere

$$\underline{v}(r,\theta) = \begin{pmatrix} v_\infty \left[1 - \dfrac{3}{2}\dfrac{R}{r} + \dfrac{1}{2}\left(\dfrac{R}{r}\right)^3 \right] \cos\theta \\[2ex] -v_\infty \left[1 - \dfrac{3}{4}\dfrac{R}{r} - \dfrac{1}{4}\left(\dfrac{R}{r}\right)^3 \right] \sin\theta \\[2ex] 0 \end{pmatrix}_{r\theta\phi}$$

(5.101)

$$p(r,\theta) = p_\infty - \rho g r \cos\theta - \dfrac{3}{2}\dfrac{\mu v_\infty}{R}\left(\dfrac{R}{r}\right)^2 \cos\theta$$

(5.102)

The quantity p_∞ is the pressure far from the sphere at the elevation of the origin of the coordinate system (i.e., at $\theta = \pi/2$, the sphere equator).

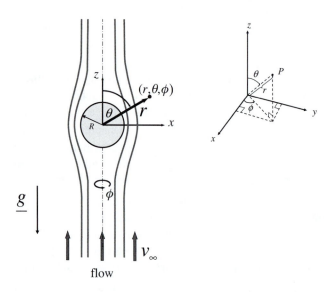

Schematic of creeping flow around a sphere. This flow is known as Stokes flow. The solution for the velocity field is discussed in Chapter 8.

Given that we were provided the solution for $\underline{v}$ for the flow around a sphere problem, we can calculate the stress field in this flow from the Newtonian constitutive equation. Because we are in spherical coordinates, we must use the correct form for the Newtonian constitutive equation in this coordinate system. The correct form is in Table B.8:

$$\underline{\underline{\tilde{\tau}}} = \mu \left(\nabla \underline{v} + \nabla \underline{v}^T \right) \tag{5.103}$$

$$\underline{v} = \begin{pmatrix} v_r(r,\theta) \\ v_\theta(r,\theta) \\ 0 \end{pmatrix}_{r\theta z} \tag{5.104}$$

$$\underline{\underline{\tilde{\tau}}}(r,\theta) = \mu \begin{pmatrix} 2\frac{\partial v_r}{\partial r} & r\frac{\partial}{\partial r}\left(\frac{v_\theta}{r}\right) + \frac{1}{r}\frac{\partial v_r}{\partial\theta} & 0 \\ r\frac{\partial}{\partial r}\left(\frac{v_\theta}{r}\right) + \frac{1}{r}\frac{\partial v_r}{\partial\theta} & 2\left(\frac{1}{r}\frac{\partial v_\theta}{\partial\theta} + \frac{v_r}{r}\right) & 0 \\ 0 & 0 & \frac{2v_r}{r} + \frac{2v_\theta \cot\theta}{r} \end{pmatrix}_{r\theta\phi} \tag{5.105}$$

$$\underline{\underline{\tilde{\Pi}}}(r,\theta) = \underline{\underline{\tilde{\tau}}} - p\underline{\underline{I}}$$

$$= \begin{pmatrix} 2\mu\frac{\partial v_r}{\partial r} - p(r,\theta) & \mu r\frac{\partial}{\partial r}\left(\frac{v_\theta}{r}\right) + \frac{\mu}{r}\frac{\partial v_r}{\partial\theta} & 0 \\ \mu r\frac{\partial}{\partial r}\left(\frac{v_\theta}{r}\right) + \frac{\mu}{r}\frac{\partial v_r}{\partial\theta} & 2\mu\left(\frac{1}{r}\frac{\partial v_\theta}{\partial\theta} + \frac{v_r}{r}\right) - p(r,\theta) & 0 \\ 0 & 0 & \frac{2\mu v_r}{r} + \frac{2\mu v_\theta \cot\theta}{r} - p(r,\theta) \end{pmatrix}_{r\theta\phi} \tag{5.106}$$

The components of $\underline{\underline{\tilde{\Pi}}}$ are calculated from the velocity and pressure solutions, Equations 5.101 and 5.102. For $\tilde{\Pi}_{rr}$ and $\tilde{\Pi}_{r\theta} = \tilde{\Pi}_{\theta r}$, we obtain:

$$\tilde{\Pi}_{rr} = 2\mu\frac{\partial v_r}{\partial r} - p(r,\theta)$$

$$= \mu v_\infty \cos\theta \left(\frac{3R}{r^2} - \frac{3R^3}{r^4}\right) - p(r,\theta)$$

$$= \mu v_\infty \cos\theta \left(\frac{3R}{r^2} - \frac{3R^3}{r^4}\right) - p_0 + \rho g r \cos\theta + \frac{3}{2}\mu v_\infty \cos\theta \left(\frac{R}{r^2}\right)$$

$$= \mu v_\infty \cos\theta \left(\frac{9R}{2r^2} - \frac{3R^3}{r^4}\right) - p_0 + \rho g r \cos\theta \tag{5.107}$$

$$\tilde{\Pi}_{r\theta} = \mu r\frac{\partial}{\partial r}\left(\frac{v_\theta}{r}\right) + \frac{\mu}{r}\frac{\partial v_r}{\partial\theta}$$

$$= -\mu v_\infty \sin\theta \frac{3R^3}{2r^4} \tag{5.108}$$

The two remaining coefficients of $\underline{\underline{\tilde{\Pi}}}$, $\tilde{\Pi}_{\theta\theta}$ and $\tilde{\Pi}_{\phi\phi}$, may be calculated similarly (see Problem 27).

EXAMPLE 5.7. *For the flow of an incompressible Newtonian fluid with the velocity field given here, calculate the extra-stress tensor $\tilde{\underline{\tau}}$. What is the force on a small plane ΔA located at coordinate point $(x, y, z) = (0, 0, l_0)$ and facing the flow ($\hat{n} = -\hat{e}_z$)?*

$$\underline{v} = \begin{pmatrix} -\frac{\dot{\varepsilon}_0 x}{2} \\ -\frac{\dot{\varepsilon}_0 y}{2} \\ \dot{\varepsilon}_0 z \end{pmatrix}_{xyz} \tag{5.109}$$

SOLUTION. The flow field in Equation 5.109 is called uniaxial elongational flow [12, 104], and it is a standard flow used in the study of non-Newtonian fluids (Figure 5.18). We calculate the stress produced by a Newtonian fluid in this flow from the velocity field by applying the Newtonian constitutive equation. The velocity field is given in a Cartesian coordinate system; thus, we begin with Equation 5.90:

$$\begin{pmatrix} \tilde{\tau}_{xx} & \tilde{\tau}_{xy} & \tilde{\tau}_{xz} \\ \tilde{\tau}_{yx} & \tilde{\tau}_{yy} & \tilde{\tau}_{yz} \\ \tilde{\tau}_{zx} & \tilde{\tau}_{zy} & \tilde{\tau}_{zz} \end{pmatrix}_{xyz} = \mu \begin{pmatrix} 2\frac{\partial v_x}{\partial x} & \frac{\partial v_y}{\partial x} + \frac{\partial v_x}{\partial y} & \frac{\partial v_z}{\partial x} + \frac{\partial v_x}{\partial z} \\ \frac{\partial v_x}{\partial y} + \frac{\partial v_y}{\partial x} & 2\frac{\partial v_y}{\partial y} & \frac{\partial v_z}{\partial y} + \frac{\partial v_y}{\partial z} \\ \frac{\partial v_x}{\partial z} + \frac{\partial v_z}{\partial x} & \frac{\partial v_y}{\partial z} + \frac{\partial v_z}{\partial y} & 2\frac{\partial v_z}{\partial z} \end{pmatrix}_{xyz} \tag{5.110}$$

Using the velocity field in Equation 5.109 to calculate these derivatives, we obtain the final result:

$$\begin{pmatrix} \tilde{\tau}_{xx} & \tilde{\tau}_{xy} & \tilde{\tau}_{xz} \\ \tilde{\tau}_{yx} & \tilde{\tau}_{yy} & \tilde{\tau}_{yz} \\ \tilde{\tau}_{zx} & \tilde{\tau}_{zy} & \tilde{\tau}_{zz} \end{pmatrix}_{xyz} = \begin{pmatrix} -\mu\dot{\varepsilon}_0 & 0 & 0 \\ 0 & -\mu\dot{\varepsilon}_0 & 0 \\ 0 & 0 & 2\mu\dot{\varepsilon}_0 \end{pmatrix}_{xyz} \tag{5.111}$$

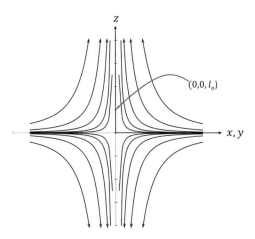

Figure 5.18 In elongational flow, the principal flow direction is in the *z*-direction. The flow stretches strongly in the *z*-direction, and contracts equally in the *x*- and *y*-directions.

The total stress tensor $\underline{\underline{\tilde{\Pi}}}$ is then:

$$\underline{\underline{\tilde{\Pi}}} = \underline{\underline{\tilde{\tau}}} - p\underline{\underline{I}} \tag{5.112}$$

$$= \begin{pmatrix} -\mu\dot{\varepsilon}_0 - p & 0 & 0 \\ 0 & -\mu\dot{\varepsilon}_0 - p & 0 \\ 0 & 0 & 2\mu\dot{\varepsilon}_0 - p \end{pmatrix}_{xyz} \tag{5.113}$$

To calculate the force on the plane of interest, we begin with the equation for the fluid force on a small plane in terms of the stress tensor, Equation 5.2:

$$\begin{array}{c} \text{Force in a fluid on} \\ \text{a surface of area } \Delta A \\ \text{with unit normal } \hat{n} \end{array} \quad \underline{f}(\Delta A, \hat{n}) = \Delta A \ [\hat{n} \cdot \underline{\underline{\tilde{\Pi}}}]_{surface} \tag{5.114}$$

We have all of the quantities needed to calculate the force:

$$\underline{f} = \Delta A \left(0 \ \ 0 \ \ -1\right)_{xyz} \cdot \begin{pmatrix} -\mu\dot{\varepsilon}_0 - p & 0 & 0 \\ 0 & -\mu\dot{\varepsilon}_0 - p & 0 \\ 0 & 0 & 2\mu\dot{\varepsilon}_0 - p \end{pmatrix}_{xyz} \Bigg|_{at \ (0,0,l_0)} \tag{5.115}$$

$$= \Delta A \left(0 \ \ 0 \ \ -2\mu\dot{\varepsilon}_0 + p\right)_{xyz} \tag{5.116}$$

$$= \begin{pmatrix} 0 \\ 0 \\ \Delta A \left(-2\mu\dot{\varepsilon}_0 + p\right) \end{pmatrix}_{xyz} \tag{5.117}$$

5.2.2 Using the constitutive equation

We return to the two flow problems we have been solving since introducing the continuum model: flow of a thin film of liquid down an inclined plane and force on a right-angle bend. We now are ready to finish these problems, and in this section, we arrive at the final velocity profile for flow down an incline (Example 5.8) and the final force on the bend (Example 5.9).

EXAMPLE 5.8 (Incline: concluded). *What is the velocity field in a wide, thin film of water that runs steadily down an inclined surface under the force of gravity? The fluid has a constant density ρ (continued from Chapters 3 and 4 and previously in this chapter).*

SOLUTION. In Chapter 3, we started this problem, which is illustrated in Figure 5.12. We performed a momentum balance on a microscopic control volume of size $\Delta x \Delta y \Delta z$; in Chapter 4, we wrote the contact forces in terms of the extra-stress tensor components and the pressure. We then incorporated the stress components into the momentum balance (see Equations 4.357–4.359), but we paused the development there because we did not know how the stress components varied with velocity. Previously in this chapter, we tried to use Newton's law of viscosity for the stress-velocity relationship, but we were unsure about how to convert Newton's law of viscosity to a new coordinate system. Now that we know how to write the stress-velocity law in any coordinate system (see Equations 5.90–5.92), we can proceed with the solution.

The stress-velocity relationship for Newtonian fluids is the Newtonian constitutive equation, which may be written in the Cartesian coordinate system as shown in the following matrix:

$$
\begin{matrix} \text{Newtonian} \\ \text{constitutive} \\ \text{equation} \end{matrix} \quad \tilde{\underline{\underline{\tau}}} = \mu \left(\nabla \underline{v} + (\nabla \underline{v})^T \right) \tag{5.118}
$$

$$
\begin{pmatrix} \tilde{\tau}_{xx} & \tilde{\tau}_{xy} & \tilde{\tau}_{xz} \\ \tilde{\tau}_{yx} & \tilde{\tau}_{yy} & \tilde{\tau}_{yz} \\ \tilde{\tau}_{zx} & \tilde{\tau}_{zy} & \tilde{\tau}_{zz} \end{pmatrix}_{xyz} = \mu \begin{pmatrix} 2\frac{\partial v_x}{\partial x} & \frac{\partial v_y}{\partial x} + \frac{\partial v_x}{\partial y} & \frac{\partial v_z}{\partial x} + \frac{\partial v_x}{\partial z} \\ \frac{\partial v_x}{\partial y} + \frac{\partial v_y}{\partial x} & 2\frac{\partial v_y}{\partial y} & \frac{\partial v_z}{\partial y} + \frac{\partial v_y}{\partial z} \\ \frac{\partial v_x}{\partial z} + \frac{\partial v_z}{\partial x} & \frac{\partial v_y}{\partial z} & 2\frac{\partial v_z}{\partial z} \end{pmatrix}_{xyz} \tag{5.119}
$$

We now simplify Equation 5.119 for the flow down an incline by considering what we know about the velocity vector in the flow. We chose our coordinate system such that the z-direction is the flow direction; thus, for our flow:

$$
\underline{v} = \begin{pmatrix} 0 \\ 0 \\ v_z \end{pmatrix}_{xyz} \tag{5.120}
$$

The components v_x and v_y are both zero. In addition, the mass balance for this problem performed in Chapter 3 gave $dv_z/dz = 0$ (see Equation 3.155), which allows us to eliminate one term on the right side of Equation 5.119. Finally, if we neglect edge effects and consider our flow to be very wide, there is no variation in the velocity in the y-direction, $dv_z/dy = 0$. With these simplifications, Equation 5.119 becomes:

$$
\tilde{\underline{\underline{\tau}}} = \mu \begin{pmatrix} 0 & 0 & \frac{\partial v_z}{\partial x} \\ 0 & 0 & 0 \\ \frac{\partial v_z}{\partial x} & 0 & 0 \end{pmatrix}_{xyz} \tag{5.121}
$$

Finally, because v_z is not a function of x or z, we can change the derivatives in Equation 5.121 from partial derivatives to total derivatives:

$$
\begin{pmatrix} \tilde{\tau}_{xx} & \tilde{\tau}_{xy} & \tilde{\tau}_{xz} \\ \tilde{\tau}_{yx} & \tilde{\tau}_{yy} & \tilde{\tau}_{yz} \\ \tilde{\tau}_{zx} & \tilde{\tau}_{zy} & \tilde{\tau}_{zz} \end{pmatrix}_{xyz} = \begin{pmatrix} 0 & 0 & \mu\frac{dv_z}{dx} \\ 0 & 0 & 0 \\ \mu\frac{dv_z}{dx} & 0 & 0 \end{pmatrix}_{xyz} \tag{5.122}
$$

This is the stress-velocity relationship for this flow problem.

Now that we know the stress-velocity relationship for the flow down an incline, we can solve the momentum-balance equations at which we arrived in Chapter 4,

Equations 4.357–4.359, repeated here:

x-Component, momentum balance on CV

$$0 = -\frac{\partial p}{\partial x} + \left[\frac{\partial \tilde{\tau}_{xx}}{\partial x} + \frac{\partial \tilde{\tau}_{yx}}{\partial y} + \frac{\partial \tilde{\tau}_{zx}}{\partial z}\right] - \rho g \sin \beta \qquad (5.123)$$

y-Component, momentum balance on CV

$$0 = -\frac{\partial p}{\partial y} + \left[\frac{\partial \tilde{\tau}_{xy}}{\partial x} + \frac{\partial \tilde{\tau}_{yy}}{\partial y} + \frac{\partial \tilde{\tau}_{zy}}{\partial z}\right] \qquad (5.124)$$

z-Component, momentum balance on CV

$$0 = -\frac{\partial p}{\partial z} + \left[\frac{\partial \tilde{\tau}_{xz}}{\partial x} + \frac{\partial \tilde{\tau}_{yz}}{\partial y} + \frac{\partial \tilde{\tau}_{zz}}{\partial z}\right] + \rho g \cos \beta \qquad (5.125)$$

These three components of the momentum balance seem complicated but, for the simple flow considered, many of the stress components are equal to zero (Equation 5.122)—for example, $\tilde{\tau}_{yx} = \tilde{\tau}_{xy} = 0$. If we incorporate the information from Equation 5.122 about which terms are zero, the three components of the momentum balance simplify considerably:

$$x\text{-Component} \quad 0 = -\frac{\partial p}{\partial x} + \frac{\partial \tilde{\tau}_{zx}}{\partial z} - \rho g \sin \beta \qquad (5.126)$$

$$y\text{-Component} \quad 0 = -\frac{\partial p}{\partial y} \qquad (5.127)$$

$$z\text{-Component} \quad 0 = -\frac{\partial p}{\partial z} + \frac{\partial \tilde{\tau}_{xz}}{\partial x} + \rho g \cos \beta \qquad (5.128)$$

We simplified the momentum balance by using the zero terms obtained by matching the tensor coefficients in the matrices on the left and right sides of the Newtonian constitutive equation, Equation 5.122. We have one more piece of information from the constitutive equation: We know that the two remaining stress components are equal to one another and are proportional to the velocity gradient with coefficient of proportionality μ, the viscosity:

$$\tilde{\tau}_{xz} = \tilde{\tau}_{zx} = \mu \frac{\partial v_z}{\partial x} \qquad (5.129)$$

Substituting Equation 5.129 into Equations 5.126–5.128, we obtain:

$$x\text{-Component} \quad 0 = -\frac{\partial p}{\partial x} + \mu \frac{\partial}{\partial z}\frac{\partial v_z}{\partial x} - \rho g \sin \beta \qquad (5.130)$$

$$y\text{-Component} \quad 0 = -\frac{\partial p}{\partial y} \qquad (5.131)$$

$$z\text{-Component} \quad 0 = -\frac{\partial p}{\partial z} + \mu \frac{\partial^2 v_z}{\partial x^2} + \rho g \cos \beta \qquad (5.132)$$

The mixed second derivative in Equation 5.130 also may be written as follows:

$$\frac{\partial}{\partial z}\frac{\partial v_z}{\partial x} = \frac{\partial}{\partial x}\frac{\partial v_z}{\partial z} \tag{5.133}$$

Because v_z is not a function of z (recall the mass balance result Equation 3.155, which gave $dv_z/dz = 0$), this derivative is equal to zero. Furthermore, because $\partial v_z/\partial y = 0$ as well (wide flow), the partial x-derivative in Equation 5.132 becomes the total derivative, $\partial^2 v_z/\partial x^2 = d^2 v_z/dx^2$. Equations 5.130–5.132 thus become:

$$x\text{-Component}\quad 0 = -\frac{\partial p}{\partial x} - \rho g \sin\beta \tag{5.134}$$

$$y\text{-Component}\quad 0 = -\frac{\partial p}{\partial y} \tag{5.135}$$

$$z\text{-Component}\quad \boxed{0 = -\frac{\partial p}{\partial z} + \mu\frac{d^2 v_z}{dx^2} + \rho g \cos\beta} \tag{5.136}$$

After all of these simplifications, the only term in the momentum balance that addresses velocity is in Equation 5.136, the z-component of the momentum balance. To proceed, we must solve Equation 5.136 for $v_z(x)$.

We turn now from questions of fluid mechanics to questions of solving differential equations. Equation 5.136 concerns two variables: v_z and pressure. The pressure appears as $\partial p/\partial z$, the pressure gradient in the z-direction. This term is nonzero if the pressure varies in the z-direction. We can be convinced that the pressure does not vary in the z-direction as follows: The flow is open to the atmosphere; thus, at the top surface, pressure does not vary with z-position. The pressure varies with depth x, which we can see from the x-component of the momentum balance. The x-variation is the same at all values of z, however; thus, at all positions, pressure is independent of z. Because p is not a function of z, then $\partial p/\partial z = 0$ everywhere, and we can omit this term from Equation 5.136 and solve the equation by direct x-integration:

$$\text{Deduce from } x\text{- and } y\text{-components}$$
$$\text{of the momentum balance}\qquad \frac{\partial p}{\partial z} = 0 \tag{5.137}$$
$$\text{and boundary conditions:}$$

$$\text{Equation 5.136 becomes:}\quad 0 = \mu\frac{d^2 v_z}{dx^2} + \rho g \cos\beta \tag{5.138}$$

$$\text{Rearrange:}\quad \frac{d^2 v_z}{dx^2} = \frac{-\rho g \cos\beta}{\mu} \tag{5.139}$$

The final velocity profile $v_z(x)$ may be obtained by integrating Equation 5.139 twice, as we now show in detail.

We define ψ to be the first derivative of the velocity relative to x and substitute ψ into Equation 5.139:

$$\psi \equiv \frac{dv_z}{dx} \tag{5.140}$$

$$\frac{d^2 v_z}{dx^2} = \frac{d\psi}{dx} = \frac{-\rho g \cos \beta}{\mu} \tag{5.141}$$

We solve this equation for ψ. Integrating once, we obtain:

$$\int d\psi = \int \left[\frac{-\rho g \cos \beta}{\mu} \right] dx \tag{5.142}$$

$$\psi = \left[\frac{-\rho g \cos \beta}{\mu} \right] x + C_1 \tag{5.143}$$

where C_1 is an arbitrary integration constant that must be determined by boundary conditions. Substituting $\frac{dv_z}{dx} \equiv \psi$ and integrating again:

$$\psi = \frac{dv_z}{dx} = \left[\frac{-\rho g \cos \beta}{\mu} \right] x + C_1 \tag{5.144}$$

$$\int dv_z = \int \left(\left[\frac{-\rho g \cos \beta}{\mu} \right] x + C_1 \right) dx \tag{5.145}$$

$$\boxed{v_z = \left[\frac{-\rho g \cos \beta}{\mu} \right] \frac{x^2}{2} + C_1 x + C_2} \tag{5.146}$$

This is the solution for $v_z(x)$ in terms of the two unknown integration constants, C_1 and C_2.

We evaluate the integration constants C_1 and C_2 by using two boundary conditions on velocity; that is, we need to know something about the velocity at two locations to evaluate C_1 and C_2. We do not know the velocity at the top surface in this problem but, at the bottom, we know that the velocity is equal to zero because the fluid sticks to the stationary wall. Substituting this information into Equation 5.146, we find the value of C_2:

$$\text{Boundary condition (BC):} \quad x = 0 \quad v_z = 0 \tag{5.147}$$

$$\Rightarrow \quad \boxed{C_2 = 0} \tag{5.148}$$

We need a second boundary condition on velocity, which is less obvious. The other boundary of the flow down the incline is the free surface; that is, the top surface where the fluid is in contact with air. At that interface, little force is transferred between the fluid and the air, which leads to the second boundary condition. At the top surface ($x = H$), the shear stress in the fluid is approximately zero (see Chapter 6 for more details):

$$\text{Boundary condition (BC):} \quad x = H \quad \tilde{\tau}_{xz} = 0 \tag{5.149}$$

From the constitutive equation, we can relate this fact about stress to the velocity derivative at the free surface:

$$\text{Newton's law of viscosity: } \tilde{\tau}_{xz} = \mu \frac{dv_z}{dx} \tag{5.150}$$

$$\tilde{\tau}_{xz}|_{x=H} = 0 \;\Rightarrow\; \left.\frac{dv_z}{dx}\right|_{x=H} = 0 \tag{5.151}$$

$$\text{Equivalent boundary condition (BC): } x = H \quad \frac{dv_z}{dx} = 0 \tag{5.152}$$

We substitute this relationship into Equation 5.143 to calculate the second integration constant C_1:

$$\text{(Equation 5.140)} \quad \frac{dv_z}{dx} = \frac{-\rho g \cos \beta}{\mu} x + C_1 \tag{5.153}$$

$$\Rightarrow \quad \boxed{C_1 = \frac{\rho g H \cos \beta}{\mu}} \tag{5.154}$$

Substituting the results from the boundary-condition calculations into Equation 5.146, we obtain the final result for the velocity profile:

$$\text{Final answer:} \quad \boxed{v_z(x) = \frac{\rho g \cos \beta}{\mu} \left[Hx - \frac{x^2}{2} \right]} \tag{5.155}$$

The result in Equation 5.155 is the information we have been seeking throughout our long consideration of this problem: the distribution of velocity with position for the steady flow of a fluid down an inclined plate. The pattern that the fluid adopts in this flow—compactly and quantitatively represented now by Equation 5.155—results from the influences of both gravity and the molecular exchange of momentum among the layers of fluid that slide down the incline. To arrive at the solution, we imposed mass and momentum conservation on a microscopic control volume and accounted for velocity-dependent molecular effects through the use of the Newtonian constitutive equation. Our ally throughout this analysis was calculus, which allowed us to account for the various local effects through derivatives, integrals, differential equations, and boundary conditions.

To make the velocity profile easier to plot, we rearrange it to be written in terms of the dimensionless variable x/H, the relative distance through the film thickness:

$$\boxed{v_z(x) = \frac{\rho g H^2 \cos \beta}{2\mu} \left[2\left(\frac{x}{H}\right) - \left(\frac{x^2}{H^2}\right) \right]} \tag{5.156}$$

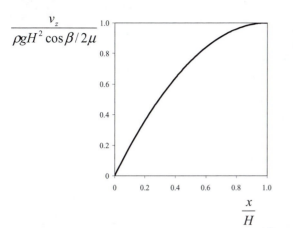

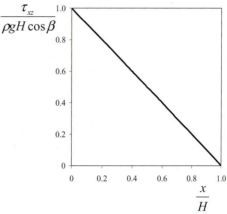

Figure 5.19 The results of microscopic balances in the falling-film example are the velocity and shear-stress profiles.

$$\underline{v} = \begin{pmatrix} 0 \\ 0 \\ \frac{\rho g H^2 \cos \beta}{2\mu} \left[2\left(\frac{x}{H}\right) - \left(\frac{x}{H}\right)^2 \right] \end{pmatrix}_{xyz}$$ (5.157)

Now that we know the velocity profile for the flow, we can evaluate the stress from Equation 5.121:

$$\underline{\underline{\tilde{\tau}}} = \mu \begin{pmatrix} 0 & 0 & \frac{\partial v_z}{\partial x} \\ 0 & 0 & 0 \\ \frac{\partial v_z}{\partial x} & 0 & 0 \end{pmatrix}_{xyz}$$ (5.158)

$$\tilde{\tau}_{xz} = \mu \frac{dv_z}{dx}$$ (5.159)

$$\tilde{\tau}_{xz} = \rho g H \cos \beta \left(1 - \frac{x}{H} \right)$$ (5.160)

$$\underline{\underline{\tilde{\tau}}} = \begin{pmatrix} 0 & 0 & \rho g H \cos \beta \left(1 - \frac{x}{H}\right) \\ 0 & 0 & 0 \\ \rho g H \cos \beta \left(1 - \frac{x}{H}\right) & 0 & 0 \end{pmatrix}_{xyz}$$ (5.161)

Note that unlike in drag shear flow between parallel plates, in flow down an incline, the shear stress is not constant but rather varies linearly with position x or dimensionless position x/H. The velocity and shear-stress profile results for the incline problem are shown in Figures 5.19 and 5.20 (compare with shear-flow

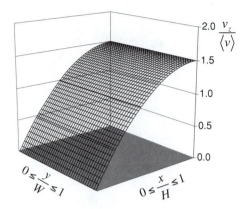

Figure 5.20 The solution for velocity profile for steady flow down an inclined plane is shown in a three-dimensional view. The velocity at the free surface is 1.5 times the average velocity of this flow. The data are plotted versus normalized coordinates x/H and y/W, which both range from 0 to 1.

results in Figures 5.5 and 5.6). The shear stress is a linear function of x/H, and the velocity profile is a parabola in the variable x/H with a maximum velocity at the free surface.

With completion of the falling-film example, we have fully demonstrated the continuum model and how it may be used to calculate velocity and stress profiles in flows. The solution began with the microscopic momentum balance on a control volume, the Reynolds transport theorem. We wrote expressions for the various forces and momentum contributions. The last step of the procedure was to quantify the relationship between molecular stress and velocity, which is the information in the Newtonian constitutive equation. In Chapter 6, we expand the method to more complex problems and standardize and simplify our solution technique. In that chapter, we also discuss additional boundary conditions used in flow problems. For complex two- and three-dimensional flows, numerical methods are used to solve the momentum-balance equations for $\underline{v}(x, y, z)$ and $p(x, y, z)$.

In Chapter 3, we started a second example in which we performed balances on a much larger control volume. In that problem, we sought the force due to flow on a 90-degree pipe bend, but we were unable to complete that problem because we lacked information on how stresses and velocity are related. Now that we have the missing information, we return to that macroscopic-balance problem and complete the calculation. Macroscopic-CV problems require the use of average or integrated properties rather than the microscopic properties used in the inclined-flow case.

EXAMPLE 5.9 (90-Degree bend: concluded). *What is the direction and magnitude of the force needed to support the 90-degree pipe bend shown in Figure 5.21 (continued from Chapters 3 and 4)? An incompressible liquid enters the pipe at volumetric flow rate Q_a and exits at volumetric flow rate Q_b. The flow is steady. The cross-sectional area of the pipe bend is πR^2.*

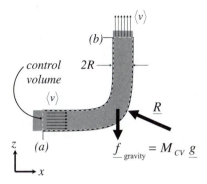

Figure 5.21

In Chapter 3, we carried out the momentum balance on a macroscopic control volume outlined here with a dotted line. In Chapter 4, we specified the molecular stresses in this problem in terms of the stress components $\tilde{\tau}_{ij}$. In this chapter, we use the constitutive equation to relate stress to velocity and complete the problem.

SOLUTION. In this problem, we used a CV that included all of the fluid in the 90-degree bend (see dotted line in Figure 5.21), and we applied and simplified the Reynolds transport theorem:

Reynolds transport theorem (momentum balance on CV)

$$\frac{d\mathbf{P}}{dt} + \iint_{CS} (\hat{n} \cdot \underline{v})\, \rho \underline{v}\, dS = \sum_{\substack{\text{on} \\ \text{CV}}} \underline{f} \tag{5.162}$$

The flow is steady; thus, $d\mathbf{P}/dt = 0$. The average velocity was used to quantify the convection of momentum into and out of the CV through surfaces (a) and (b). The average velocity through the CV was found to be constant throughout and is designated $\langle v \rangle$. The forces on the CV were identified as gravity and three molecular contact forces: the restoring force $\mathcal{R}$ for which we are solving, and the molecular forces on surfaces (a) and (b)—which we wrote in terms of integrals involving the extra-stress tensor. The momentum balance incorporating these observations was found to be as follows (see Equation 4.376):

$$\frac{d\mathbf{P}}{dt} = - \iint_S (\hat{n} \cdot \underline{v})\, \rho \underline{v}\, dS + \sum_{\substack{\text{on} \\ \text{CV}}} \underline{f} \tag{5.163}$$

$$= - \iint_S (\hat{n} \cdot \underline{v})\, \rho \underline{v}\, dS + \underline{f}_{gravity} + \underline{\mathcal{R}} + \underline{\mathcal{F}}|_a + \underline{\mathcal{F}}|_b \tag{5.164}$$

$$0 = \langle v \rangle^2 \rho \pi R^2 \begin{pmatrix} 1 \\ 0 \\ -1 \end{pmatrix}_{xyz} + M_{CV} \begin{pmatrix} 0 \\ 0 \\ -g \end{pmatrix}_{xyz} + \begin{pmatrix} \mathcal{R}_x \\ \mathcal{R}_y \\ \mathcal{R}_z \end{pmatrix}_{xyz}$$

$$+ \pi R^2 \left\langle \underline{f}_\mu \big|_a \right\rangle + \pi R^2 \begin{pmatrix} p|_a \\ 0 \\ 0 \end{pmatrix}_{xyz} + \pi R^2 \left\langle \underline{f}_\mu \big|_b \right\rangle - \pi R^2 \begin{pmatrix} 0 \\ 0 \\ p|_b \end{pmatrix}_{xyz} \tag{5.165}$$

where:

$$\left\langle \underline{f}_\mu \Big|_a \right\rangle = \frac{1}{S_a} \iint_a \left(\hat{n} \cdot \underline{\underline{\tau}} \right)\Big|_a dS \tag{5.166}$$

$$\left\langle \underline{f}_\mu \Big|_b \right\rangle = \frac{1}{S_b} \iint_b \left(\hat{n} \cdot \underline{\underline{\tau}} \right)\Big|_b dS \tag{5.167}$$

S_a and S_b are the cross-sectional areas at surfaces a and b. Now that we know the stress-constitutive equation, we can evaluate these integrals and conclude this problem.

As with the pressure integrals on this problem in Example 4.23, we carry out the integral in Equation 5.166 in a cylindrical coordinate system $r\theta\bar{z}$ with $\hat{e}_{\bar{z}} = \hat{e}_x$. The outwardly pointing unit normal at (a) therefore is $\hat{n} = -\hat{e}_x = -\hat{e}_{\bar{z}}$.

$$\left\langle \underline{f}_\mu \Big|_a \right\rangle = \frac{1}{S_a} \iint_a \left(\hat{n} \cdot \underline{\underline{\tau}} \right)\Big|_a dS \tag{5.168}$$

$$= \frac{1}{\pi R^2} \int_0^{2\pi} \int_0^R \left(-\hat{e}_{\bar{z}} \cdot \underline{\underline{\tau}}\Big|_a \right) r\, dr\, d\theta \tag{5.169}$$

For the (b) surface, the calculation is similar. The outwardly pointing unit normal at (b) is $\hat{n} = \hat{e}_z = \hat{e}_{\bar{z}}$:

$$\left\langle \underline{f}_\mu \Big|_b \right\rangle = \frac{1}{S_b} \iint_b \left(\hat{n} \cdot \underline{\underline{\tau}} \right)\Big|_b dS \tag{5.170}$$

$$= \frac{1}{\pi R^2} \int_0^{2\pi} \int_0^R \left(\hat{e}_{\bar{z}} \cdot \underline{\underline{\tau}}\Big|_b \right) r\, dr\, d\theta \tag{5.171}$$

To carry out these dot products, we need $\underline{\underline{\tau}}$ in the $r\theta\bar{z}$ cylindrical coordinate system (see Equation B.8-2):

$$\underline{\underline{\tau}} = \mu \begin{pmatrix} 2\frac{\partial v_r}{\partial r} & r\frac{\partial}{\partial r}\left(\frac{v_\theta}{r}\right) + \frac{1}{r}\frac{\partial v_r}{\partial \theta} & \frac{\partial v_r}{\partial z} + \frac{\partial v_z}{\partial r} \\ r\frac{\partial}{\partial r}\left(\frac{v_\theta}{r}\right) + \frac{1}{r}\frac{\partial v_r}{\partial \theta} & 2\left(\frac{1}{r}\frac{\partial v_\theta}{\partial \theta} + \frac{v_r}{r}\right) & \frac{1}{r}\frac{\partial v_z}{\partial \theta} + \frac{\partial v_\theta}{\partial z} \\ \frac{\partial v_r}{\partial z} + \frac{\partial v_z}{\partial r} & \frac{1}{r}\frac{\partial v_z}{\partial \theta} + \frac{\partial v_\theta}{\partial z} & 2\frac{\partial v_z}{\partial z} \end{pmatrix}_{r\theta\bar{z}} \tag{5.172}$$

$$\left(\hat{e}_{\bar{z}} \cdot \underline{\underline{\tau}} \right) = \mu \begin{pmatrix} \frac{\partial v_r}{\partial z} + \frac{\partial v_z}{\partial r} \\ \frac{1}{r}\frac{\partial v_z}{\partial \theta} + \frac{\partial v_\theta}{\partial z} \\ 2\frac{\partial v_z}{\partial z} \end{pmatrix}_{r\theta\bar{z}} \tag{5.173}$$

At both (a) and (b), we assume the velocity to be perpendicular to the surface, that is, in the $\bar{z}$-direction, with a constant speed across the cross section. This means that across the surfaces, $v_r = v_\theta = 0$ and $v_{\bar{z}}$ is independent of r and θ. Equation 5.173 thus simplifies to:

$$\left(\hat{e}_{\bar{z}} \cdot \underline{\underline{\tau}} \right) = \begin{pmatrix} 0 \\ 0 \\ 2\mu\frac{\partial v_{\bar{z}}}{\partial z} \end{pmatrix}_{r\theta\bar{z}} \tag{5.174}$$

This viscous-force contribution is not zero because the velocity $v_{\bar{z}}$ changes in the $\bar{z}$ direction. The contribution is small, however, and usually is not mentioned in discussion of flow through a bend. We can be rigorous in carrying along this calculation, however, and the derivation of the final result follows (the final result neglecting this term is in Equation 5.185).

The integrals in Equations 5.169 and 5.171 easy to carry out because the expressions they contain are constant with respect to r and θ. We can estimate the derivative $\partial v_{\bar{z}}/\partial \bar{z}$ at each surface by noting that at (a), the average $\bar{z}$-velocity goes from $\langle v \rangle$ to zero over the length of the bend L, whereas at (b), the average $\bar{z}$-velocity goes from zero to $\langle v \rangle$ over the same length:

$$\pi R^2 \left\langle \underline{f}_\mu \Big|_a \right\rangle = \int_0^{2\pi} \int_0^R \left(-\hat{e}_{\bar{z}} \cdot \underline{\tilde{\tau}} \Big|_a \right) r\, dr\, d\theta \tag{5.175}$$

$$= \int_0^{2\pi} \int_0^R -2\mu \left(\frac{\partial v_{\bar{z}}}{\partial \bar{z}} \right) \Big|_a \hat{e}_{\bar{z}}\, r\, dr\, d\theta \tag{5.176}$$

$$= \int_0^{2\pi} \int_0^R -2\mu \left(\frac{-\langle v \rangle}{L} \right) \hat{e}_{\bar{z}}\, r\, dr\, d\theta \tag{5.177}$$

$$= \frac{2\pi R^2 \mu \langle v \rangle}{L} \hat{e}_x \tag{5.178}$$

$$\pi R^2 \left\langle \underline{f}_\mu \Big|_b \right\rangle = \int_0^{2\pi} \int_0^R \left(\hat{e}_{\bar{z}} \cdot \underline{\tilde{\tau}} \Big|_b \right) r\, dr\, d\theta \tag{5.179}$$

$$= \int_0^{2\pi} \int_0^R 2\mu \left(\frac{\partial v_{\bar{z}}}{\partial \bar{z}} \right) \Big|_b \hat{e}_{\bar{z}}\, r\, dr\, d\theta \tag{5.180}$$

$$= \int_0^{2\pi} \int_0^R 2\mu \left(\frac{\langle v \rangle}{L} \right) \hat{e}_{\bar{z}}\, r\, dr\, d\theta \tag{5.181}$$

$$= \frac{2\pi R^2 \mu \langle v \rangle}{L} \hat{e}_z \tag{5.182}$$

We reverted to the overall Cartesian coordinate system in these final answers; at S_a, $\hat{e}_{\bar{z}} = \hat{e}_x$, whereas at S_b, $\hat{e}_{\bar{z}} = \hat{e}_z$. The momentum balance becomes:

$$0 = \langle v \rangle^2 \rho \pi R^2 \begin{pmatrix} 1 \\ 0 \\ -1 \end{pmatrix}_{xyz} + M_{CV} \begin{pmatrix} 0 \\ 0 \\ -g \end{pmatrix}_{xyz} + \begin{pmatrix} \mathcal{R}_x \\ \mathcal{R}_y \\ \mathcal{R}_z \end{pmatrix}_{xyz}$$

$$+ \pi R^2 \begin{pmatrix} 2\mu \frac{\langle v \rangle}{L} \\ 0 \\ 0 \end{pmatrix}_{xyz} + \pi R^2 \begin{pmatrix} 0 \\ 0 \\ 2\mu \frac{\langle v \rangle}{L} \end{pmatrix}_{xyz} + \pi R^2 \begin{pmatrix} p|_a \\ 0 \\ -p|_b \end{pmatrix}_{xyz} \tag{5.183}$$

In Example 5.11, we evaluate the relative magnitude of the two viscosity expressions in Equation 5.183 and find that the viscous terms make a negligible contribution. Neglecting these terms, the final result for restraining force vector, $\underline{R}$, is given here:

$$
\begin{pmatrix} R_x \\ R_y \\ R_z \end{pmatrix}_{xyz} = \langle v \rangle^2 \rho \pi R^2 \begin{pmatrix} -1 \\ 0 \\ 1 \end{pmatrix}_{xyz} + M_{CV} \begin{pmatrix} 0 \\ 0 \\ g \end{pmatrix}_{xyz} + \pi R^2 \begin{pmatrix} -p|_a \\ 0 \\ p|_b \end{pmatrix}_{xyz}
$$

(5.184)

$$
\underline{R} = \begin{pmatrix} -\pi R^2 \left(\rho \langle v \rangle^2 + p|_a \right) \\ 0 \\ \pi R^2 \left(\rho \langle v \rangle^2 + p|_b \right) + M_{CV} g \end{pmatrix}_{xyz}
$$

(5.185)

This completes the 90-degree bend problem. The change of direction of the flow has a profound effect on the direction of the resulting force.

The completion of these two examples signals that we have finished the development of the continuum model for flow. The essential physics of flow is momentum conservation, captured for a control volume in the Reynolds transport theorem. In the process, we developed the continuum approach, the control volume, and the molecular-stress constitutive equation to allow momentum conservation to be applied to fluids in motion. The methodology used in these two examples is general; in Chapter 6, we apply this method to an arbitrary control volume and derive the general microscopic-momentum-balance equation. Further refinements of the method in subsequent chapters lead to techniques that allow us to apply the continuum method to important flows. In Chapter 9, we develop a general macroscopic-momentum-balance equation that is useful for problems in which forces are sought and flow details are less important.

The purpose of modeling is to render mathematically the behavior of a system of interest. Once a model is complete, we can use it to calculate quantities of interest such as flow rates, forces, and other related properties. We demonstrate calculations of this type in the next two examples. In the last section of this chapter, we introduce non-Newtonian constitutive modeling.

EXAMPLE 5.10. *What are the flow rate and average velocity in the falling-film example (see Figure 5.12)? What is the effect of the incline angle β on the flow rate and average velocity achieved?*

SOLUTION. We discuss volumetric flow through a surface in Chapter 3. We show that for flow through a surface that is not necessarily oriented perpendicular to the flow, the volumetric flow rate is given by:

$$
\begin{array}{c} \text{Volumetric flow} \\ \text{of liquid through } A \\ \text{(general-orientation case;} \\ \underline{v} \text{ does not vary across } A) \end{array} \qquad Q = (\hat{n} \cdot \underline{v})A
$$

(5.186)

where Q is the volumetric flow rate through the small flat area A, $\hat{n}$ is the unit normal to A, and the velocity is given by $\underline{v}$. For an area that is finite in size and not necessarily flat and across which $\underline{v}$ may vary, we can generalize Equation 5.186

to the following (see Section 6.2.3.3):

$$\begin{array}{c} \text{Total flow rate} \\ \text{out through finite} \\ \text{surface } \mathcal{S} \end{array} \quad Q = \iint_{\mathcal{S}} [\hat{n} \cdot \underline{v}]_{\text{at surface}} \, dS \qquad (5.187)$$

To apply Equation 5.187 to the flow down an incline, we must identify $\hat{n}$, $\underline{v}$, and the surface over which we want to integrate. The flow rate in the flow-down-an-incline problem is the same at every z-position throughout the flow; therefore, we can choose as our calculation surface any plane perpendicular to the flow—we choose the cross section at the exit, $z = L$. The unit normal of our calculation surface is $\hat{n} = \hat{e}_z$, and the velocity vector is given in Equation 5.120 as $\underline{v} = v_z \hat{e}_z$. The dot product of these two vectors is $\hat{n} \cdot \underline{v} = v_z$:

$$\hat{n} \cdot \underline{v} = \begin{pmatrix} 0 \\ 0 \\ 1 \end{pmatrix}_{xyz} \cdot \begin{pmatrix} 0 \\ 0 \\ v_z \end{pmatrix}_{xyz} = v_z \qquad (5.188)$$

We solved for v_z in a previous example (see Equation 5.155); the result is given here:

$$v_z|_{z=L} = v_z = \frac{\rho g \cos \beta}{2\mu} \left(2Hx - x^2 \right) \qquad (5.189)$$

The surface $\mathcal{S}$ is a rectangle in the xy-plane; thus, $dS = dxdy$. The flow rate Q then is given by:

$$Q = \iint_{\mathcal{S}} [\hat{n} \cdot \underline{v}]_{\text{at surface}} \, dS \qquad (5.190)$$

$$= \int_0^W \int_0^H v_z|_{z=L} \, dxdy \qquad (5.191)$$

$$= \int_0^W \int_0^H \frac{\rho g \cos \beta}{2\mu} \left(2Hx - x^2 \right) \, dxdy \qquad (5.192)$$

$$= W \int_0^H \frac{\rho g \cos \beta}{2\mu} \left(2Hx - x^2 \right) \, dx \qquad (5.193)$$

The final integration is left to readers (see Problem 6 in Chapter 6). The average velocity is calculated from the flow-rate result:

$$\begin{array}{c} \text{Average velocity} \\ \text{out through} \\ \text{surface } \mathcal{S} \end{array} \quad \langle v \rangle = \frac{\iint_{\mathcal{S}} [\hat{n} \cdot \underline{v}]_{\text{at surface}} \, dS}{\iint_{\mathcal{S}} dS} \qquad (5.194)$$

$$= \frac{Q}{\int_0^W \int_0^H dxdy} \qquad (5.195)$$

$$= \frac{Q}{HW} \qquad (5.196)$$

The effect of β on the average velocity achieved is controlled by the $\cos \beta$ term in the final answer.

EXAMPLE 5.11. *What are the relative magnitudes of the various terms in the calculated force on a right-angle bend? Which forces may be neglected and under what circumstances?*

SOLUTION. In Example 5.9, we completed the solution for the force on a right-angle bend (see Figure 5.21). The result is repeated here, including the two terms accounting for viscous effects on the entry and exit surfaces of the control volume:

Force on a right-angle bend
$$\underline{\mathcal{R}} = \begin{pmatrix} -\rho \langle v \rangle^2 \pi R^2 - p|_a \pi R^2 - 2\mu \frac{\langle v \rangle}{L} \pi R^2 \\ 0 \\ \rho \langle v \rangle^2 \pi R^2 + p|_b \pi R^2 - 2\mu \frac{\langle v \rangle}{L} \pi R^2 + M_{CV} g \end{pmatrix}_{xyz}$$

(5.197)

To compare these terms in an actual situation, we choose to look at a particular right-angle bend. We choose a bend in 1-1/2-inch, Schedule 40 (cross-sectional area $A = \pi R^2 = 2.04$ in) steel pipe. We choose that the length of each arm of the bend is about 6 in $= 0.5$ ft. Water (density $= 62.25$ lb$_m$/ft^3, viscosity $= 6.005 \times 10^{-4}$ lb$_m$/(ft s)) is flowing in the pipe at 3.0 gpm.

There are four terms in the final result for force: the convective term, the pressure term, the viscous entry/exit term, and the gravity term. The convective and viscous terms are small:

$$\rho \langle v \rangle^2 A = \left(\frac{62.25 \text{ lb}_m}{\text{ft}^3} \right) \left(\frac{0.4718 \text{ ft}}{\text{s}} \right)^2 \left(\frac{2.04 \text{ in.}^2}{144 \text{ in.}^2 /\text{ft}^2} \right) \left(\frac{\text{s}^2 \text{ lb}_f}{32.174 \text{ ft lb}_m} \right)$$

$$= 6 \times 10^{-3} \text{ lb}_f = 30 \text{ mN}$$

$$2\mu \frac{\langle v \rangle}{L} A = (2) \left(\frac{6.005 \times 10^{-4} \text{ lb}_m}{\text{ft s}} \right) \left(\frac{0.4718 \text{ ft /s}}{0.5 \text{ ft}} \right) \left(\frac{2.04 \text{ in.}^2}{144 \text{ in.}^2 /\text{ft}^2} \right) \left(\frac{\text{s}^2 \text{ lb}_f}{32.174 \text{ ft lb}_m} \right)$$

$$= 5 \times 10^{-7} \text{ lb}_f = 2 \ \mu N$$

The viscous effect is negligible compared to the convective term. The impact of gravity also is modest:

$$M_{CV} g \approx \rho A (2L) g$$

$$= \left(\frac{62.25 \text{ lb}_m}{\text{ft}^3} \right) \left(\frac{2.04 \text{ in.}^2}{144 \text{ in.}^2 /\text{ft}^2} \right) (2)(0.5 \text{ ft}) \left(\frac{32.174 \text{ ft /s}^2}{32.174 \text{ ft lb}_m /\text{s}^2 \text{ lb}_f} \right)$$

$$= 0.9 \text{ lb}_f = 4 \ N$$

The dominant terms are the pressure terms. The pressures on the inlet and the outlet of the bend depend on how the bend is installed. If we imagine the pressure as due to a constant-head tank upstream of the bend (see Example 1.4), then the gauge pressure at the inlet of the bend is given by $\rho g h$. A reasonable number for the upstream head is 33.9 atm = 1 ft = 14.6 psi. For these conditions, we calculate the force due to pressure as:

$$p|_a A = \rho g h A = \left(\frac{14.696 \text{ lb}_f}{\text{in}^2} \right) \left(2.04 \text{ in}^2 \right)$$

$$= 30 \text{ lb}_f = 130 \text{ N}$$

The pressure at the outlet of the bend is slightly less due to the frictional losses within the bend, and we can calculate the pressure drop across the bend from the K_f value for a 90-degree bend (using the mechanical energy balance; see Chapter 1). This pressure drop is negligible; thus, $p|_b \approx p|_a$. Using our hypothetical installation and operation of the bend, we calculate:

$$\underline{\mathcal{R}} = \begin{pmatrix} -30 \text{ lb}_f \\ 0 \\ 31 \text{ lb}_f \end{pmatrix}_{xyz} = \begin{pmatrix} -130 \text{ } N \\ 0 \\ 140 \text{ } N \end{pmatrix}_{xyz} \tag{5.198}$$

and the only contributions that have significance in the final answer are the pressure and gravity terms. Thus, for the conditions cited, the convective and viscous terms are negligible. The numbers indicate that the viscous effect likely always will be negligible; the convective term is four orders of magnitude larger than the viscous entry/exit term and is proportional to $\langle v \rangle^2$. If the flow rate were much higher and if the pressure terms were lower, then the convective term might be a factor.

In Section 5.3, we discuss constitutive equations for fluids that do not follow the Newtonian equation. This is advanced material, but the subject is not esoteric. Rather, many common and important materials are non-Newtonian, including most foods, molten plastics, pastes, suspensions, and biological fluids. The tensor approach to stress is essential in non-Newtonian fluid mechanics. First-time readers may want to proceed to Chapter 6 and return to the discussion of non-Newtonian fluids once the techniques of Newtonian fluid mechanics are familiar.

5.3 Non-Newtonian fluids

As discussed previously in this chapter, experiments on the parallel-plate apparatus confirm the validity of Newton's law of viscosity and the Newtonian constitutive equation for many materials, including water, oil, honey, milk, and solvents. Many materials, however, do not follow the Newtonian constitutive equation. Fluids that are compressible, for example, have an additional contribution to stress other than what we described; Chapter 10 presents modifications to the Newtonian constitutive equation that account for compressibility.

Many incompressible fluids, including foods such as mayonnaise, peanut butter, and ketchup, do not follow the Newtonian constitutive equation. Many industrial materials, including molten plastics, asphalt, and concrete, are non-Newtonian. Biological fluids, including blood and mucus, are almost universally non-Newtonian. In addition, the stresses in geological flows, such as those involving soil and lava, fail to follow the Newtonian constitutive equation.

The field of study that addresses the many effects seen in non-Newtonian fluids is called *rheology*, and there is considerable literature on the subject [8, 12, 13, 90, 37, 92, 104, 164]. In this section, we introduce the basic nature of non-Newtonian flows and constitutive models for non-Newtonian fluids; in-depth information on non-Newtonian fluids is in the cited literature.

Although there is a single Newtonian constitutive equation, for non-Newtonian fluids, we have many different constitutive equations:

$$\begin{matrix} \text{Newtonian} \\ \text{constitutive equation:} \end{matrix} \qquad \tilde{\underline{\underline{\tau}}} = \mu \left(\nabla \underline{v} + (\nabla \underline{v})^T \right) \qquad (5.199)$$

$$\begin{matrix} \text{Non-Newtonian} \\ \text{constitutive equation:} \end{matrix} \qquad \tilde{\underline{\underline{\tau}}} = \text{unknown function, } f(\underline{v}) \qquad (5.200)$$

We need a variety of constitutive equations for non-Newtonian fluids because of what the constitutive equation is. Recall that the stress tensor $\tilde{\underline{\underline{\tau}}}$ accounts for the molecular-force contributions to the momentum balance. Molecular contact forces are different in every fluid because they arise from chemical interactions, and the atoms and molecules are different in every substance (see Figure 4.3). We take the continuum approach in our modeling of fluid motion but, ultimately, we must match our models to the actual chemical behavior of the fluid systems under study. The constitutive equation is the link between the continuum model and the chemical properties of the molecules that comprise the fluid.

There could be as many constitutive models as there are chemicals, but it turns out to be less complicated. For thousands of fluids, including water, oil, and even gases under most circumstances, the Newtonian constitutive equation is the stress-velocity relationship. For many materials that do not follow the Newtonian constitutive equation, simple modifications often are adequate, as discussed in Section 5.3.3. Other materials, such as polymer melts and solutions, require complex viscoelastic constitutive equations; a detailed study of such equations is beyond the scope of this book. We outline issues related to viscoelastic constitutive equations in Section 5.3.4. The discussion of non-Newtonian fluids begins with two sections in which the material functions that are used to describe non-Newtonian behavior are introduced.

5.3.1 Non-Newtonian viscosity

The description *non-Newtonian* indicates that a fluid does not follow the Newtonian constitutive equation. There are many ways to *not* follow this equation, as we now discuss.

One type of non-Newtonian behavior observed in the parallel-plate apparatus is a nonlinear relationship between F/LW and V/H in steady-drag flow (Figure 5.22). An upward curving shear-stress/velocity-gradient relationship indicates that the fluid deformed at high speeds generates more stress than expected from the Newtonian relationship. This type of behavior is known as shear-thickening. Few systems shear-thicken, but one well-known shear-thickening fluid is a concentrated suspension of cornstarch and water. This fluid is sometimes used in

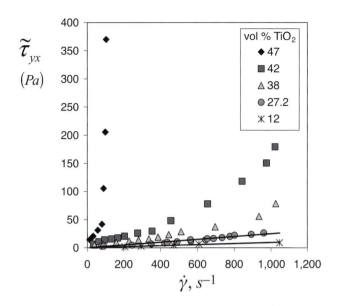

Figure 5.22   In shear flow, Newtonian fluids exhibit a linear relationship between F/LW and V/H. For the suspensions of TiO_2 in water shown here, the lowest concentrations are approximately Newtonian. Some fluids show nonlinear behavior in shear flow, such as the higher-concentration suspensions. The data are recalculated and replotted from Metzner and Whitlock [97].

science demonstrations because the effect is very striking [130]. When slowly stirring a cornstarch/water suspension, the effort required is about the same as for stirring water. When rapidly stirring the same fluid, however, the fluid develops an internal structure that thickens the fluid; this thicker fluid has a higher instantaneous viscosity and resists the stirring. When stirring ceases, the consistency of the fluid returns to its initial low value. A fun shear-thickening experiment is to run across a pool of cornstarch/water suspension (Figure 5.23). The high rate of deformation involved in running makes the viscosity increase rapidly to a very high value, and the suspension supports the runner. Stopping and standing still on the mixture reveals the fluid to be a low-viscosity liquid; under these conditions, the runner sinks [107].

Figure 5.23   Cornstarch/water suspensions are shear-thickening. If a person walks quickly across a bath of this suspension, it supports him and, in fact, provides adequate traction for the motion; it is not slippery. If he stops walking, however, he sinks into the bath and it is difficult to extricate him.

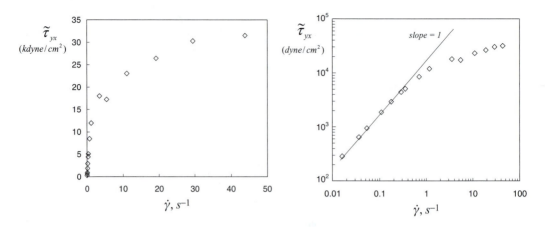

Figure 5.24 Steady shear stress as a function of shear rate plotted on a linear scale (left) and on a log-log scale (right) for a concentrated solution of a narrowly distributed polybutadiene (recalculated from Menezes and Graessley [96]). $M_w = 350$ kg/mol, $M_w/M_n < 1.05$, and concentration is 0.0676 g/cm³ in Flexon 391, a hydrocarbon oil. Note that a linear relationship between steady shear stress and shear rate is reflected in a line of slope 1 when the data are plotted log-log.

Shear-thickening is found in suspensions due to particle interactions. In the corn-starch/water suspension, solid starch granules are touching one another and water fills the spaces between the particles. When such a system is disturbed by slowly stirring it, the water fluidizes the solid particles and the mixture moves easily. When it is disturbed rapidly, however, the particles jam against one another, forming larger structures called hydroclusters [171] that strongly resist the deformation. Thus, such fluids exhibit low viscosities at low rates-of-deformation and high and increasing viscosity at high rates-of-deformation.

A downward-curving stress/velocity-gradient relationship in the steady-drag experiment (Figure 5.24) indicates that compared to a Newtonian fluid, the fluid generates less stress than expected for a given speed. This type of behavior is known as shear-thinning, which is very common. Polymer melts in particular often are shear-thinning, which may be caused in part by disentanglement of long-chain polymers (Figure 5.25). Whereas in shear-thickening a structure develops that jams the flow during high-rate flow making movement more difficult,

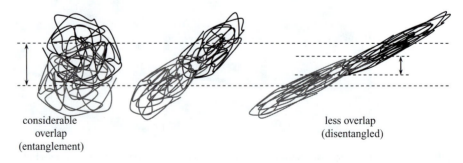

considerable   less overlap
overlap   (disentangled)
(entanglement)

Figure 5.25 Disentanglement of long-chain molecules contributes to shear-thinning. When the molecules are entangled, the viscosity is high. When the flow disentangles the chains, they flow separately and generate less stress. When the flow stops or slows, the chains reentangle due to the Brownian motion of their segments.

in shear-thinning materials, structure breaks down during high-rate flow making movement easier. Whether a fluid shear-thickens or shear-thins depends on the material's molecular structure and the details of intermolecular and interparticulate forces.

To quantify shear-thickening and shear-thinning, we define a non-Newtonian viscosity. To choose our definition, recall that for Newtonian fluids in steady-drag flow, Newton's law of viscosity states that the viscosity is the slope of the shear-stress versus velocity-gradient line (see Equation 5.61):

$$
\begin{array}{cc}
\text{Newtonian viscosity} \\
\text{(constant)}
\end{array}
\qquad
\mu = \frac{\tilde{\tau}_{yx}}{\left(\dfrac{dv_x}{dy}\right)} = \text{constant}
\qquad (5.201)
$$

where we chose the x-direction as the flow direction and the y-direction as the velocity-gradient direction. For fluids in which the graph of shear-stress versus velocity-gradient in steady-drag flow is not a line, we define the non-Newtonian viscosity η as the instantaneous ratio of shear stress and velocity gradient at each value of velocity gradient:

$$
\begin{array}{cc}
\text{Non-Newtonian viscosity} \\
\text{(variable function of shear rate)}
\end{array}
\qquad
\begin{aligned}
\eta &\equiv \frac{\tilde{\tau}_{yx}\big|_{\dot{\gamma}}}{\left(\dfrac{dv_y}{dx}\right)} = \eta(\dot{\gamma}) \\[2em]
\dot{\gamma} &= \left|\frac{dv_y}{dx}\right| \qquad \text{(shear flow)}
\end{aligned}
\qquad (5.202)
$$

The expression $\dot{\gamma}$ (read as *gamma dot*) is the rate-of-deformation. Plots of non-Newtonian viscosity versus rate-of-deformation characterize non-Newtonian fluids as shear-thinning (Figure 5.26) or shear-thickening (Figure 5.27). It is straightforward to calculate the non-Newtonian viscosity from experimental measurements in a shear apparatus.

5.3.2 Shear-induced normal stresses

Another type of non-Newtonian behavior is the generation of normal stresses in steady drag flow. In the introduction to shear flow in Section 5.1, we assumed that the force to move the top plate in the parallel-plate shear apparatus was purely in the flow (x) direction:

$$
\begin{array}{c}
\text{Force to move} \\
\text{plate in shear flow} \\
\text{(Newtonian)}
\end{array}
=
\begin{pmatrix} F_x \\ F_y \\ F_z \end{pmatrix}_{xyz}
=
\begin{pmatrix} F \\ 0 \\ 0 \end{pmatrix}_{xyz}
= F\hat{e}_x
\qquad (5.203)
$$

Experiments in the parallel-plate apparatus verify that fluids such as water, oil, and honey do not generate normal stresses in steady shear ($F_y = F_z = 0$), and the stress-velocity relationship for such materials is the Newtonian constitutive equation, as discussed previously.

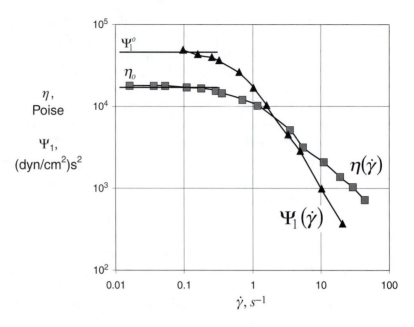

Figure 5.26 Shear-thinning is very common for polymer melts. The non-Newtonian viscosity η can vary with velocity gradient $\dot{\gamma}$ over five or more decades in magnitude. Steady-shear viscosity and first normal-stress coefficient (see Section 5.3.2) as a function of shear rate for a concentrated solution of a narrowly distributed polybutadiene (replotted from Menezes and Graessley [96]). $M_w = 350$ kg/mol, $M_w/M_n < 1.05$, and concentration is 0.0676 g/cm^3 in Flexon 391, a hydrocarbon oil.

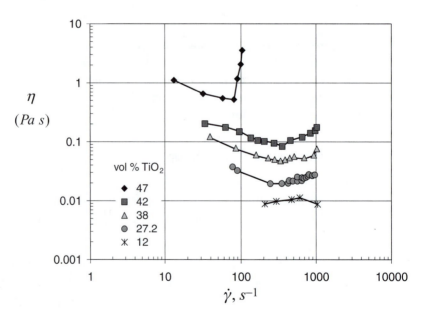

Figure 5.27 Viscosity, η, versus shear rate, $\dot{\gamma}$, for five suspensions of TiO$_2$ in water (recalculated and replotted from Metzner and Whitlock [97]). The diameters of the TiO$_2$ particles are between 0.2 and 1 microns.

Figure 5.28 The photograph at top shows how a Newtonian fluid—a dilute mixture of flour, water, and food coloring in this case—moves away from the mixing blades when it is stirred at a high rate. A non-Newtonian flour–water dough in the bottom photograph, however, climbs the mixing blades.

However, some fluids generate normal stresses in shear. We can obtain firsthand experience with shear-induced normal stresses in the kitchen (Figure 5.28). When mixing flour and water in an electric mixer, the batter is unremarkable until the flour content rises to a critical level. Flour–water dough is elastic, and the turning mixing blades create a shear flow. The dough is not constrained in the axial direction, and elasticity causes it to climb up the mixing blades in response to shear-induced normal stresses.

Shear-induced normal stresses can be detected in the parallel-plate apparatus because the normal stress manifests as an upward thrust on the top plate ($F_y \neq 0$, $F_z \neq 0$; see Section 5.1.2). For some polymeric fluids, to keep the gap between the parallel plates constant during drag flow, a downward force must be imposed on the upper plate. The upward thrust is generated by the deformation of the

Figure 5.29   One of the most striking viscoelastic effects is called rod-climbing or the Weissenberg effect [12]. In this photograph, a rod is rotated clockwise in a fluid reservoir at a slow speed of approximately 0.5 cycle/s. In response to the flow, the fluid climbs the rod. When the turning ceases, the fluid returns to the reservoir. The apparatus shown was created by John L. Schrag and Arthur S. Lodge at the University of Wisconsin; the fluid is a 2 percent aqueous polyacrylamide solution. (*Photograph by Carlos Arango Sabogal.*)

fluid; when the deformation stops, the upward thrust goes to zero. If there is no upper plate in a shearing flow, a fluid that exhibits shear normal stresses climbs up a turning rod as a way of expressing the shear-induced normal forces (Figure 5.29).

Shear-induced normal stresses appear in our continuum model as nonzero stresses $\tilde{\tau}_{xx}$, $\tilde{\tau}_{yy}$, and $\tilde{\tau}_{zz}$ in this flow. When a shear flow is assumed and the Newtonian model is used to predict stress, the diagonal normal stresses are zero ($\tilde{\tau}_{xx} = \tilde{\tau}_{yy} = \tilde{\tau}_{zz} = 0$), as Example 5.12 demonstrates. The Newtonian model, thus, is not appropriate to model $\underline{\underline{\tilde{\tau}}}$ in materials that exhibit shear normal stresses.

EXAMPLE 5.12. *What are the predicted normal stresses generated in steady drag flow for a Newtonian fluid?*

SOLUTION. We establish in Section 5.1.1 that the velocity profile for steady-shear flow (see Figure 5.5) is:

$$\underline{v} = \begin{pmatrix} v_x \\ 0 \\ 0 \end{pmatrix}_{xyz} = \begin{pmatrix} Vy/H \\ 0 \\ 0 \end{pmatrix}_{xyz} \tag{5.204}$$

We also calculated the complete stress tensor for Newtonian fluids in steady-shear flow (compare to Equation 5.87):

$$\tilde{\underline{\underline{\tau}}} = \mu \left(\nabla \underline{v} + (\nabla \underline{v})^T \right) \tag{5.205}$$

$$= \mu \begin{pmatrix} 2\frac{\partial v_x}{\partial x} & \frac{\partial v_y}{\partial x} + \frac{\partial v_x}{\partial y} & \frac{\partial v_z}{\partial x} + \frac{\partial v_x}{\partial z} \\ \frac{\partial v_x}{\partial y} + \frac{\partial v_y}{\partial x} & 2\frac{\partial v_y}{\partial y} & \frac{\partial v_z}{\partial y} + \frac{\partial v_y}{\partial z} \\ \frac{\partial v_x}{\partial z} + \frac{\partial v_z}{\partial x} & \frac{\partial v_y}{\partial z} + \frac{\partial v_z}{\partial y} & 2\frac{\partial v_z}{\partial z} \end{pmatrix}_{xyz} \tag{5.206}$$

$$= \begin{pmatrix} 0 & \mu\frac{dv_x}{dy} & 0 \\ \mu\frac{dv_x}{dy} & 0 & 0 \\ 0 & 0 & 0 \end{pmatrix}_{xyz} \tag{5.207}$$

Evaluating $\partial v_x \partial y$ from Equation 5.204, we obtain:

$$\tilde{\underline{\underline{\tau}}} = \begin{pmatrix} 0 & \mu\frac{V}{H} & 0 \\ \mu\frac{V}{H} & 0 & 0 \\ 0 & 0 & 0 \end{pmatrix}_{xyz} \tag{5.208}$$

We see from Equation 5.208 that in steady drag flow of a Newtonian fluid, the normal stresses $\tilde{\tau}_{xx}$, $\tilde{\tau}_{yy}$, and $\tilde{\tau}_{zz}$ are zero.

Shear-induced normal stresses are independent of the direction of the shearing flow: Whether the flow is in the $(+x)$- or the $(-x)$-direction, the thrust that is generated is upward. This is in contrast to the shear stress, which changes sign when the flow direction changes. Thus, rotating the mixer blades in the opposite direction still results in the dough climbing up the blades. The function defined to quantify shear normal stresses is $\Psi_1(\dot{\gamma})$, the first normal stress coefficient:

First normal stress coefficient
$$\Psi_1(\dot{\gamma}) \equiv \frac{(\tilde{\tau}_{xx} - \tilde{\tau}_{yy})}{\dot{\gamma}^2} \tag{5.209}$$

where x- is the flow direction and y- is the gradient direction of shear flow. For polymeric fluids, the first normal stress coefficient varies significantly with rate-of-deformation $\dot{\gamma}$, as shown in Figure 5.26 [104]. Normal stress effects are not insignificant in many polymeric systems.

Shear-induced normal stresses are one result of fluid elasticity. Memory is another effect associated with elasticity. Memory is exhibited by fluids such as Silly Putty [31], which stretches when pulled slowly but which also bounces. During the slow deformation, the material "forgets" its past shape, whereas in rapid deformation, it has nearly perfect "memory" of a past shape (Figure 5.30). Memory effects put a strong constraint on stress modeling. To predict flow behavior that makes reference to past shapes and deformations, a constitutive equation must refer to past shapes and deformations. Constitutive modeling that considers fluid memory can be complex [82, 104].

slow deformation:

rapid deformation:

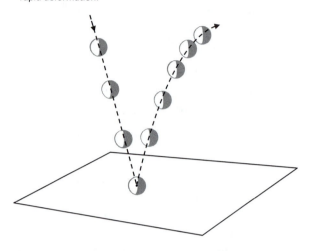

Figure 5.30   Silly Putty is a liquid and, if left on a table, it will flow into a puddle; when stretched slowly, it elongates. However, when the applied force acts rapidly, such as when a blob of Silly Putty is thrown against the floor, the fluid behaves like an elastic solid [119].

As stated at the beginning of this section, the study of non-Newtonian effects is itself an entire discipline. We discuss just two classes of non-Newtonian constitutive equations: inelastic and viscoelastic. Inelastic constitutive equations are fairly simple equations that adapt the tensor structure of the Newtonian constitutive equation and the definition of the non-Newtonian viscosity (Equation 5.202) to produce non-Newtonian constitutive equations that can be used on materials that do not exhibit memory. Inelastic constitutive equations may be useful for calculations on viscoelastic fluids, within certain limits.

Viscoelastic constitutive equations include all other types of fluid constitutive equations; Section 5.3.4 briefly describes some of the issues involved in their development and application. More information on viscoelastic constitutive equations is in the literature [37, 82, 104].

5.3.3 Inelastic constitutive equations

The variation of non-Newtonian viscosity with shear rate is a significant effect that cannot be ignored when modeling stress produced in the flow of molten polymers, foods, and many industrial and biomedical materials. For these systems, the change in the viscosity often is the most dramatic and important effect occurring in the flow. Because the viscosity is not constant, the Newtonian stress-velocity relationship $\underline{\underline{\tilde{\tau}}} = \mu(\nabla \underline{v} + (\nabla \underline{v})^T)$ is not correct for these systems and we need a new, non-Newtonian constitutive equation.

A simple way to adapt the Newtonian constitutive equation to make it non-Newtonian is to replace the constant Newtonian viscosity μ in the equation with the nonconstant, non-Newtonian viscosity $\eta(\dot{\gamma})$:

$$\begin{array}{c}\text{Proposed}\\ \text{non-Newtonian}\\ \text{constitutive equation}\end{array} \quad \underline{\underline{\tilde{\tau}}} \overset{?}{=} \eta(\dot{\gamma})\left(\nabla\underline{v} + (\nabla\underline{v})^T\right) \qquad (5.210)$$

This proposed constitutive equation is in the form of a tensor and the predicted stress is symmetric, which means that it meets two important criteria for stress. There is a problem with this equation, however, and it is related to the current definition of the quantity $\dot{\gamma}$:

$$\begin{array}{c}\text{Current definition of}\\ \text{rate-of-deformation } \dot{\gamma}:\\ \text{(shear flow)}\end{array} \quad \dot{\gamma} = \left|\frac{dv_x}{dy}\right| \qquad (5.211)$$

The definition in Equation 5.211 refers to a specific coordinate system associated with a specific flow—shear flow. For flows other than a shear flow in the x-direction that varies in the y-direction, the definition of rate-of-deformation in Equation 5.211 does not make sense.

The problem with the current definition of the rate-of-deformation $\dot{\gamma}$ is illustrated by trying to use it in flows with which we are somewhat familiar.

EXAMPLE 5.13. *For the planar-jet flow shown in Figure 5.31, what are the stress components predicted by the proposed non-Newtonian stress-velocity relationship given in Equation 5.210?*

SOLUTION. We discussed this flow previously, but this time we choose to use a coordinate system in which the direction toward the wall is the z-direction. The

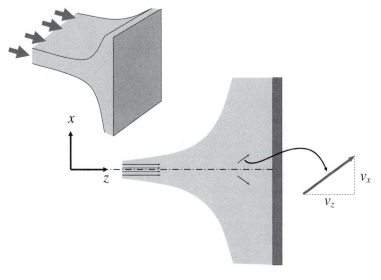

Figure 5.31   We revisit this planar-jet flow to see if the proposed non-Newtonian constitutive equation can be applied. We chose a new coordinate system, with z as the direction toward the wall and x as the direction upward.

choice of coordinate system is arbitrary; we choose z as toward the wall because this choice highlights the problem with Equation 5.210.

The flow is designed so that the y-component of velocity is zero; also, because the flow is wide, the remaining velocity components do not depend on y. We mathematically summarize these characteristics as follows:

$$\underline{v} = \begin{pmatrix} v_x \\ v_y \\ v_z \end{pmatrix}_{xyz} = \begin{pmatrix} v_x(x,z) \\ 0 \\ v_z(x,z) \end{pmatrix}_{xyz} \tag{5.212}$$

To test the proposed constitutive equation, first we write it in component form in the chosen coordinate system; and second, we simplify it using what we know about the velocity field:

$$\underline{\underline{\tilde{\tau}}} \stackrel{?}{=} \eta(\dot{\gamma}) \left(\nabla \underline{v} + (\nabla \underline{v})^T \right) \tag{5.213}$$

$$\begin{pmatrix} \tilde{\tau}_{xx} & \tilde{\tau}_{xy} & \tilde{\tau}_{xz} \\ \tilde{\tau}_{yx} & \tilde{\tau}_{yy} & \tilde{\tau}_{yz} \\ \tilde{\tau}_{zx} & \tilde{\tau}_{zy} & \tilde{\tau}_{zz} \end{pmatrix}_{xyz} \stackrel{?}{=} \eta(\dot{\gamma}) \begin{pmatrix} 2\frac{\partial v_x}{\partial x} & \frac{\partial v_y}{\partial x} + \frac{\partial v_x}{\partial y} & \frac{\partial v_z}{\partial x} + \frac{\partial v_x}{\partial z} \\ \frac{\partial v_x}{\partial y} + \frac{\partial v_y}{\partial x} & 2\frac{\partial v_y}{\partial y} & \frac{\partial v_z}{\partial y} + \frac{\partial v_y}{\partial z} \\ \frac{\partial v_x}{\partial z} + \frac{\partial v_z}{\partial x} & \frac{\partial v_y}{\partial z} + \frac{\partial v_z}{\partial y} & 2\frac{\partial v_z}{\partial z} \end{pmatrix}_{xyz} \tag{5.214}$$

$$\begin{pmatrix} \tilde{\tau}_{xx} & \tilde{\tau}_{xy} & \tilde{\tau}_{xz} \\ \tilde{\tau}_{yx} & \tilde{\tau}_{yy} & \tilde{\tau}_{yz} \\ \tilde{\tau}_{zx} & \tilde{\tau}_{zy} & \tilde{\tau}_{zz} \end{pmatrix}_{xyz} \stackrel{?}{=} \eta(\dot{\gamma}) \begin{pmatrix} 2\frac{\partial v_x}{\partial x} & 0 & \frac{\partial v_z}{\partial x} + \frac{\partial v_x}{\partial z} \\ 0 & 0 & 0 \\ \frac{\partial v_x}{\partial z} + \frac{\partial v_z}{\partial x} & 0 & 2\frac{\partial v_z}{\partial z} \end{pmatrix}_{xyz} \tag{5.215}$$

The proposed constitutive equation is intended to be non-Newtonian, with the nonconstant viscosity given by $\eta(\dot{\gamma})$. The current definition of $\dot{\gamma}$, however, refers to a particular shear coordinate system:

$$\text{Current definition of } \dot{\gamma}: \qquad \dot{\gamma} = \left| \frac{dv_x}{dy} \right| \tag{5.216}$$
$$\text{(shear flow)}$$

We are faced with the problem of adapting this definition to our current flow. In the coordinate system we are using for our flow, $dv_x/dy = 0$, which certainly does not reflect the deformation taking place in the flow. If we change x to z and y to x:

$$\text{Proposed adaptation of } \dot{\gamma}: \qquad \dot{\gamma} \stackrel{?}{=} \left| \frac{dv_z}{dx} \right| \tag{5.217}$$
$$\text{for current flow}$$

we obtain a nonzero $\dot{\gamma}$, but this value is arbitrary and does not reflect all of the deformation taking place in our flow.

The proposed constitutive equation is found to have a fundamental flaw: The expression $\eta(\dot{\gamma})$ with the current definition of $\dot{\gamma}$ refers to a coordinate system that is meaningless in our current situation and mostly meaningless in the general situation. There was a similar problem in Section 5.2.1, when we tried to adapt Newton's law of viscosity to flows other than simple shear. We cannot proceed with a prediction of the $\tilde{\tau}_{ij}$ using Equation 5.210 until we address the meaning of $\dot{\gamma}$ in flows other than shear flow.

This problem illustrates another requirement imposed on a constitutive equation: It should be valid in any coordinate system and for any flow. We must find a way to express non-Newtonian stresses that does not violate this rule.

A solution to this problem is to define $\dot{\gamma}$ generally so that it is applicable and meaningful in all coordinate systems and for all flows. The choice of definition should reduce to our original definition in simple shear flow. To derive this definition, we turn to tensor mathematics [104]. We begin by defining the rate-of-deformation tensor $\underline{\underline{\dot{\gamma}}}$:

$$\underline{\underline{\dot{\gamma}}} \equiv \nabla\underline{v} + (\nabla\underline{v})^T \tag{5.218}$$

$$= \begin{pmatrix} 2\frac{\partial v_x}{\partial x} & \frac{\partial v_x}{\partial y} + \frac{\partial v_y}{\partial x} & \frac{\partial v_x}{\partial z} + \frac{\partial v_z}{\partial x} \\ \frac{\partial v_y}{\partial x} + \frac{\partial v_x}{\partial y} & 2\frac{\partial v_y}{\partial y} & \frac{\partial v_y}{\partial z} + \frac{\partial v_z}{\partial y} \\ \frac{\partial v_z}{\partial x} + \frac{\partial v_x}{\partial z} & \frac{\partial v_z}{\partial y} + \frac{\partial v_y}{\partial z} & 2\frac{\partial v_z}{\partial z} \end{pmatrix}_{xyz} \tag{5.219}$$

This tensor appears in the Newtonian constitutive equation ($\underline{\underline{\tau}} = \mu\underline{\underline{\dot{\gamma}}}$) and in the proposed non-Newtonian constitutive equation, Equation 5.210. For our rate-of-deformation measure $\dot{\gamma}$, we choose that $\dot{\gamma}$ is given by the magnitude of the tensor $\underline{\underline{\dot{\gamma}}}$. The magnitude of a tensor is a quantity defined in tensor mathematics; tensor magnitude is independent of the coordinate system and is a measure of a tensor's size or effect [6]. The magnitude of a tensor $\underline{\underline{A}}$ is defined as:

$$\begin{array}{c} \text{Magnitude} \\ \text{of tensor } \underline{\underline{A}} \\ \text{(orthonormal coordinate system)} \end{array} \qquad |\underline{\underline{A}}| = +\sqrt{\left(\frac{1}{2}\cdot\sum_{p=1}^{3}\sum_{j=1}^{3}A_{pj}A_{jp}\right)} \tag{5.220}$$

where $\underline{\underline{A}}$ is expressed in an orthonormal coordinate system (e.g., Cartesian, cylindrical, or spherical). For a symmetric tensor (remember that $\underline{\underline{\dot{\gamma}}}$ is symmetric), the calculation is even easier because $A_{pj} = A_{jp}$:

$$\text{Symmetric tensor } \underline{\underline{A}}: \quad |\underline{\underline{A}}| = +\sqrt{\left(\frac{1}{2}\cdot\begin{array}{c}\text{sum of squares}\\\text{of each orthonormal}\\\text{component of } \underline{\underline{A}}\end{array}\right)} \tag{5.221}$$

We define the rate-of-deformation $\dot{\gamma}$ as the magnitude of the tensor $\underline{\underline{\dot{\gamma}}}$:

$$\begin{array}{c}\text{Rate-of-deformation}\\\text{(general definition)}\end{array}$$

$$\dot{\gamma} \equiv |\underline{\underline{\dot{\gamma}}}| = +\sqrt{\left(\frac{1}{2}\cdot\begin{array}{c}\text{sum of squares}\\\text{of each orthonormal}\\\text{component of } \underline{\underline{\dot{\gamma}}}\end{array}\right)}$$

$$= +\sqrt{\left(\frac{1}{2}\cdot\sum_{p=1}^{3}\sum_{j=1}^{3}\dot{\gamma}_{pj}^2\right)}$$

$$\tag{5.222}$$

Although this definition seems complex, it is straightforward to calculate from matrix components of the rate-of-deformation tensor $\dot{\underline{\underline{\gamma}}}$. We now try this new definition, beginning with shear flow.

EXAMPLE 5.14. *With the new definition for $\dot\gamma$, what is $\dot\gamma$ for steady-shear flow?*

SOLUTION. To solve this problem, we write the definition of $\dot{\underline{\underline{\gamma}}}$, simplify the expression by using the velocity field for steady-shear flow, and use the new definition of the tensor magnitude, $\dot\gamma = \left|\dot{\underline{\underline{\gamma}}}\right|$:

$$\underline{v} = \begin{pmatrix} \frac{V}{H}y \\ 0 \\ 0 \end{pmatrix}_{xyz} \tag{5.223}$$

$$\dot{\underline{\underline{\gamma}}} = \begin{pmatrix} 2\frac{\partial v_x}{\partial x} & \frac{\partial v_x}{\partial y} + \frac{\partial v_y}{\partial x} & \frac{\partial v_x}{\partial z} + \frac{\partial v_z}{\partial x} \\ \frac{\partial v_y}{\partial x} + \frac{\partial v_x}{\partial y} & 2\frac{\partial v_y}{\partial y} & \frac{\partial v_y}{\partial z} + \frac{\partial v_z}{\partial y} \\ \frac{\partial v_z}{\partial x} + \frac{\partial v_x}{\partial z} & \frac{\partial v_z}{\partial y} + \frac{\partial v_y}{\partial z} & 2\frac{\partial v_z}{\partial z} \end{pmatrix}_{xyz} \tag{5.224}$$

$$= \begin{pmatrix} 0 & \frac{V}{H} & 0 \\ \frac{V}{H} & 0 & 0 \\ 0 & 0 & 0 \end{pmatrix}_{xyz} \tag{5.225}$$

$$\dot\gamma \equiv + \left(\frac{1}{2} \cdot \begin{array}{c} \text{sum of squares} \\ \text{of each Cartesian} \\ \text{component of } \dot{\underline{\underline{\gamma}}} \end{array} \right)^{\frac{1}{2}} \tag{5.226}$$

$$= +\sqrt{\frac{1}{2}\left(0 + \frac{V^2}{H^2} + 0 + \frac{V^2}{H^2} + 0 + 0 + 0 + 0 + 0\right)} \tag{5.227}$$

$$= \left|\frac{V}{H}\right| \tag{5.228}$$

This is the same result for $\dot\gamma$ in a shear flow as would be obtained if we used the former definition of $\dot\gamma$.

Former definition of $\dot\gamma$: (shear flow only)

$$\dot\gamma = \left|\frac{dv_x}{dy}\right| \tag{5.229}$$

$$= \left|\frac{d(Vy/H)}{dy}\right| \tag{5.230}$$

$$= \left|\frac{V}{H}\right| \tag{5.231}$$

The new definition of rate-of-deformation $\dot\gamma$ has the same value in any coordinate system; this is guaranteed by tensor mathematics.[3] The new definition also

---

[3] Tensor magnitude is one of the three invariants of a second-order tensor and $\dot{\underline{\underline{\gamma}}}$ is a second-order tensor [6].

is valid for any flow: For any flow, we can write the tensor $\dot{\underline{\underline{\gamma}}} = \nabla\underline{v} + (\nabla\underline{v})^T$ and take its magnitude using Equation 5.222. We demonstrate this now by applying the new definition of $\dot{\gamma}$ to a flow with curved streamlines: the planar-jet flow.

EXAMPLE 5.15. *For the planar-jet flow shown in Figure 5.31, what are the stress components predicted by the proposed non-Newtonian constitutive equation in Equation 5.210? Use the new definition of $\dot{\gamma}$, Equation 5.222.*

SOLUTION. We choose to analyze this problem in the same Cartesian coordinate system as used before, in which z is toward the wall and the flow varies in the x- and z-directions. The solution follows the same steps as in the previous attempt, with the difference being the definition of the rate-of-deformation $\dot{\gamma}$.

$$\underline{v} = \begin{pmatrix} v_x \\ v_y \\ v_z \end{pmatrix}_{xyz} = \begin{pmatrix} v_x(x,z) \\ 0 \\ v_z(x,z) \end{pmatrix}_{xyz} \tag{5.232}$$

$$\dot{\underline{\underline{\gamma}}} = \left(\nabla\underline{v} + (\nabla\underline{v})^T \right) \tag{5.233}$$

$$= \begin{pmatrix} 2\frac{\partial v_x}{\partial x} & \frac{\partial v_y}{\partial x} + \frac{\partial v_x}{\partial y} & \frac{\partial v_z}{\partial x} + \frac{\partial v_x}{\partial z} \\ \frac{\partial v_x}{\partial y} + \frac{\partial v_y}{\partial x} & 2\frac{\partial v_y}{\partial y} & \frac{\partial v_z}{\partial y} + \frac{\partial v_y}{\partial z} \\ \frac{\partial v_x}{\partial z} + \frac{\partial v_z}{\partial x} & \frac{\partial v_y}{\partial z} + \frac{\partial v_z}{\partial y} & 2\frac{\partial v_z}{\partial z} \end{pmatrix}_{xyz} \tag{5.234}$$

$$= \begin{pmatrix} 2\frac{\partial v_x}{\partial x} & 0 & \frac{\partial v_z}{\partial x} + \frac{\partial v_x}{\partial z} \\ 0 & 0 & 0 \\ \frac{\partial v_x}{\partial z} + \frac{\partial v_z}{\partial x} & 0 & 2\frac{\partial v_z}{\partial z} \end{pmatrix}_{xyz} \tag{5.235}$$

$$\dot{\gamma} \equiv + \left(\frac{1}{2} \cdot \begin{array}{c} \text{sum of squares} \\ \text{of each Cartesian} \\ \text{component of } \dot{\underline{\underline{\gamma}}} \end{array} \right)^{\frac{1}{2}} \tag{5.236}$$

$$= + \sqrt{\begin{array}{c} \frac{1}{2}\left(4\left(\frac{\partial v_x}{\partial x}\right)^2 + 0 + \left(\frac{\partial v_z}{\partial x} + \frac{\partial v_x}{\partial z}\right)^2 + 0 + 0 \right. \\ \left. + 0 + \left(\frac{\partial v_x}{\partial z} + \frac{\partial v_z}{\partial x}\right)^2 + 0 + 4\left(\frac{\partial v_z}{\partial z}\right)^2 \right) \end{array}} \tag{5.237}$$

$$\boxed{\dot{\gamma} = +\sqrt{ 2\left(\frac{\partial v_x}{\partial x}\right)^2 + \left(\frac{\partial v_z}{\partial x} + \frac{\partial v_x}{\partial z}\right)^2 + 2\left(\frac{\partial v_z}{\partial z}\right)^2 }} \tag{5.238}$$

The final result for the rate-of-deformation is complex, and it is unlikely that we would have determined this formula on our own. Because of the formal mathematics involved in the definition of the magnitude of a tensor, however, this calculation of the rate-of-deformation $\dot{\gamma}$ correctly reflects the deformation occurring at every location in this complex flow. The proposed constitutive equation is

$\underline{\underline{\tilde{\tau}}} = \eta(\dot{\gamma})\underline{\underline{\dot{\gamma}}}$; thus, the predicted stress components are:

$$\underline{\underline{\tilde{\tau}}} = \eta(\dot{\gamma})\underline{\underline{\dot{\gamma}}} \tag{5.239}$$

$$\underline{\underline{\tilde{\tau}}} = \eta(\dot{\gamma}) \begin{pmatrix} 2\frac{\partial v_x}{\partial x} & 0 & \frac{\partial v_z}{\partial x} + \frac{\partial v_x}{\partial z} \\ 0 & 0 & 0 \\ \frac{\partial v_x}{\partial z} + \frac{\partial v_z}{\partial x} & 0 & 2\frac{\partial v_z}{\partial z} \end{pmatrix}_{xyz} \tag{5.240}$$

with $\dot{\gamma}$ given by Equation 5.238.

We have arrived at a reasonable expression for non-Newtonian stress for our planar-jet flow. The result in Equation 5.240 does not violate any rules for constitutive equations. Our final result for the stress in Equation 5.240 is incomplete, however, because we have not yet specified the details of the function $\eta(\dot{\gamma})$. We next discuss this and how well this guessed-at constitutive equation works.

The constitutive equation we are using is called the *generalized Newtonian fluid (GNF) constitutive equation*:

Generalized Newtonian constitutive equation (GNF)

$$\begin{aligned} \underline{\underline{\tilde{\tau}}} &= \eta(\dot{\gamma})\left(\nabla\underline{v} + (\nabla\underline{v})^T\right) \\ &= \eta(\dot{\gamma})\underline{\underline{\dot{\gamma}}} \\ \dot{\gamma} &= \left|\underline{\underline{\dot{\gamma}}}\right| \\ \eta(\dot{\gamma}) &= \text{specified by user} \end{aligned} \tag{5.241}$$

The function $\eta(\dot{\gamma})$ is chosen by the user to match the curve of steady-shear viscosity as a function of $\dot{\gamma}$ for a material of interest.

For many polymer melts at high rates-of-deformation, measurements of $\eta(\dot{\gamma})$ when plotted on a log-log graph result in a straight line (Figure 5.32): Data of this type can be fit with a $\eta(\dot{\gamma})$ given by a power-law function:

Power-law GNF viscosity function

$$\eta(\dot{\gamma}) = m\dot{\gamma}^{n-1} \tag{5.242}$$

where m and n are the fitting parameters of the model. The parameter m is called the consistency index and n is called the power-law index. The power-law index n is unitless, and the units of m can be worked out from Equation 5.242 (see Problem 38). We can see how m and n are related to the plot in Figure 5.32 by taking the log of both sides of Equation 5.242:

$$\eta(\dot{\gamma}) = m\dot{\gamma}^{n-1} \tag{5.243}$$

$$\log \eta = \log m + (n-1)\log \dot{\gamma} \tag{5.244}$$

$$\log \eta = (\text{intercept}) + (\text{slope})\log \dot{\gamma} \tag{5.245}$$

Thus, the slope of a $\log \eta$ versus $\log \dot{\gamma}$ plot is equal to $n-1$, and the value of $\log \eta$ at $\log \dot{\gamma} = 0$ (i.e., $\dot{\gamma} = 1$) is $\log \eta = \log m$.

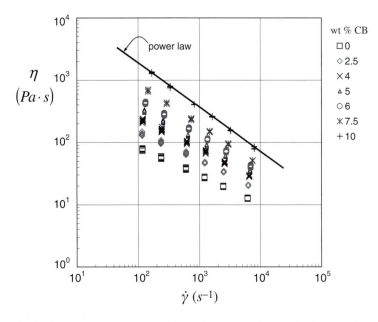

Figure 5.32 Steady-shear viscosity as a function of shear rate for composites of a liquid-crystal polymer with carbon black (CB) (concentrations indicated); the viscosity data follow a power-law relationship with shear rate. The data are for mixtures of Vectra A950RX and Ketjenblack EC-600 JD [74].

Over the entire range of $\dot{\gamma}$, many polymers exhibit a more complex shape for $\eta(\dot{\gamma})$ than what is described by the power-law model. The viscosity data in Figure 5.26 exhibit a power-law region at high rates-of-deformation; however, at low $\dot{\gamma}$, the viscosity levels off to a plateau. For some materials, viscosity also plateaus at high rates-of-deformation. A function for $\eta(\dot{\gamma})$ that can fit these more complex shapes is the Carreau–Yasuda model [181, 104] (Figure 5.33).

Carreau–Yasuda
GNF viscosity function

$$\eta(\dot{\gamma}) = \eta_\infty + (\eta_0 - \eta_\infty) \left[1 + (\lambda \dot{\gamma})^a\right]^{\frac{n-1}{a}} \qquad (5.246)$$

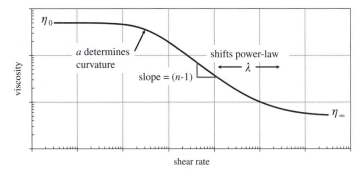

Figure 5.33 The Carreau–Yasuda model for $\eta(\dot{\gamma})$ predicts a shape that is compatible with what is observed for many polymer melts and solutions [104].

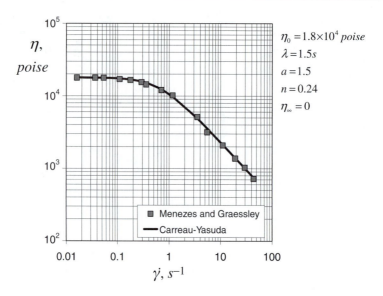

$\eta_0 = 1.8 \times 10^4\ poise$
$\lambda = 1.5 s$
$a = 1.5$
$n = 0.24$
$\eta_\infty = 0$

Menezes and Graessley
— Carreau–Yasuda

$\dot{\gamma},\ s^{-1}$

Figure 5.34 The viscosity data in Figure 5.26 can be fit to the Carreau–Yasuda model with the parameter values indicated. The infinite-shear viscosity parameter η_∞ is not needed for this fit because the data do not plateau at high shear rate.

The Carreau–Yasuda model has five parameters: η_0, which determines the level of the low shear-rate plateau; η_∞, which determines the level of the high shear-rate plateau; n, which determines the slope of the sharply decreasing region; λ, which determines when the viscosity curve begins to decrease; and a, which determines the shape of the curve as it transitions from the η_0 level to the power-law region. The Carreau–Yasuda model works well for typical polymer viscosity curves that show a low-shear-rate plateau and shear-thinning behavior (Figure 5.34).

The generalized Newtonian fluid model is popular in many fields, including plastics processing, hemorheology (i.e., blood flow), and geophysics. Numerous viscosity functions $\eta(\dot{\gamma})$ have been published, some of which are listed in Table 5.2. Making analytical calculations with GNF models is fairly simple, and complex flows are calculated readily using computer codes [5, 27, 151]). In the rheology literature there are many examples of momentum-balance calculations carried out using the power-law GNF [12, 104].

The generalized Newtonian fluid constitutive equations have limitations that they share as a group. Because of the form of the equation, $\underline{\underline{\tau}} = \eta \underline{\underline{\dot{\gamma}}}$, the GNF models predict that the normal stresses are zero in steady shearing, regardless of the form chosen for $\eta(\dot{\gamma})$ (see Problem 37). Thus, GNF models cannot predict rod-climbing (see Figure 5.29). Also, the GNF models predict stresses by considering only the rate-of-deformation at the current time, $\underline{\underline{\dot{\gamma}}}(t)$. Because they do not include the effect of past deformations, generalized Newtonian models cannot predict memory effects (see Figure 5.30). Thus, the GNF models are catagorized as inelastic models.

Table 5.2. Several choices of function $\eta(\dot{\gamma})$ used in the generalized Newtonian constitutive equation, depending on application

Name System	$\eta(\dot{\gamma})$	Parameters	Reference
Power-Law (polymer melts)	$\eta(\dot{\gamma}) = m\dot{\gamma}^{n-1}$	m – consistency index n – power-law index	[12, 131]
Carreau–Yasuda (polymer melts)	$\dfrac{\eta(\dot{\gamma}) - \eta_\infty}{(\eta_0 - \eta_\infty)} = \left[1 + (\lambda\dot{\gamma})^a\right]^{\frac{n-1}{a}}$	η_∞ – infinite-shear viscosity η_0 – zero-shear viscosity λ – relaxation time a – shape index n – power-law index	[12, 181]
Bingham (suspensions, emulsions, pastes)	$\eta(\dot{\gamma}) = \begin{cases} \infty & \tau \le \tilde{\tau}_0 \\ \mu_0 + \dfrac{\tilde{\tau}_0}{\dot{\gamma}} & \tau > \tilde{\tau}_0 \end{cases}$ $\tau = \left|\underset{=}{\tilde{\tau}}\right| = \left|\eta\underset{=}{\dot{\gamma}}\right|$	$\tilde{\tau}_0$ – yield stress μ_0 – viscosity	[12]
Ellis (polymer melts)	$\eta(\dot{\gamma}) = \dfrac{\eta_0}{1 + \left|\dfrac{\tau}{\tilde{\tau}_0}\right|^{\alpha-1}}$ $\tau = \left|\underset{=}{\tilde{\tau}}\right| = \left|\eta\underset{=}{\dot{\gamma}}\right|$	η_0 – zero-shear viscosity $\tilde{\tau}_0$ – characteristic stress α – stress power-law index	[12]
DeKee (polymer melts)	$\eta(\dot{\gamma}) = \eta_1 e^{-\lambda\dot{\gamma}} + \eta_2 e^{-0.1\lambda\dot{\gamma}} + \eta_\infty$	η_1 – first viscosity η_2 – second viscosity η_∞ – infinite-shear viscosity λ – relaxation time	[19, 41]
Casson (blood)	$\sqrt{\tau} = \sqrt{\tilde{\tau}_0} + \sqrt{\eta_0\dot{\gamma}}$ $\tau = \left|\underset{=}{\tilde{\tau}}\right| = \left|\eta\underset{=}{\dot{\gamma}}\right|$	η_0 – zero-shear viscosity $\tilde{\tau}_0$ – characteristic stress	[19]

Even with these limitations, the GNF models are useful. For flow in pipes or in which the relationship between pressure drop and flow rate is the most important aspect, GNF models perform well [12]. See the literature surrounding each equation to determine its validity for a given flow of interest.

To apply the inelastic models discussed here, we again need the momentum balance—Reynolds transport theorem applied to a microscopic control volume. The general equation that is most convenient to use for non-Newtonian calculations is a version of this balance called the Cauchy momentum equation, which is derived and discussed in Chapter 6 [12, 104].

In the following example, we see how we can use the power-law, generalized Newtonian fluid model to calculate an engineering property of interest such as force on the wall in a flow of a non-Newtonian fluid.

EXAMPLE 5.16. *A shear-thinning, power-law fluid (m and n known) is subjected to a steady-drag flow in the apparatus shown in Figure 5.35. The velocity field is given here (see also Example 6.4):*

$$\underline{v} = \begin{pmatrix} \frac{V}{H}x_2 \\ 0 \\ 0 \end{pmatrix}_{123} \tag{5.247}$$

What is the stress tensor $\underline{\tilde{\Pi}}$ for this flow? Using the stress-tensor expression found, calculate the force needed to move the top plate at the speed V.

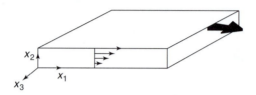

Figure 5.35   A shear-thinning, power-law fluid subjected to a drag flow.

SOLUTION. If the velocity field is known, the stress in a fluid may be calculated from the constitutive equation. For a power-law fluid, the constitutive equation is the generalized Newtonian fluid equation:

Generalized Newtonian fluid constitutive equation (GNF)

$$\underline{\tilde{\tau}}(t) = \eta(\dot{\gamma})\left(\nabla\underline{v} + (\nabla\underline{v})^T\right) \tag{5.248}$$

$$= \eta(\dot{\gamma})\underline{\dot{\gamma}} \tag{5.249}$$

The function $\eta(\dot{\gamma})$ can assume a variety of forms; for a shear-thinning power-law fluid, $\eta(\dot{\gamma})$ is given by:

Power-law GNF viscosity function:

$$\eta(\dot{\gamma}) = m\dot{\gamma}^{n-1} \tag{5.250}$$

with $n < 1$ and $\dot{\gamma}$ given by Equation 5.222 and repeated here:

$$\dot{\gamma} = +\sqrt{\left(\frac{1}{2} \cdot \sum_{p=1}^{3}\sum_{j=1}^{3} \dot{\gamma}_{pj}^2\right)} \tag{5.251}$$

For the velocity field given in Equation 5.247, we therefore calculate $\nabla \underline{v}$, $\underset{=}{\dot{\gamma}}$, $\dot{\gamma}$, and finally $\eta(\dot{\gamma})$:

$$\nabla \underline{v} = \begin{pmatrix} \frac{\partial v_1}{\partial x_1} & \frac{\partial v_2}{\partial x_1} & \frac{\partial v_3}{\partial x_1} \\ \frac{\partial v_1}{\partial x_2} & \frac{\partial v_1}{\partial x_2} & \frac{\partial v_3}{\partial x_2} \\ \frac{\partial v_1}{\partial x_3} & \frac{\partial v_2}{\partial x_3} & \frac{\partial v_3}{\partial x_3} \end{pmatrix}_{123} \tag{5.252}$$

$$= \begin{pmatrix} 0 & 0 & 0 \\ \frac{V}{H} & 0 & 0 \\ 0 & 0 & 0 \end{pmatrix}_{123} \tag{5.253}$$

$$\underset{=}{\dot{\gamma}} = \nabla \underline{v} + (\nabla \underline{v})^T \tag{5.254}$$

$$= \begin{pmatrix} 0 & \frac{V}{H} & 0 \\ \frac{V}{H} & 0 & 0 \\ 0 & 0 & 0 \end{pmatrix}_{123} \tag{5.255}$$

$$\dot{\gamma} = +\sqrt{\left(\frac{1}{2} \cdot \sum_{p=1}^{3} \sum_{j=1}^{3} \dot{\gamma}_{pj}^2 \right)} \tag{5.256}$$

$$= +\sqrt{\left(\frac{V}{H} \right)^2} = \frac{V}{H} \tag{5.257}$$

$$\eta(\dot{\gamma}) = m \dot{\gamma}^{n-1} \tag{5.258}$$

$$= m \left(\frac{V}{H} \right)^{n-1} \tag{5.259}$$

We now assemble the final expression for $\underset{=}{\tilde{\tau}}$ from the GNF constitutive equation:

$$\underset{=}{\tilde{\tau}}(t) = \eta(\dot{\gamma}) \underset{=}{\dot{\gamma}} \tag{5.260}$$

$$= m \left(\frac{V}{H} \right)^{n-1} \begin{pmatrix} 0 & \frac{V}{H} & 0 \\ \frac{V}{H} & 0 & 0 \\ 0 & 0 & 0 \end{pmatrix}_{123} \tag{5.261}$$

$$\underset{=}{\tilde{\tau}} = \begin{pmatrix} 0 & m \left(\frac{V}{H} \right)^n & 0 \\ m \left(\frac{V}{H} \right)^n & 0 & 0 \\ 0 & 0 & 0 \end{pmatrix}_{123} \tag{5.262}$$

To calculate the force on a finite surface in a fluid, we use the same integral of $\hat{n} \cdot \tilde{\underline{\Pi}}|_{surface}$ discussed in equation 4.285:

$$\begin{array}{c} \text{Total molecular fluid force} \\ \text{on a finite surface } \mathcal{S} \end{array} \quad \boxed{\mathcal{F} = \iint_{\mathcal{S}} [\hat{n} \cdot \tilde{\underline{\Pi}}]_{\text{at surface}} \, dS} \qquad (5.263)$$

For the top surface, $\hat{n} = -\hat{e}_2$ and $x_2 = H$. Therefore, the force on the top surface in this flow, $\mathcal{F}$, is given by:

$$\tilde{\underline{\Pi}} = -p\underline{I} + \tilde{\underline{\tau}} \qquad (5.264)$$

$$= \begin{pmatrix} -p & m\left(\frac{V}{H}\right)^n & 0 \\ m\left(\frac{V}{H}\right)^n & -p & 0 \\ 0 & 0 & -p \end{pmatrix}_{123} \qquad (5.265)$$

$$[\hat{n} \cdot \tilde{\underline{\Pi}}]_{x_2=H} = \begin{pmatrix} 0 & -1 & 0 \end{pmatrix}_{123} \cdot \begin{pmatrix} -p & m\left(\frac{V}{H}\right)^n & 0 \\ m\left(\frac{V}{H}\right)^n & -p & 0 \\ 0 & 0 & -p \end{pmatrix}_{123} \qquad (5.266)$$

$$= \begin{pmatrix} -m\left(\frac{V}{H}\right)^n & p & 0 \end{pmatrix}_{123} \qquad (5.267)$$

$$\mathcal{F} = \iint_{\mathcal{S}} [\hat{n} \cdot \tilde{\underline{\Pi}}]_{\text{at surface}} \, dS \qquad (5.268)$$

$$= \int_0^L \int_0^W \begin{pmatrix} -m\left(\frac{V}{H}\right)^n \\ p \\ 0 \end{pmatrix}_{123} dx_3 dx_1 \qquad (5.269)$$

$$\mathcal{F} = \begin{pmatrix} -mLW\left(\frac{V}{H}\right)^n \\ pLW \\ 0 \end{pmatrix}_{123} \qquad (5.270)$$

This is the fluid force on the top plate; the force to drive the top plate is the negative of the fluid force. The 1-component of the result gives the tangential force needed to drive the flow. The 2-component of the force result reflects the force due to the atmosphere, which acts normally to the plate.

5.3.4 Viscoelastic constitutive equations

The inelastic generalized Newtonian fluid (GNF) constitutive equation cannot predict shear normal stresses or memory effects (see Figure 5.29). To capture this kind of behavior in our models, far more sophisticated analysis is needed. This section is a brief overview of the study of constitutive equations for viscoelastic fluids. Of the two observations that inelastic constitutive equations fail to predict, fluid memory is the easier to address, and we discuss it first.

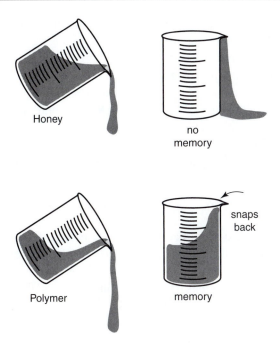

Honey

no
memory

Polymer

snaps
back

memory

Figure 5.36

Pouring honey and then stopping abruptly makes a mess (top). Pouring a polyacrylamide solution and stopping abruptly results in the fluid snapping back into the beaker from which it was poured (bottom) [112].

The difference between a fluid with memory and a fluid with no memory is shown when trying to pour two fluids from a beaker (Figure 5.36) [119]. A Newtonian fluid like honey has no memory. If we begin pouring and then attempt to stop, the honey makes a mess. No honey that had been poured returns to the beaker; rather, the pouring stream is cut off and honey dribbles down the outside of the beaker. A polyacrylamide solution, which is viscoelastic, responds qualitatively differently to the same experiment. If we begin pouring the viscoelastic fluid but then reverse direction in an attempt to stop pouring, the viscoelastic solution snaps back elastically and returns to the beaker [119].

Viscoelastic fluids move like Newtonian fluids under some circumstances, but they also have memory of past velocities, velocity gradients, and shapes. A constitutive equation that seeks to capture such complex behavior must refer to current velocities, velocity gradients, or shapes but also must refer to these quantities in the past. The generalized Newtonian constitutive equation discussed in the previous section relates stress at a time t, $\underset{\approx}{\tilde{\tau}}(t)$ to the velocity field $\underline{v}(t)$ at that same time:

GNF
constitutive equation
(no memory)

$$\underset{\approx}{\tilde{\tau}}(t) = \eta(\dot{\gamma})\underset{\approx}{\dot{\underline{\gamma}}}(t)$$

$$\underset{\approx}{\dot{\underline{\gamma}}}(t) = \nabla\underline{v}(t) + (\nabla\underline{v}(t))^{T}$$

$$\dot{\gamma} = \left|\underset{\approx}{\dot{\underline{\gamma}}}(t)\right|$$

$$\eta(\dot{\gamma}) = \text{specified by the user}$$

(5.271)

The GNF relates the instantaneous velocity gradient tensor to the instantaneous stress tensor and thus does not capture memory.

A constitutive equation that has memory of past events has terms that refer to the fluid's motion in the past. We construct a memory-constitutive equation by writing the stress as a sum of two types of contributions: (1) contributions generated as a result of the fluid's current motion; and (2) contributions generated from what the fluid's motion was sometime in the past.

If we write $t - \lambda$ as the time λ seconds before the current time t, and $t - 2\lambda$ as the time 2λ seconds before the current time t, we can write stress for one possible memory fluid as follows:

$$\underline{\underline{\tilde{\tau}}}(t) = \eta_1 \underline{\underline{\dot{\gamma}}}(t) + \eta_2 \underline{\underline{\dot{\gamma}}}(t - \lambda) + \eta_3 \underline{\underline{\dot{\gamma}}}(t - 2\lambda) \tag{5.272}$$

where η_1, η_2, η_3, and λ are scalar parameters of the model. In this equation, the stress at the current time t depends not only on $\underline{v}(t)$ but also on $\underline{v}$ at several other times in the past. These ideas lead to valid constitutive equations for viscoelastic fluids. If we consider contributions from all past times and total them such that the memory of past events fades farther back in the past, we obtain the generalized linear-viscoelastic constitutive equation [104]:

$$\begin{array}{c} \text{Generalized} \\ \text{linear-viscoelastic} \\ \text{constitutive equation} \end{array} \qquad \boxed{\underline{\underline{\tilde{\tau}}}(t) = \int_{-\infty}^{t} G(t - t') \, \underline{\underline{\dot{\gamma}}}(t') \, dt'} \tag{5.273}$$

In this equation, $G(t - t)$ is a function that describes how the fluid forgets past events. The integral represents a sum over contributions to the stress from rates-of-deformation $\underline{\underline{\dot{\gamma}}}$ at past times t'.

Incorporating memory into a constitutive equation may be accomplished by adding contributions from the past to contributions from the present. Predicting normal stresses in shear flow, however, requires a more drastic change of approach. Normal stresses in shear flow are a nonlinear effect. This means that to predict shear normal stresses, we need a constitutive equation that is more than a simple proportionality between stress and $\underline{\underline{\dot{\gamma}}}$ (e.g., the Newtonian or generalized Newtonian equations) and more than linear combinations of $\underline{\underline{\dot{\gamma}}}$ at different times (e.g., the generalized linear-viscoelastic model). To capture nonlinear effects, we must find a new nonlinear form for the constitutive equation. When considering nonlinear models, however, the possibilities are endless. To model nonlinear effects in both fluid mechanics and other fields of physics, it is rarely fruitful to guess at possible forms—which has been our approach thus far. The number of reasonable nonlinear functions we can use for $\underline{\underline{\tilde{\tau}}}(\underline{v})$ is nearly infinite.

Instead of guessing, researchers in the field looked at the rules of tensor transformations and found how those rules constrain the constitutive equation. We know that the nonlinear constitutive equation we seek must work in any coordinate system; we also know that the constitutive equation should not depend on the point of view of the observer. Thus, if we calculate stress in a flow with our constitutive equation expressed in a stationary coordinate system, or if we use a coordinate system moving at a constant velocity, we should calculate the same results.

The late 20th-century work of Oldroyd [129], Coleman and Noll [26], and others clarified the constraints that the tensor transformation rules had on constitutive modeling. Their work led to the discovery of nonlinear constitutive models that predict shear normal stresses and many other nonlinear effects. There still are many challenges in nonlinear constitutive modeling, however, because no single model has been discovered that predicts the complete range of behaviors exhibited by viscoelastic fluids. It is an ongoing research challenge to develop constitutive models that capture the behavior of nonlinear viscoelastic fluids.

Finally, there is another approach to the stress-velocity relationship in non-Newtonian fluids. We have followed the continuum approach, beginning with the velocity field and the stress tensor, and asked the question: How can we relate these two continuum field variables? An alternative approach is to forego the stress tensor, return to the momentum balance, and ask the original question: How can we quantify the molecular forces in the momentum balance?

$$
\begin{array}{l}
\text{Reynolds transport theorem} \\
\text{(momentum balance on CV)}
\end{array}
\quad
\frac{d\mathbf{P}}{dt} + \iint_{CS} (\hat{n} \cdot \underline{v}) \, \rho \underline{v} \, dS = \sum_{\substack{\text{on} \\ \text{CV}}} \underline{f}
\qquad (5.274)
$$

$$
\frac{d\mathbf{P}}{dt} + \iint_{CS} (\hat{n} \cdot \underline{v}) \, \rho \underline{v} \, dS = \underline{f}_{contact} + \underline{f}_{gravity}
$$

$$(5.275)$$

For the contact-force term, our continuum approach is to write $f_{contact}$ as an integral over $\hat{n} \cdot \underline{\underline{\tilde{\Pi}}}|_{surface}$, to define the stress tensor as $\underline{\underline{\tilde{\Pi}}} = -p\underline{\underline{I}} + \underline{\underline{\tilde{\tau}}}$, and to look for constitutive equations to obtain $\underline{\underline{\tilde{\tau}}}(\underline{v})$:

$$
\frac{d\mathbf{P}}{dt} + \iint_S (\hat{n} \cdot \underline{v}) \, \rho \underline{v} \, dS = \iint_S [\hat{n} \cdot \underline{\underline{\tilde{\Pi}}}]_{\text{at surface}} \, dS + M_{CV}\underline{g}
\qquad (5.276)
$$

$$
= \iint_S \left[\hat{n} \cdot \left(-p\underline{\underline{I}} + \underline{\underline{\tilde{\tau}}} \right) \right]_{\text{at surface}} \, dS + M_{CV} \, \underline{g} \quad (5.277)
$$

In arriving at Equation 5.277 from Equation 5.276, we substituted Equation 4.263 for the molecular-force vector and Equation 4.4 for the gravity term.

A different approach is to return to the molecules and model how they behave. For a chosen system, we know the chemistry and we often know much about the molecular forces that cause the nonlinear stress-velocity behavior observed. It is possible to begin with the molecules, their structure and their intermolecular forces, and then build up to a contact-force term that can be included in the momentum balance.

The molecular-modeling approach can be an effective strategy for complex systems. It is limited, however, in that the results are valid only for the specific fluids modeled. Also, the calculations can be time-consuming. Even with these drawbacks, however, the greatest advances in nonlinear rheological modeling in recent years have come from molecular modeling. For an introduction to molecular modeling in non-Newtonian fluids, refer to the literature [13, 83].

5.4 Summary

Our search in this chapter was for a stress-velocity relationship for molecular forces in fluids. We sought expressions that could describe all (or most) fluids. In reaching that goal, we have been reasonably successful. We arrived at the Newtonian constitutive equation, a rigorous stress-velocity relationship that captures the behavior of thousands of fluids, including the two most common: water and air. The field of fluid mechanics is the field of Newtonian fluid mechanics—that is, the study of fluids that follow the Newtonian constitutive equation.

Non-Newtonian fluids challenge the continuum approach. We can use continuum ideas to develop inelastic constitutive equations and even simple viscoelastic constitutive equations, but we find that for the most complex fluids and complex behaviors, it is more fruitful to return to a molecular approach for stress-velocity calculations.

The pattern of discovery in the quest for a proper molecular-force term in fluids follows a standard discovery pattern in science. When investigating observations, we look at the simplest explanations first, seeking to define when they are suitable. When the simplest systems are well understood, we move on to more complex cases. If the most obvious modifications fail, we move on to more complicated models, always looking for constraints that help narrow down the possible choices.

In the remaining chapters, we apply the Newtonian stress-velocity relationship and hone our skills in solving for velocity and stress fields in flowing liquids. Chapter 6 focuses on generalizing our solution methods and equations; Chapter 7 applies our techniques to flows within boundaries; and Chapter 8 focuses on unbounded flows, known as external flows, which includes the analysis of boundary layers. Chapter 9 takes forward and generalizes the macroscopic balance techniques. Advanced applications of continuum modeling are described in Chapter 10.

5.5 Problems

1. In fluid mechanics, what is a constitutive equation?
2. Is the Newtonian constitutive equation related to molecular forces in a fluid? How?
3. What does it mean to say a fluid is a "Newtonian fluid?" Give examples of non-Newtonian behavior.
4. What is stress? What is pressure? Distinguish between these two concepts in the context of fluid flow.
5. In Chapter 4, molecular force on a surface in a fluid is given by $\iint_S \hat{n} \cdot \underline{\tilde{\Pi}}|_{\text{surface}} \, dS$. How can we calculate $\underline{\tilde{\Pi}}$ for a flow?

6. The total stress tensor $\underline{\tilde{\underline{\Pi}}}$ is given by $\underline{\tilde{\underline{\Pi}}} = -p\underline{\underline{I}} + \underline{\tilde{\underline{\tau}}}$. If the velocity is zero, what is $\underline{\tilde{\underline{\Pi}}}$?

7. Sometimes we see different versions of the total-stress tensor:

$$\text{This text and [174]:} \qquad \underline{\tilde{\underline{\Pi}}} = -p\underline{\underline{I}} + \underline{\tilde{\underline{\tau}}}$$

$$\text{Bird et al. and [12, 104]:} \qquad \underline{\underline{\Pi}} = p\underline{\underline{I}} + \underline{\underline{\tau}}$$

What is the difference between these two equivalent representations?

8. Figure 5.1 and the accompanying text discusses stress at a point in terms of the different stresses that different surfaces experience at the same point in a moving fluid. If the stress tensor $\underline{\tilde{\underline{\Pi}}}$ for a flow was calculated at Point P to be (arbitrary units):

$$\underline{\tilde{\underline{\Pi}}}|_P = \begin{pmatrix} 2 & 0 & 0 \\ 0 & 0.5 & 0 \\ 0 & 0 & 1 \end{pmatrix}_{xyz}$$

calculate the force on the following six surfaces of unit area through P: The first surface is oriented with $\hat{n} = \hat{e}_x$. The five remaining surfaces are oriented with unit normals that are obtained from $\hat{e}_x$ by making successive 60-degree rotations around the z-axis. Plot these vectors.

9. For the velocity profile given here, what is $\frac{\partial v_z}{\partial z}$ along the line $x = 0$, $y = 0$? What is the value of that derivative on that line at locations $z = 1, 2,$ and 3? What is the value of the velocity vector at those locations? Comment on the connection between the two quantities, $\underline{v}$ and $\frac{\partial v_z}{\partial z}$.

$$\underline{v} = -9x\hat{e}_x - 9y\hat{e}_y + 18z\hat{e}_z$$

10. For the flow in Figure 5.18, describe the motion of a fluid particle traveling on the streamlines shown. Are the fluid particles accelerating? How do you know?

11. Drag flow is a flow with velocity vector $\underline{v} = v_x(y)\hat{e}_x$ when written in the Cartesian coordinate system (i.e., flow in the x-direction). Calculate the velocity gradient tensor $\nabla\underline{v}$ (a 3×3 matrix) for this flow.

12. For the two-dimensional flow shown in Figure 4.77, write the velocity gradient tensor $\nabla\underline{v}$ in matrix form, indicating which coefficients and derivatives are zero. Are the fluid particles accelerating?

13. For a flow (Newtonian fluid) that may be written as $\underline{v} = v_x\hat{e}_x + v_z\hat{e}_z$, calculate the 3×3 matrix $\nabla\underline{v} + (\nabla\underline{v})^T$. What is $\underline{\tilde{\underline{\tau}}}$ for this flow?

14. For a flow (Newtonian fluid) that may be written as $\underline{v} = v_\theta\hat{e}_\theta$ in a cylindrical coordinate system, calculate the 3×3 matrix $\nabla\underline{v} + (\nabla\underline{v})^T$. What is $\underline{\tilde{\underline{\tau}}}$ for this flow?

15. For a flow (Newtonian fluid) that may be written as $\underline{v} = v_r\hat{e}_r + v_\theta\hat{e}_\theta$ in a cylindrical coordinate system, calculate the 3×3 matrix $\nabla\underline{v} + (\nabla\underline{v})^T$. What is $\underline{\tilde{\underline{\tau}}}$ for this flow?

16. In Example 5.1, we use the x-component of the momentum balance in a simple shear flow to show that the shear stress is constant in simple shear

flow. What conclusions can we draw from the y- and z-components of the momentum balance?

17. In the development of Newton's law of viscosity in Section 5.1.2, we discuss experimental results that show that in steady shear flow, F/LW is proportional to V/H, where F is the tangential pulling force on the top plate, A is the area of the plate, V is the speed of the plate, and H is the gap between the plates. Use your own experience with fluids to describe how force depends on gap in shear flow for constants A and V. Also, describe a situation that illustrates how force depends on area for constants H and V.

18. A fluid is made to flow in a parallel-plate apparatus with a narrow gap of 1.0 mm. The tangential force to move the top plate at 0.012 mm/s is 13 mN. What is the viscosity of the fluid in the gap? Give the answer in centipoise and American engineering units. The plate area is 9.1 cm^2.

19. What is the vector velocity field in a simple shear flow produced in the flow between two large parallel plates? The top plate is moved at 1.2 mm/s and the gap between the plates is 0.5 mm. What is the velocity (magnitude and direction) in the plane halfway between the plates?

20. What is the shear stress if peanut butter is sheared in a narrow-gap parallel plate device? The gap is set at 0.8 mm and the velocity of the upper plate is 0.1 mm/s. What is the shear stress if the gap doubles?

21. In a narrow-gap parallel-plate device, a Newtonian fluid is made to flow in steady-drag flow. If the shear rate (i.e., shear rate = velocity/gap in this flow) is cut in half, what happens to the stress on the upper plate?

22. What is the extra-stress tensor for a Newtonian fluid undergoing the uniform flow described by the velocity profile given here? U_∞ is a constant.

$$\underline{v} = \begin{pmatrix} U_\infty \\ 0 \\ 0 \end{pmatrix}_{xyz}$$

23. What is the extra-stress tensor for a Newtonian fluid undergoing the shear flow described by the velocity profile given here?

$$\underline{v} = \begin{pmatrix} -8y \\ 0 \\ 0 \end{pmatrix}_{xyz}$$

24. What is the extra-stress tensor for a Newtonian fluid undergoing the pipe flow described by the velocity profile given here? V and R are constants.

$$\underline{v} = \begin{pmatrix} 0 \\ 0 \\ V\left(1 - \frac{r^2}{R^2}\right) \end{pmatrix}_{r\theta z}$$

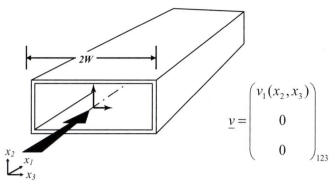

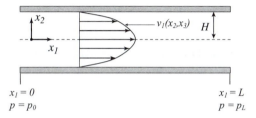

cross section:

$x_1 = 0$
$p = p_0$

$x_1 = L$
$p = p_L$

Figure 5.37 Unidirectional flow in a rectangular channel varies in three dimensions (Problem 30).

25. What is the extra-stress tensor for a Newtonian fluid undergoing the elongational flow described by the velocity profile given here?

$$\underline{v} = \begin{pmatrix} -3x \\ -3y \\ 6z \end{pmatrix}_{xyz}$$

26. What is the extra-stress tensor for a Newtonian fluid undergoing the flow in the spiral vortex tank with the velocity profile given here? K is a constant. See Chapter 6 for more details on calculating velocity profiles in flow.

$$\underline{v} = \begin{pmatrix} 0 \\ \frac{K}{r} \\ 0 \end{pmatrix}_{r\theta\phi}$$

27. When a velocity field in a flow has been calculated, it is straightforward subsequently to calculate the stress tensor. For creeping flow around a sphere, calculate $\underline{\underline{\tilde{\Pi}}}$ from the solution given in Equations 5.101 and 5.102.

28. For the problem of a film flowing down an inclined plane discussed in this chapter, what is the maximum value of the velocity? What is the average flow rate?

29. For the problem of a film flowing down an inclined plane discussed in this chapter, what is the maximum value of the shear stress? How does shear stress vary across the film thickness?

30. How do the velocity components simplify for the unidirectional flow of water through a rectangular channel (Figure 5.37)? For a Newtonian fluid, how does the extra-stress tensor simplify as a result of this?

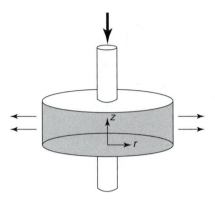

Figure 5.38 Squeeze flow between two parallel plates is a common flow test geometry (Problem 31).

31. How do the velocity components simplify for the squeezing flow shown in Figure 5.38? For a Newtonian fluid, how does the extra-stress tensor simplify as a result of this?

32. How do the velocity components simplify for the wide-slit planar-contraction flow shown in Figure 5.39? For a Newtonian fluid, how does the extra-stress tensor simplify as a result of this?

33. How do the velocity components simplify for axisymmetric-contraction flow shown in Figure 5.40? For a Newtonian fluid, how does the extra-stress tensor simplify as a result of this?

34. The problem of a thin film falling down an incline is discussed in this chapter. Figure 5.41 is a version of this problem in which a Cartesian coordinate system is proposed. Write the velocity vector in this coordinate system (i.e., which components are zero?) and the boundary conditions. How does the

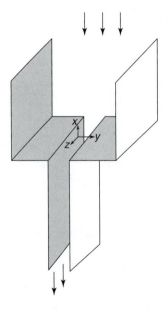

Figure 5.39 Planar contraction flow occurs when a larger reservoir drains through a slot (Problem 32).

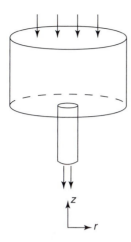

Figure 5.40   Axisymmetric-contraction flow occurs when a larger round reservoir drains through a pipe (Problem 33).

Newtonian constitutive equation simplify? Comment on the chosen coordinate system.

35. We solved the incline problem using a coordinate system that sits on the solid surface (Figure 5.12). Here, choose instead to solve the same problem with a coordinate system in which the coordinate position $x = 0$ is located at the free surface (i.e., the top of the film) (Figure 5.42). What is the gravity acceleration vector in this coordinate system? What are the flow-boundary conditions? Are there any advantages or disadvantages to this choice of coordinate system? Discuss your observations.

36. In a drag flow in the x-direction with the gradient in the y-direction, the force on the top plate is measured as $\underline{f} = 16\hat{e}_x + 2\hat{e}_y$. Is the fluid Newtonian? How do you know one way or the other?

37. The text indicates that the generalized Newtonian models cannot predict normal stresses in shear; show that this is true. (Hint: Write $\underline{\underline{\dot{\gamma}}}$ for shear flow and calculate stresses from the constitutive equation.)

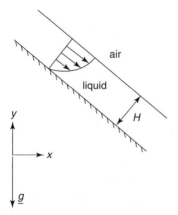

Figure 5.41   Flow problems may be solved in any coordinate system. Some choices are much better than others, however (Problem 34).

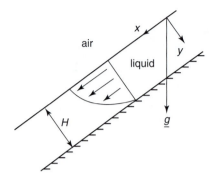

Figure 5.42 Flow problems may be solved in any coordinate system. The choice here has hidden advantages (Problem 35).

38. What are the units of the power-law parameters m and n in the power-law model for viscosity, $\eta = m\dot{\gamma}^{n-1}$ (see Equation 5.242)? Why are the units so strange? Is there an error?

39. For a high molecular-weight polymer, the stress response is modeled with the generalized Newtonian fluid constitutive equation, $\underline{\underline{\tilde{\tau}}} = \eta(\dot{\gamma})\left(\nabla\underline{v} + (\nabla\underline{v})^T\right)$. The viscosity function $\eta(\dot{\gamma})$ is measured as:

$$\eta[\text{Pa s}] = \left(3.4 \times 10^5\right)\dot{\gamma}^{-0.43} = \left(3.4 \times 10^5\right)\left|\frac{\partial v_1}{\partial x_2}\right|^{-0.43}$$

where $\dot{\gamma} = \left|\frac{\partial v_1}{\partial x_2}\right|$ is given in units of s^{-1}. Plot the viscosity function (i.e., log-log plot). In a flow with $\underline{v} = (0.90\ 1/\text{s})x_2\hat{e}_1$, what is $\tilde{\tau}_{21}$?

40. Plot the viscosity-versus-shear rate for a power-law generalized Newtonian fluid. The parameters of the model are $m = 3{,}200\ \text{Pa s}^{0.54}$ and $n = 0.54$. Choose a range of shear rate that is physically reasonable. The plot should be logarithmic on both axes.

41. Plot the viscosity versus shear rate for a Carreau–Yasuda Generalized Newtonian fluid. The parameters of the model are listed here. Choose a range of shear rate that is physically reasonable. The plot should be logarithmic on both axes:

$$\eta_0 = 3{,}500\ \text{Pa s}$$
$$\hat{n} = 0.42$$
$$a = 2.0$$
$$\eta_\infty = 0$$
$$\lambda = 11\ \text{s}$$

42. The units of the power-law model do not follow accepted rules in physics. What are the units on m in the power-law model? We can normalize the units of this model by introducing a parameter λ to nondimensionalize the shear rate $\dot{\gamma}$ [36]:

$$\text{Revised power-law model:}\quad \eta = \tilde{\eta}\left(\lambda\dot{\gamma}\right)^{n-1}$$

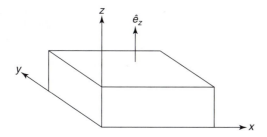

Figure 5.43   A control volume shaped like a rectangular parallelepiped (Problem 44).

For viscosity in units of Pa s, what are the units of the parameters $\tilde{\eta}$ and n of this revised power-law model? Comment on the relative desirability of the revised power-law model compared to the traditional model.

43. What is the extra-stress tensor for a non-Newtonian power-law fluid undergoing the elongational flow described by the velocity profile given here?

$$\underline{v}[\text{m/s}] = \begin{pmatrix} -4x \\ -4y \\ 8z \end{pmatrix}_{xyz}$$

44. A control volume shaped like a rectangular parallelepiped is shown in Figure 5.43. For the velocity field given in Problem 43, how would you calculate the total fluid force on the z-surface of the control volume for a power-law GNF? What additional information would you need? The dimensions of the box in the x-, y-, and z-directions are L, W, and H, respectively.

45. What is the extra-stress tensor for a non-Newtonian power-law fluid undergoing the uniform flow described by the velocity profile given here? U_∞ is a constant.

$$\underline{v} = \begin{pmatrix} U_\infty \\ 0 \\ 0 \end{pmatrix}_{xyz}$$

46. What is the extra-stress tensor for a non-Newtonian power-law fluid undergoing the shear flow described by the velocity profile given here? The parameter a is a constant.

$$\underline{v} = \begin{pmatrix} -ay \\ 0 \\ 0 \end{pmatrix}_{xyz}$$

47. What is the extra-stress tensor for a non-Newtonian power-law fluid undergoing the pipe flow described by the velocity profile given here? V and R are constants.

$$\underline{v} = \begin{pmatrix} 0 \\ 0 \\ V\left(1 - \left(\frac{r}{R}\right)^{2.85}\right) \end{pmatrix}_{r\theta z}$$

48. For Newtonian fluids, the extra-stress tensor is given by $\underset{\approx}{\tilde{\tau}} = \mu \left(\nabla \underline{v} + (\nabla \underline{v})^T \right)$, where μ is a constant—the viscosity. Fluids like "oobleck" (i.e., cornstarch and water) exhibit unusual behavior that does not follow this equation. Investigate the behavior of oobleck on the Internet and indicate the features of the Newtonian constitutive equation that prevent it from describing the behavior of oobleck.

49. For Newtonian fluids, the extra-stress tensor is given by $\underset{\approx}{\tilde{\tau}} = \mu \left(\nabla \underline{v} + (\nabla \underline{v})^T \right)$, where the viscosity μ is a constant. Fluids like pizza dough exhibit unusual behavior that does not follow this equation. Investigate and indicate what must be changed in the Newtonian constitutive equation to correctly describe the behavior of pizza dough.

50. For Newtonian fluids, the extra-stress tensor is given by $\underset{\approx}{\tilde{\tau}} = \mu \left(\nabla \underline{v} + (\nabla \underline{v})^T \right)$, where the viscosity μ is a constant. Fluids like high molecular-weight polymer melts exhibit behavior that does not follow this equation. Investigate and indicate what must be changed in the Newtonian constitutive equation to correctly describe the behavior of entangled polymer melts.

PART III

FLOW FIELD CALCULATIONS

6 Microscopic Balance Equations

In Chapter 5, we completed two example problems that we have pursued for several chapters. The first was the flow of a thin liquid film down an inclined plane. To solve that problem, we chose a microscopic control volume (CV) and performed a momentum balance on it. In the course of the solution, we adopted the use of the stress tensor $\tilde{\underline{\underline{\Pi}}} = -p\underline{\underline{I}} + \tilde{\underline{\underline{\tau}}}$ as a way to quantify the contact forces on the CV. We subsequently introduced the Newtonian constitutive equation to relate the extra-stress tensor $\tilde{\underline{\underline{\tau}}}$ to velocity. The microscopic CV solution method used in Chapter 5 is general and can be applied to other problems, although not without difficulty. Choosing a microscopic CV can be tricky; if the flow streamlines are not straight, getting all of the geometric factors correct is challenging.

The microscopic-balancing method can be made considerably easier by developing general balance equations and beginning the analysis there. We pursue this problem-solving method in this chapter. The derivations of the microscopic-mass-balance and the microscopic-momentum-balance equations are provided in Section 6.1, using vector/tensor theorems introduced there. The derivations in Section 6.1 are generalizations of the control-volume method used in Chapters 3–5 to solve the falling-film problem. The two general balance equations derived in this chapter are as follows:

$$\begin{array}{cc} \text{Continuity equation:} \\ \text{(microscopic mass balance)} \end{array} \qquad \left(\frac{\partial \rho}{\partial t} + \underline{v} \cdot \nabla \rho\right) = -\rho \nabla \cdot \underline{v} \qquad (6.1)$$

$$\begin{array}{cc} \text{Navier-Stokes equation:} \\ \text{(microscopic momentum balance)} \end{array} \qquad \rho\left(\frac{\partial \underline{v}}{\partial t} + \underline{v} \cdot \nabla \underline{v}\right) = -\nabla p + \mu \nabla^2 \underline{v} + \rho \underline{g}$$

$$(6.2)$$

These are partial differential equations (PDEs) for velocity and pressure. Using the continuity and the Navier-Stokes equations is a streamlined way to apply the continuum model and the control-volume analysis of Chapters 3–5.

The Navier-Stokes equation is the core tool used to solve problems in fluid mechanics. The tools necessary to carry out microscopic momentum balances on arbitrary flows of Newtonian fluids are (1) an understanding of the solution methodology; (2) tables of the Navier-Stokes equation written in common coordinate systems (see Appendix B); and (3) an understanding of the types of boundary conditions encountered in such problems. These topics are discussed

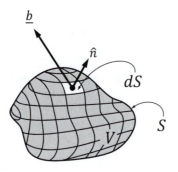

Figure 6.1   Schematic of an arbitrary volume enclosed by a surface area. Each particular piece of the surface, dS, is characterized by the direction of its unit normal, $\hat{n}$.

in Section 6.2, and it is an option for readers to proceed directly to that section. Chapters 7, 8, and 10 discuss specific applications of the Navier-Stokes equation to a variety of flow problems. In those chapters, we also sort out the various behaviors that fluids exhibit and finally learn the reasons for observed flow patterns and stresses.

The second flow problem completed in Chapter 5 is a calculation of the force needed to restrain a 90 degree pipe bend with water flowing in it. To solve the 90-degree-bend problem, we chose a macroscopic control volume and wrote a momentum balance. We generalize the macroscopic solution method in Chapter 9. The general macroscopic-balance equations also enhance our appreciation of flow-system behavior. Macroscopic balances are suited to developing practical data correlations that apply to fluid devices (see Chapter 9).

6.1 Deriving the microscopic balance equations

The solution of the flow-down-an-incline problem began with choosing the control volume, after which we applied the mass balance and the momentum balance (i.e., Reynolds transport theorem); this led to differential equations—first for stress and then for velocity—which were solved with boundary conditions. The derivation of the general microscopic balance equations follows these steps.

To calculate the momentum balance in the general case, the control volume we choose is an arbitrarily shaped microscopic volume (Figure 6.1). Because it is arbitrarily shaped, we cannot define its geometry and carry out specific calculations on this control volume; instead, we write the terms of the balances as integrals over the volume and enclosing surface. Because the choice of CV is arbitrary, we subsequently deduce that the relationships we derive must hold at every point in space.

The chosen CV has a total volume $\mathcal{V}$ and is enclosed by a surface $\mathcal{S}$. To write the amount of mass, momentum, and energy interacting with the CV, we again use the fundamental definition of an integral as a limit of a sum (see Section 1.3 and the Web appendix [108] for details). To calculate the quantity of mass, momentum, or energy in the CV, we construct a sum of small pieces of mass, momentum, or energy located in small volumes ΔV within it:

$$
\begin{array}{c}
\text{Total quantity} \\
\text{of a property within} \\
\text{volume } \mathcal{V}
\end{array}
\approx \sum_{i=1}^{N} \left(\frac{\text{quantity}}{\text{volume}} \right) \Delta V_i
\tag{6.3}
$$

$$
= \sum_{i=1}^{N} f(x_i, y_i, z_i) \Delta V_i
\tag{6.4}
$$

where f in this case is quantity per volume of mass, momentum, or energy. In the limit that ΔV goes to zero, this approximation becomes exact (see the Web appendix [108]):

$$\begin{array}{c}\text{Total quantity}\\ \text{of a property within} \\ \text{volume } \mathcal{V}\end{array} = \lim_{\Delta V \to 0} \left[\sum_{i=1}^{N} f(x_i, y_i, z_i) \Delta V_i \right] \tag{6.5}$$

This expression may be written as a triple integral of f over the volume $\mathcal{V}$ (see the Web appendix [108]):

$$\begin{array}{c}\text{Total quantity}\\ \text{of a property within} \\ \text{volume } \mathcal{V}\end{array} = \iiint_{\mathcal{V}} f(x, y, z) dV \tag{6.6}$$

For quantities that act on or pass through the control surface, we use surface integrals, much like as in Section 3.2.2 when we calculated the convection term in the Reynolds transport theorem. To calculate the total quantity of mass, momentum, or energy acting on or passing through the control surface, we construct a sum of contributions associated with small pieces of tangent plane ΔS. If we then take the limit as $\Delta A = (\hat{n} \cdot \hat{e}_z) \Delta S$ goes to zero, we obtain a surface integral:

$$\begin{array}{c}\text{Rate of transfer}\\ \text{of a property} \\ \text{through or to } \mathcal{S}\end{array} \approx \sum_{i=1}^{N} \left(\frac{\text{quantity}}{\text{area} \cdot \text{time}} \right) \Delta S_i \tag{6.7}$$

$$= \sum_{i=1}^{N} f(x_i, y_i, z_i) \Delta S_i \tag{6.8}$$

$$\begin{array}{c}\text{Rate of transfer}\\ \text{of a property} \\ \text{through or to } \mathcal{S}\end{array} = \lim_{\Delta A \to 0} \left[\sum_{i=1}^{N} f(x_i, y_i, z_i) \Delta S_i \right] \tag{6.9}$$

where, in the surface case, f is flux of a property, quantity/(area·time), and ΔA_i is the projection of the tangent-plane area ΔS_i. The two areas ΔS_i in the tangent plane and its projection ΔA_i are related through geometry (see the Web appendix [108]):

$$\Delta A_i = (\hat{n}_i \cdot \hat{e}_z) \Delta S_i \tag{6.10}$$

where $\hat{n}_i$ is the outwardly pointing unit normal to the tangent plane ΔS_i and $\hat{e}_z$ is the z-direction of a convenient Cartesian coordinate system. Substituting Equation 6.10 into Equation 6.9, we obtain:

$$\begin{array}{c}\text{Rate of transfer}\\ \text{of a property} \\ \text{through or to } \mathcal{S}\end{array} = \lim_{\Delta A \to 0} \left[\sum_{i=1}^{N} \frac{f(x_i, y_i, z_i)}{(\hat{n}_i \cdot \hat{e}_z)} \Delta A_i \right] \tag{6.11}$$

Equation 6.11 may be written as a double integral over the total projected area $\mathcal{R}$ (see the Web appendix [108]):

$$
\begin{array}{c}
\text{Rate of transfer} \\
\text{of a property} \\
\text{through or to } \mathcal{S}
\end{array}
= \iint_{\mathcal{R}} \frac{f(x, y, z)}{(\hat{n} \cdot \hat{e}_z)} \, dA
\qquad (6.12)
$$

If we define $dS \equiv dA/(\hat{n} \cdot \hat{e}_z)$, then Equation 6.12 becomes the surface integral:

$$
\begin{array}{c}
\text{Rate of transfer} \\
\text{of a property} \\
\text{through or to } \mathcal{S}
\end{array}
= \iint_{\mathcal{S}} f(x, y, z) \, dS
\qquad (6.13)
$$

The volume integral in Equation 6.6 and the surface integral in Equation 6.13 allow us to express mass, momentum, and energy terms in calculations on our arbitrary control volume. We need one more tool from vector mathematics to carry out our derivation of the microscopic balances, and we state this theorem without proof in the next section.

6.1.1 Gauss-Ostrogradskii divergence theorem

The Gauss-Ostrogradskii divergence theorem[1] relates the change of a vector or tensor property, $\underline{b}$, in a closed volume, $\mathcal{V}$, with the flux of that property through the surface $\mathcal{S}$ that encloses $\mathcal{V}$ [6, 58, 146] (Figure 6.1). The theorem is:

$$
\begin{array}{c}
\text{Gauss-Ostrogradskii} \\
\text{divergence theorem}
\end{array}
\qquad
\boxed{\iint_{\mathcal{S}} (\hat{n} \cdot \underline{b}) \, dS = \iiint_{\mathcal{V}} (\nabla \cdot \underline{b}) \, dV}
\qquad (6.14)
$$

$$
\begin{array}{c}
\text{Divergence theorem} \\
\text{(Cartesian coordinates,} \\
\underline{b} \text{ is a vector)}
\end{array}
\qquad
\begin{aligned}
&\iint_{\mathcal{S}} (n_1 b_1 + n_2 b_2 + n_3 b_3) \, dS \\
&= \iiint_{\mathcal{V}} \left(\frac{\partial b}{\partial x_1} + \frac{\partial b}{\partial x_2} + \frac{\partial b}{\partial x_3} \right) \, dV
\end{aligned}
\qquad (6.15)
$$

where $\hat{n}$ is the outwardly pointing unit normal of the differential surface element, dS. The volume $\mathcal{V}$ is not necessarily constant in time. Use of the Gauss-Ostrogradskii divergence theorem allows us to convert an integral over a volume into a surface integral (or vice versa) without loss of information. The divergence theorem can be thought of as justification for defining the divergence operator ∇ in the first place. Proof of the divergence theorem is in the literature [146].

The divergence theorem relates two different mathematical ways of expressing changes in a function with position (x, y, z) within a volume $\mathcal{V}$. The divergence operator $\nabla\cdot$ quantifies changes by considering the spatial derivatives $\partial/\partial x_i$ in three coordinate directions. Recall that in Cartesian coordinates, the gradient operator ∇ is defined as:

$$
\nabla = \begin{pmatrix} \frac{\partial}{\partial x} \\ \frac{\partial}{\partial y} \\ \frac{\partial}{\partial z} \end{pmatrix}_{xyz}
\qquad (6.16)
$$

[1] Also known as Green's theorem or simply as the divergence theorem; see Aris and Schey [6, 146].

The divergence operator $\nabla\cdot$ acts on vectors or tensors; for example, for the vector $\underline{v}$:

$$\text{Divergence of } \underline{v}: \qquad \nabla \cdot \underline{v} = \frac{\partial v_x}{\partial x} + \frac{\partial v_y}{\partial y} + \frac{\partial v_z}{\partial z} \qquad (6.17)$$

Note that the divergence of a vector is a scalar. For the tensor $\underline{\underline{A}}$, the divergence is given here. Note that the divergence of a tensor is a vector:

$$\text{Divergence of } \underline{\underline{A}}: \qquad \nabla \cdot \underline{\underline{A}} = \begin{pmatrix} [\nabla \cdot \underline{\underline{A}}]_x \\ [\nabla \cdot \underline{\underline{A}}]_y \\ [\nabla \cdot \underline{\underline{A}}]_z \end{pmatrix}_{xyz} = \begin{pmatrix} \frac{\partial A_{xx}}{\partial x} + \frac{\partial A_{yx}}{\partial y} + \frac{\partial A_{zx}}{\partial z} \\ \frac{\partial A_{xy}}{\partial x} + \frac{\partial A_{yy}}{\partial y} + \frac{\partial A_{zy}}{\partial z} \\ \frac{\partial A_{xz}}{\partial x} + \frac{\partial A_{yz}}{\partial y} + \frac{\partial A_{zz}}{\partial z} \end{pmatrix}_{xyz}$$

$$(6.18)$$

A way to characterize an amount of change occuring in a finite volume is to add up (i.e., volume integrate) the divergence of a property over the volume, giving the righthand side of Equation 6.14. When defining ∇ and choosing to add up $\nabla \cdot \underline{b}$ over a volume V, we do not have specific physics in mind—it is simply a way to characterize an amount of spatial change present in the function $\underline{b}$.

A second mathematical way to express spatial changes in a finite volume is given by the lefthand side of Equation 6.14. The lefthand side of the divergence theorem adds up (i.e., surface integrates) the outward components of a quantity acting on or crossing the bounding surface. This integral may express a quantity of interest for a physical problem such as the mass flow through a surface. The divergence theorem states that these two calculated quantities are equal.

The physical meanings of terms that resemble either the lefthand or the righthand side of the divergence theorem depend on the identity of the quantity $\underline{b}$. In the sections that follow, we use integrals of these two types. The divergence theorem provides a way to connect and to interrelate physical phenomena that naturally manifest in spatial rates-of-change in volumes (i.e., right side) or as fluxes through surfaces (i.e., left side).

We now discuss the derivation of the microscopic mass, momentum, and energy balances.

6.1.2 Mass balance

One of the fundamental laws of nature is that mass is conserved. Mathematically, this means that the time rate-of-change of mass of a body m_B is zero:

$$\text{Mass of a body is conserved} \qquad \frac{dm_B}{dt} = 0 \qquad (6.19)$$

For flow calculations, we seek to write the mass balance on a control volume rather than on a body. The general mass balance on a microscopic control volume is found by considering CV mass changes over a time interval Δt. Recall that in the Chapter 3 discussion, we consider a particular control volume and the momentum flows into and out of that volume (see Figure 3.22). That analysis, now performed for the changes in mass taking place in the arbitrary CV, yields an expression for the time rate-of-change of the mass of the CV:

$$\begin{pmatrix} \text{rate of increase} \\ \text{of mass in CV} \end{pmatrix} = \begin{pmatrix} \text{mass} \\ \text{flow in} \end{pmatrix} - \begin{pmatrix} \text{mass} \\ \text{flow out} \end{pmatrix} \tag{6.20}$$

$$\frac{dm_{CV}}{dt} = \begin{pmatrix} \text{net inward} \\ \text{mass flow} \\ \text{through } \mathcal{S} \end{pmatrix} \tag{6.21}$$

where m_{CV} is the mass of the CV and $\mathcal{S}$ is the bounding surface of the CV. This equation is the mass equivalent of Equation 3.65, the momentum balance on a CV.

To apply Equation 6.21 to the general CV in Figure 6.1, we work term by term. The net inward flux of mass through the control surface $\mathcal{S}$ can be written using Equation 6.13:

$$\begin{array}{c} \text{Rate of transfer} \\ \text{of a property} \\ \text{through or to } \mathcal{S} \end{array} = \iint_{\mathcal{S}} f(x, y, z) \, dS \tag{6.22}$$

where f is flux of a property, quantity/(area·time). Recall that this relationship is discussed in Chapter 3 (see Section 3.2.2.2). At a point on the control surface with velocity $\underline{v}$ and outward unit normal vector $\hat{n}$, we can write mass/(area · time) as follows (see Equation 3.90):

$$\frac{\text{mass}}{\text{area} \cdot \text{time}} = \frac{\left(\dfrac{\text{mass}}{\text{volume}}\right)\left(\dfrac{\text{volume}}{\text{time}}\right)}{\text{area}} \tag{6.23}$$

$$\frac{\text{mass out}}{\text{area} \cdot \text{time}} = \frac{(\rho)\begin{pmatrix} \text{outward} \\ \text{velocity} \cdot \text{area} \\ \text{component} \end{pmatrix}}{\text{area}} \tag{6.24}$$

$$= \rho \, (\hat{n} \cdot \underline{v}) \tag{6.25}$$

Recall that $\hat{n}$ is the outwardly pointing unit normal and $\hat{n} \cdot \underline{v}$ is the component of $\underline{v}$ that passes out of the CV (see Equation 3.90). Substituting Equation 6.25 into Equation 6.22 gives:

$$\begin{array}{c} \text{Net outward} \\ \text{mass flow through } \mathcal{S} \end{array} = \iint_{\mathcal{S}} \rho(\hat{n} \cdot \underline{v}) \, dS \tag{6.26}$$

Substituting Equation 6.26 into the equation for mass conservation on a CV (Equation 6.21) yields:

$$\frac{dm_{CV}}{dt} = \begin{pmatrix} \text{net inward} \\ \text{mass flow} \\ \text{through } \mathcal{S} \end{pmatrix} \tag{6.27}$$

Mass balance on a CV
$$\boxed{\frac{dm_{CV}}{dt} = \iint_{\mathcal{S}} -(\hat{n} \cdot \underline{v})\rho \, dS} \tag{6.28}$$

We include a negative sign to convert outflow to inflow. This version of the mass balance on a CV is directly analogous to the Reynolds transport theorem (see Equation 3.135), with mass substituted for momentum:

Momentum balance on a CV
(Reynolds transport theorem)
$$\boxed{\frac{d\mathbf{P}}{dt} + \iint_{CS} (\hat{n} \cdot \underline{v}) \, \rho \underline{v} \, dS = \sum_{\substack{\text{on} \\ \text{CV}}} \underline{f}} \tag{6.29}$$

The integral term in both expressions is called the convective term, and it appears when we consider a CV rather than a body. An analogous term appears in the energy balance (see Section 6.1.4).

The lefthand side of the mass balance (Equation 6.28) is evaluated for our arbitrary CV by first writing m_{CV} in terms of an integral of the local density over the CV (see Equation 6.6):

$$m_{CV} = \iiint_{\mathcal{V}} \left(\frac{\text{mass}}{\text{volume}}\right) dV \tag{6.30}$$

$$m_{CV} = \iiint_{\mathcal{V}} \rho \, dV \tag{6.31}$$

The density may be a function of time and position. The time derivative may be calculated by differentiating Equation 6.31 directly:

$$\frac{dm_{CV}}{dt} = \frac{d}{dt} \iiint_{\mathcal{V}} \rho \, dV \tag{6.32}$$

We can move the time derivative from the outside of the integral to become a partial derivative inside the volume integral by using a vector relation called the *Leibniz rule* for a constant volume $\mathcal{V}$ (see the Web appendix [108] and [58]):

Leibniz rule
(constant volume)
$$\boxed{\frac{d}{dt} \iiint_{\mathcal{V}} \rho \, dV = \iiint_{\mathcal{V}} \frac{\partial \rho}{\partial t} \, dV} \tag{6.33}$$

The mass balance is therefore:

$$\frac{dm_{CV}}{dt} = -\iint_{\mathcal{S}} (\hat{n} \cdot \underline{v})\rho \, dS \tag{6.34}$$

$$\iiint_{\mathcal{V}} \frac{\partial \rho}{\partial t} \, dV = -\iint_{\mathcal{S}} (\hat{n} \cdot \underline{v})\rho \, dS \tag{6.35}$$

To convert the righthand side of Equation 6.35 to an integral over the volume rather than over the surface, we use the divergence theorem, which relates the surface integral of $\hat{n}\cdot$ a property to the volume integral of $\nabla\cdot$ that same property. In the mass balance, we apply the divergence theorem to the property $\rho\underline{v}$. Equation 6.35 becomes:

$$\iiint_V \frac{\partial\rho}{\partial t}dV = -\iint_S (\hat{n}\cdot\underline{v})\rho\,dS \tag{6.36}$$

$$= -\iint_S \hat{n}\cdot(\rho\underline{v})\,dS \tag{6.37}$$

$$= -\iiint_V \nabla\cdot(\rho\underline{v})\,dV \tag{6.38}$$

It is important when applying the divergence theorem to note that the divergence operator $\nabla\cdot$ acts on $\rho\underline{v}$ and not on $\underline{v}$ alone because both ρ and $\underline{v}$ may vary with position. Finally, Equation 6.38 relates the integrals of two properties over the same arbitrary volume V. We now combine the two integrals under the same integration symbol:

$$\iiint_V \left[\frac{\partial\rho}{\partial t} + \nabla\cdot(\rho\underline{v})\right]dV = 0 \tag{6.39}$$

We arrive at an equation for conservation of mass over an arbitrary volume V in our flowing stream. Because that volume is arbitrary, however, this equation must hold over every volume we choose. The only way that this can be true is if the expression within the integral sign is zero at every position in space. Note that it is unusual that we may conclude that an integral being zero implies that the integrand therefore must be zero everywhere. We consider this situation in the following example.

EXAMPLE 6.1. *When an integral is zero, under what circumstances may we assume the integrand to be equal to zero?*

SOLUTION. It is highly unusual to conclude that an integral being zero implies that the integrand—the quantity being integrated over—is zero. Consider the function $y = \sin x$. Although we know that:

$$\int_0^{2\pi} \sin x\,dx = 0 \tag{6.40}$$

we also know that $\sin x$ is not equal to zero at every point (Figure 6.2). What makes the integral in Equation 6.40 equal to zero is the choice of limits. We can find any number of other limits, however, over which the integral of sine is not zero.

In the case discussed in Equation 6.39, the limits of the integral are the boundaries of V, which are arbitrary. If there is a function over which we may integrate and always obtain zero no matter which limits are chosen, that function must be

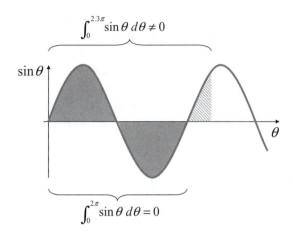

$$\int_0^{2.3\pi} \sin\theta \, d\theta \neq 0$$

$\sin\theta$

θ

$$\int_0^{2\pi} \sin\theta \, d\theta = 0$$

Figure 6.2 For many functions, an interval may be found over which the integral of the function is zero. Only the function $f(x) = 0$ integrates to zero over every possible interval.

equal to zero. It is the arbitrariness of the integration limits—$\mathcal{V}$ in the current case—that implies that the integrand must be zero at every point.

Thus, although it is highly unusual to state that when an integral is zero implies its integrand must be zero, the one time this is true is when the limits in the integral are arbitrary. The arbitrariness of $\mathcal{V}$ implies that the integrand in Equation 6.39 must be zero at every point:

Continuity equation
(microscopic mass balance)
$$0 = \frac{\partial\rho}{\partial t} + \nabla\cdot(\rho\underline{v}) \qquad (6.41)$$

The result of our calculation is the microscopic-mass-balance equation, which is known as the *continuity equation*.

We can expand Equation 6.41 to three terms by carrying out the spatial derivative $\nabla\cdot(\rho\underline{v})$, which is the divergence of the product of two variables. Manipulations of vector and tensor components may be carried out in Cartesian coordinates using matrix notation or by using a compact vector–tensor notation called *Einstein notation* [104]. We omit the details of these calculations, recording only the results (see Problem 11, the Web appendix [108], and [104]). Equation 6.41 may be shown with no additional assumptions to be equivalent to (see also Table B.1):

$$0 = \frac{\partial\rho}{\partial t} + \underline{v}\cdot\nabla\rho + \rho\nabla\cdot\underline{v} \qquad (6.42)$$

Continuity equation
(microscopic mass balance)
$$\frac{\partial\rho}{\partial t} + \underline{v}\cdot\nabla\rho = -\rho\nabla\cdot\underline{v} \qquad (6.43)$$

When the density is constant (i.e., incompressible fluid), the continuity equation becomes:

Continuity equation
(incompressible fluid)
$$0 = \nabla\cdot\underline{v} \qquad (6.44)$$

We write the equations in this section in Gibbs notation (∇, $\underline{v}$, and so on), a notation from vector–tensor mathematics that allows us to write an equation independent of the coordinate system. We can write any Gibbs expression in any coordinate system by using the defining equations for the chosen coordinate system and rigorously carrying out the indicated derivatives. The continuity equation (Equation 6.43) is worked out for the Cartesian, cylindrical, and spherical coordinate systems and appears in Section 6.2 and in Table B.5 of Appendix B. We further discuss these equations in Section 6.2.1.1.

6.1.3 Momentum balance

We turn now to the derivation of the microscopic momentum balance. First the general balance is discussed; subsequently the balance is specified to Newtonian fluids.

6.1.3.1 GENERAL FLUIDS

The derivation of the microscopic momentum balance parallels the derivation of the continuity equation, although the force terms require special attention. The momentum balance on a control volume is the Reynolds transport theorem, familiar from Chapter 3:

$$\text{Reynolds transport theorem} \atop \text{(momentum balance on CV)} \quad \frac{d\underline{P}}{dt} + \iint_{CS} (\hat{n} \cdot \underline{v})\, \rho \underline{v}\, dS = \sum_{\text{on} \atop \text{CV}} \underline{f} \quad (6.45)$$

$$\frac{d\underline{P}}{dt} = -\iint_{CS} (\hat{n} \cdot \underline{v})\, \rho \underline{v}\, dS + \underbrace{\sum \underline{f}}_{\substack{\text{contact} \\ \text{forces} \\ \text{on CV} \\ \text{(molecular)}}} + \underbrace{\sum \underline{f}}_{\substack{\text{body} \\ \text{forces} \\ \text{on CV} \\ \text{(gravity)}}} \quad (6.46)$$

As before, we apply Equation 6.46 to our arbitrary microscopic control volume and proceed term by term.

The term $d\underline{P}/dt$ on the lefthand side of Equation 6.46 is the rate-of-change of momentum in the CV. We can write the total amount of momentum in the arbitrary, microscopic CV by writing the momentum analog of Equation 6.30, the total mass in the CV. To obtain $d\underline{P}/dt$, we subsequently take the time derivative of that integral:

$$\underline{P} = \iiint_V \left(\frac{\text{momentum}}{\text{volume}} \right) dV \quad (6.47)$$

$$= \iiint_V \rho \underline{v}\, dV \quad (6.48)$$

$$\frac{d\underline{P}}{dt} = \frac{d}{dt} \iiint_V \rho \underline{v}\, dV \quad (6.49)$$

$$\boxed{\frac{d\underline{P}}{dt} = \iiint_V \frac{\partial(\rho \underline{v})}{\partial t}\, dV} \quad (6.50)$$

As before, when the time derivative moves inside the integral over a fixed volume, it becomes a partial derivative (Leibniz rule; see the Web appendix [108]).

The convection term in Equation 6.46 is already in a form that applies to the arbitrary, microscopic CV. We apply the divergence theorem to express this term as a volume integral:[2]

$$\iint_{CS} (\hat{n} \cdot \underline{v}) \, \rho \underline{v} \, dS = \iint_{S} \hat{n} \cdot (\rho \underline{v} \, \underline{v}) \, dS \tag{6.51}$$

$$\boxed{\iint_{S} (\hat{n} \cdot \underline{v}) \, \rho \underline{v} \, dS = \iiint_{V} \nabla \cdot (\rho \underline{v} \, \underline{v}) \, dV} \tag{6.52}$$

The contact-force term may be written in terms of the usual integral of $\hat{n} \cdot \underline{\underline{\tilde{\Pi}}}|_{surface}$ over a finite surface, the finite surface being the surface that bounds the CV (see Equation 4.285):

$$\begin{array}{c} \text{Total molecular fluid force} \\ \text{on a finite surface } \mathcal{S} \end{array} \qquad \underline{\mathcal{F}} = \iint_{S} [\hat{n} \cdot \underline{\underline{\tilde{\Pi}}}]_{\text{at surface}} \, dS \tag{6.53}$$

$$\underset{\substack{\text{contact} \\ \text{forces} \\ \text{on CV} \\ \text{(molecular)}}}{\sum} \underline{f} = \iint_{S} \hat{n} \cdot \left(-p\underline{\underline{I}} + \underline{\underline{\tilde{\tau}}} \right) \, dS \tag{6.54}$$

where p is pressure and $\underline{\underline{\tilde{\tau}}}$ is the extra-stress tensor. We now convert Equation 6.54 to a volume integral using the divergence theorem:

$$\boxed{\underset{\substack{\text{contact} \\ \text{forces} \\ \text{on CV} \\ \text{(molecular)}}}{\sum} \underline{f} = \iiint_{V} \nabla \cdot \left(-p\underline{\underline{I}} + \underline{\underline{\tilde{\tau}}} \right) \, dV} \tag{6.55}$$

The final term of the momentum balance on the arbitrary, microscopic CV is the gravity term, which is a straightforward volume integral:

$$\underline{f}_{gravity} = \iiint_{V} \left(\frac{\text{mass} \cdot \text{acceleration}}{\text{volume}} \right) \, dV \tag{6.56}$$

$$\boxed{\underline{f}_{gravity} = \iiint_{V} \rho \underline{g} \, dV} \tag{6.57}$$

We now put all of the pieces together in the Reynolds transport theorem:

$$\begin{array}{c} \text{Reynolds transport theorem} \\ \text{(microscopic momentum balance)} \end{array} \qquad \frac{d\mathbf{P}}{dt} = \iint_{CS} (-\hat{n} \cdot \underline{v}) \, \rho \underline{v} \, dS + \underset{\substack{\text{on} \\ \text{CV}}}{\sum} \underline{f}$$

$$\tag{6.58}$$

---

[2] The resulting term $\underline{v} \, \underline{v}$ is known as a dyadic product between two vectors, and it is a tensor. Note that there is no dot product or any other operation between these vectors. (See Sections 1.3.2.2 and 3.2.2.2.)

$$\iiint_V \frac{\partial(\rho \underline{v})}{\partial t} \, dV = -\iiint_V \nabla \cdot (\rho \underline{v} \, \underline{v}) \, dV + \iiint_V \nabla \cdot \left(-p\underline{\underline{I}} + \underline{\underline{\tilde{\tau}}}\right) \, dV$$
$$+ \iiint_V \rho \underline{g} \, dV \tag{6.59}$$

As with the equation for microscopic mass balance, we combine all of the terms under a common integral sign:

$$\iiint_V \left[\frac{\partial(\rho \underline{v})}{\partial t} + \nabla \cdot (\rho \underline{v} \, \underline{v}) - \nabla \cdot \left(-p\underline{\underline{I}} + \underline{\underline{\tilde{\tau}}}\right) - \rho \underline{g}\right] dV = 0 \tag{6.60}$$

We arrive at an equation for conservation of momentum over an arbitrary volume V in our flowing stream. Because that volume is arbitrary, however, this equation must hold over *every* volume we choose. The only way that this can be true is if the expression within the integral sign is zero at every position in space. We therefore can write the integrand of Equation 6.60 as equal to zero everywhere:

$$\frac{\partial(\rho \underline{v})}{\partial t} + \nabla \cdot (\rho \underline{v} \, \underline{v}) - \nabla \cdot \left(-p\underline{\underline{I}} + \underline{\underline{\tilde{\tau}}}\right) - \rho \underline{g} = 0 \tag{6.61}$$

Microscopic
momentum $$\frac{\partial(\rho \underline{v})}{\partial t} + \nabla \cdot (\rho \underline{v} \, \underline{v}) = \nabla \cdot \left(-p\underline{\underline{I}} + \underline{\underline{\tilde{\tau}}}\right) + \rho \underline{g} \tag{6.62}$$
balance

Equation 6.62 is a version of the microscopic momentum balance, but there are algebraic simplifications we can make by carrying out the time derivative on the product $\rho \underline{v}$ and by carrying out the divergence operations ($\nabla \cdot$). These manipulations of vector and tensor components may be made in Cartesian coordinates using either matrix notation or Einstein notation [104] (see Appendix B.1, Table B.1). We omit the details of these calculations, recording only the results (see Problem 12 and [104]). Equation 6.62 may be shown to be equivalent to:

$$\rho \frac{\partial \underline{v}}{\partial t} + \underline{v} \frac{\partial \rho}{\partial t} + \rho(\underline{v} \cdot \nabla \underline{v}) + \underline{v}\nabla \cdot (\rho \underline{v}) = -\nabla p + \nabla \cdot \underline{\underline{\tilde{\tau}}} + \rho \underline{g} \tag{6.63}$$

This equation may be simplified further if we group some of these terms:

$$\rho \left(\frac{\partial \underline{v}}{\partial t} + \underline{v} \cdot \nabla \underline{v}\right) + \underline{v} \left[\frac{\partial \rho}{\partial t} + \nabla \cdot (\rho \underline{v})\right] = -\nabla p + \nabla \cdot \underline{\underline{\tilde{\tau}}} + \rho \underline{g} \tag{6.64}$$

The terms in square brackets are simply the terms of the microscopic mass balance (see Equation 6.41), and these terms sum to zero, leaving the final result:

Cauchy momentum equation
(microscopic momentum balance, $$\boxed{\rho \left(\frac{\partial \underline{v}}{\partial t} + \underline{v} \cdot \nabla \underline{v}\right) = -\nabla p + \nabla \cdot \underline{\underline{\tilde{\tau}}} + \rho \underline{g}}$$
equation of motion [EOM])

$$\tag{6.65}$$

The microscopic momentum balance in Equation 6.65 is written in Gibbs notation, which is independent of the coordinate system. Each term in Equation 6.65 is a vector. When we write the equation of motion in a chosen coordinate system, it becomes three scalar equations—one equation for each component. The microscopic momentum equation is expressed in several coordinate systems in Table B.6 in Appendix B. We further discuss these equations in Section 6.2.1.1.

6.1.3.2 NEWTONIAN FLUIDS

The equation of motion or Cauchy momentum equation (Equation 6.65) can be specialized for incompressible Newtonian fluids by incorporating the Newtonian constitutive equation (see Equation 5.89):

$$\text{Newtonian constitutive equation (incompressible)} \qquad \boxed{\underline{\underline{\tau}} = \mu \left(\nabla \underline{v} + (\nabla \underline{v})^T \right) = \mu \underline{\underline{\dot{\gamma}}}} \qquad (6.66)$$

We proceed by substituting Equation 6.66 into the general equation of motion, Equation 6.65:

$$\rho \left(\frac{\partial \underline{v}}{\partial t} + \underline{v} \cdot \nabla \underline{v} \right) = -\nabla p + \nabla \cdot \underline{\underline{\tau}} + \rho \underline{g} \qquad (6.67)$$

$$= -\nabla p + \nabla \cdot \left(\mu \left(\nabla \underline{v} + (\nabla \underline{v})^T \right) \right) + \rho \underline{g} \qquad (6.68)$$

Equation 6.68 is the microscopic momentum balance for incompressible Newtonian fluids, but there are simplifications we can make by carrying out the spatial derivatives in the viscosity term. As stated in the previous section, manipulations of vector and tensor components may be carried out in Cartesian coordinates using matrix notation or Einstein notation [104] (see Appendix B.1, Table B.1). We omit the details of these calculations, recording only the results. Equation 6.68 may be shown to be equivalent to:

$$\rho \left(\frac{\partial \underline{v}}{\partial t} + \underline{v} \cdot \nabla \underline{v} \right) = -\nabla p + \mu \nabla \cdot \nabla \underline{v} + \mu \nabla (\nabla \cdot \underline{v}) + \rho \underline{g} \qquad (6.69)$$

$$= -\nabla p + \mu \nabla^2 \underline{v} + \mu \nabla (\nabla \cdot \underline{v}) + \rho \underline{g} \qquad (6.70)$$

Note that by definition, $\nabla^2 \underline{v} \equiv \nabla \cdot \nabla \underline{v}$. We can make a final simplification because the continuity equation for incompressible fluids indicates that $\nabla \cdot \underline{v} = 0$ (see Equation 6.44). This fact eliminates the second term containing viscosity. The final result is the microscopic momentum balance for incompressible Newtonian fluids, called the *Navier-Stokes equation*:

$$\text{Navier-Stokes equation (microscopic momentum balance, incompressible Newtonian fluids)} \qquad \boxed{\rho \left(\frac{\partial \underline{v}}{\partial t} + \underline{v} \cdot \nabla \underline{v} \right) = -\nabla p + \mu \nabla^2 \underline{v} + \rho \underline{g}}$$

$$(6.71)$$

To use the Navier-Stokes equation, we must write the vector equation in component form in a chosen coordinate system. The components in Cartesian, cylindrical, and spherical coordinate systems are given in Table B.7 in Appendix B.

The difference between the Navier-Stokes equation and the Cauchy momentum equation is that the former is specialized for incompressible Newtonian fluids. By contrast, the Cauchy momentum equation is good for all fluids, including non-Newtonian fluids that follow inelastic and viscoelastic constitutive equations (see Section 5.3).

The use of the Navier-Stokes equation is discussed in Section 6.2.

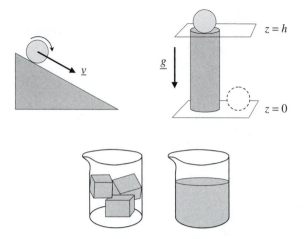

Energy is a property of a system. Energy may be stored in the state of a system—for example, as kinetic energy stored in the speed of the system, as potential energy stored in the position of the system in a potential field, or as internal energy stored in the chemical state of a system.

6.1.4 Energy balance

The third fundamental law of nature that we need that energy is conserved. The mechanical energy balance (MEB) covered in the "quick start" in Chapter 1 is based on an energy balance on a CV. We introduce energy balances here and give the complete derivation of the macroscopic energy balances in Chapter 9.

The first law of thermodynamics relates the time rate-of-change of energy of a body to the heat into the system and the work done by the system [157]:

$$\text{First law of thermodynamics:} \qquad \frac{dE_B}{dt} = Q_{in,B} - W_{by,B} \qquad (6.72)$$
$$\text{(energy conservation)}$$

The term $Q_{in,B}$ is the total rate of heat into the body; $W_{by,B}$ is the total rate of work done by the body; and E_B is the total energy of the body, consisting of internal, kinetic, and potential energy [157, 167] (Figure 6.3):

$$E_B = U + E_k + E_p \qquad (6.73)$$

The kinetic energy of a body is the energy due to the speed at which the body is moving. To calculate the kinetic energy, we first must choose a reference state; for kinetic energy, the reference state is the body at rest, $v = 0$. Relative to a body at rest, the kinetic energy of a body moving with speed v is given by:

$$\begin{pmatrix} \text{Kinetic energy} \\ \text{of a body moving} \\ \text{with speed } v \end{pmatrix} = \frac{1}{2}mv^2 = E_k \qquad (6.74)$$

where m is the mass of the body and v is the speed of the body.

Potential energy is the energy of the body by virtue of its position in a potential field. The most important potential fields are gravity and electromagnetic fields. Potential energy in Earth's gravitational field is the energy that the body has by virtue of being at a particular elevation. A ball, for example, can roll down a hill and exchange its potential energy (i.e., the energy stored in it simply by being at the top of the hill) for kinetic energy (i.e., speed). Again, energy is calculated relative to a reference state. For potential energy, we choose a reference elevation and then measure the elevation of the body relative to that reference elevation. The potential energy of a body therefore is given by:

$$\begin{pmatrix} \text{Potential energy} \\ \text{of a body at} \\ \text{elevation } z \end{pmatrix} = mg(z - z_{\text{ref}}) = E_p \qquad (6.75)$$

where m is the mass of the body, g is the acceleration due to gravity, and $(z - z_{\text{ref}})$ is the elevation of the body relative to the reference elevation z_{ref}. Often, z_{ref} is chosen to be $z = 0$ and $E_p = mgz$.

Internal energy is the energy possessed by a body internally—that is, in its molecules and atoms. The temperature of a body is one indicator of its internal energy, but a body may store internal energy in its phase (e.g., being a solid versus being a liquid) or in its chemical arrangement (e.g., being a 2:1 mixture of gases H_2 and O_2 versus being a beaker of H_2O). Internal energy is kept track of with the defined function U. Again, the value of U reported for a body is always relative to a chosen reference state:

$$\begin{pmatrix} \text{Internal energy} \\ \text{of a body} \\ \text{with respect to} \\ \text{a chosen reference} \\ \text{state} \end{pmatrix} = U \qquad (6.76)$$

The reference state for internal energy must describe fully the internal energy of the body. For example, we might choose liquid water at temperature 25°C as the reference state for a calculation involving steam. We must specify temperature (25° C in this example), phase (liquid), and chemical composition (H_2O) to fully specify the internal energy.

We seek to write the energy balance on a control volume rather than on a body. As with the mass and momentum balances on a control volume (see Equations 6.28 and 6.29), changing from the balance on a body to the balance on a control volume results in the addition of a convective term. The analysis of Chapter 3 (see Figure 3.22) that resulted in the Reynolds transport theorem also can be carried out for energy balances. The correct convective-energy term that emerges is analogous to the convective term for mass, with energy per unit volume $\rho\hat{E}$ replacing density (mass per unit volume):

Energy balance on a CV (First law of thermodynamics)
$$\boxed{\frac{dE_{CV}}{dt} = Q_{in,CV} - W_{by,CV} - \iint_S (\hat{n} \cdot \underline{v})\rho\hat{E} \, dS} \qquad (6.77)$$

In this equation, E_{CV} is the total energy of the control volume; $\hat{E}$ is the energy per unit mass of the fluid; $\hat{E} = \hat{U} + \hat{E}_k + \hat{E}_p$; and the terms $Q_{in,CV}$ and $W_{by,CV}$ are the rate of heat addition to the CV and the rate of work by the CV, respectively. The convective term (including the negative sign) represents the net flow of energy into the CV per unit time. Equation 6.77 has the same form as the mass balance on a CV (Equation 6.28) and the momentum balance on a CV (Equation 6.29).

Conservation Equations

Individual Bodies

mass
$$0 = \frac{dm_B}{dt}$$

momentum
$$\sum_{\substack{\text{on} \\ \text{body}}} \underline{f} = \frac{d(m\underline{v})_B}{dt}$$

energy
$$Q_{in,B} - W_{by,B} = \frac{d(m\hat{E})_B}{dt}$$

Control Volume Balances

mass
$$0 = \frac{dm_{CV}}{dt} + \iint_{CS} (\hat{n} \cdot \underline{v}) \rho \, dS$$

momentum
$$\sum_{\substack{\text{on} \\ \text{CV}}} \underline{f} = \frac{d(m\underline{v})_{CV}}{dt} + \iint_{CS} (\hat{n} \cdot \underline{v}) \rho \underline{v} \, dS$$

energy
$$Q_{in,CV} - W_{by,CV} = \frac{d(m\hat{E})_{CV}}{dt} + \iint_{CS} (\hat{n} \cdot \underline{v}) \rho \hat{E} \, dS$$

Figure 6.4   The mass, momentum, and energy conservation equations have their own particular physics (lefthand sides), but the transformation from the balance on a body to the balance on a control volume happens in the same way for all three relationships—that is, a convective term is added.

The three balance equations written for a body and for a CV are compared in Figure 6.4.

In fluid mechanics, the energy balance in Equation 6.77 applied to macroscopic CVs leads to important relationships among pressure, fluid velocity, and work by engineering devices such as pumps and turbines (see discussions about the mechanical energy balance and the macroscopic Bernoulli equation in Chapter 9). Equation 6.77 applied to a microscopic CV gives the microscopic energy balance, which is a fundamental relationship used to calculate properties in non-isothermal flows. In this texts we concentrate on isothermal flows; thus, the microscopic energy balance is not of central importance. The derivation of the microscopic energy balance in terms compatible with this text is available in the Web appendix [108]. The final microscopic energy balance is:

Microscopic energy balance
$$\rho \left(\frac{\partial \hat{E}}{\partial t} + \underline{v} \cdot \nabla \hat{E} \right) = -\nabla \cdot \underline{q} - \nabla \cdot (p\underline{v}) + \nabla \cdot \left(\underline{\underline{\tau}} \cdot \underline{v} \right) + S_e$$

(6.78)

Microscopic Balances

mass

$$\left(\frac{\partial \rho}{\partial t}+\underline{v}\cdot\nabla\rho\right)=-\rho(\nabla\cdot\underline{v})$$

momentum

$$\rho\left(\frac{\partial\underline{v}}{\partial t}+\underline{v}\cdot\nabla\underline{v}\right)=-\nabla p+\nabla\cdot\underline{\tilde{\tau}}+\rho\underline{g}$$

energy

$$\rho\left(\frac{\partial\hat{E}}{\partial t}+\underline{v}\cdot\nabla\hat{E}\right)=-\nabla\cdot\underline{q}-\nabla\cdot(p\underline{v})+\nabla\cdot\underline{\tilde{\tau}}\cdot\underline{v}+S_e$$

Figure 6.5   The mass, momentum, and energy conservation equations have their own particular physics (righthand sides), but the transformation from the balance on a body to the balance on a control volume happens in the same way for all three relationships—that is, a convective term is added. In the microscopic balances, this results in the substantial derivative.

where $\hat{E}$ is the energy per unit mass, $\underline{q}$ is the energy flux due to conduction, and S_e is the energy produced per time per volume by energy sources. All of the microscopic balances are summarized in Figure 6.5. Another common version of the microscopic energy balance is given in Equation 6.79:

Thermal energy equation
(no viscous dissipation,
fluid at constant p or $\rho\neq\rho(T)$)

$$\rho\hat{C}_p\left(\frac{\partial T}{\partial t}+\underline{v}\cdot\nabla T\right)=k\nabla^2 T+S_e$$

(6.79)

where T is temperature, $\hat{C}_p$ is specific heat capacity, and k is the thermal conductivity. For more information on nonisothermal flows, see the literature [15].

6.2 Using microscopic-balance equations

Key to using the microscopic balances in flow calculations is: understanding their meaning and how to employ them; having access to tables of the balances written in convenient coordinate systems; and choosing boundary conditions appropriately. In this section, we discuss the isothermal microscopic mass and momentum balances and show how to solve problems with them. The final results of such calculations are the velocity field and the pressure field for a flow. In this section, we also show how macroscopic engineering quantities (e.g., the overall flow rate or the total force on a wall) are obtained from the results of microscopic-balance calculations.

Problem-Solving Procedure –
Control-Volume Approach

1. Sketch the problem.
2. Choose a coordinate system. The coordinate system should be chosen so that the velocity vector and the boundary conditions are simplified.
3. Considering the flow, simplify $\underline{v}$ as much as possible.
4. Choose a control volume on which to perform balances. The control volume should be of infinitesimal size in any direction in which momentum is being transported.
5. Write the mass balance on the control volume. Simplify the resulting expression.
6. Write the momentum balance on the control volume. Simplify the resulting expression.
7. Take the limit as the size of the control volume vanishes. The result of taking this limit is a differential equation for the stress.
8. Solve the resulting differential equation and apply boundary conditions on stress, if known.
9. Substitute the Newtonian constitutive equation, the generalized Newtonian constitutive equation, or other appropriate constitutive equation for the stress as a function of velocity and solve the resulting differential equation for velocity field. Apply velocity boundary conditions to solve for unknown constant(s) of integration.
10. Solve for the pressure field, if applicable.
11. Calculate the stress components and engineering quantities of interest.

Figure 6.6 With the control-volume approach, the steps for solving a microscopic balance are shown above. The modified steps for solving a microscopic balance using the microscopic balance equations is shown in Figure 6.9.

6.2.1 Solution methodology

In the falling-film example of Chapters 3–5, we followed a procedure—the control-volume approach—that can be used to solve for velocity fields for Newtonian fluids. We summarize the procedure in Figure 6.6.

Implementation of the CV approach is complex. Steps 4–7 in Figure 6.6 involve manipulations of the microscopic control volume, and they can be difficult steps to get right. The control-volume choice is arbitrary and, once it is chosen, the balances must be written correctly, with all necessary terms included. If the streamlines are not straight, we face the additional challenge of getting right all of the geometric factors at this step.

A simpler approach is to use the microscopic balances derived in Section 6.1 instead of deriving the CV balances each time as in the CV approach. The microscopic-balance equations are written on an arbitrary, microscopic CV, and are applicable to any problem. The microscopic-balance equations for mass and momentum conservation contain all of the terms needed for flow calculations.[3] This preferable pathway is presented in this section.

6.2.1.1 THE EQUATIONS

The microscopic mass balance is called the continuity equation, and it is given here in Gibbs notation as well as Cartesian coordinates. This equation and the correct expressions for the continuity equation in cylindrical and spherical coordinates are listed in Table B.5 in Appendix B:

Continuity equation (microscopic mass balance)

$$0 = \frac{\partial \rho}{\partial t} + \underline{v} \cdot \nabla \rho + \rho \nabla \cdot \underline{v}$$

(6.80)

Cartesian coordinates

$$0 = \frac{\partial \rho}{\partial t} + \left(v_x \frac{\partial \rho}{\partial x} + v_y \frac{\partial \rho}{\partial y} + v_z \frac{\partial \rho}{\partial z} \right) + \rho \left(\frac{\partial v_x}{\partial x} + \frac{\partial v_y}{\partial y} + \frac{\partial v_z}{\partial z} \right)$$

(6.81)

Recall that the ∇ operator is the spatial differentiation operator; in Cartesian coordinates, ∇ is given by the following (see Section 1.3.2.2):

$$\nabla = \begin{pmatrix} \dfrac{\partial}{\partial x} \\[6pt] \dfrac{\partial}{\partial y} \\[6pt] \dfrac{\partial}{\partial z} \end{pmatrix}_{xyz}$$

(6.82)

Gibbs expressions such as $\nabla \rho$, $\nabla \underline{v}$, and $\nabla \cdot \underline{v}$ for several coordinate systems are in Table B.2 in Appendix B.

Looking at the continuity equation in Gibbs notation (Equation 6.80), we can interpret each term physically. There are three terms that sum to zero. The first two involve derivatives of the density ρ relative to time t and position x, y, z, respectively. The third term does not involve derivatives of the density ρ but rather spatial derivatives of the velocity $\underline{v}$. In words, the continuity equation states that the mass per unit volume at a point may change in one of three ways (Figure 6.7): The local density may change with time (e.g., due to reaction or the net effects of other contributions); the local density may change relative to position; or the local density may change because the velocity of fluid particles varies with position; therefore, at a given position at a particular time, more or less material may be present.

When a fluid is incompressible, the density is constant; thus, $\partial \rho / \partial t = 0$ and the partial derivatives of ρ relative to x, y, and z also are zero. For incompressible

---

[3] Modification of the microscopic balances may be necessary to account for forces in conducting fluids (see Section 2.9), where electromagnetic-force body forces are present or for materials with body couples. We do not consider such flows in this text.

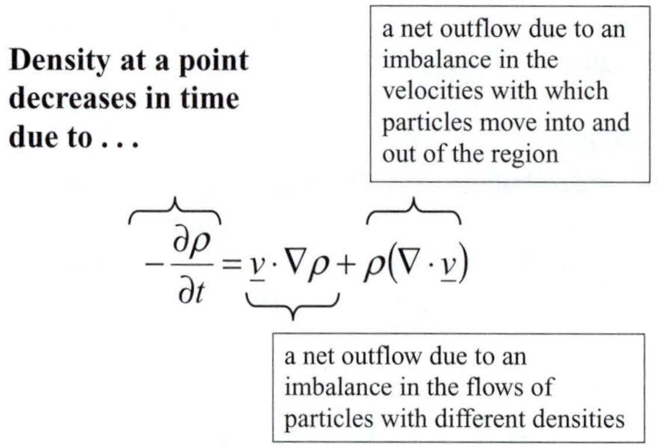

Density at a point decreases in time due to ...

a net outflow due to an imbalance in the velocities with which particles move into and out of the region

$$-\frac{\partial \rho}{\partial t} = \underline{v} \cdot \nabla \rho + \rho (\nabla \cdot \underline{v})$$

a net outflow due to an imbalance in the flows of particles with different densities

Figure 6.7   The changes in density ρ at a point are due to contributions due to density variations with position and velocity variations with position.

fluids, the continuity equation simplifies as shown here:

$$\begin{array}{ll} \text{Continuity equation} \\ \text{(incompressible fluid)} \end{array} \qquad \boxed{0 = \nabla \cdot \underline{v}} \qquad (6.83)$$

$$\begin{array}{ll} \text{Cartesian} \\ \text{coordinates} \end{array} \qquad 0 = \frac{\partial v_x}{\partial x} + \frac{\partial v_y}{\partial y} + \frac{\partial v_z}{\partial z} \qquad (6.84)$$

In the falling-film example, we had unidirectional flow of an incompressible fluid in the z-direction. For such a flow, $v_x = v_y = 0$ and the continuity equation gives:

$$\begin{array}{ll} \text{Continuity equation} \\ \text{(incompressible fluid,} \\ \text{unidirectional flow in } z\text{-direction)} \end{array} \qquad \boxed{\frac{\partial v_z}{\partial z} = 0} \qquad (6.85)$$

We arrive at the equivalent result in the CV solution to that problem (see Equation 3.155).

There are two useful versions of the microscopic momentum balance. The most general version is the Cauchy momentum equation, which is applicable for all fluids—Newtonian or non-Newtonian, compressible or incompressible:

$$\begin{array}{ll} \text{Cauchy momentum equation} \\ \text{(microscopic momentum balance)} \end{array} \qquad \boxed{\rho \left(\frac{\partial \underline{v}}{\partial t} + \underline{v} \cdot \nabla \underline{v} \right) = -\nabla p + \nabla \cdot \underline{\underline{\tau}} + \rho \underline{g}}$$

$$(6.86)$$

Note that $\underline{\underline{\tau}}$ appears in the Cauchy momentum equation. The second version of the microscopic momentum balance is specialized for incompressible Newtonian fluids; this is called the Navier-Stokes equation:

Navier-Stokes equation
(microscopic momentum
(balance for incompressible
Newtonian fluids)

$$\rho \left(\frac{\partial \underline{v}}{\partial t} + \underline{v} \cdot \nabla \underline{v} \right) = -\nabla p + \mu \nabla^2 \underline{v} + \rho \underline{g}$$

(6.87)

Because the Newtonian constitutive equation is used to write stress in terms of the velocity field, $\underline{\underline{\tilde{\tau}}}$ no longer appears in the Navier-Stokes version of the momentum balance. Both are vector equations, and we can display them as either the sum of vector terms or as three equations, one for each component in physical space (e.g., x, y, and z). Following are the two momentum balances in Cartesian coordinates, in both the vector–sum form and the three-component form. The Cauchy momentum equation and the Navier-Stokes equation written in Cartesian, cylindrical, and spherical coordinates are in Tables B.6 and B.7 (see Appendix B).

Cauchy Momentum Equation
(microscopic momentum balance)

$$\rho \begin{pmatrix} \dfrac{\partial v_x}{\partial t} \\[2mm] \dfrac{\partial v_y}{\partial t} \\[2mm] \dfrac{\partial v_z}{\partial t} \end{pmatrix}_{xyz} + \rho \begin{pmatrix} v_x \dfrac{\partial v_x}{\partial x} + v_y \dfrac{\partial v_x}{\partial y} + v_z \dfrac{\partial v_x}{\partial z} \\[2mm] v_x \dfrac{\partial v_y}{\partial x} + v_y \dfrac{\partial v_y}{\partial y} + v_z \dfrac{\partial v_y}{\partial z} \\[2mm] v_x \dfrac{\partial v_z}{\partial x} + v_y \dfrac{\partial v_z}{\partial y} + v_z \dfrac{\partial v_z}{\partial z} \end{pmatrix}_{xyz}$$

$$= - \begin{pmatrix} \dfrac{\partial p}{\partial x} \\[2mm] \dfrac{\partial p}{\partial y} \\[2mm] \dfrac{\partial p}{\partial z} \end{pmatrix}_{xyz} + \begin{pmatrix} \dfrac{\partial \tilde{\tau}_{xx}}{\partial x} + \dfrac{\partial \tilde{\tau}_{yx}}{\partial y} + \dfrac{\partial \tilde{\tau}_{zx}}{\partial z} \\[2mm] \dfrac{\partial \tilde{\tau}_{xy}}{\partial x} + \dfrac{\partial \tilde{\tau}_{yy}}{\partial y} + \dfrac{\partial \tilde{\tau}_{zy}}{\partial z} \\[2mm] \dfrac{\partial \tilde{\tau}_{xz}}{\partial x} + \dfrac{\partial \tilde{\tau}_{yz}}{\partial y} + \dfrac{\partial \tilde{\tau}_{zz}}{\partial z} \end{pmatrix}_{xyz} + \rho \begin{pmatrix} g_x \\[2mm] g_y \\[2mm] g_z \end{pmatrix}_{xyz}$$

(6.88)

x-Component Cauchy momentum equation:

$$\rho \left(\frac{\partial v_x}{\partial t} + v_x \frac{\partial v_x}{\partial x} + v_y \frac{\partial v_x}{\partial y} + v_z \frac{\partial v_x}{\partial z} \right) = -\frac{\partial p}{\partial x} + \left(\frac{\partial \tilde{\tau}_{xx}}{\partial x} + \frac{\partial \tilde{\tau}_{yx}}{\partial y} + \frac{\partial \tilde{\tau}_{zx}}{\partial z} \right) + \rho g_x$$

(6.89)

y-Component Cauchy momentum equation:

$$\rho \left(\frac{\partial v_y}{\partial t} + v_x \frac{\partial v_y}{\partial x} + v_y \frac{\partial v_y}{\partial y} + v_z \frac{\partial v_y}{\partial z} \right) = -\frac{\partial p}{\partial y} + \left(\frac{\partial \tilde{\tau}_{xy}}{\partial x} + \frac{\partial \tilde{\tau}_{yy}}{\partial y} + \frac{\partial \tilde{\tau}_{zy}}{\partial z} \right) + \rho g_y$$

(6.90)

z-Component Cauchy momentum equation:

$$\rho\left(\frac{\partial v_z}{\partial t} + v_x\frac{\partial v_z}{\partial x} + v_y\frac{\partial v_z}{\partial y} + v_z\frac{\partial v_z}{\partial z}\right) = -\frac{\partial p}{\partial z} + \left(\frac{\partial \tilde{\tau}_{xz}}{\partial x} + \frac{\partial \tilde{\tau}_{yz}}{\partial y} + \frac{\partial \tilde{\tau}_{zz}}{\partial z}\right) + \rho g_z$$

(6.91)

Navier-Stokes Equation
(microscopic momentum balance, incompressible Newtonian fluids)

$$
\rho\begin{pmatrix}\dfrac{\partial v_x}{\partial t}\\[6pt]\dfrac{\partial v_y}{\partial t}\\[6pt]\dfrac{\partial v_z}{\partial t}\end{pmatrix}_{xyz} + \rho\begin{pmatrix}v_x\dfrac{\partial v_x}{\partial x} + v_y\dfrac{\partial v_x}{\partial y} + v_z\dfrac{\partial v_x}{\partial z}\\[6pt]v_x\dfrac{\partial v_y}{\partial x} + v_y\dfrac{\partial v_y}{\partial y} + v_z\dfrac{\partial v_y}{\partial z}\\[6pt]v_x\dfrac{\partial v_z}{\partial x} + v_y\dfrac{\partial v_z}{\partial y} + v_z\dfrac{\partial v_z}{\partial z}\end{pmatrix}_{xyz}
$$

$$
= -\begin{pmatrix}\dfrac{\partial p}{\partial x}\\[6pt]\dfrac{\partial p}{\partial y}\\[6pt]\dfrac{\partial p}{\partial z}\end{pmatrix}_{xyz} + \mu\begin{pmatrix}\dfrac{\partial^2 v_x}{\partial x^2} + \dfrac{\partial^2 v_x}{\partial y^2} + \dfrac{\partial^2 v_x}{\partial z^2}\\[6pt]\dfrac{\partial^2 v_y}{\partial x^2} + \dfrac{\partial^2 v_y}{\partial y^2} + \dfrac{\partial^2 v_y}{\partial z^2}\\[6pt]\dfrac{\partial^2 v_z}{\partial x^2} + \dfrac{\partial^2 v_z}{\partial y^2} + \dfrac{\partial^2 v_z}{\partial z^2}\end{pmatrix}_{xyz} + \rho\begin{pmatrix}g_x\\g_y\\g_z\end{pmatrix}_{xyz}
$$

(6.92)

x-Component Navier-Stokes:

$$\rho\left(\frac{\partial v_x}{\partial t} + v_x\frac{\partial v_x}{\partial x} + v_y\frac{\partial v_x}{\partial y} + v_z\frac{\partial v_x}{\partial z}\right) = -\frac{\partial p}{\partial x} + \mu\left(\frac{\partial^2 v_x}{\partial x^2} + \frac{\partial^2 v_x}{\partial y^2} + \frac{\partial^2 v_x}{\partial z^2}\right) + \rho g_x$$

(6.93)

y-Component Navier-Stokes:

$$\rho\left(\frac{\partial v_y}{\partial t} + v_x\frac{\partial v_y}{\partial x} + v_y\frac{\partial v_y}{\partial y} + v_z\frac{\partial v_y}{\partial z}\right) = -\frac{\partial p}{\partial y} + \mu\left(\frac{\partial^2 v_y}{\partial x^2} + \frac{\partial^2 v_y}{\partial y^2} + \frac{\partial^2 v_y}{\partial z^2}\right) + \rho g_y$$

(6.94)

z-Component Navier-Stokes:

$$\rho\left(\frac{\partial v_z}{\partial t} + v_x\frac{\partial v_z}{\partial x} + v_y\frac{\partial v_z}{\partial y} + v_z\frac{\partial v_z}{\partial z}\right) = -\frac{\partial p}{\partial z} + \mu\left(\frac{\partial^2 v_z}{\partial x^2} + \frac{\partial^2 v_z}{\partial y^2} + \frac{\partial^2 v_z}{\partial z^2}\right) + \rho g_z$$

(6.95)

The source of each term of the Navier-Stokes equation is discussed in the derivation in Section 6.1 and summarized in Figure 6.8. The Reynolds transport theorem was the starting place, which states that the rate-of-change of momentum on a CV is equal to a convective term plus the sum of the forces on the CV

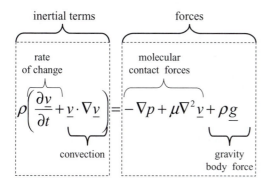

Figure 6.8 The Navier-Stokes equation is a microscopic momentum balance, and the terms are the same as the terms in the Reynolds transport theorem.

(i.e., Newton's second law applied to a control volume; see Chapter 3):

$$\text{Reynolds transport theorem} \atop \text{(momentum balance on CV)} \qquad \frac{d\underline{\mathbf{P}}}{dt} = \iint_{\mathcal{S}} -(\hat{n} \cdot \underline{v})\, \rho \underline{v}\, dS + \sum_{\substack{\text{on} \\ \text{CV}}} \underline{f} \qquad (6.96)$$

If we move the convective term to the lefthand side, the Reynolds transport theorem states that the time-rate of change of momentum in the CV, plus the net outflow of momentum, is equal to the sum of forces on the CV:

$$\text{Reynolds transport theorem} \atop \text{(momentum balance on CV)} \qquad \frac{d\underline{\mathbf{P}}}{dt} + \iint_{\mathcal{S}} (\hat{n} \cdot \underline{v})\, \rho \underline{v}\, dS = \sum_{\substack{\text{on} \\ \text{CV}}} \underline{f} \qquad (6.97)$$

In this form, the Reynolds transport theorem and the Navier-Stokes equation are parallel in construction: The Navier-Stokes equation also has the rate-of-change of momentum and a convective term on the lefthand side, and the sum of three forces: isotropic molecular stress contribution (i.e., pressure), anisotropic molecular stress contribution (i.e., viscosity), and gravity on the righthand side.

In this text we focus on isothermal problems, but nonisothermal problems can be addressed as well. The third microscopic balance needed to carry out a nonisothermal calculation is the microscopic energy balance. Energy is a scalar property; thus, the microscopic energy balance is a single equation, which may be written in the Cartesian coordinate system as shown here. This equation in Cartesian, cylindrical, and spherical coordinates is listed in Table B.9:

$$\text{Thermal energy equation} \atop \begin{array}{c} \text{(no viscous dissipation,} \\ \text{fluid at constant } p \text{ or } \rho \neq \rho(T)) \end{array} \qquad \boxed{\rho \hat{C}_p \left(\frac{\partial T}{\partial t} + \underline{v} \cdot \nabla T \right) = k \nabla^2 T + \mathcal{S}_e}$$

$$(6.98)$$

Problem-Solving Procedure –
Microscopic-Balances Approach

1. Sketch the problem.
2. Choose a coordinate system. The coordinate system should be chosen so that the velocity vector and the boundary conditions are simplified.
3. Considering the flow, simplify $\underline{v}$ as much as possible.
4. Simplify the continuity equation (i.e., microscopic mass balance).
5. Simplify the equation of motion (i.e., microscopic momentum balance, Navier-Stokes equation).
6. Solve the resulting differential equation for the velocity field. Apply velocity boundary conditions to solve for the unknown constant(s) of integration.
7. Solve for the pressure field if applicable.
8. Calculate the stress components from the Newtonian constitutive equation and for any engineering quantities of interest.

Figure 6.9   With the microscopic-balance approach, the steps for problem solving are shown.

$$\begin{aligned}\text{Cartesian} \quad & \frac{\partial T}{\partial t} + \left(v_x \frac{\partial T}{\partial x} + v_y \frac{\partial T}{\partial y} + v_z \frac{\partial T}{\partial z} \right) \\[2mm]
\text{coordinates} \quad & \\[2mm]
& = \frac{k}{\rho \hat{C}_p} \left[\frac{\partial^2 T}{\partial x^2} + \frac{\partial^2 T}{\partial y^2} + \frac{\partial^2 T}{\partial z^2} \right] + \frac{S_e}{\rho \hat{C}_p}\end{aligned}$$

$$(6.99)$$

6.2.1.2 APPLYING THE EQUATIONS

The microscopic mass, momentum, and energy equations eliminate the need to choose a control volume, write forces, and take limits. To use the microscopic balances, we follow a modified solution procedure compared to the CV approach, as shown in Figure 6.9. We are freed from choosing and manipulating a control volume in the microscopic-balance approach. Instead, we interrogate each term of the equations of change (i.e., mass, momentum, and energy balances) to determine which are zero and which must be retained in the solution.

We now demonstrate the process of using the microscopic balances by repeating a familiar problem: the flow of a thin film down an inclined plane.

EXAMPLE 6.2 (Incline, revisited). *What is the velocity field in a wide, thin film of water that runs steadily down an inclined surface under the force of gravity? The fluid has a constant density ρ.*

SOLUTION. We solve this familiar problem again, now using the procedure for problem solving with microscopic balances (see Figure 6.9).

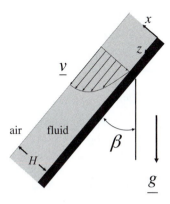

The idealized version of flow down an incline is a film of constant thickness where the velocity is unidirectional but its magnitude varies with position in the film. We want to calculate the velocity as a function of position relative to the wall, $v_z(x)$.

First, we sketch the problem as shown in Figure 6.10. We choose as our coordinate system a Cartesian system in which z is the flow direction. In thinking about the flow, we decide that it is unidirectional; that is, z is the flow direction and $v_x = v_y = 0$. The velocity vector is therefore:

$$\underline{v} = \begin{pmatrix} 0 \\ 0 \\ v_z \end{pmatrix}_{xyz} \qquad (6.100)$$

We completed Steps 1–3 of Figure 6.9. Step 4 is to simplify the continuity equation. We assume water to be an incompressible fluid; therefore, the continuity equation is the version given in Equation 6.83. In our Cartesian system, this is:

$$\begin{matrix} \text{Continuity equation} \\ \text{(incompressible fluid)} \end{matrix} \qquad \frac{\partial v_x}{\partial x} + \frac{\partial v_y}{\partial y} + \frac{\partial v_z}{\partial z} = 0 \qquad (6.101)$$

Because the velocity vector has only one nonzero component, the continuity equation simplifies to:

$$\boxed{\frac{\partial v_z}{\partial z} = 0} \qquad (6.102)$$

Step 5 is to simplify the equation of motion (EOM). Water is an incompressible Newtonian fluid; thus, we use the Navier-Stokes equation written in Cartesian coordinates (see Equation 6.92).

Navier-Stokes equation
Momentum balance for an incompressible fluid:

$$
\rho \begin{pmatrix} \dfrac{\partial v_x}{\partial t} \\[2mm] \dfrac{\partial v_y}{\partial t} \\[2mm] \dfrac{\partial v_z}{\partial t} \end{pmatrix}_{xyz} + \rho \begin{pmatrix} v_x \dfrac{\partial v_x}{\partial x} + v_y \dfrac{\partial v_x}{\partial y} + v_z \dfrac{\partial v_x}{\partial z} \\[2mm] v_x \dfrac{\partial v_y}{\partial x} + v_y \dfrac{\partial v_y}{\partial y} + v_z \dfrac{\partial v_y}{\partial z} \\[2mm] v_x \dfrac{\partial v_z}{\partial x} + v_y \dfrac{\partial v_z}{\partial y} + v_z \dfrac{\partial v_z}{\partial z} \end{pmatrix}_{xyz}
$$

$$
= - \begin{pmatrix} \dfrac{\partial p}{\partial x} \\[2mm] \dfrac{\partial p}{\partial y} \\[2mm] \dfrac{\partial p}{\partial z} \end{pmatrix}_{xyz} + \mu \begin{pmatrix} \dfrac{\partial^2 v_x}{\partial x^2} + \dfrac{\partial^2 v_x}{\partial y^2} + \dfrac{\partial^2 v_x}{\partial z^2} \\[2mm] \dfrac{\partial^2 v_y}{\partial x^2} + \dfrac{\partial^2 v_y}{\partial y^2} + \dfrac{\partial^2 v_y}{\partial z^2} \\[2mm] \dfrac{\partial^2 v_z}{\partial x^2} + \dfrac{\partial^2 v_z}{\partial y^2} + \dfrac{\partial^2 v_z}{\partial z^2} \end{pmatrix}_{xyz} + \rho \begin{pmatrix} g_x \\ g_y \\ g_z \end{pmatrix}_{xyz} \qquad (6.103)
$$

These are three equations for the x-, y-, and z-components, which we solve for the velocity and pressure fields. To begin simplifying Equation 6.103, we cancel the terms with v_x, v_y, and $\partial v_z / \partial z$ because these are zero (see Equation 6.102). Also, the flow is steady, so we eliminate the time-derivative term. This simplifies the Navier-Stokes equation as shown here:

$$
\begin{pmatrix} 0 \\ 0 \\ 0 \end{pmatrix}_{xyz} = - \begin{pmatrix} \dfrac{\partial p}{\partial x} \\[2mm] \dfrac{\partial p}{\partial y} \\[2mm] \dfrac{\partial p}{\partial z} \end{pmatrix}_{xyz} + \mu \begin{pmatrix} 0 \\ 0 \\ \dfrac{\partial^2 v_z}{\partial x^2} + \dfrac{\partial^2 v_z}{\partial y^2} \end{pmatrix}_{xyz} + \rho \begin{pmatrix} g_x \\ g_y \\ g_z \end{pmatrix}_{xyz} \qquad (6.104)
$$

We now write the gravity vector in our chosen coordinate system as follows (Figure 6.10):

$$
\underline{g} = \begin{pmatrix} -g \sin \beta \\ 0 \\ g \cos \beta \end{pmatrix}_{xyz} \qquad (6.105)
$$

Finally, because we assume the flow to be very wide, the velocity is independent of y. Thus, $\partial v_z / \partial y = 0$ and, in conjunction with Equation 6.102, we conclude that v_z is a function only of x. Because v_z is a function only of x, we can replace partial x-derivatives of v_z with regular derivatives. The simplified Navier-Stokes equation is now:

$$
\begin{pmatrix} 0 \\ 0 \\ 0 \end{pmatrix}_{xyz} = - \begin{pmatrix} \dfrac{\partial p}{\partial x} \\[2mm] \dfrac{\partial p}{\partial y} \\[2mm] \dfrac{\partial p}{\partial z} \end{pmatrix}_{xyz} + \begin{pmatrix} 0 \\ 0 \\ \mu \dfrac{d^2 v_z}{dx^2} \end{pmatrix}_{xyz} + \begin{pmatrix} -\rho g \sin \beta \\ 0 \\ \rho g \cos \beta \end{pmatrix}_{xyz} \qquad (6.106)
$$

The x- and y-components of the Navier-Stokes equation for this problem (Equation 6.106) are simple and tell us about the pressure distribution. The z-component of the Navier-Stokes equation tells us about the velocity distribution:

x-Component:
$$\frac{\partial p}{\partial x} = -\rho g \sin \beta \tag{6.107}$$

y-Component:
$$\frac{\partial p}{\partial y} = 0 \tag{6.108}$$

z-Component:
$$\frac{\partial p}{\partial z} = \mu \frac{d^2 v_z}{dx^2} + \rho g \cos \beta \tag{6.109}$$

The derivative of the pressure in the flow direction appears in the z-component of the Navier-Stokes equation. We have not attempted to eliminate any of the pressure terms on our own, and this is deliberate. As we see from the y-component of the Navier-Stokes equation, we did not need to assume that the pressure does not change in the y-direction; rather, this becomes the conclusion of the momentum balance, given the other assumptions we made. Likewise, the momentum balance in the x-direction indicates that with our assumptions thus far, the pressure varies in the x-direction due to the component of gravity in the x-direction. This is a hydrostatic pressure contribution analogous to Equation 4.64.

The momentum balance in the z-direction is more complicated, and both pressure and velocity terms appear. At this point, it is worthwhile to ask the question: Does the pressure vary in the z-direction? As when we solved this problem using the CV approach, to answer the question we must look at the boundary conditions. We know that along the top surface of the film, the pressure is atmospheric. At least at this location, $x = H$, the pressure is not a function of the z-position. The pressure varies in the x-direction, but the gradient in the x-direction is constant (Equation 6.107); and thus no z-variation is introduced between $x = H$ and $x = 0$, and pressure does not change in the z-direction anywhere in the flow. With these arguments, we deduce that $\partial p / \partial z = 0$, and the z-component of the Navier-Stokes equation becomes a simple differential equation that we can solve:

z-Component:
$$0 = \mu \frac{d^2 v_z}{dx^2} + \rho g \cos \beta \tag{6.110}$$

This is the same equation solved previously (see Equation 5.138), and it completes Step 5.

The steps to the results for v_z and $\tilde{\tau}_{xz}$ are the same from this point as in the problem solved with the CV approach (see Equation 5.155). The boundary conditions used were no-slip at the wall (see Equation 5.147) and zero shear stress at the interface with air (see Equation 5.149). The final results are:

$$v_z = \frac{\rho g \cos \beta}{\mu} \left[Hx - \frac{x^2}{2} \right] \tag{6.111}$$

$$\tilde{\tau}_{xz} = \rho g \cos \beta \, (H - x) \tag{6.112}$$

$$\underline{v} = \begin{pmatrix} 0 \\ 0 \\ \frac{\rho g \cos \beta}{\mu} \left[Hx - \frac{x^2}{2} \right] \end{pmatrix}_{xyz} \tag{6.113}$$

These results are plotted in Figures 5.19 and 5.20. The stress tensor $\tilde{\underline{\tau}}$ was calculated for this flow from its definition and the velocity result (see Equation 5.161).

To calculate the pressure distribution in the falling film, we return to the components of the Navier-Stokes equation (i.e., Equations 6.107–6.109). We argued that $\partial p / \partial z = 0$, and the y-component of the Navier-Stokes indicates that $\partial p / \partial y = 0$. We therefore conclude that pressure is a function only of x, $p = p(x)$. With this information, we now solve Equation 6.107 for the pressure as a function of x:

$$\frac{dp}{dx} = -\rho g \sin \beta \tag{6.114}$$

$$p(x) = -\rho g \sin \beta x + C_3 \tag{6.115}$$

The boundary condition on pressure is that at the top surface of the film, the pressure is atmospheric. We therefore can solve for C_3 and for the final expression for pressure:

$$\text{Boundary condition:} \quad x = H \quad p = p_{atm} \tag{6.116}$$

$$p = p_{atm} + \rho g H \sin \beta \left(1 - \frac{x}{H} \right) \tag{6.117}$$

We use this result in Section 6.2.3.1 to calculate the total force on the incline. The total molecular stress tensor $\tilde{\underline{\Pi}}$ is equal to $-p\underline{I} + \tilde{\underline{\tau}}$, which we can calculate from the Newtonian constitutive equation, $\tilde{\underline{\tau}} = \mu(\nabla \underline{v} + (\nabla \underline{v})^T)$ (see Equation 6.66), and Equations 6.113 and 6.117.

The solution to the flow-down-an-inclined-plane problem using the microscopic balances was straightforward. The Navier-Stokes equation contained all of the forces that act in this problem, and they are written correctly in our chosen coordinate system. We were free to concentrate on deciding which terms were important in the solution. We could solve for pressure without performing additional balances. Two more examples follow: a Newtonian and a non-Newtonian. In Chapters 7 and 8, we apply the equations of change to more complex flows.

EXAMPLE 6.3. *An incompressible Newtonian fluid is confined between two long, wide plates. The gap between the plates is H and the top plate moves with a constant speed V in the x_1-direction of the coordinate system shown in Figure 6.11. What is the velocity field in the flow?*

SOLUTION. To solve for the velocity field, we follow the problem-solving procedure in Figure 6.9. The flow is illustrated in Figure 6.11 and the coordinate system is specified there. The flow is in the x_1-direction; thus, $\underline{v}$ simplifies to:

$$\underline{v} = \begin{pmatrix} v_1 \\ v_2 \\ v_3 \end{pmatrix}_{123} = \begin{pmatrix} v_1 \\ 0 \\ 0 \end{pmatrix}_{123} \tag{6.118}$$

Because $v_2 = v_3 = 0$, the continuity equation simplifies as follows:

$$\begin{array}{cc} \text{Continuity equation} \\ \text{(incompressible fluid)} \end{array} \quad 0 = \nabla \cdot \underline{v} \tag{6.119}$$

$$0 = \frac{\partial v_1}{\partial x_1} + \frac{\partial v_2}{\partial x_2} + \frac{\partial v_3}{\partial x_3} \tag{6.120}$$

$$\boxed{0 = \frac{\partial v_1}{\partial x_1}} \tag{6.121}$$

We turn now to the Navier-Stokes equation. We know $v_2 = v_3 = 0$ and from the continuity equation that $\partial v_1/\partial x_1 = 0$. The flow is steady ($\partial v_1/\partial t = 0$). The Navier-Stokes equation simplifies to:

$$\begin{array}{c} \text{Navier-Stokes equation} \\ \text{(microscopic momentum balance for} \\ \text{incompressible Newtonian fluids)} \end{array} \quad \rho \left(\frac{\partial \underline{v}}{\partial t} + \underline{v} \cdot \nabla \underline{v} \right) = -\nabla p + \mu \nabla^2 \underline{v} + \rho \underline{g}$$

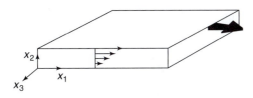

An incompressible Newtonian fluid is confined between two long, wide plates. The gap between the plates is H and the top plate moves with a constant speed V in the x_1-direction.

$$\rho \begin{pmatrix} \dfrac{\partial v_1}{\partial t} \\[8pt] \dfrac{\partial v_2}{\partial t} \\[8pt] \dfrac{\partial v_3}{\partial t} \end{pmatrix}_{123} + \rho \begin{pmatrix} v_1\dfrac{\partial v_1}{\partial x_1} + v_2\dfrac{\partial v_1}{\partial x_2} + v_3\dfrac{\partial v_1}{\partial x_3} \\[8pt] v_1\dfrac{\partial v_2}{\partial x_1} + v_2\dfrac{\partial v_2}{\partial x_2} + v_3\dfrac{\partial v_2}{\partial x_3} \\[8pt] v_1\dfrac{\partial v_3}{\partial x_1} + v_2\dfrac{\partial v_3}{\partial x_2} + v_3\dfrac{\partial v_3}{\partial x_3} \end{pmatrix}_{123}$$

$$= -\begin{pmatrix} \dfrac{\partial p}{\partial x_1} \\[8pt] \dfrac{\partial p}{\partial x_2} \\[8pt] \dfrac{\partial p}{\partial x_3} \end{pmatrix}_{123} + \mu \begin{pmatrix} \dfrac{\partial^2 v_1}{\partial x_1^2} + \dfrac{\partial^2 v_1}{\partial x_2^2} + \dfrac{\partial^2 v_1}{\partial x_3^2} \\[8pt] \dfrac{\partial^2 v_2}{\partial x_1^2} + \dfrac{\partial^2 v_2}{\partial x_2^2} + \dfrac{\partial^2 v_2}{\partial x_3^2} \\[8pt] \dfrac{\partial^2 v_3}{\partial x_1^2} + \dfrac{\partial^2 v_3}{\partial x_2^2} + \dfrac{\partial^2 v_3}{\partial x_3^2} \end{pmatrix}_{123} + \rho \begin{pmatrix} g_1 \\ g_2 \\ g_3 \end{pmatrix}_{123} \qquad (6.122)$$

$$\rho \begin{pmatrix} 0 \\ 0 \\ 0 \end{pmatrix}_{123} + \rho \begin{pmatrix} 0 \\ 0 \\ 0 \end{pmatrix}_{123} = -\begin{pmatrix} \dfrac{\partial p}{\partial x_1} \\[8pt] \dfrac{\partial p}{\partial x_2} \\[8pt] \dfrac{\partial p}{\partial x_3} \end{pmatrix}_{123} + \mu \begin{pmatrix} \dfrac{\partial^2 v_1}{\partial x_2^2} + \dfrac{\partial^2 v_1}{\partial x_3^2} \\[8pt] 0 \\[8pt] 0 \end{pmatrix}_{123} + \begin{pmatrix} 0 \\ -\rho g \\ 0 \end{pmatrix}_{123}$$

$$(6.123)$$

If we assume that the plates are very wide and if we consider only a region away from the edges, we can neglect the dependence of v_1 on x_3. The x_1-component of the momentum balance becomes:

$$\text{1-Component Navier-Stokes:} \qquad 0 = -\frac{\partial p}{\partial x_1} + \mu\frac{d^2 v_1}{dx_2^2} \qquad (6.124)$$

There is no imposed pressure gradient in the x_1-direction, and the pressure at the edges of the plate is atmospheric. We assume that there is no pressure gradient in the x_1-direction. With this final additional assumption, we can solve the x_1-component of the Navier-Stokes equation for the velocity field:

$$0 = \mu\frac{d^2 v_1}{dx_2^2} \qquad (6.125)$$

Dividing by μ and defining $\Psi \equiv dv_1/dx_2$, we integrate once:

$$\mu\frac{d^2 v_1}{dx_2^2} = 0 \qquad (6.126)$$

$$\frac{d\Psi}{dx_2} = 0 \qquad (6.127)$$

$$\Psi = C_1 \qquad (6.128)$$

where C_1 is an arbitrary integration constant. Substituting the definition of Ψ and integrating again:

$$\Psi = C_1 \tag{6.129}$$

$$\frac{dv_1}{dx_2} = C_1 \tag{6.130}$$

$$v_1 = C_1 x_2 + C_2 \tag{6.131}$$

The boundary conditions are no-slip at the top and the bottom:

$$\text{Boundary Condition 1: } x_2 = H \; v_1 = V \tag{6.132}$$

$$\text{Boundary Condition 2: } x_2 = 0 \; v_1 = 0 \tag{6.133}$$

from which we obtain the final result for the velocity field:

$$\boxed{v_1(x_2) = \frac{V}{H} x_2} \tag{6.134}$$

Having obtained the equation for the velocity field, the stress field now may be obtained from the Newtonian constitutive equation (see Equation 6.66):

$$\underline{\underline{\tilde{\tau}}} = \mu \left(\nabla \underline{v} + (\nabla \underline{v})^T \right) \tag{6.135}$$

$$= \mu \begin{pmatrix} 2\dfrac{\partial v_1}{\partial x_1} & \dfrac{\partial v_2}{\partial x_1} + \dfrac{\partial v_1}{\partial x_2} & \dfrac{\partial v_3}{\partial x_1} + \dfrac{\partial v_1}{\partial x_3} \\[2mm] \dfrac{\partial v_2}{\partial x_1} + \dfrac{\partial v_1}{\partial x_2} & 2\dfrac{\partial v_2}{\partial x_2} & \dfrac{\partial v_2}{\partial x_3} + \dfrac{\partial v_3}{\partial x_2} \\[2mm] \dfrac{\partial v_3}{\partial x_1} + \dfrac{\partial v_1}{\partial x_3} & \dfrac{\partial v_2}{\partial x_3} + \dfrac{\partial v_3}{\partial x_2} & 2\dfrac{\partial v_3}{\partial x_3} \end{pmatrix}_{123} \tag{6.136}$$

$$= \begin{pmatrix} 0 & \mu\dfrac{V}{H} & 0 \\[2mm] \mu\dfrac{V}{H} & 0 & 0 \\[2mm] 0 & 0 & 0 \end{pmatrix}_{123} \tag{6.137}$$

EXAMPLE 6.4. *An incompressible power-law fluid is confined between two long, wide plates. The gap between the plates is H and the top plate moves with a constant speed V in the z-direction of the coordinate system shown in Figure 6.11. What is the velocity field in the flow?*

SOLUTION. This example is the same as the previous except that now we are asked to solve for the case of a non-Newtonian fluid—specifically, a power-law generalized Newtonian fluid. The solution to the problem is similar to Example 6.3 at the beginning, but the momentum balance we use is the Cauchy momentum equation rather than the Navier-Stokes, which is valid only for Newtonian fluids.

Referring to the previous example, $\underline{v} = v_1\hat{e}_1$ and the mass balance is the same with $\partial v_1/\partial x_1 = 0$. We assume as we did in the Newtonian case that because the plates are wide, there is no variation of the velocity field in the width direction, $\partial v_1/\partial x_3 = 0$; we also assume that pressure is independent of position in the flow direction, yielding $\partial p/\partial x_1 = 0$ (see the Newtonian solution in Example 6.3). The momentum balance is Equation 6.86, the Cauchy momentum equation, which is valid for Newtonian and non-Newtonian fluids:

Cauchy momentum equation (Cartesian coordinates):

$$
\rho \begin{pmatrix} \dfrac{\partial v_1}{\partial t} \\[2mm] \dfrac{\partial v_2}{\partial t} \\[2mm] \dfrac{\partial v_3}{\partial t} \end{pmatrix}_{123} + \rho \begin{pmatrix} v_1\dfrac{\partial v_1}{\partial x_1} + v_2\dfrac{\partial v_1}{\partial x_2} + v_3\dfrac{\partial v_1}{\partial x_3} \\[2mm] v_1\dfrac{\partial v_2}{\partial x_1} + v_2\dfrac{\partial v_2}{\partial x_2} + v_3\dfrac{\partial v_2}{\partial x_3} \\[2mm] v_1\dfrac{\partial v_3}{\partial x_1} + v_2\dfrac{\partial v_3}{\partial x_2} + v_3\dfrac{\partial v_3}{\partial x_3} \end{pmatrix}_{123}
$$

$$
= - \begin{pmatrix} \dfrac{\partial p}{\partial x_1} \\[2mm] \dfrac{\partial p}{\partial x_2} \\[2mm] \dfrac{\partial p}{\partial x_3} \end{pmatrix}_{123} + \begin{pmatrix} \dfrac{\partial \tilde{\tau}_{11}}{\partial x_1} + \dfrac{\partial \tilde{\tau}_{21}}{\partial x_2} + \dfrac{\partial \tilde{\tau}_{31}}{\partial x_3} \\[2mm] \dfrac{\partial \tilde{\tau}_{12}}{\partial x_1} + \dfrac{\partial \tilde{\tau}_{22}}{\partial x_2} + \dfrac{\partial \tilde{\tau}_{32}}{\partial x_3} \\[2mm] \dfrac{\partial \tilde{\tau}_{13}}{\partial x_1} + \dfrac{\partial \tilde{\tau}_{23}}{\partial x_2} + \dfrac{\partial \tilde{\tau}_{33}}{\partial x_3} \end{pmatrix}_{123} + \rho \begin{pmatrix} g_1 \\ g_2 \\ g_3 \end{pmatrix}_{123} \quad (6.138)
$$

Using what we already know about the flow, the Cauchy momentum equation becomes:

$$
\begin{pmatrix} 0 \\ 0 \\ 0 \end{pmatrix}_{123} = - \begin{pmatrix} 0 \\[2mm] \dfrac{\partial p}{\partial x_2} \\[2mm] \dfrac{\partial p}{\partial x_3} \end{pmatrix}_{123} + \begin{pmatrix} \dfrac{\partial \tilde{\tau}_{11}}{\partial x_1} + \dfrac{\partial \tilde{\tau}_{21}}{\partial x_2} + \dfrac{\partial \tilde{\tau}_{31}}{\partial x_3} \\[2mm] \dfrac{\partial \tilde{\tau}_{12}}{\partial x_1} + \dfrac{\partial \tilde{\tau}_{22}}{\partial x_2} + \dfrac{\partial \tilde{\tau}_{32}}{\partial x_3} \\[2mm] \dfrac{\partial \tilde{\tau}_{13}}{\partial x_1} + \dfrac{\partial \tilde{\tau}_{23}}{\partial x_2} + \dfrac{\partial \tilde{\tau}_{33}}{\partial x_3} \end{pmatrix}_{123} + \begin{pmatrix} 0 \\ -\rho g \\ 0 \end{pmatrix}_{123}
$$

Because the coefficients of $\underline{\underline{\tilde{\tau}}}$ appear in the Cauchy momentum equation, we cannot directly simplify this equation as much as the Navier-Stokes equation, which contains pressure and the coefficients of velocity. To further simplify the Cauchy momentum equation for this problem, we must examine $\underline{\underline{\tilde{\tau}}}$. We obtain the relationship between $\underline{\underline{\tilde{\tau}}}$ and velocity from the generalized Newtonian constitutive equation with the power-law function for the viscosity:

Generalized Newtonian equation

$$
\underline{\underline{\tilde{\tau}}}(t) = \eta(\dot{\gamma}) \left(\nabla\underline{v} + (\nabla\underline{v})^T \right) \quad (6.139)
$$

$$
= \eta(\dot{\gamma})\underline{\underline{\dot{\gamma}}} \quad (6.140)
$$

The function $\eta(\dot{\gamma})$ for a shear-thinning power-law fluid, $\eta(\dot{\gamma})$ is given by:

$$\eta(\dot{\gamma}) = m\dot{\gamma}^{n-1} \tag{6.141}$$

with $n < 1$ and $\dot{\gamma}$ given by Equation 5.222, which is repeated here:

$$\dot{\gamma} = +\sqrt{\left(\frac{1}{2} \cdot \sum_{p=1}^{3}\sum_{j=1}^{3} \dot{\gamma}_{pj}^2\right)} \tag{6.142}$$

Thus, we need $\underline{\underline{\dot{\gamma}}}$, which we can construct from $\underline{v}$:

$$\underline{\underline{\dot{\gamma}}} = \nabla\underline{v} + (\nabla\underline{v})^T \tag{6.143}$$

$$\nabla\underline{v} = \begin{pmatrix} \frac{\partial v_1}{\partial x_1} & \frac{\partial v_2}{\partial x_1} & \frac{\partial v_3}{\partial x_1} \\ \frac{\partial v_1}{\partial x_2} & \frac{\partial v_1}{\partial x_2} & \frac{\partial v_3}{\partial x_2} \\ \frac{\partial v_1}{\partial x_3} & \frac{\partial v_2}{\partial x_3} & \frac{\partial v_3}{\partial x_3} \end{pmatrix}_{123} \tag{6.144}$$

For the velocity field, we know $v_2 = v_3 = 0$; from the mass balance, we obtained $\partial v_1/\partial x_1 = 0$. Therefore, we calculate:

$$\nabla\underline{v} = \begin{pmatrix} 0 & 0 & 0 \\ \frac{\partial v_1}{\partial x_2} & 0 & 0 \\ 0 & 0 & 0 \end{pmatrix}_{123} \tag{6.145}$$

$$\underline{\underline{\dot{\gamma}}} = \begin{pmatrix} 0 & \frac{\partial v_1}{\partial x_2} & 0 \\ \frac{\partial v_1}{\partial x_2} & 0 & 0 \\ 0 & 0 & 0 \end{pmatrix}_{123} \tag{6.146}$$

$$\dot{\gamma} = +\sqrt{\left(\frac{1}{2} \cdot \sum_{p=1}^{3}\sum_{j=1}^{3} \dot{\gamma}_{pj}^2\right)} \tag{6.147}$$

$$= +\sqrt{\left(\frac{\partial v_1}{\partial x_2}\right)^2} = \left|\frac{\partial v_1}{\partial x_2}\right| \tag{6.148}$$

Because v_1 increases in the x_2 direction, the derivative $\frac{\partial v_1}{\partial x_2}$ is always positive, and we choose the positive square root as $+\frac{\partial v_1}{\partial x_2}$. Now that we have $\dot{\gamma}$, we can calculate $\eta(\dot{\gamma})$ and, subsequently, $\underline{\underline{\tilde{\tau}}}$:

$$\eta(\dot{\gamma}) = m\dot{\gamma}^{n-1} \tag{6.149}$$

$$= m\left(\frac{\partial v_1}{\partial x_2}\right)^{n-1} \tag{6.150}$$

The constitutive equation for stress is given in Equation 6.140. Substituting what we know about $\underline{v}$ and η, we obtain a 3×3 matrix containing what we know about the stress tensor for the current problem:

Generalized Newtonian constitutive equation

$$\underline{\underline{\tilde{\tau}}}(t) = \eta(\dot{\gamma})\left(\nabla\underline{v} + (\nabla\underline{v})^T\right) \tag{6.151}$$

$$= \eta(\dot{\gamma})\underline{\underline{\dot{\gamma}}} \tag{6.152}$$

$$= m\left(\frac{\partial v_1}{\partial x_2}\right)^{n-1}\begin{pmatrix} 0 & \frac{\partial v_1}{\partial x_2} & 0 \\ \frac{\partial v_1}{\partial x_2} & 0 & 0 \\ 0 & 0 & 0 \end{pmatrix}_{123} \tag{6.153}$$

$$\underline{\underline{\tilde{\tau}}} = \begin{pmatrix} 0 & m\left(\frac{\partial v_1}{\partial x_2}\right)^n & 0 \\ m\left(\frac{\partial v_1}{\partial x_2}\right)^n & 0 & 0 \\ 0 & 0 & 0 \end{pmatrix}_{123} \tag{6.154}$$

From Equation 6.154, we see that all but the $\tilde{\tau}_{21}$ and $\tilde{\tau}_{12}$ components of the stress are zero; thus, returning to the momentum balance, we eliminate all of the zero stresses:

$$\begin{pmatrix} 0 \\ 0 \\ 0 \end{pmatrix}_{123} = -\begin{pmatrix} 0 \\ \frac{\partial p}{\partial x_2} \\ \frac{\partial p}{\partial x_3} \end{pmatrix}_{123} + \begin{pmatrix} \frac{\partial \tilde{\tau}_{21}}{\partial x_2} \\ \frac{\partial \tilde{\tau}_{12}}{\partial x_1} \\ 0 \end{pmatrix}_{123} + \begin{pmatrix} 0 \\ -\rho g \\ 0 \end{pmatrix}_{123}$$

We also know that $\tilde{\tau}_{21} = \tilde{\tau}_{12} = m\left(\frac{\partial v_1}{\partial x_2}\right)^n$, which is not a function of x_1; thus, the term $\partial\tilde{\tau}_{21}/\partial x_1$ drops out as well:

$$\begin{pmatrix} 0 \\ 0 \\ 0 \end{pmatrix}_{123} = -\begin{pmatrix} 0 \\ \frac{\partial p}{\partial x_2} \\ \frac{\partial p}{\partial x_3} \end{pmatrix}_{123} + \begin{pmatrix} \frac{\partial}{\partial x_2}\left(m\left(\frac{\partial v_1}{\partial x_2}\right)^n\right) \\ 0 \\ 0 \end{pmatrix}_{123} + \begin{pmatrix} 0 \\ -\rho g \\ 0 \end{pmatrix}_{123}$$

As was true in the Newtonian solution, the components of the momentum balance tell us that pressure may not vary in the x_3-direction and, in the x_2-direction, the only pressure variation is the hydrostatic effect. In the x_1-direction, we can solve for the velocity profile by integrating twice. The final result for the velocity field turns out to be the same as in the Newtonian case:

$$\frac{\partial}{\partial x_2}\left(m\left(\frac{\partial v_1}{\partial x_2}\right)^n\right) = 0 \tag{6.155}$$

$$m\left(\frac{\partial v_1}{\partial x_2}\right)^n = C_1 \tag{6.156}$$

$$\frac{\partial v_1}{\partial x_2} = \left(\frac{C_1}{m}\right)^{\frac{1}{n}} \tag{6.157}$$

$$v_1 = \left(\frac{C_1}{m}\right)^{\frac{1}{n}} x_2 + C_2 \tag{6.158}$$

The boundary conditions are given in Equations 6.132 and 6.133 and the final result is given here:

$$\text{Boundary Condition 1: } x_2 = H \ v_1 = V \tag{6.159}$$

$$\text{Boundary Condition 2: } x_2 = 0 \ v_1 = 0 \tag{6.160}$$

$$\boxed{v_1(x_2) = \frac{V}{H}x_2} \tag{6.161}$$

Although the velocity fields are the same for drag flow of Newtonian and non-Newtonian fluids, the stress fields are different, as demonstrated when we substitute the previous result into Equation 6.154 for the stress tensor $\underline{\underline{\tilde{\tau}}}$:

$$\boxed{\underline{\underline{\tilde{\tau}}} = \begin{pmatrix} 0 & m\left(\frac{V}{H}\right)^n & 0 \\ m\left(\frac{V}{H}\right)^n & 0 & 0 \\ 0 & 0 & 0 \end{pmatrix}_{123}} \tag{6.162}$$

When $n = 1$, the power-law generalized Newtonian fluid result becomes the Newtonian result with $m = \mu$.

The Navier-Stokes equations are highly complex, and the behavior they are capable of describing is extraordinary in breadth and variability, as we can see from the range of flow behavior around us: turbulent streams and rivers, tornados, and sloshing mixing tanks. The Navier-Stokes equations are nonlinear PDEs and the fact that they are nonlinear allows them to describe systems that change rapidly over short time or length scales. This is the true physics of momentum balances and the challenge to those who seek to solve them. A general solution of the Navier-Stokes equation does not yet exist:[4]

Navier-Stokes equation
(microscopic momentum balance for incompressible Newtonian fluids)
$$\rho \left(\frac{\partial \underline{v}}{\partial t} + \underline{v} \cdot \nabla \underline{v} \right) = -\nabla p + \mu \nabla^2 \underline{v} + \rho \underline{g}$$

Chapters 7 and 8 discuss solutions of the Navier-Stokes equations for simple and complex internal and external flows. For some complex flows, the Navier-Stokes equations can be solved with specialized analytical techniques. Numerical flow simulators have been developed to solve complex flow fields as well (see Section 10.2) [5, 27]. The Navier-Stokes equations are the governing equations for turbulent flow (see Section 2.4) but, in the case of turbulent flow, they must be averaged before solving (see Section 10.3).

Developing the governing equations and a solution methodology is an important first step that now must be followed with a broad study of fluid behavior and a careful and inventive application of the solution method; we show how in the remaining chapters. A critical step in solving the Navier-Stokes equation or the Cauchy momentum equation is to identify correctly the boundary conditions. We discuss commonly encountered boundary conditions in the next section. The final section of this chapter discusses how engineering quantities of interest (e.g., flow rates and forces on walls) are calculated from the solutions for velocity field $\underline{v}$ and the pressure field. The reward for the hard work of solving the Navier-Stokes equation is the ability to calculate whatever we want from the velocity and stress fields obtained.

6.2.2 Boundary conditions

An important step in the microscopic problem-solving procedure is to identify the boundary conditions. It often is helpful to think about the boundary conditions when first sketching the problem and when choosing the coordinate system. The boundary conditions hold information about how the velocity field varies and about which components of $\underline{v}$ are zero or negligible. We practice with boundary conditions in the following example.

EXAMPLE 6.5. *What are the boundary conditions for the flows shown in Figure 6.12?*

SOLUTION. To apply a boundary condition, we must write a mathematical expression that relates the variables for velocity or stress to the flow geometry. In this example, we write the boundary conditions for the flows in Figure 6.12 using the indicated coordinate systems.

---

[4]In April 2000, the Clay Mathematics Institute made the following announcement: "In order to celebrate mathematics in the new millennium, The Clay Mathematics Institute of Cambridge, Massachusetts (CMI) has named seven Prize Problems. The Scientific Advisory Board of CMI selected these problems, focusing on important classic questions that have resisted solution over the years. The Board of Directors of CMI designated a $7 million prize fund for the solution to these problems, with $1 million allocated to each." One of the millennium problems is "to make substantial progress toward a mathematical theory which will unlock the secrets hidden in the Navier-Stokes equations." (www.claymath.org/millennium)

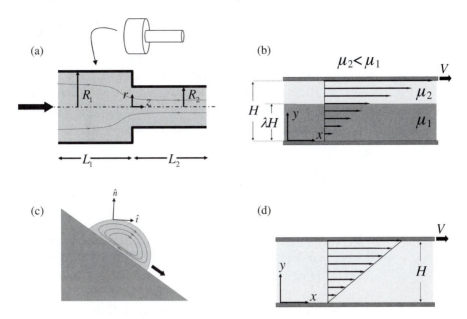

Figure 6.12 Examples of the most common boundary conditions. Clockwise from upper left: symmetry and no-slip, stress/velocity matching and no-slip, no-slip, and surface tension and no-slip.

Figure 6.12a depicts flow through a contraction from a big pipe to a smaller pipe. Both the upstream and downstream sections are circular pipes; thus, the flow is axially symmetric. For flows that are symmetric, one of the "boundaries" is the axis of symmetry. For the flow in Figure 6.12a, the centerline of the flow is the axis of symmetry. At this location, the velocity field goes through a maximum:

$$\text{Figure 6.12a}$$
$$\text{Boundary Condition 1:} \quad \text{for all } z \quad r = 0 \quad \frac{\partial v_z}{\partial r} = 0 \quad (6.163)$$
$$\text{(symmetry)}$$

The other boundary of the flow is the wall of the tube. The flow velocity goes to zero at the walls (i.e., no-slip boundary condition). The position of the wall is $r = R_1$ in the upstream portion ($-L_1 < z < 0$) and $r = R_2$ in the downstream portion ($0 < z < L_2$). At $z = 0$, there also is a vertical portion of the wall that must be assigned a no-slip boundary condition. Thus, the second boundary condition on velocity is:

$$\text{Figure 6.12a}$$
$$\text{Boundary Condition 2:} \quad \begin{cases} -L_1 < z < 0 & r = R_1 & v_z = 0 \\ z = 0 & R_1 < r < R_2 & v_z = 0 \quad (6.164) \\ 0 < z < L_2 & r = R_2 & v_z = 0 \end{cases}$$
$$\text{(no-slip)}$$

The complexity of this boundary makes it impractical to find an analytical solution to this problem (i.e., a solution arrived at by symbolic manipulation), but it is straightforward to solve this flow using numerical codes [5, 27].

Figure 6.12b depicts two fluid layers composed of different Newtonian fluids sandwiched between long, wide, parallel plates. The top fluid has a lower viscosity than the bottom fluid; the top plate moves in the x-direction with speed V and the bottom plate is stationary. The boundaries of this flow are the bottom ($y = 0$) and the top ($y = H$). Also, there is a boundary where the two fluids meet ($y = \lambda H$).

The momentum balance equation is different within the two fluids because the viscosities are different. To solve the problem, we solve separately for velocity and shear stress in Fluid 1 and Fluid 2, and then we match velocity and stress at the fluid-fluid boundary. Thus, the four boundary conditions are:

Figure 6.12b

Boundary Condition 1 (no-slip):	$y = 0$	$v_x^{(fluid\ 1)} = 0$
Boundary Condition 2 (velocity matching):	$y = \lambda H$	$v_x^{(fluid\ 1)} = v_x^{(fluid\ 2)}$
Boundary Condition 3 (stress matching):	$y = \lambda H$	$\tilde{\tau}_{yx}^{(fluid\ 1)} = \tilde{\tau}_{yx}^{(fluid\ 2)}$
Boundary Condition 4 (no-slip):	$y = 0$	$v_x^{(fluid\ 2)} = V$

$$(6.165)$$

Figure 6.12c depicts the steady flow of a drop of Newtonian fluid "rolling" down an inclined plane. Because of the complex geometry of the flow domain, this flow is best analyzed numerically, and the details of that calculation are beyond the scope of this text [70]. Although numerical methods are needed to solve the differential equations, we can arrive at the correct equations to solve by following the methods in this text (see Problem 37), and we can write the boundary conditions using the vector tools discussed.

The boundaries of the moving droplet flow are the surfaces of the drop [156]. Part of the drop surface is in contact with the wall, and the wall is not moving. Thus, for the part of the flow in contact with the wall, the boundary condition is no-slip at the wall:

$$\text{Figure 6.12c} \qquad \underline{v}\Big|_{\substack{surface \\ in\ contact \\ with\ wall}} = 0 \qquad (6.166)$$

$$\text{Boundary Condition 1:}$$

Part of the drop surface is in contact with air; at this surface, the tangential component of stress on that surface is approximately zero. The vector stress on any surface at a chosen location is shown in Equation 4.263 to be given by:

$$\begin{pmatrix} \text{stress vector} \\ \text{on a surface} \\ \text{of unit normal } \hat{n} \end{pmatrix} = [\hat{n} \cdot \underline{\underline{\tilde{\Pi}}}]_{\text{at surface}} \qquad (6.167)$$

where $\hat{n}$ is the outwardly pointing unit normal at the surface at the location of interest and $\underline{\underline{\tilde{\Pi}}} = -p\underline{\underline{I}} + \underline{\underline{\tilde{\tau}}}$ is the total stress tensor at that same location. The boundary condition on the free surface is that the tangential component of $\underline{f}$ is zero:

$$\text{Figure 6.12c} \qquad [\hat{n} \cdot \underline{\underline{\tilde{\Pi}}}] \cdot \hat{t} = 0 \qquad (6.168)$$

$$\text{Boundary Condition 2:}$$

where $\hat{t}$ is a unit vector tangent to the drop at the surface.

An important complexity in this calculation is that we do not know the shape of the drop in advance. We need to calculate the shape of the drop from the mass and momentum balances, and the shape will be affected by surface tension and gravity. Also, because the drop is moving, it is desirable to analyze this flow in a coordinate system that is moving with the drop. In a stationary coordinate system, the drop comes into view at some time and exits some time later; whereas in a coordinate system that moves with the speed of the center of gravity, the drop stays within view at all times. Discussion of a numerical solution in a moving coordinate system is in the literature [70].

Figure 6.12d is steady-drag flow of a Newtonian fluid between two very long, wide plates; the no-slip boundary conditions for this flow were discussed previously:

Figure 6.12d

Boundary Condition 1 (no-slip): $y = 0$ $v_x = 0$ (6.169)

Boundary Condition 2 (no-slip): $y = H$ $v_x = V$ (6.170)

Identifying boundary conditions is sometimes a challenge when fluid-mechanics problems are first attempted. The number of different types of boundary conditions used in fluid mechanics is relatively small, however, and the most prominent were encountered in Example 6.5. The common fluid-mechanics boundary conditions are as follows.

1. *No-slip at the wall.* This boundary condition states that the fluid in contact with a wall has the same velocity as the wall. Often, the walls are not moving, so the fluid velocity is zero at the wall. In drag flow (see Figure 6.12d and Section 5.1), the velocity of the bottom wall is zero and the velocity of the top wall is nonzero; in both cases, the fluid velocity is equal to the wall velocity:

Drag flow in x-direction, $\underline{v} = v_x(y)\hat{e}_x$:
no-slip on bottom ($y = 0$)
and top ($y = H$) plates

$$\begin{array}{l} v_x|_{y=0} = 0 \\ v_x|_{y=H} = V \end{array}$$ (6.171)

Usually, the velocity is specified in terms of a component normal to the boundary and a component tangential to the boundary. If $\hat{n}$ is the unit vector normal to the boundary and $\hat{t}$ is the unit vector tangential to the boundary, we can write the general no-slip boundary condition as:

General case: no-slip

normal component of $\underline{v}$: $(\hat{n} \cdot \underline{v})|_{\text{boundary}} = V_n$
tangential component of $\underline{v}$: $(\hat{t} \cdot \underline{v})|_{\text{boundary}} = V_t$

(6.172)

where V_n and V_t are the specified normal and tangential components of the velocity at the boundary.

2. *Symmetry.* In some flows, there is a plane or line of symmetry (see Figure 6.12a, line of cylindrical symmetry). Because the velocity field is the same on either side of the plane of symmetry or circularly symmetrical around a line of symmetry, the velocity must go through a minimum or a maximum at that location. Thus, the boundary condition to use in this

case is that the first derivative of the velocity is zero at the boundary of symmetry:

$$
\text{Axisymmetric} \atop {\text{contraction flow, } \underline{v} = v_z(r)\hat{e}_r: \atop \text{symmetry along centerline}} \qquad \boxed{\left.\frac{\partial v_z}{\partial r}\right|_{r=0} = 0} \qquad (6.173)
$$

The general case of the symmetry-boundary condition is written as:

$$
\text{General-case symmetry:} \qquad \boxed{\underline{v}\big|_{\text{symmetry boundary}} = \text{extremum}} \qquad (6.174)
$$

This expression means that the velocity must come to a maximum or minimum at a symmetry boundary.

3. *Stress continuity.* When a fluid forms one of the boundaries of the flow, the stress is continuous from one fluid to another (see Figure 6.12b,c). Thus, for a viscous fluid in contact with an inviscid (i.e., zero or very low viscosity fluid), the stress in the viscous fluid is the same as the stress in the inviscid fluid. Because the inviscid fluid cannot support any shear stress (i.e., zero viscosity), this means that the stress is zero at this interface (recall the incline problem in Section 5.2.2; see Figure 5.12).

$$
\text{Flow down an incline} \atop {(\text{see Figure 5.12}) \atop {\underline{v} = v_z(x)\hat{e}_z, \text{ open to} \atop \text{air at } x = H}} \qquad \boxed{\tilde{\tau}_{xz}\big|_{x=H} = 0} \qquad (6.175)
$$

For the general case, such as the moving droplet shown in Figure 6.12c, the boundary condition in Gibbs notation is:

$$
\text{Viscous fluid contacts} \atop {\text{inviscid fluid} \atop {\text{at boundary} \atop (\text{tangential stresses vanish})}} \qquad \boxed{\left(\hat{n}\cdot\underline{\underline{\tilde{\tau}}}\cdot\hat{t}\right)\Big|_{\text{at boundary}} = 0} \qquad (6.176)
$$

where $\hat{t}$ is a unit vector tangent to the surface.

Alternatively, if two viscous fluids meet and form a flow boundary, this same boundary condition requires that the stress in one fluid equal the stress in the other at the boundary. For example, in Figure 6.12b, the shear stresses must match at the fluid interface:

$$
\text{Drag flow } \underline{v} = v_x(y)\hat{e}_x \atop {\text{of two viscous fluid} \atop {(\text{Figure 6.12b}) \atop (\text{shear stress matches at boundary})}} \qquad \boxed{\tilde{\tau}_{yx}^{fluid1}\bigg|_{y=\lambda H} = \tilde{\tau}_{yx}^{fluid2}\bigg|_{y=\lambda H}}
$$

$$
(6.177)
$$

For the general case, this becomes:

Two viscous
fluids in contact
(shear stress
matches at
boundary)

$$\left.\left(\hat{n}\cdot\underline{\underline{\tilde{\tau}}}^{fluid1}\cdot\hat{t}\right)\right|_{\text{at interface}} = \left.\left(\hat{n}\cdot\underline{\underline{\tilde{\tau}}}^{fluid2}\cdot\hat{t}\right)\right|_{\text{at interface}}$$

(6.178)

4. *Velocity continuity.* When a fluid forms one of the boundaries of the flow as described previously and as shown in Figure 6.12b, the velocity also is continuous from one fluid to another. For the flow in Figure 6.12b, the velocity boundary condition at the interface is:

Drag flow $\underline{v} = v_x(y)\hat{e}_x$
of two viscous fluids
(Figure 6.12b)
(velocity matches
at boundary)

$$\left.v_x^{fluid1}\right|_{y=\lambda H} = \left.v_x^{fluid2}\right|_{y=\lambda H}$$

(6.179)

For the general case of two viscous fluids in contact, the boundary conditions on normal and tangential velocity components are shown here:

Two viscous fluids
in contact
(tangential and normal
velocities match)
at boundary)

$$\left.\hat{t}\cdot\underline{v}^{fluid1}\right|_{\text{at boundary}} = \left.\hat{t}\cdot\underline{v}^{fluid2}\right|_{\text{at boundary}}$$
$$\left.\hat{n}\cdot\underline{v}^{fluid1}\right|_{\text{at boundary}} = \left.\hat{n}\cdot\underline{v}^{fluid2}\right|_{\text{at boundary}}$$

(6.180)

5. *Finite velocity and stress.* Occasionally, an expression is derived that predicts infinite velocities or stresses at a point; an example is an equation that includes $1/r$, for a flow domain where $r = 0$ is included. A possible boundary condition to use in this instance is the requirement that the velocity and/or the stress be finite throughout the flow domain. This boundary condition appears occasionally in flows with cylindrical symmetry:

Finite
properties

$$\left.\underline{v}\right|_{\text{at boundary}} = \text{finite}$$
$$\left.\underline{\underline{\tilde{\tau}}}\right|_{\text{at boundary}} = \text{finite}$$

(6.181)

6. *Surface tension.* As discussed in Section 4.4, the imbalance of molecular forces at interfaces can be modeled within the continuum model using a material parameter called the surface tension. In Section 4.4, we derive an expression that relates pressures inside and outside of a spherical drop (see Equation 4.408):

$$\Delta p = p_{in} - p_{out} = \frac{2\sigma}{R}$$

(6.182)

The pressure inside is greater than the pressure outside the drop by the amount $2\sigma/R$. The unbalanced molecular forces at the surface lead to an extra inward pull on the surface molecules. The surface tension can serve as a boundary condition on pressure:

Surface tension
(spherical interface)

$$p|_{\text{at boundary}} = p_{out} + \frac{2\sigma}{R}$$ (6.183)

If the surface is not spherical, the pressure at the boundary becomes the following [5, 27]:

Surface tension
(general interface)

$$p|_{\text{at boundary}} = p_{out} + \sigma \left(\frac{1}{R_1} + \frac{1}{R_2} \right)$$

(6.184)

where R_1 and R_2 are the two principal radii of curvature of the surface at the point of interest (see Section 4.4 and the Web appendix [108] for more detail).

This list is not all-encompassing; for more information on boundary conditions, consult the literature [9, 14, 85]. We often know the most about the character of a flow near its boundaries. It is good practice to start a flow problem with consideration of the boundaries. As shown in the following example, we often can make a problem easier to solve by choosing the coordinate system such that the boundary conditions are easy to evaluate.

EXAMPLE 6.6. *For the flow of a thin film down an inclined plane, various possible choices for the coordinate system are shown in Figure 6.13. What are the boundary conditions for each choice? Discuss the relative merits of these coordinate-system choices.*

SOLUTION. By now, the problem of flow down an incline is familiar. For our problem, the flow shape is rectilinear; therefore, we choose a Cartesian coordinate system to use when solving for the velocity and stress fields. Having made this choice, we still have a number of reasonable options for the choice of coordinate system, as shown in Figure 6.13.

In general, we usually choose Cartesian systems that are aligned with gravity, such as those at the top and bottom of Figure 6.13. In this problem, however, such a choice requires us to solve for two nonzero components of the velocity vector—an unnecessary complication.

If we choose one of the two coordinate systems in the center of Figure 6.13, we reduce the number of nonzero velocity components to one:

$$\underline{v} = \begin{pmatrix} 0 \\ 0 \\ v_z \end{pmatrix}_{xyz} \qquad \begin{matrix} \text{coordinate system chosen with} \\ z \text{ in flow direction} \end{matrix}$$ (6.185)

This is an important simplification. By expressing $\underline{v}$ in a coordinate system in which there is a single nonzero component, v_z, we reduce the complexity of the problem. The two lined-up coordinate systems in the center of Figure 6.13 appear equally good for solving the incline problem; previously in Chapter 3, we

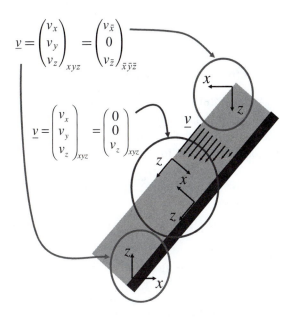

$$\underline{v} = \begin{pmatrix} v_x \\ v_y \\ v_z \end{pmatrix}_{xyz} = \begin{pmatrix} v_{\bar{x}} \\ 0 \\ v_{\bar{z}} \end{pmatrix}_{\bar{x}\bar{y}\bar{z}}$$

$$\underline{v} = \begin{pmatrix} v_x \\ v_y \\ v_z \end{pmatrix}_{xyz} = \begin{pmatrix} 0 \\ 0 \\ v_z \end{pmatrix}_{xyz}$$

Figure 6.13 The choice of a coordinate system is arbitrary but important. If we choose a horizontal Cartesian system, the velocity vector has two nonzero components, requiring us to solve simultaneously two components of the momentum balance. If we choose a Cartesian coordinate system that aligns one of the basis vectors with the direction of the flow, the velocity vector has only one nonzero component and the momentum balance is easier to solve.

chose the system that is on the surface of the incline (Figure 6.14). In our chosen coordinate system, the velocity vector is given by:

$$\underline{v} = \begin{pmatrix} 0 \\ 0 \\ v_z \end{pmatrix}_{xyz} = v_z \hat{e}_z \tag{6.186}$$

In this coordinate system, the boundaries of the problem are:

$$\begin{array}{llll} \text{Boundary conditions:} & BC1a: & x = 0 & v_z = 0 \\ \text{(origin at incline surface)} & BC2a: & x = H & \frac{dv_z}{dx} = 0 \end{array} \tag{6.187}$$

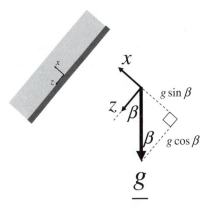

Figure 6.14 Because we chose a coordinate system that simplifies the velocity vector, the gravity vector is slightly more complicated than it might be with another choice.

If we chose instead the lined-up coordinate system with the origin at the free surface, the boundaries of the problem are:

$$
\begin{array}{llll}
\text{Boundary conditions:} & BC1b\text{:} & x = H & v_z = 0 \\
\text{(origin at free surface)} & BC2b\text{:} & x = 0 & \frac{dv_z}{dx} = 0
\end{array}
\tag{6.188}
$$

Both choices for the coordinate system are valid; there is a slight advantage to the second choice in terms of the amount of algebra to solve for C_1 and C_2. In the solution to the problem, the boundary conditions are needed once the Navier-Stokes equation is integrated to give Equation 5.146:

$$
v_z = \left[\frac{-\rho g \cos \beta}{\mu} \right] \frac{x^2}{2} + C_1 x + C_2
\tag{6.189}
$$

To apply the free-surface boundary condition, we must differentiate the result to obtain dv_z/dx:

$$
\frac{dv_z}{dx} = \left[\frac{-\rho g \cos \beta}{\mu} \right] x + C_1
\tag{6.190}
$$

If we choose $x = 0$ at the free surface, C_1 becomes zero, simplifying the algebra for determining the integration constants. If we choose $x = H$ at the free surface, we must perform more complex manipulations to obtain C_2.

Because of this advantage, it is customary to choose the origin for this problem at the free surface. When there is a symmetry plane or line of symmetry in a problem, there also is a boundary condition in terms of a derivative, and the same logic applies.

6.2.3 Engineering quantities from velocity and stress fields

A final topic that may help students is how to calculate engineering quantities from velocity and stress fields. Four important engineering quantities are force on a surface, torque to produce a rotation, flow rate, and velocity/stress maxima.

6.2.3.1 TOTAL FORCE ON A WALL

One reason that fluids are used in devices is to transfer forces. An example of this already discussed is the hydraulic lift (see Section 4.2.4.2), in which a fluid is used to amplify forces. Another example of fluids mediating forces is when a fluid is introduced between two solid parts as a lubricant to reduce the amount of force transferred (Figure 6.15). Force transfer is not always the goal—for example, when the transportation of the fluid is the engineering goal. In this case, the forces of the fluid on the wall must be overcome with a pump or another device. In fast-moving equipment, the fluid forces can be significant and the consequences of failure disastrous.

In all of these examples, the design of the apparatus depends on knowing the total force that a fluid exerts on the wall. If the stress distribution in the fluid is obtained using the microscopic momentum balance, then the total force on any surface may be calculated by evaluating the fluid stress at each point on that surface and summing the product of stress and area over the entire surface.

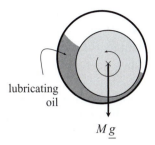

$$M\,\underline{g}$$

Figure 6.15 When two metal parts move relative to one another, such as in the journal bearing sketched here, a lubricant is used to reduce the stress transferred from one part to the other. The total force on a surface in contact with a lubricant can be calculated with the equations in this section.

We previously addressed such a sum over a surface; the result is a surface integral. The force due to the fluid at one piece of the wall tangent plane ΔS_i is given by Equation 4.263:

$$\begin{matrix}\text{Fluid force} \\ \text{on surface } \Delta S_i \\ \text{with unit normal } \hat{n} \\ \text{at point } (x_i, y_i, z_i)\end{matrix} = [\hat{n} \cdot \underline{\underline{\tilde{\Pi}}}]_{(x_i y_i z_i)}\, \Delta S_i \qquad (6.191)$$

where $\underline{\underline{\tilde{\Pi}}}$ is the total-stress tensor and $[\hat{n} \cdot \underline{\underline{\tilde{\Pi}}}]_{x_i y_i z_i}$ is the stress on ΔS_i at x_i, y_i, z_i. To obtain the force on the entire wall, we sum all of the pieces that comprise the surface and take the limit as $\Delta S = \Delta A/(\hat{n} \cdot \hat{e}_z)$ goes to zero (see the Web appendix [108]):

$$\begin{matrix}\text{Total fluid force} \\ \text{on a surface } \mathcal{S}:\end{matrix} \quad \underline{\mathcal{F}} = \lim_{\Delta A \to 0} \left[\sum_{i=1}^{N} [\hat{n} \cdot \underline{\underline{\tilde{\Pi}}}]_{(x_i y_i z_i)} \Delta S_i \right] \qquad (6.192)$$

$$= \lim_{\Delta A \to 0} \left[\sum_{i=1}^{N} \frac{[\hat{n} \cdot \underline{\underline{\tilde{\Pi}}}]_{(x_i y_i z_i)}}{\hat{n}_i \cdot \hat{e}_z} \Delta A_i \right] \qquad (6.193)$$

$$= \iint_{\mathcal{R}} \frac{[\hat{n} \cdot \underline{\underline{\tilde{\Pi}}}]_{\text{at surface}}}{\hat{n} \cdot \hat{e}_z} \, dA \qquad (6.194)$$

$$= \iint_{\mathcal{S}} [\hat{n} \cdot \underline{\underline{\tilde{\Pi}}}]_{\text{at surface}} \, dS \qquad (6.195)$$

$$\begin{matrix}\text{Total fluid force} \\ \text{on a surface } \mathcal{S}:\end{matrix} \quad \boxed{\underline{\mathcal{F}} = \iint_{\mathcal{S}} [\hat{n} \cdot \underline{\underline{\tilde{\Pi}}}]_{\text{at surface}} \, dS} \qquad (6.196)$$

We previously introduced this expression in Equation 4.285, and we use it extensively throughout the text.

We can try Equation 6.196 by calculating the total force on the incline in the falling-film example.

EXAMPLE 6.7. *What is the total vector force on the incline in the falling-film example (see Figure 6.10)?*

SOLUTION. The total force on a surface in a fluid is given by Equation 6.196. The unit normal to the incline surface written in the chosen flow coordinate system is $\hat{n} = \hat{e}_x$ (see Figure 6.10). This unit normal vector is the same at every location on the surface of the incline. The stress tensor $\tilde{\underline{\underline{\Pi}}} = -p\underline{\underline{I}} + \tilde{\underline{\underline{\tau}}}$ was solved for in pieces in previous examples; the result for $\tilde{\underline{\underline{\Pi}}}$ can be constructed from the Newtonian constitutive equation (Equation 6.66) and the shear-stress result (Equation 6.112). The final force then can be calculated with a straightforward integration of Equation 6.196.

For the flow down an incline the total-stress tensor $\tilde{\underline{\underline{\Pi}}}$ is:

$$\tilde{\underline{\underline{\Pi}}} = -p\underline{\underline{I}} + \tilde{\underline{\underline{\tau}}} \tag{6.197}$$

$$= \begin{pmatrix} -p(x) & 0 & \tilde{\tau}_{xz}(x) \\ 0 & -p(x) & 0 \\ \tilde{\tau}_{xz}(x) & 0 & -p(x) \end{pmatrix}_{xyz} \tag{6.198}$$

We solved previously for the two missing pieces of information, $p(x)$ and $\tilde{\tau}_{xz}$:

$$p(x) = p_{atm} + \rho g H \sin\beta \left(1 - \frac{x}{H}\right) \tag{6.199}$$

$$\tilde{\tau}_{xz}(x) = \rho g \cos\beta (H - x) \tag{6.200}$$

To use Equation 6.196, we need $\hat{n} \cdot \tilde{\underline{\underline{\Pi}}}$ at the incline surface. The surface of the incline is located at $x = 0$, and the unit normal to the entire surface is $\hat{n} = \hat{e}_x$:

$$[\hat{n} \cdot \tilde{\underline{\underline{\Pi}}}]_{\text{at surface}} = [\hat{e}_x \cdot \tilde{\underline{\underline{\Pi}}}]_{x=0} \tag{6.201}$$

$$= \begin{pmatrix} 1 & 0 & 0 \end{pmatrix}_{xyz} \cdot \begin{pmatrix} -p(0) & 0 & \tilde{\tau}_{xz}(0) \\ 0 & -p(0) & 0 \\ \tilde{\tau}_{xz}(0) & 0 & -p(0) \end{pmatrix}_{xyz}$$

$$= \begin{pmatrix} -p(0) & 0 & \tilde{\tau}_{xz}(0) \end{pmatrix}_{xyz} \tag{6.202}$$

$$= \begin{pmatrix} -p_{atm} - \rho g H \sin\beta \\ 0 \\ \rho g H \cos\beta \end{pmatrix}_{xyz} \tag{6.203}$$

Note that in the last step, we changed from a row vector to a column vector for convenience.

The surface integral in Equation 6.196 can be carried out by identifying dS for the surface and coordinate system. The surface of the incline is flat and rectangular with unit normal $\hat{n} = \hat{e}_x$; therefore, we write $dS = dy\,dz$. We now complete the integration:

$$\begin{matrix}\text{Total force} \\ \text{on the} \\ \text{incline surface}\end{matrix} \quad \mathcal{F} = \iint_S [\hat{n} \cdot \tilde{\underline{\underline{\Pi}}}]_{\text{at surface}} \, dS \tag{6.204}$$

$$= \int_0^L \int_0^W \begin{pmatrix} -p_{atm} - \rho g H \sin\beta \\ 0 \\ \rho g H \cos\beta \end{pmatrix}_{xyz} dy\,dz \tag{6.205}$$

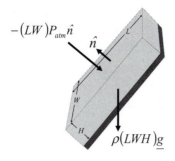

Figure 6.16

The force on the incline is a combination of the weight of the fluid and the force due to atmospheric pressure.

The limits of the integrals are chosen to cover the entire surface of the plane. The quantities L and W are the length and width of the incline. After integration, the result is:

$$
\begin{matrix}
\text{Total force} \\
\text{on the} \\
\text{incline surface}
\end{matrix}
\qquad
\underline{\mathcal{F}} = LW
\begin{pmatrix}
-p_{atm} - \rho g H \sin \beta \\
0 \\
\rho g H \cos \beta
\end{pmatrix}_{xyz}
\qquad (6.206)
$$

$$
= (-LW p_{atm} - LWH\rho g \sin \beta)\,\hat{e}_x + \rho g H \cos \beta \hat{e}_z \quad (6.207)
$$

The integration was easy because nothing in the integral varies with y or z.

Reviewing the geometry of the falling-film problem and the form of the solution in Equation 6.206, we notice that we can rewrite our solution in a form that helps to grasp its meaning:

$$
\begin{matrix}
\text{Total fluid force} \\
\text{on the} \\
\text{incline surface}
\end{matrix}
= \rho(LWH)
\begin{pmatrix}
-g \sin \beta \\
0 \\
g \cos \beta
\end{pmatrix}_{xyz}
- LW
\begin{pmatrix}
p_{atm} \\
0 \\
0
\end{pmatrix}_{xyz}
\quad (6.208)
$$

$$
= \rho(LWH)\underline{g} - (LW)p_{atm}\hat{n}
$$

We see from this final way of writing the result that the force on the incline is simply the weight of the fluid plus the force due to atmospheric pressure (Figure 6.16).

In Example 6.7 and with many calculations of this type, we need to carry out a surface integration. The surface differential dS in Equation 6.196 must be interpreted according to the specific case under consideration. For convenience, we assemble several common cases in the inside front cover of this book.

For the simple case of the flow down an incline, it appears that we could arrive at the total-force result using a straightforward force balance instead of performing the integration in Equation 6.196. The surface-integration method is general, however, and is useful in more complex situations, including those involving intricate wall shapes. We discuss a case involving spherical coordinates in Example 6.8 and we discuss more flow examples in Chapters 7 and 8.

EXAMPLE 6.8. *What is the total vector force on a sphere in creeping flow around a sphere, the flow shown in Figure 6.17?*

SOLUTION. In Chapter 8, we discuss the solution for the velocity and stress fields for flow around a sphere. We can calculate the total force on the sphere from the results for $\underline{v}$ and $\underline{\underline{\tilde{\Pi}}} = \underline{\underline{\tilde{\tau}}} - p\underline{\underline{I}}$ obtained there. We begin with Equation 6.196:

$$
\begin{array}{c}
\text{Total fluid force} \\
\text{in a fluid} \\
\text{on a surface } \mathcal{S}:
\end{array}
\qquad
\mathcal{F} = \iint_S [\hat{n} \cdot \underline{\underline{\tilde{\Pi}}}]_{\text{at surface}} \, dS
\qquad (6.209)
$$

As stated previously, to carry out this integral we need $\underline{\underline{\tilde{\Pi}}} = \underline{\underline{\tilde{\tau}}} - p\underline{\underline{I}}$ solved for with the microscopic momentum balance. If we presume that we have this result, then we can calculate the total force from Equation 6.209. The surface in which we are interested is located at $r = R$ and has an outwardly pointing unit normal vector $\hat{n} = \hat{e}_r$, where we are using the spherical coordinate system as shown in Figure 6.17. The differential surface element dS on the surface of the sphere can be written in the spherical coordinate system as $dS = R^2 \sin\theta d\theta d\phi$. The total force then is given by:

$$
\begin{array}{c}
\text{Total fluid force} \\
\text{on the sphere:}
\end{array}
\qquad
\mathcal{F} = \int_0^{2\pi} \int_0^{\pi} [\hat{e}_r \cdot \underline{\underline{\tilde{\Pi}}}]_{r=R} \, R^2 \sin\theta d\theta d\phi
\qquad (6.210)
$$

The limits on the integrations are chosen to cover the entire surface of the sphere.

The velocity has nonzero components in the r and θ directions, but there is no swirling component in the ϕ-direction:

$$
\underline{v} = \begin{pmatrix} v_r \\ v_\theta \\ v_\phi \end{pmatrix}_{r\theta\phi} = \begin{pmatrix} v_r \\ v_\theta \\ 0 \end{pmatrix}_{r\theta\phi}
\qquad (6.211)
$$

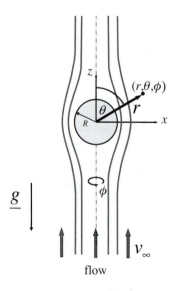

Figure 6.17   Flow around a sphere is important in droplet flow and in settling flows in suspensions.

The stress tensor in spherical coordinates is given in Table B.8 in Appendix B. In this flow, there is symmetry in the ϕ-direction, which allows us to eliminate velocity derivatives with respect to ϕ in Equation B.8-3. Also, $v_\phi = 0$; thus, four components of $\underline{\tilde{\tau}}$ are zero. With these simplifications, Equation B.8-3 becomes:

$$\underline{\tilde{\tau}} = \begin{pmatrix} \tilde{\tau}_{rr} & \tilde{\tau}_{r\theta} & \tilde{\tau}_{r\phi} \\ \tilde{\tau}_{\theta r} & \tilde{\tau}_{\theta\theta} & \tilde{\tau}_{\theta\phi} \\ \tilde{\tau}_{\phi r} & \tilde{\tau}_{\phi\theta} & \tilde{\tau}_{\phi\phi} \end{pmatrix}_{r\theta\phi} = \begin{pmatrix} \tilde{\tau}_{rr} & \tilde{\tau}_{r\theta} & 0 \\ \tilde{\tau}_{\theta r} & \tilde{\tau}_{\theta\theta} & 0 \\ 0 & 0 & \tilde{\tau}_{\phi\phi} \end{pmatrix}_{r\theta\phi} \tag{6.212}$$

We now can write $\hat{n} \cdot \underline{\tilde{\Pi}} = \hat{e}_r \cdot (\underline{\tilde{\tau}} - p\underline{I})$ as:

$$\hat{n} \cdot \underline{\tilde{\Pi}} = \hat{e}_r \cdot \underline{\tilde{\Pi}} = \begin{pmatrix} 1 & 0 & 0 \end{pmatrix}_{r\theta\phi} \cdot \begin{pmatrix} \tilde{\tau}_{rr} - p & \tilde{\tau}_{r\theta} & 0 \\ \tilde{\tau}_{\theta r} & \tilde{\tau}_{\theta\theta} - p & 0 \\ 0 & 0 & \tilde{\tau}_{\phi\phi-p} \end{pmatrix}_{r\theta\phi} \tag{6.213}$$

$$= \begin{pmatrix} \tilde{\tau}_{rr} - p & \tilde{\tau}_{r\theta} & 0 \end{pmatrix}_{r\theta\phi} \tag{6.214}$$

$$= \begin{pmatrix} \tilde{\tau}_{rr} - p \\ \tilde{\tau}_{r\theta} \\ 0 \end{pmatrix}_{r\theta\phi} \tag{6.215}$$

Substituting this result into Equation 6.210, we obtain the expression that we must evaluate to obtain the force on the sphere:

Total fluid force on the sphere:
$$\underline{\mathcal{F}} = \int_0^{2\pi} \int_0^\pi [\hat{e}_r \cdot \underline{\tilde{\Pi}}]_{r=R} \, R^2 \sin\theta \, d\theta \, d\phi \tag{6.216}$$

$$= \int_0^{2\pi} \int_0^\pi \begin{pmatrix} \tilde{\tau}_{rr}|_R - p|_R \\ \tilde{\tau}_{r\theta}|_R \\ 0 \end{pmatrix}_{r\theta\phi} R^2 \sin\theta \, d\theta \, d\phi \tag{6.217}$$

Without the microscopic-balance results for the components of $\underline{v}$, this is as far as we can go in our solution for the total force on the sphere. The solution to the flow around the sphere problem for creeping flow (i.e., slow flow) is presented in Chapter 8. It turns out that $\tilde{\tau}_{rr}$ is equal to zero at the surface $r = R$; thus, the final expression to evaluate for force is that given here:

Total fluid force on the sphere:
$$\boxed{\underline{\mathcal{F}} = \int_0^{2\pi} \int_0^\pi \begin{pmatrix} -p|_R \\ \tilde{\tau}_{r\theta}|_R \\ 0 \end{pmatrix}_{r\theta\phi} R^2 \sin\theta \, d\theta \, d\phi} \tag{6.218}$$

Note that $\hat{e}_r$ and $\hat{e}_\theta$ are both functions of θ; this must be considered in the integration. We complete this calculation in Chapter 8.

Working on a more complex problem such as calculating the forces in the flow around a sphere is made considerably easier by having the general rule, Equation 6.196, and then knowing how to apply it. More examples of the utility of Equation 6.196 appear in the remaining chapters. In complex flows, Equation 6.196 is evaluated numerically using computer code [5, 27].

6.2.3.2 TORQUE

We worked briefly with torque in Chapter 1 (see Example 1.17) and in Chapter 4 (see Example 4.20). As in Example 4.20, when machine parts turn in fluids, torque is needed or generated by the motion; thus, torque is an engineering quantity of interest in fluid mechanics.

Torque is the amount of effort to produce a rotation in a body; the definition of torque is the cross product of the lever arm and the tangential force. The lever arm vector is along the path from the axis of rotation to the point of application of the force:

$$\underline{T} = (\text{lever arm}) \times (\text{force}) \tag{6.219}$$

$$= \underline{R} \times \underline{f} \tag{6.220}$$

We calculate torque on a finite surface in a flow beginning with the fluid force on an infinitesimal surface given by Equation 4.279:

$$\begin{matrix} \text{Molecular fluid force} \\ \text{on surface } \Delta S_i \\ \text{with unit normal } \hat{n} \\ \text{at point } (x_i, y_i, z_i) \end{matrix} \qquad \underline{f}\Big|_{\Delta S_i} = [\hat{n} \cdot \underline{\tilde{\Pi}}]_{(x_i y_i z_i)} \, \Delta S_i \tag{6.221}$$

The total torque is the limit of the sum of the infinitesimal torques on small pieces of the surface tangent planes ΔS_i:

$$\begin{matrix} \text{Total torque} \\ \text{on a surface } \mathcal{S}: \end{matrix} = \lim_{\Delta A \longrightarrow 0} \left[\sum_{i=1}^{N} \underline{R}|_{\Delta S_i} \times [\hat{n} \cdot \underline{\tilde{\Pi}}]_{(x_i y_i z_i)} \, \Delta S_i \right] \tag{6.222}$$

$$= \lim_{\Delta A \longrightarrow 0} \left[\sum_{i=1}^{N} \frac{\underline{R}|_{\Delta A_i} \times [\hat{n} \cdot \underline{\tilde{\Pi}}]_{(x_i y_i z_i)}}{\hat{n}_i \cdot \hat{e}_z} \, \Delta A_i \right] \tag{6.223}$$

$$= \iint_{\mathcal{R}} \frac{[\underline{R} \times (\hat{n} \cdot \underline{\tilde{\Pi}})]_{\text{at surface}}}{\hat{n} \cdot \hat{e}_z} \, dA \tag{6.224}$$

$$= \iint_{\mathcal{S}} [\underline{R} \times (\hat{n} \cdot \underline{\tilde{\Pi}})]_{\text{at surface}} \, dS \tag{6.225}$$

$$\boxed{\begin{matrix} \text{Total torque} \\ \text{on a surface } \mathcal{S}: \end{matrix} \qquad \underline{T} = \iint_{\mathcal{S}} [\underline{R} \times (\hat{n} \cdot \underline{\tilde{\Pi}})]_{\text{at surface}} \, dS} \tag{6.226}$$

The torque thus may be calculated from the stress tensor, which (as usual) may be obtained from the solution of the momentum balance. We practice applying Equation 6.226 in Example 6.9.

EXAMPLE 6.9. *A compact device that may be used to measure viscosity and other flow properties of fluids is the parallel-plate torsional rheometer (Figure 6.18). In this device, the gap between two circular disks is filled with fluid*

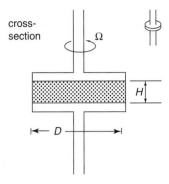

An incompressible Newtonian fluid is confined between two circular disks of diameter D. The gap between the plates is H and the top plate rotates with a constant angular velocity Ω, as shown here.

and one of the disks is turned. The design is such that the velocity field in the gap is given by:

$$\underline{v} = \begin{pmatrix} 0 \\ \frac{r\Omega z}{H} \\ 0 \end{pmatrix}_{r\theta z} = \frac{r\Omega z}{H} \hat{e}_\theta \qquad (6.227)$$

The viscosity is related to the torque required to turn the disk. For a Newtonian fluid in such an apparatus, how is the viscosity related to the total torque to turn the top disk?

SOLUTION. The total torque on a surface in a fluid is given by Equation 6.226:

Total torque on a surface $\mathcal{S}$:
$$\underline{T} = \iint_{\mathcal{S}} [\underline{R} \times (\hat{n} \cdot \underline{\tilde{\Pi}})]_{\text{at surface}} \, dS \qquad (6.228)$$

To apply this equation to calculate the torque in the current problem, we identify each quantity in the equation and carry out the integration. Torque is needed to turn the top plate because the flat circular surface at $z = H$ is in contact with the fluid. The surface in the fluid in contact with the top plate has a unit normal $\hat{n} = \hat{e}_z$. The lever-arm vector $\underline{R}$ is from the axis of rotation to a point experiencing torque. The points on the surface experiencing torque are all of the locations on the top fluid surface; thus, the lever arm is variable. We choose a small area $dS = r\,d\theta\,dr$, which is located in the plane of the plate, a distance r from the axis of rotation. For these areas, the lever-arm vectors are $\underline{R} = r\hat{e}_r$. Equation 6.228 becomes:

Total torque on the top fluid surface in the parallel-plate rheometer
$$\underline{T} = \int_0^{2\pi} \int_0^{D/2} [\underline{R} \times (\hat{n} \cdot \underline{\tilde{\Pi}})]_{\text{at } z=H} \, dS \qquad (6.229)$$

$$= \int_0^{2\pi} \int_0^{D/2} [r\hat{e}_r \times (\hat{e}_z \cdot \underline{\tilde{\Pi}})]_{\text{at } z=H} \, r\,dr\,d\theta \qquad (6.230)$$

The stress tensor $\underset{=}{\tilde{\Pi}}$ comes from the Newtonian constitutive equation evaluated for this flow. Because we know the velocity field (Equation 6.228), we can calculate the expression that we need directly from the constitutive equation, Equation 5.89:

$$\underset{=}{\tilde{\Pi}} = -p\underline{\underline{I}} + \underset{=}{\tilde{\tau}} \tag{6.231}$$

$$= -p\underline{\underline{I}} + \mu \left(\nabla \underline{v} + (\nabla \underline{v})^T \right) \tag{6.232}$$

In cylindrical coordinates, the Newtonian constitutive equation is given in Equation 5.91. We can immediately simplify Equation 5.91 because $v_r = v_z = 0$ and v_θ is not a function of θ. With these simplifications, the Newtonian constitutive equation becomes:

$$\underset{=}{\tilde{\tau}} = \begin{pmatrix} 0 & \mu \left(\frac{\partial v_\theta}{\partial r} - \frac{v_\theta}{r} \right) & 0 \\ \mu \left(\frac{\partial v_\theta}{\partial r} - \frac{v_\theta}{r} \right) & 0 & \mu \frac{\partial v_\theta}{\partial z} \\ 0 & \mu \frac{\partial v_\theta}{\partial z} & 0 \end{pmatrix}_{r\theta z} \tag{6.233}$$

Carrying out the partial derivatives of v_θ using the velocity field given and assembling $\underset{=}{\tilde{\Pi}}$, we obtain:

$$\underset{=}{\tilde{\Pi}} = -p\underline{\underline{I}} + \mu \left(\nabla \underline{v} + (\nabla \underline{v})^T \right) \tag{6.234}$$

$$= \begin{pmatrix} -p & \mu \left(\frac{\partial v_\theta}{\partial r} - \frac{v_\theta}{r} \right) & 0 \\ \mu \left(\frac{\partial v_\theta}{\partial r} - \frac{v_\theta}{r} \right) & -p & \mu \frac{\partial v_\theta}{\partial z} \\ 0 & \mu \frac{\partial v_\theta}{\partial z} & -p \end{pmatrix}_{r\theta z} \tag{6.235}$$

$$\underset{=}{\tilde{\Pi}} = \begin{pmatrix} -p & 0 & 0 \\ 0 & -p & \mu \frac{r\Omega}{H} \\ 0 & \mu \frac{r\Omega}{H} & -p \end{pmatrix}_{r\theta z} \tag{6.236}$$

The next steps are to carry out the dot product and the cross product in Equation 6.230:

Total torque on the top fluid surface in the parallel-plate rheometer

$$\underline{\mathcal{T}} = \int_0^{2\pi} \int_0^{D/2} [r\hat{e}_r \times (\hat{e}_z \cdot \underset{=}{\tilde{\Pi}})]_{\text{at } z=H} \; r\,dr\,d\theta \tag{6.237}$$

$$\hat{e}_z \cdot \underset{=}{\tilde{\Pi}} = \begin{pmatrix} 0 & 0 & 1 \end{pmatrix}_{r\theta z} \cdot \begin{pmatrix} -p & 0 & 0 \\ 0 & -p & \mu \frac{r\Omega}{H} \\ 0 & \mu \frac{r\Omega}{H} & -p \end{pmatrix}_{r\theta z} \tag{6.238}$$

$$= \begin{pmatrix} 0 & \mu \frac{r\Omega}{H} & -p \end{pmatrix}_{r\theta z} \tag{6.239}$$

$$r\hat{e}_r \times (\hat{e}_z \cdot \underset{=}{\tilde{\Pi}}) = \begin{pmatrix} r \\ 0 \\ 0 \end{pmatrix}_{r\theta z} \times \begin{pmatrix} 0 \\ \mu \frac{r\Omega}{H} \\ -p \end{pmatrix}_{r\theta z} \tag{6.240}$$

$$= \begin{pmatrix} 0 \\ rp \\ \frac{r^2 \mu \Omega}{H} \end{pmatrix}_{r\theta z} \tag{6.241}$$

where we use Equation 1.182 to evaluate the cross product. The integral to evaluate for torque thus becomes:

$$\underline{\mathcal{T}} = \int_0^{2\pi} \int_0^{D/2} [r\hat{e}_r \times (\hat{e}_z \cdot \underline{\tilde{\Pi}})]_{\text{at } z=H} \; r\,dr\,d\theta \tag{6.242}$$

$$= \int_0^{2\pi} \int_0^{D/2} \begin{pmatrix} 0 \\ rp \\ \frac{r^2\mu\Omega}{H} \end{pmatrix}_{r\theta z} r\,dr\,d\theta \tag{6.243}$$

$$= \int_0^{2\pi} \int_0^{D/2} \left[r^2 p\hat{e}_\theta + \frac{r^3\mu\Omega}{H}\hat{e}_z \right] dr\,d\theta \tag{6.244}$$

The basis vector $\hat{e}_\theta$ is a function of θ; thus, we convert to Cartesian coordinates before evaluating the integral:

$$\underline{\mathcal{T}} = \int_0^{2\pi} \int_0^{D/2} \left[r^2 p\hat{e}_\theta + \frac{r^3\mu\Omega}{H}\hat{e}_z \right] dr\,d\theta \tag{6.245}$$

$$= \int_0^{2\pi} \int_0^{D/2} \left[r^2 p\left(-\sin\theta\hat{e}_x + \cos\theta\hat{e}_y\right) + \frac{r^3\mu\Omega}{H}\hat{e}_z \right] dr\,d\theta \tag{6.246}$$

The θ-integral results in the $\hat{e}_x$ and $\hat{e}_y$ terms dropping out, leaving the $\hat{e}_z$ component as the only nonzero component of torque. The details of the remaining steps are left to readers. The final result for torque is:

Total torque on the top fluid surface in the parallel-plate rheometer (Newtonian)
$$\underline{\mathcal{T}} = \frac{\pi\Omega\mu R^4}{2H}\hat{e}_z = \begin{pmatrix} 0 \\ 0 \\ \frac{\pi\Omega\mu R^4}{2H} \end{pmatrix}_{xyz} = \begin{pmatrix} 0 \\ 0 \\ \frac{\pi\Omega\mu R^4}{2H} \end{pmatrix}_{r\theta z} \tag{6.247}$$

We therefore can calculate the viscosity μ from a measurement of the torque magnitude as:

Viscosity from torque in parallel-plate Newtonian
$$\mu = \frac{2H\mathcal{T}}{\pi\Omega R^4} \tag{6.248}$$

6.2.3.3 FLOW RATE AND AVERAGE VELOCITY

Forces and torques are two types of engineering variables; another important quantity is flow rate. Flow rate, or volume flow per unit time, may be calculated directly from a velocity profile. To calculate the flow rate through a finite surface S when the velocity varies across the surface, we again calculate a surface integral.

The flow rate through one piece of the surface tangent plane ΔS_i is given by Equation 3.90 at that point:

Flow rate through ΔS_i with unit normal $\hat{n}$ at point (x_i, y_i, z_i)
$$= [\hat{n} \cdot \underline{v}]_{(x_i y_i z_i)} \; \Delta S_i \tag{6.249}$$

To obtain the total flow rate, we sum all of the pieces that comprise the surface S, and take the limit as ΔS goes to zero (see the Web appendix [108]):

$$
\begin{array}{l}
\text{Total flow rate} \\
\text{out through} \\
\text{surface } S:
\end{array}
\quad
Q = \lim_{\Delta A \to 0} \left[\sum_{i=1}^{N} [\hat{n} \cdot \underline{v}]_{(x_i y_i z_i)} \, \Delta S_i \right]
\qquad (6.250)
$$

$$
= \lim_{\Delta A \to 0} \left[\sum_{i=1}^{N} \frac{[\hat{n} \cdot \underline{v}]_{(x_i y_i z_i)}}{\hat{n}_i \cdot \hat{e}_z} \, \Delta A_i \right]
\qquad (6.251)
$$

$$
= \iint_{\mathcal{R}} \frac{[\hat{n} \cdot \underline{v}]_{\text{at surface}}}{\hat{n} \cdot \hat{e}_z} \, dA
\qquad (6.252)
$$

$$
Q = \iint_{S} [\hat{n} \cdot \underline{v}]_{\text{at surface}} \, dS
\qquad (6.253)
$$

$$
\boxed{
\begin{array}{l}
\text{Total flow rate} \\
\text{out through} \\
\text{surface } S:
\end{array}
\quad
Q = \iint_{S} [\hat{n} \cdot \underline{v}]_{\text{at surface}} \, dS
}
\qquad (6.254)
$$

To calculate the average velocity, we divide the total volumetric flow rate by the area through which the flow passes:

$$
\boxed{
\begin{array}{l}
\text{Average velocity} \\
\text{through surface } S:
\end{array}
\quad
\langle v \rangle = \frac{Q}{\iint_{S} dS} = \frac{\iint_{S} [\hat{n} \cdot \underline{v}]_{\text{at surface}} \, dS}{\iint_{S} dS}
}
\qquad (6.255)
$$

We used this expression in Equation 5.194. We can try these expressions by calculating the total flow rate and average velocity in the steady drag-flow between parallel plates (see Example 6.3).

EXAMPLE 6.10. *What are the flow rate and average velocity in the steady drag-flow between parallel plates (see Example 6.3)?*

SOLUTION. We begin the solution with Equation 6.254:

$$
\begin{array}{l}
\text{Total flow rate} \\
\text{out through} \\
\text{surface } S:
\end{array}
\quad
Q = \iint_{S} [\hat{n} \cdot \underline{v}]_{\text{at surface:}} \, dS
\qquad (6.256)
$$

We must identify $\hat{n}$, $\underline{v}$, and the surface over which we want to integrate. The flow rate in the drag-flow-between-infinite-plates problem is the same at every x_1-position throughout the flow; therefore, we can choose as our calculation surface any plane perpendicular to the flow—we choose a plane at the exit, $x_1 = L$. The unit normal of our calculation surface is $\hat{n} = \hat{e}_1$, and the velocity vector is given in Equation 6.134 as $\underline{v} = (V/H)x_2\hat{e}_1$. The dot of these two vectors is:

$$\hat{n} \cdot \underline{v} = \begin{pmatrix} 1 \\ 0 \\ 0 \end{pmatrix}_{123} \cdot \begin{pmatrix} \frac{V}{H}x_2 \\ 0 \\ 0 \end{pmatrix}_{123} = \frac{V}{H}x_2 \tag{6.257}$$

The surface S is a rectangle in the 23-plane; thus, $dS = dx_2dx_3$ (see Figure 6.11), and the location of the surface is $x_1 = L$. The flow rate Q then is given by:

$$Q = \iint_S [\hat{n} \cdot \underline{v}]_{x_1=L} \, dS \tag{6.258}$$

$$= \int_0^W \int_0^H \left(\frac{V}{H}x_2 \right) dx_2dx_3 \tag{6.259}$$

$$= \frac{WHV}{2} \tag{6.260}$$

The average velocity is Q/HW:

Average velocity out through surface S in drag flow:
$$\langle v \rangle = \frac{\displaystyle\iint_S [\hat{n} \cdot \underline{v}]_{\text{at surface}} \, dS}{\displaystyle\iint_S dS} \tag{6.261}$$

$$= \frac{Q}{\displaystyle\int_0^W \int_0^H dx_1dx_2} \tag{6.262}$$

$$= \frac{V}{2} \tag{6.263}$$

6.2.3.4 VELOCITY AND STRESS EXTREMA

In some engineering problems, the maximum or minimum velocity or force is of interest. For example, if a fluid jet hits a surface, the maximum value of the force is important to know in designing the surface to withstand the impact. The location of the maximum force also is important when designing a bracing system for such a device.

To locate the maximum or minimum of any function (e.g., the velocity or stress component), we calculate the first derivative of the function and set it equal to zero (see Section 1.3.1) [166]:

At the maximum/minimum of $f(x)$: $\qquad \dfrac{df}{dx} = 0 \tag{6.264}$

Solving Equation 6.264 for $x_{min/max}$ gives us the location of the minimum or maximum. To determine if the extrema located is a minimum or a maximum, we calculate the second derivative [166]:

$$\frac{d^2 f}{dx^2}\bigg|_{x_{min/max}} > 0 \quad \Rightarrow \text{ minimum} \tag{6.265}$$

$$\frac{d^2 f}{dx^2}\bigg|_{x_{min/max}} < 0 \quad \Rightarrow \text{ maximum} \tag{6.266}$$

EXAMPLE 6.11. *A Newtonian fluid flows steadily between two long, wide plates under an imposed pressure difference $\Delta p = p_0 - p_L$ (Figure 6.19). In addition, the top plate moves at a velocity V. The velocity field is found by using methods described in Chapters 3–6; The solution for $v_x(y)$ in the coordinate system of Figure 6.19 is given here:*

$$\underline{v} = \left[\frac{H^2(p_L - p_0)}{2\mu L} \left(\left(\frac{y}{H} \right)^2 - 1 \right) + \frac{V}{2} \left(\frac{y}{H} + 1 \right) \right] \hat{e}_x \tag{6.267}$$

What is the location of the velocity maximum as a function of the imposed pressure difference?

SOLUTION. To find the location y_{max} at which the velocity function $v_x(y)$ attains its maximum value, we must find the location where the first derivative of v_x with respect to y goes to zero:

$$v_x(y) = \frac{H^2(p_L - p_0)}{2\mu L} \left[\left(\frac{y}{H} \right)^2 - 1 \right] + \frac{V}{2} \left(\frac{y}{H} + 1 \right) \tag{6.268}$$

Location
of $\dfrac{dv_x}{dy}\bigg|_{y=y_{max}} = 0$ $\qquad\qquad$ (6.269)
maximum:

To simplify the algebra, we define the constant B as:

$$B \equiv \frac{H^2(p_L - p_0)}{2\mu L} \tag{6.270}$$

which allows us to write the velocity as:

$$v_x(y) = B \frac{y^2}{H^2} - B + \frac{V}{2H}y + \frac{V}{2} \tag{6.271}$$

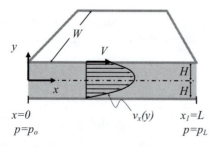

Figure 6.19 Combined pressure and drag flow of a Newtonian fluid through a wide, long slit can be modeled as shown.

We now take the first derivative of $v_x(y)$ and solve for the value of y that makes this zero:

$$\frac{dv_x}{dy} = \frac{2B}{H^2}y + \frac{V}{2H} \tag{6.272}$$

$$0 = \frac{2B}{H^2}y_{max} + \frac{V}{2H} \tag{6.273}$$

$$y_{max} = \frac{-VH}{4B} \tag{6.274}$$

$$\boxed{y_{max} = \frac{\mu L V}{2H(p_0 - p_L)}} \tag{6.275}$$

We can verify that this is, in fact, a maximum rather than a minimum by calculating the second derivative of $v_x(y)$:

$$\frac{d^2v_x}{dy^2} = \frac{2B}{H^2} \tag{6.276}$$

$$\frac{d^2v_x}{dy^2} = \frac{p_L - p_0}{\mu L} < 0 \tag{6.277}$$

Because the upstream pressure is higher than the downstream pressure ($p_0 > p_L$), Equation 6.277 indicates that the second derivative is negative throughout the flow; thus, the extremum we found is a maximum (compare to Equation 6.266).

6.3 Summary

In this chapter, we derive and use the microscopic mass and momentum balances. For general fluids, the microscopic momentum balance is the Cauchy momentum equation, Equation 6.86. For incompressible Newtonian fluids, the microscopic momentum balance is the Navier-Stokes equation, Equation 6.122. We show how to apply these equations to a problem with which we are familiar: the flow of a thin film down an inclined plane. We also discuss two topics necessary for problem solving with the microscopic balances: flow boundary conditions and methods for calculating macroscopic engineering properties from the microscopic results.

We have provided the groundwork for performing microscopic balances on a wide variety of flows. Chapters 7 and 8 discuss microscopic solutions in two important flow classes: internal and external flows, including boundary-layer flows. In those chapters, we apply the microscopic balances to simple and complicated cases, and we discuss how to use dimensional analysis to modify the microscopic analysis when a detailed microscopic solution is impractical or unnecessary. Studying the Navier-Stokes equation in various situations reveals why fluids behave the way they do.

6.4 Problems

1. Compare and contrast the solution to the flow down an incline plane pursued in Chapters 3–5 with the solution in Example 6.2.

2. What is the difference between solving for pressure with the mechanical energy balance and solving for pressure with the Navier-Stokes equation?

3. In the derivation of the continuity equation, we omit details of some vector/tensor manipulations. Using matrix notation in a Cartesian coordinate system or using Einstein notation, show that Equation 6.41 may be simplified to give Equation 6.43.

4. In the derivation of the Cauchy momentum equation, we omit details of some vector/tensor manipulations. Using matrix notation in a Cartesian coordinate system or using Einstein notation, show that Equation 6.62 may be simplified to give Equation 6.63.

5. In the derivation of the Navier-Stokes equation, we omit details of some vector/tensor manipulations. Using matrix notation in a Cartesian coordinate system or using Einstein notation, show that Equation 6.68 may be simplified to give Equation 6.70.

6. In the calculation of the total flow rate down an inclined plane, integrate Equation 5.193 to obtain the final result.

7. Show that the results for creeping flow around a sphere (see Equation 5.101) satisfy the continuity equation for incompressible fluids.

8. For each of the four coordinate systems shown in Figure 6.13, what is the vector that expresses the acceleration due to gravity?

9. In Figure 6.13, the following equality is given:

$$
\underline{v} = \begin{pmatrix} v_x \\ v_y \\ v_z \end{pmatrix}_{xyz} = \begin{pmatrix} v_{\bar{x}} \\ 0 \\ v_{\bar{z}} \end{pmatrix}_{\bar{x}\bar{y}\bar{z}}
$$

   Explain how both of these ways to express $\underline{v}$ are correct. Note that $v_x \neq v_{\bar{x}}$, $v_z \neq v_{\bar{z}}$.

10. Show with matrix operations on Cartesian coordinates that $\nabla \cdot p\underline{I} = \nabla p$. Use Table B.2 in Appendix B to obtain the Cartesian coordinates of this Gibbs expression.

11. Using matrices and the definition of the gradient of a vector (Appendix B), show that the following two expressions are equivalent:

$$
\nabla \cdot (\rho \underline{v}) = \underline{v} \cdot \nabla \rho + \rho (\nabla \cdot \underline{v})
$$

12. Using matrices and the definition of the gradient of a vector (Appendix B), show that the following two expressions are equivalent:

$$
\frac{\partial(\rho \underline{v})}{\partial t} + \nabla \cdot (\rho \underline{v}\, \underline{v}) = \rho \frac{\partial \underline{v}}{\partial t} + \underline{v} \frac{\partial \rho}{\partial t} + \rho (\underline{v} \cdot \nabla \underline{v}) + \underline{v} \nabla \cdot (\rho \underline{v})
$$

13. Using matrices and the definition of the gradient of a vector (Appendix B), show that the following two expressions are equivalent (viscosity is constant):

$$
\nabla \cdot \left(\mu \left(\nabla \underline{v} + (\nabla \underline{v})^T \right) \right) = \mu \nabla^2 \underline{v} + \mu \nabla (\nabla \cdot \underline{v})
$$

14. Using matrices and the definition of the gradient of a vector (Appendix B), show that the following two expressions are equivalent:

$$\frac{\partial \left(\rho \hat{E} \right)}{\partial t} + \nabla \cdot \left(\underline{v} \rho \hat{E} \right) = \rho \frac{\partial \hat{E}}{\partial t} + \hat{E} \left[\frac{\partial \rho}{\partial t} + \nabla \cdot (\rho \underline{v}) \right] + \rho \left(\underline{v} \cdot \nabla \hat{E} \right)$$

15. Using matrices and the definition of the gradient of a vector (Appendix B), show for a Newtonian fluid that $\underline{\underline{\tilde{\tau}}}^{T} : \nabla \underline{v}$ is always positive. This term appears in the derivation of the microscopic energy balance [108].

16. Compare and contrast the form of the momentum balance in the Navier-Stokes equation and the form of the momentum balance given by Newton's second law of motion, $\sum \underline{f} = m\underline{a}$.

17. What is the difference between the equation of motion with the extra-stress tensor $\underline{\underline{\tilde{\tau}}}$ included (i.e., the Cauchy momentum equation, Equation 6.65) and the equation of motion with viscosity μ present (i.e., the Navier-Stokes equation, Equation 6.71)? Can both equations be used for Newtonian fluids?

18. In the solution method for microscopic-momentum-balance problems outlined in Section 6.2, how would the solution steps change if the fluid under consideration were non-Newtonian rather than Newtonian?

19. A fluid flows down an inclined plane. The magnitude of the total force on the plane is $100N$. If a fluid of the same density but 10 times higher viscosity flows down the incline, what is the magnitude of the total force on the plane? Explain your answer.

20. If the velocity vector $\underline{v}$ in m/s and pressure in Pa for water flow in a pipe (radius 0.010 m, length 2.00 m) are given by the following expressions, what is the vector $\underline{\mathcal{F}}$ that indicates the magnitude and direction of the force on the walls of the pipe?

$$p[Pa] = -240z$$

$$\underline{v}[m/s] = \begin{pmatrix} 0 \\ 0 \\ 6.0 \left(1 - \left(\frac{r}{0.010} \right)^2 \right) \end{pmatrix}_{r\theta z}$$

where r and z are expressed in m.

21. If the velocity vector $\underline{v}$ in m/s for flow through a pipe of radius 0.012 m is given by the following expression, what is the volumetric flow rate Q of fluid through the pipe?

$$\underline{v}[m/s] = \begin{pmatrix} 0 \\ 0 \\ 12.0 \left(1 - \left(\frac{r}{0.012} \right)^2 \right) \end{pmatrix}_{r\theta z}$$

where r is expressed in m.

22. Fluid is trapped between two concentric cylinders and the inner cylinder (with radius $= \kappa R$) is turning, producing the velocity field $\underline{v}$ given here. What is the torque on the inner cylinder? What is the torque on the

outer cylinder (with radius $= R$)? Assume that the pressure is constant throughout.

$$\underline{v} = \begin{pmatrix} 0 \\ \left(\frac{\kappa^2 \Omega R}{\kappa^2 - 1}\right)\left(\frac{r}{R} - \frac{R}{r}\right) \\ 0 \end{pmatrix}_{r\theta z}$$

23. For a pressure-driven flow in a slit, the total stress tensor $\underline{\underline{\tilde{\Pi}}}$ is given here, where $P = Cx$ is pressure, μ is viscosity, B is the gap half-height, and A is the velocity at the centerline. The fluid is an incompressible Newtonian fluid. What is the x-component of the force due to fluid on the bottom plate?

$$v_x(z) = A\left(1 - \frac{z^2}{B^2}\right)$$

$$\underline{\underline{\tilde{\Pi}}} = \begin{pmatrix} -P & 0 & -\frac{2\mu Az}{B^2} \\ 0 & -P & 0 \\ -\frac{2\mu Az}{B^2} & 0 & -P \end{pmatrix}_{xyz}$$

24. For water in a flow with the velocity vector given here, what is the force in the fluid on a square surface with unit normal $\hat{n} = \hat{e}_z$ extending from $x = 0$, $y = 0$ to $x = 1$, $y = 1$. All distances are in meters; assume the pressure is the same everywhere.

$$\underline{v}[m/s] = \begin{pmatrix} -0.04x \\ -0.04y \\ 0.08z \end{pmatrix}_{xyz}$$

25. What is the torque on a rod turning in an infinite bath of fluid? The radius of the rod is R, the length is L, and the rod turns at angular velocity Ω in a fluid of viscosity μ. You may leave your answer in terms of the unknown velocity distribution.

26. The velocity field for squeeze flow between parallel plates is given here (This was obtained with a quasi-steady-state solution [12]), where h is the instantaneous gap height. What is the instantaneous flow rate through the circular strip of surface of height $2h$ at $r = R/2$? The area of this surface is $2\pi(R/2)2h$.

$$\underline{v} = \begin{pmatrix} v_r \\ v_\theta \\ v_z \end{pmatrix}_{r\theta z} = \begin{pmatrix} -\frac{3}{4}\frac{Vr}{h}\left(\frac{z^2}{h^2} - 1\right) \\ 0 \\ \frac{3}{2}V\left(\frac{z^3}{3h^3} - \frac{z}{h}\right) \end{pmatrix}_{r\theta z}$$

27. A fluid in a circular tank is in solid-body rotation on a turntable. The velocity field is given here. What is the fluid force on the wall due to the rotation? Explain your results.

$$\underline{v} = \begin{pmatrix} v_r \\ v_\theta \\ v_z \end{pmatrix}_{r\theta z} = \begin{pmatrix} 0 \\ r\Omega \\ 0 \end{pmatrix}_{r\theta z}$$

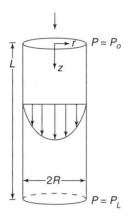

Figure 6.20

Pressure-driven flow in a tube (see Problem 30).

28. The z-direction velocity shown here results from the effect of natural convection (i.e., hot air rises) between two vertical parallel plates (the plates are long and wide). At what position in the flow does the velocity reach a maximum? In the equation, v_z is the velocity and y is the coordinate direction in a Cartesian coordinate system; all other quantities are constants related to the flow.

$$v_z(y) = \frac{\bar{\rho}\bar{\beta}(T_2 - T_1)b^2}{12\mu}\left[\left(\frac{y}{b}\right)^3 - \left(\frac{y}{b}\right)\right]$$

29. For the combined pressure-driven and wall-drag flow discussed in Example 6.11, sketch the coordinate system that was used to find the velocity solution provided in the example. What are the boundary conditions that were used?

30. What are the velocity boundary conditions in the flow shown in Figure 6.20? Express your answer in the coordinate system shown.

31. The problem of a thin film falling down an incline is discussed in this chapter. In Figure 5.41, a version of this problem is illustrated and a Cartesian coordinate system is proposed. Write the velocity vector in this coordinate system. How does the continuity equation simplify? How does the Navier-Stokes equation simplify? Comment on the chosen coordinate system.

32. For the velocity described in Figure 5.31 (i.e., the planar-jet flow), apply the microscopic mass balance (i.e., the continuity equation). What is the relationship between the velocity gradients that the mass balance requires?

33. Two different fluids with different densities and viscosities are layered between two long, parallel plates. The thickness of the bottom and more dense fluid layer is h_1; the thickness of the top and less dense fluid layer is h_2. The top plate is made to move parallel to the bottom plate at a low velocity V. The bottom plate is stationary. The flow is steady and both fluids are incompressible. The flow problem is solved in a Cartesian coordinate system with flow in the x-direction, and y is the direction perpendicular to the plates, with $y = 0$ at the surface of the bottom plate. What are the boundary conditions for this flow? Give your answer in mathematical form in the coordinate system described.

34. An incompressible fluid with density ρ and viscosity μ is placed in the region between two coaxial cylinders. The outer cylinder (radius $= R$) is stationary and the inner cylinder (radius $= \kappa R$) is moving counterclockwise at angular velocity Ω in rad/s. The flow is steady. The flow is solved in a cylindrical coordinate system with $z = 0$ at the bottom surface of the apparatus; $\hat{e}_z$ points upward. What are the boundary conditions for this flow? Give the answer in mathematical form in the coordinate system described. Check the units of your expressions.

35. In Example 6.2, we discuss the solution for the velocity field for flow down an inclined plane. The upper boundary of this flow is a free surface, meaning that there is no solid surface there. The boundary condition at the free surface where two fluids meet is that the velocity and stress should be continuous across the boundary. Consider the free surface in the flow-down-an-incline problem as the meeting point of two fluids—air and water—with the viscosity of air being much lower than the viscosity of water. Using the stress-matching boundary condition, justify the boundary condition used for the free surface in Example 6.2.

36. Sketch the flow domain for upward flow in a circular pipe inclined by a 30-degree angle to the horizontal. Pipe flow usually is analyzed in a cylindrical coordinate system. In terms of the cylindrical coordinate system for this problem, what is the gravity vector? Hint: choose gravity to be in the x-z plane. What are the implications of this complicated expression? How would this complication affect the solution of the Navier-Stokes equations for this problem?

37. Figure 6.12c depicts the steady flow of a drop of Newtonian fluid "rolling" down an inclined plane. Because of the complex geometry of the flow domain, this flow is best analyzed numerically, and the details of that calculation are beyond the scope of this text [70]. Although numerical methods are needed to solve the differential equations, we can arrive at the correct equations to solve by following the methods in this text. What is the differential equation that governs this flow and what are the appropriate boundary conditions?

38. *Flow Problem: Drag flow of a Newtonian fluid in a slit.* Calculate the velocity profile and flow rate for drag flow of an incompressible Newtonian liquid between two infinitely wide parallel plates separated by a gap of H. The pressure in the gap is uniform in the flow direction. The lower plate does not move, but the upper plate is pulled to the right at a speed V. The flow is steady and well developed.

39. *Flow Problem: Pressure-driven flow in an uphill slit.* An incompressible Newtonian fluid is made to flow between two long, wide parallel plates by a constant driving pressure gradient. The pressure at an upstream point is P_0 and a distance L downstream the pressure is P_L. The plates tilt upwards, making an angle ψ with the horizontal; do not neglect gravity. Calculate the steady state velocity profile, the flow rate, and the force on the walls. The gap between the plates is B.

40. *Flow Problem: Combined forward pressure and drag.* An incompressible Newtonian fluid is made to flow between two long, wide, horizontal parallel plates by the combined effect of a constant driving pressure gradient and

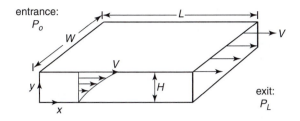

Figure 6.21   Schematic for Problem 41.

the motion of the wall. The gap between the plates is B, and the top plate is pulled to the right at a speed V while the lower plate remains stationary. The pressure at an upstream point is P_0 and a distance L downstream the pressure is P_L. Calculate the steady state velocity field and the flow rate.

41. *Flow Problem: Combined backward pressure and forward drag of a Newtonian fluid in a slit.* Calculate the velocity profile for the flow shown in Figure 6.21. The flow is steady flow of an incompressible Newtonian fluid between two wide plates. The flow is driven forward by the motion of the top plate (i.e., the top plate moves in the x-direction at speed V) and the flow is opposed by the pressure, which is slightly higher at the exit, P_L, than at the entrance, P_0, $P_L > P_0$. Neglect the effect of gravity. Use the coordinate system given in Figure 6.21.

42. *Flow Problem: Combined pressure-driven/drag flow of a Newtonian fluid in a slit that is tilted upward.* Calculate the velocity profile and flow rate for pressure-driven flow of an incompressible Newtonian liquid between two infinitely wide parallel plates separated by a gap of H. The slit is inclined to the horizontal by an angle α. The top plate moves forward at velocity V. The pressure at an upstream point is P_0; at a point a distance L downstream, the pressure is P_L. Assume that the flow between the plates is well developed and at steady state. The axial pressure gradient is constant.

43. *Flow Problem: Axial annular drag, wire coating.* An incompressible Newtonian fluid fills the annular gap between a cylindrical wire of radius κR and an outer shell of inner radius R (Figure 6.22). The wire is pulled to the

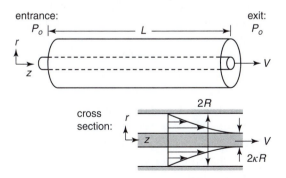

Figure 6.22   In wire-coating, a wire is drawn through a bath. This axial drag flow is addressed in Problem 43.

right at a speed V. There is no pressure variation throughout the apparatus. Calculate the steady state velocity profile, the flow rate, and the force on the wire. This geometry occurs in wire-coating. Answer: $v_z/V = \ln(r/R)/\ln\kappa$.

44. *Flow Problem: Upward, pressure-driven flow of a Newtonian fluid in a pipe.* Calculate the velocity profile and flow rate for pressure-driven flow of an incompressible Newtonian liquid in a vertical pipe of radius R. The pressure at the bottom entrance to the tube is P_0; at a point a distance L upward, the pressure is P_L. Assume that the flow is well developed and at steady state. Do not neglect gravity.

45. *Flow Problem: Two-layer drag flow between parallel plates.* Two different fluids with different densities and different viscosities are layered between two long, parallel plates. The thickness of the bottom and more dense fluid is h_1; the thickness of the top and less dense fluid is h_2. The top plate is made to move parallel to the bottom plate at a velocity V. The bottom plate is stationary. The flow is steady and both fluids are incompressible. Solve for the velocity profile in a Cartesian coordinate system with flow in the x-direction; y is the direction perpendicular to the plates, with $y = 0$ at the surface of the bottom plate.

46. *Flow Problem: Two-layer drag, pressure-driven flow between parallel plates.* Two different fluids with different densities (ρ_1, ρ_2) and viscosities (μ_1, μ_2) are layered between two long, parallel plates. The thickness of the bottom and more dense fluid layer is h_1; the thickness of the top and less dense fluid layer is h_2. Both plates are stationary and a flow is produced by the imposition of a small, constant pressure gradient such that the interface between the two fluids remains flat and parallel to the walls. The flow is steady and both fluids are incompressible. Solve for the velocity profile in a Cartesian coordinate system with flow in the x-direction; y is the direction perpendicular to the plates, with $y = 0$ at the surface of the bottom plate (copious algebra!).

47. *Flow Problem: Two-layer flow down an incline.* Two different fluids with different densities (ρ_1, ρ_2) and viscosities (μ_1, μ_2) are layered on a long plate tilted at an angle β to the horizontal. The thickness of the bottom and more dense fluid layer is h_1; the thickness of the top and less dense fluid layer is h_2. The bottom plate is stationary and a flow is produced by gravity. The flow is steady and both fluids are incompressible. Solve for the velocity profile in a Cartesian coordinate system with flow in the x-direction; y is the direction perpendicular to the plates, with $y = 0$ at the surface of the bottom plate.

48. *Flow Problem: Drag flow with viscosity varying.* Calculate the velocity profile and flow rate for drag flow of an incompressible Newtonian liquid between two infinitely wide parallel plates separated by a gap of H. The viscosity of the fluid varies linearly with position in the gap as $\mu = ay + b$. The pressure in the gap is uniform in the flow direction. The upper plate is driven such that the velocity is V and the lower plate is stationary. Assume that the flow between the plates is well developed and at steady state. Solve for the velocity profile in a Cartesian coordinate system with flow in the x-direction; y is the direction perpendicular to the plates, with $y = 0$ at the surface of the bottom plate.

49. *Flow Problem: Axial annular drag with pressure drop, wire coating.* Repeat Problem 43 with an imposed pressure gradient $= (-\Delta p/L)$ in the flow direction. Calculate the velocity field only.

50. *Flow Problem: Drag flow in a slit, power-law non-Newtonian fluid.* Repeat Problem 38 with a power-law, generalized Newtonian fluid with parameters m and n. Calculate the velocity and stress fields.

51. *Flow Problem: Pressure-driven flow in a slit tilted upward, power-law non-Newtonian fluid.* Repeat Problem 39 with a power-law, generalized Newtonian fluid with parameters m and n. Calculate the velocity field only.

52. *Flow Problem: Upward pressure-driven flow in a tube, power-law non-Newtonian fluid.* Repeat Problem 44 with a power-law, generalized Newtonian fluid with parameters m and n.

7 Internal Flows

So far, we have concentrated on developing a method for modeling fluids. We established the idea of the continuum to allow us to describe flows with continuous functions, and we introduced the use of the control volume (CV) to free us from having to follow individual fluid particles (see Chapter 3). We described a method of accounting for stresses in fluids (see Chapter 4) and showed how stress and motion are related through the constitutive equation (see Chapter 5). We developed a solution methodology that led to the microscopic balance equations and, finally, to solutions of simple flow problems (see Chapter 6). We have all of the tools necessary to solve flow problems, and we now turn to the task of modeling and understanding the flow behaviors described in Chapter 2.

In this chapter, we concentrate on internal flows, which are flows through closed conduits. Chapter 8 discusses both external flows, in which fluid moves over or around obstacles, and an important class of flows called boundary-layer flows.

In this chapter and in Chapter 8, we address complex, realistic, and practical flow problems with our modeling methods. We begin the analysis of complex problems with a microscopic analysis on an idealized system. When it is possible to solve the microscopic-analysis equations, we obtain a complete description of the flow $(\underline{v}, \underline{\tilde{\Pi}})$, from which we can calculate any engineering property of interest (e.g., flow rate Q, $\langle v \rangle$, $\mathcal{F}_{\text{drag}}$, $\mathcal{F}_{\text{wall}}$, and $\mathcal{T}_{\text{surface}}$). Numerical methods may be used to solve the microscopic balances when complex geometries are considered [49, 100]. For turbulent and other highly complex flows, we usually cannot solve microscopic balances, even with the help of computer methods; for these cases, we use dimensional analysis and data correlations to obtain results. In this chapter and the next, we see the entire progression from microscopic balance to complex flow solutions to dimensional analysis to, finally, data correlations.

7.1 Circular pipes

We begin with an important and widely applied area in fluid mechanics: flow rate and stress relationships in pipes and other closed conduits. This topic includes the issue of laminar versus turbulent flow (see Section 2.4). We start with a practical problem.

EXAMPLE 7.1. *The hose connecting the city water supply to the washing machine in a home burst while the homeowner was away (Figure 7.1). Water gushed out of the 1/2-inch Schedule 40 pipe for 48 hours before the problem was noticed by a neighbor and the water was shut off. How much water sprayed into the house over the two-day period? The water utility reports that the water pressure supplied to the house was approximately 60 psig.*

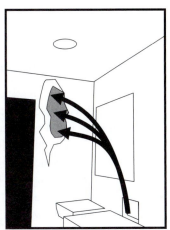

Figure 7.1   The flexible hosing that connects a cold-water shutoff valve to an automatic washing machine is the weak point in the water-delivery system of many homes. When this hose breaks (bottom), water spills into the home in a forceful, never-ending stream. The top photograph and schematic show the damage caused by the water to a nearby wall. Because the homeowner was away, the entire home flooded.

SOLUTION. Fundamentally, this problem is flow through a pipe (Figure 7.2). The pressure is the driving force for the flow, and the set driving pressure of 60 psig causes a certain flow rate for the water. That flow rate multiplied by 48 hours gives the total amount of water that sprayed into the house:

$$\begin{pmatrix} \text{volume} \\ \text{of water} \\ \text{in house} \end{pmatrix} = \left(\frac{\text{volume}}{\text{time}} \right) (\text{time interval}) \qquad (7.1)$$

$$= Q \Delta t \qquad (7.2)$$

The missing information is the relationship between driving pressure $\Delta p = 60$ psig and the volumetric flow rate Q.

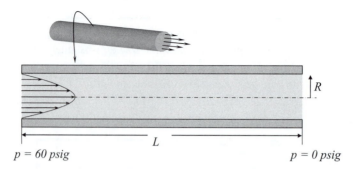

Figure 7.2 To calculate the volume of water that flows into the house over two days, we must relate the driving pressure Δp to the flow rate Q for flow of water in a pipe of circular cross section.

If we assume that the flow rate is low, then the flow is laminar. In a laminar flow, the flow is well organized and layers of fluid slide smoothly over one another (Figure 7.3). In laminar flow in pipes, the flow is unidirectional in the z-direction of a cylindrical coordinate system. Using the microscopic-momentum-balance method from Chapter 6, we can solve for the velocity and stress fields for unidirectional flow in a tube. When we have the velocity field, we can calculate the flow rate with an integral of $v_z(r)$ over the cross section of the tube, as demonstrated in Section 6.2.3 (see Equation 6.254). This integral yields the needed relationship between flow rate and pressure drop so that we can finish the calculation in Equation 7.2.

If the flow rate is not low, then Reynolds's experiments, as discussed in Section 2.4, show that the flow is not unidirectional but rather three-dimensional and disorganized (see Figure 2.19). In the case of turbulent flow, directly applying the microscopic-momentum-balance method is problematic. We cannot make enough assumptions to be able to solve the Navier-Stokes equations in the case of turbulent flow. To solve the turbulent-flow problem, we need experimental data on pressure drop as a function of flow rate.

In summary, to solve the burst-pipe problem, we must know the flow-rate/pressure-drop relationship in the pipe. The flow in the pipe may be laminar or turbulent. Using microscopic-balance techniques (see Chapter 6), we now solve the problem for laminar flow. For turbulent flow, we anticipate difficulties in solving for $Q(\Delta p)$ because the flow field is complex.

As is typical in engineering problem solving, we are presented with a complicated situation and the available information about the system and its physics is incomplete. Our strategy in these situations is to proceed where we can, beginning with the simpler case and, if necessary, building on it to learn how to address more

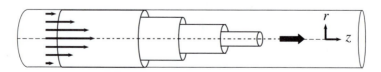

Figure 7.3 Laminar flow in a tube is a unidirectional flow in which concentric cylinders of fluid slide over one another in straight lines.

complex situations later. For the burst-pipe problem, this means we first solve the laminar pipe-flow problem and test whether that solution yields reasonable results for the problem of interest. If the laminar-flow solution is not acceptable, we redirect our efforts to the study of turbulent flow.

We suspend for the moment our work on the problem of the burst water pipe and turn to solving for $Q(\Delta p)$ for laminar pipe flow of an incompressible Newtonian fluid.

7.1.1 Laminar flow in pipes

The problem of pressure-driven flow in a tube is called *Poiseuille flow* after Jean Louis Marie Poiseuille (1797–1869), who conducted important early experiments on flow in tubes from 1838 to 1846. The final result for Q versus Δp in laminar tube flow (see Equation 7.28) is called the *Hagen-Poiseuille equation* after Poiseuille and Gotthilf Heinrich Ludwig Hagen (1797–1869), who independently deduced the law in 1839 from experimental observations. In the example that follows, we use the Navier-Stokes equations to solve for the flow-rate/pressure-drop relationship for laminar flow in a pipe.

EXAMPLE 7.2. *Calculate the velocity profile, flow rate, and shear-stress function for downward, pressure-driven flow of an incompressible Newtonian liquid in a tube of circular cross section (Figure 7.4). The pressure at an upstream point is p_0; at a point a distance L downstream, the pressure is p_L. Assume that the flow between these two points is fully developed (i.e., not affected by inlet or exit effects) and at steady state.*

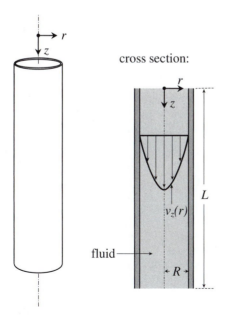

cross section:

Figure 7.4   Schematic of pressure-driven flow in a tube of circular cross section. This flow is known as Poiseuille flow in a tube.

Problem-Solving Procedure –
Microscopic-Balances Approach

1. Sketch the problem.
2. Choose a coordinate system. The coordinate system should be chosen so that the velocity vector and the boundary conditions are simplified.
3. Considering the flow, simplify $\underline{v}$ as much as possible.
4. Simplify the continuity equation (i.e., microscopic mass balance).
5. Simplify the equation of motion (i.e., microscopic momentum balance, Navier-Stokes equation).
6. Solve the resulting differential equation for the velocity field. Apply the velocity boundary conditions to solve for the unknown constant(s) of integration.
7. Solve for the pressure field, if applicable.
8. Calculate the stress components from the Newtonian constitutive equation and for any engineering quantities of interest.

Figure 7.5   With the microscopic-balance approach, the steps for solving a microscopic balance are shown here.

SOLUTION. Our task is to calculate the velocity field; thus, we proceed with the microscopic mass and momentum balances developed in Chapter 6:

$$
\begin{array}{c}
\text{Mass conservation:} \\
\text{(continuity equation,} \\
\text{constant density)}
\end{array}
\qquad
\boxed{0 = \nabla \cdot \underline{v}}
\qquad (7.3)
$$

$$
\begin{array}{c}
\text{Momentum conservation:} \\
\text{(Navier-Stokes equation)}
\end{array}
\qquad
\boxed{\rho\left(\frac{\partial \underline{v}}{\partial t} + \underline{v}\cdot\nabla\underline{v}\right) = -\nabla p + \mu\nabla^2\underline{v} + \rho\underline{g}}
\qquad (7.4)
$$

To apply these equations to the problem, we follow the procedure in Figure 6.9, which is repeated in Figure 7.5: Choose a coordinate system, simplify where possible, and solve the equations obtained. Because the tube is cylindrical, we choose to use cylindrical coordinates (see Tables B.5 and B.7 in Appendix B.2).

We assume that the velocity is only in the z-direction. This assumption permits us to cancel terms in the continuity equation and in the equation of motion (EOM) (i.e., the Navier-Stokes equation) that involve the velocity components v_θ and v_r:

$$
\underline{v} = \begin{pmatrix} v_r \\ v_\theta \\ v_z \end{pmatrix}_{r\theta z} = \begin{pmatrix} 0 \\ 0 \\ v_z \end{pmatrix}_{r\theta z}
\qquad (7.5)
$$

Mass conservation is given by the continuity equation for an incompressible fluid, which in cylindrical coordinates is given here (see Equation B.5-2):

$$
\text{Continuity equation:} \quad 0 = \nabla \cdot \underline{v}
\qquad (7.6)
$$

$$
0 = \frac{1}{r}\frac{\partial(r v_r)}{\partial r} + \frac{1}{r}\frac{\partial(v_\theta)}{\partial\theta} + \frac{\partial(v_z)}{\partial z}
\qquad (7.7)
$$

For the velocity field in Equation 7.5, $v_r = v_\theta = 0$ and the continuity equation simplifies to:

$$\text{Continuity equation, laminar pipe flow:} \quad \boxed{\frac{\partial v_z}{\partial z} = 0} \tag{7.8}$$

The microscopic-momentum balance for an incompressible Newtonian fluid is the Navier-Stokes equation. The Navier-Stokes equation in cylindrical coordinates is given in Table B.7 and here:

$$\text{Navier-Stokes equation:} \quad \rho\left(\frac{\partial \underline{v}}{\partial t} + \underline{v} \cdot \nabla \underline{v}\right) = -\nabla p + \mu \nabla^2 \underline{v} + \rho \underline{g} \tag{7.9}$$

$$\rho \begin{pmatrix} \dfrac{\partial v_r}{\partial t} \\[2mm] \dfrac{\partial v_\theta}{\partial t} \\[2mm] \dfrac{\partial v_z}{\partial t} \end{pmatrix}_{r\theta z} + \rho \begin{pmatrix} v_r\left(\dfrac{\partial v_r}{\partial r}\right) + v_\theta\left(\dfrac{1}{r}\dfrac{\partial v_r}{\partial \theta} - \dfrac{v_\theta}{r}\right) + v_z\left(\dfrac{\partial v_r}{\partial z}\right) \\[2mm] v_r\left(\dfrac{\partial v_\theta}{\partial r}\right) + v_\theta\left(\dfrac{1}{r}\dfrac{\partial v_\theta}{\partial \theta} + \dfrac{v_r}{r}\right) + v_z\left(\dfrac{\partial v_\theta}{\partial z}\right) \\[2mm] v_r\left(\dfrac{\partial v_z}{\partial r}\right) + v_\theta\left(\dfrac{1}{r}\dfrac{\partial v_z}{\partial \theta}\right) + v_z\left(\dfrac{\partial v_z}{\partial z}\right) \end{pmatrix}_{r\theta z}$$

$$= - \begin{pmatrix} \dfrac{\partial p}{\partial r} \\[2mm] \dfrac{1}{r}\dfrac{\partial p}{\partial \theta} \\[2mm] \dfrac{\partial p}{\partial z} \end{pmatrix}_{r\theta z}$$

$$+ \mu \begin{pmatrix} \dfrac{\partial}{\partial r}\left(\dfrac{1}{r}\dfrac{\partial(r v_r)}{\partial r}\right) + \dfrac{1}{r^2}\dfrac{\partial^2 v_r}{\partial \theta^2} + \dfrac{\partial^2 v_r}{\partial z^2} - \dfrac{2}{r^2}\dfrac{\partial v_\theta}{\partial \theta} \\[2mm] \dfrac{\partial}{\partial r}\left(\dfrac{1}{r}\dfrac{\partial(r v_\theta)}{\partial r}\right) + \dfrac{1}{r^2}\dfrac{\partial^2 v_\theta}{\partial \theta^2} + \dfrac{\partial^2 v_\theta}{\partial z^2} + \dfrac{2}{r^2}\dfrac{\partial v_r}{\partial \theta} \\[2mm] \dfrac{1}{r}\dfrac{\partial}{\partial r}\left(r\dfrac{\partial v_z}{\partial r}\right) + \dfrac{1}{r^2}\dfrac{\partial^2 v_z}{\partial \theta^2} + \dfrac{\partial^2 v_z}{\partial z^2} \end{pmatrix}_{r\theta z}$$

$$+ \rho \begin{pmatrix} g_r \\ g_\theta \\ g_z \end{pmatrix}_{r\theta z} \tag{7.10}$$

For the coordinate system we chose (see Figure 7.4), gravity is in the z-direction. We therefore write the gravity vector $\underline{g}$ as:

$$\underline{g} = \begin{pmatrix} g_r \\ g_\theta \\ g_z \end{pmatrix}_{r\theta z} = \begin{pmatrix} 0 \\ 0 \\ g \end{pmatrix}_{r\theta z} \tag{7.11}$$

Substituting $\underline{g}$ and what we know already about $\underline{v}$ (i.e., steady state, $v_r = v_\theta = 0$, $\partial v_z/\partial z = 0$), we obtain a simplified version of the Navier-Stokes equation for Poiseuille flow in a tube:

$$
\begin{pmatrix} 0 \\ 0 \\ 0 \end{pmatrix}_{r\theta z} = - \begin{pmatrix} \dfrac{\partial p}{\partial r} \\ \dfrac{1}{r}\dfrac{\partial p}{\partial \theta} \\ \dfrac{\partial p}{\partial z} \end{pmatrix}_{r\theta z} + \begin{pmatrix} 0 \\ 0 \\ \dfrac{\mu}{r}\dfrac{\partial}{\partial r}\left(r\dfrac{\partial v_z}{\partial r}\right) + \dfrac{\mu}{r^2}\dfrac{\partial^2 v_z}{\partial \theta^2} \end{pmatrix}_{r\theta z} + \begin{pmatrix} 0 \\ 0 \\ \rho g \end{pmatrix}_{r\theta z} \tag{7.12}
$$

The r- and θ-components of the Navier-Stokes equation indicate that pressure is a function of neither r nor θ:

$$
r\text{-Component:} \quad -\frac{\partial p}{\partial r} = 0 \tag{7.13}
$$

$$
\theta\text{-Component:} \quad -\frac{1}{r}\frac{\partial p}{\partial \theta} = 0 \tag{7.14}
$$

Thus, pressure is a function only of z; $p = p(z)$. This is a useful conclusion, which we use in solving the z-component of the Navier-Stokes equation. Note that we did not have to assume anything about the pressure distribution to arrive at Equations 7.13 and 7.14; rather, the assumptions that $v_r = v_\theta = 0$ and writing the form of $\underline{g}$, coupled with conservation of momentum, were sufficient to require that the r and θ pressure derivatives be zero.

We solve for the velocity field using the z-component of the Navier-Stokes equation, given in Equation 7.12:

$$
z\text{-Component:} \quad \frac{dp}{dz} = \frac{\mu}{r}\frac{\partial}{\partial r}\left(r\frac{\partial v_z}{\partial r}\right) + \frac{\mu}{r^2}\frac{\partial^2 v_z}{\partial \theta^2} + \rho g \tag{7.15}
$$

Notice that we changed the partial derivative $\partial p/\partial z$ to the total derivative dp/dz because pressure is a function only of z ($\partial p/\partial r = \partial p/\partial \theta = 0$).

Our task is now reduced to solving Equation 7.15. It is still a complicated equation, however, and before we try to solve it, we look to see if we can simplify it further by examining the meaning of each term.

The lefthand side of Equation 7.15 contains the term dp/dz. This is certainly not zero because the axial-pressure difference is driving the flow. The righthand side of Equation 7.15 contains θ- and r-derivatives of v_z; v_z certainly varies with r because v_z is zero at $r = R$ and nonzero in the center ($r = 0$; see Figure 7.4). Now we examine the possibility of θ-variation of v_z. We see from Equation 7.14 that the pressure does not vary in the θ-direction. Although we did not yet find any restriction on the θ-variation of the velocity v_z, with no flow in the θ-direction and no pressure variation in that direction, it is reasonable to assume that there is no variation of v_z in the θ-direction; that is, the flow should be symmetric with respect to θ. We make this assumption and, as with the previous assumption of unidirectional flow, we will check the final results against experimental observations to determine the accuracy of this assumption. On the basis of these physical arguments, we take $\partial v_z / \partial \theta = 0$. Thus, Equation 7.15 simplifies to:

$$\frac{dp(z)}{dz} - \rho g = \frac{\mu}{r} \frac{\partial}{\partial r} \left[r \frac{\partial v_z(r)}{\partial r} \right] \tag{7.16}$$

The partial differential equation (PDE) in Equation 7.16 is of a type not too difficult to solve. Although it is an equation of two variables, p and v_z, which are a function of two independent variables, z and r, it is a separable equation, meaning that we can separate the pressure and velocity parts and solve them independently (see the Web appendix [108] for more details).

Looking at Equation 7.16, we analyze it as follows. Because by the continuity equation (i.e., microscopic mass balance) v_z is not a function of z, and by previous assumption v_z also is not a function of θ, we conclude that v_z is a function of r alone. We therefore change the partial derivatives of v_z with respect to r to total derivatives:

$$\frac{dp}{dz} - \rho g = \frac{\mu}{r} \frac{d}{dr} \left(r \frac{dv_z}{dr} \right) \tag{7.17}$$

From the discussion of the r- and θ-components of the Navier-Stokes equation, we found that $p = p(z)$. Thus, the lefthand side of Equation 7.17 is a function only of z and the righthand side is a function only of r. For two functions of different, independent variables (r and z) to be equal for *all* values of the independent variables, the two functions need to be equal to the same constant. Thus, the two sides of Equation 7.17 are equal to the same constant, which we call λ (see the Web appendix [108]). Our complex, multivariable equation thus becomes two single-variable equations that we are able to solve:

$$\text{Lefthand side:} \quad \frac{dp}{dz} - \rho g = \lambda \tag{7.18}$$

$$\text{Righthand side:} \quad \frac{\mu}{r} \frac{d}{dr} \left(r \frac{dv_z}{dr} \right) = \lambda \tag{7.19}$$

The boundary conditions on velocity for this problem are no-slip at the wall and symmetry relative to the centerline of the flow (see Section 6.2.2):

$$\begin{array}{cc} \text{Velocity} & r = R \ \ v_z = 0 \\ \text{boundary conditions:} & r = 0 \ \ \dfrac{dv_z}{dr} = 0 \\ \text{(Poiseuille flow)} & \end{array} \tag{7.20}$$

For pressure we are given values at two locations, which we use as boundary conditions:

$$\begin{array}{cc} \text{Pressure} & z = 0 \ \ p = p_0 \\ \text{boundary conditions:} & z = L \ \ p = p_L \\ \text{(Poiseuille flow)} & \end{array} \tag{7.21}$$

Integrating Equations 7.18 and 7.19 and incorporating the boundary conditions, we obtain the final results; calculation details are left to readers (see Problem 6). These results are plotted in dimensionless form in Figures 7.6 and 7.7 and a

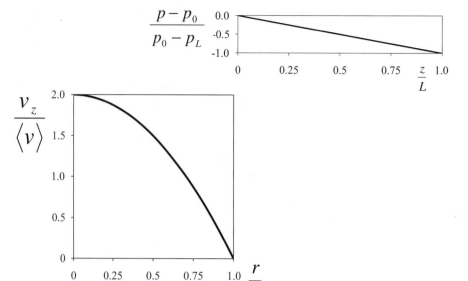

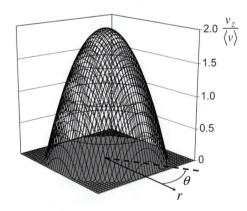

Figure 7.6 Velocity and pressure profiles calculated for Poiseuille flow (i.e., pressure-driven flow) of a Newtonian fluid in a tube. Average velocity $\langle v \rangle$ in this flow is given in Equation 7.29.

vector plot of the velocity field is in Figure 7.8:

Solution
Poiseuille flow
in a tube:

$$p(z) = -\left(\frac{p_0 - p_L}{L} \right) z + p_0$$ (7.22)

$$v_z(r) = \frac{(p_0 - p_L + \rho g L) R^2}{4 \mu L} \left[1 - \left(\frac{r}{R} \right)^2 \right]$$ (7.23)

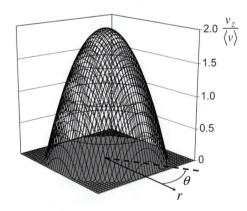

Figure 7.7 The solution for velocity profile for pressure-driven laminar flow in a tube is shown in a three-dimensional view. The centerline velocity is twice the average velocity of this flow. The data are plotted versus dimensionless cylindrical coordinates r/R and θ.

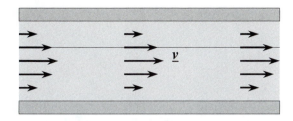

Figure 7.8 The solution for the velocity field for pressure-driven laminar flow in a tube is shown in this arrow plot. Each arrow starts at a location in the fluid and then points in the direction of the velocity at that point. The length of the arrow is proportional to the speed of the fluid at that location. Because Poiseuille flow is symmetric in the θ-direction, the arrow map shows a cross-sectional cut at a single value of θ. For a more complex flow, a three-dimensional arrow map is required; alternatively, the solution can be plotted at various cross sections.

The solution for the flow rate, Q, is calculated using Equation 6.254:

$$\begin{array}{l}\text{Total flow rate}\\ \text{out through}\\ \text{surface } \mathcal{S}:\end{array} \qquad Q = \iint_{\mathcal{S}} [\hat{n} \cdot \underline{v}]_{\text{at surface}} \, dS \qquad (7.24)$$

The surface of interest in this integral is the tube cross section at the exit. For this surface, $\hat{n} \cdot \underline{v} = \hat{e}_z \cdot \underline{v} = v_z$, and the surface-area element dS for the circular-tube cross section is $dS = rd\theta dr$ (Figure 4.61):

$$Q = \iint_{\mathcal{S}} v_z \, dS \qquad (7.25)$$

$$= \int_0^{2\pi} \int_0^R v_z(r) \, r \, dr \, d\theta \qquad (7.26)$$

Substituting the solution for $v_z(r)$ from Equation 7.23 and integrating, we obtain the relationship between pressure drop and flow rate in laminar flow in a tube (see Problem 7). This equation is known as the Hagen-Poiseuille equation, and it appeared in Chapter 2 (see Equation 2.8). Note that we neglect gravity in writing Equation 7.28 and in subsequent equations in this example:[1]

$$\text{(Gravity included)} \qquad Q = \frac{\pi(p_0 - p_L + \rho g L)R^4}{8\mu L} \qquad (7.27)$$

Hagen-Poiseuille equation
(Flow-rate/pressure-drop
for laminar tube flow)
(No gravity)

$$\boxed{Q = \frac{\pi(p_0 - p_L)R^4}{8\mu L}} \qquad (7.28)$$

For the case in which gravity is neglected, the average velocity is given by:

$$\langle v \rangle = \frac{Q}{\pi R^2} = \frac{(p_0 - p_L)R^2}{8\mu L} \qquad (7.29)$$

[1] We neglect gravity in Equation 7.28 because for most pipe flows the gravity contribution is small. It is possible to include the hydrostatic effect of the flow-direction component of gravity for all orientations of the pipe by defining a dynamic pressure that incorporates the effect of gravity as an increase in the driving pressure drop (see Equation 8.115, Problem 10, and the Glossary).

and the velocity profile (see Equation 7.23) may be written in terms of average velocity $\langle v \rangle$ as:

$$\frac{v_z}{\langle v \rangle} = 2 \left[1 - \left(\frac{r}{R} \right)^2 \right]$$ (7.30)

which is plotted in Figures 7.6 and 7.7.

Now that we know $\underline{v}$, we can calculate the stress field from the Newtonian constitutive equation. Because we are in cylindrical coordinates, we must use the correct form for the Newtonian constitutive equation in this coordinate system, found in Table B.8:

Newtonian
constitutive equation $\underline{\underline{\tilde{\tau}}}(r) = \mu \left(\nabla \underline{v} + (\nabla \underline{v})^T \right)$ (7.31)
(stress/velocity relationship)

$$\underline{v} = \begin{pmatrix} 0 \\ 0 \\ v_z(r) \end{pmatrix}_{r\theta z}$$ (7.32)

(see Table B.8) $\underline{\underline{\tilde{\tau}}}(r) = \mu \left[\begin{pmatrix} 0 & 0 & \frac{\partial v_z}{\partial r} \\ 0 & 0 & 0 \\ 0 & 0 & 0 \end{pmatrix}_{r\theta z} + \begin{pmatrix} 0 & 0 & 0 \\ 0 & 0 & 0 \\ \frac{\partial v_z}{\partial r} & 0 & 0 \end{pmatrix}_{r\theta z} \right]$ (7.33)

$$= \begin{pmatrix} 0 & 0 & \mu\frac{\partial v_z}{\partial r} \\ 0 & 0 & 0 \\ \mu\frac{\partial v_z}{\partial r} & 0 & 0 \end{pmatrix}_{r\theta z} = \begin{pmatrix} 0 & 0 & \frac{(p_L-p_0)r}{2L} \\ 0 & 0 & 0 \\ \frac{(p_L-p_0)r}{2L} & 0 & 0 \end{pmatrix}_{r\theta z}$$ (7.34)

The only nonzero stress components in $\underline{\underline{\tilde{\tau}}}$ are the shear stresses, $\tilde{\tau}_{rz} = \tilde{\tau}_{zr}$ (Figure 7.9). In this flow, a measurable quantity is the shear stress at the wall, which is given by:

Shear stress
at wall for $\tilde{\tau}_{rz}(r)|_R = \dfrac{(p_L - p_0)R}{2L}$ (7.35)
pipe flow:

The total-stress tensor $\underline{\underline{\tilde{\Pi}}}$ is given by Equation 4.302, $\underline{\underline{\tilde{\Pi}}}(r, z) = \underline{\underline{\tilde{\tau}}} - p\underline{\underline{I}}$. For Newtonian fluids in a tube $\underline{\underline{\tilde{\Pi}}}$ becomes:

$$\underline{\underline{\tilde{\Pi}}}(r, z) = \begin{pmatrix} -p(z) & 0 & \frac{(p_L-p_0)r}{2L} \\ 0 & -p(z) & 0 \\ \frac{(p_L-p_0)r}{2L} & 0 & -p(z) \end{pmatrix}_{r\theta z}$$ (7.36)

Total stress-
tensor
Poiseuille flow $\underline{\underline{\tilde{\Pi}}}(r, z) = \begin{pmatrix} \frac{\Delta p}{L}z - p_0 & 0 & \frac{-\Delta p\, r}{2L} \\ 0 & \frac{\Delta p}{L}z - p_0 & 0 \\ \frac{-\Delta p\, r}{2L} & 0 & \frac{\Delta p}{L}z - p_0 \end{pmatrix}_{r\theta z}$ (7.37)
in a tube:

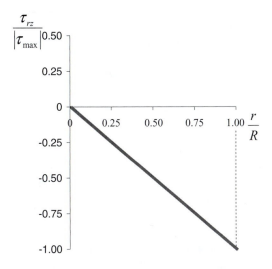

Figure 7.9 Shear-stress profile for Poiseuille flow in a tube. The magnitude of the wall stress is $|\tilde{\tau}_{max}| = \Delta p R/2L$.

where $\Delta p \equiv p_0 - p_L$. The tensor $\underline{\underline{\tilde{\Pi}}}$ is a complete description of stress in this flow, and Equation 7.37 may be used to calculate the force on *any* surface in the flow by using the usual integral of $\hat{n} \cdot \underline{\underline{\tilde{\Pi}}}|_{surface}$ (see Equation 6.196):

$$\text{Total molecular force in a fluid on a surface } \mathcal{S}: \qquad \mathcal{F} = \iint_{\mathcal{S}} [\hat{n} \cdot \underline{\underline{\tilde{\Pi}}}]_{\text{at surface}} \, dS \qquad (7.38)$$

Through this microscopic analysis, we arrive at equations for $v_z(r)$ (Equation 7.23), $Q(\Delta p)$ (Equation 7.28), and $\underline{\underline{\tilde{\Pi}}}(r, z)$ (Equation 7.37) for steady, laminar flow in a tube. With these results, we can calculate any fluid force on any surface (Equation 7.38; see, for example, Problem 20).

Example 7.2 shows the power of microscopic analysis. With a set of reasonable assumptions and applying the continuum model, we can solve the problem of steady flow for laminar pipe flow. When a flow of interest is sufficiently simple, the microscopic-balance technique is powerful.

We obtained the result for the flow-rate/pressure-drop relationship for laminar flow—that is, the Hagen-Poiseuille equation:

$$\text{Hagen-Poiseuille equation: } (Q(\Delta p) \text{ for laminar tube flow}): \qquad Q = \frac{\pi(p_0 - p_L)R^4}{8\mu L} \qquad (7.39)$$

In the burst-pipe problem that began this chapter, the missing information was the flow-rate/pressure-drop relationship. With Equation 7.39, we now can solve the burst-pipe problem for the case of laminar flow. We carry out this calculation in Example 7.3 and check the final result to see whether the laminar-flow assumption is reasonable for the case of flow in household pipes.

EXAMPLE 7.3 (Burst pipe, continued). *The hose connecting the city water supply to the washing machine in a home burst while the homeowner was away. Water gushed out of the 1/2-inch Schedule 40 pipe for 48 hours before the problem was noticed by a neighbor and the water was shut off. How much water sprayed into the house over the two-day period? The water utility reports that the water pressure supplied to the house was approximately 60 psig.*

SOLUTION. The pressure is the driving force for the flow, and the set driving pressure of 60 psig causes a certain flow rate for the water. That flow rate multiplied by 48 hours gives the total amount of water that gushed into the house:

$$\begin{pmatrix} \text{volume} \\ \text{of water} \\ \text{in house} \end{pmatrix} = \left(\frac{\text{volume}}{\text{time}} \right) (\text{time interval}) \qquad (7.40)$$

$$= Q \Delta t \qquad (7.41)$$

If we assume that flow in the pipes is laminar, then we can calculate the flow rate Q from the Hagen-Poiseuille equation, $Q = (\pi \Delta p R^4)/(8\mu L)$ (Equation 7.39).

To calculate the flow rate from the Hagen-Poiseuille equation, we must know the fluid viscosity, the radius of the pipe, the length of the pipe, and the pressure drop across the length of the pipe. We know the pressure supplied by the municipal water authority, and water is delivered to a house typically in a 1-inch main supply line from the road to the house. We investigate and find that the distance from the main to the house is 80 feet. From the terminus of the main water supply line, we also find that there is 20 feet of 1/2-inch Schedule 40 (ID = 0.622 in.) pipe carrying water to the washing machine. This is a typical arrangement in homes in the United States.

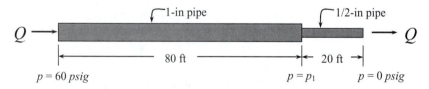

Figure 7.10 We model the piping leading to the broken hose as 80 feet of 1-inch Schedule 40 pipe connected to 20 feet of 1/2-inch Schedule 40 pipe.

Thus, we have two pipes to consider (Figure 7.10), and there also are bends, fittings, and valves in the line. For the purposes of this estimate we ignore any pressure changes caused by them[2] and consider only the pressure changes in the straight pipes. Beginning with the larger pipe (assuming this pipe also is

---

[2]In Chapter 1 (see Example 1.10), we saw that friction due to fittings often is small compared to wall drag in straight pipes.

Schedule 40, ID $= 1.049$ in.) and applying the Hagen-Poiseuille equation, we obtain:

$$Q = \frac{\pi(p_0 - p_1)R_1^4}{8\mu L_1} \tag{7.42}$$

$$= \frac{\pi(60 \text{ psig} - p_1)(0.5245 \text{ in})^4}{8\mu(80 \text{ ft})}. \tag{7.43}$$

The quantity p_1 is the pressure at the end of the 1-inch main supply line. For the smaller line to the washer, the flow rate Q is the same, and we write:

$$Q = \frac{\pi(p_1 - p_{\text{atm}})R_2^4}{8\mu L_2} \tag{7.44}$$

$$= \frac{\pi(p_1 - 0 \text{ psig})(0.311 \text{ in})^4}{8\mu(20 \text{ ft})}. \tag{7.45}$$

Note that in gauge pressure, $p_{\text{atm}} = 0$ psig. We now have two equations and two unknowns, Q and p_1, which we solve simultaneously:

$$p_1 = 40 \text{ psig}$$

$$Q = 1,200 \text{ gal/min} \tag{7.46}$$

The result we obtain for flow rate is quite high. For comparison, we learn from an Internet search that a typical 1-1/2-inch firehose delivers about 100 gal/min from a driving pressure of approximately 300 psi. The largest 4-inch supply firehose can reach 1,000 gal/min. Thus, 1,200 gal/min from a 1/2-inch line operating at 60 psi is not reasonable. Another feasibility check is to see how fast a swimming pool would fill at the calculated flow rate. An Internet search reveals that 13,000 gallons is the capacity of a typical home swimming pool; at 1,200 gal/min, it would take 11 minutes to fill it, compared to a more typical filling time of 24–34 hours, an estimate also obtained from the Web. Finally, checking the Reynolds number, we see that according to our laminar result, $\text{Re} = 7 \times 10^6$ in the small pipe. Reynolds's experiments (see Chapter 2) indicated that laminar flow persists only to $\text{Re} = 2,100$. Thus, for many reasons, this calculation of flow rate and pressure drop arrived at from laminar flow is not correct for the burst-pipe problem.

The reason for the incorrect result is that frictional losses in the pipes were underestimated by assuming laminar flow. Turbulent motions dissipate energy and slow the flow. To calculate the correct flow rate, we must study turbulent flow and how it affects flow rate.

Using the Hagen-Poiseuille equation—which is only correct for laminar flow—greatly underestimates the frictional drag present in pipes. The true estimate of flow rate in household pipes (according to the Internet, it is approximately 10 gal/min) is two orders of magnitude lower than what we calculated in the laminar example. The large discrepancy between the laminar prediction and what is observed is evidence that the flow in household pipes is not laminar. To correctly solve the burst-pipe problem, we must know more about turbulent flow.

Turbulent flow problems cannot be solved directly by following the microscopic-momentum balance from start to finish as we did for laminar flow, but we can modify our approach by incorporating experimental observations and arrive at important results, including the flow-rate/pressure-drop relationship. We discuss turbulent-flow modeling in the next section.

There are important flows for which the laminar-flow solution and the Hagen-Poiseuille equation are appropriate, such as in glass-tube viscometers that are used to find the viscosity of fluids. We show how the Hagen-Poiseuille equation applies to these instruments in Example 7.4.

EXAMPLE 7.4. *How does the measurement of efflux time Δt in a Cannon-Fenske routine viscometer (Figure 7.11) allow us to deduce the viscosity of the fluid?*

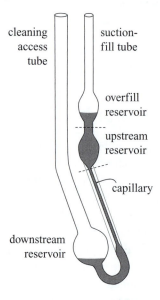

cleaning
access
tube

suction-
fill tube

overfill
reservoir

upstream
reservoir

capillary

downstream
reservoir

Figure 7.11   The Cannon-Fenske viscometer—a variation on the Ostwald viscometer invented by Wilhelm Ostwald—has two fluid reservoirs connected by a tilted capillary tube. The fluid is drawn up through the capillary to fill the second reservoir to overfull. The fluid level then is allowed to drop; the time required for the fluid meniscus to pass between the two marks shown is the efflux time Δt.

SOLUTION. The Cannon-Fenske viscometer is a glass apparatus that has two fluid reservoirs connected by a tilted capillary tube (see Figure 7.11). The capillary tube is manufactured to be straight and of uniform inside diameter (ID). An appropriate volume is charged to the downstream reservoir and, after equilibration at constant temperature, this fluid is drawn up through the capillary to fill the upstream reservoir to overfull. The fluid level then is allowed to drop under the pull of gravity. The time required for the fluid meniscus to pass between the two dashed marks shown in Figure 7.11 is the efflux time Δt.

The flow through the capillary can be analyzed as shown in Figure 7.12. Fluid in the amount that fits in the upper reservoir, volume ΔV, flows through a capillary of length L. The time required for that fluid to pass through the capillary is the efflux time Δt. Thus, the measured flow rate Q through the

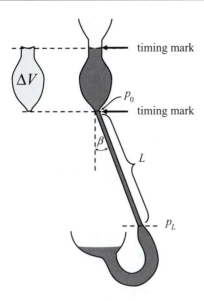

timing mark

timing mark

Figure 7.12 Schematic of the operation of a Cannon-Fenske viscometer. A volume of fluid ΔV is timed as it passes through a capillary of length L.

capillary is $Q = \Delta V / \Delta t$:

$$
\begin{array}{c}
\text{Measured flow rate} \\
\text{through capillary} \\
\text{in viscometer:}
\end{array}
\qquad
Q = \frac{\Delta V}{\Delta t}
\qquad (7.47)
$$

The flow rate through a capillary as a function of system variables was solved for in the previous discussion, and the result is the Hagen-Poiseuille equation (with gravity), Equation 7.27:

$$
\begin{array}{c}
\text{Hagen-Poiseuille} \\
\text{(gravity } g_z = g \text{ included):}
\end{array}
\qquad
Q = \frac{\pi(p_0 - p_L + \rho g L)R^4}{8\mu L}
\qquad (7.48)
$$

Because the capillary is tilted, gravity is not in the flow ($\hat{e}_z$) direction but rather is tilted from $\hat{e}_z$ by an angle β. To apply Equation 7.48 to the Cannon-Fenske system defined in Figure 7.12, we substitute the correct z-component of gravity ($g \cos \beta$) for the z-component that was used in the derivation (g):

$$
\begin{array}{c}
\text{Hagen-Poiseuille} \\
\text{(gravity } g_z = g \cos \beta \text{ included):}
\end{array}
\qquad
Q = \frac{\pi(p_0 - p_L + \rho g \cos \beta L)R^4}{8\mu L}
\qquad (7.49)
$$

We now substitute Q from Equation 7.47 and solve for Δt:

$$
Q = \frac{\Delta V}{\Delta t} = \frac{\pi(p_0 - p_L + \rho g \cos \beta L)R^4}{8\mu L}
\qquad (7.50)
$$

$$
\Delta t = \frac{8\mu \Delta V}{\pi R^4} \left(\frac{1}{\frac{p_0 - p_L}{L} + \rho g \cos \beta} \right)
\qquad (7.51)
$$

The pressure at the top of the capillary is nearly atmospheric and the pressure at the bottom also is nearly atmospheric. Taking $p_0 - p_L \approx 0$ (see Problem 18 for corrections to this assumption), we obtain the final result:

$$t_{\text{efflux}} = \Delta t = \frac{8\mu \Delta V}{\pi R^4 \rho g \cos \beta} \tag{7.52}$$

Efflux time
in Cannon-Fenske
viscometer:

$$\Delta t = \frac{8\nu \Delta V}{\pi R^4 g \cos \beta} \tag{7.53}$$

where $\nu = \mu/\rho$ is the kinematic viscosity of the fluid. We thus can write the kinematic viscosity in terms of viscometer dimensions and the measured efflux time as follows:

Kinematic viscosity
obtained using
Cannon-Fenske
viscometer
$(p_0 - p_L$ neglected):

$$\nu = \left(\frac{\pi R^4 g \cos \beta}{8 \Delta V} \right) \Delta t \tag{7.54}$$

The angle β depends on how the viscometer is mounted during experimental operation; for that reason, experimentalists use great care in vertically aligning the viscometer. The values of R and ΔV are fixed at the time of device manufacture.

Everything in parentheses in Equation 7.54 is fixed for a given viscometer, although dimensions such as ΔV and R vary slightly among instruments. As a matter of practicality, each viscometer is supplied by the manufacturer with a calibration constant that replaces the quantity in parentheses in Equation 7.54. The calibration constant, which is a function of temperature, is determined at the factory by measuring efflux time Δt for a material of known kinematic viscosity ν. This approach has the advantage of accounting for the small neglected pressure difference. Thus, the final operating equation for the Cannon-Fenske viscometer is:

$$\nu = \left[\left(\begin{array}{c} \text{correction} \\ \text{factor for} \\ p_0 - p_L \end{array} \right) \left(\frac{\pi R^4 g \cos \beta}{8 \Delta V} \right) \right] \Delta t \tag{7.55}$$

Kinematic viscosity
obtained using
Cannon-Fenske
viscometer
$(p_0 - p_L$ accounted for):

$$\nu(T) = \alpha(T) \Delta t \tag{7.56}$$

where $\alpha(T)$ is the temperature-dependent calibration constant supplied by the manufacturer for a given viscometer. Note that for accurate viscosities to be measured, each Cannon-Fenske viscometer must be mounted vertically and charged with its standardized volume of material; excess fluid alters the back pressure (p_L) and introduces variability not accounted for by the calibration (see Problem 19 for more discussion on this issue).

The solution strategy in this section is general and may be used to solve for velocity and stress fields for well-defined flows. We follow the methodology in this section to solve the Navier-Stokes equations for other flows in Sections 7.2 and 7.3 and in Chapter 8. When geometry or flow circumstances are complex, computer-implemented numerical methods may be used to solve the Navier-Stokes equations for $\underline{v}$ and p (see Section 10.2).

We return now to the burst-pipe problem and the need for information on turbulent flow.

7.1.2 Turbulent flow in pipes

In the previous section, we sought to calculate the flow rate in the burst-pipe example using a result from laminar flow. The formula used was the Hagen-Poiseulle equation, an equation that relates pressure drop to flow rate for laminar pipe flow:

$$\text{Hagen-Poiseuille equation:} \quad Q = \frac{\pi(p_0 - p_L)R^4}{8\mu L} \quad (7.57)$$
$$(Q(\Delta p) \text{ for laminar tube flow})$$

We found that the burst-pipe result predicted by this laminar-flow equation was not correct; the predicted flow rate was orders of magnitude too high. The error in that calculation was to use a laminar-flow relationship to predict turbulent flow. To correctly complete the burst-pipe calculation, we need the flow-rate/pressure-drop relationship for turbulent flow in pipes.

We can seek the missing relationship by following the same steps used to develop the laminar-flow relationship. As we attempt to follow that process, we hope to see why and how the method fails in turbulent flow. The steps leading to the Hagen-Poiseuille equation (Equation 7.57) are as follows:

Steps to Hagen-Poiseuille Equation

1. Apply the microscopic mass and momentum balances to pressure-driven tube flow (see Equations 7.7 and 7.10).
2. Simplify using reasonable assumptions (see Equations 7.8 and 7.17).
3. Calculate the velocity field, which is a function of pressure drop (see Equation 7.23).
4. Calculate the flow rate as an integral over the velocity field (see Equation 7.26).

For turbulent flow, Step 1 is easy—the microscopic mass and momentum equations for turbulent pipe flow are the same as for laminar pipe flow (see Equations 7.3 and 7.4). The difficulty arises in Step 2 as we attempt to solve these equations for the turbulent velocity field. In the laminar-flow case, the calculation was fairly simple because the flow was simple: unidirectional, steady, and symmetrical. Turbulent flow has none of these characteristics (see Section 2.4 and Figure 7.13). Turbulent flow is time-varying and three-dimensional, meaning $v_r \neq 0$ and $v_\theta \neq 0$. Because we cannot solve for the velocity field in turbulent flow, we cannot directly calculate $Q(\Delta p)$ for turbulent flow using the steps outlined here.

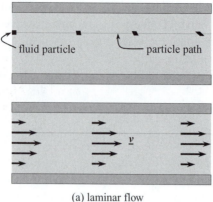

(a) laminar flow

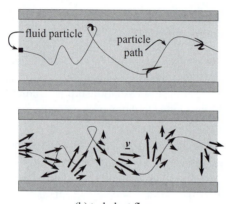

(b) turbulent flow

Figure 7.13   Laminar flow (a) is a flow in which fluid particles move in layers, one layer sliding over the other (the word *laminar* comes from the Latin word for *layer*). In laminar flow, particles move along straight paths and the velocity along them is constant at steady state. The fluid particles deform in a well-defined manner. In turbulent flow, (b) the detailed motion of fluid particles is not well defined and a great deal of mixing occurs. Particles move along tortuous trajectories, one of which is shown here, and are deformed in ways that are difficult to quantify. The velocity field, even at steady state, is a wildly fluctuating function of space and time.

A modification of the calculation procedure that works for turbulent flow is to use experiments to fill in the gaps in the theoretical approach (Figure 7.14). Combining a theoretical approach with experimental results can be an effective way to understand complex phenomena. This technique was pursued successfully for turbulent flow in pipes by Ludwig Prandtl who, in the first half of the 20th century, published a flow-rate/pressure-drop equation for turbulent flow in pipes [175, 136]. Prandtl's equation is of a mixed theoretical/experimental type, called data correlations. Prandtl's data correlation for $\Delta p(Q)$ is given here:[3]

Prandtl correlation
for flow rate/pressure drop
(turbulent pipe flow,
smooth pipe):

$$\frac{1}{\sqrt{f}} = 4.0 \log \mathrm{Re}\sqrt{f} - 0.40 \qquad (7.58)$$

[3] The Prandtl correlation is equivalent to the Colebrook equation of Chapter 1 with $\varepsilon = 0$ (see Equation 1.95).

Problem-Solving Strategies for Pipe Flow

Laminar Flow	Turbulent Flow
1. Apply microscopic mass and momentum balances to the problem.	1. Apply microscopic mass and momentum balances to the problem.
2. Solve for the velocity profile.	2. ~~Solve for the velocity profile~~ *Analyze the microscopic equations to obtain the form of $\Delta p(Q)$.*
3. Integrate the velocity profile to obtain the flow rate as a function of pressure drop.	3. ~~Integrate the velocity profile to obtain flow rate as a function of pressure drop~~ *Perform experiments to determine $\Delta p(Q)$.* Dimensional analysis

The procedure that leads to the flow-rate/pressure-drop relationship for laminar flow may be modified to give useful results in turbulent flow.

There are two dimensionless variables in this equation, f—the Fanning friction factor—which is related to pressure drop, and Re—the Reynolds number—which is related to flow rate:

$$\text{Fanning friction factor in tubes:} \qquad f = \frac{(p_0 - p_L)D}{2\rho V^2 L} \qquad (7.59)$$

$$\text{Reynolds number in tubes:} \qquad \text{Re} = \frac{\rho V D}{\mu} \qquad (7.60)$$

in which $V = \langle v \rangle = Q/\pi R^2$ is the average velocity across the pipe cross section. The Fanning friction factor and a version of the Prandtl correlation were introduced in the quick-start section in Chapter 1.

The development of Prandtl's correlation (discussed in the next section) followed the steps shown on the right side of Figure 7.14. First, the mass and momentum balances for pipe flow were written. Second, these microscopic equations were manipulated to show how the mass and momentum-conservation laws constrain $\Delta p(Q)$. This step involves a technique called *dimensional analysis*, which is discussed in detail in Section 7.1.2.2. Third, experiments were performed to determine the actual function $\Delta p(Q)$ (with all of the details filled in) for turbulent flow.

Dimensional analysis is a powerful tool of science and engineering that can be used to establish the variables that impact a given quantity. For the pipe-flow problem, Prandtl and his coworkers used dimensional analysis to determine the appropriate experiments to conduct to determine $\Delta p(Q)$ (Equation 7.58) for turbulent pipe flow. The science leading to Equation 7.58 takes time to explain and comprehend, and we turn to that effort next. First, however, we use the Prandtl correlation to solve the burst-pipe problem for turbulent flow.

EXAMPLE 7.5 (Burst pipe, concluded). *The hose connecting the city water supply to the washing machine in a home burst while the homeowner was away.*

Water gushed out of the 1/2-inch Schedule 40 pipe for 48 hours before the problem was noticed by a neighbor and the water was shut off. How much water sprayed into the house over the two-day period? The water utility reports that the water pressure supplied to the house was approximately 60 psig.

SOLUTION. Pressure is the driving force for the flow and the fixed driving pressure of 60 psig causes a certain flow rate of the water. That flow rate multiplied by 48 hours gives the total amount that sprayed into the house:

$$\begin{pmatrix} \text{volume} \\ \text{of water} \\ \text{in house} \end{pmatrix} = \left(\frac{\text{volume}}{\text{time}} \right) (\text{time interval}) \tag{7.61}$$

$$= Q \Delta t \tag{7.62}$$

If we assume that the flow in the pipes is turbulent, then we can calculate the flow rate Q from the Prandtl correlation (assuming smooth pipes).

As before (see Example 7.3), we assume that the municipal water supply is connected to the house by 80 feet of 1-inch Schedule 40 pipe, which subsequently connects in the house to the washing machine through 20-feet of 1/2-inch Schedule 40 pipe. Thus, we have two pipes to consider (Figure 7.10), and there also are fittings and valves in the line. We ignore any pressure changes due to the fittings or valves and the velocity change and consider only the pressure changes in the straight pipes (Problem 12 considers the complete case).

In principle, the solution method for this turbulent-flow problem is the same as the method used in the previous attempt, in which we considered laminar flow in the same system. The flow-rate/pressure-drop relationship (we now use the Prandtl correlation instead of the Hagen-Poiseuille equation) must be satisfied in each section, and the pressure drop over the two pipes must total 60 psig. Because the flow-rate/pressure-drop relationship for turbulent flow (i.e., the Prandtl correlation) is highly nonlinear, however, we must perform iterative calculations to solve for the pressures and flow rate.

In each pipe section, the flow-rate/pressure-drop relationship is given by the Prandtl correlation, Equation 7.58. Both the friction factor f and the Reynolds number Re are functions of flow rate through the average velocity $\langle v \rangle \equiv V = Q/\pi R^2$, which we do not know. The friction factor also depends on the pressure drop for each pipe section. There is sufficient information to solve for the two pressure drops and for the overall flow rate (i.e., two equations and two unknowns), but we cannot solve directly for these unknowns.

An iterative solution could be performed as shown in Figure 7.15.

Solution steps:

1. Guess flow rate Q.
2. Calculate the Reynolds number in the large pipe, $\text{Re}_1 = \rho \langle v \rangle_1 D_1 / \mu$.
3. Calculate f_1 for the large pipe from the friction-factor/Reynolds-number correlation for turbulent flow, the Prandtl correlation (see Equation 7.58). Note that the calculation of f_1 from Equation 7.58 must be done iteratively.
4. With the value of f_1 calculated and the assumed flow rate Q, calculate the pressure drop Δp_1 across the large pipe from the expression that gives the

Calculate Q	guess Q=	0.001972	m³/s
change units	Q=	31	gpm
density	r=	997.08	kg/m³
$V_1=4Q/pD_1^2$	$V_1=$	3.537	m/s
1" pipe given	$D_1=$	0.0266	m
viscosity	m=	8.94E-04	Pa s
calculate Re from definition	$Re_1=$	105,154	
Calculate f_1 from Prandtl correlation	**guess f_1=**	**0.004454**	
iterate to final answer	$g(f)=$	14.98	
next round $f = 1/g\,(f)^2$	$f_{final}=$	0.004454	
given	$L_1=$	80	ft
change units	$L_1=$	24.4	m
calculate Dp from friction factor	$Dp_1=$	1.02E+05	Pa
pressure drop across the first tube	$Dp_1=$	14.8	psi
total Dp (given)	$Dp_{total}=$	60	psi
pressure drop across the second tube	$Dp_2=$	45.2	psi
change units	$Dp_2=$	3.12E+05	Pa
given	$D_2=$	0.0158	m
Calculate V_2 from Prandtl correlation			
(no iteration needed)	$L_2=$	20	ft
change units	$L_2=$	6.1	m
calculate Re from correlation	LHS=	15.8006	
$f = 1/LHS^2$	$f_2=$	0.004005	
next round V_2	$V_2=$	10.0612	m/s
Next round Q=$V_2pD_2^2/4$	**Q=**	**0.001972**	**m³/s**

Prandtl correlation:

$$\underbrace{4\log(\mathrm{Re}\sqrt{f})-0.4}_{g(f)}=\frac{1}{\sqrt{f}}$$

Pipe friction factor from experiments:

$$f = \frac{\Delta p D}{2L\rho V^2}$$

Prandtl correlation:

$$\underbrace{4\log\left[\frac{\rho V D}{\mu}\sqrt{\frac{\Delta p D}{2L\rho V^2}}\right]-0.4}_{\text{LHS}}=\frac{1}{\sqrt{f}}$$

Figure 7.15 The iterative solution for the flow rate may be carried out in a spreadsheet program. The steps of the solution are described in this chapter. The values shown correspond to the final iteration.

friction factor in pipes in terms of pressure drop and geometric and material parameters (see Equation 7.59).

5. Calculate the pressure drop Δp_2 across the small tube from $\Delta p_2 = 60\ \text{psig} - \Delta p_1$.

6. Calculate velocity $\langle v\rangle_2$ in the smaller pipe through solution of Equation 7.58, the Prandtl correlation. Note that although we do not know the friction factor f_2 or the Reynolds number Re_2 for the smaller pipe, we do know the pressure drop. Both the friction factor and the Reynolds number are a function of velocity, but the product $\mathrm{Re}\sqrt{f}$ is independent of velocity and can be calculated directly from Δp_2 for the small pipe:

$$\mathrm{Re}\sqrt{f} = \frac{\rho\langle v\rangle D}{\mu}\sqrt{\frac{\Delta p D}{2L\rho\langle v\rangle^2}} \tag{7.63}$$

$$= \frac{\rho D}{\mu}\sqrt{\frac{\Delta p D}{2L\rho}} \tag{7.64}$$

$$= \sqrt{\frac{\rho\Delta p D^3}{2L\mu^2}} \tag{7.65}$$

The lefthand side of the Prandtl correlation is calculated using the expression for $Re\sqrt{f}$. This gives f, and the velocity is calculated directly from Equation 7.59. No iterations are necessary for this step.

7. Calculate flow rate from $Q = \langle v \rangle_2 \pi D_2^2/4$.
8. Compare calculated flow rate Q with the initial guess and iterate until the two values converge.

For this problem and following this procedure, we calculate that the flow rate in the burst pipe is 31 gpm. The pressure drop across the larger and smaller pipes is 15 and 45 psi, respectively. The smaller cross-sectional area of the small pipe causes a larger resistance to flow; thus, the pressure drop across the smaller pipe is much larger than across the larger pipe. Note that the flow is turbulent (i.e., $Re = 105{,}000$, as we assumed).

We now calculate the volume of water that filled the house over two days as a result of the burst water pipe:

$$\begin{pmatrix} \text{volume} \\ \text{of water} \\ \text{in house} \end{pmatrix} = \left(\frac{\text{volume}}{\text{time}} \right) (\text{time interval}) \tag{7.66}$$

$$= Q\Delta t \tag{7.67}$$

$$= \left(\frac{31 \text{ gal}}{\text{min}} \right) \left(\frac{60 \text{ min}}{\text{h}} \right) \left(\frac{24 \text{ h}}{\text{day}} \right) (2 \text{ days}) \tag{7.68}$$

$$= 89{,}280 \text{ gal} \tag{7.69}$$

$$= 89{,}000 \text{ gal} \quad (\text{two significant figures}) \tag{7.70}$$

Thus, a volume of water equal to six times the volume of a typical home swimming pool (i.e., $6 \times 13{,}000$ gal) spilled into the home over two days. If the amount of half-inch pipe were longer (i.e., the distance from the water main line to the washer is farther), the flow rate would be less because there would be more friction.

The burst-pipe example shows the usefulness of the Prandtl data correlation for calculating flow rates in pressure-driven turbulent flow in straight pipes. The Prandtl correlation also may be used to calculate flows and pressures in complex piping systems, as shown in Example 7.9. Another data correlation, the Colebrook equation (Equation 7.161; see Chapter 1 and Section 7.1.2.3), relates flow rate/pressure drop in rough pipes; there are data correlations available for $\Delta p(Q)$ in other circumstances—for example, in conduits of different shapes (see Section 7.2) [132].

The Prandtl data correlation and the others mentioned all come from experimental observations. By definition, a data correlation is a function, table, or rule of thumb that summarizes actual behavior of a real system. To arrive at data correlations, we must conduct experiments. If we have an actual apparatus of interest—a coating line or an artificial heart, for example—we can conduct experiments on the actual device and correlate the data through fitting functions, summarizing the data in tables, or establishing rules of thumb.

If we do not have the actual piece of equipment but want to make an estimate of how a device or variations on it might behave, we must exploit our knowledge of the physics that similar devices have in common. This is what Prandtl did in arriving at Equation 7.58 and what has been done by generations of physicists and engineers in the development of new technologies. Subsequent sections describe the process of dimensional analysis and how it can be applied to a complex system such as turbulent flow in a pipe. Through dimensional analysis, a process that shows the essential relationships in a system, we are able to conclude that flow-rate/pressure-drop relationships in similar pipes can be organized in a single equation if the data are written in terms of the Fanning friction factor as a function of the Reynolds number $f(\text{Re})$. Once we make this determination, experiments show the exact form of the $f(\text{Re})$ relationship, and a data correlation such as the Prandtl correlation may be produced.

The next two subsections explain how the Prandtl correlation may be developed for turbulent flow in smooth pipes with the help of dimensional analysis. Uses of data correlations in pipe-flow problems are discussed in Section 7.1.2.3. We move on to other internal flows in Section 7.2.

7.1.2.1 MOMENTUM BALANCE IN TURBULENT FLOW

We are investigating a way to quantify flow rate/pressure drop in turbulent flow. As shown previously, we calculate flow rate/pressure drop in laminar flow by solving the continuity equation and the Navier-Stokes equations in the flow:

Mass conservation:
(continuity equation,
constant density)

$$0 = \nabla \cdot \underline{v} \qquad (7.71)$$

Momentum conservation:
(Navier-Stokes equation)

$$\rho \left(\frac{\partial \underline{v}}{\partial t} + \underline{v} \cdot \nabla \underline{v} \right) = -\nabla p + \mu \nabla^2 \underline{v} + \rho \underline{g} \qquad (7.72)$$

To make the same calculation for turbulent flow, therefore, we first must solve for the velocity field and stress tensor for turbulent flow in a tube:

Turbulent pipe flow:
$$\underline{v} = \begin{pmatrix} v_r(r, \theta, z, t) \\ v_\theta(r, \theta, z, t) \\ v_z(r, \theta, z, t) \end{pmatrix}_{r\theta z} \qquad (7.73)$$

We attempt to do this now, following the same steps as in the laminar-flow calculation.

For the laminar-flow solution, we solved the mass and momentum balances for the velocity field $\underline{v}$ by making reasonable assumptions about the flow symmetry, such as unidirectional flow, steady state, and θ-symmetry. Because laminar flow is well organized and has a great deal of symmetry, we obtained the complete solutions for $\underline{v}$ and p and could calculate the flow-rate/pressure-drop relationship for laminar tube flow in the form of the Hagen-Poiseuille equation.

Using the laminar-flow case as a guide, we now think about turbulent flow to see whether we can identify any features that allow us to simplify Equations 7.71–7.73 and to proceed to calculate $\Delta p(Q)$ in turbulent flow. In the laminar-flow solution, we obtained $v_z(r)$ by solving the z-component of the equation of motion. In cylindrical coordinates, the z-component of the equation of motion for turbulent tube flow is:

$$\rho \left(\frac{\partial v_z}{\partial t} + v_r \frac{\partial v_z}{\partial r} + \frac{v_\theta}{r} \frac{\partial v_z}{\partial \theta} + v_z \frac{\partial v_z}{\partial z} \right)$$

$$= -\frac{\partial p}{\partial z} + \mu \left(\frac{1}{r} \frac{\partial}{\partial r} \left(r \frac{\partial v_z}{\partial r} \right) + \frac{1}{r^2} \frac{\partial^2 v_z}{\partial \theta^2} + \frac{\partial^2 v_z}{\partial z^2} \right) + \rho g_z \quad (7.74)$$

We see that none of the terms is zero for turbulent flow: Turbulent flow is time-varying; turbulent flow has three nonzero velocity components; v_z varies in all three coordinate directions; and pressure varies in the z-direction.

Because we cannot solve the Navier-Stokes equations for turbulent flow, we need a new approach. If we need a detailed understanding of the structure and stresses of turbulent flow, we show in Section 10.3 that a useful approach is to consider a time-averaged version of the microscopic-momentum balance. In that approach, each velocity component $v_i = v_r, v_\theta, v_z$ is written as a mean value $\overline{v_i}$ plus a time-fluctuating term v_i', and the entire momentum-balance equation is averaged over a reasonable time interval T. Statistical methods then are used to describe and model the structure and stresses associated with turbulence. The development of the time-averaging approach was critical to advances in the basic understanding of turbulence that were achieved in the second half of the 20th century [10, 165].

Fortunately, we do not need a detailed understanding of the structure and stresses of turbulence to address the current problem. We seek only to understand the flow-rate/pressure-drop relationship in turbulent flow. Because we defined the scope of the problem as the narrow issue of the flow-rate/pressure-drop relationship, we can use a combined analytical/experimental approach.

To summarize, it is not possible to obtain directly an analytical expression for $\Delta p(Q)$ in turbulent flow. It is possible to pursue a statistical exploration of turbulence, but this effort is not needed for our limited purposes. Instead, we develop a correlation for $\Delta p(Q)$ using experimental data on turbulent flow in pipes. To be sure to capture the physics of the flow, we design the $\Delta p(Q)$ experiments using the physics implicit in the mass and momentum balances (Equations 7.71–7.73). The fundamental physics can tell us the form of the correlation between Δp and Q before a single datapoint is taken.

7.1.2.2 DIMENSIONAL ANALYSIS

We seek the pressure-drop/flow-rate relationship for turbulent flow. The situation is that we know the equations that govern mass and momentum conservation for turbulent flow, Equations 7.71 and 7.72, but we cannot solve them:

Mass conservation:
(continuity equation, $0 = \nabla \cdot \underline{v}$ (7.75)
constant density)

Momentum conservation:
(Navier-Stokes equation) $\rho \left(\frac{\partial \underline{v}}{\partial t} + \underline{v} \cdot \nabla \underline{v} \right) = -\nabla p + \mu \nabla^2 \underline{v} + \rho \underline{g}$ (7.76)

We decide to conduct experiments to provide the information we need, but there are many variables and physical properties to consider. The flow is characterized by unknown fields p and $\underline{v}$ as a function of position and time, and the flow also is a function of material parameters density ρ and viscosity μ, which are different for different fluids. The physical dimensions of the system (D, L) also affect our results. Our first task is to determine how these various factors interact and influence the flow produced. We can do so by looking at how these parameters influence solutions to the Navier-Stokes equation.

The Navier-Stokes equation is the complete expression of momentum conservation for Newtonian fluids, but certain terms of the Navier-Stokes equation are more or less important in any particular case under consideration. For example, when viscosity is very large, the $\mu \nabla^2 \underline{v}$ term dominates the Navier-Stokes equation. When the time-rate-of-change is rapid, the time-derivative term $\partial \underline{v}/\partial t$ dominates. Many terms (but not all) contain velocity; some contain velocity more than once. The effect of geometry also is complex because it affects the spatial derivatives, which appear on both the lefthand ($\underline{v} \cdot \nabla \underline{v}$) and righthand ($\mu \nabla^2 \underline{v}$) sides of the equation.

To determine how each term scales for a given problem, we nondimensionalize each quantity in the equation. We used a type of nondimensionalization before when we sought a convenient way to plot complex equations. For example, the equation for the final velocity field in Poiseuille flow in a tube (see Equation 7.23), reproduced here, was plotted in Figure 7.6 in terms of the dimensionless quantity $v_z/\langle v \rangle$ as a function of the dimensionless variable r/R:

$$v_z = \frac{(p_0 - p_L)R^2}{4\mu L}\left[1 - \left(\frac{r}{R}\right)^2\right] \tag{7.77}$$

$$\langle v \rangle = \frac{Q}{\pi R^2} = \frac{(p_0 - p_L)R^2}{8\mu L} \tag{7.78}$$

$$\frac{v_z}{\langle v \rangle} = 2\left[1 - \left(\frac{r}{R}\right)^2\right] \tag{7.79}$$

We always should be able to organize an equation into nondimensional groups because every equation must be dimensionally consistent. Organizing an equation into nondimensional form serves to visualize a final solution (see also Figures 7.9, 7.28, and 7.30), find algebra mistakes, and see clearly the form of an equation without distracting details.

We extend the idea of plotting nondimensionally to our current problem of determining which quantities in the Navier-Stokes equations are important for certain flows. If we rewrite the governing equations in terms of nondimensional versions of velocity, coordinate directions r and z, velocity, time, and other variables, we see more clearly the structure of the equations. The clarity that comes from rewriting these equations in nondimensional form helps considerably in designing experiments, plotting data, and deriving correlations.

We seek to nondimensionalize the continuity equation and the z-component of the Navier-Stokes equation (Equation 7.74). To nondimensionalize a term such as $\partial v_z/\partial r$, we choose characteristic values of velocity and length. For turbulent tube flow, we choose the average velocity $V \equiv \langle v \rangle$ as the characteristic velocity and the pipe diameter $D = 2R$ as the characteristic distance. These choices are arbitrary, and how effective they are as choices can be decided only from the final results of the analysis. We define nondimensional variables v_z^* and z^* as the ratios of the dimensional variables to the characteristic values:

$$v_z^* \equiv \frac{v_z}{V} \tag{7.80}$$

$$z^* \equiv \frac{z}{D} \tag{7.81}$$

We must nondimensionalize all of the variables in the Navier-Stokes equation; thus, we define the additional dimensionless variables here:

$$v_r^* \equiv \frac{v_r}{V} \tag{7.82}$$

$$v_\theta^* \equiv \frac{v_\theta}{V} \tag{7.83}$$

$$r^* \equiv \frac{r}{D} \tag{7.84}$$

So far, we can nondimensionalize with only two characteristic scale factors, V and D. We also must nondimensionalize time; if we choose the ratio D/V as the characteristic time, we avoid introducing yet another scale factor:

$$t^* \equiv \frac{t}{D/V} \tag{7.85}$$

The time D/V represents the time that the fluid traveling at the average velocity requires to travel a distance of one tube diameter. This seems reasonable as a characteristic time for tube flow; again, the utility of these choices can be evaluated only when we see the final results of the analysis.

Having constructed dimensionless velocities, distances, and times, we now are ready to substitute these dimensionless variables into the continuity equation and the z-component of the Navier-Stokes equation. We anticipate that the resulting dimensionless equations will be easier to interpret than their dimensional analogs. Solving Equations 7.80–7.85 for the dimensional variables v_r, v_θ, v_z, z, r, and t and substituting them in the continuity equation, we obtain:

Continuity equation: (incompressible)
$$\frac{\partial v_r}{\partial r} + \frac{1}{r}\frac{\partial v_\theta}{\partial \theta} + \frac{\partial v_z}{\partial z} = 0 \tag{7.86}$$

$$\frac{\partial \left(v_r^* V \right)}{\partial r^* D} + \frac{1}{r^* D}\frac{\partial \left(v_\theta^* V \right)}{\partial \theta} + \frac{\partial \left(v_z^* V \right)}{\partial z^* D} = 0 \tag{7.87}$$

Dimensionless continuity equation:
$$\frac{\partial v_r^*}{\partial r^*} + \frac{1}{r^*}\frac{\partial v_\theta^*}{\partial \theta} + \frac{\partial v_z^*}{\partial z^*} = 0 \tag{7.88}$$

This dimensionless result does not contain any scale factors containing the characteristic dimensions V or D; thus, this equation does not yield any information about the relative importance of the dimensionless derivatives. Each term in Equation 7.88 is equally important in all flows.

Following the same steps for the z-component of the Navier-Stokes equation, we obtain:

z-component Navier-Stokes equation:

$$\rho\left(\frac{\partial v_z}{\partial t} + v_r\frac{\partial v_z}{\partial r} + \frac{v_\theta}{r}\frac{\partial v_z}{\partial \theta} + v_z\frac{\partial v_z}{\partial z}\right)$$

$$= -\frac{\partial p}{\partial z} + \mu\left(\frac{1}{r}\frac{\partial}{\partial r}\left(r\frac{\partial v_z}{\partial r}\right) + \frac{1}{r^2}\frac{\partial^2 v_z}{\partial \theta^2} + \frac{\partial^2 v_z}{\partial z^2}\right) + \rho g_z$$

Dimensionless z-component Navier-Stokes equation:

$$\rho\left(\frac{\partial(v_z^* V)}{\partial(t^* D/V)} + (v_r^* V)\frac{\partial(v_z^* V)}{\partial(r^* D)} + \frac{(v_\theta^* V)}{(r^* D)}\frac{\partial(v_z^* V)}{\partial \theta} + (v_z^* V)\frac{\partial(v_z^* V)}{\partial(z^* D)}\right)$$

$$= -\frac{\partial p}{\partial(z^* D)} + \mu\left(\frac{1}{(r^* D)}\frac{\partial}{\partial(r^* D)}\left(r^* D\frac{\partial(v_z^* V)}{\partial(r^* D)}\right) + \frac{1}{(r^* D)^2}\frac{\partial^2(v_z^* V)}{\partial \theta^2}\right.$$

$$\left. + \frac{\partial^2(v_z^* V)}{\partial(z^* D)^2}\right) + \rho g_z \tag{7.89}$$

There are two terms that have not been nondimensionalized: pressure and gravity. We nondimensionalize gravity by defining $g_z^* \equiv g_z/g$. For inspiration on how to address the pressure term, we return to a discussion of the solution of laminar flow in a tube.

We need a characteristic pressure to nondimensionalize the pressure term in Equation 7.89. We begin by calling the characteristic pressure P and nondimensionalizing as usual:

$$p^* \equiv \frac{p - p_{ref}}{P} \tag{7.90}$$

We choose to express p^* in terms of a pressure difference rather than as an absolute pressure because pressure differences cause flow. With this definition, the pressure-gradient term of the z-component of the Navier-Stokes equation becomes:

$$\begin{matrix}\text{Pressure term}\\\text{nondimensional}\\\text{Navier-Stokes:}\end{matrix} \qquad -\left(\frac{1}{D}\right)\frac{\partial p}{\partial z^*} = -\left(\frac{P}{D}\right)\frac{\partial p^*}{\partial z^*} \tag{7.91}$$

We now write P in terms of other characteristic quantities by forcing the coefficient of the pressure term to be the same as the coefficient of the $\partial v_z/\partial t$ or $\underline{v} \cdot \nabla\underline{v}$ terms. Thus:

$$\begin{pmatrix}\text{coefficient of}\\\text{pressure term}\end{pmatrix} = \begin{pmatrix}\text{coefficient of}\\\partial/\partial t \text{ term}\end{pmatrix} \tag{7.92}$$

$$\frac{P}{D} = \frac{\rho V^2}{D} \tag{7.93}$$

$$\boxed{P = \rho V^2} \tag{7.94}$$

Note that this choice of characteristic pressure is dimensionally correct.

Dimensional:

$$\rho\left(\frac{\partial \underline{v}}{\partial t} + \underline{v}\cdot\nabla\underline{v}\right) = -\nabla p + \mu\nabla^2\underline{v} + \rho\underline{g}$$

$\underbrace{\hspace{4cm}}_{\text{inertial terms}}$ $\underbrace{\hspace{4cm}}_{\text{force terms}}$

Nondimensional:

$$\left(\frac{\partial \underline{v}^*}{\partial t^*} + \underline{v}^*\cdot\nabla^*\underline{v}^*\right) = -\nabla^* p^* + \frac{1}{\mathrm{Re}}\nabla^{*2}\underline{v}^* + \frac{1}{Fr}\underline{g}^*$$

Figure 7.16   The inertial terms of the momentum balance are on the lefthand side. The righthand side of the momentum balance has the forces that act on the control volume.

The choice $P = \rho V^2$ is arbitrary, but we can argue its reasonableness. The lefthand-side terms of the Navier-Stokes equation are the rate-of-change and the convective terms, which are collectively known as the inertial terms (Figure 7.16). These terms are directly concerned with the fluid momentum; thus, choosing to link the pressure and inertial terms is the nondimensional equivalent of allowing the velocity term to be calculated in terms of the pressure. This was our approach in the laminar-flow solution, and we make this choice for nondimensionalizing pressure; it can be truly justified only by testing the final results of our analysis against what is observed in laboratory flow.

We arrive at a version of the z-component of the microscopic-momentum-balance equation written in terms of nondimensional variables:

$$\left(\frac{\rho V^2}{D}\right)\left[\frac{\partial v_z^*}{\partial t^*} + v_r^*\frac{\partial v_z^*}{\partial r^*} + \frac{v_\theta^*}{r^*}\frac{\partial v_z^*}{\partial \theta} + v_z^*\frac{\partial v_z^*}{\partial z^*}\right]$$

$$= \left(-\frac{\rho V^2}{D}\right)\frac{\partial p^*}{\partial z^*} + \left(\frac{\mu V}{D^2}\right)\left[\frac{1}{r^*}\frac{\partial}{\partial r^*}\left(r^*\frac{\partial v_z^*}{\partial r^*}\right) + \frac{1}{r^{*2}}\frac{\partial^2 v_z^*}{\partial \theta^2} + \frac{\partial^2 v_z^*}{\partial z^{*2}}\right] + (\rho g)\,g_z^*$$

$$\text{(7.95)}$$

The characteristic values V and D along with ρ and μ are grouped in front of each term of Equation 7.95, and these groupings indicate the relative magnitudes of the various terms.

$$\text{Inertial and pressure terms} \quad \frac{\rho V^2}{D} \qquad\qquad \text{(7.96)}$$

$$\text{Viscous terms} \quad \frac{\mu V}{D^2} \qquad\qquad \text{(7.97)}$$

$$\text{Gravity term} \quad \rho g \qquad\qquad \text{(7.98)}$$

We can make the front factors dimensionless if we divide through by one of them—for example, if we divide through by the inertial group $\rho V^2 / D$:

$$\frac{\partial v_z^*}{\partial t^*} + v_r^* \frac{\partial v_z^*}{\partial r^*} + \frac{v_\theta^*}{r^*} \frac{\partial v_z^*}{\partial \theta} + v_z^* \frac{\partial v_z^*}{\partial z^*}$$

$$= -\frac{\partial p^*}{\partial z^*} + \frac{1}{\mathrm{Re}} \left[\frac{1}{r^*} \frac{\partial}{\partial r^*} \left(r^* \frac{\partial v_z^*}{\partial r^*} \right) + \frac{1}{r^{*2}} \frac{\partial^2 v_z^*}{\partial \theta^2} + \frac{\partial^2 v_z^*}{\partial z^{*2}} \right] + \frac{1}{\mathrm{Fr}} g_z^* \quad (7.99)$$

The two dimensionless scale factors that appear in the nondimensional Navier-Stokes equation are called the *Reynolds number* and the *Froude*[4] *number*:

$$\text{Reynolds number:} \quad \boxed{\mathrm{Re} \equiv \frac{\rho V D}{\mu}} \quad \text{ratio of } \frac{\text{(inertial forces)}}{\text{(viscous forces)}} \quad (7.100)$$

$$\text{Froude number:} \quad \boxed{\mathrm{Fr} \equiv \frac{V^2}{gD}} \quad \text{ratio of } \frac{\text{(inertial forces)}}{\text{(gravity forces)}} \quad (7.101)$$

If we nondimensionalize the r- and θ-components of the Navier-Stokes equation, the same dimensionless groups appear. Thus, the dimensionless Navier-Stokes equation written in Gibbs notation is as follows (see Figure 7.16):

$$\text{Dimensionless Navier-Stokes:} \quad \boxed{\frac{\partial \underline{v}^*}{\partial t^*} + \underline{v}^* \cdot \nabla^* \underline{v}^* = -\nabla^* p^* + \left(\frac{1}{\mathrm{Re}} \right) \nabla^{*2} \underline{v}^* + \left(\frac{1}{\mathrm{Fr}} \right) \underline{g}^*}$$

$$(7.102)$$

where ∇^* represents the dimensionless *del* operator.

The dimensional analysis so far indicates that the Navier-Stokes equation for any given problem is specified by the values of two dimensionless groups: the Reynolds number and the Froude number. The grouping of scale factors that we performed made it easier to see how the Navier-Stokes equation would change if it were applied to different pipe-flow problems; that is, if the density or viscosity changed or if the pipe diameter changed. If Re is large, for example, the $\mathrm{Re}^{-1}\nabla^2 \underline{v}$ term drops to zero. The familiar Reynolds number is the ratio of inertial to viscous forces in the Navier-Stokes equation.

We seek the pressure-drop/flow-rate relationship for turbulent flow. We showed that for any pipe-flow problem, the mass and momentum balances are governed by the Reynolds number and the Froude number. To apply this dimensional analysis result to turbulent pipe flow, we must recognize that pressure drop and flow rate are connected through the drag on the walls of the tube. To show the relationship between wall drag and pressure drop, we calculate $\mathcal{F}_{\mathrm{drag}}$ for laminar flow next. After we identify the governing relationships for the problem of pipe flow, we return to the turbulent-flow case.

---

[4] Pronounced "Frood."

EXAMPLE 7.6. *What is the equation for drag at the wall for steady, laminar, pressure-driven flow of a Newtonian fluid in a pipe? This flow is called Poiseuille flow.*

SOLUTION. In Section 6.2.3.1, we discuss fluid forces at walls, which may be obtained from a surface integral over $\hat{n}\cdot$ the stress tensor on the surface:

$$
\begin{array}{c}
\text{Total molecular fluid force} \\
\text{on a surface } \mathcal{S}:
\end{array}
\qquad
\boxed{\underline{\mathcal{F}} = \iint_{\mathcal{S}} [\hat{n} \cdot \underline{\tilde{\Pi}}]_{\text{at surface}} \, dS}
\qquad (7.103)
$$

The stress on the inside walls of the pipe (i.e., unit normal is $-\hat{e}_r$, location is $r = R$) is $[\hat{n} \cdot \underline{\tilde{\Pi}}]_{\text{at surface}} = -\hat{e}_r \cdot \underline{\tilde{\Pi}}|_R$. In a previous example, we solved for the stress tensor in laminar pipe flow of an incompressible Newtonian fluid (see Equation 7.37). Therefore, we calculate the total stress on the pipe walls by carrying out the integration in Equation 7.103 using the known laminar-flow solution for $\underline{\tilde{\Pi}}(r, z)$:

$$
\begin{array}{c}
\text{Pressure-driven} \\
\text{laminar flow} \\
\text{in a tube} \\
\text{(Newtonian):}
\end{array}
\qquad
\underline{\tilde{\Pi}}(r,z) =
\begin{pmatrix}
-p(z) & 0 & \frac{(p_L-p_0)r}{2L} \\
0 & -p(z) & 0 \\
\frac{(p_L-p_0)r}{2L} & 0 & -p(z)
\end{pmatrix}_{r\theta z}
\qquad (7.104)
$$

$$
[\hat{n} \cdot \underline{\tilde{\Pi}}]_{\text{at surface}} = -\hat{e}_r \cdot \underline{\tilde{\Pi}}|_R
\qquad (7.105)
$$

$$
= (-1 \;\; 0 \;\; 0)_{r\theta z} \cdot
\begin{pmatrix}
-p(z) & 0 & \frac{(p_L-p_0)R}{2L} \\
0 & -p(z) & 0 \\
\frac{(p_L-p_0)R}{2L} & 0 & -p(z)
\end{pmatrix}_{r\theta z}
\qquad (7.106)
$$

$$
= \left(p(z) \;\; 0 \;\; -\frac{(p_L-p_0)R}{2L} \right)_{r\theta z}
\qquad (7.107)
$$

$$
=
\begin{pmatrix}
p(z) \\
0 \\
\frac{(p_0-p_L)R}{2L}
\end{pmatrix}_{r\theta z}
\qquad (7.108)
$$

The r-component of Equation 7.108 represents the radial force on the pipe walls, which must be counterbalanced by the material strength of the solid pipe or the pipe will burst. The z-component of Equation 7.108 represents the axial drag on the pipe walls:

$$
\mathcal{F}_{\text{drag}} =
\begin{pmatrix}
\text{axial drag} \\
\text{in laminar flow} \\
\text{in a pipe}
\end{pmatrix}
=
\begin{pmatrix}
z\text{-component of} \\
\text{total force} \\
\text{on CV surface} \\
\text{of unit normal } -\hat{e}_r
\end{pmatrix}
\qquad (7.109)
$$

$$\mathcal{F}_{\text{drag}} = \int_0^L \int_0^{2\pi} \hat{e}_z \cdot (\hat{n} \cdot \underline{\underline{\tilde{\Pi}}}|_{\text{surface}}) \ R \ d\theta dz \qquad (7.110)$$

$$= \int_0^L \int_0^{2\pi} \hat{e}_z \cdot (-\hat{e}_r \cdot \underline{\underline{\tilde{\Pi}}}|_R) \ R \ d\theta dz \qquad (7.111)$$

$$= \int_0^L \int_0^{2\pi} \begin{pmatrix} 0 & 0 & 1 \end{pmatrix}_{r\theta z} \cdot \begin{pmatrix} p(z) \\ 0 \\ \frac{\Delta p R}{2L} \end{pmatrix}_{r\theta z} \ R \ d\theta dz \qquad (7.112)$$

$$= \int_0^L \int_0^{2\pi} \frac{(p_0 - p_L)R}{2L} \ R \ d\theta dz \qquad (7.113)$$

$$= (p_0 - p_L)\pi R^2 \qquad (7.114)$$

Laminar pipe flow:
$$\boxed{\mathcal{F}_{\text{drag}} = (p_0 - p_L)\pi R^2} \qquad (7.115)$$

The inner pipe surface experiences drag in the $(+z)$-direction due to the fluid motion. This can be visualized as the stickiness of the viscous fluid grabbing onto the walls and attempting to drag the pipe forward with the fluid as it moves along.

The result in Equation 7.115 is derived here for laminar flow; in Chapter 9, we also obtain the same result for turbulent flow using a macroscopic-momentum balance (see Equation 9.236):

Laminar or turbulent
pipe flow:
$$\boxed{\mathcal{F}_{\text{drag}} = (p_0 - p_L)\pi R^2} \qquad (7.116)$$

The relationship between wall drag and pressure turns out to be independent of flow type. Our current problem of pressure drop/flow rate can be recast as a calculation of $\mathcal{F}_{\text{drag}}$ as a function of Q.

Recognizing wall drag as the cause of pressure drop in pipe flow leads to the correct governing expression for our pressure-drop/flow-rate problem for turbulent flow. The general expression for drag in pipe flow—Equation 7.110, repeated here—is the key result of Example 7.6:

Axial fluid drag
on a pipe surface
of unit normal $-\hat{e}_r$:
$$\mathcal{F}_{\text{drag}} = \int_0^L \int_0^{2\pi} \hat{e}_z \cdot (-\hat{e}_r \cdot \underline{\underline{\tilde{\Pi}}}|_R) \ R \ d\theta dz \qquad (7.117)$$

We now nondimensionalize this expression to drill down to the fundamental relationship governing pressure drop/flow rate in general pipe flow.

We can write a detailed version of Equation 7.117 for turbulent flow by using the general expression for $\tilde{\underline{\underline{\Pi}}}$:

$$\text{Stress tensor:} \quad \tilde{\underline{\underline{\Pi}}} = -p\underline{\underline{I}} + \tilde{\underline{\underline{\tau}}} \tag{7.118}$$

$$= \begin{pmatrix} \tilde{\tau}_{rr} - p & \tilde{\tau}_{r\theta} & \tilde{\tau}_{rz} \\ \tilde{\tau}_{\theta r} & \tilde{\tau}_{\theta\theta} - p & \tilde{\tau}_{\theta z} \\ \tilde{\tau}_{zr} & \tilde{\tau}_{z\theta} & \tilde{\tau}_{zz} - p \end{pmatrix}_{r\theta z} \tag{7.119}$$

Axial fluid drag on a pipe surface of unit normal $-\hat{e}_r$:
$$\mathcal{F}_{\text{drag}} = \int_0^L \int_0^{2\pi} \hat{e}_z \cdot (-\hat{e}_r \cdot \tilde{\underline{\underline{\Pi}}}|_R) \; R \, d\theta dz \tag{7.120}$$

$$= \int_0^L \int_0^{2\pi} \begin{pmatrix} 0 \\ 0 \\ 1 \end{pmatrix}_{r\theta z} \cdot \left[(-1 \; 0 \; 0)_{r\theta z} \cdot \tilde{\underline{\underline{\Pi}}}|_R \right] R \, d\theta dz \tag{7.121}$$

$$= \int_0^L \int_0^{2\pi} -\tilde{\tau}_{rz}|_R \; R \, d\theta dz \tag{7.122}$$

We arrived at the simplified expression in Equation 7.122 by using matrix calculations to carry out the dot products in Equation 7.121. We need the shear stress $\tilde{\tau}_{rz}$ to calculate $\mathcal{F}_{\text{drag}}$ for turbulent flow. For both laminar and turbulent flow, shear stress $\tilde{\tau}_{rz}$ is related to the velocity field through the Newtonian constitutive equation (see Table B.8):

$$\text{Newtonian constitutive equation:} \quad \tilde{\underline{\underline{\tau}}} = \mu(\nabla \underline{v} + (\nabla \underline{v})^T) \tag{7.123}$$

$$rz\text{-Component of } \tilde{\underline{\underline{\tau}}}: \quad \tilde{\tau}_{rz} = \mu \left(\frac{\partial v_z}{\partial r} + \frac{\partial v_r}{\partial z} \right) \tag{7.124}$$

Thus, substituting Equation 7.124 into Equation 7.122, we obtain the analytical expression for the axial drag in a pipe in turbulent flow:

Axial fluid drag in turbulent flow on a pipe surface of unit normal $-\hat{e}_r$:
$$\mathcal{F}_{\text{drag}} = \int_0^L \int_0^{2\pi} -\mu \left(\frac{\partial v_z}{\partial r} + \frac{\partial v_r}{\partial z} \right)\bigg|_R R \, d\theta dz \tag{7.125}$$

To proceed further in the calculation of $\mathcal{F}_{\text{drag}}$, we need the solution for the turbulent velocity field $\underline{v}(r, \theta, z, t)$.

We determined, in general terms, how drag is related to flow for turbulent flow. With this result, we now are ready to try dimensional analysis on our problem. As discussed previously, we cannot solve the Navier-Stokes equations for turbulent flow; thus, we cannot calculate $\mathcal{F}_{\text{drag}}$ directly. We are following the dimensional-analysis approach, however; and thus, we can nondimensionalize Equation 7.125 to see which dimensionless groups enter into wall drag and therefore into $\Delta p(Q)$.

To nondimensionalize Equation 7.125, we use the same dimensionless quantities v_z^*, r^*, and so on, as when we nondimensionalized the Navier-Stokes equation.

As in that case, we apply dimensional analysis to determine what the important factors are in our problem. Beginning with Equation 7.125, we substitute the usual dimensionless expressions and obtain:

$$\mathcal{F}_{\text{drag}} = \int_0^L \int_0^{2\pi} -\mu \left(\frac{\partial v_z}{\partial r} + \frac{\partial v_r}{\partial z} \right)\bigg|_R R \, d\theta dz \tag{7.126}$$

$$= \int_0^{L/D} \int_0^{2\pi} -\mu \left(\frac{\partial (v_z^* V)}{\partial (r^* D)} + \frac{\partial (v_r^* V)}{\partial (z^* D)} \right)\bigg|_{r^*=1/2} \frac{D}{2} d\theta (dz^* D) \tag{7.127}$$

$$\mathcal{F}_{\text{drag}} \left(\frac{2}{\mu V D} \right) = \int_0^{L/D} \int_0^{2\pi} -\left(\frac{\partial v_z^*}{\partial r^*} + \frac{\partial v_r^*}{\partial z^*} \right)\bigg|_{r^*=1/2} d\theta dz^* \tag{7.128}$$

We have not yet chosen how to nondimensionalize the wall drag $\mathcal{F}_{\text{drag}}$. All the quantities on the righthand side of Equation 7.128 are dimensionless. The dimensionless velocity gradients $\partial v_z^*/\partial r^*$ and $\partial v_r^*/\partial z^*$ come from the solution of the dimensionless Navier-Stokes equation, Equation 7.102, which is a function of the two dimensionless numbers Re and Fr. The righthand side of Equation 7.128 has an additional dimensionless group in it, L/D. We organize the lefthand side of Equation 7.128 by writing the viscosity in terms of the Reynolds number:

$$\mu = \frac{\rho V D}{\text{Re}} \tag{7.129}$$

$$\mathcal{F}_{\text{drag}} \left(\frac{1}{\frac{1}{2}\rho V^2} \frac{1}{D^2} \right) \text{Re} = -\int_0^{L/D} \int_0^{2\pi} \left(\frac{\partial v_z^*}{\partial r^*} + \frac{\partial v_r^*}{\partial z^*} \right)\bigg|_{r^*=1/2} d\theta dz^* \tag{7.130}$$

It remains to nondimensionalize the wall drag, $\mathcal{F}_{\text{drag}}$.

We could choose to define the characteristic wall drag as the quantities in the denominator on the lefthand side of Equation 7.130. The units are correct and the quantity $\rho V^2/2$ is a type of kinetic energy per unit volume characteristic of the flow; it is appealing to use in our nondimensionalization process a quantity that has physical meaning. The area D^2 is not particularly meaningful when it comes to wall drag, however, because D^2 is related to the pipe cross-sectional area $(\pi D^2/4)$. A more meaningful area would be the wetted area of the pipe—that is, the wall surface area that is in contact with the fluid. The wetted area is the actual area over which the fluid exerts drag; thus, it is preferable to nondimensionalize the wall drag with this area rather than with the pipe cross section.

Thus, we choose (arbitrarily but with some justification) to define a dimensionless drag on the wall as follows:

Dimensionless
drag on wall:
$$f \equiv \frac{\text{wall force}}{\left(\dfrac{\text{kinetic energy}}{\text{volume}} \right) (\text{characteristic area})} \tag{7.131}$$

$$= \frac{\mathcal{F}_{\text{drag}}}{\left(\frac{1}{2}\rho V^2 \right) (2\pi R L)} \tag{7.132}$$

Recall Equation 1.89; we take the wetted area to be the characteristic area. With this definition for dimensionless wall drag f, Equation 7.130 becomes:

$$f = \mathcal{F}_{\text{drag}} \left(\frac{1}{\frac{1}{2}\rho V^2} \frac{1}{2\pi RL} \right) \tag{7.133}$$

$$f = \frac{1}{\text{Re}} \frac{1}{L/D} \frac{1}{\pi} \int_0^{L/D} \int_0^{2\pi} -\left(\frac{\partial v_z^*}{\partial r^*} + \frac{\partial v_r^*}{\partial z^*} \right)\Bigg|_{r^*=1/2} d\theta\, dz^* \tag{7.134}$$

Note that the change from D^2 to $2\pi RL$ for characteristic area did not introduce any new dimensionless groups because L/D already was present in the limits of the integral. The quantity f is called the *Fanning friction factor*, which appears in Chapter 1 and in Equation 7.59 for the burst-pipe example. Equation 7.131 is the formal definition of the Fanning friction factor, which is a dimensionless force on the wall of a tube:[5]

$$\begin{array}{c} \text{Fanning} \\ \text{friction factor} \end{array} \qquad \boxed{f \equiv \frac{\mathcal{F}_{\text{drag}}}{\left(\frac{1}{2}\rho V^2\right)(2\pi RL)}} \tag{7.135}$$

If we use the macroscopic-balance result for wall drag in turbulent flow, $\mathcal{F}_{\text{drag}} = \Delta p \pi R^2$ (Equation 7.116), we can write f in terms of experimental variables—that is, in terms of Δp:

$$\begin{array}{c} \text{Fanning friction} \\ \text{factor from} \\ \text{experimental variables} \\ \text{(pipe flow):} \end{array} \qquad \boxed{f = \frac{(p_0 - p_L)D}{2\rho V^2 L}} \tag{7.136}$$

This equation is given in Equations 1.91 and 7.59. As stated previously, the relationship $\mathcal{F}_{\text{drag}} = \Delta p \pi R^2$ can be calculated directly from the velocity profile for laminar flow (Equation 7.115). In Chapter 9, we derive the identical result for turbulent flow (see Equation 9.236) from the macroscopic momentum balance.

Our nondimensionalization exercise is complete and the final result of the analysis is the determination that dimensionless wall drag f is a function of dimensionless variables (i.e., v_z^*, p^*, r^*, and so on) and three dimensionless scale factors (i.e., Re, Fr, and L/D):

$$\begin{array}{c} \text{Dimensionless} \\ \text{Navier-Stokes:} \end{array} \qquad \boxed{\frac{\partial \underline{v}^*}{\partial t^*} + \underline{v}^* \cdot \nabla^* \underline{v}^* = -\nabla^* p^* + \left(\frac{1}{\text{Re}}\right) \nabla^{*2} \underline{v}^* + \left(\frac{1}{\text{Fr}}\right) \underline{g}^*}$$

$$\tag{7.137}$$

$$\begin{array}{c} \text{Dimensionless} \\ \text{wall drag:} \end{array} \qquad \boxed{f = \frac{1}{\text{Re}} \frac{1}{L/D} \frac{1}{\pi} \int_0^{L/D} \int_0^{2\pi} -\left(\frac{\partial v_z^*}{\partial r^*} + \frac{\partial v_r^*}{\partial z^*} \right)\Bigg|_{r^*=1/2} d\theta\, dz^*}$$

$$\tag{7.138}$$

---

[5] An alternate definition of friction factor is used in engineering literature. The Darcy or Moody friction factor Λ [174] differs from the Fanning friction factor by a factor of 4: $4f = \Lambda$. Care should be exercised when reading friction-factor values in the literature; it is important to know which definition is being used.

The integral in Equation 7.138 may be evaluated for a known L/D if we know the function $\underline{v}^*(r^*, \theta, z^*)$, which is obtained by solving the dimensionless microscopic mass and momentum balances. The microscopic balances are a function of only two dimensionless groups: the Reynolds number and the Froude number. Thus, the friction factor for a particular situation is governed by three dimensionless quantities—L/D, Re, and Fr:

$$f = f\left(\text{Re, Fr, } \frac{L}{D}\right) \tag{7.139}$$

This final result is powerful and practical when combined with experimental results, as discussed in the following section.

7.1.2.3 DATA CORRELATIONS

To summarize our strategy thus far, we are looking for a way to quantify pressure drop as a function of flow rate for turbulent flows in pipes. This kind of information is needed to solve problems like the burst-pipe problem in Example 7.5. We started with laminar flow in pipes to gain familiarity with the physics of the situation. In laminar flow in pipes, the flow-rate/pressure-drop relationship is the Hagen-Poiseuille equation, Equation 7.28. However, the Hagen-Poiseuille equation is of no practical use in turbulent flow because it underestimates the frictional drag on the walls in the pipe enough to overpredict flow rates in the burst-pipe example by orders of magnitude.

Dimensional analysis of the problem of flow in pipes demonstrates that dimensionless wall drag in the form of the Fanning friction factor f is a function of at most the Reynolds number, Froude number, and L/D (Equation 7.139). The correlation $f(\text{Re, Fr, } L/D)$ is a dimensionless version of the flow-rate/pressure-drop correlation we seek, $\Delta p(Q)$. To determine the form of the function $f = f(\text{Re, Fr, } L/D)$ for turbulent flow in pipes, we must experiment.

For measurements on a known fluid (i.e., known ρ and μ), in a known apparatus, we can determine experimentally the friction factor from the pressure drop and the average velocity $\langle v \rangle = V$ as follows. Beginning with the definition of f:

$$f = \frac{\mathcal{F}_{\text{drag}}}{\left(\frac{1}{2}\rho V^2\right)(2\pi RL)} \tag{7.140}$$

we substitute Equation 7.116, which relates $\mathcal{F}_{\text{drag}}$ to pressure drop for both laminar and turbulent flow. The result is f in terms of experimental variables:

$$f = \frac{(p_0 - p_L)\pi R^2}{\left(\frac{1}{2}\rho V^2\right)(2\pi RL)} \tag{7.141}$$

Fanning friction
factor from
experimental variables
(pipe flow):

$$\boxed{f = \frac{(p_0 - p_L)}{2\rho V^2 L/D}} \tag{7.142}$$

We measure pressure drop and flow rate on an apparatus that sends fluid through a section of straight pipe. We then calculate the experimental Reynolds number and the Froude number from the data, incorporating material and geometric constants that correspond to our apparatus:

$$\text{Re} = \frac{\rho V D}{\mu} \tag{7.143}$$

$$\text{Fr} = \frac{V^2}{gD} \tag{7.144}$$

Thus, data of pressure drop versus average velocity or flow rate ($Q = \pi R^2 V$) can be plotted as friction factor versus the three dimensionless quantities: Reynolds number, Froude number, and L/D. The power of Equation 7.139 is that measurement of $f(\text{Re}, \text{Fr}, L/D)$ for a single system yields the function that describes all straight-pipe systems. This is a spectacularly powerful result.

Careful experiments on flow in straight, smooth pipes were conducted in the 1800s and 1900s [139]. Researchers found that Equation 7.139 holds for pipes of all sizes and lengths and for various fluids. They further determined that friction factor is independent of Froude number for incompressible flow in full pipes,[6] implying that gravity is not important in horizontal pipes. The hydrostatic effect of gravity in tilted or vertical pipe flow may be combined with pressure by considering a dynamic pressure (see Problem 10, Equation 8.115, and the Glossary). It also was found experimentally that the ratio L/D is not important in determining the friction factor for long pipes, $L/D > 40$ [43]. Together these experiments reveal that the important relationship for flow in long pipes is only between the friction factor and the Reynolds number. Nikuradse's data revealing this measured relationship are plotted in Figure 7.17 [126].

The friction-factor/Reynolds-number relationship for flow in pipes is a well-known data correlation, perhaps the best-known data correlation in fluid mechanics. The data show the three flow regimes observed by Osborne Reynolds in 1883 [139]: laminar flow, transitional flow, and turbulent flow. At low flow rate (i.e., low Re), the friction factor plunges steeply with increasing Reynolds number; this is the laminar-flow regime. Above $\text{Re} = 4,000$, the friction factor is a more gradually changing function of Reynolds number; this is the turbulent-flow regime. Between $\text{Re} \approx 2,100$ and $\text{Re} \approx 4,000$, the flow is neither laminar nor fully turbulent; this region is called transitional flow. In the transitional regime, the friction-factor data show a sensitivity to experimental conditions. Below $\text{Re} = 2,100$, the friction-factor follows the law $f = 16/\text{Re}$, which can be predicted from the Hagen-Poiseuille equation (see Example 7.7). Above $\text{Re} = 4,000$,

---

[6] The Froude number reflects the importance of nonhydrostatic gravity effects in the Navier-Stokes equation. Experiments show that these effects become important only when fluids of different densities are present, such as in a half-full pipe or in open channels where there are waves [178].

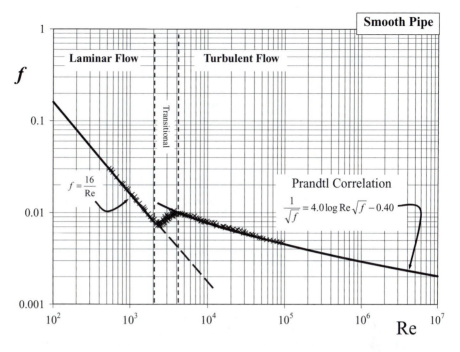

Figure 7.17 Friction factor as a function of Reynolds number for Newtonian fluids in smooth pipes (representative data from Nikuradse [126]). Three regimes are shown: laminar flow (Re < 2,100), turbulent flow (Re > 4,000), and transitional flow (2,100 < Re < 4,000). See the discussion for Equation 7.156 for more about the Prandtl friction-factor/Reynolds-number correlation for turbulent flow.

the data follow the Prandtl correlation (Equation 7.58); the data for all Reynolds numbers can be summarized by the correlations given here:

Data correlation for friction factor for pipe flow (all flow regimes):

Re	f
Re < 2,100	$\dfrac{16}{Re} = 16\,Re^{-1}$
$2,100 \le Re \le 4,000$	unstable
$4,000 \le Re \le 1 \times 10^{6}$	$\dfrac{1}{\sqrt{f}} = 4.0 \log Re\sqrt{f} - 0.40$
	or
	$f = \frac{1.02}{4} \log Re^{-2.5}$

$$(7.145)$$

See also the single-equation correlation in Figure 7.18 (see Equation 7.158).

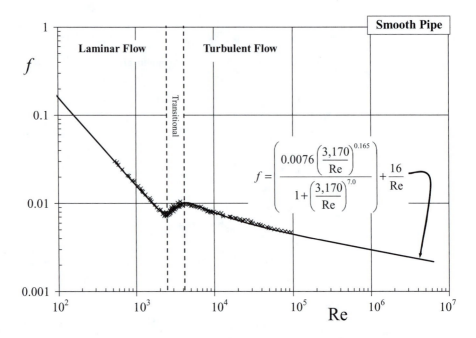

Figure 7.18 Equation 7.158 [105] captures the smooth-pipe friction factor as a function of the Reynolds number over the entire Reynolds-number range. Also shown are Nikuradse's experimental data for flow in smooth pipes [126].

EXAMPLE 7.7. *What is the predicted friction-factor/Reynolds-number relationship in steady laminar flow in a tube?*

SOLUTION. We calculated the flow-rate/pressure-drop relationship for laminar flow in a tube to be:

Hagen-Poiseuille equation
(pressure drop/flow rate
for laminar tube flow):

$$Q = \frac{\pi (p_0 - p_L) R^4}{8 \mu L} \qquad (7.146)$$

Writing the Hagen-Poiseuille equation in terms of average velocity V, we obtain:

$$V = \langle v \rangle = \frac{Q}{\pi R^2} = \frac{(p_0 - p_L) R^2}{8 \mu L} \qquad (7.147)$$

The Fanning friction factor written in terms of experimental pressure drop is given in Equation 7.142:

$$\text{Fanning friction factor} \quad f = \frac{(p_0 - p_L)}{2 \rho V^2 L / D} \qquad (7.148)$$

We now eliminate the pressure drop between these two equations and simplify:

$$f = \frac{(p_0 - p_L)}{2 \rho V^2 L / D} \qquad (7.149)$$

$$= \frac{D}{2 \rho V^2 L} \frac{8 V \mu L}{R^2} \qquad (7.150)$$

$$= \frac{16 \mu}{\rho V D} \qquad (7.151)$$

$$= \frac{16}{\text{Re}} \qquad (7.152)$$

The function $f(\text{Re})$ is usually plotted on a log-log scale. Taking the log of both sides of Equation 7.152, we obtain:

$$f = \frac{16}{\text{Re}} \tag{7.153}$$

$$\log f = \log 16 - \log \text{Re} \tag{7.154}$$

Thus, the friction-factor/Reynolds-number relationship for laminar flow in a tube, when plotted log-log, is a straight line of slope equal to -1 and an intercept of log 16. This is precisely what is on the left side of Figure 7.17:

$$\begin{array}{c}\text{Fanning friction}\\ \text{factor for}\\ \text{laminar tube flow}\end{array} \qquad \boxed{f = \frac{16}{\text{Re}}} \tag{7.155}$$

The success of dimensional analysis in pipe flow is remarkable. By nondimensionalizing the Navier-Stokes equation and the equation for drag on the tube walls, we can conclude that the dimensionless wall force (f) correlates with the dimensionless groups Re, Fr, and L/D. Experiments confirm this and tell us further that Fr is unimportant for closed full tubes and that L/D is unimportant for long pipes (see also Section 7.3.3 and Figure 7.50).

For Re > 4,000, the turbulent-flow friction-factor/Reynolds-number data are well represented by a correlation equation that arises from the work of von Kármán, Prandtl, and Nikuradse [174]. This equation was derived by Prandtl from measurements of the average velocity profile in turbulent flow. Once the velocity profile is known, we can integrate that result to obtain the flow rate Q. When combined with the definition of friction factor and adjusted for a better fit, Equation 7.156 results, as discussed in detail by White [174]:

$$\begin{array}{c}\text{Prandtl correlation}\\ \text{for } f(\text{Re})\\ \text{(smooth pipes only,}\\ \text{turbulent flow):}\end{array} \qquad \boxed{\frac{1}{\sqrt{f}} = 4.0 \log\left[\text{Re}\sqrt{f}\right] - 0.40} \tag{7.156}$$

This equation was introduced previously in the burst-pipe example, Equation 7.58. Note that f is present on both sides of Equation 7.156. A modified version of Equation 7.156 that is explicit in friction factor may be used for convenience [174]:

$$f \approx \frac{1.02}{4} \log \text{Re}^{-2.5} \tag{7.157}$$

Equations 7.157 and 7.156 differ by up to ±3 percent. A data correlation that fits smooth-pipe data over the entire range of Reynolds numbers was developed by Morrison [105]:

$$\begin{array}{c}f(\text{Re}) \text{ smooth pipes}\\ \text{(all Reynolds numbers):}\end{array} \qquad \boxed{f = \left(\frac{0.0076 \left(\frac{3{,}170}{\text{Re}}\right)^{0.165}}{1 + \left(\frac{3{,}170}{\text{Re}}\right)^{7.0}}\right) + \frac{16}{\text{Re}}} \tag{7.158}$$

This equation follows the analytical laminar result at low Reynolds numbers (i.e., $f = 16/\text{Re}$) and the Prandtl equation at high Reynolds numbers (see Figure 7.18).

It is remarkable that through dimensional analysis we can collapse pressure drop versus flow-rate data onto a single curve of f versus Re for flow in smooth tubes of any size for any Newtonian fluid. At the end of this section, we provide an example in branched piping (see Example 7.9), demonstrating the utility of our data-correlation results. The success of the friction-factor/Reynolds-number correlation is due to the powerful technique of dimensional analysis.

The first step in dimensional analysis is the key to success: identifying the correct relationships that govern the physics of the problem under consideration. In the pipe-flow case, these are the Navier-Stokes equation and the equation for force on the wall (see Equation 7.117). A second requirement is to choose reasonable characteristic values—for example, velocity V, length D, time D/V, and pressure ρV^2. If an important aspect of the physics is missed, dimensional analysis will not succeed, as shown in the next example.

EXAMPLE 7.8. *The experimental data in Figure 7.17 is for smooth pipes. Nikuradse [126] performed experiments on a rough-walled pipe to obtain the data in Figure 7.19, which are different from the smooth-wall results. Why did the curves for the different pipes not collapse to one curve, as we might expect from the previous dimensional-analysis discussion?*

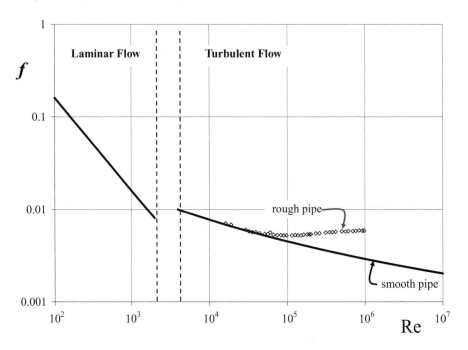

Figure 7.19   In the turbulent region, the data on f versus Re for a rough pipe are higher than the data for a smooth pipe. Data shown were taken by Nikuradse in a pipe that was roughened by gluing sand to the walls [126].

SOLUTION. Dimensional analysis on flow in a pipe indicated that wall drag should be a function of Re and Fr through the solution $\underline{v}(r^*, \theta, z^*, t^*)$ of the

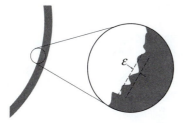

A close look at a rough pipe wall reveals that roughness introduces a new lengthscale into the problem.

dimensionless Navier-Stokes equation; it also should be a function of L/D through the boundary conditions used to evaluate wall drag (i.e., the limit on the integral; see Equation 7.138).

$$f = \frac{1}{\text{Re}} \frac{1}{L/D} \frac{1}{\pi} \int_0^{L/D} \int_0^{2\pi} -\left(\frac{\partial v_z^*}{\partial r^*} + \frac{\partial v_r^*}{\partial z^*} \right)\bigg|_{r^*=1/2} d\theta \, dz^* \quad (7.159)$$

The result $f = f(\text{Re}, \text{Fr}, L/D)$ should be good for all systems described by Equation 7.159.

For a smooth pipe, Equation 7.159 is correct because $r^* = 1/2$ or $r = D/2$ is an accurate description of the wall-surface location in this case. For a rough pipe, however, the integration in Equation 7.159 is not quite correct. The roughness

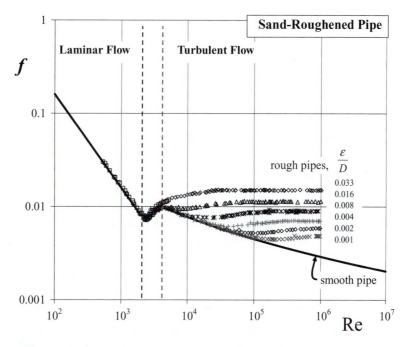

Nikuradse [126] quantified the effect of the size of wall protuberances on the friction factor by attaching a well-characterized sand to the inner walls of pipes. The data show that the friction factor is characterized by two dimensionless groups, Re and ε/D. Furthermore, f is independent of pipe roughness for laminar flow. At high Reynolds numbers, the friction factor is independent of the Reynolds number and depends on only pipe roughness.

Table 7.1. Manufactured pipes have different values of roughness ε depending on construction material

Material of construction	ε (mm)
Drawn tubing (brass, lead, glass)	1.5×10^{-3}
Commercial steel or wrought iron	0.05
Asphalted cast iron	0.12
Galvanized iron	0.15
Cast iron	0.46
Woodstove	0.2–0.9
Concrete	0.3–3.0
Riveted steel	0.9–9.0

The values are reported in the literature [132]. Colebrook [25] correlated friction-factor measurements on manufactured and sand-roughened pipes to obtain equivalent values of roughness for the manufactured pipes.

of the pipe surface introduces a new lengthscale to the problem. We have not accounted for this new lengthscale (Figure 7.20).

To solve for the friction on a rough wall, we must perform the integral in Equation 7.159, but the velocity-gradient terms must be evaluated at the rough-wall surface $r = \psi(\theta, z)$ or $r^* = \psi^*(\theta, z^*)$ rather than at the smooth wall surface of $r = D/2$ or $r^* = 1/2$:

$$
f = \frac{1}{\mathrm{Re}} \frac{1}{L/D} \frac{1}{\pi} \int_0^{L/D} \int_0^{2\pi} -\left(\frac{\partial v_z^*}{\partial r^*} + \frac{\partial v_r^*}{\partial z^*} \right)\Bigg|_{r^* = \psi^*(\theta, z^*)} d\theta dz^* \qquad (7.160)
$$

If the function $\psi(\theta, z)$ that describes the shape of the wall surface has a single characteristic dimension ε (see Figure 7.20), then nondimensionalization of this function results in a new dimensionless group ε/D that characterizes the surface shape or roughness. Pipes characterized by different values of ε/D have different curves of $f(\mathrm{Re}, \varepsilon/D)$.

In summary, the differences between rough-pipe and smooth-pipe data are predicted by dimensional analysis when the additional roughness lengthscale ε is included in the analysis. To quantify the pipe-roughness effect, Nikuradse [126] performed careful experiments on pipes artificially roughened by attaching sand to the inner walls. His results (Figure 7.21) established the validity of ε/D as the additional controlling dimensionless parameter for flow through rough pipes. This is another success for dimensional analysis.

A final point on rough pipes: Data on commercial rough pipes are similar to Nikuradse's data at larger Reynolds number but show a different shape at lower Re. Colebrook [25] gathered literature data on commercial rough pipes and deduced equivalent values of ε for actual pipes by matching the large-Re asymptotes between the commercial data and Nikuradse's data (Table 7.1). Colebrook's correlation for rough commercial pipes can be used for accurate computer

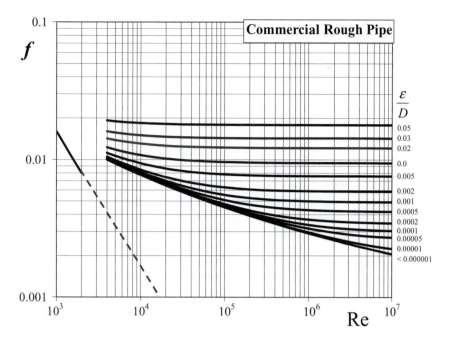

Figure 7.22 Fanning friction factor versus Reynolds number for flow in smooth and rough commercial pipes of circular cross section. After Moody [103]; calculated from the Colebrook correlation [25] (see Equation 7.161). Note that the shapes of the curves for commercial pipes are different from the data for sand-roughened pipes in the turbulent region at Reynolds numbers before f becomes independent of Re.

calculations of friction factor in commercial rough pipes (see Problem 17 for more discussion):

Colebrook correlation
for $f(\mathrm{Re})$
(smooth and rough pipes,
turbulent flow):

$$\frac{1}{\sqrt{f}} = -4.0 \log \left(\frac{\varepsilon}{D} + \frac{4.67}{\mathrm{Re}\sqrt{f}} \right) + 2.28$$

(7.161)

The Colebrook equation is introduced in Chapter 1 as Equation 1.95. A summary plot of Colebrook's correlation is shown in Figure 7.22 as plotted by Moody [103]. The Colebrook correlation also works for smooth pipes ($\varepsilon = 0$), where it reduces to the Prandtl correlation.

The previous example emphasizes that dimensional analysis works only if the physics of the problem is incorporated correctly. To determine what the correct physics is, scientists and engineers use their judgment to propose a model for a system; then, they nondimensionalize the equations, seeking predictions that can be tested. Subsequently, they perform experiments and the results indicate whether the assumptions in the model are correct or if the analysis must be modified. Data correlations can be determined from experimental results if the dimensional analysis succeeds.

The Prandtl (i.e., smooth pipe) and Colebrook (i.e., rough pipe) correlations are useful in a wide variety of practical problems involving turbulent flows in pipes, as shown in the following branched-piping example. The overall solution method used in this chapter also is widely applicable. Stated succinctly, the method is to solve practical problems using mixed analytical/empirical correlations arrived at through dimensional analysis of idealized problems:

Method for Solving Complex Engineering Problems

1. Devise a related, idealized problem that may be solved analytically.
2. Use dimensional analysis on the idealized problem to determine the governing parameters.
3. Perform experiments varying the governing parameters.
4. Obtain accurate data correlations among the governing parameters for future problem solving.

In Section 7.2, we apply this approach to the flow of Newtonian fluids in closed conduits with noncircular cross sections. In Chapter 8, we apply this same approach to external flows, obtaining correlations for drag coefficients.

EXAMPLE 7.9. *A 40-foot section of 1.0-inch ID piping branches into two pipes of the same diameter, one of which is 60.0 feet long and one of which is 85.0 feet long (Figure 7.23). The main pipe is connected to the municipal water supply, which supplies a constant water pressure of 62 psig at the pipe entrance. What are the flow rates through the two pipe exits? What is the pressure at the splitting point? Assume smooth pipes.*

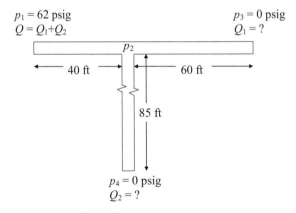

$p_1 = 62$ psig
$Q = Q_1 + Q_2$

$p_3 = 0$ psig
$Q_1 = ?$

p_2

40 ft   60 ft

85 ft

$p_4 = 0$ psig
$Q_2 = ?$

Figure 7.23 A simple pipe branch splits flow unevenly because the two branches do not resist the flow equally. The Prandtl correlation may be used to solve for the correct flow split, as discussed in this chapter.

SOLUTION. Pressure is the driving force for the flow, and the set driving pressure of 62 psig causes a certain flow rate for the water in the system. That flow rate divides unevenly into the two exit pipes, depending on resistances to flow in the pipes. The two exit pipes are the same diameter but they have different lengths. The resistance to the flow is due to the frictional drag at the walls of the pipe; thus, longer pipes have more resistance to flow. From these considerations, we

Material Properties:		
density	$\rho=$	1,000 kg/m^3
viscosity	$\mu=$	0.001 Pa s
given	$D=$	0.0254 m
At Branch Point, Guess Δp_2	$\Delta p_2=$	**20.9 psi**
change units	$\Delta p_2=$	144,100 Pa
given	$L_1=$	60 ft
change units	$L_1=$	18.3 m
calculate V_1 from Prandtl (no iteration)	LHS=	15.2199
$f = 1/\text{LHS}^2$	$f_1=$	0.004317
calculate from f	$V_1=$	4.8146 m/s
Flow Rate in Branch 1	$Q_1=$	**0.002440 m^3/s**
given	$L_2=$	85 ft
change units	$L_2=$	25.9 m
calculate V_2 from Prandtl (no iteration)	LHS=	14.9174
$f = 1/\text{LHS}^2$	$f_2=$	0.004494
calculate from f	$V_2=$	3.9647 m/s
Flow Rate in Branch 2	$Q_2=$	**0.002009 m^3/s**
main branch: total $Q=$		0.004449 m^3/s
change units	$Q=$	71 gpm
$V=4Q/\pi D^2$	$V=$	8.779 m/s
calculate Re from definition	Re=	222,995
Calculate f from Prandtl Correlation, Guess f		**0.003830**
iterate to final answer	$g(f)$	16.16
next round $f = 1/g\,(f)^2$	$f_{\text{final}}=$	0.003830
given	$L=$	40 ft
change units	$L=$	12.2 m
calculate Δp_1 from friction factor	$\Delta p_1=$	2.83E+05 Pa
pressure drop across the main tube	$\Delta p_1=$	41 psi
total Δp (given)	$\Delta p_{\text{total}}=$	62 psi
Pressure Drop Across the Branch Tubes	$\Delta p_2=$	**20.9 psi**

Figure 7.24

The iterative solution for the flow rates in branched-piping networks can be carried out in a spreadsheet program. The steps of the solution in this example are described in this chapter. The definitions of LHS and $g(f)$ are given in Figure 7.15: $LHS = 4 \log \sqrt{\Delta p D^3 \rho/(2L\mu^2)} - 0.4$ and $g(f) = 4 \log \left(\text{Re}\sqrt{f} \right) - 0.40$.

expect the proportion of the flow going to the shorter pipe to be larger than the proportion going to the longer pipe.

If we assume that the flow in the pipes is turbulent and that the pipes are smooth, then we can calculate the flow rate in each section of pipe using the Prandtl correlation. We ignore the frictional losses from any fittings or valves in the lines. Because the flow-rate/pressure-drop relationship for turbulent flow is nonlinear, we must perform iterative calculations to solve for the pressures and flow rates.

The pressure at the split p_2 determines the pressure drop across both exit pipe sections. Our iterative solution begins with a guess for p_2, which allows us to calculate the overall flow rate. Pressure drop over the main pipe, Δp_1, can be calculated from the flow-rate/pressure-drop relationship applied to that pipe; iteration yields the final solution. Details of the calculation are illustrated in Figure 7.24 and listed here. We assume turbulent flow and check that Re > 4,000 at the end of the calculation.

Solution Steps

1. Guess Δp_2, which is equal to both $p_2 - p_3$ and $p_2 - p_4$.
2. From the Prandtl correlation, calculate Q_1 and Q_2 for the two exit branches. As discussed in a previous example, calculating flow rate for a known pressure drop does not require iteration.
3. The total flow rate is the sum of the flows in the two branches, $Q = Q_1 + Q_2$.
4. From the predicted value of Q, calculate the pressure drop across the main pipe, $\Delta p_1 = p_1 - p_2$. This is an iterative calculation.
5. Calculate a revised Δp_2 from $\Delta p_2 = 62 \text{ psig} - \Delta p_1$.
6. Iterate until the values of Δp_2 converge.

From the solutions for flow rates Q_1 and Q_2 in Figure 7.24, we calculate that $Q_1/Q = 55$ percent of the flow goes toward the shorter pipe and $Q_2/Q = 45$ percent of the flow goes in the longer pipe. The pressure drop across the main pipe is $\Delta p_1 = 41$ psig, and the pressure at the split point is $62 \text{ psig} - 41 \text{ psig} = 21 \text{ psig}$.

More complex piping networks can be solved with the same equations; some amount of problem-solving strategy is needed to solve complex piping networks. More information about strategies for flow-rate/pressure-drop problems is in the literature [176].

The methods of this section may be extended to other problems. In Section 7.2, we apply these techniques to flows in noncircular conduits; in Section 7.3, we consider more complex internal flows. Chapter 8 applies the methods in this chapter to external flows.

7.2 Noncircular conduits

The flows in closed rectangular conduits (e.g., ducts and slits) or in closed conduits of other shapes (Figure 7.25) occur in engineering applications such as heat-exchanger and reactor design as well as in cutting-edge research fields such

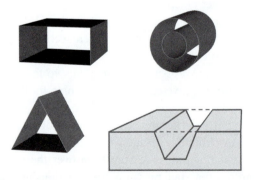

Figure 7.25   Noncircular conduits are common devices such as a rectangular duct or a double-pipe heat-exchanger (i.e., flow through an annulus); more specialized devices are pipes with a triangular or trapezoidal cross section. The trapezoidal cross section is produced in microfluidic devices as a consequence of the manufacturing process [75]. Flow through a packed bed may be modeled as flow in a closed conduit with an irregular cross section (see Section 7.2.1.2).

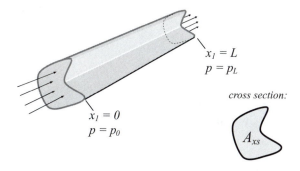

Figure 7.26 Pressure-driven laminar flow in a conduit may be analyzed with the geometry shown here.

as the development of microfluidic devices for medical research [75]. These flows are similar to the flow in pipes—an imposed pressure drop results in a flow rate that is determined by the amount of momentum lost to drag on the walls or other solid surfaces. We can extend our analysis of pipe flow to these new geometries.

As with pipe flow, the flows in noncircular conduits can be laminar or turbulent. Following the method established in this chapter, we begin with the simpler case of laminar flow. Subsequently, we perform dimensional analysis and use experiments to develop data correlations for turbulent flow in noncircular conduits.

7.2.1 Laminar flow in noncircular ducts

Steady flows through ducts of noncircular cross section share much in common with pipe flow. For laminar duct flow, we analyze the problem following the same steps as for pipe flow, beginning with the microscopic momentum balance. We begin by considering the problem generally and subsequently address specific geometries.

7.2.1.1 POISSON EQUATION

Consider the pressure-driven flow of a Newtonian fluid through a long duct of an arbitrary cross-sectional shape (Figure 7.26). The flow is assumed steady, well developed, and incompressible.

We begin with the incompressible-fluid continuity equation written in Cartesian coordinates, $x_1x_2x_3$, with the flow direction being x_1:

$$\text{Mass conservation:} \atop \text{(continuity equation,} \atop \text{constant density)} \qquad \boxed{0 = \nabla \cdot \underline{v}}$$

$$0 = \frac{\partial v_1}{\partial x_1} + \frac{\partial v_2}{\partial x_2} + \frac{\partial v_3}{\partial x_3} \qquad (7.162)$$

Because the flow is only in the x_1-direction, the 2- and 3-components of $\underline{v}$ are zero:

$$\underline{v} = \begin{pmatrix} v_1 \\ 0 \\ 0 \end{pmatrix}_{123} \qquad (7.163)$$

After making these cancellations, the continuity equation gives us:

$$\frac{\partial v_1}{\partial x_1} = 0 \qquad (7.164)$$

We also saw this result for pipe flow (compare to Equation 7.8).

Momentum conservation is given by the equation of motion for an incompressible Newtonian fluid: the Navier-Stokes equation. The components of the Navier-Stokes equation in Cartesian coordinates are in Table B.7 in Appendix B:

Momentum conservation: (Navier-Stokes equation)

$$\rho \left(\frac{\partial \underline{v}}{\partial t} + \underline{v} \cdot \nabla \underline{v} \right) = -\nabla p + \mu \nabla^2 \underline{v} + \rho \underline{g}$$

$$\rho \left(\frac{\partial v_1}{\partial t} + v_1 \frac{\partial v_1}{\partial x_1} + v_2 \frac{\partial v_1}{\partial x_2} + v_3 \frac{\partial v_1}{\partial x_3} \right)$$
$$= -\frac{\partial p}{\partial x_1} + \mu \left(\frac{\partial^2 v_1}{\partial x_1^2} + \frac{\partial^2 v_1}{\partial x_2^2} + \frac{\partial^2 v_1}{\partial x_3^2} \right) + \rho g_1 \qquad (7.165)$$

$$\rho \left(\frac{\partial v_2}{\partial t} + v_1 \frac{\partial v_2}{\partial x_1} + v_2 \frac{\partial v_2}{\partial x_2} + v_3 \frac{\partial v_2}{\partial x_3} \right)$$
$$= -\frac{\partial p}{\partial x_2} + \mu \left(\frac{\partial^2 v_2}{\partial x_1^2} + \frac{\partial^2 v_2}{\partial x_2^2} + \frac{\partial^2 v_2}{\partial x_3^2} \right) + \rho g_2 \qquad (7.166)$$

$$\rho \left(\frac{\partial v_3}{\partial t} + v_1 \frac{\partial v_3}{\partial x_1} + v_2 \frac{\partial v_3}{\partial x_2} + v_3 \frac{\partial v_3}{\partial x_3} \right)$$
$$= -\frac{\partial p}{\partial x_3} + \mu \left(\frac{\partial^2 v_3}{\partial x_1^2} + \frac{\partial^2 v_3}{\partial x_2^2} + \frac{\partial^2 v_3}{\partial x_3^2} \right) + \rho g_3 \qquad (7.167)$$

To convert the xyz-coordinate system of Table B.7 to our $x_1 x_2 x_3$ system, we write $x = x_1$, $y = x_2$, and $z = x_3$.

We now cancel all terms involving v_2, v_3, or spatial derivatives of v_1 with respect to x_1 (from the continuity-equation result). We also neglect gravity.[7] Making these substitutions, we obtain:

$$\text{1-Component:} \qquad \rho \frac{\partial v_1}{\partial t} = -\frac{\partial p}{\partial x_1} + \mu \left(\frac{\partial^2 v_1}{\partial x_2^2} + \frac{\partial^2 v_1}{\partial x_3^2} \right) \qquad (7.168)$$

$$\text{2-Component:} \qquad 0 = -\frac{\partial p}{\partial x_2} \qquad (7.169)$$

$$\text{3-Component:} \qquad 0 = -\frac{\partial p}{\partial x_3} \qquad (7.170)$$

The 2- and 3-components of the Navier-Stokes equation indicate that there are no variations of pressure in either the x_2- or x_3-direction. The 1-component of

---

[7] See Problem 10 for a method of including the effect of the flow-direction component of gravity in nonhorizontal ducts.

the Navier-Stokes equation reveals the most about the flow:

$$\text{1-Component:} \qquad \rho \frac{\partial v_1}{\partial t} = -\frac{\partial p}{\partial x_1} + \mu \left(\frac{\partial^2 v_1}{\partial x_2^2} + \frac{\partial^2 v_1}{\partial x_3^2} \right) \qquad (7.171)$$

We can simplify this expression by noting that because the flow is at steady state, the time derivative on the lefthand side is zero:

$$0 = -\frac{\partial p}{\partial x_1} + \mu \left(\frac{\partial^2 v_1}{\partial x_2^2} + \frac{\partial^2 v_1}{\partial x_3^2} \right) \qquad (7.172)$$

For the flow in a noncircular duct, v_1 is a function of two variables, x_2 and x_3. As we determined from the equation of motion, the pressure is not a function of either of these variables but rather is a function only of x_1. Placing the pressure on one side of the equation and the velocity terms on the other, we can separate the variable x_1 from x_2 and x_3:

$$\frac{\partial p(x_1)}{\partial x_1} = \mu \left(\frac{\partial^2 v_1}{\partial x_2^2} + \frac{\partial^2 v_1}{\partial x_3^2} \right) \qquad (7.173)$$

Because pressure is a function only of x_1 and velocity is a function only of x_2 and x_3 (not x_1), Equation 7.173 can hold only if the two sides are equal to the same constant. We call that constant λ, and we now have two equations to solve separately (see the Web appendix [108] for more details):

$$\text{Lefthand side:} \qquad \frac{dp}{dx_1} = \lambda \qquad (7.174)$$

$$\text{Righthand side:} \qquad \mu \left(\frac{\partial^2 v_1}{\partial x_2^2} + \frac{\partial^2 v_1}{\partial x_3^2} \right) = \lambda \qquad (7.175)$$

We changed the partial derivative symbol ∂ to the total derivative symbol d in the equation for pressure because $p = p(x_1)$ only. We cannot make the same change in the velocity equation because $v_1 = v_1(x_2, x_3)$; Equation 7.175 remains a partial differential equation (PDE).

We previously solved the pressure equation (see Equation 7.18) with the same boundary conditions ($x = 0$, $p = p_0$; $x = L$, $p = p_L$). The result is as follows (see also Equation 7.22):

$$\begin{array}{c} \text{Pressure profile} \\ \text{Poiseuille flow} \\ \text{in a duct:} \end{array} \qquad \boxed{ p(x_1) = \left(\frac{p_L - p_0}{L} \right) x_1 + p_0 } \qquad (7.176)$$

From Equations 7.176 and 7.174, we see that $\lambda = (p_L - p_0)/L$.

The PDE to solve for velocity in noncircular ducts is more complicated than the equation for tube flow because the velocity profile is three-dimensional:

$$\begin{array}{c} \text{Poisson equation:} \\ \text{Flow-direction} \\ \text{momentum balance for} \\ \text{pressure-driven flow} \\ \text{in closed conduits} \end{array} \qquad \boxed{ \frac{\partial^2 v_1}{\partial x_2^2} + \frac{\partial^2 v_1}{\partial x_3^2} = -\frac{p_0 - p_L}{\mu L} = -\frac{\Delta p}{\mu L} } \qquad (7.177)$$

This PDE has been studied for many years. It is called the Poisson equation, and it belongs to a class of equations called elliptical PDEs.[8] Once the shape of a duct is known, Equation 7.177 may be solved analytically or numerically. Numerous solutions for laminar flow are given in Shah and London [152], including solutions for ducts with rectangular, triangular, elliptical, and even limaçon[9] cross sections. Once the velocity solution is known, the stress tensor can be calculated in the usual way from the Newtonian constitutive equation (see Table B.8):

$$
\text{Newtonian constitutive equation:} \quad \tilde{\underline{\underline{\tau}}} = \mu \left(\nabla \underline{v} + (\nabla \underline{v})^T \right) \tag{7.178}
$$

$$
= \begin{pmatrix} 0 & \mu\dfrac{\partial v_1}{\partial x_2} & \mu\dfrac{\partial v_1}{\partial x_3} \\[2mm] \mu\dfrac{\partial v_1}{\partial x_2} & 0 & 0 \\[2mm] \mu\dfrac{\partial v_1}{\partial x_3} & 0 & 0 \end{pmatrix}_{123} \tag{7.179}
$$

where we simplified Equation 7.179 using $v_2 = v_3 = 0$ and the continuity equation result, $\partial v_1/\partial x_1 = 0$. The total-stress tensor then is given by:

$$
\tilde{\underline{\underline{\Pi}}} = \tilde{\underline{\underline{\tau}}} - p\underline{\underline{I}} = \begin{pmatrix} -p(x_1) & \mu\dfrac{\partial v_1}{\partial x_2} & \mu\dfrac{\partial v_1}{\partial x_3} \\[2mm] \mu\dfrac{\partial v_1}{\partial x_2} & -p(x_1) & 0 \\[2mm] \mu\dfrac{\partial v_1}{\partial x_3} & 0 & -p(x_1) \end{pmatrix}_{123} \tag{7.180}
$$

Having the stress tensor $\tilde{\underline{\underline{\Pi}}}$ and the velocity field $\underline{v}$ allows us to calculate force on any surface from the integral of $\hat{n} \cdot \tilde{\underline{\underline{\Pi}}}$ over the surface (Equation 6.53).

In Example 7.10, we solve the Poisson equation for the velocity and stress fields for a simple shape: flow between infinite parallel plates. Subsequently, we discuss a more complex geometry, a duct of rectangular cross section. For all shapes, the mathematical problem is the same—the Poisson equation with no-slip boundary conditions—but the mathematical techniques required to arrive at the final solution can be quite sophisticated for all but the most symmetric geometries.

EXAMPLE 7.10. *Calculate the velocity profile, flow rate, and shear stress for pressure-driven flow of an incompressible Newtonian liquid between two infinitely wide, parallel plates separated by a gap of 2H. The pressure at an upstream point is p_0; at a point a distance L downstream the pressure is p_L. Assume that the flow between these two points is well developed and at steady state. Gravity may be neglected.*

---

[8] See the Glossary for more on the classification of PDEs.
[9] The *limaçon*, which looks like a lima bean, is a polar curve of the form $r = a + b\cos\theta$ (a and b are constants; r and θ are polar coordinates).

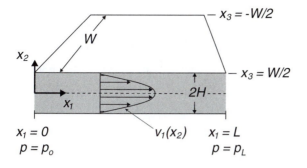

Figure 7.27 Schematic of Poiseuille flow in a wide slit.

SOLUTION. We apply the continuity equation and the Navier-Stokes equation, following the same procedure as in general noncircular ducts. The flow domain and coordinate system are shown in Figure 7.27.

Through the steps discussed in this section, we determine the differential equation for velocity and pressure fields. The resulting PDE is separable, and the pressure part is solved easily as before. The pressure profile for this problem is the same as in all steady unidirectional flows in ducts (Equation 7.176).

$$
\begin{array}{c}
\text{Pressure profile:} \\
\text{Poiseuille flow in a slit}
\end{array}
\qquad
\boxed{p(x_1) = \left(\frac{-\Delta p}{L}\right) x_1 + p_0}
\qquad (7.181)
$$

$$
\begin{array}{c}
\text{Velocity equation to solve:} \\
\text{Poiseuille flow in a slit}
\end{array}
\qquad
\boxed{\frac{\partial^2 v_1}{\partial x_2^2} + \frac{\partial^2 v_1}{\partial x_3^2} = -\frac{\Delta p}{\mu L}}
\qquad (7.182)
$$

Because the plates are infinite in width, we assume that there is no variation of any properties in the x_3-direction ($\partial v_1/\partial x_3 = 0$); therefore, for the slit, we obtain:

$$
\frac{\partial^2 v_1}{\partial x_2^2} = -\frac{\Delta p}{\mu L}
\qquad (7.183)
$$

Because v_1 is a function only of x_2, we change the partial-derivative symbol ∂ to the total-derivative symbol d and integrate Equation 7.183 twice:

$$
\frac{d^2 v_1}{dx_2^2} = -\frac{\Delta p}{\mu L}
\qquad (7.184)
$$

$$
\frac{dv_1}{dx_2} = -\frac{\Delta p}{\mu L} x_2 + C_1
\qquad (7.185)
$$

$$
v_1 = -\frac{\Delta p}{2\mu L} x_2^2 + C_1 x_2 + C_2
\qquad (7.186)
$$

where C_1 and C_2 are arbitrary integration constants.

The boundary conditions are no-slip at the two walls: $x_2 = H, -H$. Furthermore, halfway between the planes is a plane of symmetry, which means that v_1 must go through a maximum or a minimum at this plane—that is, the derivative of v_1 with respect to x_2 must be zero at this plane. These three conditions

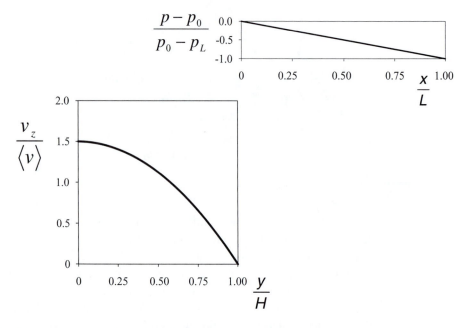

Velocity and pressure profiles calculated for Poiseuille flow (i.e., pressure-driven flow) of a Newtonian fluid in a slit.

(one is redundant) give us the needed boundary conditions on velocity:

$$\text{Boundary conditions:} \quad \begin{array}{l} x_2 = H \ \ v_1 = 0 \\[2mm] x_2 = 0 \ \ \dfrac{dv_1}{dx_2} = 0 \end{array} \tag{7.187}$$

The symmetry boundary condition is particularly desirable because it simplifies the evaluation of the integration constants. The choice of coordinate system with $x_2 = 0$ at the centerline of the channel is well matched with the boundary conditions (see Section 6.2.2). The solution for $v_1(x_2)$ is given here (the integration is left to readers and is assigned in Problem 32) and is plotted in Figures 7.28 and 7.29:

$$v_1(x_2) = \frac{H^2(p_0 - p_L)}{2\mu L} \left[1 - \left(\frac{x_2}{H} \right)^2 \right] \tag{7.188}$$

It is instructive to visually compare the two parabolic equations for the velocity profiles in tube flow (Equation 7.23) and slit flow (Equation 7.188); one is the shape of a bullet (see Figure 7.7). The other profile, also parabolic in profile, is in the shape of a rounded front (see Figure 7.29).

The solution for the flow rate Q in slit flow is calculated using Equation 6.254:

$$\text{Total flow rate out through surface } \mathcal{S}: \quad Q = \iint_{\mathcal{S}} [\hat{n} \cdot \underline{v}]_{\text{at surface}} \, dS \tag{7.189}$$

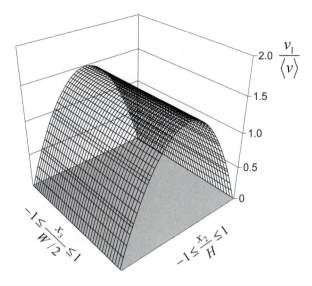

Figure 7.29 Three-dimensional representation of the velocity profile in steady Poiseuille flow in a slit. The centerline velocity is 1.5 times the average velocity of this flow. The data are plotted versus normalized coordinates x_2/H and $x_3/(W/2)$, which both range from -1 to 1.

The surface in which we are interested is the slit cross section at the exit, where $\hat{n} \cdot \underline{v} = \hat{e}_1 \cdot \underline{v} = v_1$ and $dS = dx_2 dx_3$. Thus:

$$Q = \iint_S v_1 \, dS \tag{7.190}$$

$$= \int_{-W/2}^{W/2} \int_{-H}^{H} v_1(x_2) \, dx_2 dx_3 \tag{7.191}$$

$$= W \int_{-H}^{H} v_1(x_2) \, dx_2 \tag{7.192}$$

$$= 2W \int_{0}^{H} v_1(x_2) \, dx_2 \tag{7.193}$$

Substituting $v_1(x_2)$ from Equation 7.188 and carrying out the integration (see Problem 32) yields the final result for flow rate per unit width:

Flow rate/width:
Poiseuille flow
in a slit

$$\boxed{\frac{Q}{W} = \frac{2H^3(p_0 - p_L)}{3\mu L}} \tag{7.194}$$

From $v_1(x_2)$ we can calculate the stress components from the Newtonian constitutive equation written in the Cartesian coordinate system (see Table B.8 in Appendix B):

Newtonian
constitutive
equation:

$$\underline{\underline{\tilde{\tau}}} = \mu \left(\nabla \underline{v} + (\nabla \underline{v})^T \right) \tag{7.195}$$

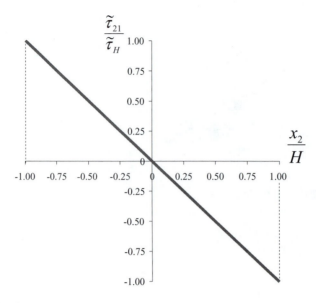

Figure 7.30 Shear stress as a function of x_2/H for Poiseuille flow in a slit.

$$\underline{v} = \begin{pmatrix} v_1(x_2) \\ 0 \\ 0 \end{pmatrix}_{123} \tag{7.196}$$

$$\underline{\underline{\tilde{\tau}}} = \begin{pmatrix} 0 & \mu\frac{dv_1}{dx_2} & 0 \\ \mu\frac{dv_1}{dx_2} & 0 & 0 \\ 0 & 0 & 0 \end{pmatrix}_{123} = \begin{pmatrix} 0 & \frac{-(p_0-p_L)x_2}{L} & 0 \\ \frac{-(p_0-p_L)x_2}{L} & 0 & 0 \\ 0 & 0 & 0 \end{pmatrix}_{123} \tag{7.197}$$

$$\underline{\underline{\tilde{\Pi}}} = \underline{\underline{\tilde{\tau}}} - p\underline{\underline{I}} = \begin{pmatrix} \frac{\Delta p}{L}x_1 - p_0 & \frac{-(p_0-p_L)x_2}{L} & 0 \\ \frac{-(p_0-p_L)x_2}{L} & \frac{\Delta p}{L}x_1 - p_0 & 0 \\ 0 & 0 & \frac{\Delta p}{L}x_1 - p_0 \end{pmatrix}_{123} \tag{7.198}$$

We see that the shear stress $\tilde{\tau}_{21}$ is a linear function of the variable x_2 in this flow (Figure 7.30). Also, the shear stress is zero at the center of the flow and is at its highest absolute values at the walls.

A quantity that can be measured in this flow is the magnitude of shear stress at the wall, which is given by:

Shear stress
at the wall: $$\boxed{|\tilde{\tau}_{21}(H)| = \tilde{\tau}_H = \frac{(p_0 - p_L)H}{L}}$$ (7.199)
(Poiseuille flow in slit)

The total drag on the walls is given by the integral of $\hat{n} \cdot \tilde{\underline{\Pi}}$ at the surface (see Equation 7.103), with the surface of interest being the wetted surfaces of the two infinite plates. In the following calculation, we evaluate drag on the bottom surface and double the results:

$$\begin{matrix} \text{Total molecular fluid force} \\ \text{on a surface } \mathcal{S}: \end{matrix} \quad \mathcal{F} = \iint_S \left[\hat{n} \cdot \tilde{\underline{\Pi}}\right]_{\text{at surface}} dS \qquad (7.200)$$

$$\mathcal{F}_{\text{drag}} = \begin{matrix} \text{Axial drag} \\ \text{in laminar flow} \\ \text{in a slit} \end{matrix} = 2 \begin{pmatrix} \text{1-component of} \\ \text{total force} \\ \text{on bottom (surface} \\ \text{of unit normal } \hat{e}_2) \end{pmatrix} \qquad (7.201)$$

$$= 2 \int_0^L \int_{-W/2}^{W/2} \begin{pmatrix} 1 \\ 0 \\ 0 \end{pmatrix}_{123} \cdot \left[(0 \;\; 1 \;\; 0)_{123} \cdot \Pi|_{x_2=-H} \right] dx_3 dx_1 \qquad (7.202)$$

$$= 2 \int_0^L \int_{-W/2}^{W/2} \tilde{\tau}_{21}|_{x_2=-H} \, dx_3 dx_1 \qquad (7.203)$$

$$= 4 \int_0^L \int_0^{W/2} \frac{H \Delta p}{L} \, dx_3 dx_1 \qquad (7.204)$$

$$= 2HW\Delta p \qquad (7.205)$$

$$\boxed{\mathcal{F}_{\text{drag}} = 2HW\Delta p} \qquad (7.206)$$

This result is analogous to the solution obtained for drag in pipes (see Equation 7.116), where $\mathcal{F}_{\text{drag}} = (\text{cross-sectional area})\Delta p$.

The calculation for flow in a slit was straightforward because the assumption of a wide slit reduced the problem to a two-dimensional flow, $\underline{v} = v_1(x_2)\hat{e}_1$. For a channel of finite width, the flow is three-dimensional, $\underline{v} = v_1(x_2, x_3)\hat{e}_1$, which makes the solution of the PDE more involved. In the case of a rectangular cross section, the differential equations are solvable using advanced mathematical techniques [61]; the solution is summarized in Example 7.11.

EXAMPLE 7.11. *Calculate the velocity profile, flow rate, and shear stress field for pressure-driven flow of an incompressible Newtonian liquid in a rectangular duct of height 2H and width 2W. The pressure at an upstream point is p_0; at a point a distance L downstream, the pressure is p_L. Assume that the flow between these two points is well developed and at steady state. Gravity may be neglected (Figure 7.31).*

SOLUTION. We apply the continuity equation and the Navier-Stokes equation to solve for $\underline{v}$ and $\tilde{\underline{\tau}}$. The continuity equation and microscopic momentum balance for pressure-driven flow in a duct simplify as before to $\partial v_1/\partial x_1 = 0$ and the

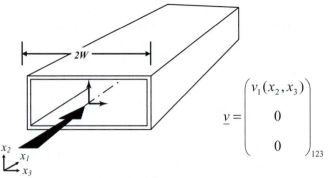

$$v = \begin{pmatrix} v_1(x_2, x_3) \\ 0 \\ 0 \end{pmatrix}_{123}$$

cross section:

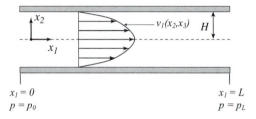

Figure 7.31   Poiseuille flow through a duct of rectangular cross section can be set up following the same procedures used for flow in a tube. The velocity is unidirectional (here, in the x_1-direction) and varies with both x_2 and x_3.

Poisson equation with pressure profile given by Equation 7.181:

$$\text{Continuity equation:} \qquad \frac{\partial v_1}{\partial x_1} = 0 \qquad (7.207)$$

$$\text{1-Component Navier-Stokes:} \qquad \frac{\partial^2 v_1}{\partial x_2^2} + \frac{\partial^2 v_1}{\partial x_3^2} = -\frac{\Delta p}{\mu L} \qquad (7.208)$$

Because the velocity profile is three-dimensional, the Poisson equation does not simplify further in the finite-duct case compared to the slit case. The solution to Equation 7.208 strongly depends on the boundary conditions, which are no-slip boundary conditions at each of the four walls:

$$x_2 = \pm H \qquad v_1 = 0 \qquad \text{for all values of } x_3 \qquad (7.209)$$

$$x_3 = \pm W \qquad v_1 = 0 \qquad \text{for all values of } x_2 \qquad (7.210)$$

Note that in the rectangular duct discussed here the width is $2W$, whereas for the infinite slit, we used width W.

The details of the solution method for the Poisson equation with these boundary conditions are in standard textbooks on solving PDEs [24, 61]. The basic method is to postulate that the solution is separable; that is, that v_1 may be written as:

$$v_1(x_2, x_3) = f(x_2)g(x_3) \qquad (7.211)$$

where $f(x_2)$ and $g(x_3)$ are unknown functions that we must determine. Once $f(x_2)$ and $g(x_3)$ are known, v_1 is reassembled from Equation 7.211, and the resulting equation contains integration constants that must be evaluated from the boundary conditions. For the rectangular-duct solution, the functions $f(x_2)$

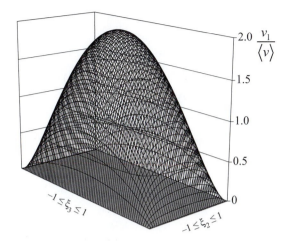

Figure 7.32 The solution for velocity profile for pressure-driven laminar flow in a rectangular duct is shown in a three-dimensional view. The centerline velocity is twice the average velocity of this flow. The data are plotted versus normalized coordinates $\xi_2 \equiv x_2/H$ and $\xi_3 \equiv x_3/W$), which both range from -1 to 1. Compare with the equivalent plot for an infinite slit in Figure 7.29.

and $g(x_3)$ are trigonometric functions, which complicates the evaluation of the integration constants; the established method for solving for these integration constants involves the use of orthogonal functions [61].

The final result for the velocity field in a rectangular duct [28] is the following infinite sum (note that the solution is given in dimensionless form):

$$\frac{v_1(\xi_2, \xi_3)}{\langle v \rangle_{slit}} = \left(\frac{48}{\pi^3} \right) \sum_{n=1,3,5,\ldots}^{\infty} (-1)^{\frac{(n-1)}{2}} \left[1 - \frac{\cosh\left(n\pi W\xi_3/2H\right)}{\cosh\left(n\pi W/2H\right)} \right] \frac{\cos\left(n\pi\xi_2/2\right)}{n^3}$$

(7.212)

where:

$$\xi_2 \equiv \frac{x_2}{H} \tag{7.213}$$

$$\xi_3 \equiv \frac{x_3}{W} \tag{7.214}$$

$$\langle v \rangle_{slit} = \frac{H^2 \Delta p}{3\mu L} \tag{7.215}$$

The result for the velocity field is plotted in Figure 7.32 (compare to the tube and slit solutions, Figures 7.7 and 7.29).

From the velocity field, we can calculate any engineering quantities of interest. The flow rate is calculated with more effort than was required in tube flow, but the calculation is again the integral of the velocity across the flow cross section as discussed in Chapter 6 (see Equation 6.254); the result is [174]:

Total flow rate
out through $$Q = \iint_{\mathcal{S}} [\hat{n} \cdot \underline{v}]_{\text{at surface}} \, dS \tag{7.216}$$
surface $\mathcal{S}$:

$$Q = \int_{-W}^{W} \int_{-H}^{H} v_1 \, dx_2 dx_3 \tag{7.217}$$

$$\boxed{Q = Q_{slit} \left(1 - \frac{192H}{\pi^5 W} \sum_{n=1,3,5,\ldots}^{\infty} \frac{\tanh\left(n\pi W/2H\right)}{n^5} \right)} \tag{7.218}$$

where:

$$Q_{slit} = \frac{4W H^3 \Delta p}{3\mu L} \tag{7.219}$$

The average velocity for the rectangular duct is $\langle v \rangle = Q/(2H)(2W)$.

The stress tensor $\underset{\approx}{\tilde{\tau}}$ is calculated from the Newtonian constitutive equation, given in rectangular coordinates in Table B.8. The velocity field is three-dimensional, $v_1 = v_1(x_2, x_3)$; thus, there are several nonzero terms in the stress-tensor expression:

Newtonian
constitutive
equation:

$$\boxed{\underset{\approx}{\tilde{\tau}} = \mu \left(\nabla \underline{v} + (\nabla \underline{v})^T \right)} \tag{7.220}$$

$$\underset{\approx}{\tilde{\tau}} = \mu \begin{pmatrix} 0 & \dfrac{\partial v_1}{\partial x_2} & \dfrac{\partial v_1}{\partial x_3} \\[2mm] \dfrac{\partial v_1}{\partial x_2} & 0 & 0 \\[2mm] \dfrac{\partial v_1}{\partial x_3} & 0 & 0 \end{pmatrix}_{123} \tag{7.221}$$

$$\underset{\approx}{\tilde{\Pi}} = \underset{\approx}{\tilde{\tau}} - p\underset{\approx}{\underline{I}} = \begin{pmatrix} -p(x_1) & \mu\dfrac{\partial v_1}{\partial x_2} & \mu\dfrac{\partial v_1}{\partial x_3} \\[2mm] \mu\dfrac{\partial v_1}{\partial x_2} & -p(x_1) & 0 \\[2mm] \mu\dfrac{\partial v_1}{\partial x_3} & 0 & -p(x_1) \end{pmatrix}_{123} \tag{7.222}$$

The velocity derivatives may be evaluated from the velocity-profile solution, Equation 7.212; $p(x_1)$ is given by the usual expression for conduits of constant cross-section, Equation 7.176.

The total force on the walls is calculated from Equation 7.103:

Total molecular fluid force
on a surface $\mathcal{S}$:

$$\underline{\mathcal{F}} = \iint_{\mathcal{S}} \left[\hat{n} \cdot \underset{\approx}{\tilde{\Pi}} \right]_{\text{at surface}} dS \tag{7.223}$$

The surfaces that experience drag are the four walls. We calculate force on the bottom and on the left side and double our result:

$$\mathcal{F}_{\text{drag}} = \begin{matrix} \text{Axial drag} \\ \text{in laminar flow} \\ \text{in a rectangular conduit} \end{matrix} \tag{7.224}$$

$$= 2 \begin{pmatrix} \text{1-component of} \\ \text{total force} \\ \text{on conduit surface} \\ \text{of unit normal } \hat{e}_2 \\ \text{(bottom)} \end{pmatrix} + 2 \begin{pmatrix} \text{1-component of} \\ \text{total force} \\ \text{on conduit surface} \\ \text{of unit normal } \hat{e}_3 \\ \text{(left side)} \end{pmatrix} \tag{7.225}$$

$$= 2 \int_0^L \int_{-W}^W \hat{e}_1 \cdot \left[\hat{e}_2 \cdot \tilde{\tilde{\Pi}} \right]_{x_2 = -H} dx_3 dx_1$$

$$+ 2 \int_0^L \int_{-H}^H \hat{e}_1 \cdot \left[\hat{e}_3 \cdot \tilde{\tilde{\Pi}} \right]_{x_3 = -W} dx_2 dx_1 \tag{7.226}$$

$$= 4\mu L \int_0^W \left. \frac{\partial v_1}{\partial x_2} \right|_{x_2 = -H} dx_3 + 4\mu L \int_0^H \left. \frac{\partial v_1}{\partial x_3} \right|_{x_3 = -W} dx_2 \tag{7.227}$$

$$= \frac{4\mu L \langle v \rangle_{slit}}{H/W} \left[\int_0^1 \left. \frac{\partial \tilde{v}_1}{\partial \xi_2} \right|_{\xi_2 = -1} d\xi_3 + \frac{H^2}{W^2} \int_0^1 \left. \frac{\partial \tilde{v}_1}{\partial \xi_3} \right|_{\xi_3 = -1} d\xi_2 \right] \tag{7.228}$$

$$= \frac{4HW \Delta p}{3} \left[\int_0^1 \left. \frac{\partial \tilde{v}_1}{\partial \xi_2} \right|_{\xi_2 = -1} d\xi_3 + \frac{H^2}{W^2} \int_0^1 \left. \frac{\partial \tilde{v}_1}{\partial \xi_3} \right|_{\xi_3 = -1} d\xi_2 \right] \tag{7.229}$$

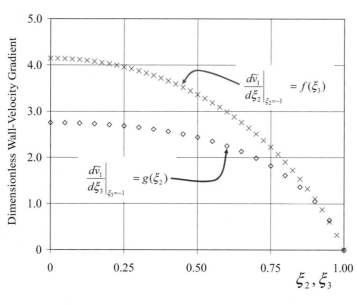

Figure 7.33 The stresses generated at the walls of a rectangular duct due to pressure-driven flow are proportional to the velocity gradients at the walls (see Equation 7.222). The velocity gradients in dimensionless form are plotted here versus $\xi_2 = x_2/H$ or $\xi_3 = x_3/W$. The shear stresses are not constant at the surfaces but rather vary with position. In the corners, the stress is zero; the stress is at maximum in the center of the faces.

$$= 4HW\Delta p \left[\frac{8}{\pi^2} \sum_{n=1,3,5,\dots}^{\infty} \frac{1}{n^2} \right] \tag{7.230}$$

$$= 4HW\Delta p \tag{7.231}$$

where $\tilde{v}_1 \equiv v_1/\langle v \rangle_{slit}$ and $\langle v \rangle_{slit} = Q_{slit}/4HW = H^2\Delta p/(3\mu L)$ (see Equation 7.194; note that W in the rectangular duct is the half-width, whereas in the slit case, W is the width). The total drag on the walls of a rectangular duct is the cross-sectional area times the pressure drop, as it was in pressure-driven slit flow and Poiseuille flow. The total-stress components used here are from Equation 7.222, and the final calculation of the expression in brackets in Equation 7.229 is carried out using the velocity result in Equation 7.212. The infinite sum in Equation 7.230 converges to $\pi^2/8$.

The dimensionless wall gradients $d\tilde{v}_1/d\xi_2$ and $d\tilde{v}_1/d\xi_3$ are proportional to wall stresses (see Equation 7.222). These quantities are plotted in Figure 7.33 for a rectangular duct that is twice as wide as it is tall ($W = 2H$). Note that unlike tube flow and slit flow, the stress is not independent of position along the perimeter in a duct of rectangular cross section.

In summary, the flow-direction momentum balance for flows through ducts of constant cross section gives the Poisson equation (Equation 7.177), which can be challenging to solve for some geometries, as in the case of the rectangular duct. More solutions for flows in ducts are in the literature [152]. In the next section, we see that even without obtaining exact solutions, we can learn much about general flow in noncircular ducts by applying dimensional analysis and the friction-factor concept to these flows. In Section 7.3, we apply the Navier-Stokes equations to unsteady flow and to flows that are not unidirectional; in Chapter 8, we model external flows—that is, flows around and over objects rather than through closed conduits.

7.2.1.2 POISEUILLE NUMBER AND HYDRAULIC DIAMETER

Our experience with flows through pipes in turbulent flow leads us to expect that dimensional analysis may result in helpful relationships for laminar and turbulent flows in noncircular conduits; this is indeed the case. Dimensional analysis of the Poisson equation leads to the definition of a general characteristic lengthscale— the hydraulic diameter—that organizes the behavior of noncircular conduits.

In the previous section, we derive the flow-direction component of the microscopic-momentum balance for Poiseuille flow in a conduit as the Poisson equation (see Equation 7.177):

Poisson equation:
Flow-direction
momentum balance for
pressure-driven flow
in closed conduits

$$\boxed{\frac{\partial^2 v_1}{\partial x_2^2} + \frac{\partial^2 v_1}{\partial x_3^2} = -\frac{\Delta p}{\mu L}} \tag{7.232}$$

The derivatives on the lefthand side of the momentum-balance Poisson equation are obtainable from the shape of the velocity profile, which varies with cross-sectional shape (see the velocity results for pressure-driven tube flow, slit flow, and rectangular-duct flow; Figures 7.7, 7.29, and 7.32). Different conduit shapes result in different pressure drops on the righthand side.

Dimensional analysis demonstrated that friction factor is a function only of Reynolds number for pipe flow; it is straightforward to show that this is true also for flows through noncircular conduits.[10] For the simple conduit flows studied thus far, the friction-factor/Reynolds-number relationships for laminar flow are:

$$\text{Friction factor in circular ducts (laminar tube flow):} \qquad \boxed{f = \frac{16}{\text{Re}}} \qquad (7.233)$$

$$\text{Friction factor (laminar slit flow):} \qquad \boxed{f = \frac{24}{\text{Re}}} \qquad (7.234)$$

For laminar flow in a general noncircular conduit, we obtain the friction-factor/ Reynolds-number relationship by a judicious rearrangement of the Poisson equation, Equation 7.232 (i.e., flow-direction momentum balance), as we now show.

For circular pipes, we obtained $\mathcal{F}_{\text{drag}} = \pi R^2 \Delta p$ for laminar pipe flow (see Section 7.1) and (width)(height)Δp for slit and rectangular duct flow. We can perform a macroscopic-momentum balance on a section of a noncircular conduit and obtain the same results for an arbitrary cross-sectional shape (assigned as Problem 18 in Chapter 9). The result is:

$$\text{Wall drag for noncircular conduits:} \qquad \boxed{\mathcal{F}_{\text{drag}} = \Delta p\, A_{xs}} \qquad (7.235)$$

where A_{xs} is the cross-sectional area of the conduit. We convert the Poisson equation (Equation 7.232) to a more general nondimensional expression by substituting $\mathcal{F}_{\text{drag}}/A_{xs}$ for Δp and defining the friction factor for a conduit of arbitrary cross section as:

$$\text{Friction factor noncircular conduit:} \qquad f \equiv \frac{\mathcal{F}_{\text{drag}}}{\frac{1}{2}\rho V^2\,(\text{wetted area})} \qquad (7.236)$$

$$= \frac{A_{xs}\,\Delta p}{\frac{1}{2}\rho V^2\,(\text{wetted area})} \qquad (7.237)$$

The wetted surface area of a noncircular conduit is the conduit perimeter p multiplied by the length L of the conduit:

$$\text{Wetted area, conduit of constant cross section} \quad = pL \qquad (7.238)$$

---

[10] An exception to this is eccentric annular flow, in which flow rate varies with eccentricity. See Shah and London [152] for solutions to laminar flow through eccentric annuli. The lengthscale introduced into the problem by placement of the center modifies the dimensional analysis, much as ε modified the dimensional analysis for rough pipes.

Using this area in Equation 7.237, we obtain:

$$f = \frac{A_{xs}\Delta p}{\frac{1}{2}\rho\langle v\rangle^2 \, pL} \tag{7.239}$$

$$= \left(\frac{\Delta p}{L}\right)\frac{A_{xs}/p}{\frac{1}{2}\rho V^2} \tag{7.240}$$

The ratio A_{xs}/p has dimensions of length; for a circular tube, it is equal to half the radius. Solving for $\Delta p/L$ gives:

$$\frac{\Delta p}{L} = \frac{\frac{1}{2}\rho V^2 f}{A_{xs}/p} \tag{7.241}$$

Combining this result with the momentum balance, Equation 7.232, gives:

$$\frac{\partial^2 v_1}{\partial x_2^2} + \frac{\partial^2 v_1}{\partial x_3^2} = -\frac{\frac{1}{2}\rho V^2 f}{(A_{xs}/p)}\frac{1}{\mu} \tag{7.242}$$

To write Equation 7.242 in a more familiar friction-factor/Reynolds-number form, we nondimensionalize the variables in the equation. Following the usual nondimensionalization techniques, we designate $V \equiv \langle v\rangle$ the characteristic velocity and D the characteristic length. It is not obvious what D should be for the noncircular conduits considered; we defer identifying D until after simplifying Equation 7.242.

We define the dimensionless velocity and positions as:

$$v_1^* \equiv \frac{v_1}{V} \tag{7.243}$$

$$x_2^* \equiv \frac{x_2}{D} \tag{7.244}$$

$$x_3^* \equiv \frac{x_3}{D} \tag{7.245}$$

Substituting these into Equation 7.242, we obtain:

$$\frac{V}{D^2}\left(\frac{\partial^2 v_1^*}{\partial x_2^{*2}} + \frac{\partial^2 v_1^*}{\partial x_3^{*2}}\right) = -\frac{\frac{1}{2}\rho V^2 f}{(A_{xs}/p)\mu} \tag{7.246}$$

$$\frac{\partial^2 v_1^*}{\partial x_2^{*2}} + \frac{\partial^2 v_1^*}{\partial x_3^{*2}} = -\frac{1}{2}\left(\frac{\rho V D}{\mu}\right)\frac{D}{(A_{xs}/p)}f \tag{7.247}$$

We see in Equation 7.247 that the Reynolds number appears ($\rho V D/\mu$), written in terms of the as-yet-unspecified characteristic length D. The form of Equation 7.247 suggests that a reasonable definition of D is A_{xs}/p. By convention, a factor of 4 is included in the definition of characteristic length, a choice that allows that length D to become pipe diameter for tube flow, the same characteristic length we used for tube flow. The characteristic length thus defined is called the *hydraulic diameter*:[11]

Hydraulic diameter :

$$D = D_H \equiv \frac{4A_{xs}}{\mathcal{P}}$$

(7.248)

Incorporating the definition of hydraulic diameter, we arrive at the nondimensional version of the flow-direction microscopic-momentum balance for laminar flow in conduits of arbitrary cross section:

Flow-direction momentum balance in laminar flow in ducts of constant cross section (Cartesian coordinates):

$$-\frac{1}{2}\left(\frac{\partial^2 v_1^*}{\partial x_2^{*2}} + \frac{\partial^2 v_1^*}{\partial x_3^{*2}}\right) = f_{D_H}\mathrm{Re}_{D_H} = \mathrm{Po}$$

(7.249)

where the friction factor and the Reynolds number are given by:

$$f_{D_H} = \left(\frac{\Delta p}{L}\right)\frac{A_{xs}/\mathcal{P}}{\frac{1}{2}\rho V^2} = \left(\frac{\Delta p}{L}\right)\frac{D_H}{2\rho V^2}$$

(7.250)

$$\mathrm{Re}_{D_H} \equiv \frac{\rho V D_H}{\mu}$$

(7.251)

The combination $f_{D_H}\mathrm{Re}_{D_H}$ is called the Poiseuille number (Po) and it is a constant (i.e., independent of the Reynolds number) for steady laminar flows in ducts. The flow-direction momentum balance also may be written in general vector–tensor (i.e., Gibbs) notation or in cylindrical coordinates, as shown here:

Gibbs notation:

$$-\frac{1}{2}\nabla^{*2}\underline{v}^* = f_{D_H}\mathrm{Re}_{D_H}$$

(7.252)

Cylindrical coordinates:

$$-\frac{1}{2}\left[\frac{1}{r^*}\frac{\partial}{\partial r^*}\left(r^*\frac{\partial v_z^*}{\partial r^*}\right) + \frac{1}{r^{*2}}\frac{\partial^2 v_z^*}{\partial \theta^2}\right] = f_{D_H}\mathrm{Re}_{D_H} = \mathrm{Po}$$

(7.253)

Equation 7.249 (or Equation 7.252 or 7.253) is a powerful general result for unidirectional, steady flow in noncircular ducts. For a duct of a chosen cross section, the velocity profile has a steady-state shape, and the lefthand side of Equation 7.249 evaluates to a numerical constant. The value obtained for the constant, the Poiseuille number, depends on only the shape of the cross section, not on flow variables such as the Reynolds number. Thus, Equation 7.249 states that $\mathrm{Po} = f_{D_H}\mathrm{Re}_{D_H} = $ constant for a given geometry. We already know this is true for tubes and slits (see Equations 7.233 and 7.234). Through this derivation,

---

[11] Note that in the literature there also is a quantity called the hydraulic radius r_H, which is equivalent to $D_H/4$ (see Problem 23). This is an unfortunate inconsistency in nomenclature.

we now see that this is a general result for laminar flow in conduits of constant cross section with f and Re based on hydraulic diameter.[12]

To illustrate the power of Equation 7.249 and to explore the concept of hydraulic diameter, we calculate the constant $f_{D_H} \mathrm{Re}_{D_H}$ for both circular tubes and conduits with triangular cross section.

EXAMPLE 7.12. *For steady laminar flow in ducts of constant cross section, the combination $f_{D_H} \mathrm{Re}_{D_H}$ is a constant. For steady laminar flow in a tube, evaluate this constant using Equation 7.253.*

SOLUTION. We calculated the velocity profile for laminar flow in a tube in Equation 7.23 and the average velocity for that flow in Equation 7.29 (results given in cylindrical coordinates):

$$v_z(r) = \frac{(p_0 - p_L)R^2}{8\mu L}\left[1 - \left(\frac{r}{R}\right)^2\right] \tag{7.254}$$

$$V = \langle v \rangle = \frac{(p_0 - p_L)R^2}{4\mu L} \tag{7.255}$$

In dimensionless form, we write the velocity profile as:

$$v_r^* = \frac{v_r}{V} = 2\left(1 - 4r^{*2}\right) \tag{7.256}$$

where $r^* = r/D_H$ and $D_H = 4(\pi R^2)/(2\pi R) = 2R$.

The flow-direction microscopic momentum balance is given by Equation 7.253:

$$-\frac{1}{2}\left[\frac{1}{r^*}\frac{\partial}{\partial r^*}\left(r^*\frac{\partial v_z^*}{\partial r^*}\right) + \frac{1}{r^{*2}}\frac{\partial v_z^*}{\partial \theta^2}\right] = f_{D_H}\mathrm{Re}_{D_H} \tag{7.257}$$

We now substitute Equation 7.256 into Equation 7.257 and simplify:

$$f_{D_H}\mathrm{Re}_{D_H} = -\frac{1}{2}\left[\frac{1}{r^*}\frac{\partial}{\partial r^*}\left(r^*\frac{\partial v_z^*}{\partial r^*}\right) + \frac{1}{r^{*2}}\frac{\partial v_z^*}{\partial \theta^2}\right] \tag{7.258}$$

$$v_r^* = \frac{v_r}{V} = 2\left(1 - 4r^{*2}\right) \tag{7.259}$$

$$\frac{\partial v_z^*}{\partial \theta} = 0 \tag{7.260}$$

$$\frac{\partial v_z^*}{\partial r^*} = -16r^* \tag{7.261}$$

$$f_{D_H}\mathrm{Re}_{D_H} = -\frac{1}{2}\left(\frac{1}{r^*}\frac{\partial}{\partial r^*}\left(-16r^{*2}\right)\right) \tag{7.262}$$

$$= 16 \tag{7.263}$$

We arrive at $f_{D_H} = 16/\mathrm{Re}_{D_H}$, the familiar result from Equation 7.152.

---

[12] A notable failure of the hydrodynamic-diameter concept is eccentric annular flow, as noted in Footnote 10.

EXAMPLE 7.13. *For steady laminar flow in ducts of constant cross section, the combination $f_{D_H}Re_{D_H}$ is a constant. If the shape of the cross section of the conduit is an equilateral triangle (Figure 7.34), evaluate this constant using Equation 7.249.*

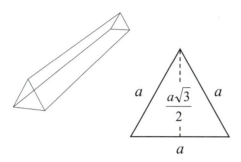

Figure 7.34 The shape of the cross section in this example is an equilateral triangle with walls of length a.

SOLUTION. The velocity profile and average velocity for laminar flow in an equilateral triangular conduit is given by White [174] to be (Cartesian coordinates):

$$v_x(y, z) = \frac{(p_0 - p_L)}{2a\sqrt{3}\mu L}\left(z - \frac{a\sqrt{3}}{2}\right)\left(3y^2 - z^2\right) \tag{7.264}$$

$$V = \langle v \rangle = \frac{a^2(p_0 - p_L)}{80\mu L} \tag{7.265}$$

To nondimensionalize the velocity profile, we need the hydraulic diameter:

$$D_H \equiv \frac{4A_{xs}}{\mathcal{P}} \tag{7.266}$$

$$= \frac{4\left(\frac{1}{2}a\frac{a\sqrt{3}}{2}\right)}{3a} \tag{7.267}$$

$$= \frac{a}{\sqrt{3}} \tag{7.268}$$

In dimensionless form, we write the velocity profile in terms of $v_x^* = v_x/V$, $y^* = y/D_H$, and $z^* = z/D_H$:

$$v_x^* = \frac{20}{9}(2z^* - 3)\left(3y^{*2} - z^{*2}\right) \tag{7.269}$$

The microscopic-momentum balance in the flow direction written in Cartesian coordinates is as follows (Equation 7.249):

$$-\frac{1}{2}\left(\frac{\partial^2 v_x^*}{\partial y^{*2}} + \frac{\partial^2 v_x^*}{\partial z^{*2}}\right) = f_{D_H}Re_{D_H} \tag{7.270}$$

where we make the substitutions $x_2^* = y^*$, $x_3^* = z^*$, and $v_1^* = v_x^*$. Calculating the required derivatives of v_x^* from Equation 7.269, we proceed to the final result for $f_{D_H}\text{Re}_{D_H}$:

$$\frac{\partial v_x^*}{\partial y^*} = \frac{20}{9}(2z^* - 3)(6y^*) \tag{7.271}$$

$$\frac{\partial^2 v_x^*}{\partial y^{*2}} = \frac{40}{3}(2z^* - 3) \tag{7.272}$$

$$\frac{\partial v_x^*}{\partial z^*} = \frac{40}{3}\left(-z^{*2} + z^* + y^{*2}\right) \tag{7.273}$$

$$\frac{\partial^2 v_x^*}{\partial z^{*2}} = \frac{40}{3}(-2z^* + 1) \tag{7.274}$$

$$f_{D_H}\text{Re}_{D_H} = -\frac{1}{2}\left(\frac{\partial^2 v_x^*}{\partial y^{*2}} + \frac{\partial^2 v_x^*}{\partial z^{*2}}\right) \tag{7.275}$$

$$= -\frac{20}{3}(2z^* - 3 - 2z^* + 1) \tag{7.276}$$

$$= \frac{40}{3} \tag{7.277}$$

The friction-factor/Reynolds-number relationship for laminar flow in a conduit with an equilateral triangular cross section is $f_{D_H} = 13.333/\text{Re}_{D_H}$.

For highly symmetric shapes, Po is a single number, whereas for more complex shapes, Po depends on geometric parameters that define the cross-sectional shape. For example, for elliptical cross sections, the shape of the cross section is defined by the lengths of the major and minor axes of the ellipse (Figure 7.35). The Poiseuille number for an elliptical cross section may be written as follows (see Problem 25):

Poiseuille number
for laminar flow
in a duct of an
elliptical cross section:

$$\text{Po}(a, b) = f_{D_H}\text{Re}_{D_H} = \frac{32\pi^2}{p}(a^2 + b^2) \tag{7.278}$$

where a and b are the semi-major and semi-minor axes of the ellipse, and p is the perimeter of the ellipse, given in terms of a and b as:

$$p = 4\int_0^{\frac{\pi}{2}}\sqrt{a^2\sin^2\psi + b^2\cos^2\psi}\,d\psi \tag{7.279}$$

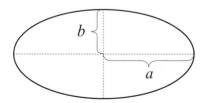

The shape of an ellipse is defined by the lengths of its major (2a) and minor (2b) axes.

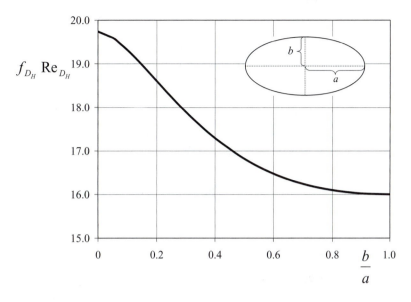

Figure 7.36 $f_{D_H}\mathrm{Re}_{D_H}$ versus b/a for elliptical ducts [152].

Note that ψ is a dummy variable of integration. The relationship $f_{D_H}\mathrm{Re}_{D_H}$ versus b/a for an ellipse is shown in Figure 7.36. The Poiseuille number for the rectangular geometry can be calculated from the laminar-flow solution presented in Equation 7.212 (this is left to readers in Problem 27). The results are given in Figure 7.37. In their text Shah and London [152] summarize and graphically display solutions for $f_{D_H}\mathrm{Re}_{D_H}$ for 40 geometries, encompassing both practical engineering shapes and those of more theoretical interest.

In summary, we can analyze steady, unidirectional, laminar flow through noncircular closed geometries as a group. For simple shapes, we can solve the

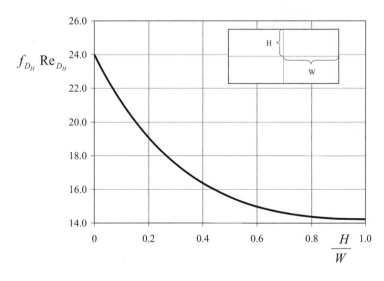

Figure 7.37 The Poiseuille number for laminar flow in a rectangular duct is a function of the ratio H/W [152], where H and W are the half-height and half-width of the duct, respectively.

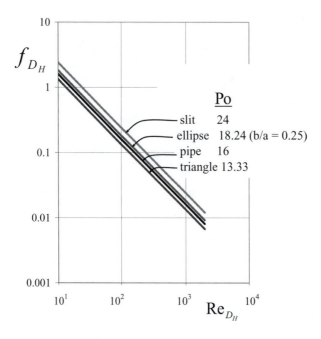

Figure 7.38 Friction-factor/Reynolds-number relationships for laminar flows in conduits of constant cross section are similar and characterized by a constant Poiseuille number: $Po = f_{D_H} Re_{D_H}$. For noncircular conduits, therefore, $f_{D_H} \propto 1/Re_{D_H}$.

problems completely—for example, slit flow (Equation 7.188) and tube flow (Example 7.2)—although the mathematics required can be sophisticated for some shapes (e.g., the rectangular duct; see Equation 7.212). For almost all shapes,[13] we show that the Poiseuille number is constant, which corresponds to the friction-factor scaling as $1/Re_{D_H}$ (Figure 7.38). The hydraulic diameter, $D_H = 4A/\mathcal{P}$, was developed as a common lengthscale for use in such flows.

We can use the laminar flow results from this section to calculate flow-rate/pressure-drop problems for noncircular conduits, as shown in the following examples. We also show in Example 7.16 how a packed bed may be analyzed as a conduit of irregular cross section. In the next subsection, we briefly discuss turbulent flows in noncircular conduits before continuing with more complex internal flows in Section 7.3.

EXAMPLE 7.14. *Water is flowing in a 15.0-m-long triangular duct (i.e., an equilateral triangle in cross section; geometry given in Figure 7.34: a = 50 mm). The upstream gauge pressure is 25.0 Pa and the downstream gauge pressure is 0 Pa. Calculate the average velocity in the conduit.*

SOLUTION. The pressure drop is only 25.0 Pa; therefore, we assume at first that the flow is laminar. We doublecheck this assumption at the end of the calculation.

---

[13] Hydrodynamic diameter does not work for eccentric annuli [152].

For laminar flow in a triangular duct, $f_{D_H} \mathrm{Re}_{D_H} = 40/3$. We therefore can write:

$$f_{D_H} \mathrm{Re}_{D_H} = 40/3 \tag{7.280}$$

$$\left(\frac{\Delta p D_H}{2 L \rho V^2} \right) \left(\frac{\rho V D_H}{\mu} \right) = \frac{40}{3} \tag{7.281}$$

Solving for $V = \langle v \rangle$, we obtain:

$$\frac{\Delta p D_H^2}{2 \mu L V} = \frac{40}{3} \tag{7.282}$$

$$V = \frac{3 D_H^2 \Delta p}{80 \mu L} \tag{7.283}$$

We substitute numerical values for the quantities in Equation 7.283 and obtain the final result for average velocity V:

$$D_H = \frac{a}{\sqrt{3}} = \frac{50 \text{ mm}}{\sqrt{3}} = 0.02887 \text{ m} \tag{7.284}$$

$$V = \frac{(3)(0.02887 \text{ m})^2 (25.0 \text{ Pa})}{(80)(15.0 \text{ m})(0.00100 \text{ Pa s})} \tag{7.285}$$

$$= 0.05209 \text{ m/s} \tag{7.286}$$

$$= 0.052 \text{ m/s} \quad \text{(2 significant figures)} \tag{7.287}$$

We check our assumption of laminar flow:

$$\mathrm{Re}_{D_H} = \frac{\rho V D_H}{\mu} = \frac{\left(1{,}000 \frac{\text{kg}}{\text{m}^3} \right) \left(0.05209 \frac{\text{m}}{\text{s}} \right) 0.02887 \text{ m}}{(0.00100 \text{ Pa s})} = 1{,}504 \tag{7.288}$$

The Reynolds number is below 2,100, which is the upper limit for laminar flow in circular pipes. We do not know a priori if this is the upper limit of Re for flow in a triangular duct, however, until we discuss turbulent flow in triangular ducts (see Section 7.2.2).

EXAMPLE 7.15. *For laminar flow in a duct at average velocity V, which geometry has more drag: a tube of diameter a or a square duct of side a?*

SOLUTION. The drag in a flow is quantified by the friction factor:

$$f_{D_H} = \frac{\mathcal{F}_{\text{drag}}}{\frac{1}{2} \rho V^2 \, \not{p} L} \tag{7.289}$$

The hydraulic diameters for the two geometries are the same:

$$\text{Circle:} \quad D_H = \frac{4 A_{xs}}{\not{p}} = \frac{4 \, \pi \left(\frac{a}{2} \right)^2}{2 \pi \frac{a}{2}} = a \tag{7.290}$$

$$\text{Square:} \quad D_H = \frac{4 A_{xs}}{\not{p}} = \frac{4 \, a^2}{4a} = a \tag{7.291}$$

For a tube, $Po = f_{D_H} Re_{D_H} = 16$; for a square duct, $Po = f_{D_H} Re_{D_H} = 14.22708$ (see Figure 7.37 and [152]). For the same fluid (ρ, μ) at the same average velocity V, we compare the two geometries to see which has higher drag:

$$f_{D_H} = \frac{\mathcal{F}_{\text{drag}}}{\frac{1}{2}\rho V^2 \, pL} = \frac{Po}{Re_{D_H}} \tag{7.292}$$

$$\mathcal{F}_{\text{drag}} = \frac{\frac{1}{2}\rho V^2}{Re_{D_H}} \, Po \, pL \tag{7.293}$$

$$\mathcal{F}_{\text{drag}}\big|_{\text{circle}} = \frac{\frac{1}{2}\rho V^2 L}{Re_{D_H}} (16)(\pi a) = 50.265 \frac{\frac{1}{2}\rho V^2 \, aL}{Re_{D_H}} \tag{7.294}$$

$$\mathcal{F}_{\text{drag}}\big|_{\text{square}} = \frac{\frac{1}{2}\rho V^2 L}{Re_{D_H}} (14.22708)(4a) = 56.908 \frac{\frac{1}{2}\rho V^2 \, aL}{Re_{D_H}} \tag{7.295}$$

A square duct has 13 percent more drag than a circular duct of the same hydraulic diameter and length.

EXAMPLE 7.16. *For pressure-driven flow through a packed bed, how can we relate pressure drop and flow rate? For example, a 1.0-cm-diameter chromatography column consists of a packing with a void fraction $\varepsilon = 0.39$ and a specific surface area (i.e., total particle surface area/particle volume) $a_v = 720 \text{ cm}^{-1}$. What pressure drop $(\Delta p / L)$ must be applied to drive toluene through the column at 1.0 ml/min?*

SOLUTION. We have succeeded thus far in relating pressure drop and flow rate in circular and noncircular ducts. We choose now to think of flow through a packed bed as flow through a duct of an extremely irregular cross section (Figure 7.39). With this picture, we can derive a functional form for the friction-factor/Reynolds-number relationship, which can be tested through comparison to experiments. We follow the development in Denn [43].

For conduits of arbitrary cross section, we show in this chapter that the Poiseuille number is constant:

$$Po \equiv f_{D_H} Re_{D_H} = \text{constant} \tag{7.296}$$

where friction factor and Reynolds number are defined in terms of the hydraulic diameter (see Equations 7.250 and 7.251):

$$f_{D_H} = \left(\frac{\Delta p}{L}\right) \frac{A_{xs}/p}{\frac{1}{2}\rho V^2} = \left(\frac{\Delta p}{L}\right) \frac{D_H}{2\rho V^2} \tag{7.297}$$

$$Re_{D_H} \equiv \frac{\rho V D_H}{\mu} \tag{7.298}$$

For circular tubes, $Po = 16$; for triangular ducts, $Po = 40/3$; and for elliptical and rectangular ducts, Po is given in Figures 7.36 and 7.37.

For simple geometries such as a tube and a triangle, we solved analytically for Po by using known solutions for the flow field. We cannot do this for the tortuous three-dimensional flow through a packed bed, but we can derive a hydraulic

Figure 7.39   The structure of a porous material is illustrated by this snapshot of an empty pore model calculated by Lev Gelb and coworkers at Washington University (St. Louis, Missouri, USA; used with permission, for related work see [144]). Flow through a packed bed follows a tortuous three-dimensional path that may be thought of as flow through a conduit of irregular cross section.

diameter and then perform experiments to see (1) if the Poiseuille number is constant; and (2) if Po is constant, to see what value is obtained for flow through packed beds. Once we know the value of Po, we can solve easily the chromatography problem posed at the beginning of this example as a pressure drop/flow rate problem.

The hydraulic diameter is defined as:

$$D_H \equiv \frac{4A_{xs}}{p} \qquad (7.299)$$

where A_{xs} is the cross-sectional area open to flow and p is the wetted perimeter. For flow in a packed bed, A_{xs} and p can be related to two properties of the bed and its packing: the void fraction ε and the specific surface area a_v.

The void fraction is a measure of how much of the bed volume is occupied by packing. The void fraction is defined as:

$$\text{Void fraction:} \qquad \varepsilon \equiv \frac{(\text{empty-bed volume})}{(\text{total-bed volume})} \qquad (7.300)$$

Thus, $1 - \varepsilon$ is the fraction of the bed volume occupied by the packing:

$$1 - \varepsilon = \frac{(\text{volume of solids})}{(\text{total-bed volume})} \qquad (7.301)$$

The cross section open to flow A_{xs} is the open volume per unit length (compare to the same calculation for a tube):

$$A_{xs} = \left(\frac{\text{empty-bed volume}}{L} \right) = \frac{\varepsilon \mathcal{V}}{L} \qquad (7.302)$$

where $\mathcal{V}$ is the total volume occupied by the bed and L is the length of the bed. The specific surface area of a packing is the total surface area of the particle per unit volume of the particle:

$$\text{Specific surface area of a packing:} \quad a_v \equiv \frac{\text{(total particle surface area)}}{\text{(particle volume)}} \tag{7.303}$$

Note that if the packing is nonporous spheres:

$$\text{Specific surface area of nonporous spheres:} \quad a_v = \frac{4\pi R^2}{\frac{4}{3}\pi R^3} = \frac{6}{D} \tag{7.304}$$

where $D = 2R$ is the diameter of the spheres. For nonspherical particles, $a_v = 6/D$ may be used to calculate an effective spherical diameter from the known a_v for the nonspherical geometry.

The wetted perimeter p is the amount of particle surface area associated with a cross section of the column. Thus, p is the total particle surface area per unit bed length. This can be related to a_v, as follows:

$$p = \left(\frac{\text{total surface area in column}}{L} \right) \tag{7.305}$$

$$= \frac{1}{L} \left(\frac{\text{surface area}}{\text{volume}} \right) \text{(volume occupied by solids)} \tag{7.306}$$

$$= \left(\frac{1}{L} \right) (a_v) \left(\mathcal{V}(1 - \varepsilon) \right) \tag{7.307}$$

$$= \frac{a_v}{L}(1 - \varepsilon)\mathcal{V} \tag{7.308}$$

We now can calculate the hydraulic diameter for the packed bed:

$$D_H \equiv \frac{4A_{xs}}{p} \tag{7.309}$$

$$= \frac{4\left(\dfrac{\varepsilon \mathcal{V}}{L}\right)}{\dfrac{a_v}{L}(1 - \varepsilon)\mathcal{V}} = \frac{4\varepsilon}{(1 - \varepsilon)a_v} \tag{7.310}$$

$$\text{Hydraulic diameter for a packed bed:} \quad \boxed{D_H = \frac{4\varepsilon}{(1 - \varepsilon)a_v}} \tag{7.311}$$

The friction factor in terms of hydraulic diameter and experimental variables for a noncircular conduit is given directly by Equation 7.297, with the velocity V interpreted for a packed bed as the average velocity through the void regions. The superficial velocity v_0 is defined as the apparent average flow velocity as if the packing were not present:

$$\text{Superficial velocity} \quad v_0 \equiv \left(\frac{\text{total flow rate}}{\text{bed cross-sectional area}} \right) = \frac{Q}{\mathcal{V}/L} \tag{7.312}$$

where Q is the volumetric flow rate of liquid through the bed. The true average velocity through the void regions V is equal to the flow rate divided by the cross section open to flow A_{xs}:

$$\text{True average velocity through void regions:} \qquad V = \frac{Q}{A_{xs}} = \frac{Q}{\varepsilon \mathcal{V}/L} = \frac{v_0}{\varepsilon} \qquad (7.313)$$

Substituting V into Equation 7.297, the friction factor for packed beds becomes:

$$f_{D_H} = \left(\frac{\Delta p}{L}\right) \frac{D_H}{2\rho V^2} \qquad (7.314)$$

$$\text{Friction factor for a packed bed: } \boxed{f_{D_H} = \left(\frac{\Delta p}{L}\right) \frac{D_H \, \varepsilon^2}{2\rho v_0^2}} \qquad (7.315)$$
(based on D_H)

The Reynolds number[14] also contains average velocity, which we replace with true average velocity, $V = v_0/\varepsilon$:

$$\text{Re}_{D_H} = \frac{\rho V D_H}{\mu} \qquad (7.316)$$

$$\text{Reynolds number for a packed bed: } \boxed{\text{Re}_{D_H} = \frac{\rho(v_0/\varepsilon)D_H}{\mu}} \qquad (7.317)$$
(based on D_H)

Our hypothesis is that we can model flow through a packed bed as flow in a highly irregular, noncircular conduit. If we are correct, then when we take measurements on packed beds, f_{D_H} should go as $1/\text{Re}_{D_H}$, at least in the laminar-flow (i.e., slow-flow) region. Experimental data by Ergun [45] and others for f_{D_H} versus Re_{D_H} are shown in Figure 7.40. The hydrodynamic-diameter model describes well flows through packed beds at small Re_{D_H}, and the Poiseuille number for flow through packed beds is found experimentally to be $100/3 = 33.33$. The data follow the curve $f_{D_H}\text{Re}_{D_H} = 33.33$ until $\text{Re}_{D_H} \approx 10$. The entire dataset is well represented by the following equation, known as the Ergun correlation [23, 45]:[15]

$$\text{Ergun correlation: friction factor/Reynolds number for flow through packed beds} \qquad \boxed{\frac{100/3}{\text{Re}_{D_H}} + \frac{1.75}{3} = f_{D_H}} \qquad (7.318)$$

The Ergun correlation trends to a constant value $(1.75/3)$ at large Reynolds numbers; the friction factor in very rough pipes also approaches to constant value as Re becomes large (see Figure 7.22). Thus, the result for packed beds shows that at low Reynolds number, the flow may be modeled as flow through a highly irregular noncircular conduit; whereas at high Reynolds number, the response is analogous to turbulent flow through extremely rough pipe (Figure 7.41).

[14] In the literature on this subject, a slightly different friction factor and Reynolds number are defined and used: the particle friction factor ($f_p = 3 f_{D_H}$) and the particle Reynolds number ($\text{Re}_p = (3/2)\text{Re}_{D_H}$).

[15] In terms of particle friction factor $f_p = 3 f_{D_H}$ and particle Reynolds number $Re_p = (3/2)Re_{D_H}$, the Ergun correlation is $150/Re_p + 1.75 = f_p$ [23].

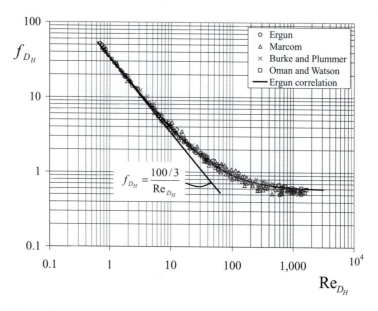

Figure 7.40 Data on friction factor versus Reynolds number at low Reynolds number validate the hypothesis that $f_{D_H} \text{Re}_{D_H}$ is constant for slow flow through packed beds. The constant $f_{D_H} \text{Re}_{D_H}$ is found to be $100/3 = 33.33$. Above $\text{Re}_{D_H} = 10$, the data deviate from the hydraulic-diameter result, following instead $f_{D_H} = \text{constant} = 1.75/3$, which is the result expected for flow in very rough pipe (compare to Figure 7.22 for large ε/D and large Re). Data are from reference [45].

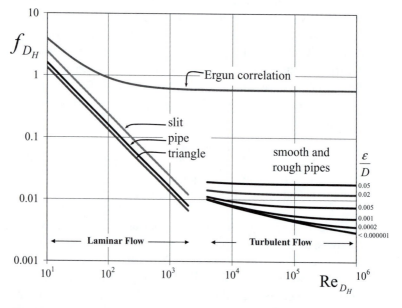

Figure 7.41 We compare the packed-bed result (i.e., Ergun correlation) with the friction-factor/Reynolds-number relationship for flows in other conduits. At low Reynolds number, the Poiseuille number, $\text{Po} = f_{D_H} \text{Re}_{D_H}$, is constant for most cross-sectional shapes and for packed beds. At high Reynolds number and high roughness in pipe flow, the friction factor becomes constant with a value that increases with increasing roughness; packed beds at high Re also have $f_{D_H} = \text{constant}$. At intermediate Reynolds number, the observed behavior of packed beds is intermediate between these two extremes.

The Ergun correlation contains the information we need to solve the question posed at the beginning of this example about flow through a chromatography column. We are asked to consider a 1.0-cm-diameter chromatography column with $\varepsilon = 0.39$ and $a_v = 720$ cm^{-1}. The desired flow rate through the column is $Q = 1.0$ ml/min, and we need to know the applied pressure drop required to achieve this flow rate.

From the literature, we obtain the density and viscosity of toluene at room temperature to be $\rho = 0.8669$ g/cm^3 and $\mu = 0.590$ cp [132]. We directly calculate the Reynolds number, and we calculate the friction factor from the Ergun correlation:

$$v_0 = \frac{Q}{\pi R^2} = \frac{\left(1.0 \, \frac{\text{cm}^3}{\text{min}}\right)\left(\frac{\text{min}}{60 \text{ s}}\right)}{(\pi)(0.5 \text{ cm})^2} = 0.02122066 \text{ cm/s} \tag{7.319}$$

$$D_H = \frac{4\varepsilon}{(1-\varepsilon)a_v} = \frac{4(0.39)}{(1-0.39)(72 \, \frac{1}{\text{cm}})} = 3.55191 \times 10^{-2} \text{ cm} \tag{7.320}$$

$$\text{Re}_{D_H} = \frac{\rho(v_0/\varepsilon)D_H}{\mu} = \frac{(0.8669 \, \frac{\text{g}}{\text{cm}^3})(0.02122066 \, \frac{\text{cm}}{\text{s}})(3.55191 \times 10^{-2} \text{ cm})}{(0.00590 \, \frac{\text{g}}{\text{cm s}})(0.39)}$$

$$= 0.28397 \tag{7.321}$$

$$f_{D_H} = \frac{100}{3\text{Re}_{D_H}} + \frac{1.75}{3} = \frac{100}{(3)(0.28397)} + \frac{1.75}{3} \tag{7.322}$$

$$= 118 \tag{7.323}$$

We now calculate $\Delta p/L$ from the definition of f_{D_H} (see Equation 7.315):

$$f_{D_H} = \left(\frac{\Delta p}{L}\right)\frac{D_H \, \varepsilon^2}{2\rho v_0^2} \tag{7.324}$$

$$\frac{\Delta p}{L} = \frac{2 f_{D_H} \rho v_0^2}{D_H \, \varepsilon^2} \tag{7.325}$$

$$= \frac{(2)(118)\left(0.8669\frac{\text{g}}{\text{cm}^3}\right)(0.02122066 \, \frac{\text{cm}}{\text{s}})^2 \left(\frac{\text{dyne s}^2}{\text{g cm}}\right)}{(3.55191 \times 10^{-2} \text{ cm})(0.39)^2} \tag{7.326}$$

$$= 17.05 \, \frac{\text{dynes}}{\text{cm}^3}\left(\frac{0.1 \text{ Pa}}{1 \text{ dyne/cm}^2}\right)\left(\frac{100 \text{ cm}}{\text{m}}\right) \tag{7.327}$$

$$= \boxed{170 \text{ Pa/m}} \tag{7.328}$$

The methods in this section provide an understanding of steady, laminar flow in noncircular conduits. As with tubes, however, flows in noncircular conduits become unstable at Reynolds numbers greater than a certain critical value ($\text{Re}_{crit} = 2,100$ for the case of tubes). A brief discussion of turbulent flow in noncircular ducts follows.

7.2.2 Turbulent flow in noncircular ducts

In circular pipes, we know that above a Reynolds number of $\mathrm{Re}_{crit} = 2{,}100$ the flow is no longer laminar. In Section 7.1.2.2, we use dimensional analysis on the expression for drag at the pipe walls to determine that the dimensionless wall drag, or friction factor, is a function of the Reynolds number only. We can perform the same calculation for noncircular conduits (assigned as Problem 18 in Chapter 9) and the result is the same: The friction factor in noncircular ducts is a function of the Reynolds number only, as long as the flow is fully developed (i.e., no entrance effects).

In laminar duct flow, we learned that the friction-factor/Reynolds-number relationships for circular and noncircular conduits were similar (i.e., $\mathrm{Po} = f_{D_H}\mathrm{Re}_{D_H} = $ constant) but not identical. In turbulent flow, therefore, we approach the problem with the expectation that the $f_{D_H}(\mathrm{Re}_{D_H})$ correlation for noncircular ducts may be similar to the pipe-flow case but not identical. In an interesting twist of physics, it turns out that the $f_{D_H}(\mathrm{Re}_{D_H})$ correlation for turbulent flow in noncircular ducts is nearly identical to the tube-flow correlations (it is within a few percentage points) [68, 126, 148]. Thus, it is a fair approximation to use the Prandtl correlation (see Equation 7.156) or the Moody chart (see Figure 7.17) for turbulent flow in noncircular ducts, with both friction factor and Reynolds number written in terms of hydraulic diameter. Experiments on rectangular [68], triangular, and annular [69] ducts show that the error in this approximation may be reduced further by adjusting the Reynolds number by a ratio of 16—the Poiseuille number for tube flow—to the laminar-flow Poiseuille number for the duct under consideration:

Modified Prandtl correlation for turbulent flow in noncircular ducts (experimental results):

$$\frac{1}{\sqrt{f_{D_H}}} = 4.0 \log \left[\frac{\mathrm{Re}_{D_H} \sqrt{f_{D_H}}}{\frac{\mathrm{Po}_{duct}}{16}} \right] - 0.40$$

(7.329)

Hydraulic diameter:

$$D_H \equiv \frac{4 A_{xs}}{\mathcal{P}}$$

(7.330)

Fanning friction factor for ducts:

$$f_{D_H} = \frac{\mathcal{F}_{drag}}{\left[\frac{1}{2}\rho \langle v \rangle^2\right](\mathcal{P}L)} = \frac{\Delta p\, A_{xs}}{\left[\frac{1}{2}\rho \langle v \rangle^2\right](\mathcal{P}L)}$$

$$= \frac{\Delta p\, D_H}{\left[\frac{1}{2}\rho \langle v \rangle^2\right](4L)}$$

(7.331)

Reynolds number for ducts:

$$\mathrm{Re}_{D_H} = \frac{\rho \langle v \rangle D_H}{\mu}$$

(7.332)

(Compare Equation 7.329 to the circular-pipe Prandtl correlation in Equation 7.156.) The combination $16 D_H/\mathrm{Po}_{duct}$ is called the effective diameter, D_{eff} [176]. Values of Po_{duct} are shown in Figure 7.38 for several cross sections.

The similarities among turbulent-flow correlations in noncircular ducts are less surprising when we examine the turbulent velocity fields in ducts of most cross

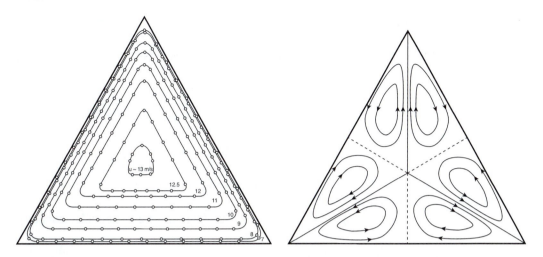

Figure 7.42 Turbulent flows in noncircular ducts are similar. The flow in the central region is plug flow, and shear stress is concentrated at the walls. A weak secondary flow brings momentum from the center of the flow to the walls [127].

sectional shapes [176]. For cross sections that are not too thin in any portion, the flow is plug flow in the core with a viscous boundary layer near the wall (Figure 7.42). This common flow profile is maintained by the presence of a weak secondary flow that exists on top of the mean unidirectional flow. The secondary flow consists of recirculating cells that bring momentum from the center of the conduit to the walls [42]. This recirculation maintains the flow structure of a large plug-like central core, with the viscous-drag effects confined to the near-wall region. Thus, the importance of the shape of the conduit is diminished, and only the amount of wall area—as quantified by the hydraulic diameter—determines the amount of friction [128]. More detail on flows through noncircular ducts is in the literature [42, 152].

The practical result of this discussion about turbulent flow in noncircular ducts is that the modified Prandtl correlation (Equation 7.329) may be used in pressure-drop/flow-rate calculations in such systems, as demonstrated in Example 7.17.

EXAMPLE 7.17. *Water at 25° is forced through a narrow slit that is 1.0 mm by 50 mm in cross section and 50.0 cm long. The driving pressure is 6.0 psi. What is the flow rate through the slit? The flow may be assumed to be turbulent.*

SOLUTION. Flow through a noncircular duct may be analyzed with the hydraulic diameter and the friction-Re correlation in Equation 7.329. For the slit, we first calculate the hydraulic diameter:

$$D_H = \frac{4A_{xs}}{\mathcal{P}} \tag{7.333}$$

$$= \frac{4(1\ \text{mm})(50\ \text{mm})}{2(50\ \text{mm}) + 2(1\ \text{mm})} \tag{7.334}$$

$$= 1.96\ \text{mm} \tag{7.335}$$

To calculate the Reynolds number or the friction factor, we need the average velocity, which is related to the flow rate as $Q = \langle v \rangle WH$, where L is the length of the slit and H is the slit height. We can guess a value of $\langle v \rangle$ and then iterate to the correct answer. Our method is as follows:

1. Guess flow rate Q; calculate $\langle v \rangle = Q/WH$.
2. Calculate Re_{D_H} from the definition of the Reynolds number for noncircular conduits, $\mathrm{Re} = \rho \langle v \rangle D_H / \mu$.
3. Following Example 7.5, calculate f_{D_H} through an iterative process. First guess f_{D_H}, calculate the right side of Equation 7.329, and then calculate a new f_{D_H} from this rightside calculation. Iterate until the value of f_{D_H} converges.
4. Calculate $\langle v \rangle$ from f_{D_H} through its definition for noncircular conduits, Equation 7.331.
5. Calculate Q from $\langle v \rangle$: $Q = \langle v \rangle WH$.
6. If Q does not match the initial guess, use the new value as the next guess and iterate.

This entire process can be carried out in spreadsheet software. The solution for the numbers in this example is:

$$Q = 96 \text{ cm}^3/\text{s} = 1.5 \text{ gpm}$$

$$\langle v \rangle = 190 \text{ cm/s}$$

$$\mathrm{Re}_{D_H} = 4,200$$

$$f_{D_H} = 1.1 \times 10^{-2}$$

7.3 More complex internal flows

In this chapter we provide an introduction to internal flows. The governing equations for all incompressible Newtonian flow problems are known:

Mass conservation:
(continuity equation, $0 = \nabla \cdot \underline{v}$ (7.336)
constant density)

Momentum conservation: $\rho \left(\dfrac{\partial \underline{v}}{\partial t} + \underline{v} \cdot \nabla \underline{v} \right) = -\nabla p + \mu \nabla^2 \underline{v} + \rho \underline{g}$
(Navier-Stokes equation)

(7.337)

Newtonian $\tilde{\underline{\underline{\tau}}} = \tilde{\underline{\underline{\Pi}}} + p\underline{\underline{I}} = \mu \left(\nabla \underline{v} + (\nabla \underline{v})^T \right)$ (7.338)
constitutive equation:

Total molecular fluid $\underline{\mathcal{F}} = \iint_S \left[\hat{n} \cdot \tilde{\underline{\underline{\Pi}}} \right]_{\text{at surface}} dS$ (7.339)
force on a surface $\mathcal{S}$:

Total flow rate out $Q = \iint_S \left[\hat{n} \cdot \underline{v} \right]_{\text{at surface}} dS$ (7.340)
through surface $\mathcal{S}$:

For complex internal flows, advanced mathematical and computational techniques are needed to find solutions to this set of equations. Here we briefly introduce solution methods common to the study of more complex internal flows.

We have solved the equations of Newtonian flow for simple flows. For steady, unidirectional flows (i.e., pipe flow and flow through noncircular ducts), the lefthand side of the Navier-Stokes equation is zero, and we solve the Poisson equation. For unsteady, undirectional flows, the $\partial \underline{v}/\partial t$ term remains, and we seek an unsteady solution (see Section 7.3.1) or a simplified, quasisteady solution (see Section 7.3.2). When flows are not unidirectional, the Navier-Stokes equations are difficult to solve, even for steady flows (discussed at length in Chapter 8 for external flows). For slowly changing geometries in internal flows, approximate solutions to steady, nonunidirectional flows are obtained with the lubrication approximation (see Section 7.3.3). Often, flows in complex geometries and unsteady flows are solved numerically (see Chapter 10). We begin with unsteady-state solutions to the Newtonian flow equations.

7.3.1 Unsteady-state solutions

The incompressible internal flows analyzed in this chapter thus far are steady flows. For steady unidirectional flows, the velocity field has a single nonzero component, and the continuity and Navier-Stokes equations simplify to:[16]

Steady, Incompressible, Unidirectional Flow

Continuity equation: $\quad 0 = \dfrac{\partial v_x}{\partial x}$

x-Component Navier-Stokes: $\quad 0 = -\dfrac{\partial p}{\partial x} + \mu \left(\dfrac{\partial^2 v_x}{\partial y^2} + \dfrac{\partial^2 v_x}{\partial z^2} \right)$ $\qquad$ (7.341)

y-Component Navier-Stokes: $\quad 0 = \dfrac{\partial p}{\partial y}$

z-Component Navier-Stokes: $\quad 0 = \dfrac{\partial p}{\partial z}$

We looked at several problems described by these equations; more solutions are in the literature [152].

For unsteady, incompressible, unidirectional flows, these equations are modified by the retention of the time derivative in the x-component of the momentum balance.

---

[16] Gravity is neglected or incorporated into pressure through the use of an equivalent or dynamic pressure $\mathcal{P}$ (see the Glossary and Problem 10).

Unsteady, Incompressible, Unidirectional Flow

Continuity equation:	$0 = \dfrac{\partial v_x}{\partial x}$
x-Component Navier-Stokes:	$\rho \dfrac{\partial v_x}{\partial t} = -\dfrac{\partial p}{\partial x} + \mu \left(\dfrac{\partial^2 v_x}{\partial y^2} + \dfrac{\partial^2 v_x}{\partial z^2} \right)$
y-Component Navier-Stokes:	$0 = \dfrac{\partial p}{\partial y}$
z-Component Navier-Stokes:	$0 = \dfrac{\partial p}{\partial z}$

$$(7.342)$$

The time derivative is one of the inertial terms (lefthand side) of the momentum balance (see Figure 7.16). So far we considered flows only influenced by viscous forces, which enter into the microscopic momentum balance through the $\mu \nabla^2 \underline{v}$ term on the righthand side. For flows in which inertia is present, there is competition between the inertial and viscous forces. We can see the effect of inertia on flow fields by considering a simple unsteady flow, the unidirectional accelerating flow at the bottom of a tall container (Figure 7.43).

EXAMPLE 7.18. *A semi-infinite fluid bounded by a wall is set in motion by the sudden acceleration of the wall (see Figure 7.43). Calculate the time-dependent velocity and stress fields. The fluid is an incompressible Newtonian fluid. The effect of gravity may be neglected.*

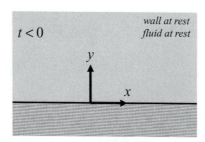

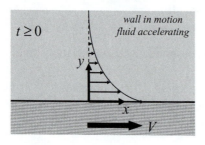

Figure 7.43   A plate forms a boundary for a semi-infinite fluid. At time $t = 0$, the plate is suddenly accelerated and then maintains a constant speed V.

SOLUTION. This flow is discussed at length by Denn [43]. In this flow, a semi-infinite fluid is set in motion by the sudden acceleration of the wall (see Figure 7.43). The fluid velocity field is unidirectional in the x-direction, and the continuity equation for unidirectional flow in the x-direction gives $dv_x/dx = 0$. For wide flow, there is no variation of v_x in the neutral direction z. Thus, v_x varies only in the y-direction. The Navier-Stokes equation reduces to:

$$x\text{-Component Navier-Stokes:} \qquad \rho\frac{\partial v_x}{\partial t} = -\frac{\partial p}{\partial x} + \mu\frac{\partial^2 v_x}{\partial y^2} \qquad (7.343)$$

$$y\text{-Component Navier-Stokes:} \qquad 0 = -\frac{\partial p}{\partial y} \qquad (7.344)$$

$$z\text{-Component Navier-Stokes:} \qquad 0 = -\frac{\partial p}{\partial z} \qquad (7.345)$$

Pressure is constant away from the wall; thus, we assume $\partial p/\partial x = 0$ everywhere. The appropriate initial and boundary conditions are that the fluid is initially quiescent, there is no-slip at the wall, and the fluid is undisturbed far from the wall. Mathematically, these conditions are:

$$t = 0 \qquad v_x = 0 \qquad \text{for all } y \geq 0 \qquad (7.346)$$

$$t > 0 \qquad v_x = V \qquad \text{for } y = 0 \qquad (7.347)$$

$$t < \infty \qquad v_x = 0 \qquad \text{for } y = \infty \qquad (7.348)$$

This flow may be solved analytically by noting that a fortuitous combination of variables, $\zeta \equiv y/\sqrt{4\mu t/\rho}$, reduces the PDE in Equation 7.343 to an ordinary differential equation (ODE). A solution that exploits such a combination of variables is called a similarity solution. For the problem of the wall suddenly set in motion, the velocity profile is as follows (the details of the solution are left to the reader) [43]:

Velocity profile
semi-infinite fluid,
wall suddenly set in motion:
$$\boxed{\frac{v_x(y,t)}{V} = 1 - \text{erf}\left(y/\sqrt{4\mu t/\rho}\right)} \qquad (7.349)$$

where erf () is the error function, defined as:

Error function defined:
$$\boxed{\text{erf } \zeta \equiv \frac{2}{\sqrt{\pi}}\int_0^\zeta e^{-\xi^2}\,d\xi} \qquad (7.350)$$

This solution for $v_x(y,t)$ is plotted in Figure 7.44 for the combined variables (left) and for $v_x(y)$ at various times (right).

The idealized problem of a wall suddenly set in motion has inertial effects due to the term $\rho\partial v_x/\partial t$ as well as viscous effects due to the term $\mu\partial^2 v_x/\partial y^2$. The competition between these two terms dictates the flow structure. The inertial term imposes the tendency of the fluid to remain at rest. The viscous term, because it describes the stress transfer between fluid layers, enforces the tendency of the fluid to match the velocity of the wall. The viscous forces dominate near the wall;

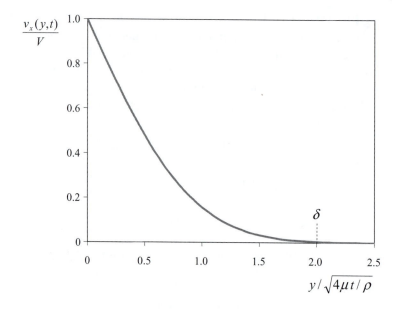

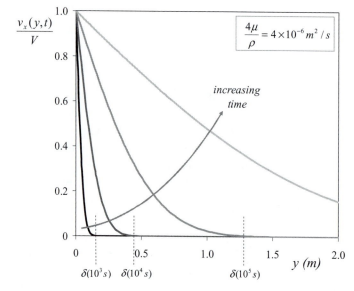

Figure 7.44   The solution to the flow caused by a wall suddenly set in motion may be represented in terms of the dimensionless similarity variable $y/\sqrt{4\mu t/\rho}$ or in terms of the physical variables y and t. In this figure, the flow is upward on the left side and the thickness of the boundary layer is represented by the location on the abscissa where the velocity goes to zero.

whereas far from the wall, the inertial forces dominate. The layer near the wall where viscous forces dominate is called the viscous boundary layer. As shown in Figure 7.44 (bottom), the boundary layer—that is, the region of flow affected by the motion of the wall—grows in thickness with time. The plot of the velocity versus the combined variable $y/\sqrt{4\mu t/\rho}$ (see Figure 7.44, top) shows that when $y/\sqrt{4\mu t/\rho} = 2$, the effect of the wall has died out (i.e., velocity goes to zero). If we consider the thickness of the boundary layer δ to be the value of y when

$y/\sqrt{4\mu t/\rho} = 2$, then δ grows with time as:

$$\text{Boundary-layer thickness, } \delta: \qquad \delta = 2\sqrt{4\mu t/\rho} \qquad (7.351)$$

The boundary layer is a natural outcome of the competition between viscous and inertial forces in this flow. Chapter 8 discusses steady flows in which viscous and inertial effects compete and boundary layers form.

More information about solutions to unsteady fluid flow problems is in the literature [9] or may be pursued numerically [49, 100].

7.3.2 Quasi-steady-state solutions

There are many important engineering flows that vary slowly with time. A modeling approach that works well in this circumstance is to solve the problem at a particular time—for example, the initial time—as if it were a steady-state problem. Subsequently, a quantity that is actually changing with time, a height, a flow rate, or a velocity, is allowed to slowly vary with time. An example of a quasi-steady-state flow solution follows.

EXAMPLE 7.19. *Water is siphoned from a tank as shown in Figure 7.45. What is the flow rate of the water in the siphon tube (ID = 2R) as a function of time? How long does it take the tank to drain?*

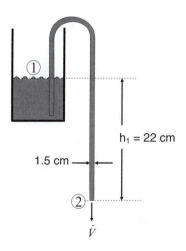

$h_1 = 22$ cm

1.5 cm

$\dot{V}$

Figure 7.45 | A siphon is inherently unsteady in its operation. To solve for the time to drain the tank, we use a quasi-steady-state approach.

SOLUTION. The system of the water flowing in the siphon is an unsteady flow of an incompressible fluid. Therefore, the mechanical energy balance (MEB), which is allowed only on steady-state problems, appears to be an inappropriate choice to use to solve the siphon problem. If, however, we assume that at any instant in time the flow is steady, perhaps we can obtain a useful result.

We seek, therefore, to use the MEB to solve for the instantaneous flow rate in the siphon as a function of the instantaneous height of the fluid in the feed tank. The system of the water in the siphon is single-input, single-output, the fluid is incompressible, and we assume that over a short period that the flow is steady (i.e., quasi-steady-state). There is no heat transfer and no chemical reaction or phase change. All of the requirements of the mechanical energy balance therefore are met over the short period considered:

$$\frac{\Delta p}{\rho} + \frac{\Delta \langle v \rangle^2}{2\alpha} + g\Delta z + F = -\frac{W_{s,by}}{m}$$

Mechanical energy balance (single-input, single-output, steady, no phase change, incompressible, $\hat{v}$ constant across inlet, outlet $\Delta T \approx 0$, no reaction)

(7.352)

To apply the MEB to our siphon, we choose the two points as (1) the free surface in the tank, and (2) the exit point of the siphon. There are no moving parts in the chosen system and therefore no shaft work. The flow in the tank and siphon is tranquil and little friction is generated. The MEB simplifies to the macroscopic Bernoulli equation. The velocity profile is flat at Points (1) and (2); therefore, $\alpha = 1$:

$$\frac{\Delta p}{\rho} + \frac{\Delta \langle v \rangle^2}{2} + g\Delta z = 0$$

Bernoulli equation (single-input, single-output, steady, no phase change, incompressible, $\Delta T \approx 0$, no reaction, no friction, no shaft work)

(7.353)

$$\frac{p_2 - p_1}{\rho} + \frac{\langle v \rangle_2^2 - \langle v \rangle_1^2}{2} + g(z_2 - z_1) = 0 \qquad (7.354)$$

At Points (1) and (2), the pressure is atmospheric; therefore, $p_2 - p_1 = 0$. The quantities z_1 and z_2 refer to the elevations of the two chosen points. We may choose the elevation of the discharge as our reference level for measuring elevation. Thus, $z_2 = 0$ and $z_1 = h(t) + h_a$, where h_a is the vertical distance from the discharge to the bottom of the tank and $h(t)$ is the height of the fluid in the tank at the instant being considered.

The average velocity of the water at the exit $\langle v \rangle_2$ is related to the discharge flow rate Q as:

$$\langle v \rangle_2 = \frac{Q}{\pi R^2} \qquad (7.355)$$

where R is the radius of the siphon tube. The expression $\langle v \rangle_1$ refers to the velocity of the tank water surface; in the solution to a related problem in Chapter 1 (Example 1.5), we assume this velocity is approximately zero. As a result, the Chapter 1 solution is limited to the early stages of draining the tank. For this problem, we must consider all times during the tank draining; thus, $\langle v \rangle$ is not constant. We can express the variable $\langle v \rangle_1$ in terms of the discharge volumetric flow rate:

$$\langle v \rangle_1 = \frac{Q}{A} \tag{7.356}$$

where A is the cross-sectional area of the tank.

Making these substitutions, the mechanical energy balance becomes:

$$\frac{p_2 - p_1}{\rho} + \frac{\langle v \rangle_2^2 - \langle v \rangle_1^2}{2} + g(z_2 - z_1) = 0 \tag{7.357}$$

$$\frac{\langle v \rangle_2^2 - \langle v \rangle_1^2}{2} - g(h(t) + h_a) = 0 \tag{7.358}$$

$$\left(\frac{Q}{\pi R^2} \right)^2 - \left(\frac{Q}{A} \right)^2 = 2g(h(t) + h_a) \tag{7.359}$$

$$Q = \sqrt{\frac{2g(h(t) + h_a)}{\left(\frac{1}{\pi R^2} \right)^2 - \left(\frac{1}{A} \right)^2}} \tag{7.360}$$

For the case of A large and $h(t)$ constant and fixed at the value of the initial height of the fluid in the tank, we recover the solution of Chapter 1 (see Equation 1.45).

To implement a quasi-steady-state solution, we note that flow rate Q is related to the average speed of the falling water level in the tank. Because $h(t)$ is the position of the water level in the tank, $-dh/dt$ is the average speed of the tank water level:

$$\langle v \rangle_1 = -\frac{dh}{dt} = \frac{Q}{A} \tag{7.361}$$

$$-\frac{dh}{dt} = \sqrt{\frac{2g(h(t) + h_a)}{\left(\frac{A}{\pi R^2} \right)^2 - 1}} \tag{7.362}$$

$$\frac{dh}{dt} = \left[-\sqrt{\frac{2g}{\left(\frac{A}{\pi R^2} \right)^2 - 1}} \right] (h(t) + h_a)^{\frac{1}{2}} \tag{7.363}$$

$$\frac{dh}{dt} = \beta (h(t) + h_a)^{\frac{1}{2}} \tag{7.364}$$

where β is the quantity in square brackets in Equation 7.363. Equation 7.364 is straightforward to integrate to obtain $h(t)$. The result is:

Level of tank
drained by a siphon:
$$\boxed{h(t) = -h_a + \left(\frac{\beta t}{2} + \sqrt{h_0 + h_a} \right)^2} \tag{7.365}$$

The flow rate now may be obtained from Equation 7.361, and the time to drain the tank t_f also may be obtained. The details are left to readers (see also Problem 35).

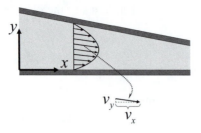

Figure 7.46 The pressure-driven flow between plates that are not parallel must be analyzed with two components of the Navier-Stokes equation.

7.3.3 Geometrically complex flows (including lubrication approximation, converging flows, and entry flows)

For steady flows that are not unidirectional, two or more components of the Navier-Stokes equation are significant. A simple example of a two-dimensional flow is pressure-driven flow between two plates that are not parallel (Figure 7.46). For such a flow (i.e., no flow in width direction $v_z = 0$, wide plates $\partial/\partial z = 0$, steady $\partial/\partial t = 0$, gravity neglected), the microscopic mass and momentum balances simplify as shown here:

Steady, Two-Dimensional, Incompressible Flow

Continuity equation:
$$0 = \frac{\partial v_x}{\partial x} + \frac{\partial v_y}{\partial y}$$

x-Component Navier-Stokes:

$$\rho \left(v_x \frac{\partial v_x}{\partial x} + v_y \frac{\partial v_x}{\partial y} \right) = -\frac{\partial p}{\partial x} + \mu \left(\frac{\partial^2 v_x}{\partial x^2} + \frac{\partial^2 v_x}{\partial y^2} \right)$$
(7.366)

y-Component Navier-Stokes:

$$\rho \left(v_x \frac{\partial v_y}{\partial x} + v_y \frac{\partial v_y}{\partial y} \right) = -\frac{\partial p}{\partial y} + \mu \left(\frac{\partial^2 v_y}{\partial x^2} + \frac{\partial^2 v_y}{\partial y^2} \right)$$

z-Component Navier-Stokes:
$$0 = \frac{\partial p}{\partial z}$$

Note the presence of inertial terms (lefthand-side terms) in both the x- and y-components of the Navier-Stokes equation.

This system of equations is extraordinarily complex and is best solved using numerical-solution techniques [27]. A highly idealized problem of this type is symmetric converging or diverging flow between two plates, called Hamel flow (Figure 7.47), which is introduced in Example 7.20 and discussed in more detail in Denn [43] and Landau and Lifshitz [80].

EXAMPLE 7.20. *Fluid enters the narrow opening between two long, wide plates that diverge as shown in Figure 7.47. Calculate the steady-state velocity and stress fields in the diverging flow. The fluid is an incompressible Newtonian fluid.*

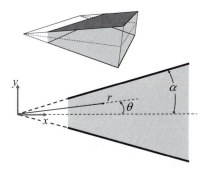

Figure 7.47 A symmetric converging or diverging flow was studied in the early 1900s by Hamel. The complete solution is given in Landau and Lifshitz [80].

SOLUTION. This problem is analyzed in cylindrical coordinates with z as the width direction. To obtain equations that we can solve analytically, we neglect the flow in the θ-direction and assume that the flow is purely radial ($v_\theta = v_z = 0$). We also assume that there are no variations of quantities in the z-direction ($\partial/\partial z = 0$) and that the flow is steady. The governing equations thus become:

Continuity equation:
$$0 = \frac{1}{r}\frac{\partial(rv_r)}{\partial r} \qquad (7.367)$$

r-Component Navier-Stokes:
$$\rho v_r \frac{\partial v_r}{\partial r} = -\frac{\partial p}{\partial r} + \mu\left(\frac{\partial}{\partial r}\left(\frac{1}{r}\frac{\partial(rv_r)}{\partial r}\right) + \frac{1}{r^2}\frac{\partial^2 v_r}{\partial\theta^2}\right) \qquad (7.368)$$

θ-Component Navier-Stokes:
$$0 = -\frac{1}{r}\frac{\partial p}{\partial\theta} + \mu\frac{2}{r^2}\frac{\partial v_r}{\partial\theta} \qquad (7.369)$$

z-Component Navier-Stokes:
$$0 = \frac{\partial p}{\partial z} \qquad (7.370)$$

This is a two-dimensional problem to be solved for $v_\theta(r, \theta)$ and $p = p(r, \theta)$. The solution process is simplified greatly by the constraint imposed by the continuity equation, Equation 7.367. Because we assume that $v_z = v_\theta = 0$, the continuity equation is a PDE that we can integrate:[17]

$$\frac{1}{r}\frac{\partial(rv_r)}{\partial r} = 0 \qquad (7.371)$$

$$(rv_r) = f(\theta) \qquad (7.372)$$

$$v_r = \frac{f(\theta)}{r} \qquad (7.373)$$

[17] Recall that when integrating a partial differential, instead of adding an arbitrary integration constant, we must add an arbitrary integration function of the other variables—in this case, θ. The partial derivative of our result, Equation 7.373, with respect to r satisfies the original equation, Equation 7.367.

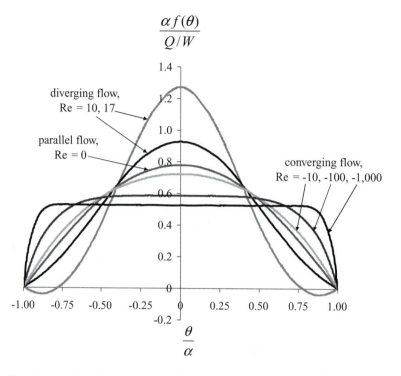

$$\frac{\alpha f(\theta)}{Q/W}$$

diverging flow, Re = 10, 17

parallel flow, Re = 0

converging flow, Re = -10, -100, -1,000

$$\frac{\theta}{\alpha}$$

Figure 7.48 The solution to Hamel flow for flow between converging or diverging plates [43]. Dimensionless variables are $\alpha f(\theta)/(Q/W)$ as a function of scaled angular position θ/α, where α is the convergence half-angle; $\alpha = 45°$ for the case shown. The Reynolds number for this flow is $2\rho Q\alpha/(\mu W)$.

Thus, the problem is simplified from solving for the two-dimensional function $v_\theta(r, \theta)$ to solving for the one-dimensional function $f(\theta)$. Pressure can be eliminated from the Navier-Stokes equations by differentiating the r-component by θ and the θ-component by r, as discussed by Denn [43] (this technique of eliminating pressure is employed in the study of potential flows in Chapter 8). In this way, the problem is reduced from solving the complex set of partial differential equations, Equations 7.367–7.370, to solving a single third-order ODE for the function $f(\theta)$. For the case of $\alpha = 45°$, the solutions for $f(\theta)$ are shown in Figure 7.48. The Reynolds number is defined for this flow as:

$$\text{Reynolds number, Hamel flow:} \qquad \text{Re} \equiv \frac{\rho(2Q\alpha/W)}{\mu} \qquad (7.374)$$

The flow rate (as quantified by Re) may be positive or negative, indicating flow in the positive r-direction (i.e., diverging flow) or in the negative r-direction (i.e., converging flow). The half-angle α also appears in Re; thus, for Re = 0, the flow is unidirectional flow between parallel plates.

The solutions to Hamel flow in Figure 7.48 show the important role of boundary layers in problems in which both inertia and viscosity are significant. Figure 7.48 demonstrates that the solutions for converging and diverging flow are quite different. For converging flow (i.e., negative Re), the velocity profile evolves with increasing |Re| toward a central plug flow with thin viscous boundary layers near the walls. This flow is stable for all |Re|. When the flow is diverging

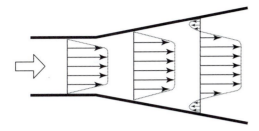

Figure 7.49 This schematic was drawn from a still from a flow-visualization experiment in a diffuser [121, 154]. The flow shows flow separation at the wall. This backflow is caused by the adverse (i.e., rising) pressure gradient in the flow.

(i.e., positive Re), there is a qualitative change to the velocity profile as |Re| increases to large positive values. With increasing |Re| in a diverging channel, the flow has increasing difficulty in maintaining the flow near the boundary where viscosity dominates. Pressure decreases near the wall and, for Re > 14, the calculated solution indicates that the velocity reverses direction near the walls. This circumstance, which is due to the pressure profiles that accompany the flow, is called backflow and is unstable in practice. In experiments, instead of producing this backflow, the flow separates from the wall and produces complex structures and, eventually, turbulence (Figure 7.49).

The discussion of Hamel flow illustrates the nature of the Navier-Stokes equations. We see in this simplest of two-dimensional flows that when both inertial and viscous forces are present, the two contributions interact, and complex flows result. Under certain circumstances—for example, the case of converging Hamel flow at large magnitude of Reynolds number—viscosity and inertia divide the flow domain in two: (1) a region dominated by the viscous effects (i.e, the boundary layer near the wall); and (2) a region dominated by the inertial effects (i.e, the core flow). In other circumstances, as exemplified by the case of diverging Hamel flow, such a division of labor is not produced effectively, and the flow becomes unstable.

Another mixed inertia–viscous flow of major importance is the entry flow in a tube; boundary layers have a role here as well (Figure 7.50). Entry flow has been studied in-depth for conduits of many shapes [152]. The flow near the entry

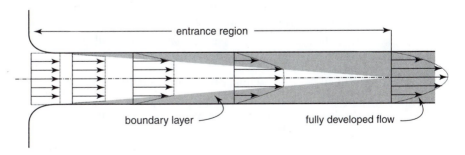

Figure 7.50 At the inlet to tube flow, the velocity profile rearranges and eventually becomes the well-developed flow studied in this chapter. The transition from the inlet to well-developed flow begins by the formation of boundary layers near the walls. These boundary layers grow and eventually merge at the center of the tube.

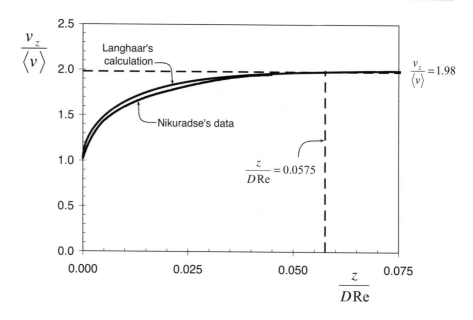

Figure 7.51 As shown in Figure 7.50, when fluid enters a pipe from a reservoir, the velocity profile in the pipe is flat and the centerline velocity $v_z(0)$ equals the average velocity $\langle v \rangle$. As the flow develops along the tube, the velocity profile rearranges and the centerline velocity increases. Calculations by Langhaar and experiments by Nikuradse agree that the entrance effect is no longer important when $L/D > 0.058$Re [81].

of a tube is different from the well-developed flow downstream. Near the entry, the flow enters as plug flow (i.e., with a flat velocity profile), but the presence of the walls causes boundary layers to form near them. Inside the boundary layers, the plug flow slows as the no-slip boundary condition is satisfied at the wall. As the flow progresses in the tube, viscous momentum-transport always is occurring, and the boundary layer grows in thickness, much as it does in the semi-infinite wall case. Eventually, the boundary layers at the wall grow in thickness to fill the entire tube, and the velocity attains its well-known parabolic profile shape (i.e., fully developed, laminar flow, and viscous-dominated). Experiments and calculations in laminar flow show precisely when the flow becomes fully developed [81] (Figure 7.51). Turbulent flows also exhibit boundary layers in which viscous effects are dominant; entry effects in turbulent flow in pipes die out for $L/D > 40$ [132]. In the modeling of blood flow in the human body, both laminar and turbulent flows exist and most flows in the body are entry flows. For this application, more precise correlations are needed [21]:

$$
\begin{array}{ll}
\text{Correlations} & \text{laminar flow: } \dfrac{L_e}{D} = 0.59 + 0.056\text{Re} \quad \text{Atkinson et al. [7]} \\[2mm]
\text{for pipe} & \\
\text{entry length:} & \text{turbulent flow: } \dfrac{L_e}{D} = 4.4\text{Re}^{\frac{1}{6}} \qquad\qquad \text{White [176]}
\end{array}
$$

(7.375)

See also Shah and London, who report an even more precise correlation for laminar entry flow [152].

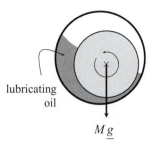

lubricating oil

$M\underline{g}$

Figure 7.52 Schematic of a lubrication flow. The central axis may be the axle of an automobile, for example, and the outer ring represents part of the housing.

The concept of the boundary layer is important in fluid mechanics and it is discussed again in Section 8.2. Boundary-layer formation is a common trait of flows for which both inertial and viscous contributions are important. For engineering applications, consideration of the boundary layer is essential because heat and mass transfer often occur through walls; thus, heat and mass must traverse the boundary layer.

When the inertial contribution to the flow momentum is slight, analytical solutions are sometimes found using a quasi-unidirectional technique known as the lubrication approximation. Lubrication flow is named for flow in narrow gaps between moving parts in which the role of the fluid is to lubricate the parts (Figure 7.52). In such gaps, the flow is only slightly different from unidirectional; the lubrication approximation takes advantage of this similarity by considering the flow to be locally unidirectional and parallel [43]. With this assumption, analytical solutions may be found. The lubrication approximation is useful in polymer-processing flow calculation. Chapter 13 in Denn [43] discusses the lubrication approximation.

This chapter demonstrates that the continuum modeling method is versatile and capable of providing insight to a wide variety of flow problems. The overall strategy is outlined in Section 7.1.2.3: When tackling a difficult flow problem, begin by identifying an idealized version of the flow that can be solved. Then, use the solution to the idealized problem to nondimensionalize the equations of change so that information in the governing equations can be accessed. Finally, solve for $\underline{v}$ and p or conduct experiments and develop data correlations so that the engineering problem may be solved.

Problems that are unidirectional and steady are not difficult to solve—the left-hand side of the Navier-Stokes equation goes to zero, eliminating the nonlinear terms. When we stray from these flows, inertia becomes increasingly important and the flow behavior becomes more complex and fascinating. Chapter 8 confronts these issues as we move on to external flows, which almost always exhibit both viscous and inertial contributions.

7.4 Problems

1. The governing equations for fluid flow are four coupled equations in four unknowns. What are these equations? What is a strategy for solving them?
2. What is the role of dimensional analysis in fluid mechanics?
3. Using the methods in this chapter, write the continuity equation (i.e., microscopic-mass balance) in dimensionless form. What can we learn from the result?
4. Figure 7.6 plots results for the velocity and pressure profiles for steady, Poiseuille flow in a tube. We choose to plot these functions using

dimensionless combinations of the variables and characteristic quantities. Why do we use dimensionless combinations? What difficulties would we encounter if we choose to plot the bare v_z versus r and p versus z?

5. In terms of the problem-solving strategy defined in Section 7.1.2.3, identify the idealized problem, the experiments, and the data correlations that were used to solve the burst-pipe problem of this chapter.

6. Complete the calculation of the velocity profile and the total-stress tensor for steady, pressure-driven flow in a tube (i.e., Poiseuille flow in a tube). In other words, show that Equations 7.22, 7.23, and 7.34 result from the integration and application of Equations 7.18 and 7.19.

7. Show that the Hagen-Poiseuille equation (Equation 7.28) for pressure drop as a function of flow rate in laminar flow follows from the integration of the velocity field across the pipe cross section (Equation 7.26).

8. In the calculation of total drag in a pipe, show using matrix calculations that the simplified expression in Equation 7.122 is equivalent to the definition of axial drag in Equation 7.120.

9. In laminar flow in a tube, calculate the axial drag by beginning with the surface integral in Equation 7.125 and incorporating the solution for the velocity profile. Neglect the effect of gravity.

10. The solution for pressure-driven laminar flow in a tube includes the effect of gravity. How does the solution change if the flow is upward instead of downward? How does the solution change if the pipe is mounted at a 30-degree angle to horizontal? Show that the effect of gravity in all cases can be accounted for by defining the dynamic pressure as given here [43] (see the Glossary):

$$\mathcal{P} \equiv p - \rho g_z Z$$

11. For the burst-pipe problem discussed in this chapter, we first attempt to solve by assuming laminar flow. For the laminar-flow result, what was the Reynolds number calculated in the small pipe? If the flow could have remained laminar up to that Reynolds number (it cannot; the flow becomes unstable), what would have been the Fanning friction factor? Compare this number and the pressure drop it implies to the actual f and Δp that we calculated. Discuss your answer.

12. We neglect the presence of fittings and the velocity change in the burst-pipe example in this chapter. What would be the effect on the burst-pipe calculation if we include the frictional loss due to velocity head, bends, fittings, and valves? Assume that there are eight 90-degree bends, two gate valves, and one globe valve half open in the smaller piping section.

13. We assume a smooth pipe in the burst-pipe example in this chapter. Repeat Example 7.5 assuming that the pipes are galvanized iron with a pipe roughness of 0.0005 foot. Was smooth pipe a good assumption?

14. An 80-foot section of 1/2-inch ID Schedule 40 piping branches into two pipes of the same diameter, one of which is 160 feet long and the other 200.0 feet long (all horizontal). The main pipe is connected to the municipal water supply, which supplies a constant 50.0 psig at the pipe entrance. What are the flow rates through the two pipe exits? What is the pressure at the

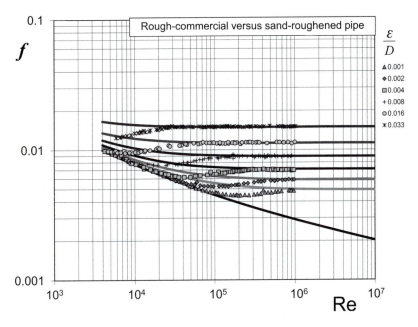

Figure 7.53 Data on rough commercial pipes, represented by the Colebrook correlation (solid lines), are compared with the data of Nikuradse [126] for sand-roughened pipes (discrete points). The two measurements agree at large Re but not at lower values (Problem 17).

splitting point? Assume smooth pipes; do not consider friction losses due to fittings.

15. For turbulent pipe flow, show that Equation 7.156—the Prandtl correlation for fluid friction—is equivalent to the case $\varepsilon = 0$ in the Colebrook correlation (Equation 7.161).

16. For steady pipe flow, repeat branched-piping, Example 7.9 for pipes with roughness $\varepsilon = 0.05$ mm.

17. The Colebrook correlation (i.e., Equation 7.161) gives friction factor as a function of Reynolds number and roughness ratio for commercial pipes. The values of roughness ε for commercial pipes were deduced by comparing the measured asymptotic values of f for real pipes, with the values for f at large Re obtained by Nikuradse [126] on pipes roughened with well-characterized sand of uniform size. The Colebrook equation and Nikuradse's data are compared in Figure 7.53. The two datasets have different shapes at Reynolds numbers below the asymptotic values. What differences can you think of between the wall surfaces on commercial pipes and those on the artificially roughened walls of Nikuradse that might account for these differences? Discuss your answer.

18. In Section 7.1.1, we initially neglect the pressure difference $p_0 - p_L$ when analyzing the Cannon-Fenske viscometer (see Figure 7.11) before ultimately resorting to experimental calibration to account for the small pressure effect (see Equation 7.56). We can account for the pressure difference $p_0 - p_L$ more formally by performing a quasi-steady-state analysis on the system.

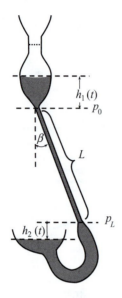

Figure 7.54

The Cannon-Fenske viscometer measures fluid viscosity by allowing the user to time the passage of a set volume of fluid through a long narrow capillary. The flow is driven primarily by gravity; the imposed pressure drop due to the changing driving fluid head $h_1(t)$ and the back pressure due to the head $h_2(t)$ may be accounted for by applying a quasi-steady-state analysis, as described in Problem 18.

Consider the expanded view of the Cannon-Fenske viscometer shown in Figure 7.54. Let $h_1(t)$ represent the time-dependent height of the upper meniscus above the second timing mark and $h_2(t)$ represent the time-dependent height difference between the fluid level in the lower reservoir and the exit of the capillary tube. In the quasi-steady-state approach, we write relationships between variables as if time were moving slowly and the system were nearly in steady state.

(a) Using the principles of fluid statics on our quasistationary system, what is the relationship among p_0, h_1, and atmospheric pressure?

(b) Using the same approach, what is the relationship among p_L, h_2, and atmospheric pressure?

(c) Writing the volumetric flow rate Q as the rate of change of the fluid volume V in the upper reservoir $-dV/dt$, integrate the appropriate equation for volume with respect to time from 0 to t_{efflux} to obtain a pressure-corrected equation for the measurement of fluid kinematic viscosity ν with the Cannon-Fenske viscometer. Assume that $h_1(t)$ and $h_2(t)$ vary linearly with time throughout the experiment:

Answer:

$$\frac{\mu}{\rho} = \left[\frac{\pi R^4 g}{8VL} \left(\frac{h_1(0)}{2} + \frac{h_2(0)}{2} + h_2(t_{\text{efflux}})/2 + L \cos \beta \right) \right] t_{\text{efflux}}$$

(d) Do $h_1(t)$ and $h_2(t)$ vary linearly with time? How important is this effect?

19. When using a calibrated Cannon-Fenske viscometer, it is necessary to employ the same fluid volume as during calibration. To achieve this, the viscometer

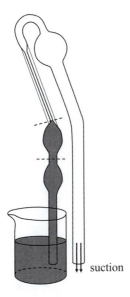

suction

Figure 7.55

Schematic of the inverted loading technique that is required when using a Cannon-Fenske viscometer (Problem 19).

is loaded with fluid as shown in Figure 7.55. The viscometer is inverted into a beaker of fluid and suction is applied to the cleaning arm. In the inverted position, when the fluid reaches the timing mark nearest the capillary, the correct volume has been loaded.

When several concentrations of solution are being measured as part of a sequence, it is convenient to dilute a concentrated solution within the viscometer to make the subsequent measurements on less concentrated solutions. This technique is used in the study of polymers [60]. The Cannon-Fenske viscometer is inappropriate for this type of measurement due to the excess, unknown back pressure that would result from adding additional solvent.

The Ubbelhode viscometer is similar to the Cannon-Fenske, but the exit of the capillary in the former is vented, preventing the back-pressure problem (Figure 7.56). Following the quasi-steady-state technique outlined in Problem 18, calculate the equation that relates kinematic viscosity and efflux time in the Ubbelhode viscometer.

20. Liquid with the physical properties of water flows in a tube in laminar flow. A researcher studying biological flows in tubes wants to conduct experiments on the apparatus and must replace part of the wall with a different solid material that is transparent to a particular kind of electromagnetic radiation. What is the force on the patch of the wall being replaced? The patch is one-eighth the circumference of the tube and is of length l.

21. What is the purpose of the concept of the hydraulic diameter?

22. The correlation between the Fanning friction factor and the Reynolds number for turbulent flow through pipes (circular cross section) is shown in the Moody plot (Figure 7.22). Which plot do we use for the correlation of $f(\text{Re})$ for noncircular conduits? Explain.

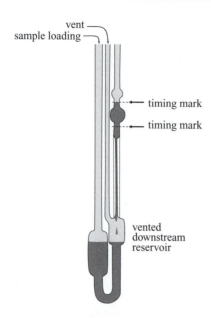

vent
sample loading

timing mark
timing mark

vented
downstream
reservoir

Figure 7.56 Schematic of a Ubbelhode viscometer. The Ubbelhode viscometer is vented at the exit of the capillary. Venting the exit ensures that the pressure at the exit is known (i.e., atmospheric) and allows the sample volume to vary (Problem 19).

23. Hydraulic radius [174] in a noncircular conduit is defined as:

$$\text{Hydraulic radius} \qquad r_H \equiv \frac{A}{\mathcal{P}}$$

where A is the cross-sectional area of the conduit and $\mathcal{P}$ is the wetted perimeter of the conduit. With this definition, how are hydraulic radius and hydraulic diameter related? Discuss your answer.

24. For steady flow in a duct of rectangular cross section, carry out the integrations in Equation 7.229 to obtain the analytical expression for the wall drag in pressure-driven flow.

25. Calculate the Poiseuille number, $f_{D_H}\text{Re}_{D_H}$, for a conduit with elliptical cross section; compare your result with Figure 7.36. The major axis of the ellipse is of length $2a$ and the minor axis is $2b$. The velocity field for laminar flow through a conduit of elliptical cross section is given by White [174] as:

$$v_x = \frac{1}{2\mu}\frac{\Delta p}{L}\frac{a^2 b^2}{a^2 + b^2}\left[1 - \frac{y^2}{a^2} - \frac{z^2}{b^2}\right]$$

The average velocity in this conduit is given by:

$$V = \langle v \rangle = \frac{\Delta p}{4\mu L}\frac{a^2 b^2}{a^2 + b^2}$$

What is the friction-factor/Reynolds-number relationship for this geometry?

26. In steady, pressure-driven, planar-slit flow of an incompressible Newtonian fluid, calculate the vector force on a plane given by the cross section at the exit (see Example 7.10).

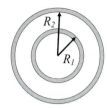

Figure 7.57 For the flow between the inner and outer surfaces of an annulus, the geometry is shown here (Problem 30).

27. Calculate the Poiseuille number, $f_{D_H} \text{Re}_{D_H}$, for a conduit the cross section of which is a rectangle of sides a and b ($b > a$). What is the friction-factor/Reynolds-number relationship for this geometry?

28. Calculate the Poiseuille number, $f_{D_H} \text{Re}_{D_H}$, for a conduit the cross section of which is a square of side a. What is the friction-factor/Reynolds-number relationship for this geometry?

29. Calculate the Poiseuille number, $f_{D_H} \text{Re}_{D_H}$, for a conduit the cross section of which is a slit of infinite width. What is the friction-factor/Reynolds-number relationship for this geometry?

30. Calculate the Poiseuille number, $f_{D_H} \text{Re}_{D_H}$, for flow between the two circular surfaces of an annulus. Let R_1 be the outside radius of the inner pipe and R_2 be the inside radius of the outer pipe (Figure 7.57). What is the friction-factor/Reynolds-number relationship for this geometry?

31. For flow through a rectangular duct, show that in the limit of infinite width, the solution for velocity (Equation 7.212) becomes the solution for velocity in steady flow through a slit.

32. In Poiseuille flow in a slit, complete the integration in Example 7.10 to obtain the final velocity profile for Poiseuille flow in a slit (Equation 7.188). Calculate the flow rate per unit width by carrying out the missing calculus/algebra to arrive at Equation 7.194.

33. Water at $25°$ is forced through an isosceles triangular duct that is 1.0 mm on a side and 5.0 cm long. The driving pressure is 6.0 psig; the exit is open to the atmosphere. What is the flow rate through the slit? Assume the flow to be turbulent.

34. Under what conditions (i.e., limits) does the solution for tangential-annular flow (see figure for Problem 37) approach the parallel-plate solution (Example 6.3)? Using the solution given here, perform a coordinate transformation to show that this is so.

$$\underline{v} = \begin{pmatrix} 0 \\ \left(\frac{\kappa^2 \Omega R}{\kappa^2 - 1}\right)\left(\frac{r}{R} - \frac{R}{r}\right) \\ 0 \end{pmatrix}_{r\theta z}$$

35. For a tank draining through an exit in the bottom, calculate the flow rate by completing a quasi-steady-state calculation like that discussed in Example 7.19. You may neglect friction.

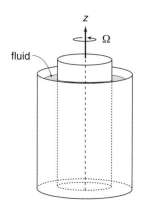
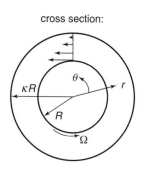

cross section:

| Figure 7.58 | Tangential annular flow of a Newtonian fluid (Problem 37). |

36. Using a numerical software package, calculate the total force on the wall for pressure-driven flow in a slit. How does your numerical result compare to the analytical result? Use the same boundary conditions in both solutions.

37. *Flow Problem: Tangential annular flow.* An incompressible Newtonian fluid fills the annular gap between a cylinder of radius κR and an outer cup of inner radius R (Figure 7.58). The inner cylinder turns counter clockwise at an angular velocity Ω radians/s. The flow may be assumed to be symmetrical in the azimuthal direction (i.e., no θ variation). A pressure gradient develops in the radial direction; the pressure at $z = L$ at the inner cylinder is p_1. Calculate the steady state velocity profile, the radial pressure distribution, and the torque needed to turn the inner cylinder.

38. *Flow Problem: Pressure-driven flow of a Newtonian fluid in an annular gap.* Calculate the velocity profile and flow rate for pressure-driven flow of an incompressible Newtonian liquid in the annular gap between two vertical cylinders. The radius of the inner cylinder is κR and the radius of the outer cylinder is R. The pressure at an upstream point is P_0; at a point a distance L downstream, the pressure is P_L. Assume that the flow is well developed and at steady state. You may neglect gravity.

39. *Flow Problem: Pressure-driven flow of a Newtonian fluid in an annular gap, numerical.* Solve Problem 38 using computer simulation software [27]. Calculate the forces on both the inner and outer surfaces.

40. *Flow Problem: Flow due to natural convection between two long plates.* The flow between the panes of glass in a double-pane window may be modeled as shown in Figure 7.59. Calculate the velocity profile at steady state. Assume the plates are infinitely long and wide (for answer, see Example 1.11). The density variation with position may be handled as follows. The density of the gas is a function of temperature as given by:

$$\rho = \bar{\rho} - \bar{\rho}\bar{\beta}(T - \bar{T})$$

where $\bar{\rho}$ is the mean density, $\bar{\beta}$ is the mean coefficient of thermal expansion, and $\bar{T}$ is the mean temperature (all constant). The temperature profile

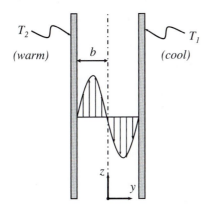

Figure 7.59   Temperature difference generates a flow between two long, wide plates (i.e., hot air rises). We obtain the velocity profile given in Equation 1.140 by using the methods in this chapter in conjunction with energy-balance equations (Problem 40).

obtained from the energy balance is:

$$T = \frac{T_1 - T_2}{2b} y + \frac{T_2 + T_1}{2}$$
$$= \frac{T_1 - T_2}{2b} y + \bar{T}$$

41. *Flow Problem: Radial flow between parallel disks.* An incompressible Newtonian fluid fills the gap between two parallel disks of radius R (Figure 7.60). Fluid is injected through a hole in the center of the top disk, and a steady radial flow occurs. The flow may be assumed to be symmetrical in the azimuthal direction (i.e., no θ variation). A pressure gradient develops in the radial direction; the pressure near the center is p_0 and the pressure at the rim is p_R. Calculate the steady state velocity profile and the radial pressure distribution.

42. *Flow Problem: Unsteady one-dimensional flow, startup.* An incompressible Newtonian fluid is in contact with a long, tall wall that initially is stationary

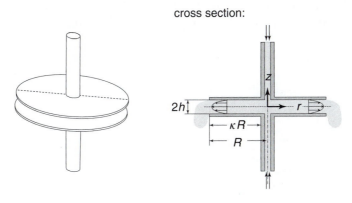

Figure 7.60   Radial flow of a Newtonian fluid from between parallel disks (Problem 41).

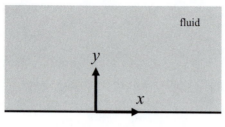

$t = 0$, stationary plate

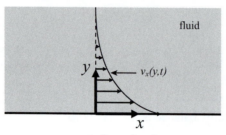

$t > 0$, moving plate

Figure 7.61   Startup flow of a plate in a semi-infinite Newtonian fluid (Problem 42).

(Figure 7.61). The wall suddenly accelerates and moves at steady velocity V. The pressure is uniform throughout the flow. Calculate the steady state velocity profile. Plot the velocity solution for various values of time.

43. *Flow Problem: Flow near an oscillating wall.* An incompressible Newtonian fluid is bounded on one side by a wall and is infinite in the y-direction (Figure 7.62). The wall is moved back and forth according to:

$$v_x(t)|_{\text{wall}} = V \cos \omega t = \mathcal{R}\{V e^{i\omega t}\}$$

What is the time-dependent velocity profile in the fluid as a function of position and time? (see also page 102 of [104]).

44. *Flow Problem: Squeeze flow.* An incompressible Newtonian fluid fills the gap between two parallel disks of radius R (Figure 7.63). The disks are subjected to axial forces that cause them to squeeze together. The fluid in the gap responds by producing a combined axial and radial flow that pushes fluid

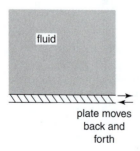

plate moves
back and
forth

Figure 7.62   A plate forms a boundary for a semi-infinite fluid. The wall is moved according to a sinusoidal function (Problem 43).

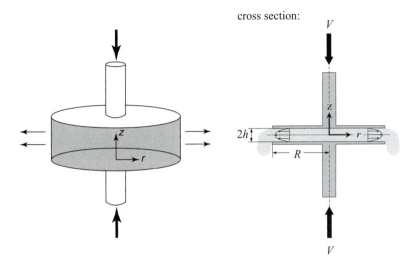

cross section:

Figure 7.63 Squeeze flow of a Newtonian fluid between parallel disks (Problem 44).

out of the gap. The flow may be assumed to be symmetrical in the azimuthal direction (i.e., no θ variation). A pressure gradient develops in the radial direction; the pressure at the center is p_0 and the pressure at the rim is p_R. Calculate the steady state velocity profile and the radial pressure distribution. If the plates are moving with speed V, calculate the force needed to maintain the motion.

45. *Flow Problem: Rod turning in an infinite fluid.* A rod rotates counterclockwise in an infinite bath of fluid. What is the velocity field in the fluid? The radius of the rod is R, the length of the rod is L, and the rod turns at angular velocity Ω in a fluid of viscosity μ. The flow is steady and the fluid is Newtonian.

46. *Flow Problem: Poiseuille flow in a rectangular duct.* An incompressible Newtonian fluid flows down the axis of a duct of rectangular cross section under the influence of a pressure gradient (Figure 7.31). The width of the duct is 2W and the height of the duct is 2H. The upstream pressure is p_0 and the pressure a distance L downstream is p_L. Calculate the steady state velocity and pressure profiles. Note: the velocity is three-dimensional and the solution involves a series of hyperbolic trigonometric functions [174].

47. *Flow Problem: Poiseuille flow in a rectangular duct, numerical.* Calculate the velocity field and flow rate for steady, well-developed, pressure-driven flow in a duct of rectangular cross section (Poiseuille flow in a duct; see Figure 7.31). Compare your result to the analytical solution [174].

48. *Flow Problem: Two-dimensional planar flow in a right-angle tee-split, numerical solution.* Flow enters a two-dimensional right-angle tee-split as shown in Figure 7.64. The flow is steady, two-dimensional flow of an incompressible Newtonian fluid (water may be used). Calculate the flow field and the force on the wall as a function of the inlet Reynolds number. Produce appropriate plots to demonstrate the characteristics of the flow.

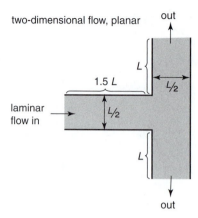

two-dimensional flow, planar

Figure 7.64   Numerical simulation software may be used to calculate the flow domain for two-dimensional planar flow in a right-angle split (Problem 48).

49. *Flow Problem: Two-dimensional axisymmetric flow into radial wall flow in a narrow gap, numerical solution.* Flow exits a pipe at the center of a disk and impinges on a wall producing a radial flow that spreads outward between parallel disks as shown in Figure 7.65. The flow is steady, two-dimensional, axisymmetric flow of an incompressible Newtonian fluid (water may be used). Calculate the flow field and the force on the wall as a function of the inlet Reynolds number. Produce appropriate plots to demonstrate characteristics of the flow.

50. *Flow Problem: Two-dimensional axisymmetric flow through an orifice, numerical solution.* Flow passes through an orifice positioned in the center of a tube as shown in Figure 7.66. The flow is steady, slow, two-dimensional flow of an incompressible Newtonian fluid (water may be used). Calculate the flow field and the pressure drop across the orifice as a function of the inlet Reynolds number. Produce appropriate plots to demonstrate characteristics of the flow.

51. *Flow Problem: Two-dimensional planar cavity flow, numerical solution.* Flow is produced in a cavity by the motion of the top wall as shown in Figure 7.67.

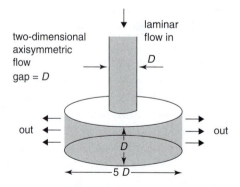

two-dimensional axisymmetric flow gap = D

Figure 7.65   Numerical simulation software may be used to calculate two-dimensional axisymmetric flow into radial wall flow in a narrow gap (Problem 49).

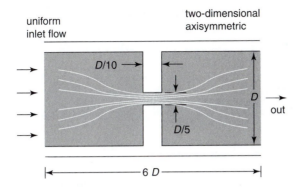

Numerical simulation software may be used to calculate the flow domain for two-dimensional axisymmetric flow through an orifice (Problem 50).

The flow is steady, two-dimensional planar flow of an incompressible Newtonian fluid (water may be used). Calculate the flow field and the force on the stationary walls as a function of a Reynolds number based on wall velocity and cavity depth. Produce appropriate plots to demonstrate the characteristics of the flow.

52. *Flow Problem: Two-dimensional planar gradual contraction near wall, numerical solution.* Flow enters a channel that gradually contracts as shown in Figure 7.68. The flow is steady, two-dimensional flow of an incompressible

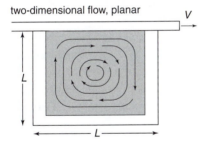

Numerical simulation software may be used to calculate the flow domain for two-dimensional planar cavity flow (Problem 51).

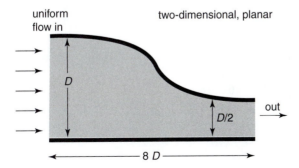

Numerical simulation software may be used to calculate the flow domain for two-dimensional planar gradual contraction near the wall (Problem 52).

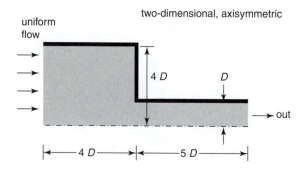

two-dimensional, axisymmetric

uniform
flow

Figure 7.69 Numerical simulation software may be used to calculate the flow domain for two-dimensional axisymmetric 4:1 contraction (Problem 53).

Newtonian fluid (water may be used). Calculate the flow field and the force on the two walls as a function of the inlet Reynolds number. Produce appropriate plots to demonstrate characteristics of the flow.

53. *Flow Problem: Two-dimensional axisymmetric 4:1 contraction, numerical solution.* Flow enters 4:1 axial contraction as shown in Figure 7.69. The flow is steady, two-dimensional, axisymmetric flow of an incompressible Newtonian fluid (water may be used). Calculate the flow field and the force on the wall as a function of the outlet Reynolds number. Produce appropriate plots to demonstrate characteristics of the flow.

54. *Flow Problem: Flow in an obstructed channel, numerical.* For the obstructed flow shown in Figure 7.70, calculate the flow field with a numerical problem solver. What is the velocity field?

55. *Flow Problem: Squeeze flow with constant force.* For the same flow as described in Problem 44, calculate the plate separation as a function of time if the applied force is constant.

56. *Flow Problem: Helical flow.* An incompressible Newtonian fluid fills the annular gap between a cylinder of radius κR and an outer shell of inner

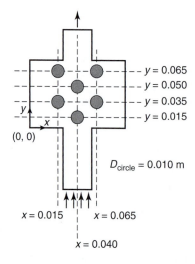

Figure 7.70 Numerical problem-solving software may be used for complex flow geometries shown here (Problem 54).

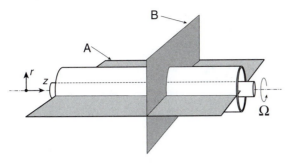

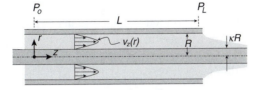

Cross section A:

Cross section B:

Figure 7.71   Helical flow of a Newtonian fluid (Problem 56).

radius R (Figure 7.71). The inner cylinder turns counter clockwise at an angular velocity Ω radians/s. In addition, the inner cylinder is pulled to the right at a velocity V. The combined effect of these two motions produces a helical flow. The flow may be assumed to be symmetrical in the azimuthal direction (i.e., no θ variation). The axial pressure gradient is constant and denoted λ, and the pressure at the inner cylinder is $P_{\kappa R}$. Calculate the steady state velocity profile, the radial pressure distribution, and the torque needed to turn the inner cylinder.

8 External Flows

In Chapter 7, we applied analysis methods to flows inside pipes and other closed conduits. We started with a practical challenge of estimating the extent of a home flood and developed our solution method by thinking about that problem from various angles (Figure 8.1). We first decided on the goal of our analysis; then, starting with the simplest models, we systematically investigated flows of increasing complexity until we found a solution to the burst-pipe problem through dimensional analysis and data correlations. This protocol is general, and it can be applied to other flows, as demonstrated in this chapter.

We turn now to external flows. External flow is a term used to describe flows over or around obstacles. The wind blowing on a skyscraper is an example of an external flow (see Example 2.5), as is an electric fan cooling a printed circuit board in a computer or a cleaning jet directed past the fender of a freshly painted automobile. Objects moving through fluids also create external flows (see Figure 2.11). Ships on the ocean, mixing blades in viscous liquids, and skydivers (Figure 8.2) are all operating in external flows. External flows are not unidirectional, steady flows; thus, both inertia and viscosity affect flow behavior.

We begin Section 8.1 with a practical problem and follow the strategy of Chapter 7 to arrive at a solution. In the process, we investigate solutions to simple, classic problems of external flow; resort to dimensional analysis; and, finally, address complex engineering problems in external flow with data correlations. The study of external flow in this chapter includes an in-depth discussion of boundary layers in Section 8.2. The creation of boundary layers is nature's way of isolating viscous effects from strong inertial effects.

Section 8.3 discusses complex external flows and introduces the use of vorticity in flow modeling. Vorticity is a flow-field property that allows us to keep track of rotational character in flows. As discussed in Section 8.3.1, a key effect of the no-slip boundary condition is to introduce the tendency to rotate into flow fields. In external flows and in complex flows of all types, it is convenient to keep track of the transport of rotational character—introduced by the wall and measured by vorticity—in addition to keeping track of the transport of velocity and the distribution of pressure as we have thus far. The definition and tracking of vorticity is a tactic devised to clarify the behavior of flows in which both inertia and viscosity are important. We turn now to introductory external-flow problems.

PROBLEM: **A pipe bursts: How much water
was wasted in the flood?**

Questions	**Answers**
What was driving the flow?	Pressure.
Under a given pressure, how much flow occurs?	Try simple laminar analysis to find out.
Is the laminar-flow solution correct?	No.
What is wrong with the laminar prediction?	Experiments show flow is turbulent.
Can we analyze turbulent flow using laminar-flow methods?	No; too difficult; flow is statistical and time-varying.
What can the governing equations tell us about turbulent flow?	Dimensional analysis; friction factor is, a function only of Reynolds number.
What is the experimental relationship f(Re) for turbulent flow?	Colebrook correlation.
How much water was wasted in the flood?	89,000 gallons

Figure 8.1   In Chapter 7 we use this process to solve the burst-pipe problem. This solution involved the analysis of an internal flow; in this chapter, the same methodology is applied to external flows.

8.1 Flow around a sphere

The topic of this section is the external flow that takes place around an obstacle in the path of a uniform flow. We begin with a simple obstacle: a single, isolated sphere. We choose to study this flow because it is an entry point to understanding flows around more complex objects such as automobiles moving through air or hurricane winds pounding a building. Investigating flows around obstacles leads to the concept of the drag coefficient C_D and development of C_D-Reynolds number correlations, which are experimentally determined relationships essential to many external-flow engineering problems. Following our usual practice, we begin with a practical problem.

EXAMPLE 8.1. *What is the maximum speed reached by a skydiver who jumps out of an airplane at 13,000 feet (see Figure 8.2)? How much can the speed of the skydiver vary depending on her body position (i.e., arms and legs flung out or pulled in tightly)?*

SOLUTION. In this example, we are asked several questions about skydiving, a problem that is fundamentally an object falling through a viscous fluid: air. According to Newton's laws of motion, a body under the pull of gravity falling through a vacuum falls with constant acceleration. A body falling in the presence

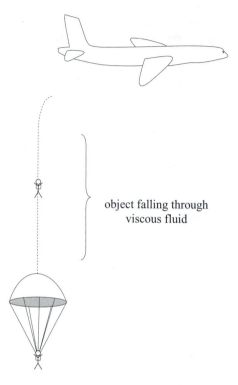

Skydiving is a flow that we can analyze through methods in this text. The skydiver is an object moving through a viscous fluid. From the point of view of the skydiver, the flow is a stationary object with a flow rushing past it.

of a viscous fluid is subject to retarding fluid forces, as discussed in Chapter 2 (see Figure 2.7). Fluid forces slow the motion of the object and, ultimately, the downward force due to gravity is balanced by the retarding fluid forces. At steady-state, the object reaches a zero-acceleration condition called the terminal speed.

Here, we seek to calculate the terminal speed of a skydiver. Our solution to the problem should consider aspects of object shape and orientation and perhaps changes in fluid viscosity due to density and temperature variations. As a first approach to this problem, we consider the skydiver to be a simple object (i.e., a sphere) falling in a fluid of constant viscosity μ (Figure 8.3). Following the same approach as in our initial problem in Chapter 7, we calculate the skydiver's terminal speed from this first analysis and see how it compares with literature values for the observed speed. Depending on the comparison, we then refine our analysis to obtain a more accurate result.

To calculate the terminal speed of a falling sphere, we apply Newton's second law (i.e., momentum conservation) to the sphere (Figure 8.4):

$$
\begin{array}{cc}
\text{Newton's second law:} \\
\text{momentum conservation} \\
\text{for a body}
\end{array}
\qquad
\boxed{\underset{\substack{\text{all forces} \\ \text{acting on body}}}{\sum} \underline{f} = m\underline{a}}
\qquad (8.1)
$$

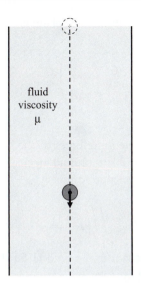

Figure 8.3 We can model a skydiver as an object falling in a vast container of a fluid of viscosity μ.

where $\underline{f}$ represents the various forces on the body, m is the mass of the body, and $\underline{a}$ is the acceleration of the body. The acceleration of the falling sphere is zero (i.e., it falls at constant terminal speed). The two forces on the skydiver are gravity and the fluid forces:

$$\begin{array}{l}\text{Momentum balance} \\ \text{on skydiver} \\ \text{at terminal speed:}\end{array} \qquad \sum_{\substack{\text{all forces} \\ \text{acting on body}}} \underline{f} = m\underline{a} = 0 \qquad (8.2)$$

$$\underline{f}_{\text{gravity}} + \underline{f}_{\text{fluid}} = 0 \qquad (8.3)$$

$$m\underline{g} + \underline{\mathcal{F}} = 0 \qquad (8.4)$$

where m is the mass of the skydiver, $\underline{g} = -g\hat{e}_z$ is the acceleration due to gravity, and $\underline{\mathcal{F}}$ is the retarding fluid force on the sphere. In the Cartesian coordinate

sphere falling at steady state:

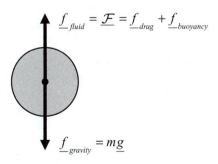

$$\underline{f}_{\text{fluid}} = \underline{\mathcal{F}} = \underline{f}_{\text{drag}} + \underline{f}_{\text{buoyancy}}$$

$$\underline{f}_{\text{gravity}} = m\underline{g}$$

Figure 8.4 The forces on the moving sphere are gravity and the forces due to the fluid. The fluid forces consist of two types: buoyancy and drag. At steady state, the upward and downward forces balance and the sphere acceleration is zero. The speed of the sphere at steady state is called the terminal speed.

system with $\hat{e}_z$ upward, the force due to gravity on the sphere is written as:

$$f_{\text{gravity}} = m\underline{g} = \begin{pmatrix} 0 \\ 0 \\ -\left(\frac{4\pi R^3}{3}\right)\rho_{\text{body}}g \end{pmatrix}_{xyz} \qquad (8.5)$$

where ρ_{body} is the density of the sphere.

The problem now becomes to calculate the fluid forces on the body $\mathcal{F}$. Calculating fluid forces on an object is a basic problem in fluid mechanics that can be solved using the microscopic-momentum-balance approach (see Chapter 6):

$$\begin{array}{c} \text{Total molecular fluid force} \\ \text{on a surface } \mathcal{S}: \end{array} \qquad \mathcal{F} = \iint_{S} \left[\hat{n} \cdot \underline{\underline{\tilde{\Pi}}}\right]_{\text{at surface}} dS \qquad (8.6)$$

To calculate $\mathcal{F}$, we need the total-stress tensor in the fluid $\underline{\underline{\tilde{\Pi}}}$ for the flow being considered; to calculate $\underline{\underline{\tilde{\Pi}}}$, which is equal to $-p\underline{\underline{I}} + \mu\left(\nabla\underline{v} + (\nabla\underline{v})^T\right)$, we need the velocity field $\underline{v}$. Thus, to proceed with our skydiver calculation, we need the velocity field around a sphere falling in a fluid of constant viscosity.

We postpone our calculation of the terminal speed of a skydiver and turn instead to learning the fundamentals of flow past a sphere in a uniform flow. From $\underline{v}$ in the sphere case, we can calculate $\underline{\underline{\tilde{\Pi}}}$; from $\underline{\underline{\tilde{\Pi}}}$, we can calculate $\mathcal{F}$ for the sphere from Equation 8.6. After absorbing the lessons of flow past a sphere, we return to the skydiver problem and continue with the calculation of the skydiver's terminal speed.

8.1.1 Creeping flow around a sphere

The problem of slow, steady flow past a sphere is called *creeping flow* or *Stokes flow*, after George Gabriel Stokes (1819–1903), the mathematician and physicist who presented groundbreaking calculations on this flow in 1851. The equation for magnitude of drag on a sphere as a function of terminal speed is called Stokes law, and we derive it in Example 8.2 (see Equation 8.62). With Claude-Louis Navier (1785–1836), Stokes also is credited with the elucidation of the microscopic-momentum balance, now called the Navier-Stokes equations. Using the microscopic-balance methodology (see Chapters 6 and 7), we arrive at the Stokes solution for the velocity and stress fields in creeping flow around a sphere.

EXAMPLE 8.2. *Calculate the velocity field, the stress field, and the force on the sphere in the steady upward flow of an incompressible Newtonian fluid around a stationary solid sphere of diameter 2R. The fluid approaches the sphere with a uniform upstream velocity v_∞ (Figure 8.5).*

SOLUTION. The flow shown in Figure 8.5 is the equivalent of a sphere falling slowly downward in a viscous fluid (see Figure 8.3). We analyze the flow in a coordinate system that is anchored to the sphere; thus, the fluid appears to rise upward at a steady speed v_∞.

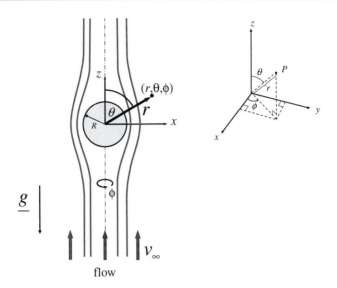

Schematic of flow around a sphere. In the creeping-flow limit (i.e., no inertia), this flow is known as Stokes flow.

The presence of the sphere makes it reasonable to analyze this problem in spherical coordinates, r, θ, and ϕ. The z-direction of the Cartesian system also is an important direction because $\hat{e}_z$ is both the far-field flow direction and is related to the direction of gravity ($\underline{g} = -g\hat{e}_z$). Therefore, both the spherical and the Cartesian systems shown in Figure 8.5 are used.

In spherical coordinates, the fluid velocity field may be written as:

$$\underline{v} = \begin{pmatrix} v_r \\ v_\theta \\ v_\phi \end{pmatrix}_{r\theta\phi} = \begin{pmatrix} v_r \\ v_\theta \\ 0 \end{pmatrix}_{r\theta\phi} \tag{8.7}$$

We assume that v_ϕ is equal to zero; that is, there is no swirling component to the flow. The flow is steady but it is not unidirectional.

Mass conservation is given by the continuity equation, and we write it in spherical coordinates (see Equation B.5-3) as follows:

$$\begin{array}{cc} \text{Continuity equation} & \nabla \cdot \underline{v} = 0 \\ \text{(Gibbs notation):} & \end{array} \tag{8.8}$$

$$\frac{1}{r^2}\frac{\partial(r^2 v_r)}{\partial r} + \frac{1}{r\sin\theta}\frac{\partial(v_\theta \sin\theta)}{\partial\theta} + \frac{1}{r\sin\theta}\frac{\partial(v_\phi)}{\partial\phi} = 0 \tag{8.9}$$

The density is constant; thus, $\partial\rho/\partial t = 0$ and ρ may be removed from the spatial derivatives and subsequently canceled out of the equation. With these considerations and $v_\phi = 0$, the continuity equation simplifies to:

$$\begin{array}{cc} \text{Continuity equation,} & \frac{1}{r^2}\frac{\partial(r^2 v_r)}{\partial r} + \frac{1}{r\sin\theta}\frac{\partial(v_\theta \sin\theta)}{\partial\theta} = 0 \\ \text{flow around a sphere:} & \end{array} \tag{8.10}$$

Comparing this result with the simplified continuity equation for Poiseuille flow in a tube (see Equation 7.8), Equation 8.10 is more complicated. Laminar flow in a tube is in the same direction at every location (i.e., $\underline{v} = v_z\hat{e}_z$ for tube flow), whereas the fluid velocity in flow around a sphere is in different directions depending on the location we choose to observe. The flow around an obstacle is more complex than pipe flow, even in the highly symmetrical case of flow around a sphere.

The microscopic-momentum balance for an incompressible Newtonian fluid is the Navier-Stokes equation. The Navier-Stokes equation written in spherical coordinates is given in Table B.7 in Appendix B and reproduced here:

$$\text{Navier-Stokes equation:} \quad \rho \left(\frac{\partial \underline{v}}{\partial t} + \underline{v} \cdot \nabla \underline{v} \right) = -\nabla p + \mu \nabla^2 \underline{v} + \rho \underline{g} \qquad (8.11)$$

$$\rho \begin{pmatrix} \frac{\partial v_r}{\partial t} \\ \frac{\partial v_\theta}{\partial t} \\ \frac{\partial v_\phi}{\partial t} \end{pmatrix}_{r\theta\phi} + \rho \begin{pmatrix} v_r \left(\frac{\partial v_r}{\partial r} \right) + v_\theta \left(\frac{1}{r} \frac{\partial v_r}{\partial \theta} - \frac{v_\theta}{r} \right) + v_\phi \left(\frac{1}{r \sin\theta} \frac{\partial v_r}{\partial \phi} - \frac{v_\phi}{r} \right) \\ v_r \left(\frac{\partial v_\theta}{\partial r} \right) + v_\theta \left(\frac{1}{r} \frac{\partial v_\theta}{\partial \theta} + \frac{v_r}{r} \right) + v_\phi \left(\frac{1}{r \sin\theta} \frac{\partial v_\theta}{\partial \phi} - \frac{v_\phi}{r} \cot\theta \right) \\ v_r \left(\frac{\partial v_\phi}{\partial r} \right) + v_\theta \left(\frac{1}{r} \frac{\partial v_\phi}{\partial \theta} \right) + v_\phi \left(\frac{1}{r \sin\theta} \frac{\partial v_\phi}{\partial \phi} + \frac{v_r}{r} + \frac{v_\theta}{r} \cot\theta \right) \end{pmatrix}_{r\theta\phi}$$

$$= - \begin{pmatrix} \frac{\partial p}{\partial r} \\ \frac{1}{r} \frac{\partial p}{\partial \theta} \\ \frac{1}{r \sin\theta} \frac{\partial p}{\partial \phi} \end{pmatrix}_{r\theta\phi}$$

$$+ \mu \begin{pmatrix} \left(\frac{\partial}{\partial r} \left(\frac{1}{r^2} \frac{\partial}{\partial r} (r^2 v_r) \right) + \frac{1}{r^2 \sin\theta} \frac{\partial}{\partial \theta} \left(\sin\theta \frac{\partial v_r}{\partial \theta} \right) + \frac{1}{r^2 \sin^2\theta} \frac{\partial^2 v_r}{\partial \phi^2} \right. \\ \left. - \frac{2}{r^2 \sin\theta} \frac{\partial}{\partial \theta} (v_\theta \sin\theta) - \frac{2}{r^2 \sin\theta} \frac{\partial v_\phi}{\partial \phi} \right) \\ \left(\frac{1}{r^2} \frac{\partial}{\partial r} \left(r^2 \frac{\partial v_\theta}{\partial r} \right) + \frac{1}{r^2} \frac{\partial}{\partial \theta} \left(\frac{1}{\sin\theta} \frac{\partial}{\partial \theta} (v_\theta \sin\theta) \right) + \frac{1}{r^2 \sin^2\theta} \frac{\partial^2 v_\theta}{\partial \phi^2} \right. \\ \left. + \frac{2}{r^2} \frac{\partial v_r}{\partial \theta} - \frac{2 \cot\theta}{r^2 \sin\theta} \frac{\partial v_\phi}{\partial \phi} \right) \\ \left(\frac{1}{r^2} \frac{\partial}{\partial r} \left(r^2 \frac{\partial v_\phi}{\partial r} \right) + \frac{1}{r^2} \frac{\partial}{\partial \theta} \left(\frac{1}{\sin\theta} \frac{\partial}{\partial \theta} (v_\phi \sin\theta) \right) + \frac{1}{r^2 \sin^2\theta} \frac{\partial^2 v_\phi}{\partial \phi^2} \right. \\ \left. + \frac{2}{r^2 \sin\theta} \frac{\partial v_r}{\partial \phi} + \frac{2 \cot\theta}{r^2 \sin\theta} \frac{\partial v_\theta}{\partial \phi} \right) \end{pmatrix}_{r\theta\phi}$$

$$+ \rho \begin{pmatrix} g_r \\ g_\theta \\ g_\phi \end{pmatrix}_{r\theta\phi} \qquad (8.12)$$

This is a complex equation but, as is true with Poiseuille flow in Chapter 7, we know much about flow past a sphere that we can use to simplify Equation 8.12. First, the flow is steady ($\partial/\partial t = 0$) and we assume that $v_\phi = 0$ and that the flow is symmetric in the ϕ-direction; therefore, we can eliminate all terms with v_ϕ or velocity derivatives with respect to ϕ. Second, gravity is in the downward direction and for the spherical coordinate system we chose (see Figure 8.5) gravity becomes:

$$\underline{g} = -g\hat{e}_z = \begin{pmatrix} g_r \\ g_\theta \\ g_\phi \end{pmatrix}_{r\theta\phi} = \begin{pmatrix} -g\cos\theta \\ g\sin\theta \\ 0 \end{pmatrix}_{r\theta\phi} \tag{8.13}$$

Substituting $\underline{g}$ and what we know already about the velocity field $\underline{v}$ (i.e., steady state, $v_\phi = 0$, symmetric in the ϕ-direction), we obtain a simplified version of the Navier-Stokes equation for steady flow around a sphere:

$$\rho \begin{pmatrix} v_r\left(\dfrac{\partial v_r}{\partial r}\right) + v_\theta\left(\dfrac{1}{r}\dfrac{\partial v_r}{\partial \theta} - \dfrac{v_\theta}{r}\right) \\[2ex] v_r\left(\dfrac{\partial v_\theta}{\partial r}\right) + v_\theta\left(\dfrac{1}{r}\dfrac{\partial v_\theta}{\partial \theta} + \dfrac{v_r}{r}\right) \\[2ex] 0 \end{pmatrix}_{r\theta\phi}$$

$$= \begin{pmatrix} \dfrac{\partial p}{\partial r} \\[2ex] \dfrac{1}{r}\dfrac{\partial p}{\partial \theta} \\[2ex] \dfrac{1}{r\sin\theta}\dfrac{\partial p}{\partial \phi} \end{pmatrix}_{r\theta\phi}$$

$$+ \mu \begin{pmatrix} \left(\dfrac{\partial}{\partial r}\left(\dfrac{1}{r^2}\dfrac{\partial}{\partial r}(r^2 v_r)\right) + \dfrac{1}{r^2\sin\theta}\dfrac{\partial}{\partial\theta}\left(\sin\theta\dfrac{\partial v_r}{\partial\theta}\right) - \dfrac{2}{r^2\sin\theta}\dfrac{\partial}{\partial\theta}(v_\theta\sin\theta)\right) \\[2ex] \left(\dfrac{1}{r^2}\dfrac{\partial}{\partial r}\left(r^2\dfrac{\partial v_\theta}{\partial r}\right) + \dfrac{1}{r^2}\dfrac{\partial}{\partial\theta}\left(\dfrac{1}{\sin\theta}\dfrac{\partial}{\partial\theta}(v_\theta\sin\theta)\right) + \dfrac{2}{r^2}\dfrac{\partial v_r}{\partial\theta}\right) \\[2ex] 0 \end{pmatrix}_{r\theta\phi}$$

$$+ \rho \begin{pmatrix} -g\cos\theta \\ g\sin\theta \\ 0 \end{pmatrix}_{r\theta\phi} \tag{8.14}$$

We have made many assumptions about the structure of the flow field and still have a complex equation that is too difficult to solve (Equation 8.14). In particular, the lefthand side of Equation 8.14 has nonlinear terms—that is, terms with velocity multiplied by a velocity derivative or multiplied by another velocity (Figure 8.6). Mathematically, these nonlinear terms make the Navier-Stokes equation intractable to us.

We appear to be blocked at this point, and we must bring something new into the problem to proceed. The terms on the lefthand side of the Navier-Stokes equation—those that are multiplied by the density—are the inertial terms (see Figure 7.16). These terms account for the tendency of a fluid to remain at rest once at rest or to remain in motion once in motion. The inertial contributions to the momentum balance are important in rapid flows; in flows in which viscosity

$$
\left.\begin{cases}
v_r\left(\dfrac{\partial v_r}{\partial r}\right) \\[2mm]
v_\theta\left(\dfrac{1}{r}\dfrac{\partial v_r}{\partial \theta}\right) \\[2mm]
-\dfrac{v_\theta^2}{r} \\[2mm]
v_r\left(\dfrac{\partial v_\theta}{\partial r}\right) \\[2mm]
\dfrac{v_\theta}{r}\left(\dfrac{\partial v_\theta}{\partial \theta}\right) \\[2mm]
\dfrac{v_\theta v_r}{r}
\end{cases}\right.
$$

Nonlinear terms in Navier-Stokes equation for flow around a sphere (lefthand side, inertial contribution)

Figure 8.6   On the left side of the Navier-Stokes equation are nonlinear terms. The nonlinear terms in Equation 8.14 are shown here.

is high or velocity is low (i.e., low-Reynolds-number flows), the inertial terms do not contribute significantly to the solution.

The external flows of interest here are sometimes slow and sometimes fast. In fact, our skydiver problem is certainly a flow in which the fluid speed is high. This is the beginning of our study of flow past objects, however, and it makes sense to study first a problem that we may be able to solve—the slow-flow problem—and see what insights we obtain. With this in mind, we now assume that the flow is slow enough so that the (nonlinear) inertial terms can be neglected. When we arrive at our solution, we can compare predictions of the analysis with actual experiments to see whether and when this assumption is valid.

Neglecting the inertial terms, the Navier-Stokes equation for flow around a sphere becomes:

$$
\begin{pmatrix} 0 \\ 0 \\ 0 \end{pmatrix}_{r\theta\phi}
=
\begin{pmatrix}
-\dfrac{\partial p}{\partial r} \\[3mm]
-\dfrac{1}{r}\dfrac{\partial p}{\partial \theta} \\[3mm]
\dfrac{1}{r\sin\theta}\dfrac{\partial p}{\partial \phi}
\end{pmatrix}_{r\theta\phi}
$$

$$
+
\begin{pmatrix}
\mu\left(\dfrac{\partial}{\partial r}\left(\dfrac{1}{r^2}\dfrac{\partial}{\partial r}(r^2 v_r)\right)+\dfrac{1}{r^2\sin\theta}\dfrac{\partial}{\partial\theta}\left(\sin\theta\dfrac{\partial v_r}{\partial\theta}\right)-\dfrac{2}{r^2\sin\theta}\dfrac{\partial}{\partial\theta}(v_\theta\sin\theta)\right) \\[3mm]
\mu\left(\dfrac{1}{r^2}\dfrac{\partial}{\partial r}\left(r^2\dfrac{\partial v_\theta}{\partial r}\right)+\dfrac{1}{r^2}\dfrac{\partial}{\partial\theta}\left(\dfrac{1}{\sin\theta}\dfrac{\partial}{\partial\theta}(v_\theta\sin\theta)\right)+\dfrac{2}{r^2}\dfrac{\partial v_r}{\partial\theta}\right) \\[3mm]
0
\end{pmatrix}_{r\theta\phi}
$$

$$
+\rho
\begin{pmatrix}
-g\cos\theta \\
g\sin\theta \\
0
\end{pmatrix}_{r\theta\phi}
\qquad (8.15)
$$

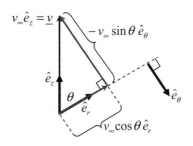

Figure 8.7

The boundary conditions on the spherical-coordinate-system velocity components can be related to the uniform velocity v_∞ through geometry.

Equation 8.15 still appears formidable; however, at this point, it is possible to find an analytical solution to the equation. The solution method [14, 43], although not obvious (especially to a beginner in the study of partial differential equations), is straightforward to understand, as we now demonstrate.

One clue that experts find useful when seeking solutions to partial differential equations (PDEs) is to look at the boundary conditions. To solve Equation 8.15, because of the second derivatives on the righthand side, we need two boundary conditions each on v_r and v_θ. Two boundary conditions that we can identify easily are no-penetration and no-slip at the surface of the sphere:

$$r = R \quad v_r = 0 \quad \text{for all } \theta, \phi \tag{8.16}$$

$$r = R \quad v_\theta = 0 \quad \text{for all } \theta, \phi \tag{8.17}$$

The other boundary conditions are that, far from the sphere, the flow must return to the uniform velocity field that exists upstream of the sphere, $\underline{v} = v_\infty \hat{e}_z$. The uniform velocity v_∞ is upward; in terms of the spherical velocity components v_r and v_θ, the uniform velocity field at infinity becomes the following (Figure 8.7):

$$r = \infty \quad v_r = \quad v_\infty \cos \theta \quad \text{for all } \phi \tag{8.18}$$

$$r = \infty \quad v_\theta = -v_\infty \sin \theta \quad \text{for all } \phi \tag{8.19}$$

The second set of boundary conditions indicates that at the edges of the flow, the θ-dependencies of the velocity components are given by $\cos \theta$ and $\sin \theta$ functions. As a first guess, therefore, it seems plausible to assume that the angular dependences of v_r and v_θ also are given by $\cos \theta$ and $\sin \theta$ functions throughout the flow. Thus, we guess that:

$$v_r = \mathcal{A}(r) \cos \theta \tag{8.20}$$

$$v_\theta = \mathcal{B}(r) \sin \theta \tag{8.21}$$

where $\mathcal{A}(r)$ and $\mathcal{B}(r)$ are functions only of r, and $\mathcal{A}$ and $\mathcal{B}$ now must be determined by solving Equations 8.10 and 8.15.

The guessed step based on boundary conditions turns out to work. Details of the solution are given in the literature [43] and they consist of substituting our guesses for the functionality of the velocity components (Equations 8.20 and 8.21) into the continuity and Navier-Stokes equations (Equations 8.10 and 8.15) and solving the resulting ordinary differential equations (ODEs) for $\mathcal{A}(r)$ and $\mathcal{B}(r)$. The final solutions for $\underline{v}(r, \theta)$ and $p(r, \theta)$ are given here; note that the velocity solution is written relative to our chosen coordinate-system basis vectors $\hat{e}_r$ and $\hat{e}_\theta$:[1]

$$\underline{v} = v_r \hat{e}_r + v_\theta \hat{e}_\theta$$

$$= \left[1 - \frac{3}{2} \frac{R}{r} + \frac{1}{2} \left(\frac{R}{r} \right)^3 \right] v_\infty \cos\theta \, \hat{e}_r - \left[1 - \frac{3}{4} \frac{R}{r} - \frac{1}{4} \left(\frac{R}{r} \right)^3 \right] v_\infty \sin\theta \, \hat{e}_\theta$$

$$(8.22)$$

Solution [43], creeping flow (Stokes flow) around a sphere:

$$\underline{v}(r,\theta) = \begin{pmatrix} v_\infty \left[1 - \frac{3}{2} \frac{R}{r} + \frac{1}{2} \left(\frac{R}{r} \right)^3 \right] \cos\theta \\ -v_\infty \left[1 - \frac{3}{4} \frac{R}{r} - \frac{1}{4} \left(\frac{R}{r} \right)^3 \right] \sin\theta \\ 0 \end{pmatrix}_{r\theta\phi}$$

$$(8.23)$$

$$p(r,\theta) = p_\infty - \rho g r \cos\theta - \frac{3}{2} \frac{\mu v_\infty}{R} \left(\frac{R}{r} \right)^2 \cos\theta$$

$$(8.24)$$

The quantity p_∞ in Equation 8.24 was introduced when a pressure boundary condition was needed; p_∞ is the pressure far from the sphere at the elevation of the origin of the coordinate system (i.e., at $\theta = \pi/2, r = \infty, p = p_\infty$).

The velocity field for creeping flow around a sphere is shown in Figure 8.8. Note that because the flow field is two-dimensional (i.e., it depends on two variables, r and θ), we cannot produce an easy one-dimensional or even three-dimensional sketch of the flow profile as we did for the Poiseuille tube flow in Figures 7.6 and 7.7. The velocity field in tube flow did not depend on the z-coordinate and was fully symmetric in the θ-direction, making it easier to plot. For the current case of creeping flow around a sphere, we render the velocity field by drawing vectors at selected points in the flow, where the length of the vector represents the magnitude of the velocity at that location. Creeping flow around a sphere is fully symmetric in the ϕ-direction by assumption; thus, Figure 8.8 is a representation of the velocity field in an arbitrary plane of constant ϕ.

An alternative way to represent the velocity field is to sketch the streamlines. Streamlines in steady flow are the equivalent of particle paths—that is, the path that a fluid particle takes as it passes through the field of view. Streamlines

---

[1] To write the solution in Cartesian coordinates, use Equation 8.22; for the two basis vectors $\hat{e}_r$ and $\hat{e}_\theta$, substitute the appropriate basis-vector transformations from Equations 1.271–1.273. The coordinates r and θ are transformed to x, y, and z in Equations 1.268–1.270.

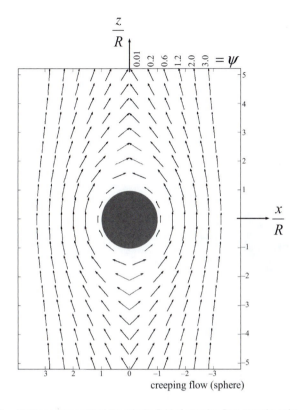

creeping flow (sphere)

Figure 8.8   Vector or arrow plot of the velocity field of creeping flow around a sphere. For points along several streamlines (i.e., particle paths), the velocity vector centered at the point is shown; the length of the arrow is proportional to the magnitude of the velocity at that point. Note that near the sphere the velocity is low due to the no-slip boundary condition.

are defined more formally as lines that are everywhere tangent to the local velocity field; this definition is appropriate for steady and unsteady flows. The streamlines for steady creeping flow around a sphere are shown in Figure 8.9.[2] For Poiseuille flow in a tube, the streamlines are straight lines of constant θ and r (see Figure 2.19).

Now that we know $\underline{v}(r, \theta)$ and $p(r, \theta)$ (Equations 8.23 and 8.24), we can calculate the stress field in creeping flow from the Newtonian constitutive equation. Because we are in spherical coordinates, we must use the correct form for the Newtonian constitutive equation in this coordinate system (see Table B.8). We

---

[2] The stream function shows the locations of the streamlines. For creeping flow, the stream function ψ is given by [85]:

$$\psi(r) = v_\infty R^2 \sin^2 \theta \left[\frac{1}{2} \left(\frac{r}{R} \right)^2 - \frac{3}{4} \left(\frac{r}{R} \right) + \frac{1}{4} \left(\frac{R}{r} \right) \right] \qquad (8.25)$$

Streamlines and the stream function ψ are discussed in Section 8.2 and in the Glossary.

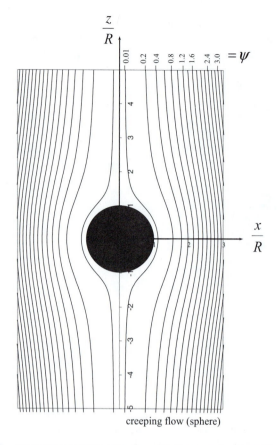

creeping flow (sphere)

Figure 8.9 Streamlines or particle paths of creeping flow around a sphere. The values of the stream function ψ for several lines are shown. For steady flows, streamlines mark the paths of fluid particles in the flow. In all flows, the local velocity vector at a point in the flow is tangent to the streamline function ψ at that point. For more on the stream function, see the literature [43, 85].

performed this calculation in Chapter 5 (see Example 5.6):

$$\tilde{\underline{\tau}} = \mu \left(\nabla \underline{v} + \nabla \underline{v}^T \right) \tag{8.26}$$

$$\underline{v} = \begin{pmatrix} v_r(r, \theta) \\ v_\theta(r, \theta) \\ 0 \end{pmatrix}_{r\theta\phi} \tag{8.27}$$

$$\tilde{\underline{\tau}}(r, \theta) = \mu \begin{pmatrix} 2\frac{\partial v_r}{\partial r} & r\frac{\partial}{\partial r}\left(\frac{v_\theta}{r}\right) + \frac{1}{r}\frac{\partial v_r}{\partial \theta} & 0 \\ r\frac{\partial}{\partial r}\left(\frac{v_\theta}{r}\right) + \frac{1}{r}\frac{\partial v_r}{\partial \theta} & 2\left(\frac{1}{r}\frac{\partial v_\theta}{\partial \theta} + \frac{v_r}{r}\right) & 0 \\ 0 & 0 & \frac{2v_r}{r} + \frac{2v_\theta \cot\theta}{r} \end{pmatrix}_{r\theta\phi} \tag{8.28}$$

$$\tilde{\underline{\Pi}}(r, \theta) = \tilde{\underline{\tau}} - p\underline{I}$$

$$= \begin{pmatrix} 2\mu\frac{\partial v_r}{\partial r} - p(r, \theta) & \mu r\frac{\partial}{\partial r}\left(\frac{v_\theta}{r}\right) + \frac{\mu}{r}\frac{\partial v_r}{\partial \theta} & 0 \\ \mu r\frac{\partial}{\partial r}\left(\frac{v_\theta}{r}\right) + \frac{\mu}{r}\frac{\partial v_r}{\partial \theta} & 2\mu\left(\frac{1}{r}\frac{\partial v_\theta}{\partial \theta} + \frac{v_r}{r}\right) - p(r, \theta) & 0 \\ 0 & 0 & \frac{2\mu v_r}{r} + \frac{2\mu v_\theta \cot\theta}{r} - p(r, \theta) \end{pmatrix}_{r\theta\phi} \tag{8.29}$$

Any components of $\underline{\underline{\tilde{\Pi}}}$ that are of interest now may be calculated from Equations 8.29, 8.23, and 8.24.

Our objective is to calculate the total amount of force that is exerted on the sphere by the fluid. The tensor $\underline{\underline{\tilde{\Pi}}}$ is a complete description of fluid stress in a flow, and the result for $\underline{\underline{\tilde{\Pi}}}$ in this flow (Equation 8.29) may be used to calculate the force on *any* surface in the flow by using Equation 6.196, repeated here:

$$\text{Total molecular fluid force on a surface } \mathcal{S}: \qquad \underline{\mathcal{F}} = \iint_{\mathcal{S}} [\hat{n} \cdot \underline{\underline{\tilde{\Pi}}}]_{\text{at surface}} \, dS \qquad (8.30)$$

Because we now have the stress tensor $\underline{\underline{\tilde{\Pi}}}$, we are ready to perform this calculation.

We want to calculate the force on the sphere in creeping flow using Equation 8.30. The surface of the sphere is the surface at $r = R$ for all values of θ and ϕ, and the unit normal to this surface is $\hat{n} = \hat{e}_r$ at every point on the sphere surface. The simplicity with which we can describe the sphere surface validates our choice of the spherical coordinate system for our calculations. The differential surface element dS for a sphere is $(R \sin \theta \, d\phi)(R \, d\theta)$. Thus, the total fluid force on the sphere $\underline{\mathcal{F}}$ becomes:

$$\underline{\mathcal{F}} = \iint_{\mathcal{S}} [\hat{n} \cdot \underline{\underline{\tilde{\Pi}}}]_{\text{at surface}} \, dS \qquad (8.31)$$

$$= \int_0^{2\pi} \int_0^{\pi} [\hat{e}_r \cdot \underline{\underline{\tilde{\Pi}}}]_{r=R} \, R^2 \sin \theta \, d\theta \, d\phi \qquad (8.32)$$

The dot product in Equation 8.32 may be evaluated using matrix manipulations as follows:

$$[\hat{e}_r \cdot \underline{\underline{\tilde{\Pi}}}] = \begin{pmatrix} 1 & 0 & 0 \end{pmatrix}_{r\theta\phi} \cdot \begin{pmatrix} \tilde{\Pi}_{rr} & \tilde{\Pi}_{r\theta} & \tilde{\Pi}_{r\phi} \\ \tilde{\Pi}_{\theta r} & \tilde{\Pi}_{\theta\theta} & \tilde{\Pi}_{\theta\phi} \\ \tilde{\Pi}_{\phi r} & \tilde{\Pi}_{\phi\theta} & \tilde{\Pi}_{\phi\phi} \end{pmatrix}_{r\theta\phi} \qquad (8.33)$$

$$= \begin{pmatrix} \tilde{\Pi}_{rr} & \tilde{\Pi}_{r\theta} & \tilde{\Pi}_{r\phi} \end{pmatrix}_{r\theta\phi} = \begin{pmatrix} \tilde{\Pi}_{rr} \\ \tilde{\Pi}_{r\theta} \\ \tilde{\Pi}_{r\phi} \end{pmatrix}_{r\theta\phi} \qquad (8.34)$$

$$= \begin{pmatrix} 2\mu \dfrac{\partial v_r}{\partial r} - p(r, \theta) \\ \mu r \dfrac{\partial}{\partial r} \left(\dfrac{v_\theta}{r} \right) + \dfrac{\mu}{r} \dfrac{\partial v_r}{\partial \theta} \\ 0 \end{pmatrix}_{r\theta\phi} \qquad (8.35)$$

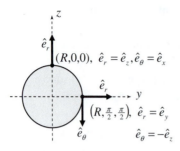

Figure 8.10 The basis vectors in the spherical coordinate system vary with position. Shown is the plane for $x = 0$. For the downstream stagnation point $(r, \theta, \phi) = (R, 0, 0)$, $\hat{e}_r = \hat{e}_z$ and $\hat{e}_\theta = \hat{e}_x$ (out of the plane of the paper). For the second point shown $(r, \theta, \phi) = (R, \pi/2, \pi/2)$, $\hat{e}_r = \hat{e}_y$ and $\hat{e}_\theta = -\hat{e}_z$.

where the coefficients of the tensor $\underset{\sim}{\tilde{\Pi}}$ were obtained from Equation 8.29. It is now straightforward to calculate $\mathcal{F}$ from the solution for $\underline{v}$:

Total fluid force on the sphere in creeping flow:

$$\underline{\mathcal{F}} = R^2 \int_0^{2\pi} \int_0^{\pi} [\hat{e}_r \cdot \underset{\sim}{\tilde{\Pi}}]_{r=R} \ \sin\theta d\theta d\phi \tag{8.36}$$

$$\underline{\mathcal{F}} = R^2 \int_0^{2\pi} \int_0^{\pi} \left[\begin{pmatrix} 2\mu\dfrac{\partial v_r}{\partial r} - p(r,\theta) \\[2mm] \mu r\dfrac{\partial}{\partial r}\left(\dfrac{v_\theta}{r}\right) + \dfrac{\mu}{r}\dfrac{\partial v_r}{\partial \theta} \\[2mm] 0 \end{pmatrix}_{r\theta\phi} \right]_{r=R} \sin\theta d\theta d\phi$$

$$\tag{8.37}$$

Carrying out the indicated derivatives using the expressions for v_r and v_θ in Equation 8.23 and evaluating the results at the surface of the sphere $r = R$, we obtain:

$$\underline{\mathcal{F}} = R^2 \int_0^{2\pi} \int_0^{\pi} \begin{pmatrix} -p(R,\theta)\sin\theta \\[2mm] \dfrac{-3\mu v_\infty \sin^2\theta}{2R} \\[2mm] 0 \end{pmatrix}_{r\theta\phi} d\theta d\phi \tag{8.38}$$

$$= R^2 \int_0^{2\pi} \int_0^{\pi} \left[(-p(R,\theta)\sin\theta)\hat{e}_r + \left(\dfrac{-3\mu v_\infty \sin^2\theta}{2R}\right)\hat{e}_\theta \right] d\theta d\phi \tag{8.39}$$

The form of the equation for $\underline{\mathcal{F}}$ in Equation 8.39 emphasizes a complicating factor in the integration. The two basis vectors $\hat{e}_r$ and $\hat{e}_\theta$ both vary with θ and ϕ. We can be convinced of this fact by considering two points on the surface of the sphere, as shown in Figure 8.10. The geometric relationships between the spherical basis vectors and the Cartesian basis vectors are discussed in Section 1.3, and the equations that relate them are given in Equations 1.271–1.273 and repeated here:

$$\hat{e}_r = (\sin\theta\cos\phi)\hat{e}_x + (\sin\theta\sin\phi)\hat{e}_y + (\cos\theta)\hat{e}_z$$

$$= \begin{pmatrix} \sin\theta\cos\phi \\ \sin\theta\sin\phi \\ \cos\theta \end{pmatrix}_{xyz} \tag{8.40}$$

$$\hat{e}_\theta = (\cos\theta\cos\phi)\hat{e}_x + (\cos\theta\sin\phi)\hat{e}_y + (-\sin\theta)\hat{e}_z$$

$$= \begin{pmatrix} \cos\theta\cos\phi \\ \cos\theta\sin\phi \\ -\sin\theta \end{pmatrix}_{xyz} \tag{8.41}$$

$$\hat{e}_\phi = (-\sin\phi)\hat{e}_x + (\cos\phi)\hat{e}_y$$

$$= \begin{pmatrix} -\sin\phi \\ \cos\phi \\ 0 \end{pmatrix}_{xyz} \tag{8.42}$$

The Cartesian basis vectors do not vary with position; that is, $\hat{e}_x$, $\hat{e}_y$, and $\hat{e}_z$ are constants. To properly consider the θ- and ϕ-dependence of the basis vectors in the integration, we substitute the expressions for $\hat{e}_r$ and $\hat{e}_\theta$ in terms of the constant Cartesian basis vectors into Equation 8.39 and proceed with the integrations:

$$\mathcal{F} = R^2 \int_0^{2\pi} \int_0^\pi (-p(R,\theta)\sin\theta)\,\hat{e}_r + \left(\frac{-3\mu v_\infty \sin^2\theta}{2R}\right)\hat{e}_\theta \, d\theta d\phi \tag{8.43}$$

$$= R^2 \int_0^{2\pi} \int_0^\pi (-p(R,\theta)\sin\theta) \begin{pmatrix} \sin\theta\cos\phi \\ \sin\theta\sin\phi \\ \cos\theta \end{pmatrix}_{xyz}$$

$$+ \left(\frac{-3\mu v_\infty \sin^2\theta}{2R}\right) \begin{pmatrix} \cos\theta\cos\phi \\ \cos\theta\sin\phi \\ -\sin\theta \end{pmatrix}_{xyz} d\theta d\phi \tag{8.44}$$

Equation 8.44 appears to be complicated, but the symmetry of the problem makes the ϕ-integration fairly simple to carry out. Note that the only ϕ-dependence in the x-component is $\cos\phi$, which appears in every term. Likewise, the only ϕ-dependence in the y-component is $\sin\phi$, and this quantity appears in every term of the y-component. The z-component of Equation 8.44 is independent of ϕ. To clarify the process of carrying out the ϕ-integration, therefore, we write Equation 8.44 as:

$$\mathcal{F} = R^2 \int_0^{2\pi} \int_0^\pi \begin{pmatrix} A(\theta)\cos\phi \\ A(\theta)\sin\phi \\ B(\theta) \end{pmatrix}_{xyz} d\theta d\phi \tag{8.45}$$

where $A(\theta)$ and $B(\theta)$ are given by:

$$A(\theta) = (-p(R,\theta)\sin\theta)\sin\theta + \left(\frac{-3\mu v_\infty \sin^2\theta}{2R}\right)\cos\theta \qquad (8.46)$$

$$B(\theta) = (-p(R,\theta)\sin\theta)\cos\theta + \left(\frac{3\mu v_\infty \sin^2\theta}{2R}\right)\sin\theta \qquad (8.47)$$

Individually integrating the x-, y-, and z-components of Equation 8.45 over ϕ, we obtain:

$$x\text{-Component of Equation 8.45:}\quad R^2 \int_0^{2\pi}\int_0^\pi (A(\theta)\cos\phi)\,d\theta d\phi$$

$$= R^2\left[\int_0^\pi A(\theta)d\theta\right]\int_0^{2\pi}\cos\phi d\phi$$

$$= 0 \qquad (8.48)$$

$$y\text{-Component of Equation 8.45:}\quad R^2 \int_0^{2\pi}\int_0^\pi (A(\theta)\sin\phi)\,d\theta d\phi$$

$$= R^2\left[\int_0^\pi A(\theta)d\theta\right]\int_0^{2\pi}\sin\phi d\phi$$

$$= 0 \qquad (8.49)$$

$$z\text{-Component of Equation 8.45:}\quad R^2 \int_0^{2\pi}\int_0^\pi B(\theta)d\theta d\phi$$

$$= R^2\left[\int_0^\pi B(\theta)d\theta\right]\int_0^{2\pi}d\phi$$

$$= R^2\left[\int_0^\pi B(\theta)d\theta\right]2\pi \qquad (8.50)$$

Incorporating these results into Equation 8.45, we obtain:

$$\underline{\mathcal{F}} = R^2 \int_0^{2\pi}\int_0^\pi \begin{pmatrix} A(\theta)\cos\phi \\ A(\theta)\sin\phi \\ B(\theta) \end{pmatrix}_{xyz} d\theta d\phi \qquad (8.51)$$

$$\underline{\mathcal{F}} = \begin{pmatrix} \mathcal{F}_x \\ \mathcal{F}_y \\ \mathcal{F}_z \end{pmatrix}_{xyz} = \begin{pmatrix} 0 \\ 0 \\ 2\pi R^2 \int_0^\pi B(\theta)d\theta \end{pmatrix}_{xyz} \qquad (8.52)$$

The expression for force on the sphere in Equation 8.52 reflects our intuition that the net fluid force on the sphere is in the far-field flow direction, $\hat{e}_z$. We calculate the remaining nonzero component of the forces on the sphere, $\mathcal{F}_z$, by substituting the equation for $B(\theta)$ (Equation 8.47) into Equation 8.52 and carrying out the θ-integration. This involves algebra and trigonometric integration, but the integrals are standard. The procedure is outlined herein. We begin with $B(\theta)$ from Equation 8.47:

$$B(\theta) = -p(R, \theta) \sin\theta \cos\theta + \left(\frac{3\mu v_\infty \sin^3 \theta}{2R} \right) \tag{8.53}$$

The creeping-flow solution for the pressure $p(r, \theta)$ is given in Equation 8.24. Taking $r = R$ in Equation 8.24 and substituting the result in the previous equation for $B(\theta)$, we obtain:

$$p(R, \theta) = p_\infty - \rho g R \cos\theta - \frac{3}{2} \frac{\mu v_\infty}{R} \cos\theta \tag{8.54}$$

$$B(\theta) = -p_\infty \sin\theta \cos\theta + \rho g R \sin\theta \cos^2\theta$$
$$+ \frac{3}{2} \frac{\mu v_\infty}{R} \sin\theta \cos^2\theta + \frac{3\mu v_\infty \sin^3\theta}{2R} \tag{8.55}$$

Substituting Equation 8.55 for $B(\theta)$ into Equation 8.52 for force on a sphere and integrating $B(\theta)$ term by term yields:

$$\mathcal{F}_z = 2\pi R^2 \int_0^\pi B(\theta) d\theta \tag{8.56}$$

$$= 2\pi R^2 \int_0^\pi -p_\infty \sin\theta \cos\theta d\theta + 2\pi R^2 \int_0^\pi \rho g R \sin\theta \cos^2\theta d\theta$$

$$+ 2\pi R^2 \int_0^\pi \frac{3}{2} \frac{\mu v_\infty}{R} \sin\theta \cos^2\theta d\theta + 2\pi R^2 \int_0^\pi \frac{3\mu v_\infty \sin^3\theta}{2R} d\theta \tag{8.57}$$

$$= 0 + \frac{4\pi R^3 \rho g}{3} + 2\pi R \mu v_\infty + 4\pi R \mu v_\infty \tag{8.58}$$

$$= \begin{pmatrix} \text{contribution} \\ \text{from far-field} \\ \text{pressure} \end{pmatrix} + \begin{pmatrix} \text{gravity contribution} \\ \text{from fluid} \\ \text{surrounding sphere} \\ \text{(buoyancy)} \end{pmatrix}$$

$$+ \begin{pmatrix} \text{contribution} \\ \text{from flow-induced} \\ \text{pressure} \\ \text{(pressure or form drag)} \end{pmatrix} + \begin{pmatrix} \text{contribution} \\ \text{from flow-induced} \\ \text{shear stress} \\ \text{(friction drag)} \end{pmatrix} \tag{8.59}$$

$$\mathcal{F}_z = \frac{4\pi R^3 \rho g}{3} + 6\pi R \mu v_\infty \tag{8.60}$$

Fluid force on a sphere in creeping flow (Stokes flow):

$$\underline{\mathcal{F}} = \begin{pmatrix} 0 \\ 0 \\ \dfrac{4\pi R^3 \rho g}{3} + 6\pi R \mu v_\infty \end{pmatrix}_{xyz} \tag{8.61}$$

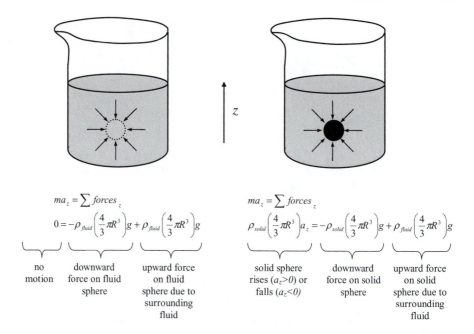

$ma_z = \sum forces_z$

$$0 = -\rho_{fluid}\left(\frac{4}{3}\pi R^3\right)g + \rho_{fluid}\left(\frac{4}{3}\pi R^3\right)g$$

| no motion | downward force on fluid sphere | upward force on fluid sphere due to surrounding fluid |

$ma_z = \sum forces_z$

$$\rho_{solid}\left(\frac{4}{3}\pi R^3\right)a_z = -\rho_{solid}\left(\frac{4}{3}\pi R^3\right)g + \rho_{fluid}\left(\frac{4}{3}\pi R^3\right)g$$

| solid sphere rises ($a_z>0$) or falls ($a_z<0$) | downward force on solid sphere | upward force on solid sphere due to surrounding fluid |

Figure 8.11   The buoyant force experienced by a sphere in a liquid can be understood as the force that the surrounding fluid exerts on the volume occupied by the sphere. In the liquid when the sphere is not present, a volume equivalent to the volume of the sphere is occupied by liquid. The surrounding liquid and this liquid sphere are in equilibrium (i.e., no motion). When that liquid volume is replaced by the sphere, if the sphere weighs more than the liquid it replaced, the sphere sinks; if the sphere weighs less than the liquid it replaced, it rises.

Equation 8.61 gives the total fluid force on a sphere in creeping flow. The first contribution in the z-component, which accounts for the effect of gravity, is the buoyancy term. The buoyancy effect is the net gravity effect of the surrounding fluid on the sphere; it is present whether or not the fluid moves (note that v_∞ does not appear in this term; see Example 4.6). A steel ball falls through water, whereas a balloon rises. In each case, the sphere experiences an upward force that is the equivalent of the force due to gravity that would have acted on the fluid that has been displaced by the sphere (Figure 8.11). If the mass of the displaced fluid is lower than the mass of the sphere (i.e., the steel case), the sphere falls—although it is retarded by the buoyancy force. If the mass of the displaced fluid is higher than the mass of the sphere, as in the case of the balloon, the sphere rises. In a force balance on the sphere without drag, there would be a downward force due to gravity on the sphere ($4\pi\rho_{sphere}g/3$), the upward buoyancy term ($4\pi\rho_{fluid}g/3$), balanced by the mass times the acceleration of the sphere ($\sum \underline{f} = m\underline{a}$).

The second contribution to the z-component of the fluid force on a sphere (Equation 8.61) comes from the motion of the sphere and is called the kinetic force. Tracing back the source of the terms (Equation 8.59), the kinetic force comes from two sources: one source is the effect of the flow-induced pressure field (i.e., the third term); and the second source is the effect of viscosity to produce shear stress on the surface of the sphere (i.e., the fourth term). Together, these two kinetic contributions—the form drag (due to the pressure distribution) and the friction drag (due to shear stress)—are the drag on the sphere:

Stokes law:
magnitude of drag
in creeping flow
around a sphere
(Stokes-Einstein-Sutherland equation)

$$\mathcal{F}_{\text{drag}} = 6\pi R \mu v_\infty \qquad (8.62)$$

The drag is the retarding fluid force due to fluid motion. Stokes law is the starting point in many important analyses of the motion of solid and liquid spheres, including the study of colloidal dispersions, suspensions, and emulsions [83].

This completes our analysis of creeping flow around a sphere. Through microscopic analysis, we arrived at equations for velocity field $\underline{v}(r, \theta)$ and pressure field $p(r, \theta)$, from which the stress field $\underline{\tilde{\Pi}}(r, \theta)$ and the force on any surface in the flow $\mathcal{F}$ (Equation 8.30) may be calculated. The assumptions of this analysis are steady flow, no ϕ-velocity, ϕ-symmetry of velocity, and no inertial effects (the left side of the Navier-Stokes equation was neglected). The drag on the sphere was calculated as $\mathcal{F}_{\text{drag}} = 6\pi R \mu v_\infty$, which is Stokes law.

This chapter begins the discussion of external flow by considering the problem of calculating the terminal speed of a skydiver. We solved the idealized case of a sphere falling through a viscous fluid or, equivalently, a viscous fluid flowing slowly around a stationary sphere. If the equation for fluid force $\mathcal{F}$ calculated for the falling-sphere case (Equation 8.61) can be applied to the skydiver, we now have enough information to calculate her terminal speed. We perform this calculation now and check our answer against observations reported in the literature to see whether the creeping-flow assumptions are justified in the skydiver case.

EXAMPLE 8.3 (Skydiver, continued). *What is the maximum speed reached by a skydiver who jumps out of an airplane at 13,000 feet? How much can the speed of the skydiver vary depending on her body position (i.e., arms and legs flung out or pulled in tightly)?*

SOLUTION. As we learned in the creeping-flow solution, a body falling in the presence of a viscous fluid such as air is subject to fluid forces of drag and buoyancy. Both effects slow the motion of the object and, ultimately, the downward force due to gravity is balanced by the retarding fluid forces; at steady state, the object reaches a zero-acceleration condition called the terminal speed.

To calculate the terminal speed as a function of the various fluid and sphere properties, we apply Newton's second law (i.e., momentum conservation) to the sphere (see Figure 8.4):

Newton's second law:

$$\sum_{\substack{\text{all forces} \\ \text{acting on body}}} \underline{f} = m\underline{a} \qquad (8.63)$$

where $\underline{f}$ represents the various forces, m is the mass of the body, and $\underline{a}$ is the acceleration of the body. When the skydiver is at terminal speed, she is moving at constant speed; therefore, the acceleration is zero. The two forces on the skydiver are gravity and the fluid force:

$$\text{Momentum balance} \atop \text{on skydiver} \atop \text{at terminal speed:} \qquad \sum_{\substack{\text{all forces} \\ \text{acting on body}}} \underline{f} = m\underline{a} = 0 \qquad (8.64)$$

$$\underline{f}_{\text{gravity}} + \underline{f}_{\text{fluid}} = 0 \qquad (8.65)$$

$$m\underline{g} + \underline{\mathcal{F}} = 0 \qquad (8.66)$$

where m is the mass of the skydiver assumed to be a sphere of density ρ_{body}; g is the acceleration due to gravity; and $\underline{\mathcal{F}}$, the fluid force on a sphere falling through a viscous liquid, is given by the result of the analysis of Stokes flow, Equation 8.61. Note that we are using the coordinate system in Figure 8.5; thus, $\underline{g} = -g\hat{e}_z$. Substituting the Stokes-flow result in Equation 8.61 into the sphere momentum-balance equation, Equation 8.66, yields:

$$m\underline{g} + \underline{\mathcal{F}} = 0 \qquad (8.67)$$

$$\begin{pmatrix} 0 \\ 0 \\ -\dfrac{4\pi R^3 \rho_{\text{body}} g}{3} \end{pmatrix}_{xyz} + \begin{pmatrix} 0 \\ 0 \\ \dfrac{4\pi R^3 \rho g}{3} + 6\pi R\mu v_\infty \end{pmatrix}_{xyz} = \begin{pmatrix} 0 \\ 0 \\ 0 \end{pmatrix}_{xyz} \qquad (8.68)$$

Solving the z-component of Equation 8.68 for the terminal speed v_∞, we obtain the final result:

$$v_\infty = \frac{(\rho_{\text{body}} - \rho)2R^2 g}{9\mu} \qquad (8.69)$$

$$\text{Terminal speed} \atop \text{of a sphere} \atop \text{(Stokes regime):} \qquad \boxed{v_\infty = \frac{(\rho_{\text{body}} - \rho)D^2 g}{18\mu}} \qquad (8.70)$$

Equation 8.70 now may be used to estimate the terminal speed of a skydiver, provided that the assumptions of Stokes flow (i.e., steady flow, no inertia, $v_\phi = 0$, ϕ-symmetry) are valid. To obtain a final answer, we must estimate the values of the physical parameters in Equation 8.70. For the skydiver, we assume sphere dimensions that approximate the size of a human being. For the physical-property data, we consult the Internet for approximate values. We can check the sensitivity of the calculation to these choices after obtaining our initial result:

$$\text{Viscosity of air:} \quad \mu = 1.7 \times 10^{-5} \text{ Pa s}$$

$$\text{Density of air:} \quad \rho = 1.3 \text{ kg/m}^3$$

$$\text{Density of human ball (water):} \quad \rho_{\text{body}} \approx 1,000 \text{ kg/m}^3$$

$$\text{Diameter of sphere:} \quad D = 0.50 \text{ m}$$

$$\text{Acceleration due to gravity:} \quad g = 9.80 \text{ m/s}^2$$

With these parameter values, we obtain the terminal speed from Equation 8.70 as:

$$\text{Stokes-flow estimate of terminal speed:} \qquad v_\infty = 8 \times 10^6 \text{ m/s} = 18 \times 10^6 \text{ mph} \qquad (8.71)$$

The result obtained, 8 million m/s, is an unphysically high number. The speed of sound, for example, is 343 m/s (770 mph), and our answer is four orders of magnitude higher. There is something seriously wrong with our analysis. If we decrease the values of the parameters in the numerator of Equation 8.70 to the lowest possible estimates and increase the viscosity of air, which appears in the denominator, to five times our previous estimate, we obtain the lowest possible estimate from this calculation method:

$$\text{Viscosity of air:} \qquad \mu \approx 8.5 \times 10^{-5} \text{ Pa s}$$

$$\text{Density of air:} \qquad \rho \approx 0.1 \text{ kg/m}^3$$

$$\text{Density of human ball:} \qquad \rho_{\text{body}} \approx 100 \text{ kg/m}^3$$

$$\text{Diameter of sphere:} \qquad D \approx 0.1 \text{ m}$$

$$\text{Acceleration due to gravity:} \qquad g = 9.80 \text{ m/s}^2$$

$$\text{New Stokes-flow estimate of terminal speed:} \qquad v_\infty \approx 6{,}400 \text{ m/s} = 14{,}000 \text{ mph} \qquad (8.72)$$

This value is still unphysically high. The problem with the analysis is not in the accuracy of the estimates of model parameters; rather, it is that the assumptions of Stokes flow are not correct in the flow regime that exists during skydiving.

The reason for the incorrect results is that highly ordered Stokes flow breaks down above a critical speed (see Section 8.2). As the velocity increases, the flow around a sphere undergoes a series of transitions (see Figure 2.12 and further discussion in this section) from creeping flow to the development of recirculation and laminar boundary layers, eventually producing a flow with a turbulent boundary layer near the object and a turbulent wake behind it.

As in the case of laminar flow in tubes in Chapter 7, the study of highly idealized flow is enlightening when we begin to study a problem, but more intensive study is needed before practical solutions can be calculated.

The creeping-flow results in Example 8.3, which were derived for very slow flows of highly viscous liquids, greatly underestimate the frictional drag present in a rapid skydiving descent. The true estimate of the terminal speed of a skydiver is about 55 m/s, two orders of magnitude lower than the lower estimate made.[3] The significant discrepancy between the creeping-flow prediction and what is observed is evidence that a skydiver is not in creeping flow. To correctly solve the skydiver problem, we must know more about rapid flows around obstacles, including the study of boundary layers, inertial forces, flow separation, and wakes—all of which were neglected in the Stokes-flow analysis.

[3] The terminal speed is approximately 55 m/s when the skydiver is in the belly-to-Earth position; in the head-first position, speeds can reach 90 m/s.

8.1.2 Noncreeping flow around a sphere

In the previous section, we calculated the terminal speed of a skydiver by using results from creeping flow around a sphere:

$$
\begin{array}{c}
\text{Fluid force on a sphere} \\
\text{in creeping flow} \\
\text{in the } z\text{-direction} \\
\text{(Stokes flow):}
\end{array}
\qquad
\underline{\underline{\mathcal{F}}} =
\begin{pmatrix}
0 \\
0 \\
\dfrac{4\pi R^3 \rho g}{3} + 6\pi R \mu v_\infty
\end{pmatrix}_{xyz}
\qquad (8.73)
$$

We find that the skydiver terminal-speed result predicted by this creeping-flow equation is not correct; the predicted flow rate is at least two orders of magnitude too high. Our error in that calculation was to use a creeping-flow relationship (i.e., inertia neglected) to make a prediction in a very rapid flow where inertia is important. To correctly complete the skydiver calculation, we need a drag/velocity relationship for rapid flow past objects.

When we addressed pipe flow in Chapter 7, we had a similar dilemma: We could calculate pressure drop/flow rate from a laminar-flow analysis (i.e., the Hagen-Poiseuille equation, Equation 7.26); however, we needed a pressure-drop/flow-rate equation for turbulent flow and we could not solve directly for that. In Chapter 7, we solved the rapid-pipe-flow problem by using a mixed analytical/experimental approach. We relied on a data correlation—either the Prandtl correlation (see Equation 7.58) or the Colebrook equation (see Equation 1.95)—to obtain a correct result for pressure drop in turbulent pipe flow.

Using the tube-flow experience as our guide, we now follow the same steps to arrive at a solution to the external-flow problem considered here. The appropriate correlation we need to solve the skydiving problem is between nondimensional drag and nondimensional flow rate in noncreeping flow. Nondimensional drag is called the drag coefficient C_D, and nondimensional flow rate is again the Reynolds number—this time defined relative to the object in the flow. We develop the drag coefficient in this section. To obtain a reasonable solution to the skydiver problem, we first present the definition of the drag coefficient without derivation:

$$
\begin{array}{c}
\text{Drag coefficient:} \\
\text{nondimensional} \\
\text{drag}
\end{array}
\qquad
C_D \equiv \frac{\mathcal{F}_{\text{drag}}}{\left(\frac{1}{2}\rho V^2\right)\left(A_p\right)}
\qquad (8.74)
$$

$$
\begin{array}{c}
\text{Reynolds number:} \\
\text{nondimensional} \\
\text{flow rate}
\end{array}
\qquad
\text{Re} \equiv \frac{\rho V D}{\mu}
\qquad (8.75)
$$

where $\mathcal{F}_{\text{drag}}$ is the drag on the object; A_p is the projected area of the object, which is $\pi R^2 = \pi D^2/4$ for the sphere; D is the diameter of the sphere; ρ and μ are the density and viscosity of the fluid, respectively; and V is a characteristic velocity of the flow, chosen to be the upstream velocity v_∞. Note the similarity between the definitions of drag coefficient and Fanning friction factor for pipes (see Equation 7.135)—this similarity is appropriate because both are nondimensional retarding fluid forces.

For creeping flow, we can calculate C_D versus Re from the Stokes-flow analysis. We demonstrate the creeping-flow calculation before discussing $C_D(\text{Re})$ in noncreeping flow.

EXAMPLE 8.4. *What is the relationship between drag coefficient C_D and Reynolds number Re for creeping flow around a sphere?*

SOLUTION. We begin with the definition of drag coefficient, Equation 8.74:

$$\text{Drag coefficient of a sphere:} \quad \text{nondimensional} \quad C_D \equiv \frac{\mathcal{F}_{\text{drag}}}{\left(\frac{1}{2}\rho V^2\right)\left(\frac{\pi D^2}{4}\right)} \tag{8.76}$$

$$\text{drag}$$

The drag $\mathcal{F}_{\text{drag}}$ appears in this definition; for creeping flow, this quantity is given by Stokes law (Equation 8.62):

$$\begin{array}{c} \text{Stokes Law:} \\ \text{drag in creeping flow} \\ \text{around a sphere} \end{array} \quad \mathcal{F}_{\text{drag}} = 6\pi R \mu v_{\infty} \tag{8.77}$$

Substituting Stokes law (Equation 8.77) into the definition of drag coefficient yields the desired relationship:

$$C_D \equiv \frac{\mathcal{F}_{\text{drag}}}{\left(\frac{1}{2}\rho V^2\right)\left(\frac{\pi D^2}{4}\right)} \tag{8.78}$$

$$= \frac{6\pi R \mu V}{\left(\frac{1}{2}\rho V^2\right)\left(\frac{\pi D^2}{4}\right)} \tag{8.79}$$

$$= \frac{24\mu}{\rho V D} = \frac{24}{\text{Re}} \tag{8.80}$$

$$\begin{array}{c} \text{Drag law for} \\ \text{creeping flow} \\ \text{around a sphere:} \end{array} \quad \boxed{C_D = \frac{24}{\text{Re}}} \tag{8.81}$$

Experiments that test the validity of Equation 8.81 were performed as long ago as the mid-1800s, and the results are shown in Figure 8.12. For Reynolds numbers below 2, the creeping-flow solution is verified. This agreement is a validation of the microscopic analysis performed in the first part of this chapter. From the experimental data, we also learn the limits of the analysis. For Re < 2, the flow is consistent with the assumptions of steady flow: $v_{\phi} = 0$, symmetry in the ϕ-direction, and negligible inertia. For Re > 2, inertia is important; and, at high Reynolds numbers, the creeping-flow equation $C_D = 24/\text{Re}$ grossly underpredicts the drag.

The data in Figure 8.12 indicate the true amount of drag observed in flow around a sphere for the velocities corresponding to the Reynolds numbers shown. For Re < 2, the drag coefficient is given by $C_D = 24/\text{Re}$, the creeping-flow solution.

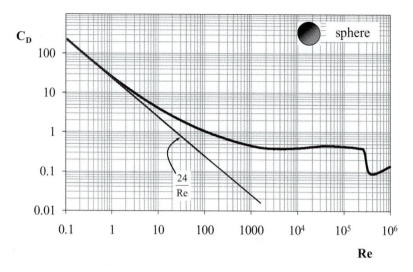

Figure 8.12 Experimental data for drag coefficient as a function of the Reynolds number for flow around a sphere [147]. Below Re = 2, the data closely follow the creeping-flow solution of $C_D = 24/$Re.

For $0.1 \leq \text{Re} \leq 1{,}000$, C_D follows a less steeply declining curve given approximately by $C_D \approx \frac{24}{\text{Re}} \left(1 + 0.14\text{Re}^{0.7}\right)$ [132]. We see that above Re = 1,000, the curve of C_D versus Re levels off, and C_D may be estimated within 15 percent by the constant $C_D \simeq 0.445$ (called Newton's drag-law regime) up to a Reynolds number of 2.6×10^5. This region is followed by a sharp drop in drag at Re = 2.6×10^5. In summary, the experimental correlation for drag coefficient versus Reynolds number for flow around a sphere is:

	Re	C_D
Data correlation for drag coefficient for flow around a sphere (all flow regimes):	Re < 2	$\dfrac{24}{\text{Re}} = 24\text{Re}^{-1}$
	$0.1 \leq \text{Re} \leq 1{,}000$	$\frac{24}{\text{Re}} \left(1 + 0.14\text{Re}^{0.7}\right)$
	$1{,}000 \leq \text{Re} \leq 2.6 \times 10^5$	≈ 0.445
	$2.8 \times 10^5 \leq \text{Re} \leq 10^6$	$\dfrac{\log C_D}{\text{Re}/10^6} = 4.43 \log \text{Re} - 27.3$

$$(8.82)$$

A single correlation equation for sphere drag coefficient as a function of Reynolds number is given here [106]; Equation 8.83 is valid from the creeping-flow limit through Re = 10^6 and is suitable for computer implementation:

Data correlation for drag coefficient for flow around a sphere (all flow regimes up to Re = 10^6) [106]:

$$f = \frac{24}{\text{Re}} + \frac{2.6 \left(\frac{\text{Re}}{5.0}\right)}{1 + \left(\frac{\text{Re}}{5.0}\right)^{1.52}} + \frac{0.411 \left(\frac{\text{Re}}{263{,}000}\right)^{-7.94}}{1 + \left(\frac{\text{Re}}{263{,}000}\right)^{-8.00}} + \frac{0.25 \left(\frac{\text{Re}}{10^6}\right)}{1 + \left(\frac{\text{Re}}{10^6}\right)} \qquad (8.83)$$

The correlation in Equation 8.83 is compared to literature data in Figure 8.13.

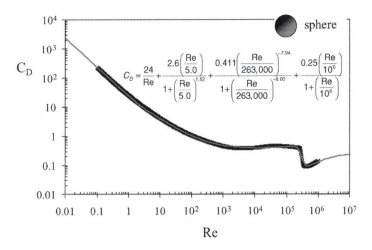

Figure 8.13 Experimental data for drag coefficient as a function of the Reynolds number for flow around a sphere [147]. Also shown are the approximations of these data that are represented in Equation 8.83 [106]. The data correlation is not to be used above Re $= 10^6$.

Now that experiments and data correlations have shown the true nature of drag in rapid flow around a sphere, we are in a better position to estimate the terminal speed of a skydiver.

EXAMPLE 8.5 (Skydiver, continued). *What is the maximum speed reached by a skydiver who jumps out of an airplane at 13,000 feet? How much can the speed of the skydiver vary depending on her body position (i.e., arms and legs flung out or pulled in tightly)?*

SOLUTION. As shown during our previous attempt at this problem, to calculate the terminal speed of a falling object, we apply Newton's second law (i.e., momentum conservation). When the skydiver is at terminal speed, the acceleration is zero. The two forces on the skydiver are gravity and the fluid forces:

$$
\begin{array}{c}
\text{Momentum balance} \\
\text{on skydiver} \\
\text{at terminal speed:}
\end{array}
\qquad
\underset{\substack{\text{all forces} \\ \text{acting on body}}}{\sum} \underline{f} = m\underline{a} = 0
\qquad (8.84)
$$

$$
\underline{f}_{\text{gravity}} + \underline{f}_{\text{fluid}} = 0 \qquad (8.85)
$$

$$
m\underline{g} + \underline{\mathcal{F}} = 0 \qquad (8.86)
$$

where m is the mass of the skydiver; the acceleration due to gravity is $\underline{g} = -g\hat{e}_z$; and for $\underline{\mathcal{F}}$, the force on a sphere falling through a viscous liquid, we previously used the result of the analysis of Stokes flow, Equation 8.61. We now know that using Stokes flow in the skydiver case is an error because Stokes flow is not produced at the high speeds attained by the skydiver. Instead, we express $\underline{\mathcal{F}}$ in terms of drag coefficient and use the experimental correlations for $C_D(\text{Re})$ in Equation 8.82 or 8.83 to obtain our final answer.

To incorporate the drag coefficient in our expressions, we review our previous solution for Stokes flow. In Stokes flow, the fluid force on the sphere $\underline{\mathcal{F}}$ was seen to be composed of two parts: a buoyancy term (contains gravity) and a drag term (contains velocity) (see Example 8.2):

$$
\begin{array}{c}
\text{Fluid force on a} \\
\text{sphere in} \\
\text{creeping flow} \\
\text{(Stokes flow)}
\end{array}
\quad
\underline{\mathcal{F}} = \begin{pmatrix} 0 \\ 0 \\ \dfrac{4\pi R^3 \rho g}{3} + 6\pi R \mu v_\infty \end{pmatrix}_{xyz}
\tag{8.87}
$$

For flow outside the Stokes regime, we replace the drag term ($6\pi R \mu v_\infty$) with drag given in Equation 8.74, the defining equation for the drag coefficient C_D:

$$
\text{Arbitrary shape:} \quad C_D \equiv \frac{\mathcal{F}_{\text{drag}}}{\left(\frac{1}{2}\rho V^2\right)(A_p)}
\tag{8.88}
$$

$$
\text{Sphere:} \quad C_D \equiv \frac{\mathcal{F}_{\text{drag}}}{\left(\frac{1}{2}\rho V^2\right)\left(\frac{\pi D^2}{4}\right)}
\tag{8.89}
$$

Solving Equation 8.88 for $\mathcal{F}_{\text{drag}}$ and substituting $\mathcal{F}_{\text{drag}}$ for the Stokes drag in the fluid-force Equation 8.87, we obtain:

$$
\mathcal{F}_{\text{drag}} = \frac{\rho V^2 A_p C_D}{2}
\tag{8.90}
$$

$$
\underline{\mathcal{F}} = \begin{pmatrix} 0 \\ 0 \\ \dfrac{4\pi R^3 \rho g}{3} + \mathcal{F}_{\text{drag}} \end{pmatrix}_{xyz}
\tag{8.91}
$$

$$
\begin{array}{c}
\text{Fluid force on a} \\
\text{sphere in} \\
\text{uniform flow} \\
\text{(all flow regimes):}
\end{array}
\quad
\underline{\mathcal{F}} = \begin{pmatrix} 0 \\ 0 \\ \dfrac{4\pi R^3 \rho g}{3} + \dfrac{\rho V^2 A_p C_D}{2} \end{pmatrix}_{xyz}
\tag{8.92}
$$

Now, returning to the momentum balance (see Equation 8.86), we substitute Equation 8.92 for the fluid force and expand the gravity term in our chosen coordinate system:

$$
\begin{pmatrix} 0 \\ 0 \\ -\mathcal{V}\rho_{\text{body}}g \end{pmatrix}_{xyz} + \begin{pmatrix} 0 \\ 0 \\ \mathcal{V}\rho g + \dfrac{\rho V^2 A_p C_D}{2} \end{pmatrix}_{xyz} = \begin{pmatrix} 0 \\ 0 \\ 0 \end{pmatrix}_{xyz}
\tag{8.93}
$$

where the mass of the sphere is given by $\mathcal{V}\rho_{\text{body}} = (4/3)\pi R^3 \rho_{\text{body}}$; ρ_{body} is the density of the body; and $\mathcal{V}$ is the volume of the body. Solving the z-component of Equation 8.93 for the drag coefficient, we obtain the equation used to generate the $C_D(\text{Re})$ correlations from experimental data on terminal speed. Solving the same equation for the terminal speed, we obtain the equation that allows us to calculate terminal speed from published experimental results for the drag coefficient:

Measured drag coefficient: (arbitrary object drop)

$$C_D = \frac{2 \mathcal{V} g \left(\rho_{\text{body}} - \rho \right)}{\rho A_p v_\infty^2} \qquad (8.94)$$

Terminal speed of an arbitrary body:

$$v_\infty = V = \sqrt{\frac{2 \mathcal{V} g(\rho_{\text{body}} - \rho)}{\rho A_p C_D}} \qquad (8.95)$$

Measured drag coefficient: (sphere drop)

$$C_D = \frac{4 g D \left(\rho_{\text{body}} - \rho \right)}{3 \rho v_\infty^2} \qquad (8.96)$$

Terminal speed of a sphere:

$$v_\infty = V = \sqrt{\frac{4(\rho_{\text{body}} - \rho) D g}{3 \rho C_D}} \qquad (8.97)$$

Equation 8.97 and the data in Figure 8.12 now may be used to estimate the terminal speed of a skydiver. Using the same original values of the physical parameters as in Example 8.3, we start by assuming that the Reynolds number for a skydiver will be high—perhaps in the $10^3 \leq \text{Re} \leq 2 \times 10^5$ region—where the drag coefficient reaches a constant value, $C_D = 0.445$:

$$\text{Density of air:} \quad \rho = 1.3 \text{ kg/m}^3 \qquad (8.98)$$

$$\text{Density of human ball (water):} \quad \rho_{\text{body}} = 1,000 \text{ kg/m}^3 \qquad (8.99)$$

$$\text{Diameter of sphere:} \quad D = 0.5 \text{ m} \qquad (8.100)$$

$$\text{Acceleration due to gravity:} \quad g = 9.80 \text{ m/s}^2 \qquad (8.101)$$

$$\text{Drag coefficient:} \quad C_D = 0.445 \qquad (8.102)$$

With these parameter values, we obtain the terminal speed as:

$$\begin{array}{c} \text{Estimate of} \\ \text{terminal speed:} \quad V = 107 \text{ m/s} \\ (C_D = 0.445) \end{array} \qquad (8.103)$$

The estimate obtained is a substantial improvement when compared to the Stokes-flow estimate. The actual speed of a skydiver in freefall is approximately 55 m/s (belly-to-Earth position) or 90 m/s (head-first position). Thus, assuming the skydiver was a sphere and that C_D was equal to a constant value of 0.445 allowed us to predict an answer within a reasonable uncertainty of the correct experimental result (i.e., within a factor of 2).

This second attempt is already good, but we have many effects to investigate as we continue our solution to this problem. Note that we only roughly modeled our skydiver, assuming that she falls like a sphere of a chosen diameter. There certainly is an effect of object shape on the terminal speed, and we explore these issues later in the chapter (see Example 8.24). In addition, we assumed a constant drag coefficient of $C_D = 0.445$; this value is valid up to a Reynolds number of about 2.6×10^5. Using the velocity result of 107 m/s, the Reynolds number is $\mathrm{Re} = 4 \times 10^6$, which is beyond this maximum Re for the Newton's drag-law regime. Finally, there is a qualitative change in the sphere $C_D(\mathrm{Re})$ correlation at high Reynolds numbers ($\mathrm{Re} > 2.6 \times 10^5$). It is reasonable to ask whether the phenomena that cause the abrupt drop in C_D seen in Figure 8.12 also might affect our skydiver. More experiments, including flow visualization, are needed to address these questions.

The reasonable estimate of terminal speed obtained in Example 8.5 can be counted as a success for the modeling methods in this text. We start with a real, practical problem; identify an idealized situation related to our real problem; investigate the idealized situation; and then use experimental results to map the idealized problem onto the real problem. This is a fundamental methodology of fluid-mechanics modeling. The answer obtained is within a factor of 2 of the observed terminal speed.

To move to the next level and improve the accuracy of our estimates, we must refine the models and broaden the experiments. We can clarify the origin of the drag coefficient (see Equation 8.74) by using dimensional analysis, and we turn to a discussion of this topic in the next section. That development shows how we knew C_D would be a function only of Reynolds number. We also discuss more thoroughly the types of flow behaviors seen in the noncreeping regime of flow past a sphere (e.g., recirculation, boundary layers, separation, and wakes), and we make a first attempt at modeling high-speed flows with the Navier-Stokes equation. In Section 8.2, we discuss boundary layers and flows past nonspherical objects and investigate the influence of shape and orientation on drag. In Section 8.3, we build on what we learned from our trials and errors and discover the importance of the rotational character of flow fields, quantified by the field property vorticity. The chapter concludes with a discussion of complex external flows.

8.1.2.1 DIMENSIONAL ANALYSIS OF NONCREEPING FLOW

In the previous section, we learn that drag coefficient C_D quantifies drag in noncreeping flow past a sphere. In this section, we discuss the origin of drag coefficient. The concept of the drag coefficient derives directly from dimensional analysis.

The situation we face with noncreeping flow around a sphere is the same situation we faced when analyzing turbulent pipe flow in Chapter 7: We know the equations that govern mass and momentum conservation for the flows of interest, but we are unable to solve them:

Mass conservation:
(continuity equation, $0 = \nabla \cdot \underline{v}$ (8.104)
constant density)

Momentum conservation:
(Navier-Stokes equation) $\rho \left(\dfrac{\partial \underline{v}}{\partial t} + \underline{v} \cdot \nabla \underline{v} \right) = -\nabla p + \mu \nabla^2 \underline{v} + \rho \underline{g}$ (8.105)

Newtonian
constitutive equation: $\tilde{\underline{\underline{\tau}}} = \tilde{\underline{\underline{\Pi}}} + p\underline{\underline{I}} = \mu \left(\nabla \underline{v} + (\nabla \underline{v})^T \right)$ (8.106)

Total fluid force
on a surface $\mathcal{S}$: $\underline{\mathcal{F}} = \iint_{\mathcal{S}} [\hat{n} \cdot \tilde{\underline{\underline{\Pi}}}]_{\text{at surface}} \, dS$ (8.107)

We will use experiments to provide the information we lack, but there are many variables and physical properties to consider. To sort out how each quantity affects a given problem, we follow the procedure in Chapter 7: We nondimensionalize each quantity in the governing equations. If we rewrite the governing equations in terms of nondimensional versions of velocity, coordinate directions, time, and other variables, we can see more clearly the structure of the equations; that clarity helps considerably in designing experiments, plotting data, and deriving correlations.

We want to nondimensionalize the governing equations for flow around a sphere: the continuity equation, the Navier-Stokes equation, and the expression for fluid force on a sphere. We choose the upstream velocity $V \equiv v_\infty$ as the characteristic velocity and the sphere diameter $D = 2R$ as the characteristic linear dimension. These choices are arbitrary; how good they are can be decided from the final results of the analysis. As in Chapter 7, we define nondimensional variables as the ratios of the dimensional variables to the characteristic values:

$$v_r^* \equiv \frac{v_r}{V} \tag{8.108}$$

$$v_\theta^* \equiv \frac{v_\theta}{V} \tag{8.109}$$

$$v_\phi^* \equiv \frac{v_\phi}{V} \tag{8.110}$$

$$r^* \equiv \frac{r}{D} \tag{8.111}$$

$$t^* \equiv \frac{t}{D/V} \tag{8.112}$$

We now solve these expressions for the dimensional variables and substitute them into the Navier-Stokes equation, continuity equation, and equation for fluid force $\mathcal{F}$ on a surface. After some algebra, we obtain new nondimensional versions of the governing equations.

For the constant-density continuity equation, we obtain:

Continuity equation: $\dfrac{1}{r^{*2}} \dfrac{\partial(r^{*2} v_r^*)}{\partial r^*} + \dfrac{1}{r^* \sin\theta} \dfrac{\partial(v_\theta^* \sin\theta)}{\partial\theta} + \dfrac{1}{r^* \sin\theta} \dfrac{\partial(v_\phi^*)}{\partial\phi} = 0$

$$\tag{8.113}$$

The nondimensional mass balance does not include any scale factors containing the characteristic dimensions V or D; thus, this version does not yield any information about the relative importance of the three terms in the continuity equation. Each term in Equation 8.113 appears to be equally important.

For the momentum balance, we use the Navier-Stokes equation—the microscopic-momentum balance:

Navier-Stokes equation: $\rho\left(\dfrac{\partial \underline{v}}{\partial t} + \underline{v} \cdot \nabla\underline{v}\right) = -\nabla p + \mu\nabla^2\underline{v} + \rho\underline{g}$

$$\tag{8.114}$$

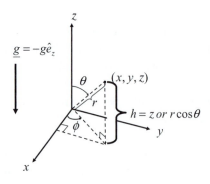

The dynamic pressure folds the hydrostatic effect of gravity into the pressure term. The height h must be expressed in whichever coordinate system is chosen for problem solving.

Before we nondimensionalize the Navier-Stokes equation, we can simplify the equation given our experience with Stokes flow. The role of gravity in the flow-around-a-sphere problem can be determined by reviewing the creeping-flow solution, Equations 8.23 and 8.24. Gravity appears only in the pressure distribution. Because the presence of gravity affects only the pressure field, we can roll the effect of gravity into the pressure function by using a *dynamic pressure* as follows—let the dynamic pressure field, $\mathcal{P}(r, \theta, \phi)$ or $\mathcal{P}(x, y, z)$, be defined as:

$$\text{Dynamic pressure} \quad \mathcal{P} \equiv p + \rho g h \tag{8.115}$$

where $p(x, y, z)$ is the pressure at a given location in the fluid, ρ is the density of the fluid, and h is the vertical height of the location (x, y, z) above an elevation chosen as the reference elevation. For example, if we choose the reference elevation for the flow-around-a-sphere problem as the horizontal plane through the origin of the coordinate systems, then in the two coordinate systems of our problem, h is given by the following (Figure 8.14):

$$h(x, y, z) = z \tag{8.116}$$

$$h(r, \theta, \phi) = r \cos \theta \tag{8.117}$$

The mathematical form of h used for a given calculation depends on the coordinate system being used.

Having defined the dynamic pressure in Equation 8.115, we can group the pressure (∇p) and gravity ($\rho \underline{g}$) terms of the Navier-Stokes equations. To see this, consider the term ∇p of the Navier-Stokes equation. In the Cartesian coordinate system, this term is a vector given by the following (Equation 1.242):

$$\nabla p = \begin{pmatrix} \frac{\partial p}{\partial x} \\ \frac{\partial p}{\partial y} \\ \frac{\partial p}{\partial z} \end{pmatrix}_{xyz} \tag{8.118}$$

In the Cartesian coordinate system of the sphere problem (see Figure 8.5), h is given by $h = z$; thus, $\mathcal{P} = p + \rho g z$. We can form $\nabla \mathcal{P}$ from the definition of ∇ (Equation 8.118), which is valid for any function, and then simplify each component by carrying out the partial derivatives:

$$\nabla \mathcal{P} = \nabla (p + \rho gh) \tag{8.119}$$

$$= \nabla (p + \rho gz) \tag{8.120}$$

$$= \begin{pmatrix} \frac{\partial(p+\rho gz)}{\partial x} \\ \frac{\partial(p+\rho gz)}{\partial y} \\ \frac{\partial(p+\rho gz)}{\partial z} \end{pmatrix}_{xyz} \tag{8.121}$$

$$= \begin{pmatrix} \frac{\partial p}{\partial x} \\ \frac{\partial p}{\partial y} \\ \frac{\partial p}{\partial z} + \rho g \end{pmatrix}_{xyz} \tag{8.122}$$

$$= \begin{pmatrix} \frac{\partial p}{\partial x} \\ \frac{\partial p}{\partial y} \\ \frac{\partial p}{\partial z} \end{pmatrix}_{xyz} + \begin{pmatrix} 0 \\ 0 \\ +\rho g \end{pmatrix}_{xyz} \tag{8.123}$$

$$\nabla \mathcal{P} = \nabla p - \rho \underline{g} \tag{8.124}$$

where we use the fact that the gravity vector in our chosen coordinate system is given by $\underline{g} = -g\hat{e}_z$. Incorporating the gradient of dynamic pressure from Equation 8.124 into the Navier-Stokes equation yields:

Navier-Stokes equation: $\quad \rho \left(\frac{\partial \underline{v}}{\partial t} + \underline{v} \cdot \nabla \underline{v} \right) = -\nabla p + \mu \nabla^2 \underline{v} + \rho \underline{g}$ (8.125)
(regular pressure term)

Navier-Stokes equation: $\quad \rho \left(\frac{\partial \underline{v}}{\partial t} + \underline{v} \cdot \nabla \underline{v} \right) = -\nabla \mathcal{P} + \mu \nabla^2 \underline{v}$ (8.126)
(dynamic pressure term)

The version of the Navier-Stokes equation in Equation 8.126 was derived in a particular Cartesian coordinate system, but it is valid in any coordinate system because vectors and tensors are independent of the coordinate system. Note that when there is no flow ($\underline{v} = 0$), the Navier-Stokes equation states that the gradient of the dynamic pressure is zero ($\nabla \mathcal{P} = 0$), which is simply the static-fluid equation (see Equation 4.52):

No-flow Navier-Stokes equation: $\qquad \nabla \mathcal{P} = 0 \tag{8.127}$

$$\begin{pmatrix} \frac{\partial p}{\partial x} \\ \frac{\partial p}{\partial y} \\ \frac{\partial p}{\partial z} + \rho g \end{pmatrix}_{xyz} = \begin{pmatrix} 0 \\ 0 \\ 0 \end{pmatrix}_{xyz} \tag{8.128}$$

We are nondimensionalizing the Navier-Stokes equation in search of the dimensionless groups that govern the flow-around-a-sphere problem. Beginning now with the Navier-Stokes equation in terms of dynamic pressure (Equation 8.126), we incorporate the nondimensional variables v_r^*, v_θ^*, and so on (i.e., Equations 8.108–8.112). For the dynamic-pressure term, we nondimensionalize

$\mathcal{P}$ with a characteristic pressure P chosen as $P = \rho V^2$; this is the same characteristic pressure we choose in tube flow.[4] The nondimensional Navier-Stokes equation becomes:

$$
\begin{pmatrix} \dfrac{\partial v_r^*}{\partial t^*} \\[2mm] \dfrac{\partial v_\theta^*}{\partial t^*} \\[2mm] \dfrac{\partial v_\phi^*}{\partial t^*} \end{pmatrix}_{r\theta\phi}
+
\begin{pmatrix} v_r^*\left(\dfrac{\partial v_r^*}{\partial r^*}\right) + v_\theta^*\left(\dfrac{1}{r^*}\dfrac{\partial v_r^*}{\partial \theta} - \dfrac{v_\theta^*}{r^*}\right) + v_\phi^*\left(\dfrac{1}{r^*\sin\theta}\dfrac{\partial v_r^*}{\partial \phi} - \dfrac{v_\phi^*}{r^*}\right) \\[3mm] v_r^*\left(\dfrac{\partial v_\theta^*}{\partial r^*}\right) + v_\theta^*\left(\dfrac{1}{r^*}\dfrac{\partial v_\theta^*}{\partial \theta} + \dfrac{v_r^*}{r^*}\right) + v_\phi^*\left(\dfrac{1}{r^*\sin\theta}\dfrac{\partial v_\theta^*}{\partial \phi} - \dfrac{v_\phi^*}{r^*}\cot\theta\right) \\[3mm] v_r^*\left(\dfrac{\partial v_\phi^*}{\partial r^*}\right) + v_\theta^*\left(\dfrac{1}{r^*}\dfrac{\partial v_\phi^*}{\partial \theta}\right) + v_\phi^*\left(\dfrac{1}{r^*\sin\theta}\dfrac{\partial v_\phi^*}{\partial \phi} + \dfrac{v_r^*}{r^*} + \dfrac{v_\theta^*}{r^*}\cot\theta\right) \end{pmatrix}_{r\theta\phi}
$$

$$
= -\begin{pmatrix} \dfrac{\partial \mathcal{P}^*}{\partial r^*} \\[3mm] \dfrac{1}{r^*}\dfrac{\partial \mathcal{P}^*}{\partial \theta} \\[3mm] \dfrac{1}{r^*\sin\theta}\dfrac{\partial \mathcal{P}^*}{\partial \phi} \end{pmatrix}_{r\theta\phi}
$$

$$
+\frac{1}{\text{Re}}
\begin{pmatrix} \left(\dfrac{\partial}{\partial r^*}\left(\dfrac{1}{r^{*2}}\dfrac{\partial}{\partial r^*}(r^{*2}v_r^*)\right) + \dfrac{1}{r^{*2}\sin\theta}\dfrac{\partial}{\partial\theta}\left(\sin\theta\dfrac{\partial v_r^*}{\partial\theta}\right) + \dfrac{1}{r^{*2}\sin^2\theta}\dfrac{\partial^2 v_r^*}{\partial\phi^2}\right. \\[2mm] \left. - \dfrac{2}{r^{*2}\sin\theta}\dfrac{\partial}{\partial\theta}(v_\theta^*\sin\theta) - \dfrac{2}{r^{*2}\sin\theta}\dfrac{\partial v_\phi^*}{\partial\phi}\right) \\[3mm] \left(\dfrac{1}{r^{*2}}\dfrac{\partial}{\partial r^*}\left(r^{*2}\dfrac{\partial v_\theta^*}{\partial r^*}\right) + \dfrac{1}{r^{*2}}\dfrac{\partial}{\partial\theta}\left(\dfrac{1}{\sin\theta}\dfrac{\partial}{\partial\theta}(v_\theta^*\sin\theta)\right) + \dfrac{1}{r^{*2}\sin^2\theta}\dfrac{\partial^2 v_\theta^*}{\partial\phi^2}\right. \\[2mm] \left. + \dfrac{2}{r^{*2}}\dfrac{\partial v_r^*}{\partial\theta} - \dfrac{2\cot\theta}{r^{*2}\sin\theta}\dfrac{\partial v_\phi^*}{\partial\phi}\right) \\[3mm] \left(\dfrac{1}{r^{*2}}\dfrac{\partial}{\partial r^*}\left(r^{*2}\dfrac{\partial v_\phi^*}{\partial r^*}\right) + \dfrac{1}{r^{*2}}\dfrac{\partial}{\partial\theta}\left(\dfrac{1}{\sin\theta}\dfrac{\partial}{\partial\theta}(v_\phi^*\sin\theta)\right) + \dfrac{1}{r^{*2}\sin^2\theta}\dfrac{\partial^2 v_\phi^*}{\partial\phi^2}\right. \\[2mm] \left. + \dfrac{2}{r^{*2}\sin\theta}\dfrac{\partial v_r^*}{\partial\phi} + \dfrac{2\cot\theta}{r^{*2}\sin\theta}\dfrac{\partial v_\theta^*}{\partial\phi}\right) \end{pmatrix}_{r\theta\phi}
\qquad (8.129)
$$

The dimensionless scale-factor that appears in the nondimensional Navier-Stokes equation is again the Reynolds number:

$$
\text{Reynolds number} \quad \boxed{\text{Re} \equiv \frac{\rho V D}{\mu}} \quad \text{ratio of } \frac{\text{(inertial forces)}}{\text{(viscous forces)}} \qquad (8.130)
$$

The nondimensional Navier-Stokes equation for flow around a sphere written in Gibbs notation is:

Nondimensional
Navier-Stokes:
(dynamic pressure)

$$
\boxed{\frac{\partial \underline{v}^*}{\partial t^*} + \underline{v}^*\cdot\nabla^*\underline{v}^* = -\nabla^*\mathcal{P}^* + \left(\frac{1}{\text{Re}}\right)\nabla^{*2}\underline{v}^*}
$$

$$(8.131)$$

---

[4] There is more discussion of this choice later in this development and in Section 8.3.2.

where ∇^* represents the nondimensional del operator. Note that compared to the nondimensional Navier-Stokes equation at which we arrived when studying Poiseuille flow in a tube (see Equation 7.99), this version does not contain the Froude number, because we write pressure in terms of dynamic pressure $\mathcal{P}$. Using the dynamic pressure is appropriate when the effect of gravity is only as a hydrostatic supplement to the pressure field. In flows with free surfaces, waves can form and gravity has a profound effect on the shape of the fluid–air interface. For such free-surface flows, the Froude number Fr is important [115], and the grouping of pressure and gravity effects into a dynamic pressure is inappropriate.

We nondimensionalized mass and momentum balances for flow around a sphere and determined that the Reynolds number is the important quantity that determines the form of solutions to these equations. To apply the mass- and momentum-balance equations to the specific problem of determining the force on the sphere, we must also nondimensionalize the expression for fluid force on a surface. To calculate fluid forces on the sphere, we use the usual expression (Equations 8.6 and 8.32):

$$
\text{Total fluid force on a surface } \mathcal{S}: \qquad \underline{\mathcal{F}} = \iint_{\mathcal{S}} [\hat{n} \cdot \underline{\underline{\tilde{\Pi}}}]_{\text{at surface}} \, dS \tag{8.132}
$$

$$
\begin{array}{c}\text{Total fluid force}\\\text{on the sphere}\\\text{in noncreeping flow:}\end{array} \qquad \underline{\mathcal{F}} = \int_0^{2\pi} \int_0^\pi [\hat{e}_r \cdot \underline{\underline{\tilde{\Pi}}}]_{r=R} \, R^2 \sin\theta d\theta d\phi \tag{8.133}
$$

Carrying out the dot product on the tensor $\underline{\underline{\tilde{\Pi}}} = \underline{\underline{\tilde{\tau}}} - p\underline{\underline{I}}$ written in the spherical coordinate system, we obtain:

$$
[\hat{e}_r \cdot \underline{\underline{\tilde{\Pi}}}] = \begin{pmatrix} 1 & 0 & 0 \end{pmatrix}_{r\theta\phi} \cdot \begin{pmatrix} \tilde{\Pi}_{rr} & \tilde{\Pi}_{r\theta} & \tilde{\Pi}_{r\phi} \\ \tilde{\Pi}_{\theta r} & \tilde{\Pi}_{\theta\theta} & \tilde{\Pi}_{\theta\phi} \\ \tilde{\Pi}_{\phi r} & \tilde{\Pi}_{\phi\theta} & \tilde{\Pi}_{\phi\phi} \end{pmatrix}_{r\theta\phi} \tag{8.134}
$$

$$
= \begin{pmatrix} \tilde{\Pi}_{rr} & \tilde{\Pi}_{r\theta} & \tilde{\Pi}_{r\phi} \end{pmatrix}_{r\theta\phi} \tag{8.135}
$$

$$
= \begin{pmatrix} \tilde{\Pi}_{rr} \\ \tilde{\Pi}_{r\theta} \\ \tilde{\Pi}_{r\phi} \end{pmatrix}_{r\theta\phi} = \begin{pmatrix} \tilde{\tau}_{rr} - p \\ \tilde{\tau}_{r\theta} \\ \tilde{\tau}_{r\phi} \end{pmatrix}_{r\theta\phi} \tag{8.136}
$$

$$
= \begin{pmatrix} 2\mu\dfrac{\partial v_r}{\partial r} - p(r,\theta,\phi) \\[2mm] \mu r\dfrac{\partial}{\partial r}\left(\dfrac{v_\theta}{r}\right) + \dfrac{\mu}{r}\dfrac{\partial v_r}{\partial\theta} \\[2mm] \dfrac{\mu}{r\sin\theta}\dfrac{\partial v_r}{\partial\phi} + \mu r\dfrac{\partial}{\partial r}\left(\dfrac{v_\phi}{r}\right) \end{pmatrix}_{r\theta\phi} \tag{8.137}
$$

where the components of $\underline{\underline{\tilde{\tau}}}$ in spherical coordinates are obtained from Table B.8 in Appendix B. Substituting this result into the equation for force on a sphere

(Equation 8.133), we obtain:

Total molecular fluid force
on the sphere
in noncreeping flow:
$$\underline{\mathcal{F}} = \int_0^{2\pi} \int_0^{\pi} [\hat{e}_r \cdot \tilde{\underline{\underline{\Pi}}}]_{r=R} \; R^2 \sin\theta \, d\theta \, d\phi \qquad (8.138)$$

$$\underline{\mathcal{F}} = R^2 \int_0^{2\pi} \int_0^{\pi} \left[\begin{pmatrix} 2\mu \dfrac{\partial v_r}{\partial r} - p(r, \theta, \phi) \\[2mm] \mu r \dfrac{\partial}{\partial r}\left(\dfrac{v_\theta}{r}\right) + \dfrac{\mu}{r}\dfrac{\partial v_r}{\partial \theta} \\[2mm] \dfrac{\mu}{r\sin\theta}\dfrac{\partial v_r}{\partial \phi} + \mu r\dfrac{\partial}{\partial r}\left(\dfrac{v_\phi}{r}\right) \end{pmatrix}_{r\theta\phi} \right]_{r=R} \sin\theta \, d\theta \, d\phi \qquad (8.139)$$

Comparing Equation 8.139 to the creeping-flow equivalent, Equation 8.37, note that we include $\tilde{\Pi}_{r\phi}$ here because in the general case we may not assume that $v_\phi = 0$ or that noncreeping flow is symmetric in the ϕ-direction. Later, we substitute the dynamic pressure $p = \mathcal{P} - \rho g h = \mathcal{P} - \rho g r \cos\theta$ into the expression for $\underline{\mathcal{F}}$ to separate the effect of buoyancy. We delay this substitution to keep the current calculation simpler.

At this point in the previous creeping-flow calculation of $\underline{\mathcal{F}}$, we used the creeping-flow solutions for velocity $\underline{v}$ and pressure p. Due to mathematical complexity, however, we are unable to solve the Navier-Stokes equations for $\underline{v}(r, \theta, \phi, t)$ and $p(r, \theta, \phi, t)$ for noncreeping flow; thus, we cannot calculate $\underline{\mathcal{F}}$ directly. It is precisely for this reason that we follow the dimensional-analysis approach.

Because we cannot solve for $\underline{\mathcal{F}}$ directly in the noncreeping-flow case, we nondimensionalize Equation 8.139 to see which dimensionless groups enter into noncreeping force on a sphere. We use the same nondimensional quantities v_z^*, p^*, r^*, and so on that we used when nondimensionalizing the Navier-Stokes equations. Making the appropriate substitutions of nondimensional quantities, we obtain:

$$\underline{\mathcal{F}} = \rho V^2 R^2 \int_0^{2\pi} \int_0^{\pi} \left[\begin{pmatrix} \dfrac{2}{\text{Re}}\dfrac{\partial v_r^*}{\partial r^*} - p^* \\[2mm] \dfrac{r^*}{\text{Re}}\dfrac{\partial}{\partial r^*}\left(\dfrac{v_\theta^*}{r^*}\right) + \dfrac{1}{\text{Re}\, r^*}\dfrac{\partial v_r^*}{\partial \theta} \\[2mm] \dfrac{1}{\text{Re}\, r^*\sin\theta}\dfrac{\partial v_r^*}{\partial \phi} + \dfrac{r^*}{\text{Re}}\dfrac{\partial}{\partial r^*}\left(\dfrac{v_\phi^*}{r^*}\right) \end{pmatrix}_{r\theta\phi} \right]_{r^*=\frac{1}{2}} \sin\theta \, d\theta \, d\phi$$

$$(8.140)$$

Note that the pressure term (prefactor $= 1$) and the velocity-gradient or viscous terms (prefactor $= 1/\text{Re}$) scale differently; this is a consequence of how we choose to scale pressure. By choosing the characteristic pressure P as ρV^2 rather than as something involving viscosity, we impose that these terms scale differently.[5]

---

[5] Note also that in the nondimensionalization of tube flow, there was no pressure term in the drag expression (see Equation 7.138) because the flow was unidirectional.

To calculate the drag from the total force $\mathcal{F}$ in Equation 8.140, we follow the steps for the analogous creeping-flow calculation and convert the vector in Equation 8.140 to Cartesian coordinates. We convert our expression to the Cartesian coordinate system so that the θ- and ϕ-dependencies of the basis vectors are accounted for appropriately prior to final integration. The basis vectors $\hat{e}_r$, $\hat{e}_\theta$, and $\hat{e}_\phi$ are written in terms of the Cartesian basis vectors in Equations 1.271–1.273:

$$\mathcal{F} = \rho V^2 R^2 \int_0^{2\pi} \int_0^\pi \left[\begin{pmatrix} \dfrac{2}{\mathrm{Re}} \dfrac{\partial v_r^*}{\partial r^*} - p^* \\[2mm] \dfrac{r^*}{\mathrm{Re}} \dfrac{\partial}{\partial r^*}\left(\dfrac{v_\theta^*}{r^*}\right) + \dfrac{1}{\mathrm{Re}\, r^*} \dfrac{\partial v_r^*}{\partial \theta} \\[2mm] \dfrac{1}{\mathrm{Re}\, r^* \sin\theta} \dfrac{\partial v_r^*}{\partial \phi} + \dfrac{r^*}{\mathrm{Re}} \dfrac{\partial}{\partial r^*}\left(\dfrac{v_\phi^*}{r^*}\right) \end{pmatrix}_{r\theta\phi} \right]_{r^*=\frac{1}{2}} \sin\theta\, d\theta\, d\phi$$

(8.141)

$$= \rho V^2 R^2 \int_0^{2\pi} \int_0^\pi \left[\left(\dfrac{2}{\mathrm{Re}} \dfrac{\partial v_r^*}{\partial r^*} - p^* \right) \hat{e}_r + \left(\dfrac{r^*}{\mathrm{Re}} \dfrac{\partial}{\partial r^*}\left(\dfrac{v_\theta^*}{r^*}\right) + \dfrac{1}{\mathrm{Re}\, r^*} \dfrac{\partial v_r^*}{\partial \theta} \right) \hat{e}_\theta \right.$$

$$\left. + \left(\dfrac{1}{\mathrm{Re}\, r^* \sin\theta} \dfrac{\partial v_r^*}{\partial \phi} + \dfrac{r^*}{\mathrm{Re}} \dfrac{\partial}{\partial r^*}\left(\dfrac{v_\phi^*}{r^*}\right) \right) \hat{e}_\phi \right]_{r^*=\frac{1}{2}} \sin\theta\, d\theta\, d\phi$$

(8.142)

Fluid force on a sphere in noncreeping flow:

$$\mathcal{F} = \rho V^2 R^2 \int_0^{2\pi} \int_0^\pi \left[\left(\dfrac{2}{\mathrm{Re}} \dfrac{\partial v_r^*}{\partial r^*} - p^* \right) \begin{pmatrix} \sin\theta \cos\phi \\ \sin\theta \sin\phi \\ \cos\theta \end{pmatrix}_{xyz} \right.$$

$$+ \left(\dfrac{r^*}{\mathrm{Re}} \dfrac{\partial}{\partial r^*}\left(\dfrac{v_\theta^*}{r^*}\right) + \dfrac{1}{\mathrm{Re}\, r^*} \dfrac{\partial v_r^*}{\partial \theta} \right) \begin{pmatrix} \cos\theta \cos\phi \\ \cos\theta \sin\phi \\ -\sin\theta \end{pmatrix}_{xyz}$$

$$\left. + \left(\dfrac{1}{\mathrm{Re}\, r^* \sin\theta} \dfrac{\partial v_r^*}{\partial \phi} + \dfrac{r^*}{\mathrm{Re}} \dfrac{\partial}{\partial r^*}\left(\dfrac{v_\phi^*}{r^*}\right) \right) \begin{pmatrix} -\sin\phi \\ \cos\phi \\ 0 \end{pmatrix}_{xyz} \right]_{r^*=\frac{1}{2}}$$

$$\times \sin\theta\, d\theta\, d\phi$$

(8.143)

The result for $\mathcal{F}$ in Equation 8.143 is more complicated than the result for $\mathcal{F}$ obtained for creeping flow (compare Equation 8.143 to Equation 8.61, repeated here):

Fluid force on a sphere in creeping flow: (Stokes flow, Equation 8.61)

$$\mathcal{F} = \begin{pmatrix} 0 \\ 0 \\ \dfrac{4\pi R^3 \rho g}{3} + 6\pi R\mu v_\infty \end{pmatrix}_{xyz}$$

(8.144)

A significant difference between these two expressions is the direction of the vector $\underline{\mathcal{F}}$. In arriving at Equation 8.144, we carried out the ϕ-integrations; from those integrations, we learned that the x- and y-components of $\underline{\mathcal{F}}$ were zero (see Equation 8.52). Thus, the force on a sphere in a uniform z-directional creeping flow is in the z-direction, the direction of the oncoming flow. In the general, noncreeping flow (Equation 8.143), we cannot carry out the ϕ-integrations, and we cannot rule out that there will be nonzero x- and y-components to the force on the sphere. Thus, for the general case of uniform flow in the z-direction around a sphere, the vector that expresses the force on the sphere likely has three nonzero components:

$$\begin{array}{c} \text{Fluid force on a sphere} \\ \text{in noncreeping flow:} \\ \text{(Equation 8.143)} \end{array} \qquad \boxed{\underline{\mathcal{F}} = \begin{pmatrix} \mathcal{F}_x \\ \mathcal{F}_y \\ \mathcal{F}_z \end{pmatrix}_{xyz}} \qquad (8.145)$$

In creeping flow, the total fluid force $\underline{\mathcal{F}}$ (see Equation 8.61) was found to be the sum of buoyancy, which is in the direction opposite to gravity, and drag, which is in the far-field flow direction $\hat{e}_z$:

$$\underline{f}_{buoyancy} = -\rho \mathcal{V}_{\text{body}}\underline{g} \qquad (8.146)$$

$$\underline{f}_{drag} = \mathcal{F}_{\text{drag}}\hat{e}_z \qquad (8.147)$$

$$\begin{array}{c} \text{Fluid force on a sphere} \\ \text{in creeping flow:} \end{array} \qquad \underline{\mathcal{F}} = \underline{f}_{buoyancy} + \underline{f}_{drag} \qquad (8.148)$$

where ρ is the density of the fluid and $\mathcal{V}_{\text{body}}$ is the volume of the body, which for a sphere is $4\pi R^3/3$. The total fluid force on a sphere in noncreeping flow is the sum of three vector contributions: the buoyancy; the drag; and, apparently, another contribution perpendicular to $\hat{e}_z$, called the *lift*, which may be arrived at by subtraction:

$$\underline{\mathcal{F}} = \underline{f}_{buoyancy} + \underline{f}_{drag} + \underline{f}_{lift} \qquad (8.149)$$

$$\begin{array}{c} \text{Fluid force on a sphere} \\ \text{in noncreeping flow} \\ \text{in terms of lift:} \\ \text{(upstream flow in } z\text{-direction)} \end{array} \qquad \boxed{\underline{\mathcal{F}} = -\rho \mathcal{V}_{\text{body}}\underline{g} + \mathcal{F}_{\text{drag}}\hat{e}_z + \underline{\mathcal{F}}_{\text{lift}}}$$

$$(8.150)$$

$$\begin{pmatrix} \mathcal{F}_x \\ \mathcal{F}_y \\ \mathcal{F}_z \end{pmatrix}_{xyz} = -\rho \mathcal{V}_{\text{body}} \begin{pmatrix} g_x \\ g_y \\ g_z \end{pmatrix}_{xyz} + \begin{pmatrix} 0 \\ 0 \\ \mathcal{F}_{\text{drag}} \end{pmatrix}_{xyz} + \begin{pmatrix} \mathcal{F}_{\text{lift},x} \\ \mathcal{F}_{\text{lift},y} \\ 0 \end{pmatrix}_{xyz} \qquad (8.151)$$

Note that because drag is defined as all of the z-direction force not due to buoyancy, the z-component of the lift is zero.

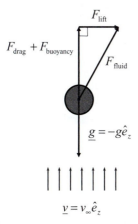

Figure 8.15 There are forces on the sphere due to gravity and due to the fluid. The fluid forces can be divided into buoyancy, which acts opposite to gravity, and two contributions due to fluid motion. The kinetic contribution to the fluid force that is parallel to the upstream flow direction is called drag; the kinetic contribution to the fluid force that is not parallel to the upstream flow direction is called lift.

In the coordinate systems we are using for flow around a sphere (see Figure 8.5), gravity is parallel to the upstream flow direction $\hat{e}_z$ but in the opposite direction, $\underline{g} = -g\hat{e}_z$. Incorporating this and the volume of the sphere into Equation 8.151, we obtain:

$$\underline{\mathcal{F}} = \begin{pmatrix} \mathcal{F}_x \\ \mathcal{F}_y \\ \mathcal{F}_z \end{pmatrix}_{xyz} = -\frac{4\pi R^3 \rho}{3} \begin{pmatrix} 0 \\ 0 \\ -g \end{pmatrix}_{xyz} + \begin{pmatrix} 0 \\ 0 \\ \mathcal{F}_{\text{drag}} \end{pmatrix}_{xyz} + \begin{pmatrix} \mathcal{F}_{\text{lift},x} \\ \mathcal{F}_{\text{lift},y} \\ 0 \end{pmatrix}_{xyz}$$

(8.152)

$$\begin{matrix} \text{Lift:} \\ \text{(gravity given} \\ \text{by } -g\hat{e}_z) \end{matrix} \qquad \underline{\mathcal{F}}_{\text{lift}} = \begin{pmatrix} \mathcal{F}_{\text{lift},x} \\ \mathcal{F}_{\text{lift},y} \\ 0 \end{pmatrix}_{xyz} = \begin{pmatrix} \mathcal{F}_x \\ \mathcal{F}_y \\ 0 \end{pmatrix}_{xyz}$$

(8.153)

Note that when $\underline{g} = -g\hat{e}_z$, the x- and y-components of $\mathcal{F}_{\text{lift}}$ are equal to the x- and y-components of $\underline{\mathcal{F}}$ overall.

Lift is an extra contribution to the force on a sphere for noncreeping flow around a sphere (Figure 8.15). We first discussed lift in Section 2.5 (see Figure 2.26). Lift is the tendency of an object in a flow to experience a kinetic fluid force in a direction other than the upstream flow direction. Lift is what allows airplanes to fly (see discussion in Example 8.6). Lift, as shown in Equation 8.150, is a portion of the fluid force on an object in a uniform flow. In creeping flow, we calculate that the lift is zero (i.e., in Equation 8.61, the x- and y-components are zero); in noncreeping flow, however, inertia is present and spheres may experience lift, meaning side-to-side forces in a uniform flow (see Figure 8.15). We discuss these effects in the next section (see also Figures 8.21 and 8.22). To make the concept of lift more intuitive, we explore the colloquial use of the term *lift* in the following example.

EXAMPLE 8.6. *How does the concept of lift, a lateral force defined for flow around a sphere, relate to lift, the upward force that allows an airplane to fly?*

SOLUTION. The two concepts of lift—lift in flow around a sphere and lift on an airplane's wings—are the same concept but are applied with flows in different directions relative to gravity (Figure 8.16). For an airplane wing moving rapidly horizontal to the ground, the flow may be analyzed from the perspective of the center of gravity of the wing. From that point of view, a horizontal flow of air in the x-direction approaches the wing, and the wing is set at some finite angle to horizontal. This angle is called the angle of attack, α. The fluid force felt by the wing has a horizontal component, called the drag, and a vertical component, the sum of the lift and a small buoyant force. The component of force in the y-direction is not considered here because the wing is wide and the flow is approximately two-dimensional.

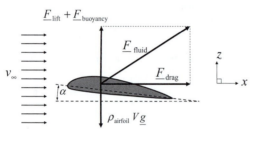

Figure 8.16 From the point of view of the airplane, the flight of an airplane is flow around an obstacle. The flow approaches the airplane wing (volume of wing = V; density of wing = $\rho_{airfoil}$) with a uniform speed v_∞. The wing experiences the downward pull of gravity, the retarding drag, buoyancy (small), and lift, which is a force in the airplane perpendicular to the flow direction and opposite to the direction of gravity.

Buoyancy, drag, and lift sum to give the total fluid force on a two-dimensional airfoil (Equation 8.149):

$$\underline{\mathcal{F}} = \mathcal{F}_{\text{fluid}} = \underline{f}_{\text{buoyancy}} + \underline{f}_{\text{drag}} + \underline{f}_{\text{lift}} \tag{8.154}$$

$$= \begin{pmatrix} 0 \\ 0 \\ \rho V g \end{pmatrix}_{xyz} + \begin{pmatrix} \mathcal{F}_{\text{drag}} \\ 0 \\ 0 \end{pmatrix}_{xyz} + \begin{pmatrix} 0 \\ 0 \\ \mathcal{F}_{\text{lift}} \end{pmatrix}_{xyz} \tag{8.155}$$

where ρ is fluid density, V is the volume of the airfoil, and g is the acceleration due to gravity. The buoyancy is negligible. Thus, flow around an airplane wing and flow around a sphere are similar, differing only in whether buoyancy is parallel to drag (i.e., sphere case) or to lift (i.e., airplane case) (compare Equations 8.152 and 8.155).

Returning to the question of drag on a sphere in noncreeping flow, we are working on nondimensionalizing the expression for $\mathcal{F}_{\text{drag}}$ on a sphere. In Equation 8.143, the component of $\underline{\mathcal{F}}$ that we need to calculate the drag is the z-component of $\underline{\mathcal{F}}$ (compare to Equation 8.152). Fortunately, the z-component of $\underline{\mathcal{F}}$ is the simplest of the three components in terms of mathematics. From Equation 8.152, we see that the drag is given by the z-direction force on a sphere, minus the buoyancy contribution:

$$\mathcal{F}_{\text{drag}} = \mathcal{F}_z - \frac{4\pi R^3 \rho g}{3} \tag{8.156}$$

The quantity $\mathcal{F}_z$ is the z-component of Equation 8.143; thus, the drag on a sphere in noncreeping flow is given by:

$$\mathcal{F}_{\text{drag}} = \mathcal{F}_z - \frac{4\pi R^3 \rho g}{3} \tag{8.157}$$

$$= \rho V^2 R^2 \int_0^{2\pi} \int_0^{\pi} \left[\left(\frac{2}{\text{Re}} \frac{\partial v_r^*}{\partial r*} - p^* \right) \cos\theta \right.$$

$$\left. + \left(\frac{r^*}{\text{Re}} \frac{\partial}{\partial r*} \left(\frac{v_\theta^*}{r^*} \right) + \frac{1}{\text{Re}\, r^*} \frac{\partial v_r^*}{\partial \theta} \right) (-\sin\theta) \right]_{r*=\frac{1}{2}} \sin\theta\, d\theta\, d\phi - \frac{4\pi R^3 \rho g}{3} \tag{8.158}$$

We can write Equation 8.158 in terms of the nondimensional dynamic pressure $\mathcal{P}^*$ as follows. The dynamic pressure was defined in Equation 8.115 as:

$$\mathcal{P} \equiv p + \rho g h \tag{8.159}$$

In the spherical coordinate system of Figure 8.5, $h = r\cos\theta$. We now nondimensionalize $\mathcal{P}$ in the usual way using the characteristic pressure $P = \rho V^2$:

$$\mathcal{P}^* = \frac{\mathcal{P} - p_{ref}}{\rho V^2} = \frac{p - p_{ref}}{\rho V^2} + \frac{\rho g r \cos\theta}{\rho V^2} \tag{8.160}$$

$$= p^* + \frac{gr \cos\theta}{V^2} \tag{8.161}$$

Solving for p^*, we obtain:

$$p^* = \mathcal{P}^* - \frac{gr \cos\theta}{V^2} \tag{8.162}$$

which we substitute into Equation 8.158. The pressure term in the integral in Equation 8.158 may be simplified by carrying out the integrations over θ and ϕ. Showing only the pressure term and carrying out the substitution for dynamic pressure, and (where possible) carrying out the integrations discussed previously, we obtain:

$$\text{Pressure term} \atop \text{(Equation 8.158)} = \rho V^2 R^2 \int_0^{2\pi} \int_0^\pi \left[-p^* \cos\theta \right]_{r^*=\frac{1}{2}} \sin\theta \, d\theta \, d\phi \tag{8.163}$$

$$= \rho V^2 R^2 \int_0^{2\pi} \int_0^\pi \left[-\mathcal{P}^* \cos\theta + \frac{gr\cos\theta}{V^2} \cos\theta \right] \sin\theta \, d\theta \, d\phi \Bigg|_{r^*=\frac{1}{2}} \tag{8.164}$$

$$= \rho V^2 R^2 \int_0^{2\pi} \int_0^\pi \left[-\mathcal{P}^* \cos\theta \right]_{r^*=\frac{1}{2}} \sin\theta \, d\theta \, d\phi$$

$$+ \rho V^2 R^2 \int_0^{2\pi} \int_0^\pi \frac{gR\cos^2\theta}{V^2} \sin\theta \, d\theta \, d\phi \tag{8.165}$$

$$= \left[\rho V^2 R^2 \int_0^{2\pi} \int_0^\pi \left[-\mathcal{P}^* \cos\theta \right]_{r^*=\frac{1}{2}} \sin\theta \, d\theta \, d\phi \right] + \frac{4\pi R^3 \rho g}{3} \tag{8.166}$$

where we carried out the second integration in Equation 8.165. Substituting Equation 8.166 into the complete Equation 8.158, the two terms containing gravity cancel, and we obtain the final expression for drag $\mathcal{F}_{\text{drag}}$ in noncreeping flow around a sphere:

$$\mathcal{F}_{\text{drag}} = \rho V^2 R^2 \int_0^{2\pi} \int_0^\pi \left[\left(\frac{2}{\text{Re}} \frac{\partial v_r^*}{\partial r^*} - \mathcal{P}^* \right) \cos\theta \right.$$

$$\left. + \left(\frac{r^*}{\text{Re}} \frac{\partial}{\partial r^*} \left(\frac{v_\theta^*}{r^*} \right) + \frac{1}{\text{Re}\, r^*} \frac{\partial v_r^*}{\partial\theta} \right) (-\sin\theta) \right]_{r^*=\frac{1}{2}} \sin\theta \, d\theta \, d\phi \tag{8.167}$$

Note that dynamic pressure $\mathcal{P}^*$, not regular pressure p^*, now appears in Equation 8.167 and that the buoyancy-subtraction term is gone. Modifying pressure to dynamic pressure has the effect of absorbing the buoyancy effect into $\mathcal{P}$, an altered pressure.

Equation 8.167 is nondimensional, except that we have not yet chosen how to nondimensionalize the wall drag, $\mathcal{F}_{\text{drag}}$. We define a nondimensional drag on the wall following the logic employed previously with tube flow. The nondimensional drag is the drag coefficient:

$$\text{Drag coefficient:} \atop \text{nondimensional drag} \atop \text{on a sphere} \quad C_D \equiv \frac{\text{wall force}}{\left(\dfrac{\text{kinetic energy}}{\text{volume}} \right) (\text{characteristic area})} \tag{8.168}$$

$$= \frac{\mathcal{F}_{\text{drag}}}{\left(\frac{1}{2}\rho V^2 \right) \left(\frac{\pi D^2}{4} \right)} \tag{8.169}$$

We choose the projected area in the flow direction $\pi D^2/4$ to be the characteristic area. With this definition for nondimensional wall drag C_D, we can write the drag in terms of the drag coefficient:

$$\mathcal{F}_{\text{drag}} = \frac{\rho V^2 \pi D^2 C_D}{8} \tag{8.170}$$

and substitute this into Equation 8.167 to obtain the general expression for non-dimensional drag for noncreeping flow around a sphere:

Nondimensional drag,
noncreeping flow around a sphere:

$$C_D = \frac{2}{\pi} \int_0^{2\pi} \int_0^\pi \left[\left(\frac{2}{\text{Re}} \frac{\partial v_r^*}{\partial r^*} - \mathcal{P}^* \right) \cos\theta \right.$$
$$\left. + \left(\frac{r^*}{\text{Re}} \frac{\partial}{\partial r^*} \left(\frac{v_\theta^*}{r^*} \right) + \frac{1}{\text{Re}\, r^*} \frac{\partial v_r^*}{\partial \theta} \right) (-\sin\theta) \right] \sin\theta\, d\theta\, d\phi \Bigg|_{r^* = \frac{1}{2}} \tag{8.171}$$

This completes our nondimensionalization exercise; the final result of our analysis (Equation 8.171) is the determination that nondimensional wall drag C_D is a function of nondimensional variables $(v_r^*, v_\theta^*, v_\phi^*, \mathcal{P}^*, r^*, \theta, \phi, t^*)$ and one dimensionless scale factor Re. Also, the microscopic balances that determine relationships among the nondimensional variables are a function of only the Reynolds number, as previously discussed for Equation 8.131. Thus, the drag coefficient for noncreeping flow around a sphere is determined by knowledge of Reynolds number alone:

Drag law,
flow around a sphere $\boxed{C_D = C_D\,(\text{Re})}$ (8.172)
(see Figure 8.13)

We saw this verified in experimental results shown in Figure 8.12 and discussed in the previous section.

The calculations of this section confirm the usefulness of dimensional analysis in understanding complex flow problems. We used dimensional analysis to determine that the drag coefficient is a function of only Reynolds number, a prediction that is confirmed by experiments (see Figure 8.12). We showed an example of how to use drag coefficient as a function of Reynolds number in the skydiver example in Section 8.1.2, and two more examples are presented here. We showed that, in general, the fluid force on a sphere is a combination of buoyancy, drag, and lift. We learned that the same procedures developed for internal flows work well for external flows (i.e., start simple, solve, nondimensionalize, and conduct experiments to correlate). However, we have not yet addressed detailed flow predictions ($\underline{v}$ and p) in noncreeping flow; we turn to this topic after the examples.

EXAMPLE 8.7. *A smooth ball the size of a baseball is dropped from the Golden Gate Bridge in San Francisco, California, USA. How fast is the ball going when it hits the water?*

SOLUTION. The dropping of a ball from a bridge is the same as the skydiver problem, Example 8.5. To calculate the terminal speed as a function of the various fluid and sphere properties, we apply Newton's second law (i.e., momentum conservation) to the ball. When the ball is at terminal speed, the acceleration is zero. The two forces on the ball are gravity and the fluid force:

$$\begin{array}{c}\text{Momentum balance}\\ \text{on ball}\\ \text{at terminal speed:}\end{array} \qquad \sum_{\substack{\text{all forces}\\ \text{acting on body}}} \underline{f} = m\underline{a} = 0 \qquad (8.173)$$

$$\underline{f}_{\text{gravity}} + \underline{f}_{\text{fluid}} = 0 \qquad (8.174)$$

$$m\underline{g} + \underline{\mathcal{F}} = 0 \qquad (8.175)$$

where m is the mass of the ball, a sphere of density ρ_{body}; $\underline{g} = -g\hat{e}_z$ is the acceleration due to gravity; and $\underline{\mathcal{F}} = \mathcal{F}\hat{e}_z$ is the fluid force on a sphere falling through a viscous liquid.

As in the solution to the skydiver problem (see Example 8.5), the fluid force on the falling object is given by:

$$\begin{array}{c}\text{Fluid force on a}\\ \text{sphere in}\\ \text{uniform flow}\\ \text{(all flow regimes):}\end{array} \qquad \underline{\mathcal{F}} = \begin{pmatrix} 0 \\ 0 \\ \dfrac{4\pi R^3 \rho g}{3} + \dfrac{\rho V^2 D^2 C_D \pi}{8} \end{pmatrix}_{xyz} \qquad (8.176)$$

where C_D is the drag coefficient, $V = v_\infty$ is the terminal velocity, $D = R/2$ is the diameter of the ball, and ρ is the density of air. Substituting this into Equation 8.175 and solving the z-component for the velocity, the final expression for terminal velocity becomes:

$$\begin{array}{c}\text{Terminal speed}\\ \text{of a sphere}\\ \text{(arbitrary regime of Re):}\end{array} \qquad \boxed{v_\infty = \sqrt{\dfrac{4(\rho_{\text{body}} - \rho)Dg}{3\rho C_D}}} \qquad (8.177)$$

We use the correlation for drag coefficient $C_D(\text{Re})$ in Equation 8.82 or 8.83 to calculate the drag.

Equation 8.177 may be used to estimate the terminal speed of a dropped baseball. Using the values of the physical parameters of air from Example 8.3, we assume that the Reynolds number will be high enough that the drag coefficient can be assumed to be a constant average value, $C_D = 0.445$. The dimensions of a baseball may be obtained from an Internet search:

$$\text{Viscosity of air:} \quad \mu = 1.7 \times 10^{-5} \text{ Pa s}$$

$$\text{Density of air:} \quad \rho = 1.23 \text{ kg/m}^3$$

$$\text{Density of baseball:} \quad \rho_{\text{body}} = 668 \text{ kg/m}^3$$

$$\text{Diameter of baseball:} \quad D = 0.0746 \text{ m}$$

$$\text{Acceleration due to gravity:} \quad g = 9.80 \text{ m/s}^2$$

$$\text{Drag coefficient (Newton's regime):} \quad C_D = 0.445$$

With these parameter values, we obtain the terminal speed as:

$$\begin{array}{l} \text{Estimate of} \\ \text{terminal speed:} \quad V = 34 \text{ m/s} = 77 \text{ mph} \\ (C_D = 0.445) \end{array} \tag{8.178}$$

To check our assumption of $C_D = 0.445$, we calculate the Reynolds number:

$$\mathrm{Re} = \frac{\rho v_\infty D}{\mu} \tag{8.179}$$

$$= 1.9 \times 10^5 \tag{8.180}$$

This value is within the range of the assumed value of drag coefficient; thus, the solution is valid.

EXAMPLE 8.8. *A smooth ball the size of a baseball is thrown with an initial velocity of 90 mph at an angle of 22 degrees from the horizontal. What is the velocity of the ball as a function of time and how far will it go? What is the path traced out by the ball?*

SOLUTION. This problem, like the dropped ball in Example 8.7, is solved with a force balance on the ball. In this case, however, the ball is accelerating throughout the time of observation:

$$\begin{array}{c} \text{Momentum balance} \\ \text{on ball:} \end{array} \qquad \sum_{\substack{\text{all forces} \\ \text{acting on body}}} \underline{f} = m\underline{a} \tag{8.181}$$

$$\underline{f}_{\text{gravity}} + \underline{f}_{\text{fluid}} = m\frac{d\underline{v}}{dt} \tag{8.182}$$

$$m\underline{g} + \underline{\mathcal{F}} = m\frac{d\underline{v}}{dt} \tag{8.183}$$

where m is the mass of the ball, a sphere of density ρ_{body}; $\underline{g}$ is the acceleration due to gravity; and $\underline{\mathcal{F}}$, the fluid force on a sphere moving through a viscous liquid, is given by the data correlations for drag coefficient and acts in the direction opposite to the motion of the ball. We use the correlation for drag coefficient $C_D(\mathrm{Re})$ in Equation 8.83 to calculate the drag. The defining equation for the drag coefficient C_D is:

$$C_D \equiv \frac{\mathcal{F}_{\text{drag}}}{\left(\frac{1}{2}\rho v^2\right)\left(\frac{\pi D^2}{4}\right)} \tag{8.184}$$

Neglecting buoyancy (a small effect) for this problem, we write the fluid force as the drag directed in the direction opposite to the ball's motion:

$$\underline{\mathcal{F}} = \underline{f}_{\text{drag}} + \underline{f}_{\text{buoyancy}} \tag{8.185}$$

$$= \mathcal{F}_{\text{drag}}\left(-\hat{v}\right) \tag{8.186}$$

$$= \frac{\rho v^2 D^2 C_D \pi}{8}\left(-\hat{v}\right) \tag{8.187}$$

Figure 8.17

From knowledge of the launch angle and the initial speed, we can calculate the entire trajectory of a thrown ball.

where $\hat{v}$ is a unit vector in the direction of the ball's motion.

We solve the problem for $\underline{v}$ written in a Cartesian coordinate system, with gravity in the $-z$-direction. The ball is thrown in the xz-plane (Figure 8.17):

$$\underline{v} = \begin{pmatrix} v_x \\ 0 \\ v_z \end{pmatrix}_{xyz} \qquad \hat{v} = \begin{pmatrix} \frac{v_x}{v} \\ 0 \\ \frac{v_z}{v} \end{pmatrix}_{xyz} \tag{8.188}$$

where $v = |\underline{v}|$. The momentum balance is thus:

$$\frac{d\underline{v}}{dt} = \underline{g} + \frac{\underline{\mathcal{F}}}{m} \tag{8.189}$$

$$\begin{pmatrix} \frac{dv_x}{dt} \\ 0 \\ \frac{dv_z}{dt} \end{pmatrix}_{xyz} = \begin{pmatrix} 0 \\ 0 \\ -g \end{pmatrix}_{xyz} + \begin{pmatrix} -\frac{\rho v^2 D^2 C_D \pi}{8m}\left(\frac{v_x}{v}\right) \\ 0 \\ -\frac{\rho v^2 D^2 C_D \pi}{8m}\left(\frac{v_z}{v}\right) \end{pmatrix}_{xyz} \tag{8.190}$$

The drag coefficient is a function of velocity; thus, we cannot use analytical techniques to solve Equation 8.190. Instead, we use numerical methods and spreadsheet software to obtain an accurate solution.

The numerical strategy for solving Equation 8.190 is to begin when the ball is thrown and calculate the location of the ball a short time later (we choose $\Delta t = 0.05$ s), assuming that the ball moves at a constant speed over the short time interval. At the new location, we recalculate the direction and speed of the ball from Equation 8.190 and use the new values to calculate another step. We continue stepping forward in time until the ball hits the ground ($z = 0$). This is known as Euler's method [24].

For our problem, the ball starts at time $t_{current} = 0$ at location $x_{current} = 0$, $z_{current} = 0$ with initial speed $v|_{current} = 90$ mph. Using geometry, the initial direction $\hat{v}|_{current}$ may be written as (see Figure 8.17):

$$\hat{v}|_{current} = \begin{pmatrix} \cos\beta \\ 0 \\ \sin\beta \end{pmatrix}_{xyz} \tag{8.191}$$

where $\beta = 22$ degrees from the horizontal. Details of the solution are given here, and a spreadsheet implementation is shown in Figure 8.18.

Subsequent Steps:

Current Values (i.e. Old)		index:	1		2	3	4	...	55	56	57
	increment with Δt	t=	0.05	s	0.1	0.15	0.2		2.75	2.8	2.85
	V_x=$V_{new,x}$ previous step	V_x=	37	m/s	37	36	36	...	20	20	20
	V_z=$V_{new,z}$ previous step	V_z=	15	m/s	14	14	13	...	-12	-13	-13
	V=sqrt($V_{new,x}{}^2$+$V_{new,z}{}^2$)	V=	40	m/s	39	39	38	...	24	24	24
	V_hat$_x$=V/V_x	V_hat$_x$=	0.927	dimensionless	0.931	0.936	0.940	...	0.858	0.849	0.839
	V_hat$_z$=V/V_z	V_hat$_z$=	0.375	dimensionless	0.364	0.353	0.342	...	-0.514	-0.529	-0.543
	Re=VD/ν	Re=	2.13E+05	SI units	2.09E+05	2.05E+05	2.01E+05	...	1.26E+05	1.26E+05	1.27E+05
	from data correlation and Re	C_D=	0.38		0.39	0.40	0.40	...	0.42	0.42	0.42
	from definition of C_D and velocity	F_x=	-1.55	Newton	-1.525	-1.497	-1.466	...	-0.552	-0.548	-0.544
	from definition of C_D and velocity	F_z=	-0.63	Newton	-0.596	-0.564	-0.533	...	0.330	0.341	0.352
New, Calculated from Old	previous location + (V_x)(Δt)	location x_{new}=	1.865	meters	3.704	5.516	7.303	...	75.228	76.242	77.248
	previous location + (V_z)(Δt)	location z_{new}=	0.754	meters	1.472	2.155	2.805	...	0.045	-0.587	-1.238
	From momentum balance	$V_{new,x}$=	37	m/s	36	36	35	...	20	20	20
	From momentum balance	$V_{new,z}$=	14	m/s	14	13	12	...	-13	-13	-13

Figure 8.18 Using spreadsheet software, we can implement the algorithm described in this example. The initial properties are set up in cells above the main calculation cells. The first step is in the column with index=1, followed by subsequent time steps in columns to the right.

Solution Steps

1. Increment the time, $t_{new} = t_{current} + \Delta t$.
2. For the current $v = |\underline{v}|$, calculate the Reynolds number $\text{Re} = \rho v D/\mu$, C_D (see Equation 8.83), and the drag on the ball.
3. From the momentum balance in Equation 8.190, calculate the new values of v_x and v_z as follows. Write the time derivatives in terms of a finite time step using the fundamental definition of derivative (see Equation 1.138):

$$\frac{dv_x}{dt} \approx \frac{v_x|_{new} - v_x|_{current}}{\Delta t} \tag{8.192}$$

$$v_x|_{new} = v_x|_{current} + \Delta t \left(\frac{dv_x}{dt} \right) \tag{8.193}$$

Substituting the time derivative from the x-component of Equation 8.190, we obtain:

$$v_x|_{new} = v_x|_{current} + \Delta t \left(-\frac{\rho v^2 D^2 C_D \pi}{8m} \left(\frac{v_x}{v} \right) \right) \Bigg|_{current} \tag{8.194}$$

Repeat these steps with the z-component:

$$\frac{dv_z}{dt} \approx \frac{v_z|_{new} - v_z|_{curent}}{\Delta t} \tag{8.195}$$

$$v_z|_{new} = v_z|_{current} + \Delta t \left(\frac{dv_z}{dt} \right) \tag{8.196}$$

$$= v_z|_{current} - g + \Delta t \left(-\frac{\rho v^2 D^2 C_D \pi}{8m} \left(\frac{v_z}{v} \right) \right) \Bigg|_{current} \tag{8.197}$$

4. Calculate the new ball speed, $v_{new} = \sqrt{v_x^2 + v_z^2}\Big|_{new}$.

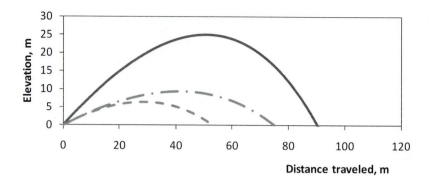

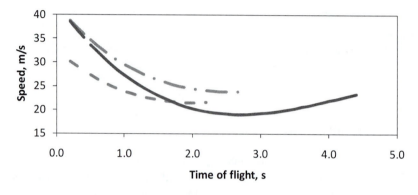

Figure 8.19 The ball trajectory is plotted for three different initial conditions: 22 degrees, 70 mph (dashed); 22 degrees, 90 mph (dash-dot); and 40 degrees, 90 mph (solid). Having set up the calculation for one set of conditions, we easily can run other initial conditions.

5. Calculate the new location of the ball assuming constant velocity at the current value of the velocity:

$$x_{new} = x_{current} + \Delta t \; v_x|_{current} \tag{8.198}$$

$$z_{new} = z_{current} + \Delta t \; z_x|_{current} \tag{8.199}$$

6. Save the dataset: t_{new}, $v_x|_{new}$, $v_z|_{new}$, v_{new}, x_{new}, and z_{new}.
7. If z_{new} is less than zero, indicating that the ball has hit the ground, stop the calculation. Otherwise, carry forward all of the *new* velocity and position data to be *current* data in the next step; then return to the first solution step.

Figure 8.18 is a spreadsheet implementation of the algorithm (i.e., Euler's method). The calculated ball trajectories for three different initial conditions are plotted in Figure 8.19 (top). For an initial angle of 22 degrees from the horizontal and an initial speed of 90 mph, the ball travels 75 m. The velocity as a function of time is obtained as part of the solution for ball position, and this is plotted in Figure 8.19 (bottom). Mathematical modeling of a process allows us to easily rerun the calculation for different initial conditions (β, $\underline{v}(0)$).

The correlation for sphere drag coefficient as a function of Reynolds number allows us to address various engineering problems related to flow past a sphere. When drag is the only issue of an external-flow problem, the knowledge of $C_D(\text{Re})$ is sufficient to find an appropriate solution.

When more than drag is an issue—when we seek the flow pattern that develops, for example, or the distribution of forces on an object—we must pursue a greater understanding of external flows than available from drag results alone. In the next section, we pursue more in-depth knowledge about such flows by exploring details of the observed flow fields that correspond to the higher-Reynolds-number results in Figure 8.12. We will learn—as expected from the fluid-force equation (Equation 8.143)—that the force on a sphere in noncreeping flow has components not only in the upstream-flow (z) direction but also in the lateral (x- and y-) directions (i.e., lift). A dramatic illustration of this effect is that a sphere dropped in a fluid, moving at a sufficiently high Reynolds number, will zigzag back and forth under the influence of lift forces arising from the interplay between inertia and fluid viscosity (see [113]; see also Section 8.2 and Figure 8.21).

As discussed in the next section, lift components of fluid force and complex flow patterns are two signature characteristics of noncreeping flow around a sphere. The flow patterns observed behind a sphere vary considerably with Reynolds numbers for Re modestly above the creeping-flow limit. At higher Reynolds numbers, the flow structure stabilizes into a pattern that reflects the dominance of inertial forces in most of the flow domain. This high-Re flow pattern is known as boundary-layer flow, and our usual methodology—start with a simple problem, proceed to dimensional analysis for the more complex case—is used once again as we seek to understand high-inertia experimental flows in the boundary-layer limit (see Section 8.2). Airplane flight and most high-speed flows fall into the realm of boundary-layer flow.

8.1.2.2 FLOW PATTERNS

The experiments that produced the drag-coefficient data in Figure 8.12 were of two types: spheres dropping in viscous fluids and viscous fluid being pushed past a stationary sphere in a wind tunnel, for example. As discussed in Section 8.1, these two experimental setups are equivalent—as long as the sphere drops in a straight line in the sphere-dropping experiments and does not move at all in the fixed-sphere experiments. Figure 8.20 illustrates that under the appropriate conditions, a sphere dropped in a liquid follows a straight path (see [113] for a video of this flow).

As mentioned at the end of the previous section, however, as the speed v_∞ increases in the dropping experiments (e.g., by using heavier spheres or lower-viscosity fluid), a curious effect is observed: The sphere no longer falls in a straight line but rather begins to weave back and forth as it falls through the fluid (Figure 8.21). The lateral motions of the sphere violate the assumption used when analyzing this flow—that is, if the sphere moves from side to side, we can no longer use a coordinate transformation to turn a sphere-dropping experiment into a flow-around-a-sphere experiment. Thus, at Reynolds numbers when the sphere weaves back and forth, we cannot compare the results of sphere-drop experiments to our fixed-sphere calculations.

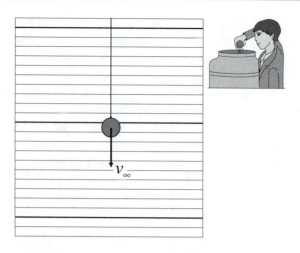

Figure 8.20 A stainless-steel sphere, dipped in ink and dropped in glycerin [155], falls in a straight line and, after a short startup period, at a constant velocity. The ink on the surface is pulled back around the sphere and colors the central trailing streamline, which is straight. A video of this experiment is available on the Web as part of the National Committee on Fluid Mechanics Films series [113].

The sideways motions in dropping-sphere experiments show that lateral forces are experienced by the sphere. These lateral forces are due to lift and were anticipated in Equation 8.143, which gives the force on a sphere under noncreeping conditions. The onset of lateral sphere motion in sphere-drop experiments marks the highest Reynolds number at which sphere-dropping experiments are useful to us; therefore, we now abandon those simple experiments. We reach higher Reynolds numbers with experiments in which the sphere is fixed in place and the flow is made to go around the sphere.

If a sphere is fixed in place and fluid is pumped around it, the flow may be visualized with the aid of smoke or reflective particles. The results of observations of this type are shown in Figure 8.22 [143]. At low Re (i.e., creeping-flow regime, not shown), the flow shows the streamlines that are predicted by the Stokes result (see Equation 8.23); this flow is equivalent to the flow in sphere-dropping experiments. For $130 < $ Re < 300 (Figure 8.22a), the streamlines near the rear of the fixed sphere (called the trailing side of the sphere) are noticeably different and vortices appear behind the sphere. With the appearance of trailing vortices, the flow pattern farther behind the sphere also changes. The central trailing streamline, which in creeping flow was a

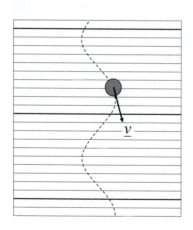

Figure 8.21 At a high Reynolds number, a sphere dropped in a viscous fluid travels back and forth under the influence of lateral lift forces. Compared to Figure 8.20, the experiment depicted here was produced by using water instead of glycerin. A video of this experiment is available on the Web as part of the National Committee on Fluid Mechanics Films series [113].

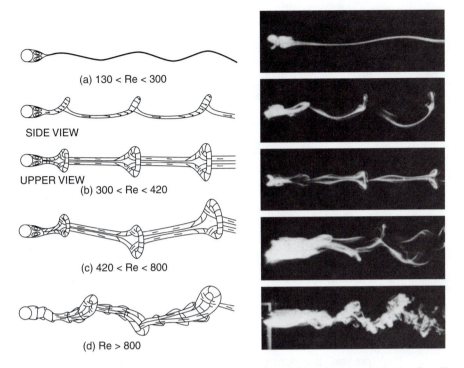

(a) 130 < Re < 300

SIDE VIEW

UPPER VIEW
(b) 300 < Re < 420

(c) 420 < Re < 800

(d) Re > 800

Figure 8.22   Schematic of sphere wakes and flow-visualization photographs from wind-tunnel experiments at various Reynolds numbers [143]. Flow is from left to right and the sphere is stationary. From top to bottom: (a) 130 < Re < 300; (b) 300 < Re < 420; (c) 420 < Re < 800; and (d) 800 < Re. Image source: H. Sakamoto and H. Haniu, Trans. ASME, vol. 112, 286 (1990), used with permission.

straight line extending from the rear stagnation point downstream (see Figure 8.20), changes character and begins to weave as the flow propagates downstream. The flow field is no longer independent of time; instead, there is a periodic character to the flow field as the central trailing streamline weaves back and forth.

At a still higher Reynolds number, the flow patterns around a fixed sphere become more interesting and complex. For 300 < Re < 420 (Figure 8.22b), the trailing vortices grow in length and their shape becomes more obviously three-dimensional and complex. These vortices, called hairpin vortices, resemble bent wires linked in a chain. For 420 < Re < 800, the chain of hairpin vortices weave back and forth; at Re > 800, the weaving vortices break up and the wake becomes more difficult to describe.

The development of vortices and wake behind the sphere is a fundamental change in the flow pattern of flow around a sphere, analogous to the appearance of transitional and turbulent flows in pipes. To correlate the effect of these pattern changes with drag, we annotate the sphere-C_D(Re) plot to match the effect of these flow-pattern changes with the measured drag coefficient (Figure 8.23). We see that the observed complex vortex patterns are associated with higher drag. The drag for Re > 120, where the vortex patterns occur, is much higher than the drag that would be obtained if the Stokes-flow solution ($C_D = 24/\text{Re}$) could be made to persist to higher Reynolds numbers. This is reminiscent of the

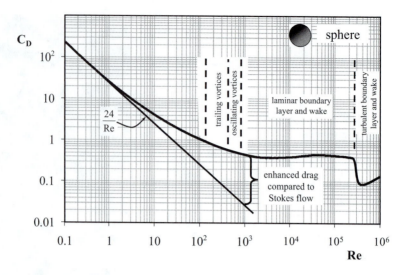

Figure 8.23 Experimental data for drag coefficient as a function of the Reynolds number for flow around a sphere [143.147]. Below Re = 0.1, the data follow the creeping-flow solution of $C_D = 24/\text{Re}$. The flow patterns associated with different Reynolds numbers are indicated.

transition that occurs in pipe flow in which higher friction factors f are observed in transitional and turbulent pipe flow (Re $> 2,100$) than expected from a laminar flow if a laminar pipe flow ($f = 16/\text{Re}$) could be made to persist to these higher Reynolds numbers (compare Figure 8.23 to Figure 7.17).

In both pipe flow and flow around a sphere, nature chooses a different flow pattern for rapid flows compared to slow flows, and the high-speed flow pattern in both cases is more complex than the slow-flow pattern. In addition, in both pipe flow and flow around a sphere, the high-speed flow pattern produces higher drag, even after all the scale factors of the flow are considered—that is, even when nondimensional friction forces f and C_D are compared. There apparently is some change in the character of the flow driving forces such that the slow-flow pattern is unattainable at high flow rates. Understanding the observed flow transitions between slow and rapid flows has been a goal of scientists and engineers for at least two centuries; considering the interest of ancient civilizations in domestic, oceanic, and meteorological flows, it likely has been of interest for considerably longer. We turn our attention now to the study of rapid-flow phenomena.

8.1.2.3 POTENTIAL FLOW

For insight into rapid external flows, we deploy our analysis techniques, which are based on the time-tested principles of mass, momentum, and energy conservation. The slow-flow and rapid-flow regimes are delineated by the value of the Reynolds number Re. We can find the governing equations for rapid flows by letting the Reynolds number go to infinity in the general governing equations. The governing equations in this limit reveal more about which forces are driving the complex behavior summarized in Figures 8.22 and 8.23.

Using dimensional analysis, we already have written nondimensional mass-balance, momentum-balance, and wall-drag equations for the general flow around a sphere:

Nondimensional continuity equation:
$$\nabla^* \cdot \underline{v}^* = 0 \tag{8.200}$$

Nondimensional Navier-Stokes (dynamic pressure):
$$\frac{\partial \underline{v}^*}{\partial t^*} + \underline{v}^* \cdot \nabla^* \underline{v}^* = -\nabla^* \mathcal{P}^* + \left(\frac{1}{\text{Re}}\right) \nabla^{*2} \underline{v}^* \tag{8.201}$$

Nondimensional drag on a sphere:
$$C_D = \frac{2}{\pi} \int_0^{2\pi} \int_0^{\pi} \left[\left(\frac{2}{\text{Re}} \frac{\partial v_r^*}{\partial r^*} - \mathcal{P}^* \right) \cos\theta \right.$$

$$+ \left(\frac{r^*}{\text{Re}} \frac{\partial}{\partial r^*} \left(\frac{v_\theta^*}{r^*} \right) \right.$$

$$\left. \left. + \frac{1}{\text{Re}\, r^*} \frac{\partial v_r^*}{\partial \theta} \right) (-\sin\theta) \right]_{r^* = \frac{1}{2}} \sin\theta\, d\theta\, d\phi \tag{8.202}$$

Beginning with these three equations, we take the limit of $\text{Re} \longrightarrow \infty$ to obtain the governing equations in the high-Re limit:

Equations of potential flow (perfect or inviscid fluid):

$$\nabla^* \cdot \underline{v}^* = 0$$

$$\frac{\partial \underline{v}^*}{\partial t^*} + \underline{v}^* \cdot \nabla^* \underline{v}^* = -\nabla^* \mathcal{P}^*$$

$$C_D = \frac{2}{\pi} \int_0^{2\pi} \int_0^{\pi} \left[-\mathcal{P}^* \cos\theta \right]_{r^* = \frac{1}{2}} \sin\theta\, d\theta\, d\phi$$

$$\tag{8.203}$$

These equations (Equation 8.203) are the governing equations for flow when the Reynolds number is high—that is, when viscous forces are not important. A fluid without viscosity is called a perfect fluid or an inviscid fluid, and a flow with no viscous effects is called an inviscid flow or *potential flow*. The use of the term potential flow comes from the observation that the system of equations in Equation 8.203 also occurs when analyzing electrical potentials [79].

The set of Equations 8.203 can be applied to flow around a sphere, and we turn now to this calculation (Example 8.9). We expect this calculation to explain the complex behavior observed in Figures 8.22 and 8.23.

EXAMPLE 8.9. *Calculate the steady-state velocity field for the flow of an incompressible, inviscid fluid around a solid sphere of diameter 2R. The fluid approaches the sphere with a uniform upstream velocity v_∞. The geometry is the same as in the viscous, creeping-flow calculation (see Figure 8.5); however, in this problem, the fluid is inviscid ($\mu = 0$) and inertia is not neglected.*

SOLUTION. The solution we seek here is called the potential-flow solution for flow around a sphere. We choose to solve the problem in spherical coordinates due to the geometry of the problem, and we begin with the microscopic mass and momentum balances (i.e., continuity equation and equation of motion) written in the chosen coordinate system.

There is no azimuthal-component of the flow ($v_\phi = 0$) and the flow is symmetric in the ϕ-direction; thus, there are only two nonzero components of velocity and all ϕ-derivatives are zero. The flow is steady and the viscosity is zero. The continuity equation becomes:

$$v = \begin{pmatrix} v_r \\ v_\theta \\ v_\phi \end{pmatrix}_{r\theta\phi} = \begin{pmatrix} v_r \\ v_\theta \\ 0 \end{pmatrix}_{r\theta\phi} \tag{8.204}$$

Continuity equation: $0 = \nabla \cdot \underline{v}$

$$0 = \frac{1}{r^2} \frac{\partial \left(r^2 v_r \right)}{\partial r} + \frac{1}{r \sin\theta} \frac{\partial \left(v_\theta \sin\theta \right)}{\partial \theta} \tag{8.205}$$

The momentum balance is next. As discussed previously, we can combine the effects of gravity $\underline{g} = -g\hat{e}_z$ and pressure p on our problem by using dynamic pressure $\mathcal{P}$ given by Equation 8.115. The microscopic-momentum balance for steady, inviscid flow becomes:

Navier-Stokes equation: $\rho \dfrac{\partial \underline{v}}{\partial t} + \rho \underline{v} \cdot \nabla \underline{v} = -\nabla \mathcal{P} + \mu \nabla^2 \underline{v}$
(dynamic pressure)

$$\rho \underline{v} \cdot \nabla \underline{v} = -\nabla \mathcal{P}$$

$$\begin{pmatrix} \rho \left(v_r \dfrac{\partial v_r}{\partial r} + \dfrac{v_\theta}{r} \dfrac{\partial v_r}{\partial \theta} - \dfrac{v_\theta^2}{r} \right) \\ \rho \left(v_r \dfrac{\partial v_\theta}{\partial r} + \dfrac{v_\theta}{r} \dfrac{\partial v_\theta}{\partial \theta} + \dfrac{v_r v_\theta}{r} \right) \\ 0 \end{pmatrix}_{r\theta\phi} = \begin{pmatrix} -\dfrac{\partial \mathcal{P}}{\partial r} \\ -\dfrac{1}{r} \dfrac{\partial \mathcal{P}}{\partial \theta} \\ -\dfrac{1}{r \sin\theta} \dfrac{\partial \mathcal{P}}{\partial \phi} \end{pmatrix}_{r\theta\phi} \tag{8.206}$$

The ϕ-component of the momentum balance confirms that there is no ϕ-variation of the pressure for this problem. The continuity equation (Equation 8.205) and the r- and θ-components of the Navier-Stokes equation (Equation 8.206) form a system of three equations in three unknowns: v_r, v_θ, and $\mathcal{P}$.

The problem is set up, but we are left with a difficult mathematical task. We are aided in solving these equations by an invention known as the *stream function* $\psi(r, \theta)$, which we develop now. For a function of two variables, the two mixed second-derivative functions are equal. For any function $\psi(r, \theta)$, therefore:

$$\frac{\partial^2 \psi}{\partial \theta \, \partial r} = \frac{\partial^2 \psi}{\partial r \, \partial \theta} \tag{8.207}$$

$$\frac{\partial}{\partial \theta} \left(\frac{\partial \psi}{\partial r} \right) = \frac{\partial}{\partial r} \left(\frac{\partial \psi}{\partial \theta} \right) \tag{8.208}$$

$$0 = \frac{\partial}{\partial r} \left(\frac{\partial \psi}{\partial \theta} \right) + \frac{\partial}{\partial \theta} \left(-\frac{\partial \psi}{\partial r} \right) \tag{8.209}$$

There are similarities between the continuity-equation result for this problem (Equation 8.205) and the mixed-partials expression in Equation 8.209. First, we rearrange the continuity equation:

$$0 = \frac{1}{r^2} \frac{\partial \left(r^2 v_r \right)}{\partial r} + \frac{1}{r \sin \theta} \frac{\partial \left(v_\theta \sin \theta \right)}{\partial \theta} \tag{8.210}$$

$$0 = \sin \theta \frac{\partial \left(r^2 v_r \right)}{\partial r} + r \frac{\partial \left(v_\theta \sin \theta \right)}{\partial \theta} \tag{8.211}$$

$$0 = \frac{\partial \left(r^2 \sin \theta v_r \right)}{\partial r} + \frac{\partial \left(r v_\theta \sin \theta \right)}{\partial \theta} \tag{8.212}$$

We can move $\sin \theta$ into the r-derivative term and r into the θ-derivative term because the derivatives are partial derivatives of the other variable. Comparing Equation 8.212 to Equation 8.209, we define the function $\psi(r, \theta)$ so that these two equations are equivalent [40]:

$$\frac{\partial \psi}{\partial \theta} \equiv r^2 \sin \theta v_r \tag{8.213}$$

$$-\frac{\partial \psi}{\partial r} \equiv r v_\theta \sin \theta \tag{8.214}$$

Solving these definitions for the velocity components v_r and v_θ in terms of the single function ψ, we obtain:

$$v_r = \frac{1}{r^2 \sin \theta} \frac{\partial \psi}{\partial \theta} \tag{8.215}$$

$$v_\theta = \frac{-1}{r \sin \theta} \frac{\partial \psi}{\partial r} \tag{8.216}$$

For the function $\psi(r, \theta)$ defined this way, the continuity equation is automatically satisfied. The problem now becomes to solve for $\psi(r, \theta)$ and $\mathcal{P}(r, \theta)$. The equations to use to solve the problem are the r- and θ-components of the momentum balance, Equation 8.206, with the appropriate expressions involving

ψ substituted for v_r and v_θ (Equations 8.215 and 8.216):

$$\rho\left(v_r\frac{\partial v_r}{\partial r} + \frac{v_\theta}{r}\frac{\partial v_r}{\partial\theta} - \frac{v_\theta^2}{r}\right) = -\frac{\partial\mathcal{P}}{\partial r} \tag{8.217}$$

$$\rho\left(v_r\frac{\partial v_\theta}{\partial r} + \frac{v_\theta}{r}\frac{\partial v_\theta}{\partial\theta} + \frac{v_r v_\theta}{r}\right) = -\frac{1}{r}\frac{\partial\mathcal{P}}{\partial\theta} \tag{8.218}$$

To obtain a single equation for $\psi(r,\theta)$, we differentiate Equation 8.217 by θ and Equation 8.218 by r, yielding the same mixed second partial derivative of dynamic pressure on the righthand side of both expressions:

$$\frac{\partial}{\partial\theta}\left(v_r\frac{\partial v_r}{\partial r} + \frac{v_\theta}{r}\frac{\partial v_r}{\partial\theta} - \frac{v_\theta^2}{r}\right) = -\frac{1}{\rho}\frac{\partial^2\mathcal{P}}{\partial\theta\,\partial r} \tag{8.219}$$

$$\frac{\partial}{\partial r}\left(r\left(v_r\frac{\partial v_\theta}{\partial r} + \frac{v_\theta}{r}\frac{\partial v_\theta}{\partial\theta} + \frac{v_r v_\theta}{r}\right)\right) = -\frac{1}{\rho}\frac{\partial^2\mathcal{P}}{\partial r\,\partial\theta} \tag{8.220}$$

Eliminating the pressure second derivative between these two equations and substituting the defining equations for the stream function $\psi(r,\theta)$ (Equations 8.215 and 8.216) produces a single, third-order, partial differential equation for $\psi(r,\theta)$. The algebra in spherical coordinates is complex but, ultimately, the equation to solve for $\psi(r,\theta)$ is an encouragingly simple expression [40]:

$$\frac{\partial^2\psi}{\partial r^2} + \frac{\sin\theta}{r^2}\frac{\partial}{\partial\theta}\left(\frac{1}{\sin\theta}\frac{\partial\psi}{\partial\theta}\right) = 0 \tag{8.221}$$

This mathematical system is studied in many areas of physics, and the entire expression is defined as the operator E^2:

$$E^2\psi \equiv \frac{\partial^2\psi}{\partial r^2} + \frac{\sin\theta}{r^2}\frac{\partial}{\partial\theta}\left(\frac{1}{\sin\theta}\frac{\partial\psi}{\partial\theta}\right) \tag{8.222}$$

The equation for the momentum balance for flow around a sphere thus can be written as:

$$\begin{array}{l}\text{Momentum-balance}\\ \text{flow around a sphere}\\ \text{(potential flow):}\end{array} \quad \boxed{E^2\psi = 0} \tag{8.223}$$

We can find a solution for ψ from Equation 8.223 (Equation 8.221) by following the same strategy used in the creeping-flow solution: Consider the boundary conditions and guess an appropriate solution. We discuss that solution now.

The boundary conditions for the flow around a sphere are: (1) the velocity goes to zero at the sphere surface, and (2) the flow is uniform in the z-direction at infinity. We first write the boundary conditions in terms of $\underline{v}$ and then convert to ψ:

$$\text{At } r = R: \quad \underline{v} = 0 \tag{8.224}$$

$$= \begin{pmatrix} v_r \\ v_\theta \\ 0 \end{pmatrix}_{r\theta\phi} = \begin{pmatrix} 0 \\ 0 \\ 0 \end{pmatrix}_{r\theta\phi} \tag{8.225}$$

In terms of ψ, these become:

$$\frac{\partial \psi}{\partial r}\bigg|_{r=R} = 0 \tag{8.226}$$

$$\frac{\partial \psi}{\partial \theta}\bigg|_{r=R} = 0 \tag{8.227}$$

Far from the sphere, the flow must return to the uniform flow at speed v_∞:

$$\text{At } r = \infty: \quad \underline{v} = v_\infty \hat{e}_z \tag{8.228}$$

$$= v_\infty \cos\theta \hat{e}_r - v_\infty \sin\theta \hat{e}_\theta \tag{8.229}$$

$$= \begin{pmatrix} v_r \\ v_\theta \\ 0 \end{pmatrix}_{r\theta\phi} = \begin{pmatrix} v_\infty \cos\theta \\ -v_\infty \sin\theta \\ 0 \end{pmatrix}_{r\theta\phi} \tag{8.230}$$

In terms of ψ, these become:

$$\frac{\partial \psi}{\partial r}\bigg|_{r=\infty} = v_\infty r \sin^2\theta \tag{8.231}$$

$$\frac{\partial \psi}{\partial \theta}\bigg|_{r=\infty} = v_\infty r^2 \sin\theta \cos\theta \tag{8.232}$$

Integrating these two boundary conditions, we obtain the same result—a single boundary condition for ψ at $r = \infty$:

$$\psi|_{r=\infty} = \frac{v_\infty r^2}{2} \sin^2\theta \tag{8.233}$$

Note that we arbitrarily set the integration constant for ψ to zero. We can do this because v_r and v_θ depend on only derivatives of ψ, not on the value of ψ (see Equations 8.215 and 8.216).

The form of the far-field boundary condition given in Equation 8.233 suggests that we might find a solution for ψ in the form:

$$\text{Guess solution:} \quad \psi(r, \theta) = f(r)\sin^2\theta \tag{8.234}$$

We can test this idea by substituting this guess into the differential equation for ψ (Equation 8.221) and solving for $f(r)$, if possible. Making this substitution, we obtain an equation for $f(r)$:

$$\frac{d^2 f}{dr} - 2f\frac{1}{r^2} = 0 \tag{8.235}$$

This is an ordinary differential equation (ODE) with a known solution, $f(r) = r^n$, where n must be solved for by substitution of the solution into the differential equation. Carrying out this substitution, we find that for Equation 8.235 to be satisfied, n must equal 2 or -1; thus, the complete solution for $f(r)$ is:

$$f(r) = Ar^2 + Br^{-1} \tag{8.236}$$

where A and B are constants that must be evaluated by using the boundary conditions.

We have two constants to evaluate, but we have three boundary conditions (i.e., Equations 8.226, 8.227, and 8.233). This occurred because our true system was second order in spatial derivatives of both v_r and v_θ (i.e., ∇^2 in the Navier-Stokes equation), requiring two boundary conditions on each velocity, component or a total of four boundary conditions. When we switched to the stream function, we obtained a second-order equation on ψ (Equation 8.221). Because of the form of ψ, the two boundary conditions at $r = \infty$ could be satisfied simultaneously; however, it is not possible to satisfy simultaneously both boundary conditions at R and we must choose to satisfy one or the other. We choose avoiding penetration of the sphere ($r = R$, $v_r = 0$) as a more important condition than no-slip at the surface ($r = R$, $v_\theta(R) = 0$).

Applying the boundary condition at infinity, we obtain $A = v_\infty/2$. At the surface of the sphere, we must avoid fluid entering the sphere; thus, $v_r = 0$ at the surface. This boundary condition combined with the result for A gives us $B = -v_\infty R^3/2$. We have no additional degrees of freedom to use to force the no-slip boundary condition to hold; therefore, *the velocity field for which we have solved will slip at the sphere surface.* The final result for the stream function $\psi(r, \theta)$ for potential flow around a sphere at high Re is given by:

$$\begin{array}{c} \text{Stream function} \\ \text{potential flow} \\ \text{around a sphere:} \end{array} \qquad \psi(r, \theta) = \frac{v_\infty R^2 \sin^2 \theta}{2} \left[\left(\frac{r}{R}\right)^2 - \left(\frac{R}{r}\right) \right] \qquad (8.237)$$

We emphasize that this solution does not respect the no-slip boundary condition at the sphere surface.

To calculate $\underline{v}$ from the stream function ψ, we return to Equations 8.215 and 8.216. Knowing the velocities, we calculate $\mathcal{P}(r, \theta)$ by integrating the pressure partial derivatives in Equations 8.217 and 8.218 and apply the boundary condition far from the sphere $r = \infty$, $\mathcal{P} = \mathcal{P}_\infty$.[6] The solutions for $\underline{v}(r, \theta)$ and $\mathcal{P}(r, \theta)$ are given here. Note that the solution is written in spherical $r\theta\phi$ coordinates:

$$\begin{array}{c} \text{Steady potential flow} \\ \text{around a sphere [85]} \\ \text{(flow in } z\text{-direction;} \\ \underline{g} = -g\hat{e}_z \text{ or neglected):} \end{array} \qquad \underline{v}(r, \theta) = \begin{pmatrix} v_\infty \left[1 - \left(\frac{R}{r}\right)^3 \right] \cos\theta \\ -v_\infty \left[1 + \frac{1}{2}\left(\frac{R}{r}\right)^3 \right] \sin\theta \\ 0 \end{pmatrix}_{r\theta\phi}$$

$$(8.238)$$

$$\mathcal{P}(r, \theta) = \mathcal{P}_\infty + \frac{1}{2}\rho v_\infty^2 \left[2\left(\frac{R}{r}\right)^3 \left(1 - \frac{3}{2}\sin^2\theta\right) - \left(\frac{R}{r}\right)^6 \left(1 - \frac{3}{4}\sin^2\theta\right) \right]$$

$$(8.239)$$

---

[6] There is a simpler equivalent way to solve for the pressure function; see Example 8.12.

These equations give velocity field $\underline{v}$ and pressure field $\mathcal{P}$ for the high-Reynolds-number problem posed, but we are unable to satisfy all of the boundary conditions.

For steady, uniform, high-Reynolds-number flow around a sphere, the velocity-field and pressure-field solutions arrived at by using the potential-flow equations (see Equation 8.203) are given here, and the streamlines $\psi(r, \theta)$ and velocity field are plotted in Figures 8.24 and 8.25 [85]. For steady flows, streamlines mark the paths of fluid particles in the flow. In all flows, the local velocity vector at a point in the flow is tangent to the streamline function ψ at that point (see Equations 8.215 and 8.216) [40, 85].

What we notice first about the potential-flow solution obtained in Example 8.9 is that it is *wrong*. The potential-flow solution does not resemble the high-Reynolds-number experimental results in Figure 8.22: There are no vortices or wake predicted. For $300 > \text{Re} > 420$, for example, experiments show a recirculating region behind the sphere and a distinct and wavy wake. The potential solution for all Reynolds numbers has no recirculation, no wake, and straight streamlines downstream of the sphere (Figure 8.26). This complete lack of agreement between prediction and measurement is an enormous surprise. Our process of obtaining the nondimensional equation of motion and the nondimensional continuity equation seemed destined to produce the correct governing equations for high-Reynolds-number flows and, therefore, the correct solutions. Yet, the predictions do not match the observations. Again, we encounter a stumbling block in our analysis, and we now must struggle to understand what is wrong with our

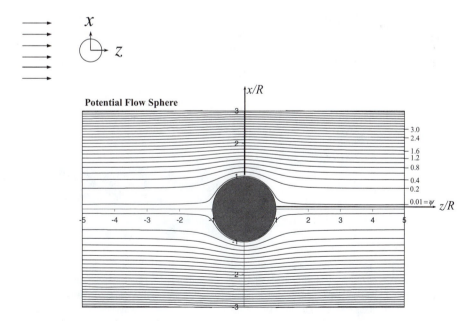

Figure 8.24   Streamlines or particle paths for potential flow around a sphere. The values of the stream function ψ for several lines are shown. For steady flows, streamlines mark the paths of fluid particles in the flow. In all flows, the local velocity vector at a point is tangent to the streamline function ψ at that point. Compare to creeping flow in Figure 8.9.

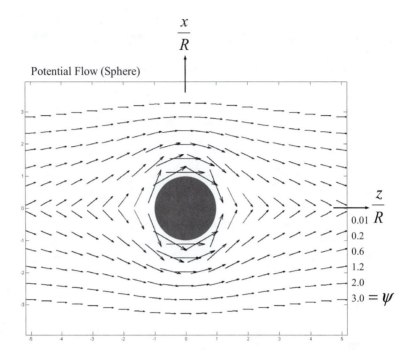

Figure 8.25 Vector or arrow plot of the velocity field of potential flow around a sphere. For points along several streamlines (i.e., particle paths), the velocity vector centered at the point is shown and the length of the arrow is proportional to the magnitude of the velocity at that point. Note that near the sphere, the velocity is very high; the potential-flow solution does not respect the no-slip boundary condition at the sphere surface.

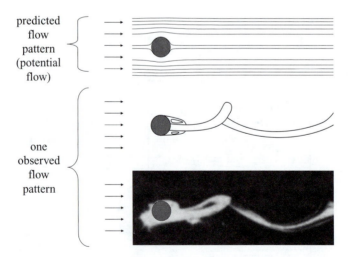

Figure 8.26 The prediction of our potential-flow calculations for rapid flow around a sphere are shown at the top. Potential flow predicts that the streamlines closely hug the sphere and are straight and parallel after the sphere. The observed flow pattern is more like the photograph at the bottom [143], which shows recirculating flow on the trailing side of the sphere and a wake that has a complex shape. A sketch of the wake shape is also shown. Image source: H. Sakamoto and H. Haniu, Trans. ASME, vol. 112, 286 (1990), used with permission.

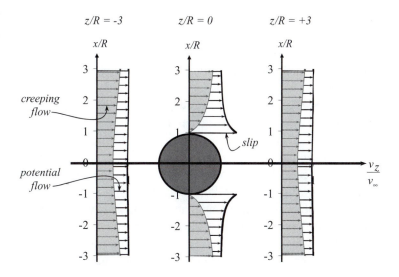

Figure 8.27 The velocity field in creeping flow around a sphere shows effects of the sphere far upstream and complies with the no-slip boundary condition. The velocity field in potential flow around a sphere is nearly indifferent to the presence of the sphere at equivalent locations upstream and downstream. It is significant that the potential-flow solution does not satisfy the no-slip boundary condition at the sphere surface.

methods. As usual, a closer examination of our results and assumptions leads us to an understanding of this situation and will get us past this roadblock.

The streamlines of potential flow do not resemble the experimental observations, but they are not entirely unfamiliar. At first glance, the streamlines predicted by the equations of potential flow (see Figure 8.24) are similar to those found in creeping flow (see Figure 8.9), a flow valid for only low Reynolds numbers. Closer examination, however, reveals that the streamlines near the centerline of the flows (i.e., $\phi = 0.01$ on both plots) are qualitatively different. In potential flow, fluid particles that follow the $\phi = 0.01$ streamline closely hug the sphere as they pass. In creeping flow, fluid particles following the $\phi = 0.01$ streamline swing away from the sphere significant a distance.

The differences in the streamline maps for potential and creeping flow around a sphere are reflected more starkly in the differences in the velocity fields. Figure 8.25 shows that the velocities near the sphere surface in potential flow are anomalously high. In creeping flow by contrast (see Figure 8.8), the fluid slows near the sphere and velocity eventually goes to zero at the sphere surface. Figure 8.27 is a closer look at creeping and potential flows, in which the velocity distributions along three vertical lines in the flow domain are shown. The three vertical lines are a line upstream of the sphere ($z/R = -3$); a line that passes through the center of the sphere ($z/R = 0$); and a downstream line ($z/R = 3$). The velocity fields of creeping and potential flows are quite different at these three locations. Upstream, potential flow is nearly plug flow; that is, the velocity is insensitive to position x/R and approximately equal to the free-stream speed v_∞ at all points along the line $z/R = 3$. By contrast, the velocity distribution in creeping flow at $z/R = -3$ shows a pronounced dip near the flow centerline

$x/R = 0$. The dip indicates that in creeping flow the presence of the sphere is strongly felt at a distance of three radii upstream of the sphere's location.

Along the line that passes through the center of the sphere ($z/R = 0$), the differences in velocity distribution between creeping and potential flow are great. In potential flow, the flow slips at the sphere surface. In addition, in potential flow, the fluid speed at the sphere surface exceeds the free-stream speed ($v_z/v_\infty > 1$) and decreases to v_∞ only at distances far from the sphere. By contrast, along this same axis in creeping flow, the velocity goes to zero at the sphere surface (i.e., the no-slip boundary condition is respected), and at no location in creeping flow does v_z/v_∞ exceed 1. Far from the sphere, v_z/v_∞ goes to 1 for both creeping and potential flow. At a position three radii downstream of the sphere $z/R = 3$, the velocity profiles in both potential and creeping flow return to the flow patterns observed at the equivalent upstream position.

We also can compare the drag on the sphere predicted by creeping and potential flow. For creeping flow, we calculated the nondimensional wall drag as follows (see Equation 8.81):

$$\text{Nondimensional wall drag for creeping flow around a sphere:} \quad C_D = \frac{24}{\text{Re}} \tag{8.240}$$

and we saw that for Re < 2, this relationship is observed experimentally. For potential flow, we can calculate the drag on the sphere from the pressure solution and the nondimensionalized fluid-force equation for infinite Reynolds number (see Equation 8.203). This calculation is shown in Example 8.10.

EXAMPLE 8.10. *Calculate the drag on a sphere in steady potential flow around a sphere (high Reynolds number, inviscid fluid).*

SOLUTION. As shown previously, the drag coefficient, which is the nondimensional drag on the sphere, may be calculated for flow of an inviscid fluid by using Equation 8.203:

$$\text{Inviscid flow:} \quad C_D = \frac{2}{\pi} \int_0^{2\pi} \int_0^\pi \left[-\mathcal{P}^* \cos\theta \right]_{r^* = \frac{1}{2}} \sin\theta \, d\theta \, d\phi \tag{8.241}$$

We carry out this integration for potential flow around a sphere using the solution for the pressure distribution in this flow (Equation 8.239):

$$\text{Pressure distribution, potential flow:} \quad \mathcal{P}^*(r, \theta) = \frac{\mathcal{P}}{\rho v_\infty^2} \tag{8.242}$$

$$= \frac{\mathcal{P}_\infty}{\rho v_\infty^2} + \frac{1}{2} \left[2 \left(\frac{R}{r} \right)^3 \left(1 - \frac{3}{2} \sin^2\theta \right) \right.$$

$$\left. - \left(\frac{R}{r} \right)^6 \left(1 - \frac{3}{4} \sin^2\theta \right) \right] \tag{8.243}$$

We must evaluate this expression at the sphere surface, $r^* = 1/2$ or $r = R$:

$$r^* = \frac{r}{D} = \frac{1}{2} \tag{8.244}$$

$$\left. P^*(r, \theta) \right|_{r^* = \frac{1}{2}} = \frac{P_\infty}{\rho v_\infty^2} + \frac{1}{2} \left[\left(2 - 3 \sin^2 \theta \right) - \left(1 - \frac{3}{4} \sin^2 \theta \right) \right] \tag{8.245}$$

$$\left. P^*(r, \theta) \right|_{r^* = \frac{1}{2}} = \frac{P_\infty}{\rho v_\infty^2} + \frac{1}{2} - \frac{9}{8} \sin^2 \theta \tag{8.246}$$

$$= \mathcal{A} + \mathcal{B} \sin^2 \theta \tag{8.247}$$

where we define constants $\mathcal{A}$ and $\mathcal{B}$ to simplify the integrations:

$$\mathcal{A} = \frac{P_\infty}{\rho v_\infty^2} + \frac{1}{2} \tag{8.248}$$

$$\mathcal{B} = -\frac{9}{8} \tag{8.249}$$

Substituting Equation 8.247 into the equation for drag coefficient (see Equation 8.241), we now finish the calculation of drag coefficient for potential flow:

$$C_D = \frac{2}{\pi} \int_0^{2\pi} \int_0^\pi \left[-P^* \cos \theta \right]_{r^* = \frac{1}{2}} \sin \theta \, d\theta \, d\phi \tag{8.250}$$

$$= -\frac{2}{\pi} \int_0^{2\pi} \int_0^\pi \left[\mathcal{A} + \mathcal{B} \sin^2 \theta \right] \cos \theta \sin \theta \, d\theta \, d\phi \tag{8.251}$$

$$= -4 \int_0^\pi \left[\mathcal{A} + \mathcal{B} \sin^2 \theta \right] \cos \theta \sin \theta \, d\theta \tag{8.252}$$

$$= -4 \int_0^\pi \mathcal{A} \cos \theta \sin \theta \, d\theta - 4 \int_0^\pi \mathcal{B} \sin^3 \theta \cos \theta \, d\theta \tag{8.253}$$

$$= -4\mathcal{A} \frac{\sin^2 \theta}{2} \Big|_0^\pi - 4\mathcal{B} \frac{\sin^4 \theta}{4} \Big|_0^\pi = 0 \tag{8.254}$$

Drag coefficient, potential flow:

$$\boxed{C_D = 0} \tag{8.255}$$

We arrive at an astonishing result. According to the high-Reynolds-number, zero-viscosity solution to the Navier-Stokes equations (i.e., the potential-flow solution), there is no drag whatsoever on a sphere in uniform flow at high speeds.

It is time to review our process. We turned to the Navier-Stokes equation at high Reynolds number to understand the unusual flow patterns observed in wind-tunnel experiments for flow around spheres, as shown in Figures 8.22 and 8.23. Instead of providing insight, however, the solution to the Navier-Stokes equations in this flow for high Reynolds numbers is unreasonable. The no-slip boundary condition is not respected at the surface of the sphere (see Figures 8.25 and 8.27), and no drag is predicted (Equation 8.255). Both of these predictions are wrong:

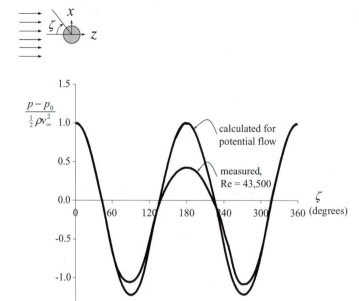

The pressure distribution very near the sphere surface as a function of the angle $\zeta = \pi - \theta$ from the forward stagnation point. Shown are the potential-flow prediction and a measurement from Flachsbart [50], as cited in Schlichting [148], at a Reynolds number of 43,500. The pressure distributions are similar.

In uniform flow, fluid adheres to the surface of the sphere and drag most definitely is measured at all Reynolds numbers (see Figure 8.23).

The failure of potential-flow calculations to predict drag is called *d'Alembert's paradox*, after Jean le Rond d'Alembert, who calculated forces in a variety of high-Re flows and found drag to be missing in each case. Researchers' initial inability to understand the problem with potential-flow solutions led to a multidecade rift between hydraulics experts, who observed phenomena that could not be explained (e.g., trailing vorticies and wake flow), and theoretical researchers, whose potential-flow solutions (exhibiting no drag and slipping at the wall) were not observed.

To understand why potential flow does not correctly capture high-speed flows, we begin with what potential flow predicts correctly. Experiments on high-Reynolds-number flows show that away from the sphere, the potential-flow predictions of the streamlines in steady flow (i.e., particle paths) are correct. More significant, the pressure distribution—as predicted by potential flow—is found to be approximately correct both far away from and near to the sphere (Figure 8.28). The problems with the potential-flow solution mostly are confined to predicting an incorrect velocity field near the sphere and predicting the total absence of drag (Figure 8.29).

Because the pressure distribution seems to be correct throughout the flow, it is particularly confusing that the drag prediction is wrong. It seems to reason that if the pressure distribution is approximately correct and since the prediction of drag coefficient C_D comes from an integration over the pressure (see Equation 8.241), the drag prediction of the potential-flow solution also should be correct. We can

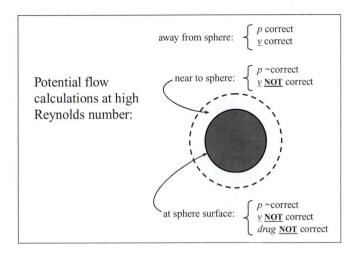

Figure 8.29 Potential-flow theory gets the pressure distribution nearly right everywhere. The velocity solution in potential flow is incorrect near the sphere; the calculation of drag at the surface of the sphere also is incorrect.

investigate this question by looking at how the shape of the pressure distribution affects the predicted drag on a sphere.

EXAMPLE 8.11. *What kind of pressure distributions lead to drag on a sphere in noncreeping flow?*

SOLUTION. The drag on a sphere in noncreeping flow is given by Equation 8.171, which contains the nondimensional dynamic pressure distribution $\mathcal{P}^*$:

$$C_D = \frac{2}{\pi} \int_0^{2\pi} \int_0^\pi \left[\left(\frac{2}{Re} \frac{\partial v_r^*}{\partial r^*} - \mathcal{P}^* \right) \cos\theta \right. $$
$$ \left. + \left(\frac{r^*}{Re} \frac{\partial}{\partial r^*} \left(\frac{v_\theta^*}{r^*} \right) + \frac{1}{Re\, r^*} \frac{\partial v_r^*}{\partial \theta} \right) (-\sin\theta) \right]_{r^*=\frac{1}{2}} \sin\theta\, d\theta\, d\phi \qquad (8.256)$$

with $\mathcal{P}^* = \mathcal{P}^*(r^*, \theta, \phi)$. When Reynolds number is large, Equation 8.256 reduces to:

Drag coefficient,
noncreeping flow
around a sphere:
(high Reynolds number)

$$C_D = \frac{2}{\pi} \int_0^{2\pi} \int_0^\pi \left[-\mathcal{P}^*(r^*, \theta, \phi) \cos\theta \right]_{r^*=\frac{1}{2}} \sin\theta\, d\theta\, d\phi$$

$$(8.257)$$

In a previous example, we carried out this integration for potential flow and obtained $C_D = 0$ (Equation 8.255).

Without assuming potential flow, if we assume that the pressure distribution is independent of ϕ, we can carry out the ϕ-integration in Equation 8.257, which results in a factor of 2π:

$$C_D = 4 \int_0^\pi \left[-\mathcal{P}^*(r^*, \theta) \right]_{r^*=\frac{1}{2}} \cos\theta \sin\theta\, d\theta \qquad (8.258)$$

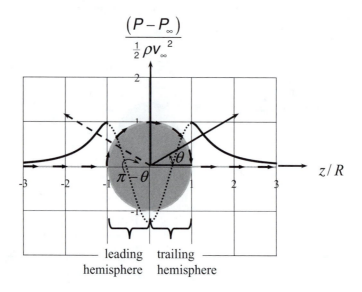

Figure 8.30 For the potential-flow solution for flow around a sphere, the pressure along the center streamline varies as shown. Upstream of the sphere, the central streamline (curve with arrows) follows the z-axis and then splits into two and flows around the sphere, rejoining as a single streamline at the rear stagnation point. For the central streamline, the pressure rises as the fluid approaches the sphere. Along the sphere surface, the pressure profile following this streamline (dotted curve) drops from a maximum at the forward stagnation point to below the mean-stream value at the sphere equator and then rises again to a maximum at the rear stagnation point. From the rear of the sphere downstream, the pressure falls again to the mean-stream value.

To explore the effect of pressure distribution on drag, we can explore other assumptions and see how they affect the predicted drag coefficient.

Pressure in the flow around a sphere typically is an important function of θ, and we do not know that dependence in general. In the potential-flow solution to flow around a sphere (i.e., the solution in which viscosity is assumed to be zero), the pressure distribution along the surface of the sphere is symmetrical in a front-to-back sense (Figure 8.30). That is, the shape of the pressure distribution on the leading hemisphere is the same as the shape on the trailing hemisphere. We write this fact mathematically as:

$$\begin{array}{c} \text{Inviscid flow} \\ \text{around a sphere;} \\ \text{pressure distribution} \\ \text{is front-to-back symmetrical:} \end{array} \qquad \mathcal{P}(\theta)|_{r=R} = \mathcal{P}(\pi - \theta)|_{r=R} \qquad (8.259)$$

To see the implications of the pressure-distribution symmetry on the prediction of drag coefficient, we can divide the integral in the drag equation, Equation 8.258, into two pieces: one over the trailing half of the sphere ($0 \le \theta \le \pi/2$) and the other over the leading half of the sphere ($\pi/2 \le \theta \le \pi$):

$$C_D = 4 \int_0^{\frac{\pi}{2}} \left[-\mathcal{P}^* \right]_{r^*=\frac{1}{2}} \cos\theta \sin\theta d\theta + 4 \int_{\frac{\pi}{2}}^{\pi} \left[-\mathcal{P}^* \right]_{r^*=\frac{1}{2}} \cos\theta \sin\theta d\theta$$

$$(8.260)$$

We define the angle ζ (zeta) to be $\zeta \equiv \pi - \theta$ and write the second integral in Equation 8.260 in terms of ζ as follows (note particularly the limits of integration on the second integral):

$$\theta \equiv \pi - \zeta$$

$$\mathcal{P}(\theta) = \mathcal{P}(\zeta) \qquad \text{(symmetry assumption)}$$

$$\sin \theta = \sin \zeta \qquad \text{(trigonometric identity)}$$

$$\cos \theta = -\cos \zeta \qquad \text{(trigonometric identity)}$$

$$d\theta = -d\zeta$$

$$\theta = \frac{\pi}{2}, \quad \zeta = \frac{\pi}{2}$$

$$\theta = \pi, \quad \zeta = 0$$

$$C_D = 4 \int_0^{\frac{\pi}{2}} \left[-\mathcal{P}^*(r^*, \theta) \right]_{r^* = \frac{1}{2}} \cos \theta \sin \theta d\theta$$

$$+ 4 \int_{\frac{\pi}{2}}^0 \left[-\mathcal{P}^*(r^*, \zeta) \right]_{r^* = \frac{1}{2}} (-\cos \zeta) \sin \zeta (-d\zeta) \qquad (8.261)$$

$$C_D = 4 \int_0^{\frac{\pi}{2}} \left[-\mathcal{P}^*(r^*, \theta) \right]_{r^* = \frac{1}{2}} \cos \theta \sin \theta d\theta$$

$$- 4 \int_0^{\frac{\pi}{2}} \left[-\mathcal{P}^*(r^*, \zeta) \right]_{r^* = \frac{1}{2}} \cos \zeta \sin \zeta d\zeta \qquad (8.262)$$

$$C_D = 0$$

We see that the pressure-distribution symmetry on the sphere surface (i.e., $\mathcal{P}^*$ independent of ϕ and $\mathcal{P}^*(\theta) = \mathcal{P}^*(\pi - \theta) = \mathcal{P}^*(\zeta)$) implies that the drag coefficient C_D is zero. Thus, because the inviscid solution of flow around a sphere predicts that the pressure distribution is symmetrical, it also predicts that there is no drag on a sphere in uniform flow.

The previous example is enlightening. From that calculation, we see that the fore–aft symmetry of the surface-pressure distribution in flow around a sphere—when coupled with zero viscosity (assumed in arriving at the equation for nondimensional force, Equation 8.257)—is associated with zero drag. In inviscid flows with such fore–aft symmetrical pressure distributions, the forces on the leading hemisphere are balanced exactly by the forces on the trailing hemisphere, and the net drag is zero.

At this point, it is helpful to review our situation. We analyzed flow around a sphere at low flow rates, neglecting inertia, and calculated the results for velocity field, pressure field, and drag on the sphere. These results match what is observed as long as the Reynolds number is less than about 2 (see Figure 8.12; $C_D = 24/\text{Re}$, creeping flow, inertia neglected).

At Reynolds numbers above 2, the Stokes drag result does not hold and flow visualization indicates that the flow around a sphere is characterized by recirculating vortices, oscillatory wake flow, and, ultimately, complex flow structure (see Figure 8.22).

To find the source of the observed flow richness, we returned to the microscopic mass and momentum balances, which were too difficult to solve when both viscous and inertial effects are included. We postulated that viscous effects are not important in rapid flows and looked at the limit in which the Reynolds number is quite large. Solving those equations, we obtained the potential-flow solutions for velocity field, pressure field, and drag. The potential-flow solutions, however, do not match what is observed: The predicted velocity field poorly matches the observed velocity field near the sphere, and the drag result is completely wrong ($\mathcal{F}_{drag} = 0$; see Figure 8.29). Potential flow does not predict wake, flow separation, vortices, or any oscillatory flow at any Reynolds number. Only the pressure field and the velocity field away from the sphere are approximately correct when the calculated potential-flow results are compared to high-Re experiments (see Figure 8.28).

The failure of potential flow to explain high-Re flow around a sphere is a setback; however, it does not need to lead to failure of our project. Instead, as shown previously, reaching a dead end on the current path simply necessitates reviewing the path, finding the wrong step, and beginning another investigation. In the next section, which is on boundary layers, we examine the flow near the surface of the sphere to track down the problem with our attempts so far to calculate flow fields and drag in high-Reynolds-number flow.

Potential flow is wrong for drag calculations, but away from walls it is right. For high-speed flow problems, we use potential flow solutions to predict pressure and flow patterns, provided the influence of the wall may be neglected (see Example 8.15). Pressure distributions are easy to calculate in potential flows because, as we show in Examples 8.12–14, the Bernoulli equation (familiar from Chapter 1) applies in potential flow. The examples here also point out the important distinction between rotational and irrotational flows and explain why this flow classification affects how the Bernoulli equation is applied to high-speed flows. After these examples we turn to boundary layers to fix what is wrong with potential flow.

EXAMPLE 8.12. *How are pressure and fluid velocity related in steady, incompressible, potential (inviscid) flows?*

SOLUTION. The governing equations for potential flow are the mass and momentum balances evaluated in the limit that the Reynolds number is very large (see Equation 8.203). We write them in dimensional form:

$$\begin{array}{cc} \text{Continuity equation} \\ \text{(incompressible):} \end{array} \qquad \nabla \cdot \underline{v} = 0 \qquad\qquad (8.263)$$

$$\begin{array}{cc} \text{Navier-Stokes equation} \\ \text{(inviscid):} \end{array} \qquad \rho \left(\frac{\partial \underline{v}}{\partial t} + \underline{v} \cdot \nabla \underline{v} \right) = -\nabla \mathcal{P} \qquad (8.264)$$

At steady state ($\partial \underline{v}/\partial t = 0$), the momentum balance simplifies further to:

$$\rho \underline{v} \cdot \nabla \underline{v} = -\nabla (p + \rho g h) \qquad (8.265)$$

Navier-Stokes
(steady, incompressible, $\underline{v} \cdot \nabla \underline{v} + \dfrac{\nabla p}{\rho} + g \nabla h = 0 \qquad (8.266)$
inviscid):

where we revert to using the bare pressure instead of the dynamic pressure $\mathcal{P} = p + \rho g h$, and h is the variable representing the vertical height of the location of a point (x, y, z) above an elevation chosen as the reference elevation (see Equation 8.115). We used the assumption of incompressible fluid (constant density) in moving ρ through the gradient operator, $\nabla(\rho g h) = \rho g \nabla h$.

Through algebraic manipulations in Cartesian coordinates, the following vector identity can be shown to hold for any vector field $\underline{v}$ [6]:

$$\text{Vector identity:} \quad \underline{v} \cdot \nabla \underline{v} = \nabla \left(\frac{1}{2} v^2 \right) - \underline{v} \times (\nabla \times \underline{v}) \qquad (8.267)$$

where $\underline{v}^2 = \underline{v} \cdot \underline{v} = v^2$ and v is the magnitude of the vector $\underline{v}$. Substituting this identity into the steady, inviscid Navier-Stokes equation (Equation 8.266), we obtain:

Navier-Stokes for
steady, incompressible,
inviscid flow
(potential flow):

$$\boxed{\underline{v} \times (\nabla \times \underline{v}) = \nabla \left(\frac{1}{2} v^2 \right) + \frac{\nabla p}{\rho} + g \nabla h}$$

$$(8.268)$$

This is known as Crocco's theorem [154], which is the relationship between pressure and velocity for steady, incompressible, inviscid flow.

We also consider an additional special case, that of *irrotational flow*—a flow for which $\nabla \times \underline{v} = 0$. Assuming irrotational flow, Equation 8.268 becomes:

$$\text{Irrotational potential flow:} \quad \nabla \left(\frac{1}{2} v^2 \right) + \frac{\nabla p}{\rho} + g \nabla h = 0 \qquad (8.269)$$

We factor out the gradient operator ∇ from each term to obtain:[7]

Navier-Stokes
(steady, incompressible, $\nabla \left(\dfrac{v^2}{2} + \dfrac{p}{\rho} + g h \right) = 0 \qquad (8.270)$
inviscid, irrotational):

The final result in Equation 8.270 indicates that the gradient of the scalar function enclosed between the parentheses is zero everywhere in a steady, incompressible, inviscid, irrotational flow. Thus, that combination of variables is constant

---

[7]In [14], $gh = \Phi$ is called the gravity potential.

throughout the flow. We recognize this final expression as the Bernoulli equation:

Bernoulli equation: integration of the Navier-Stokes equation for steady, incompressible, inviscid, irrotational flow

$$\boxed{\left(\frac{v^2}{2} + \frac{p}{\rho} + gh\right) = \text{constant}}$$

(8.271)

$$\boxed{\left(\frac{v^2}{2} + \frac{\mathcal{P}}{\rho}\right) = \text{constant}}$$

(8.272)

For *any two points* in a steady, incompressible, inviscid, *irrotational* flow:

$$\boxed{\frac{v_2^2 - v_1^2}{2} + \frac{p_2 - p_1}{\rho} + g(h_2 - h_1) = 0}$$

(8.273)

In Example 8.9, we calculate the velocity field for potential flow around a sphere; it is possible to show that uniform potential flow past an obstacle is irrotational (see Problem 53 and Example 8.25). Thus, we can use Equation 8.271 as an easy way to calculate the pressure (or dynamic pressure with Equation 8.272) from the velocity field (see Problem 27). This is a handy pressure-calculating method when the flow of interest is irrotational and inviscid.

EXAMPLE 8.13. *Show that the Bernoulli equation applies in steady, inviscid, flows in which $\nabla \times \underline{v}$ is not zero (rotational flows), if properly applied along a streamline.*

SOLUTION. For steady ($\partial \underline{v}/\partial t = 0$), inviscid ($\mu = 0$) flow, the Navier-Stokes equation simplifies to:

Navier-Stokes equation: $\rho\left(\frac{\partial \underline{v}}{\partial t} + \underline{v} \cdot \nabla \underline{v}\right) = -\nabla p + \mu \nabla^2 \underline{v} + \rho \underline{g}$ (8.274)

$$\rho \underline{v} \cdot \nabla \underline{v} = -\nabla p + \rho \underline{g}$$

(8.275)

We can write the Navier-Stokes equation in any coordinate system. We choose a coordinate system that always has one direction pointing in the direction of flow. At any point, the flow direction is $\hat{v} = \underline{v}/v$. The other two directions of the coordinate system we call $\hat{u}$ and $\hat{w}$; all three basis vectors vary with position. Thus, in the vuw-coordinate system, the velocity vector is given by:

$$\underline{v} = \begin{pmatrix} v_v \\ v_u \\ v_w \end{pmatrix}_{vuw} = \begin{pmatrix} v \\ 0 \\ 0 \end{pmatrix}_{vuw}$$

(8.276)

The simple form of the velocity vector given in the vuw-coordinate system also makes the $\hat{v}$-component of the Navier-Stokes equation quite simple:

$$\text{Navier-Stokes equation:} \qquad \rho \underline{v} \cdot \nabla \underline{v} = -\nabla p + \rho \underline{g} \qquad (8.277)$$
$$\text{(steady, inviscid)}$$

$$\hat{v}\text{-component:} \qquad \rho \left(v_v \frac{\partial v_v}{\partial x_v} + v_u \frac{\partial v_v}{\partial x_u} + v_w \frac{\partial v_v}{\partial x_w} \right) = -\frac{\partial p}{\partial x_v} + \rho g_v \qquad (8.278)$$

$$\rho v \frac{\partial v}{\partial x_v} = -\frac{\partial p}{\partial x_v} + \rho g_v \qquad (8.279)$$

It is straightforward to verify algebraically that $v \frac{\partial v}{\partial x_v} = \frac{1}{2} \frac{\partial (v^2)}{\partial x_v}$. We define h in the usual way as the vertical distance upward; thus, $g_v = -g \partial h / \partial x_v$ (see discussion with Equation 8.119). Making these two substitutions, the Navier-Stokes equation for steady, inviscid flow in our chosen coordinate system becomes:

$$\rho v \frac{\partial v}{\partial x_v} + \frac{\partial p}{\partial x_v} - \rho g_v = 0 \qquad (8.280)$$

$$\rho \frac{\partial \left(\frac{v^2}{2} \right)}{\partial x_v} + \frac{\partial p}{\partial x_v} + \rho g \frac{\partial h}{\partial x_v} = 0 \qquad (8.281)$$

$$\frac{\partial}{\partial x_v} \left(\frac{v^2}{2} + \frac{p}{\rho} + gh \right) = 0 \qquad (8.282)$$

This result is similar to the result for irrotational flow, Equation 8.270, except that in irrotational flow, the gradient operation may be taken in any direction and zero always is obtained. In the current case of inviscid but not irrotational flow, we cannot arrive at this result in any arbitrary direction; rather, we obtain the Bernoulli equation only when we integrate in the $\hat{v}$-direction—that is, along a streamline:

$$\text{Navier-Stokes} \qquad \frac{\partial}{\partial x_v} \left(\frac{v^2}{2} + \frac{p}{\rho} + gh \right) = 0 \qquad (8.283)$$
$$\text{(steady, incompressible,}$$
$$\text{inviscid, } \hat{v}\text{-component):}$$

$$\text{Bernoulli equation (again)} \qquad \boxed{\left(\frac{v^2}{2} + \frac{p}{\rho} + gh \right) = \begin{array}{c} \text{constant} \\ \text{along a} \\ \text{streamline} \end{array}}$$
$$\text{(steady, incompressible,}$$
$$\text{inviscid flow, quantity integrated}$$
$$\text{along a streamline):}$$

$$(8.284)$$

For *two points on the same streamline* in a steady, incompressible, inviscid, *rotational* flow

$$\boxed{\frac{v_2^2 - v_1^2}{2} + \frac{p_2 - p_1}{\rho} + g(h_2 - h_1) = 0}$$

$$(8.285)$$

As long as we are careful to apply this expression only in steady, inviscid flow and along a streamline, Equation 8.284 is a powerful result. It is only in irrotational flow, $\omega \equiv \nabla \times \underline{v} = 0$, that the requirement of following a streamline is not necessary.

We encountered the Bernoulli equation in Chapter 1 (see Equation 1.17). For a steady, incompressible, single-input, single-output fluid flow in which friction may be neglected and there are no shafts or reaction and little heat transfer, we can perform the macroscopic energy balance along a streamline (to ensure that we only consider a single-input, single-output case) and obtain the Bernoulli equation. On different streamlines in flows where rotational character is present ($\nabla \times \underline{v} \neq 0$), the quantity on the left side of Equation 8.271 sums to different numbers, but that number is constant along the streamline. For the case of irrotational flow, the value of this constant—the Bernoulli constant—is the same everywhere in the flow, and we do not need to confine our calculations to points on the same streamline. More discussion on rotational character in flows appears in Section 8.3, which introduces vorticity ω, a property of the velocity field. Vorticity is zero in irrotational flow ($\omega = \nabla \times \underline{v}$).

If we are careful to apply potential-flow results where they are valid, we can use these widely available solutions [9] to make useful calculations. One such application is in the wind-speed calculation in Chapter 2 (see Example 2.5), and Example 8.15 uses a potential-flow solution to calculate pressure. We apply the Bernoulli equation along a streamline in Example 9.6, in which we analyze a Pitot tube (see Chapter 9). The rules for using potential-flow solutions are summarized here.

Rules for Using Potential-Flow Solutions

1. Potential-flow solutions may be only used in rapid flows away from walls.
2. The Bernoulli equation may be used only along streamlines, not across streamlines (unless the flow is known to be irrotational).
3. In high-speed flows that have rotational character (e.g., aeronautical flows), useful potential-flow models can be constructed by superposing rotational and irrotational potential-flow solutions (see Section 10.4).

EXAMPLE 8.14. *What is the pressure distribution around a cylinder in potential flow? The flow field is irrotational.*[8]

SOLUTION. Example 8.12 shows that the Bernoulli equation holds in potential flows. Thus, we can calculate the pressure field around a cylinder from the solution for the velocity field using the Bernoulli equation.

The velocity as a function of position for potential flow around a long cylinder may be solved for following a procedure similar to that used in the sphere case in Example 8.9; the solution is given in the literature [9] and in Equation 2.44. The x-axis of the Cartesian system points in the wind direction, perpendicular to the

---

[8]Uniform potential flow past an obstacle is irrotational (see Problem 53).

cylinder; the z-axis of the $r\theta z$-system points along the cylinder axis:

Potential flow
around a long cylinder
$\underline{v} = v_\infty \hat{e}_x$:

$$\underline{v} = \begin{pmatrix} v_\infty \left(1 - \frac{R^2}{r^2}\right) \cos\theta \\ -v_\infty \left(1 + \frac{R^2}{r^2}\right) \sin\theta \\ 0 \end{pmatrix}_{r\theta z} \tag{8.286}$$

$$= v_\infty \left(1 - \frac{R^2}{r^2}\right) \cos\theta \hat{e}_r - v_\infty \left(1 + \frac{R^2}{r^2}\right) \sin\theta \hat{e}_\theta \tag{8.287}$$

To calculate the pressure distribution from this velocity field, we apply the Bernoulli equation between two points: at a point far upstream where we designate the pressure as p_∞ and at another point where it is $p(r,\theta)$. Because we are discussing an irrotational flow, we are not limited to applying the Bernoulli equation along a streamline.

The Bernoulli equation (Equation 8.271) is given by:

Bernoulli equation:
integration of the
Navier-Stokes equation for
steady, incompressible,
inviscid, irrotational flow:

$$\boxed{\left(\frac{v^2}{2} + \frac{p}{\rho} + gh\right) = \text{constant}} \tag{8.288}$$

For *any two points* in a
steady, incompressible,
inviscid, *irrotational* flow:

$$\boxed{\frac{v_2^2 - v_1^2}{2} + \frac{p_2 - p_1}{\rho} + g(h_2 - h_1) = 0} \tag{8.289}$$

Neglecting gravity and substituting the expressions for velocity and pressure at the two points, we obtain:

$$\frac{v_2^2 - v_1^2}{2} + \frac{p_2 - p_1}{\rho} + g(h_2 - h_1) = 0 \tag{8.290}$$

$$\frac{v(r,\theta)^2 - v_\infty^2}{2} + \frac{p(r,\theta) - p_\infty}{\rho} = 0 \tag{8.291}$$

$$p(r,\theta) = \rho\left(\frac{p_\infty}{\rho} - \frac{v^2}{2} + \frac{v_\infty^2}{2}\right) \tag{8.292}$$

$$= \frac{\rho v_\infty^2}{2}\left(\frac{p_\infty}{\rho v_\infty^2/2} - \frac{v^2}{v_\infty^2} + 1\right) \tag{8.293}$$

where $v = v(r,\theta)$ is the magnitude of the velocity, which may be calculated from $v = |\underline{v}| = \sqrt{\underline{v} \cdot \underline{v}}$ and the velocity result in Equation 8.286. After some algebra, we obtain:

$$\left(\frac{v}{v_\infty}\right)^2 = 1 - \left(\frac{2R^2}{r^2}\right)\cos 2\theta + \frac{R^4}{r^4} \tag{8.294}$$

Substituting this expression into Equation 8.293, we obtain:

$$\frac{p(r, \theta)}{\rho v_\infty^2/2} = \frac{p_\infty}{\rho v_\infty^2/2} - \frac{v^2}{v_\infty^2} + 1 \qquad (8.295)$$

Pressure distribution,
potential flow
around a cylinder:

$$\boxed{\frac{(p(r, \theta) - p_\infty)}{\rho v_\infty^2/2} = \left(\frac{2R^2}{r^2}\right) \cos 2\theta - \frac{R^4}{r^4}}$$

(8.296)

EXAMPLE 8.15. *A new tower hotel, cylindrical in shape and 100 feet in diameter, is built in a resort town near the sea on the windward side of an island (see Figure 2.16). Residents complain that there often are uncomfortably high winds near several entrances to the tower. In addition, the doors are sometimes difficult to open. In Chapter 2, we addressed the issue of the wind speed as a function of position around the hotel tower in Example 2.5. How does the pressure field vary at different locations around the building?*

SOLUTION. We are interested in the flow in the main stream, away from the walls; thus, we can model the flow around the tower with the potential flow around a cylinder. The pressure distribution for flow around a cylinder is calculated in Example 8.14:

Pressure distribution,
potential flow
around a cylinder:

$$\frac{(p(r, \theta) - p_\infty)}{\rho v_\infty^2/2} = \left(\frac{2R^2}{r^2}\right) \cos 2\theta - \frac{R^4}{r^4} \qquad (8.297)$$

We are interested in the values of the pressure near the doors located at points C, C', D, D', and E in Figure 2.16. The other doors are on the lee side of the hotel and therefore in the wake behind the cylinder. In the wake region, the potential-flow solution does not represent either the velocity or the pressure distribution. The pressures behind the hotel are likely to be close to the mean atmospheric pressure p_∞.

For the windward doors, we calculate the pressures from Equation 8.297 (we use air density ≈ 1.3 kg/m^3):

Location	r	θ	$v_\infty = 30$	50	70	90 mph
C, C'	60 feet	$\pm\frac{\pi}{2}$	$p - p_\infty = -0.03$	-0.09	-0.17	-0.29 psig
D, D'	60 feet	$\pm\frac{3\pi}{4}$	$p - p_\infty = -0.01$	-0.02	-0.04	-0.07 psig
E	60 feet	π	$p - p_\infty = +0.02$	$+0.04$	$+0.08$	$+0.14$ psig

Due to these forces, some of the doors tend to fly open if they are not latched (i.e., the negative pressures), whereas other doors are impossible to open if the winds are high. If we estimate that a door is 3 feet by 6 feet, then the force holding shut the door at E ranges from 40 lb$_f$ (at 30 mph) to 360 lb$_f$ (at hurricane strength of 90 mph), whereas the force sucking open the side doors at C and C' range from -80 lb$_f$ (at 30 mph) to -740 lb$_f$ (at 90 mph). Clearly, the doors should be secured when winds are high.

8.2 Boundary layers

The task remains to understand the rich flow patterns that occur at Reynolds numbers above Re $= 2$ in flow around a sphere. Examining the potential-flow results, it is striking that although the pressure distribution calculated from potential flow is approximately correct near the sphere, the calculated drag is completely wrong. This is particularly striking because the equation used to evaluate drag, Equation 8.241 (repeated here), is an integral over the surface pressure distribution only, with no influence of the (incorrect) velocity field or velocity gradients:

$$\text{Drag coefficient} \atop \text{(potential flow)} \quad C_D = \frac{2}{\pi} \int_0^{2\pi} \int_0^{\pi} \left[-\mathcal{P}^* \cos\theta \right]_{r^*=\frac{1}{2}} \sin\theta d\theta d\phi \qquad (8.298)$$

Because the pressure field of potential flow is approximately correct and the drag comes from an integration over only the pressure field, the drag calculated from potential flow is expected to be approximately correct—yet, it is not.

There are differences between the observed pressure field at high Re and the calculated potential-flow pressure field (see Figure 8.28) but, as discussed, these are not too large. The fact that we calculated drag from Equation 8.298—an equation that depends on only the pressure distribution—is worth reexamining, however. Equation 8.298 was obtained by permitting the Reynolds number to go to infinity in the more complete equation for drag on a sphere, Equation 8.171, repeated here:

$$C_D = \frac{2}{\pi} \int_0^{2\pi} \int_0^{\pi} \left[\left(\frac{2}{\text{Re}} \frac{\partial v_r^*}{\partial r^*} - \mathcal{P}^* \right) \cos\theta \right.$$

$$\left. + \left(\frac{r^*}{\text{Re}} \frac{\partial}{\partial r^*} \left(\frac{v_\theta^*}{r^*} \right) + \frac{1}{\text{Re}\, r^*} \frac{\partial v_r^*}{\partial \theta} \right) (-\sin\theta) \right]_{r^*=\frac{1}{2}} \sin\theta d\theta d\phi \qquad (8.299)$$

The potential-flow version of the equation for drag (Equation 8.298) indicates that the drag on the sphere at a high Reynolds number should be independent of velocity gradients near the sphere surface. These velocity gradients are a substantial source of drag in creeping flow, a flow dominated by viscosity (see the last term in Equations 8.58 and 8.59: friction drag $= 4\pi R\mu v_\infty = 2/3$ of total drag). Perhaps we should examine whether the viscous effects represented by these velocity gradients really are negligible, as we assume when using Equation 8.298.

In our previous analysis, we began with Equation 8.299 and took the high-Reynolds-number limit, eliminating all velocity-derivative terms (i.e., terms with the prefactor $= 1/\text{Re}$). We implicitly assumed in that analysis that all of the nondimensional velocity derivatives in Equation 8.299 would remain finite or increase slowly as the Reynolds number approached infinity; thus, terms with $1/\text{Re}$ would go to zero. Perhaps, however, at high Reynolds numbers and near the surface of the sphere, the velocity derivatives grow rapidly with Re. If the velocity

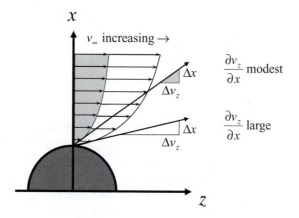

Figure 8.31 Near the sphere surface, if the no-slip velocity boundary condition is satisfied, the velocity gradients must grow as the free-stream velocity increases. The free-stream velocity is in the z-direction.

derivatives increase rapidly near the surface of the sphere, which seems likely in retrospect (Figure 8.31), it may not be possible to consider the velocity-derivative terms in Equation 8.299 to be negligible—even given the prefactor $1/\text{Re}$, which is getting very small. In other words, it is worth considering that our ad hoc scaling practices may break down near the sphere surface, causing difficulty in our calculations.

In reflecting on the flow near the surface of the sphere, we expect high-velocity derivatives because the fluid adheres to the stationary sphere—that is, the no-slip boundary condition holds at the surface. In reviewing the velocity-field results for potential flow, we are reminded that the no-slip boundary condition is not respected in that solution (see Figure 8.27). To explain why we were unable to force the potential-flow solution to respect the no-slip boundary condition, we return to the microscopic-momentum balance from which we obtained the potential-flow solution.

In the nondimensional Navier-Stokes equation, Equation 8.201:

$$
\begin{array}{c}
\text{Nondimensional} \\
\text{Navier-Stokes} \\
\text{(dynamic pressure):}
\end{array}
\quad
\frac{\partial \underline{v}^*}{\partial t^*} + \underline{v}^* \cdot \nabla^* \underline{v}^* = -\nabla^* \mathcal{P}^* + \left(\frac{1}{\text{Re}}\right) \nabla^{*2} \underline{v}^* \quad (8.300)
$$

we obtained the potential-flow equations by taking the limit of this equation as $\text{Re} \longrightarrow \infty$. In this limit, the term $\left(\frac{1}{\text{Re}}\right) \nabla^{*2} \underline{v}^*$ was eliminated. This is the only term in the momentum balance in which second derivatives of velocity appear. Because we eliminated the second derivative of velocity, leaving terms with only first derivatives, we need only a single velocity boundary condition to obtain solutions to the resulting equations. Once we specify that the normal component of the velocity is zero at the surface, the problem is completely specified and we cannot impose the additional constraint that the tangential velocity goes to zero

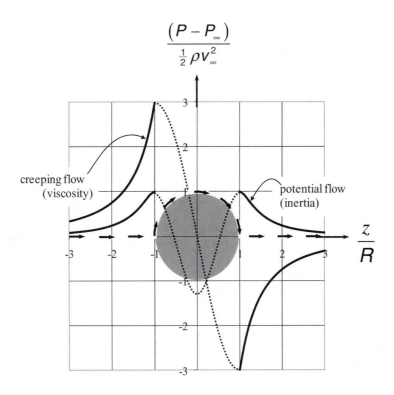

The nondimensional pressure distributions for z-direction flow around a sphere for both the inviscid-flow solution (zero viscosity) and the creeping-flow solution (zero inertia). The calculations shown follow the pressure along the negative z-axis to the upstream stagnation point, then follow the surface of the sphere (dotted-line results), rejoining the z-axis at the rear stagnation point.

at the surface of the sphere (see Equations 8.16 and 8.17):

Tangential component of the no-slip boundary condition at sphere surface *not* satisfied for potential flow:

$$
\begin{array}{lll}
r = R & v_r = 0 & \text{for all values of } \phi \\
r = R & v_\theta \neq 0 & \text{for all values of } \phi
\end{array}
$$

(8.301)

In addition to not following the no-slip boundary condition, the potential-flow solution predicts a different pressure distribution than the viscous-dominated solution, creeping flow. Figure 8.32 compares the calculated pressures of the two solutions as we follow the central streamline up to and around the sphere. For the potential-flow case (see Figure 8.30), the pressure rises as fluid approaches the forward *stagnation point*, defined as a point where the velocity approaches a wall and halts, such as when the central streamline impacts the sphere. As the stream splits and hugs the sphere, the pressure decreases to a minimum below the mean pressure $\mathcal{P}_\infty$ at the sphere equator. The pressure then rises symmetrically to its previous maximum value as the streamline reaches the rear stagnation point and subsequently decreases to the mean value of pressure as the flow continues downstream.

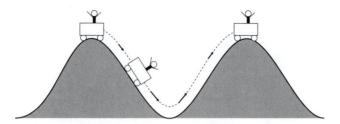

Figure 8.33   The pressure distribution caused by a purely inertial flow (potential flow) can be visualized by thinking of another inertially dominated situation, the motion of a roller coaster. A roller-coaster car at the top of a hill gains momentum as it rolls down the hill. The speed is a maximum at the bottom of the hill. As the car climbs the hill, the kinetic energy of the car, reflected in the high speed, is traded for potential energy as the car rises. If there is no friction, the car will arrive at the top of the hill just as it runs out of kinetic energy (speed goes to zero).

In creeping flow, the pressure also rises as the central streamline approaches the sphere, but the nondimensional pressure rise is higher in the creeping case. As the stream splits and hugs the surface of the sphere, the pressure in creeping flow decreases to the free-stream value at the top of the sphere. However, instead of rising back to the previous maximum value as in potential flow, the pressure in creeping flow continues to fall, reaching a negative value at the rear stagnation point equal in magnitude to the value of pressure reached at the forward stagnation point. As the flow continues downstream, the pressure rises to the mean pressure value observed away from the sphere.

The differences between the two pressure traces produced by inertia-dominated (i.e., potential) and viscous-dominated (i.e., creeping) flows can be visualized in terms of inertia and viscosity. In the inertial case, the pressure acts like a roller coaster, which also moves due to high inertia (Figure 8.33): A roller-coaster car at the top of the pressure hill at the forward stagnation point rolls downhill and, with no frictional losses (i.e., no viscosity), inertia allows it to arrive back at the original value of pressure at the rear stagnation point.

For the viscous case there is no inertia, only a sticky, gooey, viscous glue. The top of the pressure hill in the viscous case can be visualized by considering a thin cantilevered beam submerged in a viscous fluid (Figure 8.34). If a spoon is submerged in a fluid and pressed into a beam in the fluid, the beam deflects downward—the elastic energy stored in the deflected beam is like the stored energy of the high pressure at the forward stagnation point. If we try now to extract the spoon from the viscous fluid, the deflected beam will help to push the spoon upward until the beam returns to the neutral position. If we keep pulling on the spoon, the beam is sucked upward by the adhesive and cohesive fluid forces reflected in viscosity. The beam experiences a negative pressure—much as the sphere experiences a negative pressure at the rear stagnation point in creeping flow around a sphere.

We seek to understand drag in real flows at finite Reynolds numbers. Our approach has been first to study creeping flow, which generated drag from two sources: (1) an asymmetric pressure distribution (see Figure 8.32), which contributed $2\pi R\mu v_\infty$ to the drag (see Equation 8.58); and (2) viscous shear stress at the surface of the sphere, which contributed $4\pi R\mu v_\infty$ to the drag (Figure 8.35). The sum of these two quantities gave $\mathcal{F}_{\text{drag}} = 6\pi R\mu v_\infty$, the Stokes law for drag

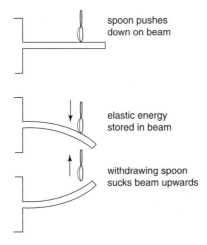

spoon pushes down on beam

elastic energy stored in beam

withdrawing spoon sucks beam upwards

Figure 8.34 The effect of viscosity on pressure in flow around a sphere may be visualized by considering another viscous-dominated scenario, the effect of viscosity on a deflecting beam submerged in a viscous liquid.

on a sphere in creeping flow. We then studied potential flow and learned that neither pressure (which was symmetric) nor viscosity (which was neglected) contributed to the drag, and no drag was produced. Based on the observation that drag indeed exists at finite Re, we now reason that in a real flow at high Reynolds numbers, viscosity may not be neglected—at least not near the sphere. The inclusion of viscosity surely introduces viscous shear stress at the surface of the sphere. Furthermore, based on the shape of the pressure trace in creeping flow, the introduction of viscosity is likely to make the surface pressure trace in real flows asymmetric and therefore drag-producing. Thus, in real high-Re flows, we expect both pressure and viscous contributions to drag. The challenge now is to reformulate our analysis of rapid flows by focusing on the surface so that we can calculate these two drag contributions: (1) pressure drag due to an asymmetric pressure distribution; and (2) viscous drag due to velocity gradients at the sphere surface. The method that facilitates these calculations is boundary-layer analysis.

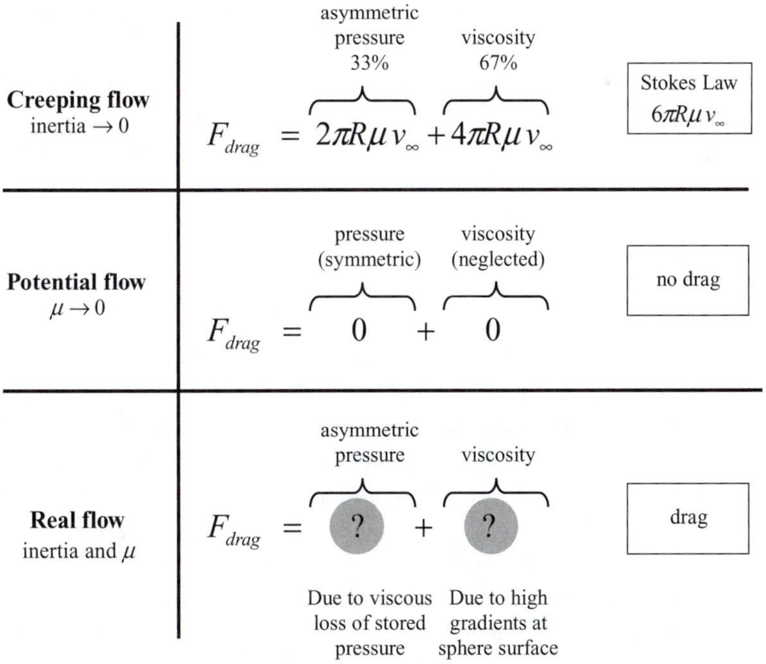

Creeping flow
inertia $\to 0$

$$F_{drag} = \overbrace{2\pi R\mu v_\infty}^{\substack{\text{asymmetric} \\ \text{pressure} \\ 33\%}} + \overbrace{4\pi R\mu v_\infty}^{\substack{\text{viscosity} \\ 67\%}}$$

Stokes Law
$6\pi R\mu v_\infty$

Potential flow
$\mu \to 0$

$$F_{drag} = \overbrace{0}^{\substack{\text{pressure} \\ \text{(symmetric)}}} + \overbrace{0}^{\substack{\text{viscosity} \\ \text{(neglected)}}}$$

no drag

Real flow
inertia and μ

$$F_{drag} = \overbrace{?}^{\substack{\text{asymmetric} \\ \text{pressure}}} + \overbrace{?}^{\text{viscosity}}$$

drag

Due to viscous loss of stored pressure   Due to high gradients at sphere surface

Figure 8.35 Drag comes from pressure asymmetry and viscosity. In the creeping-flow solution, both contributions are present (see Equation 8.58). In the potential-flow solution, neither are present. In a finite-Re solution, both are present.

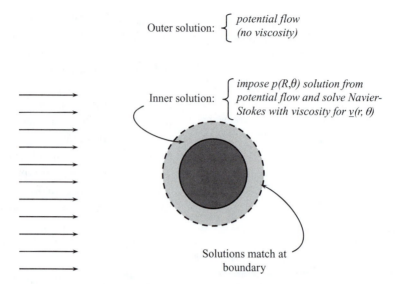

Outer solution: { *potential flow (no viscosity)*

Inner solution: { *impose p(R,θ) solution from potential flow and solve Navier-Stokes with viscosity for v(r, θ)*

Solutions match at boundary

Figure 8.36 The boundary-layer approach divides a flow domain of interest into an inner solution, near the boundary, and an outer solution, which comprises the free stream. The two solutions are matched in an overlap region. The thickness of the boundary layer is greatly exaggerated in this schematic.

8.2.1 Laminar boundary layers

An insightful solution to the problem of noncreeping flow past a surface was proposed by Ludwig Prandtl in 1904. As discussed, in flow around a sphere, both viscous and inertial effects are important. Viscosity and inertia are not equally important everywhere in the flow, however. Prandtl recognized that for computational purposes, he could divide the flow domain into two regions: a large outer region and a small, thin boundary layer near the surface of the sphere (Figure 8.36). In the outer region, viscosity is not important and inertia dominates; in the boundary-layer region, both inertia and viscosity are important. Prandtl's idea was to solve separately the momentum-balance problem in the two flow regions and to subsequently match the two solutions in an overlap region between the inner and outer flows. Any interaction between the two solutions is ignored.

For Prandtl's approach to work, we must be able to calculate the velocity and pressure fields in the inner and outer regions. Prandtl noted that the potential-flow solution for flow around a sphere correctly predicts the pressure and velocity fields far from the sphere; also, the potential-flow solution does not pose any calculation difficulties. It therefore was proposed that potential flow be the outer solution for both pressure and velocity in the boundary-layer construction for uniform flow past a sphere (see Figure 8.36).

To calculate the inner solution for both pressure and velocity fields, Prandtl reasoned that the inner-region pressure field *is largely determined by the outer solution for pressure*; the inner or boundary-layer region is very thin (a guess; verified later) and can be thought of as responsive to the pressure distribution imposed by the outer solution. If we use the outer solution for the pressure distribution as both a driving force and a boundary condition for the inner flow, we can calculate the inner flow once the outer flow is known.

The boundary-layer idea is very clever. Once this way of thinking about high-Reynolds-number flow is introduced, it is easy to be convinced of its logic and correctness. The experimental observations of high-Reynolds-number flow around spheres and other objects conform to the boundary-layer point of view: The streamlines away from the obstacle follow the inviscid, potential-flow solutions; near the obstacle, something else happens that deserves special attention. On the leading side of obstacles, thin layers of viscous flow are observed, in agreement with the boundary-layer picture. On the trailing side of obstacles, however, the flow is neither potential flow nor boundary-layer flow; thus, the boundary-layer view is not applicable on the trailing side. However, if we focus on applying the boundary-layer method to the leading side of the sphere, we can move one step closer to understanding the entire flow.

We have decided on our course of action: we will model rapid flows (e.g., flow past a sphere) as separate outer and inner flows and combine them at the boundary. For the outer flow, we know how to solve the high-Reynolds-number governing equations (see Equation 8.203) for the velocity and pressure fields. We have not yet studied the inner flow—the viscous flow near the sphere surface as fluid streams by at a rapid rate with an imposed pressure distribution. This is the problem of the laminar boundary layer.

EXAMPLE 8.16. *Calculate the steady-state velocity field for the flow of an incompressible viscous fluid near the surface of a solid sphere of diameter 2R. The fluid approaches the sphere with a uniform upstream velocity v_∞. The geometry is the same as in the creeping-flow and potential-flow calculations (see Figure 8.5) but, in this problem, the flow is not slow (i.e., the Reynolds number is finite) and viscosity may not be neglected ($\mu \neq 0$). A known pressure distribution in the flow direction is imposed at the edge of the boundary layer. The imposed pressure distribution is the pressure distribution of potential flow around a sphere (see Equation 8.239).*

SOLUTION. The solution starts the same way as our creeping-flow solution (see Example 8.2). In spherical coordinates, the fluid velocity field may be written as:

$$\underline{v} = \begin{pmatrix} v_r \\ v_\theta \\ v_\phi \end{pmatrix}_{r\theta\phi} = \begin{pmatrix} v_r \\ v_\theta \\ 0 \end{pmatrix}_{r\theta\phi} \tag{8.302}$$

We assume that v_ϕ is equal to zero—that is, there is no swirling component to the flow.

Mass conservation is written in spherical coordinates (see Equation B.5-3) as follows (i.e., constant density):

$$\text{Continuity equation} \atop \text{(Gibbs notation):} \qquad \nabla \cdot \underline{v} = 0 \qquad (8.303)$$

With $v_\phi = 0$, the continuity equation simplifies to:

$$\text{Continuity equation,} \atop \text{flow around a sphere:} \qquad \boxed{\frac{1}{r^2}\frac{\partial(r^2 v_r)}{\partial r} + \frac{1}{r\sin\theta}\frac{\partial(v_\theta \sin\theta)}{\partial\theta} = 0}$$

$$(8.304)$$

The Navier-Stokes equation written in spherical coordinates is given in Table B.7. For steady flow with no ϕ-component and with ϕ-symmetry assumed, the Navier-Stokes equation becomes:

$$\rho \begin{pmatrix} v_r\left(\dfrac{\partial v_r}{\partial r}\right) + v_\theta\left(\dfrac{1}{r}\dfrac{\partial v_r}{\partial\theta} - \dfrac{v_\theta}{r}\right) \\[2ex] v_r\left(\dfrac{\partial v_\theta}{\partial r}\right) + v_\theta\left(\dfrac{1}{r}\dfrac{\partial v_\theta}{\partial\theta} + \dfrac{v_r}{r}\right) \\[2ex] 0 \end{pmatrix}_{r\theta\phi}$$

$$= \begin{pmatrix} \dfrac{\partial \mathcal{P}}{\partial r} \\[2ex] \dfrac{1}{r}\dfrac{\partial \mathcal{P}}{\partial\theta} \\[2ex] \dfrac{1}{r\sin\theta}\dfrac{\partial \mathcal{P}}{\partial\phi} \end{pmatrix}_{r\theta\phi}$$

$$+ \mu \begin{pmatrix} \left(\frac{\partial}{\partial r}\left(\frac{1}{r^2}\frac{\partial}{\partial r}(r^2 v_r)\right) + \frac{1}{r^2\sin\theta}\frac{\partial}{\partial\theta}\left(\sin\theta\frac{\partial v_r}{\partial\theta}\right) - \frac{2}{r^2\sin\theta}\frac{\partial}{\partial\theta}(v_\theta\sin\theta)\right) \\[2ex] \left(\frac{1}{r^2}\frac{\partial}{\partial r}\left(r^2\frac{\partial v_\theta}{\partial r}\right) + \frac{1}{r^2}\frac{\partial}{\partial\theta}\left(\frac{1}{\sin\theta}\frac{\partial}{\partial\theta}(v_\theta\sin\theta)\right) + \frac{2}{r^2}\frac{\partial v_r}{\partial\theta}\right) \\[2ex] 0 \end{pmatrix}_{r\theta\phi}$$

$$(8.305)$$

The set of equations to solve (Equations 8.304 and 8.305) is the same set contemplated when we first began what became the creeping-flow problem in Example 8.2. It was daunting then, and it is daunting now. In that first problem, we reduced the complexity of the system of equations by neglecting the entire left-hand side of the Navier-Stokes equations. By neglecting these inertial terms, we simplified the problem enough to be able to solve it.

We do not want to neglect inertia in our current solution. Instead, we follow the boundary-layer approach and solve for the flow in the boundary layer with the pressure distribution from the free stream imposed as a boundary condition. Even given the assumption of the pressure boundary-condition, we still face many complexities in the current problem. Our equation contains nonzero pressure derivatives in two coordinate directions, r and θ. There also are many terms with velocity derivatives and, certainly, some are more important than others. We need to sort out which of the velocity and pressure derivatives can be neglected so that the mathematics simplifies. This is a difficult task.

This problem still may be too difficult. Some of the complexity of the problem comes from the fact that the surfaces are curved and the coordinate transformation that allowed the governing equations to be written in spherical coordinates introduced extra curvature terms. We are better off if we start with a simpler problem—one without curvature, for example, that would help determine which of the pressure and velocity derivatives are significant for the current problem. We pause, therefore, in our solution of boundary-layer flow around a sphere to address a simpler boundary-layer problem: uniform flow past a flat plate. We follow our usual protocol here; that is, we turn to simple problems to discover the fundamental issues. We return to the flow-past-a-sphere boundary-layer problem after we are more experienced with idealized flat-plate flow.

From the previous discussion, we are led to the idea to investigate the case of flow past a flat plate. By turning to this simpler case, we increase our chances of success with the sphere-boundary-layer problem. Even if it turns out that we cannot neglect curvature effects in the sphere-boundary-layer problem, solving the flat-plate boundary-layer problem is still a good first step that follows exactly our problem-solving strategy: Solve a simple related problem and then use what we learn from the simple problem to tackle the more difficult problem of interest.

EXAMPLE 8.17. *What are the velocity field and the pressure field in a viscous fluid for the flow in which a rapid, uniform flow approaches a flat plate? The flow is steady and the fluid is incompressible. Away from the plate, the flow approaches the inviscid (i.e., potential) flow solution of this same problem.*

SOLUTION. The flow is shown in Figure 8.37. On first consideration, it appears that the flow may be unidirectional throughout the flow domain. Assuming unidirectional flow, we apply the microscopic mass balance, the continuity

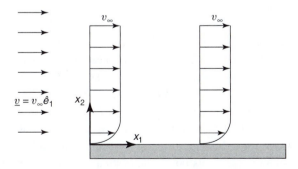

Figure 8.37 When a uniform flow meets a flat plate, the no-slip boundary condition slows the flow where it contacts the plate. Far from the plate, the flow remains the undisturbed uniform flow.

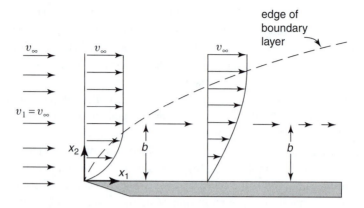

edge of
boundary
layer

Figure 8.38 A fluid particle a small distance from the plate surface initially is outside of the boundary layer but, as it moves downstream, it enters the boundary layer and slows down. Thus, the flow-direction component of fluid velocity varies with x_1, the principal direction.

equation:

$$\text{Unidirectional flow:} \quad \underline{v} = \begin{pmatrix} v_1 \\ 0 \\ 0 \end{pmatrix}_{123} \tag{8.306}$$

$$\text{Continuity equation:} \atop \text{(microscopic-mass balance)} \quad \nabla \cdot \underline{v} = 0 \tag{8.307}$$

$$\frac{\partial v_1}{\partial x_1} + \frac{\partial v_2}{\partial x_2} + \frac{\partial v_3}{\partial x_0} = 0 \tag{8.308}$$

$$\boxed{\frac{\partial v_1}{\partial x_1} = 0} \tag{8.309}$$

From this result, we immediately see a problem with our assumption that the velocity is only in the x_1-direction. If we assume unidirectional flow, then conservation of mass imposes that the flow may not vary in the flow direction. Yet, the flow does vary in the flow direction, as we can establish with a simple thought experiment. In the current flow, consider a series of locations at a constant distance b from the wall (Figure 8.38). When the fluid confronts the plate edge, the flow speed is the speed of the free stream, $v_1 = v_\infty$. As the flow progresses along the plate, however, the presence of the plate causes a slower fluid layer to form; at some x_1 position down the plate, fluid at a distance b from the wall exhibits a velocity of less than v_∞. At this point, the b-streamline is within the boundary layer. Because our reflection on the flow indicates that v_1 *is* a function of the x_1-position ($\partial v_1 / \partial x_1 \neq 0$), we must not be able to assume unidirectional flow (Equation 8.309).

Given that there is a problem with the assumption of unidirectional flow, we now relax that assumption and allow the velocity to have a nonzero x_2-component:

$$\text{Two-dimensional flow:}\quad \underline{v} = \begin{pmatrix} v_1 \\ v_2 \\ 0 \end{pmatrix}_{123} = v_1\hat{e}_1 + v_2\hat{e}_2 \qquad (8.310)$$

$$\text{Continuity equation:}\quad \nabla \cdot \underline{v} = 0 \qquad (8.311)$$

$$\boxed{\frac{\partial v_1}{\partial x_1} + \frac{\partial v_2}{\partial x_2} = 0} \qquad (8.312)$$

The continuity-equation result does not appear to be helpful, but there is nothing to be done about it because we tried the simpler case of unidirectional flow and we know that the simpler case is not correct.

We proceed now to the Navier-Stokes equation:

$$\begin{array}{c}\text{Navier-Stokes equation}\\ \text{(microscopic-momentum balance):}\end{array}\quad \rho\left(\frac{\partial \underline{v}}{\partial t} + \underline{v}\cdot\nabla\underline{v}\right) = -\nabla p + \mu\nabla^2\underline{v} + \rho\underline{g}$$

$$(8.313)$$

For steady flow, we can eliminate the time-derivative term, and we choose to combine the pressure and the gravity effects into the dynamic pressure $\mathcal{P} \equiv p + \rho g h$. Taking these steps and writing the equation in Cartesian coordinates, we obtain the simplified momentum-balance equations that govern the flow in the boundary layer near a flat plate:

$$\begin{array}{c}\text{Navier-Stokes equation}\\ \text{(steady, two-dimensional,}\\ \text{dynamic pressure):}\end{array}\quad \rho\underline{v}\cdot\nabla\underline{v} = -\nabla\mathcal{P} + \mu\nabla^2\underline{v} \qquad (8.314)$$

1-component Navier-Stokes:

$$\boxed{\rho\left(v_1\frac{\partial v_1}{\partial x_1} + v_2\frac{\partial v_1}{\partial x_2}\right) = -\frac{\partial\mathcal{P}}{\partial x_1} + \mu\left(\frac{\partial^2 v_1}{\partial x_1^2} + \frac{\partial^2 v_1}{\partial x_2^2}\right)} \qquad (8.315)$$

2-component Navier-Stokes:

$$\boxed{\rho\left(v_1\frac{\partial v_2}{\partial x_1} + v_2\frac{\partial v_2}{\partial x_2}\right) = -\frac{\partial\mathcal{P}}{\partial x_2} + \mu\left(\frac{\partial^2 v_2}{\partial x_1^2} + \frac{\partial^2 v_2}{\partial x_2^2}\right)} \qquad (8.316)$$

The problem now is to solve for v_1, v_2, and $\mathcal{P}$ from the mass- and momentum-balance equations (Equations 8.312, 8.315, and 8.316). The velocity boundary conditions are no-slip and no-penetration at the surface of the plate and matching the velocity of the free stream at the edge of the boundary layer. For pressure, our plan is to match the value at the edge of the boundary layer with the flow-direction pressure profile from the potential-flow version of this same problem.

The task is still daunting because $\mathcal{P}$ is a function of x_1 and x_2, as are v_1 and v_2. Our first solution step is to sort out which terms in Equations 8.315 and 8.316 dominate the solution and which are negligible. Turning first to determining the importance of the pressure derivatives, we must know the function $\mathcal{P}(x_1, x_2)$ to impose at the edge of the boundary layer. This function comes from the potential-flow solution for uniform flow past a flat plate. We therefore pause in our attempt to solve for the flow of a viscous fluid past a flat plate and solve instead for the inviscid case. We return to finish this problem after that solution is known.

The boundary-layer paradigm requires that we know the pressure profile for the outer flow. Once this pressure profile is known, we impose that pressure distribution on the inner flow as a boundary condition. In the next example, we pursue the pressure distribution in uniform flow of an inviscid fluid past a flat plate.

EXAMPLE 8.18. *Calculate the velocity and pressure fields for steady uniform flow of an incompressible, inviscid fluid past a flat plate.*

SOLUTION. The flow is shown in Figure 8.39. On first consideration, it appears that the flow may be unidirectional throughout the flow domain. Assuming unidirectional flow, we apply the microscopic mass balance, the continuity equation:

$$\text{Unidirectional flow:} \quad \underline{v} = \begin{pmatrix} v_1 \\ 0 \\ 0 \end{pmatrix}_{123} \tag{8.317}$$

$$\begin{array}{c} \text{Continuity equation:} \\ \text{(microscopic-mass balance)} \end{array} \quad \nabla \cdot \underline{v} = 0 \tag{8.318}$$

$$\frac{\partial v_1}{\partial x_1} + \frac{\partial v_2}{\partial x_2} + \frac{\partial v_3}{\partial x_0} = 0 \tag{8.319}$$

$$\boxed{\frac{\partial v_1}{\partial x_1} = 0} \tag{8.320}$$

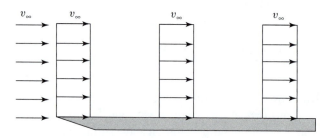

Figure 8.39   When a uniform flow of an inviscid fluid meets a flat plate, there is no mechanism for the flow to slow at the plate surface, and the no-slip boundary condition is not satisfied. The flow is the same near the plate and far from the plate.

Because we do not impose the no-slip boundary condition due to the absence of viscosity in this flow, it is possible to have a flow that does not vary in the flow direction. Thus, the conclusion of the mass balance is valid, and we continue with our assumption of unidirectional flow.

For steady ($\partial \underline{v}/\partial t = 0$), unidirectional ($\underline{v} \cdot \nabla \underline{v} = 0$) flow of an invisicid ($\mu = 0$) fluid, the Navier-Stokes equation now becomes:

$$\rho \left(\frac{\partial \underline{v}}{\partial t} + \underline{v} \cdot \nabla \underline{v} \right) = -\nabla p + \mu \nabla^2 \underline{v} + \rho \underline{g} \tag{8.321}$$

$$0 = -\nabla \mathcal{P} \tag{8.322}$$

$$0 = \begin{pmatrix} -\frac{\partial \mathcal{P}}{\partial x_1} \\ -\frac{\partial \mathcal{P}}{\partial x_2} \\ -\frac{\partial \mathcal{P}}{\partial x_3} \end{pmatrix}_{123} \tag{8.323}$$

where, as usual, $\mathcal{P} \equiv p + \rho g h$ is the dynamic pressure. Integrating each term of Equation 8.323, we see that the final solution for potential flow past a flat surface is constant dynamic pressure. In addition, the momentum balance indicates that the velocity thus cannot change. Because the incoming velocity profile is known to be $\underline{v} = v_\infty \hat{e}_1$, this must be the flow field throughout. The potential-flow solution for uniform flow over a flat plate is:

$$\begin{array}{c} \text{Potential flow} \\ \text{past a flat plate} \\ \text{(steady, incompressible, inviscid):} \end{array} \quad \begin{cases} \mathcal{P} = \text{constant} \\ \underline{v} = v_\infty \hat{e}_1 = \begin{pmatrix} v_\infty \\ 0 \\ 0 \end{pmatrix}_{123} \end{cases} \tag{8.324}$$

Without the no-slip boundary condition (associated with viscosity), the incoming uniform flow is uninterrupted by the presence of the plate. No variation of pressure or of velocity is observed. Now that we know the pressure distribution in the free stream away from the wall (i.e., constant pressure, viscosity neglected), we can continue with our solution for the velocity and pressure profiles in viscous flow near a plate with no slip at the wall.

EXAMPLE 8.19 (Flat plate, concluded). *What are the velocity field and the pressure field in a viscous fluid for the flow in which a rapid, uniform flow approaches a flat plate? The flow is steady and the fluid is incompressible. Away from the plate, the flow matches the inviscid (potential) flow solution of this same problem.*

SOLUTION. The flow is shown in Figure 8.37. From our previous treatment of this problem in Example 8.17, we know that the flow is two-dimensional. For steady, two-dimensional flow of an incompressible fluid, we previously showed that the governing equations are:

$$\text{Two-dimensional flow:} \quad \underline{v} = \begin{pmatrix} v_1 \\ v_2 \\ 0 \end{pmatrix}_{123} \tag{8.325}$$

$$\text{Continuity equation:} \quad \boxed{\nabla \cdot v = \frac{\partial v_1}{\partial x_1} + \frac{\partial v_2}{\partial x_2} = 0} \tag{8.326}$$

Navier-Stokes equation
(steady, two-dimensional, $\quad \rho \underline{v} \cdot \nabla \underline{v} = -\nabla \mathcal{P} + \mu \nabla^2 \underline{v} \tag{8.327}$
dynamic pressure):

1-component Navier-Stokes:

$$\boxed{\rho \left(v_1 \frac{\partial v_1}{\partial x_1} + v_2 \frac{\partial v_1}{\partial x_2} \right) = -\frac{\partial \mathcal{P}}{\partial x_1} + \mu \left(\frac{\partial^2 v_1}{\partial x_1^2} + \frac{\partial^2 v_1}{\partial x_2^2} \right)} \tag{8.328}$$

2-component Navier-Stokes:

$$\boxed{\rho \left(v_1 \frac{\partial v_2}{\partial x_1} + v_2 \frac{\partial v_2}{\partial x_2} \right) = -\frac{\partial \mathcal{P}}{\partial x_2} + \mu \left(\frac{\partial^2 v_2}{\partial x_1^2} + \frac{\partial^2 v_2}{\partial x_2^2} \right)} \tag{8.329}$$

Our task in this problem is to solve for v_1, v_2, and $\mathcal{P}$ from the mass- and momentum-balance equations (i.e., Equations 8.326, 8.328, and 8.329). The velocity boundary conditions are no-slip and no-penetration at the surface of the plate and matching the velocity of the free stream at the edge of the boundary layer. The boundary condition on pressure is to match the pressure at the boundary layer with the flow-direction pressure profile from the potential-flow version of this same problem. We showed in Example 8.18 that the pressure and velocity fields are uniform in the potential-flow version of the flow-past-a-plate problem.

The problem remains formidable. Our strategy is to simplify it by looking closely at each term in the three equations that we are solving for those terms that can be neglected safely. If we can eliminate some terms, the mathematical problem may become simple enough to complete.

We first consider the pressure terms. We know that at the top of the boundary layer, the pressure must become a constant. This condition arises from the need to match the pressure distribution in the outer flow. Does the pressure vary across the boundary layer? This is the equivalent to asking: Is $\partial \mathcal{P}/\partial x_2$ nonzero? The answer to this question is given by the 2-component of the Navier-Stokes equation, Equation 8.329:

$$\frac{\partial \mathcal{P}}{\partial x_2} = -\rho v_1 \frac{\partial v_2}{\partial x_1} - \rho v_2 \frac{\partial v_2}{\partial x_2} + \mu \frac{\partial^2 v_2}{\partial x_1^2} + \mu \frac{\partial^2 v_2}{\partial x_2^2} \tag{8.330}$$

We know that the pressure at the top of the boundary layer is constant (matching the potential-flow situation). If we can assume that all of the velocity gradients in Equation 8.330 are negligible, then $\partial \mathcal{P}/\partial x_2 = 0$, and the pressure must be constant throughout the boundary layer.

We consider each velocity-derivative term in Equation 8.330 separately to see whether it is negligible. We are aided by the continuity equation, which indicates that $\partial v_2 / \partial x_2 = -\partial v_1 / \partial x_1$. The magnitudes of the terms in Equation 8.330 are estimated here:

$$v_1 \frac{\partial v_2}{\partial x_1} = (\text{large})(\text{small}) \overset{?}{=} 0 \tag{8.331}$$

$$v_2 \frac{\partial v_2}{\partial x_2} = v_2 \left(-\frac{\partial v_1}{\partial x_1} \right) = (\text{small})(\text{moderate}) \overset{?}{=} 0 \tag{8.332}$$

$$\frac{\partial^2 v_2}{\partial x_1^2} = \frac{\partial}{\partial x_1} \left(\frac{\partial v_2}{\partial x_1} \right) = (\text{small change of [small quantity]}) \overset{?}{=} 0 \tag{8.333}$$

$$\frac{\partial^2 v_2}{\partial x_2^2} = \frac{\partial}{\partial x_2} \left(\frac{\partial v_2}{\partial x_2} \right) = (\text{small change of [moderate quantity]}) \overset{?}{=} 0 \tag{8.334}$$

First, if the variation of pressure in the boundary layer is to be ignored, we must neglect $\partial v_2 / \partial x_1$. This is the flow-direction change of the small component of velocity that moves fluid away from the wall. If we make this assumption, the first and the third conditions are satisfied. We are confident that the second derivative of v_2 in the direction perpendicular to the wall (Equation 8.334) also is small enough to neglect: although $\partial v_2 / \partial x_2$ may be a quantity of finite size, the rate of change of this rate of change is expected to be small.

The assumption that is the biggest stretch is the second one (Equation 8.332), in which we must assume that the product of the transverse velocity and one of the velocity derivatives $\partial v_1 / \partial x_1$ or $\partial v_2 / \partial x_2$ is negligible. The velocity derivatives in Equation 8.332 appear in the continuity equation, and neither is itself negligible, nor is v_2. We assume, however, that the product of these two quantities results in a negligible effect.

When making ordering approximations as discussed here, we necessarily rely on judgment, which may or may not hold up to reality. We proceed, however, in making these judgments and we evaluate their appropriateness based on the outcome of the analysis and comparison to experiments.

With the assumptions discussed previously, the Navier-Stokes equations become:

1-component Navier-Stokes:

$$\rho \left(v_1 \frac{\partial v_1}{\partial x_1} + v_2 \frac{\partial v_1}{\partial x_2} \right) = \frac{\partial \mathcal{P}}{\partial x_1} + \mu \left(\frac{\partial^2 v_1}{\partial x_1^2} + \frac{\partial^2 v_1}{\partial x_2^2} \right) \tag{8.335}$$

2-component Navier-Stokes: $\qquad 0 = -\dfrac{\partial \mathcal{P}}{\partial x_2}$ $\qquad$ (8.336)

We now apply our ordering judgments to the 1-component of the equation of motion as well. Examining each term of the 1-component of the Navier-Stokes, we judge that both terms on the lefthand side are significant because v_1 is the dominant velocity and its principal change is in the x_2 direction. For this latter reason, the second derivative of v_1 with respect to x_2 on the righthand side also should be retained. The second derivative of v_1 with respect to the flow direction, however, is likely to be small because it is a rate of change of the less important rate of change of v_1. Note that because $\mathcal{P}$ is constant at the edge of the boundary layer and, by the 2-component of the Navier-Stokes $\mathcal{P}$, does not vary with x_2, then $\mathcal{P}$ must not vary with x_1 anywhere ($\partial \mathcal{P}/\partial x_1 = 0$). The summary list of all of our assumptions for both components of the Navier-Stokes equation is as follows:

$$
\begin{array}{ll}
\text{Assumptions of} & \dfrac{\partial v_2}{\partial x_1} \approx 0 \\[2mm]
\text{flat-plate boundary-layer} & \dfrac{\partial^2 v_2}{\partial x_2^2} \approx 0 \\[2mm]
\text{analysis:} & \dfrac{\partial^2 v_1}{\partial x_1^2} \approx 0 \\[2mm]
& v_2 \dfrac{\partial v_2}{\partial x_2} = v_2 \left(-\dfrac{\partial v_1}{\partial x_1} \right) \approx 0
\end{array}
\tag{8.337}
$$

The final equations that remain to be solved for the velocity field for viscous flow past a flat plate are:

$$
\underline{v} = \begin{pmatrix} v_1(x_1, x_2) \\ v_2(x_2) \\ 0 \end{pmatrix}_{123}
\tag{8.338}
$$

$$
\mathcal{P} = \text{constant}
\tag{8.339}
$$

$$
\frac{\partial v_1}{\partial x_1} + \frac{d v_2}{d x_2} = 0
\tag{8.340}
$$

$$
\rho \left(v_1 \frac{\partial v_1}{\partial x_1} + v_2 \frac{\partial v_1}{\partial x_2} \right) = \mu \frac{\partial^2 v_1}{\partial x_2^2}
\tag{8.341}
$$

Having made these several assumptions discussed here, we at last arrive at a set of coupled partial differential equations that, although complex, can be solved for the velocity field $\underline{v}$. We turn now to a discussion of that solution.

The analytical solution of viscous flow past a flat plate comes from the insight that uniform flow past a flat plat is similar to the sudden acceleration of a wall in semi-infinite fluid. We discussed the solution to the wall-acceleration problem in Example 7.18. In the acceleration problem for a fixed position of observation, a boundary layer forms near the wall and grows as a function of time. In the current problem, if an observer travels to the right at a speed of v_∞, it appears as if the boundary layer grows with time (Figure 8.40). The solution for velocity in the accelerating-wall problem at time t, therefore, looks like the solution to our problem at location $x_1 = t v_\infty$. In the accelerated-wall problem, we found that we could collapse the system of partial differential equations (PDEs) into a single ordinary differential equation (ODE) by using a combined variable defined as:

$$
\text{Accelerating flow (Example 7.18): combined variable} \equiv \frac{y}{\sqrt{\nu t}}
\tag{8.342}
$$

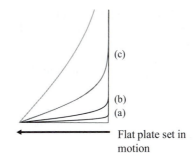

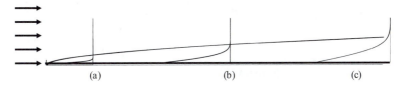

Figure 8.40 Flow near a wall suddenly set in motion (top) and uniform flow encountering a wall (bottom) are similar. The effect of the no-slip boundary condition in each case is to introduce rotational character into the flow field near the wall. The rotational character moves out from the wall in both cases, causing a characteristic boundary layer to develop.

where $\nu = \mu/\rho$ is the kinematic viscosity. Taking $t = x_1/v_\infty$, we therefore guess for the flat-plate problem that we can define a combined variable ζ as:

$$\text{Boundary layer, flat plate: } \zeta \equiv \frac{x_2}{\sqrt{\nu x_1/v_\infty}} = x_2\sqrt{\frac{\rho v_\infty}{\mu x_1}} \qquad (8.343)$$

The details of the final solution for $\underline{v}$ are in the literature [43]. The first step is to define the function $f(\zeta)$ as:

$$\frac{df(\zeta)}{d\zeta} = f' \equiv \frac{v_1}{v_\infty} \qquad (8.344)$$

$$v_1 = v_\infty f' \qquad (8.345)$$

We choose to define the function f for our problem in terms of its first derivative f' because of the way the rest of the solution develops, as we discuss herein. The second step of the solution is to use the continuity equation (Equation 8.340) to solve for v_2 in terms of f, ζ, and x_1. The result is:

$$v_2 = \frac{1}{2}\left(\frac{v_\infty\mu}{\rho x_1}\right)^{\frac{1}{2}}\left[f'\zeta - f\right] \qquad (8.346)$$

Notice that the integrated function $f = \int f'd\zeta$ appears in this expression, justifying our choice to define $f' = v_1/v_\infty$ the way we did. The integration constant for $f = \int f'd\zeta$ has been set to zero because to obtain v_1, we need only f' and not f; therefore, this constant is arbitrary.

Now that we have v_2 in terms of the combined variable ζ (although there is still an $x_1^{-\frac{1}{2}}$ in the expression), we convert the 1-component of the Navier-Stokes equation (Equation 8.341) to be in terms of f and ζ and incorporate the derived expression for v_2 (Equation 8.346). After some algebra, the result is a third-order ODE for $f(\zeta)$ with a remarkably simple although nonlinear structure:

Third-order ODE
for function f related
to velocity component v_1
for flat-plate boundary-layer flow

$$0 = 2f''' + f''f \qquad (8.347)$$

All of this effort is wasted if we cannot convert the boundary conditions on v_1 and v_2 to boundary conditions on f. The boundary conditions are no-slip at the wall ($v_1 = 0$ at the wall), no-penetration at the wall ($v_2 = 0$ at the wall), and the velocity matches the free-stream velocity field at the boundary ($\underline{v} = v_\infty \hat{e}_1$ at $x_2 = \delta(x_1)$). In terms of ζ, these become:

No-slip at the wall: $v_1|_{x_2=0} = 0 \implies f'(0) = 0$ (8.348)

No-penetration at the wall: $v_2|_{x_2=0} = 0 \implies f(0) = 0$ (8.349)

Velocity match at the boundary: $v_1|_{x_2=\delta(x_1)} = v_\infty \implies$?   (8.350)

$v_2|_{x_2=\delta(x_1)} = 0 \implies$?   (8.351)

Leading-edge velocity is

the free-stream value: $v_1|_{x_1=0} = v_\infty \implies f(\infty) = 1$ (8.352)

where $\delta(x_1)$ is the thickness of the boundary layer as a function of position.

The thickness of the boundary layer is unknown; therefore, it is awkward to apply boundary conditions that refer to the thickness of the boundary layer δ. We can eliminate reference to δ if we acknowledge that we do not really care where v_1 goes back to v_∞, only that it does reach v_∞ away from the wall. Thus, we can replace the δ boundary conditions with a velocity match in the far distance at $x_2 = \infty$. If we rewrite the boundary conditions involving the boundary-layer thickness δ as discussed here, then because the similarity variable ζ has x_2 in the numerator and x_1 in the denominator, the two uncertain boundary conditions become identical to the last boundary condition—that is, the condition at the leading edge of the plate.

Velocity match at the boundary: $v_1|_{x_2=\infty} = v_\infty \implies f(\infty) = 1$ (8.353)

$v_2|_{x_2=\infty} = 0 \implies f(\infty) = 1$ (8.354)

Leading-edge velocity

is the free-stream value: $v_1|_{x_1=0} = v_\infty \implies f(\infty) = 1$ (8.355)

The particular similarity transformation discussed for this problem, $f(\zeta)$, can satisfy all of the boundary conditions while not overconstraining the third-order ODE for f. It is this serendipitous circumstance that allows us to solve the flat-plate boundary-layer problem with the approach described here.

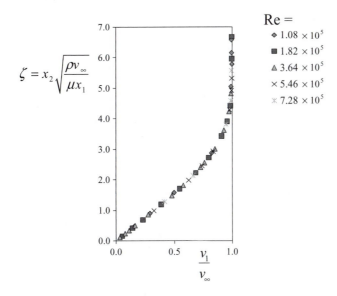

$$\zeta = x_2 \sqrt{\frac{\rho v_\infty}{\mu x_1}}$$

Re =
◆ 1.08×10^5
■ 1.82×10^5
▲ 3.64×10^5
✕ 5.46×10^5
✕ 7.28×10^5

Figure 8.41 The normalized flow-direction velocity in the boundary layer as a function of the combined variable $\zeta = x_2\sqrt{\frac{\rho v_\infty}{\mu x_1}}$; data shown; theory matches well (not shown). Note that for $\zeta = 5$, v_1/v_∞ has reached 1, indicating that the speed of the fluid in the boundary layer has reached the free-stream speed. This value of ζ marks the edge of the boundary layer. Data from J. Nikuradse, Laminare Reibungsschichten an der langsangestromten Platte. Monograph. Zentrale f. wiss. Berichtswessen, Berlin, 1942 as cited by [149].

Equation 8.347 with boundary conditions Equations 8.348, 8.349, and 8.352 can be solved numerically as discussed in Denn and in Problem 43 [43]. The solution for $f(\zeta)$ was found by Blasius in 1908 and is tabulated in [174]. The boundary-layer velocity solution $v_1/v_\infty = f'$ fits the curve-fitting function given here (maximum error $= 0.5$ percent):

Velocity profile for laminar flow past a flat plate (fit to numerical solution):

$$\frac{v_1}{v_\infty} = f'(\zeta) = 1 - (0.5434)\log\left[1 + (68.3)10^{-0.6247\zeta}\right]$$

(8.356)

Recall that $\zeta \equiv x_2\sqrt{\rho v_\infty/(\mu x_1)}$. Experiments by Nikuradse [149] and others confirm that the solution obtained here matches the actual shape of the laminar boundary layer in flow past a flat plate (Figure 8.41). This correspondence validates the ordering assumptions we made in simplifying the governing equations (see Equation 8.337).

The definition of the combined variable ζ allowed the boundary conditions to be expressed in terms of conditions at infinite distance; because of this, we did not have to assume a boundary-layer thickness to arrive at the final solution for the velocity profile. We now can calculate the boundary-layer thickness as a function of distance x_1 from the leading edge by consulting the solution for the velocity field in Figure 8.41. We see that the speed of the fluid at the surface ($\zeta = 0$) is zero due to the no-slip boundary condition at the wall. As we move to locations

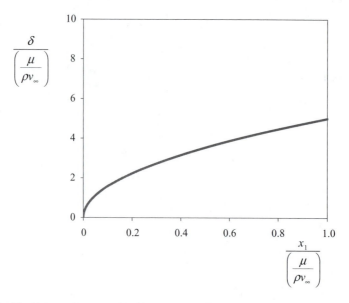

Figure 8.42   The thickness of a laminar boundary layer on a flat plate increases with the square root of the distance from the leading edge of the plate. Note that the dashed line denoting the boundary-layer edge is not a streamline but rather the locus of points where the velocity reaches the free-stream velocity.

away from the wall, we find that the v_1-speed of the fluid increases, eventually reaching the speed of the free stream ($v_1/v_\infty \approx 1$). We define the thickness of the boundary layer to be the x_2-location where the speed of the fluid in the boundary layer reaches $0.99v_\infty$; from Figure 8.41 or Equation 8.356, this happens at $\zeta \approx 5$. From the definition of ζ, we now calculate the boundary-layer thickness δ as a function of distance x_1 from the leading edge:

$$\zeta|_{BL\ edge} = \left(x_2 \sqrt{\frac{\rho v_\infty}{\mu x_1}} \right)\Bigg|_{BL\ edge} \tag{8.357}$$

$$5 = \delta \sqrt{\frac{\rho v_\infty}{\mu x_1}} \tag{8.358}$$

Boundary-layer
thickness:
$$\delta = \left(\sqrt{\frac{25\mu}{\rho v_\infty}} \right) \sqrt{x_1} \tag{8.359}$$

Note that as the free-stream velocity v_∞ increases, the boundary-layer thickness δ decreases. As expected, viscosity works to thicken the boundary layer. The boundary layer does not plateau to a constant thickness far from the leading edge (Figure 8.42); rather, boundary-layer thickness increases without bound as we look at the flow farther downstream from the leading edge.

The fluid force on the plate is calculated in the usual way (see Equation 8.6). For our flat plate, the unit normal to the surface in question is $\hat{n} = \hat{e}_2$; the surface in contact with the fluid is a rectangle with area LW located at coordinate position $x_2 = 0$:

$$\begin{matrix} \text{Total} \\ \text{molecular} \\ \text{fluid force} \\ \text{on a} \\ \text{surface } \mathcal{S}: \end{matrix} \qquad \underline{\mathcal{F}} = \iint_{\mathcal{S}} [\hat{n} \cdot \underline{\tilde{\Pi}}]_{\text{at surface}} \; dS \qquad (8.360)$$

$$= \int_0^W \int_0^L [\hat{e}_2 \cdot \underline{\tilde{\Pi}}]_{x_2=0} \; dx_1 dx_3 \qquad (8.361)$$

$$= \int_0^W \int_0^L (0\ 1\ 0)_{123} \cdot \begin{pmatrix} \tilde{\Pi}_{11} & \tilde{\Pi}_{12} & \tilde{\Pi}_{13} \\ \tilde{\Pi}_{21} & \tilde{\Pi}_{22} & \tilde{\Pi}_{23} \\ \tilde{\Pi}_{31} & \tilde{\Pi}_{32} & \tilde{\Pi}_{33} \end{pmatrix}_{123}\Bigg|_{x_2=0} dx_1 dx_3$$

$$(8.362)$$

$$= \int_0^W \int_0^L \begin{pmatrix} \tilde{\tau}_{21}|_{x_2=0} \\ (\tilde{\tau}_{22} - \mathcal{P})|_{x_2=0} \\ \tilde{\tau}_{23}|_{x_2=0} \end{pmatrix}_{123} dx_1 dx_3 \qquad (8.363)$$

The drag on the wall is equal to $\mathcal{F}_1$, the flow-direction (i.e., 1-direction) force on the plate. The 1-component of Equation 8.363 is the integration of $\tilde{\tau}_{21}$ at the wall, and we can calculate $\tilde{\tau}_{21}$ from the solution for the velocity profile $v_1(x_1, x_2)$, Equation 8.356 (see Figure 8.41):[9]

$$\tilde{\tau}_{21}|_{x_2=0} = \tilde{\tau}_w = \mu \left.\frac{\partial v_1}{\partial x_2}\right|_{x_2=0} \qquad (8.364)$$

$$= \left[\mu v_\infty f'' \frac{d\zeta}{dx_2}\right]\Bigg|_{x_2=0} \qquad (8.365)$$

$$\begin{matrix} \text{Wall shear stress} \\ \text{on flat plate} \\ \text{as a function of location } x_1 \\ \text{(laminar boundary layer):} \end{matrix} \qquad \boxed{\tilde{\tau}_w = 0.332 \mu v_\infty \left(\frac{\rho v_\infty}{\mu x_1}\right)^{\frac{1}{2}}} \qquad (8.366)$$

[9] Because of the limited accuracy of the curve-fit used to obtain Equation 8.356, if that equation is used to calculate the drag, we obtain a coefficient of 0.334 for Equation 8.366. Numerically integrating the full numerical solution for $f(\zeta)$ yields the most accurate calculation of this coefficient, a value of 0.332057 [174].

To calculate the drag, we integrate the 1-component of the fluid-force expression, Equation 8.363:

$$\mathcal{F}_{\text{drag}} = \mathcal{F}_1 = \int_0^W \int_0^L \tilde{\tau}_{21}|_{x_2=0} \, dx_1 dx_3 = \int_0^W \int_0^L \tilde{\tau}_w \, dx_1 dx_3 \quad (8.367)$$

$$= W \int_0^L 0.332 \mu v_\infty \left(\frac{\rho v_\infty}{\mu x_1} \right)^{\frac{1}{2}} dx_1 \quad (8.368)$$

$$\mathcal{F}_{\text{drag}} = 0.664 W v_\infty (\mu \rho v_\infty L)^{\frac{1}{2}} \quad (8.369)$$

In terms of drag coefficient, Equation 8.369 becomes:

$$C_D = \frac{\mathcal{F}_{\text{drag}}}{\frac{1}{2}\rho v_\infty^2 WL} \quad (8.370)$$

Drag in laminar flow past a flat plate:

$$\boxed{C_D = 1.328 \sqrt{\frac{\mu}{\rho v_\infty L}}} \quad (8.371)$$

Finally, it is possible to define a Reynolds number for the flow over a flat plate. There is no obvious characteristic lengthscale in this flow on which to base a Reynolds number, but it is the custom in boundary-layer discussions to define a Reynolds number based on the coordinate variable x_1, the distance from the leading edge:

$$\text{Re}_{x_1} \equiv \frac{\rho v_\infty x_1}{\mu} \quad (8.372)$$

$$\frac{\text{Re}_{x_1}}{x_1^2} = \frac{\rho v_\infty}{\mu x_1} \quad (8.373)$$

$$\frac{\sqrt{\text{Re}_{x_1}}}{x_1} = \sqrt{\frac{\rho v_\infty}{\mu x_1}} = \frac{\zeta}{x_2} \quad (8.374)$$

In terms of Re_{x_1}, the thickness of the boundary layer (see Equation 8.359) may be written as:

Laminar boundary-layer thickness:

$$\boxed{\delta = \frac{5x_1}{\sqrt{\text{Re}_{x_1}}}} \quad (8.375)$$

The drag coefficient for flow past a flat plate likewise can be expressed in terms of the Reynolds number as:

Drag in laminar flow past a flat plate:

$$\boxed{C_D = \frac{1.328}{\sqrt{\text{Re}_L}}} \quad (8.376)$$

where $\text{Re}_L = \text{Re}_{x_1}|_{x_1=L}$ is the Reynolds number based on the length of the plate.

The predicted velocity profile for flow past a flat plate (see Figure 8.41) has been verified experimentally [149]. Experimental agreement is obtained for the region beginning at the leading edge; however, at a location far downstream, the boundary layer is unstable and becomes turbulent. This instability occurs at values of Re_{x_1} between 2×10^5 and 6×10^5 [43]. Turbulent boundary layers are discussed in Section 8.2.2.

We can use the results of the flat-plate analysis to calculate the thickness of laminar boundary layers (see Example 8.20). If the boundary layer is found to be sufficiently thin, we can use the flat-plate solution as a stand-in for the true curved surfaces of flow around a sphere. This may allow us to proceed further on the sphere problem that we began in Example 8.16.

EXAMPLE 8.20. *What is the thickness of the boundary layer on the leading side of a baseball thrown at 90 mph? Assume that the ball is completely smooth.*

SOLUTION. If we assume that the boundary layer is very thin, we can model the flow around a baseball as flow over a flat plate; if the boundary layer is laminar, we can calculate δ from Equation 8.375. By using the flat-plate results, we are not considering the pressure effects in this flow; we suspend this concern in favor of obtaining a first answer to our question.

To apply the result of the flat-plate analysis, we must know the total distance that the fluid travels in the boundary layer; that is, we need the equivalent length of the plate. The distance that a boundary layer on a ball has traveled from the leading edge is simply the arc length on the surface of the ball up to an angle of $90° = \frac{\pi}{2}$. Beyond this angle, we know from flow observation that the flow no longer has a boundary-layer character. Thus, $x_{1,max} = R\Theta_{max} = R\left(\frac{\pi}{2}\right)$, where R is the radius of the ball (the diameter of a baseball is about 7.46 cm). Knowing this length, we now calculate the boundary-layer thickness from Equation 8.375:

$$\mathrm{Re}_{x_1} = \frac{\rho v_\infty x_1}{\mu} = \frac{v_\infty x_1}{\mu/\rho} \tag{8.377}$$

$$= \frac{\left(\dfrac{90 \text{ miles}}{\text{hr}}\right)\left(\dfrac{1{,}609.344 \text{ m}}{\text{mile}}\right)\left(\dfrac{\text{hr}}{3{,}600 \text{ s}}\right)\left(\dfrac{0.0746 \text{ m}}{2}\right)\left(\dfrac{\pi}{2}\right)}{\dfrac{1.412 \times 10^{-5} \text{ m}^2}{\text{s}}} \tag{8.378}$$

$$= 1.7 \times 10^5 \tag{8.379}$$

This Reynolds number is just within the laminar boundary-layer limit. The boundary-layer thickness is therefore:

$$\delta = \frac{5x_1}{\sqrt{\mathrm{Re}_{x_1}}} \tag{8.380}$$

$$= \frac{(5)\left(\frac{0.0746 \ m}{2}\right)\left(\frac{\pi}{2}\right)}{\sqrt{1.7 \times 10^5}} = 0.00071 \text{ m} \tag{8.381}$$

$$\boxed{\delta = 0.7 \text{ mm}} \tag{8.382}$$

The boundary layer is indeed quite thin; we are justified in neglecting curvature for such a thin boundary layer.

As shown, the region near the wall where viscosity is important is thin; thus, we can justify neglecting the effect of curvature when we return to the flow around a sphere. In subsequent sections, we look at turbulent boundary layers and the effect of object shape on the stability of boundary layers. At the end of Section 8.2, we return to the noncreeping flow around a sphere and take full account of pressure in a more complete solution to the boundary-layer flow around a sphere (see Example 8.23).

8.2.2 Turbulent boundary layers

The analysis of boundary layers discussed thus far has been successful. Because the analysis is confined to a thin region near a surface, we choose to use the problem of flow past a flat plate to obtain a solution for the velocity profile in the boundary layer (see Equation 8.356). Experiments performed on flows past flat plates confirm that the calculated velocity profile is observed, at least near the leading edge of the plate [149]. As we observe the flow downstream from the leading edge, however, the velocity eventually becomes unstable and the boundary layer changes from the calculated laminar-flow solution to a new, turbulent boundary layer.

The transition to turbulence in the boundary layer occurs at Reynolds numbers in the range of $2 \times 10^5 < \mathrm{Re}_{x_1} < 6 \times 10^5$; the value of the Reynolds number at the transition depends on the smoothness of the surface and the uniformity of the upstream flow. Turbulent boundary layers are thicker than laminar boundary layers and they do not grow with $x_1^{0.5}$ like laminar boundary layers; rather, they grow more steeply as $x_1^{0.8}$ [147]:

$$\begin{array}{c} \text{Turbulent} \\ \text{boundary-layer growth} \\ \text{with distance along the plate:} \end{array} \qquad \delta = 0.37 x_1^{0.8} \left(\frac{\rho v_\infty}{\mu} \right)^{-\frac{1}{5}} \qquad (8.383)$$

In terms of Reynolds number Re_{x_1}, this is:

$$\begin{array}{c} \text{Turbulent} \\ \text{boundary-layer growth} \\ \text{with Reynolds number} \\ \text{(due to Prandtl [147]):} \end{array} \qquad \boxed{\delta = \frac{0.37 x_1}{\mathrm{Re}_{x_1}^{\frac{1}{5}}}} \qquad (8.384)$$

White [174] gives a slightly different correlation that is more accurate:

$$\begin{array}{c} \text{Turbulent} \\ \text{boundary-layer growth} \\ \text{with Reynolds number} \end{array} \qquad \boxed{\delta = \frac{0.16 x_1}{\mathrm{Re}_{x_1}^{\frac{1}{7}}}} \qquad (8.385)$$

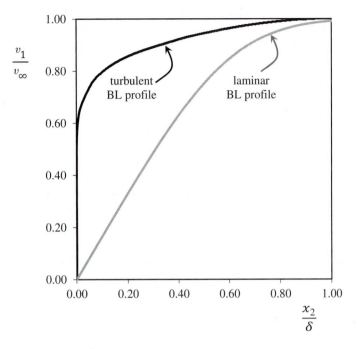

Figure 8.43 The velocity profile in laminar boundary layers can be obtained from the solution to Blasius's equation (see Equation 8.347). For turbulent boundary layers, the average velocity profile in the boundary layer may be obtained from experimental observations, and the shape of the turbulent profile is approximately $v_1/v_\infty = (x_2/\delta)^{\frac{1}{7}}$ [174].

The two types of boundary layers also differ in the distribution of velocity within the boundary layer. Because of the nature of turbulent flow, the velocity in a turbulent boundary layer fluctuates in three dimensions. Thus, to compare the turbulent velocity profile with the velocity profile in laminar flow, we consider a time-averaged, 1-direction turbulent velocity profile (Figure 8.43). Compared to laminar boundary layers, the averaged velocity profile in a turbulent boundary layer has a much higher velocity gradient (slope) close to the surface than a laminar boundary layer (see Figure 8.43). The time-averaged velocity $\overline{v_1}$ of a turbulent boundary layer is found to follow a power-law shape with an exponent of 1/7.

$$\begin{array}{c}\text{Time-averaged} \\ \text{turbulent velocity profile in} \\ \text{flow past a flat plate} \\ \text{(experimental result):}\end{array} \qquad \frac{\overline{v_1}}{v_\infty} = \left(\frac{x_2}{\delta}\right)^{\frac{1}{7}} \qquad (8.386)$$

We cannot solve the governing equations for turbulent flow past a flat plate, but we know from dimensional analysis that the drag coefficient is a function only of the Reynolds number. Therefore, as in pipe flow and in flow past a sphere, we can use measurements on actual flat plates to determine the empirical correlations for drag on a flat plate in turbulent flow. The data for the flat-plate drag coefficient are shown in Figure 8.44. The experimental $C_D(\text{Re})$ correlations for various flow

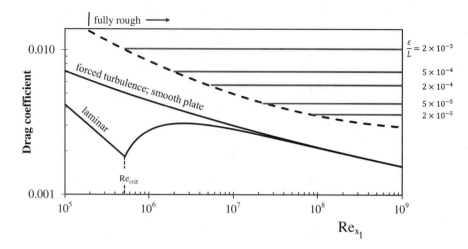

Figure 8.44 Drag coefficient for flow past a flat plate as a function of plate Reynolds number [63, 176]. Laminar boundary-layer flow is unstable above $\text{Re}_{x_1} = 5 \times 10^5$, and a turbulent boundary larger forms. For rough plates C_D is much larger than for smooth plates.

regimes are given here [147, 183]:

$$\text{Laminar flow:} \quad C_D = \frac{1.328}{\text{Re}_L^{0.5}} \tag{8.387}$$

$$\text{Transitional flow, } \text{Re}_L < 5 \times 10^5: \quad C_D = \frac{0.455}{(\log \text{Re}_L)^{2.58}} - \frac{1,700}{\text{Re}_L} \tag{8.388}$$

$$\text{Turbulent, smooth plate:} \quad C_D = \frac{0.455}{(\log \text{Re}_L)^{2.58}} \tag{8.389}$$

$$\text{Turbulent, surface roughness } \epsilon: \quad C_D = \left[1.89 - 1.62 \log \left(\frac{\epsilon}{L} \right) \right]^{-2.5} \tag{8.390}$$

The result for the laminar regime was derived in the previous section. For turbulent boundary layers on smooth surfaces, the drag coefficient is roughly double the laminar drag coefficient. For rough plates, the drag increases significantly with surface roughness, but the drag coefficient for rough plates is independent of Reynolds number (see Figure 8.44).

Because the drag on a flat plate is higher when the boundary layer is turbulent than when it is laminar, we always can reduce the force on the plate or the energy of pumping fluid past a flat plate by designing the flow to produce laminar boundary layers. To do this, we must keep the Reynolds number below the critical value for transition to turbulence. For objects other than flat plates, however, it is not always true that producing laminar boundary layers results in less drag on the object. This counterintuitive result is due to the outer-flow pressure distribution present for objects other than flat plates. The pressure distribution in the outer flow causes flow separation—a drastic change in the flow pattern in flow past nonflat objects. Flow past blunt objects and flow separation are discussed in Section 8.2.3.

The following example demonstrates an application of the flat-plate analysis.

EXAMPLE 8.21. *The flow over an airplane wing is modeled, in a first attempt, as the flow over a flat plate of length equal to the wing's chord length. For the*

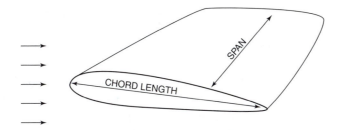

An airplane wing is a type of airfoil. The shape of an airfoil minimizes drag by gradually tapering off after the thickest portion. This type of shape change promotes attachment of the boundary layer, thereby reducing the dominant form of drag—the pressure drag due to boundary-layer detatchment. The dimensions of an airfoil are chord length (related to the length in the flow direction) and span (width in the neutral direction).

airplane wing in Figure 8.45, what is the maximum thickness of the boundary layer? The wing is very wide and chord length is 2.00 m. The airplane is moving at a cruising speed of 9.00×10^2 km/hour.

SOLUTION. The thickness of the boundary layer in flow over a flat plate is a function of the Reynolds number and the distance traveled along the flat plate (see Equations 8.375 and 8.384 for laminar and turbulent flow, respectively). We have all of the information needed to calculate the boundary-layer thickness δ for the situation under consideration. We begin by calculating the Reynolds number:

$$\mathrm{Re}_{x_1} = \frac{\rho v_\infty x_1}{\mu} = \frac{v_\infty x_1}{\mu/\rho} \tag{8.391}$$

$$= \frac{\left(\dfrac{900\ \mathrm{km}}{\mathrm{hr}}\right)(2\ \mathrm{m})\left(\dfrac{10^3\ \mathrm{m}}{\mathrm{km}}\right)\left(\dfrac{\mathrm{hr}}{3{,}600\ \mathrm{s}}\right)}{\dfrac{1.412 \times 10^{-5}\ \mathrm{m^2}}{\mathrm{s}}} \tag{8.392}$$

$$= 3.541 \times 10^7 \tag{8.393}$$

This Reynolds number is above the laminar boundary-layer limit. We therefore calculate δ from the equation that gives boundary-layer thickness for turbulent boundary layers, Equation 8.384:

$$\delta = 0.37 \left(\frac{x_1}{\mathrm{Re}_{x_1}^{0.2}}\right) \tag{8.394}$$

$$= \frac{(0.37)(2\ \mathrm{m})}{\left(3.541 \times 10^7\right)^{0.2}} = 0.023\ \mathrm{m} \tag{8.395}$$

$$\boxed{\delta = 2.3\ \mathrm{cm}} \tag{8.396}$$

At these high Reynolds numbers, the boundary layer is less than a few centimeters thick (about an inch); thus, we are justified in ignoring curvature effects in the boundary-layer calculations.

The boundary-layer problems discussed thus far are for flow over a flat plate; the pressure is constant throughout this flow. When boundary layers form on nonflat objects, a pressure distribution develops throughout the flow field, and this has a profound effect on the velocity field and the drag produced by the object. Flow past nonflat objects is discussed in the next section. The pressure distribution developed in the flow past nonflat objects causes boundary-layer separation, which occurs when pressure *rises* in the direction of flow. In addition, because rising pressure affects laminar and turbulent boundary layers differently, under some circumstances flow separation leads to counterintuitive results, such as lower drag in turbulent flow than in laminar flow.

This section concludes with an example in which we set up the study of flow past nonflat objects by carrying out a formal dimensional analysis on a boundary layer with a nonconstant pressure distribution.

EXAMPLE 8.22. *What are the boundary-layer equations for flows in which the pressure field is not constant (i.e., objects other than a flat plate)?*

SOLUTION. In the discussion of boundary-layer flow past a flat plate, we were dealing with a simple flow and could make the ordering judgments needed to simplify the problem. When the obstacle is not a flat plate, the pressure field will not be uniform; thus, to determine the governing equations, we must proceed more formally. The discussion presented here follows that of Denn [43].

We consider the case of an obstacle of arbitrary shape. Based on our experience with flat plates, we assume that the boundary layer on the obstacle is thin. Thus, we can model flow around the nonflat obstacle as equivalent to flow over a flat plate with a nonuniform pressure distribution imposed at the location where the boundary layer meets the outer flow. The imposed pressure distribution is obtained from the potential-flow solution for flow around the object.

We showed previously that for steady, two-dimensional flow of an incompressible fluid past a flat plate, the governing equations are given by Equations 8.312, 8.315, and 8.316, repeated here:

Continuity equation:
$$\frac{\partial v_1}{\partial x_1} + \frac{\partial v_2}{\partial x_2} = 0$$

(8.397)

1-component Navier-Stokes:
$$\rho\left(v_1\frac{\partial v_1}{\partial x_1} + v_2\frac{\partial v_1}{\partial x_2}\right) = -\frac{\partial \mathcal{P}}{\partial x_1} + \mu\left(\frac{\partial^2 v_1}{\partial x_1^2} + \frac{\partial^2 v_1}{\partial x_2^2}\right)$$

(8.398)

2-component Navier-Stokes:
$$\rho\left(v_1\frac{\partial v_2}{\partial x_1} + v_2\frac{\partial v_2}{\partial x_2}\right) = -\frac{\partial \mathcal{P}}{\partial x_2} + \mu\left(\frac{\partial^2 v_2}{\partial x_1^2} + \frac{\partial^2 v_2}{\partial x_2^2}\right)$$

(8.399)

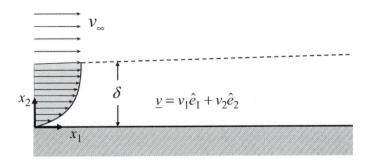

Figure 8.46 Characteristic values should be chosen so that the nondimensional derivatives are $O(1)$. Thus, for the 1-velocity, for $\partial v_1^* / \partial x^*$ to be $O(1)$, we choose U to be the maximum value of v_1. Then, we choose the characteristic length to be the distance over which v_1 undergoes a change of magnitude U.

Thus far we make no assumptions about the pressure distribution. To see which terms can be neglected, we use dimensional analysis.

We begin with the continuity equation. The first step in dimensional analysis is to choose the characteristic values. We choose these values so that the resulting nondimensional derivatives in Equation 8.397 will be scaled to $O(1)$ (order of magnitude 1) because both derivatives are important. Thus, to scale the x_1-component of velocity, we choose the maximum 1-direction velocity in the boundary layer, which is the free-stream velocity $U = v_\infty$ (Figure 8.46). The distance over which the 1-velocity changes in the flow direction is unknown; we designate this distance L. The distance over which this velocity changes in the x_2-direction is the height of the boundary layer δ. We choose, therefore, two characteristic lengths—δ for x_2 and L for x_1—because the velocity varies differently in the two directions.

The two velocity components of flow near a surface are very different in size; therefore, they merit separate characteristic values. For the x_1-direction velocity, we choose a characteristic velocity $U = v_\infty$ as discussed previously; for the x_2-direction velocity, we designate V as the characteristic velocity. We refer to the governing equations for guidance on how the various lengthscales and velocity scales interrelate. The characteristic pressure is designated P.

Beginning with the continuity equation (see Equation 8.397), we nondimensionalize as usual:

$$v_1^* \equiv \frac{v_1}{U} \tag{8.400}$$

$$v_2^* \equiv \frac{v_2}{V} \tag{8.401}$$

$$\frac{\partial v_1}{\partial x_1} + \frac{\partial v_2}{\partial x_2} = 0 \tag{8.402}$$

$$\frac{U}{L}\frac{\partial v_1^*}{\partial x_1^*} + \frac{V}{\delta}\frac{\partial v_2^*}{\partial x_2^*} = 0 \tag{8.403}$$

$$\left(\frac{U\delta}{VL}\right)\frac{\partial v_1^*}{\partial x_1^*} + \frac{\partial v_2^*}{\partial x_2^*} = 0 \tag{8.404}$$

A dimensionless group appeared in the continuity equation: $U\delta/VL$. If this dimensionless group is small, then v_2 is not a function of x_2. Because v_2 is zero at the boundary, this means that v_2 is zero everywhere. This is not the result anticipated because boundary layers are known to grow in the flow direction. If $U\delta/VL$ is large, then v_1 is not a function of x_1—the position variable in the flow direction. Again, we believe that this is incorrect, as discussed previously. Our conclusion, therefore, is that $U\delta/VL$ must be of order 1 ($O(1)$), and we can use this fact to define V with respect to the other variables:

$$\text{Choose:} \quad \frac{U\delta}{VL} = 1 \tag{8.405}$$

$$\begin{array}{cc} \text{Scaled continuity equation} \\ \text{in boundary layer:} \end{array} \quad \boxed{\frac{\partial v_1^*}{\partial x_1^*} + \frac{\partial v_2^*}{\partial x_2^*} = 0} \tag{8.406}$$

$$\begin{array}{cc} \text{Characteristic} \\ \text{velocity for } v_2: \end{array} \quad \boxed{V = \frac{U\delta}{L}} \tag{8.407}$$

The result for V is a characteristic velocity in the x_2-direction that is proportional to the characteristic velocity in the main (x_1-) direction but much smaller ($V = U\delta/L$).

The 1-component of the Navier-Stokes (see Equation 8.399) likewise can be nondimensionalized:

$$\rho \left(v_1 \frac{\partial v_1}{\partial x_1} + v_2 \frac{\partial v_1}{\partial x_2} \right) = -\frac{\partial \mathcal{P}}{\partial x_1} + \mu \left(\frac{\partial^2 v_1}{\partial x_1^2} + \frac{\partial^2 v_1}{\partial x_2^2} \right) \tag{8.408}$$

$$\frac{\rho U^2}{L} v_1^* \frac{\partial v_1^*}{\partial x_1^*} + \frac{\rho V U}{\delta} v_2^* \frac{\partial v_1^*}{\partial x_2^*} = \frac{P}{L} \left(-\frac{\partial \mathcal{P}^*}{\partial x_1^*} \right) + \frac{\mu U}{L^2} \frac{\partial^2 v_1^*}{\partial x_1^{*2}} + \frac{\mu U}{\delta^2} \frac{\partial^2 v_1^*}{\partial x_2^{*2}} \tag{8.409}$$

$$\frac{\rho U^2}{L} \left(v_1^* \frac{\partial v_1^*}{\partial x_1^*} + v_2^* \frac{\partial v_1^*}{\partial x_2^*} \right) = \frac{P}{L} \left(-\frac{\partial \mathcal{P}^*}{\partial x_1^*} \right) + \frac{\mu U}{L^2} \frac{\partial^2 v_1^*}{\partial x_1^{*2}} + \frac{\mu U}{\delta^2} \frac{\partial^2 v_1^*}{\partial x_2^{*2}} \tag{8.410}$$

We know that v_1^* is a strong function of x_2^* and the $\partial^2 v_1^*/\partial x_2^{*2}$ term is therefore a significant term; thus, we divide through by the coefficient of this term, leaving this term with a coefficient of 1:

$$\frac{\rho U \delta^2}{\mu L} \left(v_1^* \frac{\partial v_1^*}{\partial x_1^*} + v_2^* \frac{\partial v_1^*}{\partial x_2^*} \right) = \frac{P\delta^2}{\mu UL} \left(-\frac{\partial \mathcal{P}^*}{\partial x_1^*} \right) + \left(\frac{\delta}{L} \right)^2 \frac{\partial^2 v_1^*}{\partial x_1^{*2}} + \frac{\partial^2 v_1^*}{\partial x_2^{*2}} \tag{8.411}$$

The coefficient of the inertial terms is $\rho U\delta^2/\mu L$. If this quantity is large, the inertial terms dominate; if this quantity is small, the viscous terms dominate. Because we seek a solution in a regime where neither inertial nor viscous terms dominate, we set this coefficient to 1 and use it to define the behavior of δ:

$$\text{Choose:} \quad \frac{\rho U \delta^2}{\mu L} = 1 \tag{8.412}$$

$$
\begin{array}{c}
\text{Characteristic} \\
\text{lengthscale} \\
\text{in } x_2\text{-direction} \\
\text{(boundary-layer height):}
\end{array}
\qquad
\boxed{\delta = \left(\frac{\mu L}{\rho U}\right)^{\frac{1}{2}}}
\tag{8.413}
$$

The boundary-layer thickness δ thus defines the region of the flow in which neither inertial nor viscous forces dominate the flow physics. If we define a Reynolds number based on the principal velocity v_1 and the flow-direction lengthscale L:

$$\mathrm{Re}_L \equiv \frac{\rho U L}{\mu} \tag{8.414}$$

we can write the boundary-layer thickness as:

$$\delta = \frac{L}{(\mathrm{Re}_L)^{\frac{1}{2}}} \tag{8.415}$$

Returning to Equation 8.411, the coefficient of the pressure-gradient term contains δ^2; substituting the scaling of δ (Equation 8.413) into the pressure coefficient in Equation 8.411, we obtain:

$$
\begin{array}{c}
\text{Coefficient of} \\
\text{pressure-gradient term:}
\end{array}
\qquad
\frac{P\delta^2}{\mu U L} = \frac{P}{\rho U^2}
\tag{8.416}
$$

Following our usual practice when we want to retain a term, we set this coefficient equal to 1 and define $P = \rho U^2$, a characteristic pressure based on the inertia of the free stream. The 1-component of the Navier-Stokes now becomes:

$$v_1^* \frac{\partial v_1^*}{\partial x_1^*} + v_2^* \frac{\partial v_1^*}{\partial x_2^*} = -\frac{\partial \mathcal{P}^*}{\partial x_1^*} + \left(\frac{\delta}{L}\right)^2 \frac{\partial^2 v_1^*}{\partial x_1^{*2}} + \frac{\partial^2 v_1^*}{\partial x_2^{*2}} \tag{8.417}$$

The boundary-layer thickness δ is known to be very small compared to any macroscopic lengthscale in the flow direction; in addition, the nondimensional second derivative $\partial^2 v_1^*/\partial x_1^{*2}$ was scaled to be $O(1)$—thus, the $\partial^2 v_1^*/\partial x_1^{*2}$ term may be neglected. Omitting this term yields the properly scaled 1-component of the Navier-Stokes equation for boundary layers:

$$
\begin{array}{c}
\text{Scaled Navier-Stokes} \\
\text{in boundary layer} \\
\text{(1-component):}
\end{array}
\qquad
\boxed{v_1^* \frac{\partial v_1^*}{\partial x_1^*} + v_2^* \frac{\partial v_1^*}{\partial x_2^*} = -\frac{\partial \mathcal{P}^*}{\partial x_1^*} + \frac{\partial^2 v_1^*}{\partial x_2^{*2}}}
\tag{8.418}
$$

This result is consistent with the result obtained in Example 8.19 for a flat-plate. Compared to the flat-plate result, however, in this general solution the pressure gradient is retained.

We see from these calculations that from basic knowledge about how the flow in the boundary layer behaves, we can establish the correct scaling factors for the flow. Equation 8.404 states that v_2 must vary with x_2 and v_1 must vary with x_1; therefore, scale factors V and δ must be related as given in Equation 8.407. Equation 8.411 states that neither inertia nor viscosity dominate the momentum balance in the boundary layer; thus, δ and Re_L are related as given in Equation 8.415. The scaled x_1-component of the Navier-Stokes equation in Equation 8.418 comes together with no additional assumptions.

We also can nondimensionalize the x_2-component of the Navier-Stokes equation (see Equation 8.399), using the same characteristic values that we have established for the problem. Recall that $V = U\delta/L$ and $P = \rho U^2$:

$$\rho \left(v_1 \frac{\partial v_2}{\partial x_1} + v_2 \frac{\partial v_2}{\partial x_2} \right) = -\frac{\partial P}{\partial x_2} + \mu \left(\frac{\partial^2 v_2}{\partial x_1^2} + \frac{\partial^2 v_2}{\partial x_2^2} \right) \tag{8.419}$$

$$\frac{\rho U V}{L} v_1^* \frac{\partial v_2^*}{\partial x_1^*} + \frac{\rho V^2}{\delta} v_2^* \frac{\partial v_2^*}{\partial x_2^*} = \frac{P}{\delta} \left(-\frac{\partial P^*}{\partial x_2^*} \right) + \frac{\mu V}{L^2} \frac{\partial^2 v_2^*}{\partial x_1^2} + \frac{\mu V}{\delta^2} \frac{\partial^2 v_2^*}{\partial x_2^{*2}} \tag{8.420}$$

$$\frac{\rho U^2 \delta}{L^2} \left(v_1^* \frac{\partial v_2^*}{\partial x_1^*} + v_2^* \frac{\partial v_2^*}{\partial x_2^*} \right) = \frac{\rho U^2}{\delta} \left(-\frac{\partial P^*}{\partial x_2^*} \right) + \frac{\mu U \delta}{L^3} \frac{\partial^2 v_2^*}{\partial x_1^{*2}} + \frac{\mu U}{\delta L} \frac{\partial^2 v_2^*}{\partial x_2^{*2}} \tag{8.421}$$

Dividing through by the coefficient of the pressure-gradient term and eliminating all viscosities μ in favor of δ ($\mu = \delta^2 \rho U/L$), we obtain:

$$\frac{\delta^2}{L^2} \left(v_1^* \frac{\partial v_2^*}{\partial x_1^*} + v_2^* \frac{\partial v_2^*}{\partial x_2^*} \right) = -\frac{\partial P^*}{\partial x_2^*} + \frac{\delta^4}{L^4} \left(\frac{\partial^2 v_2^*}{\partial x_1^{*2}} \right) + \frac{\delta^2}{L^2} \left(\frac{\partial^2 v_2^*}{\partial x_2^{*2}} \right) \tag{8.422}$$

Solving for nondimensional pressure gradient:

Scaled
Navier-Stokes
in boundary layer
(2-component):

$$\boxed{\frac{\partial P^*}{\partial x_2^*} = \frac{\delta^2}{L^2} \left(\frac{\delta^2}{L^2} \frac{\partial^2 v_2^*}{\partial x_1^{*2}} + \frac{\partial^2 v_2^*}{\partial x_2^{*2}} - v_1^* \frac{\partial v_2^*}{\partial x_1^*} - v_2^* \frac{\partial v_2^*}{\partial x_2^*} \right)}$$

$$\tag{8.423}$$

Equation 8.423 shows an important result. Each of the nondimensional derivative terms was scaled carefully to be of $O(1)$. Every term is multiplied by δ^2/L^2 (or δ^4/L^4), which is a very small number. Thus, we may conclude that the nondimensional pressure gradient in the x_2-direction is zero; that is, there is no variation of dynamic pressure P in the direction normal to the surface:

Scaled Navier-Stokes
in boundary layer
(x-component, all shapes):

$$\boxed{\frac{\partial P^*}{\partial x_2^*} = 0} \tag{8.424}$$

This is the same assumption we made in Example 8.19 for a flat plate, and we find it again here for any obstacle shape. This is a key component of boundary-layer theory. Combining dimensional-analysis results, we obtain the simplified nondimensional equations that govern the flow in the viscous boundary layer.

Governing Equations in Boundary Layer (Arbitrary Shape, Thin Layer)

Scaled continuity equation
in boundary layer:

$$\frac{\partial v_1^*}{\partial x_1^*} + \frac{\partial v_2^*}{\partial x_2^*} = 0$$

(8.425)

Scaled Navier-Stokes
in boundary layer
(1-component):

$$v_1^*\frac{\partial v_1^*}{\partial x_1^*} + v_2^*\frac{\partial v_1^*}{\partial x_2^*} = -\frac{\partial \mathcal{P}^*}{\partial x_1^*} + \frac{\partial^2 v_1^*}{\partial x_2^{*2}}$$

(8.426)

Scaled Navier-Stokes
in boundary layer
(1-component):

$$\frac{\partial \mathcal{P}^*}{\partial x_2^*} = 0$$

(8.427)

The formal dimensional analysis confirms the ordering performed on the flat-plate problem in Example 8.19. For the flow past obstacles other than a flat plate, $\mathcal{P}$ is a function of x_1 outside the boundary layer and the term $\partial \mathcal{P}/\partial x_1$ becomes prominent. We obtain $\partial \mathcal{P}/\partial x_1$ from the potential-flow solution of the flow past an obstacle of interest. We use these governing equations in Example 8.23 in a solution for boundary-layer flow past a sphere.

8.2.3 Flow past blunt objects

In this chapter, we discuss external flows. We began with the skydiver problem, which was modeled as a sphere falling in a viscous liquid. We found that we could solve the flow-past-a-sphere problem in the creeping-flow limit (no inertia), but the terminal speed we calculate for the skydiver assuming creeping flow is wildly incorrect. For further insight on rapid flows past objects, we turned to dimensional analysis and correlations from experimental sphere data, and we arrived at a value of the terminal velocity for the skydiver that was within a factor of 2 of the correct speed. This is excellent agreement considering that we modeled the shape of a skydiver as a sphere, which is a rough approximation.

To further improve our calculations on flow past an obstacle, we moved on to investigate the different flow regimes seen in flow past a sphere. Experiments outside the creeping-flow limit show that with increasing Reynolds number, recirculation appears on the trailing side of the sphere, followed by development of an oscillatory wake. Ultimately, boundary-layer flow and fully turbulent wake are observed behind the sphere. We cannot calculate this behavior with either creeping-flow or potential-flow models.

Following a different approach, called the boundary-layer method, we investigated rapid flows near surfaces and found that we could predict velocity fields in laminar boundary layers on flat plates. Furthermore, through dimensional analysis and experiments, we found that we could correlate flat-plate boundary-layer data outside of the laminar regime. The resulting plot of drag coefficient versus Reynolds number for flat plates provides a complete picture of the drag/flow-rate relationship in that flow. Figure 8.44 is a flat-plate analog to the pipe-flow Moody chart for the Fanning friction factor versus Reynolds number. The pressure is constant in flow past a flat plate.

Flow past a three-dimensional object is more complex than flow past a plate, however, because there is a distribution of pressure in such flows. The boundary-layer method is still applicable in these cases; the only adjustment is that the outer-flow pressure distribution must be included when the boundary-layer equations are solved. The pressure distribution in flow past an obstacle is obtained from the potential-flow solution for the situation of interest. In uniform potential flow around an obstacle, the flow is irrotational (see Problem 53, Equation 8.480, and discussion in Example 8.25), and the pressure distribution may be calculated with the Bernoulli equation. The Bernoulli states that where flow speeds increase, pressure must decrease, and when flow speeds decrease, pressure must rise.

We now have the tools to address the high-Re sphere problem with the boundary-layer method. The pressure distribution for potential flow around a sphere was calculated in Example 8.9 (see Equation 8.239). In Example 8.23, this pressure distribution is imposed on the boundary layer near a sphere to obtain results for the velocity field within the sphere's boundary layer. Example 8.23 is a first step toward calculating the effects of object shape on external flows. In that example, we see striking effects that demonstrate that rising pressure in the flow direction has a profound effect on boundary layers. As mentioned previously, it turns out that positive pressure gradients from the outer flow—which are created by the shape of the object—are the source of much of the remaining unexplained flow complexity in flows around obstacles.

EXAMPLE 8.23 (Sphere, concluded). *Calculate the steady-state velocity field for the flow of an incompressible viscous fluid near the surface of a solid sphere of diameter 2R. The fluid approaches the sphere with a uniform upstream velocity* v_∞. *The geometry is the same as in the creeping-flow and potential-flow calculations (see Figure 8.5), but in this problem, the flow is not slow (i.e. the Reynolds number is finite) and viscosity may not be neglected ($\mu \neq 0$). A known pressure distribution in the flow direction is imposed at the edge of the boundary layer. The imposed pressure distribution is the pressure distribution of potential flow around a sphere (see Equation 8.239).*

SOLUTION. We started this problem in Example 8.16. There, we assumed that density is constant, the ϕ-component of the velocity is zero, and the flow is steady and symmetrical in the ϕ-direction:

$$\underline{v} = \begin{pmatrix} v_r(r, \theta) \\ v_\theta(r, \theta) \\ 0 \end{pmatrix}_{r\theta\phi} \tag{8.428}$$

We also assumed that we could confine our calculations to the boundary layer at the surface, with the pressure from the potential-flow solution imposed as a boundary condition. The spherical geometry is a complication, and we since have learned that because the boundary layer is thin, we can model the flow domain as a flat plane using local Cartesian coordinates xyz (Figure 8.47). For these local

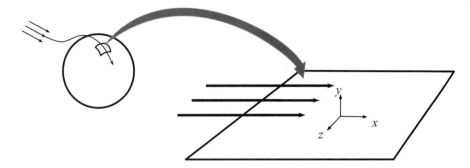

Figure 8.47

For the very thin layer of fluid near the surface of the sphere, the curvature can be neglected and the equations analyzed in Cartesian coordinates. This is analogous to using Cartesian coordinates on Earth's surface rather than spherical coordinates.

coordinates and incorporating our assumptions thus far, the governing equations become:

Continuity equation:
$$\frac{\partial v_x}{\partial x} + \frac{\partial v_y}{\partial y} = 0 \qquad (8.429)$$

Navier-Stokes:
$$\rho \left(\frac{\partial \underline{v}}{\partial t} + \underline{v} \cdot \nabla \underline{v} \right) = -\nabla \mathcal{P} + \mu \nabla^2 \underline{v} \qquad (8.430)$$

$$\rho \underline{v} \cdot \nabla \underline{v} = -\nabla \mathcal{P} + \mu \nabla^2 \underline{v} \qquad (8.431)$$

$$\rho \begin{pmatrix} v_x \dfrac{\partial v_x}{\partial x} + v_y \dfrac{\partial v_x}{\partial y} \\[2ex] v_x \dfrac{\partial v_y}{\partial x} + v_y \dfrac{\partial v_y}{\partial y} \\[2ex] 0 \end{pmatrix}_{xyz} = \begin{pmatrix} -\dfrac{\partial \mathcal{P}}{\partial x} \\[2ex] -\dfrac{\partial \mathcal{P}}{\partial y} \\[2ex] -\dfrac{\partial \mathcal{P}}{\partial z} \end{pmatrix}_{xyz} + \mu \begin{pmatrix} \dfrac{\partial^2 v_x}{\partial x^2} + \dfrac{\partial^2 v_x}{\partial y^2} \\[2ex] \dfrac{\partial^2 v_y}{\partial x^2} + \dfrac{\partial^2 v_y}{\partial y^2} \\[2ex] 0 \end{pmatrix}_{xyz}$$

$$(8.432)$$

Incorporating the usual boundary-layer approximations (see Equation 8.337), all of the velocity gradients in the y-component of the Navier-Stokes equation are neglected, leaving zero pressure gradient in the y-direction ($\partial \mathcal{P}/\partial y$) and the x-component simplifies by the omission of one term, $\partial^2 v_x/\partial x^2$. The system of equations to solve for $v_x(x, y)$ and $v_y(y)$ becomes:

$$\frac{\partial v_x}{\partial x} + \frac{\partial v_y}{\partial y} = 0 \qquad (8.433)$$

$$\rho \left(v_x \frac{\partial v_x}{\partial x} + v_y \frac{\partial v_x}{\partial y} \right) = \xi(x) + \mu \frac{\partial^2 v_x}{\partial y^2} \qquad (8.434)$$

where $\xi(x) = -\frac{\partial \mathcal{P}}{\partial x}$ is a known function obtained from $\mathcal{P}(r, \theta)|_{r=R}$ in the potential-flow solution to flow around a sphere, as discussed now.

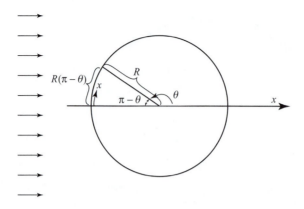

Figure 8.48 We can relate the spherical coordinate system to the near-surface Cartesian system through geometry.

The outer-flow pressure distribution for potential flow around a sphere is given in Equation 8.239. Note that the solution is expressed in the spherical coordinate system used in the potential-flow problem:

$$P(r, \theta) = P_\infty + \frac{1}{2}\rho v_\infty^2 \left[2 \left(\frac{R}{r} \right)^3 \left(1 - \frac{3}{2}\sin^2 \theta \right) - \left(\frac{R}{r} \right)^6 \left(1 - \frac{3}{4}\sin^2 \theta \right) \right]$$

(8.435)

At the surface of the sphere ($r = R$), this becomes:

$$P(r, \theta)|_{r=R} = P_\infty + \frac{1}{2}\rho v_\infty^2 \left[1 - \frac{9}{4}\sin^2 \theta \right]$$

(8.436)

To use this result in our near-surface solution, we translate the spherical coordinate-system variable θ into the near-surface Cartesian coordinates we are using. Figure 8.48 shows that we can write:

$$\text{arc length} = (\text{radius})(\text{included angle})$$

(8.437)

$$x = R(\pi - \theta)$$

(8.438)

$$\sin \left(\frac{x}{R} \right) = \sin(\pi - \theta) = \sin \theta$$

(8.439)

Substituting this into the pressure distribution, we now calculate $\xi(x)$ (Figure 8.49).

$$P(x) = P_\infty + \frac{1}{2}\rho v_\infty^2 \left[1 - \frac{9}{4}\sin^2 \left(\frac{x}{R} \right) \right]$$

(8.440)

$$\xi(x) = -\frac{\partial P}{\partial x}$$

(8.441)

$$\xi(x) = \frac{9\rho v_\infty^2}{4R} \sin \left(\frac{x}{R} \right) \cos \left(\frac{x}{R} \right)$$

(8.442)

The governing equations can be solved using a mathematical software package [94, 180]. The boundary conditions on velocity are the usual ones: no-slip and no-penetration at the sphere surface and matching the free-stream velocity at

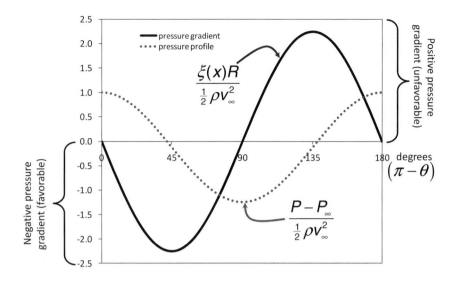

Figure 8.49   The calculated velocity profile for potential flow around a sphere has regions of negative and positive pressure gradient. When the pressure gradient is negative, conditions are favorable to boundary-layer attachment because the pressure gradient helps to push the flow forward. When the pressure gradient is positive, the conditions are unfavorable to boundary-layer attachment.

large distances from the sphere and as the fluid enters the boundary layer:

$$\text{No-slip:} \quad y = 0 \quad v_x = 0 \tag{8.443}$$

$$\text{Match outer flow:} \quad y = \infty \quad v_x = v_\infty \tag{8.444}$$

$$\text{No-penetration:} \quad y = 0 \quad v_y = 0 \tag{8.445}$$

$$\text{Match at entrance:} \quad x = 0 \quad v_x = v_\infty \tag{8.446}$$

The solutions for $v_x(y, \tilde{\theta})$ [147, 149] are plotted in Figure 8.50, where $\tilde{\theta} = (\pi - \theta) = (x/R)$ is the angle measured clockwise around the origin as the fluid proceeds around the sphere (see Figure 8.48). The shape of the velocity profile at the upstream stagnation point ($\tilde{\theta} = 0$, bottom curve) is familiar from our discussion of flow in the boundary layer along a flat plate: The no-slip boundary condition is respected at $y = 0$, and the velocity increases as we move away from the wall (increasing y), reaching the free-stream velocity v_∞ at a value of the dimensionless scaled variable $\tilde{\zeta} = y/R\sqrt{v_\infty R/\nu}$ of about 1.5.

Looking at the solution $v_x(y, \tilde{\theta})$ at various positions $\tilde{\theta}$ on the sphere surface, we see that the shape of the velocity profile changes as the flow proceeds around the sphere. The value of $\tilde{\zeta}$ at which the velocity reaches the free-stream value increases with increasing $\tilde{\theta}$, reaching $\tilde{\zeta} = 3$ at the largest angle computed, $\tilde{\theta} = 109.6$ degrees. Close to the wall, the slope of the velocity profile dv_x/dy decreases steadily as the flow moves around the sphere. The slope begins at a finite value of about 1.6 but, at the last value of $\tilde{\theta}$ shown, the velocity gradient near the wall approaches zero $\left(\frac{dv_x}{dy} \Big|_{\tilde{\theta}=109.6°} = 0.1 \right)$.

The variable pressure gradient imposed from the outer flow is the cause of the position-dependent changes in velocity-profile shape (Figure 8.51). The

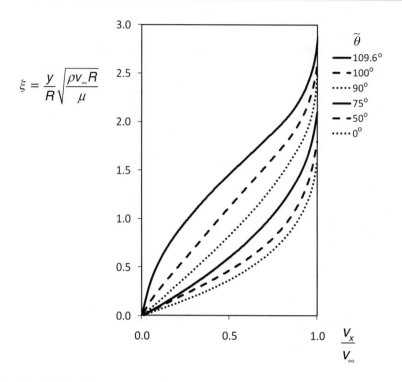

$$\xi = \frac{y}{R}\sqrt{\frac{\rho v_\infty R}{\mu}}$$

Figure 8.50 The solution in the boundary layer for flow around a sphere is reported in Schlichting [148, page 238].

pressure gradient (Equation 8.442) is negative on the leading side of the sphere and therefore drives the flow forward and adds to the inertia of the fluid. As usual, the presence of the wall retards the fluid near the wall, but the imposed high upstream pressure helps to replace some of this lost fluid momentum and pushes the fluid forward.

The magnitude of the favorable pressure gradient falls with increasing $\tilde{\theta}$, however; finally, at $\tilde{\theta} = 90° = \pi/2$, the imposed pressure gradient is zero

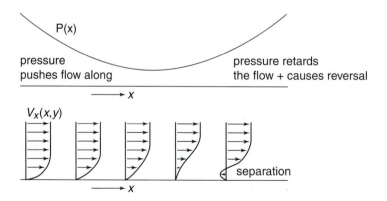

Figure 8.51 Boundary-layer separation is caused by rising pressure in the direction of flow in the boundary layer. In modeling calculations of this flow, the pressure distribution is imposed on the boundary layer by the outer flow (inviscid). The shape of the object determines the pressure profile, which can fall or rise. Falling pressure does not cause any difficulties; rising pressure, however, slows the flow in the boundary layer and can cause flow reversal near the surface.

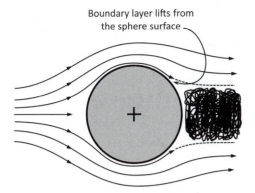

Boundary layer lifts from
the sphere surface

Figure 8.52 When the imposed pressure from the free stream rises in the flow direction (adverse pressure gradient), the boundary-layer character of the flow ends at a point called the separation point. Beyond this point, the boundary layer is no longer attached to the wall and a stagnant region with recirculation forms.

$(-\partial \mathcal{P}/\partial x = \xi(R\pi/2) = 0)$. From this point forward ($\tilde{\theta} > 90°$), the pressure gradient is positive, and the forces from the imposed pressure now are pushing the flow in the opposite direction, working against the fluid inertia. Eventually, the inertia runs out, and the pressure wins. As a consequence, the flow near the wall—which was always the slowest flow in the boundary layer—stops. The location where the flow in the boundary layer stops and subsequently reverses direction is called the separation point.

When the flow near the wall stops and reverses, the flow pattern looks like the boundary layer has lifted off of the surface and joined the outer stream (Figure 8.52). The calculations described here cannot tell us what happens in the flow after flow separation occurs because the assumptions used to simplify the governing equations are no longer valid in separated flow. In particular, once the flow has separated, the pressure distribution near the sphere is no longer given by the potential-flow result, as we discuss now.

Although we cannot calculate the velocity profile near the sphere past the separation point, this does not prevent us from calculating drag on the sphere from our boundary-layer analysis. The drag on the sphere is calculated in the usual way, from an integration over the wetted surface of $\hat{n} \cdot \underline{\tilde{\Pi}}$ at the surface (Equation 8.6):

Total molecular fluid
force on a surface $\mathcal{S}$:
$$\underline{\mathcal{F}} = \iint_{\mathcal{S}} [\hat{n} \cdot \underline{\tilde{\Pi}}]_{\text{at surface}} \, dS \tag{8.447}$$

Total fluid force
on the sphere in
noncreeping flow:
$$\underline{\mathcal{F}} = \int_0^{2\pi} \int_0^\pi [\hat{e}_r \cdot \underline{\tilde{\Pi}}]_{r=R} \, R^2 \sin\theta \, d\theta \, d\phi \tag{8.448}$$

$$= \int_0^{2\pi} \int_0^\pi \begin{pmatrix} (\tilde{\tau}_{rr} - \mathcal{P})|_{r=R} \\ \tilde{\tau}_{r\theta}|_{r=R} \\ \tilde{\tau}_{r\phi}|_{r=R} \end{pmatrix}_{r\theta\phi} R^2 \sin\theta \, d\theta \, d\phi$$

$$\tag{8.449}$$

We include the effect of gravity by substituting $\mathcal{P}$ for p in the expression for the stress components. The drag is given by the z-component of this expression, which we obtained previously (see Example 8.2) by converting all three basis vectors $\hat{e}_r$, $\hat{e}_\theta$, and $\hat{e}_\phi$ to Cartesian coordinates and then taking the z-component. We also can obtain the z-component by writing $\hat{e}_z$ in the $r\theta\phi$-coordinate system and carrying out the dot product with the fluid force written in the same coordinate system. We follow this approach here:

$$\mathcal{F}_{\text{drag}} = \mathcal{F}_z = \hat{e}_z \cdot \underline{\mathcal{F}} \tag{8.450}$$

$$= \begin{pmatrix} \cos\theta \\ -\sin\theta \\ 0 \end{pmatrix}_{r\theta\phi} \cdot \begin{pmatrix} \mathcal{F}_r \\ \mathcal{F}_\theta \\ \mathcal{F}_\phi \end{pmatrix}_{r\theta\phi} \tag{8.451}$$

$$= \cos\theta \, \mathcal{F}_r - \sin\theta \, \mathcal{F}_\theta \tag{8.452}$$

$$\mathcal{F}_{\text{drag}} = \int_0^{2\pi} \int_0^\pi (\tilde{\tau}_{rr} - \mathcal{P})|_{r=R} \cos\theta \; R^2 \sin\theta \, d\theta d\phi$$

$$+ \int_0^{2\pi} \int_0^\pi \tilde{\tau}_{r\theta}|_{r=R} (-\sin\theta) \; R^2 \sin\theta \, d\theta d\phi \tag{8.453}$$

$$\mathcal{F}_{\text{drag}} = \int_0^{2\pi} \int_0^\pi -\mathcal{P}|_{r=R} \cos\theta \; R^2 \sin\theta \, d\theta d\phi$$

$$+ \int_0^{2\pi} \int_0^\pi [\tilde{\tau}_{rr} \cos\theta - \tilde{\tau}_{r\theta} \sin\theta]_{r=R} \; R^2 \sin\theta \, d\theta d\phi \tag{8.454}$$

$$= \begin{pmatrix} \text{pressure} \\ \text{contribution} \\ \text{to drag} \end{pmatrix} + \begin{pmatrix} \text{viscous} \\ \text{contribution} \\ \text{to drag} \end{pmatrix} \tag{8.455}$$

$$= \begin{pmatrix} \text{form} \\ \text{drag} \end{pmatrix} + \begin{pmatrix} \text{skin-friction} \\ \text{drag} \end{pmatrix} \tag{8.456}$$

Equation 8.454 shows that we can calculate the drag on the sphere from two contributions: (1) the pressure distribution around the sphere (i.e., the true, flow-separated pressure distribution, not the potential-flow distribution, which gives zero pressure drag); and (2) the stress distribution near the surface, which is calculable from the boundary-layer velocity results.

The pressure contribution to drag on a sphere sometimes is called the *form drag* because the asymmetry in the pressure distribution caused by flow separation is a function of the shape or form of an obstacle. Form drag cannot be calculated from the current results because the true pressure profile in the

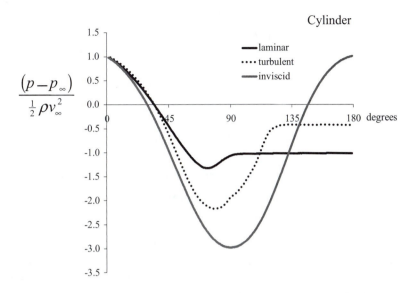

Cylinder

$$\frac{(p-p_\infty)}{\frac{1}{2}\rho v_\infty^2}$$

— laminar
···· turbulent
— inviscid

Figure 8.53

Pressure profiles in the flow past a cylinder. Boundary-layer separation leads to a low-pressure pocket behind blunt objects [178]. The overall pressure distribution in separated flow is asymmetric, and this configuration leads to drag through the pressure contribution to drag described in Equation 8.454.

separated flow is not obtained. If, however, experiments are conducted that give the pressure distribution (see Figure 8.53 for flow around a cylinder), the form drag may be calculated from the data by evaluating the first term in Equation 8.454.

The viscous or skin-friction contribution to drag on a sphere can be calculated from our velocity results. Focusing then on the second term of Equation 8.454, we first note that the flow is assumed to be symmetrical in the ϕ-direction, allowing us to carry out the ϕ-integration:

$$\left(\begin{array}{c}\text{skin-friction}\\\text{drag}\end{array}\right) = \int_0^{2\pi} \int_0^{\pi} [\tilde{\tau}_{rr}\cos\theta - \tilde{\tau}_{r\theta}\sin\theta]_{r=R} \; R^2 \sin\theta d\theta d\phi \quad (8.457)$$

$$= 2\pi R^2 \int_0^{\pi} [\tilde{\tau}_{rr}\cos\theta - \tilde{\tau}_{r\theta}\sin\theta]_{r=R} \; \sin\theta d\theta \quad (8.458)$$

The remaining θ-integration is over two stress components, $\tilde{\tau}_{rr}$ and $\tilde{\tau}_{r\theta}$; the stress $\tilde{\tau}_{rr}$ is zero at the surface because the boundary-layer solution for v_y shows that $\tilde{\tau}_{rr}(\theta)|_{r=R} = \tilde{\tau}_{yy}(x)|_{y=0} = 0$. The stress $\tilde{\tau}_{r\theta}$ is nonzero at the sphere surface and must be integrated. Finally, the limits of the integration should be modified to take into account flow separation. When the boundary layer separates from the surface, a slow recirculating region is observed on the lee side of the sphere, and the skin friction is assumed to be approximately zero in this region. Thus, to calculate the skin-friction drag, we integrate the shear stress at the surface from the separation point ($\theta = \pi - \tilde{\theta}_{max}$) to where the flow first impacts the sphere, the forward stagnation point ($\theta = \pi$):

$$\begin{pmatrix} \text{skin-friction} \\ \text{drag} \end{pmatrix} = 2\pi R^2 \int_0^\pi [\tilde{\tau}_{rr} \cos\theta - \tilde{\tau}_{r\theta} \sin\theta]_{r=R} \; \sin\theta \, d\theta \qquad (8.459)$$

$$= 2\pi R^2 \int_{\pi - \tilde{\theta}_{max}}^\pi - \tilde{\tau}_{r\theta}(\theta)|_{r=R} \; \sin^2\theta \, d\theta \qquad (8.460)$$

$$= 2\pi R^2 \int_{\pi - \tilde{\theta}_{max}}^\pi - \tilde{\tau}_{yx}\Big|_{\substack{y=0 \\ x=R(\pi-\theta)}} \; \sin^2\theta \, d\theta \qquad (8.461)$$

The required stress components may be obtained from the constitutive equation and the velocity solution. The integration may be performed numerically.

Although not without limitations, Prandtl's basic boundary-layer analysis has brought us a long way. From the results of this analysis, we become aware of an important aspect of flow past obstacles: Adverse (rising) pressure gradients cause fluid-layer separation. Fluid-layer separation has the important effect of altering the pressure distribution in the flow from what would be expected from potential-flow theory: The low pressures on the lee side (i.e., separated side) of obstacles fail to balance the high pressures on the windward side of the obstacle, producing an asymmetric pressure profile and a large drag due to the unbalanced forces. Measured pressure distributions for laminar and turbulent flow past a cylinder are shown in Figure 8.53. A pressure minimum is observed ahead of the sphere equator ($\theta = 90°$) for both flow regimes, and—although there is some pressure recovery after the minimum—the pressure downstream of the sphere never returns to upstream pressure levels, and the pressure profile is asymmetric.

The result in Equation 8.454 confirms that drag has two contributions: viscous and pressure drag (see Figure 8.35). The amount of friction drag produced by an obstacle versus how much pressure drag is produced depends on the object's shape. For example, the drag on a flat plate is pure friction drag because the pressure is everywhere constant. The drag on a cylinder is almost purely pressure drag (drag on a cylinder is about 3 percent friction drag [178]). We can see the effect of object shape on drag in Figure 8.54, which portrays the relative amount of friction drag versus pressure drag produced by an object called a streamlined cylinder, which is a cylinder altered to have a tapered shape on its lee side. When the thickness t of the object is small compared to the length of the tail measured by its chord length c, the object is essentially a flat plate and all the drag is friction drag. When t/c is about 1, the streamlined cylinder is approximately a cylinder, and the pressure drag dominates the drag. Between these two shapes, the ratio of the friction drag to the pressure drag drops gradually and smoothly from 1 to zero as the ratio t/c goes from zero to 1.

The process of modifying body shape to reduce pressure drag is called *streamlining*, which works by delaying and eliminating boundary-layer separation. Boundary-layer separation, we now know, is affected by the shape of the body; specifically, long flat bodies do not exhibit flow separation and blunt objects always exhibit flow separation. By changing the shape of an obstacle, we change

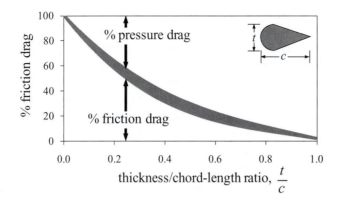

Figure 8.54 A cylinder modified to have a tapered back section is known as a streamlined cylinder. As the shape of the streamlined cylinder changes from essentially a flat plate to a conventional cylinder, the drag produced by the object switches from pure friction drag to essentially pure pressure drag [178]. The scatter in the data is represented by the line thickness.

the pressure distribution in the outer flow. Shape changes that eliminate adverse pressure gradients or that make more gradual the rising pressure on the backside of the obstacle eliminate or delay boundary-layer separation.

Streamlining does not reduce drag due to viscosity, however; in fact, streamlining typically increases the surface area in contact with fluid, thereby increasing viscous drag. Streamlining nevertheless is beneficial because the enormous reduction of pressure drag that results when boundary-layer separation is avoided more than compensates for the slightly increased viscous drag associated with making an object more streamlined. Figure 8.55 shows two objects, drawn to scale, that have the same amount of drag. Although the streamlined shape is much larger than the small blunt circular shape, the streamlined object has relatively little drag for its size because in the flow around it boundary-layer separation is avoided.

Another curious observation is that turbulent boundary layers are better able to resist boundary-layer separation than laminar boundary layers. The extra stability of turbulent boundary layers relative to flow separation means that the wake behind a sphere is more narrow when a turbulent boundary layer has separated

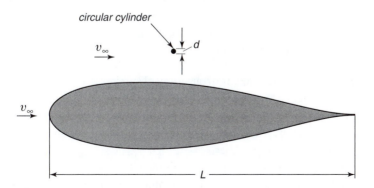

Figure 8.55 A small wire, which is a blunt object, generates as much drag as a much larger streamlined object (airfoil NACA $63_4 - 021$; $L = 167d$) [147, 149]. See [113] for a demonstration of this effect.

than when a laminar boundary layer separates. The turbulent boundary layer separates farther back along the sphere surface because the turbulent flow in the boundary layer is better able to accommodate itself to the adverse pressure gradient that causes flow separation. Two reasons are cited most commonly as accounting for the turbulent boundary layer's robustness in the face of adverse pressure gradients. First, there is more momentum near the wall in a turbulent boundary layer due to the mean shape of turbulent boundary layers (see Figure 8.43). This configuration protects the boundary-layer momentum, in a sense, from the influence of the outer flow. Second, the three-dimensional fluctuations that occur in turbulent boundary layers contain significant energy that helps the boundary layer to resist the decelerating effects of an adverse pressure gradient.

For blunt objects, the drag coefficient is insensitive to the effect of Reynolds number above a certain threshold (i.e., Newton's drag law regime; see Figure 8.12). Why this is so is easily understood in terms of the flow-separation concepts discussed previously. Total drag is generated from two sources: pressure asymmetry and viscosity. Viscous drag is calculated from an integration of velocity gradients near the surface of an object; therefore, viscous drag is a function of the Reynolds number: The speed of the flow affects the magnitude of the velocity gradients near the surface, which increases viscous drag. Pressure drag does not depend on the Reynolds number, however. Once the boundary layer has separated, the pressure profile does not change any more with increasing Reynolds number. The overall drag for a blunt object is primarily pressure drag, which is independent of the Reynolds number. This is true provided that the boundary layer has separated. Because most objects are blunt, a single value of drag coefficient is all that is needed to calculate the drag in flow past most objects. Tables of (constant) drag coefficients for a variety of blunt objects are in the literature [63].

A skydiver falling through the air is an approximately streamlined object when her arms and legs are pulled in and is more blunt when her arms and legs are flung out. In the following example, we return to the skydiver problem to settle a final question that was asked in that problem: What is the effect of body position on the drag on a skydiver?

EXAMPLE 8.24 (Skydiver, concluded). *What is the maximum speed reached by a skydiver who jumps out of an airplane at 13,000 feet? How much can the speed of the skydiver vary depending on her body position (i.e., arms and legs flung out or pulled in tightly)?*

SOLUTION. We learned from our study of spheres that flow around a blunt object forms a boundary layer on the upstream side of the object, and the boundary layer separates from the surface when the pressure begins to rise at the widest part the object. The skydiver is a blunt object moving through a fluid, and we expect the same type of flow around her.

The drag on a blunt object is dominated by the pressure drag due to flow separation. Because the pressure distribution around a blunt object is insensitive to Reynolds number, pressure drag is independent of Reynolds number and a single value of drag coefficient is sufficient to describe the drag on blunt objects.

We can estimate the effect of the skydiver's posture on her terminal speed by making terminal-speed calculations with drag coefficients from two different shapes: one that mimics the pulled-in-tight posture (a rectangular solid in the shape of a person, $C_D = 0.5$, $A_p = 0.111$ m^2) and one that mimics the arms-flung-out posture (a flat disk, $C_D = 1.17$, $A_p = 0.84$ m^2) [63].

We previously obtained the terminal velocity of an object of arbitrary shape as follows (Equation 8.95):

$$\begin{array}{c} \text{Terminal speed} \\ \text{of a blunt object} \\ \text{(arbitrary regime of Re):} \end{array} \qquad \boxed{v_\infty = V = \sqrt{\frac{2\mathcal{V}(\rho_{\text{body}} - \rho)g}{\rho A_p C_D}}} \qquad (8.462)$$

where v_∞ is the terminal speed, $\mathcal{V}$ is the volume of the object, ρ_{body} is the density of the object, g is the acceleration due to gravity, ρ is the density of air, A_p is the reference area for the chosen shape (projected area in the direction of motion), and C_D is the drag coefficient of the chosen shape. For the volume of the body, we assume our skydiver to be $h = 1.78$ m tall, and we calculate her volume as $\mathcal{V} = A_p h$ based on the A_p of the rectangular solid shape.

Using the same original values of the physical parameters as in Example 8.3, we estimate the terminal speed:

$$\text{Density of air:} \quad \rho = 1.3 \text{ kg/m}^3 \qquad (8.463)$$

$$\text{Density of human (water):} \quad \rho_{\text{body}} = 1{,}000 \text{ kg/m}^3 \qquad (8.464)$$

$$\text{Acceleration due to gravity:} \quad g = 9.80 \text{ m/s}^2 \qquad (8.465)$$

$$\text{Bullet shape:} \quad C_D = 0.5, A_p = 0.111 \text{ m}^2 \qquad (8.466)$$

$$\text{Disk shape:} \quad C_D = 1.17, A_p = 0.84 \text{ m}^2 \qquad (8.467)$$

With these parameter values, we obtain the terminal speeds as:

$$\begin{array}{c} \text{Estimate of fastest terminal speed:} \\ \text{(arms and legs pulled in; rectangular solid)} \end{array} \quad V = 170 \text{ m/s} \quad (8.468)$$

$$\begin{array}{c} \text{Estimate of slowest terminal speed:} \\ \text{(arms and legs flung out; disk)} \end{array} \quad V = 55 \text{ m/s} \quad (8.469)$$

The estimate using the drag coefficient of the disk is the same as the belly-to-Earth estimates of terminal skydiver speeds from the Internet. The terminal speed estimate for the pulled-in-tight posture is too high (i.e., terminal speed ≈ 90 m/s for this position). The two estimates bracket the original estimate (107 m/s), which used a sphere as the shape of the skydiver. By varying how spread out she is, the skydiver can vary her speed of descent by a factor of 2 or so.

This concludes our introduction to external flow. The physics of fluid flow is captured by the continuity equation and the Navier-Stokes equation, but these equations are so mathematically complex that we can solve them analytically only in certain cases. In this and in the previous chapter, we studied several cases of relatively simple flows: flows in which viscosity dominates (i.e., laminar flow and creeping flow) and in which inertia dominates (i.e., turbulent pipe flow and potential flow). The boundary-layer method of Prandtl is a first foray into the study of mixed flows. Boundary-layer flow is a simplified mixed flow in which viscous effects are confined to a small region, whereas the majority of the flow is free from the effects of viscosity. More advanced fluid-mechanics study continues from this point, seeking methods to obtain solutions to the Navier-Stokes equations for mixed viscous and inertial flows. An important method for advanced-flow analyses is tracking vorticity in mixed flows. Vorticity, which is introduced in the next section, is a property of a flow field that helps us to study viscous and inertial interactions in flows.

8.3 More complex external flows

We completed our studies of flow basics in external flows and now move on to study more complex flows. In Section 8.3.1, we introduce and motivate the use of vorticity to study complex flows. In Section 8.3.2, we revisit dimensional analysis and discuss how the proper use of dimensional analysis guides us to sophisticated techniques for the study of complex flow.

8.3.1 Vorticity

To make predictions in complex flows, we begin as usual with the governing equations of fluid motion—the mass and momentum balances—and we make follow-up calculations with the appropriate expressions for forces or for whatever is of interest:

$$\text{Continuity equation:} \quad \nabla \cdot \underline{v} = 0 \tag{8.470}$$

$$\text{Navier-Stokes equation:} \quad \rho \left(\frac{\partial \underline{v}}{\partial t} + \underline{v} \cdot \nabla \underline{v} \right) = -\nabla \mathcal{P} + \mu \nabla^2 \underline{v} \tag{8.471}$$

$$\begin{array}{c} \text{Total molecular fluid force} \\ \text{on a finite surface } \mathcal{S}: \end{array} \quad \underline{\mathcal{F}} = \iint_{\mathcal{S}:} [\hat{n} \cdot \underline{\tilde{\Pi}}]_{\text{at surface}} \, dS \tag{8.472}$$

With complex flows, it is difficult to solve the governing equations, especially when both viscous and inertial effects are important. Thus far, we can solve the Navier-Stokes equations only when we make strong simplifying assumptions, such as creeping flow or steady, unidirectional flow (no inertia); and potential flow (no viscosity).

We have succeeded with a flow that is somewhat complex: boundary-layer flow. In boundary-layer flows, both viscous and inertial effects are present, but Prandtl's idea of incorporating viscous effects only near the wall enabled the solution of significant problems in flows around obstacles. Many flows with mixed viscous and inertial effects are not boundary-layer flows, however, and these problems require a different approach. To devise a new approach, we begin by reflecting on the types of flow patterns that have been overlooked so far. By examining the as-yet-unsolved flow problems, we hope to discover methods and techniques that provide understanding of them.

Figure 8.56 shows some of the flow behavior that we currently cannot predict. In Figure 8.56a, smoke is used to visualize the flow behind a model of an

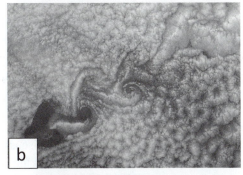

Figure 8.56 There is still a significant amount of flow behavior that is difficult to predict due to the mathematical complexity of the governing equations (the Navier-Stokes equations). In (a), a complex wake is produced immediately behind an airplane. In (b) clouds show that a passing aircraft leaves a complex pattern over long distances; in (c) flood water passing under a low bridge produces a whirlpool. Image credits: (a) Courtesy of NASA; (b) NASA/GSFC/JPL, MISR Team; (c) Courtesy USGS photographer Mark Landers.

airplane in a wind tunnel. The flow is curly, complex, and three dimensional. In Figure 8.56b, an airplane moves through clouds that show the flow pattern behind the moving aircraft. The trail behind the plane shows that curly vortices form in an alternating pattern behind the plane (called a von Karman vortex street) and the vortices slowly move away from the plane's path as they die out. In the flooded river upstream of a bridge (see Figure 8.56c), a whirlpool forms. The whirlpool is the top of a three-dimensional vortex caused by the water passing under the bridge. The vortex is a flow pattern in which there is a concentrated rotational character in a small region of fluid. Vortices form in idealized flows as well, such as the clockwise vortex that forms on the leeward side of a sphere at intermediate Reynolds numbers (see Figure 8.22). These vortices are not shed from the sphere but rather stay in place.

The flows in Figure 8.56 have in common that they exhibit a degree of rotary motion. In flows that produce a vortex, the rotary motion present in the flow is intense and localized. We succeeded in modeling flow around obstacles when we (meaning Ludwig Prandtl) noticed that the effect of viscosity was localized at solid surfaces—we could ignore viscosity away from surfaces. In flows that produce vortices, there is a different property that is localized: rotational character. Perhaps we can devise a boundary-layer–like approach to rotational character and divide flows into regions that are *rotational* and *irrotational*. Following the boundary-layer idea, we could model the rotational and irrotational parts separately and combine the solutions where they overlap.

This is the motivation for using vorticity to model complex flows. *Vorticity* is a property of a velocity field that is a measure of rotational character as a function of position and time. Vorticity is associated with a time and a location in space: The direction of the vorticity vector indicates the direction of the axis around which the local velocity field tends to rotate a particle, and the magnitude of the vorticity indicates the intensity of the rotational character of the local velocity field. Understanding the role of vorticity and visualizing the meaning of *rotational character* of a flow field can be difficult. In this section, we discuss why vorticity is worth tracking in flows that are driven by both viscous and inertial forces.

Vorticity $\underline{\omega}$ is defined as the *curl*[10] of the fluid velocity field. Vorticity is a vector field:

$$
\begin{array}{c}
\text{Vorticity defined} \\
\text{(a measure} \\
\text{of rotational character} \\
\text{of the flow field } \underline{v}\text{):}
\end{array}
\qquad
\boxed{\underline{\omega} \equiv \nabla \times \underline{v}}
\qquad (8.473)
$$

The vector cross product is reviewed in Chapter 1 (see Equation 1.183); the curl of the vector $\underline{v}$ is given by:

$$
\underline{\omega} = \nabla \times v = \det
\begin{vmatrix}
\hat{e}_1 & \hat{e}_2 & \hat{e}_3 \\
\dfrac{\partial}{\partial x_1} & \dfrac{\partial}{\partial x_2} & \dfrac{\partial}{\partial x_3} \\
v_1 & v_2 & v_3
\end{vmatrix}
\qquad (8.474)
$$

$$
\underline{\omega} =
\begin{pmatrix}
\dfrac{\partial v_3}{\partial x_2} - \dfrac{\partial v_2}{\partial x_3} \\[2ex]
\dfrac{\partial v_1}{\partial x_3} - \dfrac{\partial v_3}{\partial x_1} \\[2ex]
\dfrac{\partial v_2}{\partial x_1} - \dfrac{\partial v_1}{\partial x_2}
\end{pmatrix}_{123}
\qquad (8.475)
$$

---

[10] The curl is a vector-field operator defined as $\nabla \times$ a field; see [146].

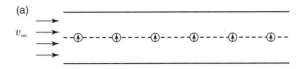

(a)

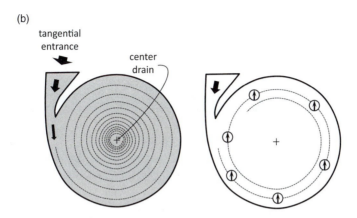

(b)

In uniform flow in a channel (a), a ping-pong ball placed in the center of the flow away from the wall does not rotate; the flow there is free of vorticity. A spiral vortex tank (b) is a flow that is irrotational away from the drain; however, as the fluid approaches the drain, the flow becomes intensely rotational [114, 154].

It can be shown that the magnitude of the vorticity is equal to twice the potential rate of spin caused by the flow field [114]. The direction of the vorticity is the local axis of potential spin; the relationship between the rotation direction of the potential spin and the direction of the vorticity vector is dictated by the righthand rule.

To familiarize ourselves with vorticity, we first look for flows or regions of flow that have no vorticity (i.e., no rotational character) and those that have rotational character. An example of a flow with no vorticity is a uniform flow away from a wall as shown in Figure 8.57a. We can use a ping-pong ball as a rough vorticity meter; we paint lines on the ball so that we can track its orientation. If we then place this marked ball in uniform flow away from walls, it floats downstream in pure translation and does not rotate. Uniform flow away from walls has no vorticity. A surprising flow that also has no vorticity is the flow in a spiral vortex tank (see Figure 8.57b [114]), which is a large tank full of fluid that has a tangential inlet at the top and an outlet hole in the center of the bottom. The fluid follows nearly circular paths as it moves toward the drain. If we place in the flow a marked ping-pong ball or another device that is sensitive to the local rate of rotation, we see that the ball translates in a circular path as it follows the streamlines but the orientation remains constant (i.e., the ball does not rotate). The stillness of the ball in terms of rotation is reminiscent of the stillness of the needle of a compass placed on a turntable: As the compass translates in a circle, the needle always points north. From these examples, we see that vorticity can be absent in flows with both straight and curved streamlines.

An example of a flow with vorticity throughout is the "flow" of water in a tank when it is located on a rotating turntable (Figure 8.58b). In a body of water moving in rigid-body rotation, the lines on a vorticity meter (i.e., the marked

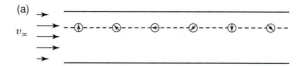

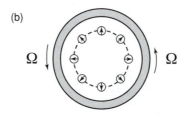

Figure 8.58   In uniform flow in a channel (a), a ping-pong ball placed in the flow near the wall rotates due to the vorticity in the boundary layer. Fluid in a tank that is on a rotating turntable moves in solid-body rotation (b). A marked ping-pong ball shows that the flow has vorticity; that is, the orientation of the ball rotates as it moves along with the fluid [114, 154].

ping-pong ball) rotate with the same angular rotational speed as the tank. The streamlines in rigid-body flow are circles, as are the approximate streamlines in the spiral vortex tank. However, in rigid-body rotation, there is vorticity (i.e., a local tendency to rotate), whereas in the spiral vortex tank, there is none. Another flow with vorticity is the flow in the straight-channel boundary layer near a wall (Figure 8.58a). Within the boundary layer, the flow is shear flow, and the faster speed of the layers away from the wall compared to the layers near the wall causes the vorticity meter to spin. The near-wall-flow example indicates that having straight-line flow does not guarantee that there will be no vorticity—flow in a boundary layer has nearly straight streamlines, yet there is vorticity. The example in Figure 8.57b of a vorticity-free flow, the spiral vortex tank, has vorticity in one location—the center: If we place the vorticity meter near the center of the spiral-vortex-tank, the meter spins intensely. The spiral-vortex-tank flow has no vorticity away from the drain and concentrated vorticity near the drain.

In these descriptions, it is striking that the unidirectional flow past a surface was invoked when describing flows with vorticity (i.e., the flow in the boundary layer) as well as flows without vorticity (i.e., the flow in the free stream). This ideal flow seems like one that we should examine more closely if we want to understand rotational character, including how rotational character is generated in a flow and how it propagates throughout the flow field.

In flow past a flat plate, an irrotational free-stream flow approaches a flat plate. When the flow meets the flat plate, fluid elements that pass near the wall move in approximately straight lines, but the flow field near the wall is shear and has rotational character (Figure 8.59). A marked ping-pong ball placed upstream in this flow would not begin to rotate until it encountered the boundary layer. The question is: What is the source of this rotational character? The answer is that it is coming from the wall. The no-slip boundary condition slows the fluid near the wall and sets up a shear flow near the wall. This shear flow is rotational and within the boundary layer there is vorticity.

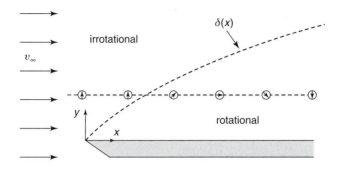

A marked ping-pong ball in a free stream that approaches a flat plat initially does not rotate because the uniform upstream flow is without vorticity. As the ball reaches the plate, it eventually begins to spin due to the vorticity produced by the no-slip boundary condition at the wall. The vorticity diffuses from the wall and also is convected downstream by the flow.

The production and transport of vorticity in a flow can be calculated from the Navier-Stokes equation, which is an equation for the velocity $\underline{v}$. If we form the cross product of ∇ with each term of the Navier-Stokes equation, we produce a transport equation that concerns the vorticity $\underline{\omega} = \nabla \times \underline{v}$.

We begin by taking the curl of the Navier-Stokes equation:

$$\text{Navier-Stokes equation} \atop \text{(microscopic-momentum balance):} \qquad \rho \left(\frac{\partial \underline{v}}{\partial t} + \underline{v} \cdot \nabla \underline{v} \right) = -\nabla \mathcal{P} + \mu \nabla^2 \underline{v}$$

(8.476)

$$\nabla \times \left[\rho \left(\frac{\partial \underline{v}}{\partial t} + \underline{v} \cdot \nabla \underline{v} \right) \right] = \nabla \times \left[-\nabla \mathcal{P} + \mu \nabla^2 \underline{v} \right] \qquad (8.477)$$

The first term on the left becomes the time-derivative of the vorticity. The second term is the curl of $\underline{v} \cdot \nabla \underline{v}$, which may be shown to satisfy the following identity [6, 146]:

$$\text{Vector identity:} \qquad \nabla \times (\underline{v} \cdot \nabla \underline{v}) = \underline{v} \cdot \nabla \underline{\omega} + \underline{\omega} (\nabla \cdot \underline{v}) - \underline{\omega} \cdot \nabla \underline{v} \qquad (8.478)$$

Assuming incompressible fluid ($\nabla \cdot \underline{v} = 0$) and substituting Equation 8.478 into Equation 8.477, we obtain:

$$\rho \left(\frac{\partial \underline{\omega}}{\partial t} + \underline{v} \cdot \nabla \underline{\omega} - \underline{\omega} \cdot \nabla \underline{v} \right) = \nabla \times \left[-\nabla \mathcal{P} + \mu \nabla^2 \underline{v} \right] \qquad (8.479)$$

Working now on the right-hand side, the pressure term is zero because the curl of the gradient of a scalar function is zero ($\nabla \times \nabla f = 0$) [6, 146]; in the last term, we can show that $\nabla \times \nabla^2 \underline{v} = \nabla^2 (\nabla \times \underline{v}) = \nabla^2 \underline{\omega}$. Both of these simplifications result from the fact that we can carry out the spatial derivatives in any order. The final result for the momentum balance in terms of vorticity is:

$$\text{Vorticity-transport equation} \atop \text{(curl of Navier-Stokes;} \atop \text{incompressible fluid):} \qquad \boxed{\frac{\partial \underline{\omega}}{\partial t} + \underline{v} \cdot \nabla \underline{\omega} = \underline{\omega} \cdot \nabla \underline{v} + \nu \nabla^2 \underline{\omega}} \qquad (8.480)$$

where $\nu = \mu/\rho$ is the kinematic viscosity.

In two-dimensional flows—flows that can be written as $\underline{v}(x_1, x_2) = v_1\hat{e}_1 + v_2\hat{e}_2$—the vorticity-transport equation has an even simpler form, as demonstrated in the following example.

EXAMPLE 8.25. *How does the vorticity-transport equation simplify in two-dimensional flow? Comment on your results.*

SOLUTION. A two-dimensional flow is one that in a Cartesian coordinate system can be written with only two components:

$$\underline{v}(x_1, x_2) = \begin{pmatrix} v_1 \\ v_2 \\ 0 \end{pmatrix}_{123} \tag{8.481}$$

Vorticity is the curl of the velocity vector; for the components given here, we calculate $\underline{\omega}$ as:

$$\underline{\omega} \equiv \nabla \times \underline{v} \tag{8.482}$$

$$\underline{\omega} = \begin{pmatrix} \dfrac{\partial v_3}{\partial x_2} - \dfrac{\partial v_2}{\partial x_3} \\[2mm] \dfrac{\partial v_1}{\partial x_3} - \dfrac{\partial v_3}{\partial x_1} \\[2mm] \dfrac{\partial v_2}{\partial x_1} - \dfrac{\partial v_1}{\partial x_2} \end{pmatrix}_{123} = \begin{pmatrix} 0 \\[2mm] 0 \\[2mm] \dfrac{\partial v_2}{\partial x_1} - \dfrac{\partial v_1}{\partial x_2} \end{pmatrix}_{123} \tag{8.483}$$

This equation indicates that for any two-dimensional flow, the vorticity at every location is perpendicular to the plane of the flow. This means that at every point, the tendency of the flow field to rotate fluid particles always produces rotation in the x_1-x_2-plane (i.e., around the x_3 axis).

The vorticity-transport equation is given by Equation 8.480:

$$\text{Vorticity-transport equation:} \qquad \frac{\partial \underline{\omega}}{\partial t} + \underline{v} \cdot \nabla \underline{\omega} = \underline{\omega} \cdot \nabla \underline{v} + \nu \nabla^2 \underline{\omega} \tag{8.484}$$

For steady flow, the time-derivative is zero. For two-dimensional flow, we can use matrix calculations to show that the first term on the right-hand side also is zero (see Problem 55). The vorticity-transport equation for steady two-dimensional flow past a flat plate becomes:

$$\text{Vorticity-transport equation (steady, two-dimensional, incompressible):} \qquad \underline{v} \cdot \nabla \underline{\omega} = \nu \nabla^2 \underline{\omega} \tag{8.485}$$

$$\underline{v} \cdot \nabla \omega = \nu \nabla^2 \omega \tag{8.486}$$

The second equation for the scalar ω is the 3-component and only nonzero component of Equation 8.485, where $\omega = \omega_3$ is the magnitude of the vorticity.

The vorticity-transport equation indicates how vorticity moves around in a flow. There are two terms in the vorticity-transport equation for steady two-dimensional flow (Equation 8.485). On the left is the convective term. *Vorticity can be transported by flow* from one location to another much like momentum,

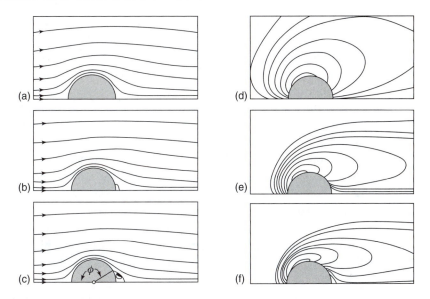

Figure 8.60   Plotting vorticity contour (right) gives more of the physical sense of what is happening in flow around a sphere than what is obtained from the streamline plots (left). (a) Re = 5, no separation; (b) Re = 20, separation at 171 degrees; (c) Re = 40, separation at 148 degrees. After calculations done by V. G. Jenson (Proc. Roy. Soc. London A, vol. 249, 346 (1959) reproduced in [147].

energy, or mass (see Figure 6.4; the convective terms contain $\underline{v} \cdot \nabla \rho$ [mass], $\underline{v} \cdot \nabla \underline{v}$ [momentum], and $\underline{v} \cdot \nabla \hat{E}$ [energy]). On the right of Equation 8.485 is the diffusive term. *Vorticity can diffuse*, and the transport coefficient for the diffusion of vorticity is the kinematic viscosity $\nu = \mu / \rho$. In steady, incompressible, two-dimensional flow, vorticity produced by the no-slip boundary condition at solid surfaces moves away from the wall by diffusion and convects and diffuses throughout the flow.

Vorticity contours are often more effective than streamline plots in showing the character of complex flows. Figure 8.60 plots streamlines and isovorticity contours for a familiar flow: flow around a sphere at finite Reynolds numbers. The streamlines for the flows at three different Reynolds numbers are shown on the left, and the differences with Reynolds number are rather subtle in terms of the streamline pattern. The isovorticity contours on the right, however, distinctly and intuitively show the effect on the flow of the rising Reynolds number. The vorticity originates at the sphere due to the no-slip boundary condition and spreads into the flow by diffusion. We also can see the effect of convection: The upstream flow carries the vorticity downstream.

The vorticity-transport equation has important implications in uniform, two-dimensional flows around obstacles. Note also that Equation 8.485 has no vorticity production term. If applied along a streamline, we see that this equation indicates that there is no vorticity production along a streamline. If the vorticity is zero at any point on the streamline, then it is zero at all points along the streamline. This relationship explains why uniform potential flow around an obstacle is irrotational.

The vorticity-transport equation (Equation 8.480) indicates how momentum conservation governs the transport of vorticity. We understand the convection and diffusion terms of that equation. There is an additional term that appears, however, which has interesting implications in three-dimensional flows. The term $\underline{\omega} \cdot \nabla \underline{v}$ dropped out of the vorticity-transport equation when we considered two-dimensional flow in Example 8.25. This term captures a physics that has no analogy in other transport laws. It is associated with *vorticity intensification*, which is a three-dimensional effect related to the conservation of angular momentum. Briefly, the term $\underline{\omega} \cdot \nabla \underline{v}$ tracks the flow-field analogy of the acceleration that happens when a twirling iceskater draws his arms inward to spin faster. The skater's acceleration occurs because he has reduced his moment of inertia by drawing in his arms. For the skater's motion to contain the same amount of momentum with a smaller moment of inertia, his angular speed must increase. The motion of tornados is governed by the $\underline{\omega} \cdot \nabla \underline{v}$ term of the vorticity transport equation (see Section 10.5). The upward motion of the fluid in the center of a tornado stretches the vortex; conservation of momentum then causes the spin of the storm to intensify, sometimes to disastrous effect. This is discussed in Chapter 10.

This concludes our brief introduction to vorticity and its use in complex-flow modeling. Chapter 10 discusses the role of vorticity in the production of lift and its usefulness when considering flows with curved streamlines. Readers who are interested in learning more about vorticity, circulation, and lift are encouraged to view the two-part National Committee on Fluid Mechanics Films film on the subject [114].

In the next section, we return to the topic of dimensional analysis and explore in detail how nondimensionalization techniques misled us in our study of external flows (i.e., d'Alembert's paradox, $\mathcal{F}_{\text{drag}} = 0$). We will see that not all nondimensionalization choices are equal when carrying out dimensional analysis: Care must be taken to choose characteristic values that guarantee meaningful results throughout a complex flow. As discussed in the next section, when we simplify the governing equations based on dimensional analysis, we must be rigorous in checking the magnitudes of all terms in the equations and be prepared to use different scalings in different flow regions. The section on dimensional analysis concludes our discussion of external flow. In Chapter 9, we return to the topic of macroscopic balances and their application to complex engineering flows. In Chapter 10, we revisit the flow behaviors described in Chapter 2, assess our progress, and point the way to advanced study in fluid mechanics.

8.3.2 Dimensional analysis redux

In Chapter 7, we introduced the concept of dimensional analysis and first applied that technique to problems in fluid mechanics. Dimensional analysis is the method in which system variables such as velocity, distance, and pressure are scaled relative to characteristic values of those quantities. At a fundamental level, scaling the variables has the effect of eliminating units from the calculations. We used dimensional analysis for this purpose in Section 7.1, and we saw that dimensional analysis also had the practical effect of checking our calculations for dimensional consistency. A second practical application of dimensional analysis is that it allows us to plot functions without having to detail the values of too many parameters (see, e.g., Figure 7.6).

The powerful use of dimensional analysis, however, is in tailoring the equations of change: the microscopic mass, momentum, and energy balances. As demonstrated in Section 8.1, most fluid-mechanics problems require the solution of the full, complex, nonlinear versions of the continuity equation and the Navier-Stokes equation. We can simplify this task if we identify terms of these equations that have little effect in a chosen problem. This is the value of dimensional analysis: Properly applied, it demonstrates which terms in an equation are important for a given flow situation and which may be eliminated safely (see Example 8.22 on the topic of dimensional analysis on boundary-layer flows).

Sometimes formal dimensional analysis appears to be superfluous. In the analysis of creeping flow around a sphere (Section 8.1.1), we casually simplified the governing equations, choosing to eliminate the inertial terms from the dimensional Navier-Stokes equation without conducting any dimensional analysis. We arrived at solutions for velocity and pressure fields using these ad hoc methods. Given that the results of the analysis were borne out by experiments ($C_D = 24/\text{Re}$), that methodology seems justified.

We encountered trouble with this casual approach when we went to higher Reynolds numbers, however. When we considered the more complex behavior seen at higher Reynolds numbers (Section 8.1.2.2), ad hoc dimensional analysis produced the following nondimensional equations that we used to explore noncreeping flow around a sphere:

For all Reynolds numbers:

Continuity equation: $\nabla^* \cdot \underline{v}^* = 0$

Navier-Stokes equation: $\dfrac{\partial \underline{v}^*}{\partial t^*} + \underline{v}^* \cdot \nabla^* \underline{v}^* = -\nabla^* \mathcal{P}^* + \left(\dfrac{1}{\text{Re}}\right) \nabla^{*2} \underline{v}^*$

$$C_D = \frac{2}{\pi} \int_0^{2\pi} \int_0^{\pi} \left[\left(\frac{2}{\text{Re}} \frac{\partial v_r^*}{\partial r^*} - \mathcal{P}^*\right) \cos\theta\right.$$

Drag on a sphere:

$$+ \left(\frac{r^*}{\text{Re}} \frac{\partial}{\partial r^*}\left(\frac{v_\theta^*}{r^*}\right)\right.$$

$$\left.\left. + \frac{1}{\text{Re}\, r^*} \frac{\partial v_r^*}{\partial \theta}\right)(-\sin\theta)\right]_{r^*=\frac{1}{2}} \sin\theta\, d\theta\, d\phi$$

(8.487)

For rapid flows, we let $\text{Re} \longrightarrow \infty$, and these three equations became:

For Re $\longrightarrow \infty$:

Continuity equation: $\nabla^* \cdot \underline{v}^* = 0$

Navier-Stokes equation: $\dfrac{\partial \underline{v}^*}{\partial t^*} + \underline{v}^* \cdot \nabla^* \underline{v}^* = -\nabla^* \mathcal{P}^*$

Drag on a sphere: $C_D = \dfrac{2}{\pi} \displaystyle\int_0^{2\pi} \int_0^{\pi} \left[-\mathcal{P}^* \cos\theta\right]_{r^*=\frac{1}{2}} \sin\theta\, d\theta\, d\phi$

(8.488)

These are the potential-flow equations, and we discuss their solution for uniform flow around a sphere in Section 8.1.2.2.

As discussed in Section 8.1.2.2, the predicted behavior in the potential-flow solution is not what we expected. It did not correctly calculate drag, predicting that drag was zero. Potential flow also predicted slip at the sphere surface. The source of the confusion can be traced to our initial attitude toward dimensional analysis. When introducing dimensional analysis, we correctly recognized that if scaled properly, the equations of change would be displayed in a way that organizes and systematizes the effects of various quantities, such as density, viscosity, velocity, time, and distance. The missing rigor in our method was that we failed to examine what it meant to scale properly our equations; as we discuss here, different choices for characteristic quantities can lead us in different directions. Sometimes the choice does not have a major impact and sometimes it does. We were fortunate when we scaled the equations for pressure-driven flow in a tube, and our results led to friction-factor/Reynolds-number correlations that were correct and helpful. We were less fortunate in noncreeping flow around a sphere; in that case, our scaling was improper and our analysis failed.

Although we hit some dead ends before arriving at our current state of understanding, we are now in an excellent position to revisit dimensional analysis and to appreciate the true merits and power of this type of analysis. With the benefit of hindsight, we can clarify the requirements that must be met when choosing characteristic lengths, times, and other quantities. With the appropriate choices for characteristic quantities, we are guided by dimensional analysis to the proper equations that govern different regions of complex flows. With careful attention to method, we turn dimensional analysis into a powerful and effective tool for solving the most difficult problems in fluid mechanics.

We now revisit dimensional analysis, determined to be more rigorous. Two problems on which we used dimensional analysis are turbulent pipe flow and noncreeping flow past a sphere. The procedure is straightforward: Beginning with the governing equations, choose characteristic values of the variables in the problem, define nondimensional versions of those variables, and substitute the new nondimensional variables in the governing equations. For the problems mentioned, we made similar choices: For characteristic length, we chose the pipe or sphere diameter; for characteristic velocity, we chose the dominate velocity in the flow; and for characteristic time and pressure, we built a characteristic value from the previous choices for D and V (Table 8.1). Our choices for characteristic values were based on capturing the magnitude of the changes taking place in the system variables.

With the nondimensional variables thus defined, we substitute these expressions into the Navier-Stokes equation and factor out the dimensions:

$$
\frac{\rho V}{T}\left(\frac{\partial v_z^*}{\partial t^*}\right) + \frac{\rho V^2}{D}\left(v_x^*\frac{\partial v_z^*}{\partial x^*} + v_y^*\frac{\partial v_z^*}{\partial y^*} + v_z^*\frac{\partial v_z^*}{\partial x_1^*}\right)
$$

$$
= \frac{P}{D}\left(-\frac{\partial \mathcal{P}^*}{\partial z^*}\right) + \frac{\mu V}{D^2}\left(\frac{\partial^2 v_z^*}{\partial x^{*2}} + \frac{\partial^2 v_z^*}{\partial y^{*2}} + \frac{\partial^2 v_z^*}{\partial z^{*2}}\right) \qquad (8.489)
$$

Dividing Equation 8.489 by the coefficient of the second term on the left-hand side (i.e., the convective term), the inverse of the Reynolds number appears as

Table 8.1. Characteristic values in dimensional analysis of turbulent pipe flow and noncreeping flow around a sphere

Dimension	Symbol	Pipe flow	Flow around sphere	Nondimensional variable
Length	D	Pipe diameter	Sphere diameter	$r^* = \dfrac{r}{D}$
				$z^* = \dfrac{z}{D}$
Velocity	V	$\langle v_z \rangle$	v_∞	$v_r^* = \dfrac{v_r}{V}$
				$v_\theta^* = \dfrac{v_\theta}{V}$
				$v_z^* = \dfrac{v_z}{V}$
Time	T	$\dfrac{D}{V}$	$\dfrac{D}{V}$	$t^* = \dfrac{t}{T} = \dfrac{tV}{D}$
Pressure	P	ρV^2	ρV^2	$p^* = \dfrac{p}{P} = \dfrac{p}{\rho V^2}$

the coefficient of the viscous momentum term on the right-hand side:

$$\frac{D}{TV}\left(\frac{\partial v_z^*}{\partial t^*}\right) + \left(v_x^*\frac{\partial v_z^*}{\partial x^*} + v_y^*\frac{\partial v_z^*}{\partial y^*} + v_z^*\frac{\partial v_z^*}{\partial z^*}\right)$$

$$= \frac{P}{\rho V^2}\left(-\frac{\partial \mathcal{P}^*}{\partial z^*}\right) + \frac{\mu}{\rho V D}\left(\frac{\partial^2 v_z^*}{\partial x^{*2}} + \frac{\partial^2 v_z^*}{\partial y^{*2}} + \frac{\partial^2 v_z^*}{\partial z^{*2}}\right) \qquad (8.490)$$

From Equation 8.490, we see that our standard choices for T and P correspond to choosing that the coefficients of the velocity-time-derivative term and the pressure-derivative term both be equal to 1. By choosing these coefficients to be 1, we force the terms to be as important to the solution as the other terms with coefficient 1—that is, as important as the convective term.

$$Choose: \qquad T \equiv \frac{D}{V} \qquad P \equiv \rho V^2$$

Time changes, pressure gradient, and convection of comparable importance:

$$\left(\frac{\partial v_z^*}{\partial t^*}\right) + \left(v_x^*\frac{\partial v_z^*}{\partial x^*} + v_y^*\frac{\partial v_z^*}{\partial y^*} + v_z^*\frac{\partial v_z^*}{\partial z^*}\right)$$

$$= \left(-\frac{\partial \mathcal{P}^*}{\partial z^*}\right) + \frac{1}{Re}\left(\frac{\partial^2 v_z^*}{\partial x^{*2}} + \frac{\partial^2 v_z^*}{\partial y^{*2}} + \frac{\partial^2 v_z^*}{\partial z^{*2}}\right)$$

$$(8.491)$$

This is the version of the nondimensional Navier-Stokes equation discussed in Sections 7.1.2.2 and 8.1.2.1. The characteristic values are chosen so that the nondimensional derivatives—$\partial v_z^*/\partial t^*$, $\partial v_z^*/\partial x^*$, and others—are expected to be independent of the characteristic values. This scaling may be described as producing terms of order one, written $O(1)$. The nomenclature $y = O(x)$ means in general that the quantity y is proportional to x; $O(1)$ means that y is independent of the parameters of the model [85].

Several of the steps reviewed here were taken arbitrarily under the presumption that the choice would make no difference. For example, we choose in Equation 8.489 to divide the entire equation by the coefficient of the convective inertial term. If we had chosen instead to divide the entire equation by the coefficient of the viscous term $\mu V/D^2$, the result would be:

$$\frac{\rho D^2}{T\mu}\left(\frac{\partial v_z^*}{\partial t^*}\right) + \frac{\rho V D}{\mu}\left(v_x^*\frac{\partial v_z^*}{\partial x^*} + v_y^*\frac{\partial v_z^*}{\partial y^*} + v_z^*\frac{\partial v_z^*}{\partial z^*}\right)$$
$$= \frac{PD}{\mu V}\left(-\frac{\partial \mathcal{P}^*}{\partial z^*}\right) + \left(\frac{\partial^2 v_z^*}{\partial x^{*2}} + \frac{\partial^2 v_z^*}{\partial y^{*2}} + \frac{\partial^2 v_z^*}{\partial z^{*2}}\right) \qquad (8.492)$$

Again, the Reynolds number appears—albeit in a different location—but equally significant, the coefficients of the pressure- and time-derivative terms have changed:

$$\frac{\rho D^2}{T\mu}\left(\frac{\partial v_z^*}{\partial t^*}\right) + \text{Re}\left(v_x^*\frac{\partial v_z^*}{\partial x^*} + v_y^*\frac{\partial v_z^*}{\partial y^*} + v_z^*\frac{\partial v_z^*}{\partial z^*}\right)$$
$$= \frac{PD}{\mu V}\left(-\frac{\partial \mathcal{P}^*}{\partial z^*}\right) + \left(\frac{\partial^2 v_z^*}{\partial x^{*2}} + \frac{\partial^2 v_z^*}{\partial y^{*2}} + \frac{\partial^2 v_z^*}{\partial z^{*2}}\right) \qquad (8.493)$$

If we choose the same characteristic time $T = D/V$ and pressure $P = \rho V^2$ as before, the result for nondimensional Navier-Stokes will be the same (Equation 8.491). If, however, we follow the same inspiration as we did previously and set the coefficients of convenient terms in Equation 8.493 to 1, we are led to different choices for T and P:

$$\text{Choose:} \quad T \equiv \frac{\rho D^2}{\mu} \qquad P \equiv \frac{\mu V}{D}$$

$$\left(\frac{\partial v_z^*}{\partial t^*}\right) + \text{Re}\left(v_x^*\frac{\partial v_z^*}{\partial x^*} + v_y^*\frac{\partial v_z^*}{\partial y^*} + v_z^*\frac{\partial v_z^*}{\partial z}\right)$$
$$= \left(-\frac{\partial \mathcal{P}^*}{\partial z^*}\right) + \left(\frac{\partial^2 v_z^*}{\partial x^{*2}} + \frac{\partial^2 v_z^*}{\partial y^{*2}} + \frac{\partial^2 v_z^*}{\partial z^{*2}}\right) \qquad (8.494)$$

Time changes, viscosity, and pressure gradient of comparable importance:

$$\frac{1}{\text{Re}}\left(\frac{\partial v_z^*}{\partial t^*}\right) + \left(v_x^*\frac{\partial v_z^*}{\partial x^*} + v_y^*\frac{\partial v_z^*}{\partial y^*} + v_z^*\frac{\partial v_z^*}{\partial z^*}\right)$$
$$= \frac{1}{\text{Re}}\left(-\frac{\partial \mathcal{P}^*}{\partial z^*}\right) + \frac{1}{\text{Re}}\left(\frac{\partial^2 v_z^*}{\partial x^{*2}} + \frac{\partial^2 v_z^*}{\partial y^{*2}} + \frac{\partial^2 v_z^*}{\partial z^{*2}}\right)$$

$$(8.495)$$

Equations 8.491 and 8.495 are different versions of the nondimensional Navier-Stokes equation, and we arrived at them by following similar steps. The differences between the two versions are in the coefficients of the time-derivative and pressure-gradient terms. In Equation 8.491, the time-derivative and pressure-gradient coefficients are chosen to be 1; in Equation 8.495, the coefficients turn out to be $1/\text{Re}$.

To a certain extent, the differences between Equations 8.491 and 8.495 are insignificant. If our goal was, as in Chapter 7, to determine which dimensionless groups are important for developing friction correlations, both versions of the nondimensionalization are satisfactory, and they both identify the Reynolds number as the important dimensionless parameter in the flow.

We now are interested more in the second use of dimensional analysis, which is to guide us as we attempt to solve the nonlinear Navier-Stokes equations for complex flows. For this purpose, we need to know the order of magnitude of each term in the equation so that we may judge which terms are dominant and which may be neglected. When applied to this purpose, the two versions of the nondimensional Navier-Stokes equation give different answers, as demonstrated in the following examples.

EXAMPLE 8.26. *For steady creeping flow around a sphere (Re → 0), which version of the nondimensional Navier-Stokes equation (Equation 8.491 or Equation 8.495) predicts the better approximate momentum balance?*

SOLUTION. We are asked to consider two versions of the steady-state Navier-Stokes equation: one nondimensionalized with a characteristic pressure $P = \rho V^2$ and one nondimensionalized with a characteristic pressure $P = \mu D/V$. Written in Gibbs notation, the two versions are as follows:

$$P \equiv \rho V^2 \Rightarrow \text{Re}\,(\underline{v}^* \cdot \nabla^* \underline{v}^*) = \text{Re}\,(-\nabla^* \mathcal{P}^*) + \nabla^{*2} \underline{v}^* \qquad (8.496)$$

$$P \equiv \frac{\mu V}{D} \Rightarrow \text{Re}\,(\underline{v}^* \cdot \nabla^* \underline{v}^*) = \quad (-\nabla^* \mathcal{P}^*) + \nabla^{*2} \underline{v}^* \qquad (8.497)$$

Creeping flow around a sphere is flow at a vanishingly low Reynolds number. Taking the limit that Re ⟶ 0 in the two versions of the Navier-Stokes equation, we obtain two different predictions for the governing momentum equation:

$$P \equiv \rho V^2 \Rightarrow 0 = \nabla^{*2} \underline{v}^* \qquad (8.498)$$

$$P \equiv \frac{\mu V}{D} \Rightarrow 0 = -\nabla^* \mathcal{P}^* + \nabla^{*2} \underline{v}^* \qquad (8.499)$$

We see that the first choice for the characteristic pressure leads to a governing momentum equation that does not contain the pressure gradient; in the second version, the pressure is retained. When we solve creeping flow in Section 8.1.1, we retain the pressure term and find that the pressure distribution is a significant feature of creeping flow. Thus, it appears that for creeping flow around a sphere, Equation 8.499 is the correct nondimensionalization, and we should choose $P = \mu V/D$ for the characteristic pressure.

Creeping flow around a sphere:	Characteristic pressure:	$P \equiv \dfrac{\mu V}{D}$	
	Navier-Stokes:	$0 = -\nabla^* \mathcal{P}^* + \nabla^{*2} \underline{v}^*$	(8.500)

We see from this example that the two different choices for P do not give equivalent simplified Navier-Stokes equations.

In the previous example, we carefully consider steady flow around a sphere and find that there are at least two ways to nondimensionalize the pressure. To choose between the two possible characteristic pressures, we consider creeping flow and find that only the characteristic pressure based on viscosity is appropriate. This seems to settle the issue: We should choose $P = \mu V/D$. Perhaps, however, we should check another limit to verify that the characteristic pressure based on viscosity is always the correct choice.

EXAMPLE 8.27. *For steady rapid flow around a sphere, which version of the nondimensional Navier-Stokes equation (Equation 8.491 or Equation 8.495) predicts the better approximate momentum balance?*

SOLUTION. We are asked to consider two versions of the steady-state Navier-Stokes equation, one nondimensionalized with a characteristic pressure $P = \rho V^2$ and one nondimensionalized with a characteristic pressure $P = \mu D/V$. Written in Gibbs notation, the two versions are as follows:

$$P \equiv \rho V^2 \Rightarrow \text{Re}\,(\underline{v}^* \cdot \nabla^* \underline{v}^*) = \text{Re}\,(-\nabla^* \mathcal{P}^*) + \nabla^{*2} \underline{v}^* \qquad (8.501)$$

$$P \equiv \frac{\mu V}{D} \Rightarrow \text{Re}\,(\underline{v}^* \cdot \nabla^* \underline{v}^*) = \quad (-\nabla^* \mathcal{P}^*) + \nabla^{*2} \underline{v}^* \qquad (8.502)$$

Rapid flow around a sphere is represented by a high Reynolds number. Taking the limit that $\text{Re} \to \infty$ in the two versions of the Navier-Stokes equation, we obtain two different predictions for the governing momentum equation:

$$P \equiv \rho V^2 \Rightarrow (\underline{v}^* \cdot \nabla^* \underline{v}^*) = -\nabla^* \mathcal{P}^* \qquad (8.503)$$

$$P \equiv \frac{\mu V}{D} \Rightarrow (\underline{v}^* \cdot \nabla^* \underline{v}^*) = 0 \qquad (8.504)$$

We see that in the rapid-flow case, the characteristic pressure based on viscosity leads to a governing momentum equation that does not contain the pressure gradient. The inertial choice, with $P = \rho V^2$, correctly retains the pressure. In both cases, the effect of viscosity is completely lost. Equation 8.503 is the momentum balance for potential flow, and the pros and cons of potential flow are discussed in the previous section. Thus, it appears that for rapid flow around a sphere, Equation 8.503 is the least detrimental nondimensionalization; therefore, for rapid flow, we should choose $P = \rho V^2$ for the characteristic pressure:

$$
\begin{array}{ll}
\text{Potential flow} \\
\text{around a sphere:}
\end{array}
\boxed{
\begin{array}{ll}
\text{Characteristic} & P \equiv \rho V^2 \\
\text{pressure:} & \\
\text{Navier-Stokes:} & (\underline{v}^* \cdot \nabla^* \underline{v}^*) = -\nabla^* \mathcal{P}^*
\end{array}
}
\qquad (8.505)
$$

We see from this example that again the two choices for P are not equivalent and which choice is preferable depends on the flow situation: $P = \rho V^2$ is better in the case of potential flow, whereas $P = \mu V/D$ is preferred for creeping flow.

To summarize, we can think of at least two choices for characteristic pressure in flow around a sphere. We see that the viscosity-based P is preferable for the creeping-flow case, in which viscosity dominates; choosing the viscosity-based pressure leads to governing equations at zero Reynolds number that correctly include the pressure effects. The inertia-based P is more meaningful where inertia dominates; choosing the inertia-based characteristic pressure leads to governing equations at an infinite Reynolds number that correctly include the pressure effects. It appears that if we want to simplify the governing equations, it matters what we choose for characteristic values—if we make the wrong choice, we are led to the wrong simplified equations.

In addition to the confusing issue of which characteristic pressure to choose, the conundrum of potential flow remains. Dimensional analysis, even with the correct characteristic pressure chosen, leads in the high-Reynolds-number limit to results that simply do not match what is observed. It appears that dimensional analysis has failed for the case of rapid flow around a sphere: It has not led to simplified equations that predict the rich flow behavior observed (see Figure 8.22).

The failure of dimensional analysis in the case of rapid flow around a sphere is due to the choice of the sphere diameter D for the lengthscale for nondimensionalization [85]. In the boundary layer, the lengthscale over which the velocity changes is not the large lengthscale D but rather the much smaller lengthscale δ (see Example 8.22). Thus, the flow around a sphere has the property that the characteristic dimensions over which properties change are different for different regions of the flow. The choice of D as the single dimension over which to nondimensionalize leads to the difficulties experienced with the potential-flow solution [85]. When we recognize that a problem has regimes with different characteristic lengths, we can build our solution methods around the correct lengthscales. This is a technique of advanced fluid mechanics (i.e., matched asyptotic expansion). For an indepth treatment of scaling issues in fluid mechanics, see Leal [85]; see also Problem 57.

This concludes our discussion of the continuum model. The continuum model is a successful model of fluid behavior. For simple flows, with the help of calculus, we solve for the velocity and stress fields. For complex flows, with the help of dimensional analysis and advanced methods (Chapters 7, 8, and [85]), we also solve for the velocity and stress fields. In this text, we have seen how to calculate flow quantities of interest from the velocity and stress fields. In the remaining chapters of this book, we explore the origins of the macroscopic balance equations and apply these balances to more complex situations (Chapter 9) and we revisit our Chapter 2 tour of fluid behavior and see how much of that behavior is now within our modeling means.

8.4 Problems

1. The classic internal flow is pipe flow; the classic external flow is uniform flow past a sphere. Using these two examples, compare and contrast internal and external flows.
2. Compare and contrast the Fanning friction factor and the drag coefficient. What is the purpose of each?
3. Why does a skydiver reach terminal velocity? Why does the skydiver not accelerate continuously as she falls?

4. Spherical coordinates are used to solve for the velocity profile in a flow. The result is given here. Convert $\underline{v}$ from spherical coordinates to Cartesian coordinates.

$$\underline{v} = \begin{pmatrix} \left[a + b\frac{1}{r} + c\frac{1}{r^3}\right]\cos\theta \\ -\left[a + \frac{b}{2}\frac{1}{r} - \frac{c}{2}\frac{1}{r^3}\right]\sin\theta \\ 0 \end{pmatrix}_{r\theta\phi}$$

5. In Example 8.2, gravity is given by $\underline{g} = -g\hat{e}_z$. Using Equations 1.271–1.273, calculate this $\underline{g}$ in the spherical coordinate system (the answer is given in Equation 8.13).

6. From intuition, sketch the velocity field for flow around a sphere at modest flow rates. Make your arrows proportional to what you believe the velocity magnitude should be at each point.

7. In the creeping flow around a sphere problem (see Example 8.1), which terms of the Navier-Stokes equation are neglected? How is this justified?

8. In creeping flow around a sphere, we calculate the final velocity profile beginning with the guess for the velocity components in Equations 8.20 and 8.21. Carry out the detailed calculation of the final velocity profile. [Lengthy]

9. In calculating forces on the sphere in creeping flow around a sphere, we use the fact that $\tau_{rr}|_{r=R}$ at the surface of the sphere is equal to zero. Confirm this.

10. Using plotting software, plot the pressure distribution in creeping flow around a sphere. Comment on the results.

11. For a 1.0-mm-diameter polystyrene bead falling in water, what is the expected terminal speed? Assume creeping flow. What is the Reynolds number of this flow? Would this flow represent creeping flow?

12. For a 1.0-mm-diameter ball made of stainless steel falling in glycerol, what is the expected terminal speed? Assume Stokes flow. Will the ball fall in the Stokes regime?

13. What is the largest Reynolds number that we can explore with sphere-dropping experiments? What limits this experimental technique?

14. For stainless-steel spheres of reasonable sizes, in reasonable fluids, what is the minimum fluid viscosity you may use in a terminal velocity experiment? What sizes of steel ball would you use to obtain these measurements of terminal velocity?

15. When we nondimensionalize the Navier-Stokes equation in pipe flow, two dimensionless groups appear: the Reynolds number, Re, and the Froude number, Fr. When the Navier-Stokes equation was nondimensionalized for flow around a sphere, the Froude number did not appear. Explain the difference.

16. The force on a sphere in creeping flow was found to be unidirectional: $\underline{F}|_{\text{creeping}} = F_z\hat{e}_z$, whereas for noncreeping flow, the force is not unidirectional. Why?

17. What is lift? What are the dimensions of lift?

18. A cricket ball is thrown with an initial speed of 52 mph straight up in the air. How long until the ball hits the ground? With what speed will the ball hit the ground? Do not neglect air resistance.

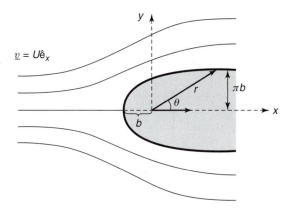

Figure 8.61 A Rankine half-body (Problem 29).

19. A smooth ball the size of a soccer ball is dropped from a bridge to a river 140 m below. Calculate the speed of the ball both with and without drag. How much error is there in the calculation if air resistance is neglected?

20. A smooth ball 4.0 cm in diameter weighing 0.25 kg is launched at an initial velocity of 40.0 mph at an angle of 34 degrees from the horizontal. What is the speed of the ball as a function of time and how far will the ball go? What is the path traced out by the ball?

21. Calculate the true pressure drag on a cylinder by numerically integrating the experimental pressure data in Figure 8.53.

22. The flow patterns behind a sphere at high Reynolds numbers are shown in Figure 8.22. Compare these flow patterns to what is observed behind a long cylinder. Discuss the comparison.

23. What is the definition and meaning of *stream function*?

24. What are the governing equations for potential flow around a sphere? Where do these equations originate?

25. Using plotting software, plot the pressure distribution in potential flow around a sphere. Comment on the results.

26. What is d'Alembert's paradox? Why is this observation important?

27. For potential flow around a sphere, calculate the pressure distribution. Start from the velocity solution given in Equation 8.238.

28. Demonstrate the error involved when the Bernoulli equation is applied inappropriately by carrying out the following calculation and comparison: Beginning with the correct velocity profile result for creeping flow around a sphere, use the Bernoulli equation (incorrectly) to calculate the pressure profile. Compare the incorrect profile obtained from the Bernoulli equation to the correct pressure profile for creeping flow around a sphere. Comment on your results.

29. The velocity field for uniform upstream flow $\underline{v} = U\hat{e}_x$ flowing in potential flow around an obstacle called a Rankine half-body is sketched in Figure 8.61. The shape of the obstacle follows the equation $r_{body}(\theta)$ given here. What is the pressure field for this flow? You may neglect gravity. The quantities b and U are constants. Plot the results for a half-body with $b = 1.0$ m and

upstream flow speed $U = 1.0$ m/s.

$$r_{\text{body}}(\theta) = \frac{(\pi - \theta)b}{\sin \theta}$$

$$\underline{v} = \begin{pmatrix} -U \cos \theta - \dfrac{Ub}{r} \\ -U \sin \theta \\ 0 \end{pmatrix}_{r\theta z}$$

30. Calculate the extra-stress tensor $\underline{\underline{\tilde{\tau}}}$ for potential flow around a sphere of an inviscid Newtonian fluid. Calculate the total-stress tensor $\underline{\underline{\tilde{\Pi}}}$. Comment on what is obtained.

31. At first glance, the streamlines for creeping flow and potential flow around a sphere (see Figures 8.9 and 8.9) seem similar. The arrow plots of the velocity fields for these two flow solutions, however, show the striking differences between the two scenarios (see Figures 8.8 and 8.25). Summarize the differences in velocity fields. Why do the streamline plots look similar? When looking at streamline plots, how can a viewer perceive the differences in flows?

32. Are boundary layers important in both internal and external flows? Explain your answer.

33. What type of forces dominate in the boundary layer? What type of forces dominate outside the boundary layer?

34. To solve the microscopic mass and momentum balances in the boundary layer, we make many assumptions. List the assumptions that go into developing the simplified equations of change for the boundary layer. Comment on each.

35. For a laminar boundary layer on a flat plate, how does the boundary-layer thickness vary with viscosity? How does the thickness vary with distance from the leading edge?

36. The flow in a boundary layer near a flat plate has two components: one that is large (v_1), and one that is much smaller but nonzero (v_2). For several locations x_1, plot $v_2(x_1)$. Comment on your results.

37. The solution for the boundary-layer flow near a flat plate is given by Equation 8.356. Plot the velocity v_1 as a function of the distance away from the plate x_2 for various distances from the leading edge (i.e., various x_1 values).

38. For water flow at 1.5 m/s over a flat plate, at what distance downstream will the boundary-layer thickness be 1 inch? Assume laminar boundary layer.

39. A boundary layer is considered thin if $\delta/x < 0.1$. For these conditions, calculate whether the boundary layer is thin for the following system: water flowing over a 1.0-m-long flat plate with a free-stream velocity of 0.010 m/s.

40. What is the force (i.e., drag) on a thin plate given the following conditions? The fluid is water, the plate is 0.52 m long and 6.3 m wide, and the free-stream velocity is 1.3×10^{-2} m/s.

41. What is the thickness of the boundary layer on a golf ball driven from the tee at 145 mph? Assume that the ball is completely smooth and therefore has a laminar boundary layer. For a real golf ball, the dimples on the surface

ensure that the boundary layer is everywhere turbulent. What is the thickness of a turbulent boundary layer under these conditions?

42. For the flow in the boundary layer near a flat plate, derive the third-order, ordinary differential equation that governs the spatial variation of the principal velocity component. Begin with the continuity equation (see Equation 8.340) and the Navier-Stokes equation (see Equation 8.341) and incorporate the coordinate transformations defined in this chapter (see Equation 8.343). The final result is Equation 8.347.

43. Example 8.19 addresses the solution for the velocity field in the problem of boundary-layer flow past a flat plate. To obtain the velocity field, we need the solution to the third-order, nonlinear ODE in Equation 8.347. Solve Equation subject to the boundary conditions in Equations 8.348, 8.349, and 8.352. This can be done using Mathematica [203] or equivalent software and by using a shooting algorithm, whereby the initial value of the second derivative of the function is guessed and adjusted until the boundary condition at infinity is satisfied. The correct guess for f"(0) is 0.332 [48].

44. Derive the expression for wall shear stress on a flat plate as a function of location (see Equation 8.366). Use the empirical curve fit (see Equation 8.356) for the velocity profile.

45. What is streamlining? Why does it work?

46. Blunt objects experience drag from two sources: pressure drag and friction drag. Explain these two types of drag. Which type is eliminated through streamlining?

47. How much faster will a cyclist traveling at 40 mph go if he buys a recumbent bicycle compared to an upright posture on a standard bicycle?

48. When riding downhill on a bicycle, a cyclist can slow down by sitting up straight rather than crouching over. How much deceleration can be expected from this posture change? Make reasonable assumptions in your calculations and indicate those assumptions.

49. What would the drag coefficient have to be to obtain the correct value of the terminal speed of a skydiver? Calculate for both the head-first and the belly-to-Earth positions.

50. If a coin falls flat-side-down through water versus edge-side-down, what is the speed difference at terminal speed?

51. What is vorticity? It is mentioned only in the advanced study of fluid mechanics, yet it is a fundamental property of flow fields. Discuss the utility of vorticity.

52. The isovorticity lines in Figure 8.60 appear to be pushed downstream by the flow. Describe what is happening in the flow that results in this effect.

53. Show that uniform potential flow past an obstacle is an irrotational flow. Hint: Far upstream of the obstacle, the flow is irrotational.

54. A vector identity useful in vorticity calculation is given in Equation 8.267. Writing the vectors in Cartesian coordinates, verify this vector identity.

55. For two-dimensional flow, use matrix calculations to show that $\underline{\omega} \cdot \nabla \underline{v} = 0$, where $\underline{\omega} = \nabla \times \underline{v}$ is the vorticity.

56. Show that $\nabla \times \nabla f = 0$, where f is a scalar function.

57. In this chapter, we always nondimensionalize time with a characteristic time $T = D/V$. For this characteristic time to be appropriate, the scaled time-derivative should be $O(1)$. This is true if characteristic changes in the velocity take place over an amount of time equal to T. A second characteristic time what we could construct from various quantities in the flow is based on the viscosity:

$$\tilde{T} = \frac{\rho D^2}{\mu} = \frac{D^2}{\nu}$$

where ν is the kinematic viscosity, which takes the role of a momentum-diffusion coefficient. Also, if the flow has its own imposed characteristic time—such as an imposed frequency of oscillation—this is another potential characteristic time to adopt.

(a) Using the definition of characteristic time $\tilde{T}$ introduced in this problem, what are the two forms of the nondimensional Navier-Stokes equation that result from choosing characteristic pressure to be first $P = \rho V^2$ and then $P = \mu V/D$?

(b) The *Strouhal number Str* is defined as the dimensionless ratio of time scales in the flow:

$$Str = \frac{T}{D/V}$$

$$Str = \frac{T}{\rho D^2/\mu}$$

Incorporate the Strouhal number into the two forms of the nondimensional Navier-Stokes equation found in (a).

(c) For the nondimensional Navier-Stokes equation discussed in this chapter, what value do we implicitly assume for the Strouhal number? In unsteady and oscillating flows, the Strouhal number assumes a prominent role [85].

PART IV

ADVANCED FLOW CALCULATIONS

9 Macroscopic Balance Equations

The mass- and momentum-balance techniques described in Chapters 3–5 are general and apply to any control volume (CV). We apply those techniques to a general microscopic control volume in Chapter 6 and use the microscopic balances in Chapters 7 and 8. Microscopic-control-volume calculations yield the equations that govern three-dimensional velocity and stress fields. If the equations can be solved, the information that microscopic balances provide is complete. Solving the microscopic balances is difficult, however, because the continuity equation and the Navier-Stokes equation are a set of four nonlinear, coupled, partial differential equations (PDEs).

For many fluids problems, the information sought is relatively large scale and flow details are not very important. For these problems—such as the calculation of the total force on a wall; overall flow rate in a device; and the total work performed by a pump, a turbine, or a mixer—balancing on a larger CV can be a fast and simple way to arrive at quantities of interest. Macroscopic CV balances are mathematically easier to calculate than microscopic CV balances, although they generally require information that must be determined experimentally.

In this chapter, we derive and learn to use the macroscopic mass, momentum, and energy balances, including the mechanical energy balance (MEB), which is discussed in Chapter 1. The macroscopic momentum balance is introduced here; it is a generalization of the problem solving methods we used in Chapters 3–5 to calculate the force caused by fluid moving through a 90-degree pipe bend (see Figure 5.21). Setting up effective macroscopic balances requires ingenuity because to obtain useful information, we must choose the control volume carefully.

We begin with derivations in Section 9.1. Section 9.2 shows how macroscopic balances can be applied to complex engineering problems.

9.1 Deriving the macroscopic balance equations

In the subsections that follow, we derive the macroscopic-balance equations. These three equations represent fundamental laws of physics: Mass is conserved, momentum is conserved, and energy is conserved. They are called the *macroscopic balances* because we write them on large systems, as contrasted with the *microscopic balances* of Chapters 6–8, which were applied to infinitesimally small control volumes and resulted in detailed differential equations. The equations derived here are coarser but easier to solve. Macroscopic balances

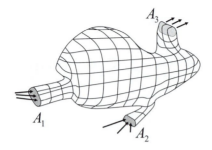

Figure 9.1   To derive the general macroscopic balances, we imagine a device of arbitrary shape and size with the input and output surfaces oriented at arbitrary angles. The flow enters and exits the device through these surfaces but is not necessarily perpendicular to the surfaces.

incorporate experimentally determined parameters to improve accuracy. Both the microscopic and macroscopic balances have a place in fluid-mechanics modeling.

9.1.1 Macroscopic mass-balance equation

In Chapter 6, we arrived at a general expression for the mass balance on a control volume (CV) of any size (see Equation 6.28, repeated here):

$$\text{Mass balance on a CV:} \qquad \boxed{\frac{dm_{CV}}{dt} + \iint_{CS} \rho(\hat{n} \cdot \underline{v})\, dS = 0} \qquad (9.1)$$

where m_{CV} is the total mass of fluid in the CV and the integral represents the net outflow of mass from the CV. The integral must be carried out on the entire control surface, CS, that forms the boundary of the CV.

In Chapters 7 and 8, we perform balances on microscopic CVs because we wanted detailed information on the velocity and stress fields. Here, we choose macroscopic CVs for our balances. Macroscopic CVs can be as large as an entire piece of equipment, and they also can be complex in shape. The complexity in CV shape is not a hindrance in macroscopic balances because the quantities we want to calculate also are macroscopic. A macroscopic CV should be chosen to incorporate the entire flow or force of interest.

To derive a general macroscopic balance, we draw a nonspecific macroscopic control volume (Figure 9.1). The device in Figure 9.1 is drawn with an arbitrary shape and with input and output surfaces at unusual angles so that we can determine the general equations for macroscopic mass, momentum, and energy balances.

To apply the mass balance in Equation 9.1 to our macroscopic CV, we first work on the integral in that equation. For the CV in Figure 9.1, the fluid velocity $\underline{v}$ is zero on all parts of the boundary except on the surfaces A_i—the input and output surfaces for the CV. We therefore can write:

$$\iint_{CS} \rho(\hat{n} \cdot \underline{v})\, dS = \sum_{A_i} \iint_{A_i} \rho(\hat{n} \cdot \underline{v})\, dS \qquad (9.2)$$

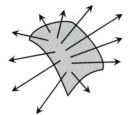

velocity varies in magnitude and direction across the surface

velocity varies in magnitude, but not in direction across the surface

Figure 9.2   Vector quantities have magnitude and direction. In our macroscopic-balance derivations we allow magnitude of velocity to vary but we hold the direction $\hat{v}$ constant.

The fluid density ρ appears in these integrals and although the fluid density may vary across the A_i, for many engineering problems this variation is insignificant. If we assume ρ_i does not vary across the surfaces A_i, we may move the density outside the integrals in Equation 9.2. With this change, the mass balance in Equation 9.1 becomes:

$$\frac{dm_{CV}}{dt} + \iint_{CS} \rho(\hat{n} \cdot \underline{v})\, dS = 0 \tag{9.3}$$

$$\frac{dm_{CV}}{dt} + \sum_{A_i} \rho_i \iint_{A_i} (\hat{n} \cdot \underline{v})\, dS = 0 \tag{9.4}$$

The integrals in this mass balance may be simplified further if we assume that the velocity through the inlet and outlet surfaces varies in magnitude across the surface but not in direction (Figure 9.2). In terms of the variables of the problem, this means that the magnitude v of the fluid velocity is a variable but the direction $\hat{v}$ of the fluid velocity is a constant (Figure 9.3).

$$\underline{v} = v\hat{v} \tag{9.5}$$

Modeling assumption $\Longrightarrow$ $\hat{v}|_{A_i}$ is constant across surface A_i (9.6)

With this assumption, Equation 9.4 becomes:

$$\frac{dm_{CV}}{dt} + \sum_{A_i} \rho_i \iint_{A_i} v(\hat{n} \cdot \hat{v})\, dS = 0 \tag{9.7}$$

$$\frac{dm_{CV}}{dt} + \sum_{A_i} \rho_i \cos\theta_i \iint_{A_i} v\, dS = 0 \tag{9.8}$$

where the θ_i are the angles between $\underline{v}|_{A_i}$ and $\hat{n}|_{A_i}$ at the input and output surfaces and, therefore, $(\hat{n} \cdot \hat{v})|_{A_i} = \cos\theta_i$ (Figure 9.4). Note that for inlet surfaces, $\hat{n}$ and $\hat{v}$ form an angle greater than 90 degrees and $\cos\theta < 0$; for outlet surfaces, the angle is less than 90 degrees and $\cos\theta > 0$.

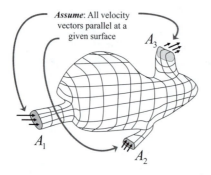

For many flows, the direction of the velocity is a constant across the inlet and outlet surfaces, whereas the magnitude of the velocity may vary.

The integrals that remain in Equation 9.8 are related to the average value of the fluid speed across the inlet and outlet surfaces:

$$\text{Average fluid speed across } A_i \qquad \langle v \rangle|_{A_i} = \frac{\iint_{A_i} v \, dS}{\iint_{A_i} dS} = \frac{1}{A_i} \iint_{A_i} v \, dS \qquad (9.9)$$

The mass balance on the macroscopic CV thus becomes:

$$\frac{dm_{CV}}{dt} + \sum_{A_i} \rho_i \cos \theta_i \iint_{A_i} v \, dS = 0 \qquad (9.10)$$

$$\frac{dm_{CV}}{dt} + \sum_{A_i} \rho_i \cos \theta_i \, A_i \, \langle v \rangle|_{A_i} = 0 \qquad (9.11)$$

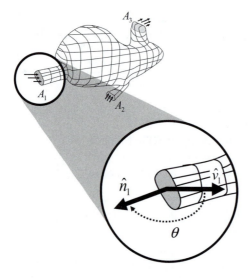

The vector $\hat{v}$ is the direction of the velocity and $\hat{n}$ denotes the outwardly pointing unit normal to a surface through which fluid passes; the angle between these two directions is θ. For inlet surfaces $\hat{n}$ and $\hat{v}$ form an angle greater than 90 degrees and $\cos \theta < 0$, whereas for outlet surfaces, $\theta < 90$ degrees and $\cos \theta > 0$.

Macroscopic
mass balance
($\rho, \hat{v}$ constant across A_i,
θ_i = angle between $\hat{n}_i$ and $\hat{v}_i$):

$$\frac{dm_{CV}}{dt} + \sum_{i=1}^{\text{\# streams}} \left[\rho A \cos\theta\langle v\rangle\right]\Big|_{A_i} = 0$$

(9.12)

Single-input,
single-output:

$$\frac{dm_{CV}}{dt} + \rho_1 A_1 \cos\theta_1 \langle v\rangle_1 + \rho_2 A_2 \cos\theta_2 \langle v\rangle_2 = 0 \qquad (9.13)$$

where N is the number of inlet and outlet surfaces; if $N = 2$ (i.e., one inlet and one outlet), the macroscopic mass balance becomes Equation 9.13. Note that $\hat{n}$ is an outwardly pointing unit normal. If we further assume that $\hat{v}$ is parallel to $\hat{n}$ at both the inlet and outlet surfaces, then the macroscopic mass balance simplifies further to ($\cos\theta_1 = -1$; $\cos\theta_2 = 1$):

Single-input,
single-output
$\hat{v}$ parallel to $\hat{n}$
(velocity perpendicular
to surface)

$$\frac{dm_{CV}}{dt} + (\rho A\langle v\rangle)|_{\text{out}} - (\rho A\langle v\rangle)|_{\text{in}} = 0 \qquad (9.14)$$

In Section 9.2, we practice using the macroscopic-mass-balance equation by applying Equation 9.13 to a Venturi meter.

9.1.2 Macroscopic momentum-balance equation

We turn now to the derivation of the macroscopic momentum balance. This balance is useful when calculating total fluid force on an apparatus. The momentum balance on a control volume of any size is given by the Reynolds transport theorem (see Equation 3.135), which is Newton's second law $\left(\sum f = m\underline{a}\right)$ written on a CV:

Reynolds transport theorem
(momentum balance on CV):

$$\frac{d\mathbf{P}}{dt} + \iint_{CS} (\hat{n} \cdot \underline{v})\,\rho\underline{v}\,dS = \sum_{\substack{\text{on} \\ \text{CV}}} \underline{f} \qquad (9.15)$$

The convective integral in Equation 9.15 may be simplified for a macroscopic CV by making the same assumptions made for developing the macroscopic mass balance (Equation 9.12)—namely, that the density is constant across surfaces A_i and the direction of the fluid velocity $\hat{v}$ does not vary across A_i:

$$\frac{d\mathbf{P}}{dt} - \sum_{\substack{\text{on} \\ \text{CV}}} \underline{f} = -\iint_{CS} (\hat{n} \cdot \underline{v})\,\rho\underline{v}\,dS \qquad (9.16)$$

$$= -\sum_{A_i} \rho_i \iint_{A_i} v^2(\hat{n} \cdot \hat{v})\hat{v}\,dS \qquad (9.17)$$

$$= -\sum_{A_i} \rho_i \cos\theta_i\, \hat{v}|_{A_i} \iint_{A_i} v^2\,dS \qquad (9.18)$$

The remaining integrals in Equation 9.18 are related to the average values across the surfaces A_i of the *square* of the fluid speed:

$$\text{Average across } A_i \text{ of fluid speed squared} \qquad \langle v^2 \rangle \big|_{A_i} = \frac{\iint_{A_i} v^2 \, dS}{\iint_{A_i} dS} = \frac{1}{A_i} \iint_{A_i} v^2 \, dS \quad (9.19)$$

The momentum balance on the macroscopic CV thus becomes:

$$\frac{d\mathbf{P}}{dt} - \sum_{\substack{\text{on} \\ \text{CV}}} \underline{f} = - \sum_{A_i} \rho_i \cos\theta_i \, \hat{v}\big|_{A_i} \iint_{A_i} v^2 \, dS \qquad (9.20)$$

$$= - \sum_{A_i} \rho_i \cos\theta_i \, \hat{v}\big|_{A_i} A_i \, \langle v^2 \rangle \big|_{A_i} \qquad (9.21)$$

Macroscopic momentum balance ($\rho, \hat{v}$ constant across A_i):
$$\frac{d\mathbf{P}}{dt} + \sum_{i=1}^{\# \text{ streams}} \left[\rho A \cos\theta \langle v^2 \rangle \hat{v} \right]\Big|_{A_i} = \sum_{\substack{\text{on} \\ \text{CV}}} \underline{f} \qquad (9.22)$$

Single-input, single-output:
$$\frac{d\mathbf{P}}{dt} + \rho_1 A_1 \cos\theta_1 \langle v^2 \rangle_1 \, \hat{v}\big|_{A_1} + \rho_2 A_2 \cos\theta_2 \langle v^2 \rangle_2 \, \hat{v}\big|_{A_2} = \sum_{\substack{\text{on} \\ \text{CV}}} \underline{f}$$
$$(9.23)$$

where N is the number of inlet and outlet surfaces; if $N = 2$ (i.e., one inlet and one outlet), the macroscopic momentum balance becomes Equation 9.23. As noted for the macroscopic mass-balance equation, $\cos\theta_i$ is negative for inlet surfaces.

Equation 9.22 contains an expression for the average of the square of the speed $\langle v^2 \rangle$. For turbulent flows, because the velocity profile is flat (or uniform), this quantity is approximately equal to the square of the average speed. For laminar flows, the velocity profile is not flat but rather parabolic (see Equation 7.23) and $\langle v^2 \rangle$ is appreciably larger than the square of the average speed (see Example 9.1). To account for these two situations, we define the parameter β as:

$$\text{Momentum velocity profile parameter} \qquad \boxed{\beta \equiv \frac{\langle v \rangle^2}{\langle v^2 \rangle}} \qquad (9.24)$$

Incorporating $\langle v^2 \rangle = \langle v \rangle^2 / \beta$ into Equations 9.22 and 9.23, we obtain:

Macroscopic momentum balance ($\rho, \hat{v}$ constant across A_i, $\theta_i = $ angle between $\hat{n}_i$ and $\hat{v}_i$):
$$\boxed{\frac{d\mathbf{P}}{dt} + \sum_{i=1}^{\# \text{ streams}} \left[\frac{\rho A \cos\theta \langle v \rangle^2}{\beta} \hat{v} \right]\Bigg|_{A_i} = \sum_{\substack{\text{on} \\ \text{CV}}} \underline{f}}$$
$$(9.25)$$

$$\frac{d\mathbf{P}}{dt} + \sum_{i=1}^{\# \ streams} \left[\frac{\rho A \cos\theta \langle v \rangle^2}{\beta} \begin{pmatrix} \hat{v}_1 \\ \hat{v}_2 \\ \hat{v}_3 \end{pmatrix}_{123} \right]_{A_i} = \sum_{\substack{on \\ CV}} \underline{f} \qquad (9.26)$$

Single-input, single-output:
$$\frac{d\mathbf{P}}{dt} + \frac{\rho_1 A_1 \cos\theta_1 \langle v \rangle_1^2}{\beta_1} \hat{v}|_{A_1} + \frac{\rho_2 A_2 \cos\theta_2 \langle v \rangle_2^2}{\beta_2} \hat{v}|_{A_2} = \sum_{\substack{on \\ CV}} \underline{f}$$

$$(9.27)$$

In the second form of the macroscopic momentum-balance equation (Equation 9.26), we write the unit vectors $\hat{v}|_{A_i}$ in matrix form to emphasize the vector direction associated with those terms. Because momentum is a vector, the vector directions $\hat{v}|_{A_i}$ have an important role in problems that involve the macroscopic momentum balance.

To determine the values of β for laminar and turbulent flow, we carry out the integrations of the velocity (see Equation 9.9) and the square of the velocity (see Equation 9.19) for these two flows.

EXAMPLE 9.1. *For laminar and turbulent pipe flow, what are the correct values of the momentum velocity-profile parameter β?*

SOLUTION. The momentum velocity-profile parameter β is defined in Equation 9.24:

$$\beta \equiv \frac{\langle v \rangle^2}{\langle v^2 \rangle} \qquad (9.28)$$

If the speed $v = |\underline{v}|$ is constant across the cross-sectional area A_1, we calculate β to be:

$$v = v_0 = \text{constant} \qquad (9.29)$$

$$\beta = \frac{\langle v \rangle^2}{\langle v^2 \rangle} \qquad (9.30)$$

$$= \frac{\left(\frac{1}{A_1} \iint_{A_1} v \, dS \right)^2}{\left(\frac{1}{A_1} \iint_{A_1} v^2 \, dS \right)} \qquad (9.31)$$

$$= \frac{\left(\frac{1}{A_1} \iint_{A_1} v_0 \, dS \right)^2}{\left(\frac{1}{A_1} \iint_{A_1} v_0^2 \, dS \right)} = \frac{\left(\frac{(v_0 A_1)}{A_1} \right)^2}{v_0^2 \frac{A_1}{A_1}} \qquad (9.32)$$

$$\beta = 1 \qquad (9.33)$$

A value of $\beta = 1$ indicates that there is no variation in the velocity across the cross section.

For laminar pipe flow, we calculate in Chapter 7 that the velocity is not constant across the cross section of the pipe. The velocity in laminar pipe flow was found to be as follows (see Equation 7.23):

$$\text{Laminar pipe flow:} \quad \underline{v} = v\hat{v} = \begin{pmatrix} 0 \\ 0 \\ v_z \end{pmatrix}_{r\theta z} = v_z \hat{e}_z \tag{9.34}$$

$$v = v_z(r) = \frac{(p_0 - p_L + \rho g L)R^2}{4\mu L}\left[1 - \left(\frac{r}{R}\right)^2\right] \tag{9.35}$$

$$= v_{z,max}\left[1 - \left(\frac{r}{R}\right)^2\right] \tag{9.36}$$

where $v_{z,max} = (p_0 - p_L + \rho g L)R^2/4\mu L$. We can use Equation 9.36 to carry out the calculation of β for laminar pipe flow:

$$\beta = \frac{\langle v \rangle^2}{\langle v^2 \rangle} = \frac{\left(\dfrac{1}{A_1}\displaystyle\iint_{A_1} v\, dS\right)^2}{\left(\dfrac{1}{A_1}\displaystyle\iint_{A_1} v^2\, dS\right)} \tag{9.37}$$

$$= \frac{\left(\dfrac{1}{\pi R^2}\displaystyle\int_0^{2\pi}\int_0^R v_{z,max}\left[1 - \left(\frac{r}{R}\right)^2\right]r\,dr\,d\theta\right)^2}{\left(\dfrac{1}{\pi R^2}\displaystyle\int_0^{2\pi}\int_0^R \left(v_{z,max}\left[1 - \left(\frac{r}{R}\right)^2\right]\right)^2 r\,dr\,d\theta\right)} \tag{9.38}$$

$$= \frac{\left(\dfrac{2v_{z,max}}{R^2}\displaystyle\int_0^R\left[r - \frac{r^3}{R^2}\right]dr\right)^2}{\dfrac{2v_{z,max}^2}{R^2}\displaystyle\int_0^R\left[r - \frac{2r^3}{R^2} + \frac{r^5}{R^4}\right]dr} \tag{9.39}$$

$$= \frac{\dfrac{2}{R^2}\left(\left[\dfrac{r^2}{2} - \dfrac{r^4}{4R^2}\right]\Big|_0^R\right)^2}{\left[\dfrac{r^2}{2} - \dfrac{r^4}{2R^2} + \dfrac{r^6}{6R^4}\right]\Big|_0^R} = \frac{\dfrac{2}{R^2}\left(\dfrac{R^2}{4}\right)^2}{\dfrac{R^2}{6}} = \frac{3}{4} \tag{9.40}$$

$$\boxed{\beta = \frac{3}{4} = 0.75} \qquad \begin{array}{l}\text{Momentum} \\ \text{velocity profile} \\ \text{parameter for} \\ \text{laminar flow}\end{array} \tag{9.41}$$

We find that $\beta = \langle v \rangle^2/\langle v^2 \rangle = 0.75$ for laminar flow. Thus, the quantity that appears in the macroscopic momentum balance, $\langle v^2 \rangle$, in laminar flow is equal to:

$$\text{Laminar flow:} \quad \langle v^2 \rangle = (4/3)\langle v \rangle^2 = 1.33\langle v \rangle^2 \tag{9.42}$$

We can carry out the same calculation for turbulent pipe flow if we have an expression for $v_z(r)$ for that case. Experimental measurements show that the velocity profile for turbulent flow may be written as [14, 137]:

$$\begin{array}{c}\text{Velocity profile}\\\text{for turbulent}\\\text{pipe flow}\\\text{(experimental result):}\end{array} \qquad v_z(r) = v_{z,max}\left(1 - \frac{r}{R}\right)^{\frac{1}{7}} \qquad (9.43)$$

where $v_{z,max}$ is the maximum value of the velocity. Substituting Equation 9.43 into Equation 9.37, we arrive at the value of β for turbulent flow (the details are left to readers; Problem 6):

$$\begin{array}{c}\text{Momentum}\\\text{velocity-profile}\\\text{parameter for}\\\text{turbulent flow:}\end{array} \qquad \boxed{\beta = 0.98} \qquad (9.44)$$

The momentum velocity-profile parameter β thus varies from $\beta = 3/4$ for laminar flow to $\beta = 0.98$ or $\beta = 1$ for turbulent flow or plug flow, respectively.

We arrive at the macroscopic momentum balance (see Equation 9.25, repeated here), which we obtained by adapting the general momentum balance on a CV (see Equation 9.15) to the case of a macroscopic CV:

$$\begin{array}{c}\text{Macroscopic}\\\text{momentum}\\\text{balance}\\(\rho, \hat{v} \text{ constant across } A_i):\end{array} \qquad \boxed{\frac{d\mathbf{P}}{dt} + \sum_{i=1}^{\# \text{ streams}}\left[\frac{\rho A \cos\theta \langle v \rangle^2}{\beta}\hat{v}\right]\Bigg|_{A_i} = \sum_{\substack{\text{on}\\\text{CV}}} \underline{f}}$$

$$(9.45)$$

In this adaptation, we specified that there is no change in either the density or the direction of the velocity across the input and output surfaces. The modifications to the general momentum balance in Equation 9.15 are only in the convective term (i.e., the integral).

The sum-of-the-forces term and the rate-of-change term are unaltered for macroscopic versus microscopic CVs. In Chapters 3–5, the forces on a macroscopic CV were body forces (i.e., gravity) and the molecular surface forces on the bounding control surfaces. In a macroscopic CV, the bounding surface may be solid (i.e., walls) or fluid:

$$\sum_{\substack{\text{on}\\\text{CV}}} \underline{f} = \underline{f}_{\text{gravity}} + \underline{f}_{\text{surface}} \qquad (9.46)$$

$$= \underline{f}_{\text{gravity}} + \underline{f}_{\text{inlet}} + \underline{f}_{\text{outlet}} + \underline{f}_{\text{walls}} \qquad (9.47)$$

The force due to gravity on the CV may be written (as usual) as follows:

$$\underline{f}_{\text{gravity}} = M_{CV}\underline{g} \qquad (9.48)$$

The fluid surface forces, including pressure, are expressed using the stress tensor $\underset{=}{\tilde{\Pi}}$ (see Equation 4.221):

$$\begin{array}{ll} \text{Total molecular fluid force} \\ \text{on a surface } \mathcal{S}: \end{array} \qquad \mathcal{F} = \iint_S [\hat{n} \cdot \underset{=}{\tilde{\Pi}}]_{\text{at surface}} \, dS \qquad (9.49)$$

$$= \iint_S \left[\hat{n} \cdot \left(-p\underset{=}{I} + \underset{=}{\tilde{\tau}}\right)\right]_{\text{at surface}} \, dS \quad (9.50)$$

$$= \iint_S \left[-p\hat{n} + \hat{n} \cdot \underset{=}{\tilde{\tau}}\right]_{\text{at surface}} \, dS \qquad (9.51)$$

The simplification $\hat{n} \cdot p\underset{=}{I} = p\hat{n}$ in Equation 9.51 is due to vector/tensor relationships discussed in Section 1.3.2.2. The pressure contribution $\iint -p\hat{n} \, dS$ on fluid surfaces of macroscopic CVs is almost always important; the extra-stress $\iint \hat{n} \cdot \underset{=}{\tilde{\tau}} \, dS$ contribution on fluid surfaces of CVs is almost never important, as demonstrated in the examples in this chapter. This particular viscous term (viscous stresses on fluid surfaces) may be omitted; we formally address this term in this chapter to demonstrate that omitting it as regular practice is justified. Viscous stresses on solid surfaces are very important and are included in f_{walls}.

In Section 9.2, we apply the macroscopic momentum balance (Equation 9.45) to engineering problems, and we present examples that guide readers to evaluate the various force terms. The derivation of the macroscopic energy balance is discussed next.

9.1.3 Energy balance

The third macroscopic balance is on energy. When we describe microscopic balances in Chapter 6, little was discussed about microscopic energy balances; microscopic energy balances are mostly important in flows in which the temperature varies or in which there are large thermal energy flows. Macroscopic energy balances, conversely, are widely used, even in flows in which the temperature is constant. In fact, a particular version of the macroscopic energy balance, the mechanical energy balance (MEB) (see Equation 1.3), is possibly the most widely used equation in fluid mechanics, as discussed in Chapter 1. Later in this section, we derive the MEB from the general macroscopic energy balance.

The energy balance on a CV is given in Chapter 6 in Equation 6.77 and repeated here:

$$\begin{array}{ll} \text{Energy balance} \\ \text{on a CV:} \end{array} \qquad \boxed{\frac{dE_{CV}}{dt} + \iint_{CS} (\hat{n} \cdot \underline{v})\rho \hat{E} \, dS = Q_{in,CV} - W_{by,CV}} \qquad (9.52)$$

The energy balance states that the rate-of-change of total energy in a CV ($E_{CV} = U + E_k + E_p$), plus the net convective outflow of total energy (i.e., the integral), is balanced by the rate of heat into the CV $Q_{in,CV}$ minus the rate of work done by the fluid in the CV $W_{by,CV}$. This is the first law of thermodynamics written on a constant CV.

Energy balances are an important tool in the field of thermodynamics, and a wide body of literature exists that describes how energy balances are applied to physical systems [99, 157, 167]. It is conventional in thermodynamics to

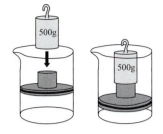

Closed System: gas
trapped by a piston

Open System: fluid in
a centrifugal pump

Figure 9.5   Examples of a closed system and an open system. The gas trapped by a piston is a closed system that may be heated or worked on, but the mass of the system does not change. The liquid in a centrifugal pump also may be heated and worked on, but it also changes in mass as fluid flows in and out of the pump; this is an example of an open system.

consider two classes of problems: *closed systems* and *open systems*. Applying energy balances on closed systems is analogous to applying mass, momentum, and energy balances on a body—the mass of a closed system does not change. A system of unchanging mass is *closed* in the sense that mass does not cross the boundaries of the system (Figure 9.5, top). By contrast, mass crosses the boundaries of an open system (Figure 9.5, bottom). Applying energy balances on open systems is analogous to applying mass, momentum, and energy balances on a control volume. In fluid mechanics, we are concerned with open-system energy balances.

We derive two versions of Equation 9.52, the *macroscopic closed-system energy balance* and the *macroscopic open-system energy balance*. From the open-system energy balance, we can obtain the mechanical energy balance (MEB) and its specialized version, the macroscopic Bernoulli equation. These equations are the principal tools used for analyzing pumps, piping networks, and other flow machinery such as turbines and mixers. We present all three energy balances in the next sections to compile all of the energy balances on liquids within the same organizational structure. The discussion of fluid-mechanics problem solving with the MEB is in Chapter 1 for elementary problems in Section 9.2 and for more complex problems.

9.1.3.1 CLOSED SYSTEMS

We begin with the energy balance on a control volume, Equation 9.52:

Energy balance
on a CV
(First law of
(thermodynamics)

$$\frac{dE_{CV}}{dt} + \iint_{CS} (\hat{n} \cdot \underline{v}) \rho \hat{E} \, dS = Q_{in,CV} - W_{by,CV}$$

(9.53)

For a closed system, no mass flows into or out of the system; thus, the convective term (i.e., the integral) is zero:

$$\text{Energy balance on a closed system} \qquad \frac{dE_{CV}}{dt} = Q_{in,CV} - W_{by,CV} \qquad (9.54)$$

When performing balances on a closed system, we usually are interested in the changes that take place between an initial time t_0 and a final time t_f. To write Equation 9.54 in terms of these two times, we integrate between these two limits:[1]

$$\int_{t_0}^{t_f} dE_{CV} = \oint_{t_0}^{t_f} Q_{in,CV}\, dt - \oint_{t_0}^{t_f} W_{by,CV}\, dt \quad (9.55)$$

$$E_{CV}|_{t_f} - E_{CV}|_{t_0} = \underset{f-i}{\Delta} E_{CV} = \oint_{t_0}^{t_f} Q_{in,CV}\, dt - \oint_{t_0}^{t_f} W_{by,CV}\, dt \quad (9.56)$$

Here, we introduce the symbol $\underset{f-i}{\Delta}$ to indicate the difference between the final and the initial values of the property. The integrals on the righthand side of Equation 9.56 are the total amount of heat and work associated with the energy change $\underset{f-i}{\Delta} E_{CV}$. We use the line integral symbol for these integrations to remind us that heat and work in a process depend on the path taken in the process. These quantities are not state functions (see the thermodynamics literature [157]). We define the heat and work integrals as $\mathcal{Q}_{in}$ and $\mathcal{W}_{by}$:

$$\mathcal{Q}_{in} \equiv \oint_{t_0}^{t_f} Q_{in,CV}\, dt \qquad (9.57)$$

$$\mathcal{W}_{by} \equiv \oint_{t_0}^{t_f} W_{by,CV}\, dt \qquad (9.58)$$

The closed-system energy balance thus becomes:

$$\text{Energy balance, closed system:} \qquad \underset{f-i}{\Delta} E_{CV} = \mathcal{Q}_{in} - \mathcal{W}_{by} \qquad (9.59)$$

As discussed in Chapter 6, the total energy of a system is composed of the sum of three contributions to energy: internal, kinetic, and potential energy, $E_{CV} = U + E_k + E_p$. Incorporating this sum into Equation 9.59, we arrive at the final version of the macroscopic closed-system energy balance:

$$\text{Macroscopic closed-system energy balance:} \qquad \boxed{\underset{f-i}{\Delta} U + \underset{f-i}{\Delta} E_k + \underset{f-i}{\Delta} E_p = \mathcal{Q}_{in} - \mathcal{W}_{by}} \qquad (9.60)$$

where the $\underset{f-i}{\Delta}$ are the changes in that property in the sense *final–initial*, and $\mathcal{Q}_{in}$ and $\mathcal{W}_{by}$ are given by Equations 9.57 and 9.58.

---

[1] The energy of a system is a state function; that is, its value depends on only the state of a system and not on the particular path that a system takes to arrive at that state. State variables integrate straightforwardly. The heat and work associated with a system are *not* state variables. The work and heat that go into changing the energy of a system are different depending on the path chosen. For that reason, we cannot evaluate directly the integrals in Equations 9.57 and 9.58 unless the path is known. For more on this aspect of thermodynamics, see the literature [99, 157].

9.1.3.2 OPEN SYSTEMS

As with the macroscopic closed-system balance, the derivation of the macroscopic open-system balance begins with the energy balance on a CV, Equation 9.52. For an open system, mass flows into and out of the system; thus, the convective term (i.e., the integral) is an important part of this balance:

$$
\begin{array}{c} \text{Energy balance} \\ \text{on a CV} \\ \text{(First law of} \\ \text{thermodynamics):} \end{array} \qquad \frac{dE_{CV}}{dt} + \iint_{CS} (\hat{n} \cdot \underline{v})\rho \hat{E} \, dS = Q_{in,CV} - W_{by,CV}
$$

$$(9.61)$$

The convective integral in Equation 9.61 may be simplified for a macroscopic CV by making the same assumptions as previously in developing the macroscopic mass and momentum balances—namely, that the density ρ_i is constant across control surfaces A_i and the direction of the fluid velocity $\hat{v}_i$ does not vary across A_i. With these assumptions and writing $\underline{v} = v\hat{v}$, the convective integral becomes:

$$
\iint_{CS} (\hat{n} \cdot \underline{v})\rho \hat{E} \, dS = \sum_{A_i} \iint_{A_i} (\hat{n} \cdot \underline{v})\rho \hat{E} \, dS \tag{9.62}
$$

$$
= \sum_{A_i} \cos \theta_i \rho_i \iint_{A_i} \hat{E} v \, dS \tag{9.63}
$$

Note that we incorporate the fact that $\hat{n} \cdot \underline{v} = v|_{A_i} \cos \theta_i$. If we further assume that $\hat{E}_i$ does not vary across surface A_i, Equation 9.63 becomes (see the caveat in Example 9.2):

$$
\iint_{CS} (\hat{n} \cdot \underline{v})\rho \hat{E} \, dS = \sum_{A_i} \cos \theta_i \rho_i \hat{E}_i \iint_{A_i} v \, dS \tag{9.64}
$$

With the help of Equation 9.9, we recognize the remaining integral as related to the average speed of the fluid across A_i:

$$
\begin{array}{c} \text{Average fluid speed} \\ \text{across } A_i: \end{array} \qquad \langle v \rangle|_{A_i} = \frac{\iint_{A_i} v \, dS}{\iint_{A_i} dS} = \frac{1}{A_i} \iint_{A_i} v \, dS \tag{9.65}
$$

The convective term in the energy balance becomes:

$$
\iint_{CS} (\hat{n} \cdot \underline{v})\rho \hat{E} \, dS = \sum_{A_i} \cos \theta_i \rho_i \hat{E}_i A_i \, \langle v \rangle|_{A_i} \tag{9.66}
$$

For many open-system energy-balance problems, we can further assume that $\hat{n}$ and $\hat{v}$ are parallel (Figure 9.6). This is the case if the fluid enters and exits the CV perpendicular to the inlet and outlet control surfaces. If $\hat{n}$ and $\underline{v}$ are parallel and in the same direction ($\hat{n} = \hat{v}$), then $\hat{n} \cdot \hat{v} = \cos \theta = 1$; this is the case for an outflow surface. If $\hat{n}$ and $\underline{v}$ are parallel but in opposite directions ($\hat{n} = -\hat{v}$), then $\hat{n} \cdot \hat{v} = \cos \theta = -1$; this is the case for an inflow surface. We can break up the summation in Equation 9.66 into outflow and inflow surfaces. We further

inlet surfaces   *outlet surfaces*

Figure 9.6   For many systems analyzed by the macroscopic balances, the velocity is perpendicular to the inlet and outlet control surfaces.

simplify the convective expression by writing the mass flow rate through A_i as $m_i = \rho_i A_i \, \langle v \rangle|_{A_i}$:

$$\iint_{CS} (\hat{n} \cdot \underline{v}) \rho \hat{E} \, dS = \sum_{A_i} \cos \theta_i \, \rho_i \hat{E}_i A_i \, \langle v \rangle|_{A_i} \tag{9.67}$$

$$= \sum_{\substack{A_j \\ \text{out}}} \left(\rho_j A_j \, \langle v \rangle|_{A_j} \right) \hat{E}_j - \sum_{\substack{A_i \\ \text{in}}} \left(\rho_i A_i \, \langle v \rangle|_{A_i} \right) \hat{E}_i \tag{9.68}$$

$$= \sum_{\substack{A_j \\ \text{out}}} m_j \hat{E}_j - \sum_{\substack{A_i \\ \text{in}}} m_i \hat{E}_i \tag{9.69}$$

The macroscopic energy balance becomes:

$$\frac{dE_{CV}}{dt} + \sum_{\substack{A_j \\ \text{out}}} m_j \hat{E}_j - \sum_{\substack{A_i \\ \text{in}}} m_i \hat{E}_i = Q_{in,CV} - W_{by,CV} \tag{9.70}$$

The total energy of a system is composed of the sum of three contributions to energy: internal, kinetic, and potential, $E_{CV} = U + E_k + E_p$, or on a per-unit-mass basis, $\hat{E}_{CV} = \hat{U} + \hat{E}_k + \hat{E}_p$. Incorporating these expressions into Equation 9.69, we arrive at the final version of the convective terms of the macroscopic open-system energy balance:

$$\sum_{\substack{A_j \\ \text{out}}} m_j \hat{E}_j - \sum_{\substack{A_i \\ \text{in}}} m_i \hat{E}_i$$

$$= \sum_{\substack{A_j \\ \text{out}}} m_j \left[\hat{U} + \hat{E}_k + \hat{E}_p \right]_j - \sum_{\substack{A_i \\ \text{in}}} m_i \left[\hat{U} + \hat{E}_k + \hat{E}_p \right]_i \tag{9.71}$$

$$= \left[\sum_{\substack{A_j \\ \text{out}}} m_j \hat{U}_j - \sum_{\substack{A_i \\ \text{in}}} m_i \hat{U}_i \right] + \left[\sum_{\substack{A_j \\ \text{out}}} m_j \hat{E}_{k,j} - \sum_{\substack{A_i \\ \text{in}}} m_i \hat{E}_{k,i} \right]$$

$$+ \left[\sum_{\substack{A_j \\ \text{out}}} m_j \hat{E}_{p,j} - \sum_{\substack{A_i \\ \text{in}}} m_i \hat{E}_{p,i} \right] \tag{9.72}$$

$$= \underset{o-i}{\Delta U} + \underset{o-i}{\Delta E_k} + \underset{o-i}{\Delta E_p} \tag{9.73}$$

where the expression $\underset{o-i}{\Delta}$ means change $\sum$ out $- \sum$ in. The $\underset{o-i}{\Delta}$ terms in Equation 9.73 are defined here:

$$\underset{o-i}{\Delta} U \equiv \sum_{out} m_j \hat{U}_j - \sum_{in} m_i \hat{U}_i \tag{9.74}$$

$$\underset{o-i}{\Delta} E_k \equiv \sum_{out} m_j \hat{E}_{k,j} - \sum_{in} m_i \hat{E}_{k,i} \tag{9.75}$$

$$\underset{o-i}{\Delta} E_p \equiv \sum_{out} m_j \hat{E}_{p,j} - \sum_{in} m_i \hat{E}_{p,i} \tag{9.76}$$

Substituting these results into Equation 9.70 gives the final version of the microscopic open-system energy balance:

Macroscopic open-system energy balance:

$$\boxed{\frac{dE_{CV}}{dt} + \underset{o-i}{\Delta} U + \underset{o-i}{\Delta} E_k + \underset{o-i}{\Delta} E_p = Q_{in,CV} - W_{by,CV}}$$

$$\tag{9.77}$$

$$\frac{d}{dt}\left(U + E_k + E_p\right) + \underset{o-i}{\Delta} U + \underset{o-i}{\Delta} E_k + \underset{o-i}{\Delta} E_p = Q_{in,CV} - W_{by,CV} \tag{9.78}$$

Again, the $\underset{o-i}{\Delta} E_k$, $\underset{o-i}{\Delta} E_p$, and $\underset{o-i}{\Delta} U$ in Equation 9.77 refer to the differences between the sum of contributions from the outlet streams minus the sum of contributions from the inlet streams ($\sum$ out $- \sum$ in).[2]

At steady state, the time derivative in Equation 9.77 is zero. The steady-state, macroscopic, open-system energy balance is shown here:

$$\boxed{\Delta E_p + \Delta E_k + \Delta U = Q_{in,CV} - W_{by,CV}}$$

Steady-state, macroscopic, open-system, energy balance (preliminary form)

$$\tag{9.79}$$

In this equation, we revert to using Δ for the change-in-energy terms in preference to $\underset{o-i}{\Delta}$ because using the bare symbol Δ is more standard. In this and all related open-system equations, the symbol Δ signifies $\sum$ out $- \sum$ in.

The preliminary form of the open-system balance shown in Equation 9.79 is correct, but this equation may be written in a more convenient form with a few adaptations. In open systems, the work term $W_{by,CV}$ contains two contributions. The first contribution is due to moving parts that intrude into the system, such as mixing shafts, turbines, and the internal workings of pumps (Figure 9.7). The work performed by the fluid associated with moving shafts is called *shaft work* and is given the symbol $W_{s,by}$. The other contribution to $W_{by,CV}$ in an open system is the work done by the fluid as it enters or leaves the system (Figure 9.8); this contribution is called *flow work*. Flow work usually is combined with the convective terms, as follows.

---

[2]Contrast the $\underset{o-i}{\Delta}$ here with $\underset{f-i}{\Delta}$ in Equation 9.60.

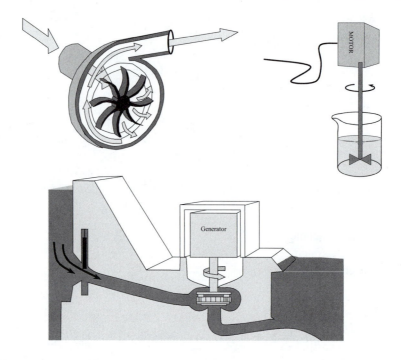

Figure 9.7 Work is force times displacement; thus, moving parts are one source of work. Work performed by the fluid associated with moving parts is called shaft work. Examples of systems with shaft work are centrifugal pumps, mixers, and turbines used in hydropower generation.

A stream entering a chosen open system flows with a pressure p_i and at a volumetric flow rate of $Q_i = \langle v \rangle_i A_i$, where $\langle v \rangle_i$ is the average speed of the fluid in the ith inlet stream and A_i is the cross-sectional area of the ith inlet stream. Pressure is force per unit area, and rate-of-work is force multiplied by velocity;

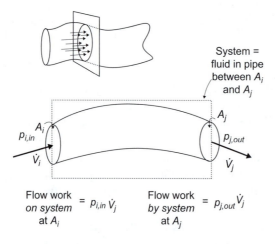

$$\begin{matrix} \text{Flow work} \\ \textit{on system} \\ \text{at } A_i \end{matrix} = p_{i,in}\,\dot{V}_j \qquad \begin{matrix} \text{Flow work} \\ \textit{by system} \\ \text{at } A_j \end{matrix} = p_{j,out}\,\dot{V}_j$$

Figure 9.8 Work is force times displacement; thus, moving fluid is a source of work. Work done by or on the fluid as it enters or leaves the system is called flow work. The work on the boundaries of a flow system is done by fluid outside the boundary on the fluid inside the system. If the system works on its surroundings, such as at the exit shown here, then the work on the system is negative.

thus, just at the system boundary as the fluid enters, the pressure times the cross-sectional area of the pipe is a force acting on the fluid in the CV, doing work on the fluid as it crosses into the system (see Figure 9.8). The rate-of-work *by* the fluid is the negative of this:

$$\left(\begin{array}{c} \text{rate of flow work} \\ by \text{ fluid system at entrance} \\ \text{for } i\text{th input stream} \end{array} \right) = (\text{force}) \left(\frac{\text{displacement}}{\text{time}} \right)$$

$$= \left[\left(\frac{\text{force}}{\text{area}} \right) (\text{area}) \right] \left(\frac{\text{displacement}}{\text{time}} \right)$$

$$= -p_i A_i \langle v \rangle_i \qquad (9.80)$$

$$= -p_i Q_i \qquad (9.81)$$

A stream exiting a chosen open system flows with a pressure p_j and at a volumetric flow rate of $Q_j = \langle v \rangle_j A_j$, where $\langle v \rangle_j$ is the average speed of the jth exit stream and A_j is the cross-sectional area of the jth exit stream. As before, just at the system boundary as the fluid exits, the pressure times the cross-sectional area of the pipe is a force acting on fluid. However, because this stream is an exiting stream, the work is done by the fluid in the CV on fluid that is outside of the CV. Thus, the work done *by* the chosen system at the exit is the force times the fluid displacement at the exit (i.e., no sign change):

$$\left(\begin{array}{c} \text{rate of flow work} \\ by \text{ fluid system at exit} \\ \text{for } j\text{th stream} \end{array} \right) = p_j A_j \langle v \rangle_j \qquad (9.82)$$

$$= p_j Q_j \qquad (9.83)$$

We now sum all of the flow-work contributions and rearrange the open-system energy balance to include the separation of shaft work and flow work into the different expressions derived previously:

$$\Delta E_p + \Delta E_k + \Delta U = Q_{in,CV} - W_{by,CV} \qquad (9.84)$$

$$= Q_{in,CV} - W_{s,by} + \sum_{in} p_i Q_i - \sum_{out} p_j Q_j \qquad (9.85)$$

$$\Delta E_p + \Delta E_k + \left[\Delta U + \sum_{out} p_j Q_j - \sum_{in} p_i Q_i \right] = Q_{in,CV} - W_{s,by,CV} \qquad (9.86)$$

The two flow-work terms commonly are combined with the internal-energy term and expressed in terms of the change in the thermodynamic function enthalpy, as we now show. *Specific enthalpy* or enthalpy per unit mass $\hat{H}$ is defined as:

Specific enthalpy $\boxed{\hat{H} \equiv \hat{U} + p\hat{V}}$ $\qquad (9.87)$

where $\hat{V} = 1/\rho$ is the specific volume and, therefore, $m\hat{V} = Q$:

$$m \qquad \hat{V} \qquad = \qquad Q \qquad (9.88)$$

$$\left(\frac{\text{mass}}{\text{time}}\right)\left(\frac{\text{volume}}{\text{mass}}\right) = \left(\frac{\text{volume}}{\text{time}}\right) \qquad (9.89)$$

For each flow stream in our system, we can calculate the amount of enthalpy brought in or taken out; summing as previously we calculate an overall change in enthalpy for our system:

$$\begin{pmatrix} \text{net rate} \\ \text{of enthalpy} \\ \text{flow out of} \\ \text{open system} \end{pmatrix} = \underset{o-i}{\Delta} H = \Delta H = \sum_{out} m_j \hat{H}_j - \sum_{in} m_i \hat{H}_i \qquad (9.90)$$

$$= \sum_{out}\left(m_j \hat{U}_j + m_j p_j \hat{V}_j\right) - \sum_{in}\left(m_i \hat{U}_i + m_i p_i \hat{V}_i\right) \qquad (9.91)$$

The $mp\hat{V} = pQ$ terms can be recognized as the flow-work terms that appeared in Equation 9.86 (see also Equation 9.81):

$$\begin{pmatrix} \text{net rate of} \\ \text{enthalpy flow out} \\ \text{of an open system} \end{pmatrix} = \Delta H \qquad (9.92)$$

$$= \sum_{out}\left(m_j \hat{U}_j + p_j Q_j\right) - \sum_{in}\left(m_i \hat{U}_i + p_i Q_i\right) \qquad (9.93)$$

$$= \left(\sum_{out} m_j \hat{U}_j - \sum_{in} m_i \hat{U}_i\right) + \sum_{out} p_j Q_j - \sum_{in} p_i Q_i \qquad (9.94)$$

$$= \left[\Delta U + \sum_{out} p_j Q_j - \sum_{in} p_i Q_i\right] \qquad (9.95)$$

Equation 9.95 matches the bracketed terms in Equation 9.86. Returning to Equation 9.86 and combining with Equation 9.95, we obtain the conventional form of the macroscopic, open-system energy balance:

Macroscopic open-system energy balance (steady state):

$$\boxed{\Delta E_p + \Delta E_k + \Delta H = Q_{in,CV} - W_{s,by,CV}} \qquad (9.96)$$

where here Δ refers to $\sum$ out $- \sum$ in and ΔH is given by the following expression:

$$\Delta H = \left[\sum_{\substack{A_j \\ out}} m_j \hat{H}_j - \sum_{\substack{A_i \\ in}} m_i \hat{H}_i\right] \qquad (9.97)$$

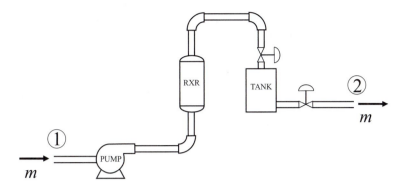

A common system often is one with a single input stream and a single output stream, and in which an incompressible $(1/\hat{V} = \rho = \text{constant})$, nonreacting, nearly isothermal (U small) fluid is flowing.

For many heat-transfer systems, separation systems, and reactors, the kinetic and potential energy changes are not important and there is no shaft work (i.e., no mixers, no turbines, and no pumps). Under these conditions, the open-system energy balance reduces to:

$$\Delta H = Q_{in,CV}$$

Open-system energy balance
when ΔE_p, ΔE_k, $W_{s,by,CV} \approx 0$ (9.98)
(steady state)

A way to think about enthalpy, therefore, is as the energy function that changes when heat is added to an open system (i.e., mass flows in and out) under the fairly common conditions listed with Equation 9.98.

Note that for all of the Δ terms in the open-system balances, Δ refers to $\sum$ out $- \sum$ in. Techniques for applying the steady-state, open-system energy balance are discussed in introductory engineering textbooks [47].

9.1.3.3 MECHANICAL ENERGY BALANCE

The simple form of the steady-state, macroscopic, open-system energy balance discussed previously, $\Delta H = Q_{in,CV}$ (Equation 9.98), is common in heat exchangers and reactors; however, in the flow of liquids and gases through conduits, the kinetic energy, potential energy, and shaft work dominate the energy balance. This circumstance is so common, in fact, that a simplified version of the steady-state, macroscopic, open-system energy balance is given a name: the *mechanical energy balance* (MEB). We derive the MEB in this section.

We consider the special case of a single-input, single-output system such as a liquid pushed through a piping system by a pump (Figure 9.9), and we apply the steady-state, open-system energy balance (Equation 9.96).

Macroscopic
open-system
energy balance $\Delta E_p + \Delta E_k + \Delta H = Q_{in,CV} - W_{s,by,CV}$ (9.99)
(steady state):

For such a system, there is only a single mass flow rate, m; thus, all of the $\sum$ out $- \sum$ in summations implicit in the Δ terms of the open-system energy balance become simple differences, as we now show. We label the outlet of our system (i.e., the control volume) as Position 2 and the inlet as Position 1. We further substitute $\hat{E}_k = E_k/m = v^2/2 = \langle v \rangle^2/2$ (see Equation 6.74) and $\hat{E}_p = E_p/m = gz$ (see Equation 6.75). For the case of single-input, single-output, steady flow, each term in the open-system energy balance simplifies as shown here:

$$\Delta E_p = \sum_{out} m_j \hat{E}_{p,j} - \sum_{in} m_i \hat{E}_{p,i} \tag{9.100}$$

$$= m\hat{E}_{p,2} - m\hat{E}_{p,1} \tag{9.101}$$

$$= m\left(\hat{E}_{p,2} - \hat{E}_{p,1}\right) \tag{9.102}$$

$$= mg(z_2 - z_1) \tag{9.103}$$

$$\Delta E_k \equiv \sum_{out} m_j \hat{E}_{k,j} - \sum_{in} m_i \hat{E}_{k,i} \tag{9.104}$$

$$= m\hat{E}_{k,2} - m\hat{E}_{k,1} \tag{9.105}$$

$$= m\left(\hat{E}_{k,2} - \hat{E}_{k,1}\right) \tag{9.106}$$

$$= m\left(\frac{\langle v \rangle_2^2}{2} - \frac{\langle v \rangle_1^2}{2}\right) \tag{9.107}$$

$$\Delta H = \left(\sum_{out} m_j \hat{U}_j - \sum_{in} m_i \hat{U}_i\right) + \sum_{out} m_j p_j \hat{V}_j - \sum_i m_j p_i \hat{V}_i \tag{9.108}$$

$$= m\hat{U}_2 - m\hat{U}_1 + mp_2\hat{V}_2 - mp_1\hat{V}_1 \tag{9.109}$$

$$= m\left(\hat{U}_2 - \hat{U}_1 + p_2\hat{V}_2 - p_1\hat{V}_1\right) \tag{9.110}$$

$$= m\left(\hat{U}_2 - \hat{U}_1 + \frac{p_2}{\rho_2} - \frac{p_1}{\rho_1}\right) \tag{9.111}$$

In Equation 9.111 we used the fact that $\hat{V} = 1/\rho$, where ρ is fluid density. For an incompressible fluid, the density is constant and $\rho_1 = \rho_2 = \rho$. We now substitute all of these results into the steady-state, open-system energy balance and simplify:

$$\Delta E_k + \Delta E_p + \Delta H = Q_{in,CV} - W_{s,by,CV} \tag{9.112}$$

$$m\left(\frac{\langle v \rangle_2^2}{2} - \frac{\langle v \rangle_1^2}{2}\right) + mg(z_2 - z_1) + m\left(\hat{U}_2 - \hat{U}_1 + \frac{p_2}{\rho} - \frac{p_1}{\rho}\right)$$
$$= Q_{in,CV} - W_{s,by,CV} \tag{9.113}$$

$$\left(\frac{\langle v \rangle_2^2}{2} - \frac{\langle v \rangle_1^2}{2}\right) + g(z_2 - z_1) + \left(\hat{U}_2 - \hat{U}_1\right) + \left(\frac{p_2}{\rho} - \frac{p_1}{\rho}\right)$$
$$= \frac{Q_{in,CV}}{m} - \frac{W_{s,by,CV}}{m} \tag{9.114}$$

$$\frac{\Delta p}{\rho} + \frac{\Delta \langle v \rangle^2}{2} + g\Delta z + \left[\Delta \hat{U} - \frac{Q_{in,CV}}{m} \right] = -\frac{W_{s,by,CV}}{m} \qquad (9.115)$$

In Equation 9.115, Δ again means *out–in*.

The terms in square brackets in Equation 9.115 are small for the flow of incompressible fluids in pipes because (1) temperature is approximately constant; (2) no phase or other chemical changes take place; and (3) only modest amounts of heat are transferred. We group these terms together and call them the friction term, F:

$$\boxed{\frac{\Delta p}{\rho} + \frac{\Delta \langle v \rangle^2}{2} + g\Delta z + F = -\frac{W_{s,by}}{m}}$$

Mechanical energy balance
(single-input, single-output,
steady, no phase change,
incompressible,
turbulent flow,   (9.116)
$\hat{v}$ constant across
cross section
$\Delta T \approx 0$, no reaction)
(turbulent flow only)

This is the mechanical energy balance. For compactness and following convention, we write the shaft work on the CV in the mechanical energy balance as $W_{s,by}$ rather than $W_{s,by,CV}$.

There is one subtlety that we ignored in our derivation of the mechanical energy balance. In the kinetic-energy term, although velocity varies across the cross section, we assume that $\hat{E}_k = \langle v \rangle^2 / 2$; that is, kinetic energy per unit mass equals half the square of the *average* velocity. By making this assumption, we assume the velocity to be constant across the cross section of the input and outlet flows. We make this assumption in Equation 9.64 when we assume that $\hat{E} = \hat{E}_k + \hat{E}_p + \hat{U}$ was independent of position and could be moved out of the integral. This is a good assumption for turbulent flow because velocity does not vary much with position in turbulent flow, but it is incorrect for laminar flow. In the next example, we consider this effect and add a correction to Equation 9.116 so that it may be used in laminar flow.

EXAMPLE 9.2. *What is the mechanical energy balance for laminar flows and other flows where the velocity varies across the tube cross section (Figure 9.10)?*

SOLUTION. In our development of the steady-state, open-system energy balance, we customized the energy balance on a CV, beginning with the general energy-balance in Equation 9.61:

Energy balance
on a CV
(First law of
thermodynamics):

$$\frac{dE_{CV}}{dt} + \iint_{CS} (\hat{n} \cdot \underline{v}) \rho \hat{E}\, dS = Q_{in,CV} - W_{by,CV}$$

$$(9.117)$$

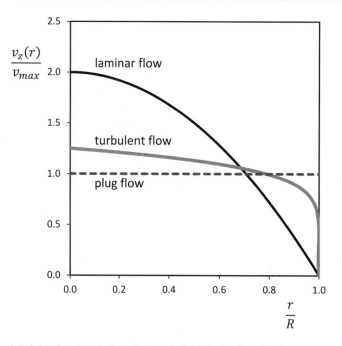

Figure 9.10 In turbulent flow, the velocity profile is mostly flat; in laminar flow, there is a pronounced variation of velocity across the tube cross section. This must be considered in the energy balances.

In our customization, we set the time-derivative to zero (steady state), and we worked on the convective integral, making it specific to a single-input, single-output system. In that development, we assumed that the specific energy $\hat{E}$ was independent of position on an input or output control surface. This allowed us to bring several terms, including $\hat{E}_i$, out of the double integral:

$$\iint_{CS} (\hat{n} \cdot \underline{v}) \rho \hat{E} \, dS = \sum_{A_i} \cos \theta_i \, \rho_i \, \hat{E}_i \iint_{A_i} v \, dS \qquad (9.118)$$

where $\hat{E} = \hat{U} + \hat{E}_k + \hat{E}_p$.

For internal energy $\hat{U}$ and potential energy $\hat{E}_p$, the assumption is correct and Equation 9.118 holds. For the kinetic energy $\hat{E}_k$, however, we must be more careful. The kinetic-energy-per-unit-mass term $\hat{E}_k$ is given by $\hat{E}_k = v^2/2$ (no average brackets $\langle \rangle$; see Equation 6.74). For turbulent flow, the velocity is approximately constant across the inlet and outlet surfaces, and the assumption $v \approx \langle v \rangle$ is valid. For steady laminar flow in a tube, however, the velocity has a parabolic profile across the cross section (see Equation 7.23):

$$\begin{array}{c} \text{Velocity for} \\ \text{laminar tube flow:} \end{array} \qquad v_z(r) = \frac{(p_0 - p_L + \rho g L) R^2}{4 \mu L} \left[1 - \left(\frac{r}{R} \right)^2 \right] \qquad (9.119)$$

and we may not ignore the variation of $\hat{E}_k$ across the flow cross section. As a result, in laminar flow, the kinetic-energy expression cannot be taken out of the integral in Equation 9.117.

For laminar flow and other flows where v varies across the cross section, we thus take a slightly different path to the mechanical energy balance. Beginning with the convective integral from Equation 9.117, we write the energy as its three parts:

$$\iint_{CS} (\hat{n} \cdot \underline{v}) \rho \hat{E} \, dS = \sum_{A_i} \cos \theta_i \rho_i \iint_{A_i} \left(\hat{U} + \hat{E}_k + \hat{E}_p \right) v \, dS \tag{9.120}$$

$$= \sum_{A_i} \cos \theta_i \rho_i \iint_{A_i} \hat{U} v \, dS + \sum_{A_i} \cos \theta_i \rho_i \iint_{A_i} \hat{E}_k v \, dS$$

$$+ \sum_{A_i} \cos \theta_i \rho_i \iint_{A_i} \hat{E}_p v \, dS \tag{9.121}$$

The internal- and potential-energy terms may be simplified as they were in the turbulent case; the kinetic-energy term needs additional attention:

$$\iint_{CS} (\hat{n} \cdot \underline{v}) \rho \hat{E} \, dS = \underset{o-i}{\Delta} U + \sum_{A_i} \cos \theta_i \rho_i \iint_{A_i} \hat{E}_k v \, dS + \underset{o-i}{\Delta} E_p \tag{9.122}$$

Concentrating now on the kinetic-energy term, we write $\hat{E}_k = v^2/2$:

$$\sum_{A_i} \cos \theta_i \rho_i \iint_{A_i} \hat{E}_k v \, dS = \sum_{A_i} \cos \theta_i \rho_i \iint_{A_i} \frac{v^2}{2} v \, dS \tag{9.123}$$

$$= \sum_{A_i} \frac{\cos \theta_i \rho_i}{2} \iint_{A_i} v^3 \, dS \tag{9.124}$$

The remaining integral in Equation 9.124 may be recognized as related to the average of the *cube* of the fluid speed across A_i:

$$\begin{array}{c} \text{Average across } A_i \text{ of} \\ \text{fluid speed cubed:} \end{array} \quad \langle v^3 \rangle \Big|_{A_i} = \frac{\iint_{A_i} v^3 \, dS}{\iint_{A_i} dS} = \frac{1}{A_i} \iint_{A_i} v^3 \, dS \tag{9.125}$$

Thus:

$$\sum_{A_i} \cos \theta_i \rho_i \iint_{A_i} \hat{E}_k v \, dS = \sum_{A_i} \frac{\cos \theta_i \rho_i}{2} \iint_{A_i} v^3 \, dS \tag{9.126}$$

$$= \sum_{A_i} \frac{\cos \theta_i \rho_i A_i \langle v^3 \rangle \big|_{A_i}}{2} \tag{9.127}$$

Equation 9.127 contains an expression for the average of the cube of the speed $\langle v^3 \rangle$. For turbulent flows, this quantity is approximately equal to the cube of the average speed, whereas for laminar flows, it is appreciably larger than the cube of the average speed (Figure 9.11). To account for these two situations, we define the parameter α as:

$$\begin{array}{c} \text{Energy} \\ \text{velocity-profile} \\ \text{parameter:} \end{array} \quad \boxed{\alpha \equiv \frac{\langle v \rangle^3}{\langle v^3 \rangle}} \tag{9.128}$$

$$v_z(r) = 2\langle v \rangle \left(1 - \left(\frac{r}{R} \right)^2 \right)$$

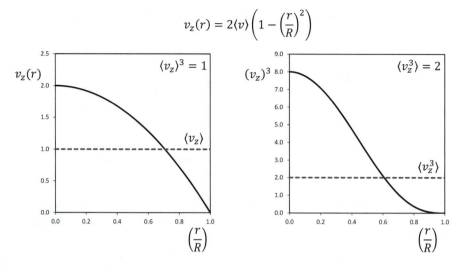

Figure 9.11 For a function that varies with position, the cube of the average of the function is not generally equal to the average of the cube of the function.

Incorporating $\langle v^3 \rangle = \langle v \rangle^3 / \alpha$ into Equation 9.127, we obtain:

$$\sum_{A_i} \cos \theta_i \rho_i \iint_{A_i} \hat{E}_k v \, dS = \sum_{A_i} \frac{\cos \theta_i \rho_i A_i \, \langle v^3 \rangle \big|_{A_i}}{2} \qquad (9.129)$$

$$= \sum_{A_i} \cos \theta_i \rho_i A_i \left(\frac{\langle v \rangle^3 \big|_{A_i}}{2\alpha} \right) \qquad (9.130)$$

We further simplify the kinetic-energy expression by writing the mass flow rate through A_i as $m_i = \rho_i A_i \, \langle v \rangle |_{A_i}$:

$$\sum_{A_i} \cos \theta_i \rho_i \iint_{A_i} \hat{E}_k v \, dS = \sum_{A_i} \cos \theta_i \rho_i A_i \, \langle v \rangle |_{A_i} \left(\frac{\langle v \rangle^2 \big|_{A_i}}{2\alpha} \right) \qquad (9.131)$$

$$= \sum_{A_i} \cos \theta_i m_i \left(\frac{\langle v \rangle^2 \big|_{A_i}}{2\alpha} \right) \qquad (9.132)$$

The development of the MEB now proceeds in the same manner as before. We assume that $\hat{n}$ and $\hat{v}$ are parallel, which is the case if the fluid enters and exits the CV perpendicular to the inlet and outlet control surfaces. If $\hat{n} = \hat{v}$, then $\hat{n} \cdot \hat{v} = \cos\theta = 1$; this is the case for an outflow surface. If $\hat{n} = -\hat{v}$, then $\hat{n} \cdot \hat{v} = \cos\theta = -1$; this is the case for an inflow surface. We break up the summation in Equation 9.132 into outflow and inflow surfaces:

$$\sum_{A_i} \cos \theta_i \rho_i \iint_{A_i} \hat{E}_k v \, dS = \sum_{\substack{A_j \\ \text{outflow}}} m_j \left(\frac{\langle v \rangle^2 \big|_{A_j}}{2\alpha} \right) - \sum_{\substack{A_i \\ \text{inflow}}} m_i \left(\frac{\langle v \rangle^2 \big|_{A_i}}{2\alpha} \right) \qquad (9.133)$$

We thus can redefine $\underset{o-i}{\Delta} E_k$ to apply to both turbulent and laminar flows as:

$$\underset{o-i}{\Delta} E_k \equiv \sum_{out} m_j \left(\frac{\langle v \rangle^2 |_{A_j}}{2\alpha} \right) - \sum_{in} m_i \left(\frac{\langle v \rangle^2 |_{A_i}}{2\alpha} \right) \tag{9.134}$$

(Compare to Equation 9.75.)

The mechanical energy balance applies to the special case of a single-input, single-output system such as a liquid pushed through a piping system by a pump (see Figure 9.9). For such a system, there is only a single mass flow rate, m; thus, the $\sum$ out $- \sum$ in summations become simple differences. Proceeding as before, we label the outlet as Position 2 and the inlet as Position 1 and the general kinetic-energy term becomes:

$$\underset{o-i}{\Delta} E_k = \Delta E_k = \sum_{out} m_j \left(\frac{\langle v \rangle^2 |_{A_j}}{2\alpha} \right) - \sum_{in} m_i \left(\frac{\langle v \rangle^2 |_{A_i}}{2\alpha} \right) \tag{9.135}$$

$$= m \left(\frac{\langle v \rangle_2^2}{2\alpha} - \frac{\langle v \rangle_1^2}{2\alpha} \right) \tag{9.136}$$

Compare this result to Equation 9.107. The inclusion of α in the denominator makes the expression correct for both laminar and turbulent flow.

We see that the only change in this laminar-flow analysis compared to the original turbulent-flow analysis is the inclusion of α in the denominator of the kinetic-energy term. The final version of the mechanical energy balance that is applicable for both laminar and turbulent flows is thus:

$$\boxed{\frac{\Delta p}{\rho} + \frac{\Delta \langle v \rangle^2}{2\alpha} + g \Delta z + F = -\frac{W_{s,by}}{m}}$$

Mechanical energy balance
(single-input, single-output,
steady, no phase change,
incompressible,   (9.137)
$\hat{v}$ constant across
cross section
$\Delta T \approx 0$, no reaction)

(Compare to Equation 9.116.) The values of $\alpha = \langle v \rangle^3 / \langle v^3 \rangle$ for laminar and turbulent flows may be calculated formally (as we did with β in the macroscopic momentum balance) from the definition of α in Equation 9.128 and the velocity profiles $v_z(r)$ for laminar and turbulent flow (see Problem 6). The values of α and β for plug flow (i.e., constant velocity across the cross section), laminar flow, and turbulent flow are summarized in Table 9.1.

The mechanical energy balance gives the relationship among pressure, velocity, elevation, frictional losses, and shaft work for the steady flow of incompressible

Table 9.1. Parameter α from the MEB and parameter β from the macroscopic momentum balance

Name	Flow:	Plug	Turbulent	Laminar
energy velocity-profile parameter	$\alpha = \dfrac{\langle v \rangle^3}{\langle v^3 \rangle}$	1	0.90–0.99	$\dfrac{1}{2}$
momentum velocity-profile parameter	$\beta = \dfrac{\langle v \rangle^2}{\langle v^2 \rangle}$	1	0.98	$\dfrac{3}{4}$

Note: Both parameters reflect the deviation of the velocity profile from plug flow.

fluids where there is little heat transfer, no phase changes, no chemical changes, and little change in temperature:

Mechanical energy balance
(single-input, single-output,
steady, no phase change,
incompressible,
$\hat{v}$ constant across
cross section
$\Delta T \approx 0$, no reaction):

$$\frac{\Delta p}{\rho} + \frac{\Delta \langle v \rangle^2}{2\alpha} + g \Delta z + F = -\frac{W_{s,by}}{m} \quad (9.138)$$

The values of α and β for various flow types are summarized in Table 9.1. Application of the mechanical energy balance is limited to single-input, single-output systems. Pressure, fluid velocity, and elevation are easily measured in experimental systems, and shaft work is often the quantity to be calculated with the MEB. The friction term sometimes may be neglected; when it cannot be neglected, it must be calculated from experimental results—that is, from data correlations (see Section 1.2).

We discuss applications of the mechanical energy balance in Section 9.2, including the analysis of valves and fittings that leads to the concept of K_f introduced in Chapter 1. Other significant applications of the MEB include the analyses of pumping systems and open-channel flows, which are also discussed in the next section.

9.2 Using the macroscopic balance equations

The macroscopic balances on mass, momentum, and energy, derived in the previous section, are used widely to calculate flow information in industrial problems. The balances are written on an arbitrarily shaped macroscopic control volume (CV) (see Section 9.1 for derivations).

The macroscopic mass balance is given by:

Macroscopic
mass balance
($\rho, \hat{v}$ constant across A_i):

$$\frac{dm_{CV}}{dt} + \sum_{i=1}^{\# \, streams} \left[\rho A \cos \theta \langle v \rangle \right]\Big|_{A_i} = 0 \quad (9.139)$$

Macroscopic
mass balance
single-input,
single-output:

$$\frac{dm_{CV}}{dt} + \rho_1 A_1 \cos\theta_1 \langle v \rangle_1 + \rho_2 A_2 \cos\theta_2 \langle v \rangle_2 = 0 \qquad (9.140)$$

where m_{CV} is the mass of the control volume, ρ is the density, A is the cross-sectional area of an inlet or outlet surface, θ is the angle between the unit normal $\hat{n}$ to the surface A and the direction of the velocity through the surface, and $\langle v \rangle$ is the average velocity through the surface. If there is one inlet and one outlet stream, the macroscopic mass balance becomes Equation 9.140. Note that $\hat{n}$ is an outwardly pointing unit normal; thus, for inlet surfaces, $\hat{n}$ and $\hat{v}$ form an angle greater than 90 degrees (see Figure 9.4) and $\cos\theta$ is negative. If we further assume that $\hat{v}$ is parallel to $\hat{n}$ at both inlet and outlet surfaces ($\cos\theta_1 = -1$; $\cos\theta_2 = 1$), then the macroscopic mass balance simplifies to:

Macroscopic mass balance
single-input,
single-output
$\hat{v}$ parallel to $\hat{n}$
(velocity perpendicular
to surface)

$$\frac{dm_{CV}}{dt} + (\rho A \langle v \rangle)|_{\text{out}} - (\rho A \langle v \rangle)|_{\text{in}} = 0$$

$$(9.141)$$

We show how to use the macroscopic mass balance (Equations 9.139–9.141) in this section.

The macroscopic momentum balance is given by:

Macroscopic
momentum balance
(ρ, $\hat{v}$ constant across A_i)

$$\frac{d\mathbf{P}}{dt} + \sum_{i=1}^{\#\,streams} \left[\frac{\rho A \cos\theta \langle v \rangle^2}{\beta} \hat{v} \right]\Bigg|_{A_i} = \sum_{\substack{\text{on} \\ \text{CV}}} \underline{f}$$

$$(9.142)$$

where $\mathbf{P}$ is the momentum in the control volume; $\underline{f}$ is a force on the control volume; $\beta = \langle v \rangle^2 / \langle v^2 \rangle$ is the momentum velocity-profile parameter; $\hat{v}$ is a unit vector in the direction of the velocity through surface A; and ρ, A, $\langle v \rangle$, and θ are as described previously for the macroscopic mass balance. The values of β for various flow types are summarized in Table 9.1. In Equation 9.142, we specify that neither the density nor the direction of the velocity changes across the input and output surfaces. The forces in the summation in Equation 9.142 are body forces (i.e., gravity) and the molecular surface forces on the bounding control surfaces:

$$\sum_{\substack{\text{on} \\ \text{CV}}} \underline{f} = \underline{f}_{\text{gravity}} + \underline{f}_{\text{surface}} \qquad (9.143)$$

$$= \underline{f}_{\text{gravity}} + \underline{f}_{\text{inlet}} + \underline{f}_{\text{outlet}} + \underline{f}_{\text{walls}} \qquad (9.144)$$

The molecular surface forces, including pressure, are expressed using the stress tensor $\underset{=}{\tilde{\Pi}}$ (see Equation 4.221):

Total molecular
fluid force
on a surface $\mathcal{S}$:

$$\mathcal{F} = \iint_S [\hat{n} \cdot \underset{=}{\tilde{\Pi}}]_{\text{at surface}} \, dS \qquad (9.145)$$

$$= \iint_S \left[\hat{n} \cdot \left(-p\underline{I} + \underset{=}{\tilde{\tau}}\right)\right]_{\text{at surface}} \, dS \qquad (9.146)$$

$$= \iint_S \left[-p\hat{n} + \hat{n} \cdot \underset{=}{\tilde{\tau}}\right]_{\text{at surface}} \, dS \qquad (9.147)$$

The simplification $\hat{n} \cdot p\underline{I} = p\hat{n}$ in Equation 9.147 is due to vector/tensor relationships discussed in Section 1.3.2.2. We show how to use the macroscopic momentum balance (Equations 9.142– 9.147) in this section.

The third macroscopic balance is the macroscopic energy balance. In Section 9.1.3, we derive three macroscopic energy balances: (1) the closed-system energy balance (see Equation 9.60), which is useful for systems where no flow occurs; (2) the open-system energy balance (see Equation 9.96), which is widely used in chemical engineering for the analysis of heat exchangers, evaporators, and other devices in which flow occurs and temperature change dominates; and (3) the mechanical energy balance (see Equation 9.138), which is a useful energy balance for fluid systems in which the fluid motion is dominant and thermal and chemical effects are negligible. Here we focus on the mechanical energy balance (MEB).

The mechanical energy balance gives the relationship among pressure, velocity, elevation, frictional losses, and shaft work for the steady flow of incompressible fluids in which there is little heat transfer, no phase changes, no chemical changes, and minimal change in temperature:

Mechanical energy balance
(single-input, single-output,
steady, no phase change,
incompressible, $\hat{v}$ constant
across cross section
$\Delta T \approx 0$, no reaction)

$$\frac{\Delta p}{\rho} + \frac{\Delta\langle v\rangle^2}{2\alpha} + g\Delta z + F = -\frac{W_{s,by}}{m}$$

$$(9.148)$$

The symbol Δ indicates "*out–in*." The values of the energy velocity-profile parameter α for various flow types are summarized in Table 9.1. Application of the MEB is limited to single-input, single-output systems. Pressure, fluid velocity, and elevation are measured easily in experimental systems, and shaft work often is the quantity to be calculated with the MEB. The friction term sometimes may be neglected; when the friction term cannot be neglected, it must be calculated from experimental results—that is, from data correlations, as discussed in Section 1.2 and examined here.

The examples in this section demonstrate the techniques used to apply the macroscopic mass, momentum, and energy balances in fluid mechanics. The MEB problems discussed here are more complex than those in Chapter 1. In addition, we demonstrate how the macroscopic momentum balance can be useful in fluid-mechanics analysis, particularly when forces on devices are of interest. The discussion is organized around the types of devices we consider: pressure measurement, flow-rate measurement, valves and fittings, and pumps. We also show how the MEB may be used in the analysis of open-channel flows.

9.2.1 Pressure-measurement devices

Pressure is an important variable in fluids engineering, and many devices have been developed to measure pressure. Here, we carry out two examples of applying the macroscopic balances to measure pressure in flows.

The results in this section are expressed in terms of *head*. The concept of head—that is, energy per unit weight—is common in fluids engineering, and we can understand it by examining the mechanical energy balance. As discussed in Sections 1.2 and 9.1, each term of the mechanical energy balance has units of energy per unit mass, either J/kg or ft lb_f/lb_m:

Mechanical energy balance
(single-input, single-output,
steady, no phase change,
incompressible,
$\hat{v}$ constant across cross section
$\Delta T \approx 0$, no reaction)

$$\frac{\Delta P}{\rho} + \frac{\Delta \langle v \rangle^2}{2\alpha} + g\Delta z + F = -\frac{W_{s,by}}{m}$$

(9.149)

If we divide the MEB by the acceleration due to gravity g, each term becomes energy per unit weight, which has units of length, feet or meters. Energy per unit weight of flowing liquid is called fluid head:

Mechanical
energy balance
(units of head)

$$\frac{\Delta P}{\rho g} + \frac{\Delta \langle v \rangle^2}{2g\alpha} + \Delta z + \frac{F}{g} = -\frac{W_{s,by}}{mg}$$

(9.150)

It is common for hydraulic engineers to discuss flow energies in terms of head. The practice of using head units can be traced to the still-current practice of using manometer tubes and Pitot tubes to measure flow pressures and flow rates [118]. The first example shows how head is related to the reading of a manometer tube; the second example discusses Pitot tubes, which report the stagnation head—a quantity related to flow rate.

EXAMPLE 9.3. *A vertical manometer tube is attached to the wall of a flow channel as shown in Figure 9.12. The flowing liquid rises in the manometer tube to a height h_{static}. Relate the height h_{static} to a property of the flow.*

SOLUTION. We choose our system to be the fluid in the manometer between Point 1 at the base of the manometer in contact with the moving fluid and Point 2 at the top free surface of the fluid in the manometer. The system of water in the vertical manometer tube in Figure 9.12 is a single-input, single-output, steady flow of an incompressible fluid. There is no heat transfer and no chemical reaction or phase change. All of the requirements of the mechanical energy

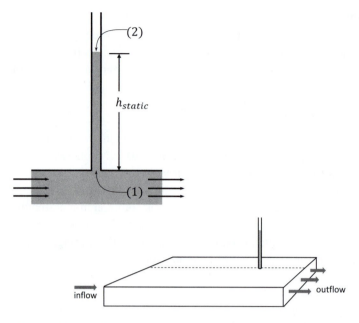

Figure 9.12 Manometer tubes are clear vertical tubes of arbitrary cross section installed in a flow such that the fluid can enter and rise into the tube. The height of the fluid in the manometer tube may be shown to represent the gauge pressure in the flow.

balance therefore are met. We begin with the MEB in terms of head:

$$\text{Mechanical energy balance (units of head):} \quad \frac{\Delta p}{\rho g} + \frac{\Delta \langle v \rangle^2}{2 g \alpha} + \Delta z + \frac{F}{g} = -\frac{W_{s,by}}{mg} \qquad (9.151)$$

In our chosen system, there are no shafts and therefore no shaft work; the velocity in the manometer is zero at both points and therefore $\Delta \langle v \rangle^2 = 0$. With no velocity, there is no friction. The pressure at Point 1 is the pressure in the flow, and the pressure at Point 2 is atmospheric. The mechanical energy balance reduces to:

$$\frac{p_2 - p_1}{\rho g} + \frac{\langle v \rangle_2^2 - \langle v \rangle_1^2}{2 g \alpha} + (z_2 - z_1) + \frac{F_{2,1}}{g} = -\frac{W_{s,by.21}}{mg} \qquad (9.152)$$

$$\frac{p_{atm} - p_1}{\rho g} + (z_2 - z_1) = 0 \qquad (9.153)$$

$$\boxed{h_{\text{static}} = z_2 - z_1 = \frac{p - p_{atm}}{\rho g}} \qquad (9.154)$$

where $p_1 = p$ is the flow pressure where the manometer is installed. We see that h_{static} is the gauge pressure (see the Glossary) at the point of manometer installation, expressed in head units. Note that we can arrive at the same result by analyzing the manometer with the static fluid equations discussed in Section 4.2.4.

EXAMPLE 9.4. *A Pitot tube is a "J"-shaped manometer tube installed through the wall of a flow such that the curved end directly faces the oncoming flow (Figure 9.13). The flowing liquid rises in the Pitot tube to a height h_{stag}. Relate the height h_{stag} to a property of the flow.*

SOLUTION. Our system is the fluid in the Pitot tube. We choose Point 1 to be the surface of fluid that faces the flow. This surface intercepts the flow and produces a stagnation point (see the Glossary). The flow decelerates and the velocity comes to zero on this surface. Point 2 is chosen to be the top free surface of the fluid in the vertical portion of the tube. The system of fluid in the installed Pitot tube is a single-input, single-output, steady flow of an incompressible fluid. There is no heat transfer and no chemical reaction or phase change. All of the requirements of the mechanical energy balance therefore are met. We begin with the MEB in terms of head:

$$\text{Mechanical energy balance (units of head):} \quad \frac{\Delta p}{\rho g} + \frac{\Delta \langle v \rangle^2}{2 g \alpha} + \Delta z + \frac{F}{g} = -\frac{W_{s,by}}{mg} \qquad (9.155)$$

In our chosen system, there are no shafts and therefore no shaft work; the velocity is zero at both Points 1 and 2; thus, $\Delta \langle v \rangle^2 = 0$. With no velocity, there is no friction. The pressure at Point 1 is the pressure at the stagnation point, and

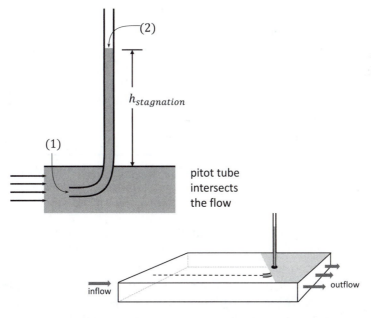

Pitot tubes are clear J-shaped tubes of arbitrary cross section installed in a flow such that the curved bottom portion directly faces the incoming flow. The height of the fluid in the Pitot tube may be shown to represent the stagnation pressure at the point where the bottom portion of the tube stops the flow. Stagnation pressure, in turn, may be shown to be related to the velocity in the flow.

the pressure at Point 2 is atmospheric. The MEB reduces to:

$$\frac{p_2 - p_1}{\rho g} + \frac{\langle v \rangle_2^2 - \langle v \rangle_1^2}{2g\alpha} + (z_2 - z_1) + \frac{F_{2,1}}{g} = -\frac{W_{s,by.21}}{mg} \qquad (9.156)$$

$$\frac{p_{atm} - p_1}{\rho g} + (z_2 - z_1) = 0 \qquad (9.157)$$

$$\boxed{h_{\text{stag}} = z_2 - z_1 = \frac{p_{\text{stag}} - p_{atm}}{\rho g}} \qquad (9.158)$$

where the pressure at the stagnation point p_1 now is called p_{stag}.

We see that h_{stag} is the gauge flow pressure at the stagnation point, expressed in head units. The expression h_{stag} is called the stagnation head. A Pitot tube allows us to measure the gauge pressure at the stagnation point where the Pitot tube intercepts the flow. This is a useful quantity for flow measurements (see Example 9.6).

Static manometer tubes and Pitot tubes are used widely in demonstrations of fluid phenomena (see, e.g., the NCFMF film on pressure [112]), and the combination of a static tube with a Pitot tube makes a device that can measure flow rate. In the second example in the next section, we discuss Pitot-static tubes, which are used on airplanes to determine flight speed.

9.2.2 Flow-rate-measurement devices

In Chapter 1, we discuss the need for flow measurement in process streams. Although the pail-and-scale method for measuring flow rate is accurate, it is disruptive to the system; therefore, many devices have been developed to allow for accurate flow-rate measurement without breaking into the flow loop. One such flow-measurement device is discussed in Chapter 1: the Venturi meter. To illustrate the application of the macroscopic mass balance (see Equation 9.140), we repeat the (simple) mass balance from that example to see how to apply the formal macroscopic mass-balance equation.

EXAMPLE 9.5. *What is the relationship between measured pressure drop and flow rate through a Venturi meter? The flow may be assumed to be steady and the fluid is incompressible.*

SOLUTION. We solved this problem in Chapter 1 and the figure is repeated here (Figure 9.14). To determine the relationship between pressure drop and flow rate, we apply the mechanical energy balance as well as mass conservation (see Equation 1.15). To see how our general macroscopic mass balance works, we formally apply Equation 9.140 to the Venturi problem.

The first step in applying the macroscopic mass balance is to choose our control volume. We choose a shape that encloses all of the fluid between Planes 1 and 2 in Figure 9.14. Point 1 is the point of the upstream pressure measurement and Point 2 is at the throat, the location of the other pressure measurement. For this

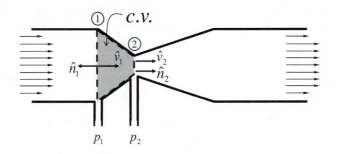

Figure 9.14 In a Venturi meter, flow is directed through a gently tapering tube. Pressure is measured before the contraction (1) and at the point of smallest diameter (2) (i.e., the throat). The relationship between the measured pressures and the fluid velocity may be deduced from the mechanical energy balance with assistance of the macroscopic mass balance.

CV, there is one inlet surface and one outlet surface (single-input, single-output). The macroscopic mass balance on this CV is given by Equation 9.140:

$$\begin{array}{l}\text{Macroscopic} \\ \text{mass balance,} \\ \text{single-input,} \\ \text{single-output}\end{array} \qquad \frac{dm_{CV}}{dt} + \rho_1 A_1 \cos\theta_1 \langle v \rangle_1 + \rho_2 A_2 \cos\theta_2 \langle v \rangle_2 = 0 \qquad (9.159)$$

For the Venturi meter, the flow is steady; thus, the rate-of-change term is zero. The other terms may be identified readily as:

$$\rho_1 = \rho_2 = \rho \quad \text{(incompressible fluid)} \qquad (9.160)$$

$$\theta_1 = 180° \quad \cos\theta_1 = -1 \quad \text{(input surface)} \qquad (9.161)$$

$$\theta_2 = 0° \quad \cos\theta_2 = 1 \quad \text{(output surface)} \qquad (9.162)$$

The mass balance becomes:

$$\frac{dm_{CV}}{dt} + \rho_1 A_1 \cos\theta_1 \langle v \rangle_1 + \rho_2 A_2 \cos\theta_2 \langle v \rangle_2 = 0 \qquad (9.163)$$

$$\rho A_2 \langle v \rangle_2 - \rho A_1 \langle v \rangle_1 = 0 \qquad (9.164)$$

For a circular pipe of diameter D_i, the area is given by $A_i = \pi D_i^2/4$ and we obtain:

$$A_1 \langle v \rangle_1 = A_2 \langle v \rangle_2 \qquad (9.165)$$

$$\frac{\pi D_1^2 \langle v \rangle_1}{4} = \frac{\pi D_2^2 \langle v \rangle_2}{4} \qquad (9.166)$$

$$\begin{array}{l}\text{Velocity relationship} \\ \text{in Venturi meter:}\end{array} \qquad \boxed{\langle v \rangle_1 = \left(\frac{D_2}{D_1}\right)^2 \langle v \rangle_2} \qquad (9.167)$$

The complete solution for the friction in a Venturi meter is in Chapter 1. In that solution, Equation 9.167 appears as Equation 1.25.

The use of Equation 9.159 for the problem in the previous example is more rigorous than necessary, but it is useful to see how the formal equation works in a problem for which we know the solution. In the next example, we show how the combination of a static-pressure manometer tube and a Pitot tube can be used to measure flow speed. The Pitot-static tube is a common sight on the nose of commercial jets.

EXAMPLE 9.6. *A static-pressure manometer and a Pitot tube are installed in the flow as shown in Figure 9.15. Note that the static-pressure manometer is installed upstream of the Pitot tube along the same streamline. How are the heights of the fluid in each tube related to flow variables?*

SOLUTION. To see the usefulness of the combination of sensors shown in Figure 9.15, consider the system of the streamline that connects Points 1 and 2 in Figure 9.15. A steady-flow streamline is the path followed by a series of fluid particles (Figure 9.16; see also the Glossary). Because the only mass particles traveling along the streamline are those that start at the upstream point of the streamline, a streamline is a single-input, single-output system, even without any walls to enclose it.

The system of the streamline between Points 1 and 2 in Figure 9.15 is a single-input, single-output, steady flow of an incompressible fluid. There is no heat transfer and no chemical reaction or phase change. All of the requirements of the mechanical energy balance therefore are met. We begin with the

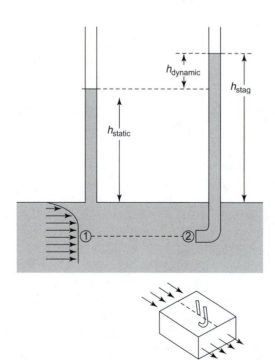

Figure 9.15   Schematic of velocity measurement with a combination of a Pitot table and a static tube.

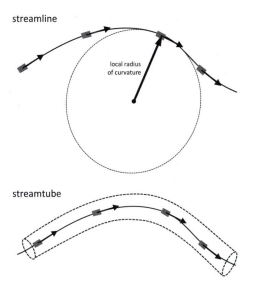

streamline

local radius
of curvature

streamtube

A streamline in steady flow is the path that fluid particles take as they advance through the flow. Packets of fluid enter the streamline at an upstream point and stay on the streamline throughout. Thus, a streamline is a single-input, single-output system and we can use the mechanical energy balance along streamlines.

MEB in terms of head:

$$\begin{array}{c}\text{Mechanical} \\ \text{energy balance} \\ \text{(units of head):}\end{array} \qquad \frac{\Delta p}{\rho g} + \frac{\Delta \langle v \rangle^2}{2g\alpha} + \Delta z + \frac{F}{g} = -\frac{W_{s,by}}{mg} \qquad (9.168)$$

We choose Point 2 to be the stagnation point of the Pitot tube. The streamline that ends at the stagnation point originates upstream; we choose Point 1 to be on this streamline, at a point directly below the vertical manometer tube (see Figure 9.15).

In our chosen system there are no shafts and therefore no shaft work; the velocity is zero at Point 2, and at Point 1 it is equal to the free-stream average velocity $\langle v \rangle_1$. The pressures at Points 1 and 2 are different and can be written as p_1 and p_2. The elevations of Points 1 and 2 are the same; thus, $\Delta z = 0$. We choose to neglect any frictional losses in the flow between Points 2 and 1. Finally, we assume turbulent flow ($\alpha = 1$). The MEB reduces to:

$$\frac{p_2 - p_1}{\rho g} + \frac{\langle v \rangle_2^2 - \langle v \rangle_1^2}{2g\alpha} + (z_2 - z_1) + \frac{F_{2,1}}{g} = -\frac{W_{s,by.21}}{mg} \qquad (9.169)$$

$$\frac{p_2 - p_1}{\rho g} + \frac{0 - \langle v \rangle_1^2}{2g} = 0 \qquad (9.170)$$

$$\frac{\langle v \rangle_1^2}{2g} = \frac{(p_2 - p_1)}{\rho g} \qquad (9.171)$$

The result in Equation 9.171 demonstrates that the average fluid velocity near the point of installation of the manometer tube is related to the pressure difference between the two chosen points used in our balance: (2) a downstream point where the flow was halted by the Pitot tube; and (1) an upstream point along the same streamline.

From Examples 9.3 and 9.4, we may write the pressures p_1 and p_2 in terms of the heads at the two points. The head reflected by the manometer tube (Point 1) is called the static head h_{static}, and it displays the gauge pressure of the flowing liquid at Point 1. The head at the Pitot tube (Point 2) is called the stagnation head h_{stag}, and it reflects the pressure at the stagnation point, which includes a contribution due to the deceleration of the fluid that is halted at the tip of the Pitot tube:

$$
h_{static} = \frac{p_{static} - p_{atm}}{\rho g} = \frac{p_1 - p_{atm}}{\rho g} \tag{9.172}
$$

$$
h_{stag} = \frac{p_{stag} - p_{atm}}{\rho g} = \frac{p_2 - p_{atm}}{\rho g} \tag{9.173}
$$

We now substitute these relationships into our MEB results (Equation 9.171) and simplify:

$$
\frac{\langle v \rangle_1^2}{2g} = \frac{(p_2 - p_1)}{\rho g} \tag{9.174}
$$

$$
= \frac{1}{\rho g} \left((\rho g h_{stag} + p_{atm}) - (\rho g h_{static} + p_{atm}) \right) \tag{9.175}
$$

$$
\frac{\langle v \rangle_1^2}{2g} = \left(h_{stag} - h_{static} \right) \tag{9.176}
$$

Our analysis shows that the difference between the stagnation head and the static head provides a measurement of the average velocity at Point 1:

Fluid velocity head, $h_{velocity}$, equals difference between stagnation and static head:

$$
\boxed{h_{velocity} = \frac{\langle v \rangle^2}{2g} = \left(h_{stag} - h_{static} \right)} \tag{9.177}
$$

$$
\boxed{\langle v \rangle = \sqrt{2g \left(h_{stag} - h_{static} \right)}} \tag{9.178}
$$

where $\langle v \rangle = \langle v \rangle_1$ is the average fluid velocity near the installation of the vertical manometer tube. The flow energy due to kinetic energy is termed the velocity head and is given by $h_{velocity} = \langle v \rangle^2 / 2g$. The velocity head is given by the difference between the stagnation head and the static head in a Pitot-static device (Equation 9.177).

Our analysis shows that the readings on a vertical manometer tube and on a Pitot tube installed as shown in Figure 9.15 give readings of static head and stagnation head. The difference between these two quantities is the velocity head, which also is called the dynamic head. The measurement of this head difference gives the average velocity in a turbulent flow through Equation 9.178 (friction neglected). Calibrated Pitot-static tubes (friction accounted for) are used on aircraft (Figure 9.17).

Figure 9.17   Pitot-static tubes are used on aircraft to measure the speed of the air rushing past the vessel in flight. The Pitot-static tube measures the air velocity relative to the position of the device on the hull of the airplane. The exposed sensor tubes seen on the front of a jet airplane have the classic "J" shape of a Pitot tube.

Head appears naturally in the discussion of some devices, and it is particularly prominent in the analysis of centrifugal pumps, which we discuss in Section 9.2.4. Head has another advantage that is obscured in the recent examples by the fact that no numbers were used. To understand another reason why engineers sometimes prefer head units, we examine the pressure head produced by a pressure difference of 50 lb_f/ft^2 in a system pumping water ($\rho = 62.43\ lb_m/ft^3$). For the pressure term in the mechanical energy balance, we obtain:

$$\begin{array}{l} \text{Energy} \\ \text{per unit } \textit{mass} \\ \text{due to pressure:} \end{array} = \frac{\Delta P}{\rho}$$

$$= \left(50\frac{lb_f}{ft^2}\right)\left(\frac{ft^3}{62.43\ lb_m}\right)$$

$$= 0.80\frac{ft\ lb_f}{lb_m} \qquad (9.179)$$

$$\begin{array}{l} \text{Energy} \\ \text{per unit } \textit{weight} \\ \text{due to pressure:} \end{array} = \frac{\Delta P}{\rho g} = \left(\frac{\Delta P}{\rho}\right)\left(\frac{1}{g}\right)$$

$$= \left(0.80\frac{ft\ lb_f}{lb_m}\right)\left(\frac{s^2}{32.174\ ft}\right)\left(\frac{32.174\ ft\ lb_m}{s^2 lb_f}\right) \qquad (9.180)$$

$$= 0.80\ ft \qquad (9.181)$$

The numbers for the two different quantities in Equations 9.179 and 9.181 are the same. As we see in Equation 9.180, when converting energy/mass to head and making subsequent unit conversions, the number 32.174 appears in two places: (1) in the numerator as part of the unit conversion from lb_f to ft lb_m/s^2; and (2) in the denominator as g, the acceleration due to gravity. Because these two factors are numerically the same, the two quantities of magnitude 32.174 cancel (i.e., the numbers cancel, not the units), and expressions with units of ft lb_f/lb_m and ft of head are numerically the same. We made this same observation in the final example in Chapter 1: The numbers for energy/mass in ft lb_f/lb_m are the same as the head results, energy/weight in ft. Thus, we can take a shortcut when performing calculations in American engineering units and calculate all types of energy/mass (i.e., pressure-based, velocity-based, and friction loss) in ft lb_f/lb_m, and the numbers obtained can be recognized immediately as equal to the corresponding head numbers (i.e., ft of head, energy per unit weight).

In metric units, when we repeat the analogous calculations, there is no such numerical serendipity:

$$\begin{matrix}\text{Energy}\\\text{per unit } mass\\\text{due to pressure:}\end{matrix} = \frac{\Delta P}{\rho}$$

$$= \left(50\frac{lb_f}{ft^2}\right)\left(\frac{ft^3}{62.43\ lb_m}\right)$$

$$= \left(0.800897\frac{ft\ lb_f}{lb_m}\right)\left(\frac{J/s}{0.7376ft\ lb_f/s}\right)\left(\frac{lb_m}{0.453593\ kg}\right)$$

$$= 2.3938\ Nm/kg$$

$$= \boxed{2.4\ Nm/kg} \tag{9.182}$$

$$\begin{matrix}\text{Energy}\\\text{per unit } weight\\\text{due to pressure:}\end{matrix} = \frac{\Delta P}{\rho g} = \left(\frac{\Delta P}{\rho}\right)\left(\frac{1}{mg}\right)$$

$$= \left(\frac{2.3938\ Nm}{kg}\right)\left(\frac{s^2}{9.8066\ m}\right)\left(\frac{kg\ m/s^2}{1\ N}\right)$$

$$= 0.24410\ m$$

$$= \boxed{0.24\ m} \tag{9.183}$$

In the metric system, the numerical value of energy per unit mass is not the same as the numerical value of energy per unit weight. There is no arithmetic advantage to using head in the metric system.

In summary, ft of head is a unit of length. It expresses the energy per unit weight in a flowing system. It may be converted to other systems by using the conversion factors for length, and some engineers find it intuitive to compare energies in a flow in terms of head. In the American engineering system of units, ft of head (energy per unit weight) has exactly the same numerical value as ft lb$_f$/lb$_m$ (energy per unit mass). In the SI system, N m/kg (the analogous expression to ft lb$_f$/lb$_m$) is not numerically equal to m of head.

9.2.3 Valves and fittings

In this section, we show several calculations using the macroscopic mass, momentum, and energy balances to determine forces as well as frictional losses in valves and fittings. We used some of these results in Chapter 1, but now we apply our formal modeling methods to obtain them directly.

EXAMPLE 9.7 (90-Degree bend, revisited). *What is the direction and magnitude of the force needed to support the 90-degree pipe bend shown in Figure 9.18? This problem was solved previously in Chapters 3 and 5. The flow is steady and turbulent and the cross section of the pipe bend is πR^2.*

SOLUTION. We seek a macroscopic force caused by a flow; density and velocity direction are constant across the inlet and outlet surfaces. The macroscopic momentum balance for a single-input, single-output system applies to this flow situation (see Equation 9.142, $N = 2$):

Macroscopic momentum balance on a CV, $\rho, \hat{v}$ constant across A_i:

$$\frac{d\underline{\mathbf{P}}}{dt} + \frac{\rho_1 A_1 \cos\theta_1 \langle v\rangle_1^2}{\beta_1}\, \hat{\underline{v}}|_{A_1} + \frac{\rho_2 A_2 \cos\theta_2 \langle v\rangle_2^2}{\beta_2}\, \hat{\underline{v}}|_{A_2} = \sum_{\substack{\text{on}\\ \text{CV}}} \underline{f}$$

(9.184)

We choose as our control volume all of the fluid in the bend, as outlined by the dashed curve in Figure 9.18. For our analysis, we choose a rectangular coordinate

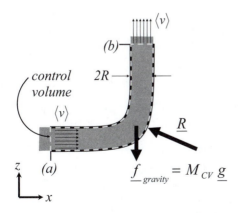

Figure 9.18 In Chapters 3–5, we worked on this problem as we developed the continuum model for fluids. Here, we solve it again using the macroscopic momentum balance.

system, which is desirable because the inlet and outlet velocities are expressed most conveniently in a Cartesian system. The forces in this problem are discussed in Chapter 7. The part that is different here is the convective term, which we now can treat formally with the macroscopic momentum-balance equation.

The flow is steady; thus, the rate of change of momentum in the CV $\partial \underline{P}/dt$ is zero. The forces on the bend are gravity and the molecular surface forces on the CV: inlet, outlet, and walls:

$$\sum_{\substack{\text{on} \\ \text{CV}}} \underline{f} = \underline{f}_{\text{gravity}} + \underline{f}_{\text{surface}} \tag{9.185}$$

$$= \underline{f}_{\text{gravity}} + \underline{f}_{\text{inlet}} + \underline{f}_{\text{outlet}} + \underline{f}_{\text{walls}} \tag{9.186}$$

The two convective terms are evaluated at the inlet and outlet surfaces of the bend. The momentum balance becomes:

$$\frac{d\underline{P}}{dt} + \frac{\rho_1 A_1 \cos \theta_1 \langle v \rangle_1^2}{\beta_1} \, \hat{v}|_{A_1} + \frac{\rho_2 A_2 \cos \theta_2 \langle v \rangle_2^2}{\beta_2} \, \hat{v}|_{A_2} = \sum_{\substack{\text{on} \\ \text{CV}}} \underline{f} \tag{9.187}$$

$$\frac{\rho_1 A_1 \cos \theta_1 \langle v \rangle_1^2}{\beta_1} \, \hat{v}|_{A_1} + \frac{\rho_2 A_2 \cos \theta_2 \langle v \rangle_2^2}{\beta_2} \, \hat{v}|_{A_2}$$
$$= \underline{f}_{\text{gravity}} + \underline{f}_{\text{inlet}} + \underline{f}_{\text{outlet}} + \underline{f}_{\text{walls}} \tag{9.188}$$

From Figure 9.18 and the chosen coordinate system, we can identify the following quantities in the convective terms:

$$\rho_1 = \rho_2 = \rho \qquad \text{(incompressible fluid)}$$

$$A_1 = A_2 = \pi R^2 \qquad \text{(constant cross section)}$$

$$\beta_1 = \beta_2 \approx 1 \qquad \text{(turbulent flow)}$$

$$\hat{n}_1 = -\hat{e}_x = \begin{pmatrix} -1 \\ 0 \\ 0 \end{pmatrix}_{xyz}$$

$$\hat{n}_2 = \hat{e}_z = \begin{pmatrix} 0 \\ 0 \\ 1 \end{pmatrix}_{xyz}$$

$$\hat{v}|_1 = \hat{e}_x = \begin{pmatrix} 1 \\ 0 \\ 0 \end{pmatrix}_{xyz}$$

$$\hat{v}|_2 = \hat{e}_z = \begin{pmatrix} 0 \\ 0 \\ 1 \end{pmatrix}_{xyz}$$

$$\cos \theta_1 = \hat{n}_1 \cdot \hat{v}|_1 = -1$$

$$\cos \theta_2 = \hat{n}_2 \cdot \hat{v}|_2 = 1$$

Substituting these values into the macroscopic momentum balance yields:

$$\rho \pi R^2 \langle v \rangle_1^2 \begin{pmatrix} -1 \\ 0 \\ 0 \end{pmatrix}_{xyz} + \rho \pi R^2 \langle v \rangle_2^2 \begin{pmatrix} 0 \\ 0 \\ 1 \end{pmatrix}_{xyz}$$

$$= \underline{f}_{\text{gravity}} + \underline{f}_{\text{inlet}} + \underline{f}_{\text{outlet}} + \underline{f}_{\text{walls}} \qquad (9.189)$$

We can relate $\langle v \rangle_1$ and $\langle v \rangle_2$ through the macroscopic mass balance (see Equation 9.140). Substituting the values of various parameters discussed previously into the macroscopic mass balance yields:

Macroscopic
mass balance
($\rho, \hat{v}$ constant $\dfrac{dm_{CV}}{dt} + \rho_1 A_1 \cos \theta_1 \langle v \rangle_1 + \rho_2 A_2 \cos \theta_2 \langle v \rangle_2 = 0 \qquad (9.190)$
across A_i),
single-input,
single-output

$$-\rho \pi R^2 \langle v \rangle_1 + \rho \pi R^2 \langle v \rangle_2 = 0 \qquad (9.191)$$

$$\langle v \rangle_1 = \langle v \rangle_2 \qquad (9.192)$$

Thus, the average velocity is the same at the inlet and the outlet, and we write the average velocity simply as $\langle v \rangle$. With this notation, the macroscopic momentum balance becomes:

$$\rho \pi R^2 \langle v \rangle^2 (-\hat{e}_x + \hat{e}_z) = \underline{f}_{\text{gravity}} + \underline{f}_{\text{inlet}} + \underline{f}_{\text{outlet}} + \underline{f}_{\text{walls}} \qquad (9.193)$$

$$\rho \pi R^2 \langle v \rangle^2 \begin{pmatrix} -1 \\ 0 \\ 1 \end{pmatrix}_{xyz} = \underline{f}_{\text{gravity}} + \underline{f}_{\text{inlet}} + \underline{f}_{\text{outlet}} + \underline{f}_{\text{walls}} \qquad (9.194)$$

We obtained this same result for the convective term in Chapter 3 (see Equation 3.181). From this point on, the force terms are calculated as discussed previously, and the final result is obtained in the same manner as in Chapter 5 (see Equation 5.185).

We now present a second example, to calculate the effect of pressure on flow in a horizontal pipe. We calculated this quantity in Chapter 7 for laminar flow, beginning with the microscopic-balance results for $v_z(r)$ (see Equation 7.103). Now we perform this calculation for general flow—laminar and turbulent—using the macroscopic momentum-balance equation.

EXAMPLE 9.8. *What is the total force on the wall for a Newtonian fluid of viscosity μ flowing in a long circular pipe under pressure? Over a length L, the pressure drops from p_0 to p_L; the flow may be laminar or turbulent.*

SOLUTION. We seek a macroscopic force caused by a flow; density and velocity direction are constant across the inlet and outlet surfaces. The macroscopic

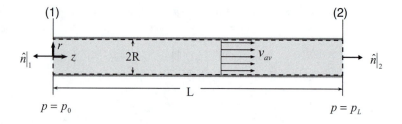

Figure 9.19 A macroscopic balance on a section of straight pipe allows us to relate the total force on the pipe to the pressure drop. The macroscopic control volume is indicated by a dotted line.

momentum balance applies to this flow situation (see Equation 9.142):

Macroscopic
momentum
balance
on a CV,
$\rho, \hat{v}$ constant
across A_i:

$$\frac{d\underline{\mathbf{P}}}{dt} + \frac{\rho_1 A_1 \cos\theta_1 \langle v \rangle_1^2}{\beta_1} \hat{v}|_{A_1} + \frac{\rho_2 A_2 \cos\theta_2 \langle v \rangle_2^2}{\beta_2} \hat{v}|_{A_2} = \sum_{\substack{\text{on} \\ \text{CV}}} \underline{f} \qquad (9.195)$$

We choose as our CV all of the fluid inside the pipe, as outlined by the dotted line in Figure 9.19. A horizontal cylindrical coordinate system is a reasonable choice for this problem because it is easy to express the incoming and exiting velocities in such a coordinate system. We choose the $(r\theta z)$-coordinate system shown in Figure 9.19.[3]

Because the flow is steady, the time derivative on the lefthand side of Equation 9.195 is zero. The convective term can be simplified if we realize that for the current problem, the density is constant ($\rho_1 = \rho_2 = \rho$), the cross-sectional area is constant ($A_1 = A_2 = \pi R^2$), and the quantities $\cos\theta_1$ and $\cos\theta_2$ are given by:

$$\theta_1 = 180° \quad \cos\theta_1 = -1 \qquad (9.196)$$

$$\theta_2 = 0° \quad \cos\theta_2 = 1 \qquad (9.197)$$

The forces on the CV are body forces (i.e., gravity) and surface forces (i.e., wall forces and molecular forces on the fluid ends of the CV). We assume that the flow is either laminar or turbulent (i.e., does not switch between laminar and turbulent); thus, β is constant. The direction of the velocity at Surfaces (1) and (2) is $\hat{v}|_{A_1} = \hat{v}|_{A_2} = \hat{e}_z$. The macroscopic momentum-balance equation becomes:

$$\frac{d\underline{\mathbf{P}}}{dt} + \frac{\rho_1 A_1 \cos\theta_1 \langle v \rangle_1^2}{\beta_1} \hat{v}|_{A_1} + \frac{\rho_2 A_2 \cos\theta_2 \langle v \rangle_2^2}{\beta_2} \hat{v}|_{A_2} = \sum_{\substack{\text{on} \\ \text{CV}}} \underline{f} \qquad (9.198)$$

$$-\frac{\rho \pi R^2 \langle v \rangle_1^2}{\beta} \hat{e}_z + \frac{\rho \pi R^2 \langle v \rangle_2^2}{\beta} \hat{e}_z = \underline{f}_{\text{gravity}} + \underline{f}_{\text{surface}} \qquad (9.199)$$

where $\underline{f}_{\text{surface}} = \underline{f}_{\text{inlet}} + \underline{f}_{\text{outlet}} + \underline{f}_{\text{walls}}$.

[3] In Chapter 5 and in the previous problem, we use a Cartesian system in our calculations. We can do the same here; readers are encouraged to carry out such a calculation for practice.

We can relate $\langle v \rangle_1$ and $\langle v \rangle_2$ through the mass balance. The macroscopic mass balance is given by Equation 9.140, repeated here:

Macroscopic mass balance ($\rho, \hat{v}$ constant across A_i), single-input, single-output

$$\frac{dm_{CV}}{dt} + \rho_1 A_1 \cos\theta_1 \langle v \rangle_1 + \rho_2 A_2 \cos\theta_2 \langle v \rangle_2 = 0 \qquad (9.200)$$

Making the same substitutions, we obtain:

$$\frac{dm_{CV}}{dt} + \rho_1 A_1 \cos\theta_1 \langle v \rangle_1 + \rho_2 A_2 \cos\theta_2 \langle v \rangle_2 = 0 \qquad (9.201)$$

$$-\langle v \rangle_1 + \langle v \rangle_2 = 0 \qquad (9.202)$$

$$\boxed{\langle v \rangle_1 = \langle v \rangle_2 = \langle v \rangle} \qquad (9.203)$$

Returning to the momentum balance (see Equation 9.199) and incorporating the mass-balance result, we see that the convective terms cancel for both laminar ($\beta = 0.75$) and turbulent ($\beta \approx 1$) flow. The macroscopic momentum balance for this problem now simplifies to:

$$-\frac{\rho \pi R^2 \langle v \rangle^2}{\beta} \begin{pmatrix} 0 \\ 0 \\ 1 \end{pmatrix}_{r\theta z} + \frac{\rho \pi R^2 \langle v \rangle^2}{\beta} \begin{pmatrix} 0 \\ 0 \\ 1 \end{pmatrix}_{xyz} = \underline{f}_{\text{gravity}} + \underline{f}_{\text{surface}} \qquad (9.204)$$

$$0 = \underline{f}_{\text{gravity}} + \underline{f}_{\text{surface}} \qquad (9.205)$$

All that remains is to write the four forces on the CV: gravity and molecular forces on the inlet, outlet, and the walls. The gravity force in the chosen coordinate system is:

Force on CV due to gravity:

$$\underline{f}_{\text{gravity}} = m_{CV}\underline{g} = m_{CV} \begin{pmatrix} g_r \\ g_\theta \\ 0 \end{pmatrix}_{r\theta z} \qquad (9.206)$$

where m_{CV} is the mass of fluid in the CV. Although the use of cylindrical coordinates is awkward for the gravity vector, it does not cause any difficulty because our main concern is with the z-component (i.e., flow-direction component) of the momentum balance.

The surface force consists of forces on the inlet, outlet, and walls:

$$\underline{f}_{\text{surface}} = \underline{f}_{\text{inlet}} + \underline{f}_{\text{outlet}} + \underline{f}_{\text{walls}} \qquad (9.207)$$

The force exerted on the fluid by the walls of the pipe is the force we seek. We can write this force as:

$$\begin{pmatrix} \text{force on CV} \\ \text{due to contact} \\ \text{with walls} \end{pmatrix} = - \begin{pmatrix} \text{force on walls} \\ \text{due to contact} \\ \text{with fluid} \end{pmatrix} = \begin{pmatrix} R_r \\ R_\theta \\ R_z \end{pmatrix}_{r\theta z} = \underline{R} \qquad (9.208)$$

The source of this term is the molecular contact between the fluid and the walls. The no-slip boundary condition imposes molecular stresses on the walls, and to remain stationary the walls exert an equal and opposite force. The remaining two forces are the molecular forces on the inlet and outlet surfaces, $\underline{f}\big|_{A_1}$ and $\underline{f}\big|_{A_2}$, which we discuss now.

The macroscopic momentum balance for this problem thus far is shown here:

$$0 = \underline{f}_{\text{gravity}} + \underline{f}_{\text{surface}} \tag{9.209}$$

$$0 = m_{CV}\underline{g} + \underline{f}\big|_{A_1} + \underline{f}\big|_{A_2} + \underline{R} \tag{9.210}$$

In Equation 9.210, the gravity term is known, and $\underline{R}$ is the quantity we seek. The two terms $\underline{f}|_{A_1}$ and $\underline{f}|_{A_2}$ are the molecular forces on the end surfaces of the CV, including pressure and viscous forces. To calculate these terms, we turn to the stress tensor.

The molecular fluid forces on any surface may be expressed using the stress tensor $\tilde{\underline{\underline{\Pi}}}$ (see Equation 4.221):

Total molecular fluid force on a surface $\mathcal{S}$:

$$\underline{\mathcal{F}} = \iint_{\mathcal{S}} [\hat{n} \cdot \tilde{\underline{\underline{\Pi}}}]_{\text{at surface}} \, dS \tag{9.211}$$

$$= \iint_{\mathcal{S}} \left[\hat{n} \cdot \left(-p\underline{\underline{I}} + \tilde{\underline{\underline{\tau}}}\right)\right]_{\text{at surface}} \, dS \tag{9.212}$$

$$= \iint_{\mathcal{S}} \left[-p\hat{n} + \hat{n} \cdot \tilde{\underline{\underline{\tau}}}\right]_{\text{at surface}} \, dS \tag{9.213}$$

When we apply Equation 9.213 to the inlet A_1 and outlet A_2 surfaces, we obtain $\underline{f}|_{A_1}$ and $\underline{f}|_{A_2}$:

$$\underline{f}\big|_{A_1} = \iint_{A_1} \left[-p\hat{n} + \hat{n} \cdot \tilde{\underline{\underline{\tau}}}\right]_{A_1} \, dA \tag{9.214}$$

$$= \iint_{A_1} [-p\hat{n}]_{A_1} \, dA + \iint_{A_1} \left[\hat{n} \cdot \tilde{\underline{\underline{\tau}}}\right]_{A_1} \, dA \tag{9.215}$$

$$\underline{f}\big|_{A_2} = \iint_{A_2} \left[-p\hat{n} + \hat{n} \cdot \tilde{\underline{\underline{\tau}}}\right]_{A_2} \, dA \tag{9.216}$$

$$= \iint_{A_2} [-p\hat{n}]_{A_2} \, dA + \iint_{A_2} \left[\hat{n} \cdot \tilde{\underline{\underline{\tau}}}\right]_{A_2} \, dA \tag{9.217}$$

The pressure contribution to each force is straightforward to calculate. The unit normal vectors of Surfaces (1) and (2) are $\hat{n}|_{A_1} = -\hat{e}_z$ and $\hat{n}|_{A_2} = \hat{e}_z$.

We therefore can write:

$$\iint_{A_1} [-p\hat{n}]_{A_1} \, dA = \int_0^{2\theta} \int_0^R (-p_0) \begin{pmatrix} 0 \\ 0 \\ -1 \end{pmatrix}_{r\theta z} r\,dr\,d\theta \qquad (9.218)$$

$$= \begin{pmatrix} 0 \\ 0 \\ \pi R^2 p_0 \end{pmatrix}_{r\theta z} \qquad (9.219)$$

$$\iint_{A_2} [-p\hat{n}]_{A_2} \, dA = \int_0^{2\theta} \int_0^R (-p_L) \begin{pmatrix} 0 \\ 0 \\ 1 \end{pmatrix}_{r\theta z} r\,dr\,d\theta \qquad (9.220)$$

$$= \begin{pmatrix} 0 \\ 0 \\ -\pi R^2 p_L \end{pmatrix}_{r\theta z} \qquad (9.221)$$

The contributions of the extra-stress tensor $\underline{\underline{\tilde{\tau}}}$ to $\underline{f}|_{A_1}$ and $\underline{f}|_{A_2}$ are subtler; thus, to calculate these terms, it is best to proceed formally. This contribution cancels out of the momentum balance equation, as we show here. Some readers may prefer to skip this discussion of the extra-stress contribution on the fluid ends, accept that the contributions are both zero, and proceed to the assembled momentum balance, Equation 9.231.

To calculate the extra-stress contribution to the force on the ends of our CV, we proceed as follows. For laminar flow, we calculate $\underline{\underline{\tilde{\tau}}}$ from the constitutive equation for Newtonian fluids (see Equation 5.89). For the current problem, the velocity is only in the z-direction and varies only with r; the density is constant. Therefore, using Table B.8, in Appendix B we calculate the stress tensor $\underline{\underline{\tilde{\tau}}}$ to be:

Extra-stress tensor
laminar flow
in a tube:
$$\underline{\underline{\tilde{\tau}}} = \mu \left(\nabla \underline{v} + (\nabla \underline{v})^T \right) \qquad (9.222)$$

$$= \begin{pmatrix} 0 & 0 & \mu\dfrac{\partial v_z}{\partial r} \\ 0 & 0 & 0 \\ \mu\dfrac{\partial v_z}{\partial r} & 0 & 0 \end{pmatrix}_{r\theta z} \qquad (9.223)$$

For turbulent flow, we calculate $\underline{\underline{\tilde{\tau}}}$ from the time-averaged result for $\underline{\underline{\bar{\tilde{\tau}}}}$, developed in the Web appendix [108]. Both results simplify to the same form; thus, we can develop both cases together in subsequent calculations:

Extra-stress tensor
turbulent flow
in a tube:
$$\underline{\underline{\tilde{\tau}}} = \begin{pmatrix} 0 & 0 & \mu\dfrac{\partial \overline{v_z}}{\partial r} \\ 0 & 0 & 0 \\ \mu\dfrac{\partial \overline{v_z}}{\partial r} & 0 & 0 \end{pmatrix}_{r\theta z} \qquad (9.224)$$

Having obtained $\underline{\underline{\tilde{\tau}}}$, the extra-stress tensor, we now substitute Equation 9.224 into the second integrals of Equations 9.215 and 9.217 and calculate the $\underline{\underline{\tilde{\tau}}}$ contributions:

$$\left[\hat{n}\cdot\underline{\underline{\tilde{\tau}}}\right]\Big|_{A_1} = \begin{pmatrix}0 & 0 & -1\end{pmatrix}_{r\theta z}\cdot\begin{pmatrix} 0 & 0 & \mu\dfrac{\partial\overline{v}_z}{\partial r} \\ 0 & 0 & 0 \\ \mu\dfrac{\partial\overline{v}_z}{\partial r} & 0 & 0 \end{pmatrix}_{r\theta z}\Bigg|_{A_1} \tag{9.225}$$

$$= \begin{pmatrix}-\mu\dfrac{\partial\overline{v}_z}{\partial r} & 0 & 0\end{pmatrix}_{r\theta z}\Bigg|_{A_1} = -\mu\dfrac{\partial\overline{v}_z}{\partial r}\Bigg|_{A_1}\hat{e}_r \tag{9.226}$$

$$\int_0^{2\pi}\int_0^R\left[\hat{n}\cdot\underline{\underline{\tilde{\tau}}}\right]\Big|_{A_1} r\,dr\,d\theta = \int_0^{2\pi}\int_0^R -\mu\dfrac{\partial\overline{v}_z}{\partial r}\Bigg|_{A_1}\hat{e}_r\, r\,dr\,d\theta \tag{9.227}$$

$$= \int_0^{2\pi}\int_0^R -\mu\dfrac{\partial\overline{v}_z}{\partial r}\Bigg|_{A_1}\begin{pmatrix}\cos\theta \\ \sin\theta \\ 0\end{pmatrix}_{xyz} r\,dr\,d\theta \tag{9.228}$$

$$= 0 \tag{9.229}$$

Note that we converted $\hat{e}_r$ to Cartesian coordinates, which allows us to carry out the θ-integration. The θ-integral in both the x- and y-components is zero. A similar calculation leads to the same result for the $\underline{\underline{\tilde{\tau}}}$-related molecular force on A_2.

Substituting all of the calculated forces on the CV into the macroscopic momentum-balance equation, we obtain an equation that we can solve for the wall force $\underline{R}$:

$$0 = \sum_{\substack{\text{on}\\\text{CV}}}\underline{f} = m_{CV}\underline{g} + \underline{f}\Big|_{A_1} + \underline{f}\Big|_{A_2} + \underline{R} \tag{9.230}$$

$$\begin{pmatrix}0\\0\\0\end{pmatrix}_{r\theta z} = m_{CV}\begin{pmatrix}g_r\\g_\theta\\0\end{pmatrix}_{r\theta z} + \begin{pmatrix}0\\0\\\pi R^2\left(p_0 - p_L\right)\end{pmatrix}_{r\theta z} + \begin{pmatrix}R_r\\R_\theta\\R_z\end{pmatrix}_{r\theta z} \tag{9.231}$$

The momentum-balance result in Equation 9.231 is a vector equation, and we obtain information from all three components. The r- and θ-components of Equation 9.231 show that a portion of $\underline{R}$ must counter gravity:

$$r\text{-component:} \quad 0 = m_{CV}g_r + R_r \tag{9.232}$$

$$\theta\text{-component:} \quad 0 = m_{CV}g_\theta + R_\theta \tag{9.233}$$

The z-component of Equation 9.231 relates axial drag and pressure drop:

$$z\text{-component:} \quad 0 = \pi R^2\left(p_0 - p_L\right) + R_z \tag{9.234}$$

$$\mathcal{F}_{\text{drag}} = -R_z = \pi R^2(p_0 - p_L) \tag{9.235}$$

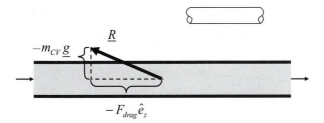

Figure 9.20 The calculated result for the net force on a straight pipe is a combination of two forces: an upward component that counteracts gravity, and a second component, the drag, that is in the direction opposite to the flow direction.

Note that this is the same answer obtained for laminar flow from the microscopic calculation (see Equation 7.115):

Axial drag on walls
in tube flow,
laminar or turbulent:
$$\mathcal{F}_{\text{drag}} = (p_0 - p_L)\pi R^2 \qquad (9.236)$$

We can rewrite Equation 9.231 as shown here, a form that emphasizes the two parts of $\underline{R}$—axial drag and vertical gravity (Figure 9.20):

Wall force on fluid
in tube flow,
laminar or turbulent:
$$\underline{R} = -\mathcal{F}_{\text{drag}}\hat{e}_z - m_{CV}\underline{g} \qquad (9.237)$$

Force on walls
in tube flow,
laminar or turbulent:
$$-\underline{R} = \mathcal{F}_{\text{drag}}\hat{e}_z + m_{CV}\underline{g} \qquad (9.238)$$

In Example 9.8, we arrive at the molecular force on the ends of the CV by using the fundamental equation $\mathcal{F} = \iint_S [\hat{n} \cdot \underline{\underline{\tilde{\Pi}}}]_{\text{at surface}} \, dS$. We did not have to draw on intuition about the effect of molecular stresses (i.e., pressure and viscosity) because we completely accounted for them with the stress tensor $\underline{\underline{\tilde{\Pi}}}$. Although we may have been able to use intuition to arrive at pressure as the only relevant surface force, proceeding formally is helpful in more complex problems in which intuition fails. The formal macroscopic momentum-balance equation is particularly helpful in sorting out the effect of direction changes in flows, as shown in the next example.

EXAMPLE 9.9. *Calculate the net force on the horizontal U-shaped pipe bend shown in Figure 9.21. The pipe is circular in cross section, water is flowing in the pipe, and the flow is steady.*

SOLUTION. We seek a macroscopic force caused by a flow; density and velocity direction are constant across the inlet and outlet surfaces. The macroscopic

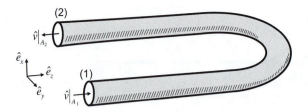

When fluid flows in a U-shaped tube, the momentum changes direction and forces are required to restrain the tube.

momentum balance applies to this flow situation.

Macroscopic
momentum
balance
on a CV,
$\rho, \hat{v}$ constant
across A_i:

$$\frac{d\underline{P}}{dt} + \frac{\rho_1 A_1 \cos\theta_1 \langle v \rangle_1^2}{\beta_1} \, \hat{v}|_{A_1} + \frac{\rho_2 A_2 \cos\theta_2 \langle v \rangle_2^2}{\beta_2} \, \hat{v}|_{A_2} = \sum_{\substack{\text{on} \\ \text{CV}}} \underline{f}$$

(9.239)

We choose a macroscopic CV that encloses all of the fluid inside the pipe section. A horizontal rectangular coordinate system is a reasonable choice for expressing the momentum in this problem because it is easy to express the incoming and exiting velocities and the effect of gravity in such a coordinate system.

Having chosen the CV and the coordinate system, we proceed with writing the terms of the macroscopic mass and momentum balances on the CV as they apply to these choices. The mass balance is given by Equation 9.140, repeated here:

Macroscopic
mass balance on CV
$(\rho, \hat{v}$ constant across $A_i)$,
single-input, single-output

$$\frac{dm_{CV}}{dt} + \rho_1 A_1 \cos\theta_1 \langle v \rangle_1 + \rho_2 A_2 \cos\theta_2 \langle v \rangle_2 = 0$$

(9.240)

Because our flow is steady, the time-derivative on the lefthand side of Equation 9.240 is zero. The convective terms can be simplified if we note that for the current problem, density is constant ($\rho_1 = \rho_2 = \rho$), the cross-sectional area is constant ($A_1 = A_2 = \pi R^2$), and the quantities $\cos\theta_1$ and $\cos\theta_2$ are given by:

$$\theta_1 = 180° \quad \hat{n}|_{A_1} \cdot \hat{v}|_{A_1} = \cos\theta_1 = -1$$

(9.241)

$$\theta_2 = 0° \quad \hat{n}|_{A_1} \cdot \hat{v}|_{A_2} = \cos\theta_2 = 1$$

(9.242)

Making these substitutions, we find:

$$\frac{dm_{CV}}{dt} + \rho_1 A_1 \cos\theta_1 \langle v \rangle_1 + \rho_2 A_2 \cos\theta_2 \langle v \rangle_2 = 0$$

(9.243)

$$-\langle v \rangle_1 + \langle v \rangle_2 = 0$$

(9.244)

$$\boxed{\langle v \rangle_1 = \langle v \rangle_2 = \langle v \rangle}$$

(9.245)

Turning now to the macroscopic momentum balance, again the flow is steady; therefore, the rate of change of momentum of the CV is zero, $d\underline{P}/dt = 0$. The convective terms can be simplified using the same information used before to simplify the mass balance (ρ, A constant) as well as the result of the mass balance ($\langle v \rangle$ constant). We assume that the flow is either laminar or turbulent (it does not switch between laminar and turbulent); thus, β is constant. The forces on the CV are body forces (i.e., gravity) and surface forces (i.e., wall forces and forces on the inlet and outlet). The macroscopic-momentum-balance equation becomes:

$$\frac{d\underline{P}}{dt} + \frac{\rho_1 A_1 \cos\theta_1 \langle v \rangle_1^2}{\beta_1}\, \hat{v}|_{A_1} + \frac{\rho_2 A_2 \cos\theta_2 \langle v \rangle_2^2}{\beta_2}\, \hat{v}|_{A_2} = \sum_{\substack{\text{on} \\ \text{CV}}} \underline{f} \tag{9.246}$$

$$-\frac{\rho\pi R^2 \langle v \rangle^2}{\beta}\, \hat{v}|_{A_1} + \frac{\rho\pi R^2 \langle v \rangle^2}{\beta}\, \hat{v}|_{A_2} = \underline{f}_{\text{gravity}} + \underline{f}_{\text{surface}} \tag{9.247}$$

The direction of the velocity at Surface (1) is $\hat{v}|_{A_1} = \hat{e}_z$ and the direction of the velocity at Surface (2) is $\hat{v}|_{A_2} = -\hat{e}_z$; thus, the convective terms do not cancel. The macroscopic momentum balance for this problem simplifies to the following (compare to the straight-pipe Equations 9.204 and 9.205):

$$-\frac{\rho\pi R^2 \langle v \rangle^2}{\beta}\hat{e}_z + \frac{\rho\pi R^2 \langle v \rangle^2}{\beta}(-\hat{e}_z) = \underline{f}_{\text{gravity}} + \underline{f}_{\text{surface}} \tag{9.248}$$

$$\frac{-2\rho\pi R^2 \langle v \rangle^2}{\beta}\hat{e}_z = \underline{f}_{\text{gravity}} + \underline{f}_{\text{surface}} \tag{9.249}$$

The gravity force in the chosen coordinate system is:

$$\begin{matrix}\text{Force on CV} \\ \text{due to gravity:}\end{matrix} \quad \underline{f}_{\text{gravity}} = m_{CV}\underline{g} = \begin{pmatrix} -m_{CV}g \\ 0 \\ 0 \end{pmatrix}_{xyz} \tag{9.250}$$

where m_{CV} is the mass of fluid in the CV. The surface force is the force on the inlet, outlet, and walls:

$$\underline{f}_{\text{surface}} = \underline{f}_{\text{inlet}} + \underline{f}_{\text{outlet}} + \underline{f}_{\text{walls}} \tag{9.251}$$

$$= \underline{f}\big|_{A_1} + \underline{f}\big|_{A_2} + \underline{R} \tag{9.252}$$

The momentum balance becomes:

$$-\frac{2\rho\pi R^2 \langle v \rangle^2}{\beta}\hat{e}_z = -m_{CV}g\hat{e}_x + \underline{f}\big|_{A_1} + \underline{f}\big|_{A_2} + \underline{R} \tag{9.253}$$

The molecular forces $\underline{f}|_{A_i}$ may be expressed in terms of the usual expression for molecular forces, $\hat{n} \cdot \underline{\tilde{\Pi}}$, integrated over the surface of interest (see Equation 4.221):

Total molecular
fluid force $\qquad \underline{\mathcal{F}} = \iint_{\mathcal{S}} [\hat{n} \cdot \underline{\tilde{\Pi}}]_{\text{at surface}} \, dS \qquad (9.254)$
on a surface $\mathcal{S}$:

$$= \iint_{\mathcal{S}} \left[\hat{n} \cdot \left(-p\underline{I} + \underline{\tilde{\tau}} \right) \right]_{\text{at surface}} \, dS \qquad (9.255)$$

$$= \iint_{\mathcal{S}} \left[-p\hat{n} + \hat{n} \cdot \underline{\tilde{\tau}} \right]_{\text{at surface}} \, dS \qquad (9.256)$$

For the inlet and outlet fluid surfaces, we write:

$$\underline{f}\Big|_{A_1} = \iint_{A_1} \left[-p\hat{n} + \hat{n} \cdot \underline{\tilde{\tau}} \right]_{A_1} \, dA \qquad (9.257)$$

$$= \iint_{A_1} [-p\hat{n}]_{A_1} \, dA + \iint_{A_1} \left[\hat{n} \cdot \underline{\tilde{\tau}} \right]_{A_1} \, dA \qquad (9.258)$$

$$\underline{f}\Big|_{A_2} = \iint_{A_2} \left[-p\hat{n} + \hat{n} \cdot \underline{\tilde{\tau}} \right]_{A_2} \, dA \qquad (9.259)$$

$$= \iint_{A_2} [-p\hat{n}]_{A_2} \, dA + \iint_{A_2} \left[\hat{n} \cdot \underline{\tilde{\tau}} \right]_{A_2} \, dA \qquad (9.260)$$

The pressure contribution to each force is straightforward to calculate. The unit normal vectors of Surfaces (1) and (2) are the same, $\hat{n}|_{A_1} = \hat{n}|_{A_2} = -\hat{e}_z$. We therefore can write:

$$\iint_{A_1} [-p\hat{n}]_{A_1} \, dA = \int_0^{2\theta} \int_0^R (-p_1) \begin{pmatrix} 0 \\ 0 \\ -1 \end{pmatrix}_{r\theta z} r \, dr \, d\theta \qquad (9.261)$$

$$= \begin{pmatrix} 0 \\ 0 \\ \pi R^2 p_1 \end{pmatrix}_{r\theta z} \qquad (9.262)$$

$$\iint_{A_2} [-p\hat{n}]_{A_2} \, dA = \int_0^{2\theta} \int_0^R (-p_2) \begin{pmatrix} 0 \\ 0 \\ -1 \end{pmatrix}_{r\theta z} r \, dr \, d\theta \qquad (9.263)$$

$$= \begin{pmatrix} 0 \\ 0 \\ \pi R^2 p_2 \end{pmatrix}_{r\theta z} \qquad (9.264)$$

The two pressure terms combine and reinforce one another. The contribution of the extra-stress tensor $\underline{\tilde{\tau}}$ to $\underline{f}|_{A_1}$ and $\underline{f}|_{A_2}$ may be calculated formally from the constitutive equation as before, and the calculation is similar to that in Example 5.9. In this case, the small viscous contributions on the two surfaces due to the fluid deceleration dv_z/dz cancel out (the detailed calculation is left to readers; see Problem 23).

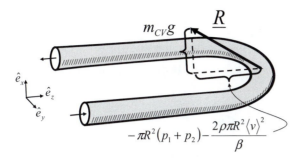

$$-\pi R^2 \left(p_1 + p_2 \right) - \frac{2\rho\pi R^2 \langle v \rangle^2}{\beta}$$

Figure 9.22   The solution we calculated can be understood directly as a backward force countering the fluid motion and an upward force countering gravity.

Putting all of the forces on the control volume into the macroscopic momentum-balance equation (Equation 9.253), we obtain an equation that we can solve for the wall force $\underline{R}$:

$$0 = -m_{CV}g\hat{e}_x + \underline{f}\Big|_{A_1} + \underline{f}\Big|_{A_2} + \underline{R} + \frac{2\rho\pi R^2 \langle v \rangle^2}{\beta}\hat{e}_z \qquad (9.265)$$

$$0 = \begin{pmatrix} -m_{CV}g \\ 0 \\ 0 \end{pmatrix}_{xyz} + \begin{pmatrix} 0 \\ 0 \\ \pi R^2 \left(p_1 + p_2 \right) \end{pmatrix}_{xyz} + \begin{pmatrix} R_x \\ R_y \\ R_z \end{pmatrix}_{xyz} + \begin{pmatrix} 0 \\ 0 \\ \frac{2\rho\pi R^2 \langle v \rangle^2}{\beta} \end{pmatrix}_{xyz} \qquad (9.266)$$

The x-component of the macroscopic momentum balance states that R_x balances gravity. The y-component states that there is no need for any y-restraining force (i.e., no side-to-side force). The z-component gives the required horizontal force (Figure 9.22):

Force on CV from walls:
$$\boxed{\underline{R} = \begin{pmatrix} R_x \\ R_y \\ R_z \end{pmatrix}_{xyz} = \begin{pmatrix} m_{CV}g \\ 0 \\ -\pi R^2 \left(p_1 + p_2 \right) - \frac{2\rho\pi R^2 \langle v \rangle^2}{\beta} \end{pmatrix}_{xyz}}$$
$$(9.267)$$

Force on walls from fluid:
$$\boxed{-\underline{R}} \qquad (9.268)$$

The two remaining examples use macroscopic balances to calculate the friction of an expansion fitting and a valve.

EXAMPLE 9.10. *How does friction in a sudden expansion (Figure 9.23) depend on fluid velocity?*

SOLUTION. To solve this problem, we need all three of the macroscopic balances. First, we perform the macroscopic mass balance. Our chosen control volume, shown in Figure 9.23, consists of the fluid just inside the expansion;

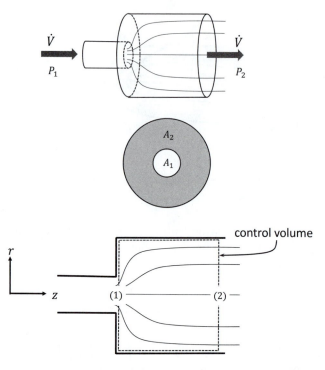

Figure 9.23 The friction in a sudden expansion may be calculated directly from a combination of macroscopic mass, momentum, and energy balances. The flow cross section changes from circle A_1 to circle A_2.

Inlet-Surface 1 is just at the plane where the cross-sectional area changes, and Outlet-Surface 2 is slightly downstream where the flow has straightened out.

For a single-input, single-output system, the macroscopic mass balance is given by Equation 9.140:

Macroscopic
mass balance
Single-input,
single-output
$$\frac{dm_{CV}}{dt} + \rho_1 A_1 \cos\theta_1 \langle v \rangle_1 + \rho_2 A_2 \cos\theta_2 \langle v \rangle_2 = 0 \qquad (9.269)$$

For our system, the time-derivative is zero (i.e., steady state), the density is constant $\rho_1 = \rho_2 = \rho$, θ_1 is equal to 180 degrees, and θ_2 is equal to 0 degrees. The mass balance becomes:

$$\rho A_1 (-1) \langle v \rangle_1 + \rho A_2 \langle v \rangle_2 = 0 \qquad (9.270)$$

$$\boxed{\langle v \rangle_2 = \frac{A_1}{A_2} \langle v \rangle_1} \qquad (9.271)$$

We seek a macroscopic force caused by a flow; density and velocity direction are constant across the inlet and outlet surfaces. The macroscopic momentum balance applies to this flow situation:

Macroscopic momentum balance on a CV, ρ, $\hat{v}$ constant across A_i:

$$\frac{d\mathbf{P}}{dt} + \frac{\rho_1 A_1 \cos\theta_1 \langle v\rangle_1^2}{\beta_1}\,\hat{v}|_{A_1} + \frac{\rho_2 A_2 \cos\theta_2 \langle v\rangle_2^2}{\beta_2}\,\hat{v}|_{A_2} = \sum_{\substack{\text{on}\\\text{CV}}} \underline{f} \tag{9.272}$$

We choose a Cartesian coordinate system with flow in the z-direction; thus, $\hat{v}|_{A_1} = \hat{v}|_{A_2} = \hat{e}_z$. Assuming steady, turbulent flow ($\beta_1 = \beta_2 = 1$) and incorporating the same simplifications as for the mass balance, we obtain:

$$\frac{d\mathbf{P}}{dt} + \frac{\rho_1 A_1 \cos\theta_1 \langle v\rangle_1^2}{\beta_1}\,\hat{v}|_{A_1} + \frac{\rho_2 A_2 \cos\theta_2 \langle v\rangle_2^2}{\beta_2}\,\hat{v}|_{A_2} = \sum_{\substack{\text{on}\\\text{CV}}} \underline{f} \tag{9.273}$$

$$-\rho A_1 \langle v\rangle_1^2 \begin{pmatrix} 0 \\ 0 \\ 1 \end{pmatrix}_{xyz} + \rho A_2 \langle v\rangle_2^2 \begin{pmatrix} 0 \\ 0 \\ 1 \end{pmatrix}_{xyz} = \sum_{\substack{\text{on}\\\text{CV}}} \underline{f} \tag{9.274}$$

The forces on the expansion are gravity, which we neglect, and the surface forces on the inlet, outlet, and walls:

$$\sum_{\substack{\text{on}\\\text{CV}}} \underline{f} = \underline{f}_{\text{gravity}} + \underline{f}_{\text{surface}} \tag{9.275}$$

$$= \underline{f}_{\text{surface}} \tag{9.276}$$

$$\sum_{\substack{\text{on}\\\text{CV}}} \underline{f} = \underline{f}_{\text{inlet}} + \underline{f}_{\text{outlet}} + \underline{f}_{\text{walls}} \tag{9.277}$$

The molecular surface forces, including pressure, are expressed using the stress tensor $\underline{\underline{\tilde{\Pi}}}$ (see Equation 4.221):

Total molecular fluid force on a surface $\mathcal{S}$:

$$\mathcal{F} = \iint_{\mathcal{S}} [\hat{n} \cdot \underline{\underline{\tilde{\Pi}}}]_{\text{at surface}}\, dS \tag{9.278}$$

$$= \iint_{\mathcal{S}} \left[\hat{n} \cdot \left(-p\underline{\underline{I}} + \underline{\underline{\tilde{\tau}}}\right)\right]_{\text{at surface}}\, dS \tag{9.279}$$

$$= \iint_{\mathcal{S}} \left[-p\hat{n} + \hat{n} \cdot \underline{\underline{\tilde{\tau}}}\right]_{\text{at surface}}\, dS \tag{9.280}$$

The surface forces on the expansion are the fluid forces on the horizontal and vertical solid walls and the fluid forces on the inlet and outlet surfaces, which are fluid surfaces. Thus, the macroscopic momentum balance becomes:

$$
\rho A_2 \langle v \rangle_2^2 \begin{pmatrix} 0 \\ 0 \\ 1 \end{pmatrix}_{xyz} - \rho A_1 \langle v \rangle_1^2 \begin{pmatrix} 0 \\ 0 \\ 1 \end{pmatrix}_{xyz} = \sum_{\substack{\text{on} \\ \text{CV}}} \underline{f} \tag{9.281}
$$

$$
+ \rho \left(A_2 \langle v \rangle_2^2 - A_1 \langle v \rangle_1^2 \right) \begin{pmatrix} 0 \\ 0 \\ 1 \end{pmatrix}_{xyz} = \underline{f}_{\text{inlet}} + \underline{f}_{\text{outlet}} + \underline{f}_{\text{walls}} \tag{9.282}
$$

$$
\rho \left(A_2 \langle v \rangle_2^2 - A_1 \langle v \rangle_1^2 \right) \begin{pmatrix} 0 \\ 0 \\ 1 \end{pmatrix}_{xyz} = \underline{f}\Big|_{A_1} + \underline{f}\Big|_{A_2} + \underline{f}_{\text{walls}} \tag{9.283}
$$

The forces on the fluid boundaries are calculated just as in the two previous examples. Surface A_1 has outwardly pointing unit normal $\hat{n}|_{A_1} = -\hat{e}_z$, whereas Surface A_2 has unit normal $= \hat{e}_z$. Using local cylindrical coordinates for the integrations, we obtain:

$$
\underline{f}\Big|_{A_1} + \underline{f}\Big|_{A_2} = \iint_S \left[-p\hat{n} + \hat{n} \cdot \underline{\underline{\tilde{\tau}}} \right]_{\text{at surface}} dS \tag{9.284}
$$

$$
= \int_0^{2\pi} \int_0^{R_1} -p_1 \begin{pmatrix} 0 \\ 0 \\ -1 \end{pmatrix}_{r\theta z} r\,dr\,d\theta + \int_0^{2\pi} \int_0^{R_1} -\hat{e}_z \cdot \underline{\underline{\tilde{\tau}}}\Big|_{A_1} r\,dr\,d\theta
$$

$$
+ \int_0^{2\pi} \int_0^{R_2} -p_2 \begin{pmatrix} 0 \\ 0 \\ 1 \end{pmatrix}_{r\theta z} r\,dr\,d\theta + \int_0^{2\pi} \int_0^{R_2} \hat{e}_z \cdot \underline{\underline{\tilde{\tau}}}\Big|_{A_2} r\,dr\,d\theta
$$

$$
\tag{9.285}
$$

From our solution of the 90-degree-bend and the U-bend problems, we know that the two $\underline{\underline{\tilde{\tau}}}$ integrals have a contribution when $\partial v_z / \partial z$ is important. Although there is a change of v_z in the flow direction due to the expansion, this viscous-stress contribution, always small, is very small in this fitting. We neglect the $\hat{n} \cdot \underline{\underline{\tilde{\tau}}}$ integrals. The momentum balance now becomes (Equation 9.283):

$$
\begin{pmatrix} 0 \\ 0 \\ \rho \left(A_2 \langle v \rangle_2^2 - A_1 \langle v \rangle_1^2 \right) \end{pmatrix}_{xyz} = \begin{pmatrix} 0 \\ 0 \\ p_1 \pi R_1^2 - p_2 \pi R_2^2 \end{pmatrix}_{r\theta z} + \underline{f}_{\text{walls}} \tag{9.286}
$$

The last force we need is the z-direction force on the solid-wall surfaces. The z-directional force on the solid surfaces includes the fluid force on the horizontal tube wall and on the vertical washer-shaped wall at Surface 1. The force on the horizontal wall should be small and we neglect this term:

$$
f_{\text{walls},z} = f_{\substack{\text{vertical} \\ \text{wall, } z}} + f_{\substack{\text{horizontal} \\ \text{wall, } z}} \tag{9.287}
$$

$$
\approx f_{\substack{\text{vertical} \\ \text{wall, } z}} \tag{9.288}
$$

The force on the vertical washer-shaped surface near the inlet is not negligible because the pressure exerts a z-direction force. We calculate this force in the usual way with an integral over the stress tensor $\tilde{\tilde{\Pi}}$. We neglect the $\hat{n} \cdot \tilde{\tilde{\tau}}$ integral as before due to the small $\partial v_z / dz$ gradients. Using cylindrical coordinates for the integrations, we obtain:

$$\underline{f}_{\substack{\text{vertical} \\ \text{wall}}} = \iint_S \left[-p\hat{n} + \hat{n} \cdot \tilde{\tilde{\tau}} \right]_{\text{at surface}} dS \tag{9.289}$$

$$= \int_0^{2\pi} \int_{R_1}^{R_2} -p_1 \begin{pmatrix} 0 \\ 0 \\ -1 \end{pmatrix}_{r\theta z} r \, dr \, d\theta \tag{9.290}$$

$$= p_1 \pi \left(R_2^2 - R_1^2 \right) \begin{pmatrix} 0 \\ 0 \\ 1 \end{pmatrix}_{r\theta z} \tag{9.291}$$

$$f_{\substack{\text{vertical} \\ \text{wall, } z}} = p_1 \pi \left(R_2^2 - R_1^2 \right) \tag{9.292}$$

Combining this result with the z-component of the momentum balance in Equation 9.286, we obtain:

$$\rho \left(A_2 \langle v \rangle_2^2 - A_1 \langle v \rangle_1^2 \right) = p_1 \pi R_1^2 - p_2 \pi R_2^2 + f_{\text{walls},z} \tag{9.293}$$

$$= p_1 \pi R_1^2 - p_2 \pi R_2^2 + p_1 \pi \left(R_2^2 - R_1^2 \right) \tag{9.294}$$

$$= p_1 A_1 - p_2 A_2 + p_1 \left(A_2 - A_1 \right) \tag{9.295}$$

where we write $\pi R_1^2 = A_1$ and $\pi R_2^2 = A_2$. We now incorporate the mass-balance result and rearrange:

$$\rho A_2 \langle v \rangle_2^2 - \rho A_1 \langle v \rangle_1^2 = p_1 A_1 - p_2 A_2 + p_1 (A_2 - A_1) \tag{9.296}$$

$$\rho \left(A_2 \langle v \rangle_2^2 - A_1 \langle v \rangle_1^2 \right) = p_1 A_2 - p_2 A_2 \tag{9.297}$$

$$\frac{p_2 - p_1}{\rho} = \frac{A_1}{A_2} \langle v \rangle_1^2 - \langle v \rangle_2^2 \tag{9.298}$$

$$\boxed{\frac{p_2 - p_1}{\rho} = \frac{A_1}{A_2} \langle v \rangle_1^2 \left(1 - \frac{A_1}{A_2} \right)} \tag{9.299}$$

The z-direction momentum balance and the mass balance yield a relationship between pressure and velocity. To relate the pressure difference to friction, we now apply the mechanical energy balance for turbulent flow. The system of water in the expansion between Points 1 and 2 is a single-input, single-output, steady flow of an incompressible fluid. There is no heat transfer and no chemical reaction or phase change. All the requirements of the mechanical energy balance therefore are met:

Mechanical energy balance
(single-input, single-output,
steady, no phase change,
incompressible,
turbulent flow ($\alpha = 1$),
$\hat{v}$ constant across cross section
$\Delta T \approx 0$, no reaction)

$$\boxed{\frac{\Delta p}{\rho} + \frac{\Delta \langle v \rangle^2}{2} + g\Delta z + F = -\frac{W_{s,by}}{m}}$$

(9.300)

There is no shaft work and there is no elevation change. The MEB thus becomes:

$$\frac{p_2 - p_1}{\rho} + \frac{\langle v \rangle_2^2 - \langle v \rangle_1^2}{2} + g(z_2 - z_1) + F_{2,1} = -\frac{W_{s,by,21}}{m} \qquad (9.301)$$

$$\frac{p_2 - p_1}{\rho} + \frac{\langle v \rangle_2^2 - \langle v \rangle_1^2}{2} + F_{2,1} = 0 \qquad (9.302)$$

$$F_{2,1} = -\frac{p_2 - p_1}{\rho} - \frac{\langle v \rangle_2^2 - \langle v \rangle_1^2}{2} \qquad (9.303)$$

To complete the calculation of $F_{2,1}$ for an expansion, we substitute the pressure term from the z-momentum balance (see Equation 9.299), eliminate $\langle v \rangle_2^2$ with the mass balance (see Equation 9.271), and solve for the friction $F_{2,1}$. The result (after some algebra) is:

$$F_{\text{expansion}} = F_{2,1} = -\frac{p_2 - p_1}{\rho} - \frac{\langle v \rangle_2^2 - \langle v \rangle_1^2}{2} \qquad (9.304)$$

Friction in
an expansion

$$\boxed{F_{\text{expansion}} = \frac{\langle v \rangle_1^2}{2} \left(1 - \frac{A_1}{A_2}\right)^2} \qquad (9.305)$$

This expression was given in Chapter 1 (see Equation 1.121) as the formula to calculate the frictional contributions from expansion fittings.

EXAMPLE 9.11. *A new valve has been invented and manufactured. Show how to account for the valve's friction as a function of the number of handle turns open.*

SOLUTION. When manufacturers sell valves, they provide a specifications sheet that includes a quantity known as the valve flow-coefficient C_V. The valve coefficient C_V usually is provided in a plot as a function of the setting of the valve in terms of the number of turns the valve is opened (Figure 9.24). This information indicates how much friction is produced by the valve in all of its operating positions. The C_V plot also is useful information for determining how the valve opens and closes; that is, it distinguishes between valves that close suddenly after

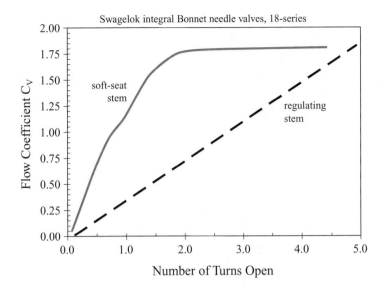

Swagelok integral Bonnet needle valves, 18-series

soft-seat stem

regulating stem

Figure 9.24 For a particular valve, a manufacturer supplies information on the amount of friction the valve produces in terms of the valve flow coefficient C_V as a function of the number of turns open. The flow characteristics shown are for the Swagelok integral Bonnet needle valves, 18-series [162].

many turns and valves that gradually (linearly) close as the valve stem is turned (i.e., the trim of the valve; see the Glossary).

The valve flow coefficient C_V is defined in terms of data recorded in specified units. The definition of C_V is [132]:

$$\text{Valve flow coefficient:} \quad \boxed{C_V \equiv Q(\text{gpm})\sqrt{\frac{SG}{p_1(psi) - p_2(psi)}}} \quad (9.306)$$

where Q is the volumetric flow rate through the valve in units of gallons per minute (gpm), $SG = \rho/\rho_{\text{ref}}$ is the specific gravity of the fluid, ρ_{ref} is the density of water at $4°C$ and 1 atm pressure, and $p_1 - p_2$ is the pressure drop across the valve in units of $lb_f/in.^2$ (psi). In piping or pumping discussions, volumetric flow rate Q often is called capacity. From Equation 9.306, we see that the units of C_V are $gpm/psi^{0.5}$.

The valve flow coefficient is related to pressure drop versus flow rate for the valve, but the relationship is somewhat convoluted. We can sort out the role of C_V and understand its unusual definition by first applying the mechanical energy balance to the valve (Figure 9.25). We choose as our system the water flowing between Points 1 and 2, as shown.

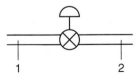

1   2

Figure 9.25 A valve is a single-input, single-output system with few thermal energy effects; thus, it is an excellent candidate for the mechanical energy balance.

The system of water in the valve is a single-input, single-output, steady flow of an incompressible fluid. There is no heat transfer and no chemical reaction or phase change. All the requirements of the mechanical energy balance are therefore met. The mechanical energy balance is:

Mechanical energy balance
(single-input, single-output,
steady, no phase change,
incompressible,
turbulent flow ($\alpha = 1$)
$\hat{v}$ constant across cross section
$\Delta T \approx 0$, no reaction)

$$\frac{\Delta p}{\rho} + \frac{\Delta \langle v \rangle^2}{2} + g\Delta z + F = -\frac{W_{s,by}}{m}$$

(9.307)

Because we assume turbulent flow, $\alpha = 1$.

Our chosen system is fluid between Points 1 and 2 in Figure 9.25. For this system, there is no shaft work, no change in elevation, and no change in velocity. Thus, the MEB reduces to:

$$\frac{p_2 - p_1}{\rho} + \frac{\langle v \rangle_2^2 - \langle v \rangle_1^2}{2} + g(z_2 - z_1) + F_{2,1} = -\frac{W_{s,by,21}}{m} \qquad (9.308)$$

$$\frac{p_2 - p_1}{\rho} + F_{2,1} = 0 \qquad (9.309)$$

$$F_{\text{valve}} = F_{2,1} = \frac{(p_1 - p_2)}{\rho} \qquad (9.310)$$

To relate F_{valve} to C_V, we begin with the definition of C_V in Equation 9.306. First, we square both sides of Equation 9.306 and incorporate the definition of specific gravity, $SG = \rho/\rho_{ref}$:

$$C_V^2 = Q^2 \left(\frac{SG}{p_1 - p_2} \right) = \frac{Q^2 \rho}{\rho_{ref}(p_1 - p_2)} \qquad (9.311)$$

For a pipe of diameter D, we can write the volumetric flow rate Q in terms of the average velocity $\langle v \rangle$ in the usual way:

$$Q = \frac{\pi D^2 \langle v \rangle}{4} \qquad (9.312)$$

Substituting this into Equation 9.311 and rearranging, we arrive at a ratio that is related to the friction term we seek:

$$C_V^2 = \frac{\rho}{(p_1 - p_2)} \frac{Q^2}{\rho_{ref}} \qquad (9.313)$$

$$= \frac{\rho}{(p_1 - p_2)} \frac{\pi^2 D^4 \langle v \rangle^2}{16 \rho_{ref}} \qquad (9.314)$$

$$\frac{(p_1 - p_2)}{\rho} = \frac{\pi^2 D^4 \langle v \rangle^2}{C_V^2 16 \rho_{ref}} \qquad (9.315)$$

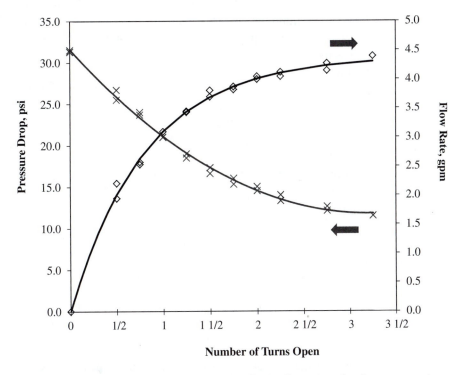

Figure 9.26 To characterize the performance of a valve and to quantify the frictional losses due to the valve, measurements are taken of pressure drop and flow rate as a function of number of turns open. The data can then be cast as C_V as a function of number of turns open. Data shown are for a laboratory metering valve.

We now calculate F_{valve} as:

$$F_{\text{valve}} = F_{2,1} = \frac{p_1 - p_2}{\rho} \qquad (9.316)$$

$$F_{\text{valve}} = \left(\frac{\pi^2 D^4}{8\rho_{\text{ref}}C_V^2}\right)\frac{\langle v \rangle^2}{2} \qquad (9.317)$$

We see that the friction term for the valve is proportional to $\langle v \rangle^2$. The friction coefficient K_f is defined by Equation 1.120:

Valve friction coefficient, K_f defined:

$$F_{\text{valve}} = K_f \frac{\langle v \rangle^2}{2} \qquad (9.318)$$

Thus, we find that K_f is given by:

Valve friction coefficient:

$$K_f \equiv \left(\frac{\pi^2 D^4}{8\rho_{\text{ref}}C_V^2}\right) \qquad (9.319)$$

For the C_V and K_f of the new valve, we take measurements of pressure drop and flow rate across the valve as a function of valve position (Figure 9.26). From the pressure drop and flow rate measurements, we calculate the function C_V (number of turns) using Equation 9.306; from this function, we calculate K_f (number of turns) from Equation 9.319. Valve characteristics are reported as either C_V as a function of valve position or K_f as a function of valve position.

In this section, we have worked several examples with the macroscopic momentum balance. There is a pattern to how these problems come together, which can be summarized in a more specific version of the macroscopic momentum-balance equation. Beginning with the usual form of the balance (see Equation 9.45), we explicitly write the force terms, following the pattern set in the examples of this section.

$$
\frac{d\underline{P}}{dt} + \sum_{i=1}^{\# \ streams} \left[\frac{\rho A \cos\theta \langle v \rangle^2}{\beta} \ \hat{v} \right]\Bigg|_{A_i}
$$

$$
= \sum_{i=1}^{\# \ streams} [-pA\hat{n}]_{A_i} + \left[\sum_{i=1}^{\# \ streams} \iint_{A_i} [\hat{n} \cdot \underline{\tilde{\tau}}]_{A_i} \ dS \right] + \underline{R} + M_{CV}\underline{g} \quad (9.320)
$$

where $\underline{R}$ is the viscous force on the solid walls of the control volume. We assume that the pressure does not vary across A_i, which allows us to carry out the pressure integral. We note that the term on the righthand side in square brackets is small or zero. Omitting this term, our final equation for the macroscopic momentum balance has the same form as the microscopic momentum balance or Navier-Stokes equation: Rate of change plus convective term equals a pressure term, a term dealing with viscous forces, and a gravity term.

Macroscopic Momentum Balance on a Control Volume (ρ, $\hat{v}$, p constant across A_i):

$$
\boxed{\frac{d\underline{P}}{dt} + \sum_{i=1}^{\# \ streams} \left[\frac{\rho A \cos\theta \langle v \rangle^2}{\beta} \ \hat{v} \right]\Bigg|_{A_i} = \sum_{i=1}^{\# \ streams} [-pA\hat{n}]_{A_i} + \underline{R} + M_{CV}\underline{g}} \quad \begin{cases} \beta_{laminar} = 0.75 \\ \beta_{turbulent} \approx 1 \end{cases}
$$

$$(9.321)$$

This version of the macroscopic momentum balance is suitable for most problems of interest.

9.2.4 Pumps

Moving liquids through piping systems usually is accomplished with a pump, and the most common and robust pump design is the centrifugal pump. In the sections that follow, we show how the mechanical energy balance may be used to choose the right pump for a given application. The MEB also shows why low-suction pressures are harmful to pumps; this limitation must be considered when installing pumps.

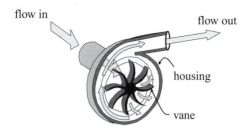

flow in

flow out

housing

vane

Figure 9.27   In a centrifugal pump, a motor spins a vane within a housing. The spinning vane draws fluid in along a central pipe, and the fluid spins outward with centrifugal acceleration. The fluid is collected along the outer part of the housing and is discharged at an accelerated speed or higher pressure.

9.2.4.1 PUMP SIZING

The centrifugal pump is one of the most effective and economical machines for transporting water-like liquids (Figure 9.27). Centrifugal pumps move fluid along by spinning an appropriately shaped vane within a housing and hurling fluid outward with centrifugal force. The vane is driven by an electrical motor, and the speed of rotation of the vane (i.e., revolutions per minute [RPM]) and the design of the pump determine how much shaft work is delivered by the pump to the fluid. The energy balance governs how the shaft work delivered by the pump affects the speed, pressure, and other characteristics of the fluid. In the following examples, we explore the operation of a centrifugal pump in a variety of common systems.

EXAMPLE 9.12. *A new pump has been invented and manufactured. Potential owners of the pump need to know how it operates under various conditions. Show how to account for the operation of the pump.*

SOLUTION. When manufacturers sell pumps, they provide specifications that indicate how the pump operates under various conditions. The key chart is a plot of pumping head (discussed here) versus capacity (volumetric flow rate). This curve is called the *pumping-head curve*. We show here the meaning and utility of the pumping-head curve.

Consider a pump that is fit with pressure taps just upstream and just downstream of the pump (Figure 9.28). The upstream tap is called the suction tap, and the downstream tap is called the discharge tap. The system of the fluid between Points s and d is a single-input, single-output, steady flow of an incompressible fluid.

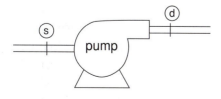

Figure 9.28   Schematic of pump suction-discharge system.

There is no heat transfer and no chemical reaction or phase change. All of the requirements of the mechanical energy balance therefore are met:

$$\frac{\Delta p}{\rho} + \frac{\Delta \langle v \rangle^2}{2\alpha} + g\Delta z + F = -\frac{W_{s,by}}{m}$$

Mechanical energy balance
(single-input, single-output,
steady, no phase change,
incompressible,
$\hat{v}$ constant across cross section
$\Delta T \approx 0$, no reaction)

$$(9.322)$$

Applied to the suction-discharge system, the MEB becomes:

$$\frac{p_d - p_s}{\rho} + \frac{\langle v \rangle_d^2 - \langle v \rangle_s^2}{2\alpha} + g(z_d - z_s) + F_{d,s} = -\frac{W_{s,by,ds}}{m} \qquad (9.323)$$

where d denotes the property at the discharge point and s denotes the property at the suction point. The work done by the fluid as it passes through the pump is negative ($W_{s,by,ds} < 0$). The total work delivered by the pump is positive and is known as the brake horsepower (bhp).

Pump
Brake horsepower, bhp
(total work delivered by pump):

$$bhp = -W_{s,by,ds} \qquad (9.324)$$

$$bhp = \omega T \qquad (9.325)$$

where ω is the angular velocity of the vane and T is the magnitude of the torque on the vane. The friction term $F_{d,s}$ in Equation 9.323 is the fluid friction between Points d and s, which almost entirely is the friction in the pump. Following the procedure we used when considering turbines in Chapter 1, we group the pump losses with the pump shaft work and define a pump efficiency. The pump efficiency reflects the fraction of the energy delivered by the pump that is actually converted to usable work on the fluid in the form of pressure change, velocity change, or elevation change, rather than being dissipated as friction:

Pump
efficiency:

$$\eta = \frac{\left(\dfrac{\text{useful energy}}{\text{mass fluid}} \right)}{\left(\dfrac{\text{total energy input}}{\text{mass fluid}} \right)} \qquad (9.326)$$

$$\eta \equiv \frac{\left(\dfrac{-W_{s,by,ds}}{m} - F_{d,s} \right)}{\left(\dfrac{-W_{s,by,ds}}{m} \right)} \qquad (9.327)$$

Substituting pump efficiency η from Equation 9.327 into Equation 9.323, we obtain:

$$\frac{p_d - p_s}{\rho} + \frac{\langle v \rangle_d^2 - \langle v \rangle_s^2}{2\alpha} + g(z_d - z_s) + F_{d,s} = -\frac{W_{s,by,ds}}{m} \tag{9.328}$$

$$\frac{p_d - p_s}{\rho} + \frac{\langle v \rangle_d^2 - \langle v \rangle_s^2}{2\alpha} + g(z_d - z_s) = -\frac{W_{s,by,ds}}{m} - F_{d,s} \tag{9.329}$$

$$\frac{p_d - p_s}{\rho} + \frac{\langle v \rangle_d^2 - \langle v \rangle_s^2}{2\alpha} + g(z_d - z_s) = \eta \left(-\frac{W_{s,by,ds}}{m} \right) \tag{9.330}$$

$$= \frac{\eta \, bhp}{m} \tag{9.331}$$

All of the quantities on the lefthand side of Equation 9.330 are readily measurable on the pump instrumented as in Figure 9.28, and they provide a measure of the performance of the pump in terms of the useful energy/mass provided by the pump (i.e., the righthand side of the equation). The plot of the lefthand side of Equation 9.330 versus flow rate is the pumping-head curve we seek.

To write the pump-head curve in its standard form, we now divide Equation 9.330 by gravity to convert the equation to head and regroup the terms into discharge and suction terms. Taking $\alpha = 1$ (turbulent flow), we obtain:

$$\frac{p_d - p_s}{\rho} + \frac{\langle v \rangle_d^2 - \langle v \rangle_s^2}{2\alpha} + g(z_d - z_s) = \eta \left(-\frac{W_{s,by,ds}}{m} \right) \tag{9.332}$$

$$\left(\frac{p_d}{\rho g} + \frac{\langle v \rangle_d^2}{2g} + z_d \right) - \left(\frac{p_s}{\rho g} + \frac{\langle v \rangle_s^2}{2g} + z_s \right) = \eta \left(-\frac{W_{s,by,ds}}{mg} \right) \tag{9.333}$$

The expressions in brackets are called the discharge head and the suction head:

Discharge head:
$$\boxed{H_d \equiv \frac{p_d}{\rho g} + \frac{\langle v \rangle_d^2}{2g} + z_d} \tag{9.334}$$

Suction head:
$$\boxed{H_s \equiv \frac{p_s}{\rho g} + \frac{\langle v \rangle_s^2}{2g} + z_s} \tag{9.335}$$

Using these quantities, Equation 9.333 becomes:

$$\left(\frac{p_d}{\rho g} + \frac{\langle v \rangle_d^2}{2g} + z_d \right) - \left(\frac{p_s}{\rho g} + \frac{\langle v \rangle_s^2}{2g} + z_s \right) = \eta \left(-\frac{W_{s,by,ds}}{mg} \right) \tag{9.336}$$

$$H_d - H_s = \eta \left(\frac{-W_{s,by,ds}}{mg} \right) \tag{9.337}$$

Each pumping-head term is a combination of pressure head, velocity head, and elevation head. These three types of head represent the three ways that energy from the pump may be transferred to the fluid. Plots of $H_d - H_s$ measured on a

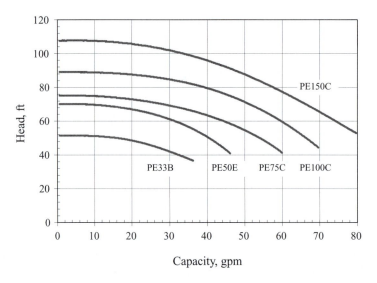

Figure 9.29   Actual pump characteristic curves, $H_{d,s}$ versus Q, for several commercial centrifugal pumps. Each curve represents the performance capabilities of a particular pump. (*Source:* Krum Pump Company, Kalamazoo, MI)

pump as a function of capacity Q in gpm are known as pumping-head curves:

$$\text{Pumping-head curve:} \qquad \boxed{H_d - H_s = H_{d,s} = \eta\left(-\frac{bhp}{mg}\right)} \qquad (9.338)$$

Examples of pumping-head curves for a family of commercial pumps are shown in Figure 9.29. Each curve represents the performance of a particular centrifugal pump operating at a fixed angular velocity (i.e., fixed RPM). These curves are determined experimentally by measuring pressure, average velocity, and elevation at the suction and discharge locations on a pump (Equations 9.334 and 9.335).

The pumping ability of a centrifugal pump depends on shaft rotational speed, shape of the vanes in the pump, flow path within the housing, and other pump-design variables that are fixed at the time of manufacture. At a given RPM, the amount of pumping head $H_{d,s}$ (i.e., energy per unit weight of fluid) that a pump can deliver depends on flow rate Q. When the amount of head delivered is high, the flow rate is low; as the flow rate increases, the amount of head delivered decreases. For an application that requires a pump, the details of the actual system—how high the fluid is to be raised in elevation and is there a pressure increase, for example—determine how much head is needed from the pump. The right pump is one that can deliver the amount of pumping head needed at the desired flow rate. We discuss pump sizing in the examples that follow.

The previous example shows how we use pumping-head curves to supply information about pump performance. We saw that a pump operating at a chosen head is limited in the amount of flow rate it can produce, and it produces more flow when pumping against a system that presents less resistance in the form of head. When installing a pump, we must know how much head is needed to choose the pump wisely.

To understand these issues of head deliverable by a pump versus the amount of head a system requires, we consider a simple example.

EXAMPLE 9.13. *After the local river flooded, 3.0 feet of water filled the basement of a university building. A PE33B pump (see Figure 9.29) is available to pump the water out of the basement. The vertical distance from the water surface to the nearest drain is 60 feet. Will the pump work?*

SOLUTION. The proposed operation of the pump is shown in Figure 9.30. The system of the water between Points 1 and 2 is a single-input, single-output, steady flow of an incompressible fluid. There is no heat transfer and no chemical reaction or phase change. All of the requirements of the mechanical energy balance therefore are met. Assuming turbulent flow, we write:

Mechanical energy balance
(single-input, single-output,
steady, no phase change,
incompressible,
turbulent flow ($\alpha = 1$),
$\hat{v}$ constant across cross section
$\Delta T \approx 0$, no reaction):

$$\frac{\Delta p}{\rho g} + \frac{\Delta \langle v \rangle^2}{2g} + \Delta z + \frac{F}{g} = -\frac{W_{s,by}}{mg}$$

(9.339)

In our chosen system, the pressures at Points 1 and 2 are both atmospheric; thus, there is no pressure change. The velocity at Point 1 is zero and the velocity at Point 2 is nonzero and unknown. The elevation change $z_2 - z_1$ is roughly 60 feet. There also are frictional losses in the system, but they should be minor except

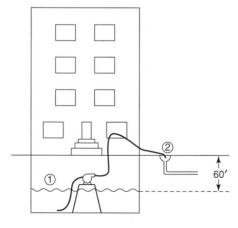

Figure 9.30   We want to install a pump to move water up 60 feet from a flooded basement. The mechanical energy balance tells us whether this scheme will work.

for the losses in the pump, which we include through pump efficiency. The MEB becomes:

$$\frac{p_2 - p_1}{\rho g} + \frac{\langle v \rangle_2^2 - \langle v \rangle_1^2}{2g} + (z_2 - z_1) + \frac{F_{2,1}}{g} = -\frac{W_{s,by,21}}{mg} \tag{9.340}$$

$$\frac{\langle v \rangle_2^2}{2g} + (z_2 - z_1) = \eta \left(-\frac{W_{s,by,21}}{mg} \right) \tag{9.341}$$

$$\frac{\langle v \rangle_2^2}{2g} + 60 \text{ ft} = \eta \left(-\frac{W_{s,by,21}}{mg} \right) = H_{d,s} \tag{9.342}$$

We do not know what the velocity through the hose will be, but the elevation change is at least 60 feet, and the velocity head will make the lefthand side of Equation 9.342 larger. The pump's performance, represented by the righthand side of Equation 9.342, must be powerful enough to match the demands of the system at the flow rate achieved. The pumping head curve $H(Q)$ is shown in Figure 9.29 (see also Equation 9.338).

Consulting the characteristic curve of the PE33B pump in Figure 9.29, we see that the maximum head deliverable is about 50 feet; this maximum value occurs at $Q = 0$. The demands of the system are at least 60 feet of head due to the elevation rise from the basement to the street drain. Although we do not know the velocity head in the hose, we know enough about the system to conclude that the PE33B pump is inadequate. This pump will not perform the task we are asking of it because it cannot deliver the minimum of 60 feet of head.

The pumping-head curve is useful in the previous example, and we easily determine that it was pointless to ask the available pump to lift the flood waters by 60 feet. It is clear that a pump must be able to deliver at least the elevation head required by the system. This type of verification is quick and easy to make.

To choose the appropriate pump for a task that involves velocity head, friction loss, and pressure head in addition to elevation head, we need a technique that allows us to account for all of the demands a particular system makes on a pump. The MEB applied to a system can be used to construct a characteristic curve for the system that can provide the needed information. We demonstrate this technique next.

EXAMPLE 9.14. *A family built a small vacation home on a hill near a lake. They plan to obtain wash water directly from the lake. To bring the water from the lake, they plan to install the piping/tank system shown schematically in Figure 9.31. When the pump is running to fill the tank, they desire a flow rate of at least 2 gpm. Which pump should they install? Choose your answer from among those with pumping-head curves appearing in Figure 9.29.*

SOLUTION. The size of the pump needed depends on the load placed on the pump. We analyze the load on the pump using the mechanical energy balance.

The details of the piping system for the cottage are shown in Figure 9.32. Nominal 1/2-inch, type K, copper water tubing is used throughout the installation.

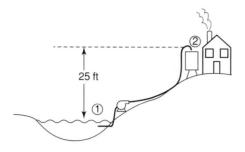

The flow rate achieved in the piping system shown depends on the energy demands of the system compared to the ability of the pump as reflected in the pumping-head curve.

The system of fluid between Points 1 and 2 is a single-input, single-output, steady flow of an incompressible fluid. There is no heat transfer and no chemical reaction or phase change. All of the requirements of the mechanical energy balance are therefore are met:

Mechanical energy balance
(single-input, single-output,
steady, no phase change,
incompressible,
turbulent flow ($\alpha = 1$),
$\hat{v}$ constant across cross section
$\Delta T \approx 0$, no reaction):

$$\frac{\Delta p}{\rho g} + \frac{\Delta \langle v \rangle^2}{2g} + \Delta z + \frac{F}{g} = -\frac{W_{s,by}}{mg}$$

(9.343)

In our chosen system, the pressure at Points 1 and 2 are both atmospheric; thus, there is no pressure change. The velocity at Point 1 is zero and the velocity at Point 2 is nonzero and unknown; at a minimum, it should be 2 gpm. The elevation change is roughly 25 feet. There also are frictional losses, which include the losses

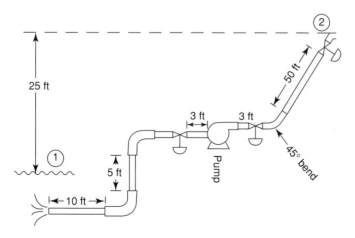

To calculate the friction loss in the system, we need detailed information about the lengths of runs of piping and the number of bends and fittings.

in the piping F_{piping} and the losses in the pump, which we include through a pump efficiency. The MEB becomes:

$$\frac{p_2 - p_1}{\rho g} + \frac{\langle v \rangle_2^2 - \langle v \rangle_1^2}{2g} + (z_2 - z_1) + \frac{F_{2,1}}{g} = -\frac{W_{s,by,21}}{mg} \qquad (9.344)$$

$$\frac{\langle v \rangle_2^2}{2g} + (z_2 - z_1) + \frac{F_{\text{piping}}}{g} = \eta\left(-\frac{W_{s,by,21}}{mg}\right) \qquad (9.345)$$

$$\frac{\langle v \rangle_2^2}{2g} + 25 \text{ ft} + \frac{F_{\text{piping}}}{g} = \eta\left(-\frac{W_{s,by,21}}{mg}\right) \qquad (9.346)$$

Thus far, the problem is similar to the previous example. To perform a first calculation, we neglect the frictional losses in the piping and assume that the velocity is the maximum of 2 gpm. If the problem turns out to be impossible with these assumptions, we avoid the unnecessary work of more detailed calculations. With these assumptions, we calculate the required head as shown here:

$$\frac{\langle v \rangle_2^2}{2g} + 25 \text{ ft} + \frac{F_{\text{piping}}}{g} = \eta\left(-\frac{W_{s,by,21}}{mg}\right) \qquad (9.347)$$

$$\frac{\langle v \rangle_2^2}{2g} + 25 \text{ ft} = \eta\left(-\frac{W_{s,by,21}}{mg}\right) \qquad (9.348)$$

We calculate the average velocity in the tubing from the volumetric flow rate (i.e., 2 gpm) and the true inner diameter (ID) of the copper tubing. We obtain the true ID from the literature [132]:

$$\text{Copper water tubing}\atop{1/2\text{-inch nominal}\atop\text{type K:}}\left\{\begin{array}{lll}\text{outer diameter:} & 0.625 \text{ in.} \\ \text{wall thickness:} & 0.049 \text{ in.} \\ \text{inner diameter:} & 0.625 \text{ in.} - (2)(0.049 \text{ in.}) \\ & = 0.527 \text{ in.}\end{array}\right. \qquad (9.349)$$

$$\langle v \rangle_2 = \frac{4Q}{\pi D^2} \qquad (9.350)$$

$$= (2.0 \text{ gpm})\left(\frac{4}{\pi\left(\frac{0.527}{12}\text{ ft}\right)^2}\right)\left(\frac{2.22802 \times 10^{-3} \text{ ft}^3/\text{s}}{\text{gpm}}\right) \qquad (9.351)$$

$$= 2.94171 \text{ ft/s} = \boxed{2.9 \text{ ft/s}} \qquad (9.352)$$

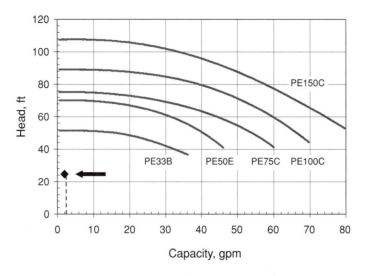

Figure 9.33 The calculated operating point for 2 gpm is shown with the available pumping-head curves, $H_{d,s}$. All of the pumps are capable of supplying this capacity. (*Source:* Krum Pump Company, Kalamazoo, MI)

Using this value for velocity in Equation 9.348, we calculate the work needed from the pump:

$$\begin{pmatrix} \text{work supplied} \\ \text{by pump} \end{pmatrix} = \begin{pmatrix} \text{work needed} \\ \text{by system} \end{pmatrix} \tag{9.353}$$

$$H_{d,s} = \eta \left(-\frac{W_{s,by,21}}{mg} \right) \tag{9.354}$$

$$= \frac{\langle v \rangle_2^2}{2g} + 25 \text{ ft} \tag{9.355}$$

$$= \left(\frac{1}{2} \right) \left(\frac{s^2}{32.174 \text{ ft}} \right) \left(2.94171 \frac{\text{ft}}{\text{s}} \right)^2 + 25 \text{ ft}$$

$$= 25.13448 \text{ ft}$$

$$\approx \boxed{25.1 \text{ ft}} \tag{9.356}$$

The estimate of 25.1 feet is reachable by all of the pumps in Figure 9.29, which we can determine by plotting the point (i.e., 2 gpm, 25.1 ft) on the same axes as the pumping-head curves (Figure 9.33). Because this point is below the pumping-head curves for all of the pumps, this indicates that they are capable of producing this amount of head at the desired flow rate. We neglected the friction, however, and it is possible that with the addition of the friction head, the smallest pump may not be able to meet the need. To add in the effect of friction in our system-head calculation, we use the correlations for friction in pipes and fittings discussed in Chapter 1 to calculate piping friction loss F_{piping}.

In Chapter 1 and earlier in this chapter, we show that friction in piping can be calculated from data correlations (Equation 1.124):

$$F_{\text{piping}} = \sum \left(\begin{array}{c} \text{friction of} \\ \text{straight-pipe sections} \end{array} \right) + \sum \left(\begin{array}{c} \text{friction of} \\ \text{fittings and valves} \end{array} \right) \quad (9.357)$$

Friction in a piping system:

$$F_{\text{piping}} = \sum_{\substack{j,\,straight \\ pipe \\ segments}} \left[4 f_j \frac{L_j}{D_j} \frac{\langle v \rangle_j^2}{2} \right] + \sum_{i,\,fittings} \left[n_i K_{f,i} \frac{\langle v \rangle_i^2}{2} \right]$$

(9.358)

For our piping system, we have the same average velocity throughout, and Equation 9.358 becomes:

Friction in a constant-velocity piping system

$$F_{\text{piping}} = \left[4f \frac{L}{D} + \sum_{i,\,fittings} n_i K_{f,i} \right] \frac{\langle v \rangle^2}{2} \quad (9.359)$$

where L is the total length of piping of diameter D.

For the piping loop under consideration in this problem, there are two 90-degree bends, one 45-degree bend, three gate valves, a contraction as the flow enters the pipe entrance, and a total of 71 feet of 1/2-inch, type K, copper water tubing. The friction coefficients for the fittings are listed in Table 1.4 and summarized here:

Fitting	n	K_f	nK_f
90° bend	2	0.75	1.50
45° bend	1	0.35	0.35
gate valve	3	0.17	0.51
contraction $\infty \rightarrow 0.527$ in	1	0.55	0.55
$\sum nK_f =$			2.91

(9.360)

For the friction due to the straight-pipe sections, we need the friction factor f, which we obtain from the Colebrook equation (see Equation 1.95) and the Reynolds number:

$$\frac{1}{\sqrt{f}} = -4.0 \log \left(\frac{4.67}{\text{Re}\sqrt{f}} \right) + 2.28 \qquad \begin{array}{c} \text{Colebrook formula,} \\ \text{Fanning friction factor} \\ \text{in steady turbulent flow} \\ \text{in smooth pipes} \end{array} \quad (9.361)$$

$$\text{Re}_{2\,\text{gpm}} = \frac{\rho \langle v \rangle D}{\mu} = \frac{\left(\frac{62.25 \text{lb}_m}{\text{ft}^3} \right) (2.94171 \text{ ft/s}) \left(\frac{0.527 \text{ in.}}{12 \text{ in./ft.}} \right)}{\left(6.005 \times 10^{-4} \frac{\text{lb}_m}{\text{ft s}} \right)} \quad (9.362)$$

$$= 13,392 = 1.3 \times 10^4 \Rightarrow f = 0.007158 = \boxed{f = 0.007} \quad (9.363)$$

where we calculate the friction factor iteratively with the Colebrook Equation (see Chapter 1).

We calculate F_{piping} in units of head from Equation 9.359:

$$\frac{F_{piping}}{g} = \frac{1}{g}\left[4f\frac{L}{D} + \sum_{i,\,fittings} n_i K_{f,i}\right]\frac{\langle v\rangle^2}{2} \tag{9.364}$$

$$= \frac{1}{32.174 \text{ ft/s}^2}\left[4(0.007158)\frac{71 \text{ ft}}{0.527/12 \text{ ft}} + 2.91\right]\frac{(2.94171 \text{ ft/s})^2}{2}$$

$$= 6.61643 \text{ ft} \tag{9.365}$$

We substitute the piping friction result into Equation 9.347 and recalculate the system head, H_{21}:

$$H_{21} = \eta\left(-\frac{W_{s,by,21}}{mg}\right) = \frac{\langle v\rangle_2^2}{2g} + 25 \text{ ft} + \frac{F_{piping}}{g} \tag{9.366}$$

$$= \frac{(2.94171 \text{ ft/s})^2}{(2)(32.174 \text{ ft/s}^2)} + 25 \text{ ft} + 6.61643 \text{ ft} \tag{9.367}$$

$$= 31.7509 \text{ ft} \tag{9.368}$$

$$= \boxed{32 \text{ ft}} \tag{9.369}$$

We calculate that we need a pump that can operate at the point 2.0 gpm and 32 feet of head. Reviewing the pumping head curves in Figure 9.29, we see that even with the friction losses added in, all of the pumps can operate at this value of head and capacity. We therefore recommend that the smallest pump (i.e., PE33B) be purchased for this application. However, if there is a likelihood that elements will be added to the system that raise the needed head by 18 or more feet (e.g., if many valves or an orifice plate flow meter are added), this pump will be inadequate because the PE33B develops only 50 feet of head. If system modification is likely, the next largest pump (i.e., PE50E) should be selected.

The calculation in the previous example allowed us to choose a pump that could perform at our minimum specification of 2 gpm. If we are interested in knowing what the actual flow rate will be on a system of interest for a chosen pump, we can calculate the system head H_{21} at a variety of flow rates and plot a system head curve. We demonstrate this procedure in the final example of this section.

EXAMPLE 9.15. *A family built a small vacation home on a hill near a lake. They plan to provide water for washing directly from the lake. To bring the water from the lake, they plan to install the piping/tank system shown schematically in Figure 9.31. They want to install a PI33B pump. What will be the flow rate through the piping?*

$$\begin{pmatrix} \text{mechanical} \\ \text{work required} \\ \text{by the system} \end{pmatrix} = \begin{pmatrix} \text{mechanical} \\ \text{work supplied} \\ \text{by the pump} \end{pmatrix}$$

$$H_{2,1} = H_{d,s}$$

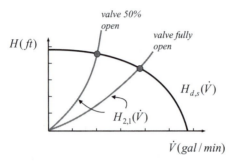

Figure 9.34   To calculate the actual flow rate that a pump will produce on a chosen system, we plot system head as a function of capacity on the same axes as the pump head deliverable by the pump. The intersection of the two curves indicates where the pump will operate on the chosen system.

SOLUTION. The problem is again approached using the mechanical energy balance, and most of the information we need is discussed in the previous example. In that example, we calculate the friction in the piping for a single chosen flow rate of 2 gpm. We also calculate the velocity-head term $\langle v \rangle^2/2g$ and frictional losses at that single value of capacity. To calculate the actual operating flow rate, we need to guess various other flow rates and calculate for each new flow rate the terms in the MEB on the system. The several values of flow rate for which we calculate head form a curve that we can plot on the same axes as the pumping-head curve. The intersection of the calculated system requirements (i.e., the system curve) and the pumping-head curve indicate the actual operating point of the system (Figure 9.34).

We formally define the system-head curve as follows. We begin with the mechanical energy balance on our system, the water between Points 1 and 2 in Figure 9.32:

Mechanical energy balance (single-input, single-output, steady, no phase change, incompressible, turbulent flow ($\alpha = 1$), $\hat{v}$ constant across cross section $\Delta T \approx 0$, no reaction):

$$\frac{\Delta p}{\rho g} + \frac{\Delta \langle v \rangle^2}{2g} + \Delta z + \frac{F}{g} = -\frac{W_{s,by}}{mg} \qquad (9.370)$$

$$\frac{p_2 - p_1}{\rho g} + \frac{\langle v \rangle_2^2 - \langle v \rangle_1^2}{2g} + (z_2 - z_1) + \frac{F_{2,1}}{g} = -\frac{W_{s,by,21}}{mg} \qquad (9.371)$$

Table 9.2. System-head calculations for the lake house in Example 9.15

Q (gpm)	$\langle v \rangle_2$ (ft/s)	$\dfrac{\langle v \rangle_2^2}{2g}$ (ft)	$\dfrac{Re}{10^4}$	$10^4 f$	$\dfrac{F_{\text{piping}}}{g}$ (ft)	$H_{2,1}$ (ft)
2.0	2.9	0.1	1.34	71.6	6.5	31.7
4.0	5.9	0.5	2.68	60.3	22.3	47.8
6.0	8.8	1.2	4.02	54.9	45.8	72.0
8.0	11.8	2.2	5.36	51.5	76.7	103.8
10.0	14.7	3.4	6.70	49.0	114.5	142.8

We separate the friction from the pump as before, introducing the pump efficiency (see Equation 9.327):

$$\frac{p_2 - p_1}{\rho g} + \frac{\langle v \rangle_2^2 - \langle v \rangle_1^2}{2g} + (z_2 - z_1) + \frac{F_{\text{piping}}}{g} = \eta \left(-\frac{W_{s,by,21}}{mg} \right) \quad (9.372)$$

The lefthand side of Equation 9.372 is all of the information that addresses the amount of pumping head presented by our system. We define $H_{2,1}$ as the system head for the system defined as the fluid between Points 1 and 2. Note that the friction of the pump is not included in the system head because it already is accounted for on the righthand side through the pump efficiency:

System-head curve for system between Points 1 and 2:

$$H_{2,1} \equiv \frac{p_2 - p_1}{\rho g} + \frac{\langle v \rangle_2^2 - \langle v \rangle_1^2}{2g} + (z_2 - z_1) + \frac{F_{\text{piping}}}{g}$$

$$(9.373)$$

F_{piping} is given by Equation 9.358. The right-hand side of Equation 9.372 indicates that the system head will be provided by a pump. The pump's performance is given by its pumping-head curve, $H_{d,s}$:

$$\eta \left(\frac{-W_{s,by,21}}{mg} \right) = H_{d,s} \quad (9.374)$$

The mechanical energy balance becomes:

$$\frac{p_2 - p_1}{\rho g} + \frac{\langle v \rangle_2^2 - \langle v \rangle_1^2}{2g} + (z_2 - z_1) + \frac{F_{\text{piping}}}{g} = \eta \left(-\frac{W_{s,by,21}}{mg} \right) \quad (9.375)$$

Operating point: $\quad H_{2,1} = H_{d,s} \quad (9.376)$

We use spreadsheet software to carry out the calculation of $H_{2,1}$ for the lake house. We choose to calculate the system losses for flow rates of 2.0, 4.0, 6.0, 8.0, and 10.0 gpm. Setting up the same calculations as in the previous example, we obtain the results shown in Table 9.2 and plotted as the system curve in Figure 9.35.

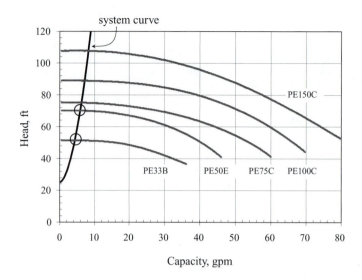

Figure 9.35 To calculate the operating point, the system curve is plotted with the pumping-head curve. The intersection of the system curve with the pumping-head curve indicates the flow rate and head that will be produced with a given pump inserted in the loop. For the system in the chapter example, the operating points for the PE33B pump (i.e., 4 gpm, 50 feet) and the PE50E pump (i.e., 6 gpm, 70 feet) are indicated. (*Source:* Krum Pump Company, Kalamazoo, MI)

$$H_{2,1} = \frac{p_2 - p_1}{\rho g} + \frac{\langle v \rangle_2^2 - \langle v \rangle_1^2}{2g} + (z_2 - z_1) + \frac{F_{piping}}{g} \qquad (9.377)$$

$$= \frac{\langle v \rangle_2^2}{2g} + 25 \text{ ft} + \frac{F_{piping}}{g} \qquad (9.378)$$

$$\frac{F_{piping}}{g} = \frac{1}{g} \left[4f\frac{L}{D} + \sum_{i, fittings} n_i K_{f,i} \right] \frac{\langle v \rangle^2}{2} \qquad (9.379)$$

The intersection of the system-head curve with the PE33B pumping-head curve is at approximately 4 gpm, 50 feet; the intersection with the head curve for the PE50 pump is 6 gpm, 70 feet. Thus, if the smaller pump is installed, the operating flow rate will be 4 gpm, whereas if the larger pump is installed, the flow will be 6 gpm.

9.2.4.2 NET POSITIVE SUCTION HEAD

The operation of a pump is captured by the mechanical energy balance, as discussed in the previous section. The MEB indicates that knowing differences in pressure, velocity, and elevation between two points allows us to calculate frictional losses and shaft work in a pumping system:

Mechanical
energy balance
(units of head):
$$\frac{\Delta p}{\rho g} + \frac{\Delta \langle v \rangle^2}{2g\alpha} + \Delta z + \frac{F}{g} = -\frac{W_{s,by}}{mg} \qquad (9.380)$$

$$\frac{p_2 - p_1}{\rho g} + \frac{\langle v \rangle_2^2 - \langle v \rangle_1^2}{2g} + z_2 - z_1 + \frac{F_{2,1}}{g} = -\frac{W_{s,by,21}}{mg} \qquad (9.381)$$

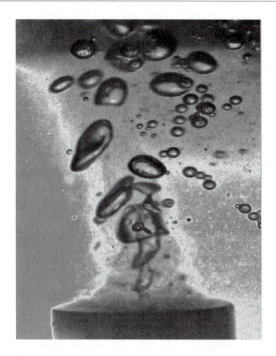

Figure 9.36 Cavitation is the formation of vapor bubbles when the pressure drops to the value of the fluid's vapor pressure. In turbo equipment, the formation of vapor bubbles and their subsequent collapse causes serious damage to propellers and vanes. In this figure a sonic wand generates cavitation at a surface. Image credit: K. S. Susick and J. K. Kolbeck, University of Illinois.

The absolute magnitudes of pressure, velocity, and elevation are not significant for the MEB calculation because only the differences *out* minus *in* appear. There is one circumstance, however, when the bare magnitude of pressure is important: when cavitation is possible.

Cavitation is the formation of vapor bubbles within a flow system, and it occurs when the local pressure level drops to and below the vapor pressure of the fluid being pumped (Figure 9.36). At pressures below the vapor pressure p_v^*, fluid flashes and forms vapor. The formation and collapse of vapor bubbles in turbomachinery can cause serious damage to propellers, drive shafts, and vanes. Also, a cavitating pump delivers far less head and flow rate than a pump operating properly. Cavitation is highly undesirable and may be avoided through proper design and installation of pumping systems.

To guard against cavitation, designers configure flow systems so that at no time does the absolute pressure in the machinery fall below the fluid's vapor pressure. In a pumping system, the likely low-pressure point is the suction side of the pump (see Figure 9.28). The magnitude of pressure at the suction side of a pump depends on how the pump is operated; that is, it depends on the system against which the pump is working. For example, if a pump is fed by a tank that is elevated (Figure 9.37), the pressure at the suction side will be higher than if the pump is drawing fluid from a reservoir at a lower elevation. The raised elevation of the feed tank increases the absolute magnitude of head at the suction point and may prevent or eliminate cavitation.

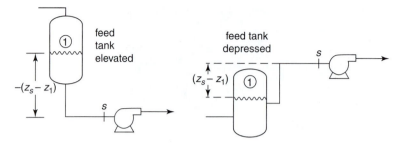

Figure 9.37 The elevation of the feed tank affects the head on the suction side of a pump. If the suction head is too low, the pump will cavitate.

Suction pressure also depends on the pump's internal workings; thus, for different pumps operating on the same system, suction pressure varies. Manufacturers test their pumps to identify the operating conditions when cavitation occurs, and they report the pump's cavitation performance in terms of required *net positive suction head* (NPSH), which we describe here [111]. NPSH as a function of flow rate is plotted on the pump-characteristic curves obtained from the manufacturer (Figure 9.38). Net positive suction head is a measure of the head required at the pump inlet to keep the liquid from cavitating or boiling. To use NPSH in designing flow loops, we must relate its definition to the flow configuration of interest.

Net positive suction head is a measure of how much the suction head of a pump exceeds the vapor pressure. The suction head (see Equation 9.335) is:

$$\text{Suction head:} \quad H_s \equiv \frac{p_s}{\rho g} + \frac{\langle v \rangle_s^2}{2g} + z_s \qquad (9.382)$$

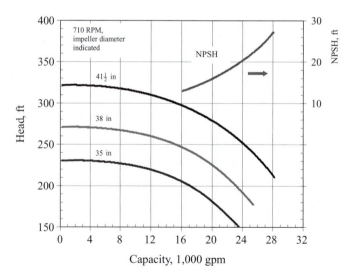

Figure 9.38 Characteristic curves of pumping head versus capacity and NPSH versus capacity for the pump in the example. All three curves are for the same pump at the same rotational speed, but different sized impellers are installed (*Source*: Ingersoll-Rand Corporation, Cameron Pump Division; from White [174]).

where H_s is the head at pump suction; p_s is the pressure; ρ is the fluid density; g is the acceleration due to gravity; $\langle v \rangle_s$ is the average velocity; and $z_s = 0$ is the elevation of the suction point, which is chosen to be the reference datum-level. In the expression for suction head, the suction pressure p_s and the average velocity at suction $\langle v \rangle_s$ depend on the design of the pump, whereas the suction elevation depends on how the pump is installed.

To give the installing engineer information about pump performance, manufacturers report the safe (i.e., cavitation-free) operating values of suction head. This is reported through the required net positive suction head defined here:

$$\text{Required net positive suction head (NPSH}_R): \quad \text{NPSH}_R = H_s - \frac{p_v^*}{\rho g} \tag{9.383}$$

$$= \frac{p_s}{\rho g} + \frac{\langle v \rangle_s^2}{2g} - \frac{p_v^*}{\rho g} \tag{9.384}$$

where p_v^* is the vapor pressure of the flowing liquid at the operating temperature and z_s is assumed to be zero. NPSH_R represents the increment above vapor pressure head $(p_v^*/\rho g)$ that the suction head $p_s/\rho g + \langle v \rangle_s^2/2g$ must meet to avoid cavitation.[4] NPSH_R is a quantity, therefore, that represents the intrinsic performance of the pump in the limit of cavitation occurring. To learn how to use NPSH_R, we consider a pump-installation problem in the next example.

EXAMPLE 9.16. *A pump manufacturer needs to measure NPSH$_R$ on a new pump. What tests need to be run?*

SOLUTION. To determine the NPSH_R for the new pump, we need to operate it at incipient cavitation. A possible apparatus for measuring NPSH_R is shown in Figure 9.39. A pump running at fixed RPM moves water in a closed cycle from an open reservoir through some piping and back to the reservoir. The elevation of the reservoir with respect to the pump may be varied by use of a lift. The pump is outfitted with a suction pressure tap, a flow meter, and a metering valve (i.e., a valve to adjust the flow rate). Measurements are taken of suction pressure p_s and flow rate. Average velocity, which is constant throughout the loop, is calculated from the flow rate, $\langle v \rangle = 4Q/\pi D^2$, where D is the pipe diameter.

To measure NPSH_R, a slow flow first is established in the loop with the reservoir at a high elevation; no cavitation should be observed. The elevation of the reservoir subsequently is reduced below the suction level until cavitation occurs. At cavitation or somewhat before, NPSH_R is calculated from its definition and the measured values of p_s and $\langle v \rangle = \langle v \rangle_s$ near or just below cavitation:

$$\text{NPSH}_R \equiv \left[\frac{p_s}{\rho g} + \frac{\langle v \rangle_s^2}{2g} - \frac{p_v^*}{\rho g} \right]\bigg|_{\substack{\text{at incipient} \\ \text{cavitation}}} \tag{9.385}$$

---

[4]Because vapor pressure is an absolute pressure, p_s also must be written in absolute pressure rather than gauge pressure.

orifice meter

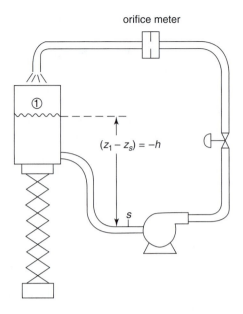

$(z_1 - z_s) = -h$

s

Schematic of an apparatus for measuring NPSH_R.

The measurement of NPSH_R as described produces a single datapoint of (Q, NPSH_R). To obtain a second datapoint at a higher flow rate, the tank is raised and the flow rate is increased by opening the valve by an increment. The tank then is lowered until just before cavitation occurs. The data of NPSH_R as a function of flow rate Q is recorded and supplied to the customer. NPSH_R represents a minimum value for suction head to exceed vapor-pressure head. When designing a flow loop, the minimum value must be exceeded:

$$\left[\frac{p_s}{\rho g} + \frac{\langle v \rangle_s^2}{2g} - \frac{p_v^*}{\rho g} \right]\Bigg|_{\substack{\text{available} \\ \text{as installed}}} \geq \text{NPSH}_R \qquad (9.386)$$

$$\text{NPSH}_A \geq \text{NPSH}_R \qquad (9.387)$$

where NPSH_A is the available net positive suction head. This quantity is calculated for an installation and the height of the suction above the feed and/or the pressure of the feed is chosen so that NPSH_A exceeds the pump's NPSH_R.

The required net positive suction head is a measure of pump performance. To use this quantity in designing a pump installation, we apply the mechanical energy balance to the proposed installation and compare the NPSH_A to the reported NPSH_R to ensure that the design avoids cavitation. We demonstrate this type of calculation in the following two examples.

EXAMPLE 9.17. *The pump with pumping-head curves given in Figure 9.38 (35 inch impeller) is installed as shown in Figure 9.40. The pressure in the tank is 760 mmHg and the desired flow rate is 2.0×10^4 gpm. If frictional head loss*

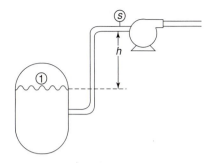

Figure 9.40 The pump in Example 9.17 is installed as shown here.

$F_{s,1}/g$ from the reservoir to the pump inlet is 7.0 feet, at what elevation should the pump suction be placed relative to the feed tank to avoid cavitation for water at 77 degrees F (25 degrees C)? If the water temperature changes to 200°F (93.3°C), what is the new location of the pump?

SOLUTION. The system of water in the pumping loop in Figure 9.40 is a single-input, single-output, steady flow of an incompressible fluid. There is no heat transfer and no chemical reaction or phase change. All of the requirements of the mechanical energy balance therefore are met. We begin with the MEB in terms of head:

Mechanical energy balance (units of head):

$$\frac{\Delta p}{\rho g} + \frac{\Delta \langle v \rangle^2}{2g\alpha} + \Delta z + \frac{F}{g} = -\frac{W_{s,by}}{mg} \qquad (9.388)$$

We choose as our system the water between the free surface in the tank and the suction point of the pump. We assume turbulent flow ($\alpha = 1$). There is no pump in our chosen system (s, 1), and the velocity at Point 1 is approximately zero. The MEB becomes:

$$\frac{p_s - p_1}{\rho g} + \frac{\langle v \rangle_s^2 - \langle v \rangle_1^2}{2g} + z_s - z_1 + \frac{F_{s,1}}{g} = -\frac{W_{s,by,s1}}{mg} \qquad (9.389)$$

$$\frac{p_s - p_1}{\rho g} + \frac{\langle v \rangle_s^2}{2g} + z_s - z_1 + \frac{F_{s,1}}{g} = 0 \qquad (9.390)$$

We take z_s as zero (i.e., reference elevation); thus, $z_1 = -h$, where h is the positive vertical height between the feed-tank fluid level and the suction elevation. We obtain:

$$\frac{p_s - p_1}{\rho g} + \frac{\langle v \rangle_s^2}{2g} + h + \frac{F_{s,1}}{g} = 0 \qquad (9.391)$$

$$\left[\frac{p_s}{\rho g} + \frac{\langle v \rangle_s^2}{2g} \right] = \frac{p_1}{\rho g} - h - \frac{F_{s,1}}{g} \qquad (9.392)$$

We can calculate NPSH_A from this equation by subtracting vapor-pressure head $p_v^*/\rho g$ from both sides:

$$\left[\frac{p_s}{\rho g} + \frac{\langle v\rangle_s^2}{2g} - \frac{p_v^*}{\rho g}\right] = \frac{p_1}{\rho g} - h - \frac{F_{s,1}}{g} - \frac{p_v^*}{\rho g} \qquad (9.393)$$

$$\text{NPSH}_A = \frac{p_1}{\rho g} - h - \frac{F_{s,1}}{g} - \frac{p_v^*}{\rho g} \qquad (9.394)$$

To avoid cavitation, the pump should be installed so that the NPSH_A is greater than the manufacturer-reported NPSH_R of the pump:

$$\text{NPSH}_A \geq \text{NPSH}_R \qquad (9.395)$$

$$\frac{p_1}{\rho g} - h - \frac{F_{s,1}}{g} - \frac{p_v^*}{\rho g} \geq \text{NPSH}_R \qquad (9.396)$$

For our current problem, we now calculate the required pump location. According to Figure 9.38, at 20,000 gpm, the NPSH_R is 16 feet (note the scale on the righthand side). Vapor pressures and densities for various fluids may be found in the literature [132]; water at 25°C has a vapor pressure of 23.756 mmHg, and the density of water at that temperature is 62.25 lb$_\text{m}$/ft^3. We calculate the needed $(z_1 - z_s)$ from Equation 9.394:

$$\frac{p_1}{\rho g} - h - \frac{F_{s,1}}{g} - \frac{p_v^*}{\rho g} \geq \text{NPSH}_R \qquad (9.397)$$

$$h \leq \frac{(p_1 - p_v^*)}{\rho g} - \frac{F_{s,1}}{g} - \text{NPSH}_R$$

$$\leq \frac{((760 - 23.756)\ \text{mmHg}) \left(\frac{(14.696\ \text{lb}_\text{f}/\text{in.}^2)(144\ \text{in.}^2/\text{ft}^2)}{760\ \text{mmHg}}\right)\left(\frac{32.174\ \text{ft lb}_\text{m}/\text{s}^2}{\text{lb}_\text{f}}\right)}{(62.25\ \text{lb}_\text{m}/\text{ft}^3)(32.174\ \text{ft/s}^2)}$$

$$- 7.0\ \text{ft} - 16\ \text{ft}$$

$$\leq 32.93\ \text{ft} - 7.0\ \text{ft} - 16\ \text{ft}$$

$$\leq 9.93\ \text{ft}$$

$$\boxed{h \leq 9.5\ \text{ft}} \qquad (9.398)$$

We calculate at this temperature that the pump may be located up to about 9.5 feet above the feed reservoir without cavitating.

At higher temperatures, the vapor pressure and density change. Consulting Perry's [132], we find that for water $p_v^*(200°\ \text{F}) = 595.21$ mmHg and

$\rho(200°F) = 0.96308 \text{ g/cm}^3 = 60.125 \text{ lb}_m/\text{ft}^3$. For these new conditions, we obtain:

$$h \le \frac{(p_1 - p_v^*)}{\rho g} - \frac{F_{s,1}}{g} - \text{NPSH}_R$$

$$\le \frac{((760 - 595.21) \text{ mmHg})\left(\frac{(14.696 \text{ lb}_f/\text{in.}^2)(144 \text{ in.}^2/\text{ft}^2)}{760 \text{ mmHg}}\right)\left(\frac{32.174 \text{ ft lb}_m/\text{s}^2}{\text{lb}_f}\right)}{(60.125 \text{ lb}_m/\text{ft}^3)(32.174 \text{ ft/s}^2)}$$

$$- 7.0 \text{ ft} - 16 \text{ ft}$$

$$\le 7.63 \text{ ft} - 7.0 \text{ ft} - 16 \text{ ft} = -15 \text{ ft}$$

$$\boxed{(-h) \ge 15 \text{ ft}} \tag{9.399}$$

At the higher temperature of 200°F, we calculate that the feed tank must be located at least 15 feet *higher* than the pump to avoid cavitation in the pump.

In Example 9.17 we see that when cavitation is likely to occur, we need to elevate the feed tank to avoid cavitation. If raising the feed tank is not practical, another solution is to pressurize it. We can calculate the feed-tank pressure needed to avoid cavitation by using the MEB.

EXAMPLE 9.18. *The pump with characteristic curves given in Figure 9.38 is operated as shown in Figure 9.41. The 35-inch impeller is installed. The desired flow rate is 2.0×10^4 gpm. Frictional head loss $F_{s,1}/g$ from the reservoir to the pump inlet is 7.0 feet, and the fluid is water at 200°F (93.3°C). The location of the pump is fixed at 3.0 feet below the level of liquid in the tank. To what pressure must the feed tank be raised to avoid cavitation in the pump?*

SOLUTION. The system of water in the pumping loop in Figure 9.41 is a single-input, single-output, steady flow of an incompressible fluid. There is no heat transfer and no chemical reaction or phase change. All of the requirements of the mechanical energy balance therefore are met. We begin with the mechanical energy balance in terms of head:

Mechanical energy balance (units of head):

$$\frac{\Delta p}{\rho g} + \frac{\Delta \langle v \rangle^2}{2 g \alpha} + \Delta z + \frac{F}{g} = -\frac{W_{s,by}}{mg} \tag{9.400}$$

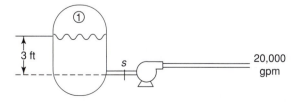

Figure 9.41 The pump in this example is installed with the suction port 3 feet below the tank water level.

Again, we choose as our system all of the water between the open free surface in the tank and the suction point of the pump. We assume turbulent flow ($\alpha = 1$). There is no pump in our chosen calculation system, and the velocity at Point 1 is approximately zero. The MEB becomes:

$$\frac{p_s - p_1}{\rho g} + \frac{\langle v \rangle_s^2 - \langle v \rangle_1^2}{2g} + z_s - z_1 + \frac{F_{s,1}}{g} = -\frac{W_{s,by,s1}}{mg} \tag{9.401}$$

$$\frac{p_s - p_1}{\rho g} + \frac{\langle v \rangle_s^2}{2g} + (z_s - z_1) + \frac{F_{s,1}}{g} = 0 \tag{9.402}$$

As usual, we assume $z_s = 0$ and, relative to this datum, $z_1 = h = 3$. We now rearrange Equation 9.402 and calculate the NPSH_A:

$$\frac{p_s - p_1}{\rho g} + \frac{\langle v \rangle_s^2}{2g} + (z_s - z_1) + \frac{F_{s,1}}{g} = 0 \tag{9.403}$$

$$\left[\frac{p_s}{\rho g} + \frac{\langle v \rangle_s^2}{2g} \right] = \frac{p_1}{\rho g} + h - \frac{F_{s,1}}{g} \tag{9.404}$$

$$\text{NPSH}_A = \frac{p_1}{\rho g} + h - \frac{F_{s,1}}{g} - \frac{p_v^*}{\rho g} \tag{9.405}$$

To avoid cavitation, NPSH_A must exceed the manufacturer's reported NPSH_R for the pump. For the current problem, we use this to devise a condition on the feed-tank pressure p_1:

$$\text{NPSH}_A \geq \text{NPSH}_R \tag{9.406}$$

$$\frac{p_1}{\rho g} + h - \frac{F_{s,1}}{g} - \frac{p_v^*}{\rho g} \geq \text{NPSH}_R \tag{9.407}$$

$$\frac{p_1}{\rho g} \geq -h + \frac{F_{s,1}}{g} + \frac{p_v^*}{\rho g} + \text{NPSH}_R \tag{9.408}$$

We now can evaluate the minimum pressure to avoid cavitation for this problem. Using the values of vapor pressure and density obtained in the previous example, ($p_v^*(200°F) = 595.21$ mmHg and $\rho(200° F) = 60.125$ lb$_m$/ft^3 [132]), we calculate:

$$\frac{p_1}{\rho g} \geq -h + \frac{F_{s,1}}{g} + \frac{p_v^*}{\rho g} + \text{NPSH}_R$$

$$\geq (-3 \text{ ft}) + 7.0 \text{ ft}$$

$$+ \frac{(595.21 \text{ mmHg}) \left(\frac{(14.696 \text{ lb}_f/\text{in.}^2)(144 \text{ in.}^2/\text{ft}^2)}{760 \text{ mmHg}} \right) \left(\frac{32.174 \text{ ft lb}_m/\text{s}^2}{\text{lb}_f} \right)}{(60.125 \text{ lb}_m/\text{ft}^3)(32.174 \text{ ft/s}^2)} + 16 \text{ ft}$$

$$\geq -3 \text{ ft} + 7 \text{ ft} + 27.57 \text{ ft} + 16 \text{ ft} = 47.57 \text{ ft}$$

$$\geq \boxed{48 \text{ ft}} \tag{9.409}$$

The pressure in the tank, in units of head, must be at least 48 feet of water at 200°F. To convert this pressure to psi, we solve the head expression for p_1:

$$\frac{p_1}{\rho g} = 47.57 \text{ ft} \tag{9.410}$$

$$p_1 = 47.57 \text{ ft} \left(\frac{60.125 \text{ lb}_m}{\text{ft}^3} \right) \left(\frac{32.174 \text{ ft}}{\text{s}^2} \right) \left(\frac{\text{lb}_f}{32.174 \text{ ft lb}_m/\text{s}^2} \right) \left(\frac{\text{ft}^2}{144 \text{ in.}^2} \right)$$

$$= 19.86 \text{ psi}$$

$$\boxed{p_1 = 20 \text{ psia (absolute)}} \tag{9.411}$$

We calculate that the tank must be held approximately 6 psi above atmospheric to avoid cavitation at the pump.

The mechanical energy balance is the most widely used relationship in engineering fluid mechanics, and this chapter explains why. From siphons to pumps, much engineering equipment can be analyzed using the MEB, and these algebraic calculations provide useful information. The key to properly using the MEB is to always check that its assumptions are met: single-input, single-output, steady flow of an incompressible fluid with no heat transfer, no chemical reaction, and no phase change. Neither the MEB nor the macroscopic momentum balance provides information about flow patterns, flow stresses, or velocity distributions. If more detail about flow patterns is required, then the microscopic balancing-methods in this book should be pursued (see Chapters 6–8) rather than macroscopic control-volume calculations.

9.2.5 Open-channel flow

Open-channel flow refers to flows open to the atmosphere. Water flowing in open culverts or gutters is a common example of open-channel flow. The presence of the air–water free surface and the fact that the free-surface shape may change in response to imposed forces gives the flow a unique character that separates open-channel flows from the internal and external flows discussed thus far. Open-channel flows most often are turbulent.

Gravity is the main driving force in open-channel flow. Because open-channel flows are open to the atmosphere, pressure does not vary in the flow direction and is not a flow driving force. When a fixed volumetric flow rate of water is flowing in a channel, the height of the water in the channel is that which minimizes the amount of wall drag produced in the flow. For a fixed flow rate (Figure 9.42), if the flow chooses a high water level, the fixed volume of fluid moves at a slower average velocity, but more of the channel is wetted by the water (thereby producing more drag). If the flow chooses a low water level, the fixed volume of fluid moves at a higher average velocity, but a rapid flow has a higher velocity-gradient near the wall. High wall velocity-gradients produce higher wall drag

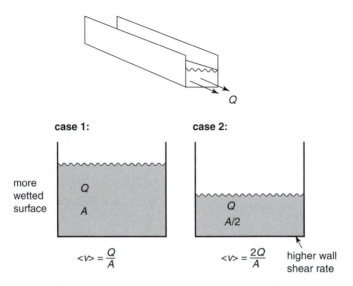

case 1:

more
wetted
surface

Q
A

$<v> = \dfrac{Q}{A}$

case 2:

Q
$A/2$

$<v> = \dfrac{2Q}{A}$ higher wall
shear rate

Figure 9.42   The tradeoffs of open-channel flow can be visualized by considering the cross-sectional area of the flow. For a fixed flow rate, if a low flow speed is chosen, the cross-sectional area must be large, increasing the wetted surface of the channel, thereby increasing the drag. If a higher flow speed is chosen for the same flow rate, the cross-sectional area is lower; however, the velocity gradients near the wall also will be higher, which increases drag. The observed channel depth balances these two effects.

than low velocity-gradients due to Newton's law of viscosity:

Shear flow near wall:
($\delta =$ boundary-layer thickness)

$$\left. v \right|_{wall} \approx \left. \frac{\langle v \rangle}{\delta} y \right|_{wall} \hat{e}_x \tag{9.412}$$

Drag at wall:
$$\mathcal{F}_{drag} \propto \left. \tau_{yx} \right|_{wall} = \left. \mu \frac{\partial v_x}{\partial y} \right|_{wall} \approx \mu \frac{\langle v \rangle}{\delta} \tag{9.413}$$

Therefore, an optimum fluid height balances the tradeoff between too-high gradients near the wall and too much wetted surface. In Example 9.19 we calculate the height of water in a channel by applying the mechanical energy balance.

EXAMPLE 9.19. *A rough cement rectangular channel (roughness $\varepsilon = 0.0080$ feet) slopes downward at an angle of 0.6 degree carrying water. The channel is 8.0 feet wide and the water depth is 4.5 feet. What is the flow rate in the channel? Assume that the depth is constant.*

SOLUTION. The system of water in the channel is a single-input, single-output, steady flow of an incompressible fluid. There is no heat transfer and no chemical reaction or phase change. All of the requirements of the mechanical energy balance therefore are met. We begin with the MEB in terms of head:

Mechanical
energy balance
(units of head):

$$\frac{\Delta p}{\rho g} + \frac{\Delta \langle v \rangle^2}{2 g \alpha} + \Delta z + \frac{F}{g} = -\frac{W_{s,by}}{mg} \tag{9.414}$$

We choose our system to be the fluid in the conduit between Points 1 and 2, which are a distance L apart (Figure 9.43).

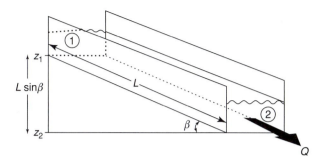

Figure 9.43 The depth of a gravity-driven flow depends on the pitch of the channel and the flow rate. The depth observed is the depth that minimizes drag.

In our chosen system, there are no shafts and therefore no shaft work. The velocity in the channel is the same at both points; thus, $\Delta\langle v\rangle^2 = 0$. The pressure is atmospheric throughout the flow. We assume the flow is turbulent (this is the usual case; $\alpha = 1$). The MEB reduces to:

$$\frac{p_2 - p_1}{\rho g} + \frac{\langle v\rangle_2^2 - \langle v\rangle_1^2}{2g\alpha} + (z_2 - z_1) + \frac{F_{2,1}}{g} = -\frac{W_{s,by.21}}{mg} \tag{9.415}$$

$$(z_2 - z_1) + \frac{F_{2,1}}{g} = 0 \tag{9.416}$$

$$h_f = \frac{F_{2,1}}{g} = (z_1 - z_2) = L\sin\beta \tag{9.417}$$

where $h_f = F_{2,1}/g$ is the head loss in the system and $\beta = 0.6$ degree is the angle between the channel bed and horizontal. We can relate $F_{2,1}$ to the Fanning friction factor (see Equation 9.358) and use the Colebrook equation (see Equations 1.95 and 9.361) for f, substituting the hydraulic diameter D_H (see Equation 7.248) for D because the cross section is noncircular (see Chapter 7):

$$\begin{array}{c} \text{Friction in pipe} \\ \text{(Equation 9.358):} \end{array} \qquad \frac{F_{2,1}}{g} = 4f\frac{L}{D_H}\frac{\langle v\rangle^2}{2g} \tag{9.418}$$

$$= L\sin\beta \tag{9.419}$$

Rearranging, we obtain:

$$\langle v\rangle = \sqrt{\frac{\sin\beta D_H g}{2f}} \tag{9.420}$$

The hydraulic diameter is given by Equation 7.248:

$$D_H = \frac{4A_{xs}}{\not{p}} = \frac{(4)(4.5\text{ ft})(8.0\text{ ft})}{((2)(4.5\text{ ft}) + 8\text{ ft})} \tag{9.421}$$

$$= \boxed{8.47\text{ ft}} \tag{9.422}$$

For the Fanning friction factor, we use the Colebrook equation (see Equation 1.95) at high Reynolds numbers:

Colebrook formula,
Fanning friction factor
in steady turbulent flow
in rough conduits

$$\frac{1}{\sqrt{f}} = -4.0 \log \left(\frac{\varepsilon}{D_H} + \frac{4.67}{\mathrm{Re}\sqrt{f}} \right) + 2.28 \qquad (9.423)$$

At high Reynolds numbers:

$$\frac{1}{\sqrt{f}} = -4.0 \log \left(\frac{\varepsilon}{D_H} \right) + 2.28 \qquad (9.424)$$

We have all of the numerical values necessary to perform our final calculations. We leave the detailed solution to readers (see Problem 42). The results are:

$$\langle v \rangle = 17.176 = \boxed{17 \text{ ft/s}} \qquad (9.425)$$

$$Q = \langle v \rangle A_{xs} \qquad (9.426)$$

$$= 618.3 \text{ ft}^3/\text{s} = \boxed{620 \text{ ft}^3/\text{s} = 280{,}000 \text{ gpm}} \qquad (9.427)$$

An interesting effect can occur in open-channel flow when there is a slope change in the flow path or a flow cross section change (Figure 9.44). If water is flowing in a steep open channel, it establishes its fluid height h_a. For the same volumetric flow rate, if the downward pitch of the channel is smaller, the equilibrium fluid height h_b in the less-steep section is higher than it was for the steep portion, $h_b > h_a$. This follows from the relationships used in Example 9.19. In a channel

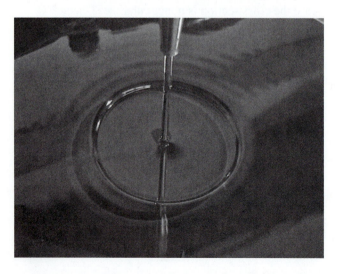

Figure 9.44 A hydraulic jump occurs when a rapid upstream flow can no longer be accommodated by the conditions downstream. This happens, for example, when the pitch of a channel decreases or as shown in the figure, when the cross section of flow increases. Image/Photo courtesy of John W. M. Bush, Jeffrey M. Aristoff, and Jeff Leblanc of the Massachusetts Institute of Technology.

in which such a slope-decrease occurs—that is, when the rapidly moving, thinner, upstream flow meets the slower-moving, thick, downstream flow—a discontinuity occurs that is called a *hydraulic jump* (Figure 9.44). Hydraulic jumps are seen commonly at dam spillways where such a slope change occurs. In the following example, we show how the macroscopic balances may be used to calculate flow depth after a hydraulic jump.

EXAMPLE 9.20. *Water flows over a dam spillway and produces a hydraulic jump on the horizontal apron onto which the spillway empties. The spillway is 150 feet wide and the fluid velocity in this region is 22 ft/s. The depth of the water on the apron is 6.0 inches. What is the fluid depth after the hydraulic jump?*

SOLUTION. We choose our macroscopic control volume as that shown in Figure 9.45 [183]. For a single-input, single-output system, the macroscopic mass balance is given by Equation 9.140:

Macroscopic
mass balance
Single-input,
single-output:
$$\frac{dm_{CV}}{dt} + \rho_1 A_1 \cos\theta_1 \langle v \rangle_1 + \rho_2 A_2 \cos\theta_2 \langle v \rangle_2 = 0 \qquad (9.428)$$

For our system, the time derivative is zero (i.e., steady state), the density is constant $\rho_1 = \rho_2 = \rho$, θ_1 is equal to 180 degrees, and θ_2 is equal to 0 degrees. The mass balance becomes:

$$\rho A_1 (-1)\langle v \rangle_1 + \rho A_2 \langle v \rangle_2 = 0 \qquad (9.429)$$

$$W y_1 \langle v \rangle_1 = W y_2 \langle v \rangle_2 = Q \qquad (9.430)$$

Macroscopic
mass balance result:
$$\boxed{y_1 \langle v \rangle_1 = y_2 \langle v \rangle_2 = \frac{Q}{W}} \qquad (9.431)$$

where y_1 is the upstream depth, y_2 is the downstream depth, and W is the width of the channel.

Momentum must be conserved across the jump. We use the macroscopic momentum balance with density and velocity direction constant across the inlet

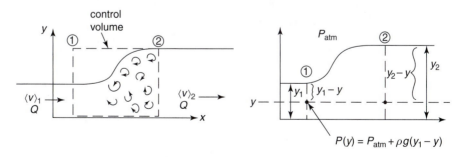

To calculate the height of a hydraulic jump, we choose the control volume shown here. When we analyze the hydraulic jump, we do not need to consider the pitch of either section; macroscopic balances applied to a horizontal jump yield the needed relationships.

and outlet surfaces:

Macroscopic
momentum
balance
on a CV,
ρ, $\hat{v}$ constant
across A_i:

$$\frac{d\mathbf{P}}{dt} + \frac{\rho_1 A_1 \cos\theta_1 \langle v \rangle_1^2}{\beta_1} \hat{v}|_{A_1} + \frac{\rho_2 A_2 \cos\theta_2 \langle v \rangle_2^2}{\beta_2} \hat{v}|_{A_2} = \sum_{\substack{\text{on} \\ \text{CV}}} f \qquad (9.432)$$

We choose a Cartesian coordinate system with flow in the x-direction; thus, $\hat{v}|_{A_1} = \hat{v}|_{A_2} = \hat{e}_x$. Assuming steady, turbulent flow ($\beta_1 = \beta_2 = 1$) and incorporating the same simplifications as for the mass balance, we obtain:

$$\frac{d\mathbf{P}}{dt} + \frac{\rho_1 A_1 \cos\theta_1 \langle v \rangle_1^2}{\beta_1} \hat{v}|_{A_1} + \frac{\rho_2 A_2 \cos\theta_2 \langle v \rangle_2^2}{\beta_2} \hat{v}|_{A_2} = \sum_{\substack{\text{on} \\ \text{CV}}} f \qquad (9.433)$$

$$-\rho A_1 \langle v \rangle_1^2 \begin{pmatrix} 1 \\ 0 \\ 0 \end{pmatrix}_{xyz} + \rho A_2 \langle v \rangle_2^2 \begin{pmatrix} 1 \\ 0 \\ 0 \end{pmatrix}_{xyz} = \sum_{\substack{\text{on} \\ \text{CV}}} f \qquad (9.434)$$

The forces on the control volume are gravity, which we neglect, and the surface forces on the bounding surfaces of our chosen control volume:

$$\sum_{\substack{\text{on} \\ \text{CV}}} f = f_{\text{gravity}} + f_{\text{surface}} \qquad (9.435)$$

$$= f_{\text{surface}} \qquad (9.436)$$

$$\sum_{\substack{\text{on} \\ \text{CV}}} f = f_{\text{inlet}} + f_{\text{outlet}} + f_{\text{top}} + f_{\text{bottom}} \qquad (9.437)$$

The top surface is air and the forces there are zero. The hydraulic jump will not be very long; thus, we neglect the viscous wall forces on the bottom. The remaining surface forces are on the inlet and the outlet.

The molecular surface forces are expressed using the usual integral of $\hat{n} \cdot \tilde{\underline{\Pi}}$ (see Equation 4.221). We learned in previous examples that the viscous $\hat{n} \cdot \tilde{\underline{\tau}}$ forces on the inlet and outlet surfaces are zero or small; thus, we omit the $\tilde{\underline{\tau}}$ term entirely for the calculation of surface forces on the inlet and outlet surfaces:

Total molecular fluid force
on a surface $\mathcal{S}$:

$$\mathcal{F} = \iint_{\mathcal{S}} [\hat{n} \cdot \tilde{\underline{\Pi}}]_{\text{at surface}} \, dS \qquad (9.438)$$

$$= \iint_{\mathcal{S}} \left[\hat{n} \cdot \left(-p\underline{I} + \tilde{\underline{\tau}}\right)\right]_{\text{at surface}} dS \qquad (9.439)$$

$$= \iint_{\mathcal{S}} [-p\hat{n}]_{\text{at surface}} \, dS \qquad (9.440)$$

$$f_{\text{inlet}} = \int_0^W \int_0^{y_1} -p(y)(-\hat{e}_x) \, dy dz \qquad (9.441)$$

We can use the static-pressure equation to write the variable pressure $p(y)$ at the inlet of our control volume: Pressure at the bottom of a column of fluid equals pressure at the top plus (density)(gravity)(height) (see Equation 4.136).

We neglect atmospheric pressure in this calculation:

$$\underline{f}_{inlet} = W \int_0^{y_1} [p_{atm} + \rho g(y_1 - y)]\,\hat{e}_x\,dy \tag{9.442}$$

$$= W \int_0^{y_1} [\rho g(y_1 - y)]\,\hat{e}_x\,dy \tag{9.443}$$

$$= W\rho g\hat{e}_x\,(y_1 y - \frac{y^2}{2}\Big|_0^{y_1} \tag{9.444}$$

$$= \frac{W\rho g y_1^2}{2}\hat{e}_x \tag{9.445}$$

A similar calculation for the outlet surface yields:

$$\underline{f}_{outlet} = -\frac{W\rho g y_2^2}{2}\hat{e}_x \tag{9.446}$$

We now assemble the momentum balance:

$$\frac{d\underline{P}}{dt} + \frac{\rho_1 A_1 \cos\theta_1 \langle v\rangle_1^2}{\beta_1}\,\hat{v}|_{A_1} + \frac{\rho_2 A_2 \cos\theta_2 \langle v\rangle_2^2}{\beta_2}\,\hat{v}|_{A_2} = \sum_{\substack{on \\ CV}} \underline{f} \tag{9.447}$$

$$-\rho A_1\langle v\rangle_1^2 \begin{pmatrix} 1 \\ 0 \\ 0 \end{pmatrix}_{xyz} + \rho A_2\langle v\rangle_2^2 \begin{pmatrix} 1 \\ 0 \\ 0 \end{pmatrix}_{xyz} = \underline{f}_{inlet} + \underline{f}_{outlet} \tag{9.448}$$

$$-\rho A_1\langle v\rangle_1^2 \begin{pmatrix} 1 \\ 0 \\ 0 \end{pmatrix}_{xyz} + \rho A_2\langle v\rangle_2^2 \begin{pmatrix} 1 \\ 0 \\ 0 \end{pmatrix}_{xyz} = \frac{W\rho g y_1^2}{2}\begin{pmatrix} 1 \\ 0 \\ 0 \end{pmatrix}_{xyz} - \frac{W\rho g y_2^2}{2}\begin{pmatrix} 1 \\ 0 \\ 0 \end{pmatrix}_{xyz}$$
$$\tag{9.449}$$

The x-component of the macroscopic momentum balance gives the following (note that the cross-sectional area $A_i = W y_i$):

$$-W y_1\langle v\rangle_1^2 + W y_2\langle v\rangle_2^2 = \frac{W g y_1^2}{2} - \frac{W g y_2^2}{2} \tag{9.450}$$

$$y_2\langle v\rangle_2^2 - y_1\langle v\rangle_1^2 = \frac{g}{2}\left(y_1^2 - y_2^2\right) \tag{9.451}$$

Macroscopic momentum-balance result:

$$\boxed{\frac{y_1\langle v\rangle_1}{g}\left(\langle v\rangle_2 - \langle v\rangle_1\right) = \frac{1}{2}\left(y_1^2 - y_2^2\right)} \tag{9.452}$$

where we used the mass balance results to arrive at the final equation.

The two equations obtained from the macroscopic mass- and momentum-balances (see Equations 9.431 and 9.452) form a set of two nonlinear algebraic equations with two unknowns (i.e., $\langle v\rangle_1$ and y_1 are given; $\langle v\rangle_2$ and y_2 are unknown). We use Equation 9.431 to eliminate $\langle v \rangle_2$ from Equation 9.452:

$$\frac{y_1 \langle v \rangle_1}{g} \left(\frac{\langle v \rangle_1 y_1}{y_2} - \langle v \rangle_1 \right) = \frac{1}{2} \left(y_1^2 - y_2^2 \right) \tag{9.453}$$

$$\frac{y_1 \langle v \rangle_1^2}{g y_2} (y_1 - y_2) = \frac{1}{2} (y_1 + y_2)(y_1 - y_2) \tag{9.454}$$

Canceling $(y_1 - y_2)$, multiplying through by y_2/y_1^2, and simplifying, we obtain:

$$\frac{2 \langle v \rangle_1^2}{g y_1} = \left(\frac{y_2}{y_1} \right) + \left(\frac{y_2}{y_1} \right)^2 \tag{9.455}$$

This is a quadratic equation for y_2/y_1, which we now solve:

$$\begin{array}{c} \text{Height of a} \\ \text{hydraulic jump:} \end{array} \quad \frac{y_2}{y_1} = -\frac{1}{2} + \frac{1}{2} \sqrt{1 + 8 \left(\frac{\langle v \rangle_1^2}{g y_1} \right)} \tag{9.456}$$

We choose the root that gives a positive y_2/y_1. The quantity $\frac{\langle v \rangle_1^2}{g y_1}$ is the Froude number for this flow (see Chapter 7):

$$\text{Froude number:} \quad \boxed{\text{Fr} \equiv \frac{V^2}{gD}} \quad \text{ratio of} \quad \frac{\text{(inertial forces)}}{\text{(gravity forces)}} \tag{9.457}$$

where $D = y_1$ is the characteristic lengthscale of this flow. We can calculate the height of the hydraulic jump for our problem using the final result in Equation 9.456:

$$\begin{array}{c} \text{Height of the} \\ \text{spillway jump:} \end{array} \quad y_2 = -\frac{y_1}{2} + \frac{y_1}{2} \sqrt{1 + 8 \left(\frac{\langle v \rangle_1^2}{g y_1} \right)} \tag{9.458}$$

$$= \boxed{3.6 \text{ ft}} \tag{9.459}$$

The Froude number for this flow is 30.

For more information on open-channel flows, consult the literature [178, 183].

9.3 Problems

1. The name "mechanical energy balance" implies, perhaps, that "mechanical energy" exists. Does it? Does it balance? Explain the meaning of the name of this important equation.

2. What is "macroscopic" about the macroscopic mass, momentum, and energy balances? How are these balances different compared to their microscopic analogs?

3. In the mechanical energy balance, the symbol Δ signifies "out" minus "in." If we mistakenly write "in" minus "out," what are the consequences to the final results? Discuss various scenarios. Is this distinction important in all cases?

4. What is the difference between solving for the vector force on a surface with the macroscopic momentum balance and solving for the vector force on a surface with the following equation?

$$\underline{F} = \int\int_S [\hat{n} \cdot \underline{\tilde{\Pi}}]_{\text{surface}} \, dS$$

5. Show that the energy velocity-profile parameter α is approximately equal to 0.5 for laminar tube flow.

6. Show that the energy velocity-profile parameter α is equal to 0.99 for turbulent flow. Assume that the velocity profile in turbulent flow is given by Equation 9.43.

7. For a slit flow with the velocity profile given here, what are the correct values of the momentum velocity-profile parameter β and the energy velocity-profile parameter α?

$$\underline{v} = v_x \hat{e}_x$$

$$v_x = v_{max} \left(1 - \frac{y}{H}\right)^{0.24}$$

8. For laminar flow of water (25°C) in a pipe that is 200.0 km long, what are the frictional losses? The pipe inner diameter is 40.0 cm and the flow rate is the highest it can be and still be laminar flow.

9. Derive the rule of thumb that the losses in turbulent flow in a pipe that is 50 diameters long are approximately equal to one velocity head [43].

10. Three 10-foot horizontal sections of pipe are connected and water at room temperature is pushed through by an upstream pressure of 55 psig. The three sections are 1/2-inch, 3/8-inch, and 1/4-inch, nominal type L, copper tubing. What is the flow rate through this series of tubes? At the exit, the 1/4-inch tubing is open to the atmosphere. Do not neglect velocity head changes.

11. Apply the appropriate balances to the fittings shown in Figure 9.46 and show that the expressions obtained are consistent with Equation 1.120.

12. For the flow loop in Figure 1.17, a colleague states that it is not necessary to do the detailed calculation. She suggests that you consider only the straight-pipe friction and neglect all of the fittings. What error would be associated with adopting her suggestion? Is it a good idea?

13. A vertical manometer tube is attached to the wall of a closed-channel water flow as shown in Figure 9.12. The flowing liquid rises in the manometer tube to a height of 12 cm. What is the pressure at that location in the flow? Give your answer in psig and psia.

14. A Pitot-static tube is installed through the wall of a water flow (27°C) such that the curved end (i.e., the Pitot tube) directly faces the oncoming flow at the center of the pipe. The liquid rises in the Pitot tube to a height of 34 cm. In the static tube, the fluid height is 12 cm. What is the fluid velocity in the pipe?

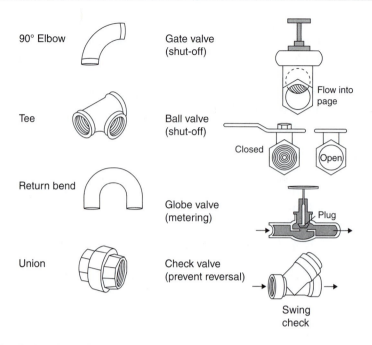

90° Elbow

Gate valve (shut-off)

Flow into page

Tee

Ball valve (shut-off)

Closed

Open

Return bend

Globe valve (metering)

Plug

Union

Check valve (prevent reversal)

Swing check

Figure 9.46   For these fittings, Problem 11 asks for the frictional losses.

15. The friction generated by straight piping can be written in terms of head. How many meters of 20 mm inner-diameter smooth tubing does it take to generate 1 meter of friction head? The Reynolds number of the flow is 20,000 and the fluid is water.

16. How does friction in a sudden contraction depend on fluid velocity? Can you derive your answer from fundamental relationships (i.e., mass, energy, and momentum balances)?

17. Devise a way to redefine the valve flow coefficient that is independent of the units of pressure drop and flow rate.

18. Show that for a noncircular conduit, the drag is given by:

$$\mathcal{F}_{\text{drag}} = \Delta p \; A_{xs}$$

where A_{xs} is the cross-sectional area of the conduit.

19. What is the direction and magnitude of the force needed to support the 90-degree expanding pipe bend shown in Figure 9.47? The water flow is steady and turbulent, the cross section of the inlet of the pipe bend is πR_1^2, and the cross section of the outlet of the bend is πR_2^2, where $R_2 > R_1$. Evaluate your answer for $R_1 = 0.545$ inch and $R_2 = 0.834$ inch and various values of flow parameters.

20. What is the direction and magnitude of the force needed to support the 60-degree expanding pipe bend shown in Figure 9.48? The water flow is steady and turbulent, the cross section of the inlet of the pipe bend is πR_1^2, and the cross section of the outlet of the bend is πR_2^2, where $R_2 > R_1$. Evaluate your answer for $R_1 = 0.545$ inch and $R_2 = 0.834$ inch.

21. What is the direction and magnitude of the force needed to support the 90-degreee contracting pipe bend shown in Figure 9.49? The water flow is steady and turbulent, the cross section of the outlet of the pipe bend is πR_1^2, and the

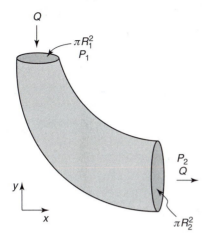

Figure 9.47   Force on a 90-degree expanding bend (Problem 19).

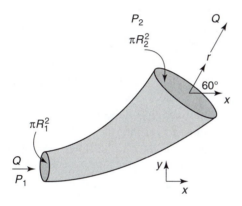

Figure 9.48   Force on a 60-degree expanding bend (Problem 20).

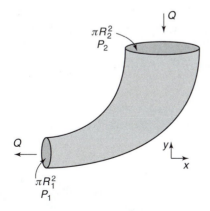

Figure 9.49   Force on a 90-degree contracting bend (Problem 21).

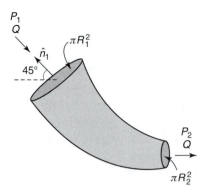

Figure 9.50   Force on a 45-degree contracting bend (Problem 22).

cross section of the inlet of the bend is πR_2^2, where $R_2 > R_1$. Evaluate your answer for $R_2 = 0.834$ inch and $R_1 = 0.430$ inch.

22. What is the direction and magnitude of the force needed to support the 45-degree contracting pipe bend shown in Figure 9.50? The water flow is steady and turbulent, the cross section of the inlet of the pipe bend is πR_1^2, and the cross section of the outlet of the bend is πR_2^2, where $R_2 < R_1$.

23. For the flow in a U-tube bend, carry out the integration in Equations 9.258 and 9.260 to show that the viscous contribution due to the rate-of-change of velocity in the flow direction cancels out in this problem.

24. Consider two different sections of pipe of the same diameter and same length. One is straight and the other is bent into the U shape. What are the forces on these two sections when 3.0 gpm of water flows through them (installed horizontally)? Assume the inlet and outlet pressures in the two cases are the same. Discuss the effect of the shape of the fitting on the force to which the fitting is subjected.

25. A Y-shaped piping installation for water flow is extended as shown in Figure 9.51 by the addition of a 70-foot section of the same type of pipe. The flow rate into the piping is a constant value of 4.0 gpm. What is the flow-rate split before the modification? What is the flow-rate split after the modification? All of the piping is 1/2-inch nominal Schedule 40 steel pipe. You may neglect the losses in the fittings.

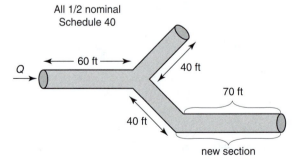

Figure 9.51   A modified Y-shaped piping installation with a split (Problem 25).

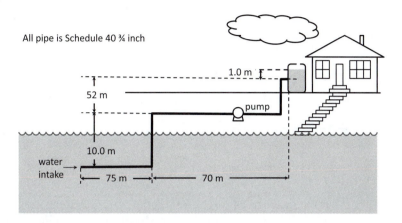

All pipe is Schedule 40 ¾ inch

1.0 m

52 m

pump

water intake

10.0 m

75 m   70 m

Figure 9.52   Designing a lawn-irrigation system (Problem 30).

26. For the installation described in Problem 25, if the inlet flow pressure is held constant (instead of the flow rate) at 60 psig, what is the flow-rate split before and after the modification?

27. What is the difference between a centrifugal pump and a positive-displacement pump?

28. What is net positive suction head? What is the danger involved in ignoring NPSH when installing a pump?

29. What are the signs and implications of cavitation?

30. A lawn-irrigation system is to be built next to a natural pond. The installers plan to obtain water for irrigation directly from the lake. To bring the water from the lake, they plan to install the piping system shown schematically in Figure 9.52. When the pump is running to fill the storage tank, they desire a flow rate of at least 14 liters/min. Which pump should they install? If possible choose your suggestion from among those with pumping-head curves in Figure 9.29.

31. For the pump installation shown in Figure 9.53, calculate and plot the system curve. The tank water levels are 6 feet apart in elevation. What is the minimum head needed to pump water in this loop at low flow rates? All pipe is Schedule 40.

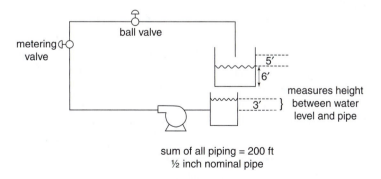

ball valve

metering valve

5′
6′

3′   } measures height between water level and pipe

sum of all piping = 200 ft
½ inch nominal pipe

Figure 9.53   To choose a pump for the system shown (Problem 31), a system-head curve is constructed.

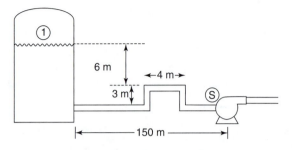

Figure 9.54   Avoiding cavitation in an installed pump (Problem 35).

32. After the local river flooded, 3.0 feet of water filled the basement of a university building. Several pumps are available to help empty the basement (see Figure 9.29). Which pump do you recommend for this operation? Justify your answer. Make reasonable assumptions.

33. Careful calculation of frictional losses in a system led to the following equation for system head in feet as a function of flow rate Q in gpm. At the last minute, the final tank in the installation was raised 5.0 feet. What is the new curve for system head?

$$H_{\text{system}} = 0.023 Q^2 + 35.2 Q + 34$$

34. The frictional losses in a system are represented accurately by the following equation for system head in feet as a function of flow rate Q in gpm.

$$H_{\text{system}} = 0.023 Q^2 + 34$$

Included in the frictional losses are those for a metering valve. The equation was calculated for the valve fullopen. Plot the system-head equation. Sketch qualitatively how the curve will shift to if the metering valve is closed halfway.

35. The pump with characteristic curves given in Figure 9.38 is installed as shown in Figure 9.54. The pressure in the tank headspace is 1,660 mmHg, and the desired flow rate is 2.2×10^4 gpm. If frictional head loss $h_f = F_{s,1}/g$ from the reservoir to the pump inlet is 11.0 feet, at what elevation should the pump suction be placed relative to the feed-tank fluid level to avoid cavitation for water at 122°F? All piping is 6.0 inches ID.

36. What data must be collected to determine the efficiency of a pump?

37. What is a hydraulic jump?

38. When heavy rains cause flooding, the water moves downstream under the pull of gravity. For a given flow rate, if the water moved very fast, the depth of the moving water could be shallow. For the same flow rate if the water moved less rapidly, the moving water would be very deep. What physics determines which of these two states nature chooses?

39. A rough cement rectangular channel (roughness $\varepsilon = 0.0085$ feet) slopes downward at an angle of 1.2 degrees carrying water. The channel is 20.0 m wide and the water depth is 1.2 m. What is the flow rate in the channel? Assume that the depth is constant.

40. A rough cement rectangular channel (roughness $\varepsilon = 0.0080$ feet) slopes downward at an angle of 3.2 degrees carrying water at 2.0 million gpm. The

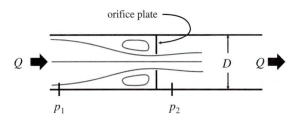

orifice plate

Q D Q

p_1 p_2

Figure 9.55 An orifice meter measures flow rate from the pressure drop across a plate with a hole in the center (Problem 43).

channel is 20.0 m wide. How deep is the water? Assume that the depth is constant.

41. A dilute aqueous solution flowing in a drying operation over a tilted surface encounters a slope change in the middle of its passage down the surface. The initial slope of the surface is 20 degrees relative to horizontal, and the new slope is 10 degrees relative to horizontal. The slope change produces a hydraulic jump. The flow is 15 m wide and the fluid velocity in the upstream region is 0.52 m/s. The fluid depth in the upstream region is 1.0 cm. What is the fluid depth after the hydraulic jump?

42. Calculate the flow rate for the circumstances described in Example 9.19. Show all of your work.

43. An orifice meter (see Figure 9.55 and Glossary) is a device that is used to measure flow rates of liquids and gases. The flow in a pipe of inner diameter D is obstructed by a plate with an orifice of diameter D_0. The flow streamlines contract from an upstream cross-sectional area of $\pi D^2/4$ to a jet of approximate cross-sectional area $\pi D_0^2/4$. The pressure in the pipe is measured upstream and slightly downstream of the orifice plate as shown in Figure 9.55. Show that the volumetric flow rate may be obtained from the following equation:

$$Q = \frac{\pi D^2}{4} \sqrt{\frac{\frac{2(p_2 - p_1)}{\rho}}{\left(1 - \frac{D^4}{D_0^4}\right)}}$$

You may neglect friction; friction in an orifice meter is accounted for by including a prefactor C_0, which is determined experimentally (For $\frac{D_0}{D} < 0.5$ and $Re = \frac{\rho v_2 D_0}{\mu} > 2 \times 10^4$, $C_0 \approx 0.61$ [132].

10 How Fluids Behave (Redux)

In this text, our goal is to explain flow; this chapter surveys how far we have come. With the completion of nine chapters of study, we find that we can make sense of much of the fluid behavior we observe. A reexamination of those behaviors helps consolidate our knowledge.

Section 10.1 is an integrated summary of the concepts of viscosity, drag, and boundary layers. Section 10.2 provides guidance on how numerical tools are used to pursue advanced flow-field models. In Section 10.3, we turn to turbulent flow, which until now was addressed only through data correlations for friction factor and drag coefficient. Sophisticated applications involving turbulent flow (e.g., airplane flight, mixing, and reactor design) require more detailed understanding of turbulent flow structure than discussed so far. We introduce the statistical study of turbulence in Section 10.3. Lift—briefly introduced in Chapter 8—is studied most effectively with advanced tools such as vorticity and circulation (Section 10.4). Section 10.5 continues with vorticity to show how this tool improves our understanding of curvy flow. Compressible fluid flow is discussed in Section 10.6. The flow behaviors not addressed in this text are accessible through advanced study based on the introductory methods in Chapters 1–9.

10.1 Viscosity, drag, and boundary layers

Our first topic in Chapter 2 was viscosity, and there we stated only that viscosity is a measure of a fluid's ability to resist flow. Chapter 5 formally defined viscosity μ: Viscosity is a material property that enters into the stress constitutive equation of the continuum model of fluids. In simple shear flow, the stress constitutive equation is:

$$\begin{array}{c}\text{Newton's law of viscosity}\\\text{(flow in } x_1\text{-direction,}\\\text{gradient in } x_2\text{-direction):}\end{array} \qquad \tilde{\tau}_{21} = \mu \frac{\partial v_1}{\partial x_2} \qquad (10.1)$$

In an arbitrary flow, the stress constitutive equation for a Newtonian fluid is a tensor equation:

$$\text{Newtonian constitutive equation:} \quad \underline{\underline{\tilde{\tau}}} = \mu \left(\nabla \underline{v} + (\nabla \underline{v})^T \right) \qquad (10.2)$$

We learned how to use the Newtonian constitutive equation in Chapters 6–8.

The stress constitutive equation relates the stresses generated by a flow to the velocity field. The Newtonian constitutive equation is the correct constitutive equation for a wide variety of fluids including water, air, and oil. To customize the Newtonian constitutive equation for a particular fluid, all that changes is the value of the viscosity. This is a striking simplicity! It is almost unbelievable that we can relate a complex flow field to its complex stress field with a single equation that has a single, scalar material parameter.

Actually, it *is* a bit too good to be true, and we learn in Section 5.3 that fluids that follow the Newtonian constitutive law are but one type of fluid. All other fluids, called non-Newtonian [104], follow more complex constitutive laws that vary in their mathematical structure and in the number of material parameters needed to fully specify the model.

The Navier-Stokes equation captures the physics of the flow of Newtonian fluids and is a nonlinear equation of great mathematical complexity—a complexity that matches the fluid behavior it describes:

Navier-Stokes Equation
(Microscopic momentum balance, continuum model):

$$\rho \left(\frac{\partial \underline{v}}{\partial t} + \underline{v} \cdot \nabla \underline{v} \right) = -\nabla p + \mu \nabla^2 \underline{v} + \rho \underline{g} \qquad (10.3)$$

The terms on the lefthand side of the Navier-Stokes are the inertial terms. When flows are rapid, these terms and the physics they represent dominate in the flow. The inertial terms are nonlinear. The terms on the righthand side of the Navier-Stokes equation represent the forces on fluid particles: pressure, viscous, and gravity; the viscous term is linear in velocity.

Because of the nonlinearity of the inertial terms of the Navier-Stokes, a general solution has not been found to this equation; rather, the history of fluid mechanics is of applying the Navier-Stokes equation to select problems, simplifying the governing equations, and solving the simplified equations. When the lefthand terms are zero (e.g., unidirectional or creeping flows), solutions of the Navier-Stokes equation are found (see Sections 7.1 and 8.1). When viscosity can be neglected (away from boundaries in rapid flow; see Section 8.1.2.3), the potential-flow solutions of the Navier-Stokes equation are found. In addition, analytical solutions to the Navier-Stokes equation are found for a limited number of flows in which both inertia and viscosity contribute; we discuss one such solution: boundary-layer flow (see Section 8.2). For most real flows, all terms of the Navier-Stokes equation must be retained.

The future of research into the effect of viscosity is numerical simulation of the Navier-Stokes equation for flows in which both viscosity and inertia are important. This field, which is active and growing, is called computational fluid dynamics (CFD). Commercial CFD codes are available, permitting contemporary researchers to benefit from decades of software development in the field. In many industries (e.g., aeronautics, meteorology, and reactor design), CFD modeling is essential to engineering design and decision making. We briefly introduce numerical approaches to the Navier-Stokes equation in the next section.

10.2 Numerical solution methods

The modeling process described in this text leads us to the differential equations that govern fluid flow—the continuity equation and the Navier-Stokes equation—which are nonlinear, multicomponent, coupled equations for the velocity and pressure fields. These equations are difficult to solve. We are familiar with analytical solutions to the governing equations that are obtained when the inertial terms are zero and when the viscous term is zero. For flows in which the equations may not be simplified, there are strategies that allow us to obtain useful results (e.g., boundary-layer approximation and circulation/lift calculations; see Section 10.4). In all cases, analytical solutions are limited to flows with simple geometries, and usually we are restricted to flows that are steady. For flows in complex geometries, for flows in which viscous and inertial contributions are both important, and for unsteady flows, the governing equations must be solved numerically.

The state of the art in numerical problem solving is to use software designed and written by coding experts rather than to develop standalone code. To obtain and interpret correctly solutions with such codes, we need some background in computational methods. In this section, we introduce numerical problem solving in fluid mechanics and discuss issues that affect the accuracy of numerical solutions. Numerical problem solving of nonlinear, coupled, partial differential equations (PDEs) is a complex topic; more on computational methods in fluid mechanics is available in the literature [49, 57, 86, 100].

10.2.1 Strategy

Many methods have been developed for numerically solving differential equations; all begin with the idea of dividing the flow domain into many small pieces. The map of the divided flow domain is called the mesh. To see why dividing the flow domain helps solve the differential equation, consider the value of our functions $\underline{v}$ and p at an arbitrary location in the mesh (Figure 10.1). If the mesh is fine, the values of the functions $\underline{v}|_{i,j}$ and $p|_{i,j}$ at an arbitrary mesh point i, j will be not very different from the values of these functions at neighboring points. Because the neighboring values are not so different, we can propose approximate methods (e.g., linear interpolation) based on the values of $\underline{v}$ and p at i, j to estimate the values of $\underline{v}$ and p and their derivatives at the neighboring points. The formulas used in this step vary from algorithm to algorithm, and the method choice affects accuracy and computational efficiency. The result of the estimization step is the generation of a large number of simple and interrelated equations that approximate the values of the functions $\underline{v}$ and p and their derivatives at all points in the mesh.

Having approximated the values of the function at every point in terms of its neighbors, we then must find a way to include in the problem the constraint of

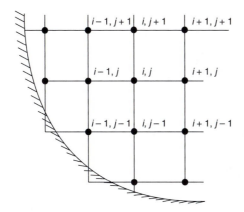

Numerical schemes divide a flow domain into small pieces that are numbered sequentially. Individual methods differ in how they estimate properties at a grid point; in all cases, the idea is to approximate the value based on the values of quantities such as velocity and pressure at neighboring grid locations.

the differential equation that we are trying to solve—for example, the Navier-Stokes equation. For two neighboring points i, j and $i + 1, j + 1$ (considering a two-dimensional flow domain and a single component of the Navier-Stokes), we can write:

$$\text{Differential equation to solve:} \quad f(v_x, v_y, p) = 0 \qquad (10.4)$$

If we now write the differential equation at each mesh point and substitute the simple approximate expressions developed, we obtain:

$$
\begin{aligned}
\text{Algebraic equations at every} \qquad f(v_x, v_y, p)\big|_{i,j} &= R_{i,j} \\
\text{mesh point, obtained from substituting} \qquad f(v_x, v_y, p)\big|_{i+1,j+1} &= R_{i+1,j+1} \\
\underline{v}, p \text{ estimates into differential equation:} \qquad & etc.
\end{aligned}
$$

$$(10.5)$$

where $R_{i,j}$ is the residual calculated when the approximations for $\underline{v}$ and p at i, j are substituted into the differential equation and $R_{i+1,j+1}$ is the residual at the location $i + 1, j + 1$. The residuals appear because the mesh-point approximations for $\underline{v}$ and p are not exact; therefore, the differential equation is satisfied only approximately at every point. We can write equations for residuals at every point in the mesh. The substitution of the estimates, which are different at every mesh point, changes the single differential equation to a set of many coupled algebraic equations. The best solution of the problem is obtained when all of the residuals are minimized.

With these steps, we transform our single differential equation into hundreds, thousands, or even tens of thousands of much simpler (the form depends on how the functions at the neighboring points are estimated from their neighbors), coupled, algebraic equations that are inconceivable to solve manually but which a computer easily solves using techniques from linear algebra. The final result is a database of the values of $\underline{v}$ and p for every point on the mesh. Figure 10.2 shows a finite-element mesh with a triangular grid that can be used for flow calculations in a complex geometry.

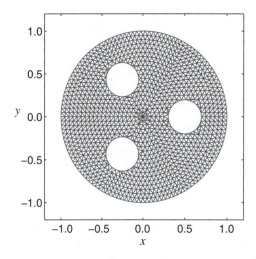

Figure 10.2 To numerically solve a differential equation, the first step is to divide the flow domain into a mesh. The discrete points of the mesh are chosen so that the functions we seek do not vary much between mesh points. The mesh shown allows us to calculate the velocity field in a cross section of complex shape—the free space around three tubes encased in a much larger outer pipe. (Finite-element mesh provided by Tomas Co [24].)

10.2.2 Software packages

The most likely way that a contemporary engineer approaches numerical solutions to problems is by using commercial software packages. Many packages are available that require little or no user programming. The introduction here describes enough about the process to allow us to begin to use a flow software package. Discussion of individual numerical methods (e.g., finite difference, finite volume, and finite-element methods) is found in the literature [24, 49, 57, 86, 100]. We outline the steps that lead to a numerical solution when using a software package.

Steps for Using a Numerical Software Package to Solve Flow Problems

1. *Choose the flow geometry.* If possible, take advantage of symmetry to reduce the size of the computational domain. It may be necessary to add sections before and/or after the flow section of interest (i.e., the test section) to eliminate edge effects or other boundary issues.
2. *Design and generate the mesh.* The mesh may be uniform or nonuniform; a nonuniform mesh with smaller elements near points of higher interest or higher rates of change is the norm because the approximations in the method assume that the function does not vary much between neighboring points.
3. *Choose the physics.* For flow problems, this is the equation of motion, the continuity equation (compressible or incompressible), the energy equation (for nonisothermal flow), or other equations as appropriate.
4. *Define the boundary conditions.* The fluid behavior at all boundaries is specified; the initial conditions are defined in nonsteady-state problems. If the desired boundary condition is not available in the code, it may be necessary to add a section to the flow domain to generate the desired boundary conditions on the test section (Figure 10.3).

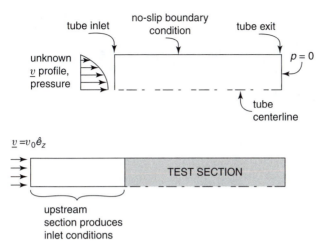

Figure 10.3 It often is necessary to add sections before or after flow that can be used to allow flows to come to steady state or to eliminate exit effects. The portion of the flow of interest is called the test section. The added upstream section shown here allows the uniform inlet velocity profile to rearrange to the steady-state velocity profile before entering the test section.

5. *Solve the problem.* To solve the problem, a specific solution method is applied (i.e., finite elements, finite volume, finite difference, or other), algebraic versions of the differential equations are generated for each mesh location, and the residuals are minimized by the computer code. If the code fails to find a solution in this step, common causes include: the Reynolds number is too high (i.e., turbulence appears); the mesh is too refined (i.e., rounding errors become the same size as some values of velocity and pressure); the boundary conditions are inappropriate; some assumptions were violated (e.g., incompressibility, isothermal, flow or steady state); or a numerical instability occurred due to roundoff or other numerical issues.

6. *Calculate and plot the engineering quantities of interest.* This step is performed in a postprocessor that can access the database of $\underline{v}$, p results at each mesh point. The postprocessing code makes the appropriate interpolations and integrations/differentiations to calculate the requested quantities. If the plotting capabilities of the code are inadequate, datasets of engineering quantities can be output for plotting with graphing software.

10.2.3 Accuracy

For various reasons, neither analytical nor numerical solutions to flow problems are 100 percent accurate to a real flow. When a flow is modeled, we make assumptions throughout the modeling process, from the assumption that the fluid is a continuum to assumptions about boundary conditions or flow symmetry. For analytical solutions, we often make additional assumptions, such as about the importance to the problem of certain terms in the differential equations; we do this to make the equations more tractable to analytical solution. With

Table 10.1. Summary of issues affecting accuracy of solutions of flow problems using the Navier-Stokes equations

Issue	Analytical	Numerical
continuum hypothesis	X	X
symmetry assumptions	X	X
approximate geometry	X	X
approximate boundary conditions	X	X
steady-state assumption	X	X
incompressible fluid	X	X
Newtonian fluid	X	X
isothermal, single-phase flow	X	X
finite domain size	X	X
neglect inconvenient terms (creeping flow, inviscid flow)	X	
linearization and other approximate analytical solution methods	X	
final solution series truncation error	X	
discretization of the flow domain (finite grid size, resolution)		X
derivative approximation errors		X
roundoff error		X
interpolation error in final calculations of engineering properties of interest		X
numerical instability induced by accumulation of error		X
inappropriate implementation of comercial code		X

numerical solutions, there also are assumptions made to allow the solutions to be obtained, and these are different in character from those in analytical methodologies. Because making assumptions affects the accuracy of the solutions obtained, we always must be aware of the assumptions we are making and how those assumptions affect our answers. Table 10.1 summarizes accuracy issues that we face in solving flow problems.

When using code on a new problem, it is essential to solve a related problem for which we know the solution and to compare the numerical and analytical results. This allows us to verify whether we understand how to make the code function. The design of the mesh is important: If the mesh is too fine, the computer may be slow or may not be able to find a solution; if the mesh is too coarse, the approximation errors may be unacceptably large. When the mesh is refined between calculations (i.e., made finer), the result should become more accurate. If refining the mesh results in a less reasonable solution, the reasons for this failure must be determined before proceeding.

Although most computer programming has been eliminated from flow simulation due to the availability of effective codes and fast computers, the task of obtaining reliable results is still challenging. As we discuss in Section 10.3.2, flows can become unstable due to their inherent physics. Numerical instabilities also can

Figure 10.4 Turbulent flow, such as in a swollen creek, is full of eddies, which are curvy regions with significant vorticity. The flow structure is three-dimensional and chaotic.

be observed, caused by roundoff error or other reasons. Some flow problems are more compatible with a particular type of numerical analysis, and this may be discovered only through long investigation. Finally, as with any modeling, the solution obtained from a numerical model is only as good as the assumptions incorporated into the model. The responsibility for the accuracy and relevancy of a numerical result lies entirely with the individual carrying out the analysis.

10.3 Laminar flow, turbulent flow

Laminar flow and related orderly flows are described in depth in this text. Turbulent flow, however, was examined only briefly. It is worthwhile to further discuss turbulent flow, which has a structure and complexity that requires a different approach than what is used in the study of laminar flow.

Turbulent flow is a highly disordered flow that in many ways is the opposite of laminar flow (Figure 10.4 and [117]). Turbulent flow is a three-dimensional flow characterized by swirling eddies and efficient mixing. It is observed at higher flow speeds than laminar flow, and it takes place in a wide variety of situations. Turbulent flow is characterized by having vorticity (i.e., rotational character of the velocity field) continuously but irregularly distributed throughout the flow in all three dimensions. Laminar flow also has vorticity—for example, a marked ping-pong ball set down in a shear flow spins continuously—but the vorticity field in laminar flow is rather tame and organized, not the disheveled riot that it is in turbulent flow. Turbulent flow occurs spontaneously and unavoidably at high flow speeds as a result of instabilities in laminar flow.

The continuum approach that provides velocity and stress fields in laminar flow also is applicable to turbulent flow; the only obstacle to using the Navier-Stokes equations in a turbulent flow is that turbulent flow is inherently three-dimensional and complex. The complexity of turbulent flow makes the continuum-mechanics equations impossible to solve for the detailed motions in these flows. In fact, the disorder of turbulent flow is such that not all aspects are even reproducible: Experiments and simulations on turbulent flows in which every aspect is scrupulously repeated unfortunately result in flows that are not identical in their details. The averages of many results in these repeated experiments are the same, but the precise location of individual eddies or swirls is not reproduced every time the experiment or simulation is run.

What *is* reproducible in turbulent flow is the statistically averaged variables for the velocity and stress fields. The averages used in the study of turbulent flow are ensemble averages such as those used in statistical thermodynamics. These quantities are averages of many implementations—carried out through modeling or experimentation—of a flow under identical circumstances. The modern study of turbulence uses statistical models and high-performance computers to obtain weather predictions, the shape of the wake behind boats and airplanes, and the mixing characteristics of chemical reactors.

In this section, we present the statistically averaged equations of change (i.e., continuity and Navier-Stokes) and discuss issues that arise in using them. Numerical solutions to the statistically averaged equations of change give researchers the tools they need to study turbulent-flow behavior in detail. We also introduce the concept of instability as related to flow studies, including the transition from laminar to turbulent flow. Complex flow structure often can be understood by tracking the evolution of the flow from a well-understood stable flow to the more complex state. Tools such as frequency-response analysis are helpful in studies of this type. More in-depth understanding of turbulent flow may be pursued through additional reading [10, 165].

10.3.1 Statistical modeling of turbulence

Turbulent flow is a rapidly fluctuating flow with many fine structures throughout. For steady turbulent flows, we can monitor the velocity as a function of time at a fixed location, and when we look at the data, we see a rapidly varying signal that is characteristic of turbulence. Fluid arrives continuously at the velocity probe from upstream positions; thus, the value of the signal at any one time represents the value of the velocity at that location for a single, instantaneous implementation of the flow. The time-average of that signal, therefore, is comparable to an ensemble-average velocity for that location in the flow [10].

The velocity components for one implementation of a chosen turbulent flow may be written as a time-averaged (i.e., ensemble-averaged) velocity contribution $\overline{v_i}$ plus a deviation from the average $\overline{v_i}'$ that corresponds to that particular flow implementation. In Cartesian coordinates, this becomes:

$$v_x = \overline{v_x} + v'_x \tag{10.6}$$

$$v_y = \overline{v_y} + v'_y \tag{10.7}$$

$$v_z = \overline{v_z} + v'_z \tag{10.8}$$

where v_x, v_y, and v_z are the velocity components for a single implementation of the flow. The flow must satisfy the continuity equation and the Navier-Stokes equation. Our strategy then is to write the Navier-Stokes and continuity equations for v_x, v_y, and v_z and subsequently carry out a time-average of the entire equation. If the Navier-Stokes equations were linear, after averaging we would obtain them again with every term replaced with its time-average. However, the Navier-Stokes equations are not linear; thus, interesting fluctuation terms appear when we do the time-average, and these terms help to quantify turbulence.

The most convenient forms of the equations of change to use for the time-averaging calculation are those given here (incompressible fluid assumed; see Equation 6.62):

$$\text{Continuity equation:} \quad \nabla \cdot \underline{v} = 0 \tag{10.9}$$

$$\text{Cauchy momentum equation:} \quad \rho \frac{\partial \underline{v}}{\partial t} + \nabla \cdot (\rho \underline{v}\,\underline{v}) = -\nabla p + \nabla \cdot \underline{\tilde{\tau}} + \rho \underline{g} \tag{10.10}$$

$$\text{Newtonian constitutive equation:} \quad \underline{\tilde{\tau}} = \mu \left(\nabla \underline{v} + (\nabla \underline{v})^T \right) \tag{10.11}$$

The time-averaging of the equations of change is performed over a short time interval t_0; the time interval is chosen to be long enough to eliminate the random fluctuations of turbulent flow but short enough not to eliminate changes with time that are part of the average character of the flow. The time-averaging of velocity components, pressure, and other flow variables is carried out using the following integral:

$$\overline{(\text{quantity})} = \frac{1}{t_0} \int_t^{t+t_0} (\text{quantity}) \, dt' \tag{10.12}$$

To calculate the fluctuation-averaged equations of change, we substitute the velocity expressions in Equations 10.6–10.8 into the equations of change and carry out the time-averaging integrations.

The averaging process described here produces three types of terms: those that yield average properties, those that average to zero, and those that quantify fluctuations. An example of a term that yields average properties is the following expression, which occurs in the x-component of the momentum-balance equation:

$$\overline{\rho \frac{\partial v_x}{\partial t}} = \frac{1}{t_0} \int_t^{t+t_0} \left(\rho \frac{\partial \overline{v_x}}{\partial t} \right) dt' \tag{10.13}$$

$$= \frac{\rho}{t_0} \frac{\partial}{\partial t} \left[\int_t^{t+t_0} \overline{v_x} \, dt' \right] \tag{10.14}$$

$$= \rho \frac{\partial \overline{v_x}}{\partial t} \tag{10.15}$$

For terms like this, the averaging process returns the average of the same quantities as when we started. For terms with a fluctuating quantity multiplying an average quantity, we obtain:

$$\frac{\partial(\overline{v_x' \overline{v_y}})}{\partial y} = \frac{1}{t_0} \int_t^{t+t_0} \frac{\partial(v_x' \overline{v_y})}{\partial y} \, dt' \tag{10.16}$$

$$= \frac{1}{t_0} \frac{\partial}{\partial y} \left[\int_t^{t+t_0} v_x' \overline{v_y} \, dt' \right] \tag{10.17}$$

$$= \frac{1}{t_0} \frac{\partial}{\partial y} \left[\overline{v_y} \int_t^{t+t_0} v_x' \, dt' \right] \tag{10.18}$$

$$= 0 \tag{10.19}$$

This integral gives zero because v_x' has a random value with zero mean; thus, the integration of this random value gives zero.

The significant terms are those in which two fluctuating terms multiply each other. The fluctuating terms may be positive and negative with equal probability. When both are negative, they produce a positive product, which skews the value of the integral toward a positive, nonzero value. We must retain terms of this type as an average of the product with no simplification:

$$\frac{\partial(\overline{v_x' v_y'})}{\partial y} = \frac{1}{t_0} \int_t^{t+t_0} \frac{\partial(v_x' v_y')}{\partial y} \, dt' \tag{10.20}$$

$$= \frac{\partial}{\partial y} \left[\frac{1}{t_0} \int_t^{t+t_0} v_x' v_y' \, dt' \right] \tag{10.21}$$

$$= \frac{\partial \overline{v_x' v_y'}}{\partial y} \tag{10.22}$$

Following the time-averaging procedure for each term in the equations of change, we obtain the ensemble-averaged equations of change for turbulent flow.

Fluctuation-averaged equations of change for turbulent flow:

Continuity equation: $$0 = \frac{\partial \overline{v_x}}{\partial x} + \frac{\partial \overline{v_y}}{\partial y} + \frac{\partial \overline{v_z}}{\partial z} \tag{10.23}$$

Cauchy
momentum equation:

(x-component) $$\rho \frac{\partial \overline{v_x}}{\partial t} + \left(\rho \frac{\partial (\overline{v_x} \, \overline{v_x})}{\partial x} + \rho \frac{\partial (\overline{v_y} \, \overline{v_x})}{\partial y} + \rho \frac{\partial (\overline{v_z} \, \overline{v_x})}{\partial z} \right)$$

$$+ \left[\rho \frac{\partial \overline{v_x' v_x'}}{\partial x} + \rho \frac{\partial \overline{v_y' v_x'}}{\partial y} + \rho \frac{\partial \overline{v_z' v_x'}}{\partial z} \right]$$

$$= -\frac{\partial \overline{p}}{\partial x} + \nabla \cdot \underline{\underline{\overline{\tau}}} + \rho g_x \tag{10.24}$$

Newtonian
constitutive equation: $$\underline{\underline{\overline{\tau}}} = \mu \left(\nabla \underline{\overline{v}} + (\nabla \underline{\overline{v}})^T \right) \tag{10.25}$$

$$
\begin{pmatrix} \overline{\tau}_{xx} & \overline{\tau}_{xy} & \overline{\tau}_{xz} \\ \overline{\tau}_{yx} & \overline{\tau}_{yy} & \overline{\tau}_{yz} \\ \overline{\tau}_{zx} & \overline{\tau}_{zy} & \overline{\tau}_{zz} \end{pmatrix}_{xyz} = \mu \begin{pmatrix} 2\dfrac{\partial \overline{v}_x}{\partial x} & \left(\dfrac{\partial \overline{v}_x}{\partial y} + \dfrac{\partial \overline{v}_y}{\partial x}\right) & \left(\dfrac{\partial \overline{v}_x}{\partial z} + \dfrac{\partial \overline{v}_z}{\partial x}\right) \\ \left(\dfrac{\partial \overline{v}_y}{\partial x} + \dfrac{\partial \overline{v}_x}{\partial y}\right) & 2\dfrac{\partial \overline{v}_y}{\partial y} & \left(\dfrac{\partial \overline{v}_y}{\partial z} + \dfrac{\partial \overline{v}_z}{\partial y}\right) \\ \left(\dfrac{\partial \overline{v}_z}{\partial x} + \dfrac{\partial \overline{v}_x}{\partial z}\right) & \left(\dfrac{\partial \overline{v}_z}{\partial y} + \dfrac{\partial \overline{v}_y}{\partial z}\right) & 2\dfrac{\partial \overline{v}_z}{\partial z} \end{pmatrix}_{xyz}
$$

$$(10.26)$$

Similar y- and z-components to the momentum equation were omitted.

Comparing the x-momentum equation (Equation 10.24) to the original Cauchy momentum equation, we see that they are identical except that the velocity has become the time-averaged velocity and there are three extra terms (boxed in Equation 10.24). The boxed terms contain convective momentum contributions due to turbulent fluctuations, and these are called the *Reynolds stresses*:

Reynolds stresses
(x-component of momentum):

$$
\overline{\tau}_{xx}^{\text{turb}} \equiv -\rho \overline{v_x' v_x'}
$$
$$
\overline{\tau}_{yx}^{\text{turb}} \equiv -\rho \overline{v_y' v_x'} \qquad (10.27)
$$
$$
\overline{\tau}_{zx}^{\text{turb}} \equiv -\rho \overline{v_z' v_x'}
$$

The negative sign is introduced into the definition of Reynolds stresses by convention to move them from the left side of the momentum balance (i.e., inertial terms) to the right side (i.e., forces)—we think of them as an additional force/area acting within the fluid. Writing the x-component of the momentum balance equation this way, we obtain:

Momentum equation:

$$
(x\text{-component}) \quad \rho \frac{\partial \overline{v}_x}{\partial t} + \left(\rho \frac{\partial (\overline{v}_x \, \overline{v}_x)}{\partial x} + \rho \frac{\partial (\overline{v}_y \, \overline{v}_x)}{\partial y} + \rho \frac{\partial (\overline{v}_z \, \overline{v}_x)}{\partial z} \right)
$$

$$
= -\frac{\partial \overline{p}}{\partial x} + \nabla \cdot \underline{\underline{\overline{\tau}}} - \left[\frac{\partial \rho \overline{v_x' v_x'}}{\partial x} + \frac{\partial \rho \overline{v_y' v_x'}}{\partial y} + \frac{\partial \rho \overline{v_z' v_x'}}{\partial z} \right] + \rho g_x
$$

$$(10.28)$$

$$
= -\frac{\partial \overline{p}}{\partial x} + \nabla \cdot \left[\underline{\underline{\overline{\tau}}}^{\text{lam}} + \underline{\underline{\overline{\tau}}}^{\text{turb}} \right] + \rho g_x \qquad (10.29)
$$

where we adopt the nomenclature $\underline{\underline{\overline{\tau}}}^{\text{lam}}$ for the averaged, viscosity-based stresses in Equation 10.26 to distinguish material-based stresses present in both laminar and turbulent flows from the convective turbulent stresses discussed here.

The final form of the fluctuation-averaged momentum balance is appealing in its simplicity:

Fluctuation-averaged momentum balance for turbulent flow:

$$\rho \left(\frac{\partial \overline{\underline{v}}}{\partial t} + \nabla \cdot \overline{\underline{v}\,\underline{v}} \right) = -\nabla \overline{p} + \nabla \cdot \left[\underline{\underline{\tau}}^{\text{lam}} + \underline{\underline{\tau}}^{\text{turb}} \right] + \rho \underline{g}$$

(10.30)

$$\rho \left(\frac{\partial \overline{\underline{v}}}{\partial t} + \overline{\underline{v}} \cdot \nabla \overline{\underline{v}} \right) = -\nabla \overline{p} + \nabla \cdot \left[\underline{\underline{\tau}}^{\text{lam}} + \underline{\underline{\tau}}^{\text{turb}} \right] + \rho \underline{g}$$

(10.31)

The equivalence of the first and second versions may be shown using the fluctuation-averaged continuity equation, $\nabla \cdot \overline{\underline{v}} = 0$. When $\underline{\underline{\tau}}^{\text{turb}} = 0$, we recover the usual momentum-balance equation, which—when combined with the Newtonian constitutive equation (Equation 10.31)—may be solved for the averaged properties $\overline{\underline{v}}$ and $\overline{p}$ using the methods in this text.

When $\underline{\underline{\tau}}^{\text{turb}} \neq 0$, we cannot solve the equations of motion unless we know the relationship between $\underline{\underline{\tau}}^{\text{turb}}$ and the averaged velocity field $\overline{\underline{v}}$. In viscous flow, we use the constitutive equation to express the stress tensor in terms of the velocity components. Making an analogy to viscous flow, for the turbulent stresses we need some type of "turbulent constitutive equation" to relate the turbulent stresses to the velocity field.

We seek a turbulent constitutive equation to allow us to solve Equation 10.31 for $\overline{\underline{v}}$ and $\overline{p}$. For viscous flow, we obtained the material-stress constitutive equation by observing material behavior and guessing an appropriate equation; for Newtonian fluids, this equation is well known (Equation 10.25). For turbulent flow, many turbulent-stress constitutive equations have been proposed, and we give two such equations here. Note that the relationship between $\underline{\underline{\tau}}^{\text{turb}}$ and the averaged velocity field $\overline{\underline{v}}$ is not a material relationship: Instead, $\underline{\underline{\tau}}^{\text{turb}}(\overline{\underline{v}})$ is a characteristic of a flow or rather of the particular flow implementation under discussion.

A reasonable first guess at a turbulent constitutive equation is to pattern the form of the equation on the Newtonian material-stress constitutive equation. This was suggested by Boussinesq in 1877 [16]:

Eddy viscosity, defined:

$$\underline{\underline{\tau}}^{\text{turb}} = \mu_{\text{eddy}} \left(\nabla \overline{\underline{v}} + (\nabla \overline{\underline{v}})^T \right)$$

(10.32)

Eddy viscosity (x-direction shear flow):

$$\overline{\tau}_{yx}^{\text{turb}} = \mu_{\text{eddy}} \frac{\partial \overline{v_x}}{\partial y}$$

(10.33)

The eddy viscosity is not a constant; if it were, the effect of turbulence would be simply to increase the viscosity of the fluid, but this is not the case. Eddy viscosity is a function of position, which we can deduce by considering the turbulent stresses at the wall. Because the fluctuations v_x' and v_y' must go to zero at the wall, then $\overline{\tau}_{yx}^{\text{turb}} = -\rho\overline{v_x'v_y'}$ also must go to zero at the wall. Because the velocity derivative is not zero at the wall, Equation 10.33 implies that the eddy viscosity must go to zero. Although these considerations mean that the eddy-viscosity concept is not very helpful for quantitative modeling, the concept is useful in discussions and in visualization of turbulent flow. The eddies of turbulent flow can be visualized as organized and fluctuating structures that bump into one another and introduce extra dissipation into the flow. In this sense, they have a viscosity.

A more useful model for turbulent stresses was developed by Prandtl [135]. Using an analogy to how molecules move about in a gas (an analogy, unfortunately, that does not accurately describe eddy motion), Prandtl proposed for shear flow that the turbulent shear stress should be given by the following function of the velocity field:

$$\text{Mixing length, defined:}\quad \overline{\tau}_{yx}^{\text{turb}} = \rho l^2 \left|\frac{\partial \overline{v}_x}{\partial y}\right| \left(\frac{\partial \overline{v}_x}{\partial y}\right) \tag{10.34}$$

where the length l is called the mixing length. Experiments on two types of turbulent flows—shear flow near a wall and free-jet flow—show that the mixing length increases with distance from the wall in the shear case and increases with the width of the mixing zone in the free-jet case:

$$\text{Shear near a wall (wall turbulence):}\quad l = \kappa_1 y \tag{10.35}$$

$$\text{Free jet (free turbulence):}\quad l = \kappa_2 b \tag{10.36}$$

where κ_1 and κ_2 are constants, y is the distance from the wall in wall shear flow, and b is the width of the mixing zone in the free jet [15]. The mixing length can be thought of as the distance over which a turbulent eddy retains its identity.

Once the model for the Reynolds stresses is chosen, a turbulence problem is solved like a problem in nonturbulent flow: The equations of change are set up for the flow geometry, and solution methods (i.e., analytical and numerical) for partial differential equations are used. Flow simulators have various turbulence models built in [27]. For more on this and other approaches to turbulence modeling (e.g., K-ε models), refer to the literature [40, 165].

10.3.2 Flow instability

We can broaden our understanding of turbulent flow by reviewing the origins of turbulence. Reynolds's tube-flow experiments showed that organized laminar flow becomes disorganized turbulent flow when the same flow is operated at higher Reynolds numbers (see Figure 7.17). In other words, laminar flow is unstable above Re = 2,100. In a laminar flow in which the Reynolds number gradually increases, instability is the first step in the chain of events that produce turbulence (paraphrased from [116]).

An accurate flow model should capture flow instability. However, models must be exceptionally accurate if they are to predict correctly flow instabilities. To see why this is true, consider the task of creating a model of Reynolds's pipe-flow experiment. Our goal is to find within the model the origins of the laminar–turbulent transition.

To model flow in a pipe, we know that the governing equations are the continuity equation and the equation of motion in the form of the Navier-Stokes equation. In identifying these equations as the governing equations for our problem, however, we already have made (or soon will make) a series of assumptions. The Navier-Stokes equations, for example, are valid only for fluids with a constant viscosity. Additional usual assumptions are that the flow is isothermal and incompressible. If we make these assumptions, however, we technically are constraining our model predictions to the behavior of a fluid that is perfectly Newtonian (i.e., viscosity is unwaveringly constant), perfectly incompressible (i.e., the density is absolutely constant), and perfectly isothermal (i.e., the temperature does not change under any circumstances). Mathematics allows for the consideration of such perfect systems; however, real-life experiments are not perfect.

Strangely enough, we obtain reasonable solutions in the steady, laminar pipe-flow problem making all three of these assumptions (see Chapter 7). The laminar-flow modeling result matches experimental measurements (i.e., the Hagen-Poiseuille equation and the parabolic velocity profile) either because the modeling assumptions are true (highly unlikely because no experimental system can be "perfect") or possibly because slight deviations from what is assumed in the model simply do not matter much. The latter statement is another way of stating that the flow turns out to be stable relative to small variations in viscosity, density, and temperature.

In an experimental flow, flow stability can be investigated by introducing slight changes to various flow conditions and then observing their effect on the flow produced. For example, we can investigate the experimental stability of laminar flow to changes in fluid density by comparing runs made with fluids of slightly different density.

Experiments are costly and difficult, however, and we prefer to investigate mathematically a flow's stability by using a model. We can understand stability of laminar flow by subjecting an existing flow solution to a perturbation of a variable that we are investigating. If the flow returns to the starting steady state after the perturbation dies out, then the flow is stable relative to that perturbation. If the flow moves to a new operating condition as a result of the perturbation, or if the perturbation grows without bound, the flow is unstable to that type of perturbation.

The predictions of a stability analysis are necessarily limited by the physics included in the model. If the model with which we are working assumes constant density, then the mathematics indicates the behavior of a system that is constrained to have a constant density. A real system is not constrained in this way, however; therefore, the real system may respond to the perturbation in a way that includes a change in the density. Any variable of the flow may be perturbed in this way, and the model will reveal the stability of the flow to perturbations of the individual variables. Stability analyses can teach us much about flows, although they are limited to the physics that we input.

The National Committee for Fluid Mechanics Films (NCFMF) film on flow instabilities [116] illustrates how we can quantify the character of flow instabilities

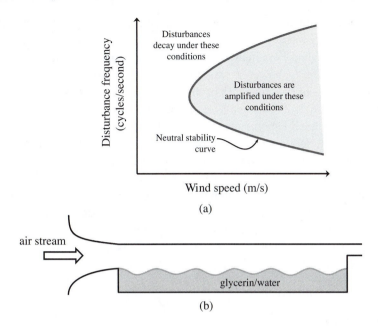

Figure 10.5 An open-channel flow (b) is perturbed at different frequencies to map the character of the instabilities that occur. The data obtained are summarized in the neutral stability curve (a) [116, 155].

(Figure 10.5). The flow examined in that film is air of various speeds blown over water. The flow is perturbed by tapping on the water surface in a controlled manner. At low air speeds, with no disturbances, there are no waves on the water and the flow is stable. When disturbances are added at low air speeds, waves appear, but they damp out rapidly. At higher air speeds with still no waves produced by the air flow alone, added disturbances produce waves that do not damp out but that, instead, grow as they move along in the flow. Finally, at an even higher air speed, waves are produced on the water without any need for external disturbances. The instabilities observed when no forced disturbance is applied are caused by accidental disturbances present in the incoming airstream.

The result of such a frequency analysis can be plotted to produce the neutral stability curve, which shows when amplification begins to be observed as a function of frequency and wave speed. The neutral stability curve (Figure 10.5a) indicates which frequency is amplified in a flow exposed to perturbations containing all frequencies; these are the conditions present in most real systems. For more on stability analysis and the origins of turbulence, refer to the literature [22, 40, 165].

10.4 Lift, circulation

In Chapter 2, we introduced lift and briefly discuss the fluid physics of airplane flight. Airplane flight is complicated and only a simplified version of the flow mechanics of lift is accessible to students who do not have a fluid-mechanics background. When a wing moves rapidly through air, a particular flow pattern (i.e., velocity field) develops, along with an accompanying pressure field. If we assume a flow pattern (e.g. a particular flow on the upper wing surface), we can present arguments to justify lift and drag on airplanes; likewise, if we assume a pressure field (i.e., higher pressure on the lower wing surface), we can justify flight. In reality, however, we do not specify the flow field or the pressure field: They are both consequences of the horizontal motion of the obstacle (i.e., wing) through the fluid (i.e., air). The simple but unsatisfying answer to why airplanes can fly is that the momentum balance allows it!

Aeronautical engineers need a more thorough explanation of the physics of flight, including how wing shape, aircraft speed, angle of attack, and other factors affect flight. Now that we understand basic fluid flow, we can examine more carefully the mechanics of flight and point out important concepts for future study. As is true for drag, the complete picture of lift comes from a fluid-force calculation on an object in a free stream using solutions to the Navier-Stokes equation to obtain the stress tensor, $\tilde{\underline{\Pi}} = -p\underline{I} + \tilde{\underline{\tau}}$. Lift is the fluid force perpendicular to drag in a uniform flow:

$$\text{Total molecular fluid force} \atop \text{on a finite surface } \mathcal{S}: \qquad \underline{\mathcal{F}} = \iint_{\mathcal{S}} [\hat{n} \cdot \tilde{\underline{\Pi}}]_{\text{at surface}} \, dS \qquad (10.37)$$

$$\text{Uniform flow in } z\text{-direction} \atop \text{past an obstacle}: \qquad \underline{\mathcal{F}} = \begin{pmatrix} 0 \\ 0 \\ \mathcal{F}_{\text{drag}} \end{pmatrix}_{xyz} + \begin{pmatrix} \mathcal{F}_{\text{lift},x} \\ \mathcal{F}_{\text{lift},y} \\ 0 \end{pmatrix}_{xyz} \qquad (10.38)$$

A wing moving through air may be studied as the two-dimensional flow of an airstream moving past an airfoil cross section. The flow is rapid; thus, we assume the Reynolds number to be high. Chapter 8 explains that we expect the flow around an airfoil at high Reynolds numbers to form a thin boundary layer with the rest of the flow being inviscid. The flow in the boundary layer is viscous, and vorticity is produced in the boundary layer. For each potential airfoil shape, designers can carry out a complete boundary-layer calculation or CFD analysis to see what the drag and lift are on the airfoil and then use this information to modify and perfect their designs. This method is sufficient but needlessly complex; a much simpler method was discovered almost 100 years ago.

As noted previously, most of the flow past an airplane wing is a very rapid flow in which viscous effects are negligible; thus, it is tempting to use potential flow to model airplane flight. Unfortunately, calculations on potential flow past a cylinder (the cylinder is a stand-in for the airplane wing) predict incorrectly that there is neither lift nor drag on the wing. This is another instance of the D'Alembert paradox discussed in Chapter 8. However, researchers who were solving potential-flow problems in the early 1900s found that there are potential-flow models that generate lift: The potential flow that represents flow past a cylinder superimposed with a vortex flow of a certain strength (i.e., free vortex; more discussion of this follows) produces a lift that is proportional to the strength of the vortex. This observation became a modeling strategy that is the basis of modern aircraft design; we explain this aeronautical fluid-mechanics model here.

The Navier-Stokes equations for inviscid flows (i.e., potential-flow equations; see Equation 8.203) are nonlinear; however, for steady, two-dimensional flows, they can be transformed to a single linear equation by using the stream function ψ

(see Equation 8.223). The advantage of the transformation to a linear differential equation is that linear combinations of solutions to linear equations also are solutions of the equations. Rather than choosing a flow and seeing if we can model it, we can turn around the process and find new solutions of the equations by simply adding up known solutions. The new potential flow described by such a superposed solution is not guaranteed to be useful or even physically realizable; it turns out, however, that this tactic produced a model that is useful for the study of lift.

The potential-flow solution to flow around a long cylinder is known [176]:

$$
\begin{array}{c}
\text{Velocity field for} \\
\text{potential flow} \\
\text{around a long cylinder} \\
\text{(viscosity neglected):}
\end{array}
\quad
\underline{v} =
\begin{pmatrix}
v_\infty \cos\theta \left(1 - \frac{R^2}{r^2}\right) \\
-v_\infty \sin\theta \left(1 + \frac{R^2}{r^2}\right) \\
0
\end{pmatrix}_{r\theta z}
\qquad (10.39)
$$

This flow is irrotational. Another known solution to the potential-flow equations is the free vortex, which is a theoretical flow that approximates the flow in a spiral vortex tank (see Figure 8.57). The streamlines in a free vortex are concentric circles. The velocity field for a free vortex satisfies the potential-flow equations and is given here:

$$
\begin{array}{c}
\text{Velocity field for} \\
\text{a free vortex flow} \\
\text{(potential flow):}
\end{array}
\quad
\underline{v} =
\begin{pmatrix}
0 \\
\frac{K}{r} \\
0
\end{pmatrix}_{r\theta z}
\qquad (10.40)
$$

K is called the strength of the vortex [76, 176], and the value of K determines the speed of the flow. This flow has rotational character. We can create a new solution to the potential-flow equations by adding the free-vortex solution to that for potential flow past a cylinder. The addition of these two potential-flow solutions creates a new flow that is similar to the flow-past-a-cylinder problem, but the new combined flow is different in that it has a degree of rotational character. It is the added rotational character throughout the flow field—introduced by the addition of the free-vortex solution—that produces lift in the superposed solution.

We might wonder why adding additional rotational character improves a potential-flow model—that is, makes the flawed potential-flow solution more like a real flow. Missing from the potential-flow solution to flow around a cylinder is any effect of the boundary layer and viscosity. When we neglect viscosity, we take out of the flow solution the rotational character associated with viscosity; recall that vorticity is strongly generated by viscosity in the boundary layer. The superposition described here—adding a free-vortex solution to the solution for flow around an obstacle—is a way of returning to the problem the rotational character taken out when we neglected viscosity in the momentum balance. As we see in the discussion that follows if we return just the right amount of rotational character we can make a model that accurately predicts the flow field, the pressure field, and the lift in real flow around an object.

We already defined a function that quantifies local rotational character of a flow field: the vorticity. To quantify the rotational character of an entire region of a flow, we can integrate the vorticity over an area $\mathcal{S}$:

$$\begin{array}{c}\text{Net vorticity flux} \\ \text{through } \mathcal{S}: \end{array} = \iint_{\mathcal{S}} (\hat{n} \cdot \underline{\omega}) \, dS \qquad (10.41)$$

where $\mathcal{S}$ is any chosen area in the flow. Using Stokes's theorem [146], we can relate this integral to a line integral around the perimeter of $\mathcal{S}$. Stokes's theorem for a general vector field $\underline{f}$ is given by:

$$\begin{array}{c}\text{Stokes's theorem} \\ \text{for vector field } \underline{f}: \end{array} \quad \boxed{\iint_{\mathcal{S}} \hat{n} \cdot \left(\nabla \times \underline{f} \right) dS = \oint_{C} \left(\hat{t} \cdot \underline{f} \right) dl} \qquad (10.42)$$

where C is the curve that encloses the surface $\mathcal{S}$, $\hat{t}$ is a unit vector locally tangent to C, and dl is a small displacement counterclockwise around C. Choosing $\underline{f}$ as the velocity $\underline{v}$, we can calculate the net flux of vorticity $\underline{\omega} = \nabla \times \underline{v}$ through $\mathcal{S}$ as the following line integral:

$$\begin{array}{c}\text{Net vorticity flux} \\ \text{through } \mathcal{S}: \end{array} = \iint_{\mathcal{S}} (\hat{n} \cdot \underline{\omega}) \, dS = \iint_{\mathcal{S}} \hat{n} \cdot (\nabla \times \underline{v}) \, dS \qquad (10.43)$$

$$= \oint_{C} (\hat{t} \cdot \underline{v}) \, dl \equiv \Gamma \qquad (10.44)$$

The line integral in the previous equation defines the circulation, Γ, which is a property of the flow field and depends on the choice of C (or, equivalently, $\mathcal{S}$). The circulation is the counterclockwise integration of the tangential component of the velocity field around a closed curve:

$$\text{Circulation on } C: \quad \boxed{\Gamma \equiv \oint_{C} (\hat{t} \cdot \underline{v}) \, dl} \qquad (10.45)$$

The definition of circulation allows us to quantify how the strength of an added free vortex is related to lift in the superposed models discussed here. In the early 20th century, Kutta and Joukowski separately studied the theory behind the production of lift in superposed flow solutions [71, 77]. The addition of the free-vortex solution to the initially irrotational flow past an obstacle introduced rotational character into the flow field. Kutta and Joukowski established that the observed lift on an object in superposed uniform and vortex solutions can be calculated readily from the circulation associated with the superposed flow:

$$\begin{array}{c}\text{Kutta–Joukowski theorem} \\ \text{for production of lift} \\ \text{on a two-dimensional object of any shape:} \end{array} \quad \boxed{\frac{\mathcal{F}_{\text{lift}}}{\text{(width)}} = -\rho v_{\infty} \Gamma} \qquad (10.46)$$

where $\mathcal{F}_{\text{lift}}$ is the lift, Γ is the circulation, ρ is the density of the fluid, and v_{∞} is the speed of the oncoming flow. Upward lift is associated with clockwise circulation (i.e., negative Γ). The Kutta–Joukowski result applies to inviscid flow for objects of any shape. For the wing model discussed here (i.e., free vortex plus flow past

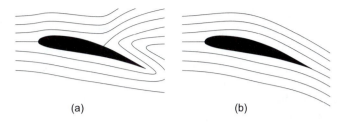

(a)   (b)

Figure 10.6 Calculated potential-flow streamline patterns for flow around an airfoil. In (a), no circulation is added to the model and unrealistic streamlines result. In (b), the correct amount of circulation is added by superposing a vortex flow; thus, the rear stagnation point is located at the trailing edge of the airfoil (i.e., Kutta condition).

a cylinder), the calculation of circulation using any curve C that encloses the cylinder results in the same value of circulation Γ.[1]

Proving the Kutta–Joukowski theorem is beyond the scope of this text [76, 136]. In Example 10.1, we verify that the Kutta–Joukowski theorem holds for flow around a cylinder with a superposed free vortex of strength K. Within the Kutta–Joukowski theorem is the genesis of a clever simplification to the airfoil-design problem. The Kutta–Joukowski result applies to inviscid-flow calculations. Real objects cannot be analyzed with inviscid-flow theories because they neglect the strong effect of viscosity in the boundary layer. In lift calculations on real airfoils, however, the main role of viscosity is to produce vorticity in the boundary layer and, hence, circulation on circuits C that are drawn around the airfoil. The clever idea is this: We can analyze real airfoils using inviscid-flow calculations if we insert into the inviscid model an amount of circulation that is equal to the circulation actually produced by the viscous boundary layer of the airfoil (Figure 10.6). In this strategy, we neglect the viscosity in the calculation of velocity and pressure from the momentum balance, but we return an important effect of the viscosity to our solution by imposing a particular value of circulation on the inviscid, irrotational flow field past the airfoil.

The circulation added to the flow in this method is the "vortex bound in the wing" referred to in the aerodynamics literature [114]. The strategy described here is similar to the boundary-layer strategy. In the study of boundary layers, we solved the potential-flow equations for the pressure field and then imposed that pressure field on the boundary layer to obtain the boundary-layer velocity profile, the drag on the surface, and the location where the boundary layer separates from the surface. For airfoil study, we solve the inviscid-flow problem around the airfoil with various values of circulation imposed (i.e., adding free vortices of various strengths K). The correct value of the circulation for the real airfoil is the value that produces flow-field streamlines that best match experimental results on the airfoil. Experiments show that the real streamlines are those in which the flows from the upper and lower surfaces meet smoothly at the trailing edge of the airfoil (Figure 10.6b). Another way to say this is that the rear stagnation point of the flow around the airfoil should be located at the sharp tip at the end of the airfoil. This requirement is known as the Kutta condition [76, 176].

---

[1] This is true because all of the vorticity in the free vortex is located at the center; the rest of the free-vortex flow (i.e., spiral-vortex tank) is irrotational [176].

In summary, the flow of air past an airplane wing is a complex flow in which both inertia and viscosity are important. The inertia dominates the flow but the viscous boundary layer has the important effect of producing vorticity at the wing surface. The vorticity produced by the no-slip boundary condition adds circulation to the flow, which can be modeled as the addition to the purely inertial flow field of a free vortex of the appropriate strength; this vortex has its core or center inside the airplane wing. With this clever strategy, we can use purely inertial solutions to the Navier-Stokes equations (i.e., the potential-flow solutions) to make meaningful calculations of lift in aeronautical flows. Pressure drag on an airfoil also may be calculated from these solutions (i.e., drag caused by asymmetric pressure profiles); viscous drag is zero in potential-flow solutions and must be calculated from a boundary-layer approach.

For more on lift and aeronautics, refer to the literature [11, 76].

EXAMPLE 10.1. *An airplane wing is modeled as a long cylinder placed in an oncoming stream. We assume that the flow may be represented by the potential-flow solution to flow around a cylinder with an added free vortex of strength K supplying circulation. The added free vortex accounts for the effect of viscosity in the boundary layer, which otherwise is neglected. What is the circulation in the superposed flow in terms of the vortex strength? What is the lift experienced by the wing? Is the Kutta–Joukowski theorem satisfied?*

SOLUTION. We are asked to model a cylindrical airplane wing using the velocity field obtained by adding the potential-flow velocity solution (Equation 10.39) to the free-vortex velocity solution (Equation 10.40). Thus, the velocity field for flow around the wing is given by:

$$
\begin{array}{c}
\text{Velocity field for} \\
\text{potential flow} \\
\text{around a long cylinder} \\
\text{with circulation:}
\end{array}
\quad
\underline{v} =
\begin{pmatrix}
v_\infty \cos\theta \left(1 - \frac{R^2}{r^2}\right) \\
-v_\infty \sin\theta \left(1 + \frac{R^2}{r^2}\right) + \frac{K}{r} \\
0
\end{pmatrix}_{r\theta z}
\quad (10.47)
$$

For $K = 0$ (i.e., no circulation), the streamlines for this flow are given in Figure 10.7 (left). When the free-vortex contribution is added, the streamlines become those shown in Figure 10.7 (right).

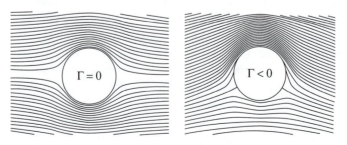

The streamlines for flow around a cylinder somewhat resemble the streamlines of flow around a sphere. When circulation is added to the solution by superposing a vortex flow, the streamlines compress on one side of the cylinder. The direction of lift is toward the compressed streamlines [178].

To calculate the circulation in our model flow, we begin with the definition of circulation, Equation 10.45:

$$\text{Circulation on } C: \quad \Gamma \equiv \oint_C (\hat{t} \cdot \underline{v})\, dl \tag{10.48}$$

Because all curves C that surround the cylinder center result in the same value of circulation Γ, we choose the simplest curve for our calculation: the circumference of the cylinder; thus, $dl = R\, d\theta$ and θ goes from 0 to 2π. The tangential unit vector $\underline{t}$ is equal to $\hat{e}_\theta$:

$$\begin{matrix}\text{Circulation} \\ \text{around the wing:}\end{matrix} \quad \Gamma = \oint_C (\hat{t} \cdot \underline{v})\, dl \tag{10.49}$$

$$= \int_0^{2\pi} (\hat{e}_\theta \cdot \underline{v})|_{r=R}\, R\, d\theta \tag{10.50}$$

$$= \int_0^{2\pi} (-2R v_\infty \sin\theta + K)\, d\theta \tag{10.51}$$

$$= (2R v_\infty \cos\theta + K\theta)\Big|_0^{2\pi} = 2\pi K \tag{10.52}$$

$$\boxed{\Gamma = 2\pi K} \tag{10.53}$$

The lift is calculated as the upward component of fluid force on the wing. Choosing the flow to be in the x-direction and the lift to be in the y-direction, we write:

$$\begin{matrix}\text{Force on an obstacle} \\ \text{in a free stream:}\end{matrix} \quad \underline{\mathcal{F}} = \iint_S [\hat{n} \cdot \underline{\tilde{\Pi}}]_{\text{at surface}}\, dS \tag{10.54}$$

$$= \begin{pmatrix} \mathcal{F}_{\text{drag}} \\ \mathcal{F}_{\text{lift}} \\ 0 \end{pmatrix}_{xyz} \tag{10.55}$$

The normal vector on the cylinder surface is $\hat{n} = \hat{e}_r$ and, because viscosity has been neglected, $\underline{\tilde{\tau}} = \mu \left(\nabla \underline{v} + (\nabla \underline{v})^T \right) = 0$:

$$\underline{\mathcal{F}} = \iint_S \left[\hat{n} \cdot (-p\underline{I} + \underline{\tilde{\tau}}) \right]_{\text{at surface}}\, dS \tag{10.56}$$

$$= \int_0^L \int_0^{2\pi} \left[\hat{e}_r \cdot -p\underline{I} \right]_{r=R}\, R\, d\theta\, dz \tag{10.57}$$

$$= RL \int_0^{2\pi} (-p\hat{e}_r)|_{r=R}\, d\theta = RL \int_0^{2\pi} \begin{pmatrix} -pR \\ 0 \\ 0 \end{pmatrix}_{r\theta z}\, d\theta \tag{10.58}$$

$$= RL \int_0^\pi p_R \left[\cos\theta \hat{e}_x + \sin\theta \hat{e}_y \right] d\theta = RL \int_0^{2\pi} \begin{pmatrix} -p_R \cos\theta \\ -p_R \sin\theta \\ 0 \end{pmatrix}_{xyz}\, d\theta \tag{10.59}$$

where $p_R(\theta)$ is the pressure at the surface of the cylinder. To calculate $p_R(\theta)$, we write the Bernoulli equation along the central streamline. Recall that the Bernoulli equation applies along streamlines in potential flow. Along the central streamline but far upstream of the cylinder, the pressure is p_∞ and the fluid speed is v_∞. On the same streamline but at $r = R$ and various angles θ, we write the pressure as $p_R(\theta)$. The velocity is given by the expression in Equation 10.47 with $r = R$, as given here:

$$\underline{v}|_{r=R} = \begin{pmatrix} 0 \\ -2v_\infty \sin\theta + \frac{K}{R} \\ 0 \end{pmatrix}_{r\theta z} \tag{10.60}$$

$$v_R = \left| \underline{v}|_{r=R} \right| = -2v_\infty \sin\theta + \frac{K}{R} \tag{10.61}$$

Writing the Bernoulli equation (i.e., integrated momentum balance along a streamline[2]) at the points indicated and neglecting the effect of gravity on the pressure field, we obtain:

Bernoulli equation:
$$0 = \frac{p_2 - p_1}{\rho} + \frac{v_2^2 - v_1^2}{2} + g(z_2 - z_1) \tag{10.62}$$

$$0 = \frac{p_R - p_\infty}{\rho} + \frac{v_R^2 - v_\infty^2}{2} \tag{10.63}$$

$$p_R = p_\infty + \frac{\rho}{2}\left(v_\infty^2 - \left(-2v_\infty \sin\theta + \frac{K}{R}\right)^2\right) \tag{10.64}$$

$$= p_\infty + \frac{\rho}{2}\left(v_\infty^2 - 4v_\infty^2 \sin^2\theta + \frac{4K}{R}v_\infty \sin\theta - \frac{K^2}{R^2}\right) \tag{10.65}$$

The lift now may be calculated from the y-component of Equation 10.59:

$$\mathcal{F}_{\text{lift}} = RL \int_0^{2\pi} -p_R \sin\theta\, d\theta \tag{10.66}$$

$$\mathcal{F}_{\text{lift}} = RL \int_0^{2\pi} \left[-p_\infty - \frac{\rho}{2}\left(v_\infty^2 - 4v_\infty^2 \sin^2\theta + \frac{4K}{R}v_\infty \sin\theta - \frac{K^2}{R^2}\right) \right] \sin\theta\, d\theta \tag{10.67}$$

$$= RL\left(-p_\infty - \frac{\rho}{2}v_\infty^2 + \frac{\rho K^2}{2R^2}\right) \int_0^{2\pi} \sin\theta\, d\theta + \left(2RL\rho v_\infty^2\right) \int_0^{2\pi} \sin^3\theta\, d\theta$$
$$- (2\rho L K v_\infty) \int_0^{\pi} \sin^2\theta\, d\theta \tag{10.68}$$

Evaluating the integrals, we obtain zero for the first two and π for the third, yielding the final result for lift:

$$\mathcal{F}_{\text{lift}} = -2\pi\rho K L v_\infty \tag{10.69}$$

---

[2] We must follow a streamline because the flow is not irrotational; see Example 8.13.

We calculated that the circulation $\Gamma = 2\pi K$ (see Equation 10.53); therefore, the lift per unit length of cylinder becomes:

$$
\begin{array}{c}
\text{Lift per unit length} \\
\text{on a transverse cylinder} \\
\text{in a uniform flow:}
\end{array}
\quad
\boxed{\frac{\mathcal{F}_{\text{lift}}}{L} = -\rho v_{\infty}\Gamma}
\qquad (10.70)
$$

This is the result predicted by the Kutta–Joukowski theorem for lift on a long obstacle of arbitrary cross section.

10.5 Flows with curved streamlines

In Chapter 2, we discuss the unconventional behavior of flows with curved streamlines, including smoke rings, the violent flow of a tornado, and the twisty flow of rivers and streams. Because of the curvature of the streamlines, there is acceleration in these flows, which means that inertia is important and the lefthand side of the Navier-Stokes equation does not go to zero (as it does in unidirectional flow). A possible strategy in analyzing such flows is to simplify the momentum balance by making appropriate assumptions. A model simplification investigated in Chapter 8 neglected viscous effects and therefore droped the $\mu\nabla^2\underline{v}$-term from the righthand side of the Navier-Stokes equation. As we learned by studying boundary layers, however, this approximation will not be satisfactory when the flow is near a surface because viscosity is important in the shear flow near surfaces. When we attempt to apply the Navier-Stokes equation to curly, curving flows, we face the same problem encountered with flow around a sphere (see Example 8.16): We are unable to simplify the equations enough to allow us to solve them analytically.

As a result of these difficulties, flows with curved streamlines usually must be modeled with numerical methods. If we confine our studies to two-dimensional flows, numerical simulators [5, 27] often can find the velocity and stress fields. For three-dimensional problems, numerical solutions are more complex, although still possible. Computer-based problem solving for both two- and three-dimensional flow is time-consuming and requires learning numerical solution techniques and strategies (see Section 10.2). For both two- and three-dimensional flows, it often is better first to think intuitively about the flow and to reason out how the fluid is likely to behave. With that in mind, we discuss in this section general ways in which the momentum balance can help us to understand the behavior of two- and three-dimensional curvy flows.

For two-dimensional flows, basic information about rapid, curved flows may be deduced from applying the momentum balance normal to the streamlines. Recall that the momentum balance along a streamline in potential flow (see Example 8.13) results in the Bernoulli equation, which relates flow-direction pressure variations and velocity. As we see in the following example, if we consider the component of the momentum balance that is perpendicular to the streamlines, we can state something about the cross-stream pressure distribution in two-dimensional potential flows with curved streamlines.

EXAMPLE 10.2. *Example 8.13 shows that in potential flow, the pressure and velocity variations along a streamline are related by the Bernoulli equation. How does pressure vary across streamlines for two-dimensional potential flow?*

SOLUTION. We begin in the same way as in Example 8.13. For steady ($\partial \underline{v}/\partial t = 0$), inviscid ($\mu = 0$) flow, the Navier-Stokes equation simplifies to:

$$\text{Navier-Stokes equation:} \quad \rho\left(\frac{\partial \underline{v}}{\partial t} + \underline{v} \cdot \nabla \underline{v}\right) = -\nabla p + \mu \nabla^2 \underline{v} + \rho\underline{g} \quad (10.71)$$

$$\rho\underline{v} \cdot \nabla \underline{v} = -\nabla p + \rho\underline{g} \quad (10.72)$$

We can write the Navier-Stokes equation in any coordinate system. To address our problem, we choose a spatially varying coordinate system that always has one direction pointing in the (varying) direction of flow. At any point, the flow direction is $\hat{v} = \underline{v}/v$. Previously, when we used a streamwise coordinate system (see Example 8.13), we defined the other two directions of our coordinate system, $\hat{u}$ and $\hat{w}$, to be mutually perpendicular and perpendicular to $\hat{v}$, an orthonormal coordinate system. This choice simplified the evaluation of the vuw-components of the Navier-Stokes equation. In the current problem, we are investigating pressure distributions normal to the streamlines. In this case, there is an advantage in choosing the second coordinate direction to be along $\hat{t}$, a spatially varying unit vector pointing toward the local center of streamline curvature (Figure 10.8). Because we are considering two-dimensional flow, the third coordinate direction, $\hat{z} = \hat{e}_z$, is chosen to be perpendicular to the horizontal plane of the flow. This is a spatially varying, orthonormal coordinate system.

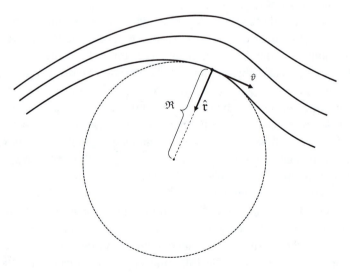

Figure 10.8 We choose our normal coordinate direction to be toward the local center of curvature. This allows us to express fluid acceleration in terms of the radius of curvature $\Re$.

In the $v\tau z$-coordinate system, the velocity vector is given by:

$$\underline{v} = \begin{pmatrix} v_v \\ v_\tau \\ v_z \end{pmatrix}_{v\tau z} = \begin{pmatrix} v \\ 0 \\ 0 \end{pmatrix}_{v\tau z} \tag{10.73}$$

We previously used the v-component of the Navier-Stokes equation to show that the Bernoulli equation holds along a streamline in potential flow. We now want to consider the τ-component of the equation of motion to deduce the pressure variation normal to the streamlines:

$$\text{Navier-Stokes equation:} \qquad \rho \underline{v} \cdot \nabla \underline{v} = -\nabla p + \rho \underline{g} \tag{10.74}$$
$$\text{(steady, inviscid)}$$

$$\begin{array}{c} \tau\text{-component:} \\ \text{(toward local center of curvature)} \end{array} \qquad \rho \left[\underline{v} \cdot \nabla \underline{v} \right]_\tau = -\frac{\partial p}{\partial x_\tau} + \rho g_\tau \tag{10.75}$$

Because of our choice of the vertical vector $\hat{e}_z$ as one of the basis vectors of our coordinate system, we know that $\underline{g} = -g\hat{e}_z$; thus, $g_\tau = 0$. Using the appropriate geometric arguments to relate the local fluid acceleration to $\mathfrak{R}$, the local radius of curvature ([111], page 167), we can write:

$$\left[\underline{v} \cdot \nabla \underline{v} \right]_\tau = \frac{v^2}{\mathfrak{R}} \tag{10.76}$$

$$\tau\text{-component Navier-Stokes:} \qquad \rho \left[\underline{v} \cdot \nabla \underline{v} \right]_\tau = -\frac{\partial p}{\partial x_\tau} + \rho g_\tau \tag{10.77}$$

$$\rho \left(\frac{v^2}{\mathfrak{R}} \right) = -\frac{\partial p}{\partial x_\tau} \tag{10.78}$$

$$\begin{array}{c} \text{Cross-stream pressure gradient} \\ \text{(steady, incompressible,} \\ \text{inviscid, } \tau\text{-component):} \end{array} \qquad \boxed{\frac{\partial p}{\partial x_\tau} = -\rho \left(\frac{v^2}{\mathfrak{R}} \right)} \tag{10.79}$$

The form of $[\underline{v} \cdot \nabla \underline{v}]_\tau$ is the familiar expression for centrifugal acceleration [167]. The result in Equation 10.79 shows that the pressure in a rapid, inviscid flow decreases in the direction of the local center of streamline curvature [123]. We can use this information to interpret, for example, the secondary flow that occurs along the bottom of a riverbed. In that flow, when the water goes around a curve, Equation 10.79 indicates that the pressure is higher along the outside of the curve than on the inside. For the main part of the channel where the flow may be modeled as inviscid (i.e., away from the walls), this pressure distribution exists. Along the bottom of the channel, however, the fluid has been slowed in the boundary layer. This slower fluid cannot maintain the pressure gradient of the fully inertial flow; instead, it is pushed by the pressure gradient toward the center of curvature. We see evidence of this in the motion of tea leaves in a stirred mug (see Figure 2.44). The cross-stream secondary flow in a river tends to scour material from the outer walls of the riverbed and deposit that material on the inner walls; this action causes slightly curvy streams to take on exaggerated curves and bends.

For an excellent discussion of the effect of streamline curvature, we recommend the NCFMF film on pressure and acceleration [118], which illustrates the effects of inertia in flows with curved streamlines. The experiments shown in the film are designed to minimize viscous effects.

Example 10.2 demonstrated that for two-dimensional curved flows, the cross-stream momentum balance simplifies to a useful expression that relates the flow field to the pressure field. The momentum balance is more difficult to apply to three-dimensional flow, however, because there are uniquely three-dimensional effects to consider. Tornados, wing-tip vortices on airplanes, and the details of turbulent-flow structure are examples of flows in which three-dimensional effects are essential to a fluid's behavior. Three-dimensional-curvature effects are discernible in the momentum balance through vorticity, as we discuss briefly here.

We can develop insight into the behavior of three-dimensional rotational flows by considering the physics of vorticity production in a flow. The mechanisms of vorticity production were studied in the late 19th century, and a major contribution from that period is *Kelvin's circulation theorem* [76, 114], which evolves from the momentum balance. Kelvin's theorem identifies three sources of vorticity generation: torques due to pressure; torques due to viscosity; and torques due to nonconservative body forces such as the Coriolis force, a pseudo force that appears in rotating reference frames [9, 161].

Kelvin's circulation theorem (circulation on a closed material curve $\mathfrak{C}$):

$$\frac{\partial \Gamma_\mathfrak{C}}{\partial t} + \underline{v} \cdot \nabla \Gamma_\mathfrak{C} = -\oint_\mathfrak{C} \frac{dp}{\rho} + \oint_\mathfrak{C} \hat{t} \cdot \left(\nu \nabla^2 \underline{v}\right) \, dl + \oint_\mathfrak{C} \hat{t} \cdot \underline{G} \, dl \qquad (10.80)$$

$$\frac{D\Gamma_\mathfrak{C}}{Dt} = -\oint_\mathfrak{C} \frac{dp}{\rho} + \oint_\mathfrak{C} \hat{t} \cdot \left(\nu \nabla^2 \underline{v}\right) \, dl + \oint_\mathfrak{C} \hat{t} \cdot \underline{G} \, dl \qquad (10.81)$$

The circulation in Kelvin's theorem is calculated around $\mathfrak{C}$, a closed curve in the fluid that always is composed of the same fluid particles; $\underline{G}$ is a nonconservative body force. We do not discuss Kelvin's theorem in detail here [76] except to examine the common case in which the pressure is a function only of density (thus, the pressure term integrates to zero), viscosity is negligible (potential flow), and the conservative force gravity is the only body force present ($\underline{G} = 0$).

Under these conditions, Kelvin's theorem states that the circulation around a material curve is a constant:

Kelvin's circulation theorem (barotropic, inviscid, conservative body forces):

$$\frac{\partial \Gamma_\mathfrak{C}}{\partial t} + \underline{v} \cdot \nabla \Gamma_\mathfrak{C} = \frac{D\Gamma_\mathfrak{C}}{Dt} = 0 \qquad (10.82)$$

The constancy of the circulation on a material curve explains some fluid behavior seen in tornados and in turbulent flow, for example. We explore this effect in the following example.

EXAMPLE 10.3. *A vortex tube is a region of a flow bounded by vortex lines, lines that are everywhere tangent to the vorticity. Consider a volume consisting of a portion of a vortex tube in the shape of a cylinder of height l_1 and radius R_1. If the flow stretches this volume to a length l_2, what is the effect of this stretch on the vorticity in the tube? We are interested in the barotropic, inviscid, conservative-body-force case.*

SOLUTION. The volumes under consideration are shown in Figure 10.9. We first can relate the two volumes by imposing conservation of mass on the two states. The vortex tube always consists of the same material points; this is another consequence of Kelvin's theorem [76]:

$$\text{Mass conservation:} \quad \rho_1 \left(l_1 \pi R_1^2 \right) = \rho_2 \left(l_2 \pi R_2^2 \right) \tag{10.83}$$

$$R_1^2 = \frac{\rho_2 l_2 R_2^2}{l_1 \rho_1} \tag{10.84}$$

where ρ_1 and ρ_2 are the densities of the fluid in the two states and the geometrical parameters are defined in Figure 10.9.

From Kelvin's theorem (i.e., barotropic, inviscid, and conservative body forces), the circulation is constant throughout the motion:

$$\begin{array}{c} \text{Kelvin's circulation theorem} \\ \text{(barotropic, inviscid,} \qquad \Gamma_1 = \Gamma_2 \\ \text{and conservative body forces):} \end{array} \tag{10.85}$$

From the discussion in Section 10.4 (see Equation 10.44), we now can write the circulations in the two states, Γ_1 and Γ_2, in terms of the vorticity flux. The top surfaces S_1 and S_2 are vortex-tube cross-sectional areas associated with the same material points for the two states, 1 and 2. The net vorticity fluxes through S_1 and

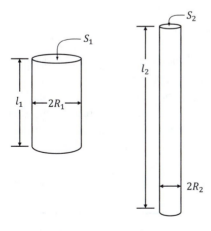

Figure 10.9   We consider a portion of a vortex tube that is the shape of a cylinder of proscribed length and radius. The motion of the fluid stretches the cylinder, and the dynamical laws of fluid mechanics indicate how this stretching affects the vorticity in the tube.

S_2 are equal to Γ_1 and Γ_2, as follows:

$$\text{Net vorticity flux through } S: \qquad \iint_S (\hat{n} \cdot \underline{\omega})\, dS = \oint_C (\hat{t} \cdot \underline{v})\, dl = \Gamma \qquad (10.86)$$

Applying this to our two states and substituting into Kelvin's theorem, we obtain:

$$\Gamma_1 = \Gamma_2 \qquad (10.87)$$

$$\iint_{S_1} (\hat{n} \cdot \underline{\omega})|_1\, dS = \iint_{S_2} (\hat{n} \cdot \underline{\omega})|_2\, dS \qquad (10.88)$$

$$\int_0^{2\pi} \int_0^{R_1} (\hat{e}_z \cdot \omega_1 \hat{e}_z) r\, dr\, d\theta = \int_0^{2\pi} \int_0^{R_2} (\hat{e}_z \cdot \omega_2 \hat{e}_z) r\, dr\, d\theta \qquad (10.89)$$

$$\omega_1 2\pi R_1^2 = \omega_2 2\pi R_2^2 \qquad (10.90)$$

Combining this result with the mass balance, yields:

$$\omega_1 R_1^2 = \omega_2 R_2^2 \qquad (10.91)$$

$$\omega_1 \frac{\rho_2 l_2 R_2^2}{l_1 \rho_1} = \omega_2 R_2^2 \qquad (10.92)$$

$$\boxed{\frac{\omega_1}{l_1 \rho_1} = \frac{\omega_2}{l_2 \rho_2}} \qquad (10.93)$$

Because the choice of which material curve to follow is arbitrary, the result applies to all material curves bounding the vortex tube:

$$\text{On a vortex line (barotropic, inviscid, and conservative body forces):} \qquad \boxed{\frac{\omega}{l\rho} = \text{constant}} \qquad (10.94)$$

Equation 10.94 and Kelvin's theorem help us to understand three-dimensional flow effects such as the creation and intensification of tornados. If an updraft of air occurs and if Coriolis forces in the rotating frame of reference of Earth initiate a circulatory motion, a vortex core is produced and lengthened by the continuing updraft. The lengthening of the vortex core increases the vorticity in the core according to Equation 10.94 because the value of the ratio $\omega/l\rho$ must be preserved. The intensification of the vorticity by this three-dimensional mechanism causes the flow speed to increase, creating the damaging storms that occur. The flow speeds in tornados can be so rapid that extremely low pressures are produced in the core. The Bernoulli equation applies along the circular streamlines in this inviscid flow; thus, high speed is accompanied by low pressure. The extreme low pressure at the core of a tornado pulls off roofs and can cause buildings to explode.

Vorticity, circulation, and vortex mapping all are tools that can help us understand the highly curly flows that dominate meteorology, magnetohydrodynamics, and turbulent flow dynamics. For more information on applications of Kelvin's circulation theorem and the related Helmholtz vortex theorems, see the literature [11, 76].

10.6 Compressible flow and supersonic flow

We consider only incompressible fluids thus far (i.e., constant density). Because water, oil, and most liquids are approximately incompressible, this is a reasonable starting point for the study of fluid mechanics. Gases may be modeled as incompressible fluids in both microscopic and macroscopic calculations, as long as the pressure changes are less than about 20 percent of the mean pressure [43, 55]; this is true for friction-factor/Reynolds-number correlations, for example [43, 132].

There are circumstances, however, in which fluid compressibility must be considered, such as in gas flows with high pressure drops, for example. For both compressible and incompressible flows, density variations are captured by the continuity equation. The Newtonian constitutive equation does not, however, take compressibility into account, and it must be modified for use in high-pressure-drop flows.

For incompressible Newtonian fluids, we discuss in Chapter 5 that the extra-stress tensor $\underset{=}{\tilde{\tau}}$ is given by a simple expression that works for all flows:

$$\underset{=}{\tilde{\tau}} = \mu \underset{=}{\dot{\gamma}}$$

Newtonian constitutive equation for incompressible fluids   (10.95)

For compressible Newtonian fluids, the relationship between stress and deformation is found to be [12]:

$$\underset{=}{\tilde{\tau}} = \mu \underset{=}{\dot{\gamma}} + \left(\frac{2}{3}\mu - \kappa \right) (\nabla \cdot \underline{v}) \underset{=}{I}$$

Newtonian constitutive equation for compressible fluids   (10.96)

The parameter κ is the dilational or bulk viscosity, a coefficient that expresses viscous momentum transport that occurs when density changes; κ is zero for ideal, monatomic gases [12, 43]. The bulk viscosity is only important when very large expansions take place, and it can usually be neglected [43]. Note that for incompressible fluids, $\nabla \cdot \underline{v} = 0$ and Equation 10.96 reduces to Equation 10.95. Problem solving with Equation 10.96 follows the same methods as described in this text. For more on modeling of high-pressure-gradient gas flows, see the literature [3, 124].

There is an important group of flows for which compressibility changes the physics—high-speed gas flows, including sonic flows. These flows occur in space flight, through nozzles, in turbines, and in relief valves (Figure 10.10). Relief valves exhibit a condition called choked flow that limits the flow rate at which a gas can escape. Properly understanding choked flow in a relief valve can be a life-and-death safety issue. In high-speed gas flows, the physics changes because the flow moves faster than does the information about the pressure field. In this section we discuss information travel in flows and its effects on high-speed gas flows.

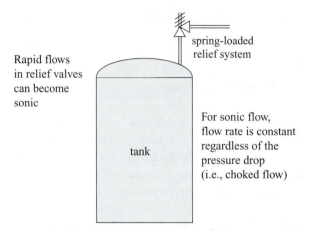

Rapid flows
in relief valves
can become
sonic

spring-loaded
relief system

For sonic flow,
flow rate is constant
regardless of the
pressure drop
(i.e., choked flow)

tank

Figure 10.10   When an explosion or runaway reaction occurs in a storage tank, the pressure is designed to be relieved by flow through valves known as pressure-relief valves. Because the pressures are high in the tank, the flow through the relief valves may be supersonic and may therefore become choked. Lack of understanding of supersonic flow can lead to the tragic underdesign of relief systems.

All fluids are compressible—liquids and gases both change in density if sufficient pressure is applied. The distinction between compressible- and incompressible-fluid modeling is a question of whether the fluid compressibility needs to be considered in a given modeling situation.

For water and other liquids, when pressure is applied at one location, the information about that applied pressure appears to travel instantaneously to other parts of the flow. Consider the flow of water in a straw (Figure 10.11). When a person applies suction to one end of a straw submerged in water, the water moves instantaneously throughout straw. Likewise, in a long, completely filled piping system, if a pump is turned on at one end, an incompressible fluid immediately begins to flow out of the other end of the pipe. In a hydraulic lift (see Figure 4.37), pushing down on fluid on one side of the apparatus immediately causes the fluid

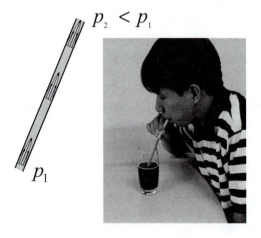

$$p_2 < p_1$$

$$p_1$$

Figure 10.11   Drinking water from a straw works because a lower pressure is created in the mouth, and this imposes a pressure gradient on the liquid in the straw. The liquid flows under the pressure gradient.

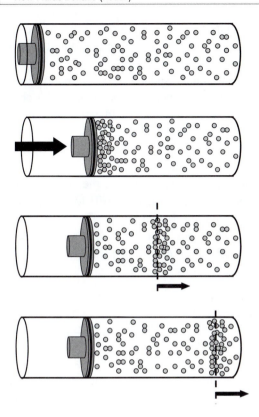

Figure 10.12 In a compressible fluid, pressure applied at one end is not transmitted instantaneously to the rest of the fluid. The pressure pulse takes a finite amount of time to travel and affects the density of the fluid along the way.

to push up on the other side. No significant density changes take place in liquids under these circumstances, and they may be modeled as incompressible flows.

In gases at modest flow speeds, the situation is the same. If we draw air into a straw, the air moves seemingly instantaneously and may be modeled as an incompressible fluid. Likewise, for air flowing in pipes and ducts at modest speeds there is no significant density change, and incompressible-flow equations apply.

The speed of pressure waves is not actually instantaneous in either liquids or gases, however. For both types of fluid the imposition of a force at one end of a system actually does not result in an immediate motion throughout the system. Instead, the fluid near where the force was applied compresses (Figure 10.12); that is, its density increases locally in response to the force. The locally compressed fluid expands against neighboring fluid, causing the neighboring fluid to compress and setting in motion a wave pulse that travels through the system. The pulse of higher-density fluid takes time to travel from the source of the disturbance through the pipe to the far end of the system. Sound travels through fluid with this same mechanism: Sound is matter's response to a compression wave. The speed of a compression wave through a medium is called the speed of sound.

The speed of sound is not important in fluid mechanics, except when the speed of the fluid begins to approach the speed of sound. When this happens, the information about the flow travels at a speed comparable to the fluid speed, and things work a bit differently. The physics changes, because the speed of sound is the speed of information travel in a fluid. Once a fluid approaches the speed of information travel, the fluid cannot respond to the information about the pressure in the same way as it does at slower speeds. To account for this change of physics in rapidly flowing systems, we must include fluid compressibility in our models, as we now discuss.

For rapid gas flows, we can account for compressibility with the mechanical energy balance. In calculations with the MEB, we must make modifications to allow it to apply to compressible fluids because the derivation of the MEB in Chapter 9 assumes incompressible fluid. The MEB for compressible fluids can be obtained by writing it on a differential length of straight pipe and assuming ideal gas ($pV = nRT$). The details are in the literature [55] (See Problem 19). The final result for the MEB in isothermal compressible flow is given here:

Mechanical energy balance (isothermal, compressible flow):

$$p_2 - p_1 = \frac{4fLG^2}{2D\rho_{ave}} + \frac{G^2}{\rho_{ave}} \ln \left(\frac{p_1}{p_2} \right)$$

(10.97)

where L is the pipe length and $G = \langle v \rangle \rho$. Note that because $\langle v \rangle \rho A$ is the mass flow rate (A is the pipe cross-sectional area), $G = \langle v \rangle \rho$ is constant for a pipe of constant cross section.

A striking aspect of the compressible-fluid mechanical energy balance result is that it predicts a maximum velocity at high pressure drops. This is choked flow. Taking the derivative of Equation 10.97 and setting it to zero, we calculate that:

Condition during choked flow: (compressible fluids)
$$v_{max} = \sqrt{\frac{p_2}{\rho_2}}$$
(10.98)

The reason for choked-flow behavior in compressible fluids can be traced to the previously discussed limiting speed of information travel in a medium. In flows at modest speeds, the instantaneous transmission of forces through the fluid (compared to the speed of the fluid) allows flow around an obstacle, for example, to rearrange to allow for the smooth passage of flow around the obstacle (Figure 10.13). In incompressible flows, forces are applied and the fluid, moving much slower than the information, responds directly to the forces by moving, displacing neighboring particles, and thereby establishing the appropriate flow field. The fact of the presence of an obstacle in a flow is transmitted upstream through pressure; the fluid pattern reflects the presence of the obstacle even though the fluid particles have not arrived yet at the obstacle. Because information transmission happens very rapidly compared to the time-scale of the flow, incompressible flows easily transmit stress information throughout the flow field. Information travel is not instantaneous for compressible fluids; hence, choked flow is observed.

In Examples 10.4 and 10.5 we relate the choked-flow speed in compressible flow predicted by the MEB (Equation 10.98) to the speed of sound in the medium.

Information about
the presence of the
obstacle travels upstream

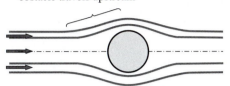

Flow in an obstructed channel

Figure 10.13

In most flows, pressure information travels very rapidly compared to flow speeds. Thus, information about the presence of an obstacle downstream can transmit upstream by way of rapid pressure waves, and the flow rearranges quickly. In supersonic flows—in which the speed of propagation of the pressure wave is close to or slower than the flow speed—this type of rearrangement is not possible and the flow develops differently.

EXAMPLE 10.4. *The compressible-fluid mechanical energy balance (Equation 10.97) predicts that there is a maximum speed for flows for which it applies (single-input, single-output, steady, no reaction, no phase change, little temperature change, compressible fluid). How is the predicted maximum flow speed related to the speed of sound in the fluid? You may assume that the sound travels under isothermal conditions.*

SOLUTION. If the pressure variation is not too large, the speed of sound in a medium is given by [167]:

$$v = \sqrt{\frac{B}{\rho_0}} \tag{10.99}$$

where ρ_0 is density and B is the bulk modulus, defined as the ratio of the change in pressure to the fractional decrease in volume:

$$B = \frac{\Delta p}{-\Delta V / V} \tag{10.100}$$

The bulk modulus relates the change in volume to changes in pressure. This information is found in the equation of state for a material. For example, for an ideal gas, pressure and volume are related by the ideal gas law:

$$\text{Ideal gas law}: \quad pV = nRT \tag{10.101}$$

If we differentiate the ideal gas law and assume that temperature is constant, we obtain:

$$p\,dV + V\,dp = 0 \tag{10.102}$$

which may be rearranged to give the bulk modulus for an ideal gas under isothermal conditions:

$$\boxed{B = \frac{dp}{-dV/V} = p} \quad \begin{array}{l}\text{Bulk modulus} \\ \text{of an ideal gas} \\ \text{at constant T}\end{array} \tag{10.103}$$

The speed of sound, therefore, is given by:

$$v = \sqrt{\frac{B}{\rho_o}} \tag{10.104}$$

$$= \sqrt{\frac{p}{\rho_o}} = \sqrt{\frac{RT}{M}} \tag{10.105}$$

where we substitute for p the expression given by the ideal gas law $p = nRT/V = \rho_o RT/M$, and M is the molecular weight of the medium. Comparing to the maximum velocity calculated in Equation 10.98 for nonisothermal compressible flow, we see that the maximum fluid velocity in single-input, single output-flow is the speed of sound in the medium, Equation 10.99.

EXAMPLE 10.5. *The compressible-fluid mechanical energy balance (Equation 10.97) predicts that there is a maximum speed for flows for which it applies (single-input, single-output, steady, no reaction, no phase change, little temperature change, compressible fluid). How is the predicted maximum flow speed related to the speed of sound in the fluid? You may assume that the sound travels under adiabatic conditions.*

SOLUTION. The expression for the speed of sound in Equation 10.105 is found to be 20 percent too small when compared with experimental results, as assumed [167] because the passage of sound through a medium does not occur isothermally. The compressions and decompressions that take place tend to change locally the temperature of the gas (consider the relationship between volume and temperature in the ideal gas law) and, because these temperature changes occur rapidly, there is no time for much heat transfer to take place. Rather than assume isothermal passage, it is better to assume that the movement of sound is adiabatic—that is, no heat transfer occurs.

Application of the first law of thermodynamics under quasistatic, adiabatic conditions results in an expression that relates pressure and volume when a gas undergoes volume changes under adiabatic conditions (see the Web appendix [108]):

$$pV^\gamma = \text{constant} \tag{10.106}$$

where $\gamma \equiv C_p/C_v$, the ratio of the heat capacity at constant pressure to the heat capacity at constant volume. We can derive B for the quasistatic, adiabatic case by taking the derivative of Equation 10.106 with respect to pressure and rearranging:

$$\frac{d(pV^\gamma)}{dp} = 0 \tag{10.107}$$

$$p\gamma V^{\gamma-1}\frac{dV}{dp} + V^\gamma = 0 \tag{10.108}$$

$$p\gamma dV + V dp = 0 \tag{10.109}$$

$$B = \frac{dp}{-dV/V} = \gamma p \tag{10.110}$$

Substituting this into Equation 10.99 for the speed of sound in terms of B yields a more accurate expression for the speed of sound in an ideal gas:

$$v = \sqrt{\frac{B}{\rho_o}} \tag{10.111}$$

$$= \sqrt{\frac{\gamma p}{\rho_o}} \tag{10.112}$$

$$\boxed{v = \sqrt{\frac{\gamma RT}{M}}} \quad \begin{array}{c} \text{Speed of sound} \\ \text{of an ideal gas} \\ \text{(adiabatic)} \end{array} \tag{10.113}$$

This result makes predictions that are close to experimental observations. Note that v_{max} for adiabatic flow is the adiabatic speed of sound [55].

The relief-valve problem occurs because there is a fixed, constant speed that pressure waves travel in matter—that is, the speed of sound in that material. The fixed speed of pressure waves also creates interesting phenomena when sonic or supersonic speeds are achieved by objects moving through matter. Objects moving through air push the air aside as they move forward. At subsonic speeds, the object pushing aside the air creates a pressure wave in the air that propagates forward of the object at a wave speed (i.e., the speed of sound) that greatly exceeds the speed of the object. When the speed of the object exceeds the speed of the pressure wave, however (i.e., supersonic flow), the object essentially outruns the pressure wave. This is how a shock wave is produced (Figure 10.14).

For more information on compressible flow, refer to the literature [3].

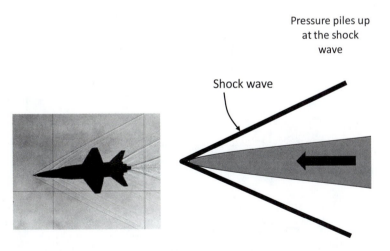

Pressure piles up
at the shock
wave

Shock wave

Figure 10.14 A shock wave is produced when an object moves faster than the pressure waves in the medium. In subsonic flows, the information about the object moves ahead of the object in the flow. In supersonic flows, the object outpaces the information about its motion. The information about the motion—the pressure—piles up in a characteristic pattern that fans out from the moving object (Photo courtesy NASA).

10.7 Summary

The modeling discussed in this text is the shared background of all flow modeling. From this starting point, we now can pursue more complex flow issues, from plastics processing to aeronautics. In this text we also focused on revealing a process of problem solving, in which a complex problem is approached by first turning to a solvable but necessarily simpler problem, followed by dimensional analysis, experimentation, and data correlation. If we look around in other fields of science and engineering, we find the same pattern of investigation as discussed here for fluid mechanics. This problem-solving methodology offers a place to start when we face a new and challenging problem.

10.8 Problems

1. Under what flow conditions are the Newtonian constitutive equation (Equation 10.2) and Newton's law of viscosity (Equation 10.1) equivalent?
2. Design a flow domain for numerical analysis on which to calculate the results from Hamel flow (see Figure 7.47). Indicate the boundary conditions on every boundary. Indicate regions where the mesh should be refined to improve accuracy.
3. Figure 10.3 is a proposed flow domain for the numerical evaluation of pressure-driven flow through a tube. Why has a section been added before the beginning of the test section?
4. A flow domain for the analysis of the steady flow through a two-dimensional contraction is shown in Figure 10.15. The flow rate is constant, the fluid is incompressible, and the fluid exits into air. What are the boundary conditions for this flow?

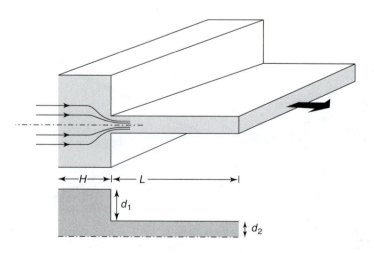

Figure 10.15 A flow domain for numerical evaluation of a flow (Problem 4).

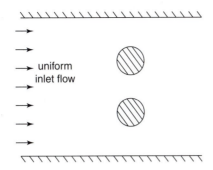

Figure 10.16 Flow field for Problem 5.

5. For the two-dimensional flow domain in Figure 10.16, indicate the regions where viscous effects are important and indicate the regions where inertial effects are expected to be important. The flow is rapid (i.e., Re is high) and the fluid is incompressible. The hashed regions are solid walls or obstacles.

6. For a pipe with laminar flow established, an increase in flow rate eventually causes the flow to become unstable and for turbulent flow to appear. Why is the flow unstable? Why is it that we cannot achieve arbitrarily high flow rates with laminar flow?

7. What are the $K-\epsilon$ models of turbulence? To answer this, research information in the literature.

8. What is the physical origin of the stresses described by the $\underset{\approx}{\tilde{\tau}}^{\text{turb}}$ term in the fluctuation-averaged Navier-Stokes equation for turbulent flow?

9. What is circulation?

10. Calculate the circulation around the center point of the free-vortex flow given in Equation 10.40.

11. Sketch the calculated flow field around an airfoil if too little and too much circulation are added. Sketch how the flow field around an airfoil should look if the correct amount of circulation (produced by the viscous flow in the boundary layer) is included in the model. What is the characteristic of the streamlines that shows that the correct amount of circulation has been added?

12. For the gradually contracting flow in Figure 10.17, where is the pressure highest? Indicate your answer on the a sketch of the flow. Explain.

2:1 contraction

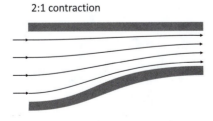

Figure 10.17 A flow with streamline curvature (Problem 12).

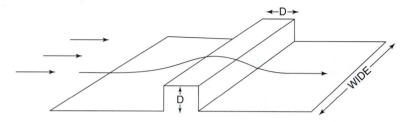

Figure 10.18   Flow over a bump (Problem 16).

13. Calculate the pressure distribution in a spiral-vortex tank. The velocity field is given in Equation 10.40.

14. Using a numerical simulator [5, 27], calculate the velocity field for Poiseuille flow in a tube. Assume axisymmetric steady flow of an incompressible Newtonian fluid. Compare your result for the velocity field with the analytical solution obtained in Chapter 7.

15. Using a numerical simulator [5, 27], calculate the velocity field for flow in a 2:1 axisymmetric contraction. Assume two-dimensional steady flow of an incompressible Newtonian fluid. Choose a steady flow rate in the creeping-flow regime. What is the highest Reynolds number you can simulate with the available code?

16. Using a numerical simulator [5, 27], calculate the velocity field for flow over the bump shown in Figure 10.18. Assume two-dimensional steady flow of an incompressible Newtonian fluid. Make your calculations in the creeping-flow regime.

17. Using a numerical simulator [5, 27], calculate the velocity field for axial unidirectional flow in the void space around three tubes mounted inside a cylindrical shell (Figure 10.19). Assume steady flow of an incompressible Newtonian fluid. Make your calculations in the creeping-flow regime. Review the mesh in Figure 10.2.

18. Using a numerical simulator [5, 27], calculate the drag coefficient as a function of the Reynolds number for flow around a sphere. The flow is solved most easily in axisymmetric mode with a flow domain resembling that in Figure 10.20. Compare your results with experimental data.

19. Derive the mechanical energy balance for isothermal, compressible flow, Equation 10.97 [55].

rod diameter = D
outer tube diameter =8D

Figure 10.19   Shell-and-tube heat exchangers have flow in the outer region similar to that shown here (Problem 17).

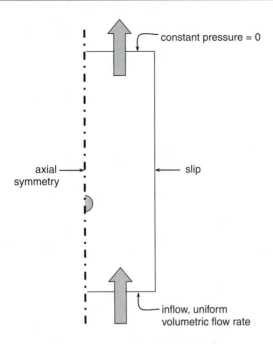

constant pressure = 0

axial — symmetry

slip

inflow, uniform
volumetric flow rate

Figure 10.20 The steady, noncreeping flow around a sphere may be calculated using finite-element analysis or other numerical techniques. A reasonable formulation of the problem is to use the flow domain and boundary conditions shown here. The flow domain is chosen to be longer after the sphere to capture the wake structure expected to form there. The mesh chosen for the calculation should be refined near the sphere (Problem 18).

20. What is the speed of sound in room-temperature air? What is the speed of sound in air at $-20°C$?

21. What causes a shock wave? Investigate in the literature and provide citations.

22. How do the Navier-Stokes equations change if we want to study magnetohydrodynamics?

23. How do the Navier-Stokes equations change if we want to study flows with a rotating reference frame?

24. What is a Taylor column? Investigate in the literature and provide citations.

25. What is the Coriolis force? Investigate in the literature and provide citations.

26. Describe a tornado in fluid-mechanics terms.

27. Explain a boundary layer in terms suitable for an elementary-school student.

28. Explain the use of dimensionless numbers in fluid mechanics.

29. Discuss the utility of dividing flows into internal and external flows.

30. What are the forces accounted for in the Navier-Stokes equation? Which forces dominate? Explain.

PART V

APPENDICES

Glossary

API gravity: This is the American Petroleum Institute standard for grading the densities of petroleum crude oils. The definition of API gravity is:

$$API\ gravity = \frac{141.5}{SG} - 131.5 \tag{A.1}$$

where $SG = \rho/\rho_{ref}$ is the specific gravity of the fluid, and ρ_{ref} is chosen as the density of water at the triple point (4°C). A petroleum with a specific gravity of 1.0 has an API gravity of:

$$API\ gravity = \frac{141.5}{1} - 131.5 = 10 \tag{A.2}$$

An API gravity of 10 is written as 10° API. Petroleum grades are defined as follows:

$$
\begin{array}{rl}
\text{light crude} & API > 31.1° \\
\text{medium oil} & 22.3° < API < 31.1° \\
\text{heavy oil} & API < 22.3° \\
\text{extra-heavy oil (bitumen)} & API < 10°
\end{array} \tag{A.3}
$$

body forces: Forces exerted on fluid particles from a distance—that is, without contact. These forces include gravity and electomagnetic forces [167]. These also are called field forces or noncontact forces.

bulk deformation: This is a deformation that results in a volume change; also called volume deformation.

calibration: This term refers to taking experimental data on a device and adjusting so that an expected relationship becomes more accurate. Some devices have internal calibrations in which a procedure is followed and the adjustment is made within the instrument (usually using software). In this text, we refer to calibration in which experimental data are collected on an instrument and the results are plotted to create a calibration curve that can be used to operate the device.

capacity: This term in the discussion of turbo machinery is synonymous with volumetric flow rate.

check valve: See **valves and fittings**.

coefficient of sliding friction: When two solids rub together, there is a frictional force that opposes the relative motion of the solids. The frictional force in solids is proportional to the normal force with which the solids are pressed together.

The constant of proportionality between the magnitudes of the frictional retarding force $\mathcal{F}$ and the normal force N is the coefficient of sliding friction μ_k:

$$\mathcal{F} = \mu_k N \qquad (A.4)$$

conservative force: A force is conservative if the work it does on a particle sums to zero over any path that returns the particle to the initial position [167]. Gravity is a conservative force (the work required to lift a mass to a certain elevation is independent of the path followed). Friction is not a conservative force (nonzero work is required to move a mass along a closed path).

constitutive equation: A constitutive equation in fluid mechanics refers to a stress constitutive equation, such as the Newtonian constitutive equation (see Equation 5.89). These equations give the relationship between the motion of the fluid ($\underline{v}$) and the nine components of stress generated by the motion ($\underline{\underline{\tau}}$). For non-Newtonian fluids, there is a wide variety of constitutive equations that capture the nonlinear and time-dependent effects not present in Newtonian flows (see Section 5.3 and [104]).

contact forces: Contact forces act on a surface of a system under consideration; contrast with noncontact (or body) forces.

control surface: This is the surface that encloses all or part of a control volume.

control volume (CV): This is an arbitrary volume in a flow used for performing balances such as on mass, momentum, and energy. In general, control volumes may move and change in shape, but the simplest case of a fixed, rigid control volume is quite useful in fluid mechanics (see Section 3.2.2).

convection: This is another term for flow.

convective terms: Mass can flow into and out of the control volume. The terms in mass, momentum, and energy balances that consider changes in control-volume mass, momentum, and energy due to this material flow are called the convective terms.

correlations: A data correlation is an experimentally determined relationship between two or more variables. For example, the friction factor and the Reynolds number are correlated, and we can measure the correlation by setting the Reynolds number, measuring the friction factor, and plotting the friction factor versus the Reynolds number. The resulting curve (often fit to a mathematical expression) is called a data correlation.

coupling: See **valves and fittings**.

data correlation: See **correlation**.

differential equations: See **partial differential equations**.

divergence of a vector, tensor: The divergence of a vector, $\underline{v}$, is $\nabla \cdot \underline{v}$; the divergence of a tensor, $\underline{\underline{A}}$, is $\nabla \cdot \underline{\underline{A}}$.

drag coefficient C_D: This is a dimensionless measure of drag defined as:

$$C_D \equiv \frac{\text{drag force}}{\frac{1}{2}\rho v^2 A} \qquad (A.5)$$

where ρ is the fluid density, v is the fluid velocity, and A is a characteristic area facing the flow.

dynamic pressure: The dynamic pressure, $\mathcal{P}$, also called equivalent pressure, is a function that combines the effects of flow pressure, p, and gravity, $\underline{g}$. The force due to gravity, $\rho\underline{g}$, is a conservative force; that is, the work done by the force due to gravity is independent of the path taken in the course of doing the work [167]. All conservative forces may be written as the negative of the gradient of a potential-energy function:

$$\rho\underline{g} = -\nabla\Phi$$

where Φ is potential energy due to gravity. Φ is equal to ρgh, where g is the gravitational force constant and h is the height of a particle of interest above a reference plane. Thus:

$$\rho\underline{g} = -\nabla(\rho gh)$$

and, for constant density:

$$\underline{g} = -g\nabla h$$

The dynamic pressure appears when the pressure and gravity terms of the Navier-Stokes equation are combined [43, 168]:

$$\rho\left(\frac{\partial\underline{v}}{\partial t} + \underline{v}\cdot\nabla\underline{v}\right) = -\nabla p + \mu\nabla^2\underline{v} + \rho\underline{g}$$

$$= -\nabla p + \mu\nabla^2\underline{v} - \nabla\Phi$$

$$= -\nabla(p + \Phi) + \mu\nabla^2\underline{v}$$

$$= -\nabla\mathcal{P} + \mu\nabla^2\underline{v}$$

where $\mathcal{P} \equiv p + \Phi = p + \rho gh$. To evaluate $\mathcal{P}$, the function h must be expressed correctly in the coordinate system of interest. See also Equation 8.115.

Einstein notation: This is a way of writing vectors and tensors in an orthonormal coordinate system. A vector or tensor written in Einstein notation is written as a sum of coefficients and basis vectors (in the case of vectors) or of coefficients and diads of basis vectors (in the case of tensors). In Einstein notation the summation signs are not written explicitly. Every pair of repeated indices indicates that the terms must be summed. Examples of Einstein notation are given here:

$$\underline{v} = \sum_{p=1}^{3} v_p\hat{e}_p = v_p\hat{e}_p$$

$$\underline{\underline{A}} = \sum_{s=1}^{3}\sum_{m=1}^{3} A_{sm}\hat{e}_s\hat{e}_m = A_{sm}\hat{e}_s\hat{e}_m$$

$$\underline{v}\cdot\underline{\underline{A}} = v_i\hat{e}_i \cdot A_{kn}\hat{e}_k\hat{e}_n = v_i A_{in}\hat{e}_n$$

empirical relation: This results from experimental observation rather than theoretical derivation.

Eulerian description of fluid mechanics: Traditionally, there are two ways to describe a flow: (1) from the point of view of individual fluid packets (i.e., a body approach, Newton's second law formulation); and (2) from the point of view of a fixed observer (i.e., control-volume approach, Reynolds-transport-theorem formulation). Both methods are equivalent but one or the other may be easier for certain problems. When a flow is described from the point of view of individual fluid packets, this is called an *Lagrangian description*. When a flow is described from the point of view of a fixed observer, this is called an *Eulerian description*. The Eulerian approach is common in Newtonian fluid mechanics, although for some atmospheric, oceanographic, and groundwater problems, the Lagrangian approach is favored. In polymer fluid dynamics, many calculations are made from a Lagrangian point of view because elastic behavior of the fluid necessitates keeping track of the histories of individual packets.

external flow: This is a flow around an object, such as a sphere or particles in a packed bed. Objects moving through a fluid (e.g., balls and airplanes) may be analyzed from the point of view of the moving object; therefore, the flow field is considered an external flow. Contrast this with internal flow.

extra-stress tensor: The total-stress tensor, $\underline{\underline{\Pi}}$, expresses the molecular surface forces that produce stress at a point in a flowing fluid. The extra-stress tensor, $\underline{\underline{\tilde{\tau}}}$, is the part of the total stress tensor, $\underline{\underline{\Pi}}$, that is anisotropic; that is, $\underline{\underline{\Pi}} = -p\underline{\underline{I}} + \underline{\underline{\tilde{\tau}}}$.

fluid: This is defined as a substance that cannot withstand a shear force without continuously deforming. This formal definition is consistent with our intuitive understanding that a fluid is a type of matter that moves and deforms easily. The fact that fluids continuously deform under shear distinguish them from elastic solids (e.g., gelatin), which deform under shear forces but stop deforming when their equilibrium shape is reached. A fluid will not stop deforming as long as a shear force is applied. Therefore, if a fluid is at rest, the shear forces must be everywhere zero throughout the fluid.

free surface: This refers to a fluid surface open to the atmosphere. Thus, in a film of fluid flowing down an incline, the top surface in contact with air is the free surface. In a fluid emerging from a nozzle and forming a jet, the entire surface of the jet is a free surface.

fully developed flow: A flow is termed fully developed if the velocity and stress fields no longer exhibit any entrance effects. For example, fully developed tube flow takes place sufficiently downstream from the entrance that the geometry of the tube entrance has no effect. The velocity and stress fields of a fully developed unidirectional flow do not vary in the flow direction.

gate valve: See **valves and fittings**.

gauge pressure: This is the term for pressure measurement relative to atmospheric pressure. A gauge reads zero when exposed to atmospheric pressure. Thus, pressures measured by gauges are relative to atmospheric pressure. To obtain absolute pressure from gauge pressure, we add 1 atm.

Gauss's integral theorem: Also known as the divergence theorem, this is a key theorem in vector mathematics that allows us to convert surface integrals to volume integrals:

$$\iint_{S} \hat{n} \cdot \underline{w} \, dS = \iiint_{V} \nabla \cdot \underline{w} \, dV$$

The two-dimensional version is known as Stokes's theorem (see **Stokes's theorem**).

g_c: This is the symbol given for the conversion factor that converts units of (mass · acceleration) to force units. The equivalence of these two quantities is a result of Newton's second law:

$$\underline{F} = m\underline{a}$$

$$N = [kg] \left[\frac{m}{s^2} \right]$$

where the force unit N (Newton) in the SI unit system (from French: Système international d'unités, also known as the metric system) is defined as $1\,kg \cdot m/s^2$. In American engineering units, Newton's second law is

$$[force] = [lb_m] \left[\frac{ft}{s^2} \right]$$

The customary unit for force in American engineering units is the *pound-force* (lb_f); there are $32.174 ft \cdot \frac{lb_m}{s^2} / lb_f$. This conversion factor is defined as g_c.

$$[lb_f] = \left(\frac{lb_f}{32.174\,ft\frac{lb_m}{s^2}} \right) [lb_m] \left[\frac{ft}{s^2} \right]$$

$$[lb_f] = \left(\frac{1}{g_c} \right) [lb_m] \left[\frac{ft}{s^2} \right]$$

Although technically $g_c = \frac{1N}{kg \cdot m/s^2}$, this nomenclature is not used in SI unit calculation. It is preferable not to write equations explicitly with unit conversion factors; we must check all calculations for dimensional consistency.

Gibbs notation: This is the vector–tensor notation used in this book (e.g., $\underline{v}$, $\nabla \underline{u}$ $\underline{\underline{B}}$, $\nabla \cdot \underline{v}$). Gibbs notation makes no reference to coordinate system.

globe valve: See **valves and fittings**.

head: This is energy per unit weight of fluid in a flow loop.

head loss: This is the frictional loss in a piping system expressed in units of head.

indeterminate vector product: This product between two vectors produces a tensor dyad. This product is not evaluated; rather, the two vectors are written side by side. When dot- or cross-multiplied, either from the left or the right, the multiplication is carried out with the nearest dyad vector:

$$\underline{v} \cdot \underline{a}\,\underline{b} = (\underline{v} \cdot \underline{a})\underline{b}$$

$$\underline{a}\,\underline{b} \cdot \underline{v} = \underline{a}(\underline{b} \cdot \underline{v}) = (\underline{b} \cdot \underline{v})\underline{a}$$

Note that in the first expression, the result is parallel to $\underline{b}$, whereas in the second expression, the result is parallel to $\underline{a}$. For more information on tensors, see Section 1.3.2.2 or [6].

internal flow: This is a flow that is bounded by a surface or surfaces. The principal example of an internal flow is flow in a pipe or other conduit but internal flow also includes flow between parallel plates (drag flow) or concentric cylinders (Couette flow). Contrast this with external flow.

inviscid fluid: A fluid is inviscid if its viscosity is zero or negligible. The flow of an inviscid fluid also is called potential flow. Because shear stress is proportional to viscosity, inviscid fluids may not transfer shear stresses.

irrotational flow: A flow is irrotational if the circulation around every point is zero. Another way of expressing the same condition is that the vorticity $\underline{\omega}$ is zero everywhere. In an irrotational flow, there is no tendency of fluid particles to rotate.

isotropic stress: Stress is isotropic at a point if the magnitude of the force per unit area on any plane through the point is the same. Pressure is an isotropic stress. The stress tensor for an isotropic stress is diagonal (i.e., all of the off-diagonal coefficients are zero), and all the diagonal components are equal.

kinematic viscosity: See **viscosity, kinematic**.

kinematics: This term refers to all information about the flow's motion. The flow kinematics consist of the velocity field, $\underline{v}$.

Kronecker delta: The Kronecker delta function, δ_{pk}, is defined here:

$$\delta_{pk} \equiv \begin{cases} 1 & \text{for } p = k \\ 0 & \text{for } p \neq k \end{cases}$$

Lagrangian description of fluid mechanics: See **Eulerian description.**

manometer: This is a device that measures pressure by inducing changes in the heights of columns of fluid. A U-tube manometer is a U-shaped tube that contains a measuring fluid (e.g., mercury, water, or oil). The tops of the two sides of the manometer are exposed to fluids at different pressures, causing the level of the measuring fluid to change. The height difference between the measuring fluid on the two sides of the U-tube manometer can be related to the difference between the two pressures (see Section 4.2.4.1).

molecular forces: These are forces in a fluid due to chemical or physical interactions between molecules. These forces include electrostatic attraction or repulsion between molecules, van der Waals forces, hydrogen bonding, and intermolecular entanglement, as well as other forces specific to the materials under consideration.

Newton, Sir Isaac: Isaac Newton (1643–1727) was an English mathematician who made many seminal contributions to the fields of mathematics, mechanics, and optics. In 1687, Newton published the *Philosophiae Naturalis Principia Mathematica* in which—from three postulates now known as Newton's laws of motion—he derived how bodies move in both the heavens (i.e., planetary motion) and on Earth. Newton's laws of motion form the core of any current introduction to physics. Newton made a significant contribution to the invention of calculus.

Newton's second law: This law states that momentum is conserved and can be written mathematically as:

$$\sum_{\substack{\text{all forces} \\ \text{acting on body}}} \underline{f} = \frac{d(m\underline{v})}{dt} \tag{A.6}$$

where f are the various forces on the body, m is the mass of the body, $\underline{v}$ is the velocity of the body, and t is time. The derivative $d(m\underline{v})/dt$ is the rate of accumulation of momentum for the body. Note that if the mass of the system is constant and there is a single force, Newton's second law becomes the familiar $\underline{f} = m\underline{a}$, where $\underline{a} = d\underline{v}/dt$ is the acceleration of the body of constant mass m.

noncontact forces: See **body forces**.

orifice plate or orifice meter: This is a device used to measure volumetric flow rate in a pipe (Figure 7.66). The device consists of a plate that obstructs the flow except for a small hole (orifice) in the plate center. The blockage of the flow increases the pressure upstream of the plate. A measurement of the pressure difference across the orifice plate may be related to the volumetric flow rate through application of the mechanical energy balance. In flow through an orifice, the streamlines are observed to converge axisymmetrically and reach their narrowest cross section slightly downstream of the orifice; the location of this narrowest jet cross section is called the vena contracta.

orthonormal basis vectors: Any three nonzero, noncoplanar vectors may form a basis in physical (i.e., three-dimensional) space. Basis vectors that are mutually perpendicular and of unit length are called orthonormal basis vectors.

partial differential equations (PDEs): The microscopic momentum balance (see Equation 6.71) is an example of a partial differential equation, which is a differential equation that involves partial derivatives. Sommerfeld [158, 174] determined that second-order PDEs (i.e., those that involve second derivatives) may be divided into three classes according to the following scheme. For the equation:

$$A\frac{\partial^2 \phi}{\partial x^2} + B\frac{\partial^2 \phi}{\partial x \partial y} + C\frac{\partial^2 \phi}{\partial y^2} = D \qquad (A.7)$$

where A, B, C, and D may be nonlinear functions of x, y, ϕ, $\partial \phi/\partial x$, and $\partial \phi/\partial y$ but not of the second derivatives of ϕ, the three classes are divided according to the value of the discriminant function $B^2 - 4AC$:

$$
\begin{array}{ll}
B^2 - 4AC < 0 & \text{elliptic PDE} \\
B^2 - 4AC = 0 & \text{parabolic PDE} \\
B^2 - 4AC > 0 & \text{hyperbolic PDE}
\end{array}
\qquad (A.8)
$$

If the equation is elliptic, the problem is a boundary-value problem, and it can be solved only by specifying the boundary conditions on a complete contour enclosing the region. Parabolic equations are mixed initial- and boundary-value problems, and the boundary conditions must be closed in one direction but remain open at one end of the other direction. Hyperbolic equations are initial-value problems, and they can be solved in a given region by specifying the conditions at only one portion of the boundary with the other boundaries remaining open [174]. Numerical schemes to solve PDEs usually are specialized for one of the three types; thus, problems that mix or change type during the solution process may present difficulties during solution.

Pascal's principle: This states that the pressure exerted on an enclosed liquid is transmitted equally to every part of the liquid and to the walls of the container.

Pascal's principle is the reason for the functioning of manometers and hydraulic devices such as automotive brakes.

pressure, equivalent: See **dynamic pressure**.

return bend: See **valves and fittings**.

Reynolds, Osborne: Osborne Reynolds (1842–1912) was a British engineer and professor of engineering at Owens College in Manchester (now the University of Manchester) who studied the transition from laminar to turbulent flow in pipes. The Reynolds number is named after him. In the field of turbulent flow, Reynolds contributed the technique of Reynolds-averaging, in which the local fluid velocity is divided into the average velocity and a superposed fluctuating component. His work on dimensional analysis and dimensional similarity made important contributions to naval architecture, enabling accurate ship-design inferences from experiments on small-scale models.

Reynolds transport theorem: This is the equivalent of Newton's second law (i.e., momentum is conserved) written for a control volume:

Reynolds transport theorem:

$$\frac{d\underline{\mathbf{P}}}{dt} + \iint_{CS} (\hat{n} \cdot \underline{v}) \, \rho \underline{v} \, dS = \sum_{\substack{\text{on} \\ \text{CV}}} \underline{f} \tag{A.9}$$

where $\sum \underline{f}$ is the sum of forces acting on the control volume, $\frac{d\underline{\mathbf{P}}}{dt}$ is the rate of change of momentum in the control volume, $\hat{n}$ is the outwardly pointing unit normal of dS, $\underline{v}$ is the velocity of fluid passing through surface element dS, and ρ is the density of fluid passing through dS. The integral is taken over CS, the surface that bounds the control volume.

rotameter: This is a device used to measure flow rate. Flow is directed through the rotameter chamber, causing a float to rise. The internal shape of the rotameter is designed so that the vertical height of the float (as measured on an inscribed scale) is proportional to flow rate. Rotameter readings are reported in units of percent full scale; they must be calibrated to give absolute units (see **calibration**).

roundoff error: This term refers to the loss in accuracy that happens in digital devices when a number is truncated, or rounded off. For example, the fraction one-third (1/3) in digital form is 0.333333 with a never-ending number of digits. By necessity, calculators and computers must truncate this number to a finite number (8 or 16 or 32 are typical choices). When calculations use this truncated number, there is a loss of accuracy in the calculation. The effect of roundoff error may be minimized by keeping a large excess of digits in intermediate calculations. Final answers then may be truncated (or rounded off) to reflect the known precision to which the calculation may be reported.

scalar: This is a quantity that has magnitude only. Examples of scalars include mass, energy, density, volume, and the number of automobiles in a parking lot. In ordinary arithmetic, we are using scalars.

shear force: The force on an arbitrary surface in a fluid can be resolved into two components: (1) the normal force, which is perpendicular to the surface; and (2) the shear force, which tangential to the surface.

specific gravity (SG): This is the ratio of a substance density to its density at a reference temperature and pressure. The usual reference temperature for water is the triple point at atmospheric pressure, which is 4°C. The source that publishes specific gravity should indicate the reference conditions.

stagnation point: This refers to a point on a solid surface where a streamline terminates. The classic stagnation point is the point in the center of a planar jet where the flow comes to a halt (Figure 4.77). Streamlines to the right and left of the central streamline are diverted by the wall to the right or left. There is one streamline, the stagnation streamline, that terminates in the surface at the stagnation point.

Stokes's theorem: The two-dimensional version of the Gauss-Ostrogradskii divergence theorem (see Equation 6.14) is Stokes's theorem. This theorem allows us to convert a line integral along a path to an area integral:

Stokes's theorem for vector field $\underline{F}$:
$$\oint_C \left(\hat{t} \cdot \underline{F} \right) dl = \iint_S \hat{n} \cdot (\nabla \times \underline{F}) \, dS$$

where $\underline{F}$ is any vector field, C is a contour in the field, dl is a differential length along the contour C, $\hat{n}$ is an outward-pointing unit vector from a surface S, $\hat{t}$ is a tangential unit vector along C, and dS is a differential piece of S [142].

stream function (ψ): This is a function that is everywhere tangent to the velocity field. In two-dimensional flow expressed in Cartesian coordinates:

$$v_x = \frac{\partial \psi}{\partial y}$$

$$v_y = \frac{\partial \psi}{\partial x}$$

ψ is a function of x and y and may be a function of t. Lines of $\psi =$ constant are called *streamlines*. In steady flows, streamlines are coincident with particle paths; thus, we often visualize steady-state streamlines—by using tracer particles, for example—to show how the fluid deforms. When flows are not steady, streamlines are not equivalent to pathlines and more sophisticated fluid tracing methods must be adopted.

streamlines: See **stream function**.

tensors: A second-order tensor or simply a tensor is an ordered pair of coordinate directions or the indeterminate vector product of two vectors. The simplest tensor is called a dyad or dyadic product and is written as two vectors side by side. For example, the tensor $\underline{\underline{A}}$ is the indeterminate vector product of the vectors $\underline{a}$ and $\underline{b}$:

$$\underline{\underline{A}} = \underline{a} \, \underline{b}$$

Tensors are operators that express linear vector functions. For example, in the following equation in which a tensor $\underline{\underline{A}}$ dot-multiplies the vector $\underline{v}$ giving the vector $\underline{w}$:

$$\underline{\underline{A}} \cdot \underline{v} = \underline{w}$$

the tensor $\underline{\underline{A}}$ transforms the vector $\underline{v}$ into the vector $\underline{w}$. This is the action of a vector function. Tensors are linear vector functions [6, 104]. Any second-order tensor may be expressed in a coordinate system as the linear combination of nine dyads formed from the basis vectors of the coordinate system. For example, in the Cartesian coordinate system $\hat{e}_1$, $\hat{e}_2$, $\hat{e}_3$, a tensor $\underline{\underline{A}}$ can be written as:

$$\underline{\underline{A}} = \sum_{n=1}^{3} \sum_{j=1}^{3} A_{nj} \hat{e}_n \hat{e}_j$$

The coefficients A_{nj} are called the scalar coefficients of $\underline{\underline{A}}$ relative to the basis $\hat{e}_1 \hat{e}_2 \hat{e}_3$, and they may be written for convenience in a 3×3 matrix:

$$\underline{\underline{A}} = \begin{pmatrix} A_{11} & A_{12} & A_{13} \\ A_{21} & A_{22} & A_{23} \\ A_{31} & A_{32} & A_{33} \end{pmatrix}_{123}$$

The subscript 123 indicates that the entries in the matrix represent the coefficients of a tensor written relative to the $\hat{e}_1 \hat{e}_2 \hat{e}_3$ coordinate system.

trim of a valve: See **valve trim**.

union: See **valves and fittings**.

unit vector: This is a vector the magnitude of which is 1. For example, if for a vector $\underline{m}$:

$$|m| = \sqrt{m \cdot m} = 1$$

then $\underline{m}$ is a unit vector. In this text, unit vectors are written with a carat ($\hat{}$) over the symbol:

$$|\hat{m}| = 1$$

valve trim: This is a curve that shows how the flow rate from the valve varies with the valve's position. A metering valve is designed to have a linear trim. With a linear trim, when the valve is twice as open, there is twice the flow. Valves less precisely designed can require many turns to slow the flow, and the flow may stop suddenly with a small amount of additional turning.

valves and fittings: Several common valves and fittings are defined as follows:

1. *Coupling.* A fitting that unites tubing of the same or different diameter through soldered or other permanent connections.
2. *Return bend.* A fitting in the shape of a U that reverses the direction of the flow.
3. *Gate valve.* A valve that varies the passing flow by raising and lowering a gate that blocks the flow. Gate valves are used when minimum flow restriction is desired through an open valve.
4. *Globe valve.* A valve that varies the flow by seating a plug to close off the flow. The flow in a globe valve is obstructed by the seating structure even when the valve is open; thus, these valves are not appropriate when minimum flow restriction is desired through an open valve.
5. *Check valve.* A valve that allows flow in one direction but prevents flow in the reverse direction (i.e., no back flow).

6. *Union.* A fitting that unites tubing of the same or different diameter through screws or other temporary connections.

vapor lock: This term refers to a device failing to operate because fluid within it has vaporized, breaking liquid continuity and causing the device to cease operating. When pressures within a device fall below the vapor pressure of a fluid within the device, the fluid vaporizes. Devices often are designed to have liquids rather than gases; the presence of gases causes device failure.

vectors: These are quantities that have associated magnitude and direction. Examples that appear in fluid mechanics are fluid velocity, $\underline{v}$, and force, $\underline{f}$.

velocity field: In fluids, different portions of the flow move in different directions and at different speeds. To describe the motion of fluids, we use the velocity field, a two- or three-dimensional continuous function that describes the fluid motion as a function of position and time.

viscosity, kinematic: This is the ratio of viscosity μ and density ρ:

$$\text{Kinematic viscosity:} \quad \nu \equiv \frac{\mu}{\rho}$$

The kinematic viscosity is measured in a Cannon-Fenske viscometer (see Figure 7.11). Kinematic viscosity also appears in the Reynolds number, $Re = \rho \langle v \rangle D/\mu = \langle v \rangle D/\nu$.

vorticity $\underline{\omega}$: This is the curl of the velocity vector:

$$\underline{\omega} \equiv \nabla \times \underline{v}$$

Vorticity is a flow-field property that allows us to keep track of rotational character in flows.

weir: This is a low overflow dam placed across a waterway to raise its level or to divert the flow. The presence of a weir makes it straightforward to calculate flow rate in the stream [183].

Mathematics

B.1 Differential operations on vectors and tensors

In fluid mechanics we are concerned with variables such as density, velocity, and stress that take on different values at different positions in a field. The modeling we describe in this text relies on our ability to keep track of spatial variations of these functions.

We are familiar with taking derivatives of scalar functions with respect to spatial variables x, y, and z in the Cartesian coordinate system. Less familiar, perhaps, is the idea of taking spatial derivatives of vectors and tensors. When we write a vector with respect to a basis and then take its derivative, we must treat both the vector coefficients and the basis vectors as variables. Consider a vector $\underline{v}$ written with respect to the arbitrary coordinate system $\tilde{\underline{e}}_1, \tilde{\underline{e}}_2, \tilde{\underline{e}}_3$ (not the Cartesian system):

$$\underline{v} = \tilde{v}_1 \tilde{\underline{e}}_1 + \tilde{v}_2 \tilde{\underline{e}}_2 + \tilde{v}_3 \tilde{\underline{e}}_3 \tag{B.1}$$

When calculating a spatial derivative of the vector $\underline{v}$ with respect to, for example x_2, we differentiate the three terms on the right of Equation B.1, applying the product rule of differentiation to each term:

$$\frac{\partial \underline{v}}{\partial x_2} = \frac{\partial}{\partial x_2}(\tilde{v}_1 \tilde{\underline{e}}_1 + \tilde{v}_2 \tilde{\underline{e}}_2 + \tilde{v}_3 \tilde{\underline{e}}_3) \tag{B.2}$$

$$= \frac{\partial}{\partial x_2}(\tilde{v}_1 \tilde{\underline{e}}_1) + \frac{\partial}{\partial x_2}(\tilde{v}_2 \tilde{\underline{e}}_2) + \frac{\partial}{\partial x_2}(\tilde{v}_3 \tilde{\underline{e}}_3) \tag{B.3}$$

$$= \tilde{v}_1 \frac{\partial \tilde{\underline{e}}_1}{\partial x_2} + \tilde{\underline{e}}_1 \frac{\partial \tilde{v}_1}{\partial x_2} + \tilde{v}_2 \frac{\partial \tilde{\underline{e}}_2}{\partial x_2} + \tilde{\underline{e}}_2 \frac{\partial \tilde{v}_2}{\partial x_2} + \tilde{v}_3 \frac{\partial \tilde{\underline{e}}_3}{\partial x_2} + \tilde{\underline{e}}_3 \frac{\partial \tilde{v}_3}{\partial x_2} \tag{B.4}$$

The derivatives of the basis vectors $\frac{\partial \tilde{\underline{e}}_3}{\partial x_1}, \frac{\partial \tilde{\underline{e}}_3}{\partial x_2}, \frac{\partial \tilde{\underline{e}}_3}{\partial x_3}$ are not zero because in the general case, the basis vectors vary with position, and the expression in Equation B.4 cannot be simplified.

There is a special coordinate system in which the spatial derivatives of the basis vectors are zero: The Cartesian coordinate system. In the Cartesian system, for every location in space, the basis vectors $\hat{e}_1$, $\hat{e}_1$, and $\hat{e}_3$ point in the same directions. Because they do not vary with spatial position, the spatial derivatives of the Cartesian basis vectors are zero, and Equation B.4 becomes:

$$\frac{\partial \underline{v}}{\partial x_2} = v_1 \frac{\partial \hat{e}_1}{\partial x_2} + \hat{e}_1 \frac{\partial v_1}{\partial x_2} + v_2 \frac{\partial \hat{e}_2}{\partial x_2} + \hat{e}_2 \frac{\partial v_2}{\partial x_2} + v_3 \frac{\partial \hat{e}_3}{\partial x_2} + \hat{e}_3 \frac{\partial v_3}{\partial x_2} \qquad (B.5)$$

$$= \hat{e}_1 \frac{\partial v_1}{\partial x_2} + \hat{e}_2 \frac{\partial v_2}{\partial x_2} + \hat{e}_3 \frac{\partial v_3}{\partial x_2} \qquad (B.6)$$

Note that we have removed the tilde from all symbols to indicate that we are now considering the Cartesian components of $\underline{v}$. There are similar terms for the x_1 and x_3 derivatives of $\underline{v}$.

The differentiation of a vector expressed in Cartesian coordinates can be communicated with the spatial differentiation operator ∇, called *del*:

$$\text{Del operator: } \nabla \equiv \hat{e}_1 \frac{\partial}{\partial x_1} + \hat{e}_2 \frac{\partial}{\partial x_2} + \hat{e}_3 \frac{\partial}{\partial x_3} \qquad (B.7)$$

Del is a vector operator. We verify here that the derivative in Equation B.6 can be written in terms of del as follows:

$$\frac{\partial \underline{v}}{\partial x_2} = \hat{e}_2 \cdot \nabla \underline{v} \qquad (B.8)$$

$$= \hat{e}_2 \cdot \left[\left(\hat{e}_1 \frac{\partial}{\partial x_1} + \hat{e}_2 \frac{\partial}{\partial x_2} + \hat{e}_3 \frac{\partial}{\partial x_3} \right) (v_1 \hat{e}_1 + v_2 \hat{e}_2 + v_3 \hat{e}_3) \right] \qquad (B.9)$$

$$= \hat{e}_2 \cdot \left[\hat{e}_1 \frac{\partial v_1}{\partial x_1} \hat{e}_1 + \hat{e}_1 \frac{\partial v_2}{\partial x_1} \hat{e}_2 + \hat{e}_1 \frac{\partial v_3}{\partial x_1} \hat{e}_3 \right.$$

$$+ \hat{e}_2 \frac{\partial v_1}{\partial x_2} \hat{e}_1 + \hat{e}_2 \frac{\partial v_2}{\partial x_2} \hat{e}_2 + \hat{e}_2 \frac{\partial v_3}{\partial x_2} \hat{e}_3$$

$$\left. + \hat{e}_3 \frac{\partial v_1}{\partial x_3} \hat{e}_1 + \hat{e}_3 \frac{\partial v_2}{\partial x_3} \hat{e}_2 + \hat{e}_3 \frac{\partial v_3}{\partial x_3} \hat{e}_3 \right] \qquad (B.10)$$

$$= \frac{\partial v_1}{\partial x_2} \hat{e}_1 + \frac{\partial v_2}{\partial x_2} \hat{e}_2 + \frac{\partial v_3}{\partial x_2} \hat{e}_3 \qquad (B.11)$$

To simplify the expressions here, we have used the fact that the Cartesian coordinate system is orthonormal. The del operator follows the distributive law of algebra, as shown; also, the differentiation operations of del operate on everything to their right.

The del operator is useful in expressing many spatially changing quantities in physics and engineering. Del operates on scalars, vectors, and tensors. When operating on a scalar (e.g., α), $\nabla\alpha$ becomes:

$$\nabla\alpha = \left(\hat{e}_1\frac{\partial}{\partial x_1} + \hat{e}_2\frac{\partial}{\partial x_2} + \hat{e}_3\frac{\partial}{\partial x_3}\right)\alpha \tag{B.12}$$

$$= \hat{e}_1\frac{\partial\alpha}{\partial x_1} + \hat{e}_2\frac{\partial\alpha}{\partial x_2} + \hat{e}_3\frac{\partial\alpha}{\partial x_3} \tag{B.13}$$

$$= \begin{pmatrix} \frac{\partial\alpha}{\partial x_1} \\[4pt] \frac{\partial\alpha}{\partial x_2} \\[4pt] \frac{\partial\alpha}{\partial x_3} \end{pmatrix}_{123} \quad \text{(a vector)} \tag{B.14}$$

When del operates on a scalar (order 0), it produces a vector (order 1). The laws of algebra for del operating on a scalar are given here:[1]

$$\begin{array}{ll}
\text{Laws of algebra} & \left\{\begin{array}{ll} \textit{Not} \text{ commutative} & \nabla\alpha \neq \alpha\nabla \\ \textit{Not} \text{ associative} & \nabla(\zeta\alpha) \neq (\nabla\zeta)\alpha \\ \text{distributive} & \nabla(\zeta + \alpha) = \nabla\zeta + \nabla\alpha \end{array}\right.
\end{array}$$

for del operating on scalars:

The associative law does not hold with the del operator because we must follow the rules of differentiating a product when faced with a term such as $\nabla(\zeta\alpha)$.

$$\frac{\partial(\gamma\alpha)}{\partial x} = \gamma\frac{\partial\alpha}{\partial x} + \alpha\frac{\partial\gamma}{\partial x} \tag{B.15}$$

Del operates on vectors and tensors in addition to scalars [6, 12, 104, 160]. When del operates on a vector (order 1) it produces a tensor (order 2). The laws of algebra for del operating on a vector are given here:

$$\begin{array}{ll}
\text{Laws of algebra} & \left\{\begin{array}{ll} \textit{Not} \text{ commutative} & \nabla\underline{w} \neq \underline{w}\nabla \\ \textit{Not} \text{ associative} & \nabla(\underline{a}\cdot\underline{b}) \neq (\nabla\underline{a})\cdot\underline{b} \\ & \nabla(\underline{a}\times\underline{b}) \neq (\nabla\underline{a})\times\underline{b} \\ \text{distributive} & \nabla(\underline{w}+\underline{b}) = \nabla\underline{w} + \nabla\underline{b} \end{array}\right.
\end{array}$$

for del operating on vectors:

The quantities $\nabla\alpha$ and $\nabla\underline{w}$ are called the *gradients* of α and $\underline{w}$, respectively. Two other operations involving ∇ are commonly used: the *divergence*, written as $(\nabla\cdot)$; and the *Laplacian*, written as $(\nabla\cdot\nabla) = (\nabla^2)$. The divergence reduces the

---

[1] Scalars, vectors, and tensors can all be classified as tensors of different orders. Scalars are zero-order tensors, vectors are first-order tensors, and the usual tensors encountered in fluid mechanics are second-order tensors. What changes when del operates on a scalar or vector is the order of the quantity on which it acts [6, 12, 104, 160].

order of a quantity on which it operates, and thus we cannot take the divergence of a scalar. The Laplacian does not change the order of a quantity on which it operates.

A complete discussion of vector operators is beyond the scope of this text. We include here examples of the divergence and Laplacian operations; these calculations are valid in Cartesian coordinates. When we wish to express these operators in non-Cartesian coordinates, the differentiation operators $\partial/\partial x_i$ must act on the (non-constant) basis vectors as well as the scalar coefficients. The tables at the end of this chapter contain vector-tensor components of various quantities written in cylindrical and spherical coordinates.

EXAMPLE B.1. *What are the following quantities:* $(\nabla \underline{w})$; $(\nabla \cdot \underline{w})$; $\left(\nabla \cdot \underline{\underline{B}}\right)$; $(\nabla^2 \alpha)$; $(\nabla^2 \underline{w})$?

SOLUTION.
The *gradient* of a vector $\underline{w}$:

$$\nabla \underline{w} = \left(\hat{e}_1 \frac{\partial}{\partial x_1} + \hat{e}_2 \frac{\partial}{\partial x_2} + \hat{e}_3 \frac{\partial}{\partial x_3} \right) (w_1 \hat{e}_1 + w_2 \hat{e}_2 + w_3 \hat{e}_3) \tag{B.16}$$

$$= \frac{\partial w_1}{\partial x_1} \hat{e}_1 \hat{e}_1 + \frac{\partial w_2}{\partial x_1} \hat{e}_1 \hat{e}_2 + \frac{\partial w_3}{\partial x_1} \hat{e}_1 \hat{e}_3 + \frac{\partial w_1}{\partial x_2} \hat{e}_2 \hat{e}_1 + \frac{\partial w_2}{\partial x_2} \hat{e}_2 \hat{e}_2$$

$$+ \frac{\partial w_3}{\partial x_2} \hat{e}_2 \hat{e}_3 + \frac{\partial w_1}{\partial x_3} \hat{e}_3 \hat{e}_1 + \frac{\partial w_2}{\partial x_3} \hat{e}_3 \hat{e}_2 + \frac{\partial w_3}{\partial x_3} \hat{e}_3 \hat{e}_3 \tag{B.17}$$

$$= \sum_{p=1}^{3} \sum_{k=1}^{3} \hat{e}_p \hat{e}_k \frac{\partial w_k}{\partial x_p} \tag{B.18}$$

$$= \begin{pmatrix} \frac{\partial w_1}{\partial x_1} & \frac{\partial w_2}{\partial x_1} & \frac{\partial w_3}{\partial x_1} \\ \frac{\partial w_1}{\partial x_2} & \frac{\partial w_2}{\partial x_2} & \frac{\partial w_3}{\partial x_2} \\ \frac{\partial w_1}{\partial x_3} & \frac{\partial w_2}{\partial x_3} & \frac{\partial w_3}{\partial x_3} \end{pmatrix}_{123} \quad \text{(a tensor)} \tag{B.19}$$

where the matrix holds the coefficients of the expressions $\hat{e}_1 \hat{e}_1$, $\hat{e}_1 \hat{e}_2$, and so on. These expressions $\hat{e}_i \hat{e}_j$ are called indeterminate vector products and are themselves simple second-order tensors (see Chapter 1 and [104]).

The *divergence* of a vector $\underline{w}$:

$$\nabla \cdot \underline{w} = \left(\hat{e}_1 \frac{\partial}{\partial x_1} + \hat{e}_2 \frac{\partial}{\partial x_2} + \hat{e}_3 \frac{\partial}{\partial x_3} \right) \cdot (w_1 \hat{e}_1 + w_2 \hat{e}_2 + w_3 \hat{e}_3) \tag{B.20}$$

$$= \frac{\partial w_1}{\partial x_1} + \frac{\partial w_2}{\partial x_2} + \frac{\partial w_3}{\partial x_3} \quad \text{(a scalar)} \tag{B.21}$$

The *divergence* of a tensor $\underline{\underline{B}}$:

$$\nabla \cdot \underline{\underline{B}} = \left(\hat{e}_1 \frac{\partial}{\partial x_1} + \hat{e}_2 \frac{\partial}{\partial x_2} + \hat{e}_3 \frac{\partial}{\partial x_3} \right) \cdot \sum_{m=1}^{3} \sum_{n=1}^{3} B_{mn} \hat{e}_m \hat{e}_n \tag{B.22}$$

$$= \sum_{m=1}^{3} \sum_{n=1}^{3} \frac{\partial}{\partial x_1} B_{mn} (\hat{e}_1 \cdot \hat{e}_m) \hat{e}_n + \sum_{m=1}^{3} \sum_{n=1}^{3} \frac{\partial}{\partial x_2} B_{mn} (\hat{e}_2 \cdot \hat{e}_m) \hat{e}_n$$

$$+ \sum_{m=1}^{3} \sum_{n=1}^{3} \frac{\partial}{\partial x_3} B_{mn} (\hat{e}_3 \cdot \hat{e}_m) \hat{e}_n \tag{B.23}$$

$$= \left(\frac{\partial B_{11}}{\partial x_1} + \frac{\partial B_{21}}{\partial x_2} + \frac{\partial B_{31}}{\partial x_3} \right) \hat{e}_1 + \left(\frac{\partial B_{12}}{\partial x_2} + \frac{\partial B_{22}}{\partial x_2} + \frac{\partial B_{32}}{\partial x_3} \right) \hat{e}_2$$

$$+ \left(\frac{\partial B_{13}}{\partial x_3} + \frac{\partial B_{23}}{\partial x_2} + \frac{\partial B_{33}}{\partial x_3} \right) \hat{e}_3 \tag{B.24}$$

$$= \begin{pmatrix} \frac{\partial B_{11}}{\partial x_1} + \frac{\partial B_{21}}{\partial x_2} + \frac{\partial B_{31}}{\partial x_3} \\ \frac{\partial B_{12}}{\partial x_2} + \frac{\partial B_{22}}{\partial x_2} + \frac{\partial B_{32}}{\partial x_3} \\ \frac{\partial B_{13}}{\partial x_3} + \frac{\partial B_{23}}{\partial x_2} + \frac{\partial B_{33}}{\partial x_3} \end{pmatrix}_{123} \quad \text{(a vector)} \tag{B.25}$$

The *Laplacian* of a scalar α:

$$\nabla \cdot \nabla \alpha = \left(\hat{e}_1 \frac{\partial}{\partial x_1} + \hat{e}_2 \frac{\partial}{\partial x_2} + \hat{e}_3 \frac{\partial}{\partial x_3} \right) \cdot \left(\hat{e}_1 \frac{\partial}{\partial x_1} + \hat{e}_2 \frac{\partial}{\partial x_2} + \hat{e}_3 \frac{\partial}{\partial x_3} \right) \alpha \tag{B.26}$$

$$= \frac{\partial^2 a}{\partial x_1^2} + \frac{\partial^2 a}{\partial x_2^2} + \frac{\partial^2 a}{\partial x_3^2} \quad \text{(a scalar)} \tag{B.27}$$

The *Laplacian* of a vector $\underline{w}$:

$$\nabla \cdot \nabla \underline{w} = \left(\hat{e}_1 \frac{\partial}{\partial x_1} + \hat{e}_2 \frac{\partial}{\partial x_2} + \hat{e}_3 \frac{\partial}{\partial x_3} \right)$$

$$\cdot \left(\hat{e}_1 \frac{\partial}{\partial x_1} + \hat{e}_2 \frac{\partial}{\partial x_2} + \hat{e}_3 \frac{\partial}{\partial x_3} \right) (w_1 \hat{e}_1 + w_2 \hat{e}_2 + w_3 \hat{e}_3) \tag{B.28}$$

$$= \left(\frac{\partial^2 w_1}{\partial x_1^2} + \frac{\partial^2 w_1}{\partial x_2^2} + \frac{\partial^2 w_1}{\partial x_3^2} \right) \hat{e}_1 + \left(\frac{\partial^2 w_2}{\partial x_1^2} + \frac{\partial^2 w_2}{\partial x_2^2} + \frac{\partial^2 w_2}{\partial x_3^2} \right) \hat{e}_2$$

$$+ \left(\frac{\partial^2 w_3}{\partial x_1^2} + \frac{\partial^2 w_3}{\partial x_2^2} + \frac{\partial^2 w_3}{\partial x_3^2} \right) \hat{e}_3 \tag{B.29}$$

$$= \begin{pmatrix} \frac{\partial^2 w_1}{\partial x_1^2} + \frac{\partial^2 w_1}{\partial x_2^2} + \frac{\partial^2 w_1}{\partial x_3^2} \\[2mm] \frac{\partial^2 w_2}{\partial x_1^2} + \frac{\partial^2 w_2}{\partial x_2^2} + \frac{\partial^2 w_2}{\partial x_3^2} \\[2mm] \frac{\partial^2 w_3}{\partial x_1^2} + \frac{\partial^2 w_3}{\partial x_2^2} + \frac{\partial^2 w_3}{\partial x_3^2} \end{pmatrix}_{123} \qquad \text{(a vector)} \qquad \text{(B.30)}$$

EXAMPLE B.2. *What is* $\nabla \cdot \alpha\, \underline{b}$?

SOLUTION. We begin by writing $\nabla \cdot \alpha\, \underline{b}$ in a Cartesian coordinate system:

$$\nabla \cdot \alpha\, \underline{b} = \left(\hat{e}_1 \frac{\partial}{\partial x_1} + \hat{e}_2 \frac{\partial}{\partial x_2} + \hat{e}_3 \frac{\partial}{\partial x_3} \right) \cdot \alpha(b_1\hat{e}_1 + b_2\hat{e}_2 + b_3\hat{e}_3) \qquad \text{(B.31)}$$

$$= \left(\hat{e}_1 \frac{\partial}{\partial x_1} + \hat{e}_2 \frac{\partial}{\partial x_2} + \hat{e}_3 \frac{\partial}{\partial x_3} \right) \cdot (\alpha b_1\hat{e}_1 + \alpha b_2\hat{e}_2 + \alpha b_3\hat{e}_3) \qquad \text{(B.32)}$$

Because both α and the coefficients of $\underline{b}$ are to the right of the del operator, they all are acted on by the differentiation action. The Cartesian unit vectors also are affected, but these are constant. Now we carry out the dot product, using the distributive law. Because the basis vectors are orthogonal and of unit length, most of the dot products are zero:

$$\nabla \cdot \alpha\, \underline{b} = \hat{e}_1 \frac{\partial}{\partial x_1} \cdot (\alpha b_1\hat{e}_1 + \alpha b_2\hat{e}_2 + \alpha b_3\hat{e}_3)$$

$$+ \hat{e}_2 \frac{\partial}{\partial x_2} \cdot (\alpha b_1\hat{e}_1 + \alpha b_2\hat{e}_2 + \alpha b_3\hat{e}_3)$$

$$+ \hat{e}_3 \frac{\partial}{\partial x_3} \cdot (\alpha b_1\hat{e}_1 + \alpha b_2\hat{e}_2 + \alpha b_3\hat{e}_3) \qquad \text{(B.33)}$$

$$= \hat{e}_1 \cdot \hat{e}_1 \frac{\partial(\alpha b_1)}{\partial x_1} + \hat{e}_1 \cdot \hat{e}_2 \frac{\partial(\alpha b_2)}{\partial x_1} + \hat{e}_1 \cdot \hat{e}_3 \frac{\partial(\alpha b_3)}{\partial x_1}$$

$$+ \hat{e}_2 \cdot \hat{e}_1 \frac{\partial(\alpha b_1)}{\partial x_2} + \hat{e}_2 \cdot \hat{e}_2 \frac{\partial(\alpha b_2)}{\partial x_2} + \hat{e}_2 \cdot \hat{e}_3 \frac{\partial(\alpha b_3)}{\partial x_2}$$

$$+ \hat{e}_3 \cdot \hat{e}_1 \frac{\partial(\alpha b_1)}{\partial x_3} + \hat{e}_3 \cdot \hat{e}_2 \frac{\partial(\alpha b_2)}{\partial x_3} + \hat{e}_3 \cdot \hat{e}_3 \frac{\partial(\alpha b_3)}{\partial x_3} \qquad \text{(B.34)}$$

$$= \frac{\partial(\alpha b_1)}{\partial x_1} + \frac{\partial(\alpha b_2)}{\partial x_2} + \frac{\partial(\alpha b_3)}{\partial x_3} \qquad \text{(B.35)}$$

To further expand this expression, we use the product rule of differentiation on the quantities in parentheses:

$$\nabla \cdot \alpha \, \underline{b} = \alpha \frac{\partial b_1}{\partial x_1} + b_1 \frac{\partial \alpha}{\partial x_1} + \alpha \frac{\partial b_2}{\partial x_2} + b_2 \frac{\partial \alpha}{\partial x_2} + \alpha \frac{\partial b_3}{\partial x_3} + b_3 \frac{\partial \alpha}{\partial x_3} \tag{B.36}$$

$$= \alpha \left(\frac{\partial b_1}{\partial x_1} + \frac{\partial b_2}{\partial x_2} + \frac{\partial b_3}{\partial x_3} \right) + \left(b_1 \frac{\partial \alpha}{\partial x_1} + b_2 \frac{\partial \alpha}{\partial x_2} + b_3 \frac{\partial \alpha}{\partial x_3} \right) \tag{B.37}$$

This is as far as we can proceed. It is possible to write this final result in vector (also called Gibbs) notation:

$$\nabla \cdot \alpha \, \underline{b} = \alpha \nabla \cdot \underline{b} + \underline{b} \cdot \nabla \alpha \tag{B.38}$$

The equivalency of Equations B.38 and B.39 may be verified by writing the terms in Equation B.39 and carrying out the dot products. If the differentiation of the product is not carried out correctly, the second term on the righthand side would be omitted (incorrectly).

A summary of vector identities involving the ∇ operator is in Table B.1 and the inside front cover of this text.

Table B.1. Additional vector identities involving the ∇ operator

$\nabla(\underline{v} \cdot \underline{f})$	$=$	$\nabla \underline{f} \cdot \underline{v} + \nabla \underline{v} \cdot \underline{f}$	B-1.1
$\nabla \cdot (\underline{\underline{A}} \cdot \underline{v})$	$=$	$\underline{\underline{A}}^T : \nabla \underline{v} + \underline{v} \cdot (\nabla \cdot \underline{\underline{A}})$	B-1.2
$\nabla \cdot (\underline{v} \cdot \underline{\underline{A}})$	$=$	$\underline{\underline{A}} : \nabla \underline{v} + \underline{v} \cdot (\nabla \cdot \underline{\underline{A}}^T)$	B-1.3
$\nabla \cdot p\underline{\underline{I}}$	$=$	∇p	B-1.4
$\nabla \cdot \nabla \underline{v}$	$=$	$\nabla^2 \underline{v}$	B-1.5
$\nabla \cdot (\nabla \underline{v})^T$	$=$	$\nabla(\nabla \cdot \underline{v})$	B-1.6
$\nabla \cdot (\rho \underline{v} \, \underline{f})$	$=$	$\rho(\underline{v} \cdot \nabla \underline{f}) + \underline{f} \nabla \cdot (\rho \underline{v})$	B-1.7

See also the table in the inside cover of this text.

B.2 Differential operations in rectangular and curvilinear coordinates

Calculations of flow fields require that the governing equations be written in chosen coordinate systems. In this section, we list the governing equations in Cartesian, cylindrical, and spherical coordinates.

Table B.2. Differential operations in the rectangular coordinate system (x, y, z)

$$\underline{w} = \begin{pmatrix} w_x \\ w_y \\ w_z \end{pmatrix}_{xyz} \tag{B.2-1}$$

$$\nabla = \hat{e}_x \frac{\partial}{\partial x} + \hat{e}_y \frac{\partial}{\partial y} + \hat{e}_z \frac{\partial}{\partial z} \tag{B.2-2}$$

$$\nabla a = \begin{pmatrix} \frac{\partial a}{\partial x} \\ \frac{\partial a}{\partial y} \\ \frac{\partial a}{\partial z} \end{pmatrix}_{xyz} \tag{B.2-3}$$

$$\nabla \cdot \nabla a = \nabla^2 a = \frac{\partial^2 a}{\partial x^2} + \frac{\partial^2 a}{\partial y^2} + \frac{\partial^2 a}{\partial z^2} \tag{B.2-4}$$

$$\nabla \cdot \underline{w} = \frac{\partial w_x}{\partial x} + \frac{\partial w_y}{\partial y} + \frac{\partial w_z}{\partial z} \tag{B.2-5}$$

$$\nabla \times \underline{w} = \begin{pmatrix} \frac{\partial w_z}{\partial y} - \frac{\partial w_y}{\partial z} \\ \frac{\partial w_x}{\partial z} - \frac{\partial w_z}{\partial x} \\ \frac{\partial w_y}{\partial x} - \frac{\partial w_x}{\partial y} \end{pmatrix}_{xyz} \tag{B.2-6}$$

$$\underline{\underline{A}} = \begin{pmatrix} A_{xx} & A_{xy} & A_{xz} \\ A_{yx} & A_{yy} & A_{yz} \\ A_{zx} & A_{zy} & A_{zz} \end{pmatrix}_{xyz} \tag{B.2-7}$$

$$\nabla \underline{w} = \begin{pmatrix} \frac{\partial w_x}{\partial x} & \frac{\partial w_y}{\partial x} & \frac{\partial w_z}{\partial x} \\ \frac{\partial w_x}{\partial y} & \frac{\partial w_y}{\partial y} & \frac{\partial w_z}{\partial y} \\ \frac{\partial w_x}{\partial z} & \frac{\partial w_y}{\partial z} & \frac{\partial w_z}{\partial z} \end{pmatrix}_{xyz} \tag{B.2-8}$$

$$\nabla^2 \underline{w} = \begin{pmatrix} \frac{\partial^2 w_x}{\partial x^2} + \frac{\partial^2 w_x}{\partial y^2} + \frac{\partial^2 w_x}{\partial z^2} \\ \frac{\partial^2 w_y}{\partial x^2} + \frac{\partial^2 w_y}{\partial y^2} + \frac{\partial^2 w_y}{\partial z^2} \\ \frac{\partial^2 w_z}{\partial x^2} + \frac{\partial^2 w_z}{\partial y^2} + \frac{\partial^2 w_z}{\partial z^2} \end{pmatrix}_{xyz} \tag{B.2-9}$$

$$\nabla \cdot \underline{\underline{A}} = \begin{pmatrix} \frac{\partial A_{xx}}{\partial x} + \frac{\partial A_{yx}}{\partial y} + \frac{\partial A_{zx}}{\partial z} \\ \frac{\partial A_{xy}}{\partial x} + \frac{\partial A_{yy}}{\partial y} + \frac{\partial A_{zy}}{\partial z} \\ \frac{\partial A_{xz}}{\partial x} + \frac{\partial A_{yz}}{\partial y} + \frac{\partial A_{zz}}{\partial z} \end{pmatrix}_{xyz} \tag{B.2-10}$$

$$\underline{u} \cdot \nabla \underline{w} = \begin{pmatrix} u_x \frac{\partial w_x}{\partial x} + u_y \frac{\partial w_x}{\partial y} + u_z \frac{\partial w_x}{\partial z} \\ u_x \frac{\partial w_y}{\partial x} + u_y \frac{\partial w_y}{\partial y} + u_z \frac{\partial w_y}{\partial z} \\ u_x \frac{\partial w_z}{\partial x} + u_y \frac{\partial w_z}{\partial y} + u_z \frac{\partial w_z}{\partial z} \end{pmatrix}_{xyz} \tag{B.2-11}$$

Table B.3. Differential operations in the cylindrical coordinate system (r, θ, z)

$$\underline{w} = \begin{pmatrix} w_r \\ w_\theta \\ w_z \end{pmatrix}_{r\theta z} \tag{B.3-1}$$

$$\nabla = \hat{e}_r \frac{\partial}{\partial r} + \hat{e}_\theta \frac{1}{r} \frac{\partial}{\partial \theta} + \hat{e}_z \frac{\partial}{\partial z} \tag{B.3-2}$$

$$\nabla a = \begin{pmatrix} \frac{\partial a}{\partial r} \\ \frac{1}{r} \frac{\partial a}{\partial \theta} \\ \frac{\partial a}{\partial z} \end{pmatrix}_{r\theta z} \tag{B.3-3}$$

$$\nabla \cdot \nabla a = \nabla^2 a = \frac{1}{r} \frac{\partial}{\partial r} \left(r \frac{\partial a}{\partial r} \right) + \frac{1}{r^2} \frac{\partial^2 a}{\partial \theta^2} + \frac{\partial^2 a}{\partial z^2} \tag{B.3-4}$$

$$\nabla \cdot \underline{w} = \frac{1}{r} \frac{\partial}{\partial r} (r w_r) + \frac{1}{r} \frac{\partial w_\theta}{\partial \theta} + \frac{\partial w_z}{\partial z} \tag{B.3-5}$$

$$\nabla \times \underline{w} = \begin{pmatrix} \frac{1}{r} \frac{\partial w_z}{\partial \theta} - \frac{\partial w_\theta}{\partial z} \\ \frac{\partial w_r}{\partial z} - \frac{\partial w_z}{\partial r} \\ \frac{1}{r} \frac{\partial (r w_\theta)}{\partial r} - \frac{1}{r} \frac{\partial w_r}{\partial \theta} \end{pmatrix}_{r\theta z} \tag{B.3-6}$$

$$\underline{\underline{A}} = \begin{pmatrix} A_{rr} & A_{r\theta} & A_{rz} \\ A_{\theta r} & A_{\theta\theta} & A_{\theta z} \\ A_{zr} & A_{z\theta} & A_{zz} \end{pmatrix}_{r\theta z} \tag{B.3-7}$$

$$\nabla \underline{w} = \begin{pmatrix} \frac{\partial w_r}{\partial r} & \frac{\partial w_\theta}{\partial r} & \frac{\partial w_z}{\partial r} \\ \frac{1}{r} \frac{\partial w_r}{\partial \theta} - \frac{w_\theta}{r} & \frac{1}{r} \frac{\partial w_\theta}{\partial \theta} + \frac{w_r}{r} & \frac{1}{r} \frac{\partial w_z}{\partial \theta} \\ \frac{\partial w_r}{\partial z} & \frac{\partial w_\theta}{\partial z} & \frac{\partial w_z}{\partial z} \end{pmatrix}_{r\theta z} \tag{B.3-8}$$

$$\nabla^2 \underline{w} = \begin{pmatrix} \frac{\partial}{\partial r} \left(\frac{1}{r} \frac{\partial (r w_r)}{\partial r} \right) + \frac{1}{r^2} \frac{\partial^2 w_r}{\partial \theta^2} + \frac{\partial^2 w_r}{\partial z^2} - \frac{2}{r^2} \frac{\partial w_\theta}{\partial \theta} \\ \frac{\partial}{\partial r} \left(\frac{1}{r} \frac{\partial (r w_\theta)}{\partial r} \right) + \frac{1}{r^2} \frac{\partial^2 w_\theta}{\partial \theta^2} + \frac{\partial^2 w_\theta}{\partial z^2} + \frac{2}{r^2} \frac{\partial w_r}{\partial \theta} \\ \frac{1}{r} \frac{\partial}{\partial r} \left(r \frac{\partial w_z}{\partial r} \right) + \frac{1}{r^2} \frac{\partial^2 w_z}{\partial \theta^2} + \frac{\partial^2 w_z}{\partial z^2} \end{pmatrix}_{r\theta z} \tag{B.3-9}$$

$$\nabla \cdot \underline{\underline{A}} = \begin{pmatrix} \frac{1}{r} \frac{\partial}{\partial r} (r A_{rr}) + \frac{1}{r} \frac{\partial A_{\theta r}}{\partial \theta} + \frac{\partial A_{zr}}{\partial z} - \frac{A_{\theta\theta}}{r} \\ \frac{1}{r^2} \frac{\partial}{\partial r} (r^2 A_{r\theta}) + \frac{1}{r} \frac{\partial A_{\theta\theta}}{\partial \theta} + \frac{\partial A_{z\theta}}{\partial z} + \frac{A_{\theta r} - A_{r\theta}}{r} \\ \frac{1}{r} \frac{\partial}{\partial r} (r A_{rz}) + \frac{1}{r} \frac{\partial A_{\theta z}}{\partial \theta} + \frac{\partial A_{zz}}{\partial z} \end{pmatrix}_{r\theta z} \tag{B.3-10}$$

$$\underline{u} \cdot \nabla \underline{w} = \begin{pmatrix} u_r \left(\frac{\partial w_r}{\partial r} \right) + u_\theta \left(\frac{1}{r} \frac{\partial w_r}{\partial \theta} - \frac{w_\theta}{r} \right) + u_z \left(\frac{\partial w_r}{\partial z} \right) \\ u_r \left(\frac{\partial w_\theta}{\partial r} \right) + u_\theta \left(\frac{1}{r} \frac{\partial w_\theta}{\partial \theta} + \frac{w_r}{r} \right) + u_z \left(\frac{\partial w_\theta}{\partial z} \right) \\ u_r \left(\frac{\partial w_z}{\partial r} \right) + u_\theta \left(\frac{1}{r} \frac{\partial w_z}{\partial \theta} \right) + u_z \left(\frac{\partial w_z}{\partial z} \right) \end{pmatrix}_{r\theta z} \tag{B.3-11}$$

Table B.4. Differential operations in the spherical coordinate system (r, θ, ϕ)

$$\underline{w} = \begin{pmatrix} w_r \\ w_\theta \\ w_\phi \end{pmatrix}_{r\theta\phi}$$
    B.4-1

$$\nabla = \hat{e}_r \frac{\partial}{\partial r} + \hat{e}_\theta \frac{1}{r} \frac{\partial}{\partial \theta} + \hat{e}_\phi \frac{1}{r\sin\theta} \frac{\partial}{\partial \phi}$$
    B.4-2

$$\nabla a = \begin{pmatrix} \frac{\partial a}{\partial r} \\ \frac{1}{r} \frac{\partial a}{\partial \theta} \\ \frac{1}{r\sin\theta} \frac{\partial a}{\partial \phi} \end{pmatrix}_{r\theta\phi}$$
    B.4-3

$$\nabla \cdot \nabla a = \nabla^2 a = \frac{1}{r^2} \frac{\partial}{\partial r}\left(r^2 \frac{\partial a}{\partial r}\right) + \frac{1}{r^2\sin\theta} \frac{\partial}{\partial \theta}\left(\sin\theta \frac{\partial a}{\partial \theta}\right) + \frac{1}{r^2\sin^2\theta} \frac{\partial^2 a}{\partial \phi^2}$$
    B.4-4

$$\nabla \cdot \underline{w} = \frac{1}{r^2} \frac{\partial}{\partial r}\left(r^2 w_r\right) + \frac{1}{r\sin\theta} \frac{\partial}{\partial \theta}\left(w_\theta \sin\theta\right) + \frac{1}{r\sin\theta} \frac{\partial w_\phi}{\partial \phi}$$
    B.4-5

$$\nabla \times \underline{w} = \begin{pmatrix} \frac{1}{r\sin\theta} \frac{\partial}{\partial \theta}\left(w_\phi \sin\theta\right) - \frac{1}{r\sin\theta} \frac{\partial w_\theta}{\partial \phi} \\ \frac{1}{r\sin\theta} \frac{\partial w_r}{\partial \phi} - \frac{1}{r} \frac{\partial}{\partial r}\left(r w_\phi\right) \\ \frac{1}{r} \frac{\partial}{\partial r}\left(r w_\theta\right) - \frac{1}{r} \frac{\partial w_r}{\partial \theta} \end{pmatrix}_{r\theta\phi}$$
    B.4-6

$$\underline{\underline{A}} = \begin{pmatrix} A_{rr} & A_{r\theta} & A_{r\phi} \\ A_{\theta r} & A_{\theta\theta} & A_{\theta\phi} \\ A_{\phi r} & A_{\phi\theta} & A_{\phi\phi} \end{pmatrix}_{r\theta\phi}$$
    B.4-7

$$\nabla \underline{w} = \begin{pmatrix} \frac{\partial w_r}{\partial r} & \frac{\partial w_\theta}{\partial r} & \frac{\partial w_\phi}{\partial r} \\ \frac{1}{r} \frac{\partial w_r}{\partial \theta} - \frac{w_\theta}{r} & \frac{1}{r} \frac{\partial w_\theta}{\partial \theta} + \frac{w_r}{r} & \frac{1}{r} \frac{\partial w_\phi}{\partial \theta} \\ \frac{1}{r\sin\theta} \frac{\partial w_r}{\partial \phi} - \frac{w_\phi}{r} & \frac{1}{r\sin\theta} \frac{\partial w_\theta}{\partial \phi} - \frac{w_\phi}{r}\cot\theta & \frac{1}{r\sin\theta} \frac{\partial w_\phi}{\partial \phi} + \frac{w_r}{r} + \frac{w_\theta}{r}\cot\theta \end{pmatrix}_{r\theta\phi}$$
    B.4-8

$$\nabla^2 \underline{w} = \begin{pmatrix} \left(\frac{\partial}{\partial r}\left(\frac{1}{r^2}\frac{\partial}{\partial r}(r^2 w_r)\right) + \frac{1}{r^2\sin\theta}\frac{\partial}{\partial\theta}\left(\sin\theta\frac{\partial w_r}{\partial\theta}\right) + \frac{1}{r^2\sin^2\theta}\frac{\partial^2 w_r}{\partial\phi^2}\right. \\ \left. - \frac{2}{r^2\sin\theta}\frac{\partial}{\partial\theta}(w_\theta\sin\theta) - \frac{2}{r^2\sin\theta}\frac{\partial w_\phi}{\partial\phi}\right) \\ \left(\frac{1}{r^2}\frac{\partial}{\partial r}\left(r^2\frac{\partial w_\theta}{\partial r}\right) + \frac{1}{r^2}\frac{\partial}{\partial\theta}\left(\frac{1}{\sin\theta}\frac{\partial}{\partial\theta}(w_\theta\sin\theta)\right) + \frac{1}{r^2\sin^2\theta}\frac{\partial^2 w_\theta}{\partial\phi^2}\right. \\ \left. + \frac{2}{r^2}\frac{\partial w_r}{\partial\theta} - \frac{2\cot\theta}{r^2\sin\theta}\frac{\partial w_\phi}{\partial\phi}\right) \\ \left(\frac{1}{r^2}\frac{\partial}{\partial r}\left(r^2\frac{\partial w_\phi}{\partial r}\right) + \frac{1}{r^2}\frac{\partial}{\partial\theta}\left(\frac{1}{\sin\theta}\frac{\partial}{\partial\theta}(w_\phi\sin\theta)\right) + \frac{1}{r^2\sin^2\theta}\frac{\partial^2 w_\phi}{\partial\phi^2}\right. \\ \left. + \frac{2}{r^2\sin\theta}\frac{\partial w_r}{\partial\phi} + \frac{2\cot\theta}{r^2\sin\theta}\frac{\partial w_\theta}{\partial\phi}\right) \end{pmatrix}_{r\theta\phi}$$
    B.4-9

$$\nabla \cdot \underline{\underline{A}} = \begin{pmatrix} \frac{1}{r^2}\frac{\partial}{\partial r}(r^2 A_{rr}) + \frac{1}{r\sin\theta}\frac{\partial}{\partial\theta}(A_{\theta r}\sin\theta) + \frac{1}{r\sin\theta}\frac{\partial A_{\phi r}}{\partial\phi} - \frac{A_{\theta\theta}+A_{\phi\phi}}{r} \\ \frac{1}{r^3}\frac{\partial}{\partial r}(r^3 A_{r\theta}) + \frac{1}{r\sin\theta}\frac{\partial}{\partial\theta}(A_{\theta\theta}\sin\theta) + \frac{1}{r\sin\theta}\frac{\partial A_{\phi\theta}}{\partial\phi} + \frac{(A_{\theta r}-A_{r\theta})-A_{\phi\phi}\cot\theta}{r} \\ \frac{1}{r^3}\frac{\partial}{\partial r}(r^3 A_{r\phi}) + \frac{1}{r\sin\theta}\frac{\partial}{\partial\theta}(A_{\theta\phi}\sin\theta) + \frac{1}{r\sin\theta}\frac{\partial A_{\phi\phi}}{\partial\phi} + \frac{(A_{\phi r}-A_{r\phi})+A_{\phi\theta}\cot\theta}{r} \end{pmatrix}_{r\theta\phi}$$
    B.4-10

$$\underline{u} \cdot \nabla \underline{w} = \begin{pmatrix} u_r\left(\frac{\partial w_r}{\partial r}\right) + u_\theta\left(\frac{1}{r}\frac{\partial w_r}{\partial\theta} - \frac{w_\theta}{r}\right) + u_\phi\left(\frac{1}{r\sin\theta}\frac{\partial w_r}{\partial\phi} - \frac{w_\phi}{r}\right) \\ u_r\left(\frac{\partial w_\theta}{\partial r}\right) + u_\theta\left(\frac{1}{r}\frac{\partial w_\theta}{\partial\theta} + \frac{w_r}{r}\right) + u_\phi\left(\frac{1}{r\sin\theta}\frac{\partial w_\theta}{\partial\phi} - \frac{w_\phi}{r}\cot\theta\right) \\ u_r\left(\frac{\partial w_\phi}{\partial r}\right) + u_\theta\left(\frac{1}{r}\frac{\partial w_\phi}{\partial\theta}\right) + u_\phi\left(\frac{1}{r\sin\theta}\frac{\partial w_\phi}{\partial\phi} + \frac{w_r}{r} + \frac{w_\theta}{r}\cot\theta\right) \end{pmatrix}_{r\theta\phi}$$
    B.4-11

Table B.5. Continuity equation in three coordinate systems

Cartesian coordinates:

$$\frac{\partial \rho}{\partial t} + \left(v_x \frac{\partial \rho}{\partial x} + v_y \frac{\partial \rho}{\partial y} + v_z \frac{\partial \rho}{\partial z} \right) + \rho \left(\frac{\partial v_x}{\partial x} + \frac{\partial v_y}{\partial y} + \frac{\partial v_z}{\partial z} \right) = 0 \quad \text{B.5-1}$$

Cylindrical coordinates:

$$\frac{\partial \rho}{\partial t} + \frac{1}{r} \frac{\partial (\rho r v_r)}{\partial r} + \frac{1}{r} \frac{\partial (\rho v_\theta)}{\partial \theta} + \frac{\partial (\rho v_z)}{\partial z} = 0 \quad\quad\quad \text{B.5-2}$$

Spherical coordinates:

$$\frac{\partial \rho}{\partial t} + \frac{1}{r^2} \frac{\partial (\rho r^2 v_r)}{\partial r} + \frac{1}{r \sin \theta} \frac{\partial (\rho v_\theta \sin \theta)}{\partial \theta} + \frac{1}{r \sin \theta} \frac{\partial (\rho v_\phi)}{\partial \phi} = 0 \quad \text{B.5-3}$$

Table B.6. Equation of motion for incompressible fluids in three coordinate systems

Cartesian coordinates:

$$\rho \left(\frac{\partial v_x}{\partial t} + v_x \frac{\partial v_x}{\partial x} + v_y \frac{\partial v_x}{\partial y} + v_z \frac{\partial v_x}{\partial z} \right) = -\frac{\partial p}{\partial x} + \left(\frac{\partial \tau_{xx}}{\partial x} + \frac{\partial \tau_{yx}}{\partial y} + \frac{\partial \tau_{zx}}{\partial z} \right) + \rho g_x \quad \text{B.6-1}$$

$$\rho \left(\frac{\partial v_y}{\partial t} + v_x \frac{\partial v_y}{\partial x} + v_y \frac{\partial v_y}{\partial y} + v_z \frac{\partial v_y}{\partial z} \right) = -\frac{\partial p}{\partial y} + \left(\frac{\partial \tau_{xy}}{\partial x} + \frac{\partial \tau_{yy}}{\partial y} + \frac{\partial \tau_{zy}}{\partial z} \right) + \rho g_y \quad \text{B.6-2}$$

$$\rho \left(\frac{\partial v_z}{\partial t} + v_x \frac{\partial v_z}{\partial x} + v_y \frac{\partial v_z}{\partial y} + v_z \frac{\partial v_z}{\partial z} \right) = -\frac{\partial p}{\partial z} + \left(\frac{\partial \tau_{xz}}{\partial x} + \frac{\partial \tau_{yz}}{\partial y} + \frac{\partial \tau_{zz}}{\partial z} \right) + \rho g_z \quad \text{B.6-3}$$

Cylindrical coordinates:

$$\rho \left(\frac{\partial v_r}{\partial t} + v_r \frac{\partial v_r}{\partial r} + \frac{v_\theta}{r} \frac{\partial v_r}{\partial \theta} - \frac{v_\theta^2}{r} + v_z \frac{\partial v_r}{\partial z} \right) = -\frac{\partial p}{\partial r} + \left(\frac{1}{r} \frac{\partial (r \tau_{rr})}{\partial r} + \frac{1}{r} \frac{\partial \tau_{\theta r}}{\partial \theta} - \frac{\tau_{\theta\theta}}{r} + \frac{\partial \tau_{rz}}{\partial z} \right) + \rho g_r \quad \text{B.6-4}$$

$$\rho \left(\frac{\partial v_\theta}{\partial t} + v_r \frac{\partial v_\theta}{\partial r} + \frac{v_\theta}{r} \frac{\partial v_\theta}{\partial \theta} + \frac{v_\theta v_r}{r} + v_z \frac{\partial v_\theta}{\partial z} \right) = -\frac{1}{r} \frac{\partial p}{\partial \theta} + \left(\frac{1}{r^2} \frac{\partial (r^2 \tau_{r\theta})}{\partial r} + \frac{1}{r} \frac{\partial \tau_{\theta\theta}}{\partial \theta} + \frac{\partial \tau_{\theta z}}{\partial z} \right) + \rho g_\theta \quad \text{B.6-5}$$

$$\rho \left(\frac{\partial v_z}{\partial t} + v_r \frac{\partial v_z}{\partial r} + \frac{v_\theta}{r} \frac{\partial v_z}{\partial \theta} + v_z \frac{\partial v_z}{\partial z} \right) = -\frac{\partial p}{\partial z} + \left(\frac{1}{r} \frac{\partial (r \tau_{rz})}{\partial r} + \frac{1}{r} \frac{\partial \tau_{\theta z}}{\partial \theta} + \frac{\partial \tau_{zz}}{\partial z} \right) + \rho g_z \quad \text{B.6-6}$$

Spherical coordinates:

$$\rho \left(\frac{\partial v_r}{\partial t} + v_r \frac{\partial v_r}{\partial r} + \frac{v_\theta}{r} \frac{\partial v_r}{\partial \theta} + \frac{v_\phi}{r \sin \theta} \frac{\partial v_r}{\partial \phi} - \frac{v_\theta^2 + v_\phi^2}{r} \right)$$

$$= -\frac{\partial p}{\partial r} + \left(\frac{1}{r^2} \frac{\partial (r^2 \tau_{rr})}{\partial r} + \frac{1}{r \sin \theta} \frac{\partial (\tau_{\theta r} \sin \theta)}{\partial \theta} + \frac{1}{r \sin \theta} \frac{\partial \tau_{\phi r}}{\partial \phi} - \frac{\tau_{\theta\theta} + \tau_{\phi\phi}}{r} \right) + \rho g_r \quad \text{B.6-7}$$

$$\rho \left(\frac{\partial v_\theta}{\partial t} + v_r \frac{\partial v_\theta}{\partial r} + \frac{v_\theta}{r} \frac{\partial v_\theta}{\partial \theta} + \frac{v_\phi}{r \sin \theta} \frac{\partial v_\theta}{\partial \phi} + \frac{v_r v_\theta}{r} - \frac{v_\phi^2 \cot \theta}{r} \right)$$

$$= -\frac{1}{r} \frac{\partial p}{\partial \theta} + \left(\frac{1}{r^3} \frac{\partial (r^3 \tau_{r\theta})}{\partial r} + \frac{1}{r \sin \theta} \frac{\partial (\tau_{\theta\theta} \sin \theta)}{\partial \theta} + \frac{1}{r \sin \theta} \frac{\partial \tau_{\phi\theta}}{\partial \phi} + \frac{(\tau_{\theta r} - \tau_{r\theta})}{r} - \frac{(\cot \theta) \tau_{\phi\phi}}{r} \right)$$

$$+ \rho g_\theta \quad \text{B.6-8}$$

$$\rho \left(\frac{\partial v_\phi}{\partial t} + v_r \frac{\partial v_\phi}{\partial r} + \frac{v_\theta}{r} \frac{\partial v_\phi}{\partial \theta} + \frac{v_\phi}{r \sin \theta} \frac{\partial v_\phi}{\partial \phi} + \frac{v_r v_\phi}{r} + \frac{v_\phi v_\theta \cot \theta}{r} \right)$$

$$= -\frac{1}{r \sin \theta} \frac{\partial p}{\partial \phi} + \left(\frac{1}{r^3} \frac{\partial (r^3 \tau_{r\phi})}{\partial r} + \frac{1}{r \sin \theta} \frac{\partial (\tau_{\theta\phi} \sin \theta)}{\partial \theta} + \frac{1}{r \sin \theta} \frac{\partial \tau_{\phi\phi}}{\partial \phi} + \frac{\tau_{\phi r} - \tau_{r\phi}}{r} + \frac{(\cot \theta) \tau_{\phi\theta}}{r} \right)$$

$$+ \rho g_\phi \quad \text{B.6-9}$$

Table B.7. Equation of motion for incompressible Newtonian fluids: Navier-Stokes equations in three coordinate systems

Cartesian coordinates:

$$\rho\left(\frac{\partial v_x}{\partial t} + v_x\frac{\partial v_x}{\partial x} + v_y\frac{\partial v_x}{\partial y} + v_z\frac{\partial v_x}{\partial z}\right) = -\frac{\partial p}{\partial x} + \mu\left(\frac{\partial^2 v_x}{\partial x^2} + \frac{\partial^2 v_x}{\partial y^2} + \frac{\partial^2 v_x}{\partial z^2}\right) + \rho g_x \qquad \text{B.7-1}$$

$$\rho\left(\frac{\partial v_y}{\partial t} + v_x\frac{\partial v_y}{\partial x} + v_y\frac{\partial v_y}{\partial y} + v_z\frac{\partial v_y}{\partial z}\right) = -\frac{\partial p}{\partial y} + \mu\left(\frac{\partial^2 v_y}{\partial x^2} + \frac{\partial^2 v_y}{\partial y^2} + \frac{\partial^2 v_y}{\partial z^2}\right) + \rho g_y \qquad \text{B.7-2}$$

$$\rho\left(\frac{\partial v_z}{\partial t} + v_x\frac{\partial v_z}{\partial x} + v_y\frac{\partial v_z}{\partial y} + v_z\frac{\partial v_z}{\partial z}\right) = -\frac{\partial p}{\partial z} + \mu\left(\frac{\partial^2 v_z}{\partial x^2} + \frac{\partial^2 v_z}{\partial y^2} + \frac{\partial^2 v_z}{\partial z^2}\right) + \rho g_z \qquad \text{B.7-3}$$

Cylindrical coordinates:

$$\rho\left(\frac{\partial v_r}{\partial t} + v_r\frac{\partial v_r}{\partial r} + \frac{v_\theta}{r}\frac{\partial v_r}{\partial \theta} - \frac{v_\theta^2}{r} + v_z\frac{\partial v_r}{\partial z}\right)$$
$$= -\frac{\partial p}{\partial r} + \mu\left(\frac{\partial}{\partial r}\left(\frac{1}{r}\frac{\partial(r v_r)}{\partial r}\right) + \frac{1}{r^2}\frac{\partial^2 v_r}{\partial \theta^2} - \frac{2}{r^2}\frac{\partial v_\theta}{\partial \theta} + \frac{\partial^2 v_r}{\partial z^2}\right) + \rho g_r \qquad \text{B.7-4}$$

$$\rho\left(\frac{\partial v_\theta}{\partial t} + v_r\frac{\partial v_\theta}{\partial r} + \frac{v_\theta}{r}\frac{\partial v_\theta}{\partial \theta} + \frac{v_r v_\theta}{r} + v_z\frac{\partial v_\theta}{\partial z}\right)$$
$$= -\frac{1}{r}\frac{\partial p}{\partial \theta} + \mu\left(\frac{\partial}{\partial r}\left(\frac{1}{r}\frac{\partial(r v_\theta)}{\partial r}\right) + \frac{1}{r^2}\frac{\partial^2 v_\theta}{\partial \theta^2} + \frac{2}{r^2}\frac{\partial v_r}{\partial \theta} + \frac{\partial^2 v_\theta}{\partial z^2}\right) + \rho g_\theta \qquad \text{B.7-5}$$

$$\rho\left(\frac{\partial v_z}{\partial t} + v_r\frac{\partial v_z}{\partial r} + \frac{v_\theta}{r}\frac{\partial v_z}{\partial \theta} + v_z\frac{\partial v_z}{\partial z}\right) = -\frac{\partial p}{\partial z} + \mu\left(\frac{1}{r}\frac{\partial}{\partial r}\left(r\frac{\partial v_z}{\partial r}\right) + \frac{1}{r^2}\frac{\partial^2 v_z}{\partial \theta^2} + \frac{\partial^2 v_z}{\partial z^2}\right) + \rho g_z \qquad \text{B.7-6}$$

Spherical coordinates:

$$\rho\left(\frac{\partial v_r}{\partial t} + v_r\frac{\partial v_r}{\partial r} + \frac{v_\theta}{r}\frac{\partial v_r}{\partial \theta} + \frac{v_\phi}{r\sin\theta}\frac{\partial v_r}{\partial \phi} - \frac{v_\theta^2 + v_\phi^2}{r}\right)$$
$$= -\frac{\partial p}{\partial r} + \mu\left[\frac{1}{r^2}\frac{\partial^2}{\partial r^2}(r^2 v_r) + \frac{1}{r^2\sin\theta}\frac{\partial}{\partial \theta}\left(\sin\theta\frac{\partial v_r}{\partial \theta}\right) + \frac{1}{r^2\sin^2\theta}\frac{\partial^2 v_r}{\partial \phi^2}\right] + \rho g_r \qquad \text{B.7-7}$$

$$\rho\left(\frac{\partial v_\theta}{\partial t} + v_r\frac{\partial v_\theta}{\partial r} + \frac{v_\theta}{r}\frac{\partial v_\theta}{\partial \theta} + \frac{v_\phi}{r\sin\theta}\frac{\partial v_\theta}{\partial \phi} + \frac{v_r v_\theta}{r} - \frac{v_\phi^2\cot\theta}{r}\right)$$
$$= -\frac{1}{r}\frac{\partial p}{\partial \theta} + \mu\left[\frac{1}{r^2}\frac{\partial}{\partial r}\left(r^2\frac{\partial v_\theta}{\partial r}\right) + \frac{1}{r^2}\frac{\partial}{\partial \theta}\left(\frac{1}{\sin\theta}\frac{\partial}{\partial \theta}(v_\theta\sin\theta)\right)\right.$$
$$\left. + \frac{1}{r^2\sin^2\theta}\frac{\partial^2 v_\theta}{\partial \phi^2} + \frac{2}{r^2}\frac{\partial v_r}{\partial \theta} - \frac{2\cot\theta}{r^2\sin\theta}\frac{\partial v_\phi}{\partial \phi}\right] + \rho g_\theta \qquad \text{B.7-8}$$

$$\rho\left(\frac{\partial v_\phi}{\partial t} + v_r\frac{\partial v_\phi}{\partial r} + \frac{v_\theta}{r}\frac{\partial v_\phi}{\partial \theta} + \frac{v_\phi}{r\sin\theta}\frac{\partial v_\phi}{\partial \phi} + \frac{v_r v_\phi}{r} + \frac{v_\phi v_\theta\cot\theta}{r}\right)$$
$$= -\frac{1}{r\sin\theta}\frac{\partial p}{\partial \phi} + \mu\left[\frac{1}{r^2}\frac{\partial}{\partial r}\left(r^2\frac{\partial v_\phi}{\partial r}\right) + \frac{1}{r^2}\frac{\partial}{\partial \theta}\left(\frac{1}{\sin\theta}\frac{\partial}{\partial \theta}(v_\phi\sin\theta)\right) + \frac{1}{r^2\sin^2\theta}\frac{\partial^2 v_\phi}{\partial \phi^2}\right.$$
$$\left. + \frac{2}{r^2\sin\theta}\frac{\partial v_r}{\partial \phi} + \frac{2\cot\theta}{r^2\sin\theta}\frac{\partial v_\theta}{\partial \phi}\right] + \rho g_\phi \qquad \text{B.7-9}$$

Table B.8. Newtonian constitutive equation for incompressible fluids in rectangular, cylindrical, and spherical coordinates

Cartesian coordinates:

$$
\begin{pmatrix}
\tau_{xx} & \tau_{xy} & \tau_{xz} \\
\tau_{yx} & \tau_{yy} & \tau_{yz} \\
\tau_{zx} & \tau_{zy} & \tau_{zz}
\end{pmatrix}_{xyz}
= \mu
\begin{pmatrix}
2\frac{\partial v_x}{\partial x} & \frac{\partial v_x}{\partial y} + \frac{\partial v_y}{\partial x} & \frac{\partial v_x}{\partial z} + \frac{\partial v_z}{\partial x} \\
\frac{\partial v_y}{\partial x} + \frac{\partial v_x}{\partial y} & 2\frac{\partial v_y}{\partial y} & \frac{\partial v_y}{\partial z} + \frac{\partial v_z}{\partial y} \\
\frac{\partial v_z}{\partial x} + \frac{\partial v_x}{\partial z} & \frac{\partial v_z}{\partial y} + \frac{\partial v_y}{\partial z} & 2\frac{\partial v_z}{\partial z}
\end{pmatrix}_{xyz}
$$

B.8-1

Cylindrical coordinates:

$$
\begin{pmatrix}
\tau_{rr} & \tau_{r\theta} & \tau_{rz} \\
\tau_{\theta r} & \tau_{\theta\theta} & \tau_{\theta z} \\
\tau_{zr} & \tau_{z\theta} & \tau_{zz}
\end{pmatrix}_{r\theta z}
= \mu
\begin{pmatrix}
2\frac{\partial v_r}{\partial r} & r\frac{\partial}{\partial r}\left(\frac{v_\theta}{r}\right) + \frac{1}{r}\frac{\partial v_r}{\partial \theta} & \frac{\partial v_r}{\partial z} + \frac{\partial v_z}{\partial r} \\
r\frac{\partial}{\partial r}\left(\frac{v_\theta}{r}\right) + \frac{1}{r}\frac{\partial v_r}{\partial \theta} & 2\left(\frac{1}{r}\frac{\partial v_\theta}{\partial \theta} + \frac{v_r}{r}\right) & \frac{1}{r}\frac{\partial v_z}{\partial \theta} + \frac{\partial v_\theta}{\partial z} \\
\frac{\partial v_r}{\partial z} + \frac{\partial v_z}{\partial r} & \frac{1}{r}\frac{\partial v_z}{\partial \theta} + \frac{\partial v_\theta}{\partial z} & 2\frac{\partial v_z}{\partial z}
\end{pmatrix}_{r\theta z}
$$

B.8-2

Spherical coordinates:

$$
\begin{pmatrix}
\tau_{rr} & \tau_{r\theta} & \tau_{r\phi} \\
\tau_{\theta r} & \tau_{\theta\theta} & \tau_{\theta\phi} \\
\tau_{\phi r} & \tau_{\phi\theta} & \tau_{\phi\phi}
\end{pmatrix}_{r\theta\phi}
$$

$$
= \mu
\begin{pmatrix}
2\frac{\partial v_r}{\partial r} & r\frac{\partial}{\partial r}\left(\frac{v_\theta}{r}\right) + \frac{1}{r}\frac{\partial v_r}{\partial \theta} & \frac{1}{r\sin\theta}\frac{\partial v_r}{\partial \phi} + r\frac{\partial}{\partial r}\left(\frac{v_\phi}{r}\right) \\
r\frac{\partial}{\partial r}\left(\frac{v_\theta}{r}\right) + \frac{1}{r}\frac{\partial v_r}{\partial \theta} & 2\left(\frac{1}{r}\frac{\partial v_\theta}{\partial \theta} + \frac{v_r}{r}\right) & \frac{\sin\theta}{r}\frac{\partial}{\partial \theta}\left(\frac{v_\phi}{\sin\theta}\right) + \frac{1}{r\sin\theta}\frac{\partial v_\theta}{\partial \phi} \\
\frac{1}{r\sin\theta}\frac{\partial v_r}{\partial \phi} + r\frac{\partial}{\partial r}\left(\frac{v_\phi}{r}\right) & \frac{\sin\theta}{r}\frac{\partial}{\partial \theta}\left(\frac{v_\phi}{\sin\theta}\right) + \frac{1}{r\sin\theta}\frac{\partial v_\theta}{\partial \phi} & 2\left(\frac{1}{r\sin\theta}\frac{\partial v_\phi}{\partial \phi} + \frac{v_r}{r} + \frac{v_\theta\cot\theta}{r}\right)
\end{pmatrix}_{r\theta\phi}
$$

B.8-3

Note: These expressions are general and applicable to three-dimensional flows. For unidirectional flows, they reduce to Newton's law of viscosity (see Chapter 5).

Table B.9. The microscopic energy equation in rectangular, cylindrical, and spherical coordinates

Cartesian coordinates:

$$
\frac{\partial T}{\partial t} + \left(v_x\frac{\partial T}{\partial x} + v_y\frac{\partial T}{\partial y} + v_z\frac{\partial T}{\partial z}\right) = \frac{k}{\rho\hat{C}_p}\left[\frac{\partial^2 T}{\partial x^2} + \frac{\partial^2 T}{\partial y^2} + \frac{\partial^2 T}{\partial z^2}\right] + \frac{S}{\rho\hat{C}_p}
$$

B.9.1

Cylindrical coordinates:

$$
\frac{\partial T}{\partial t} + \left(v_r\frac{\partial T}{\partial r} + \frac{v_\theta}{r}\frac{\partial T}{\partial \theta} + v_z\frac{\partial T}{\partial z}\right) = \frac{k}{\rho\hat{C}_p}\left[\frac{1}{r}\frac{\partial}{\partial r}\left(r\frac{\partial T}{\partial r}\right) + \frac{1}{r^2}\frac{\partial^2 T}{\partial \theta^2} + \frac{\partial^2 T}{\partial z^2}\right] + \frac{S}{\rho\hat{C}_p}
$$

B.9.2

Spherical coordinates:

$$
\frac{\partial T}{\partial t} + \left(v_r\frac{\partial T}{\partial r} + \frac{v_\theta}{r}\frac{\partial T}{\partial \theta} + \frac{v_\phi}{r\sin\theta}\frac{\partial T}{\partial \phi}\right)
$$

$$
= \frac{k}{\rho\hat{C}_p}\left[\frac{1}{r^2}\frac{\partial}{\partial r}\left(r^2\frac{\partial T}{\partial r}\right) + \frac{1}{r^2\sin\theta}\frac{\partial}{\partial \theta}\left(\sin\theta\frac{\partial T}{\partial \theta}\right) + \frac{1}{r^2\sin^2\theta}\frac{\partial^2 T}{\partial \phi^2}\right] + \frac{S}{\rho\hat{C}_p}
$$

B.9.3

Table B.10. Power-law, generalized Newtonian constitutive equation for incompressible fluids in rectangular, cylindrical, and spherical coordinates

Cartesian coordinates:

$$\begin{pmatrix} \tau_{xx} & \tau_{xy} & \tau_{xz} \\ \tau_{yx} & \tau_{yy} & \tau_{yz} \\ \tau_{zx} & \tau_{zy} & \tau_{zz} \end{pmatrix}_{xyz} = \eta \underline{\underline{\dot{\gamma}}} \qquad \text{B.10-1}$$

$$\eta \equiv m \dot{\gamma}^{n-1} = m \left| \frac{1}{2} \cdot \begin{array}{c} \text{sum of squares} \\ \text{of each term in } \underline{\underline{\dot{\gamma}}} \end{array} \right|^{\frac{n-1}{2}} = m \left| \frac{1}{2} \cdot \sum_{p=1}^{3} \sum_{j=1}^{3} \dot{\gamma}_{pj}^{2} \right|^{\frac{n-1}{2}} \qquad \text{B.10-2}$$

$$\underline{\underline{\dot{\gamma}}} = \begin{pmatrix} 2\frac{\partial v_x}{\partial x} & \frac{\partial v_x}{\partial y} + \frac{\partial v_y}{\partial x} & \frac{\partial v_x}{\partial z} + \frac{\partial v_z}{\partial x} \\ \frac{\partial v_y}{\partial x} + \frac{\partial v_x}{\partial y} & 2\frac{\partial v_y}{\partial y} & \frac{\partial v_y}{\partial z} + \frac{\partial v_z}{\partial y} \\ \frac{\partial v_z}{\partial x} + \frac{\partial v_x}{\partial z} & \frac{\partial v_z}{\partial y} + \frac{\partial v_y}{\partial z} & 2\frac{\partial v_z}{\partial z} \end{pmatrix}_{xyz} \qquad \text{B.10-3}$$

Cylindrical coordinates:

$$\begin{pmatrix} \tau_{rr} & \tau_{r\theta} & \tau_{rz} \\ \tau_{\theta r} & \tau_{\theta\theta} & \tau_{\theta z} \\ \tau_{zr} & \tau_{z\theta} & \tau_{zz} \end{pmatrix}_{r\theta z} = \eta \underline{\underline{\dot{\gamma}}} \qquad \text{B.10-4}$$

$$\eta \equiv m \dot{\gamma}^{n-1} = m \left| \frac{1}{2} \cdot \begin{array}{c} \text{sum of squares} \\ \text{of each term in } \underline{\underline{\dot{\gamma}}} \end{array} \right|^{\frac{n-1}{2}} = m \left| \frac{1}{2} \cdot \sum_{p=1}^{3} \sum_{j=1}^{3} \dot{\gamma}_{pj}^{2} \right|^{\frac{n-1}{2}} \qquad \text{B.10-5}$$

$$\underline{\underline{\dot{\gamma}}} = \begin{pmatrix} 2\frac{\partial v_r}{\partial r} & r\frac{\partial}{\partial r}\left(\frac{v_\theta}{r}\right) + \frac{1}{r}\frac{\partial v_r}{\partial \theta} & \frac{\partial v_r}{\partial z} + \frac{\partial v_z}{\partial r} \\ r\frac{\partial}{\partial r}\left(\frac{v_\theta}{r}\right) + \frac{1}{r}\frac{\partial v_r}{\partial \theta} & 2\left(\frac{1}{r}\frac{\partial v_\theta}{\partial \theta} + \frac{v_r}{r}\right) & \frac{1}{r}\frac{\partial v_z}{\partial \theta} + \frac{\partial v_\theta}{\partial z} \\ \frac{\partial v_r}{\partial z} + \frac{\partial v_z}{\partial r} & \frac{1}{r}\frac{\partial v_z}{\partial \theta} + \frac{\partial v_\theta}{\partial z} & 2\frac{\partial v_z}{\partial z} \end{pmatrix}_{r\theta z} \qquad \text{B.10-6}$$

Spherical coordinates:

$$\begin{pmatrix} \tau_{rr} & \tau_{r\theta} & \tau_{r\phi} \\ \tau_{\theta r} & \tau_{\theta\theta} & \tau_{\theta\phi} \\ \tau_{\phi r} & \tau_{\phi\theta} & \tau_{\phi\phi} \end{pmatrix}_{r\theta\phi} = \eta \underline{\underline{\dot{\gamma}}} \qquad \text{B.10-7}$$

$$\eta \equiv m \dot{\gamma}^{n-1} = m \left| \frac{1}{2} \cdot \begin{array}{c} \text{sum of squares} \\ \text{of each term in } \underline{\underline{\dot{\gamma}}} \end{array} \right|^{\frac{n-1}{2}} = m \left| \frac{1}{2} \cdot \sum_{p=1}^{3} \sum_{j=1}^{3} \dot{\gamma}_{pj}^{2} \right|^{\frac{n-1}{2}} \qquad \text{B.10-8}$$

$$\underline{\underline{\dot{\gamma}}} = \begin{pmatrix} 2\frac{\partial v_r}{\partial r} & r\frac{\partial}{\partial r}\left(\frac{v_\theta}{r}\right) + \frac{1}{r}\frac{\partial v_r}{\partial \theta} & \frac{1}{r\sin\theta}\frac{\partial v_r}{\partial \phi} + r\frac{\partial}{\partial r}\left(\frac{v_\phi}{r}\right) \\ r\frac{\partial}{\partial r}\left(\frac{v_\theta}{r}\right) + \frac{1}{r}\frac{\partial v_r}{\partial \theta} & 2\left(\frac{1}{r}\frac{\partial v_\theta}{\partial \theta} + \frac{v_r}{r}\right) & \frac{\sin\theta}{r}\frac{\partial}{\partial \theta}\left(\frac{v_\phi}{\sin\theta}\right) + \frac{1}{r\sin\theta}\frac{\partial v_\theta}{\partial \phi} \\ \frac{1}{r\sin\theta}\frac{\partial v_r}{\partial \phi} + r\frac{\partial}{\partial r}\left(\frac{v_\phi}{r}\right) & \frac{\sin\theta}{r}\frac{\partial}{\partial \theta}\left(\frac{v_\phi}{\sin\theta}\right) + \frac{1}{r\sin\theta}\frac{\partial v_\theta}{\partial \phi} & 2\left(\frac{1}{r\sin\theta}\frac{\partial v_\phi}{\partial \phi} + \frac{v_r}{r} + \frac{v_\theta\cot\theta}{r}\right) \end{pmatrix}_{r\theta\phi} \qquad \text{B.10-9}$$

Note: These expressions are general and applicable to three-dimensional flows. For unidirectional flows, they reduce to the simple power-law expression (see Chapter 5).

Bibliography

[1] R. J. Adrian, "Particle-imaging technique for experimental fluid mechanics." *Annual Reviews in Fluid Mechanics*, 23, 261–304 (1991).

[2] Agilent Technologies, makers of Lab-on-a-Chip, a technology that employs microfluidics; www.agilent.com.

[3] J. D. Anderson, *Modern Compressible Flow: With Historical Perspective* (McGraw-Hill: Boston, 2003).

[4] Shelley Anna, Departments of Chemical and Mechanical Engineering, Carnegie Mellon University, Pittsburgh, PA.

[5] ANSYS, Inc., 10 Cavendish Court, Lebanon, NH, Web site: www.ansys.com.

[6] R. Aris, *Vectors, Tensors, and the Basic Equations of Fluid Mechanics* (Dover Publications: New York, 1989) unabridged and corrected republication of the work first published by Prentice-Hall, Inc., Englewood Cliffs, NJ, 1962.

[7] B. Atkinson, M. P. Brocklebank, C. C. H. Card, and J. M. Smith, "Low Reynolds number developing flows," *AIChE Journal*, 15, 548–53 (1969).

[8] H. A. Barnes, J. F. Hutton, and K. Walters, *An Introduction to Rheology* (Elsevier Science Publishers: New York, 1989).

[9] G. K. Batchelor, *An Introduction to Fluid Dynamics* (Cambridge University Press: New York, 1967).

[10] G. K. Batchelor, *The Theory of Homogeneous Turbulence* (Cambridge University Press: New York, 1953).

[11] J. J. Bertin, *Aerodynamics for Engineers*, 4th edition (Prentice Hall: Upper Saddle River, NJ, 2002).

[12] R. B. Bird, R. C. Armstrong, and O. Hassager, *Dynamics of Polymeric Liquids, Volume 1: Fluid Mechanics*, 2nd edition (John Wiley & Sons: New York, 1987).

[13] R. B. Bird, C. F. Curtiss, R. C. Armstrong, and O. Hassager, *Dynamics of Polymeric Liquids, Volume 2: Kinetic Theory*, 2nd edition (John Wiley & Sons: New York, 1987).

[14] R. B. Bird, W. Stewart, and E. Lightfoot, *Transport Phenomena* (John Wiley & Sons: New York, 1960).

[15] R. B. Bird, W. Stewart, and E. Lightfoot, *Transport Phenomena*, 2nd edition (John Wiley & Sons: New York, 2002).

[16] T. V. Boussinesq, *Mém. prés. Acad. Sci.*, 3rd edition, Paris, XXIII, 46(1877), as cited in [14, p. 160].

[17] W. E. Boyce and R. C. DiPrima, *Elementary Differential Equations and Boundary Value Problems,* 3rd edition (John Wiley & Sons: New York, 1977).

[18] California Institute of Technology, "Caltech scientists develop new cell sorter," Press Release November 8, 1999; pr.caltech.edu/media/Press_Releases/PR12016.html.

[19] P. J. Carreau, D. C. R. De Kee, and R. P. Chhabra, *Rheology of Polymeric Systems: Principles and Applications* (Hanser Publishers: New York, 1997).

[20] H. S. Carslaw and J. C. Yeager, *Conduction of Heat in Solids*, 2nd edition (Oxford University Press: New York, 1959).

[21] K. B. Chandran, A. Y. Yoganathan, and S. E. Rittgers, *Biofluid Mechanics: The Human Circulation* (CRC Press, Taylor & Francis Group: Boca Raton, FL, 2007).

[22] S. Chandrasekhar, *Hydrodynamic and Hydromagnetic Stability* (Clarendon Press: Oxford, 1961).

[23] S. W. Churchill, *Viscous Flows: The Practical Use of Theory* (Butterworths: Stoneham, MA, 1988).

[24] T. B. Co, *Applied Mathematics for Chemical Engineers* (Cambridge University Press: New York, 2013, forthcoming).

[25] C. F. Colebrook, "Turbulent flow in pipes, with particular reference to the transition between the smooth and rough pipe laws," *Journal Institution of Civil Engineering London*, 11, 133–56 (1938–1939).

[26] B. D. Coleman and W. Noll, "Recent results in the continuum theory of viscoelastic fluids," *Annals of the New York Academy of Science*, 89, 672–714 (1961).

[27] COMSOL Multiphysics Software, COMSOL, Inc., 1 New England Executive Park, Suite 350, Burlington, MA; www.comsol.com.

[28] R. J. Cornish, "Flow in a pipe of rectangular cross-section," *Proceedings of the Royal Society of London*, 120(A), 691–700 (1928).

[29] M. M. Couette, "Distinction de deux regimes dans le mouvement des fluides," *Journal de Physique* [Ser. 2], 9, 414–424 (1890), as cited in Macosko [90], p. 107.

[30] J. M. Coulson, J. F. Richardson, J. R. Backhurst, and J. H. Harker, *Chemical Engineering Volume 2: Particle Technology and Separation Processes*, 4th edition (Pergamon Press: New York, 1991).

[31] Crayola LLC manufactures Silly Putty; www.sillyputty.com.

[32] D. A. Crowl and J. F. Levar, *Chemical Process Safety: Fundamentals with Applications*, 3rd edition (Prentice Hall: Englewood Cliffs, NJ, 2011).

[33] E. L. Cussler, *Diffusion: Mass Transfer in Fluid Systems* (Cambridge University Press: Cambridge, 1984).

[34] R. L. Daugherty, J. B. Franzini, and E. J. Finnemore, *Fluid Mechanics with Applications* (McGraw Hill: Singapore, 1989).

[35] P. A. Davidson, *An Introduction to Magnetohydrodynamics* (Cambridge University Press: New York, 2001).

[36] J. M. Dealy, "Weissenberg and Deborah numbers: Their definition and use," *Rheology Bulletin*, 2(July), 14–18 (2010).

[37] J. M. Dealy and K. F. Wissbrun, *Melt Rheology and Its Role in Plastics Processing* (Van Nostrand Reinhold: New York, 1990).

[38] W. R. Dean, "Note on the motion of fluid in a curved pipe," *Philosophical Magazine*, 4, 208–23 (1927).

[39] W. R. Dean, "The stream line motion of fluid in a curved pipe," *Philosophical Magazine*, 5, 674–95 (1928).

[40] W. M. Deen, *Analysis of Transport Phenomena* (Oxford University Press: New York, 1998).

[41] D. De Kee, "Equations rheologique pour decrier le comportent des fluids polymeriques," Ph.D. thesis, Ecole Polytechnique, Montreal, QC, Canada, 1977, as cited in [19].

[42] A. O. Demuren and W. Rodi, "Calculations of turbulence-driven secondary motion in non-circular ducts," *Journal of Fluid Mechanics*, 140, 189–222 (1984).

[43] M. M. Denn, *Process Fluid Mechanics* (Prentice-Hall: Englewood Cliffs, NJ, 1980).

[44] K. E. Drexler, *Nanosystems—Molecular Machinery, Manufacturing, and Computation* (John Wiley & Sons: New York, 1992).

[45] S. Ergun, "Fluid flow through packed columns," *Chemical Engineering Progress*, 48, 89–94 (1952).

[46] J. B. Evett and C. Liu, *2,500 Solved Problems in Fluid Mechanics and Hydraulics* (McGraw-Hill: New York, 1989).

[47] R. M. Felder and R. W. Rousseau, *Elementary Principles of Chemical Processes*, 3rd edition (John Wiley & Sons: New York, 1999).

[48] J. D. Ferry, *Viscoelastic Properties of Polymers* (John Wiley & Sons: New York, 1980).

[49] J. H. Ferziger and M. Peric, *Computational Methods for Fluid Mechanics*, 3rd revised edition (Springer: Heidelberg, Germany, 2002).

[50] O. Flachsbart, "Neuere Untersuchungen über den Luftwiderstand von Kugeln," *Phys. Z.*, 28, 461–9 (1927). Cited in Schlichting [148], p. 21, ref. 3.

[51] A. Y. Fu, C. Spence, A. Scherer, F. H. Arnold, and S. R. Quake, "A microfabricated fluorescence-activated cell sorter," *Nature Biotechnology* 17, 1109–11 (1999).

[52] Y. C. Fung, *Biodynamics: Circulation* (Springer-Verlag: New York, 1984).

[53] Y. C. Fung, *Biomechanics: Circulation*, 2nd edition (Springer: New York, 1997).

[54] J. D. S. Gaylor, "Membrane oxygenators: Current developments in design and application," *Journal of Biomedical Engineering* 10, 541–4 (1988).

[55] C. J. Geankoplis, *Transport Processes and Unit Operations*, 3rd edition (Prentice-Hall: Englewood Clifs, NJ, 1993).

[56] R. V. Giles, C. Liu, and J. B. Evett, *Schaum's Outline of Fluid Mechanics and Hydraulics,* 3rd edition (McGraw-Hill: New York, 1994).

[57] V. Girault and P.-A. Raviart, *Finite Element Methods for Navier-Stokes Equations: Theory and Algorithms* (Springer: New York, 1986).

[58] M. D. Greenberg, *Foundations of Applied Mathematics* (Prentice-Hall: Englewood Cliffs, NJ, 1978).

[59] J. Herivel, *Joseph Fourier: The Man and the Physicist* (Clarendon Press: Oxford, 1975).

[60] P. C. Hiemenz and T. P. Lodge, *Polymer Chemistry*, 2nd edition (CRC Press: Boca Raton, FL, 2007).

[61] F. B. Hildebrand, *Advanced Calculus for Applications,* 2nd edition (Prentice-Hall: Englewood Cliffs, NJ, 1976).

[62] J. O. Hirschfelder, C. F. Curtiss, and R. B. Bird, *Molecular Theory of Gases and Liquids* (John Wiley & Sons: New York, 1954).

[63] S. F. Hoerner, *Fluid-Dynamic Drag: Practical Information on Aerodynamic Drag and Hydrodynamic Resistance* (published by the author: Midland Park, NJ, 1965).

[64] S. F. Hoerner and H. V. Borst, *Fluid-Dynamic Lift: Practical Information on Aerodynamic and Hydrodynamic Lift* (published by Liselotte A. Hoerner: Hoerner Fluid Dynamics, Brick Town, NJ, 1975).

[65] G. M. Homey, H. Aref, K. S. Breuer, S. Hochgreb, J. R. Koseff, B. R. Munson, K. G. Powell, C. R. Robertson, and S. T. Thoroddsen, *Multi-Media Fluid Mechanics,* CD-ROM (Cambridge University Press: Cambridge, 2000).

[66] A. A. Johnson, "Airflow Past an Automobile," Army HPC Research Center, Network Computing Services, Inc., Minneapolis, MN; www.arc.umn.edu/johnson/autoex.html.

[67] E. Jones and R. Childers, *Contemporary College Physics,* 3rd edition (WCB McGraw-Hill: New York, 1999).

[68] O. C. Jones, Jr., "An improvement in the calculations of turbulent friction in rectangular ducts," *Journal of Fluids Engineering*, 173–81 (June 1976).

[69] O. C. Jones, Jr., and J. C. M. Leung, "An improvement in the calculation of turbulent friction in smooth concentric annuli," *Journal of Fluids Engineering*, 103, 615–23 (1981).

[70] D. D. Joseph and Y. Y. Renardy, *Fundamentals of Two-Fluid Dynamics* (Springer-Verlag: New York, 1993).

[71] N. Joukowski, *Zeitschr. f. Flugt. u. Motorluftsch.* Vol. 1, p. 281 (1910). As cited (p. 204) in Prandtl [136].

[72] J. M. Kay, *An Introduction to Fluid Mechanics & Heat Transfer*, 2nd edition (Cambridge University Press: New York, 1963).

[73] Khan Academy is a not-for-profit educational organization focussed on Internet-based educational approaches; www.khanacademy.org.

[74] J. A. King, T. M. Tambling, F. A. Morrison, J. M. Keith, A. J. Cole, and R. M. Pagel, "Effects of carbon fillers on the rheology of highly filled liquid-crystal polymer based resins," *Journal of Applied Polymer Science*, 108(3), 1646–56 (2008).

[75] B. J. Kirby, *Micro- and Nanoscale Fluid Mechanics: Transport in Microfluidic Devices* (Cambridge University Press: New York, 2010).

[76] A. M. Kuethe and J. D. Schetzer, *Foundations of Aerodynamics*, 2nd edition (John Wiley & Sons: New York, 1959).

[77] W. M. Kutta, *Sitzungsber. d. Bayr. Akad. d. Wiss., M.-Ph. Kl.*, 1910, 1911. As cited (p. 204) in Prandtl [136].

[78] W. M. Lai, D. Rubin, and E. Krempl, *Introduction to Continuum Mechanics* (Pergamon Press: New York, 1978).

[79] H. Lamb, *Hydrodynamics*, 6th edition (Dover: New York, 1945).

[80] L. D. Landau and E. M. Lifshitz, *Fluid Mechanics* (Addison-Wesley: Reading, MA, 1959).

[81] H. L. Langhaar, "Steady flow in the transition length of a straight tube," *Journal of Applied Mechanics*, 64, A55–A58 (1942).

[82] R. G. Larson, *Constitutive Equations for Polymer Melts and Solutions* (Butterworths: Boston, 1988).

[83] R. G. Larson, *The Structure and Rheology of Complex Fluids* (Oxford University Press: New York, 1999).

[84] R. G. Larson and T. J. Rehg, "Spin coating," in *Liquid Film Coating*, S. F. Kistler and P. M. Schweizer (eds.) (Chapman & Hall: London, 1997).

[85] G. L. Leal, *Laminar Flow and Convective Transport Processes: Scaling Principles and Asymptotic Analysis* (Butterworth-Heinemann: Boston, 1992).

[86] R. W. Lewis, P. Nithiarasu, and K. N. Seetharamu, *Fundamentals of the Finite Element Method for Heat and Fluid Flow* (John Wiley & Sons: West Sussex, UK, 2004).

[87] D. R. Lide, *CRC Handbook of Chemistry and Physics,* 71st edition (CRC Press: Boston, 1990).

[88] A. S. Lodge, *Body Tensor Fields in Continuum Mechanics* (Academic Press: New York, 1974).

[89] A. S. Lodge, *Elastic Liquids* (Academic Press: New York, 1964).

[90] C. W. Macosko, *Rheology Principles, Measurements, and Applications* (VCH Publishers, Inc: New York, 1994).

[91] H. Markovitz, "The emergence of rheology," *Physics Today*, 21, 23 (1968).

[92] H. Markovitz, *Rheological Behavior of Fluids* (Educational Services: Watertown, MA, 1965). Also cited as [119].

[93] G. E. Mase, *Schaum's Outline of Theory and Problems of Continuum Mechanics* (McGraw-Hill: New York, 1970).

[94] Mathsoft, Inc., Cambridge, MA; www.ptc.com/products/mathcad/.

[95] J. McMurry and R. C. Fay, *Chemistry*, 4th edition (Prentice-Hall: Upper Saddle River, NJ, 2004).

[96] E. V. Menezes and W. W. Graessley, "Nonlinear rheological behavior of polymer systems for several shear-flow histories," *Journal of Polymer Science, Polymer Physics*, 20, 1817–33 (1982).

[97] A. B. Metzner and M. Whitlock, "Flow behavior of concentrated (dilatant) suspensions," *Transactions of the Society of Rheology*, 2, 239–54 (1958).

[98] S. Middleman, *Fundamentals of Polymer Processing* (McGraw-Hill: New York, 1977).

[99] M. Modell and R. C. Reid, *Thermodynamics and Its Applications in Chemical Engineering* (Prentice-Hall: Englewood Cliffs, NJ, 1974).

[100] P. Moin, *Fundamentals of Engineering Numerical Analysis*, 2nd edition (Cambridge University Press: New York, 2010).

[101] E. Möller, "Luftwiderstandsmessungen am Volkswagen-Lieferwagen," *Automobiltechnische Z.*, 53, 1–4 (1951).

[102] A. Monterosso and N. Wimmer, "Pumping head curve at lab bench 5 in the junior chemical engineering laboratory," report submitted in CM3215 Fundamentals of Chemical Engineering Laboratory, November 10, 2010, Michigan Technological University, Houghton, MI.

[103] Lewis F. Moody, "Friction factors for pipe flow," *ASME Transactions*, 66, 671–84 (1944).

[104] F. A. Morrison, *Understanding Rheology* (Oxford University Press: New York, 2001).

[105] F. A. Morrison, "Data correlation for friction factor in smooth pipes," Department of Chemical Engineering, Michigan Technological University, Houghton, MI; www.chem.mtu.edu/fmorriso/DataCorrelationForSmooth Pipes2010.pdf.

[106] F. A. Morrison, "Data correlation for drag coefficient for a sphere," Department of Chemical Engineering, Michigan Technological University, Houghton, MI; www.chem.mtu.edu/fmorriso/DataCorrelationForSphereDrag2010.pdf.

[107] F. A. Morrison, "Michigan Tech Oobleck Run 2008," Michigan Technological University, Houghton, MI; http://youtu.be/ Lz8VWZY0iOE, accessed November 14, 2012.

[108] F. A. Morrison "Mathematics Appendix," Michigan Technological University, Houghton, MI; www.chem.mtu.edu/~fmorriso/MathematicsAppendix2012.pdf.

[109] R. L. Mott, *Applied Fluid Mechanics*, 5th edition (Prentice-Hall: Upper Saddle River, NJ, 2000).

[110] P. Moulin, J. C. Rouch, C. Serra, M. J. Clifton, and P. Aptel, "Mass transfer improvement by secondary flows: Dean vortices in coiled tubular membranes," *Journal of Membrane Science* 114, 235–44 (1996).

[111] B. R. Munson, D. F. Young, and T. H. Okiishi, *Fundamentals of Fluid Mechanics*, 5th edition (John Wiley & Sons: New York, 2006).

[112] National Committee for Fluid Mechanics Films (NCFMF), 26 films illustrating key topics in fluid mechanics made between 1961 and 1969. Distributed as VHS videos by Encyclopaedia Britanica Educational Corporation, Chicago, IL. The accompanying film notes were published by MIT Press [154]. Videos may be streamed at web.mit.edu/fluids/www/Shapiro/ncfmf.html.

[113] National Committee for Fluid Mechanics Films (NCFMF), *The Fluid Dynamics of Drag*, a four-part film on drag, part of the NCFMF series [112]. The dropping sphere appears in Part IV.

[114] National Committee for Fluid Mechanics Films (NCFMF), *Vorticity*, a two-part film on vorticity, part of the NCFMF series [112].

[115] National Committee for Fluid Mechanics Films (NCFMF), *Waves in Fluids*, part of the NCFMF series [112].

[116] National Committee for Fluid Mechanics Films (NCFMF), *Flow Instabilities*, part of the NCFMF series [112].

[117] National Committee for Fluid Mechanics Films (NCFMF), *Turbulence*, part of the NCFMF series [112].

[118] National Committee for Fluid Mechanics Films (NCFMF), *Pressure Fields and Fluid Acceleration*, part of the NCFMF series [112].

[119] National Committee for Fluid Mechanics Films (NCFMF), *Rheological Behavior of Fluids*, part of the NCFMF series [112].

[120] National Committee for Fluid Mechanics Films (NCFMF), *Eulerian Lagrangian Description*, part of the NCFMF series [112].

[121] National Committee for Fluid Mechanics Films (NCFMF), *Fundamentals of Boundary Layers*, part of the NCFMF series [112].

[122] National Committee for Fluid Mechanics Films (NCFMF), *Surface Tension in Fluid Mechanics*, part of the NCFMF series [112].

[123] National Committee for Fluid Mechanics Films (NCFMF), *Secondary Flow*, part of the NCFMF series [112].

[124] National Committee for Fluid Mechanics Films (NCFMF), *Channel Flow of a Compressible Fluid*, part of the NCFMF series [112].

[125] I. S. Newton, *PhilosophiæNaturalis Principia Mathematica* (1687). Available in English online as G. E. Smith, "Newton's PhilosophiæNaturalis Principia Mathematica," *The Stanford Encyclopedia of Philosophy* (Winter 2008), E. N. Zalta (ed.).

[126] J. Nikuradse, "Stromungsgesetze in Rauhen Rohren," *VDI Forschungsh*, 361 (1933); English translation, *NACA Technical Memorandum 1292*.

[127] J. Nikuradse, "Untersuchungen über turbulente Strömungen in nicht kreisförmigen Rohren," Archive of Applied Mechanics (Ingenieur Archiv), 1(3), 306–32 (1930).

[128] N. T. Obot, "Determination of incompressible flow friction in smooth circular and noncircular passages: A generalized approach including validation of the nearly century old hydraulic diameter concept," *Journal of Fluids Engineering* 110, 431–40 (1988).

[129] J. G. Oldroyd, "On the formulation of rheological equations of state," *Proceedings of the Royal Society*, A200, 523–41 (1950).

[130] "Oobleck," available at online encyclopedia entry from Wikipedia, the free encyclopedia, en.wikipedia.org/wiki/Oobleck. A Spanish television program "El Hormiguero" achieved international attention in October and December of 2006 for having people walk across a swimming pool filled with a cornstarch and water suspension; Youtube.com (search for "oobleck").

[131] W. Ostwald, "Ueber die Geschwindigkeitsfunktion der Viskositat Disperser Systeme, I.," *Kolloid Z.*, 36, 99–117 (1925), as cited in [19].

[132] R. H. Perry and D. W. Green, *Perry's Chemical Engineers' Handbook*, 8th edition (McGraw Hill: New York, 2008). Online version available at www.knovel.com/.

[133] "Physics of Fluids: The Gallery of Fluid Motion," a selection of award-winning photographs and videos from Physics of Fluids. The entries were chosen from among entries to the Annual Picture Gallery of Fluid Motion exhibit, held at the annual meeting of the American Physical Society, Division of Fluid Dynamics; ojps.aip.org/phf/.

[134] L. Prandtl, *Verhandlungen des III Internationalen Mathematiker-Kongresses* (Heidelberg, Germany, 1904), published in Leipzig, pp. 484–91.

[135] L. Prandtl, "Bericht uber Untersuchungen zur Ausgebideten Turbulenz," *Zeits. f. angew. Math. u. Mech.*, 5, 136–9 (1925).

[136] L. Prandtl, *Essentials of Fluid Dyanamics* (Hafner Publishing Company: New York, 1952).

[137] R. S. Prengle and R. R. Rothfus, "Fluid mechanics studies – Transition phenomena in pipes and annular cross sections," *Industrial and Engineering Chemistry*, 47, 379–86 (1955).

[138] M. Raffel, C. Willert, and J. Kompenhans, *Particle Image Velocimetry: A Practical Guide* (Springer: New York, 2002).

[139] O. Reynolds, "An experimental investigation of the circumstances which determine whether the motion of water shall be direct or sinuous and of the law of resistance in parallel channels," *Philosophical Transactions of the Royal Society*, 174, 935–82 (1883).

[140] A. M. Robertson and S. J. Muller, "Flow of Oldroyd-B fluids in curved pipes of circular and annular cross section," *International Journal of Non-Linear Mechanics*, 31, 1–20 (1996).

[141] R. Rowland, "Patient gets first totally implanted artificial heart," July 3, 2001; www.cnn.com/2001/HEALTH/conditions/07/03/artificial.heart/.

[142] R. H. Sabersky, A. J. Acosta, and E. G. Hauptmann, *Fluid Flow: A First Course in Fluid Mechanics*, 3rd edition (Macmillan Publishing Company: New York, 1989).

[143] H. Sakamoto and H. Haniu, "A study on vortex shedding from spheres in a uniform flow," *Transactions of American Society of Mechanical Engineers*, 112, 386–92, December 1990.

[144] R. Salazar and L. D. Gelb, "A computational study of the reconstruction of amorphous mesoporous materials from gas adsorption isotherms and structure factors via evolutionary optimization," *Langmuir*, 23, 530–541 (2007).

[145] W. M. Saltzman, *Biomedical Engineering: Bridging Medicine and Technology* (Cambridge University Press: New York, 2009).

[146] H. M. Schey, *Div, Grad, Curl, and All That: An Informal Test on Vector Calculus*, 3rd edition (W. W. Norton & Company: New York, 1996).

[147] H. Schlichting, *Boundary Layer Theory* (McGraw-Hill: New York, 1955).

[148] H. Schlichting, *Boundary Layer Theory*, 7th edition (McGraw-Hill: New York, 1979).

[149] H. Schlichting, K. Gersten, and K. Gersten, *Boundary Layer Theory*, 8th edition (Springer: New York, 2000).

[150] W. R. Schowalter, *Mechanics of Non-Newtonian fluids* (Oxford University Press: New York, 1978).

[151] SGI, Inc., Sunnyvale, CA; www.sgi.com.

[152] R. K. Shah and A. L. London, *Laminar Flow Forced Convection in Ducts* (Academic Press: New York, 1978).

[153] P. A. Shamlou, *Handling of Bulk Solids* (Butterworths: London, 1988).

[154] A. H. Shapiro, *Illustrated Experiments in Fluid Mechanics: The NCFMF Book of Film Notes* (MIT Press: Cambridge, MA, 1972).

[155] A. H. Shapiro, *Shape and Flow: The Fluid Dynamics of Drag* (Anchor Books, Doubleday & Company: New York, 1961).

[156] W. A. Sirignano, *Fluid Dynamics and Transport of Droplets and Sprays*, 2nd edition (Cambridge University Press: New York, 2010).

[157] J. M. Smith, H. C. Van Hess, and M. M. Abbott, *Introduction to Chemical Engineering Thermodynamics*, 6th edition (McGraw Hill: New York, 2001).

[158] A. J. W. Sommerfeld, *Partial Differential Equations in Physics*, (Academic Press: New York, 1949).

[159] "Space Educators Handbook, NASA Lunar Feather Drop Home Page." This page contains a video of the astronaut David Scott dropping two objects on the moon, Jerry Woodfill, author (jared.woodfill1@jsc.nasa.gov), Cecilia Breigh, Curator, NASA JSC ER; http://er.jsc.nasa.gov/seh/feather.html.

[160] M. R. Spiegel, *Schaum's Outline of Theory and Problems of Advanced Mathematics for Engineers and Scientists* (McGraw-Hill: New York, 1971).

[161] H. M. Stommel and D. W. Moore, *An Introduction to the Coriolis Force* (Columbia University Press: New York, 1989).

[162] Swagelok, Incorporated; www.swagelok.com.

[163] Z. Tadmor and C. G. Gogos, *Principles of Polymer Processing* (John Wiley & Sons: New York, 1979).

[164] R. I. Tanner, *Engineering Rheology,* revised edition (Clarendon Press: Oxford, 1988).

[165] H. Tennekes and J. L. Lumley, *A First Course in Turbulence* (MIT Press, Cambridge, MA, 1972).

[166] G. B. Thomas and R. L. Finney, *Calculus and Analytic Geometry*, 6th edition (Addison-Wesley: Reading, MA, 1984).

[167] P. A. Tipler, *Physics* (Worth Publishers: New York, 1976).

[168] D. J. Tritton, *Physical Fluid Dynamics*, 2nd edition (Oxford University Press: Oxford, England, 1988).

[169] C. Truesdell and W. Noll, *The Nonlinear Field Theories of Mechanics* (Springer-Verlag: Berlin, 1965).

[170] M. Van Dyke, *An Album of Fluid Motion* (Parabolic Press: Stanford, CA, 1982).

[171] N. J. Wagner and J. F. Brady, "Shear thickening in colloidal dispersions," *Physics Today*, October 2009, 27–32.

[172] C. M. Walker, "Device to prop open arteries may prove major heart treatment advance," Cardiovascular Institute of the South, Houma; www.chelationtherapyonline.com/articles/p79.htm, accessed November 14, 2012.

[173] J. Walker, "The amateur scientist," *Scientific American*, 240, 180–9, April 1979.

[174] Frank M. White, *Viscous Fluid Flow* (McGraw-Hill, Inc.: New York, 1974).

[175] Frank M. White, *Viscous Fluid Flow*, 3rd edition (McGraw-Hill, Inc.: New York, 2006).

[176] Frank M. White, *Fluid Mechanics*, 3rd edition (McGraw-Hill, Inc.: New York, 1994).

[177] Frank M. White, *Fluid Mechanics*, 4th edition (McGraw-Hill, Inc.: New York, 1999).

[178] Frank M. White, *Fluid Mechanics*, 5th edition (McGraw-Hill, Inc.: New York, 2002).

[179] R. E. Williamson and H. F. Trotter, *Multivariable Mathematics: Linear Algebra, Calculus, Differential Equations,* 2nd edition (Prentice-Hall: Englewood Cliffs, NJ, 1979).

[180] Wolfram Research, Inc., Champaign, IL; www.wolfram.com.

[181] K. Yasuda, "Investigation of the analogies between viscometric and linear viscoelastic properties of polystyrene fluids," Ph.D. Thesis, Massachusetts Institute of Technology, Cambridge, MA, 1979, as cited in [19].

[182] You Tube; www.youtube.com. Flow-visualization videos may be found using keyword searches such as flow visualization, fluid mechanics, aerodynamics, and non-Newtonian fluids. The author's You Tube channel is DrMorrisonMTU.

[183] D. F. Young, B. R. Munson, and T. H. Okiishi, *A Brief Introduction to Fluid Mechanics* (John Wiley & Sons: New York, 1997).

[184] D. Zwillinger, *Handbook of Differential Equations*, 3rd edition (Academic Press: New York, 1989).

Index

Note to index: An *n* following a page number indicates a note on that page; an *f* following a page number indicates a figure on that page; a *t* following a page number indicates a table on that page.